AF352844

Special Issue in Honor of Dr. Michael Weber's 70th Birthday: Photodynamic Therapy: Rising Star in Pharmaceutical Applications

Special Issue in Honor of Dr. Michael Weber's 70th Birthday: Photodynamic Therapy: Rising Star in Pharmaceutical Applications

Editors

Eduard Preis
Matthias Wojcik
Gerhard Litscher
Udo Bakowsky

MDPI • Basel • Beijing • Wuhan • Barcelona • Belgrade • Manchester • Tokyo • Cluj • Tianjin

Editors

Eduard Preis
Department of Pharmaceutics
and Biopharmaceutics
University of Marburg
Marburg
Germany

Matthias Wojcik
Department of Pharmaceutics
and Biopharmaceutics
University of Marburg
Marburg
Germany

Gerhard Litscher
Research Unit of Biomedical
Engineering in Anesthesia
and Intensive Care Medicine
Medical University of Graz
Graz
Austria

Udo Bakowsky
Department of Pharmaceutics
and Biopharmaceutics
University of Marburg
Marburg
Germany

Editorial Office
MDPI
St. Alban-Anlage 66
4052 Basel, Switzerland

This is a reprint of articles from the Special Issue published online in the open access journal *Pharmaceutics* (ISSN 1999-4923) (available at: www.mdpi.com/journal/pharmaceutics/special_issues/Honor_Michael).

For citation purposes, cite each article independently as indicated on the article page online and as indicated below:

LastName, A.A.; LastName, B.B.; LastName, C.C. Article Title. *Journal Name* **Year**, *Volume Number*, Page Range.

ISBN 978-3-0365-5960-5 (Hbk)
ISBN 978-3-0365-5959-9 (PDF)

Contents

 *pharmaceutics*

 MDPI

Editorial

Editorial on the "Special Issue in Honor of Dr. Michael Weber's 70th Birthday: Photodynamic Therapy: Rising Star in Pharmaceutical Applications"

Eduard Preis [1,*], Matthias Wojcik [1,*], Gerhard Litscher [2,*] and Udo Bakowsky [1,*]

1 Department of Pharmaceutics and Biopharmaceutics, University of Marburg, Robert-Koch-Str. 4, 35037 Marburg, Germany
2 President of ISLA (International Society for Medical Laser Applications), Research Unit of Biomedical Engineering in Anesthesia and Intensive Care Medicine, Research Unit for Complementary and Integrative Laser Medicine, Traditional Chinese Medicine (TCM) Research Center Graz, Department of Anesthesiology and Intensive Care Medicine, Medical University of Graz, Auenbruggerplatz 39, 8036 Graz, Austria
* Correspondence: eduard.preis@pharmazie.uni-marburg.de (E.P.); matthias.wojcik@pharmazie.uni-marburg.de (M.W.); gerhard.litscher@medunigraz.at (G.L.); ubakowsky@aol.com (U.B.)

Citation: Preis, E.; Wojcik, M.; Litscher, G.; Bakowsky, U. Editorial on the "Special Issue in Honor of Dr. Michael Weber's 70th Birthday: Photodynamic Therapy: Rising Star in Pharmaceutical Applications". *Pharmaceutics* 2022, 14, 1786. https://doi.org/10.3390/pharmaceutics14091786

Received: 23 August 2022
Accepted: 24 August 2022
Published: 26 August 2022

Publisher's Note: MDPI stays neutral with regard to jurisdictional claims in published maps and institutional affiliations.

Thousands of years ago, phototherapy or heliotherapy was performed by ancient Egyptians, Greeks, and Romans. However, from the mid-19th century onward, names such as Arnold Rikli, Niels Ryberg Finsen, Downes and Blunt, Oscar Raab, and Hermann von Tappeiner started to appear and paved the way for current experts in the field of photodynamic therapy (PDT), such as Wainwright, Maisch, and Hamblin. Still, only a tiny fraction of PDT's potential has been realized in the clinical practice guidelines [1].

This Special Issue of *Pharmaceutics* commemorates the tremendous influence Michael Weber's work had on the use and practical applications of PDT. Looking back after his 70th birthday, in addition to his inventions, patents, and publications, he always found room for hands-on experience as a practicing doctor. He pioneered the clinical use of lasers and photodynamic therapy for nearly 25 years in Germany and several other countries.

Multiple contributions from all over the world emphasize the importance of PDT and signal that there is much more to expect. The mechanism of PDT generally relies on three main components, i.e., light, a photosensitizer (PS), and molecular oxygen; however, it can be subdivided into different applications, which sometimes leads to confusing abbreviations and definitions. When applied against bacteria and fungi, it is often referred to as antimicrobial photodynamic therapy (aPDT) or photodynamic antimicrobial chemotherapy (PACT); against viruses, it is called antiviral photodynamic therapy (aPDT); against cancer cells, most researchers use the unmodified term photodynamic therapy (PDT). To simplify this matter, we divide photodynamic therapy into two parts, i.e., cancer treatment (PDT) or antimicrobial and antiviral therapy (aPDT).

The most extensive section in this Special Issue is aPDT, which comprises five articles and one communication covering bacteria treatment, three articles focusing on wound healing, and one article each looking at antifungal and antiviral therapy. González et al. and Núñez et al. used transition metal complexes, i.e., a homo-bimetallic Re(I) complex [2] and a polypyridine Ir(III) complex [3]. Both proved that these complexes could be effectively used in aPDT. Whereas the first showed an enhanced effect by combining the PS with cefotaxime, the latter successfully used imipenem. Garcia et al. applied Fotoenticine®, a new PS derived from chlorin e-6, on a microcosm biofilm [4]. In this ex vivo model, which is closer to the complex in vivo conditions, a qualitative and quantitative reduction in bacterial viability was shown. Regarding biofilms, Battisti et al. highlighted the new fluorescence lifetime imaging, which might offer more insights into the biofilm dynamics and facilitate treatment optimization [5].

Cuadrado et al. and Ayoub et al. used a more technological approach and processed the PS. They incorporated their PS (i.e., zinc menthol-phthalocyanine or parietin) into magnetic nanocomposites [6] or cyclodextrin-inclusion complexes [7].

Additionally, three articles took bacteria treatment one step further and focused more on wound healing. They bridge the gap between in vitro and in vivo conditions [8] and show new in vivo quantification methods [9,10].

Pérez-Laguna et al. assessed the effect of combination therapy against different strains of *Candida* spp. using methylene blue as PS and chlorhexidine [11]. They achieved a reduction in methylene blue concentration while maintaining the same photodynamic efficacy. Another study by Sadraeian et al. compared the effects of UV-C light and aPDT with photodithazine against SARS-CoV-2 pseudovirus [12].

The PDT section consists of five articles covering different aspects. Chai et al. synthesized a new PS that showed beneficial properties (i.e., tracking and ablation) against HepG2 human hepatocellular carcinoma cells [13]. Dobre et al. investigated the gene expression pattern of HT29 cells treated with a new porphyrin derivate that they had previously synthesized and analyzed [14]. Nanosized drug delivery systems are vital when applying the most PS in PDT. Thus, Lehmann et al. and Yeh et al. incorporated their PS in liposomes and lipid-calcium phosphate nanoparticles, respectively [15,16]. While the first group reported the feasibility of liposome nebulization and pulmonary drug delivery, the second successfully treated SCC4 and SAS cells in vitro and in a xenograft model with a combination therapy using EGFR siRNA and PDT. Bartosińska et al. compared three different forms of 5-aminolevulinic acid in treating actinic keratosis and showed that 5-aminolevulinic acid phosphate was superior to the other forms at present due to its higher tolerability and lesser pain [17].

Two articles covering immunomodulatory therapy extended the scope of the two defined sections, PDT and aPDT. Dorst et al. assessed the efficacy of IRDye700DX-loaded liposomes in the treatment of arthritis and provided insights into the difficulties of these treatment regimens [18]. Christensen et al. presented the first-in-human study of 5-aminolevulinic acid against chronic graft-versus-host disease [19]. They used extracorporeal photopheresis combined with photoactivation of the generated protoporphyrin IX and proved the tolerability and safety of the procedure. Apart from these captivating research articles, seven profound review articles deal with different fields in PDT and aPDT. A broad overview of PDT, from history to future perspectives, was given by Correia et al. [20]. They summarized the essential parameters, discussed advantages and limitations, and emphasized that PDT is a promising therapeutic option. Lange et al. and Ailioaie et al. focused on the use of cyanine-derived dyes and curcumin in PDT, respectively [21,22]. Whereas the first group thoroughly described all used dyes, the second included all technological advances and presented the use of curcumin against different cancer types in detail. Fahmy et al. focused their article on liposomal formulations in PDT, highlighted the versatility of liposomes, and listed an astonishing number of the most recent state-of-the-art studies [23]. The comprehensive review by Piaserico et al. deals with the possible applications of PDT against actinic keratoses. They provided important information on the benefits of post- and pre-treatment strategies that improve the therapeutic efficiency of PDT [24].

A systematic review by Dalvi et al. on using aPDT against periodontitis demanded more robust and well-designed studies due to the substantial flaws limiting their reproducibility [25]. Starting from single PS, along with their different classes and drug-delivery systems, Youf et al. continued to thoroughly describe all possible combinations with aPDT [26].

Overall, photodynamic therapy is a lively topic that can be used in various fields. In 2017, despite some breakthroughs, the world did not seem prepared for PDT and aPDT. We look forward to the changes in clinical practice in the upcoming years.

We thank *Pharmaceutics* (MDPI) and their excellent team for permanently staying by our side and helping us to produce this Special Issue.

Funding: This research received no external funding.

Conflicts of Interest: The authors declare no conflict of interest.

References

1. Wainwright, M.; Maisch, T.; Nonell, S.; Plaetzer, K.; Almeida, A.; Tegos, G.P.; Hamblin, M.R. Photoantimicrobials—Are we afraid of the light? *Lancet Infect. Dis.* **2017**, *17*, e49–e55. [CrossRef]
2. González, I.A.; Palavecino, A.; Núñez, C.; Dreyse, P.; Melo-González, F.; Bueno, S.M.; Palavecino, C.E. Effective Treatment against ESBL-Producing Klebsiella pneumoniae through Synergism of the Photodynamic Activity of Re (I) Compounds with Beta-Lactams. *Pharmaceutics* **2021**, *13*, 1889. [CrossRef] [PubMed]
3. Núñez, C.; Palavecino, A.; González, I.A.; Dreyse, P.; Palavecino, C.E. Effective Photodynamic Therapy with Ir(III) for Virulent Clinical Isolates of Extended-Spectrum Beta-Lactamase Klebsiella pneumoniae. *Pharmaceutics* **2021**, *13*, 603. [CrossRef] [PubMed]
4. Garcia, M.T.; Da Ward, R.A.C.; Gonçalves, N.M.F.; Pedroso, L.L.C.; Da Neto, J.V.S.; Strixino, J.F.; Junqueira, J.C. Susceptibility of Dental Caries Microcosm Biofilms to Photodynamic Therapy Mediated by Fotoenticine. *Pharmaceutics* **2021**, *13*, 1907. [CrossRef]
5. Battisti, A.; Morici, P.; Sgarbossa, A. Fluorescence Lifetime Imaging Microscopy of Porphyrins in Helicobacter pylori Biofilms. *Pharmaceutics* **2021**, *13*, 1674. [CrossRef]
6. Cuadrado, C.F.; Díaz-Barrios, A.; Campaña, K.O.; Romani, E.C.; Quiroz, F.; Nardecchia, S.; Debut, A.; Vizuete, K.; Niebieskikwiat, D.; Ávila, C.E.; et al. Broad-Spectrum Antimicrobial ZnMintPc Encapsulated in Magnetic-Nanocomposites with Graphene Oxide/MWCNTs Based on Bimodal Action of Photodynamic and Photothermal Effects. *Pharmaceutics* **2022**, *14*, 705. [CrossRef]
7. Ayoub, A.M.; Gutberlet, B.; Preis, E.; Abdelsalam, A.M.; Abu Dayyih, A.; Abdelkader, A.; Balash, A.; Schäfer, J.; Bakowsky, U. Parietin Cyclodextrin-Inclusion Complex as an Effective Formulation for Bacterial Photoinactivation. *Pharmaceutics* **2022**, *14*, 357. [CrossRef]
8. Chan, B.C.L.; Dharmaratne, P.; Wang, B.; Lau, K.M.; Lee, C.C.; Cheung, D.W.S.; Chan, J.Y.W.; Yue, G.G.L.; Lau, C.B.S.; Wong, C.K.; et al. Hypericin and Pheophorbide a Mediated Photodynamic Therapy Fighting MRSA Wound Infections: A Translational Study from In Vitro to In Vivo. *Pharmaceutics* **2021**, *13*, 1399. [CrossRef]
9. Zuhayri, H.; Nikolaev, V.V.; Knyazkova, A.I.; Lepekhina, T.B.; Krivova, N.A.; Tuchin, V.V.; Kistenev, Y.V. In Vivo Quantification of the Effectiveness of Topical Low-Dose Photodynamic Therapy in Wound Healing Using Two-Photon Microscopy. *Pharmaceutics* **2022**, *14*, 287. [CrossRef]
10. Zuhayri, H.; Nikolaev, V.V.; Lepekhina, T.B.; Sandykova, E.A.; Krivova, N.A.; Kistenev, Y.V. The In Vivo Quantitative Assessment of the Effectiveness of Low-Dose Photodynamic Therapy on Wound Healing Using Optical Coherence Tomography. *Pharmaceutics* **2022**, *14*, 399. [CrossRef]
11. Pérez-Laguna, V.; Barrena-López, Y.; Gilaberte, Y.; Rezusta, A. In Vitro Effect of Photodynamic Therapy with Different Lights and Combined or Uncombined with Chlorhexidine on *Candida* spp. *Pharmaceutics* **2021**, *13*, 1176. [CrossRef] [PubMed]
12. Sadraeian, M.; Junior, F.F.P.; Miranda, M.; Galinskas, J.; Fernandes, R.S.; da Cruz, E.F.; Fu, L.; Zhang, L.; Diaz, R.S.; Cabral-Miranda, G.; et al. Study of Viral Photoinactivation by UV-C Light and Photosensitizer Using a Pseudotyped Model. *Pharmaceutics* **2022**, *14*, 683. [CrossRef] [PubMed]
13. Chai, C.; Zhou, T.; Zhu, J.; Tang, Y.; Xiong, J.; Min, X.; Qin, Q.; Li, M.; Zhao, N.; Wan, C. Multiple Light-Activated Photodynamic Therapy of Tetraphenylethylene Derivative with AIE Characteristics for Hepatocellular Carcinoma via Dual-Organelles Targeting. *Pharmaceutics* **2022**, *14*, 459. [CrossRef] [PubMed]
14. Dobre, M.; Boscencu, R.; Neagoe, I.V.; Surcel, M.; Milanesi, E.; Manda, G. Insight into the Web of Stress Responses Triggered at Gene Expression Level by Porphyrin-PDT in HT29 Human Colon Carcinoma Cells. *Pharmaceutics* **2021**, *13*, 1032. [CrossRef] [PubMed]
15. Lehmann, J.; Agel, M.R.; Engelhardt, K.H.; Pinnapireddy, S.R.; Agel, S.; Duse, L.; Preis, E.; Wojcik, M.; Bakowsky, U. Improvement of Pulmonary Photodynamic Therapy: Nebulisation of Curcumin-Loaded Tetraether Liposomes. *Pharmaceutics* **2021**, *13*, 1243. [CrossRef]
16. Yeh, C.-H.; Chen, J.; Zheng, G.; Huang, L.; Hsu, Y.-C. Novel Pyropheophorbide Phosphatydic Acids Photosensitizer Combined EGFR siRNA Gene Therapy for Head and Neck Cancer Treatment. *Pharmaceutics* **2021**, *13*, 1435. [CrossRef]
17. Bartosińska, J.; Szczepanik-Kułak, P.; Raczkiewicz, D.; Niewiedzioł, M.; Gerkowicz, A.; Kowalczuk, D.; Kwaśny, M.; Krasowska, D. Topical Photodynamic Therapy with Different Forms of 5-Aminolevulinic Acid in the Treatment of Actinic Keratosis. *Pharmaceutics* **2022**, *14*, 346. [CrossRef]
18. Dorst, D.N.; Boss, M.; Rijpkema, M.; Walgreen, B.; Helsen, M.M.A.; Bos, D.L.; van Bloois, L.; Storm, G.; Brom, M.; Laverman, P.; et al. Photodynamic Therapy Targeting Macrophages Using IRDye700DX-Liposomes Decreases Experimental Arthritis Development. *Pharmaceutics* **2021**, *13*, 1868. [CrossRef]
19. Christensen, E.; Foss, O.A.; Quist-Paulsen, P.; Staur, I.; Pettersen, F.; Holien, T.; Juzenas, P.; Peng, Q. Application of Photodynamic Therapy with 5-Aminolevulinic Acid to Extracorporeal Photopheresis in the Treatment of Patients with Chronic Graft-versus-Host Disease: A First-in-Human Study. *Pharmaceutics* **2021**, *13*, 1558. [CrossRef]
20. Correia, J.H.; Rodrigues, J.A.; Pimenta, S.; Dong, T.; Yang, Z. Photodynamic Therapy Review: Principles, Photosensitizers, Applications, and Future Directions. *Pharmaceutics* **2021**, *13*, 1332. [CrossRef]

21. Lange, N.; Szlasa, W.; Saczko, J.; Chwiłkowska, A. Potential of Cyanine Derived Dyes in Photodynamic Therapy. *Pharmaceutics* **2021**, *13*, 818. [CrossRef] [PubMed]
22. Ailioaie, L.M.; Ailioaie, C.; Litscher, G. Latest Innovations and Nanotechnologies with Curcumin as a Nature-Inspired Photosensitizer Applied in the Photodynamic Therapy of Cancer. *Pharmaceutics* **2021**, *13*, 1562. [CrossRef] [PubMed]
23. Fahmy, S.A.; Azzazy, H.M.E.-S.; Schaefer, J. Liposome Photosensitizer Formulations for Effective Cancer Photodynamic Therapy. *Pharmaceutics* **2021**, *13*, 1345. [CrossRef] [PubMed]
24. Piaserico, S.; Mazzetto, R.; Sartor, E.; Bortoletti, C. Combination-Based Strategies for the Treatment of Actinic Keratoses with Photodynamic Therapy: An Evidence-Based Review. *Pharmaceutics* **2022**, *14*, 1726. [CrossRef]
25. Dalvi, S.; Benedicenti, S.; Sălăgean, T.; Bordea, I.R.; Hanna, R. Effectiveness of Antimicrobial Photodynamic Therapy in the Treatment of Periodontitis: A Systematic Review and Meta-Analysis of In Vivo Human Randomized Controlled Clinical Trials. *Pharmaceutics* **2021**, *13*, 836. [CrossRef]
26. Youf, R.; Müller, M.; Balasini, A.; Thétiot, F.; Müller, M.; Hascoët, A.; Jonas, U.; Schönherr, H.; Lemercier, G.; Montier, T.; et al. Antimicrobial Photodynamic Therapy: Latest Developments with a Focus on Combinatory Strategies. *Pharmaceutics* **2021**, *13*, 1995. [CrossRef]

Article

Effective Treatment against ESBL-Producing *Klebsiella pneumoniae* through Synergism of the Photodynamic Activity of Re (I) Compounds with Beta-Lactams

Iván A. González [1], Annegrett Palavecino [2], Constanza Núñez [2], Paulina Dreyse [3], Felipe Melo-González [4], Susan M. Bueno [4] and Christian Erick Palavecino [2,*]

[1] Departamento de Química, Facultad de Ciencias Naturales, Matemática y del Medio Ambiente, Universidad Tecnológica Metropolitana, Las Palmeras 3360, Ñuñoa, Santiago 7800003, Chile; igonzalezp@utem.cl

[2] Laboratorio de Microbiología Celular, Instituto de Investigación e Innovación en Salud, Facultad de Ciencias de la Salud, Universidad Central de Chile, Lord Cochrane 418, Santiago 8330546, Chile; annegrett.palavecino@alumnos.ucentral.cl (A.P.); constanza.nunezc@alumnos.ucentral.cl (C.N.)

[3] Departamento de Química, Universidad Técnica Federico Santa María, Av. España 1680, Casilla, Valparaíso 2390123, Chile; paulina.dreyse@usm.cl

[4] Departamento de Genética Molecular y Microbiología, Millennium Institute on Immunology and Immunotherapy, Facultad de Ciencias Biológicas, Pontificia Universidad Católica de Chile, Santiago 8330025, Chile; famelo@bio.puc.cl (F.M.-G.); sbueno@bio.puc.cl (S.M.B.)

* Correspondence: christian.palavecino@ucentral.cl; Tel.: +56-225851334

Citation: González, I.A.; Palavecino, A.; Núñez, C.; Dreyse, P.; Melo-González, F.; Bueno, S.M.; Palavecino, C.E. Effective Treatment against ESBL-Producing *Klebsiella pneumoniae* through Synergism of the Photodynamic Activity of Re (I) Compounds with Beta-Lactams. *Pharmaceutics* **2021**, *13*, 1889. https://doi.org/10.3390/pharmaceutics13111889

Academic Editors: Udo Bakowsky, Matthias Wojcik, Eduard Preis and Gerhard Litscher

Received: 14 September 2021
Accepted: 1 November 2021
Published: 8 November 2021

Publisher's Note: MDPI stays neutral with regard to jurisdictional claims in published maps and institutional affiliations.

Abstract: Background: Extended-spectrum beta-lactamase (ESBL) and carbapenemase (KPC^+) producing *Klebsiella pneumoniae* are multidrug-resistant bacteria (MDR) with the highest risk to human health. The significant reduction of new antibiotics development can be overcome by complementing with alternative therapies, such as antimicrobial photodynamic therapy (aPDI). Through photosensitizer (PS) compounds, aPDI produces local oxidative stress-activated by light (photooxidative stress), nonspecifically killing bacteria. Methodology: Bimetallic Re(I)-based compounds, PSRe-μL1 and PSRe-μL2, were tested in aPDI and compared with a Ru(II)-based PS positive control. The ability of PSRe-μL1 and PSRe-μL2 to inhibit *K. pneumoniae* was evaluated under a photon flux of 17 μW/cm². In addition, an improved aPDI effect with imipenem on KPC^+ bacteria and a synergistic effect with cefotaxime on ESBL producers of a collection of 118 clinical isolates of *K. pneumoniae* was determined. Furthermore, trypan blue exclusion assays determined the PS cytotoxicity on mammalian cells. Results: At a minimum dose of 4 μg/mL, both the PSRe-μL1 and PSRe-μL2 significantly inhibited in $3\log_{10}$ (>99.9%) the bacterial growth and showed a lethality of 60 and 30 min of light exposure, respectively. Furthermore, they were active on clinical isolates of *K. pneumoniae* at 3–$6 \log_{10}$. Additionally, a remarkably increased effectiveness of aPDI was observed over KPC^+ bacteria when mixed with imipenem, and a synergistic effect from 3 to $6\log_{10}$ over ESBL producers of *K. pneumoniae* clinic isolates when mixed with cefotaxime was determined for both PSs. Furthermore, the compounds show no dark toxicity and low light-dependent toxicity in vitro to mammalian HEp-2 and HEK293 cells. Conclusion: Compounds PSRe-μL1 and PSRe-μL2 produce an effective and synergistic aPDI effect on KPC^+, ESBL, and clinical isolates of *K. pneumoniae* and have low cytotoxicity in mammalian cells.

Keywords: photodynamic therapy; multi-drug resistance; antibiotic synergy; *Klebsiella pneumoniae*

1. Introduction

Due to the emergence of multi-drug resistance (MDR) pathogenic bacteria, the deficit of new antibiotics is one of the most pressing threats to human health in the 21st century [1]. The world health organization has presented a ranking of the most relevant MDR bacteria that require the urgent development of new antimicrobial therapies. Strains of

Klebsiella pneumoniae producing extended-spectrum β-lactamase (ESBL) and carbapenemase (KPC) are among the most relevant [2,3]. *K. pneumoniae* is a Gram-negative bacillus associated with pneumonia and urinary tract infections (UTI) [4,5]. Additionally, *K. pneumoniae* is one of the most relevant agents of healthcare-associated infections (HAIs) [6]. The HAIs produced by *K. pneumoniae* can be severe, producing mortalities as high as 30 to 70% [7,8]. The use of polymyxins (colistin) or tigecycline antibiotics are the only therapeutic options to treat severe KPC^+ infections [9]. The global increase in pan-resistant Enterobacteriaceae has resulted in increased use of colistin, which has accelerated the onset of resistance to polymyxins; the emergence of the mcr-1 gene is a good example [1,10]. Therefore, MDR-*K. pneumoniae* strains have a great potential to become a "superbug"; therefore, they are an excellent model for discovering new antimicrobial treatments [8].

In this scenario, non-antibiotic therapeutic options with antimicrobial properties should be explored. An alternative is the antimicrobial photodynamic inactivation (aPDI) based on light-activated photosensitizer compounds (PS) [11–13]. The PSs are chemical compounds that absorb and accumulate the quantized energy of a specific wavelength accessing a triplet excited state by intersystem crossing processes [14]. The accumulated energy is transferred to the molecular oxygen commonly present in biological solutions through two mechanisms of action to produce reactive oxygen species (ROS): The Type I effect transfers energetic electrons that produce superoxide ($O_2^{\bullet-}$); the $O_2^{\bullet-}$ produces other ROS, such as hydrogen peroxide (H_2O_2) and hydroxyl radical ($HO^{\bullet}$) [15,16]. The Type II effect transfers the energy (with no electrons) to generate singlet oxygen ($^1O_2$) [17,18]. ROS such as $^1O_2$ produces photooxidative stress by concerted addition reactions of alkene groups in closer organic macromolecules such as protein alkylation, lipid carboxylation, and DNA degradation [12,19,20]. Hence, photooxidative stress results in non-specific bacterial cell death produced by damage over bacterial structures such as plasma membranes or DNA [17,21]. Many initiatives have developed PS compounds with aPDI properties for bacteria such as *K. pneumoniae* [13,20,22–24]. PS with a longer-lived excited state lifetime improve the probability of interacting with triplet oxygen and must produce more $^1O_2$ [25–28]. Moreover, it has been identified that cationic PSs produce more significant inhibition in bacterial growth [29], probably due to a more intimate interaction with the negatively charged bacterial envelope [30]. Therefore, cationic PSs may show a better photodynamic effect on *K. pneumoniae* than anionic PSs [29–31]. Our laboratory has tested various cationic Ir(III) organometallic PSs with antimicrobial properties against *K. pneumoniae* [23,29,32]. Other authors have also developed PS for *K. pneumoniae* [31,33–35], where some of these are organic molecules able to inhibit bacterial growth in vitro [33]. The PS molecule must show low levels of cytotoxicity to reduce the probability of adverse pharmacological effects. In this regard, the PSs based on organic molecules should be less toxic [19]. Other coordination compounds based on transition metals into tetrapyrrole structures or 5-aminolevulinic acids (ALA) have been successfully used for photodynamic treatment of cancer [36,37]. The Ru(II) complex containing three phenanthroline ligands has shown highlighted antimicrobial activity [38]. Additionally, the Re(I) complexes have been used in vitro against a broad spectrum of bacteria [39]. In this sense, the complexes with transition metals such as Re(I) can be considered good options according to their photophysical properties to produce reactive oxygen species useful for antimicrobial treatment [40].

Here we verify that PS compounds that absorb in a wide range of the visible spectrum, such as bimetallic Re (I) bimetallic complexes with polypyridine bridging ligand, may be helpful in aPDI over *K. pneumoniae*. These complexes have (Re(CO)$_3$Cl)$_2$μ-N^N general formula and here were evaluated with the following N^N ligands: 2,3-Dicarboxypyrazino [2,3-f] [4,7] phenanthrolinedicarboxylic (L1) and 2,3-Diethoxycarbonylypyrazino [2,3-f] [4,7] phenanthroline (L2) to obtain the corresponding PSRe-μL1 and PSRe-μL2 compounds [41]. The photodynamic effect of the PSRe-μL1 and PSRe-μL2 compounds was tested in vitro in two laboratory strains of *K. pneumoniae*: the carbapenem susceptible (KPC^-) KPPR-1 and the carbapenem-resistant (KPC^+) ST258 strain, and also in a previously characterized population of 118 clinical isolates of *K. pneumoniae*, including 66 ESBL-producers [42]. In addition, the PSs capacity to inhibit bacterial growth was verified, as well as the pharmacological properties, such as minimum effective dose (MEC), lethality time, and synergy with imipenem (Imp) and cefotaxime (Cfx) antibiotics. Finally, the low cytotoxicity in mammalian cells determined in vitro makes these PSs a promising alternative to complement the treatment of complicated infections.

2. Materials and Methods

2.1. Synthesis of the Photosensitizer Compounds

Our group had previously synthesized and characterized the structure, photophysics, and purity of the PSRe-μL1 and PSRe-μL2 compounds, published in González et al., 2020 [41]. The characterizations included nuclear magnetic resonance (NMR), Fourier transforms infrared spectroscopy (FT-IR), elemental analysis TD-DFT calculations, and cyclic voltammetry. Additionally, the absorption spectra measured in acetonitrile solution were performed in a Shimadzu UV–Vis-NIR 3101-PC spectrophotometer. Finally, the molar extinction coefficients of the characteristic absorption bands and the area under the curve between 500–700 nm were determined. The complexes can be described using the following general formula: $(Re(CO)_3Cl)_2\mu$-N^N, where N^N is 2,3-Dicarboxypyrazino [2,3-f] [4,7] phenanthrolinedicarboxylic (L1) or 2,3-Diethoxycarbonylypyrazino [2,3-f] [4,7] phenanthroline (L2). The Re(I) PSs obtained with both L1 and L2 are then designated as PSRe-μL1 and PSRe-μL2, respectively.

2.2. Antimicrobial Activity of Photosensitizers Compounds

Stock solutions of 2 mg/mL of each photosensitizer compound were prepared in dimethyl sulfoxide DMSO (Sigma-Aldrich, Saint louis, MO, USA), from which suitable work concentrations were obtained by dilution in an aqueous medium. The imipenem susceptible (KPC-) strain KPPR1 and the imipenem resistant strain ST258 (KPC⁺) of *K. pneumoniae* were used for the antimicrobial assay. Additionally, 118 clinical isolates of *K. pneumoniae* were used. Those clinical isolates were previously characterized for sensitivity and pathogenicity patterns [42]. For photodynamic experiments, all bacteria were grown as axenic culture in Luria Bertani medium and suspended at 1×10^7 colony forming units (CFU)/mL cation-adjusted Muller Hinton broth (ca-MH). Bacteria were mixed with each PS in 24-well plates, in a final volume of 500 μL and light-irradiated immediately after adding the PS in a chamber with a white LED lamp at a photon flux of 17 μW/cm². Controls plates with bacteria but no PS and with PS and no light were also included. All plates, including controls, were incubated for 1 h or the indicated time, and broth-micro dilution and sub-cultured on ca-MH agar plates were used to determine the CFUs of the viable bacteria. The agar plates were incubated at 37 °C, and colony count was recorded using a stereoscopic microscope after 16–20 h of incubation in the dark, following the recommendations of the Clinical and Laboratory Standards Institute (CLSI 2017) [43]. Control wells with bacteria, with or without photosensitizer but not exposed to light, were also included.

2.3. Determination of the Synergy between PSs and Cfx

To determine the fractional inhibitory concentration index (FIC) value, the following formula was used [44,45]:

$$\text{FIC Index} = \frac{\text{MICac}}{\text{MIC}_a} + \frac{\text{MICbc}}{\text{MIC}_b}$$

MICac is the MIC of compound A, combined with compound B, and MICbc is the MIC of compound B combined with compound A. The MICa and MICb are the MIC of the A and B compounds alone, respectively. Values in the FIC index ≤ 0.5 are considered synergistic, and values > 4 are considered antagonistic [44]. To determine the MIC- Cfx in combination with each PS, 1×10^7 UFC/mL of ESBL-producing bacteria were aPDI treated for 30 min with 4 μg/mL of each PSRe mixed with serial dilution (32–0.125 μg/mL) of

Cfx in ca-MH broth, as stated above. To determine the modification in the PS-MEC when combined with Cfx, 1×10^7 UFC/mL of ESBL-producing bacteria were added to 4 μg/mL of Cfx and mixed with a serial dilution of each PS (32–0.125 μg/mL) in ca-MH broth, as mentioned previously.

2.4. Cell Culture

The human cell lines from the American-type culture collection (ATCC), HEp-2 (CCL-23), and HEK293T (CRL-3216) were grown in DMEM with a mix of 1% streptomycin/penicillin antibiotics. The medium was supplemented with 10% fetal bovine serum (FBS), and the cultures were incubated in a 5% CO_2 atmosphere. Initial cultures of 5×10^5 cells per well were incubated for 24 to 48 h to a 70–90% confluence in triplicate in 24-well plates.

2.5. Cytotoxicity Tests in Eukaryotic Cells

The cytotoxic effect of PS compounds on HEp-2 and HEK293T cells was determined by evaluating cell death by the exclusion of trypan blue. The 24-well plates with cells at 70–90% confluence were mixed with 4 μg/mL of PSRe-μL1 and PSRe-μL2 compounds or control with D-PBS and incubated for 1 h in the dark or light-activated in a white LED light chamber with 17 μW/ cm^2 of photon flux. Subsequently, the PSs were removed by washing twice with D-PBS, and the cells were incubated in the dark for an additional 24 h in DMEM supplemented with antibiotics + 10% FBS in a 5% CO_2 atmosphere. After incubation, the cells were trypsinized, and the cell death was determined by trypan blue exclusion in a hemocytometer chamber.

2.6. Statistical Analysis

The Prism version 9.0 Software (GraphPad Software, LLC, San Diego, CA, USA) was used to perform the statistical analysis. Statistical significance was assessed using a two-tail *T*-test for parametric pairing groups or one-way ANOVA followed by the Tukey post-test for the lethality curves.

3. Results

3.1. Absorption Properties of the PSRe-μL1 and PSRe-μL2 Compounds

We previously showed that cationic Ir(III) are effective photosensitizers against *K. pneumoniae* [29]. However, these compounds present absorption peaks below 400 nm, limiting their excitation with wavelengths that better penetrate tissues [11]. Two bimetallic Re (I) compounds have been tested to enhance the PS activation in deeper tissue infections. Those bimetallic Re(I) complexes present polypyridine ligands (Figure 1A,B), they are fully structurally characterized, and they are thermodynamically stable in acetonitrile solution [41,46]. As shown in Table 1 and Figure 1D, the PSRe-μL1 and PSRe-μL2 present similar absorption processes at 358 and 360 nm, respectively. However, they show dark coloration attributed to a wide range of absorption bands in the visible range, with the maximums at 587 and 588 nm (Figure 1C,D). As shown in Figure 1D, both PSRe-μL1 and PSRe-μL2 compounds have an intense dark color, which shows their characteristic of absorbing light with a significant molar extinction coefficient in a wide visible range. The broad range absorption (500–700 nm) shows a significant area under the curve of 16.28 and 14.39 for PSRe-μL1 and PSRe-μL2, respectively. The compounds also present significantly high molar absorption of 3384 and 5600 $M^{-1}cm^{-1}$ for the PSRe-μL1 and PSRe-μL2, respectively (Table 1). The absorptions are attributable to the singlet metal-to-ligand (^{1}MLCT) and singlet ligand-to-ligand (^{1}LLCT) charge transfer transitions from the metal and chloride ligand to ligand bridge (L1 or L2), following the behavior of analogous compounds and the trends of cyclic voltammetry experiments and TD-DFT calculations [41]. In addition, the cyclic voltammetry showed a first oxidation process at 1.51 and 1.70 for PSRe-μL1 and PSRe-μL2, respectively. A second oxidation process was obtained only for PSRe-μL1 at 1.65 V (Table 1). The HOMO-LUMO energy gaps calculated for the PSRe-μL1 and PSRe-μL2 complexes were of similar magnitude, 2.73 and 2.84 eV, respectively (Table 1).

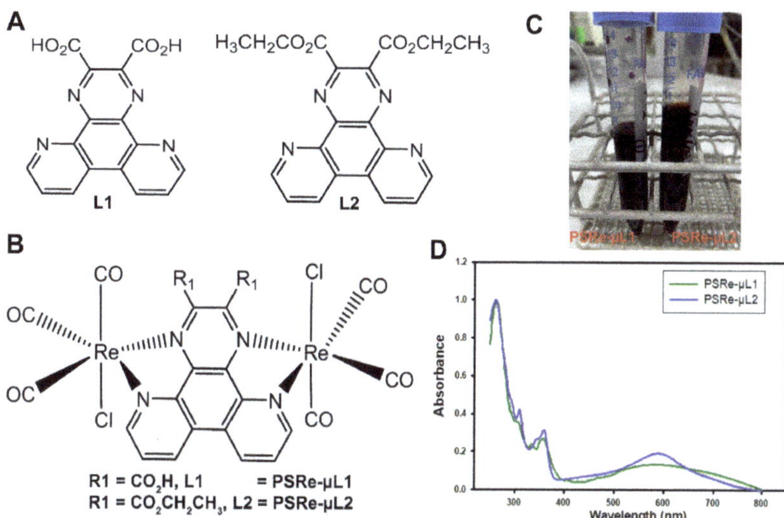

Figure 1. Chemical structure of the Re(I) bimetallic photosensitizer compounds and their L ligands and absorption spectra. The chemical structure of Re(I) complexes show the L1 and L2 ligands (**A**) and the (Re(CO)$_3$Cl)$_2$μ-N^N, whose replacement of R_1 results in the PSRe-μL1 and PSRe-μL2 compounds (**B**). Macroscopic appearance of PSRe-μL1 and PSRe-μL2 compounds concentrated in solution at 2 mg/mL (**C**). The absorption spectra for the PSRe-μL1 and PSRe-μL2 compounds in acetonitrile solution (**D**).

Table 1. Photophysical and electrochemical evaluation of photosensitizers.

Compounds	λ_{abs}/nm	Area (500–700 nm)	$E_{ox}1$/V	$E_{ox}2$/V	(ε/M^{-1}cm^{-1})
PSRe-μL1	358/587	16.28	1.51	1.65	3384
PSRe-μL2	360/588	14.39	1.70		5600
PS-Ru	550	0.83	1.29 *		600 *

* Extracted from Campagna et al., 2007 [47]. λ_{abs}, wavelength. Area, the area under the curve. E_{ox}, oxidation state. ε, molar absorption.

The PS-Ru [Ru(bpy)$_3$](PF$_6$)$_2$ (bpy = 2,2′-bipyridine) compound was used as a control to compare the aPDI activity of the PSRe-μL1 and PSRe-μL2 compounds [38]. According to the literature, in acetonitrile, the PS-Ru shows a charge-transfer absorption process at 450 nm [48] and a significantly low molar extinction coefficient at 550 nm of ~600 M^{-1}cm^{-1} [47]. In addition, we determine its area under the curve of 0.83 between 500–700 nm wavelength, which is significantly smaller than the PS-Re compounds (Table 1).

3.2. Antimicrobial Photodynamic Inactivation Activity of Compounds PSRe-μL1 and PSRe-μL2

The photodynamic antimicrobial capacity of the two new Re (I) compounds was determined by inhibiting the bacterial growth of *K. pneumoniae*. PSRe-μL1 and PSRe-μL2 at 16 μg/mL were then mixed with each *K. pneumoniae* strain, KPPR1 (KPC$^-$) and ST258 (KPC$^+$) at 1×10^7 CFU/ml. The aPDI activity of the PSRe-μL1 and PSRe-μL2 compounds was compared with the reference PS-Ru antimicrobial activity as a positive control [34,38,49]. Initial tests were performed with 16 μg/mL of each compound in an aqueous solution (ca-MH). Figure 2 shows the photodynamic treatment with PSRe-μL1 and PSRe-μL2 (Green bars), which inhibits in 3 log$_{10}$ (<99.9%) the growth of both strains of *K. pneumoniae* (* $p < 0.05$) compared to untreated control bacteria (Blue bars). The results show that the bactericidal effect produced by both PSs is dependent on light (Red bars) (ns $p > 0.05$, compared to untreated control) because growth inhibition is observed after light activation. Therefore, the PSRe-µL1 and the PSRe-µL2 compounds must be activated by light to exhibit their bactericidal effect. Comparable results were obtained with the 16 µg/mL PS-Ru control, as the bacterial growth inhibition was observed only after light activation ($p < 0.05$).

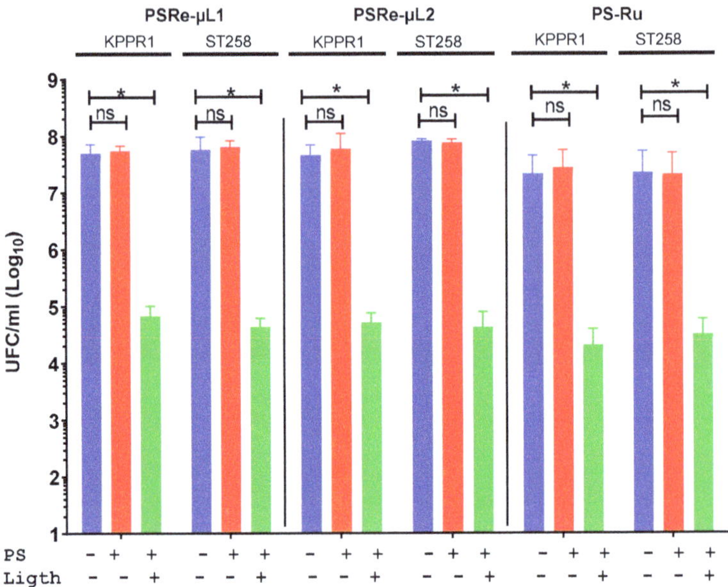

Figure 2. The Photodynamic antimicrobial inhibition of PSRe-µL1 and PSRe-µL2 compounds. The imipenem-sensitive strain (KPPR1) and the imipenem resistant strain (ST258) of *K. pneumoniae* were used at a 1×10^7 CFU/mL and mixed in triplicate with 16 µg/mL of PSRe-µL1, PSRe-µL2, or PS-Ru compounds. The mix was incubated for 1 h at 17 µW/cm^2 with light for aPDI (green bars) or in the dark for controls (red bars). The results were compared to control of bacteria not combined with the PSs (blue bars). Colony count enumerated viable bacteria on ca-MH agar after serial micro-dilution. The CFU/mL values are presented as means $\pm$ SD on a $\log_{10}$ scale. The + and − signs indicate the presence or absence of a compound or condition. Not significant (ns) $p > 0.05$ by Student's *t*-test among bacteria treated with PS without light compared to untreated control bacteria; * $p < 0.05$ by Student's *t*-test among bacteria treated with PS exposed to light compared to untreated control bacteria.

3.3. Determination of the Minimum Effective Concentration and Lethality for the Compounds PSRe-µL1 and PSRe-µL2

Because the PSRe-µL1 and PSRe-µL2 compounds showed a significant aPDI effect on the two *K. pneumoniae* strains, a further two pharmacologic parameters were determined: the minimum effective concentration (MEC) and the minimum light exposition time (lethality) on the KPPR1 and the ST258 *K. pneumoniae* strains. The PS-MEC was determined by exposition of 1×10^7 CFU/mL of each *K. pneumoniae* strain to serial dilutions (ranging from 0 to 32 µg/mL) of each PS in ca-MH broth. The aPDI treatment was performed for 1 h at a fluence rate of 17 µW/cm^2. After treatment, viable bacteria were enumerated as above by serial micro-dilution. We established the MEC as the concentration where the bacterial viability decreased in 3 $\log_{10}$ (99.9%). As shown in Figure 3, the MEC was determined at 4 µg/mL for both PSRe-µL1 (Figure 3A) and PSRe-µL2 (Figure 3B) (** $p < 0.01$, *** $p < 0.001$, Student's *t*-test comparing to control with 0 µg/mL). The MEC of 4 µg/mL of PSRe-µL1 or PSRe-µL2 were used to determine the lethality time on 1×10^7 CFU/mL of each bacterial strain in ca-MH broth. The mix was exposed to 17 μW/cm^2 of a white LED light for 5, 15, 30, 60, and 120 min. Control wells with bacteria without PS were also included. For each time, viable bacteria were enumerated by serial micro-dilution and colony counted in ca-MH agar. We established the lethality when the bacterial viability decreased in 3 log$_{10}$ (99.9%). As seen in Figure 3C, although PSRe-μL1 produced a significant reduction after 30 min of incubation (p = 0.013, compared to time 0), a 3log$_{10}$ reduction was observed after 60 min of light exposure (p < 0.01, compared to time 0). In comparison, the PSRe-μL2 significantly (p < 0.01) reduced in 3log$_{10}$ the bacterial viability after 30 min (Figure 3D).

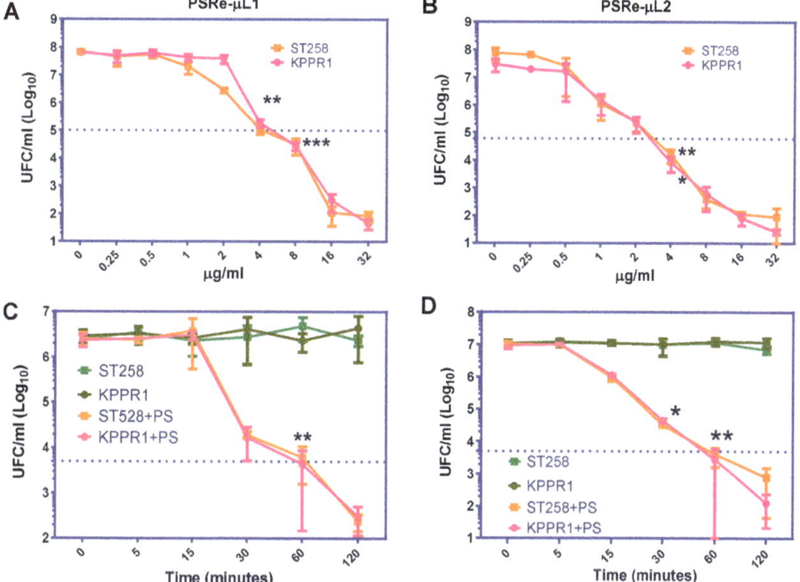

Figure 3. Determination of minimum effective concentration and time lethality. The minimum effective concentration (MEC) and lethality of the PSRe-μL1 and PSRe-μL2 were determined using the imipenem-sensitive (KPPR1) and the imipenem resistant (ST258) strains of *K. pneumoniae*. For MEC determination, bacteria at 1×10^7 CFU/mL were incubated with increasing concentrations (0–32 μg/mL) of the compounds PSRe-μL1 (**A**) or PSRe-μL2 (**B**) and exposed for 1 h to 17 μW/cm^2 of white LED light. The time lethality was determined, mixing the bacteria with 4 μg/mL of PSRe-μL1 (**C**) or PSRe-μL2 (**D**), and exposure for increasing times (5, 15, 30, 60, and 120 min) to 17 μW/cm^2 of white LED light. Colony count enumerated viable bacteria on ca-MH agar after serial-microdilution. The CFU/mL values are presented as means $\pm$ SD on a log$_{10}$ scale (* p < 0.05, ** p < 0.01, *** p < 0.001 by Student's *t*-test among bacteria treated with PS exposed to light compared to untreated control bacteria).

3.4. Increased Photodynamic Effect of PSRe-µL1 and PSRe-µL2 in Combination with Imipenem

The desired quality for photosensitizer compounds is to be used as adjunctive therapy with antibiotics. The improvement in combined therapy of the compounds PSRe-μL1 and PSRe-μL2 with imipenem was used to verify their usefulness in eradicating carbapenemase-producing *K. pneumoniae*. The strains of *K. pneumoniae* susceptible to carbapenem KPC$^-$ (KPPR1), and the resistant strain KPC$^+$ (ST258), were exposed to the preparation of 4 μg/mL of imipenem mixed in aqueous solution with 4 μg/mL of each PSRe-μL1, PSRe-μL2 (corresponding to their MECs), or the control compound PS-Ru [29]. Additionally, control bacteria exposure to light but without imipenem were included. As expected, when mixed with imipenem, the PSRe-μL1 and PSRe-μL2 compounds showed a significantly (*** p < 0.001) increased effect over bacterial viability, increasing from 3 to

$6\log_{10}$ the bactericidal effect for the KPC$^+$ strain (Figure 4). As seen previously, this behavior was not observed when combining the imipenem with the PS-Ru control compound, keeping the $3\log_{10}$ inhibitory effect ($p < 0.05$) [29].

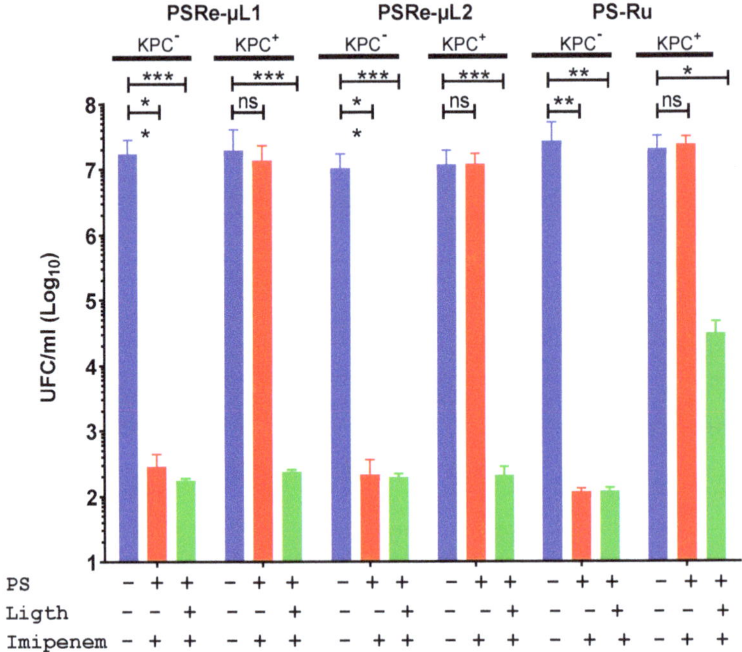

Figure 4. Increased photodynamic effect with imipenem. The effect of carbapenem on the PSRe-µL1 and PSRe-µL2 activity was determined in the strains of *K. pneumoniae* sensitive (KPPR1) or resistant (ST258) to imipenem. The bacteria at 1×10^7 UFC/mL were exposed to a mixture of 4 µg/mL of imipenem and with 4 µg/mL of each PS or the control PS-Ru. For aPDI treatment, the mixes were incubated for 1 h at 17 µW/cm^2 of light (green bars) or in the dark (red bars). Control also includes bacteria not treated (blue bars). Colony count enumerated viable bacteria on ca-MH agar after serial microdilution. The CFU/mL values are presented as means $\pm$ SD on a $\log_{10}$ scale. The + and - signs indicate the presence or absence of a compound or condition. Not significant (ns) $p > 0.05$ by Student's *t*-test among bacteria treated with PS + imipenem without light compared to untreated control bacteria; * $p < 0.05$, ** $p < 0.01$ *** $p < 0.001$ by Student's *t*-test among bacteria treated with PS + imipenem exposed to light compared to untreated control bacteria.

3.5. Antimicrobial Photodynamic Inhibition of the PSRe-µL1 and PSRe-µL2 Compounds over Clinical Isolates

We have in our laboratory a collection of 118 clinical isolates of *K. pneumoniae* from patients who had an active infection [42]. We used these isolates to verify the photodynamic activity of the PSRe-µL1 and PSRe-µL2 compounds on community bacteria and compared them with PS-Ru positive control [49]. As seen in Figure 5, photodynamic treatment with 4 µg/mL PSRe-µL1 or PSRe-µL2 significantly (*** $p < 0.001$; compared to untreated control) inhibits bacterial growth $> 3\log_{10}$ (>99.9%) of clinical isolates of *K. pneumoniae*. The results show that the bactericidal effect produced by PSRe-µL1 and PSRe-µL2 is light-dependent (ns = $p > 0.05$; compared to the untreated control). Those results are comparable with the positive control PS-Ru (4 µg/mL) (*** $p < 0.001$). Because this collection of strains has been characterized by its resistance profile and presents 66 ESBL-producing isolates, we analyzed whether a synergic effect occurred with cefotaxime. It was first determined whether the combined treatment with Cfx increases the inhibition of bacterial growth of aPDI with

PSRe-µL1 or PSRe-µL2. The clinical isolates were exposed to the preparation of 4 µg/mL of cefotaxime with 4 µg/mL of PSRe-µL1 or PSRe-µL2. As expected, the PSs compounds mixed with cefotaxime significantly (*** $p < 0.001$) increased the bactericidal effect on the clinical isolate population from 3 to 6log$_{10}$ reduction (Figure 5). No significantly increased inhibitory effect was observed for the PS-Ru control combined with cefotaxime (ns $p > 0.05$).

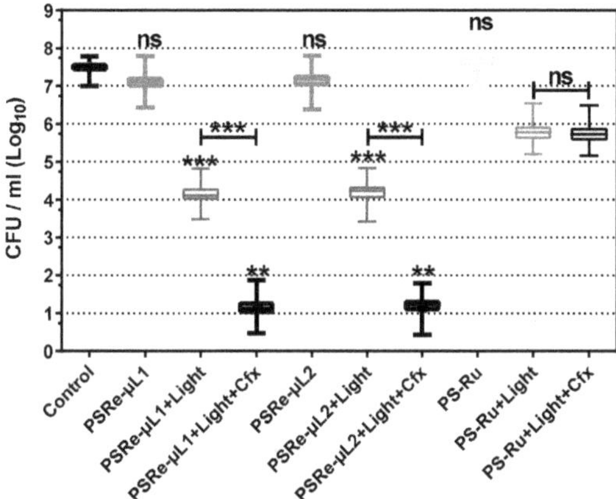

Figure 5. Antimicrobial photodynamic inactivation of clinical isolates of *K. pneumoniae*. (**A**) Growth inhibition of 118 clinical isolates of *K. pneumoniae* subjected to antimicrobial photodynamic inactivation (aPDI) with PSRe-µL1 and PSRe-µL2 compounds compared to control, PS-Ru. The bacteria were utilized at 1×10^7 CFU/mL and mixed in triplicate with 4 µg/mL of PSRe-µL1 and PSRe-µL2 compounds. For the aPDI, the mixture of bacteria with the PSs was exposed for 1 h at 17 µW/cm^2 of white light. As a control, bacteria combined with the PSs not exposed to light and bacteria not combined with the control PS-Ru were included. Colony count enumerated of viable bacteria on ca-MH agar after serial micro-dilution. The CFUs/mL values are presented as means ± SD on a log$_{10}$ scale. Not significant (ns) $p > 0.05$ by Student's *t*-test among treated bacteria compared to control; ** $p < 0.01$, *** $p < 0.001$ by Student's *t*-test among treated bacteria compared to control.

3.6. Synergism between aPDI with Cefotaxime and FIC Index Determination

The fractional inhibitory concentration (FIC) index was determined to verify when the PS and Cfx combination increases the bactericidal activity synergistically or additively. We used the MEC determination for the photosensitizer compounds, and the results of the mixture with antibiotics were tabulated as the MIC for simplicity. The set of 66 ESBL-producing clinical isolates of *K. pneumoniae* [42] were mixed with 4 µg/mL PSRe-µL1, PSRe-µL2, or PS-Ru and added to a serial log2 dilutions of Cfx (from 0 to 32 µg/mL). The mixes were incubated for 1 h for aPDI or in the dark, and the Cfx-MIC was determined 16–20 h after. As seen in Figure 6A, compared to the aPDI untreated group, a significant reduction (*** $p < 0.001$) from 8 µg/mL (8–8) to 0.17 µg/mL (0.09–0.34) with PSRe-µL1 and 0.23 µg/mL (0.15–0.41) with PSRe-µL2 on Cfx-MIC was observed (Table 2). In addition, the combined treatment also reduced the PSRe-µL1-MEC from 4 to 0.5 µg/mL and the PSRe-µL2 MEC from 4 to 0.5 µg/mL (Figure 6B, Table 2). Both PSRe-µL1 and PSRe-µL2 compounds produced a significant change in Cfx-susceptibility with an FIC index of 0.15 for (Table 2). Figure 6B and Table 2 shows that the control compound, PS-Ru, did not significantly change Cfx-susceptibility and shows an FIC index of 1.58. Given that synergy is defined with an FIC index ≤ 0.5 [44], the increase in the inhibitory effect shown by the combination of PSRe-µL1 or PSRe-µL2 with Cfx is synergistic and not additive.

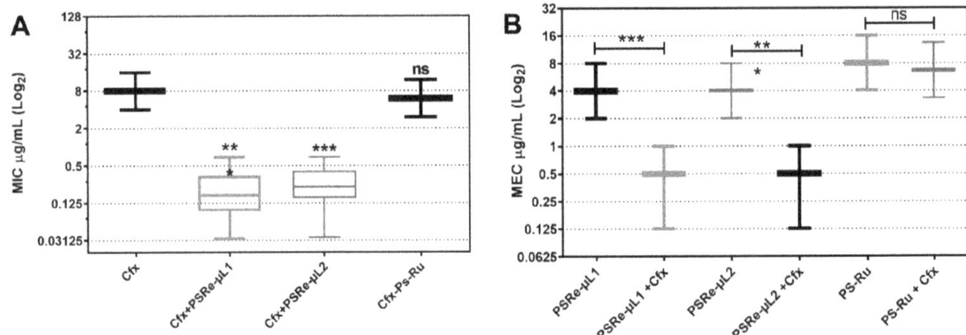

Figure 6. Determination of FIC index using combined photodynamic inactivation and cefotaxime antibiotic. A population of 66 ESBL-producing clinical isolates of *K. pneumoniae* was used to determine the modification on Cfx-MIC by the PSs and the modification in PSs-MECs by cefotaxime. To determine the Cfx-MIC modification, 1×10^7 CFU/mL of bacteria were mixed in triplicate with 4 µg/mL of PSRe-µL1, PSRe-µL2, or PS-Ru. The mix was added to serial dilutions of Cfx and incubated for 1 h with 17 µW/cm^2 of white light or in the dark. The Cfx-MIC was then determined after 16–20 h at 37 °C in the dark (**A**). The MEC for PSRe-µL1 and PSRe-µL2 were determined in combination with 4 µg/mL of Cfx, performed in triplicate in ca-MH agar for 16–20 h (**B**). The MIC values are presented as median ± SD of µg/mL on a log$_2$ scale. Not significant (ns) $p > 0.05$ by Student's *t*-test among treated bacteria compared to control; * $p < 0.05$, ** $p < 0.01$, *** $p < 0.001$ by Student's *t*-test among treated bacteria compared to control.

Table 2. FIC index calculation.

Compounds	[a] MIC (µg/mL)	MIC Combined (µg/mL)	FIC	FIC Index
Cfx	8.00			
PSRe-µL1	4.00	0.17	0.02	0.15
PSRe-µL1 *		0.50	0.13	
PSRe-µL2	4.00	0.23	0.03	0.15
PSRe-µL2 *		0.50	0.13	
PS-Ru	8.00	6.00	0.75	1.58
PS-Ru *		6.67	0.83	

[a] MIC values are the median for the ESBL-producing *K. pneumoniae*, *n* = 66. * The PS-MEC values modified by Cfx.

3.7. The Cytotoxic Effect of PSRe-µL1 and PSRe-µL2 on Mammalian Cells

The PSRe-µL1 and PSRe-µL2 compounds were tested in vitro and found to be secure for mammalian cells. This work tested the intrinsic cytotoxicity (dark cytotoxicity) and light-dependent cytotoxicity of the PSs compounds in the human HEp-2 and HEK293 cell lines. The trypan blue exclusion technique allowed us to determine cell death of 500,000 cells in the presence of 4 µg/mL PSRe-µL1 or PSRe-µL2. The cells were incubated with PSs for 1 h in the dark or activated with 17 µW/cm^2 of white LED light when the PSs were removed. Fresh medium DMEM with 10% FBS without PS was replaced, and cells were incubated 24 h more in the dark at 37 °C in a 5% CO$_2$ atmosphere. As shown in Figure 7, when HEp-2 and HEK293T cells were exposed to 4 µg/mL of PSRe-µL1 or PSRe-µL2 in the dark, no significant reduction in the cell survival was observed (ns = $p > 0.05$, Student's *t*-test, comparing treated cells with the control). Similarly, the light exposition induced no significant cell death for HEp-2 cells, although a slight (12.5 ± 5%) but significant ($p > 0.05$) reduction in cell viability of PSRe-µL1 over HEK293T cells were shown.

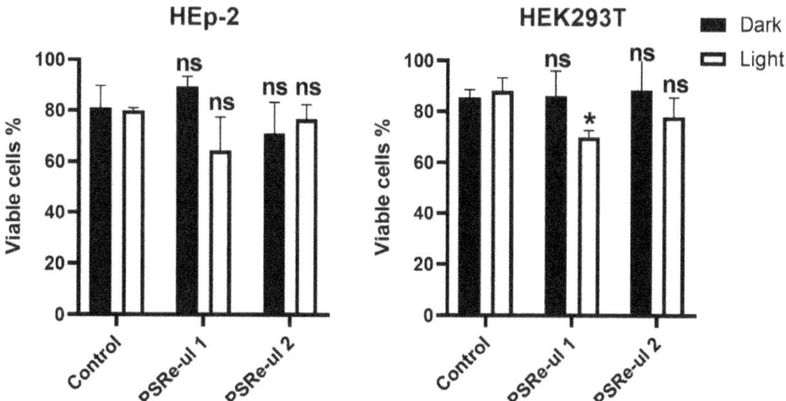

Figure 7. Cytotoxicity of the PSRe-µL1 and PSRe-µL2 compounds. Survival of 5×10^5 cells of HEp-2 and HEK293 human cell lines exposed to 4µg/mL of each PSRe-µL1 and PSRe-µL2 compounds were normalized to control untreated cells and expressed as a percentage of dead cells. Dark cytotoxicity was evaluated with cells exposed to the compound but not activated with light. Light-dependent cytotoxicity was evaluated, exposing the cells for 1 h at 17 µW/cm^2 of white LED light. Not significant (ns) $p > 0.05$, by Student's *t*-test between cells exposed to each PS compared to control cells; * $p < 0.05$ by Student's *t*-test cells exposed to each PS compared to control cells.

4. Discussion

Previously, we reported photosensitizing compounds based on Ir (III), which exhibited absorption processes close to 400 nm wavelength [29]. Therefore, although those compounds are microbicide in vitro, they may be more challenging to access exciting light into the tissues [11]. In this sense, this work tested Re(I) PS compounds that exhibit absorption processes at longer wavelengths, improving the possibility to access excited states into the tissues. Considering the sustained decrease in antibiotic options, these SPs could greatly complement the treatment of infections with MDR microorganisms such as *K. pneumoniac* [1,3]. In this context, using aPDI as complementary therapy becomes viable due to their usefulness as a rescue therapy for infections with MDR bacteria and reverse resistance to antibiotics of choice, reducing the spread of MDR strains [50].

The effective microbial activity of the PSRe-µL1 and PSRe-µL2 photosensitizers rests on the photophysical properties provided by their chemical structure. Although these compounds are neutral (do not have positive charges in the coordination sphere), they can easily polarize by absorption of light, as described in a previous report [41]. This characteristic may improve their molecular proximity to the *K. pneumoniae* cell envelope. Additionally, by exciting these high-rate light-absorbing dark compounds, they can produce high-energy triplet states, which promote energy transfer to molecular oxygen [51]. At first glance, the antimicrobial effect observed with PSRe-µL1 and PSRe-µL2 can be explained by access to a triplet state, as reported [41]. It will then depend on the nature of the Re (I) complex; the excited state responsible for the generation of the ROS producing photooxidative stress could be states ^{3}MLCT, ^{3}LC, or a mixture of both states [52]. Although aPDI performance is similar for both PSs, the PSRe-µL2 requires less time exposition than PSRe-µL1 to get the same effect. This difference could be associated with the higher coefficients of molar extinction presented for the PSRe-µL2 compound. The difference in the chemical structure of the L2 polypyridine ligand involving more complex hydrocarbon chains may be related [52]. A more extensive photophysical characterization should be carried out to corroborate these hypotheses, such as determining the bacterial cell envelope damage by TEM or the production of reactive oxygen species.

At concentrations as low as 4 µg/mL, both the PSRe-µL1 and the PSRe-µL2 compounds showed an effective aPDI activity, inhibiting the growth of *K. pneumoniae* (>3log$_{10}$).

These results are comparable to using cyclometalated Ir (III) complexes as aPDI, as previously reported [23,29]. Similar values have been reported for other PS compounds against Gram-negative bacteria [53–56]. When combined with imipenem, the PSRe-μL1 and PSRe-μL2 compounds produced a similar effect, shown previously by Ir(III) based cyclometalated compounds [29] and other photosensitizers such as rose bengal for *Acinetobacter baumannii* [57]. Similarly, alternative compounds such as anti-biofilm peptides mixed with conventional antibiotics reported good antimicrobial activity in an in vivo model [58]. This behavior could be related, as was mentioned before, to the external chemical structures of the polypyridine ligands. In the Re(I)-PSRs, polypyridine ligands have heteroatoms, allowing dipole–dipole interactions that may weaken the bacterial cell envelope facilitating the β-lactam action. The synergism shown by these compounds with β-lactams suggests that they damage the bacterial cell wall structure. Oxidative stress is known to modify the membrane permeability of *K. pneumoniae* [59]. However, at the moment, we have not carried out experiments, such as transmission microscopy, to confirm this idea. Our results also show the capacity of the Re(I)-PSRs compounds to inhibit the bacterial growth of a collection of 118 clinical isolates of *K. pneumoniae*. We used this collection of strains because its antibiotic resistance is characterized and presents a well-established sub-population of ESBL-producing bacteria [42]. The aPDI was not only effective; the FIC also demonstrates the synergic combination with Cfx over the total population and a significant reduction in Cfx-MIC in an ESBL-producers subpopulation. The synergistic effect exhibited by these compounds is one of their most remarkable qualities. The use of photodynamic therapy may resolve the lost susceptibility of MDR bacteria. Therefore, photodynamic therapy combined with more conventional antibiotics avoids the use of rescue therapy [50,57]. Similar to Cfx-MIC being reduced in combination, PSs-MEC was also reduced, indicating that lower concentration antibiotic/PS regimens could be used effectively.

The antimicrobial activity of the PSRe-μL1 and PSRe-μL2 is photodynamic and, therefore, its activation is dependent on light. The dependence on light suggests that the PS compounds themselves and in the dark are not toxic, as the results with mammalian cells show us. The slight but significant light-dependent cytotoxicity exhibited by PSRe-μL1 on HEK293T cells reinforces the low intrinsic toxicity. However, it may imply potential damage to the host tissues that must be considered and prevented with a better characterization of the exposure times and concentration of the compound. Furthermore, the cytotoxicity shown by the PSRe-μL1 and PSRe-μL2 compounds is of low significance compared to that shown by the antitumor photosensitizer compounds [60]. However, these compounds must be tested in vivo to establish whether a significant cytotoxic effect may arise, such as an anaphylactic reaction [61]. A murine model may then probe if these compounds could treat infectious diseases in vivo.

For now, it is difficult to accurately calculate the dose of light necessary to activate these PSRe compounds fully; however, we observed in vitro that with photon flux as low as 17 μW/cm² of white light, they are bactericidal. Compounds with optimum absorbance at higher wavelengths ranging from 450–750 nm would improve the exposure of PS to light into the tissues, but being less energetic, the PSs must have a triplet excited state of easy access to promote energy transfer [62]. The dark color of PSRe-μL1 and PSRe-μL2 may imply that light of different wavelengths can also be absorbed by this Re-PS, increasing the possibility for it to be excited into the tissues [51]. Therefore, we need to characterize its antimicrobial activity better when activated with defined wavelengths before starting in vivo studies in infection models [63].

5. Conclusions

The present study shows that the increased MICs of resistant bacteria could be reverted by aPDI, turning resistant strains into susceptible ones. Thus, aPDI would effectively treat MDR bacteria, greatly complementing antibiotic therapy. Furthermore, therapeutic regimens with lower doses of antibiotics and PS can be used effectively due to the synergistic effect. The requirement for lower doses of antibiotics will help reduce the generation of resistance. In this work, two bimetallic Re(I) compounds were tested as PSs to be used in the aPDI. According to their UV-vis absorption characterization, it was identified that the absorptions at lower energies occur at wavelengths higher than 450 nm, improving its use in infections compared to the PSs based on Ir(III), previously studied. The dark quality in both powder solids and liquid solutions of the Re(I) compounds may imply that light of a wide wavelength range can be absorbed [51], increasing the possibility of it being excited into the tissues. Although its low excitation at more penetrating wavelengths, during infections of internal organs such as UTI, optical fibers can be used through a catheter to deliver the required light dose [64,65]. Furthermore, the bactericidal activity of these PSs compounds occurred at similar concentrations to those used in antibiotic therapies (CLSI 2017). At these concentrations, the PSs showed no dark cytotoxicity or low light-dependent toxicity on mammalian cells.

Author Contributions: Conceptualization, C.E.P. and I.A.G.; methodology, C.E.P. and I.A.G.; software, C.E.P. and I.A.G.; validation, C.E.P., P.D., S.M.B. and I.A.G.; formal analysis, C.E.P., P.D., S.M.B. and I.A.G.; investigation, C.N., A.P., and F.M.-G.; resources, C.E.P., P.D., S.M.B. and I.A.G.; data curation, C.E.P., S.M.B. and I.A.G.; writing—original draft preparation, C.E.P., P.D. and I.A.G.; writing—review and editing, C.E.P., P.D. and I.A.G.; visualization, C.E.P. and I.A.G.; supervision, C.E.P., S.M.B. and I.A.G.; project administration, C.E.P.; funding acquisition, C.E.P., S.M.B., P.D. and I.A.G. All authors have read and agreed to the published version of the manuscript.

Funding: Authors are supported by grants: UCEN grants CIP2019007 (awarded to C.E.P.), ANID-Fondecyt grant N° 11180185 (awarded to I.A.G.), 11200764 (awarded to F.M.-G.), 1201173 (awarded to P.D.), PI_M_2020_31 USM project (awarded to P.D.), Millennium Institute of Immunology and Immunotherapy P09/016-F, ICN09_016, (awarded to S.B.).

Institutional Review Board Statement: The study was conducted according to the guidelines of the Declaration of Helsinki and approved by the Institutional Ethics Committee of the Faculty of Health Sciences of the Universidad Central de Chile N°: 01/2017 approved on 01-04-2017.

Informed Consent Statement: The informed consent form was approved by the ethics committee of the Faculty of Health Sciences of the Central University of Chile and by the Bioethics Committee of the Central Metropolitan Health Service of Chile (MHSC) act number: N° 124/07.

Data Availability Statement: The data presented in this study are available on request from the corresponding author. The data are not publicly available because they are confidential data of patients protected by the informed consent protocol.

Acknowledgments: We would like to thank Myriam Pizarro, head of the "Hospital El Carmen de Maipú" laboratory, for her collaboration and provision of the clinical isolates of Klebsiella pneumoniae. We also would like to thank Susan Bueno from Pontificia Universidad Católica de Chile for providing the control strains of *K. pneumoniae*, KPPR1, and the typo strain ST258.

Conflicts of Interest: The authors declare no conflict of interest.

Ethical Approval: The study protocol was approved by the ethics committee of the Faculty of Health Sciences of the Central University of Chile and by the Bioethics Committee of the Central Metropolitan Health Service of Chile (MHSC), Act number: N° 124/07.

References

1. Liu, Y.-Y.; Wang, Y.; Walsh, T.; Yi, L.-X.; Zhang, R.; Spencer, J.; Doi, Y.; Tian, G.; Dong, B.; Huang, X.; et al. Emergence of plasmid-mediated colistin resistance mechanism MCR-1 in animals and human beings in China: A microbiological and molecular biological study. *Lancet Infect. Dis.* **2016**, *16*, 161–168. [CrossRef]
2. WHO. WHO Publishes List of Bacteria for which New Antibiotics are Urgently Needed. Available online: https://www.who.int/news/item/27-02-2017-who-publishes-list-of-bacteria-for-which-new-antibiotics-are-urgently-needed (accessed on 19 April 2021).
3. Willyard, C. The drug-resistant bacteria that pose the greatest health threats. *Nat. Cell Biol.* **2017**, *543*, 15. [CrossRef] [PubMed]
4. Podschun, R.; Ullmann, U. Klebsiella spp. as Nosocomial Pathogens: Epidemiology, Taxonomy, Typing Methods, and Pathogenicity Factors. *Clin. Microbiol. Rev.* **1998**, *11*, 589–603. [CrossRef] [PubMed]

5. Martin, R.M.; Bachman, M.A. Colonization, Infection, and the Accessory Genome of Klebsiella pneumoniae. *Front. Cell. Infect. Microbiol.* **2018**, *8*, 4. [CrossRef]

6. Paczosa, M.K.; Mecsas, J. Klebsiella pneumoniae: Going on the Offense with a Strong Defense. *Microbiol. Mol. Biol. Rev.* **2016**, *80*, 629–661. [CrossRef]

7. Tofas, P.; Skiada, A.; Angelopoulou, M.; Sipsas, N.; Pavlopoulou, I.; Tsaousi, S.; Pagoni, M.; Kotsopoulou, M.; Perlorentzou, S.; Antoniadou, A.; et al. Carbapenemase-producing Klebsiella pneumoniae bloodstream infections in neutropenic patients with haematological malignancies or aplastic anaemia: Analysis of 50 cases. *Int. J. Antimicrob. Agents* **2016**, *47*, 335–339. [CrossRef]

8. Daikos, G.L.; Markogiannakis, A.; Souli, M.; Tzouvelekis, L.S. Bloodstream infections caused by carbapenemase-producingKlebsiella pneumoniae: A clinical perspective. *Expert Rev. Anti-Infect. Ther.* **2012**, *10*, 1393–1404. [CrossRef] [PubMed]

9. Pragasam, A.K.; Shankar, C.; Veeraraghavan, B.; Biswas, I.; Nabarro, L.E.B.; Inbanathan, F.Y.; George, B.; Verghese, S. Molecular Mechanisms of Colistin Resistance in Klebsiella pneumoniae Causing Bacteremia from India—A First Report. *Front. Microbiol.* **2017**, *7*, 2135. [CrossRef]

10. Jiang, Y.; Zhang, Y.; Lu, J.; Wang, Q.; Cui, Y.; Wang, Y.; Quan, J.; Zhao, D.; Du, X.; Liu, H.; et al. Clinical relevance and plasmid dynamics of mcr-1-positive Escherichia coli in China: A multicentre case-control and molecular epidemiological study. *Lancet Microbe* **2020**, *1*, e24–e33. [CrossRef]

11. Agostinis, P.; Berg, K.; Cengel, K.A.; Foster, T.H.; Girotti, A.W.; Gollnick, S.O.; Hahn, S.M.; Hamblin, M.R.; Juzeniene, A.; Kessel, D.; et al. Photodynamic therapy of cancer: An update. *CA Cancer J. Clin.* **2011**, *61*, 250–281. [CrossRef] [PubMed]

12. Trindade, A.C.; De Figueiredo, J.A.P.; De Oliveira, S.D.; Barth, V.C., Jr.; Gallo, S.W.; Follmann, C.; Wolle, C.F.B.; Steier, L.; Morgental, R.D.; Weber, J.B.B. Histopathological, Microbiological, and Radiographic Analysis of Antimicrobial Photodynamic Therapy for the Treatment of Teeth with Apical Periodontitis: A Study in Rats' Molars. *Photomed. Laser Surg.* **2017**, *35*, 364–371. [CrossRef] [PubMed]

13. Valenzuela-Valderrama, M.; González, I.A.; Palavecino, C.E. Photodynamic treatment for multidrug-resistant Gram-negative bacteria: Perspectives for the treatment of Klebsiella pneumoniae infections. *Photodiagnosis Photodyn. Ther.* **2019**, *28*, 256–264. [CrossRef] [PubMed]

14. Cieplik, F.; Deng, D.; Crielaard, W.; Buchalla, W.; Hellwig, E.; Al-Ahmad, A.; Maisch, T. Antimicrobial photodynamic therapy—what we know and what we don't. *Crit. Rev. Microbiol.* **2018**, *44*, 571–589. [CrossRef]

15. Winterbourn, C.C.; Kettle, A.J. Redox Reactions and Microbial Killing in the Neutrophil Phagosome. *Antioxid. Redox Signal.* **2013**, *18*, 642–660. [CrossRef]

16. Alvarez, L.; Kovačič, L.; Rodriguez, J.; Gosemann, J.-H.; Kubica, M.; Pircalabioru, G.G.; Friedmacher, F.; Cean, A.; Ghişe, A.; Sărăndan, M.B.; et al. NADPH oxidase-derived H2O2 subverts pathogen signaling by oxidative phosphotyrosine conversion to PB-DOPA. *Proc. Natl. Acad. Sci. USA* **2016**, *113*, 10406–10411. [CrossRef]

17. Briviba, K.; Klotz, L.-O.; Sies, H. Toxic and signaling effects of photochemically or chemically generated singlet oxygen in biological systems. *Biol. Chem.* **1997**, *378*, 1259–1265.

18. Tim, M. Strategies to optimize photosensitizers for photodynamic inactivation of bacteria. *J. Photochem. Photobiol. B Biol.* **2015**, *150*, 2–10. [CrossRef] [PubMed]

19. Di Mascio, P.; Martinez, G.R.; Miyamoto, S.; Ronsein, G.E.; Medeiros, M.H.G.; Cadet, J. Singlet Molecular Oxygen Reactions with Nucleic Acids, Lipids, and Proteins. *Chem. Rev.* **2019**, *119*, 2043–2086. [CrossRef]

20. Hu, X.; Huang, Y.-Y.; Wang, Y.; Wang, X.; Hamblin, M.R. Antimicrobial Photodynamic Therapy to Control Clinically Relevant Biofilm Infections. *Front. Microbiol.* **2018**, *9*, 1299. [CrossRef]

21. Agnez-Lima, L.; Melo, J.T.; Silva, A.E.; Oliveira, A.H.S.; Timoteo, A.R.S.; Lima-Bessa, K.M.; Martinez, G.R.; de Medeiros, M.H.G.; Di Mascio, P.; Galhardo, R.; et al. DNA damage by singlet oxygen and cellular protective mechanisms. *Mutat. Res. Mutat. Res.* **2012**, *751*, 15–28. [CrossRef]

22. Mulani, M.S.; Kamble, E.; Kumkar, S.N.; Tawre, M.S.; Pardesi, K.R. Emerging Strategies to Combat ESKAPE Pathogens in the Era of Antimicrobial Resistance: A Review. *Front. Microbiol.* **2019**, *10*, 539. [CrossRef]

23. Valenzuela-Valderrama, M.; Carrasco-Véliz, N.; González, I.A.; Dreyse, P.; Palavecino, C.E. Synergistic effect of combined imipenem and photodynamic treatment with the cationic Ir(III) complexes to polypyridine ligand on carbapenem-resistant Klebsiella pneumoniae. *Photodiagnosis Photodyn. Ther.* **2020**, *31*, 101882. [CrossRef]

24. Pérez, C.; Zúñiga, T.; Palavecino, C.E. Photodynamic therapy for treatment of Staphylococcus aureus infections. *Photodiagnosis Photodyn. Ther.* **2021**, *34*, 102285. [CrossRef] [PubMed]

25. DeRosa, M. Photosensitized singlet oxygen and its applications. *Coord. Chem. Rev.* **2002**, *233–234*, 351–371. [CrossRef]

26. Wilkinson, F.; Helman, W.P.; Ross, A.B. Quantum Yields for the Photosensitized Formation of the Lowest Electronically Excited Singlet State of Molecular Oxygen in Solution. *J. Phys. Chem. Ref. Data* **1993**, *22*, 113–262. [CrossRef]

27. Ogilby, P.R. Singlet oxygen: There is indeed something new under the sun. *Chem. Soc. Rev.* **2010**, *39*, 3181–3209. [CrossRef] [PubMed]

28. Foote, C.S. Definition of Type I and Type Ii Photosensitized Oxidation. *Photochem. Photobiol.* **1991**, *54*, 659. [CrossRef]

29. Valenzuela-Valderrama, M.; Bustamante, V.; Carrasco, N.; González, I.A.; Dreyse, P.; Palavecino, C.E. Photodynamic treatment with cationic Ir(III) complexes induces a synergistic antimicrobial effect with imipenem over carbapenem-resistant Klebsiella pneumoniae. *Photodiagnosis Photodyn. Ther.* **2020**, *30*, 101662. [CrossRef]

30. Dai, T.; Huang, Y.-Y.; Hamblin, M.R. Photodynamic therapy for localized infections—State of the art. *Photodiagn. Photodyn. Ther.* **2009**, *6*, 170–188. [CrossRef]
31. Miretti, M.; Clementi, R.; Tempesti, T.C.; Baumgartner, M.T. Photodynamic inactivation of multiresistant bacteria (KPC) using zinc(II)phthalocyanines. *Bioorganic Med. Chem. Lett.* **2017**, *27*, 4341–4344. [CrossRef] [PubMed]
32. Bustamante, V.; González, I.A.; Dreyse, P.; Palavecino, C.E. The mode of action of the PSIR-3 photosensitizer in the photodynamic inactivation of Klebsiella pneumoniae is by the production of type II ROS which activate RpoE-regulated extracytoplasmic factors. *Photodiagnosis Photodyn. Ther.* **2020**, *32*, 102020. [CrossRef] [PubMed]
33. Liu, C.; Zhou, Y.; Wang, L.; Han, L.; Lei, J.; Ishaq, H.M.; Nair, S.P.; Xu, J. Photodynamic inactivation of Klebsiella pneumoniae biofilms and planktonic cells by 5-aminolevulinic acid and 5-aminolevulinic acid methyl ester. *Lasers Med. Sci.* **2016**, *31*, 557–565. [CrossRef]
34. Edwards, L.; Turner, D.; Champion, C.; Khandelwal, M.; Zingler, K.; Stone, C.; Rajapaksha, R.; Yang, J.; Ranasinghe, M.; Kornienko, A.; et al. Photoactivated 2,3-distyrylindoles kill multi-drug resistant bacteria. *Bioorganic Med. Chem. Lett.* **2018**, *28*, 1879–1886. [CrossRef]
35. Marković, Z.M.; Kováčová, M.; Humpolíček, P.; Budimir, M.D.; Vajd'ák, J.; Kubát, P.; Mičušík, M.; Švajdlenková, H.; Danko, M.; Capáková, Z.; et al. Antibacterial photodynamic activity of carbon quantum dots/polydimethylsiloxane nanocomposites against Staphylococcus aureus, Escherichia coli and Klebsiella pneumoniae. *Photodiagnosis Photodyn. Ther.* **2019**, *26*, 342–349. [CrossRef] [PubMed]
36. Abrahamse, H.; Hamblin, M.R. New photosensitizers for photodynamic therapy. *Biochem. J.* **2016**, *473*, 347–364. [CrossRef]
37. Wang, L.; Yin, H.; Jabed, M.A.; Hetu, M.; Wang, C.; Monro, S.; Zhu, X.; Kilina, S.; McFarland, S.A.; Sun, W. π-Expansive Heteroleptic Ruthenium(II) Complexes as Reverse Saturable Absorbers and Photosensitizers for Photodynamic Therapy. *Inorg. Chem.* **2017**, *56*, 3245–3259. [CrossRef]
38. Li, F.; Collins, J.G.; Keene, F.R. Ruthenium complexes as antimicrobial agents. *Chem. Soc. Rev.* **2015**, *44*, 2529–2542. [CrossRef] [PubMed]
39. Viganor, L.; Howe, O.; McCarron, P.; McCann, M.; Devereux, M. The Antibacterial Activity of Metal Complexes Containing 1,10- phenanthroline: Potential as Alternative Therapeutics in the Era of Antibiotic Resistance. *Curr. Top. Med. Chem.* **2017**, *17*, 1280–1302. [CrossRef]
40. Fernández-Moreira, V.; Thorp-Greenwood, F.L.; Coogan, M.P. Application of d6 transition metal complexes in fluorescence cell imaging. *Chem. Commun.* **2010**, *46*, 186–202. [CrossRef] [PubMed]
41. González, I.; Cortés-Arriagada, D.; Dreyse, P.; Sanhueza, L.; Crivelli, I.; Ngo, H.M.; Ledoux-Rak, I.; Toro-Labbe, A.; Maze, J.; Loeb, B. Studies on the solvatochromic effect and NLO response in new symmetric bimetallic Rhenium compounds. *Polyhedron* **2020**, *187*, 114679. [CrossRef]
42. Núñez, C.; Palavecino, A.; González, I.; Dreyse, P.; Palavecino, C. Effective Photodynamic Therapy with Ir(III) for Virulent Clinical Isolates of Extended-Spectrum Beta-Lactamase *Klebsiella pneumoniae*. *Pharmaceutics* **2021**, *13*, 603. [CrossRef]
43. CLSI. *Performance Standards for Antimicrobial Susceptibility Testing*, 27th ed.; Clinical and Laboratory Standards Institute: Wayne, PA, USA, 2017.
44. Stokes, J.M.; MacNair, C.R.; Ilyas, B.; French, S.; Côté, J. P.; Bouwman, C.; Farha, M.A.; Sieron, A.O.; Whitfield, C., Coombes, B.K.; et al. Pentamidine sensitizes Gram-negative pathogens to antibiotics and overcomes acquired colistin resistance. *Nat. Microbiol.* **2017**, *2*, 1–8. [CrossRef]
45. Hall, M.J.; Middleton, R.F.; Westmacott, D. The fractional inhibitory concentration (FIC) index as a measure of synergy. *J. Antimicrob. Chemother.* **1983**, *11*, 427–433. [CrossRef]
46. González, I.; Natali, M.; Cabrera, A.R.; Loeb, B.; Maze, J.; Dreyse, P. Substituent influence in phenanthroline-derived ancillary ligands on the excited state nature of novel cationic Ir(iii) complexes. *N. J. Chem.* **2018**, *42*, 6644–6654. [CrossRef]
47. Campagna, S.P.F.; Nastasi, F.; Bergamini, G.; Balzani, V. Photochemistry and Photophysics of Coordination Compounds: Ruthenium. In *Photochemistry and Photophysics of Coordination Compounds I*; Balzani, V., Campagna, C., Eds.; Springer: Berlin/Heidelberg, Germany, 2007; Volume 280, pp. 117–214.
48. Ishida, H.; Tobita, S.; Hasegawa, Y.; Katoh, R.; Nozaki, K. Recent advances in instrumentation for absolute emission quantum yield measurements. *Coord. Chem. Rev.* **2010**, *254*, 2449–2458. [CrossRef]
49. Wachter, E.; Heidary, D.K.; Howerton, B.S.; Parkin, S.; Glazer, E.C. Light-activated ruthenium complexes photobind DNA and are cytotoxic in the photodynamic therapy window. *Chem. Commun.* **2012**, *48*, 9649–9651. [CrossRef] [PubMed]
50. Bassetti, M.; Righi, E. New antibiotics and antimicrobial combination therapy for the treatment of gram-negative bacterial infections. *Curr. Opin. Crit. Care* **2015**, *21*, 402–411. [CrossRef] [PubMed]
51. Braun, J.D.; Lozada, I.B.; Herbert, D.E. In Pursuit of Panchromatic Absorption in Metal Coordination Complexes: Experimental Delineation of the HOMO Inversion Model Using Pseudo-Octahedral Complexes of Diarylamido Ligands. *Inorg. Chem.* **2020**, *59*, 17746–17757. [CrossRef]
52. Peng, X.; Chen, H.; Draney, D.R.; Volcheck, W.; Schutz-Geschwender, A.; Olive, D.M. A nonfluorescent, broad-range quencher dye for Förster resonance energy transfer assays. *Anal. Biochem.* **2009**, *388*, 220–228. [CrossRef] [PubMed]
53. Maisch, T.; Eichner, A.; Späth, A.; Gollmer, A.; Koenig, B.; Regensburger, J.; Bäumler, W. Fast and Effective Photodynamic Inactivation of Multiresistant Bacteria by Cationic Riboflavin Derivatives. *PLoS ONE* **2014**, *9*, e111792. [CrossRef]

54. Tortora, G.; Orsini, B.; Pecile, P.; Menciassi, A.; Fusi, F.; Romano, G. An Ingestible Capsule for the Photodynamic Therapy of Helicobacter Pylori Infection. *IEEE/ASME Trans. Mechatron.* **2016**, *21*, 1935–1942. [CrossRef]
55. Wikene, K.O.; Rukke, H.V.; Bruzell, E.; Tønnesen, H.H. Investigation of the antimicrobial effect of natural deep eutectic solvents (NADES) as solvents in antimicrobial photodynamic therapy. *J. Photochem. Photobiol. B Biol.* **2017**, *171*, 27–33. [CrossRef]
56. De Mello, M.M.; De Barros, P.P.; Bernardes, R.D.C.; Alves, S.R.; Ramanzini, N.P.; Figueiredo-Godoi, L.M.A.; Prado, A.C.C.; Jorge, A.O.C.; Junqueira, J.C. Antimicrobial photodynamic therapy against clinical isolates of carbapenem-susceptible and carbapenem-resistant Acinetobacter baumannii. *Lasers Med. Sci.* **2019**, *34*, 1755–1761. [CrossRef]
57. Woźniak, A.; Rapacka-Zdonczyk, A.; Mutters, N.T.; Grinholc, M. Antimicrobials Are a Photodynamic Inactivation Adjuvant for the Eradication of Extensively Drug-Resistant Acinetobacter baumannii. *Front. Microbiol.* **2019**, *10*, 229. [CrossRef]
58. Pletzer, D.; Mansour, S.C.; Hancock, R.E.W. Synergy between conventional antibiotics and anti-biofilm peptides in a murine, sub-cutaneous abscess model caused by recalcitrant ESKAPE pathogens. *PLoS Pathog.* **2018**, *14*, e1007084. [CrossRef]
59. Yang, S.-K.; Yusoff, K.; Thomas, W.; Akseer, R.; Alhosani, M.S.; Abushelaibi, A.; Lim, E.; Lai, K.-S. Lavender essential oil induces oxidative stress which modifies the bacterial membrane permeability of carbapenemase producing Klebsiella pneumoniae. *Sci. Rep.* **2020**, *10*, 819. [CrossRef]
60. Yang, Y.-J.; Qi, S.-N.; Shi, R.-Y.; Yao, J.; Wang, L.-S.; Yuan, H.-Q.; Jing, Y.-X. Induction of apoptotic DNA fragmentation mediated by mitochondrial pathway with caspase-3-dependent BID cleavage in human gastric cancer cells by a new nitroxyl spin-labeled derivative of podophyllotoxin. *Biomed. Pharmacother.* **2017**, *90*, 131–138. [CrossRef]
61. Anand, S.; Yasinchak, A.; Bullock, T.; Govande, M.; Maytin, E.V. A non-toxic approach for treatment of breast cancer and its metastases: Capecitabine enhanced photodynamic therapy in a murine breast tumor model. *J. Cancer Metastasis Treat.* **2019**, *5*, 6. [CrossRef] [PubMed]
62. Kachynski, A.V.; Pliss, A.; Kuzmin, A.N.; Ohulchanskyy, T.; Baev, A.; Qu, J.; Prasad, P.N. Photodynamic therapy by in situ nonlinear photon conversion. *Nat. Photon.* **2014**, *8*, 455–461. [CrossRef]
63. Páez, P.L.; Bazán, C.M.; Bongiovanni, M.E.; Toneatto, J.; Albesa, I.; Becerra, M.C.; Argüello, G.A. Oxidative Stress and Antimicrobial Activity of Chromium(III) and Ruthenium(II) Complexes onStaphylococcus aureusandEscherichia coli. *BioMed Res. Int.* **2013**, *2013*, 906912. [CrossRef] [PubMed]
64. Bendsoe, N.; Svanberg, K.; Axelsson, J.; Andersson-Engels, S.; Svanberg, S. Photodynamic therapy: Superficial and interstitial illumination. *J. Biomed. Opt.* **2010**, *15*, 041502. [CrossRef] [PubMed]
65. Baran, T.M.; Foster, T. Comparison of flat cleaved and cylindrical diffusing fibers as treatment sources for interstitial photodynamic therapy. *Med. Phys.* **2014**, *41*, 22701. [CrossRef] [PubMed]

pharmaceutics

Article

Effective Photodynamic Therapy with Ir(III) for Virulent Clinical Isolates of Extended-Spectrum Beta-Lactamase *Klebsiella pneumoniae*

Constanza Núñez [1], Annegrett Palavecino [1], Iván A. González [2], Paulina Dreyse [3] and Christian Erick Palavecino [1,*]

[1] Laboratorio de Microbiología Celular, Instituto de Investigación y Postgrado, Facultad de Ciencias de la Salud, Universidad Central de Chile, Santiago 8330546, Chile; constanza.nunezc@alumnos.ucentral.cl (C.N.); Annegrett.palavecino@alumnos.ucentral.cl (A.P.)

[2] Laboratorio de Química Aplicada, Instituto de Investigación y Postgrado, Facultad de Ciencias de la Salud, Universidad Central de Chile, Santiago 8330546, Chile; ivan.gonzalez@ucentral.cl

[3] Departamento de Química, Universidad Técnica Federico Santa María, Av. España 1680, Casilla 2390123, Valparaíso 2390123, Chile; paulina.dreyse@usm.cl

* Correspondence: christian.palavecino@ucentral.cl; Tel.: +56-225851334

Abstract: Background: The extended-spectrum beta-lactamase (ESBL) *Klebsiella pneumoniae* is one of the leading causes of health-associated infections (HAIs), whose antibiotic treatments have been severely reduced. Moreover, HAI bacteria may harbor pathogenic factors such as siderophores, enzymes, or capsules, which increase the virulence of these strains. Thus, new therapies, such as antimicrobial photodynamic inactivation (aPDI), are needed. Method: A collection of 118 clinical isolates of *K. pneumoniae* was characterized by susceptibility and virulence through the determination of the minimum inhibitory concentration (MIC) of amikacin (Amk), cefotaxime (Cfx), ceftazidime (Cfz), imipenem (Imp), meropenem (Mer), and piperacillin–tazobactam (Pip–Taz); and, by PCR, the frequency of the virulence genes K2, magA, rmpA, entB, ybtS, and allS. Susceptibility to innate immunity, such as human serum, macrophages, and polymorphonuclear cells, was tested. All the strains were tested for sensitivity to the photosensitizer PSIR-3 (4 µg/mL) in a 17 µW/cm^2 for 30 min aPDI. Results: A significantly higher frequency of virulence genes in ESBL than non-ESBL bacteria was observed. The isolates of the genotype K2+, ybtS+, and allS+ display enhanced virulence, since they showed higher resistance to human serum, as well as to phagocytosis. All strains are susceptible to the aPDI with PSIR-3 decreasing viability in 3log10. The combined treatment with Cfx improved the aPDI to 6log10 for the ESBL strains. The combined treatment is synergistic, as it showed a fractional inhibitory concentration (FIC) index value of 0.15. Conclusions: The aPDI effectively inhibits clinical isolates of *K. pneumoniae*, including the riskier strains of ESBL-producing bacteria and the K2+, ybtS+, and allS+ genotype. The aPDI with PSIR-3 is synergistic with Cfx.

Keywords: antibiotic resistance; virulence factors; *Klebsiella pneumoniae*; photodynamic therapy

Citation: Núñez, C.; Palavecino, A.; González, I.A.; Dreyse, P.; Palavecino, C.E. Effective Photodynamic Therapy with Ir(III) for Virulent Clinical Isolates of Extended-Spectrum Beta-Lactamase *Klebsiella pneumoniae*. *Pharmaceutics* **2021**, *13*, 603. https://doi.org/10.3390/pharmaceutics13050603

Academic Editor: Udo Bakowsky

Received: 6 April 2021
Accepted: 15 April 2021
Published: 22 April 2021

Publisher's Note: MDPI stays neutral with regard to jurisdictional claims in published maps and institutional affiliations.

1. Introduction

Klebsiella pneumoniae is one of the major health-associated infection (HAIs) producers worldwide, including pneumonia, urinary tract, and bloodstream infections [1,2]. Moreover, HAI-producing *K. pneumoniae* strains progressively accumulate more multiple-drugs resistance (MDR), including extended-spectrum β-lactamases (ESBL) and carbapenemases, such as KPC [3–5]. Treatment options for infections caused by MDR strains of *K. pneumoniae* are severely reduced to colistin and tigecycline [6–8]. The high MDR shown by strains of *K. pneumoniae* does not fully explain its notable success as one of the most important agents of HAIs, suggesting the participation of other factors [9]. An increasing number of authors suggest that increased bacterial survival during infections is related to virulence factors [1,3,9–14].

Virulence factors shown by *Klebsiella pneumoniae* strains are common to enterobacteria, such as siderophores, enzymes, and capsules [9]. These factors may be part of genes conserved amongst all *K. pneumoniae*, called the core set of genes, but other genes vary in frequency between strains, and they are part of the accessory set of genes [9]. During colonization and infection, *K. pneumoniae* requires the expression of several core and accessory genes to deal with, for example, the nutritional stress due to nutrient sequestration or the immune response of the host [15]. For instance, during lung infection, the acquisition of iron is facilitated by the siderophores enterobactin (*ent*B) and yersiniabactin (*ybt*S) genes encoded on the bacterial chromosome [1,16]. Like iron, during infections of the lungs, urinary tract, and bloodstream, access to nitrogen is limited. The *all*S gene allows *K. pneumoniae* to use allantoin degradation as an alternative nitrogen source [17,18]. The polysaccharide capsule is involved, among others, in resistance to death by complement-mediated opsonophagocytosis [9], and inhibition of macrophages [19,20]. Some capsular varieties (such as K2) may produce hypervirulent strains with characteristic phenotypes such as the hypermucoviscous [19]. Some hypervirulent phenotypes are associated with the expression of the mucoviscosity-associated gene A (*mag*A) and the regulator of the mucoid phenotype A gene (*rmp*A) [21]. Moreover, the *rmp*A gene increases the ability of ESBL-producing strains to resist the bactericidal activity of serum and phagocytosis [22]. Since the virulence factors increase bacterial survival and may influence antibiotic resistance expression, the composition of the pool of the virulence gene of *K. pneumoniae* strains may be associated with the antibiotic susceptibility pattern [9,13].

Due to the emergence of multi-drug resistance (MDR), the deficit of new antibiotics is one of the most pressing threats to human health in the 21st century [23]. Given the high risk to public health caused by the deficiency of new antibiotics, the use of complementary antimicrobial therapies non-antibiotic-based, such as antibacterial photodynamic inactivation (aPDI), emerges as a promising alternative [24]. The aPDI may support the lack of antibiotic therapies against MDR and KPC strains [24,25]. The aPDI is a procedure based on the use of photosensitizer (PS) compounds to produce light-activated local cytotoxicity (photooxidative stress) [26]. The PSs work by energy absorbs of a specific wavelength in the UV–Vis range, which is then transferred to molecular oxygen in solution to produce reactive oxygen species (ROS) [27]. Molecular O_2 can accept this energy together with electrons, or it can only undergo a one-electron reduction to produce superoxide anion radical ($O_2^{\bullet-}$), hydrogen peroxide (H_2O_2), and hydroxyl radical ($HO^\bullet$) [28,29]. The energy transferred to the O_2 without one-electron reduction produces singlet oxygen ($^1O_2$) [27]. The ROS production generates photooxidative stress induced by aPDI, which occurs mainly due to the action of $^1O_2$. The oxidative action of $^1O_2$ occurs on organic molecules close to the PS through concerted addition reactions of alkene groups [30]. These organic molecules can be structural proteins or lipids of the bacterial envelope; therefore, the damage occurs to the cell wall plasma membrane or other bacterial structures, leading to nonspecific cell death [27,31]. Previously, a PS compound based on a polypyridine Ir(III) complex (PSIR-3, see Figure 4) demonstrated aPDI activity, inhibiting the bacterial growth of CKP and synergism with imipenem [32,33]. In this work, we determine the frequency of the K2, *ent*B, *ybt*S, *all*S, *rmp*A, and *mag*A virulence genes, in a population of 118 clinical isolates of *K. pneumoniae* and evaluate their association to the susceptibility pattern to several antibiotics. Our data show that virulence factors K2$^+$, *ybt*S$^+$, and *all*S$^+$ were associated with a modification in the bacterial minimum inhibitory concentrations (MICs) and increased ESBL production probability. These virulence genes significantly increased the survival of the bacteria, to innate immunity. These more virulent ESBL bacteria were all susceptible to the aPDI treatment with PSIR-3 and demonstrate synergism with cephotaxime.

2. Materials and Methods

2.1. Bacterial Isolates

The ethics committee of the Faculty of Health Sciences, Central University of Chile, and the Central Metropolitan Health Service of Chile (MHSC), Act number: N 124/07, approved the study protocol and the informed consent form. The clinical isolates of *K. pneumoniae* were obtained from 122 samples from unrelated patients, received in the bacteriology laboratory of "Hospital el Carmen" (HEC) for a period of six months (2017/2018). The HEC is a complex hospital with 412 beds and serving a population of more than 600,000 inhabitants. All clinical isolates were identified as *K. pneumoniae* following the protocols of the Institute of Clinical and Laboratory Standards (CLSI) [34]. As controls, the virulent strain (ATCC 43816 KPPR1) of susceptible *K. pneumoniae* (K2$^+$, *ybt*S$^+$, and *all*S$^+$) [35] and the MDR KPC$^+$ ST258 (KP35) (K2$^-$, *ybt*S$^-$ and *all*S$^-$) [36] strains were also included.

2.2. Antimicrobial Susceptibility Testing

The MIC of the antimicrobial agents were determined in 96-well plates by micro-dilution methodology in cations-adjusted Mueller-Hinton (ca-MHB) broth. Inoculum of 1×10^6 colony forming units (CFUs)/mL of each clinical isolate was mixed with decreasing concentrations of each antibiotic and incubated overnight at 37 °C following the CLSI recommendations. The MIC for each antibiotic was determined as the last dilution in which no bacterial growth occurred, and the susceptibility intervals were assigned based on the cut-off points established by the CLSI (2018) for amikacin (Amk), cefotaxime (Cfx), ceftazidime (Cfz), imipenem (Imp), meropenem (Mer), and piperacillin–tazobactam (Pip–Taz). According to the CLSI protocols [34], clinical isolates were strains considered ESBL-producing when resistant to cefotaxime but susceptible to the combination of Cfx/clavulanic acid. Values are presented as the median in mg/L and interquartile range (IQR).

2.3. DNA Extraction and PCR Amplification

The total DNA was obtained from stationary bacterial cultures in LB broth using the phenol-chloroform methodology. In brief, pelleted bacteria were suspended in 300 μL of PBS and mixed with 300 μL of phenol: chloroform (25:24 vol:vol); the final mix was stirred well in a vortex and centrifuged at 13,000× *g* at 4 °C for 15 min. The aqueous phase was mixed 1:1 with chloroform and centrifuged as above. Genomic and plasmidial DNA contained in the aqueous phase was precipitated in 0.6 vol of 2-propanol and sedimented at 13,000× *g* at 4 °C for 20 min. The nucleic acids were washed twice with 70% ethanol and resuspended in Tris-EDTA buffer (10 mM Tris-HCl, 1 mM disodium EDTA, pH 8.0). PCR reactions were performed, using 0.5 μM of each specific primer pair listed in Table 1, in 1× master mix GoTaq (Promega). The amplification was carried out at a final volume of 20 μL in a Veriti (Applied Biosystem) PCR machine with an initial denaturation step of 10 s at 95 °C, followed by 35 cycles of 10 s at 95 °C, 15 s at 58 °C, and 30 s of extension at 72 °C. A final extension step of 7 min at 72 °C was included, and PCR products were visualized on a 1.7% agarose gel.

2.4. K. Pneumoniae Survival to Innate Immunity

To evaluate the survival of *K. pneumoniae* to the innate immunity, the bactericidal activity of normal human serum (NHS), as well as phagocytosis by human macrophages (MΦ), and polymorphonuclear (PMN) cells was determined. Both serum and leukocytes were obtained from blood samples of healthy voluntary donors who had not taken any antibiotic or anti-inflammatory medication for at least ten days before the day of sampling.

2.4.1. Susceptibility to Normal Human Serum

Serum susceptibility was carried out as before [37]; in brief, 75 μL of pooled NHS were mixed with 25 μL suspension containing 2×10^6 CFUs of each isolate in a 96-well plate. As a control, bacterial cultures of each isolate were mixed with PBS. The mixtures were incubated at 37 °C for 3 h, and viable bacteria were enumerated by serial micro-dilution and colony counting on ca-MH agar plates. Serum resistance is expressed as viable bacteria in CFUs/mL compared to untreated isolates controls.

Table 1. Primers for genes encoding virulence factors of *Klebsiella pneumoniae*.

Gene	Primers	Gene Type	Amplicon Size
*ybt*S	GACGGAAACAGCACGGTAAA GAGCATAATAAGGCGAAAGA	Siderophores	242
*ent*B	GTCAACTGGGCCTTTGAGCCGTC TATGGGCGTAAACGCCGGTGAT	Siderophores	400
*mag*A	GGTGCTCTTTACATCATTGC GCAATGGCCATTTGCGTTAG	Capsular serotype K1 and hypermucoviscosity phenotype	128
*rmp*A	CATAAGAGTATTGGTTGACAG CTTGCATGAGCCATCTTTCA	Regulator of mucoid phenotype A	461
K2	CAACCATGGTGGTCGATTAG TGGTAGCCATATCCCTTTGG	Capsular serotype K2 and hypermucoviscosity phenotype	531
*all*S	CATTACGCACCTTTGTCAGC GAATGTGTCGGCGATCAGCTT	Allantoin metabolism	764
16S	ATTTGAAGAGGTTGCAAACGAT TTCACTCTGAAGTTTTCTTGTGTTC	Gene encoding the 16S ribosomal RNA	133

2.4.2. Susceptibility to Phagocytosis by Macrophages and Polymorphonuclear Cells

The separation of mononuclear and polymorphonuclear cells was performed by centrifugation in a gradient density column of Histopaque (Sigma-Aldrich, St. Louis, MO, USA), following the manufacturer instructions. Monocytes were selected from other mononuclear cells by incubation in RPMI-1640 medium (Sigma-Aldrich) without fetal bovine serum (FBS) and differentiated into macrophages incubating during 7–9 days in RPMI-1640 10% FBS at 37 °C with 5% CO_2 [38]. The macrophage monolayer was infected with 2.5×10^7 CFUs with a multiplicity of infection (MOI) of 50:1 of each bacterial isolate and centrifuged at $200 \times g$ for 5 min, to synchronize phagocytosis. After 2 h of incubation, cells were washed and incubated for an additional 60 min in RPMI-1640 + 100 µg/mL gentamicin. Macrophages were lysed with 0.1% saponin for 10 min at room temperature, and viable bacteria were enumerated as above. For PMN assays [39], 5×10^5 cells were combined with 5×10^6 CFUs of each isolate (MOI 10:1) in serum-free RPMI-1640 and synchronize by centrifugation at $524 \times g$ for 8 min at 4 °C. After 3 h, PMNs were lysed with 0.1% saponin, and viable bacteria were enumerated. Control groups of non-phagocyted bacteria were included.

2.5. Synthesis of the PSIR-3 Compound

The structural and photophysical characterization of the PSIR-3 compound was described previously [40]. The complex synthesized can be described by using the following general formula: [Ir(C^N)$_2$(N^N)](PF$_6$), where N^N is the ancillary ligand; and C^N corresponds to a cyclometalating ligand. In this study PSIR-3 is [Ir(ppy)$_2$(ppdh)]PF$_6$ where ppdh is pteridino(7,6-f)(1,10)phenanthroline-1,13(10H,12H)-dihydroxy and ppy is 2-phenylpyridine [41]. The structure and purities of the compound were confirmed by nuclear magnetic resonance (NMR), Fourier-transform infrared spectroscopy (FTIR), and mass spectroscopy (MALDI–MS) measurements. The absorption spectra were measured in acetonitrile (ACN) solutions using a Shimadzu UV–Vis Spectrophotometer UV-1900. The molar extinction coefficients of the characteristic bands were determined from the absorption spectra. Photoluminescence spectra were taken on an Edinburgh Instrument spectrofluorimeter using ACN solutions of the compounds previously degassed with N_2 for approximately 20 min. The emission quantum yields (Φ_{em}) were calculated according to the description of the literature [42]. Fluorescence lifetimes were measured by using a time-correlated single-photon counting (TC-SPC) apparatus (PicoQuant Picoharp 300) equipped with a sub-nanosecond LED source (excitation at 380 nm) powered by a PicoQuant PDL 800-B variable (2.5–40 MHz) pulsed power supply.

2.6. Antimicrobial Activity of Photosensitizer Compounds

Stock solutions of 2 g/L of the PSIR-3 compound solubilized in ACN were used to prepare working solutions in distilled water. For the antimicrobial assay, the collection of 118 clinical isolates of *K. pneumoniae* was used, and the control strains of susceptible *K. pneumoniae* KPPR1 and the MDR strain ST258 were also included. All bacteria were grown as axenic culture in Luria Bertani broth or agar medium as appropriate. PSIR-3 was mixed in 24-well plates at a final concentration of 4 mg/L for photodynamic experiments, with suspensions of 1×10^7 colony forming units (CFUs)/mL of each bacterium, in a final volume of 500 µL of cation-adjusted Mueller-Hinton (ca-MH) broth. Exposure to light was performed for 30 min in a chamber with a white LED lamp at a photon flux of 17 µW/cm^2. After exposure to light, the CFUs of the viable bacteria were determined by broth-micro dilution and sub-cultured on ca-MH agar plates. Following the recommendations of the Clinical and Laboratory Standards Institute (CLSI 2017) [34], the agar plates were incubated during 16–20 h at 37 °C, in the dark, and colony count was recorded, using a stereoscopic microscope. Control wells with bacteria culture with no photosensitizer or photosensitizer but not exposed to light were also included.

2.7. Determination of the Synergy between PSIR-3 and Cfx

The fractional inhibitory concentration index (FIC) value was determined using the following formula [43,44]. MICac is the MIC of a compound A, combined with a compound B, and MICbc is the MIC of the compound B combined with the compound A. The MICa and MICb are the MIC of the A and B compounds alone, respectively. Values in the FIC index ≤ 0.5 are considered synergistic, and values > 4 are considered antagonistic [44].

$$\text{FIC Index} = \frac{\text{MICac}}{\text{MIC}_a} + \frac{\text{MICbc}}{\text{MIC}_b}$$

To determine the MIC-Cfx combined with each PSs, 1×10^7 UFC/mL of ESBL-producing bacteria were aPDI treated for 30 min with 4 mg/L of each PS and mixed with serial dilution (32–0.125 mg/L) of Cfx in ca-MH broth, as above. To determine the MIC-PSs combined with Cfx, 1×10^7 UFC/mL of ESBL-producing bacteria was mixed with serial dilution of each PSs (32–0.125 mg/L) and a fixed concentration of 4 mg/L of Cfx, and then it was subjected to aPDI, as above.

2.8. Statistical Analyses

Statistical analyses were performed by using the Systat 13.2 software (Systat Software, Inc., San Jose, CA, USA) and GraphPad v6.01 (Prism) software. The X^2 test or Fisher's exact test for categorical variables and the Mann–Whitney *U* test for continuous non-parametric variables were used. The risk of virulence genes to modify the MIC of the antibiotic was established by determining the odds ratio with a CI: 95%.

3. Results

3.1. Demographic Characterization

This work seeks to demonstrate the capacity of aPDI to inhibiting the growth of clinical isolates of *K. pneumoniae*, which are diverse in antimicrobial susceptibility, genotype, and virulence. Phenotypic and genotypic characterization was conducted to determine the frequency of genes encoding virulence factors, the MIC values determined for various antibiotics, and susceptibility to innate immune components. From the clinical isolates of *K. pneumoniae* received in the laboratory, 118 were selected from different unrelated patients. The isolates were recovered mainly from urine samples, 113 (95.8%), and only five from respiratory samples (three endotracheal aspirates (2.5%) and two expectorations (1.7%)). As shown in Table 2, the samples were obtained from 78 (66%) females and 40 (34%) males, from 12 clinical services, with a higher contribution from the emergency room 41 (34.75%), secondly the unit of medicine 24 (20.34%), third ambulatory 16 (13.56%) and in the fourth place urology 14 (11.86%). Of the 118 clinical isolates, 114 came from adults and 4 from pediatric patients. As shown in Figure 1A, the patient's age fluctuated between 7 months and 94 years, with a median (IQR: 25–75%) of 69.5 (54.8–84) years for females, and between 10 months and 92 years, with a median of 68.5 (60.3–80.8) years for males. There are no significant differences in age between genders ($p = 0.279$ Mann–Whitney U test). The results confirm the infections with MDR and not MDR *K. pneumoniae* mainly affects the elderly population. The elderly are the most susceptible population to present complications derived from infectious diseases [45].

Table 2. Frequency of *K. pneumoniae* isolation by hospital unit and genre.

Service	Female	Male	Total	Percent
Outpatient	12	4	16	13.56%
surgery	5	5	10	8.47%
Endocrinology	1	0	1	0.85%
Geriatrics	1	2	3	2.54%
Gynecology	2	0	2	1.69%
Home hospitalization	1	0	1	0.85%
Medicine	2	0	24	20.34%
Medical–surgical service	13	11	2	1.69%
Pediatrics	1	0	1	0.85%
Critical patient unit	2	1	3	2.54%
Emergency room	31	10	41	34.75%
Urology	7	7	14	11.86%
TOTAL	78	40	118	100%

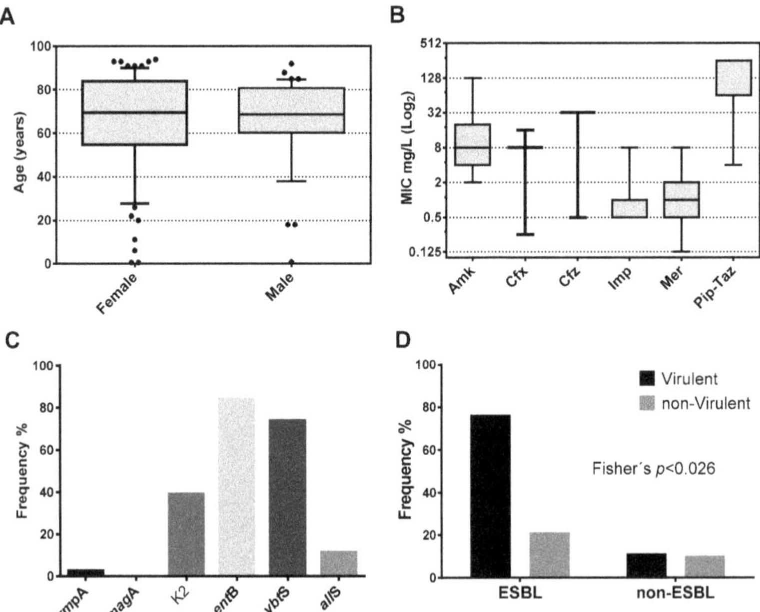

Figure 1. Characterization of patients and clinical isolates of *K. pneumoniae*. (**A**) Box plot for the age of patients stratified by gender. Patients show a median of 69.5 years for females and 68.5 years for males. (**B**) Box plot showing the median in the population of the minimum inhibitory concentration (MIC) of the seven antibiotics commonly used to treat infections by Gram-negative bacteria. (**C**) Frequency of carrying of the six genes evaluated in the study population. (**D**) Frequency distribution of carriers of three or more virulence factors stratified by extended-spectrum beta-lactamase (ESBL) production.

3.2. Antibiotic Susceptibility

In the *K. pneumoniae* population, the median and IQR (25–75%) of the MIC were determined and expressed in mg/L in a $\log_2$ box plot. As shown in Figure 1B, the bacterial population was mainly susceptible to Amk with a median (IQR) of 8 (4–20). Similar to Amk, the population was mainly susceptible to carbapenem, Imp, and Mer with a median of 1 (0.5–1) and 1 (0.5–2), respectively. In contrast, most of the population was resistant to the cephalosporins, Cfx, and Cfz, with medians of 8 (8–8) and 32 (32–32). Finally, similar to cephalosporin, almost the entire population was resistant to the Pip–Taz combination with a median of 256 (64–256). From the population, 66 isolates resistant to Cfx were susceptible to the combination Cfx/clavulanic acid. Those isolates were identified as ESBL-producing bacteria.

3.3. Virulence Gene Frequency

In this work, we select certain virulence genes belonging to families with different mechanisms of action. PCR determined the frequency of carrying genes encoding the virulence factors *rmp*A, *mag*A, K2, *ent*B, *all*S, and *ybt*S for each isolate using specific primers (Table 1). These genes were selected for being representatives of different families of virulence factors. As shown in Figure 1C, the harboring frequency was; *rmp*A [2.54% (3/118)], *mag*A [0% (0/118)], K2 [39% (46/118)], *ent*B [83.9% (99)], *all*S [11.2% 13/118)] and *ybt*S [73.8% (87/118)]. The most frequent virulence factor was the *ent*B gene, followed by the *ybt*S gene. There were no isolates with the *mag*A gene, and only three isolates harbor the *rmp*A gene. No isolates showed the hypermucoviscosity phenotype (by string test) in agar plates in the population of 118 unrelated isolates.

3.4. Correlation of Virulence Factors with Antibiotic Resistance

As shown in Figure 1D, there is a significantly higher frequency of ESBL strains that harbor three or more virulence factors compared to non-ESBL strains (Fisher's; $p < 0.026$). The non-parametric Mann–Whitney U test (Systat 13 software) was used to determine the association that each virulence gene has on the median MIC value assuming a null hypothesis $\alpha = 5\%$. As shown in Table 3, the *rmp*A gene was not significantly associated with a modification of the median-MIC of any of the antibiotics analyzed in this study ($p > 0.05$). On the other hand, the *all*S gene significantly influenced the median-MIC of Cfx, Cfz, and Pip–Taz ($p < 0.05$). Similarly, the K2 gene significantly influenced the median-MIC of Cfx and Pip–Taz ($p < 0.05$). The *ent*B gene significantly affected the median-MIC of Amk and Imp ($p < 0.05$). The *ybt*S gene significantly influenced the median-MIC of Cfx ($p < 0.006$). Finally, the *all*S gene significantly influenced the median-MIC of Cfx and Cfz antibiotics ($p < 0.05$). The MIC of the antibiotics more sensitive to the presence of virulence factors were the Cfx (sensitive to K2, *ybt*S, and *all*S genes) and Pip–Taz (sensitive to the K2 and *all*S genes). These data show that, regardless of the patient's conditions, the multi-virulence is effectively an independent risk factor that promotes ESBL-production of clinical populations of *K. pneumoniae* with a $p < 0.026$ Fisher's exact test.

Table 3. Mann–Whitney U test p-values for modification of antibiotic MIC by virulence genes.

	Amikacin	Cefotaxime	Ceftazidime	Imipenem	Meropenem	Pip–Taz
*rmp*A	0.285	0.051	0.432	0.363	0.527	0.411
K2	0.158	**0.002**	0.919	0.102	0.588	**0.001**
*ent*B	**0.009**	0.918	0.117	**0.015**	0.291	0.072
*ybt*S	0.572	**0.006**	0.722	0.340	0.502	0.067
*all*S	0.326	**0.024**	**0.001**	0.086	0.537	**0.013**

Values in bold represent a $p < 0.05$, below the null hypothesis value with $\alpha = 5\%$, which means a significant difference in MIC due to the presence of the virulence factor. Pip–Taz, piperacillin–tazobactam.

As shown in Figure 2, the box plot for each antibiotic stratified by the presence or absence of virulence genes was constructed to verify if the influence is to increase or decrease the median of the MIC. The presence of the *ent*B gene is associated with a decrease in the median-MIC for Amk and Imp. On the other hand, the K2 gene is associated with an increase in the median-MIC for Cfx and Pip–Taz. Similarly, the *ybt*S gene is associated with an increase in the median-MIC for Cfx. Finally, the *all*S gene is associated with an increased median-MIC for Cfz and decreased median-MIC for Cfx and Pip–Taz. Some of the virulence factors studied here were associated with the median- MIC modification for several antibiotics. In the *K. pneumoniae* population tested, the most influencing virulence factors were *ent*B, *ybt*S, and *all*S genes. The stratified MICs-box plot made it possible to distinguish how virulence genes modulate antibiotic susceptibility by increasing or decreasing. Remarkably, the sum of the genes of the $K2^+$, $ybtS^+$, and $allS^+$ genes contributes to increasing the median-MICs to values higher than the clinical susceptibility breakpoint established by the CLSI. These increased values occur similarly when efflux pumps are activated in other Gram-negative bacteria [46]. For example, in a population of *Pseudomonas aeruginosa*, the combined overexpression of the *mex*A and *mex*X efflux pump increased the median MIC for ciprofloxacin and cefepime above the cutoff points [46,47].

3.5. Association of $K2^+$, $ybtS^+$, and $allS^+$ Virulence Genes to Survive the Innate Immunity

Our results show that the K2 and *ybt*S virulence genes are risk factors for the production of ESBL and that the *all*S gene acts as a protective factor. Then, we selected two groups of clinical isolates of *K. pneumoniae*, which harbor or not these three genes, to assess their susceptibility to innate immunity components. Only three isolates, identified by the numbers 81, 92, and 111, are of the $K2^+$, $ybtS^+$, and $allS^+$ genotype, and unexpectedly all of them are ESBL-producers. From the genotype $K2^-$, $ybtS^-$ and $allS^-$, the isolates 13, 21, and 22 were selected as being non-ESBL. As controls, the susceptible but highly virulent *K. pneumoniae* KPPR1 ($K2^+$, $ybtS^+$, and $allS^+$) and the MDR but less virulent ST258 ($K2^-$, $ybtS^-$ and $allS^-$) strains were also included.

To determine the bacterial susceptibility to normal human serum (NHS), they were exposed for 3 h, and then the number of viable bacteria was determined by microdilution and plate counting. As shown in Figure 3A, the presence of all three virulence genes, $K2^+$, $ybtS^+$, and $allS^+$, significantly (** = $p < 0.003$; two-way ANOVA) increases the survival of ESBL-producing bacteria compared to non-ESBL bacteria lacking all three virulence genes. On average, the serum resistance was improved by four orders of magnitude ($4\log_{10}$). Similar to that observed with serum, ESBL-producing bacteria of the $K2^+$, $ybtS^+$, and $allS^+$ genotype show a significant (**** = $p < 0.0001$) increase, $4\log_{10}$, in the survival to the MΦ activity, compared to no- ESBL bacteria that lacks these virulence genes (Figure 3B). Finally, comparable to MΦ, survival to the activity of PMNs of the bacteria of genotype $K2^+$, $ybtS^+$, and $allS^+$ increased significantly (* = $p < 0.02$), at least one time on average, compared to non-ESBL strains that lack the virulence genes (Figure 3C). As shown in Figure 3, the virulent control KPPR1 strain was more resistant to serum (Figure 3D), macrophages (Figure 3E), and PMN (Figure 3F) compared to the MDR ST258 *K. pneumoniae* strain. Consistently the clinical control isolates have shown to be more susceptible to the bactericidal activity of the serum (3D) and the phagocytic activity of MΦ (3E) and PMN (3F). The co-occurrence of harboring multiple genes that encode virulence factors and the ESBL-production leads to enhanced virulence. The ESBL-producing strains of *K. pneumoniae* of the genotype $K2^+$, $ybtS^+$, and $allS^+$ were more resistant to innate immunity, consistently with studies over MDR populations of *K. pneumoniae* that increased their 30-day mortality over patients undergoing bloodstream infections [1,48]. Moreover, virulence genes, such as siderophores, which have an essential role in bacterial survival and virulence [49,50], have been previously associated with the MDR-*K. pneumoniae* [16,51]. In this study, ESBL-producer *K. pneumoniae* of the $K2^+$, $ybtS^+$, and $allS^+$ genotype shown a survival improvement for killing by PMNs. Previously, the PMN has demonstrated a limited binding and uptake capacity for MDR-*K. pneumoniae* [39]. The activity of the PMN is the most important cellular component of the innate immune response, essential as the first line of defense against bacterial infections [52].

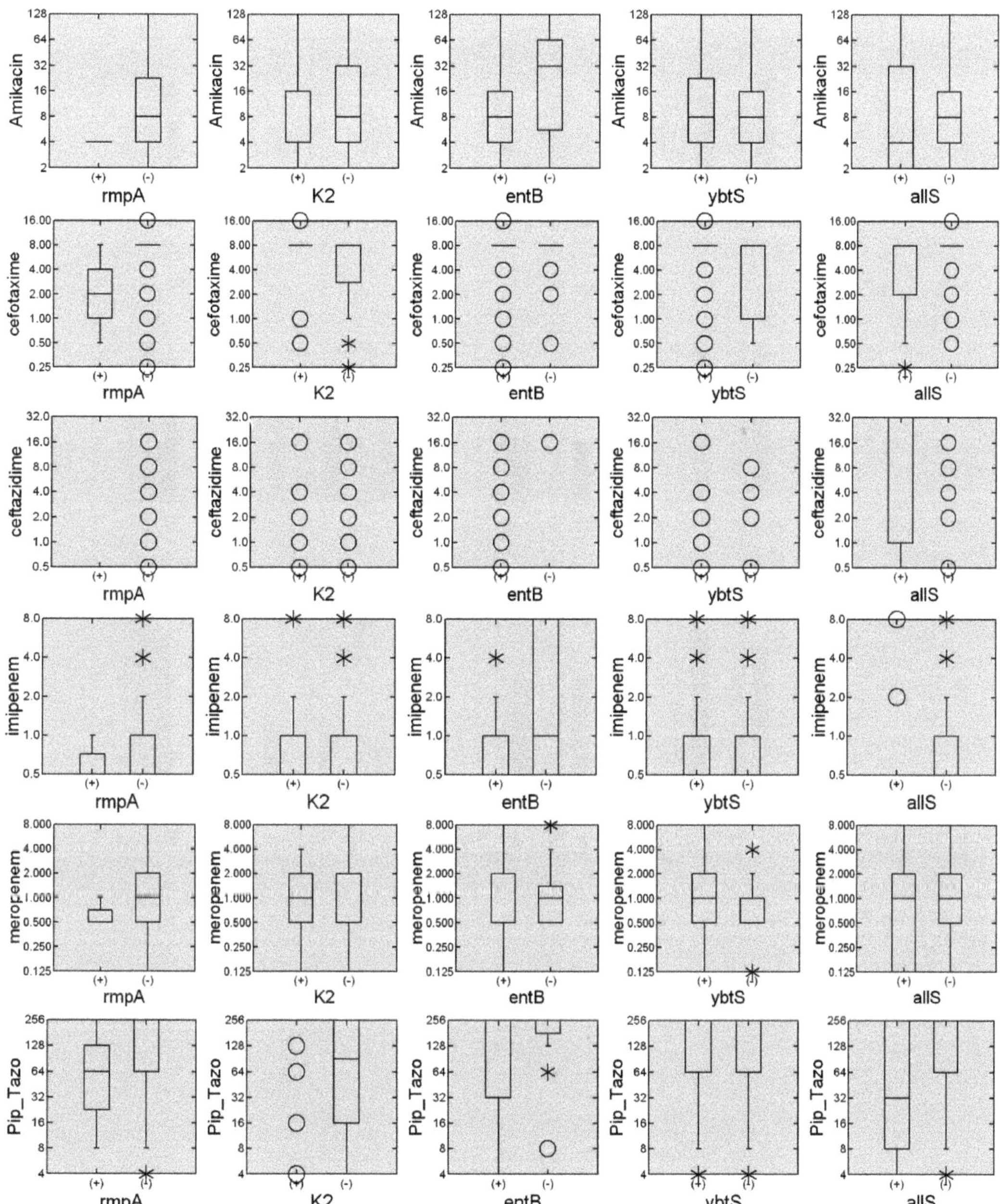

Figure 2. Association of virulence genes in the measurement of median MIC. The box diagrams created with the Systat 13 software summarize the MIC measures for six commonly used antibiotics, stratified by the presence or absence of each of the six virulence genes. The data are presented as mg/L in a Log_2 scale of MICs for each antibiotic stratified by the presence (+) or absence (−) of each virulence gene. The horizontal lines represent the median and the frequency limits between 25 and 75% of the individuals, the hollow dots represent single individuals, and the asterisks represent small groups of individuals.

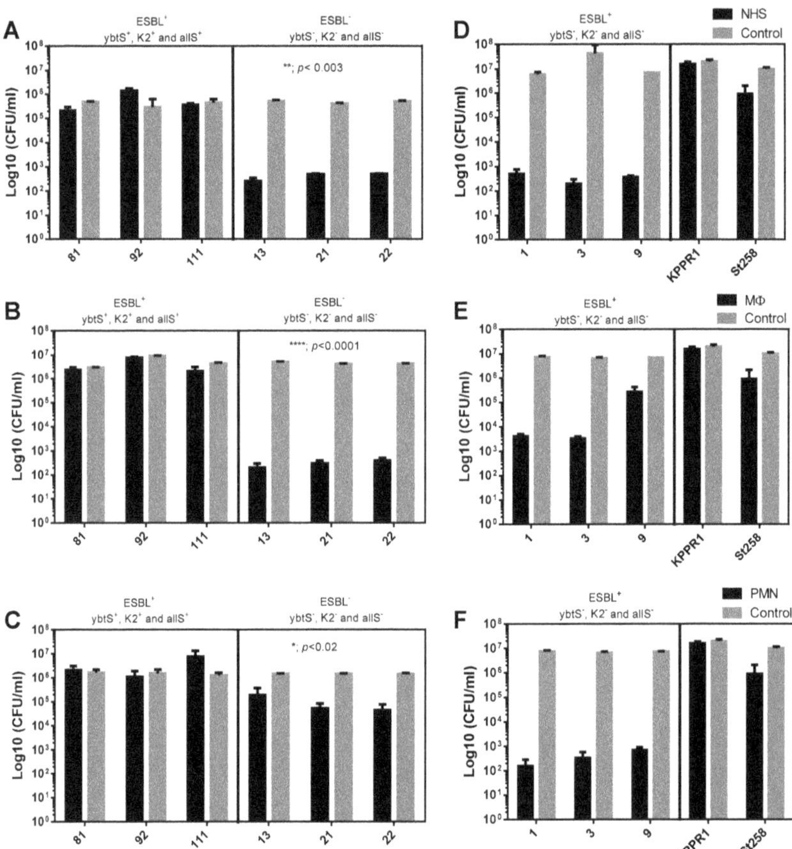

Figure 3. Effect of virulence genes on the susceptibility of ESBL or non-ESBL strains to components of innate immunity. The susceptibility of ESBL-producing *K. pneumoniae* clinical isolates bearing the virulence genes *ybtS*+, K2+, and *allS*+ was determined and compared with non-ESBL producing bacteria that lack the virulence genes. (**A,D**) Susceptibility to normal human serum (NHS), (**B,E**) susceptibility to phagocytosis by macrophages (MΦ), and (**C,F**) susceptibility to phagocytosis by polymorphonuclear cells (PMN). The results of two independent experiments performed in triplicate are shown (n = 6). Viable bacteria were enumerated by colony count on ca-MH agar after serial micro-dilution. The colony forming units (CFUs)/mL values are presented as means +/− SD, on a $\log_{10}$ scale of treated bacteria (black bars) compared to untreated control bacteria (gray bars). **** = $p < 0.0001$, ** = $p < 0.003$ and * = $p < 0.02$ of two-way ANOVA comparing the proportion of treated/untreated ESBL bacteria with non-ESBL bacteria.

3.6. Susceptibility of Clinical Isolates to aPDI with PSIR-3

3.6.1. Photophysical Properties of the PSIR-3 Compound

We have previously shown that Ir(III)-based compounds, such as PSIR-3, have photodynamic antimicrobial activity against imipenem-resistant *Klebsiella pneumoniae* [32,33]. In this work, we tested a coordination compound characterized by a positive charge in the first coordination sphere (Figure 4B). The photophysical evaluation of the PSIR-3 performed in acetonitrile solution [40] revealed absorption processes at 375 and 392 nm attributable at the first instance of charge-transfer transitions (Figure 4C,D) [32]. When the PSIR-3 compound was excited with a wavelength corresponding to the lowest charge-transfer absorption energy, 375 nm, it showed maximum emission at 598 nm (Figure 4C,D). Figure 4C shows the recorded lifetimes of excited states in 0.32 µs and the calculated quantum yield (Φ_{em}) in

0.011 [40]. The aPDI activity of the PSIR-3 compound was compared with the positive PS control [Ru(bpy)$_3$](PF$_6$)$_2$ (bpy = 2,2'-bipyridine) called PS-Ru. According to the literature, the PS-Ru shows a charge-transfer absorption process at 450 nm with maximum emission at 600 nm (excited in 450 nm) in acetonitrile [42], with Φ_{em} of 0.095 [42], and a lifetime registered of its excited state of 0.855 µs (Figure 4C) [53]. The maximum absorption of PSIR-3 occurs at wavelengths below 400 nm, that although it is more energetic, it penetrates the tissues poorly. Therefore, this compound will activate better if it is directly irradiated, such as in superficial wounds. However, UTI is one of the most common diseases caused by *K. pneumoniae*, where probes that deliver the light dose within internal surfaces can irradiate the epithelial lining of the bladder [54]. Certain kinds of fiber arrays or inflatable balloons may provide a homogenous light power delivery [55,56]. This catheterization can be applied for inpatients suffering UTI that does not respond to antibiotic treatment.

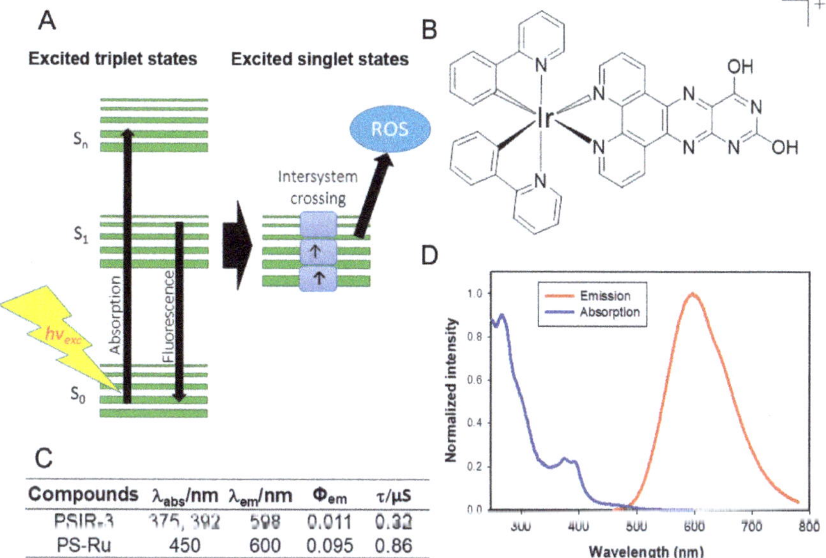

C

Compounds	λ_{abs}/nm	λ_{em}/nm	Φ_{em}	τ/µS
PSIR-3	375, 392	598	0.011	0.32
PS-Ru	450	600	0.095	0.86

Figure 4. The photophysical properties of the PSIR-3 photosensitizer. (**A**) schematic representation of the absorption–emission process of photosensitizer molecules. Light excites external electrons to accede from a basal state S_0 to a higher energetic state S_1–S_n. Suppose the electrons return to its ground state; the energy is then released as fluorescence. However, when the excited electron enters an intersystem crossing process, the released energy excites molecular oxygen and converts it to reactive oxygen species (ROS). (**B**) chemical structure of the Ir(III) compound (PSIR-3, [Ir(ppy)$_2$(ppdh)]PF$_6$). (**C**) Summary of the photophysical properties of the PSIR-3 and PS-Ru compounds, where λ_{abs} = wavelength of absorbance, λ_{em} = wavelength of emission, Φ_{em} = emission quantum yield, and τ = time in the excited state. Data for PS-Ru were obtained from the literature. Adapted from [42], Elsevier, 2010; Adapted from [53], American Chemical Society, 1983. (**D**) The absorption and the emission spectra of the PSIR-3 in acetonitrile (ACN).

3.6.2. Antimicrobial Photodynamic Inhibition of the PSIR-3 over Clinical Isolates

Photodynamic treatment was verified to produce the observed growth inhibition of the 118 clinical isolates of *K. pneumoniae* compared to the untreated bacteria. The photodynamic activity of the PSIR-3 compound was compared to the activity of the PS-Ru reference compound as a positive control [41,57–59]. As seen in Figure 5, compared to the control of untreated bacteria, photodynamic treatment with 4 µg/mL PSIR-3 inhibits bacterial growth > 3 $\log_{10}$ (>99.9%) of clinical isolates of *K. pneumoniae* (**** $p < 0.0001$; compared to untreated control). The results show that the bactericidal effect produced by PSIR-3 is light-dependent (ns = $p > 0.05$; compared to the untreated control). These results are comparable with those obtained with the positive control compound PS-Ru, which has shown that bacterial growth inhibition is light-dependent (**** $p < 0.0001$; compared to the untreated control).

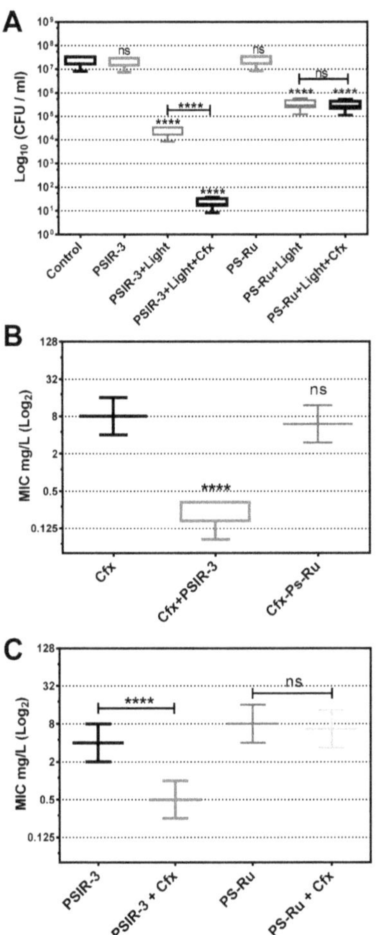

Figure 5. Antimicrobial photodynamic inactivation of clinical isolates of *K. pneumoniae*. (**A**) Growth inhibition of 118 clinical isolates of *K. pneumoniae* subjected to antimicrobial photodynamic inactivation (aPDI) with PSIR-3. The bacteria were used at a concentration of 1×10^7 CFUs/mL and mixed in triplicate with 4 mg/L of PSIR-3 or PSIR-4 compounds. For the aPDI, the mixture of bacteria with PS was exposed for 1 h at 17 μW/cm^2 of white light. As a control, bacteria combined with the photosensitizers (PSs) not exposed to light (PSIR-3 or PS-Ru) and bacteria not combined with the PSs (control) were included. Colony count enumerated of viable bacteria on ca-MH agar after serial micro-dilution. The CFUs/mL values are presented as means ± SD on a log$_{10}$ scale. (**B**) From the clinical isolates, 66 ESBL-producing bacteria were exposed to aPDI, using PSIR-3 or PS-Ru, and MIC for cefotaxime (Cfx) was performed in triplicate, in ca-MH broth, for 16–20 h. (**C**) For ESBL-producing bacteria, the MIC for PSIR-3 or PS-Ru, were determined in combination with 4 mg/L of Cfx, performed in triplicate in ca-MH agar for 16–20 h. The MIC values are presented as median ± SD of mg/mL on a log$_2$ scale. Not significant (ns) $p > 0.05$ by Student's *t*-test among treated bacteria compared to control; **** $p < 0.0001$ by Student's *t*-test among treated bacteria compared to control.

3.6.3. Synergism between aPDI with PSIR-3 and Cefotaxime

Since PSIR-3 showed synergism combined with imipenem [32], we analyzed whether it shows synergism with cefotaxime in the population of clinical isolates. First, it was determined whether the combined treatment with Cfx increases the inhibition of bacterial growth of aPDI with PSIR-3. The 118 clinical isolates of *K. pneumoniae* were exposed to the preparation of 4 μg/mL of cefotaxime with 4 μg/mL of PSIR-3 (its MIC). Control bacteria without Cfx and exposure to light were included. As expected, the PSIR-3 compound mixed with cefotaxime significantly (**** $p < 0.0001$) increased the bactericidal effect on the clinical isolates population from 3 to 6 $\log_{10}$ reduction (Figure 5A). As seen before, a significantly increased inhibitory effect was not observed when combining the cefotaxime with the PS-Ru control compound (ns $p > 0.05$). Secondly, the set of 66 clinical isolates characterized as ESBL-producing *K. pneumoniae* were treated with PSIR-3 aPDIfor 1 h, and serial dilutions determined the MIC for Cfx in ca-MH broth. As seen in Figure 5B, a significant reduction (**** $p < 0.0001$) from 8 μg/mL (8–8) to 0.17 μg/mL (0.17–0.333) on Cfx-MIC was observed compared to the untreated group. The combined treatment also reduced the PSIR-3-MIC, from 4 to 0.5 mg/L (Figure 5C). Similar to previously shown with imipenem [32], compound PSIR-3 produced a significant change in Cfx-susceptibility with a fractional inhibitory concentration (FIC) index of 0.15 (Table 4). Figure 5C shows the control compound, PS-Ru, did not significantly change Cfx-susceptibility with an FIC Index 1.58 (Table 4). Because synergy is defined as an FIC index of ≤ 0.5 [44], the increased inhibitory effect observed when combining PSIR-3 with Cfx is not summative but synergistic.

Table 4. Fractional inhibitory concentration (FIC) index calculation.

Compounds	MIC	MIC Combined	FIC	FIC Index
Cfx	8.00			
PSIR-3	4.00	0.17	0.02	0.15
PSIR-3 *		0.50	0.13	
PS-Ru	8.00	6.67	0.83	1.58
PS-Ru *		6.00	0.75	

MIC values are de median for the ESBL-producing *K. pneumoniae*, n = 66. * The MIC value modified by Cfx.

The behavior exhibited by PSIR-3 must be related, as mentioned in previous reports [32,33], to the external substituents bonded to polypyridine ligand structures and affinity to bacterial envelope [41]. This synergism is also comparable to other photosensitizers; for example, rose bengal showed an increase in the susceptibility of *Acinetobacter baumannii* for a range of antibiotics used along with aPDI [60]. In another example, conventional antibiotics and alternative compounds reported synergism in a murine model for pathogens of the ESKAPE group (*Enterococcus faecium*, *Staphylococcus aureus*, *Klebsiella pneumoniae*, *Acinetobacter baumannii*, *Pseudomonas aeruginosa*, and *Enterobacter* species) [61], using anti-biofilm peptides [62]. For now, it is difficult to accurately calculate the dose of light necessary to activate PSIR-3 effectively; however, we observed in vitro that with low doses of energy (17 μW/cm^2 of white light), it is bactericidal. Compounds with optimum absorbance at higher wavelengths, bordering the 630–750 nm, would improve the exposure of PS to light into the tissues, but being less energetic, the PSs must have a triplet excited state of easy access to promote energy transfer [63]. We, therefore, need to perform a better characterization of its antimicrobial activity when activated with defined wavelengths, before starting in vivo studies, in urinary infection models and to evaluate the need to deliver light through intraurethral catheters [64].

4. Conclusions

In this work, we saw that multi-drug resistance and virulence are significant factors in clinical isolates of *K. pneumoniae*. However, the increase in MICs can be neutralized by aPDI, turning resistant strains susceptible. APDI is effective in treating multidrug-resistant bacteria and more virulent strains, as well as strains that combine both characteristics. The aPDI then becomes a great support to antimicrobial therapy in a shortage of new effective antibiotics. The photophysical characterization of PS indicates that its maximum absorption occurs at wavelengths lower than 400 nm, which could constitute a problem for its use in infections of internal organs due to low penetration. In UTI, optical fibers can be used through a catheter to deliver the dose of light [54,55]. Moreover, compounds with optimum absorbance at higher wavelengths, bordering the 630–750 nm, would improve the exposure of PS to light in the tissues.

Author Contributions: Conceptualization, C.E.P. and I.A.G.; methodology, C.E.P. and I.A.G.; software, C.E.P. and I.A.G.; validation, C.E.P., P.D. and I.A.G.; formal analysis, C.E.P., P.D. and I.A.G.; investigation, C.N. and A.P.; resources, C.E.P., P.D. and I.A.G.; data curation, C.E.P. and I.A.G.; writing—original draft preparation, C.E.P.; writing—review and editing, C.E.P.; visualization, C.E.P. and I.A.G.; supervision, C.E.P. and I.A.G.; project administration, C.E.P.; funding acquisition, C.E.P., P.D. and I.A.G. All authors have read and agreed to the published version of the manuscript.

Funding: The authors are supported by UCEN grants CIP2019007 (awarded to C.E.P.), Fondecyt grants 1180673 and 1201173 (awarded to I.A.G. and P.D., respectively), Office of Naval Research (ONR) grant N62909-18-1-2180 (awarded to I.A.G.) and USM PM_I_2020_31 project of P.D.

Institutional Review Board Statement: The study was conducted according to the guidelines of the Declaration of Helsinki, and approved by the Institutional Ethics Committee of the Faculty of Health Sciences, of the Universidad Central de Chile N: 01/2017 approved on 01-04-2017.

Informed Consent Statement: The informed consent form was approved by the ethics committee of the Faculty of Health Sciences of the Central University of Chile and by the Bioethics Committee of the Central Metropolitan Health Service of Chile (MHSC), Act number: N 124/07.

Data Availability Statement: The data presented in this study are available on request from the corresponding author. The data are not publicly available because they are confidential data of patients protected by the informed consent protocol.

Acknowledgments: We would like to thank Myriam Pizarro, head of the "Hospital El Carmen de Maipú" laboratory, for her collaboration and provision of the clinical isolates of Klebsiella pneumoniae. We also would like to thank Susan Bueno from Pontificia Universidad Católica de Chile for providing the control strains of K. pneumoniae, KPPR1, and the typo strain ST258.

Conflicts of Interest: The authors declare no conflict of interest.

References

1. Kim, D.; Park, B.Y.; Choi, M.H.; Yoon, E.J.; Lee, H.; Lee, K.J.; Park, Y.S.; Shin, J.H.; Uh, Y.; Shin, K.S.; et al. Antimicrobial resistance and virulence factors of *Klebsiella pneumoniae* affecting 30 day mortality in patients with bloodstream infection. *J. Antimicrob. Chemother.* **2019**, *74*, 190–199. [CrossRef] [PubMed]
2. Zhen, X.; Lundborg, C.S.; Sun, X.; Hu, X.; Dong, H. Economic burden of antibiotic resistance in ESKAPE organisms: A systematic review. *Antimicrob. Resist. Infect. Control* **2019**, *8*, 137. [CrossRef]
3. Lev, A.I.; Astashkin, E.I.; Kislichkina, A.A.; Solovieva, E.V.; Kombarova, T.I.; Korobova, O.V.; Ershova, O.N.; Alexandrova, I.A.; Malikov, V.E.; Bogun, A.G.; et al. Comparative analysis of *Klebsiella pneumoniae* strains isolated in 2012–2016 that differ by antibiotic resistance genes and virulence genes profiles. *Pathog. Glob. Health* **2018**, *112*, 142–151. [CrossRef]
4. Tofas, P.; Skiada, A.; Angelopoulou, M.; Sipsas, N.; Pavlopoulou, I.; Tsaousi, S.; Pagoni, M.; Kotsopoulou, M.; Perlorentzou, S.; Antoniadou, A.; et al. Carbapenemase-producing *Klebsiella pneumoniae* bloodstream infections in neutropenic patients with haematological malignancies or aplastic anaemia: Analysis of 50 cases. *Int. J. Antimicrob. Agents* **2016**, *47*, 335–339. [CrossRef]
5. Li, Y.; Sun, Q.L.; Shen, Y.; Zhang, Y.; Yang, J.W.; Shu, L.B.; Zhou, H.W.; Wang, Y.; Wang, B.; Zhang, R.; et al. Rapid Increase in Prevalence of Carbapenem-Resistant Enterobacteriaceae (CRE) and Emergence of Colistin Resistance Gene mcr-1 in CRE in a Hospital in Henan, China. *J. Clin. Microbiol.* **2018**, *56*. [CrossRef]
6. WHO. *WHO Publishes List of Bacteria for Which New Antibiotics Are Urgently Needed*; WHO: Geneva, Switzerland, 2017.
7. Willyard, C. The drug-resistant bacteria that pose the greatest health threats. *Nature* **2017**, *543*, 15. [CrossRef]
8. Pragasam, A.K.; Shankar, C.; Veeraraghavan, B.; Biswas, I.; Nabarro, L.E.; Inbanathan, F.Y.; George, B.; Verghese, S. Molecular Mechanisms of Colistin Resistance in *Klebsiella pneumoniae* Causing Bacteremia from India-A First Report. *Front. Microbiol.* **2016**, *7*, 2135. [CrossRef] [PubMed]
9. Martin, R.M.; Bachman, M.A. Colonization, Infection, and the Accessory Genome of *Klebsiella pneumoniae*. *Front. Cell. Infect. Microbiol.* **2018**, *8*, 4. [CrossRef]

10. Sonda, T.; Kumburu, H.; van Zwetselaar, M.; Alifrangis, M.; Mmbaga, B.T.; Lund, O.; Kibiki, G.S.; Aarestrup, F.M. Molecular epidemiology of virulence and antimicrobial resistance determinants in *Klebsiella pneumoniae* from hospitalised patients in Kilimanjaro, Tanzania. *Eur. J. Clin. Microbiol. Infect. Dis. Off. Publ. Eur. Soc. Clin. Microbiol.* **2018**, *37*, 1901–1914. [CrossRef]

11. Khaertynov, K.S.; Anokhin, V.A.; Rizvanov, A.A.; Davidyuk, Y.N.; Semyenova, D.R.; Lubin, S.A.; Skvortsova, N.N. Virulence Factors and Antibiotic Resistance of *Klebsiella pneumoniae* Strains Isolated from Neonates with Sepsis. *Front. Med.* **2018**, *5*, 225. [CrossRef] [PubMed]

12. Rastegar, S.; Moradi, M.; Kalantar-Neyestanaki, D.; Ali Golabi, D.; Hosseini-Nave, H. Virulence Factors, Capsular Serotypes and Antimicrobial Resistance of Hypervirulent *Klebsiella pneumoniae* and Classical *Klebsiella pneumoniae* in Southeast Iran. *Infect. Chemother.* **2019**. [CrossRef]

13. Derakhshan, S.; Hosseinzadeh, A. Resistant Pseudomonas aeruginosa carrying virulence genes in hospitalized patients with urinary tract infection from Sanandaj, west of Iran. *Gene Rep.* **2020**, 100675. [CrossRef]

14. Imtiaz, W.; Syed, Z.; Rafaque, Z.; Andrews, S.C.; Dasti, J.I. Analysis of Antibiotic Resistance and Virulence Traits (Genetic and Phenotypic) in *Klebsiella pneumoniae* Clinical Isolates from Pakistan: Identification of Significant Levels of Carbapenem and Colistin Resistance. *Infect. Drug Resist.* **2021**, *14*, 227–236. [CrossRef] [PubMed]

15. Paczosa, M.K.; Mecsas, J. *Klebsiella pneumoniae*: Going on the Offense with a Strong Defense. *Microbiol. Mol. Biol. Rev. MMBR* **2016**, *80*, 629–661. [CrossRef] [PubMed]

16. Bachman, M.A.; Oyler, J.E.; Burns, S.H.; Caza, M.; Lepine, F.; Dozois, C.M.; Weiser, J.N. *Klebsiella pneumoniae* yersiniabactin promotes respiratory tract infection through evasion of lipocalin 2. *Infect. Immun.* **2011**, *79*, 3309–3316. [CrossRef]

17. Kim, Y.J.; Kim, S.I.; Kim, Y.R.; Wie, S.H.; Lee, H.K.; Kim, S.Y.; Park, Y.J. Virulence factors and clinical patterns of hypermucoviscous *Klebsiella pneumoniae* isolated from urine. *Infect. Dis.* **2017**, *49*, 178–184. [CrossRef]

18. French, J.B.; Neau, D.B.; Ealick, S.E. Characterization of the structure and function of *Klebsiella pneumoniae* allantoin racemase. *J. Mol. Biol.* **2011**, *410*, 447–460. [CrossRef] [PubMed]

19. Nadasy, K.A.; Domiati-Saad, R.; Tribble, M.A. Invasive *Klebsiella pneumoniae* syndrome in North America. *Clin. Infect. Dis.* **2007**, *45*, e25–e28. [CrossRef] [PubMed]

20. Chuang, Y.P.; Fang, C.T.; Lai, S.Y.; Chang, S.C.; Wang, J.T. Genetic determinants of capsular serotype K1 of *Klebsiella pneumoniae* causing primary pyogenic liver abscess. *J. Infect. Dis.* **2006**, *193*, 645–654. [CrossRef] [PubMed]

21. Cubero, M.; Grau, I.; Tubau, F.; Pallares, R.; Dominguez, M.A.; Linares, J.; Ardanuy, C. Hypervirulent *Klebsiella pneumoniae* clones causing bacteraemia in adults in a teaching hospital in Barcelona, Spain (2007–2013). *Clin. Microbiol. Infect. Off. Publ. Eur. Soc. Clin. Microbiol. Infect. Dis.* **2016**, *22*, 154–160. [CrossRef]

22. Lin, H.A.; Huang, Y.L.; Yeh, K.M.; Siu, L.K.; Lin, J.C.; Chang, F.Y. Regulator of the mucoid phenotype A gene increases the virulent ability of extended-spectrum beta-lactamase-producing serotype non-K1/K2 *Klebsiella pneumonia*. *J. Microbiol. Immunol. Infect. Wei Mian Yu Gan Ran Za Zhi* **2016**, *49*, 494–501. [CrossRef]

23. Liu, Y.Y.; Wang, Y.; Walsh, T.R.; Yi, L.X.; Zhang, R.; Spencer, J.; Doi, Y.; Tian, G.; Dong, B.; Huang, X.; et al. Emergence of plasmid-mediated colistin resistance mechanism MCR-1 in animals and human beings in China: A microbiological and molecular biological study. *Lancet Infect. Dis.* **2016**, *16*, 161–168. [CrossRef]

24. Valenzuela-Valderrama, M.; Gonzalez, I.A.; Palavecino, C.E. Photodynamic treatment for multidrug-resistant Gram-negative bacteria: Perspectives for the treatment of *Klebsiella pneumoniae* infections. *Photodiagn. Photodyn. Ther.* **2019**, *28*, 256–264. [CrossRef]

25. Tosato, M.G.; Schilardi, P.; Lorenzo de Mele, M.F.; Thomas, A.H.; Lorente, C.; Minan, A. Synergistic effect of carboxypterin and methylene blue applied to antimicrobial photodynamic therapy against mature biofilm of *Klebsiella pneumoniae*. *Heliyon* **2020**, *6*, e03522. [CrossRef] [PubMed]

26. Agostinis, P.; Berg, K.; Cengel, K.A.; Foster, T.H.; Girotti, A.W.; Gollnick, S.O.; Hahn, S.M.; Hamblin, M.R.; Juzeniene, A.; Kessel, D.; et al. Photodynamic therapy of cancer: An update. *CA Cancer J. Clin.* **2011**, *61*, 250–281. [CrossRef]

27. Briviba, K.; Klotz, L.O.; Sies, H. Toxic and signaling effects of photochemically or chemically generated singlet oxygen in biological systems. *Biol. Chem.* **1997**, *378*, 1259–1265. [PubMed]

28. Winterbourn, C.C.; Kettle, A.J. Redox reactions and microbial killing in the neutrophil phagosome. *Antioxid. Redox Signal.* **2013**, *18*, 642–660. [CrossRef]

29. Alvarez, L.A.; Kovacic, L.; Rodriguez, J.; Gosemann, J.H.; Kubica, M.; Pircalabioru, G.G.; Friedmacher, F.; Cean, A.; Ghise, A.; Sarandan, M.B.; et al. NADPH oxidase-derived H_2O_2 subverts pathogen signaling by oxidative phosphotyrosine conversion to PB-DOPA. *Proc. Natl. Acad. Sci. USA* **2016**, *113*, 10406–10411. [CrossRef]

30. Ziegelhoffer, E.C.; Donohue, T.J. Bacterial responses to photo-oxidative stress. *Nat. Rev. Microbiol.* **2009**, *7*, 856–863. [CrossRef]

31. Trindade, A.C.; de Figueiredo, J.A.; de Oliveira, S.D.; Barth Junior, V.C.; Gallo, S.W.; Follmann, C.; Wolle, C.F.; Steier, L.; Morgental, R.D.; Weber, J.B. Histopathological, Microbiological, and Radiographic Analysis of Antimicrobial Photodynamic Therapy for the Treatment of Teeth with Apical Periodontitis: A Study in Rats' Molars. *Photomed. Laser Surg.* **2017**, *35*, 364–371. [CrossRef] [PubMed]

32. Valenzuela-Valderrama, M.; Carrasco-Véliz, N.; González, I.A.; Dreyse, P.; Palavecino, C.E. Synergistic effect of combined imipenem and photodynamic treatment with the cationic Ir(III) complexes to polypyridine ligand on carbapenem-resistant *Klebsiella pneumoniae*. *Photodiagn. Photodyn. Ther.* **2020**, *31*, 101882. [CrossRef] [PubMed]

33. Bustamante, V.; Gonzalez, I.A.; Dreyse, P.; Palavecino, C.E. The mode of action of the PSIR-3 photosensitizer in the photodynamic inactivation of *Klebsiella pneumoniae* is by the production of type II ROS which activate RpoE-regulated extracytoplasmic factors. *Photodiagn. Photodyn. Ther.* **2020**, *32*, 102020. [CrossRef]

34. CLSI. *Performance Standards for Antimicrobial Susceptibility Testing*, 27th ed.; 2017. Available online: https://clsi.org/media/3481/m100ed30_sample.pdf (accessed on 2 January 2021).

35. Broberg, C.A.; Wu, W.; Cavalcoli, J.D.; Miller, V.L.; Bachman, M.A. Complete Genome Sequence of *Klebsiella pneumoniae* Strain ATCC 43816 KPPR1, a Rifampin-Resistant Mutant Commonly Used in Animal, Genetic, and Molecular Biology Studies. *Genome Announc.* **2014**, *2*. [CrossRef]

36. Ahn, D.; Penaloza, H.; Wang, Z.; Wickersham, M.; Parker, D.; Patel, P.; Koller, A.; Chen, E.I.; Bueno, S.M.; Uhlemann, A.C.; et al. Acquired resistance to innate immune clearance promotes *Klebsiella pneumoniae* ST258 pulmonary infection. *JCI Insight* **2016**, *1*, e89704. [CrossRef] [PubMed]

37. Sahly, H.; Aucken, H.; Benedi, V.J.; Forestier, C.; Fussing, V.; Hansen, D.S.; Ofek, I.; Podschun, R.; Sirot, D.; Tomas, J.M.; et al. Increased serum resistance in *Klebsiella pneumoniae* strains producing extended-spectrum beta-lactamases. *Antimicrob. Agents Chemother.* **2004**, *48*, 3477–3482. [CrossRef] [PubMed]

38. Ares, M.A.; Fernandez-Vazquez, J.L.; Rosales-Reyes, R.; Jarillo-Quijada, M.D.; von Bargen, K.; Torres, J.; Gonzalez-y-Merchand, J.A.; Alcantar-Curiel, M.D.; De la Cruz, M.A. H-NS Nucleoid Protein Controls Virulence Features of *Klebsiella pneumoniae* by Regulating the Expression of Type 3 Pili and the Capsule Polysaccharide. *Front. Cell. Infect. Microbiol.* **2016**, *6*, 13. [CrossRef]

39. Kobayashi, S.D.; Porter, A.R.; Dorward, D.W.; Brinkworth, A.J.; Chen, L.; Kreiswirth, B.N.; DeLeo, F.R. Phagocytosis and Killing of Carbapenem-Resistant ST258 *Klebsiella pneumoniae* by Human Neutrophils. *J. Infect. Dis.* **2016**, *213*, 1615–1622. [CrossRef]

40. González, I.; Gómez, J.; Santander-Nelli, M.; Natali, M.; Cortés-Arriagada, D.; Dreyse, P. Synthesis and photophysical characterization of novel Ir(III) complexes with a dipyridophenazine analogue (ppdh) as ancillary ligand. *Polyhedron* **2020**, *186*, 114621. [CrossRef]

41. González, I.; Natali, M.; Cabrera, A.R.; Loeb, B.; Maze, J.; Dreyse, P. Substituent influence in phenanthroline-derived ancillary ligands on the excited state nature of novel cationic Ir(iii) complexes. *New J. Chem.* **2018**, *42*, 6644–6654. [CrossRef]

42. Ishida, H.; Tobita, S.; Hasegawa, Y.; Katoh, R.; Nozaki, K. Recent advances in instrumentation for absolute emission quantum yield measurements. *Coord. Chem. Rev.* **2010**, *254*, 2449–2458. [CrossRef]

43. MacNair, C.R.; Stokes, J.M.; Carfrae, L.A.; Fiebig-Comyn, A.A.; Coombes, B.K.; Mulvey, M.R.; Brown, E.D. Overcoming mcr-1 mediated colistin resistance with colistin in combination with other antibiotics. *Nat. Commun.* **2018**, *9*, 458. [CrossRef] [PubMed]

44. Stokes, J.M.; MacNair, C.R.; Ilyas, B.; French, S.; Cote, J.P.; Bouwman, C.; Farha, M.A.; Sieron, A.O.; Whitfield, C.; Coombes, B.K.; et al. Pentamidine sensitizes Gram-negative pathogens to antibiotics and overcomes acquired colistin resistance. *Nat. Microbiol.* **2017**, *2*, 17028. [CrossRef]

45. Weng, N.P. Aging of the immune system: How much can the adaptive immune system adapt? *Immunity* **2006**, *24*, 495–499. [CrossRef] [PubMed]

46. Riou, M.; Avrain, L.; Carbonnelle, S.; El Garch, F.; Pirnay, J.P.; De Vos, D.; Plesiat, P.; Tulkens, P.M.; Van Bambeke, F. Increase of efflux-mediated resistance in *Pseudomonas aeruginosa* during antibiotic treatment in patients suffering from nosocomial pneumonia. *Int. J. Antimicrob. Agents* **2016**, *47*, 77–83. [CrossRef]

47. EUCAST. EUCAST Guidelines for Detection of Resistance Mechanisms and Specific Resistances of Clinical and/or Epidemiological Importance. 2016. Available online: https://www.eucast.org/resistance_mechanisms/ (accessed on 25 December 2020).

48. Brady, M.; Cunney, R.; Murchan, S.; Oza, A.; Burns, K. *Klebsiella pneumoniae* bloodstream infection, antimicrobial resistance and consumption trends in Ireland: 2008 to 2013. *Eur. J. Clin. Microbiol. Infect. Dis. Off. Publ. Eur. Soc. Clin. Microbiol.* **2016**, *35*, 1777–1785. [CrossRef] [PubMed]

49. Marques, C.; Menezes, J.; Belas, A.; Aboim, C.; Cavaco-Silva, P.; Trigueiro, G.; Telo Gama, L.; Pomba, C. *Klebsiella pneumoniae* causing urinary tract infections in companion animals and humans: Population structure, antimicrobial resistance and virulence genes. *J. Antimicrob. Chemother.* **2019**, *74*, 594–602. [CrossRef]

50. Russo, T.A.; Olson, R.; MacDonald, U.; Beanan, J.; Davidson, B.A. Aerobactin, but not yersiniabactin, salmochelin, or enterobactin, enables the growth/survival of hypervirulent (hypermucoviscous) *Klebsiella pneumoniae ex vivo* and *in vivo*. *Infect. Immun.* **2015**, *83*, 3325–3333. [CrossRef]

51. Vargas, J.M.; Moreno Mochi, M.P.; Nunez, J.M.; Caceres, M.; Mochi, S.; Del Campo Moreno, R.; Jure, M.A. Virulence factors and clinical patterns of multiple-clone hypermucoviscous KPC-2 producing *K. pneumoniae*. *Heliyon* **2019**, *5*, e01829. [CrossRef]

52. DeLeo, F.R.; Nauseef, W.M. Granulocytic Phagocytes. In *Mandell, Douglas, and Bennett's Principles and Practice of Infectious Diseases*; Bennett, J.E., Dolin, R., Blaser, M.J., Eds.; Elsevier: Philedelphia, PA, USA, 2014; Volume 1, pp. 78–92.

53. Rodgers, M.A.J. Solvent-induced deactivation of singlet oxygen: Additivity relationships in nonaromatic solvents. *J. Am. Chem. Soc.* **1983**, *105*, 6201–6205. [CrossRef]

54. Svanberg, K.; Bendsoe, N.; Axelsson, J.; Andersson-Engels, S.; Svanberg, S. Photodynamic therapy: Superficial and interstitial illumination. *J. Biomed. Opt.* **2010**, *15*, 041502. [CrossRef] [PubMed]

55. Baran, T.M.; Foster, T.H. Comparison of flat cleaved and cylindrical diffusing fibers as treatment sources for interstitial photodynamic therapy. *Med. Phys.* **2014**, *41*, 022701. [CrossRef] [PubMed]

56. Beyer, W. Systems for light application and dosimetry in photodynamic therapy. *J. Photochem. Photobiol. B Biol.* **1996**, *36*, 153–156. [CrossRef]

57. Wachter, E.; Heidary, D.K.; Howerton, B.S.; Parkin, S.; Glazer, E.C. Light-activated ruthenium complexes photobind DNA and are cytotoxic in the photodynamic therapy window. *Chem. Commun.* **2012**, *48*, 9649–9651. [CrossRef]

58. Edwards, L.; Turner, D.; Champion, C.; Khandelwal, M.; Zingler, K.; Stone, C.; Rajapaksha, R.D.; Yang, J.; Ranasinghe, M.I.; Kornienko, A.; et al. Photoactivated 2,3-distyrylindoles kill multi-drug resistant bacteria. *Bioorganic Med. Chem. Lett.* **2018**, *28*, 1879–1886. [CrossRef]

59. Li, F.; Collins, J.G.; Keene, F.R. Ruthenium complexes as antimicrobial agents. *Chem. Soc. Rev.* **2015**, *44*, 2529–2542. [CrossRef] [PubMed]

60. Hachey, A.C.; Havrylyuk, D.; Glazer, E.C. Biological activities of polypyridyl-type ligands: Implications for bioinorganic chemistry and light-activated metal complexes. *Curr. Opin. Chem. Biol.* **2021**, *61*, 191–202. [CrossRef]

61. Wozniak, A.; Rapacka-Zdonczyk, A.; Mutters, N.T.; Grinholc, M. Antimicrobials Are a Photodynamic Inactivation Adjuvant for the Eradication of Extensively Drug-Resistant *Acinetobacter baumannii*. *Front. Microbiol.* **2019**, *10*, 229. [CrossRef] [PubMed]

62. Pletzer, D.; Mansour, S.C.; Hancock, R.E.W. Synergy between conventional antibiotics and anti-biofilm peptides in a murine, sub-cutaneous abscess model caused by recalcitrant ESKAPE pathogens. *PLoS Pathog.* **2018**, *14*, e1007084. [CrossRef]

63. Kachynski, A.V.; Pliss, A.; Kuzmin, A.N.; Ohulchanskyy, T.Y.; Baev, A.; Qu, J.; Prasad, P.N. Photodynamic therapy by in situ nonlinear photon conversion. *Nat. Photonics* **2014**, *8*, 455–461. [CrossRef]

64. Paez, P.L.; Bazan, C.M.; Bongiovanni, M.E.; Toneatto, J.; Albesa, I.; Becerra, M.C.; Arguello, G.A. Oxidative stress and antimicrobial activity of chromium(III) and ruthenium(II) complexes on *Staphylococcus aureus* and *Escherichia coli*. *BioMed Res. Int.* **2013**, *2013*, 906912. [CrossRef]

 *pharmaceutics*

Article

Susceptibility of Dental Caries Microcosm Biofilms to Photodynamic Therapy Mediated by Fotoenticine

Maíra Terra Garcia [1], Rafael Araújo da Costa Ward [1], Nathália Maria Ferreira Gonçalves [1], Lara Luise Castro Pedroso [1], José Vieira da Silva Neto [2], Juliana Ferreira Strixino [3] and Juliana Campos Junqueira [1,*]

[1] Department of Biosciences and Oral Diagnosis, Institute of Science and Technology/ICT, São Paulo State University/UNESP, São José dos Campos 12245-000, Brazil; maira.garcia@unesp.br (M.T.G.); r.ward@unesp.br (R.A.d.C.W.); nmf.goncalves@unesp.br (N.M.F.G.); lara.luise@unesp.br (L.L.C.P.)

[2] Associate Laboratory of Sensors and Materials/LABAS, National Institute for Space Research, São José dos Campos 12227-010, Brazil; jose.v.silva@inpe.br

[3] Photobiology Applied to Health, Research and Development Institute IP&D, University of Vale do Paraiba/UNIVAP, São José dos Campos 12244-390, Brazil; juferreira@univap.br

* Correspondence: juliana.junqueira@unesp.br

Abstract: Photodynamic therapy (PDT) mediated by Fotoenticine® (FTC), a new photosensitizer derived from chlorin e-6, has shown in vitro inhibitory activity against the cariogenic bacterium *Streptococcus mutans*. However, its antimicrobial effects must be investigated on biofilm models that represent the microbial complexity of caries. Thus, we evaluated the efficacy of FTC-mediated PDT on microcosm biofilms of dental caries. Decayed dentin samples were collected from different patients to form in vitro biofilms. Biofilms were treated with FTC associated with LED irradiation and analyzed by counting the colony forming units (log10 CFU) in selective and non-selective culture media. Furthermore, the biofilm structure and acid production by microorganisms were analyzed using microscopic and spectrophotometric analysis, respectively. The biofilms from different patients showed variations in microbial composition, being formed by streptococci, lactobacilli and yeasts. Altogether, PDT decreased up to 3.7 log10 CFU of total microorganisms, 2.8 log10 CFU of streptococci, 3.2 log10 CFU of lactobacilli and 3.2 log10 CFU of yeasts, and reached eradication of *mutans* streptococci. PDT was also capable of disaggregating the biofilms and reducing acid concentration in 1.1 to 1.9 mmol lactate/L. It was concluded that FTC was effective in PDT against the heterogeneous biofilms of dental caries.

Keywords: photodynamic therapy; dental caries; oral biofilms; Fotoenticine; chlorine; methylene blue

Citation: Garcia, M.T.; Ward, R.A.d.C.; Gonçalves, N.M.F.; Pedroso, L.L.C.; Neto, J.V.d.S.; Strixino, J.F.; Junqueira, J.C. Susceptibility of Dental Caries Microcosm Biofilms to Photodynamic Therapy Mediated by Fotoenticine. *Pharmaceutics* **2021**, *13*, 1907. https://doi.org/10.3390/pharmaceutics13111907

Academic Editor: Maria Nowakowska

Received: 9 October 2021
Accepted: 5 November 2021
Published: 10 November 2021

Publisher's Note: MDPI stays neutral with regard to jurisdictional claims in published maps and institutional affiliations.

1. Introduction

Dental caries is a multifactorial biofilm-mediated disease characterized by the mineral loss of tooth hard tissues [1,2]. Caries development is a result of dental biofilm acidification from carbohydrate consumption that alters the oral ecology, favoring the growth of acidogenic and acid-tolerating species [2–4]. Despite the restorative treatments available for dental caries, dentists commonly encounter failures in restorations associated with secondary caries [5]. Therefore, a proper disinfection of the dental cavity before the restorative procedure is essential to avoid the growth of bacteria under restorative material and caries recurrence [6,7]. In this context, alternative adjuvant methods to restorative treatment have been extensively investigated, such as antimicrobial photodynamic therapy (PDT).

The antimicrobial effect of PDT is based on a photophysical reaction through the association of a photosensitizer, light at an appropriate wavelength (usually from the visible to near infrared spectrum) and molecular oxygen [8–12]. In this process, the photosensitizer is activated by light to produce reactive oxygen species (ROS) that kill the microbial cells via an oxidative burst [8,13]. Therefore, the efficacy of PDT depends on the type of photosensitizer, parameters of irradiation, and application site [14]. In the last decades, different photosensitizers have been investigated, including phenothiazines and chlorine derivatives [8,9,15,16]. Phenothiazines, such as methylene blue (MB), are widely studied and known, being already approved for clinical dental application in several countries around the world [17]. Chlorines are tetrapyrrolic substances derived from chlorophyll that have gained attention due to their photodynamic effects on a large number of pathogens [18]. Some derivatives of chlorines have been described as potential photosensitizers with high absorption in the red region of the visible spectrum and mechanism of action predominantly via type II reaction, producing singlet oxygen [19]. Among them, Fotoenticine (FTC), a new photosensitizer derived from chlorin e-6, has demonstrated strong photodynamic activity against the cariogenic bacterium *Streptococcus mutans* [9,19,20].

Recently, Terra-Garcia et al. [9] and Nie et al. [19] verified that PDT mediated by FTC was able to reduce the viability of biofilms formed by a reference *S. mutans* strain (UA159). In both studies, the antimicrobial effects of PDT with FTC were greater than MB, suggesting that FTC can be a useful photosensitizer in the control of *S. mutans*. Later, our research group reported that photodynamic inactivation by FTC was extended to several clinical strains of *S. mutans* isolated from patients with the active disease [20]. In all of these clinical strains, FTC-mediated PDT was able to eliminate the viable cells and disrupt the dense structure of *S. mutans* biofilms [20].

Although previous studies have suggested that PDT mediated by FTC can be a promising strategy to prevent or treat caries, these studies were limited to monospecies biofilms that do not represent the microbial complexity of dental biofilms. It is known that dental biofilms are composed of different microbial species that form a highly complex and organized structure on the tooth tissues, in which each microorganism has specific positions and functions. More than 700 bacterial species have been identified from dental biofilm, and approximately 40 species were already associated with caries [21]. In addition to *S. mutans*, other acidogenic and aciduric microorganisms in the biofilm may play an important role in dental caries, such as *Streptococcus sobrinus*, *Lactobacillus* spp., *Veillonella* spp., *Actinomyces* spp. and *Candida* spp. [22].

To overcome the limitations of in vitro biofilm studies with one or three selected species, many researchers have introduced different models of microcosm biofilms. Microcosms are more reliable models to simulate the in situ conditions of the oral cavity. Microcosm biofilms consist of a set of microorganisms from the natural oral microbiota that are formed in vitro using samples collected from the oral cavity, such as saliva or biofilms [23]. Thus, microcosm biofilms are capable of reproducing the biodiversity, ecological interactions, acid production and pH conditions of dental biofilms in the human oral cavity [24]. Using a microcosm model, the objective of this study was to evaluate the antimicrobial activity of PDT mediated by FTC on dental caries biofilm, investigating its effects on the microbial cells, biofilm structure and acid production.

2. Materials and Methods

2.1. Collection of Dental Caries Samples

This study was approved by the Ethics Committee for Research in Human Beings of the Institute of Science and Technology (ICT/UNESP) under protocol number 158617/2019, date of approval: 13 November 2020). Three patients of the dental clinics in the city of São José dos Campos were selected. Inclusion criteria were age between 20 and 40 years and presence of at least one dentin caries lesion in a molar tooth. The non-inclusion criteria were the use of antibiotics in the last 90 days and presence of caries lesions with pulp exposure.

The collection of samples from infected dentin was carried out after anesthesia and absolute isolation with a rubber dam. The material was collected with a sterile curette, transferred to brain heart infusion (BHI) broth with 20% glycerol and stored in a freezer at $-80\,^{\circ}\mathrm{C}$. Teeth were then restored according to pre-determined treatment for each patient.

2.2. Preparation of Photosensitizer and Light Source

The photosensitizer Fotoenticine® (Nuevas Tecnologias Cientificas, NTC, Lianera Asturias, Spain) obtained at a concentration of 6.89 mg/mL was sterilized by filtration on membranes with pores of 0.22 μm (MFS, Dublin, Ireland) and stored in dark conditions. The chemical structure and absorption spectrum of the Fotoenticine® are showed in Figure 1. The light source was a LED at 660 nm (IrradLED®, Biopdi, São Carlos, Brazil) with a power density of 42.8 mW/cm², energy density of 30 J/cm² and exposure time of 700 s.

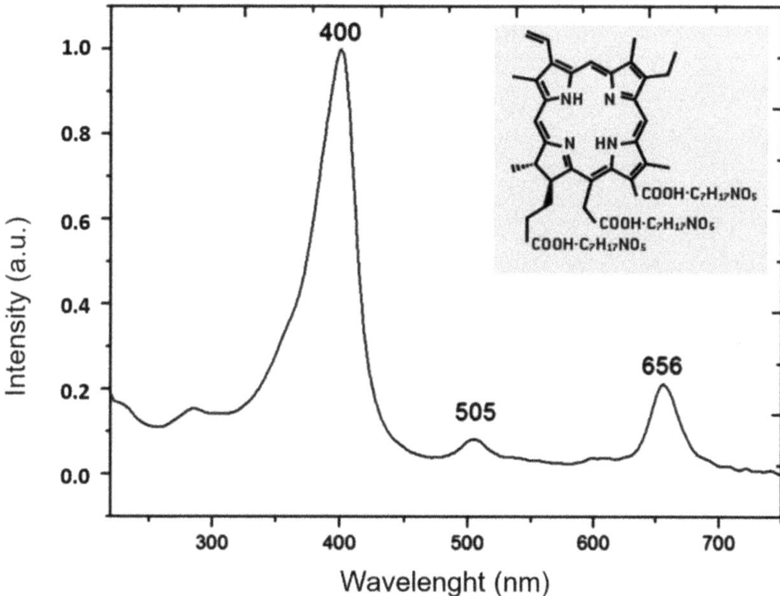

Figure 1. Chemical structure [25] and absorption spectrum of Fotoenticine (200 μg/mL in 0.9% NaCl) determined with a spectrophotometer (DS-11, Denovix, DE, USA).

2.3. Preparation of Bovine Tooth Specimens

Bovine teeth were employed as substrate to form in vitro biofilms following the methodology described by Garcia et al. [9] with some modifications. Mandibular incisor teeth were used to prepare specimens measuring 4 mm in diameter and 2 mm in thickness. In summary, the crowns were separated from the roots using a straight handpiece with a diamond disc. The crowns were attached to a circular sample cutter with the buccal side facing upwards and cut with a trephine drill. Then, specimens were inserted into a metal matrix (with the dentin surface positioned in its interior) and polished with sandpaper discs of decreasing grain size in a low-speed water-cooling system (DP-10, Panambra Industrial e Técnica, São Paulo, Brazil) (Figure 2). The final thickness was confirmed with a digital caliper (Starrett, São Paulo, Brazil). Finally, the specimens were stored in a 0.1% (*v/v*) thymol solution at 4 °C.

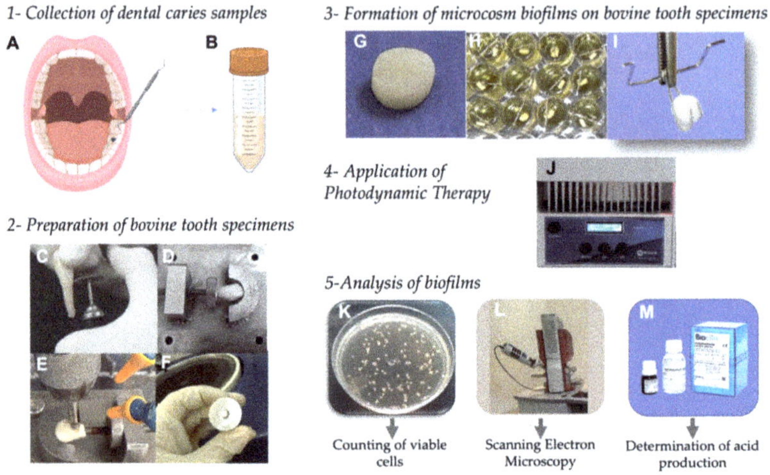

Figure 2. (**A**) Representation of sample collection from infected dentin of patient. (**B**) Sample transferred to brain heart infusion (BHI) broth with 20% glycerol. (**C**) Separation of crown from the root of bovine tooth using a straight handpiece with a diamond disc. (**D**) Crown attached to a circular sample cutter with the buccal side facing upwards. (**E**) Tooth cut with a trephine drill. (**F**) Specimen inserted into a metal matrix for polishing. (**G**) Bovine specimen prepared. (**H**) Specimen positioned in the metal wire and submerged in BHI broth to form the biofilms. (**I**) Biofilm formed on bovine specimen. (**J**) Irradiation of biofilms with LED at 660 nm. (**K**) Colonies formed on BHI agar. (**L**) Scanning electron microscope. (**M**) Lactic dehydrogenase LDH UV K014 (Bioclin) to determine the acid production.

2.4. Formation of Microcosm Biofilms on Bovine Tooth Specimens

Microcosm biofilms were formed according to the methodology of Méndez et al. with some modifications [24]. Dental caries samples stored in freezer were thawed and homogenized. Then, 400 µL were transferred to 20 mL of BHI broth supplemented with 5% sucrose and incubated at 37 °C in 5% CO_2 for 48 h. The growth was centrifuged and washed twice in phosphate buffered saline (PBS). The pellet was resuspended in 20 mL of PBS. An inoculum of 225 µL of this microbial suspension was added in each well of 24-well microplates, containing the bovine tooth specimen suspended by a metal wire and submerged in 2 mL of BHI broth with 5% sucrose. The plates were incubated at 37 °C for 120 h to form the biofilms, under 5% CO_2 pressure since most cariogenic bacteria are capnophilic [26]. Every 24 h, the wells were washed (2×) with PBS, and 2 mL of a fresh BHI broth with 5% sucrose was replaced.

2.5. Application of Photodynamic Therapy on Microcosm Biofilms

After biofilm formation, the specimens were removed from the metal wire support, transferred to a new well and submerged in 200 µL of photosensitizer or PBS. To obtain an antimicrobial effect on biofilms, the photosensitizer was used at a concentration of 0.6 mg/mL that is considered able to diffuse through *S. mutans* biofilms [20]. The plates were shaken for 15 min (pre-irradiation time) in dark conditions. After that, the plates were taken to the IrradLED® and irradiated according to the parameters described above. The control groups not irradiated remained in dark conditions for the same period. According to these procedures, four groups of microcosm biofilms were established: treated with FTC and irradiated (FTC + L+), treated with PBS and irradiated (P − L+), treated with FTC in the dark (FTC + L−) and treated with PBS in the dark (P − L−).

2.6. Analysis of Biofilms by Counting the Number of Viable Cells

The specimens containing the treated biofilms (five per group) were transferred to Falcon tubes containing 4 mL of PBS. The adhered biofilms were detached using an ultrasonic homogenizer (Sonopuls HD2200, Bandelin Eletronic, Berlin, Germany) with a power of 7 W for 30 s. Serial dilutions of the biofilm suspension were performed and plated on BHI agar for counting the total number of microorganisms, as well as on the following selective culture media: Mitis Salivarius for counting streptococci, Mitis Salivarius Bacitracin Sucrose (MSBS) agar for mutans streptococci, Rogosa agar for lactobacilli, and Sabouraud Dextrose agar with chloramphenicol for yeasts. Next, the plates were incubated at 37 °C for 48 h to determine the number of colony forming units (CFU).

2.7. Analysis of Biofilms by Scanning Electron Microscopy (SEM)

After the treatments, the specimens with biofilms (two per group) were submerged in 1 mL of 2.5% glutaraldehyde (1 h) for cell fixation. After that, they were dehydrated with 1 mL of increasing ethanol concentrations (10, 25, 50, 75 and 90%), remaining for 20 min at each concentration. Finally, the biofilms were added to absolute ethanol for 1 h. The dehydrated biofilms were incubated at 37 °C for 24 h, and then placed in aluminum stubs and covered with gold for 160 s at 40 mA (Denton Vacuum Desk II, Denton Vacuum, Moorestown, NJ, USA). After metallization, the biofilms were analyzed in the scanning electron microscope (JEOL JSM-7900, JEOL, Peabody, MA, USA).

2.8. Determination of Lactic Acid Production by the Microbial Cells in Biofilms

The specimens with the treated biofilms (5 per group) were transferred to Falcon tubes containing 5 mL of PBS and incubated at 37 °C for 3 h. Then, 1 mL of each tube was placed in a freezer at −80 °C for 5 min to stop the acid production. To verify the concentrations of lactate, the enzymatic method of lactic dehydrogenase was used according to the manufacturer's instructions (Lactic dehydrogenase LDH UV K014, Bioclin, Belo Horizonte, Brazil). Absorbance was measured at 340 nm using a microplate spectrophotometer (Epoch, Biotek, Winooski, VT, USA), and the obtained values were expressed as mmol lactate/L.

2.9. Statistical Analysis

The results obtained in the viable cell counts (log10 CFU) and acid production were analyzed by ANOVA and Tukey test using the GraphPad Prism 8.4.3 program (GraphPad Software, Inc., San Diego, CA, USA). For statistical analysis, five biofilms per group were used. The assays were repeated twice. Level of significance of 5% was adopted in all the analyses.

3. Results

3.1. Effects of FTC-Mediated PDT on Viable Cells of Microcosm Biofilms

In the viable cell counts of the non-treated biofilms (control group P − L−), the microbial growth in non-selective culture media (BHI) was 7.21 log10 CFU for patient 1, 8.09 log10 CFU for patient 2 and 10.31 log10 CFU for patient 3. These data showed that the microcosm biofilm of patient 3 had the highest quantity of microorganisms, followed by patient 2 and 1. In the selective culture media, all patients showed cariogenic biofilms formed by streptococci (5 to 8 log10 CFU), mutans streptococci (5 to 7 log10 CFU) and lactobacilli (7 to 8 log10 CFU). However, the presence of yeasts was found only in microcosm biofilms of patients 2 and 3 (9 log10 CFU) (Figure 3).

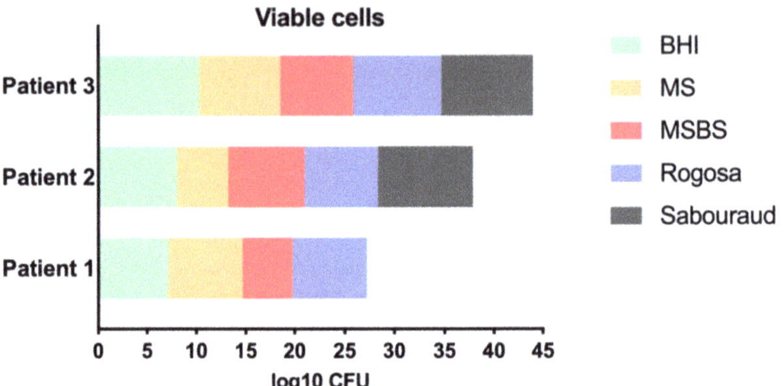

Figure 3. Viable cell counts (log10 CFU) of microcosm biofilms from patients 1, 2 and 3 according to the following culture mediums: brain heart infusion (total number of microorganisms), Mitis Salivarius (streptococci), Mitis Salivarius Bacitracin Sucrose (mutans streptococci), Rogosa (lactobacilli), and Sabouraud Dextrose with chloramphenicol (yeasts).

In relation to the viable cell counts of the biofilms treated with PDT (FTC + L+), we observed statistically significant microbial reductions relative to the groups P − L− and P − L+. These microbial reductions occurred in both selective and non−selective culture media for all the patients studied. In the non-selective medium, the FTC + L+ caused reductions from 1.8 to 3.7 log10 CFU for the total microorganisms compared to the P − L− group. Significant microbial reductions were also found in the FTC + L+ group in all the selective culture media, indicating that the FTC-mediated PDT was effective in inhibiting the growth of different microorganisms associated with cariogenic biofilms, including streptococci, mutans streptococci, lactobacilli and yeasts. In addition, the group treated with FTC alone (FTC + L−) showed a reduction from 0.2 to 1.2 log10 CFU in comparison to the P − L− group, suggesting a slightly toxic effect of FTC in dark conditions (Figure 4).

Considering the biofilms of all patients, the mean of microbial reductions achieved by FTC-mediated PDT (in comparison to the P − L− group) ranged from 2.1 to 2.8 log10 CFU for streptococci, 1.8 to 3.2 log10 CFU for lactobacilli and 1.7 to 3.2 log10 CFU for yeasts. Interestingly, mutans streptococci had a greater susceptibility to PDT relative to the other microorganisms. FTC-mediated PDT caused a mutans streptococci reduction of 6.02 log10 CFU for the biofilm of patient 3 and a total eradication for patients 1 and 2 (Figure 5).

3.2. Effects of FTC-Mediated PDT on Microcosm Biofilm Structures

The SEM images confirmed that all the caries samples collected from patients were able to provide an in vitro formation of microcosm biofilms on the surface of bovine dentin specimens. However, the morphology and structure of the biofilms varied according to the experimental groups. In all the patient samples, the P − L− and P − L+ groups presented a dense and compact biofilm, forming a complex three-dimensional structure. These biofilms were composed of a large quantity of microbial aggregates embedded in an extracellular matrix, covering the entire dentin surface. Different morphologies of microorganisms were found, including coccus, coccobacilli, filamentous bacilli, fusiform bacilli and fungal hyphae, which represented the microbial diversity of cariogenic biofilms.

The FTC + L− group also showed a dense biofilm formed by a large number of microorganisms; however, the biofilm density was lower than that in the P − L− and P − L+ groups. When the microcosm biofilms were treated with PDT (FTC + L+ group), a substantial reduction in the amount of microorganisms adhered to tooth surface was clearly observed. The PDT was capable of disaggregating the biofilms, exposing the dentin surface formed by several dentinal tubules. The microorganisms on dentin surface exhibited few cellular aggregates or isolated cells, allowing a more precise observation of microbial morphologies (Figure 6).

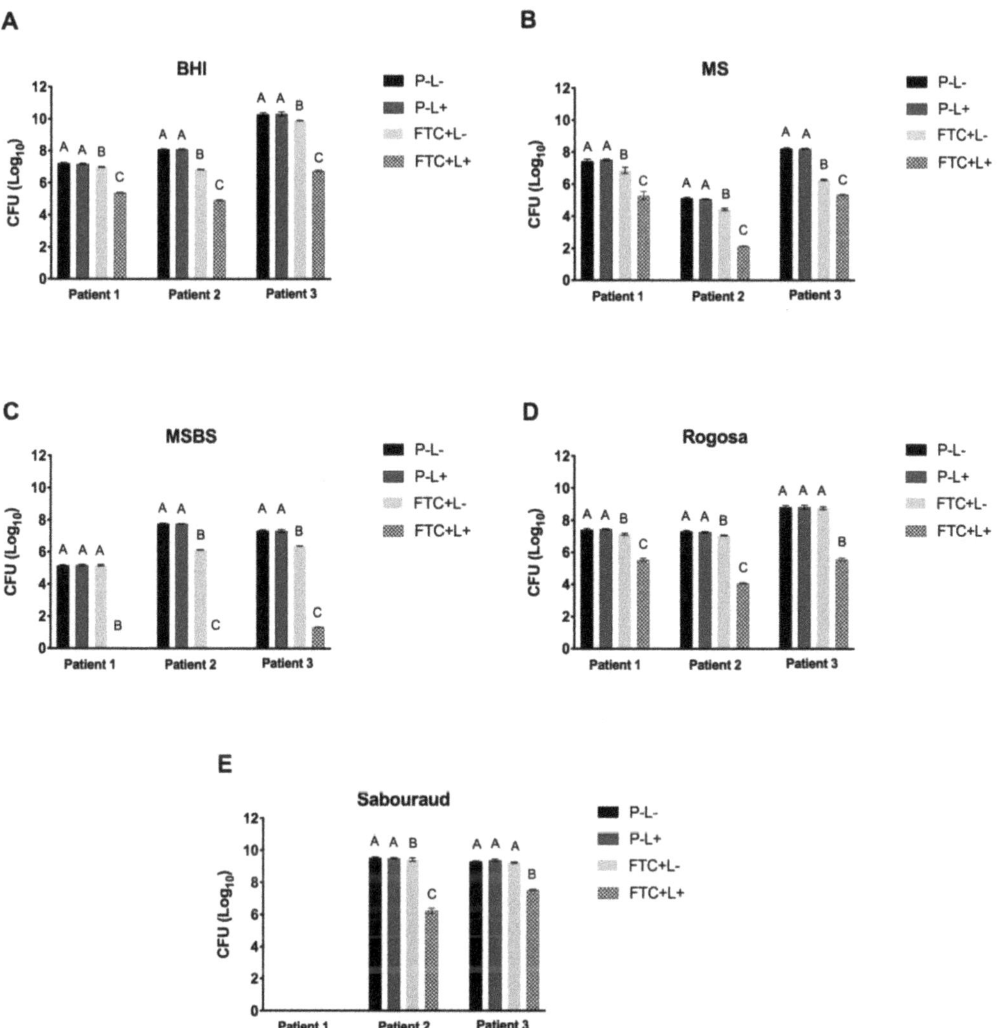

Figure 4. Means of log10 CFU and SD of microcosm biofilms (patients 1, 2 and 3) obtained in the following groups: treated with PBS in the dark (P − L−), treated with PBS and irradiated (P − L+), treated with FTC in the dark (FTC + L−) and treated with FTC and irradiated (FTC + L+). (**A**) Total microorganisms in brain heart infusion agar. (**B**) Streptococci in Mitis Salivarius agar. (**C**) Mutans streptococci in Mitis Salivarius Bacitracin Sucrose (MSBS) agar. (**D**) Lactobacilli in Rogosa agar. (**E**) Yeasts in Sabouraud Dextrose agar with chloramphenicol.

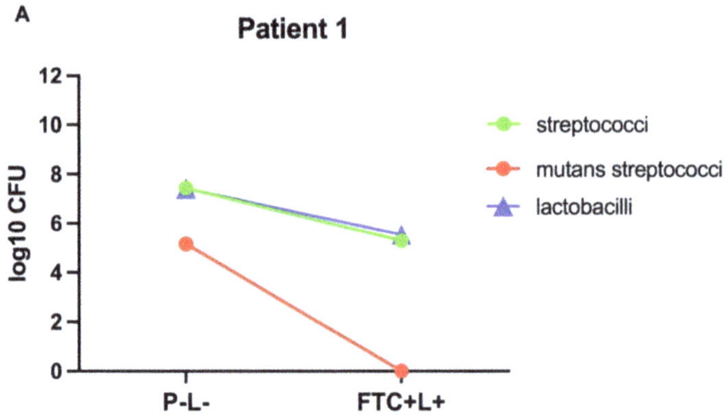

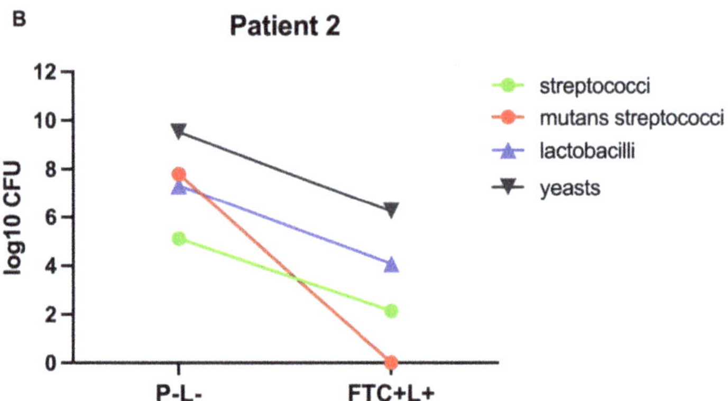

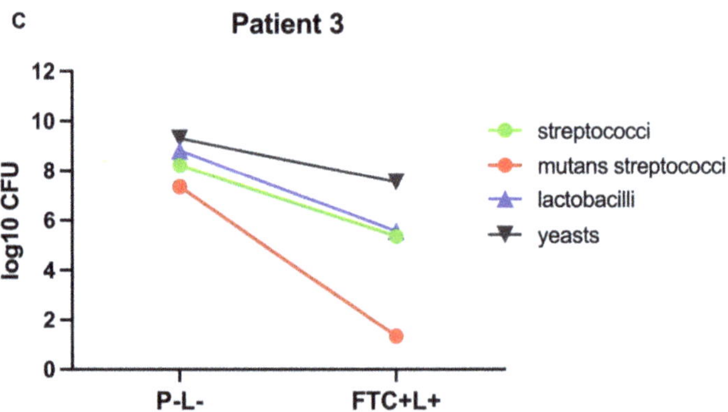

Figure 5. Microbial reductions of streptococci, mutans streptococci, lactobacilli and yeasts expressed in log10 CFU obtained in the group treated with FTC and irradiated (FTC + L+) in comparison to the control group treated with PBS in the dark (P − L−). (**A**) Microcosm biofilm from patient 1. (**B**) Microcosm biofilm from patient 2. (**C**) Microcosm biofilm from patient 3.

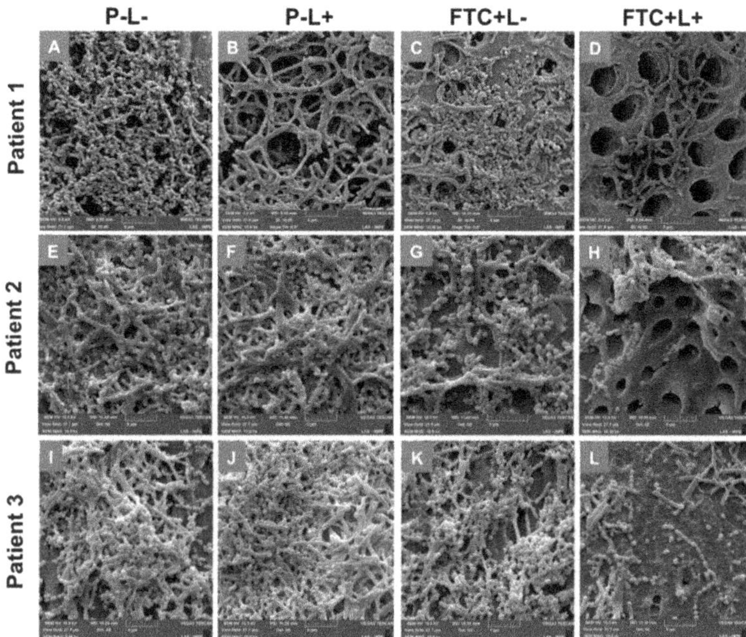

Figure 6. Scanning electron microscopy images according to the groups: treated with PBS in the dark (P − L−), treated with PBS and irradiated (P − L+), treated with FTC in the dark (FTC + L−) and treated with FTC and irradiated (FTC + L+). (**A–D**) Microcosm biofilm from patient 1. (**E–H**) Microcosm biofilm from patient 2. (**I–L**) Microcosm biofilm from patient 3.

3.3. Influence of FTC-Mediated PDT on Acid Production by the Microbial Cells of Microcosm Biofilms

In the biofilms not treated with PDT (P − L−, P − L+ and FTC + L− groups), the results of acid production ranged from 1.95 to 1.96 mmol lactate/L for patient 1, from 2.79 to 2.82 mmol lactate/L for patient 2, and from 3.80 to 3.84 mmol lactate/L for patient 3. As with the results obtained in the viable cell counts, patient 3 showed the highest quantity of acid production, followed by patient 2 and 1. These results indicated that the acidogenic capacity of caries microcosm biofilms was related to the number of microbial cells.

The biofilms treated with PDT (FTC + L+ group) showed statistically significant reductions of acid production when compared to the P − L− group for all patients studied. The acid reductions caused by PDT were 1.28, 1.92 and 0.99 mmol lactate/L, respectively, for patients 1, 2 and 3 (Figure 7). In general, it was possible to observe that PDT mediated by FTC was able to kill the microbial cells with consequent disruption of biofilm structure and decreased acidogenicity.

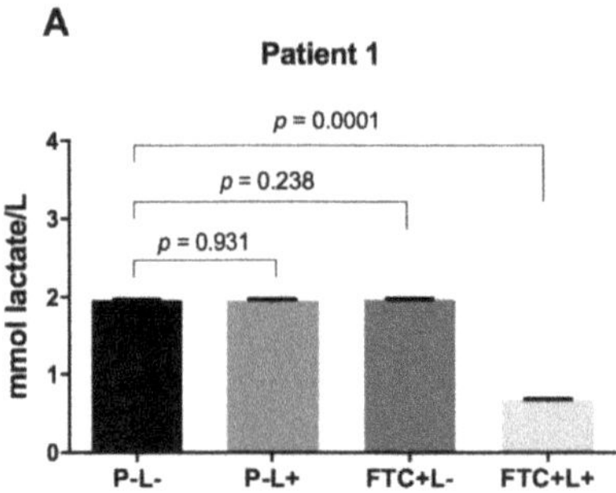

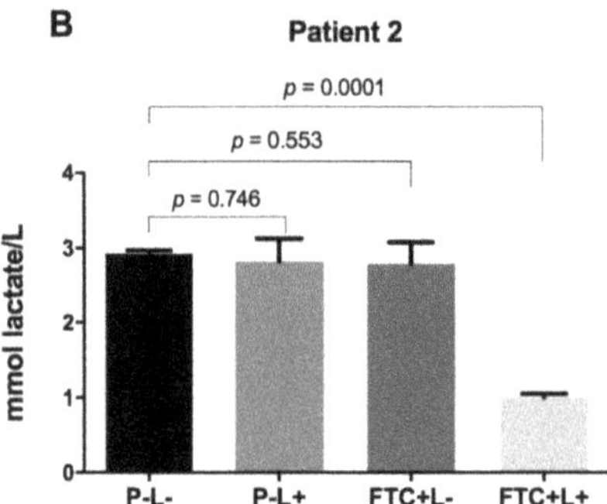

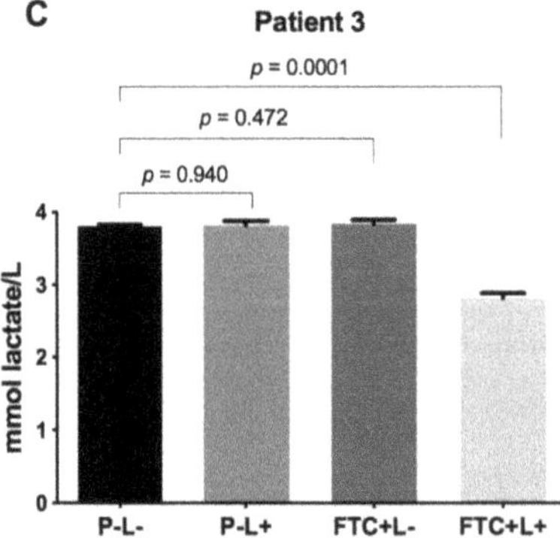

Figure 7. Means and SD of lactic acid concentration (mmol lactate/L) for the groups: treated with PBS in the dark (P − L−), treated with PBS and irradiated (P − L+), treated with FTC in the dark (FTC + L−) and treated with FTC and irradiated (FTC + L+). (**A**) Microcosm biofilm from patient 1. (**B**) Microcosm biofilm from patient 2. (**C**) Microcosm biofilm from patient 3.

4. Discussion

Cariogenic biofilms are active and complex ecosystems rich in organic acid [27], which form a spatial and ordered organization of microbial communities on tooth surfaces [28]. The dental surfaces and the development of caries lesions provide different microenvironments that allow the colonization of adapted microbial communities, leading to variations in the bacterial amount and microbial composition of cariogenic biofilms according to the caries stage [29,30]. In addition, several individual factors can influence the biofilm diversity, such as the host's diet, oral hygiene habits, tooth morphology, saliva composition, and discrepancies in the immune system [27,31]. Considering the variations in cariogenic biofilms among different individuals, in this study we collected samples from three patients with dentin caries to form in vitro microcosm biofilms and to investigate the potential of FTC-mediated PDT for the control of dental caries.

The microcosm biofilm obtained from patient 3 (10.31 log10 CFU) showed the highest total microbial count, followed by patient 2 (8.09 log10 CFU) and 1 (7.21 log10 CFU). In the three patients analyzed, the results were superior to those found by Mendéz et al. [24], who obtained a total microbial count of approximately 6.5 log10 CFU in the microcosm biofilms from dentin caries of children. Possibly, these differences can be attributed to the age of patients because in our study, only adults were included. However, Rudney et al. [32] did not observe differences in the quantity of microorganisms between adult and pediatric patients when microcosm biofilms were formed from saliva or dental plaque. The total microbial count ranged from 8 to 10 log10 CFU, independently of the inoculum source. Although the quantity of microorganisms formed in microcosm biofilms can vary depending on the study in the literature, it is evident that the microcosm model is able to provide very dense biofilms that can be useful for a large number of in vitro studies.

In this study, the microcosm biofilms from all patients were composed of streptococci, mutans streptococci and lactobacilli. In addition, the microcosm biofilms of patients 2 and 3 presented yeasts as well. Indeed, caries is closely associated with high counts of streptococci (particularly *Streptococcus mutans*), lactobacilli and yeasts of the genus *Candida* [2,33]. *Streptococcus* spp. and *Lactobacillus* spp. exhibit special properties that make them cariogenic bacteria, including the ability to adhere to dental surfaces, to produce a high quantity of acids from fermentable sugars, and to survive in an acid environment [34]. *Candida* species have been frequently isolated from dental caries lesions and also display acidogenic and aciduric properties. The presence of *Candida albicans* in cariogenic biofilms seems to enhance the virulence of *S. mutans* and, consequently, the severity of dental caries [35,36]. Many other bacterial genera have also been associated with cariogenic biofilms, such as *Actionomyces*, *Bifidobacteria*, *Veilonella*, *Prevotella* [2,34], and more recently, *Scardovia* [2]. However, the present study was limited to the identification of streptococci, mutans streptococci, lactobacilli and yeasts in the microcosm biofilms. Therefore, additional studies with a focus on other microbial groups are required, especially those that use molecular tools.

The viable cell counts in the microcosm biofilms of patients ranged from 5 to 8 log10 CFU for streptococci, 5 to 7 log10 CFU for mutans streptococci, 7 to 8 log10 CFU for lactobacilli and 0 to 9 log10 CFU for yeasts. Similar results were found by Vertuan et al. [37], Câmara et al. [38], and Mendéz et al. [24], who obtained bacterial counts between 6 to 7 log10 CFU for streptococci, 5 to 7 log10 CFU for mutans streptococci and 6 to 7 log10 CFU for lactobacilli in microcosm biofilms from oral samples. Nonetheless, these authors did not investigate the presence of yeasts in the microcosm biofilms formed in their studies.

When the FTC-mediated PDT was applied on microcosm biofilms, the viable cell numbers of all microbial groups studied were significantly decreased. In general, the microcosm biofilms from patients showed reductions between 2.1 to 2.8 log10 CFU for streptococci, 6.0 log10 CFU to total inhibition for mutans streptococci, 1.8 to 3.2 log10 CFU for lactobacilli and 1.7 to 3.2 log10 CFU for yeasts after the PDT treatment. Interestingly, the antimicrobial effects of FTC observed in our study were more pronounced than other photosensitizers cited in the literature. Using microcosm biofilms from dentin caries of patients, some authors found microbial reductions of 1.5, 1.3 and 1.9 log10 CFU with the photosensitizer methylene blue [24] and 1.5, 1.7 and 2.3 log10 CFU when curcumin was used as photosensitizer [39]. Probably, the greater photodynamic activity of FTC can be related to the positive charges of chlorins, action predominantly via type II mechanism with singlet oxygen quantum yields from 0.5 to 0.8 [8], and high capacity to penetrate into the bacterial cells [13,19].

To our knowledge, this is the first study to investigate the effects of FTC-mediated PDT on microcosm biofilms. Until now, this new photosensitizer has been studied in single biofilms formed by *S. mutans*. Terra-Garcia [20] applied the FTC-mediated PDT on single biofilms of *S. mutans* formed by the reference strain UA159 or clinical strains isolated from dental caries, verifying microbial reductions between 5 to 6 log10 CFU for the clinical strains and total inhibition for the reference strain. Using a similar photosensitizer also derived from chlorin-e6, Nie et al. [19] found a bacterial reduction of 5 to 6 log10 CFU of *S. mutans* (UA159) in single biofilms treated with PDT. Promisingly, the results of the present study using microcosm models showed that *S. mutans* remained susceptible to PDT mediated by FTC even when mixed in a complex polymicrobial biofilm.

In addition to remaining susceptible to FTC-mediated PDT in microcosm biofilms, mutans streptococci were the microorganisms most susceptible to this therapy, being totally eradicated in the microcosm biofilms from patients 1 and 2. In relation to the biofilm of patient 3, mutans streptococci were also more susceptible to PDT than lactobacilli and yeasts. We hypothesized that the highest susceptibility of *S. mutans* to PDT mediated by FTC could be associated with the growth rate, microbial cell composition, or spatial organization of biofilms. The growth rates and generation times of *S. mutans* cells were more elevated than those of *Lactobacillus* cells [40], which could influence the structure and permeability of cell membrane. The cell composition of bacteria and yeast is also considered an important factor in the susceptibility to PDT. Due to the presence of a cell wall with porous layer of peptidoglycan, Gram-positive bacteria such as *S. mutans* are more susceptible to PDT than fungal cells such as *Candida* spp. that present a cell wall of chitin with a less porous layer of beta-glucan [41]. In addition, the spatial organization of *S. mutans* cells with other microorganisms in polymicrobial biofilms can affect their susceptibility to antimicrobial treatments [42]. Using dual-species model biofilms, previous studies showed that *S. mutans* use *Candida* cells as support for their adherence [43,44]. According to these authors, *S. mutans* cells have affinity for the *C. albicans* hyphae, appearing positioned over the hyphal surface in SEM images [43,44]. This fact may leave *S. mutans* more exposed to PDT action. However, future studies are required to unveil the spatial structure of cariogenic biofilms and to develop therapies targeting the biogeography of polymicrobial infections [28].

The SEM analysis performed in the present study confirmed the capacity of microcosm biofilms in reproducing the dental caries microbiome. The biofilms of non-treated groups showed a spatial structure formed by heterogeneous and complex communities embedded in an extracellular matrix. The FTC-mediated PDT was able to decrease the number of microbial cells and to disaggregate the biofilm structure, exposing the dentin surface. In a review article, Hu et al. [13] reported that PDT not only kills the bacterial and fungal cells present in biofilms, but the ROS generated can concomitantly degrade the matrix structure by attacking several biomolecules. During the PDT reaction, the burst of ROS causes oxidative damage in multiple non-specific targets, such as amino acids, nucleic acid bases, lipids, etc. Therefore, PDT can disaggregate the biofilms by a synergistic action on microbial cells and extracellular matrix. These factors make PDT an attractive approach for chronic biofilm infections [13].

Since the production of organic acids by microbial cells in biofilms is a determinant of the development of dental caries [34], in this study we also evaluated the influence of PDT on the acidogenicity of biofilms. The PDT mediated by FTC caused a substantial reduction in lactic acid production for the three biofilms analyzed in this study, probably as a consequence of the decrease in the number of acidogenic bacteria. On the other hand, Mendez et al. [24,39] did not observe reductions in lactic acid production when the dentin microcosm biofilms were treated with PDT mediated by methylene blue [24] or curcumin [39]. These authors did not find correlations between the reductions in CFU counts and lactic acid production, raising the hypothesis that the surviving cells could have increased their acid production capacity, which resulted in maintaining the acidogenicity of the biofilms. However, we think that the bacterial reductions caused by PDT with MB [24] or curcumin [39] may not have been enough to result in a decrease in acid production because the CFU reductions were lower than the microbial reductions found in our study using FTC.

In summary, we concluded that PDT mediated by FTC, a photosensitizer derived from chlorin e-6, has antimicrobial activity against dental caries microcosm biofilms. FTC-mediated PDT led to significant reductions in total microorganisms, streptococci, mutans group streptococci, lactobacilli and yeasts. Mutans group streptococci were the microorganisms most susceptible to PDT, showing total eradication after this therapy. In addition, PDT with FTC was able to disaggregate the biofilm structure and to reduce the lactic acid concentration. Promisingly, FTC can be an attractive photosensitizer for PDT targeted to the control of cariogenic biofilms and dental caries.

Author Contributions: J.C.J. and J.F.S. conceived and designed the experiments. M.T.G., R.A.d.C.W., N.M.F.G. and L.L.C.P. performed the experiments. J.V.d.S.N. contributed to SEM analysis. M.T.G. and J.C.J. analyzed the data. M.T.G., J.V.d.S.N. and J.C.J. wrote and revised the manuscript. All authors have read and agreed to the published version of the manuscript.

Funding: National Council for Scientific Development (CNPq): Researcher on Productivity PQ-1C (306330/2018-0) and research grant (408369/2018-3) for JCJ. Coordination for the Improvement of Higher Education Personnel (CAPES): scholarship for MTG. São Paulo Research Foundation (FAPESP): scholarships for RACW (2020/08345-0) and NMFG (2020/06909-4).

Institutional Review Board Statement: The study was conducted according to the guidelines of the Declaration of Helsinki, and approved by the Ethics Committee for Research in Human Beings of the Institute of Science and Technology (ICT/UNESP) (protocol code: 158617/2019, date of approval: 11/13/2020).

Informed Consent Statement: Informed consent was obtained from all subjects involved in the study.

Conflicts of Interest: The authors declare no conflict of interest.

References

1. Mathur, V.P.; Dhillon, J.K. Dental Caries: A Disease Which Needs Attention. *Indian J. Pediatrics* **2018**, *85*, 202–206. [CrossRef] [PubMed]
2. Chen, X.; Daliri, E.B.; Tyagi, A.; Oh, D.H. Cariogenic Biofilm: Pathology-Related Phenotypes and Targeted Therapy. *Microorganisms* **2021**, *9*, 1311. [CrossRef] [PubMed]
3. Esberg, A.; Sheng, N.; Marell, L.; Claesson, R.; Persson, K.; Boren, T.; Stromberg, N. *Streptococcus mutans* Adhesin Biotypes that Match and Predict Individual Caries Development. *EBioMedicine* **2017**, *24*, 205–215. [CrossRef] [PubMed]
4. Zhang, Q.; Ma, Q.; Wang, Y.; Wu, H.; Zou, J. Molecular mechanisms of inhibiting glucosyltransferases for biofilm formation in *Streptococcus mutans*. *Int. J. Oral Sci.* **2021**, *13*, 30. [CrossRef]
5. Avoaka-Boni, M.C.; Djolé, S.X.; Kaboré, W.A.D.; Gnagne-Koffi, Y.N.; Koffi, A.F. The causes of failure and the longevity of direct coronal restorations: A survey among dental surgeons of the town of Abidjan, Cote d'Ivoire. *J. Conserv. Dent.* **2019**, *22*, 270–274. [CrossRef]
6. Godbole, E.; Tyagi, S.; Kulkarni, P.; Singla, S.; Mali, S.; Helge, S. Efficacy of Liquorice and Propolis Extract Used as Cavity Cleaning Agents against *Streptococcus mutans* in Deciduous Molars Using Confocal Microscopy: An In Vitro Study. *Int. J. Clin. Pediatric Dent.* **2019**, *12*, 194–200. [CrossRef]
7. Galdino, D.Y.T.; da Rocha Leodido, G.; Pavani, C.; Goncalves, L.M.; Bussadori, S.K.; Benini Paschoal, M.A. Photodynamic optimization by combination of xanthene dyes on different forms of *Streptococcus mutans*: An in vitro study. *Photodiagnosis Photodyn. Ther.* **2021**, *33*, 102191. [CrossRef]
8. Cieplik, F.; Deng, D.; Crielaard, W.; Buchalla, W.; Hellwig, E.; Al-Ahmad, A.; Maisch, T. Antimicrobial photodynamic therapy—What we know and what we don't. *Crit. Rev. Microbiol.* **2018**, *44*, 571–589. [CrossRef]
9. Terra Garcia, M.; Correia Pereira, A.H.; Figueiredo-Godoi, L.M.A.; Jorge, A.O.C.; Strixino, J.F.; Junqueira, J.C. Photodynamic therapy mediated by chlorin-type photosensitizers against *Streptococcus mutans* biofilms. *Photodiagnosis Photodyn. Ther.* **2018**, *24*, 256–261. [CrossRef]
10. Marcolan De Mello, M.; De Barros, P.P.; de Cassia Bernardes, R.; Alves, S.R.; Ramanzini, N.P.; Figueiredo-Godoi, L.M.A.; Prado, A.C.C.; Jorge, A.O.C.; Junqueira, J.C. Antimicrobial photodynamic therapy against clinical isolates of carbapenem-susceptible and carbapenem-resistant *Acinetobacter baumannii*. *Lasers Med. Sci.* **2019**, *34*, 1755–1761. [CrossRef]

11. De Souza, C.M.; Garcia, M.T.; de Barros, P.P.; Pedroso, L.L.C.; da Costa Ward, R.A.; Strixino, J.F.; Melo, V.M.M.; Junqueira, J.C. Chitosan enhances the Antimicrobial Photodynamic Inactivation mediated by Photoditazine(R) against *Streptococcus mutans*. *Photodiagnosis Photodyn. Ther.* **2020**, *32*, 102001. [CrossRef]

12. Martins Antunes de Melo, W.C.; Celiesiute-Germaniene, R.; Simonis, P.; Stirke, A. Antimicrobial photodynamic therapy (aPDT) for biofilm treatments. Possible synergy between aPDT and pulsed electric fields. *Virulence* **2021**, *12*, 2247–2272. [CrossRef]

13. Hu, X.; Huang, Y.Y.; Wang, Y.; Wang, X.; Hamblin, M.R. Antimicrobial Photodynamic Therapy to Control Clinically Relevant Biofilm Infections. *Front. Microbiol.* **2018**, *9*, 1299. [CrossRef]

14. Santin, G.C.; Oliveira, D.S.; Galo, R.; Borsatto, M.C.; Corona, S.A. Antimicrobial photodynamic therapy and dental plaque: A systematic review of the literature. *Sci. World J.* **2014**, *2014*, 824538. [CrossRef]

15. Lacerda Rangel Esper, M.A.; Junqueira, J.C.; Uchoa, A.F.; Bresciani, E.; Nara de Souza Rastelli, A.; Navarro, R.S.; de Paiva Goncalves, S.E. Photodynamic inactivation of planktonic cultures and *Streptococcus mutans* biofilms for prevention of white spot lesions during orthodontic treatment: An in vitro investigation. *Am. J. Orthod. Dentofac. Orthop.* **2019**, *155*, 243–253. [CrossRef]

16. Chan, B.C.L.; Dharmaratne, P.; Wang, B.; Lau, K.M.; Lee, C.C.; Cheung, D.W.S.; Chan, J.Y.W.; Yue, G.G.L.; Lau, C.B.S.; Wong, C.K.; et al. Hypericin and Pheophorbide a Mediated Photodynamic Therapy Fighting MRSA Wound Infections: A Translational Study from In Vitro to In Vivo. *Pharmaceutics* **2021**, *13*, 1399. [CrossRef]

17. Hamblin, M.R. Antimicrobial photodynamic inactivation: A bright new technique to kill resistant microbes. *Curr. Opin. Microbiol.* **2016**, *33*, 67–73. [CrossRef]

18. Abrahamse, H.; Hamblin, M.R. New photosensitizers for photodynamic therapy. *Biochem. J.* **2016**, *473*, 347–364. [CrossRef]

19. Nie, M.; Deng, D.M.; Wu, Y.; de Oliveira, K.T.; Bagnato, V.S.; Crielaard, W.; Rastelli, A.N.S. Photodynamic inactivation mediated by methylene blue or chlorin e6 against Streptococcus mutans biofilm. *Photodiagnosis Photodyn. Ther.* **2020**, *31*, 101817. [CrossRef]

20. Terra-Garcia, M.; de Souza, C.M.; Ferreira Goncalves, N.M.; Pereira, A.H.C.; de Barros, P.P.; Borges, A.B.; Miyakawa, W.; Strixino, J.F.; Junqueira, J.C. Antimicrobial effects of photodynamic therapy with Fotoenticine on Streptococcus mutans isolated from dental caries. *Photodiagnosis Photodyn. Ther.* **2021**, *34*, 102303. [CrossRef]

21. Salli, K.M.; Ouwehand, A.C. The use of in vitro model systems to study dental biofilms associated with caries: A short review. *J. Oral Microbiol.* **2015**, *7*, 26149. [CrossRef] [PubMed]

22. Pires, J.G.; Braga, A.S.; Andrade, F.B.; Saldanha, L.L.; Dokkedal, A.L.; Oliveira, R.C.; Magalhaes, A.C. Effect of hydroalcoholic extract of *Myracrodruon urundeuva* All. and *Qualea grandiflora* Mart. leaves on the viability and activity of microcosm biofilm and on enamel demineralization. *J. Appl. Oral Sci.* **2019**, *27*, e20180514. [CrossRef] [PubMed]

23. Chevalier, M.; Ranque, S.; Precheur, I. Oral fungal-bacterial biofilm models in vitro: A review. *Med. Mycol.* **2018**, *56*, 653–667. [CrossRef] [PubMed]

24. Mendez, D.A.C.; Gutierrez, E.; Dionisio, E.J.; Oliveira, T.M.; Buzalaf, M.A.R.; Rios, D.; Machado, M.; Cruvinel, T. Effect of methylene blue-mediated antimicrobial photodynamic therapy on dentin caries microcosms. *Lasers Med. Sci.* **2018**, *33*, 479–487. [CrossRef]

25. De Almeida, R.M.S.; Fontana, L.C.; Dos Santos Vitorio, G.; Pereira, A.H.C.; Soares, C.P.; Pinto, J.G.; Ferreira-Strixino, J. Analysis of the effect of photodynamic therapy with Fotoenticine on gliosarcoma cells. *Photodiagnosis Photodyn. Ther.* **2020**, *30*, 101685. [CrossRef]

26. Fysal, N.; Jose, S.; Kulshrestha, R.; Arora, D.; Hafiz, K.A.; Vasudevan, S. Antibiogram pattern of oral microflora in periodontic children of age group 6 to 12 years: A clinicomicrobiological study. *J. Contemp. Dent. Pract.* **2013**, *14*, 595–600. [CrossRef]

27. Mosaddad, S.A.; Tahmasebi, E.; Yazdanian, A.; Rezvani, M.B.; Seifalian, A.; Yazdanian, M.; Tebyanian, H. Oral microbial biofilms: An update. *Eur. J. Clin. Microbiol. Infect. Dis.* **2019**, *38*, 2005–2019. [CrossRef]

28. Kim, D.; Koo, H. Spatial Design of Polymicrobial Oral Biofilm in Its Native Disease State. *J. Dent. Res.* **2020**, *99*, 597–603. [CrossRef]

29. Bhaumik, D.; Manikandan, D.; Foxman, B. Cariogenic and oral health taxa in the oral cavity among children and adults: A scoping review. *Arch. Oral Biol.* **2021**, *129*, 105204. [CrossRef]

30. De Jesus, V.C.; Khan, M.W.; Mittermuller, B.A.; Duan, K.; Hu, P.; Schroth, R.J.; Chelikani, P. Characterization of Supragingival Plaque and Oral Swab Microbiomes in Children with Severe Early Childhood Caries. *Front. Microbiol.* **2021**, *12*, 683685. [CrossRef]

31. Astasov-Frauenhoffer, M.; Kulik, E.M. Cariogenic Biofilms and Caries from Birth to Old Age. *Monogr. Oral Sci.* **2021**, *29*, 53–64. [CrossRef]

32. Rudney, J.D.; Chen, R.; Lenton, P.; Li, J.; Li, Y.; Jones, R.S.; Reilly, C.; Fok, A.S.; Aparicio, C. A reproducible oral microcosm biofilm model for testing dental materials. *J. Appl. Microbiol.* **2012**, *113*, 1540–1553. [CrossRef]

33. Klinke, T.; Urban, M.; Luck, C.; Hannig, C.; Kuhn, M.; Kramer, N. Changes in *Candida* spp., mutans streptococci and lactobacilli following treatment of early childhood caries: A 1-year follow-up. *Caries Res.* **2014**, *48*, 24–31. [CrossRef]

34. Meyer, F.; Enax, J.; Epple, M.; Amaechi, B.T.; Simader, B. Cariogenic Biofilms: Development, Properties, and Biomimetic Preventive Agents. *Dent. J.* **2021**, *9*, 88. [CrossRef]

35. Alkhars, N.; Zeng, Y.; Alomeir, N.; Al Jallad, N.; Wu, T.T.; Aboelmagd, S.; Youssef, M.; Jang, H.; Fogarty, C.; Xiao, J. Oral Candida Predicts *Streptococcus mutans* Emergence in Underserved US Infants. *J. Dent. Res.* **2021**, 220345211012385. [CrossRef]

36. Zhang, Q.; Qin, S.; Xu, X.; Zhao, J.; Zhang, H.; Liu, Z.; Chen, W. Inhibitory Effect of Lactobacillus plantarum CCFM8724 towards *Streptococcus mutans*- and *Candida albicans*-Induced Caries in Rats. *Oxidative Med. Cell. Longev.* **2020**, *2020*, 4345804. [CrossRef]

37. Vertuan, M.; Machado, P.F.; de Souza, B.M.; Braga, A.S.; Magalhaes, A.C. Effect of TiF4/NaF and chitosan solutions on the development of enamel caries under a microcosm biofilm model. *J. Dent.* **2021**, *111*, 103732. [CrossRef]

38. Frazao Camara, J.V.; Araujo, T.T.; Mendez, D.A.C.; da Silva, N.D.G.; de Medeiros, F.F.; Santos, L.A.; de Souza Carvalho, T.; Reis, F.N.; Martini, T.; Moraes, S.M.; et al. Effect of a sugarcane cystatin on the profile and viability of microcosm biofilm and on dentin demineralization. *Arch. Microbiol.* **2021**, *203*, 4133–4139. [CrossRef]

39. Cusicanqui Mendez, D.A.; Gutierres, E.; Jose Dionisio, E.; Afonso Rabelo Buzalaf, M.; Cardoso Oliveira, R.; Andrade Moreira Machado, M.A.; Cruvinel, T. Curcumin-mediated antimicrobial photodynamic therapy reduces the viability and vitality of infected dentin caries microcosms. *Photodiagnosis Photodyn. Ther.* **2018**, *24*, 102–108. [CrossRef]

40. Llena, C.; Almarche, A.; Mira, A.; Lopez, M.A. Antimicrobial efficacy of the supernatant of *Streptococcus dentisani* against microorganisms implicated in root canal infections. *J. Oral Sci.* **2019**, *61*, 184–194. [CrossRef]

41. Sharma, S.K.; Dai, T.; Kharkwal, G.B.; Huang, Y.Y.; Huang, L.; De Arce, V.J.; Tegos, G.P.; Hamblin, M.R. Drug discovery of antimicrobial photosensitizers using animal models. *Curr. Pharm. Des.* **2011**, *17*, 1303–1319. [CrossRef]

42. Kim, D.; Barraza, J.P.; Arthur, R.A.; Hara, A.; Lewis, K.; Liu, Y.; Scisci, E.L.; Hajishengallis, E.; Whiteley, M.; Koo, H. Spatial mapping of polymicrobial communities reveals a precise biogeography associated with human dental caries. *Proc. Natl. Acad. Sci. USA* **2020**, *117*, 12375–12386. [CrossRef]

43. Metwalli, K.H.; Khan, S.A.; Krom, B.P.; Jabra-Rizk, M.A. *Streptococcus mutans*, *Candida albicans*, and the human mouth: A sticky situation. *PLoS Pathog.* **2013**, *9*, e1003616. [CrossRef]

44. Barbosa, J.O.; Rossoni, R.D.; Vilela, S.F.; de Alvarenga, J.A.; Velloso Mdos, S.; Prata, M.C.; Jorge, A.O.; Junqueira, J.C. *Streptococcus mutans* Can Modulate Biofilm Formation and Attenuate the Virulence of *Candida albicans*. *PLoS ONE* **2016**, *11*, e0150457. [CrossRef]

 *pharmaceutics*

Communication

Fluorescence Lifetime Imaging Microscopy of Porphyrins in *Helicobacter pylori* Biofilms

Antonella Battisti [1], **Paola Morici** [1,2] **and Antonella Sgarbossa** [1,*]

[1] Istituto Nanoscienze—CNR and NEST—Scuola Normale Superiore, Piazza S. Silvestro 12, I-56127 Pisa, Italy; antonella.battisti@nano.cnr.it (A.B.); paola.morici@hsanmartino.it (P.M.)

[2] Ospedale Policlinico San Martino, Largo Rosanna Benzi 10, I-16132 Genova, Italy

* Correspondence: antonella.sgarbossa@nano.cnr.it

Abstract: Bacterial biofilm constitutes a strong barrier against the penetration of drugs and against the action of the host immune system causing persistent infections hardly treatable by antibiotic therapy. *Helicobacter pylori* (Hp), the main causative agent for gastritis, peptic ulcer and gastric adenocarcinoma, can form a biofilm composed by an exopolysaccharide matrix layer covering the gastric surface where the bacterial cells become resistant and tolerant to the commonly used antibiotics clarithromycin, amoxicillin and metronidazole. Antimicrobial PhotoDynamic Therapy (aPDT) was proposed as an alternative treatment strategy for eradicating bacterial infections, particularly effective for Hp since this microorganism produces and stores up photosensitizing porphyrins. The knowledge of the photophysical characteristics of Hp porphyrins in their physiological biofilm microenvironment is crucial to implement and optimize the photodynamic treatment. Fluorescence lifetime imaging microscopy (FLIM) of intrinsic bacterial porphyrins was performed and data were analyzed by the 'fit-free' phasor approach in order to map the distribution of the different fluorescent species within Hp biofilm. Porphyrins inside bacteria were easily distinguished from those dispersed in the matrix suggesting FLIM-phasor technique as a sensitive and rapid tool to monitor the photosensitizer distribution inside bacterial biofilms and to better orientate the phototherapeutic strategy.

Keywords: porphyrins; fluorescence lifetime imaging miscroscopy (FLIM); antimicrobial photodynamic therapy (aPDT); *Helicobacter pylori* biofilm

Citation: Battisti, A.; Morici, P.; Sgarbossa, A. Fluorescence Lifetime Imaging Microscopy of Porphyrins in *Helicobacter pylori* Biofilms. *Pharmaceutics* **2021**, *13*, 1674. https://doi.org/10.3390/pharmaceutics13101674

Academic Editors: Udo Bakowsky, Matthias Wojcik, Eduard Preis and Gerhard Litscher

Received: 3 September 2021
Accepted: 11 October 2021
Published: 13 October 2021

Publisher's Note: MDPI stays neutral with regard to jurisdictional claims in published maps and institutional affiliations.

1. Introduction

Helicobacter pylori (Hp) can chronically colonize the human stomach where it plays an essential role in the development of gastritis, gastroduodenal ulcers and gastric cancer [1]. A key contribution to the persistence and the recurrence of Hp infections is provided by the bacterial ability to efficiently form a biofilm. Biofilms are matrix-enclosed bacterial populations in close proximity to each other and adherent to surfaces or interfaces. Hp biofilms are characterized by an extensive 3D network of highly hydrated exopolysaccharides where extracellular DNA, extracellular proteins, and outer membrane vesicles are embedded [2]. This extracellular matrix (ECM) protects the bacterial community from immune system cells and antimicrobial drugs. As a matter of fact, conventional pharmacological therapies have become less effective and new alternative strategies to fight bacterial infections are emerging. In this regard, one of the most promising approaches to overcome the antibiotic resistance problems is the photodynamic therapy (PDT). PDT combines the use of light irradiation with the presence of a photosensitizer molecule, able to generate cytotoxic reactive oxygen species (ROS) upon illumination. When this strategy is directed against microorganisms, the process is called antimicrobial photodynamic therapy (aPDT). The in situ-generated ROS are able to destroy biomacromolecules thus causing cell death. Typically, a photosensitizing drug is delivered to the area of interest before the treatment, but PDT can be particularly effective when the target microorganism naturally produces endogenous photosensitizers such as in the case of Hp porphyrins [3]. In previous studies we spectroscopically characterized the composition of Hp porphyrin extracts, showing that protoporphyrin IX (PPIX) and coproporphyrin I (CPI) are two of the main endogenous photosensitizing species [4,5]. Thanks to their presence, visible light irradiation conveyed by an intragastric fiberoptic laser system or by a LED-based illuminator has been shown to efficiently cause Hp photokilling without the administration of exogenous photosensitizers [6–8]. The correlation between photophysical properties and photosensitizing ability of porphyrins makes it mandatory to carefully analyze their fluorescence properties when they are closely associated with the biological target [9,10]. It has long been known that the vast majority of bacterial pathogens produce fluorescing porphyrins emitting in the red spectral region when excited by 405 nm violet light [11]. Thus, the assessment of red fluorescence in biological samples can be a signal of bacterial infection. Fluorescence imaging has recently emerged as a rapid and non-invasive technique for real-time visualization of the occurrence of bacteria and of their spatial distribution in several biological tissues and matrixes. The detection of the presence, location, and extent of fluorescent bacteria in wounds, for example, can be exploited by clinicians for targeted sampling during biopsies or to selectively treat the infected area [12,13]. Among the several fluorescence-based techniques, FLIM brings about several advantages because, in contrast to fluorescence intensity, the fluorescence decay time is independent of the local concentration of fluorophores and of the experimental set up. The fluorescence lifetime is a specific molecular feature of the fluorophore that can be sensitive to its surroundings. Fluorophores exhibiting the same spectral distribution in emission and in excitation as well may have different lifetimes, indicative of different molecular species or different conformations of the same molecule [14]. The analysis of FLIM data can be simplified by the phasor approach, a frequency domain technique that allows the transformation of the signal from every pixel in the image to a point in the phasor plot [15]. The FLIM-phasor approach has been used to create a metabolic fingerprint of individual bacteria and pop-ulations by monitoring the lifetime of the autofluorescent molecule reduced nicotinamide adenine dinucleotide phosphate (NAD(P)H). While the fluorescence spectra of free and bound NAD(P)H are very similar, their lifetimes differ significantly and depend on the metabolic state of bacteria. Phasor fingerprints were generated for *Lactobacillus acidophilus* [16], *Escherichia coli*, *Salmonella enterica* serovar Typhimurium, *Pseudomonas aeruginosa*, *Bacillus subtilis*, and *Staphylococcus epidermidis* [17] in different metabolic state during the growth phase or under antibiotic stress as well.

In this work, the phasor analysis coupled to FLIM was applied to Hp biofilms in order to map the distribution of the bacterial porphyrins in different microenvironments. Our findings show that it is possible to discriminate and image different average lifetimes for porphyrins corresponding to different molecular states inside bacteria or dispersed in the ECM. Therefore, this method can be used to provide a real-time assessment and a rapid visualization of the photosensitizer distribution along with its degree of molecular packing not only in Hp biofilms but also in all the situations where it is important to know the uptake, localization pattern and interactions of the exogenously added dye to optimize the PDT.

2. Materials and Methods

2.1. Bacterial Strains and Culture Conditions

Bacterial strains Hp ATCC® 43504™ (laboratory-adapted) and Hp ATCC® 700824™ (J99, virulent, cagA+ and vacA+), supplied by LGC Standards S.r.l. (Milan, Italy), were selected for this study. Both strains were kept at −80 °C in Brucella Broth (BB, Thermo Fisher Scientific Remel Products, Lenexa, KS, USA) supplemented with 20% (*v/v*) glycerol and 10% (*v/v*) heat-inactivated fetal bovine serum (FBS, Gibco, Life Technologies, Carlsbad, CA, USA). Thawed bacteria were grown in BB supplemented with 10% (*v/v*) FBS and incubated in the dark at 37 °C in a microaerophilic atmosphere (CampyGen Compact, Oxoid Hampshire, UK) under shaking (170 rpm).

2.2. Biofilm Formation Assay

Hp strains from frozen stocks were cultured on Brucella agar plates supplemented with 7% laked horse blood (Oxoid, Hampshire, UK) and incubated for 3 days at 37 °C under microaerophilic conditions in accordance with previously published protocols [18]. A sterile cotton swab was used to harvest bacteria after incubation. Bacteria were then suspended in brain heart infusion (BHI) broth (Oxoid, Hampshire, UK) additioned with 0.5% β-cyclodextrin (Sigma-Aldrich, St. Louis, MO, USA) and 0.4% yeast extract (Oxoid, Hampshire, UK). Aliquots of 2 mL of this suspension, adjusted to 5×10^6 CFU/mL concentration, were inoculated in the wells of 12-well culture plates (BD Falcon, Franklin Lakes, NJ, USA), where sterilized round glass coverslips were placed vertically in order to promote the adhesion of Hp cells at the air-liquid interface. BHI broth (without bacteria) was the negative control. Plates were incubated for 4 days at 37 °C under microaerophilic conditions. Biofilms grown on the glass coverslips were washed twice with phosphate-buffered saline (PBS, Gibco, Life Technologies, Carlsbad, CA, USA) to remove excess bacterial cells and immediately set for imaging at room temperature.

2.3. Extraction of Porphyrins from Hp Strains

According to a previously published protocol [19] with minor modifications, Hp cells of different ages (1–4 days) were centrifugated at 4 °C ($7000 \times g$ for 10 min), washed in 20 mL pre-cooled buffer (0.05 M Tris, 2 mM EDTA, pH 8.2) and resuspended in 10 mL of the same buffer. An aliquot of 1.5 mL of a mixture of ethyl acetate and acetic acid (3:1, v/v) was added to the suspension and bacterial cells were lysed by sonication in ice (6 cycles as in the following: 30 s, stop, 60 s). Bacterial debris was removed by centrifugation at 4 °C ($7000 \times g$ for 10 min), and the organic layer was washed twice with distilled water. Por-phyrins were extracted from the organic phase by adding 150 µL of HCl 3M. After fast vortexing, this solution was centrifuged for 5 min at $7000 \times g$ and then the bottom aqueous layer was collected for chromatographic analyses.

2.4. High Performance Liquid Chromatography

The HPLC used for analyses was a Dionex Ultimate 3000 HPLC (ThermoFisher Scientific Inc., Waltham, MA, USA), equipped with a fraction collector, PDA detector and a (3 × 150 mm) Kinetex PFP column (Phenomenex Inc., Torrance, CA, USA) using water/formic acid 100/0.1 v/v (A) and acetonitrile/formic acid 100/0.1 v/v (B) as mobile phases, with a flow of 0.8 mL/min. Runs were performed in four steps as follows: 2 min at 25% B, 26 min with a linear gradient up to 95% B, 4 min purge step at 95% B and 8 min re-equilibration step.

2.5. Fluorescence Lifetime Imaging (FLIM)

Standard porphyrins PPIX and CPI were either analyzed in methanol solution (about 1×10^{-5} M) or crystallized from solution on a glass coverslip by slow evaporation of the solvent. The measurements were performed using a Leica TCS SP5 inverted confocal microscope (Leica Microsystems, Wetzlar, Germany). An external pulsed diode laser provided excitation at 405 nm and 640 nm, while a TCSPC acquisition card (PicoHarp 300, PicoQuant, Berlin, Germany) connected to internal spectral photomultipliers allowed for detection. On the basis on the lifetime values, laser repetition rate was fixed at 20 or 40 Hz. Image size was set to 256 × 256 pixels and scan speed was modulated between 200 and 400 Hz (lines per second). The detection wavelength range was set between 580 and 720 nm for λ_{ex} = 405 nm and between 650 and 750 nm for λ_{ex} = 640 nm thanks to the built-in AOBS detection system. About 200–300 photons per pixel were collected for each measurement, at a photon counting rate of 100–200 kHz. Data collected from different ROIs in the biofilms images were averaged. Globals for Images—SimFCS v.2 (Globals Software by LFD-UCI, Irvine, CA, USA, available at www.lfd.uci.edu, accessed on 1 September 2021) was used to calculate phasors from the FLIM images.

3. Results

3.1. Chromatographic Analysis of Hp Porphyrin Production

Bacterial extracts were obtained from Hp cultures of different ages (1–4 days) by means of a specific protocol and characterized by HPLC analyses as described in Materials and Methods. The main peaks in the chromatograms were attributed to fluorescent porphyrins and flavins produced by bacteria, according to previously published results [5,6]. HPLC allowed for a comparison between the concentrations of the fluorescent species at different ages of the bacterial cultures in order to follow the production of these molecules with time (Figure 1A). As shown by the increasing height of the peaks, the fluorophores accumulate during the observed time span, suggesting a continuous production and accumulation. To visualize changes in the global fluorophores concentration, the main known peaks in the chromatograms (7.0–8.5 min and 15.7–16.7 min) were integrated and the total calculated area was reported in Figure 1B as a function of culture age. After a 2-day lag phase, where no significant change can be noticed, the global concentration of fluorophores starts to grow reaching almost a tenfold increase after four days.

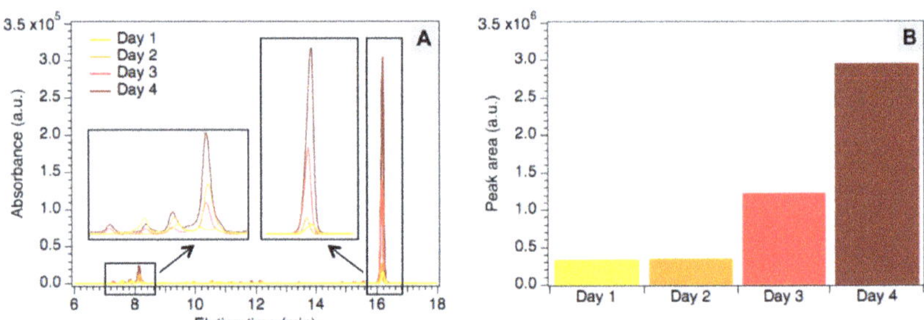

Figure 1. (**A**) Chromatograms of the extracts after 1 up to 4 days. Insets: expansion of peaks between 7.0 and 8.5 min and between 15.7 and 16.7 min (rescaled for better visualization). (**B**) Total area under the main peaks in the chromatograms (7.0–8.5 min and 15.7–16.7 min), corresponding to the global concentration of fluorescent species in the extracts.

3.2. FLIM-Phasor Analysis of Porphyrins in Hp Biofilms

Fluorescence microscopy of Hp biofilms revealed that bacterial porphyrins are present both inside bacterial cells and in the ECM [4], as they confer red fluorescence to the whole system. Biofilms obtained from Hp after four days of incubation were then observed by confocal laser scanning microscopy and analyzed by the FLIM-phasor technique developed by Jameson et al. [15]. Figure 2A shows a false-color image of a fluorescent Hp biofilm where bacteria can be easily distinguished from the matrix thanks to the higher fluorescence intensity. Conversion of the FLIM data to the phasor plot produced a cloud of pixels inside the universal circle (Figure 2B), revealing a multiexponential fluorescence lifetime decay. The corresponding phasor image in Figure 2C clearly shows that different areas of the biofilm (bacteria/ECM) are associated with different positions in the phasor plot. To inspect the behavior of porphyrins lifetime in different contexts, PPIX and CPI were also analyzed as monomers dissolved in methanol solutions and in their crystalline form. All these samples gave strictly monoexponential decays, which fall on the edge of the universal circle in the phasor plot (Figure 2B). As expected, crystalline porphyrins gave pixels clouds falling in the bottom-right area of the plot, corresponding to very short lifetimes values (<1 ns), while monomeric porphyrins in methanol solution produced pixels in the left part of the phasor plot, with lifetimes exceeding 10 ns.

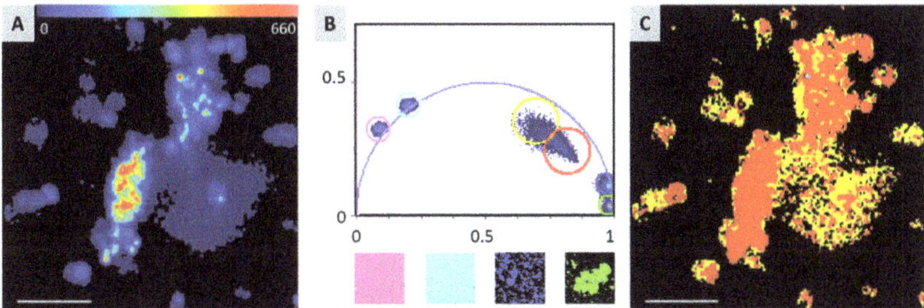

Figure 2. (**A**) False-color intensity-based image of bacterial biofilm. (**B** top) Overlap of the phasor plots of standard porphyrins PPIX (pink circle: methanol solution, green circle: crystal) and CPI (cyan circle: methanol solution, blue circle: crystal) and biofilm sample (red circle: bacteria, yellow circle: biofilm matrix). (**B** bottom) Phasor images of the standards, color code as in the plot. (**C**) Corresponding phasor image of the biofilm in A. Scale bar in A and C: 25 μm.

4. Discussion

.Bacterial biofilms are composed of living cells embedded in an extracellular polymeric matrix along with several exocellular molecules excreted by the cells or captured from the environment. When bacterial cells die, they release substances in the ECM, mainly DNA along with other endogenous cytoplasmic components. Dead cells can be a supplier for the ECM because the leakage of cytoplasmic components due to bacterial cell lysis has been shown to favor biofilm formation [20]. On this basis, it is not surprising to find endogenous Hp porphyrins both inside the bacterial cells and dispersed in the ECM of Hp biofilms.

The chromatographic analysis performed on Hp extracts after different periods of culture revealed that porphyrins are produced from the first day, but their concentration drastically increases after 4 days of incubation. Although the planktonic and biofilm growing conditions are rather different, after 4 days fully formed biofilms likewise provided a sufficient concentration of porphyrins for analysis. Accordingly, Hp biofilms were grown for 4 days and analyzed by FLIM coupled to the phasor approach.

It is known that intracellular lifetime measurements on porphyrins usually produce three lifetime values, generally attributed to the monomeric (>10 ns), dimeric (1.5–3.0 ns) and highly aggregated (<1 ns) forms [21], but these Hp biofilms do not show lifetimes >10 ns, as if no porphyrin monomers were present [4]. Nevertheless, the positioning of pixels inside the universal circle in the phasor plot matches with the presence of a combination of different lifetimes (Figure 2B). Monoexponential decays fall on the edge of the universal circle, as in the case of monomeric (pink and cyan circles) or crystalline (blue and green circles) porphyrins. The area where the biofilm pixels fall suggests that porphyrins are mostly present in several aggregated forms, likely in a mixture of dimers and higher oligomers, whose lifetimes are combined to produce the observed pixel cloud. Notably, the phasor analysis applied to FLIM acquisitions of Hp biofilms reveals different average lifetimes for porphyrins inside bacteria or dispersed in the ECM, as expressed by the slightly different position of the red and yellow circles in the plot of Figure 2B, corresponding to the red (bacteria) and yellow (matrix) areas in the phasor image (Figure 2C). This could be attributed either to a different composition between the intracellular and extracellular porphyrins mixture or to their different aggregation grade inside and outside the cells, along with the different environmental interactions. It is worth noting how the phasor approach allows for easy discrimination of the two different fluorescent areas.

5. Conclusions

To the best of our knowledge, in this work the FLIM-phasor approach was employed for the first time to easily discriminate porphyrins located in different fluorescent areas of Hp biofilms. This could be exploited to assess the extent of bacterial contamination and the presence of a biofilm inside an infected tissue, suggestive of an advanced stage of infection, thus simplifying the choice of the best treatment. Our findings indicate that porphyrins inside the cells and dispersed in the ECM show a slightly different global lifetime; likely, on the basis of the position of the relevant pixels cloud in the phasor plot, intracellular porphyrins may be organized in more packed structures. From the phototherapeutical point of view, this knowledge can be of great importance because, due to the short diffusion path of singlet oxygen and other cytotoxic reactive oxygen species, the more strictly the photosensitizer is bound to the cellular target, the more effective its photokilling activity will be. In addition, as the intracellular pigment concentration varies with the age of bacteria, the detection and the spatial distribution pattern of porphyrin fluorescence in Hp biofilm could provide some hints for the choice of the most appropriate therapeutical time window to perform aPDT. In general, this kind of analysis could be considered to be a useful and rapid tool for the exploration of fluorescent bacterial biofilms properties not only to obtain new insights into the biofilm structure and dynamics but also to develop and optimize novel antimicrobial strategies.

Author Contributions: Conceptualization, A.B., P.M. and A.S.; methodology, A.B., P.M. and A.S.; investigation, A.B., P.M. and A.S.; writing—review & editing, A.B., P.M. and A.S. All authors have read and agreed to the published version of the manuscript.

Funding: This work was supported by Regione Toscana Bando FAS Salute 2014 Italy (Grant No. CUP B52I14005760002).

Institutional Review Board Statement: Not applicable.

Informed Consent Statement: Not applicable.

Data Availability Statement: Data are available from the authors upon request.

Conflicts of Interest: The authors declare no conflict of interest.

References

1. Burkitt, M.D.; Duckworth, C.A.; Williams, J.M.; Pritchard, D.M. *Helicobacter pylori*-induced gastric pathology: Insights from in vivo and ex vivo models. *Dis. Model. Mech.* **2017**, *10*, 89–104. [CrossRef]
2. Krzyżek, P.; Grande, R.; Migdał, P.; Paluch, E.; Gościniak, G. Biofilm Formation as a Complex Result of Virulence and Adaptive Responses of *Helicobacter pylori*. *Pathogens* **2020**, *9*, 1062. [CrossRef] [PubMed]
3. Hamblin, M.R.; Viveiros, J.; Yang, C.; Ahmadi, A.; Ganz, R.A.; Tolkoff, M.J. *Helicobacter pylori* accumulates photoactive porphyrins and is killed by visible light. *Antimicrob. Agents Chemother.* **2005**, *49*, 2822–2827. [CrossRef] [PubMed]
4. Battisti, A.; Morici, P.; Ghetti, F.; Sgarbossa, A. Spectroscopic characterization and fluorescence imaging of *Helicobacter pylori* endogenous porphyrins. *Biophys. Chem.* **2017**, *229*, 19–24. [CrossRef] [PubMed]
5. Battisti, A.; Morici, P.; Signore, G.; Ghetti, F.; Sgarbossa, A. Compositional analysis of endogenous porphyrins from *Helicobacter pylori*. *Biophys. Chem.* **2017**, *229*, 25–30. [CrossRef]
6. Morici, P.; Battisti, A.; Tortora, G.; Menciassi, A.; Checcucci, G.; Ghetti, F.; Sgarbossa, A. The in vitro photoinactivation of *Helicobacter pylori* by a novel LED-based device. *Front. Microbiol.* **2020**, *11*, 283. [CrossRef]
7. Ganz, R.A.; Viveiros, J.; Ahmad, A.; Ahmadi, A.; Khalil, A.; Tolkoff, M.J.; Nishioka, N.S.; Hamblin, M.R. *Helicobacter pylori* in patients can be killed by visible light. *Lasers Surg. Med.* **2005**, *36*, 260–265. [CrossRef]
8. Lembo, A.J.; Ganz, R.A.; Sheth, S.; Cave, D.; Kelly, C.; Levin, P.; Kazlas, P.T.; Baldwin, P.C.; Lindmark, W.R.; Hamlin, M.R.; et al. Treatment of *Helicobacter pylori* infection with intra-gastric violet light phototherapy: A pilot clinical trial. *Lasers Surg. Med.* **2009**, *41*, 337–344. [CrossRef] [PubMed]
9. Rennie, M.Y.; Dunham, D.; Lindvere-Teene, L.; Raizman, R.; Hill, R.; Linden, R. Understanding Real-Time Fluorescence Signals from Bacteria and Wound Tissues Observed with the MolecuLight i:XTM. *Diagnostics* **2019**, *9*, 22. [CrossRef]
10. Blackshaw, E.L.; Jeffery, S.L.A. Efficacy of an imaging device at identifying the presence of bacteria in wounds at a plastic surgery outpatients clinic. *J. Wound Care* **2018**, *27*, 20–26. [CrossRef] [PubMed]
11. Philipp-Dormston, W.K.; Doss, M. Comparison of porphyrin and heme biosynthesis in various heterotrophic bacteria. *Enzyme* **1973**, *16*, 57–64. [CrossRef]
12. Lopez, A.J.; Jones, L.M.; Reynolds, L.; Diaz, R.C.; George, I.K.; Little, W.; Fleming, D.; D'Souza, A.; Rennie, M.Y.; Rumbaugh, K.P.; et al. Detection of bacterial fluorescence from in vivo wound biofilms using a point-of-care fluorescence imaging device. *Int. Wound J.* **2021**, *18*, 1–13. [CrossRef]
13. Jones, L.M.; Dunham, D.; Rennie, M.Y.; Kirman, J.; Lopez, A.J.; Keim, K.C.; Little, W.; Gomez, A.; Bourke, J.; Ng, H.; et al. In vitro detection of porphyrin-producing wound bacteria with real-time fluorescence imaging. *Future Microbiol.* **2020**, *15*, 319–332. [CrossRef]

14. Suhling, K.; French, P.M.W.; Phillips, D. Time-resolved fluorescence microscopy. *Photochem. Photobiol. Sci.* **2005**, *4*, 13–22. [CrossRef] [PubMed]
15. Jameson, D.M.; Gratton, E.; Hall, R.D. The measurement and analysis of heterogeneous emissions by multifrequency phase and modulation fluorometry. *Appl. Spectrosc. Rev.* **1984**, *20*, 55–106. [CrossRef]
16. Torno, K.; Wright, B.K.; Jones, M.R.; Digman, M.A.; Gratton, E.; Phillips, M. Real-time Analysis of metabolic activity within *Lactobacillus acidophilus* by phasor fluorescence lifetime imaging microscopy of NADH. *Curr. Microbiol.* **2013**, *66*, 365–367. [CrossRef]
17. Bhattacharjee, A.; Datta, R.; Gratton, E.; Hochbaum, A.I. Metabolic fingerprinting of bacteria by fluorescence lifetime imaging microscopy. *Sci. Rep.* **2017**, *7*, 3743. [CrossRef]
18. Yang, F.L.; Hassanbhai, A.M.; Chen, H.Y.; Huang, Z.Y.; Lin, T.L.; Wu, S.H.; Ho, B. Proteomannans in biofilm of *Helicobacter pylori* ATCC 43504. *Helicobacter* **2011**, *16*, 89–98. [CrossRef] [PubMed]
19. Mancini, S.; Imlay, J.A. Bacterial porphyrin extraction and quantification by LC/MS/MS analysis. *Bio-protocol* **2015**, *5*, e1616. [CrossRef]
20. Flemming, H.C.; Wingender, J. The biofilm matrix. *Nat. Rev. Microbiol.* **2010**, *8*, 623–633. [CrossRef]
21. Ricchelli, F. Photophysical properties of porphyrins in biological membranes. *J. Photochem. Photobiol. B Biol.* **1995**, *29*, 109–118. [CrossRef]

 *pharmaceutics*

Article

Broad-Spectrum Antimicrobial ZnMintPc Encapsulated in Magnetic-Nanocomposites with Graphene Oxide/MWCNTs Based on Bimodal Action of Photodynamic and Photothermal Effects

Coralia Fabiola Cuadrado [1],*, Antonio Díaz-Barrios [2], Kleber Orlando Campaña [1], Eric Cardona Romani [3], Francisco Quiroz [4], Stefania Nardecchia [5], Alexis Debut [6], Karla Vizuete [6], Dario Niebieskikwiat [7], Camilo Ernesto Ávila [8], Mateo Alejandro Salazar [8], Cristina Garzón-Romero [8], Ailín Blasco-Zúñiga [8], Miryan Rosita Rivera [8],* and María Paulina Romero [1]

[1] Laboratorio de Nuevos Materiales, Departamento de Materiales, Facultad de Ingeniería Mecánica, Escuela Politécnica Nacional, Quito 170525, Ecuador; kleber.campana@epn.edu.ec (K.O.C.); maria.romerom@epn.edu.ec (M.P.R.)
[2] School of Chemical Sciences and Engineering, Yachay Tech University, Urcuquí 100119, Ecuador; adiaz@yachaytech.edu.ec
[3] Instituto SENAI de Inovação, Serviço Nacional de Aprendizagem Industrial (Firjan SENAI), Rio de Janeiro 999074, Brazil; eromani@firjan.com.br
[4] Departamento de Ciencia de Alimentos y Biotecnología DECAB, Escuela Politécnica Nacional, Quito 170525, Ecuador; francisco.quiroz@epn.edu.ec
[5] Magnetic Soft Matter Group, Department of Applied Physics, Faculty of Sciences, University of Granada, 18071 Granada, Spain; stefania@ugr.es
[6] Centro de Nanociencia y Nanotecnología, Universidad de Las Fuerzas Armadas ESPE, Sangolquí 171103, Ecuador; apdebut@espe.edu.ec (A.D.); ksvizuete@gmail.com (K.V.)
[7] Departamento de Física, Colegio de Ciencias e Ingenierías, Universidad San Francisco de Quito, Quito 170901, Ecuador; dniebieskikwiat@usfq.edu.ec
[8] Laboratorio de Investigación en Citogenética y Biomoléculas de Anfibios (LICBA), Centro de Investigación para la Salud en América Latina—CISeAL, Facultad de Ciencias Exactas y Naturales, Pontificia Universidad Católica del Ecuador, Quito 170143, Ecuador; camoavila@gmail.com (C.E.Á.); massuno@hotmail.com (M.A.S.); ccgarzonrom@outlook.com (C.G.-R.); ailin.blasco.z@gmail.com (A.B.-Z.)
* Correspondence: coralia.cuadrado@epn.edu.ec (C.F.C.); mriverai@puce.edu.ec (M.R.R.); Tel.: +593-987526539 (C.F.C.)

Citation: Cuadrado, C.F.; Díaz-Barrios, A.; Campaña, K.O.; Romani, E.C.; Quiroz, F.; Nardecchia, S.; Debut, A.; Vizuete, K.; Niebieskikwiat, D.; Ávila, C.E.; et al. Broad-Spectrum Antimicrobial ZnMintPc Encapsulated in Magnetic-Nanocomposites with Graphene Oxide/MWCNTs Based on Bimodal Action of Photodynamic and Photothermal Effects. *Pharmaceutics* **2022**, *14*, 705. https://doi.org/10.3390/pharmaceutics14040705

Academic Editors: Udo Bakowsky, Matthias Wojcik, Eduard Preis and Gerhard Litscher

Received: 31 December 2021
Accepted: 24 February 2022
Published: 26 March 2022

Publisher's Note: MDPI stays neutral with regard to jurisdictional claims in published maps and institutional affiliations.

Abstract: Microbial diseases have been declared one of the main threats to humanity, which is why, in recent years, great interest has been generated in the development of nanocomposites with antimicrobial capacity. The present work studied two magnetic nanocomposites based on graphene oxide (GO) and multiwall carbon nanotubes (MWCNTs). The synthesis of these magnetic nanocomposites consisted of three phases: first, the synthesis of iron magnetic nanoparticles (MNPs), second, the adsorption of the photosensitizer menthol-Zinc phthalocyanine (ZnMintPc) into MWCNTs and GO, and the third phase, encapsulation in poly (N-vinylcaprolactam-co-poly(ethylene glycol diacrylate)) poly (VCL-co-PEGDA) polymer VCL/PEGDA a biocompatible hydrogel, to obtain the magnetic nanocomposites VCL/PEGDA-MNPs-MWCNTs-ZnMintPc and VCL/PEGDA-MNPs-GO-ZnMintPc. In vitro studies were carried out using *Escherichia coli* and *Staphylococcus aureus* bacteria and the *Candida albicans* yeast based on the Photodynamic/Photothermal (PTT/PDT) effect. This research describes the nanocomposites' optical, morphological, magnetic, and photophysical characteristics and their application as antimicrobial agents. The antimicrobial effect of magnetics nanocomposites was evaluated based on the PDT/PTT effect. For this purpose, doses of 65 mW·cm^{-2} with 630 nm light were used. The VCL/PEGDA-MNPs-GO-ZnMintPc nanocomposite eliminated *E. coli* and *S. aureus* colonies, while the VCL/PEGDA-MNPs-MWCNTs-ZnMintPc nanocomposite was able to kill the three types of microorganisms. Consequently, the latter is considered a broad-spectrum antimicrobial agent in PDT and PTT.

Keywords: antimicrobial nanomaterials; carbon nanotubes; graphene; magnetic nanoparticles; hydrogel; photodynamic therapy; photothermal therapy; nanocarrier

---

1. Introduction

In recent years, the study of microorganisms has increased substantially due to their ability to spread rapidly and adapt to different environments. For this reason, the World Health Organization has declared microbial diseases as one of the main threats for humanity [1]. Currently, there are antibiotic therapies that help fight infections caused by microorganisms such as *Klebsiella pneumoniae*, *Escherichia coli*, *Staphylococcus aureus* and *Neisseria gonorrhoeae*, among others. However, these have been losing their efficacy, which is why the development of new treatments is necessary [2]. The generation of new nanocomposites that help eliminate microorganisms provides a new alternative for the fight against infections caused by these pathogens [3,4].

Nanotechnology for the control of infectious diseases includes several strategies, such as the use of metal oxides for the generation of reactive oxygen species (ROS), nanocomposites capable of damaging membrane integrity or causing physical damage to the bacterial wall [5], inhibiting DNA replication and adenosine triphosphate (ATP) production in cells [6], use of graphene-family materials for the microbial elimination [7–10], and others. In this regard, Romero et al., 2020 showed the ability of GO as an antimicrobial agent to eliminate *E. coli* and *S. aureus* bacteria through the combination of Photodynamic Therapy (PDT) and Photothermic Therapy (PTT) [11]. Mei et al., 2021, synthesized a ZnPc-TEGMME@GO nanocomposite, which has a thickness between 1.47 and 2.61 nm, using concentrations ≥ 25 $\mu g \cdot mL^{-1}$ of the nanocomposite and irradiation with dual light of 450 nm and 680 nm for 10 min. The nanocomposite increases its temperature to $100\,^{\circ}C$ and rapidly promotes singlet oxygen generation, causing cuts in the membranes of bacterial cells and, consequently, the death of Gram-positive and Gram-negative bacteria [12]. Yang et al., 2018, obtained the nonchemotherapeutic nanoagent Fe_3O_4-CNT-PNIPAM with a diameter of around 200 to 400 nm, which at a concentration of 0.1 $mg \cdot mL^{-1}$, and under 808 nm laser irradiation, 3 $W \cdot cm^{-2}$, for 5 min is capable of killing *S. aureus* and *E. coli* by PTT [13].

Photodynamic Therapy (PDT) is well known and studied for its use in cancer treatments, producing minimum side effects compared to other therapies used for cancer. PDT is an attractive operating modality based on the interaction of light with a photosensitizer (PS) and oxygen [14,15], thus producing ROS and free radicals capable of causing cell and microorganism death with high cytotoxity [16–19]. This therapy has been used to treat skin diseases and cancers such as prostate, neck, lung, breasts and bladder cancers [20–23].

PS generally are classified by their activation wavelength, duration, and tissue penetration. One of the most widely used PS in PDT is the second generation PS, due to its photophysical and photochemical properties. Within this group are phthalocyanines, chlorins, and benzoporphyrins [24,25]. Phthalocyanines (Pcs) are hydrophobic compounds whose activation wavelengths of 650 and 700 nm allow deep tissue penetration. Pc reaches a high concentration in tumor tissue after 1 to 3 h of administration, and they are generally used to treat skin and subcutaneous lesions [26].

Zinc Menthol-Phthalocyanine (ZnMintPc) is a hydrophobic drug derived from phthalocyanine, whose structure is based on porphyrins but with a central Zinc atom with four methoxy groups around it that allow PS to be soluble in certain organic solvents [27]. The use of ZnMintPc has been limited due to its hydrophobic nature. When encapsulated by hydrogels, Pcs is soluble in aqueous media [28–30]. Pcs can also be combined with nanoparticles (Np) to create hybrid nanostructures that increase the quantum yield of singlet oxygen, cell uptake, and their therapeutic effectiveness [31,32].

Due to the hydrophobic nature of PS, several types of nanocarriers have been studied that prevent them from adding to each other and losing their physicochemical characteristics [20,33,34]. Among these nanocarriers are carbon nanostructures such as carbon nanotubes [35–37], graphene [33] and fullerenes [20,38,39], among others. These functionalized nanocarriers have excellent optical and mechanical properties. These compounds have cytotoxicity in biological systems, depending on their concentration, size, surface property and functional groups [40]. The surface modification of these nanocarriers with biomolecules such as polyethyleneimine, polyethylene glycol, and human proteins, improves the cytotoxicity and biocompatibility for biological applications [41], in addition to having other applications such as immunotherapy, imaging, and the development of vaccines and antimicrobial agents [42–46].

Photothermal treatment (PTT) is a type of phototherapy that works by turning light energy into heat through the use of photothermal agents (PTAs) [47]. There has been great interest in applying PTT to eliminate pathogenic bacteria in recent years. PTAs, when irradiated with near-infrared (NIR) light, generate much heat, causing protein denaturation, rupture of the bacterial cell membrane, and death of microorganisms [13,48,49]. An ideal PTA must meet specific requirements, such as high photothermal conversion efficiency, biocompatibility, ease of synthesis, photostability, and rapid elimination [50,51]. The best known PTAs are carbon-based nanomaterials, conjugated polymer-based nanomaterials, inorganic, and small molecule-based nanomaterials [48,52].

PTT/PTT synergies are considered among the most effective methods for microbe killing due to high specificity, minimal invasiveness, low risk of developing drug resistance, and selectivity [53,54]. These phototherapy techniques have a dual mode of action in which the PTT facilitates the absorption of PS because the increase in temperature will causes an increase in the permeability of the microorganism. At the same time, by PDT, the PS produces ROS, which can destroy proteins, lipids, and microbial DNA, causing the death of the bacteria, and increasing the effectiveness of the PTT since they are capable of reducing the heat resistance of bacteria [12,48].

CNTs were discovered by Iijima in 1991 [55] and consist of quasi-one-dimensional structures formed by several layers of graphene rolled up coaxially to form tubes. They are classified into two types: those with a single layer known as single-wall carbon nanotubes (SWCNTs), and those with several layers known as multiwall carbon nanotubes (MWCNTs) [56,57]. CNTs possess excellent optical, electrical, thermal, physical, and kinetic properties, and excellent cell permeability [58,59]. For this reason, there is much interest in their application as drug transport systems, electrochemical biosensors, and biological markers, among others [60,61].

CNTs have been used as biosensors, biomarkers, and drug transporters with a high carrying capacity [62,63]. Hybrid compounds have been created to improve the effectiveness of their action. Proteins, polymers, cell recognition ligands, nanoparticles, hydrophilic coatings have been incorporated into these structures, which provide them with new functions such as cell recognition, and controlled drug release, avoiding aggregations in aqueous media during targeted transport [64–66].

Graphene is a two-dimensional nanostructure with sp^2 hybridization and strongly cohesive carbons, which gives the structure excellent optical, electronic, mechanical, and chemical properties, which vary depending on its lateral size [67]. Graphene is a material with very high resistance and hardness. It is light and has low toxicity and increased flexibility, making it an innovative material in construction, technology, medicine, and other industries [43,68,69]. Because of its excellent physicochemical characteristics, graphene has been used in the biomedical field to make biosensors, drug delivery systems, antibacterial agents for early detection of cancer, gene therapy, and for cancer cell imaging/mapping, among other uses [70].

In recent years, great interest has arisen in the synthesis of magnetic nanoparticles due to their unique physicochemical properties. They can be used in PPT/PDT [71], drug delivery, theranostics, and others [72–74]. Magnetic nanoparticles (MNPs), such as magnetite and hematite obtained from iron oxides, are widely used in biomedicine because they are biocompatible and have no cytotoxic effects at concentrations below

$100~\mu g \cdot mL^{-1}$. For breast, glia, and normal cells with cancer [75], at higher concentrations, an interaction between cell membrane phospholipids and iron nanoparticles occurs, resulting in membrane failure [76–78]. MNPs can be anchored to carbon structures to obtain drug nanocarriers and directed by an external magnetic field [79,80].

VCL/PEGDA is a biocompatible hydrogel that can be obtained by emulsion polymerization. This hydrogel is very promising since it responds to physiological changes in the temperature of the human body [81,82]. Therefore, it can be used as a drug delivery system to encapsulate hydrophobic and hydrophilic agents. A study carried out by Romero et al., 2021 showed the sustained release capacity of the drug colchicine encapsulated in VCL/PEGDA [83].

PS must meet specific requirements for PDT strategy. They must be selective, disperse well in tissue, and their photostability time must be adequate for the treatment. Due to these needs, we developed two magnetic nanocarriers based on MWCNTs and GO materials in this work. Both nanocarriers are decorated with Fe-MNPs, giving the compounds superparamagnetic properties. These nanocarriers were functionalized with PS ZnMintPc, a drug used in Photodynamic Therapy. A VCL/PEGDA hydrogel was used to help with dispersion of the hydrophobic compounds ZnMintPc and MWCNTs in aqueous media by providing them with a lipophilic envelope. Finally, the antimicrobial effect of these nanocomposites was evaluated to eliminate the bacteria *S. aureus*, *E. coli*, *C. albicans* using the PDT/PTT strategy.

2. Materials and Methods

2.1. Materials

For the synthesis and functionalization process, MWCNTs and GO were provided by the Van de Graff Laboratory, Department of Physics PUC-RIO, Rio de Janeiro, Brazil. Zinc Menthol-Phthalocyanine was provided by the Federal University of São Carlos, São Carlos, Brazil; $Fe(SO_4)$, $H_2SO_4 \cdot 7H_2O$ from MERCK, Rio de Janeiro, Brazil; NH_4OH, 14.8N and N,N Dimethylformamide from Fisher Scientific, Princeton, NJ, USA; $Fe_2(SO_4)_3 \cdot H_2O$ from Fisher Scientific. HNO_3 from the Fermont, Monterrey, Mexico; saline solution Fisiol UB (pH = 7) from Lamosan, Quito, Ecuador and Tween80 from La casa del Químico, Quito, Ecuador.

For VCL/PEGDA hydrogel synthesis by emulsion polymerization, the following reagents were used. N-vinylcaprolactam (VCL; Sigma Aldrich, Darmstadt, Germany, 98%), and Poly(ethylene glycol) diacrylate (PEGDA; Sigma Aldrich, Mn 250), the initiator ammonium persulfate (APS; FMC Corporation, Philadelphia, PA, USA, >99%), the emulsifier sodium dodecyl sulfate (SDS; STEOL®CS-230 Stepan, Northbrook, IL, USA) and a buffer of sodium hydrogen carbonate (Sigma Aldrich, ≥99.7%) were used as provided.

Characterizations of MNPs-MWCNTs and MNPs-GO were carried out by FT-IR spectroscopy analysis in a JASCO FT/IR-4100 spectrometer, JASCO International Co., Ltd., Tokyo, Japan. ES (wavenumber range 7800 to 350 cm^{-1}, resolution of 0.7 cm^{-1}) and Raman spectroscopy analysis in a HORIBA Raman spectrometer LabRAM HR Evolution (Horiba, Kyoto, Japan), where the samples were excited with a 2.33 eV (532 nm).

Magnetization (M) measurements were carried out using a Quantum Design Versalab VSM, Quantum Design, Darmstadt, Germany. FR, magnetometer in the temperature range between -210 and $+60~^\circ$C with applied magnetic fields, $\mu_0 H$, up to 3 T.

Stability over time of the PS magnetic nanocomposites was characterized by UV-VIS spectroscopy analysis at a wavelength range of 280 to 780 nm (resolution better than 1.8 Å), using a UV-VIS Spectrophotometer model Evolution 220 from Thermo Fisher Scientific, Waltham, MA, USA. DPBF photobleaching and thermic studies were carried out with homemade equipment using an LED red light at 635 nm, 65.5 $mW \cdot cm^{-2}$. A DPBF photobleaching study was characterized in a UV-VIS Specord 210 Plus. XRD analysis was performed on a PANalytical brand EMPYREAN Diffractometer (Malvern Panalytical, Malvern, UK) operating in a θ–2θ configuration (Bragg-Brentano geometry) equipped with a copper X-ray tube ($K\alpha$ radiation $\lambda = 1.54056$ Å) at 45 kV and 40 mA.

2.2. Methods

2.2.1. Morphological Studies

Scanning Electron Microscopy (SEM) and Energy Dispersive X-ray Spectroscopy (EDS) evaluated morphology and semiquantitative elemental composition. For this, an aliquot of the sample was fixed to an aluminum sample holder using a double layer of carbon tape. The analyses were carried out on an SEM Tescan Mira 3 (Tescan, Brno, Czech Republic). equipped with a Schottky field emitter (JEOL Ltd., Tokyo, Japan). EDS was performed in the SEM chamber using a Bruker, X-Flash 6 | 30 detector (Bruker, Billerica, MA, USA) with a resolution of 123 eV at Mn Kα. Tapping-mode atomic force microscopy was used to determine the shape and thickness of GO (Bruker Dimension Icon AFM).

For Transmission Electron Microscopy (TEM), the samples were dispersed in a BRAN-SON 1510 ultrasonic (Transcat, Rochester, NY, USA) bath for 30 min. Next, approximately 5 µL of the sample was placed on a TEM grid (formvar/carbon, 300 mesh), and the solvent was removed with filter paper. The samples were observed in a TEM FEI, Tecnai G2 Spirit Twin (ELECMI, Zaragoza, Spain) equipped with an Eagle 4k HR camera (Tronic Extreme, Coventry, UK) at 80 kV.

For scanning transmission electron microscopy (STEM), 5 µL of the sample was placed on a TEM grid (formvar/carbon, 300 mesh), and the solvent was removed with filter paper. Staining was performed with 1% Phosphotungstic Acid for 1 s for *S. aureus* and 1 min for *E. coli*, the solvent was removed, and the sample was observed in an SEM Tescan, Mira 3 in transmission mode.

2.2.2. Synthesis of Graphene Oxide

GO was prepared according to Hummers' method [84]. Graphite powder with 99% purity was used for the synthesis of GO. Chemical products, including $NaNO_3$, $KMnO_4$, H_2SO_4, and aqueous solutions of H_2O_2, and HCl, were purchased from Sigma Aldrich. GO powder was obtained after lyophilization of the suspected GO in deionized water.

2.2.3. Purification of MWCNTs

The purification of MWCNTs was carried out using an acid attack. A 5.3 mg sample of MWCNTs was dispersed with 5% Tween 80 in 10 mL of distilled water stirred at 200 rpm for 24 h to ensure that the MWCNTs were well dispersed. Then, in a solution of 4 mL of HNO_3 and 12 mL of H_2SO_4 (1:3 ratio), the solution of MWCNTs previously treated was allowed to cool, and then stirred magnetically for 3 h. Several washes of the MWCNTs were performed using 0.22 µm pore micropore filtration until a neutral pH was obtained.

2.2.4. Synthesis of MNPs-MWCNTs and MNPs-GO

In 18 mL of pure water, the following were dispersed: purified MWCNTs, 225 mg of $FeSO_4$, 450 mg of $(Fe_2 (SO_4)_3)$, and 5% Tween 80. The sample was placed in a magnetic stirrer for 3 h, then was added carefully to 150 mL of NH_4OH. The mixture was exposed to magnetic stirring in an inert atmosphere for 1 h at 200 rpm. Finally, several magnetic purifications of the MNPs-MWCNTs nanocarrier were carried out, until the pH was neutralized, and allowed to dry.

The same process was used to synthesize of MNPs-GO, using 5.3 mg of GO instead of MWCNTs.

As a control sample for the magnetic measurements, free-standing iron nanoparticles were prepared using the same co-precipitation method but without GO or MWCNTs.

2.2.5. Synthesis of Polyethylene Glycol Diacrylate-Vinylcaprolactam (VCL/PEGDA)

Hydrogel synthesis was carried out by emulsion polymerization of 2 g of VCL, 0.08 g of PEGDA, STEOL CS-230, and 0.08 of $NaHCO_3$ dispersed in 235 mL deionized water. The mixture was slowly placed in the chemical reactor with stirring at 350 rpm and a temperature of 70 °C, maintaining a stream of Nitrogen for one hour. Then, the initiator (0.03 g of Ammonium Persulfate dissolved in 15 mL of distilled water) was added to the solution, and the reaction was kept at 70 °C for 7 h. After this time, the mixture was allowed to cool under stirring at 200 rpm and 25 °C to avoid aggregation for 12 h. Finally, the hydrogel was dialyzed against DDI water to remove impurities and unreacted reagents [83].

2.2.6. Functionalization of MNPs-MWCNTs; MNPs-GO with ZnMintPc in the Presence of VCL/PEGDA

In 10 mL of pure water, 2 mg of MNPs-MWCNTs were dispersed with 5% Tween, and the sample stirred magnetically for 24 h at 200 rpm, then sonicated for 30 min. A volume of 10 mL of VCL/PEGDA was added to the mixture, and it was stirred magnetically for 4 h at 200 rpm.

To 20 mL of the VCL/PEGDA-MNPs-MWCNTs solution, 0.67 mL of ZnMintPc solution (0.25 mM) were added, after which the mixture was sonicated for 4 h at 250 rpm. The solution was covered with aluminum foil to avoid ZnMintPc photodegradation.

The same process was carried out with MNPs-GO, obtaining the VCL/PEGDA-MNPs-GO nanocomposite.

2.2.7. Quenching Experiment of 1,3-Diphenyl Isobenzofuran

The fluorescence and absorption characteristics of 1,3-diphenylisobenzofuran (DPBF), a singlet oxygen trapping chemical, are widely employed to detect and quantify singlet oxygen [85]. This agent has an absorption range between 410–420 nm, emitting blue fluorescence. When DPBF interacts with oxygen, it produces o-dibenzoylbenzene, without absorbing visible light. The amount of $^1O_2$ produced is shown by a reduction in DPBF absorbance [11]. A sample of 5 mg of DPBF was dispersed in 1 mL of DMF to distribute in the solutions with nanocomposites later.

For this experiment, a UV-VIS SPECORD 210 Plus spectrophotometer (resolution of 2.3–2.5 nm) in a range of 300 to 800 nm was used. A reference solution was prepared with 10 mL of deionized water and 10% Tween 80. Then, in 3 mL of this reference solution, 100 µL of VCL/PEGDA-MNPs-GO (0.1 mg/mL) was added. The same process was repeated with the nanocomposites VCL/PEGDA-MNPs-MWCNTs (0.1 mg/mL), VCL/PEGDA-MNPs-MWCNTs-ZnMintPc (0.1 mg/mL) and VCL/PEGDA-MNPs-GO-ZnMintPc (0.1 mg/mL). After that, 20 µL of the DPBF (18.5 mM) solution was placed in each of them, and they were irradiated at different times while observing a decrease in absorbance of 418 nm due to DPBF.

2.2.8. Thermal Studies

Thermal studies were carried out by irradiating, in deionized water, VCL/PEGDA, MNPs, GO, MWCNTs, VCL/PEGDA-MNPs-GO, and VCL/PEGDA-MNPs-MWCNTs, with a red light of 630 nm and 65.5 mW·cm^{-2} and taking the temperature every 5 to 10 min until it did not change.

2.2.9. Antimicrobial Studies

The antimicrobial study used cryovials with the microorganisms *Staphylococcus aureus* ATCC: 25923, *Escherichia coli* ATCC: 25922, and *Candida albicans* ATCC: 10231 as test subjects. Cryovials were thawed at room temperature, and the contents were inoculated in Mueller-Hinton broth (Difco™). The culture medium was incubated overnight at 37 °C. After this period, the absorbance of each medium was determined by spectrophotometry (Thermo Scientific ™ Orion ™ AquaMate 8000 UV-Vis, Thermo Fisher Scientific, Waltham, MA, USA) and diluted in Mueller-Hinton broth to reach the concentration established for the bioassays: 10^7 CFUs mL^{-1} for *E. coli*, 10^6 CFUs mL^{-1} for *S. aureus*, and 10^5 CFUs mL^{-1} for *C. albicans*.

Upon reaching the indicated concentration, each medium was dispensed into microtubes: six aliquots of 1 mL per magnetic nanocomposites (VCL/PEGDA-MNPs-GO-ZnMintPc, VCL/PEGDA-MNPs-MWCNTs-ZnMintPc, VCL/PEGDA-MNPs-GO, and VCL/PEGDA-MNPs-MWCNTs) and six aliquots of 1 mL as control. Half of the aliquots were used to evaluate the activity of the magnetic nanocomposites under light irradiation, while the other half was not exposed to these conditions.

The microtubes were centrifuged at 3000 rpm for 10 min. The supernatant was discarded, and 1 mL of PBS was added to wash the cells and remove the remains of the culture medium. The pellet was resuspended and recentrifuged under the same conditions. PBS was discarded and 1 mL of the compound was dispensed into each microtube. In the control microtubes, the cells were resuspended again with PBS. Each tube was vortexed to dissolve the pellet again and incubated at 37 °C in the dark for 45 min.

At the end of the incubation period, three aliquots of each compound and three controls were subjected to light irradiation using red light of 630 nm, 65.5 mW·cm^{-2}. The remaining tubes were not irradiated.

Serial dilutions of each aliquot were made in PBS. A 4 µL aliquot of each dilution was inoculated into an 8-part Petri dish with Mueller-Hinton agar (Difco^{TM}). Each inoculum was streaked for colony isolation. Petri dishes were incubated for 24 h at 37 °C, and after this period the number of colonies was counted per dilution.

3. Results

3.1. Composition and Structure Characterization of Magnetic Nanocomposites

FT-IR spectroscopy was performed on the magnetic nanocomposites to study the difference in oxygen-related functional groups (Figure 1). Figure 1a shows the FT-IR spectra of MNPs-MWCNTs, which presents characteristic bands. As the literature indicates, the band at 3373 cm^{-1} is attributed to the vibration of the hydroxyl group (OH), the 1636 cm^{-1} band corresponding to the FeOO and shows the successful decoration of MWCNTs with iron nanoparticles through hydrogen bonds [86–88]. Other bands appear in the range 1400–1730 cm^{-1}, corresponding to the OH–C=O, –COO, –COOH groups added due to acid treatment and functionalization with MNPs. Finally, the characteristic bands of the MNPs appear at 709 and 623 cm^{-1}, which indicate the stretching vibration of Fe–O–Fe characteristic of Fe-MNPs, which agrees with Abrinaei, Kimiagar and Zolghadr 2019 [89].

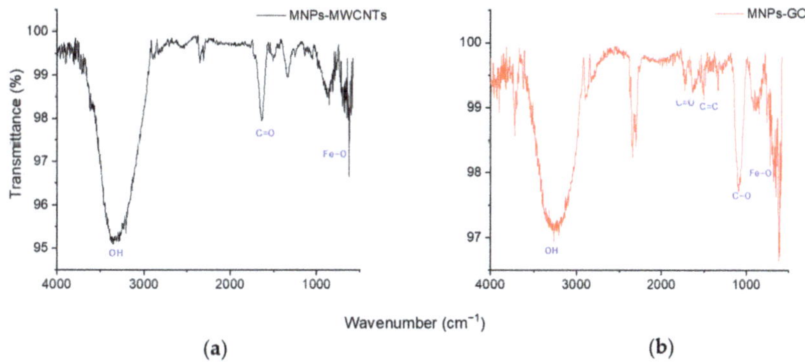

Figure 1. FT-IR spectra: (**a**) MNPs-MWCNTs; (**b**) MNPs-GO.

Figure 1b presents the FT-IR spectra of the MNPs-GO nanocarrier where characteristic bands of GO are identified. There is a band around 3327 cm^{-1} corresponding to the bending and vibration of the OH stretching of the C-groups, the band at 1636 cm^{-1} corresponds to the vibration of the C=C, the bands around 800 cm^{-1} and 1269 cm^{-1} represent the epoxy group, and bands at 1342 cm^{-1}, 1247 cm^{-1}, 1049 cm^{-1} show the stretching vibration of the carboxy groups C−O, C−C−O, and alkoxy C−O, respectively; as shown in the work of Al-Ruqeishi et al., 2020 [90].

Figure 2a shows the Raman spectra of non-purified MWCNTs (Figure 2a), purified MWCNTs (Figure 2b), and MNPs-MWCNTs (Figure 2c). These spectra show characteristic bands of the MWCNTs: band D or band of defects and the first and second-order bands G and 2D, respectively. The D-band is at 1337 cm^{-1} and indicates the presence of defects in the graphite, resulting from the presence of multiple carbon sheets that are not directly aligned sheet to sheet, which induces a loss of translational symmetry in the two-dimensional network. Due to the same effect, a secondary phonon is produced that gives rise to the presence of the G-band at 1566 cm^{-1}. The fundamental band G is a tangential elongation band attributed to the in-plane vibration of the C-C bond, is typical of carbon-derived materials, and is consistent with reports in the literature [91–94].

Figure 2d shows the Raman spectra of Iron-MNPs with its characteristic bands, two A_{1g} modes at 226, 502 cm^{-1}, five E_g modes at 248, 291, 300, 407 and 615 cm^{-1}, and the characteristic two-magnon scattering band at 1320 cm^{-1}, which agree with the results presented by Soler and Qu 2012 [95]. Some of these bands can be observed in the spectra shown in Figure 2c,f.

Figure 2a,b show Raman spectra of unpurified MWCNTs and purified MWCNTs. The $\frac{I_D}{I_G}$ ratio of the purified MWCNTs is higher than that of unpurified MWCNTs because treatment with acids causes some bonds to break and form functional groups, generating defects in the structure of MWCNTs.

In the Raman spectrum of MNPs-MWCNTs (Figure 2c), the increase in the $\frac{I_D}{I_G}$ ratio is due to charge transfer effects between MNPs and MWCNTs; the result of the functionalization is a structure of MWCNTs with defects. In addition, extra bands that correspond to the MNPs are observed.

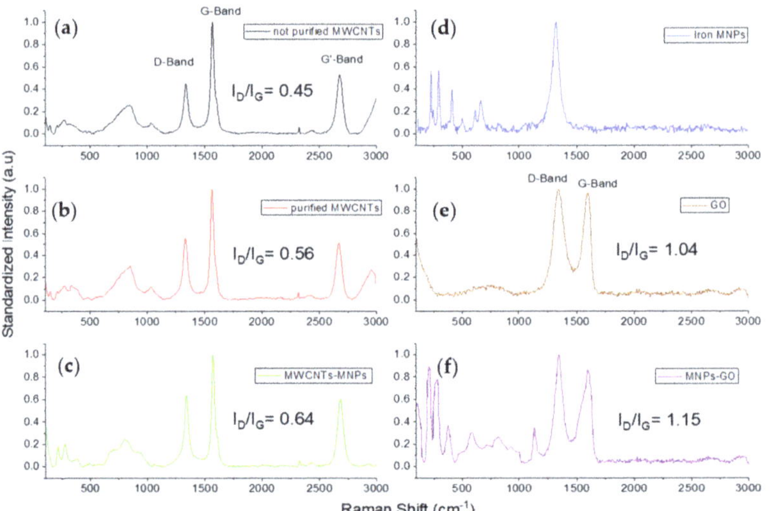

Figure 2. Raman Spectra: (**a**) non-purified MWCNTs, (**b**) purified MWCNTs, (**c**) MNPs-MWCNTs; (**d**) Fe-MNPs; (**e**) GO; (**f**) MNPs-GO; $E_{láser}$ = 2.33 eV, λ = 532 nm.

Figure 2e shows the characteristic bands of GO, an intense D band at 1340 cm^{-1}, a G-phase vibration band at 1589 cm^{-1}, and the D + D′ band located around 2900 cm^{-1}, which is activated by defects and appears with a combination of phonons with different linear moments around points K and Γ in the Brillöuin zone and agrees with Cançado et al., 2011, and Muhammad Hafiz et al., 2014 [96,97]. The ratio of the bands $\frac{I_D}{I_G}$ = 1.036 results from the degree of disorder in GO due to the presence of Carboxylic acid functional groups at its ends.

Figure 2f shows the spectra of the MNPs-GO nanocomposite where there is a shift of D band, since the nature of iron-MNPs when combined with carbon nanostructures affects the spectral amplification of the phonon peaks, agreeing with the information presented by Ramirez et al., 2017 and Satheesh et al., 2018 [98,99].

3.2. Magnetic Properties

We studied the magnetic properties of three samples: the nanocomposites VCL/PEGDA-MNPs-MWCNTs and VCL/PEGDA-MNPs-GO and, for comparison, the free-standing iron nanoparticles (MNPs). All the studied samples showed Fe nanoparticles' typical ferromagnetic (FM) behavior, with a very small coercivity of less than 2 mT at room temperature (less than 5 mT at T = −210 °C). This small coercivity indicates that the MNPs were not directly attached to the carbon structures, since directly attached MNPs have large coercivities of hundreds of mT [100,101]. This was expected because the hydrogen bonds present between MNPs and carbon structures mentioned in the results of the FT-IR spectroscopy in Section 3.1. are weak and therefore generate a reduction of the Fe–C magnetic coupling. Figure 3a shows some representative M vs. H loops measured at two different temperatures, −210 °C and +20 °C, for the three samples. From these curves we obtained saturation magnetization (M_S) as a function of temperature using the usual law of approach to saturation (LAS) [102–104] given by:

$$M(H) = M_S\left(1 - \frac{a}{H} - \frac{b}{H^2}\right) + \chi H, \tag{1}$$

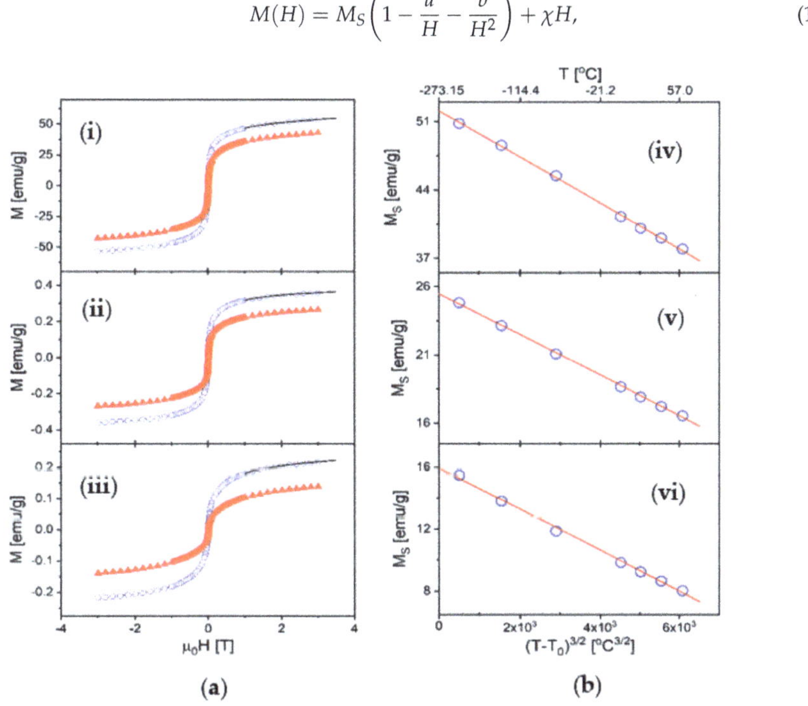

Figure 3. Magnetic properties: (**a**) Magnetization vs. applied magnetic field: (**i**) the free-standing nanoparticles, (**ii**) VCL-PEGDA-MNPs-MWCNTs, and (**iii**) VCL-PEGDA-MNPs-GO. The curves shown reflect measurements at two different temperatures, T = −210 °C (circles) and T = 20 °C (triangles). The solid lines are examples of the fit of the data using the law of approach to saturation (LAS) of Equation (1). (**b**) Saturation magnetization as a function of temperature: (**iv**) the free-standing nanoparticles, (**v**) VCL-PEGDA-MNPs-MWCNTs and (**vi**) VCL-PEGDA-MNPs-GO. The solid lines match the data showing that MS follows Bloch's law given Equation (2).

In this equation, a and b are constants, and the last term accounts for the non-ferromagnetic contributions, such as the disordered shell of the nanoparticles. The latter is valid only at high fields, close to saturation; thus, we used it to fit our $M(H)$ curves for applied fields between 1 and 3 T, as shown by the solid lines in Figure 3a for T = −210 °C. From these fits we obtained M_S at several temperatures between −210 °C and +60 °C (see

Figure 3b). In the case of the VCL-PEGDA-MNPs-MWCNTs and VCL-PEGDA-MNPs-GO samples, the magnetization value appeared substantially reduced since most of the mass (>98%) corresponds to the hydrogel. Therefore, to directly compare the measured magnetic moment with that of pure iron, the saturation magnetization value was corrected by the mass fraction of nanoparticles, x = 1.43% for the VCL-PEGDA-MNPs-MWCNTs and x = 1.37% for the VCL-PEGDA-MNPs-GO sample.

The saturation magnetization as a function of temperature is presented in Figure 3b, where we also show that, for all three samples, M_S closely follows the well-known Bloch's law for FM nanoparticles [105–107]:

$$M_S(T) = M_0 \times \left[1 - B\,(T - T_0)^{\frac{3}{2}}\right],\tag{2}$$

where M_0 is the saturation magnetization at the absolute zero degrees Kelvin ($T_0 = -273.15\,^\circ\mathrm{C}$), and B is the so-called spin-wave constant. The results for these magnetic parameters are presented in Table 1. The values obtained for B are very similar to those measured in other Fe-nanoparticles systems [104,105] ($B \sim 10^{-5}$–$10^{-4}\,^\circ\mathrm{C}^{-3/2}$) implying the existence of comparable thermally induced magnetic field excitations within the FM volume. On the other hand, our results show that the saturation magnetization is, in all cases, much smaller than that of pure iron, $M_{Fe} = 222\ \mathrm{emu\cdot g^{-1}}$. However, in the case of the free-standing nanoparticles, M_S remains larger than for the nanoparticles submerged within the hydrogel, indicating that the non-FM (disordered) shell increases in the presence of the gel medium. Moreover, the difference between the samples becomes even greater at room temperature, where the saturation magnetization decreases due to thermal effects (see Table 1). From this analysis, the effective FM volume (the size of the FM core of the nanoparticles) can be estimated by comparing M_0 with M_{Fe}, such that the fraction of the nanoparticle volume that remains, FM, can be estimated as:

$$f = \frac{M_0}{M_{Fe}},\tag{3}$$

The very small FM volume fraction in the case of the VCL-PEGDA-MNPs-MWCNTs and VCL-PEGDA-MNPs-GO samples likely indicates that the presence of the hydrogel induces strong oxidation of the surface of the nanoparticles or some interdiffusion that corrodes the surface and reduces the magnetic moment. Moreover, the larger value of the spin-wave constant in these two samples is consistent with a more significant influence of surface effects [107], which produce a faster decrease in saturation magnetization when the temperature increases.

Table 1. Magnetic parameters for the three studied samples: saturation magnetization at $T_0 = -273.15\,^\circ\mathrm{C}$ (M_0), saturation magnetization M_S at 20 °C, a fraction of the volume of the nanoparticles that remain ferromagnetic (f), and spin-wave constant (B).

Sample	M_0 (emu·g^{-1})	M_S @ 20 °C (emu·g^{-1})	f (%)	B (°C$^{-3/2}$)
MNPs	52.1	40.1	23.5	4.5×10^{-5}
VCL-PEGDA-MNPs-MWCNTs	25.5	17.9	11.5	5.8×10^{-5}
VCL-PEGDA-MNPs-GO	15.9	9.3	7.2	8.3×10^{-5}

3.3. Optical Properties of Magnetic Nanocomposites

UV-VIS absorbance spectra and photoluminescence emission were obtained. In the Supplementary Material (Figure S1a,b,d), we show the ZnMintPc-DMF and the nanocomposites VCL/PEGDA-ZnMintPc and VCL/PEGDA-MNPs-MWCNTs-ZnMintPc, respectively. We observed the presence of bands at 354 nm (Band of B or Soret), 616 nm, and 684 nm (B and Q), characteristic of the absorption spectra of ZnMintPc.

Furthermore, it can be seen that as the concentration of ZnMintPc increases, the absorption of the B and Q bands increases proportionally without their dislocation. This allows us to conclude that the Beer-Lambert law is fulfilled and there is no drug aggregation, indicating that N,N-Dimetilformamida (DMF) is a suitable solvent for this PS [108,109]. Furthermore, it is also concluded that the VCL-PEGDA hydrogel disperses PS similarly to DMF. In the Supporting Information (Figure S1c,d), the UV-VIS spectra of nanocomposites based on MNPs-MWCNTs (VCL-PEGDA-MNPs-MWCNTs and VCL-PEGDA-MNPs-MWCNTs-ZnMintPc) are presented, showing the band corresponding to MWCNTS at 265 nm, which is mentioned in the research by Wang et al., 2012 [110].

Figure S2a in the Supporting Information shows UV-VIS spectra of MNPs-GO, where GO shows a $\pi\rightarrow\pi^*$ transition of aromatic C–C bonds in the 230 nm band, which cannot be displayed, and a shoulder around 315 nm attributed to the $n\rightarrow\pi^*$ transitions of C=O bonds, as in Bera et al., 2017 [111]. The last band has undergone a hypsochromic shift due to functionalization with both MWCNTs and GO in the presence of an organic solvent (ammonium hydroxide used for MNPs syntheses), as described in the literature [112,113].

The calibration curve shown in Figure S3a was obtained from the Q band at 684 nm related to ZnMintPc and described in Figures S1 and S2, which is in a region of the spectra of interest for treatment in PDT. It can be observed that the higher the concentration of ZnMintPc in the nanocomposites (VCL-PEGDA-ZnMintPc, VCL-PEGDA-MNPs-GO-ZnMintPc, and VCL- PEGDA-MNPs-MWCNTs-ZnMintPc), the greater the intensity of the 684 nm band; but when comparing with the ZnMintPc dispersed in DMF, the latter shows the highest intensity in relation to all nanocomposites. This indicates that the VCL-PEGDA hydrogel adequately disperses the hydrophobic ZnMintPc PS in an aqueous solution, as shown in the literature [114–116]. It can be seen that the 684 nm absorption band of ZnMintPc (0.52 µM) decreases as the number of functionalized components increases, as presented in Figure S3b. The percentage of decrease for VCL/PEGDA-MNPs-MWCNTs-ZnMintPc nanocomposites is 38.2%, and for VCL/PEGDA-MNPs-GO-ZnMintPc it is 35.54%.

The temporal stability study of the VCL/PEGDA-ZnMintPc nanocomposites VCL/PEGDA-MNPs-MWCNTs-ZnMintPc and VCL/PEGDA-MNPs-GO-ZnMintPc, observed in Figure 4, was conducted at one, two and 24 h. In all three cases, as time passed, the intensity of the B band increased while the Q band decreased. In all three cases, as time passes, the intensity of the B band increased while the Q band decreased due to PS photobleaching without aggregation in 24 h.

The curves showing stability of the magnetics nanocomposites vs. time (Figure 5) indicates that the intensity of PS decays exponentially. The nanocomposites for 24 h were evaluated, and the nanocomposites based on GO and MWCNTs decreased the photobleaching rate of the nanocomposite. Table 2 shows the percentage decay of PS over time that is related to PS release. After 24 h, nanocomposites based on GO and MWCNT only allowed PS decays of 45.67% and 43.33% while, in the hydrogel, the PS decayed 56.24%. Photobleaching in VCL/PEGDA-MNPs-GO-ZnMintPc and VCL/PEGDA-MNPs-MWCNTs-ZnMintPc nanocomposites can be considered very successful.

3.4. Photodynamic Analyses

DPBF photobleaching was performed in VCL/PEGDA-MNPs-GO, VCL/PEGDA-MNPs-MWCNTs, VCL/PEGDA-MNPs-GO-ZnMintPc, and VCL/PEGDA-MNPs-MWCNTs-ZnMintPc, to evaluate the efficiency of singlet oxygen production of the nanocomposites using 630 nm light with an intensity of 65.5 mW·cm^{-2}. In the Supporting Information (Figures S4 and S5), the decrease of each nanocomposite's 418 nm DPBF band as a function of time is shown. For the VCL/PEGDA-MNPs-GO (Figure S4a) and VCL/PEGDA-MNPs-MWCNTs (Figure S4b) nanocomposites, a slight decrease in the 418 nm band at an irradiation time of 29 and 35 min, respectively, was observed. On the other hand, for nanocomposites with the presence of PS (VCL/PEGDA-MNPs-GO-ZnMintPc, Figure S5a and VCL/PEGDA-MNPs-MWCNTs-ZnMintPc Figure S5b), rapid decay of the 418 nm band was observed until all the DPBF present was photobleached in solution. Total photobleaching after 9 and 11 min of irradiation, respectively, was observed.

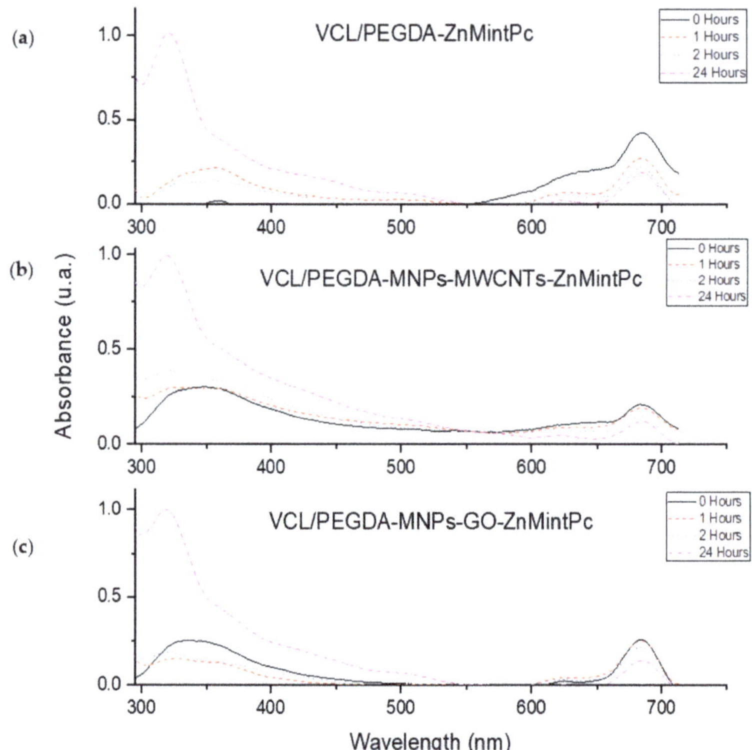

Figure 4. Stability curves over time for (**a**) VCL-PEGDA-ZnMintPc; (**b**) VCL/PEGDA-MNPs-MWCNTs-ZnMintPc and (**c**) VCL/PEGDA-MNPs-GO-ZnMintPc; ZnMintPc = 0.27 μM.

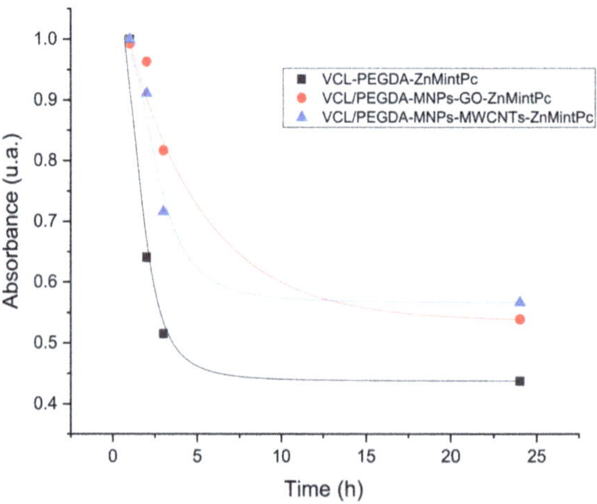

Figure 5. Decay curves of: VCL-PEGDA-ZnMintPc, VCL/PEGDA-MNPs-MWCNTs-ZnMintPc and VCL/PEGDA-MNPs-GO-ZnMintPc. ZnMintPc = 0.27 μM.

Table 2. Decay of ZnMintPc in nanocomposites in 24 h.

Time (h)	VCL-PEGDA-ZnMintPc		VCL/PEGDA-MNPs-GO-ZnMintPc		VCL/PEGDA-MNPs-MWCNTs-ZnMintPc	
	Absorbance	PS Released	Absorbance	PS Released	Absorbance	PS Released
0	100%	-	100%	-	100%	-
1	64.13%	35.87%	97.02%	2.98%	91.17%	8.83%
2	51.55%	48.55%	82.30%	17.7%	71.68%	28.32%
24	43.76%	56.24%	54.33%	45.67%	56.67%	43.33%

For each nanocomposite, absorbance vs. photoirradiation curves were constructed. Taking the 418 nm band of DPBF, an exponential decay fit to obtain the decay time presented in Figure 6a,b was made.

Figure 6a shows that the nanocomposites without PS (VCL/PEGDA-MNPs-GO and VCL/PEGDA-MNPs-MWCNTs) have long DPBF photobleaching times. The VCL/PEGDA-MNPs-GO photobleaching time was less than in VCL/PEGDA-MNPs-MWCNTs, indicating that GO has a better capacity to generate $^1O_2$ by photooxidation of DBPF, as corroborated by the work of Romero et al., 2020 [11]. For nanocomposites with PS (VCL/PEGDA-MNPs-GO-ZnMintPc and VCL/PEGDA-MNPs-MWCNTs-ZnMintPc), curve fitting show photobleaching over two times. The first one indicates a rapid decay in the first 60 s of photoirradiation, and the second one indicates a slow decay from 60 s to 700 s of photoirradiation. Once again, slow decay results in longer photobleaching time, as seen in the purple curve in Table 3 for the MWCNT-based nanocomposite (1140 ± 260 s).

The presence of PS in the nanocomposites generates a rapid photooxidation of DPBF. Therefore, the results indicate that the photobleaching of DPBF was mainly due to PDT effects mediated by VCL/PEGDA-MNPs-GO-ZnMintPc and VCL/PEGDA-MNPs-MWCNTs-ZnMintPc nanocomposites, with the nanocomposite containing GO and the photosensitizing ZnMintPc exhibiting more significant singlet oxygen generation.

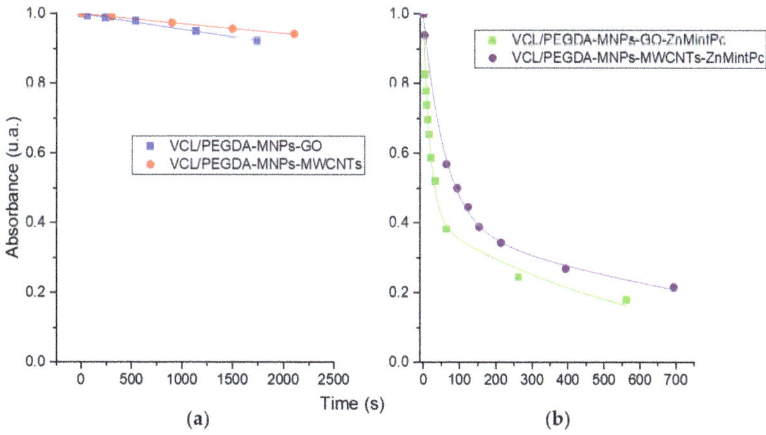

Figure 6. Decay curve of DPBF from (**a**) VCL/PEGDA-MNPs-GO and VCL/PEGDA-MNPs-MWCNTs; (**b**) VCL/PEGDA-MNPs-GO-ZnMintPc and VCL/PEGDA-MNPs-MWCNTs-ZnMintPc. DPBF = 18.5 mM, GO = 3.47 µg·mL^{-1}, MWCNTs = 3.47 µg·mL^{-1}, MNPs = 93.3 µg·mL^{-1} and ZnMintPc = 8.1 µM.

Table 3. Decay times of nanocomposites.

VCL/PEGDA-MNPs-MWCNTs	**VCL/PEGDA-MNPs-MWCNTs-ZnMintPc**
34700 ± 700 s	1140 ± 260 s
VCL/PEGDA-MNPs-GO	**VCL/PEGDA-MNPs-GO-ZnMintPc**
22300 ± 900 s	630 ± 190 s

3.5. Thermal Studies

The thermal study curves in Figure 7 show that the solutions of MNPs, GO and MWCNTs (blue, violet and light blue, respectively) act as photothermal materials when irradiated with red light for about 100 min. They can reach temperatures between 50.8 and 54.8 °C, making them suitable for use in Photothermal Therapy due to their ability to convert red and near-infrared (NIR) light into heat, and to transport drugs, as mentioned in the literature [71,117–120]. In the thermal curves of Figure 7, it can also be observed that VCL/PEGDA, VCL/PEGDA-MNP-GO, and VCL/PEGDA-MNP-MWCNTs nanocomposites irradiated with red light for about 100 min show a slight decrease in temperature compared to the thermal curve of deionized water (reference), and around 80 min of irradiation, curves representing the presence of hydrogel reach temperatures close to deionized water (the light doses are shown in the Table 4).

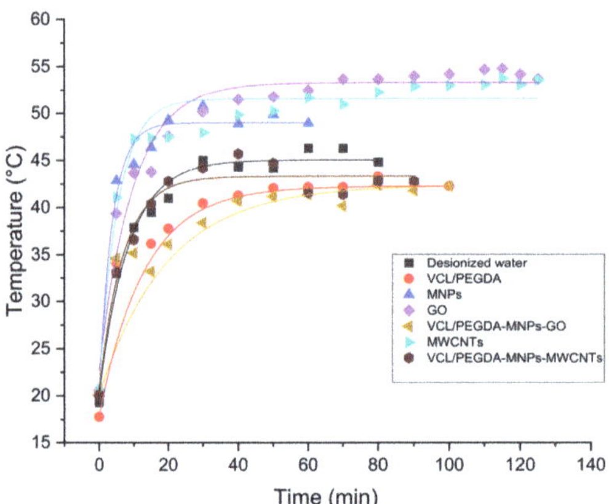

Figure 7. Thermal studies. deionized water (control black line), VCL/PEGDA (red line), MNPs (blue line), GO (violet line), MWCNTs (light blue line), VCL/PEGDA-MNPs-GO (yellow line) and VCL/PEGDA-MNPs-MWCNTs (brown line). GO = 3.47 μg·mL^{-1}, MWCNTs = 3.47 μg·mL^{-1}, MNPs = 93.3 μg·mL^{-1}.

MNPs, GO, and MWCNTs solutions raised their temperature around 10 °C, relative to the control sample (deionized water) and VCL/PEGDA-MNP-GO and VCL/PEGDA-MNP-MWCNTs nanocomposites. It can be concluded that the VCL/PEGDA hydrogel can absorb a large amount of energy without increasing its temperature. This would be expected because its main components, VCL and PEGDA, have good calorific capacities [121–123]. Therefore the hydrogel would inherit this property, which explains why the nanocomposites covered by hydrogel maintain a temperature similar to the temperature of the control sample (deionized water) at an irradiation time of more than 100 min with red light. Around 30 to 45 min, all the nanocomposites reach a threshold temperature that did not change when irradiated for a longer time.

Table 4. Light dose applied in the nanocomposites.

Time (min)	Light Dose (J·cm^{-2})
0	0
10	39.3
20	78.6
30	117.9
40	157.2
50	196.5
60	235.8
70	275.1
80	314.4
90	353.7
100	393

3.6. Morphological Studies of Magnetic Nanocomposites

The functionalized VCL/PEGDA-MNPs-GO-ZnMintPc and VCL/PEGDA-MNPs-MWCNTs-ZnMintPc nanocomposites were characterized by SEM, TEM, EDS, and XRD, and the results are presented in Figures 8b,e,f , S6, S9, S11 and S13 for GO-based nanocomposites, in Figures 8a,d, S7, S10, S12 and S14 for MWCNTs-based nanocomposites and in Figures 8c and S8 for free-standing MNPs.

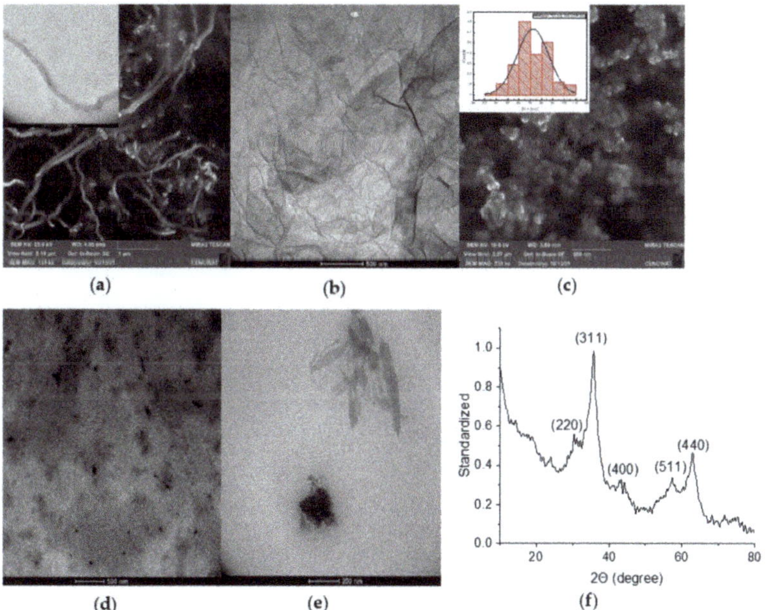

Figure 8. (**a**) SEM and TEM images of purified MWCNTs; (**b**) TEM of GO; (**c**) SEM of MNPs. (**d**) TEM of VCL/PEGDA-MNPs-MWCNTs-ZnMintPc. (**e**) TEM of VCL/PEGDA-MNPs-GO-ZnMintPc; (**f**) XRD analysis of MNPs-GO.

The SEM and TEM images in Figure 8 show the morphology of (a) carbon nanotubes, that have the shape of fibers and in TEM their internal structure and walls, (b) the structure of large sheets of GO. Figure S6d presents the height profile of GO and indicates that the thickness of the GO is roughly 2.8 nm. According to Sun et al., 2010 study [124], this indicates that the sheet is four-layered. Various GO sheet sizes are depicted, but the most common is roughly 2 μm. In Figure 8c, we can observe the MNPs with a spherical shape with an average size of ~72 nm. Figure 8d,e shows the morphology of the hydrogel coating MNPs-GO-ZnMintPc and MNPs-MWCNTs-ZnMintPc, so it is difficult to differentiate the structures covered by the hydrogel. In the Figure 8f, we present the XRD pattern for the MNPs-GO sample, where the diffraction peaks of the iron nanoparticles decorating the GO are observed at $2\theta = 30.27°, 35.6°, 43.3°, 53.7°, 57.1°$ and $63.0°$, indicating that the MNPs retain their original crystalline structure after functionalization, agreeing with the results of Amiri, Baghayeri, and Sedighi 2018; Cao et al., 2016 [125,126].

3.7. Morphological Studies of Bacteria in Magnetic Nanocomposites

To understand the interaction of magnetic nanocomposites with microorganisms, STEM images of *S. aureus* and *E. coli* bacteria were obtained in the presence of VCL/PEGDA-MNPs-GO-ZnMintPc and VCL/PEGDA-MNPs-MWCNT-ZnMintPc nanocomposites (Figure 9b,c,e,f). In the STEM image of *S. aureus* and *E. coli* pure (Figure 9a,d), their structure and morphology are not altered, and the membrane covers them without ruptures. When interacting with the VCL/PEGDA-MNPs-GO-ZnMintPc and VCL/PEGDA-MNPs-MWCNT-ZnMintPc nanocomposites, no changes were observed in the morphology of the bacteria (images taken 24 h after the bacterial-nanocomposite solutions were prepared) [127–129]. It was possible to observe how the nanocomposite can completely cover the microorganism, allowing the nanocomposites dispersed in the hydrogel to be photoexcited with the red-light source and cause microbial elimination.

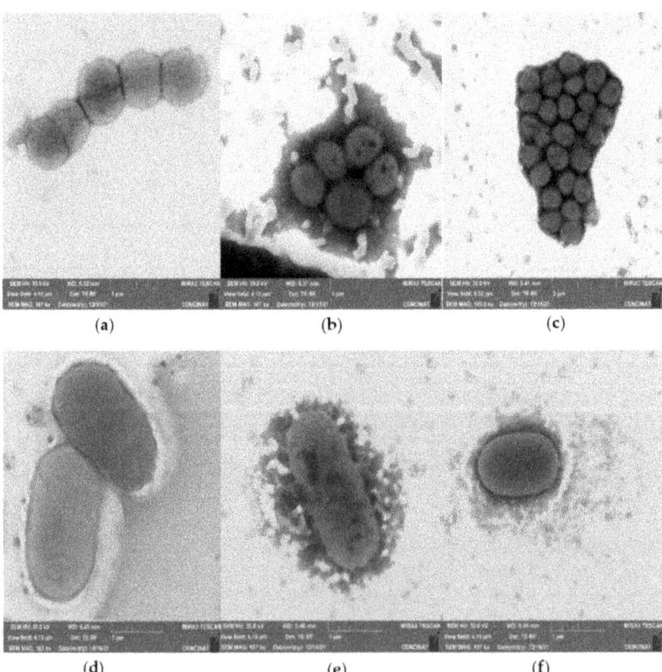

Figure 9. STEM of (**a**) *S. aureus*; (**b**) *S. aureus* + C1 (VCL/PEGDA-MNPs-GO-ZnMintPc); (**c**) *S. aureus* + C2(VCL/PEGDA-MNPs-MWCNTs-ZnMintPc); (**d**) *E. coli*; (**e**) *E. coli* + C1; (**f**) *E. coli* + C2.

3.8. Antimicrobial Effect of Magnetic Nanocomposites

The antimicrobial effect of the VCL/PEGDA-MNPs-GO-ZnMintPc (C1), VCL/PEGDA-MNPs-MWCNTs-ZnMintPc (C2), MNPs-GO (C3), and MNPs-MWCNTs (C4) nanocomposites were evaluated while irradiating with 630 nm red light at 65 mW cm^{-2} in the presence of the microorganisms *S. aureus*, *E. coli* and *C. albicans*.

The results obtained for each colony are shown in Figure 10. The histograms present the LOG (CFU mL^{-1}) evaluation standardized to 1. A concentration of 10^7 CFU mL^{-1} for *S. aureus*, 10^6 CFU mL^{-1} for *E. coli*, and 10^5 CFU mL^{-1} for *C. albicans* was used. Results show that light alone cannot eliminate microorganisms (control+light samples), as in Figure 11 with *C. albicans*. To eliminate the microorganism, it was necessary to irradiate it in the presence of a nanocomposite (C1 + light or C2 + light).

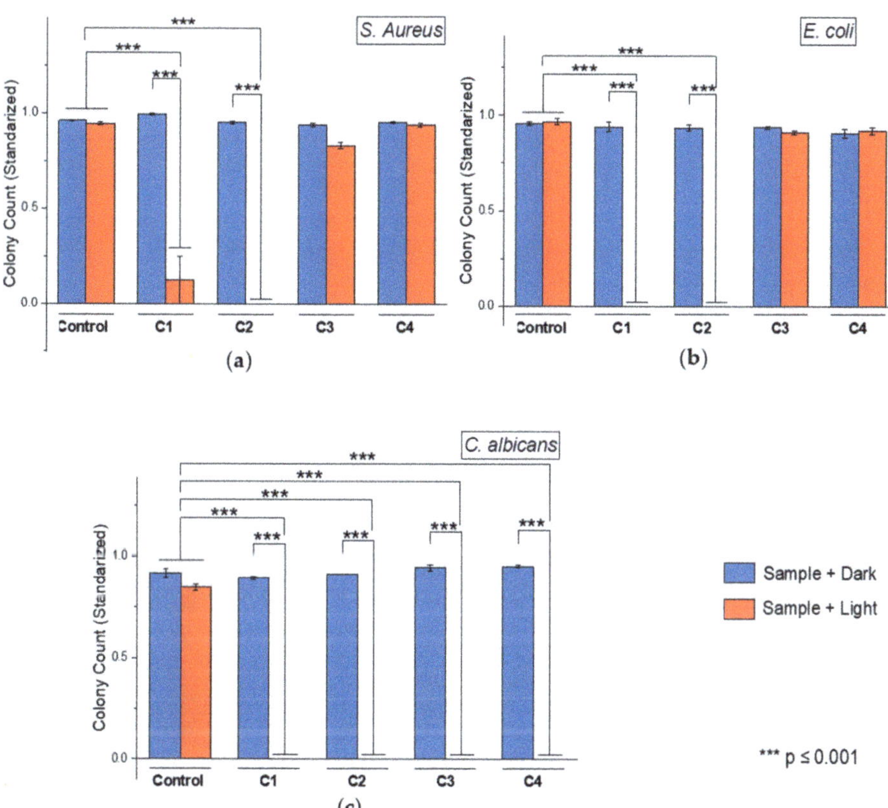

Figure 10. PDT/PTT antimicrobial effect. Standardized result of LOG(UFC/mL) by (**a**) *S. aureus*; (**b**) *E. coli* and (**c**) *C. albicans* based in C1, C2, C3 and C4 nanocomposites. The concentration of MNPs in C1, C2, C3 and C4 was 93.3 µg·mL^{-1}. The concentration of ZnMintPc in C1 and C2 was 8.1 µM. The concentration of GO in C1 and C3 was 3.47 µg·mL^{-1}. The concentration of MWCNTs in C2 and C4 was 3.47 µg·mL^{-1}. Time of irradiation of C1 and C2 nanocomposites was 30 min and for C3 and C4 was 40 min. Significant differences in means according to the Tukey test (*** $p \leq 0.001$).

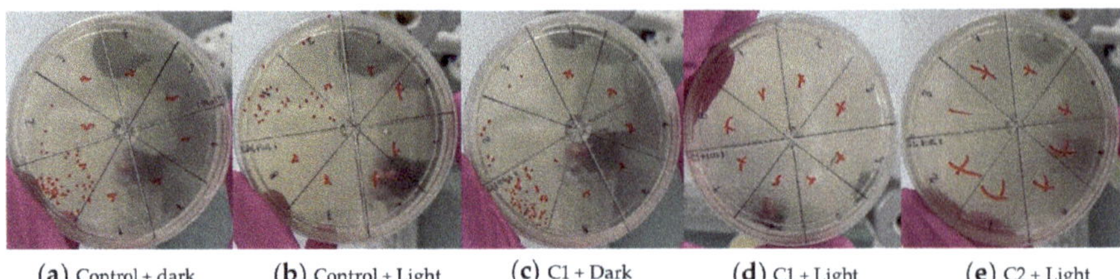

Figure 11. Antimicrobial effect of C1 and C2 nanocomposites at *C. albicans* in vitro. C1: GO-MNPs-VGLPEGDA- ZnMintPc, C2: MWCNT-MNPs-VGLPEGDA- ZnMintPc, C3: GO-MNPs-VGLPEGDA, C4: MWCNT-MNPs-VGLPEGDA. The concentration of MNPs in C1, C2, C3 and C4 was 93.3 $\mu g \cdot mL^{-1}$. The concentration of ZnMintPc in C1 and C2 was 8.1 μM. The concentration of GO in C1 and C3 was 3.47 $\mu g \cdot mL^{-1}$. The concentration of MWCNTs in C2 and C4 was 3.47 $\mu g \cdot mL^{-1}$.

The quantitative result of colony counts of *S. aureus*, *E. coli*, and *C. albicans* after being irradiated with a red LED (630 nm 65 mW cm^{-2}) indicates that for the colonies of *S. aureus* and *E. coli* (Figure 10a,b), samples C3, C4 in light and dark, were not different from the control group. In contrast, the count in samples C1 + light and C2 + light was significantly lower than the control group (*** $p \leq 0.001$). The colony count of *C. albicans* in the samples + dark was not different from the control group, but the count in the samples C1 + light, C2 + light, C3 + light, and C4 + light was significantly lower than the control group (*** $p \leq 0.001$). This means that after irradiation, the C1 nanocomposite eliminated all *E. coli*, and *C. albicans*, and some *S. aureus*. C2 nanocomposite eliminated *S. aureus*, *E. coli*, and *C. albicans*. C3 and C4 nanocomposites eliminated all *C. albicans*. Therefore, all nanocomposites can eliminate some of the microorganisms used in this study, with C2 being the best due to its ability to eliminate the three types of microorganisms. This agrees with the results obtained by Huo et al., 2021; Liu et al., 2021 and Ren et al., 2020 [129–131].

In Figure 10c, the samples without hydrogel C3 + light and C4 + light only had a complete response with *C. albicans*, and it could be considered that the photothermic effect allowed microbial elimination. This is in agreement with dos Santos et al., 2019 [132]. This conclusion is based on the results shown in Figure 7 of the photothermal effect for compounds based on MWCNT, MNP, GO, where it is observed that their temperature increased as the irradiation time increased. For the samples with hydrogel and photosensitizer (C1 + light and C2 + light), it can be concluded that the elimination of the microorganisms *S. aureus* (Figure 10a), *E. coli* (Figure 10b), and *C. albicans* (Figure 10c) was produced by the photodynamic effect, as supported by the research of Mei et al., 2021 [12] as well as the data reported by Xu, Yao and Xu 2019 [52]. This conclusion is based on the results in Table 3 that show that compounds containing ZnMintPc have a longer decay period than those which do not.

According to the results obtained, the VCL/PEGDA-MNPs-GO-ZnMintPc and VCL/PEGDA-MNPs-MWCNTs-ZnMintPc nanocomposites have a ferromagnetic character, typical of nanocomposites with iron nanoparticles and with low saturation magnetization, due to being covered by a diamagnetic hydrogel layer, in agreement with the studies carried out by Donadel et al., 2008, Mahdavi et al., 2013, and Qu et al., 2010 [133–135]; in which MNPs were synthesized for bioapplications and it was demonstrated that the surface modification caused a reduction in saturation magnetization, with values between 67 to 22 emu·g^{-1} depending on the type of biopolymer to be used.

Due to the physical and magnetic properties of these nanocomposites, it was shown that they could avoid the early extinction of the fluorescence of PS ZnMintPc, thus improving their photodynamic effect, as mentioned in the work of Huang et al., 2011 and Xiao et al., 2021. The thermodynamic studies carried out by Kou et al., 2019, Srivastava and Kumar, 2010, and Tager et al., 1993 indicate that VCL and PEGDA have good heat capacities of around 258 and 94 J·mol^{-1}·K^{-1}, which is why they are usually used to synthesize cryogels. This explains that the nanocomposites covered by the VCL/PEGDA hydrogel are capable of being maintained up to a maximum temperature of 40 °C receiving doses of red light of up to 393 J·cm^{-2}, which in comparison to nanocomposites without hydrogel, raise their temperature to around 55 °C, as described in the literature [71,117–120].

The internalization mechanism of these nanocomposites occurs through direct contact between the microorganism-nanocomposite and the PTT/PDT effect, as seen in Figure 9. Several authors have discussed the mechanism of programmed cell death due to the PDT/PTT effect, and as mentioned by Buzzá et al., 2021, Patil et al., 2021, among others in their scientific articles, the ability of PS, GO, and MWCNT to locate in various organelles and the action of the PTT/PDT effect promotes ROS generation followed by physical damage to the membrane. Oxidative stress leads to changes in calcium and lipid metabolism, generating cytokines and stress response mediators that lead to induction of apoptosis by the mitochondrial pathway and specific protein oxidation [136–140]. It can be concluded that this mechanism is the cause of the death of the microorganisms *S. aureus*, *E. coli*, and *C. albicans* in the present study. The results shown in Figure 10c indicate the elimination of *C. albicans* by the PTT effect and, as mentioned by Mocan et al., 2014, and Pérez-Hernández et al., 2015 in their research, PTT causes apoptosis, or programmed cell death, rather than necrotic cell death, by activating the intrinsic route. Inflammatory reactions are triggered by necrotic cell death, carbon-based nanomaterials in the photothermal treatment activate the flux of free radicals within the cell, and the oxidative state mediates cellular damage in PC cells via apoptotic pathway [141,142]. Therefore, it can be concluded that this is the mechanism that causes the death of *C. albicans* when there is no presence of PS.

The excretion pathway of these nanocomposites in living systems has been investigated by authors such as Dias et al., 2021 and Liu et al., 2008, who indicate that these nanocomposites based on carbon materials and MNPs can be excreted through the biliary and urinary tracts [143,144].

4. Conclusions

The present study involved synthesis of magnetic nanocomposites VCL/PEGDA-MNPs-MWCNTs-ZnMintPc and VCL/PEGDA-MNPs-GO-ZnMintPc to eliminate three types of microbial colonies: *S. aureus*, *E. coli* and *C. albicans*.

After these nanocomposites were synthesized, optical, magnetic, and morphological characterizations showed that GO, MWCNTS, iron MNPS and ZnMintPc are covered by VCL/PEGDA hydrogel.

The optical properties of these nanocomposites allow them to prevent the rapid disintegration of PS, which is essential in PDT.

Using photodynamic analysis, the nanocomposites the applicability in PDT of VCL/PEGDA-MNPs-MWCNTs-ZnMintPc and VCL/PEGDA-MNPs-GO-ZnMintPc were with low dose red light. It was observed that the nanocomposite VCL/PEGDA-MNPs-GO-ZnMintPc had higher efficiency than the nanocomposite based on MWCNTs since it produced faster photobleaching of DPBF because it is capable of transporting a more significant amount of ZnMintPc and MNPs due to its large specific area. In addition, GO contributed to PS in the formation of $^1O_2$. Since the nanocomposites are coated with a hydrogel, they are also suitable for controlled PS release systems, with promising applications for PDT.

The nanocomposite that contains GO and PS ZnMintPc had higher efficiency since it produced faster photobleaching of DPBF because it is capable of transporting a more significant amount of ZnMintPc and MNPs due to its large specific area. In addition, the GO contributed to PS in the formation of $^1O_2$.

Finally, we demonstrated that the VCL/PEGDA-MNPs-GO-ZnMintPc nanocomposite was able to eliminate colonies of *E. coli* and *C. albicans*, and the VCL/PEGDA-MNPs-MWCNTs-ZnMintPc nanocomposite eliminated the three types of microorganisms, which can therefore be considered as a broad-spectrum antimicrobial agent in PDT and PTT.

Supplementary Materials: The following supporting information can be downloaded at: https://www.mdpi.com/article/10.3390/pharmaceutics14040705/s1, Figure S1. Optical properties: (a) UV-VIS spectra of DMF(3mL) with different concentrations of ZnMintPc, (b) UV-VIS spectra of VCL/PEGDA(3mL) with different concentrations of ZnMintPc, (c). UV-VIS spectra of VCL/PEGDA (3mL) with MNPs-MWCNTs at different concentrations. (d). VCL/PEGDA with MNPs-MWCNTs and ZnMintPc at different concentrations; Figure S2. Optical properties of Magnetic Nanocomposites: (a) UV-VIS spectra of VCL/PEGDA (3mL) with different concentrations of MNPs-GO, (b) UV-VIS spectra of VCL/PEGDA (3mL) with MNPs-GO-ZnMintPc at different concentrations; Figure S3. (a) Composites calibration curve of ZnMintPc (0.52 µM) from: PS in DMF, VCL/PEGDA, VCL/PEGDA-MNPs-GO and VCL/PEGDA-MNPs-MWCNTs; (b) Absorbance of ZnMintPc-DMF and nanocomposites; Figure S4. Photodynamic Analyses of: (a) UV-VIS spectra of VCL/PEGDA-MNPs-GO and (b) UV-VIS spectra of VCL/PEGDA-MNPs-MWCNTs; Figure S5. Photodynamic Analyses of: (a) UV-VIS spectra of VCL/PEGDA-MNPs-GO-ZnMintPc and (b) UV-VIS spectra of VCL/PEGDA-MNPs-MWCNTs-ZnMintPc; Figure S6: (a) SEM analysis, (b) TEM analysis, (c) Atomic force microscopy (AFM) image (d)Height profile along the line of the panel from GO (e) EDS analysis and (f) XRD of GO; Figure S7. (a) SEM, (b) TEM, (c) EDS and (d) XRD analysis of MWCNTs; Figure S8. (a) SEM, (b) TEM, (c) EDS and (d) XRD analysis of MNPs; Figure S9. (a) SEM, (b) TEM, (c) EDS and (d) XRD analysis of MNPs-GO; Figure S10. (a) SEM, (b) TEM, (c) EDS and (d) XRD analysis of MNPs-MWCNTs; Figure S11: (a) SEM, (b) TEM, (c) EDS analysis of VCL/PEGDA-MNPs-GO; Figure S12: (a) SEM, (b) TEM, (c) EDS analysis of VCL/PEGDA-MNPs-MWCNTs; Figure S13. (a) SEM, (b) TEM, (c) EDS and (d) XRD analysis of VCL/PEGDA-MNPs-GO-ZnMintPc; Figure S14. (a) SEM, (b) TEM, (c) EDS and (d) XRD analysis of VCL/PEGDA-MNPs-MWCNTs-ZnMintPc.

Author Contributions: Validation, C.F.C.; formal analysis, C.F.C. and M.P.R.; investigation, C.F.C., A.D.-B., K.O.C., E.C.R., F.Q., S.N., A.D., K.V., D.N., C.E.Á., M.A.S., C.G.-R., A.B.-Z., M.R.R. and M.P.R.; resources, F.Q., M.R.R. and M.P.R.; data curation, C.F.C.; writing—original draft preparation, C.F.C.; writing—review and editing, C.F.C., A.D.-B., E.C.R., S.N., A.D., D.N., C.G.-R. and M.P.R.; visualization, C.F.C.; supervision, M.P.R.; project administration, M.P.R.; funding acquisition, M.R.R. and M.P.R. All authors have read and agreed to the published version of the manuscript.

Funding: National Polytechnic School through the PIIF-20-05 project. Research Office of the Pontificia Universidad Católica del Ecuador project code QINV0324-IINV529010100.

Institutional Review Board Statement: Not applicable.

Informed Consent Statement: Not applicable.

Acknowledgments: Zinc Menthol-Phthalocyanine was given by the Federal University of São Carlos-Brazil. To student Carlos Moscoso for helping with experiments in micro-organisms. To the National Polytechnic School through the PIIF-20-05 project. To the Research Office of the Pontificia Universidad Católica del Ecuador project code QINV0324-IINV529010100.

Conflicts of Interest: The authors declare no conflict of interest.

References

1. World Health Organization. *Managing Epidemics: Key Facts About Major Deadly Diseases*; World Health Organization: Geneva, Switzerland, 2018; Volume 1, ISBN 9789241565530. Available online: https://apps.who.int/iris/handle/10665/272442 (accessed on 15 December 2021).
2. World Health Organization. Antimicrobial Resistance. Available online: https://www.who.int/news-room/fact-sheets/detail/antimicrobial-resistance (accessed on 19 December 2021).
3. Ali, A.; Ovais, M.; Cui, X.; Rui, Y.K.; Chen, C. Safety Assessment of Nanomaterials for Antimicrobial Applications. *Chem. Res. Toxicol.* **2020**, *33*, 1082–1109. [CrossRef] [PubMed]
4. Zhao, Y.; Zhang, Z.; Pan, Z.; Liu, Y. Advanced bioactive nanomaterials for biomedical applications. *Exploration* **2021**, *1*, 20210089. [CrossRef]
5. Akhavan, O.; Ghaderi, E. Toxicity of Graphene and Graphene Oxide Nanowalls Against Bacteria. *ACS Nano* **2010**, *4*, 5731–5736. [CrossRef] [PubMed]
6. Gunawan, C.; Teoh, W.Y.; Marquis, C.P.; Amal, R. Cytotoxic origin of copper(II) oxide nanoparticles: Comparative studies with micron-sized particles, leachate, and metal salts. *ACS Nano* **2011**, *5*, 7214–7225. [CrossRef] [PubMed]
7. Zou, X.; Zhang, L.; Wang, Z.; Luo, Y. Mechanisms of the Antimicrobial Activities of Graphene Materials. *J. Am. Chem. Soc.* **2016**, *138*, 2064–2077. [CrossRef]

8.   Al-Jumaili, A.; Alancherry, S.; Bazaka, K.; Jacob, M.V. Review on the Antimicrobial Properties of Carbon Nanostructures. *Materials* **2017**, *10*, 1066. [CrossRef]
9.   Karahan, H.E.; Wiraja, C.; Xu, C.; Wei, J.; Wang, Y.; Wang, L.; Liu, F.; Chen, Y. Graphene Materials in Antimicrobial Nanomedicine: Current Status and Future Perspectives. *Adv. Healthc. Mater.* **2018**, *7*, 1701406. [CrossRef]
10.   Matharu, R.K.; Tabish, T.A.; Trakoolwilaiwan, T.; Mansfield, J.; Moger, J.; Wu, T.; Lourenço, C.; Chen, B.; Ciric, L.; Parkin, I.P.; et al. Microstructure and antibacterial efficacy of graphene oxide nanocomposite fibres. *J. Colloid Interface Sci.* **2020**, *571*, 239–252. [CrossRef]
11.   Romero, M.P.; Marangoni, V.S.; de Faria, C.G.; Leite, I.S.; de Carvalho Castro e Silva, C.; Maroneze, C.M.; Pereira-da-Silva, M.A.; Bagnato, V.S.; Inada, N.M. Graphene Oxide Mediated Broad-Spectrum Antibacterial Based on Bimodal Action of Photodynamic and Photothermal Effects. *Front. Microbiol.* **2020**, *10*, 2995. [CrossRef]
12.   Mei, L.; Shi, Y.; Cao, F.; Liu, X.; Li, X.M.; Xu, Z.; Miao, Z. PEGylated Phthalocyanine-Functionalized Graphene Oxide with Ultrahigh-Efficient Photothermal Performance for Triple-Mode Antibacterial Therapy. *ACS Biomater. Sci. Eng.* **2021**, *7*, 2638–2648. [CrossRef]
13.   Yang, Y.; Ma, L.; Cheng, C.; Deng, Y.; Huang, J.; Fan, X.; Nie, C.; Zhao, W.; Zhao, C. Nonchemotherapic and Robust Dual-Responsive Nanoagents with On-Demand Bacterial Trapping, Ablation, and Release for Efficient Wound Disinfection. *Adv. Funct. Mater.* **2018**, *28*, 1705708. [CrossRef]
14.   Tabish, T.A.; Scotton, C.J.; JFerguson, D.C.; Lin, L.; der Veen, A.V.; Lowry, S.; Ali, M.; Jabeen, F.; Ali, M.; Winyard, P.G.; et al. Biocompatibility and toxicity of graphene quantum dots for potential application in photodynamic therapy. *Nanomedicine* **2018**, *13*, 1923–1937. [CrossRef] [PubMed]
15.   Tabish, T.A.; Zhang, S.; Winyard, P.G. Developing the next generation of graphene-based platforms for cancer therapeutics: The potential role of reactive oxygen species. *Redox Biol.* **2018**, *15*, 34–40. [CrossRef]
16.   Dąbrowski, J.M. Reactive Oxygen Species in Photodynamic Therapy: Mechanisms of Their Generation and Potentiation. In *Advances in Inorganic Chemistry*; Elsevier: Amsterdam, The Netherlands, 2017; Volume 70, pp. 343–394.
17.   Henderson, B.; Dougherty, T. How does photodynamic therapy work? *Photochem. Photobiol.* **1992**, *55*, 145–157. [CrossRef] [PubMed]
18.   Kwiatkowski, S.; Knap, B.; Przystupski, D.; Saczko, J.; Kędzierska, E.; Knap-Czop, K.; Kotlińska, J.; Michel, O.; Kotowski, K.; Kulbacka, J. Photodynamic therapy—mechanisms, photosensitizers and combinations. *Biomed. Pharmacother.* **2018**, *106*, 1098–1107. [CrossRef] [PubMed]
19.   Klausen, M.; Ucuncu, M.; Bradley, M. Design of Photosensitizing Agents for Targeted Antimicrobial Photodynamic Therapy. *Molecules* **2020**, *25*, 5239. [CrossRef]
20.   Chen, J.; Fan, T.; Xie, Z.; Zeng, Q.; Xue, P.; Zheng, T.; Chen, Y.; Luo, X.; Zhang, H. Advances in nanomaterials for photodynamic therapy applications: Status and challenges. *Biomaterials* **2020**, *237*, 119827. [CrossRef]
21.   Banerjee, S.M.; MacRobert, A.J.; Mosse, C.A.; Periera, B.; Bown, S.G.; Keshtgar, M.R.S. Photodynamic therapy: Inception to application in breast cancer. *Breast* **2017**, *31*, 105–113. [CrossRef]
22.   Wu, D.; Fan, Y.; Yan, H.; Li, D.; Zhao, Z.; Chen, X.; Yang, X.; Liu, X. Oxidation-sensitive polymeric nanocarrier-mediated cascade PDT chemotherapy for synergistic cancer therapy and potentiated checkpoint blockade immunotherapy. *Chem. Eng. J.* **2020**, *404*, 126481. [CrossRef]
23.   Railkar, R.; Agarwal, P.K. Photodynamic Therapy in the Treatment of Bladder Cancer: Past Challenges and Current Innovations. *Eur. Urol. Focus* **2018**, *4*, 509–511. [CrossRef]
24.   Allison, R.R.; Downie, G.H.; Cuenca, R.; Hu, X.-H.; Childs, C.J.; Sibata, C.H. Photosensitizers in clinical PDT. *Photodiagnosis Photodynosis Ther.* **2004**, *1*, 27–42. [CrossRef]
25.   Wharton, T.; Gali, H.; Hamblin, M. Photosensitizers for Targeted Photodynamic Therapy. *J. Nanomed. Nanotechnol.* **2009**, *8*, 1–23.
26.   Bonnett, R.; Diamond, I.; Granelli, S.; McDonagh, A.F.; Nielsen, S.; Wilson, C.B.; Jaenicke, R.; Böhm, F.; Marston, G.; Truscott, T.G.; et al. Photosensitizers of the porphyrin and phthalocyanine series for photodynamic therapy. *Chem. Soc. Rev.* **1995**, *24*, 19. [CrossRef]
27.   Romero, M.P. *Estudo da Interação de Fotossensibilizantes Derivados de Ftalocianinas com Sistemas de Distribuição de Fármacos para Terapia Fotodinâmica*; Pontifícia Universidade Católica do Rio de Janeiro: Rio de Janeiro, Brazil, 2012.
28.   Zhou, L.; Liu, J.H.; Zhang, J.; Wei, S.H.; Feng, Y.Y.; Zhou, J.H.; Yu, B.Y.; Shen, J. A new sol-gel silica nanovehicle preparation for photodynamic therapy in vitro. *Int. J. Pharm.* **2010**, *386*, 131–137. [CrossRef] [PubMed]
29.   Chowdhary, R.; Dolphin, D.; Donnelly, R.; McCarron, P.; Chowdhary, R.; Dolphin, D. Drug delivery systems for photodynamic therapy. *Recent Pat. Drug Deliv. Formul.* **2009**, *3*, 1–7.
30.   Calixto, G.; Bernegossi, J.; de Freitas, L.; Fontana, C.; Chorilli, M. Nanotechnology-Based Drug Delivery Systems for Photodynamic Therapy of Cancer: A Review. *Molecules* **2016**, *21*, 342. [CrossRef] [PubMed]
31.   Nombona, N.; Antunes, E.; Chidawanyika, W.; Kleyi, P.; Tshentu, Z.; Nyokong, T. Synthesis, photophysics and photochemistry of phthalocyanine-ε-polylysine conjugates in the presence of metal nanoparticles against *Staphylococcus aureus*. *J. Photochem. Photobiol. A Chem.* **2012**, *233*, 24–33. [CrossRef]
32.   Jia, X.; Jia, L. Nanoparticles Improve Biological Functions of Phthalocyanine Photosensitizers Used for Photodynamic Therapy. *Curr. Drug Metab.* **2012**, *13*, 1119–1122. [CrossRef]

33. Li, Y.; Dong, H.; Li, Y.; Shi, D. Graphene-based nanovehicles for photodynamic medical therapy. *Int. J. Nanomed.* **2015**, *10*, 2451–2459.

34. Chatterjee, D.K.; Fong, L.S.; Zhang, Y. Nanoparticles in photodynamic therapy: An emerging paradigm. *Adv. Drug Deliv. Rev.* **2008**, *60*, 1627–1637. [CrossRef]

35. Sanginario, A.; Miccoli, B.; Demarchi, D. Carbon Nanotubes as an Effective Opportunity for Cancer Diagnosis and Treatment. *Biosensors* **2017**, *7*, 9. [CrossRef]

36. Guo, Q.; Shen, X.-T.; Li, Y.-Y.; Xu, S.-Q. Carbon nanotubes-based drug delivery to cancer and brain. *Curr. Med. Sci.* **2017**, *37*, 635–641. [CrossRef] [PubMed]

37. Sundaram, P.; Abrahamse, H. Effective Photodynamic Therapy for Colon Cancer Cells Using Chlorin e6 Coated Hyaluronic Acid-Based Carbon Nanotubes. *Int. J. Mol. Sci.* **2020**, *21*, 4745. [CrossRef]

38. Augustine, S.; Singh, J.; Srivastava, M.; Sharma, M.; Das, A.; Malhotra, B.D. Recent advances in carbon based nanosystems for cancer theranostics. *Biomater. Sci.* **2017**, *5*, 901–952. [CrossRef] [PubMed]

39. Jovanović, S.; Marković, Z.; Marković, B.T. Carbon-based nanomaterials as agents for photodynamic therapy. *Int. J. Cancer Res. Prev.* **2017**, *10*, 125–172.

40. Maiti, D.; Tong, X.; Mou, X.; Yang, K. Carbon-Based Nanomaterials for Biomedical Applications: A Recent Study. *Front. Pharmacol.* **2019**, *9*, 1401. [CrossRef] [PubMed]

41. Saleem, J.; Wang, L.; Chen, C. Carbon-Based Nanomaterials for Cancer Therapy via Targeting Tumor Microenvironment. *Adv. Healthc. Mater.* **2018**, *7*, e1800525. [CrossRef]

42. Lee, J.; Kim, J.; Kim, S.; Min, D.H. Biosensors based on graphene oxide and its biomedical application. *Adv. Drug Deliv. Rev.* **2016**, *105*, 275–287. [CrossRef]

43. Chauhan, N.; Maekawa, T.; Kumar, D.N.S. Graphene based biosensors—Accelerating medical diagnostics to new-dimensions. *J. Mater. Res.* **2017**, *32*, 2860–2882. [CrossRef]

44. Farrera, C.; Torres Andón, F.; Feliu, N. Carbon Nanotubes as Optical Sensors in Biomedicine. *ACS Nano* **2017**, *11*, 10637–10643. [CrossRef]

45. Yapici, M.; Alkhidir, T. Intelligent Medical Garments with Graphene-Functionalized Smart-Cloth ECG Sensors. *Sensors* **2017**, *17*, 875. [CrossRef] [PubMed]

46. Bekmukhametova, A.; Ruprai, H.; Hook, J.M.; Mawad, D.; Houang, J.; Lauto, A. Photodynamic therapy with nanoparticles to combat microbial infection and resistance. *Nanoscale* **2020**, *12*, 21034–21059. [CrossRef] [PubMed]

47. Yang, K.; Hu, L.; Ma, X.; Ye, S.; Cheng, L.; Shi, X.; Li, C.; Li, Y.; Liu, Z. Multimodal imaging guided photothermal therapy using functionalized graphene nanosheets anchored with magnetic nanoparticles. *Adv. Mater.* **2012**, *24*, 1868–1872. [CrossRef] [PubMed]

48. Chen, Y.Y.; Gao, Y.; Chen, Y.Y.; Liu, L.; Mo, A.; Peng, Q. Nanomaterials-based photothermal therapy and its potentials in antibacterial treatment. *J. Control Release* **2020**, *328*, 251–262. [CrossRef] [PubMed]

49. Li, J.; Liu, X.; Tan, L.; Cui, Z.; Yang, X.; Liang, Y.; Li, Z.; Zhu, S.; Zheng, Y.; Yeung, K.W.K.; et al. Zinc-doped Prussian blue enhances photothermal clearance of Staphylococcus aureus and promotes tissue repair in infected wounds. *Nat. Commun.* **2019**, *10*, 4490. [CrossRef]

50. Zhou, J.; Lu, Z.; Zhu, X.; Wang, X.; Liao, Y.; Ma, Z.; Li, F. NIR photothermal therapy using polyaniline nanoparticles. *Biomaterials* **2013**, *34*, 9584–9592. [CrossRef]

51. Shi, Y.; Liu, M.; Deng, F.; Zeng, G.; Wan, Q.; Zhang, X.; Wei, Y. Recent progress and development on polymeric nanomaterials for photothermal therapy: A brief overview. *J. Mater. Chem. B* **2017**, *5*, 194–206. [CrossRef]

52. Xu, J.W.; Yao, K.; Xu, Z.K. Nanomaterials with a photothermal effect for antibacterial activities: An overview. *Nanoscale* **2019**, *11*, 8680–8691. [CrossRef]

53. Zhu, H.; Cheng, P.; Chen, P.; Pu, K. Recent progress in the development of near-infrared organic photothermal and photodynamic nanotherapeutics. *Biomater. Sci.* **2018**, *6*, 746–765. [CrossRef]

54. Wang, Y.; Wei, T.; Qu, Y.; Zhou, Y.; Zheng, Y.; Huang, C.; Zhang, Y.; Yu, Q.; Chen, H. Smart, Photothermally Activated, Antibacterial Surfaces with Thermally Triggered Bacteria-Releasing Properties. *ACS Appl. Mater. Interfaces* **2020**, *12*, 21283–21291. [CrossRef]

55. Iijima, S. Helical microtubules of graphitic carbon. *Nature* **1991**, *354*, 56–58. [CrossRef]

56. Cheung, W.; Pontoriero, F.; Taratula, O.; Chen, A.M.; He, H. DNA and carbon nanotubes as medicine. *Adv. Drug Deliv. Rev.* **2010**, *62*, 633–649. [CrossRef]

57. Reilly, R.M. Carbon nanotubes: Potential benefits and risks of nanotechnology in nuclear medicine. *J. Nucl. Med.* **2007**, *48*, 1039–1042. [CrossRef] [PubMed]

58. Usrey, M.; Heller, D.; Tour, J.; Collier, C.; Freitag, M.; Dyke, C.; Doorn, S.; Mann, D.; Chae, H.; Liu, J.; et al. *Carbon Nanotubes: Properties and Applications*; CRC Press: London, UK, 2006.

59. Plombon, J.J.; O'Brien, K.P.; Gstrein, F.; Dubin, V.M.; Jiao, Y. High-frequency electrical properties of individual and bundled carbon nanotubes. *Appl. Phys. Lett.* **2007**, *90*, 063106. [CrossRef]

60. Menezes, B.R.C.; De Rodrigues, K.F.; Fonseca, B.C.D.S.; Ribas, R.G.; Montanheiro, T.L.D.A.; Thim, G.P. Recent advances in the use of carbon nanotubes as smart biomaterials. *J. Mater. Chem. B* **2019**, *7*, 1343–1360. [CrossRef]

61. Herlem, G.; Picaud, F.; Girardet, C.; Micheau, O. *Carbon Nanotubes: Synthesis, Characterization, and Applications in Drug-Delivery Systems*; Elsevier Inc.: Amsterdam, The Netherlands, 2019; ISBN 9780128140338.

62. Muthoosamy, K.; G Bai, R.; Manickam, S. Graphene and Graphene Oxide as a Docking Station for Modern Drug Delivery System. *Curr. Drug Deliv.* **2014**, *11*, 701–718. [CrossRef]
63. Wei, Y.; Zhou, F.; Zhang, D.; Chen, Q.; Xing, D. A graphene oxide based smart drug delivery system for tumor mitochondria-targeting photodynamic therapy. *Nanoscale* **2016**, *8*, 3530–3538. [CrossRef]
64. Sajjadi, M.; Nasrollahzadeh, M.; Jaleh, B.; Jamalipour Soufi, G.; Iravani, S. Carbon-based nanomaterials for targeted cancer nanotherapy: Recent trends and future prospects. *J. Drug Target.* **2021**, 716–741. [CrossRef]
65. Maleki, R.; Afrouzi, H.H.; Hosseini, M.; Toghraie, D.; Rostami, S. Molecular dynamics simulation of Doxorubicin loading with N-isopropyl acrylamide carbon nanotube in a drug delivery system. *Comput. Methods Programs Biomed.* **2020**, *184*, 105303. [CrossRef]
66. Qin, X.; Li, Y. Strategies To Design and Synthesize Polymer-Based Stimuli-Responsive Drug-Delivery Nanosystems. *ChemBioChem* **2020**, *21*, 1236–1253. [CrossRef]
67. Geim, A.K.; Novoselov, K.S. The rise of graphene. *Nat. Mater.* **2007**, *6*, 183–191. [CrossRef] [PubMed]
68. Avouris, P.; Xia, F. Graphene applications in electronics and photonics. *MRS Bull.* **2012**, *37*, 1225–1234. [CrossRef]
69. Perreault, F.; Fonseca de Faria, A.; Elimelech, M. Environmental applications of graphene-based nanomaterials. *Chem. Soc. Rev.* **2015**, *44*, 5861–5896. [CrossRef] [PubMed]
70. Singh, D.P.; Herrera, C.E.; Singh, B.; Singh, S.; Singh, R.K.; Kumar, R. Graphene oxide: An efficient material and recent approach for biotechnological and biomedical applications. *Mater. Sci. Eng. C* **2018**, *86*, 173–197. [CrossRef]
71. Estelrich, J.; Busquets, M.A. Iron Oxide Nanoparticles in Photothermal Therapy. *Molecules* **2018**, *23*, 1567. [CrossRef]
72. Tapeinos, C. Magnetic Nanoparticles and Their Bioapplications. In *Smart Nanoparticles for Biomedicine*; Elsevier: Amsterdam, The Netherland, 2018; pp. 131–142. ISBN 9780128141571.
73. Cardoso, V.F.; Francesko, A.; Ribeiro, C.; Bañobre-López, M.; Martins, P.; Lanceros-Mendez, S. Advances in Magnetic Nanoparticles for Biomedical Applications. *Adv. Healthc. Mater.* **2018**, *7*, 1700845. [CrossRef]
74. Li, X.; Wei, J.; Aifantis, K.E.; Fan, Y.; Feng, Q.; Cui, F.Z.; Watari, F. Current investigations into magnetic nanoparticles for biomedical applications. *J. Biomed. Mater. Res. Part A* **2016**, *104*, 1285–1296. [CrossRef]
75. Ankamwar, B.; Lai, T.C.; Huang, J.H.; Liu, R.S.; Hsiao, M.; Chen, C.H.; Hwu, Y.K. Biocompatibility of Fe_3O_4 nanoparticles evaluated by in vitro cytotoxicity assays using normal, glia and breast cancer cells. *Nanotechnology* **2010**, *21*, 75102. [CrossRef]
76. Miri, A.; Khatami, M.; Sarani, M. Biosynthesis, Magnetic and Cytotoxic Studies of Hematite Nanoparticles. *J. Inorg. Organomet. Polym. Mater.* **2020**, *30*, 767–774. [CrossRef]
77. Rajendran, K.; Sen, S.; Suja, G.; Senthil, S.L.; Kumar, T.V. Evaluation of cytotoxicity of hematite nanoparticles in bacteria and human cell lines. *Colloids Surf. B Biointerfaces* **2017**, *157*, 101–109. [CrossRef]
78. Yu, Q.; Xiong, X.-Q.; Zhao, L.; Xu, T.-T.; Bi, H.; Fu, R.; Wang, Q.-H. Biodistribution and Toxicity Assessment of Superparamagnetic Iron Oxide Nanoparticles In Vitro and In Vivo. *Curr. Med Sci.* **2018**, *38*, 1096–1102. [CrossRef] [PubMed]
79. Avval, Z.M.; Malekpour, L.; Raeisi, F.; Babapoor, A.; Mousavi, S.M.; Hashemi, S.A.; Salari, M. Introduction of magnetic and supermagnetic nanoparticles in new approach of targeting drug delivery and cancer therapy application. *Drug Metab. Rev.* **2020**, *52*, 157–184. [CrossRef]
80. Ramasamy, M.; Kumar, P.S.; Varadan, V.K. Magnetic nanotubes for drug delivery. In Proceedings of the Nanosensors, Biosensors, Info-Tech Sensors and 3D Systems 2017, Portland, OR, USA, 26–29 March 2017; Volume 10167, p. 1016715.
81. Garcia, J.F.; Makazaga, A.I.; Sáez, A.S.; Julián, J.R.; Van Herk, A.M.; Heuts, J.A. Cationic Nanogels for Biotechnological Uses. U.S. Patent 13/635,759, 21 March 2013.
82. Ramos, J.; Imaz, A.; Forcada, J.; Oh, J.K.; Drumright, R.; Siegwart, D.J.; Matyjaszewski, K.; Lyon, L.A.; Meng, Z.; Singh, N.; et al. Temperature-sensitive nanogels: Poly(N-vinylcaprolactam) versus poly(N-isopropylacrylamide). *Polym. Chem.* **2012**, *3*, 852–856. [CrossRef]
83. Romero, J.F.; Diaz-Barrios, A.; Gonzalez, G. Biocompatible thermo-responsive N-vinylcaprolactam based hydrogels for controlled drug delivery systems. *Rev. Bionatura* **2021**, *6*, 1712–1719. [CrossRef]
84. Hummers, W.S.; Offeman, R.E. Preparation of Graphitic Oxide. *J. Am. Chem. Soc.* **1958**, *80*, 1339. [CrossRef]
85. Zhang, X.-F.; Li, X. The photostability and fluorescence properties of diphenylisobenzofuran. *J. Lumin* **2011**, *131*, 2263–2266. [CrossRef]
86. Cunha, C.; Panseri, S.; Iannazzo, D.; Piperno, A.; Pistone, A.; Fazio, M.; Russo, A.; Marcacci, M.; Galvagno, S. Hybrid composites made of multiwalled carbon nanotubes functionalized with Fe_3O_4 nanoparticles for tissue engineering applications. *Nanotechnology* **2012**, *23*, 465102. [CrossRef]
87. Navamani, J. Development of Nanoprobe for the Determination of Blood Cholesterol. *J. Biosens. Bioelectron.* **2012**, *3*, 15. [CrossRef]
88. Pourzamani, H.; Hajizadeh, Y.; Fadaei, S. Efficiency enhancement of multi-walled carbon nanotubes by ozone for benzene removal from aqueous solution. *Int. J. Environ. Health Eng.* **2015**, *4*, 29. [CrossRef]
89. Abrinaei, F.; Kimiagar, S.; Zolghadr, S. Hydrothermal synthesis of hematite-GO nanocomposites at different GO contents and potential application in nonlinear optics. *Opt. Mater.* **2019**, *96*, 109285. [CrossRef]
90. Al-Ruqeishi, M.S.; Mohiuddin, T.; Al-Moqbali, M.; Al-Shukaili, H.; Al-Mamari, S.; Al-Rashdi, H.; Al-Busaidi, R.; Sreepal, V.; Nair, R.R. Graphene Oxide Synthesis: Optimizing the Hummers and Marcano Methods. *Nanosci. Nanotechnol. Lett.* **2020**, *12*, 88–95. [CrossRef]
91. Santoro, G.; Domingo, C. Espectroscopía Raman de nanotubos de carbono. *Opt. Pura Apl.* **2007**, *40*, 175–186.

92. Dresselhaus, M.S.; Dresselhaus, G.; Saito, R.; Jorio, A. Raman spectroscopy of carbon nanotubes. *Phys. Rep.* **2005**, *409*, 47–99. [CrossRef]

93. Marković, Z.; Jovanović, S.; Kleut, D.; Romčević, N.; Jokanović, V.; Trajković, V.; Todorović-Marković, B. Comparative study on modification of single wall carbon nanotubes by sodium dodecylbenzene sulfonate and melamine sulfonate superplasticiser. *Appl. Surf. Sci.* **2009**, *255*, 6359–6366. [CrossRef]

94. Bokobza, L.; Zhang, J. Raman spectroscopic characterization of multiwall carbon nanotubes and of composites. *Express Polym. Lett.* **2012**, *6*, 601–608. [CrossRef]

95. Soler, M.A.G.; Qu, F. Raman spectroscopy of iron oxide nanoparticles. In *Raman Spectroscopy for Nanomaterials Characterization*; Springer: Berlin/Heidelberg, Germany, 2012; pp. 379–416. ISBN 9783642206207. [CrossRef]

96. Cançado, L.G.; Jorio, A.; Ferreira, E.H.M.; Stavale, F.; Achete, C.A.; Capaz, R.B.; Moutinho, M.V.O.; Lombardo, A.; Kulmala, T.S.; Ferrari, A.C. Quantifying defects in graphene via Raman spectroscopy at different excitation energies. *Nano Lett.* **2011**, *11*, 3190–3196. [CrossRef]

97. Hafiz, S.M.; Ritikos, R.; Whitcher, T.J.; Razib, N.M.; Bien, D.C.S.; Chanlek, N.; Nakajima, H.; Saisopa, T.; Songsiriritthigul, P.; Huang, N.M.; et al. A practical carbon dioxide gas sensor using room-temperature hydrogen plasma reduced graphene oxide. *Sens. Actuators B Chem.* **2014**, *193*, 692–700. [CrossRef]

98. Ramirez, S.; Chan, K.; Hernandez, R.; Recinos, E.; Hernandez, E.; Salgado, R.; Khitun, A.G.; Garay, J.E.; Balandin, A.A. Thermal and magnetic properties of nanostructured densified ferrimagnetic composites with graphene-graphite fillers. *Mater. Des.* **2017**, *118*, 75–80. [CrossRef]

99. Satheesh, M.; Paloly, A.R.; Krishna Sagar, C.K.; Suresh, K.G.; Bushiri, M.J. Improved Coercivity of Solvothermally Grown Hematite (α-Fe$_2$O$_3$) and Hematite/Graphene Oxide Nanocomposites (α-Fe$_2$O$_3$/GO) at Low Temperature. *Phys. Status Solidi Appl. Mater. Sci.* **2018**, *215*, 1700705. [CrossRef]

100. Sajitha, E.P.; Prasad, V.; Subramanyam, S.V.; Mishra, A.K.; Sarkar, S.; Bansal, C. Size-dependent magnetic properties of iron carbide nanoparticles embedded in a carbonmatrix. *J. Phys. Condens. Matter* **2007**, *19*, 046214. [CrossRef]

101. Lipert, K.; Kazmierezak, J.; Pelech, I.; Narkiewicz, U.; Slawska-Waniewska, A.; Lachowicz, H.K. Magnetic properties of cementite (Fe~3C) nanoparticle agglomerates in a carbon matrix. *Mater. Sci.* **2007**, *25*, 399.

102. Grössinger, R. Correlation between the inhomogeneity and the magnetic anisotropy in polycrystalline ferromagnetic materials. *J. Magn. Magn. Mater.* **1982**, *28*, 137–142. [CrossRef]

103. Hauser, H.; Jiles, D.C.; Melikhov, Y.; Li, L.; Grössinger, R. An approach to modeling the dependence of magnetization on magnetic field in the high field regime. *J. Magn. Magn. Mater.* **2006**, *300*, 273–283. [CrossRef]

104. Batlle, X.; García Del Muro, M.; Tejada, J.; Pfeiffer, H.; Görnert, P.; Sinn, E. Magnetic study of M-type doped barium ferrite nanocrystalline powders. *J. Appl. Phys.* **1993**, *74*, 3333–3340. [CrossRef]

105. Kodama, R.H. Magnetic nanoparticles. *J. Magn. Magn. Mater.* **1999**, *200*, 359–372. [CrossRef]

106. Zhang, D.; Klabunde, K.; Sorensen, C.; Hadjipanayis, G. Magnetization temperature dependence in iron nanoparticles. *Phys. Rev. B Condens. Matter Mater. Phys.* **1998**, *58*, 14167–14170. [CrossRef]

107. Xiao, G.; Chien, C.L. Temperature dependence of spontaneous magnetization of ultrafine Fe particles in Fe-SiO$_2$ granular solids. *J. Appl. Phys.* **1987**, *61*, 3308–3310. [CrossRef]

108. Romero, M.P.; Gobo, N.R.S.; de Oliveira, K.T.; Iamamoto, Y.; Serra, O.A.; Louro, S.R.W. Photophysical properties and photodynamic activity of a novel menthol–zinc phthalocyanine conjugate incorporated in micelles. *J. Photochem. Photobiol. A Chem.* **2013**, *253*, 22–29. [CrossRef]

109. Nunes, S.M.T.M.T.; Sguilla, F.S.S.; Tedesco, A.C.C. Photophysical studies of zinc phthalocyanine and chloroaluminum phthalocyanine incorporated into liposomes in the presence of additives. *Brazil. J. Med. Biol. Res.* **2004**, *37*, 273–284. [CrossRef]

110. Wang, Y.H.; Song, X.; Shao, S.; Zhong, H.; Lin, F. An efficient, soluble, and recyclable multiwalled carbon nanotubes-supported TEMPO for oxidation of alcohols. *RSC Adv.* **2012**, *2*, 7693. [CrossRef]

111. Bera, M.; Gupta, P.; Maji, P.K. Facile One-Pot Synthesis of Graphene Oxide by Sonication Assisted Mechanochemical Approach and Its Surface Chemistry. *J. Nanosci. Nanotechnol.* **2018**, *18*, 902–912. [CrossRef] [PubMed]

112. Chen, P.; Marshall, A.S.; Chi, S.-H.; Yin, X.; Perry, J.W.; Jäkle, F. Luminescent Quadrupolar Borazine Oligomers: Synthesis, Photophysics, and Two-Photon Absorption Properties. *Chem. Eur. J.* **2015**, *21*, 18237–18247. [CrossRef] [PubMed]

113. Meier, H.; Gerold, J.; Kolshorn, H.; Mühling, B. Extension of Conjugation Leading to Bathochromic or Hypsochromic Effects in OPV Series. *Chem. Eur. J.* **2004**, *10*, 360–370. [CrossRef] [PubMed]

114. Wu, W.; Shen, J.; Gai, Z.; Hong, K.; Banerjee, P.; Zhou, S. Multi-functional core-shell hybrid nanogels for pH-dependent magnetic manipulation, fluorescent pH-sensing, and drug delivery. *Biomaterials* **2011**, *32*, 9876–9887. [CrossRef]

115. Wu, W.; Shen, J.; Banerjee, P.; Zhou, S. Core-shell hybrid nanogels for integration of optical temperature-sensing, targeted tumor cell imaging, and combined chemo-photothermal treatment. *Biomaterials* **2010**, *31*, 7555–7566. [CrossRef]

116. Sánchez-Iglesias, A.; Grzelczak, M.; Rodríguez-González, B.; Guardia-Girós, P.; Pastoriza-Santos, I.; Pérez-Juste, J.; Prato, M.; Liz-Marzán, L.M. Synthesis of multifunctional composite microgels via in situ Ni growth on pNIPAM-coated au nanoparticles. *ACS Nano* **2009**, *3*, 3184–3190. [CrossRef]

117. Dong, X.; Sun, Z.; Wang, X.; Leng, X. An innovative MWCNTs/DOX/TC nanosystem for chemo-photothermal combination therapy of cancer. *Nanomed. Nanotechnol. Biol. Med.* **2017**, *13*, 2271–2280. [CrossRef]

118. Burke, A.R.; Singh, R.N.; Carroll, D.L.; Wood, J.C.S.; D'Agostino, R.B.; Ajayan, P.M.; Torti, F.M.; Torti, S.V. The resistance of breast cancer stem cells to conventional hyperthermia and their sensitivity to nanoparticle-mediated photothermal therapy. *Biomaterials* **2012**, *33*, 2961–2970. [CrossRef]
119. Lim, J.H.; Kim, D.E.; Kim, E.-J.; Ahrberg, C.D.; Chung, B.G. Functional Graphene Oxide-Based Nanosheets for Photothermal Therapy. *Macromol. Res.* **2018**, *26*, 557–565. [CrossRef]
120. Huang, C.; Hu, X.; Hou, Z.; Ji, J.; Li, Z.; Luan, Y. Tailored graphene oxide-doxorubicin nanovehicles via near-infrared dye-lactobionic acid conjugates for chemo-photothermal therapy. *J. Colloid Interface Sci.* **2019**, *545*, 172–183. [CrossRef] [PubMed]
121. Tager, A.A.; Safronov, A.P.; Sharina, S.V.; Galaev, I.Y. Thermodynamic study of poly(N-vinyl caprolactam) hydration at temperatures close to lower critical solution temperature. *Colloid Polym. Sci.* **1993**, *271*, 868–872. [CrossRef]
122. Srivastava, A.; Kumar, A. Thermoresponsive poly(N-vinylcaprolactam) cryogels: Synthesis and its biophysical evaluation for tissue engineering applications. *J. Mater. Sci. Mater. Med.* **2010**, *21*, 2937–2945. [CrossRef]
123. Kou, Y.; Wang, S.; Luo, J.; Sun, K.; Zhang, J.; Tan, Z.; Shi, Q. Thermal analysis and heat capacity study of polyethylene glycol (PEG) phase change materials for thermal energy storage applications. *J. Chem. Thermodyn.* **2019**, *128*, 259–274. [CrossRef]
124. Sun, X.; Luo, D.; Liu, J.; Evans, D.G. Monodisperse chemically modified graphene obtained by density gradient ultracentrifugal rate separation. *ACS Nano* **2010**, *4*, 3381–3389. [CrossRef] [PubMed]
125. Cao, X.T.; Showkat, A.M.; Kang, I.; Gal, Y.S.; Lim, K.T. Beta-Cyclodextrin Multi-Conjugated Magnetic Graphene Oxide as a Nano-Adsorbent for Methylene Blue Removal. *J. Nanosci. Nanotechnol.* **2016**, *16*, 1521–1525. [CrossRef]
126. Amiri, A.; Baghayeri, M.; Sedighi, M. Magnetic solid-phase extraction of polycyclic aromatic hydrocarbons using a graphene oxide/Fe$_3$O$_4$@polystyrene nanocomposite. *Microchim. Acta* **2018**, *185*, 393. [CrossRef]
127. Yougbaré, S.; Mutalik, C.; Krisnawati, D.I.; Kristanto, H.; Jazidie, A.; Nuh, M.; Cheng, T.M.; Kuo, T.R. Nanomaterials for the Photothermal Killing of Bacteria. *Nanomaterials* **2020**, *10*, 1123. [CrossRef]
128. Ma, G.; Qi, J.; Cui, Q.; Bao, X.; Gao, D.; Xing, C. Graphene Oxide Composite for Selective Recognition, Capturing, Photothermal Killing of Bacteria over Mammalian Cells. *Polymers* **2020**, *12*, 1116. [CrossRef]
129. Liu, Z.; Zhao, X.; Yu, B.; Zhao, N.; Zhang, C.; Xu, F.J. Rough Carbon-Iron Oxide Nanohybrids for Near-Infrared-II Light-Responsive Synergistic Antibacterial Therapy. *ACS Nano* **2021**, *15*, 7482–7490. [CrossRef]
130. Ren, Y.; Liu, H.; Liu, X.; Zheng, Y.; Li, Z.; Li, C.; Yeung, K.W.K.; Zhu, S.; Liang, Y.; Cui, Z.; et al. Photoresponsive Materials for Antibacterial Applications. *Cell Rep. Phys. Sci.* **2020**, *1*, 100245. [CrossRef]
131. Huo, J.; Jia, Q.; Huang, H.; Zhang, J.; Li, P.; Dong, X.; Huang, W. Emerging photothermal-derived multimodal synergistic therapy in combating bacterial infections. *Chem. Soc. Rev.* **2021**, *50*, 8762–8789. [CrossRef] [PubMed]
132. dos Santos, K.F.; Sousa, M.S.; Valverde, J.V.P.; Olivati, C.A.; Souto, P.C.S.; Silva, J.R.; de Souza, N.C. Fractal analysis and mathematical models for the investigation of photothermal inactivation of Candida albicans using carbon nanotubes. *Colloids Surf. B Biointerfaces* **2019**, *180*, 393–400. [CrossRef] [PubMed]
133. Mahdavi, M.; Bin Ahmad, M.; Haron, J.; Namvar, F.; Nadi, B.; Ab Rahman, M.Z.; Amin, J. Synthesis, Surface Modification and Characterisation of Biocompatible Magnetic Iron Oxide Nanoparticles for Biomedical Applications. *Molecules* **2013**, *18*, 7533–7548. [CrossRef]
134. Donadel, K.; Felisberto, M.D.V.; Fávere, V.T.; Rigoni, M.; Batistela, N.J.; Laranjeira, M.C.M. Synthesis and characterization of the iron oxide magnetic particles coated with chitosan biopolymer. *Mater. Sci. Eng. C* **2008**, *28*, 509–514. [CrossRef]
135. Qu, J.; Liu, G.; Wang, Y.; Hong, R. Preparation of Fe$_3$O$_4$–chitosan nanoparticles used for hyperthermia. *Adv. Powder Technol.* **2010**, *21*, 461–467. [CrossRef]
136. Lagos, K.J.; Buzzá, H.H.; Bagnato, V.S.; Romero, M.P. Carbon-Based Materials in Photodynamic and Photothermal Therapies Applied to Tumor Destruction. *Int. J. Mol. Sci.* **2021**, *23*, 22. [CrossRef]
137. Patil, T.V.; Patel, D.K.; Dutta, S.D.; Ganguly, K.; Randhawa, A.; Lim, K.T. Carbon Nanotubes-Based Hydrogels for Bacterial Eradiation and Wound-Healing Applications. *Appl. Sci.* **2021**, *11*, 9550. [CrossRef]
138. Marangon, I.; Ménard-Moyon, C.; Silva, A.K.A.; Bianco, A.; Luciani, N.; Gazeau, F. Synergic mechanisms of photothermal and photodynamic therapies mediated by photosensitizer/carbon nanotube complexes. *Carbon N. Y.* **2016**, *97*, 110–123. [CrossRef]
139. Mroz, P.; Yaroslavsky, A.; Kharkwal, G.B.; Hamblin, M.R. Cell Death Pathways in Photodynamic Therapy of Cancer. *Cancers* **2011**, *3*, 2516–2539. [CrossRef]
140. Naserzadeh, P.; Ansari Esfeh, F.; Kaviani, M.; Ashtari, K.; Kheirbakhsh, R.; Salimi, A.; Pourahmad, J. Single-walled carbon nanotube, multi-walled carbon nanotube and Fe$_2$O$_3$ nanoparticles induced mitochondria mediated apoptosis in melanoma cells. *Cutan. Ocul. Toxicol.* **2017**, *37*, 157–166. [CrossRef] [PubMed]
141. Mocan, T.; Matea, C.T.; Cojocaru, I.; Ilie, I.; Tabaran, F.A.; Zaharie, F.; Iancu, C.; Bartos, D.; Mocan, L. Photothermal Treatment of Human Pancreatic Cancer Using PEGylated Multi-Walled Carbon Nanotubes Induces Apoptosis by Triggering Mitochondrial Membrane Depolarization Mechanism. *J. Cancer* **2014**, *5*, 679. [CrossRef] [PubMed]
142. Pérez-Hernández, M.; Del Pino, P.; Mitchell, S.G.; Moros, M.; Stepien, G.; Pelaz, B.; Parak, W.J.; Gálvez, E.M.; Pardo, J.; De La Fuente, J.M. Dissecting the molecular mechanism of apoptosis during photothermal therapy using gold nanoprisms. *ACS Nano* **2015**, *9*, 52–61. [CrossRef] [PubMed]

143. Dias, L.D.; Buzzá, H.H.; Stringasci, M.D.; Bagnato, V.S. Recent Advances in Combined Photothermal and Photodynamic Therapies against Cancer Using Carbon Nanomaterial Platforms for In Vivo Studies. *Photochem* **2021**, *1*, 434–447. [CrossRef]
144. Liu, Z.; Davis, C.; Cai, W.; He, L.; Chen, X.; Dai, H. Circulation and long-term fate of functionalized, biocompatible single-walled carbon nanotubes in mice probed by Raman spectroscopy. *Proc. Natl. Acad. Sci. USA* **2008**, *105*, 1410–1415. [CrossRef]

 *pharmaceutics*

Article

Parietin Cyclodextrin-Inclusion Complex as an Effective Formulation for Bacterial Photoinactivation

Abdallah Mohamed Ayoub [1,2,†], **Bernd Gutberlet** [1,†], **Eduard Preis** [1], **Ahmed Mohamed Abdelsalam** [1,3], **Alice Abu Dayyih** [1], **Ayat Abdelkader** [1,4], **Amir Balash** [5], **Jens Schäfer** [1] and **Udo Bakowsky** [1,*]

1 Department of Pharmaceutics and Biopharmaceutics, University of Marburg, Robert-Koch-Str. 4, 35037 Marburg, Germany; abdallah.ayoub@pharmazie.uni-marburg.de (A.M.A.); bernd.gutberlet@pharmazie.uni-marburg.de (B.G.); eduard.preis@pharmazie.uni-marburg.de (E.P.); ahmed.abdelsalam@pharmazie.uni-marburg.de (A.M.A.); alice.abudayyih@pharmazie.uni-marburg.de (A.A.D.); aboelela@staff.uni-marburg.de (A.A.); j.schaefer@staff.uni-marburg.de (J.S.)

2 Department of Pharmaceutics and Industrial Pharmacy, Faculty of Pharmacy, Zagazig University, Zagazig 44519, Egypt

3 Department of Pharmaceutics and Industrial Pharmacy, Faculty of Pharmacy, Al-Azhar University, Assiut 71524, Egypt

4 Assiut International Center of Nanomedicine, Al-Rajhy Liver Hospital, Assiut University, Assiut 71515, Egypt

5 Department of Pharmaceutical Chemistry, University of Marburg, Marbacher Weg 10, 35032 Marburg, Germany; amir.balash@pharmazie.uni-marburg.de

* Correspondence: ubakowsky@aol.com

† These authors contributed equally to this work.

Citation: Ayoub, A.M.; Gutberlet, B.; Preis, E.; Abdelsalam, A.M.; Abu Dayyih, A.; Abdelkader, A.; Balash, A.; Schäfer, J.; Bakowsky, U. Parietin Cyclodextrin-Inclusion Complex as an Effective Formulation for Bacterial Photoinactivation. *Pharmaceutics* **2022**, *14*, 357. https://doi.org/10.3390/pharmaceutics14020357

Academic Editor: Nejat Düzgüneş

Received: 20 December 2021
Accepted: 1 February 2022
Published: 4 February 2022

Publisher's Note: MDPI stays neutral with regard to jurisdictional claims in published maps and institutional affiliations.

Abstract: Multidrug resistance in pathogenic bacteria has become a significant public health concern. As an alternative therapeutic option, antimicrobial photodynamic therapy (aPDT) can successfully eradicate antibiotic-resistant bacteria with a lower probability of developing resistance or systemic toxicity commonly associated with the standard antibiotic treatment. Parietin (PTN), also termed physcion, a natural anthraquinone, is a promising photosensitizer somewhat underrepresented in aPDT because of its poor water solubility and potential to aggregate in the biological environment. This study investigated whether the complexation of PTN with (2-hydroxypropyl)-β-cyclodextrin (HP-β-CD) could increase its solubility, enhance its photophysical properties, and improve its phototoxicity against bacteria. At first, the solubilization behavior and complexation constant of the PTN/HP-β-CD inclusion complexes were evaluated by the phase solubility method. Then, the formation and physicochemical properties of PTN/HP-β-CD complexes were analyzed and confirmed in various ways. At the same time, the photodynamic activity was assessed by the uric acid method. The blue light-mediated photodegradation of PTN in its free and complexed forms were compared. Complexation of PTN increased the aqueous solubility 28-fold and the photostability compared to free PTN. PTN/HP-β-CD complexes reduce the bacterial viability of *Staphylococcus saprophyticus* and *Escherichia coli* by > 4.8 log and > 1.0 log after irradiation, respectively. Overall, the low solubility, aggregation potential, and photoinstability of PTN were overcome by its complexation in HP-β-CD, potentially opening up new opportunities for treating infections caused by multidrug-resistant bacteria.

Keywords: physcion; hydroxypropyl-β-cyclodextrin; photosensitizer; antimicrobial photodynamic therapy

1. Introduction

The overuse of antibiotics has resulted in inexorable antibiotic resistance, which worsens day by day, extending hospital admissions, raising healthcare costs, and eventually leading to increased death rates. In 2020, more than 670,000 people were infected in Europe with antibiotic-resistant bacteria, with approximately 33,000 deaths. The total economic burden of antibiotic resistance in Europe is expected to be 1.1 billion Euros [1]. The loss in economic output due to illness or death-related infection was estimated to surpass 1 trillion USD per year after 2030 and approach 2 trillion USD annually by 2050, accounting for a total of 100 trillion USD decline in global production between 2014 and 2050 [2]. Even worse, the number of deaths caused by worldwide antibiotic resistance is expected to increase to 10 million per year by 2050, up from 700,000 in 2014 [3]. Development of antibiotic resistance results in higher costs because of switching to more expensive antibiotics [4]. Multiple reasons, scientific, economic, and regulatory, hinder the development of new antibiotics [5]. Moreover, other drugs for chronic diseases are more appealing to pharmaceutical companies than short-course antibiotics. Consequently, approvals of new systemic antibiotics by the US Food and Drug Administration have dropped by 90% in the last 30 years [5]. Thus, alternative treatment strategies against resistant bacteria are urgently sought.

Photodynamic therapy (PDT) against cancer cells or antimicrobial photodynamic therapy (aPDT) have advantages over traditional treatments, e.g., minimal invasiveness and excellent safety profiles [6]. PDT and aPDT for selective damage to the area of interest depend on three nontoxic components: a photosensitizer (PS), light, and oxygen [7]. They are two-step procedures in which a photosensitizer (PS) is administered to patients, followed by light irradiation at a specific wavelength matching the absorbance spectra of the PS. Once exposed to light, the PS is excited from its low-energy ground state into a high-energy singlet state which undergoes intersystem crossing to the longer-lived triplet state [6]. Via type I and type II mechanisms, the PS in the triplet state can generate reactive oxygen species (ROS) such as singlet oxygen superoxide anions and hydroxyl radicals, which are the leading cause of cell death [8].

Unlike traditional antibiotics, which have a particular target known as the key-hole principle, this mechanism singles out aPDT as a multi-target procedure leaving no room for resistance development [9]. Additionally, aPDT does not necessitate a specific extracellular or intracellular localization of the PS to exert its damaging effect [9]. As a result, bacteria are unlikely to establish antibiotic resistance against this multi-targeted approach even after repeated administration [8]. Furthermore, such a broad-spectrum activity against gram-positive and gram-negative bacteria, fungi, viruses, and protozoa is beneficial in the empirical therapy of undiagnosed infections. Another advantage of aPDT is that fast-growing cells such as bacteria can accumulate PSs at a higher rate, resulting in increased selectivity [10]. This is especially harmful to bacteria, as their DNA is not isolated from the cytoplasm like that of eukaryotic cells [11]. Additionally, the photodestructive effect of aPDT can be physically controlled in practice by local light applications. The photoantimicrobial effect is much faster than traditional therapies that can take days to be effective [12]. Moreover, the generated ROS during PDT can chemically oxidize the virulence factors such as lipopolysaccharide, protein toxins, proteases, and α-hemolysin [11]. Although extensive work has been performed in aPDT, only three photosensitizers (methylene blue, toluidine blue O, and indocyanine green) have received clinical approval in dentistry so far [12].

Parietin (PTN), also termed physcion, is an anthraquinone naturally present as a secondary metabolite in lichens (e.g., *Xanthoria parietina* [13]), in other fungi (e.g., *Aspergillus*, *Penicillium* [14]), and also in plants (e.g., *Rheum, Rumex,* and *Ventilago* [14]). The UV protectant role of PTN has been detected in lichens adapted to UV stressed environments, whereby the mycobiont synthesizes PTN to protect the photobiont against oxidation mediated by excessive solar radiation [14,15]. *Xanthoria parietina* has been used in traditional medicine for kidney problems and menstrual disorders and as a painkiller [16]. Although PTN offers several activities, including antibacterial and antifungal [14], and has antitumoral [17] and other photosensitizing properties [18,19], it was studied only in organic solvents and is not applicable in clinical use. Like many lipophilic PSs, PTN aggregates unduly in the biological environment, losing its singlet oxygen quantum yield and limiting its routine use. To solve these drawbacks, we recently reported the encapsulation of PTN into liposomes to promote its aqueous solubility and stability in the biological milieu and enhance selectivity and delivery to the targeted cells [19]. However, its high hydrophobicity does not favor entrapping a higher amount in the lipid bilayer. Such a problem can be addressed with cyclodextrin (CD) complexation to increase the inclusion capacity of PTN in the delivery system.

CD is widely utilized as a typical solubilizer for hydrophobic drugs to enhance their aqueous solubility and chemical stability with a high loading capacity and a straightforward procedure [20]. CD can enclose the hydrophobic drug in its central cavity to form a host-guest complex or supramolecular species without altering its framework structure, while the outer surface is still hydrophilic to ensure water solubility [11]. CD is a cyclic biocompatible oligosaccharide structurally composed of 6–8 D-glucopyranose monomers connected by α-1,4-glucose bonds [11]. According to the number of glucose units, the natural CD may differ in water solubility and hydrophobic cone dimension available for drug accommodation. There are different varieties of CD, such as α-CD (six glucose units with a cavity volume of 0.174 nm^3), β-CD (seven glucose units with a cavity volume of 0.262 nm^3), or γ-CD (eight glucose units with a cavity volume of 0.427 nm^3) [21]. However, natural CD, particularly β-CD, can damage the renal tubule either by microcrystalline precipitation in the kidney because of lower water solubility or as a cyclodextrin/cholesterol complex [20]. Therefore, the natural CDs were chemically modified to enhance their aqueous solubility and decrease nephrotoxicity. Among them, hydroxypropyl-β-CD (HP-β-CD) is the most notable derivative, with a greater water solubility of 60% (*w/w*) compared to 2% for its parent form, β-CD [20]. Owing to its biocompatibility and minimal toxicity, HP-β-CD has long been used as a delivery system in PDT for many photosensitizers [11] such as hypericin [10], chlorophyll a [22], aminolevulinic acid [23], temoporfin [24], chlorin e6 [25], and curcumin [26]. These studies showed enhanced water solubility and delivery to the targeted tissues without significantly changing their photophysical properties. To the best of our knowledge, no work has investigated the inclusion of PTN in any cyclodextrin complex.

Various mechanisms of aPDT have already been investigated, but its success depends mainly on the radiation strength and radiant exposure, or the overall dosimetry [27]. Differences among the used photosensitizers require a multifactorial concept in dosimetry that should always be considered [28]. PTN by itself shows a concentration-dependent antimicrobial effect during irradiation and in the dark [17]. Free PTN is not suitable for therapy because of its low water solubility and poor bioavailability, which can be overcome by water-soluble PTN/HP-β-CD complexes. Hegge et al. showed an advantage of their cyclodextrin conjugate in terms of thermal stability, photostability, and easier solubilization than the ethanolic solution [29]. The conjugation to cyclodextrin results in a slightly reduced affinity of photosensitizers for gram-negative bacteria, which is compensated by the improved solubility, availability, and efficiency of singlet oxygen generation [30]. Furthermore, targeting moieties can easily be incorporated into the complex to improve selectivity [31,32].

This study aimed to improve the aqueous solubility of PTN, simultaneously maintaining its photoactivity, to obtain a promising candidate for further use in PDT. The solubility behavior was studied at different concentrations of HP-β-CD. Additionally, the complex formation (PTN/HP-β-CD) was evidenced by various characterizations, such as proton nuclear magnetic resonance (^{1}H NMR), Fourier-transform infrared spectroscopy (FT-IR), powder x-ray diffraction (PXRD), scanning electron microscopy (SEM), and differential scanning calorimetry (DSC). PTN delivery with minimal loss of photoactivity was evaluated by the uric acid method assessing the photodynamic activity of PTN/HP-β-CD complexes in water. Furthermore, the photodegradation profile of PTN/HP-β-CD complexes upon exposure to blue LED irradiation was monitored and compared to that of free PTN in ethanol. The photodynamic activity of PTN/HP-β-CD complexes was tested for the first time against gram-positive and gram-negative bacterial strains.

2. Materials and Methods

2.1. Materials

Parietin (PTN, purity >98%) was purchased from Cayman (Hamburg, Germany). 2-hydroxypropyl-β-cyclodextrin (HP-β-CD, average MW = 1483 g/mol) was obtained from Sigma-Aldrich (Taufkirchen, Germany). Ultrapure water was generated by PURELAB flex 4 (ELGA LabWater, High Wycombe, UK) and used for all experiments in this study. All other chemicals and solvents were of analytical grades and used as received.

2.2. Bacterial Strains and Media

Glycerol stock cultures of *Staphylococcus saprophyticus* subsp. *bovis* (*S. saprophyticus*, DSM 18669, DSMZ, Braunschweig, Germany) and *Escherichia coli* DH5 alpha (*E. coli*, DSM 6897, DSMZ, Braunschweig, Germany) were prepared and stored at −80 °C. The stocks were thawed one day before the bacterial viability assay and cultured in Mueller Hinton broth (MHB, Sigma Aldrich Chemie) on an orbital shaker (Compact Shaker KS 15 A, equipped with Incubator Hood TH 15, Edmund Bühler, Bodelshausen, Germany) set at 200 rpm and 37 °C.

2.3. Light Source

The irradiation experiment was performed with a prototype low-power LED device consisting of an array of light-emitting diodes custom-made by Lumundus GmbH (Eisenach, Germany). The device can emit light at wavelengths of 457 nm and 652 nm for blue and red regions, respectively. The actual light dose (J/cm^2) = Irradiance (W/cm^2) × irradiation time (in sec) [33].

2.4. Stoichiometry: Job's Plot

The continuous variation technique (Job's plot) was employed to determine the complex stoichiometry [34]. Briefly, equimolar stock solutions of PTN (100 μM in ethanol) and HP-β-CD (100 μM in water) were mixed at different ratios (1:9; 2:8; 3:7, and so on), maintaining a final volume of 10 mL to get different mole fractions of PTN from 0 to 1, while the total concentration of PTN and HP-β-CD remained constant. The suspensions were then stirred overnight on a magnetic stirrer at 200 rpm (IKA RT 15, IKA-Werke, Staufen, Germany). Any insoluble materials were removed by centrifugation of the suspension (16,800× *g* for 15 min) (Centrifuge 5418, Eppendorf, Hamburg, Germany) and filtration of the supernatant (0.45 μm nylon filter, Pall Corporation, New York, NY, USA). The amount of PTN solubilized in the complex was estimated spectrophotometrically at λ = 434 nm and was plotted (Δ*A* × *R*) against the mole fraction (*R*) of PTN, where Δ*A* denotes the difference in absorbance in the absence and presence of HP-β-CD:

$$R = \frac{[PTN]}{[PTN] + [HP\text{-}\beta\text{-}CD]} \tag{1}$$

2.5. Phase Solubility Study

A phase solubility study was performed in water at 25 °C according to the method described by Higuchi and Connors [35]. A known excess of PTN was added to vials containing different concentrations of HP-β-CD (0.28 mM-35.7 mM), and the suspensions were stirred at 25 °C for 48 h on a magnetic stirrer at 200 rpm (IKA RT 15, IKA-Werke, Staufen, Germany). Uncomplexed PTN was removed by centrifugation at 16,800× *g* for 15 min (Centrifuge 5418, Eppendorf, Hamburg, Germany), and PTN concentration was measured spectrophotometrically at λ = 434 nm (UV mini-1240, Shimadzu, Kyōto, Japan). The phase solubility graph was constructed by plotting PTN concentration (mM) against HP-β-CD concentration, and the apparent stability constant (K_S) was calculated according to the following equation.

$$K_S = \frac{Slope}{[S_0(1 - Slope)]} \tag{2}$$

S_0 is the intrinsic solubility of PTN in water as measured in our study to be 0.17 μg/mL.

2.6. Preparation of PTN/HP-β-CD Complexes and Physical Mixture

PTN/HP-β-CD inclusion complexes were prepared by the freeze-drying method [10,34]. Briefly, PTN and HP-β-CD (at a molar rate of 1:1) were dissolved in ethanol and ultrapure water, respectively. PTN was added stepwise to the HP-β-CD aqueous solution placed on a magnetic stirrer at 200 rpm and 25 °C for 48 h (IKA RT 15, IKA-Werke, Staufen, Germany). Insoluble PTN was removed by centrifugation at $16,800 \times g$ for 15 min and filtration (0.45 μm nylon filter, Pall Corporation, New York, USA). The obtained solution was then lyophilized in a freeze-dryer (Christ Alpha 1-4 LSC, Martin Christ Gefriertrocknungsanlagen, Osterode am Harz, Germany). The physical mixture was prepared by mixing PTN and HP-β-CD (in a 1:1 molar ratio) in a porcelain mortar with grinding for 15 min and then stored in a desiccator until further analysis.

2.7. Characterization of PTN/HP-β-CD Complexes

2.7.1. ^{1}H NMR Spectroscopy and 2D ROESY

^{1}H NMR spectroscopic measurements were recorded by an NMR spectrometer equipped with an auto-tune sample head (JEOL ECX-400 Nuclear Magnetic Resonance Instrument, JEOL, Akishima, Japan). Samples were prepared by dissolving an equivalent amount of free and complexed PTN in a suitable NMR solvent (DMSO-d6). Samples were then transferred into NMR tubes and assessed through several scanning cycles fixed to a minimum of 64 scans. The results were processed by MNOVA software (version 14.2.1, Mestrelab Research S.L., Santiago de Compostela, Spain).

2.7.2. FT-IR Analysis

The FT-IR measurements were performed by a Brucker α-alpha FT-IR instrument (Bruker Optic, Ettlingen, Germany) with an attenuated total internal reflectance diamond crystal. The neat solid samples were placed directly on the diamond crystal, after which the adjustable pressure arm was positioned over the sample to press it gently. FT-IR spectra were recorded in the transmission mode from 4000 to 400 cm^{-1} [36].

2.7.3. Powder X-ray Diffraction (PXRD)

The formation of PTN/HP-β-CD complexes was confirmed by XRD, determining its crystallinity after inclusion. XRD patterns of PTN/HP-β-CD complexes, the corresponding physical mixture, and the individual solid components were recorded with CuK α radiation (λ = 1.7903 Å) at a voltage of 40 kV and 35 mA current (X'Pert Pro MDP X-ray powder diffractometer, PANalytical, Almelo, Netherlands). Samples were scanned at room temperature from $2\theta = 10°$ to $2\theta = 60°$ with a step of 0.03°/min.

2.7.4. Scanning Electron Microscopy (SEM)

The surface morphology of the PTN/HP-β-CD inclusion complexes was investigated by SEM (Hitachi S-510, Hitachi-High Technologies Europe, Krefeld, Germany) equipped with a secondary electron detector. Briefly, the individual components (pure PTN, HP-β-CD), their physical mixture, and PTN/HP-β-CD inclusion complexes were mounted on an aluminum pin stub by conductive double-sided adhesive carbon tabs. The powders were sputter-coated thrice with a thin layer of gold (10 mA for 1 min) in an Edwards S150 Sputter Coater (Edwards Vacuum, Crawley, UK). The samples were visualized at an acceleration voltage of 10 kV.

2.7.5. Differential Scanning Calorimetry (DSC)

The thermal behavior of the obtained complex and its pure substances were assessed by DSC measurements (DSC-7, Perkin Elmer, Rodgau, Germany). Briefly, accurately weighted amounts of the solid materials were filled into aluminum pans and heated over the temperature range 20–300 °C at a rate of 10 °C min^{-1} under a nitrogen purge for cooling. In parallel, an empty pan sealed in the same way was served as a reference.

2.7.6. UV/Vis Absorption Spectroscopy

The absorption spectrum of PTN/HP-β-CD inclusion complexes (80 µg/mL) was obtained from λ = 200 to 700 nm by a UV/Vis spectrophotometer (Multiskan GO, Thermo Scientific, Waltham, MA, USA) and compared with that of free PTN in ethanol. The absorbance spectra were recorded using an aqueous solution of HP-β-CD or pure ethanol as background references for PTN/HP-β-CD complexes and free PTN, respectively.

2.7.7. Singlet Oxygen Quantum Yield

The photodynamic activity of PTN/HP-β-CD complexes was evaluated by analyzing the singlet oxygen generation ($^1O_2$) using uric acid (UA), and as previously mentioned, using rose bengal (RB) as a standard photosensitizer with a reported singlet oxygen quantum yield of 0.75 in water [19]. Briefly, uric acid (100 µM) was mixed with PTN (10 µg/mL) and irradiated 5 times for 1 min by a blue LED (λ$_{irr}$ = 457 nm, irradiance = 220.2 W/m^2, Lumundus, Eisenach, Germany). The absorption spectra of UA were recorded after each irradiation procedure by a UV-Vis spectrophotometer (Multiskan GO, Thermo Scientific, Waltham, MA, USA). The normalized UA absorbance at λ = 296 nm was plotted versus the irradiation time (in seconds), and the decomposition rate constant of uric acid was calculated to quantify the singlet oxygen quantum yield using the following equation.

$$\Phi[^1O_2]_{PTN} = \Phi[^1O_2]_{RB} \frac{K_{PTN} \, F_{RB}}{K_{RB} \, F_{PTN}} \tag{3}$$

$\Phi[^1O_2]_{PTN}$ and $\Phi[^1O_2]_{RB}$ are the singlet oxygen quantum yields of PTN and RB, respectively. k$_{PTN}$ and k$_{RB}$ are the rate constants of uric acid degradation by PTN and RB, respectively. F is the absorption correction factor given by F = 1 − 10^{-OD} (OD at the irradiation wavelength).

2.7.8. Photostability of Inclusion Complexes

The effect of complexation on photodegradation of PTN was studied, and the degradation profile was compared with that of free PTN in ethanol. Solutions of either PTN/HP-β-CD complexes in water or free PTN in ethanol (100 µg/mL) were irradiated by an LED device (λ$_{irr}$ = 457 nm, irradiance = 220.2 W/m^2) at 5 min interval with a total irradiation time of 30 min. After each irradiation time, the absorbance spectra were measured at λ$_{max}$ = 434 nm (Multiskan GO, Thermo Scientific, Waltham, MA, USA). The absorbance spectra were normalized for better comparison, and PTN's remaining percentage was calculated and compared with free PTN [37].

2.8. Bacterial Viability Assay

The antibacterial activity was determined by incubating the formulations with the bacterial suspensions and irradiating the samples. Both microorganisms (*S. saprophyticus* and *E. coli*) were treated equally, and similar bacterial densities were used. The overnight cultures were diluted to an optical density (OD$_{600}$) of 0.025 measured by a spectrophotometer (Shimadzu UV mini-1240, Kyōto, Japan). These suspensions were placed in an orbital shaker (Compact Shaker KS 15 A, equipped with Incubator Hood TH 15, Edmund Bühler) set at 300 rpm and 37 °C. The growth of the bacterial suspensions was stopped by placing them on ice at an OD$_{600}$ of 0.4. A total of 150 uL of each bacterial suspension was incubated with an equal volume of PTN/HP-β-CD complexes (containing 200 µM PTN) in 12-well cell culture plates (TC plate standard, Sarstedt, Nümbrecht, Germany) for 30 min at 37 °C and 100 rpm, so that the final PTN concentration was set to 100 µM. These solutions were irradiated with blue-LED (λ_{irr} = 457 nm) for 30 min at a radiant exposure of 39.6 J/cm^2. After irradiation, the suspensions were serially diluted with MHB and plated onto Mueller Hinton II Agar plates (BD, Heidelberg, Germany). After incubating the plates for 18 h at 37 °C and 90% relative humidity, the viable colonies were counted and the colony-forming units per milliliter were calculated [CFU/mL]. Filter-sterilized phosphate-buffered saline (PBS) (pH 7.4) was used as a control. The experiments were performed in triplicates.

2.9. Statistical Analysis

Unless otherwise stated, all experiments were performed in triplicates, and the results are expressed as means ± standard deviations. A two-tailed Student's t-test was performed to identify statistically significant differences. Probability values of $p < 0.05$ were considered significant.

3. Results and Discussion

3.1. Stoichiometry: Job's Plot

Job's method was employed to determine the stoichiometry of PTN and HP-β-CD, which was indicated by a maximum on Job's plot at a specific molar ratio. As shown in Figure 1A, (ΔAbs x R) is greatest at a mole fraction of 0.5, indicating that the stoichiometry between PTN and HP-β-CD is 1:1, where a single PTN molecule can be included in the cavity of one HP-β-CD molecule. Qiu et al. dealt with the complexation of anthraquinone, emodin, in HP-β-CD [34], and the authors presented a similar profile with 1:1 stoichiometry between emodin and HP-β-CD.

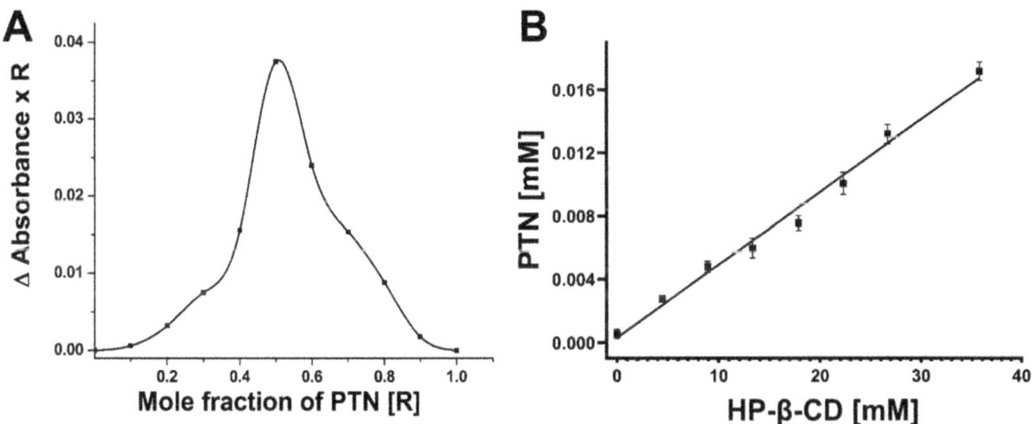

Figure 1. (**A**) Job's plot for different mole fractions of PTN (R). (**B**) Phase solubility diagram of PTN/HP-β-CD complex system at 25 °C. The concentration of HP-β-CD was in the range of 0–35 mM. All measurements were performed in triplicate, and the values were expressed as means ± SDs (n = 3).

3.2. Phase Solubility Study

The phase solubility was conducted by evaluating the change in PTN solubility as a function of HP-β-CD concentration to determine the stoichiometric ratio and stability constant of PTN in HP-β-CD. The inclusion complex formation was visually observed by the typical yellow color of PTN, indicating its solubilization in HP-β-CD solution. As depicted in Figure 1B, PTN solubility increased proportionally with an increase in HP-β-CD concentration, resulting in a correlation coefficient of 0.975 over the studied concentration range. PTN solubility in water rose considerably from 0.00059 mM in the absence of HP-β-CD to 0.017 mM in the presence of 35 mM HP-β-CD because of the inclusion of PTN within the hydrophobic cavity of HP-β-CD. This host-guest system is a typical A_L type revealing soluble complex formation according to Higuchi and Connors [35] and showing a 1:1 stoichiometry between PTN and HP-β-CD. The slope in Figure 1B is below unity, implying the formation of an inclusion complex at a molar ratio of 1:1 consistent with Job's plot. The stability constant is K_S = 733.46 M^{-1}, and according to the literature, it is considered optimal in the range of 50 to 2000 M^{-1} [34]. Smaller values suggest poor interactions between the guest and the CD, whereas higher values indicate difficult complex dissociation, leading to incomplete guest release from the inclusion complex. As a result, a 1:1 molar ratio of PTN and HP-β-CD was employed for the inclusion complex formation for further characterization.

3.3. Characterization of PTN/HP-β-CD Complexes

3.3.1. ^{1}H NMR Spectroscopy and 2D ROESY

The ^{1}H NMR investigations for free and complexed PTN were performed to explore the possible inclusion mode of the PTN/HP-β-CD complexes [38]. Accordingly, we compared the NMR spectra of free PTN, pure HP-β-CD, and the PTN/HP-β-CD complexes. As illustrated in Figure 2A, pure PTN in DMSO-d6 showed peaks corresponding to the methyl and methoxy protons at 2.4 and 3.91 ppm, respectively. In addition, the aromatic ring protons exhibited their corresponding peaks at 6.8, 7.1, and 7.5 ppm. The intense peaks at 11.9 and 12.1 ppm corresponded to the phenolic protons on each side of the anthraquinone structure [39]. Figure 2B indicated the prominent peaks of pure HP-β-CD in DMSO-d6, showing the primary chemical shifts for H-1 to H-6 protons between 3 and 6 ppm except for the methyl protons that appeared at 1 ppm. The inclusion of PTN into HP-β-CD resulted in the disappearance of the PTN's aromatic and hydroxyl protons between 6.5 and 12 ppm, as annotated in Figure 2C. Moreover, the methyl and methoxy protons entirely vanished at 2.4 and 3.9 ppm.

For further explanation of the inclusion pattern, the chemical shifts in the presence and absence of PTN were recorded (Table 1). The inclusion of PTN into the HP-β-CD had a negligible influence on the H-4, H-5, and H-6 protons (0.01 ppm). In contrast, the values for H-2 and H-3 indicated a slightly significant change (0.02–0.05 ppm), which might be brought about by the spatial interaction of these protons with the hydroxyl protons of PTN. It is worth noting that the H-3 and H-5 protons are on the interior side of the CD cavity, with the H-3 near the wider opening and the H-5 protons near the narrower side of HP-β-CD [40]. Because of the minor change in the chemical shift of the H-5 (about 0.01 ppm) compared to the more considerable change in the H-3 (about 0.02 ppm), we could presume that PTN might interact with the HP-β-CD through its wider side. According to these findings, PTN is suspected of penetrating the HP-β-CD cavity by its methyl or methoxy substituted rings. The disappearance of their corresponding protons supported this after inclusion with HP-β-CD. Therefore, we hypothesized that PTN could have two inclusion possibilities, as illustrated in Figure 3A,B.

Furthermore, the 2D ROESY of the PTN/HP-β-CD complexes was obtained to attain more conformational details of the spatial configuration of the complexed PTN [41]. The ROESY spectrum of the PTN/HP-β-CD complexes (Figure 3C) showed a relative correlation between the hydroxyl protons of PTN and the H-3 protons of the CD molecule, which is likely driven by hydrogen bonding forces. Moreover, the aromatic protons of PTN at positions 2, 4, 8, and 10 had a spatial correlation with the H-2 and H-5 protons of HP-β-CD, which explains the significant proton shifting of the H-2 (0.05 ppm) and strengthens the hypothesis that PTN was introduced through the broader side of the CD cavity.

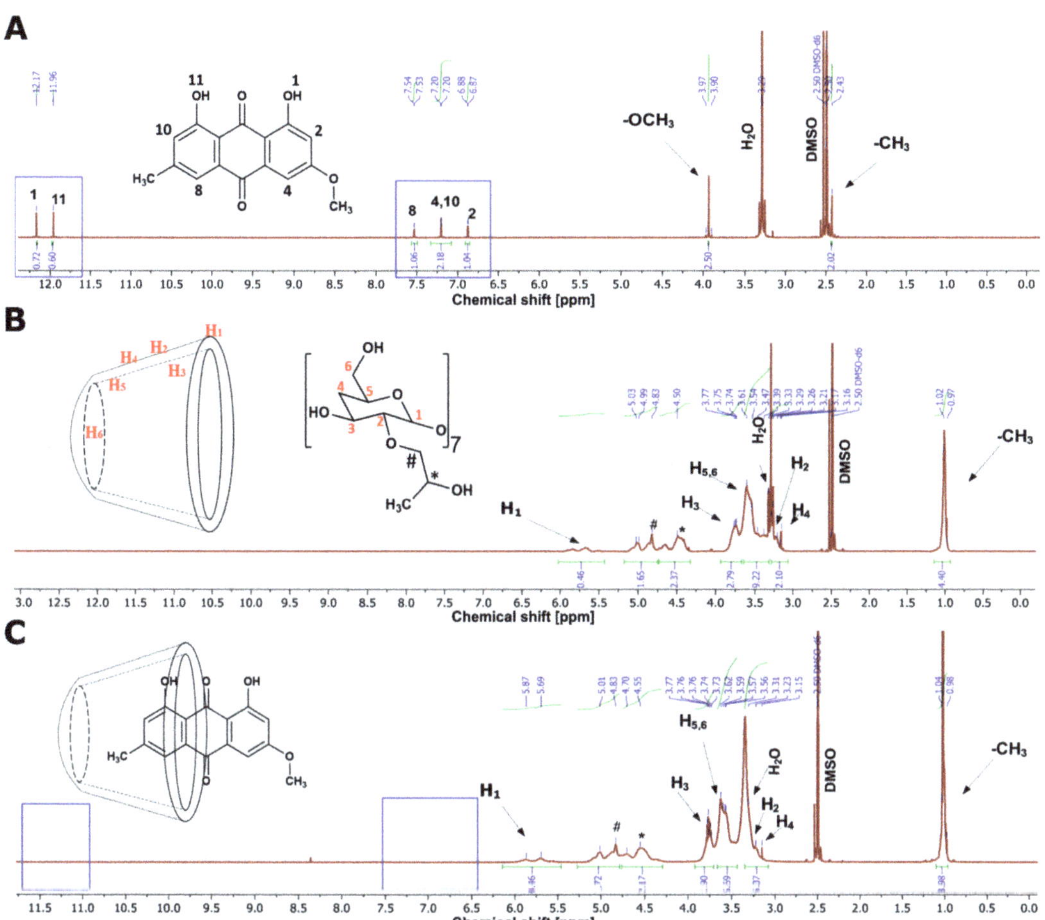

Figure 2. ^{1}H NMR spectra of (**A**) PTN, (**B**) HP-β-CD, and (**C**) the PTN/HP-β-CD complexes in DMSO-d$_{6}$.

Table 1. Selected chemical shifts (δ in ppm) of HP-β-CD and PTN/HP-β-CD complexes in DMSO-d6.

Protons	δ (ppm)	
	HP-β-CD	**PTN/HP-β-CD Complexes**
H-1	5.64	5.66
H-2	3.27	3.32
H-3	3.72	3.74
H-4	3.13	3.12
H-5	3.54	3.53
H-6	3.58	3.59
CH$_3$	1.00	1.00

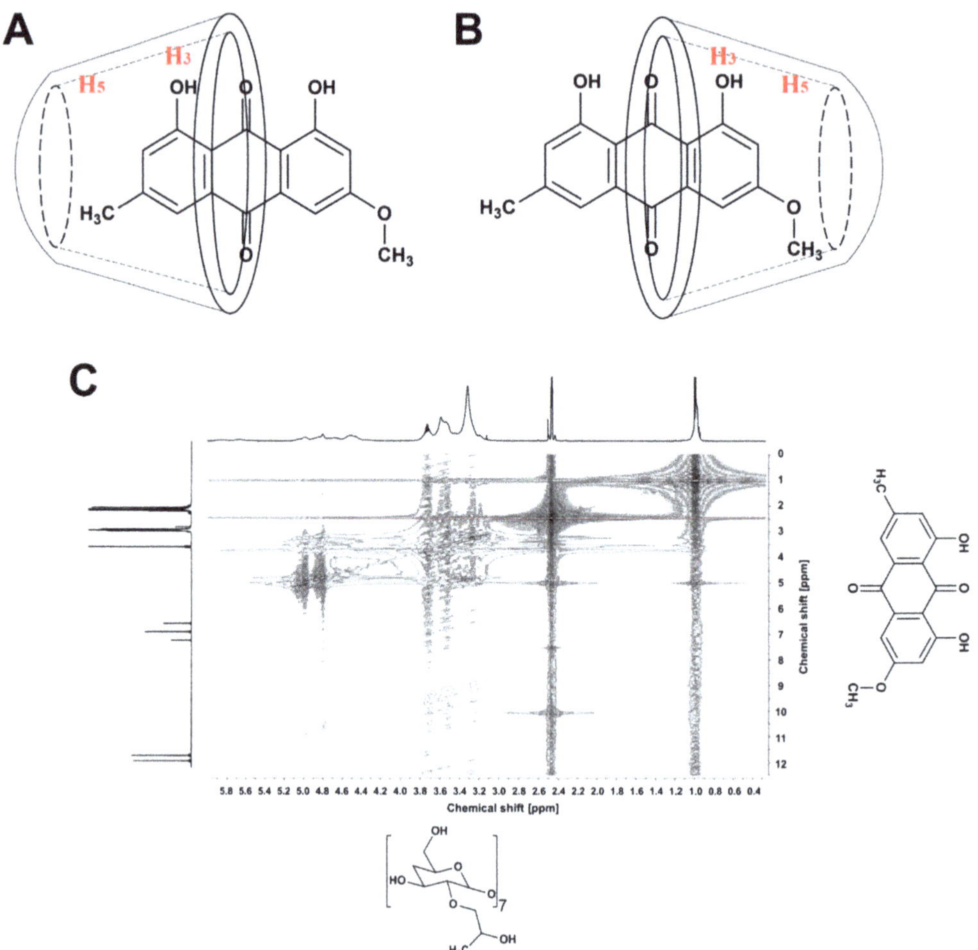

Figure 3. The possible penetration patterns of PTN through the HP-β-CD cavity by either the methyl (**A**) or the methoxy (**B**) substituted rings. (**C**) ROESY spectrum of the PTN/HP-β-CD complexes in DMSO-d6.

3.3.2. FT-IR Analysis

The possible interactions between PTN and HP-β-CD in the solid-state were assessed by comparing the FT-IR spectra of pure PTN, HP-β-CD, their physical mixture, and PTN/HP-β-CD complexes. The FT-IR spectrum of pure PTN (Figure 4A) shows many absorption bands: at 2936 cm^{-1} (CH$_3$ asymmetric), 2844 cm^{-1} (CH$_3$ symmetric), 1674–1613 cm^{-1} (C=O free and conjugated), 1557 cm^{-1} (C=C aromatic), and 1383–1364 cm^{-1} (C–O phenyl) as previously reported [39]. The pure HP-β-CD spectrum (Figure 4B) showed a broad band at 3331 cm^{-1} corresponding to OH stretching vibrations of various hydroxyl groups. The absorption bands at 2922 cm^{-1}, 1640 cm^{-1}, and 1149 cm^{-1} are also related to CH$_2$ stretching vibrations, O-H bending vibrations, and C-O-C stretching vibrations, respectively. The absorption bands of the valence vibrations of the C–O bonds in the ether and hydroxyl groups of HP-β-CD (1082 and 1035 cm^{-1}) were noticed as previously reported [42]. Figure 4C depicts that the physical mixture spectrum was equivalent to a simple combination of PTN and HP-β-CD. However, the characteristic carbonyl peaks of pure PTN were present in the physical mixture but absent in PTN/HP-β-CD complexes (Figure 4D). Also, there is neither a band at 1557 cm^{-1} nor at 1383–1364 cm^{-1} assigned to C=C aromatic and C–O phenyl, respectively. The band of the valence vibration of the O–H bond in PTN/HP-β-CD complexes was shifted to 3353 cm$^{-1,}$ and this was related to water release upon host-guest interaction as observed previously in the literature [42]. Overall, the inclusion complex did not display any new IR peaks signifying that no chemical bonds are formed with the obtained complex.

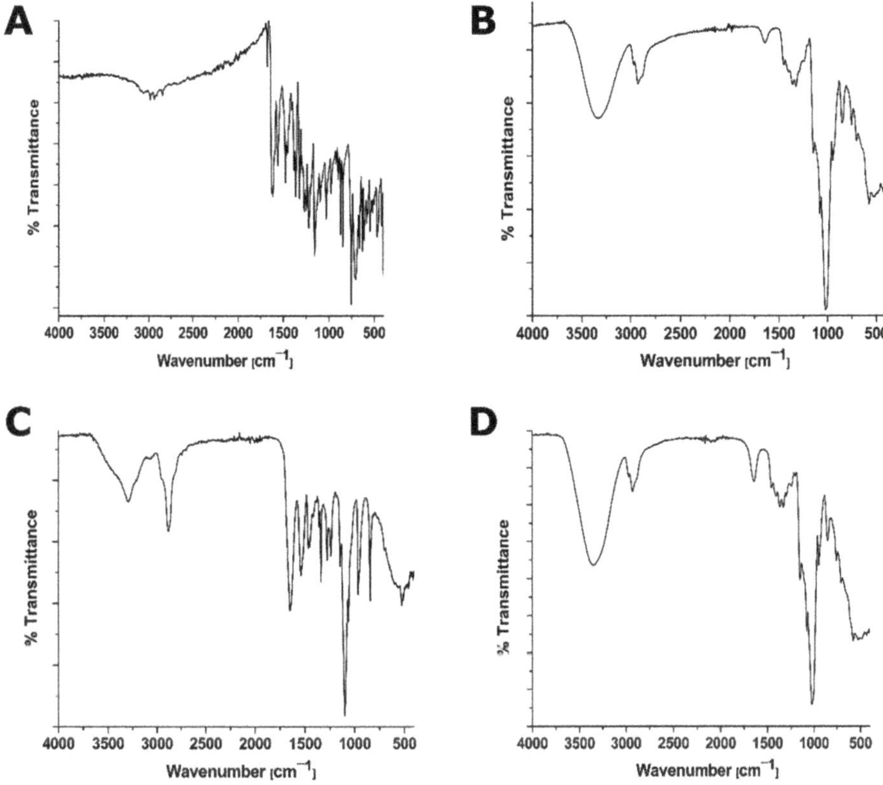

Figure 4. FT-IR spectra of (**A**) PTN, (**B**) HP-β-CD, (**C**) physical mixture (1:1 molar ratio of PTN and HP-β-CD), and (**D**) PTN/HP-β-CD complexes over the wavenumber range 4000–400 cm^{-1}.

3.3.3. Powder X-ray Diffraction (PXRD)

PXRD was used to identify any change in the crystallinity of PTN upon complexation. The PXRD pattern of pure PTN (Figure 5A) displayed well-resolved characteristic diffraction peaks at different 2θ, revealing its crystalline character. The position of peaks corresponds to what has previously been described in the literature [43]. In the case of HP-β-CD (Figure 5B), a diffuse pattern without any sharp peaks was obtained, indicating its amorphous nature. The physical mixture (Figure 5C) exhibited almost all of the distinctive peaks of PTN and HP-β-CD but with lower intensity confirming the presence of both components as isolated species while preserving the crystalline nature of PTN.

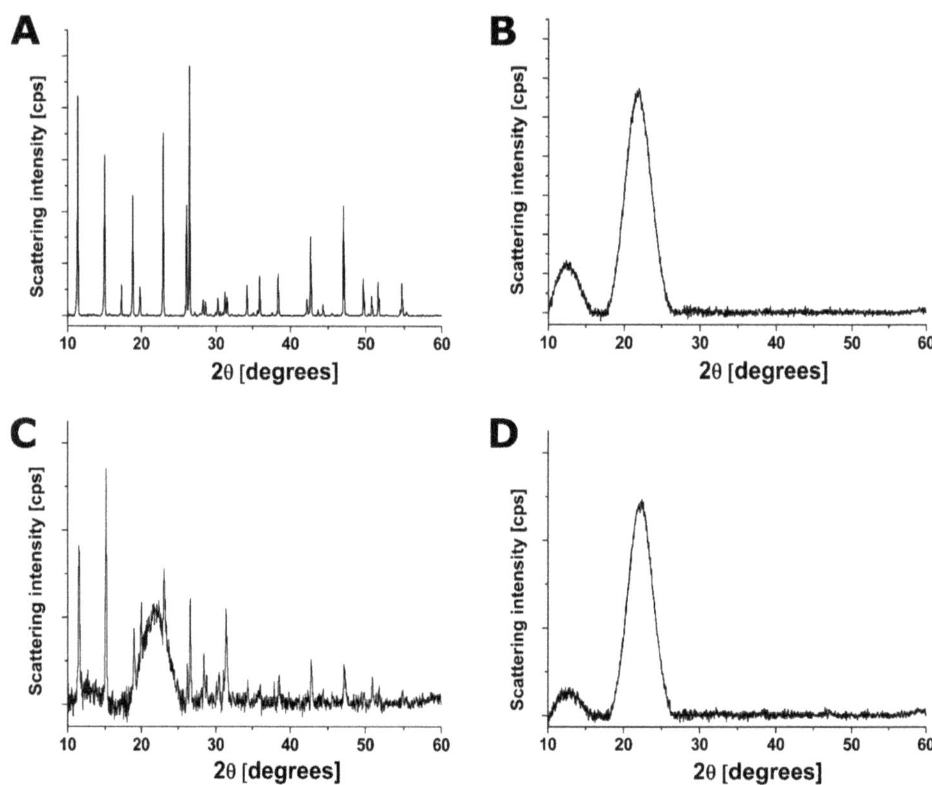

Figure 5. Powder X-ray diffraction patterns of (**A**) PTN, (**B**) HP-β-CD, (**C**) physical mixture (1:1 molar ratio of PTN and HP-β-CD), and (**D**) PTN/HP-β-CD complexes. Scanning angle of 2θ = 10–60; step width of 0.03°/min.

On the other hand, the diffractogram of PTN/HP-β-CD complexes (Figure 5D) exhibited the broad peak of pure HP-β-CD with the absence of peaks assigned to pure PTN, which may be due to the presence of the drug in a molecularly dispersed form in HP-β-CD. According to the literature, the amorphous structure is attributed in part to the HP-β-CD structure and in part to the lyophilization step during preparation [44]. These results confirm those obtained by FT-IR, which imply the disappearance of drug crystallinity in the obtained complex.

3.3.4. Scanning Electron Microscopy (SEM)

SEM was used to demonstrate the morphological changes in PTN upon formation of PTN/HP-β-CD complexes, and the results are presented in Figure 6. Pure PTN appeared as irregularly sized needle-shaped crystals (Figure 6A), while pure HP-β-CD existed as spherical crystals, well separated from each other with a porous surface (Figure 6B). The shape of the physical mixture and the inclusion complex was completely different. SEM pictures of the physical mixture indicated no change in the crystal state regarding the original morphology of individual components. The needle-shaped PTN crystals are distributed between HP-β-CD crystals (Figure 6C). However, PTN/HP-β-CD complexes appeared as roughly rectangular homogeneous particles without the spherical shape of HP-β-CD or the needle shape of PTN. This morphological alteration to a single solid phase can indicate a successful formation of PTN/HP-β-CD complexes and is consistent with the data obtained in the FT-IR and XRPD studies.

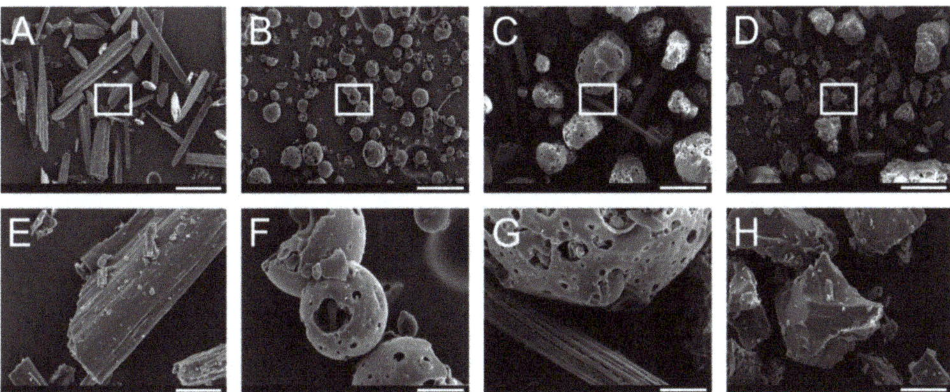

Figure 6. SEM micrographs recorded at an acceleration voltage of 10 kV of (**A**) PTN, (**B**) HP-β-CD, (**C**) physical mixture (1:1 molar ratio of PTN and HP-β-CD), and (**D**) PTN/HP-β-CD complexes with the respective magnification (**E–H**) corresponding to the white rectangle. Scale bars represent 200 μm in (**A–D**) and 30 μm in (**E–H**).

3.3.5. Differential Scanning Calorimetry (DSC)

DSC is widely used to characterize the inclusion of active moieties in the CD cavity by a shift or disappearance in the melting point of the guest molecules [45]. Figure 7 displays that pure PTN exhibits a sharp endothermic peak at 210 °C corresponding to its melting point, indicating its crystalline nature, which is well-matched with previous reports [39,46]. As reported before, the pure HP-β-CD showed a broad peak at 187 °C because of water loss from the cyclodextrin cavity at higher temperatures [47]. As clearly evidenced in Figure 7, the physical mixture thermogram was quite different from that of the pure components. In the physical mixture, the PTN peak was reduced in intensity and shifted to a higher temperature (218 °C), while HP-β-CD shifted to a lower temperature (183 °C) in comparison to the pure HP-β-CD (187 °C). This may be due to interactions during the DSC run, at least within the ~100–180 °C temperature range, and agrees with a previous study attributing these slight peak shifts to the weak interaction between the cyclodextrin and drug during the physical mixture preparation [48].

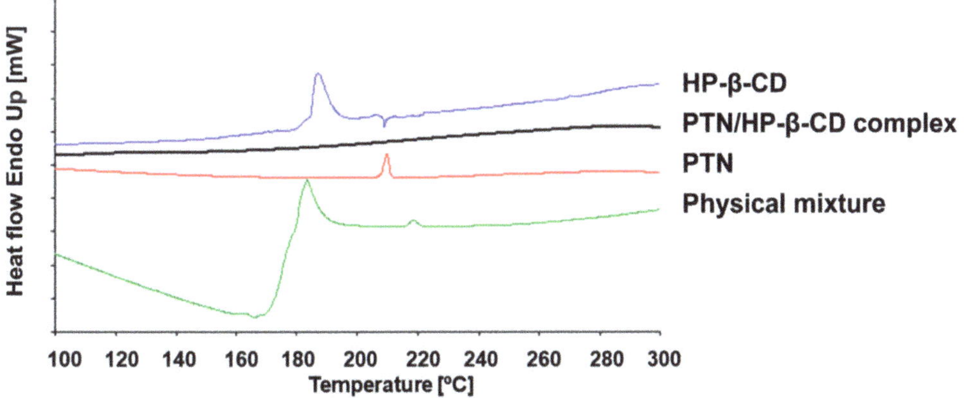

Figure 7. DSC curves of HP-β-CD, PTN, physical mixture (1:1 molar ratio of PTN and HP-β-CD), and PTN/HP-β-CD complexes. The heating rate was 10 °C min^{-1}. The thermograms were adjusted for better visualization.

On the contrary, Figure 7 depicts a notable difference in the thermal profile of PTN/HP-β-CD complexes in comparison to the parent material and their physical mixture. The endothermic peak of PTN melting was no longer visible, revealing the amorphous state of the drug because of an interaction between PTN and HP-β-CD as a result of guest–host inclusion complex formation. The absence of drug and cyclodextrin peaks in the complex thermogram indicated successful inclusion complex formation. Moreover, the dehydration peak of HP-β-CD appeared only in the pure HP-β-CD and in the physical mixture thermograms but was unseen in those of PTN/HP-β-CD complexes, confirming the complex formation. The plausible explanation of this phenomenon may be that the cyclodextrin cavity was occupied with the hydrophobic drug and no space was available for water molecules [48].

3.3.6. UV/Vis Absorption Spectroscopy

Figure 8A shows that HP-β-CD has no absorption peak within the recorded spectrum as it does not have any double bond (π-electrons) to absorb UV energy, as previously reported [49]. As depicted in Figure 8A, the UV-Vis absorption spectra of free and complexed PTN have two characteristic absorption peaks at λ_{max} = 290 nm and 434 nm, respectively, similar to those reported before [17]. The UV-Vis absorption spectra of free PTN and PTN/HP-β-CD complexes were comparable along the scanned wavelength. The peak position is similar to that of free PTN in ethanol, revealing that the complexation had no significant effect on the PTN absorption spectrum. Moreover, it indicates that the solubility of PTN in water is enhanced without aggregations. However, a slightly hypochromic shift was observed in PTN/HP-β-CD complexes, which could be attributed to the more hydrophilic microenvironment surrounding PTN [50]. Similar behavior was reported by Cannavà et al. [51], who studied the effect of sulfobutyl ether β-cyclodextrin on idebenone and reported a significant hypochromic shift because of the complexation within the cyclodextrin cavity [51]. The absorption spectrum of nifedipine was also almost identical to the free form but with lower absorbance intensity at some wavelengths [52].

3.3.7. Singlet Oxygen Quantum Yield

Since the phototoxicity of the photosensitizer involves the production of cytotoxic species such as singlet oxygen, it is useful to investigate the singlet oxygen generation efficiency. This can be performed by monitoring the decay curves of UA absorbance as a function of irradiation time. Following the established measuring procedure [19], the decomposition of UA upon exposure to generated singlet oxygen was employed to measure singlet oxygen generation. UA was selected because it does not absorb light in the blue region, specifically at the PTN irradiation wavelength. A similar irradiation experiment was also performed in parallel as a negative control, where UA solution was irradiated with a blue LED. In the absence of PTN, no photobleaching of UA was observed after irradiation (data not shown), and therefore, any loss in UA absorbance upon irradiation is due to the photodynamic activity of PTN. Figure 8B,C depicts that the decrease in UA absorbance is a function of light exposure time, indicating the irradiation time-dependent singlet oxygen production, with the lowest UA absorbance at 5 min irradiation time. A good linear relation between Ln normalized absorbance of UA at λ = 296 nm and the irradiation time denotes that the decay kinetics is first order.

A comparison of UA decays in Figure 8B revealed that free PTN induced significantly faster UA degradation than observed in the case of complexed PTN (Figure 8C). The rate constants for UA decomposition were 7.04×10^{-4} and 2.28×10^{-4} s^{-1}, while the singlet oxygen quantum yields were 0.49 and 0.35 for free and complexed PTN, respectively. This decrease in singlet oxygen generation upon complexation with HP-β-CD may be due to the photosensitizers' environment affecting the quantum yield. It is well-established that singlet oxygen has a longer lifetime in organic solvents than in water, where fast quenching of singlet oxygen between water molecules reduces the singlet oxygen lifetime [53]. Moreover, PTN/HP-β-CD complexes may be more stable and less susceptible to photodegradation than ethanolic PTN solutions. According to the literature, cyclodextrin complexation changes the photochemical characteristics of the guest molecules and has been long used to protect drugs against photodegradation [54]. Previous research suggests that β-cyclodextrin complexation has a negligible effect on the UV spectra, but it causes significant changes in the emissions spectra and fluorescence quantum yields as it usually impacts the ground and the excited states [50]. These results further confirm that in water, HP-β-CD maintains the PTN molecules mainly as monomeric species, which are the photoactive form for ROS generation.

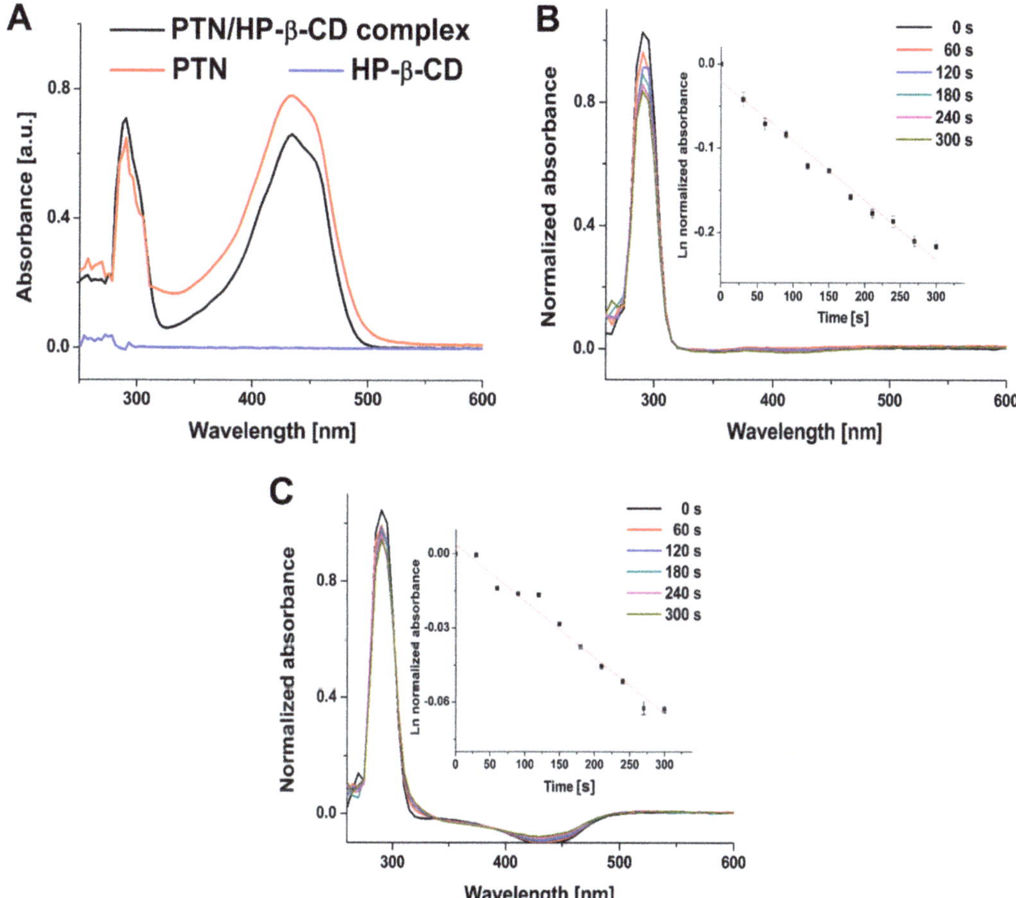

Figure 8. (**A**) UV/Vis spectra of PTN in ethanol (80 µg/mL), HP-β-CD in water (0.5 mg/mL) and PTN/HP-β-CD inclusion complexes in water (80 µg/mL). (**B,C**) Absorption spectra of uric acid upon irradiation for different times in the presence of (**B**) free PTN and (**C**) PTN/HP-β-CD complexes. The insets represent the plot of Ln normalized absorbance of uric acid at λ = 296 nm versus irradiation time in s (λ_{irr} = 457 nm, irradiance = 220.2 W/m^2). All measurements were performed in triplicate, and the values were expressed as means $\pm$ SDs (n = 3).

3.3.8. Photostability of Inclusion Complex

Considering the practical use of PTN/HP-β-CD complexes in aPDT, it is better to investigate the influence of irradiation time on the degradation of free and complexed PTN. Figure 9A,B show that the decrease in absorbance measured at λ_{max} = 434 nm is light-dependent, with the lowest absorbance detected after 30 min irradiation. This may be due to several photoactivations of the photosensitizer with successive irradiation, leading to photodecomposition or photobleaching. As shown in Figure 9C, less than 28% of free PTN remained after 30 min irradiation compared to about 70% in the case of the complexed form, indicating a better photostability of HP-β-CD. Figure 9D presents that complexed PTN could retain its characteristic yellow color even after 30 min irradiation. The slight color change of free PTN to yellowish-orange is most probably due to the formation of photodegradation products evidenced by a decrease in the peak intensity at 434 and 287 nm and a simultaneous increase of the new small peak at 488 nm, which appeared only after irradiation with the highest intensity after 30 min (Figure 9A). This photostability can be beneficial in PDT since PS molecules are still available for further photoactivation leading to more ROS generation with additional irradiation.

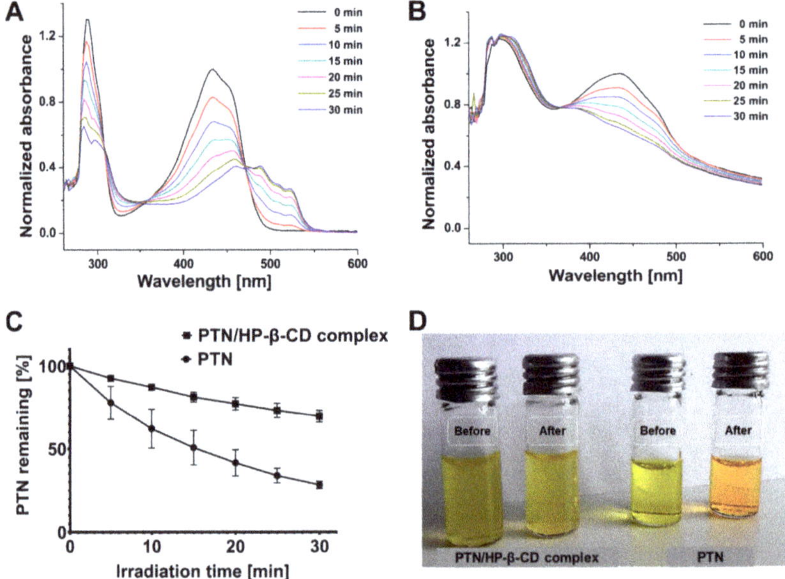

Figure 9. Effect of HP-β-CD on the photodegradation of PTN after blue LED irradiation at different time intervals (λ_{irr} = 457 nm, irradiance = 220.2 W/m^2). (**A**) The absorption spectra of PTN in ethanol and (**B**) the absorption spectra of PTN/HP-β-CD complexes in ultrapure water. (**C**) Decrease in the PTN content of free and complexed PTN at different irradiation intervals with a blue LED. (**D**) Photomicrograph of PTN/HP-β-CD complexes and PTN before and after 30 min irradiation. All measurements were performed in triplicate, and the values were expressed as means ± SDs (*n* = 3).

Additionally, the photobleaching of free PTN could limit the irradiation time and reduce its photosensitizing efficacy, especially when a longer irradiation time is required. This is consistent with a previous study, where HP-β-CD complexed cilnidipine exhibited lower photodegradation than free cilnidipine in ethanol, and the authors related this to the protective effect conferred by the host-guest system [55]. This effect was also reported before for curcumin after complexation in cyclodextrin [56]. According to the literature [57], the host-guest structure can reduce the hydrolytic or photolytic degradation of a drug in inclusion complexes by providing a molecular shield against reactive substances. This shielding effect can presumably result in lower availability and interaction of complexed PTN molecules with the light source, leading to lower singlet oxygen generation as stated before compared to the free PTN.

3.4. Bacterial Viability Assay

In this study, antibacterial activity was examined by irradiating bacteria treated with the prepared PTN/HP-β-CD complexes, and the results were presented in Figure 10. For the used concentration (100 µM PTN), no significant dark toxicity could be observed toward either microorganism when comparing the non-irradiated control sample and the non-irradiated formulation. In addition, the effect of the light source was negligible as it did not influence bacterial viability as seen in the irradiated control sample [9,58,59]. Furthermore, the unirradiated and irradiated HP-β-CD also showed no significant change in bacterial viability. In contrast, the irradiated PTN/HP-β-CD complexes significantly reduced the bacterial viability of the gram-positive *S. saprophyticus* by > 4.8 log. Therefore, it can be assumed that the antibacterial effect originated exclusively from the irradiated PTN/HP-β-CD complexes. According to the American Society for Microbiology, a bacterial reduction of 99.9986% qualifies the formulation as antibacterial [60]. The viability of the gram-negative *E. coli* was reduced by > 1.0 log after incubation with PTN/HP-β-CD complexes and irradiation. As described in the literature, the antibacterial effect of PTN is lower with gram-negative germs, and higher concentrations are needed for the same effect [14]. In addition, the affinity of the cyclodextrin complex to gram-negative bacteria can be slightly reduced because of their unique cellular structure. The difference in the cell wall structure between gram-negative and gram-positive bacteria may account for their different response to aPDT. The accumulation of PSs inside gram-negative bacteria is limited because of two mechanisms. Firstly, the lipid-rich membrane bilayer enclosing the cell wall reduces the inward penetration of PTN/HP-β-CD complexes into gram-negative bacteria [5]. Secondly, gram-negative bacteria widely use porins and efflux pumps to regulate nutrient and toxin influx and efflux, and hence they act as natural resistance mechanisms [5]. However, compared with antibacterial studies on pure PTN, the concentration used here is reduced almost tenfold. Comini et al. showed no dark toxicity of PTN against *E. coli* up to 320 µg/mL, and an antibacterial activity of irradiated PTN only above a concentration of 250 µg/mL [17]. This underlines the superiority of the cyclodextrin complex at a used PTN concentration of just 28.45 µg/mL. Overall, the complexation of PTN not only improved the water solubility and photostability of PTN but also increased the antibacterial effect, enabling the therapeutic application of PTN.

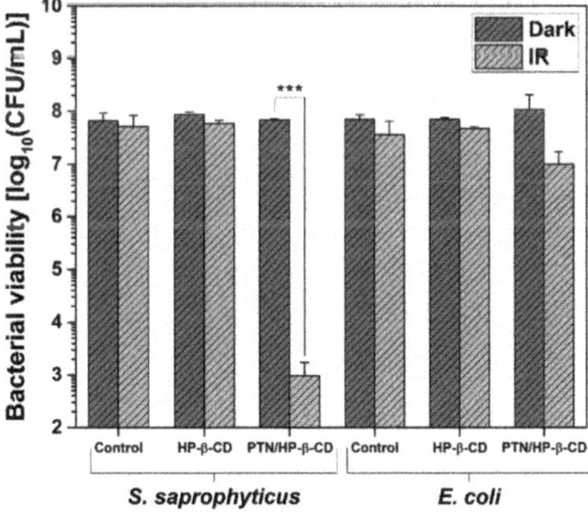

Figure 10. Bacterial viability of *Staphylococcus saprophyticus* subsp. *bovis* (*S. saprophyticus*) and *Escherichia coli* DH5 alpha (*E. coli*) treated with PTN/HP-β-CD complexes (100 µM PTN) or only with HP-β-CD for 30 min at 37 °C and then irradiated (IR) with a blue LED (λ_{irr} = 457 nm, 39.6 J/cm^2) or kept in the dark. Control represents bacterial suspension treated with filter-sterilized phosphate-buffered saline (PBS) (pH 7.4). The results are expressed as means ± SDs (*n* = 3). Statistical differences are denoted as "***" *p* < 0.001.

4. Conclusions

The efficacy of parietin (PTN) as a natural photosensitizer is poor because of its hydrophobic structure and π–π stacking, which consequently results in aggregation-induced quenched fluorescence, limited ROS production, and a diminished photodynamic effect. In this study, HP-β-CD was employed to enhance the water solubility of PTN and decrease its aggregation in aqueous vehicles. The water solubility of PTN was enhanced up to 28-fold after complexation with HP-β-CD. Various characterization methods confirmed that PTN was adequately included in the cyclodextrin cage. Even though PTN/HP-β-CD complexes have a polar environment, they are still photoactive because HP-β-CD provides a protecting hydrophobic cavity for PTN, resulting in a shielding effect from water molecules. This can be a critical factor in maintaining its photodynamic activity, as confirmed by uric acid degradation due to singlet oxygen generation. Additionally, the complexation was shown to slow down the photodegradation of PTN. Although PTN is reported to have a dark antimicrobial effect, it is presumed that aPDT could considerably reduce the required dose to give a comparable effect and lower toxic effects. In the long-term prospect, the limited efficacy of PTN/HP-β-CD against gram-negative bacteria may be overcome by the delivery of PTN in cationic formulations.

Author Contributions: Conceptualization, A.M.A. (Abdallah Mohamed Ayoub), B.G. and U.B.; investigation, A.M.A. (Abdallah Mohamed Ayoub), B.G., A.M.A. (Ahmed Mohamed Abdelsalam), E.P., A.A.D., A.A. and A.B.; writing—original draft preparation, A.M.A. (Abdallah Mohamed Ayoub), B.G. and E.P.; writing—review and editing, A.M.A. (Abdallah Mohamed Ayoub), B.G., E.P., J.S. and U.B.; supervision, U.B.; project administration, U.B. All authors have read and agreed to the published version of the manuscript.

Funding: This research received no external funding.

Institutional Review Board Statement: Not applicable.

Informed Consent Statement: Not applicable.

Data Availability Statement: The data presented in this study are available on request from the corresponding author.

Acknowledgments: Abdallah Mohamed Ayoub and Ahmed Mohamed Abdelsalam would like to express their gratitude to Yousef Jameel Foundation, the German Academic Exchange Service and the Egyptian Ministry of Higher Education and Scientific Research (DAAD/MHESR), respectively for providing the scholarship funding for the doctoral scholarships.

Conflicts of Interest: The authors declare no conflict of interest.

References

1. Organisation for Economic Co-operation and Development (OECD) and European Centers for Disease Control and Prevention. *Antimicrobial Resistance: Trackling the Burden in the European Union*; Briefing Note for EU/EEA Countries; OECD Publications: Paris, France, 2019.
2. O'Neill, J. *Tackling Drug-Resistant Infections Globally: Final Report and Recommendations. The Review on Antimicrobial Resistance*; Government of the United Kingdom: London, UK, 2016.
3. Renwick, M.J.; Simpkin, V.; Mossialos, E. *Targeting Innovation in Antibiotic Drug Discovery and Development: The Need for a One Health—One Europe—One World Framework*; European Observatory on Health Systems and Policies: Copenhagen, Denmark, 2016; ISBN 9789289050401.
4. Prestinaci, F.; Pezzotti, P.; Pantosti, A. Antimicrobial resistance: A global multifaceted phenomenon. *Pathog. Glob. Health* **2015**, *109*, 309–318. [CrossRef] [PubMed]
5. Spellberg, B. The Future of Antibiotics. *Crit. Care* **2014**, *18*, 228. [CrossRef] [PubMed]
6. Polat, E.; Kang, K. Natural Photosensitizers in Antimicrobial Photodynamic Therapy. *Biomedicines* **2021**, *9*, 584. [CrossRef] [PubMed]
7. Hamblin, M.R. Antimicrobial photodynamic inactivation: A bright new technique to kill resistant microbes. *Curr. Opin. Microbiol.* **2016**, *33*, 67–73. [CrossRef] [PubMed]
8. Cieplik, F.; Deng, D.; Crielaard, W.; Buchalla, W.; Hellwig, E.; Al-Ahmad, A.; Maisch, T. Antimicrobial photodynamic therapy—What we know and what we don't. *Crit. Rev. Microbiol.* **2018**, *44*, 571–589. [CrossRef] [PubMed]

9. Agel, M.R.; Baghdan, E.; Pinnapireddy, S.R.; Lehmann, J.; Schäfer, J.; Bakowsky, U. Curcumin loaded nanoparticles as efficient photoactive formulations against gram-positive and gram-negative bacteria. *Colloids Surf. B Biointerfaces* **2019**, *178*, 460–468. [CrossRef]
10. Plenagl, N.; Seitz, B.S.; Duse, L.; Pinnapireddy, S.R.; Jedelska, J.; Brüßler, J.; Bakowsky, U. Hypericin Inclusion Complexes Encapsulated in Liposomes for Antimicrobial Photodynamic Therapy. *Int. J. Pharm.* **2019**, *570*, 118666. [CrossRef]
11. Maldonado-Carmona, N.; Ouk, T.S.; Calvete, M.J.F.; Pereira, M.M.; Villandier, N.; Leroy-Lhez, S. Conjugating biomaterials with photosensitizers: Advances and perspectives for photodynamic antimicrobial chemotherapy. *Photochem. Photobiol. Sci.* **2020**, *19*, 445–461. [CrossRef]
12. Wainwright, M.; Maisch, T.; Nonell, S.; Plaetzer, K.; Almeida, A.; Tegos, G.P.; Hamblin, M.R. Photoantimicrobials-Are We Afraid of the Light? *Lancet Infect. Dis.* **2017**, *17*, e49–e55. [CrossRef]
13. Solhaug, K.A.; Gauslaa, Y. Parietin, a Photoprotective Secondary Product of the Lichen Xanthoria Parietina. *Oecologia* **1996**, *108*, 412–418. [CrossRef]
14. Basile, A.; Rigano, D.; Loppi, S.; di Santi, A.; Nebbioso, A.; Sorbo, S.; Conte, B.; Paoli, L.; de Ruberto, F.; Molinari, A.; et al. Antiproliferative, Antibacterial and Antifungal Activity of the Lichen Xanthoria Parietina and Its Secondary Metabolite Parietin. *Int. J. Mol. Sci.* **2015**, *16*, 7861–7875. [CrossRef] [PubMed]
15. Nguyen, K.-H.; Chollet-Krugler, M.; Gouault, N.; Tomasi, S. UV-Protectant Metabolites from Lichens and Their Symbiotic Partners. *Nat. Prod. Rep.* **2013**, *30*, 1490–1508. [CrossRef] [PubMed]
16. Boustie, J.; Grube, M. Lichens—A Promising Source of Bioactive Secondary Metabolites. *Plant Genet. Resour.* **2005**, *3*, 273–287. [CrossRef]
17. Comini, L.R.; Morán Vieyra, F.E.; Mignone, R.A.; Páez, P.L.; Laura Mugas, M.; Konigheim, B.S.; Cabrera, J.L.; Núñez Montoya, S.C.; Borsarelli, C.D. Parietin: An Efficient Photo-Screening Pigment in Vivo with Good Photosensitizing and Photodynamic Antibacterial Effects in Vitro. *Photochem. Photobiol. Sci.* **2017**, *16*, 201–210. [CrossRef] [PubMed]
18. Cogno, I.S.; Gilardi, P.; Comini, L.; Núñez-Montoya, S.C.; Cabrera, J.L.; Rivarola, V.A. Natural Photosensitizers in Photodynamic Therapy: In Vitro Activity against Monolayers and Spheroids of Human Colorectal Adenocarcinoma SW480 Cells. *Photodiagnosis Photodyn. Ther.* **2020**, *31*, 101852. [CrossRef]
19. Ayoub, A.M.; Amin, M.U.; Ambreen, G.; Dayyih, A.A.; Abdelsalam, A.M.; Somaida, A.; Engelhardt, K.; Wojcik, M.; Schäfer, J.; Bakowsky, U. Photodynamic and Antiangiogenic Activities of Parietin Liposomes in Triple Negative Breast Cancer. *Mater. Sci. Eng. C* **2021**, 112543. [CrossRef]
20. Davis, M.E.; Brewster, M.E. Cyclodextrin-Based Pharmaceutics: Past, Present and Future. *Nat. Rev. Drug Discov.* **2004**, *3*, 1023–1035. [CrossRef]
21. Szejtli, J. Introduction and General Overview of Cyclodextrin Chemistry. *Chem. Rev.* **1998**, *98*, 1743–1754. [CrossRef] [PubMed]
22. Dentuto, P.L.; Catucci, L.; Cosma, P.; Fini, P.; Agostiano, A.; Hackbarth, S.; Rancan, F.; Roeder, B. Cyclodextrin/Chlorophyll a Complexes as Supramolecular Photosensitizers. *Bioelectrochemistry* **2007**, *70*, 39–43. [CrossRef]
23. Aggelidou, C.; Theodossiou, T.A.; Gonçalves, A.R.; Lampropoulou, M.; Yannakopoulou, K. A Versatile δ-Aminolevulinic Acid (ALA)-Cyclodextrin Bimodal Conjugate-Prodrug for PDT Applications with the Help of Intracellular Chemistry. *Beilstein J. Org. Chem.* **2014**, *10*, 2414–2420. [CrossRef]
24. Yakavets, I.; Lassalle, H.-P.; Scheglmann, D.; Wiehe, A.; Zorin, V.; Bezdetnaya, L. Temoporfin-in-Cyclodextrin-in-Liposome—A New Approach for Anticancer Drug Delivery. The Optimization of Composition. *Nanomaterials* **2018**, *8*, 847. [CrossRef] [PubMed]
25. Paul, S.; Heng, P.W.S.; Chan, L.W. PH-Dependent Complexation of Hydroxypropyl-Beta-Cyclodextrin with Chlorin E6: Effect on Solubility and Aggregation in Relation to Photodynamic Efficacy. *J. Pharm. Pharmacol.* **2016**, *68*, 439–449. [CrossRef] [PubMed]
26. Wikene, K.O.; Hegge, A.B.; Bruzell, E.; Tønnesen, H.H. Formulation and Characterization of Lyophilized Curcumin Solid Dispersions for Antimicrobial Photodynamic Therapy (APDT): Studies on Curcumin and Curcuminoids LII. *Drug Dev. Ind. Pham.* **2015**, *41*, 969–977. [CrossRef] [PubMed]
27. Allison, R.R.; Downie, G.H.; Cuenca, R.; Hu, X.-H.; Childs, C.J.; Sibata, C.H. Photosensitizers in Clinical PDT. *Photodiagnosis Photodyn. Ther.* **2004**, *1*, 27–42. [CrossRef]
28. Henderson, B.W.; Busch, T.M.; Snyder, J.W. Fluence Rate as a Modulator of PDT Mechanisms. *Lasers Surg. Med.* **2006**, *38*, 489–493. [CrossRef]
29. Hegge, A.B.; Vukicevic, M.; Bruzell, E.; Kristensen, S.; Tønnesen, H.H. Solid Dispersions for Preparation of Phototoxic Supersaturated Solutions for Antimicrobial Photodynamic Therapy (APDT): Studies on Curcumin and Curcuminoides L. *Eur. J. Pharm. Biopharm.* **2013**, *83*, 95–105. [CrossRef]
30. Ribeiro, C.P.S.; Gamelas, S.R.D.; Faustino, M.A.F.; Gomes, A.T.P.C.; Tomé, J.P.C.; Almeida, A.; Lourenço, L.M.O. Unsymmetrical Cationic Porphyrin-Cyclodextrin Bioconjugates for Photoinactivation of Escherichia Coli. *Photodiagnosis Photodyn. Ther.* **2020**, *31*, 101788. [CrossRef]
31. Mora, S.J.; Cormick, M.P.; Milanesio, M.E.; Durantini, E.N. The Photodynamic Activity of a Novel Porphyrin Derivative Bearing a Fluconazole Structure in Different Media and against Candida Albicans. *Dye. Pigment.* **2010**, *87*, 234–240. [CrossRef]
32. Gao, Y.; Wang, J.; Hu, D.; Deng, Y.; Chen, T.; Jin, Q.; Ji, J. Bacteria-Targeted Supramolecular Photosensitizer Delivery Vehicles for Photodynamic Ablation against Biofilms. *Macromol. Rapid Commun.* **2019**, *40*, e1800763. [CrossRef] [PubMed]

33. Abu Dayyih, A.; Alawak, M.; Ayoub, A.M.; Amin, M.U.; Abu Dayyih, W.; Engelhardt, K.; Duse, L.; Preis, E.; Brüßler, J.; Bakowsky, U. Thermosensitive liposomes encapsulating hypericin: Characterization and photodynamic efficiency. *Int. J. Pharm.* **2021**, *609*, 121195. [CrossRef] [PubMed]

34. Qiu, N.; Zhao, X.; Liu, Q.; Shen, B.; Liu, J.; Li, X.; An, L. Inclusion Complex of Emodin with Hydroxypropyl-β-Cyclodextrin: Preparation, Physicochemical and Biological Properties. *J. Mol. Liq.* **2019**, *289*, 111151. [CrossRef]

35. Higuchi, T.; Connors, K.A. Phase Solubility Techniques. *Adv. Anal. Chem. Instrum.* **1965**, *4*, 117–212.

36. Abdelsalam, A.M.; Somaida, A.; Ambreen, G.; Ayoub, A.M.; Tariq, I.; Engelhardt, K.; Garidel, P.; Fawaz, I.; Amin, M.U.; Wojcik, M.; et al. Surface Tailored Zein as a Novel Delivery System for Hypericin: Application in Photodynamic Therapy. *Mater. Sci. Eng. C* **2021**, *129*, 112420. [CrossRef]

37. Raschpichler, M.; Preis, E.; Pinnapireddy, S.R.; Baghdan, E.; Pourasghar, M.; Schneider, M.; Bakowsky, U. Photodynamic Inactivation of Circulating Tumor Cells: An Innovative Approach against Metastatic Cancer. *Eur. J. Pharm. Biopharm.* **2020**, *157*, 38–46. [CrossRef]

38. Zhao, R.; Sandström, C.; Zhang, H.; Tan, T. NMR Study on the Inclusion Complexes of β-Cyclodextrin with Isoflavones. *Molecules* **2016**, *21*, 372. [CrossRef] [PubMed]

39. Edwards, H.G.M.; Newton, E.M.; Wynn-Williams, D.D.; Coombes, S.R. Molecular Spectroscopic Studies of Lichen Substances 1: Parietin and Emodin. *J. Mol. Struct.* **2003**, *648*, 49–59. [CrossRef]

40. Zhou, S.-Y.; Ma, S.-X.; Cheng, H.-L.; Yang, L.-J.; Chen, W.; Yin, Y.-Q.; Shi, Y.-M.; Yang, X.-D. Host–Guest Interaction between Pinocembrin and Cyclodextrins: Characterization, Solubilization and Stability. *J. Mol. Struct.* **2014**, *1058*, 181–188. [CrossRef]

41. Jahed, V.; Zarrabi, A.; Bordbar, A.; Hafezi, M.S. NMR (1H, ROESY) Spectroscopic and Molecular Modelling Investigations of Supramolecular Complex of β-Cyclodextrin and Curcumin. *Food Chem.* **2014**, *165*, 241–246. [CrossRef]

42. Savic-Gajic, I.; Savic, I.M.; Nikolic, V.D.; Nikolic, L.B.; Popsavin, M.M.; Kapor, A.J. Study of the Solubility, Photostability and Structure of Inclusion Complexes of Carvedilol with β-Cyclodextrin and (2-Hydroxypropyl)-β-Cyclodextrin. *J. Incl. Phenom. Macrocycl. Chem.* **2016**, *86*, 7–17. [CrossRef]

43. Soomro, N.A.; Wu, Q.; Amur, S.A.; Liang, H.; Ur Rahman, A.; Yuan, Q.; Wei, Y. Natural Drug Physcion Encapsulated Zeolitic Imidazolate Framework, and Their Application as Antimicrobial Agent. *Colloids Surf. B Biointerfaces* **2019**, *182*, 110364. [CrossRef]

44. Yang, B.; Lin, J.; Chen, Y.; Liu, Y. Artemether/Hydroxypropyl-Beta-Cyclodextrin Host-Guest System: Characterization, Phase-Solubility and Inclusion Mode. *Bioorganic Med. Chem.* **2009**, *17*, 6311–6317. [CrossRef] [PubMed]

45. Sid, D.; Baitiche, M.; Elbahri, Z.; Djerboua, F.; Boutahala, M.; Bouaziz, Z.; Le Borgne, M. Solubility Enhancement of Mefenamic Acid by Inclusion Complex with β-Cyclodextrin: In Silico Modelling, Formulation, Characterisation, and in Vitro Studies. *J. Enzyme Inhib. Med. Chem.* **2021**, *36*, 605–617. [CrossRef] [PubMed]

46. Zhang, L.-M.; Xie, W.-G.; Wen, T.-T.; Zhao, X. Thermal Behavior of Five Free Anthraquinones from Rhubarb. *J. Therm. Anal. Calorim.* **2010**, *100*, 215–218. [CrossRef]

47. Păduraru, O.M.; Bosînceanu, A.; Ţântaru, G.; Vasile, C. Effect of Hydroxypropyl-β-Cyclodextrin on the Solubility of an Antiarrhythmic Agent. *Ind. Eng. Chem. Res.* **2013**, *52*, 2174–2181. [CrossRef]

48. Parvathaneni, V.; Elbatanony, R.S.; Goyal, M.; Chavan, T.; Vega, N.; Kolluru, S.; Muth, A.; Gupta, V.; Kunda, N.K. Repurposing Bedaquiline for Effective Non-Small Cell Lung Cancer (NSCLC) Therapy as Inhalable Cyclodextrin-Based Molecular Inclusion Complexes. *Int. J. Mol. Sci.* **2021**, *22*, 4783. [CrossRef] [PubMed]

49. Kim, J.-S. Study of Flavonoid/Hydroxypropyl-β-Cyclodextrin Inclusion Complexes by UV-Vis, FT-IR, DSC, and X-Ray Diffraction Analysis. *Prev. Nutr. Food Sci.* **2020**, *25*, 449–456. [CrossRef] [PubMed]

50. Sanramé, C.N.; De Rossi, R.H.; Argüello, G.A. Effect of β-Cyclodextrin on the Excited State Properties of 3-Substituted Indole Derivatives. *J. Phys. Chem.* **1996**, *100*, 8151–8156. [CrossRef]

51. Cannavà, C.; Crupi, V.; Guardo, M.; Majolino, D.; Stancanelli, R.; Tommasini, S.; Ventura, C.A.; Venuti, V. Phase Solubility and FTIR-ATR Studies of Idebenone/Sulfobutyl Ether β-Cyclodextrin Inclusion Complex. *J. Incl. Phenom. Macrocycl. Chem.* **2013**, *75*, 255–262. [CrossRef]

52. Heydari, A.; Iranmanesh, M.; Doostan, F.; Sheibani, H. Preparation of Inclusion Complex Between Nifedipine and Ethylenediamine-β-Cyclodextrin as Nanocarrier Agent. *Pharm. Chem. J.* **2015**, *49*, 605–612. [CrossRef]

53. Wang, Y.; Cohen, B.; Aykaç, A.; Vargas-Berenguel, A.; Douhal, A. Femto- to Micro-Second Photobehavior of Photosensitizer Drug Trapped within a Cyclodextrin Dimer. *Photochem. Photobiol. Sci.* **2013**, *12*, 2119–2129. [CrossRef]

54. Monti, S.; Sortino, S. Photoprocesses of Photosensitizing Drugs within Cyclodextrin Cavities. *Chem. Soc. Rev.* **2002**, *31*, 287–300. [CrossRef] [PubMed]

55. Hu, L.; Zhang, H.; Song, W.; Gu, D.; Hu, Q. Investigation of Inclusion Complex of Cilnidipine with Hydroxypropyl-β-Cyclodextrin. *Carbohydr. Polym.* **2012**, *90*, 1719–1724. [CrossRef]

56. Del Castillo, M.L.R.; López-Tobar, E.; Sanchez-Cortes, S.; Flores, G.; Blanch, G.P. Stabilization of Curcumin against Photodegradation by Encapsulation in Gamma-Cyclodextrin: A Study Based on Chromatographic and Spectroscopic (Raman and UV-Visible) Data. *Vib. Spectrosc.* **2015**, *81*, 106–111. [CrossRef]

57. Loftsson, T.; Brewster, M.E. Pharmaceutical Applications of Cyclodextrins. 1. Drug Solubilization and Stabilization. *J. Pharm. Sci.* **1996**, *85*, 1017–1025. [CrossRef]

58. Preis, E.; Baghdan, E.; Agel, M.R.; Anders, T.; Pourasghar, M.; Schneider, M.; Bakowsky, U. Spray Dried Curcumin Loaded Nanoparticles for Antimicrobial Photodynamic Therapy. *Eur. J. Pharm. Biopharm.* **2019**, *142*, 531–539. [CrossRef] [PubMed]

59. Preis, E.; Anders, T.; Širc, J.; Hobzova, R.; Cocarta, A.-I.; Bakowsky, U.; Jedelská, J. Biocompatible Indocyanine Green Loaded PLA Nanofibers for in Situ Antimicrobial Photodynamic Therapy. *Mater. Sci. Eng. C* **2020**, *115*, 111068. [CrossRef] [PubMed]
60. Kiesslich, T.; Gollmer, A.; Maisch, T.; Berneburg, M.; Plaetzer, K. A Comprehensive Tutorial on in Vitro Characterization of New Photosensitizers for Photodynamic Antitumor Therapy and Photodynamic Inactivation of Microorganisms. *BioMed. Res. Int.* **2013**, *2013*, 840417. [CrossRef]

Article

Hypericin and Pheophorbide a Mediated Photodynamic Therapy Fighting MRSA Wound Infections: A Translational Study from In Vitro to In Vivo

Ben Chung Lap Chan [1,†], Priyanga Dharmaratne [2,†], Baiyan Wang [3], Kit Man Lau [1], Ching Ching Lee [1,2], David Wing Shing Cheung [1,3], Judy Yuet Wa Chan [1], Grace Gar Lee Yue [1], Clara Bik San Lau [1], Chun Kwok Wong [1,4], Kwok Pui Fung [1,3] and Margaret Ip [2,*]

[1] Institute of Chinese Medicine and State Key Laboratory of Research on Bioactivities and Clinical Applications of Medicinal Plants, The Chinese University of Hong Kong, Shatin, N.T, Hong Kong 999077, China; benchan99@cuhk.edu.hk (B.C.L.C.); virginialau@cuhk.edu.hk (K.M.L.); cclee316@gmail.com (C.C.L.); dcwshk@gmail.com (D.W.S.C.); judychanyw@cuhk.edu.hk (J.Y.W.C.); graceyue@cuhk.edu.hk (G.G.L.Y.); claralau@cuhk.edu.hk (C.B.S.L.); ck-wong@cuhk.edu.hk (C.K.W.); kpfung@cuhk.edu.hk (K.P.F.)

[2] Department of Microbiology, Faculty of Medicine, The Chinese University of Hong Kong, Shatin, N.T, Hong Kong 999077, China; priyanga@cuhk.edu.hk

[3] School of Biomedical Sciences, Faculty of Medicine, The Chinese University of Hong Kong, Shatin, N.T, Hong Kong 999077, China; tinawang@cuhk.edu.hk

[4] Department of Chemical Pathology, Faculty of Medicine, The Chinese University of Hong Kong, Shatin, N.T, Hong Kong 999077, China

* Correspondence: margaretip@cuhk.edu.hk

† These authors contributed equally to this work.

Citation: Chan, B.C.L.; Dharmaratne, P.; Wang, B.; Lau, K.M.; Lee, C.C.; Cheung, D.W.S.; Chan, J.Y.W.; Yue, G.G.L.; Lau, C.B.S.; Wong, C.K.; et al. Hypericin and Pheophorbide a Mediated Photodynamic Therapy Fighting MRSA Wound Infections: A Translational Study from In Vitro to In Vivo. *Pharmaceutics* **2021**, *13*, 1399. https://doi.org/10.3390/pharmaceutics13091399

Academic Editors: Udo Bakowsky, Matthias Wojcik, Eduard Preis, Gerhard Litscher and Maria Nowakowska

Received: 20 July 2021
Accepted: 27 August 2021
Published: 3 September 2021

Abstract: High prevalence rates of methicillin-resistant *Staphylococcus aureus* (MRSA) and lack of effective antibacterial treatments urge discovery of alternative therapeutic modalities. The advent of antibacterial photodynamic therapy (aPDT) is a promising alternative, composing rapid, nonselective cell destruction without generating resistance. We used a panel of clinically relevant MRSA to evaluate hypericin (Hy) and pheophobide a (Pa)-mediated PDT with clinically approved methylene blue (MB). We translated the promising in vitro anti-MRSA activity of selected compounds to a full-thick MRSA wound infection model in mice (in vivo) and the interaction of aPDT innate immune system (cytotoxicity towards neutrophils). Hy-PDT consistently displayed lower minimum bactericidal concentration (MBC) values (0.625–10 μM) against ATCC RN4220/pUL5054 and a whole panel of community-associated (CA)-MRSA compared to Pa or MB. Interestingly, Pa-PDT and Hy-PDT topical application demonstrated encouraging in vivo anti-MRSA activity (>1 $\log_{10}$ CFU reduction). Furthermore, histological analysis showed wound healing via re-epithelization was best in the Hy-PDT group. Importantly, the dark toxicity of Hy was significantly lower ($p < 0.05$) on neutrophils compared to Pa or MB. Overall, Hy-mediated PDT is a promising alternative to treat MRSA wound infections, and further rigorous mechanistic studies are warranted.

Keywords: photodynamic therapy; methicillin-resistant *Staphylococcus aureus*; hypericin; wound infection model

1. Introduction

Infections caused by antimicrobial resistant (AMR) bacteria are serious global health concerns and are exacerbated with prior asymptomatic carriage [1–3]. Methicillin-resistant *Staphylococcus aureus* (MRSA) is one of the commonest AMR bacteria that confers illnesses ranging from localized skin infections to systemic diseases, including toxic shock syndrome [4]. The prevalence of hospital-associated MRSA (HA-MRSA) infection varies geographically, and Hong Kong is one of the high-prevalence regions in Asia. According to the Asian Network for Surveillance of Resistant Pathogens (ANSORP) study, 57% of all inpatient isolates of *S. aureus* from Hong Kong hospitals were confirmed as methicillin-resistant [5]. Additionally, in Hong Kong, the prevalence of nasal carriage of *S. aureus* and MRSA were 27.6% and 1.3%, respectively, among children in daycare centers and kindergartens [6,7].

Microbes by their nature continually adapt to survive the antimicrobial treatments we use to combat them, resulting in an ever increasing level of antimicrobial resistance [8], and the development of nonantimicrobial treatments may be beneficial concerning resistance development. Photodynamic therapy (PDT) consists of the administration of a nontoxic drug or dye known as a photosensitizer (PS) either systemically, locally, or topically applied to a patient, followed by illumination with visible or near-infrared (NIR) light in the presence of oxygen, leading to the generation of cytotoxic reactive oxygen species (ROS) in the proximate environment causing cell death/ tissue damage [9,10]. The advantages of PDT over conventional therapies include rapid bacterial killing, applicability over a broad spectrum (Gram-negative or Gram-positive) [11,12], and efficacy against biofilms [13,14], fungi [15,16], parasites [17] and viruses [18]. To date, the clinical applications of PDT have been confined mainly to localized infections in dermatology and dentistry [19,20], wound healing [21,22], and for surface disinfection including medical devices [23].

It was reported that PDT for localized microbial infections exerts its therapeutic effect both by direct bacterial killing and the activation of the host immune response, particularly innate immunity [24]. Neutrophils are among the first line of defence recruited to the site of infection to release enzymes for killing infectious organisms and to secrete cytokines that promote inflammation. The importance of neutrophils against microbial infections is reflected by the observation that Photofrin®-PDT exhibited significant cytotoxicity for cultured MRSA, but the therapy had a low efficacy in a murine model of MRSA arthritis, even though Photofrin® accumulated well in the infected joint. It was discovered that 30% of intra-articular leukocytes, mainly neutrophils, were killed immediately during or following Photofrin-PDT [25]. Therefore, we assume it is important to examine specific PS-PDTs cytotoxicity towards human neutrophils.

Hypericin (Hy) is a naturally occurring polycyclic quinine (Figure 1a) extracted from plant species of the genus *Hypericium* including the species *Hypericum perforatum* L. (St John's Wort) [26]. Recent reports showed that Hy has the potential to treat several types of cancer and some benign skin disorders [27,28]. Interestingly, Yow et al. reported Hy could induce a significant cytotoxic effect on clinically isolated methicillin-sensitive *S. aureus* (MSSA) and MRSA [29]. In the aspect of wound healing, *H. perforatum*, which is a popular folk remedy for the treatment of wounds in Turkey, has been shown to possess remarkable in vivo wound healing activity, and Hy was found in the active fractions [30].

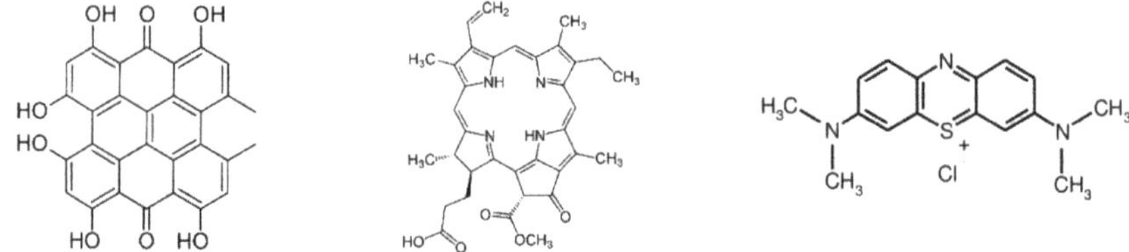

(**a**) Structure of Hypericin   (**b**) Structure of Pheophobide a   (**c**) Structure of Methylene blue

Figure 1. Chemical structures of 3 PSs used in the current investigation.

Pheobhobide a (Pa) is also a natural compound, derived from the breakdown of chlorophyll a [31]. The extended π-π conjugated system (Figure 1b) and stability of the compound in various solvents make it suitable as a photosensitizing agent. Studies have revealed that Pa-PDT is effective in eradicating a variety of tumors, including pigmented melanoma, colonic cancer, Jurkat leukemia, and pancreatic carcinoma [32–35]. Besides the anticancer activity of Pa, it has also been tested for its photodynamic activity against MRSA with modification to its structure (Na salt of Pa) [36].

Photophysical properties are of paramount importance when selecting a photosensitizer. The absorption spectrum of the compound plays a pivotal role during in vivo applications, and it has to be within the therapeutic window (550–950 nm) [37]. The key photophysical properties of Hy and Pa are summarized in Table 1 along with the gold standard of PDT studies (Methylene blue, MB, Figure 1c).

Table 1. Electronic absorption and basic photophysical data for 3 photosensitizers used.

Compound	λ_{max}/nm	λ_{em}/nm	Φ_F [a]	Φ_Δ [b]	Ref
Methylene Blue	664 (monomer in aqueous medium)	709	0.04	0.5	[38,39]
Hypericin	598 (DMSO)	651	0.2	0.73	[40,41]
Pheophobide a	667 (DMSO)	677	0.26	0.62	[42,43]

[a] Fluorescence quantum yield; [b] Singlet oxygen quantum yield.

The compelling evidence of Hy and Pa led us to investigate their PDT effects in vitro and in vivo against a broad spectrum of clinically relevant MRSA panels along with their toxicity towards neutrophils, in view of depicting their overall anti-MRSA efficacy.

2. Materials and Methods

2.1. General

Pheophorbide a was purchased from Frontier Scientific Inc. (Logan, UT, USA) and hypericin and methylene blue were purchased from Sigma-Aldrich Co. (St Louis, MA, USA). The PS solution for in vitro PDT study was prepared freshly by dissolving Pa and Hy in DMSO to make a 10 mM stock solution. It was then diluted in Tween 80 and MHB to set the desired stock solution. A serial two-fold dilution procedure was employed to obtain final working concentrations. Tween 80 and DMSO concentrations were maintained $\leq 0.1\%$ and $\leq 1\%$ (v/v), respectively, in each test group.

The bacterial strains MRSA, ATCC 43300, ATCC BAA-42, ATCC BAA-43, ATCC BAA-44, two mutant strains [AAC(6)′ APH(2)″ and RN4220/pUL5054], five community-acquired (CA-MRSA) and five hospital-acquired MRSA (HA-MRSA) clinical strains were obtained from the Department of Microbiology, Faculty of Medicine, The Chinese University of Hong Kong.

2.2. In Vitro Photodynamic Minimal Bactericidal Concentration (PD-MBC) Studies

Minimal bactericidal concentrations (MBCs) of Pa-PDT, Hy-PDT and MB-PDT for sixteen MRSA strains were determined according to the modified method adopted by Clinical and Laboratory Standards Institute (CLSI) guidelines [12,44,45]. Briefly, an overnight bacterial culture suspension was adjusted to McFarland Standard 0.5 and suspended in Mueller Hinton Broth (MHB) to make a final concentration of 1.0×10^6 colony forming unit (CFU)/mL. Photosensitizers at different concentrations (100 μL) and MRSA suspension (100 μL) were added into 96-well plate and incubated at 37°C for 2 h under dark condition as a pre-irradiation step. After incubation, the mixed solutions were irradiated from above at a light intensity of 40 mW/cm^2 using a 300 W quartz-halogen lamp attenuated by a 5 cm layer of water as a heat buffer and a color filter cut-on at 610 nm (for MB and Pa, $\lambda \geq 610$ nm) or 590 nm (for Hy, $\lambda \geq 590$ nm) for 20 min, i.e., 48 J/cm^2. Dark control group and a solvent control group were included. All experiments were repeated three times. The MBCs were determined as the minimum concentration of the photosensitizers required for complete inhibition of bacterial growth on a blood agar plate.

2.3. Animal Studies-Mouse Model of MRSA-Infected Wound

2.3.1. Animal Model

Previously published murine skin infection models [44–47] were used to validate the in vivo efficacy of the PS-PDT treatment against MRSA. All animal experiments were conformed to the university guidelines and approved by the Animal Experimentation Ethics Committee (Ref. no.12/076/MIS, 8 February 2013) of The Chinese University of Hong Kong. Female Balb/c mice (25–30 g) were supplied by Laboratory Animal Services Centre (LASEC), The Chinese University of Hong Kong. They were housed in individually ventilated cages (IVC) under the conditions of 22–25 °C and a 12-h light-dark cycle, with free access to chow and tap water.

It is apparent from the in vitro results that, MRSA ATCC RN4220/pUL5054 strain was susceptible for all three PSs with comparatively lower MBC values. Hence, this selected to establish infection on a full-thick wound in mice. Mice were anesthetized by an intraperitoneal (i.p.) injection of ketamine (40 mg/kg) and xylazine (8 mg/kg), with the hair of the back shaved and the skin cleansed with 10% povidone-iodine solution. A circular full-thickness wound (4 mm in diameter) was established through a disposable skin puncher on the back subcutaneous tissue of each animal. The lesion, overlaid with gauze, was dressed with an adhesive bandage.

For the in vivo studies, Pa and Hy were prepared according to our previously published protocol [48]. Briefly, 1% DMSO, 4% ethanol and 95% PBS constituted the final test solution.

2.3.2. Intravenous Treatment

Our research group previously investigated Pa-PDT-mediated anticancer activity (against MCF-7 tumors) in vivo [48]. So, the dosage and optimum therapeutic window for these kinds of compounds were established (2.5 mg/Kg) for intravenous injection. Three days after wound induction, mice were anesthetized with a ketamine/ xylazine cocktail, and 50 µL of MRSA (1×10^8 CFU/ mL) was inoculated onto the wound. One day later, 20 µL of photosensitizers (Pa or Hy) at 2.5 mg/kg were intravenously injected into the mice via the tail vein as for the stratified groups. Ten minutes after the application of photosensitizers, PDT illumination at 1 W was performed for either 30 s or 10 min for all PDT groups, corresponding to 30 J/wound and 600 J/wound, respectively. A continuous-wave laser was generated from the Ceralas medical laser system with excitation at 670 nm (Biolitec group, Bonn, Germany). The treatments were repeated every other day and lasted until Day 8 (three treatment cycles, Figure 2). Wound sizes were recorded before and after treatment. At the end of the experiment, animals were euthanized with an overdose of terminal pentobarbital solution. The wound (5×10 mm) was then excised aseptically.

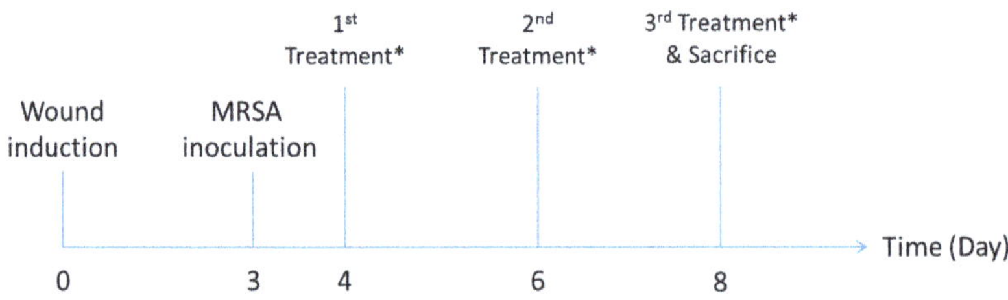

Figure 2. Timeline for the intravenous treatment.

2.3.3. Topical Treatment

Three days after wound induction, mice were anesthetized with a ketamine/ xylazine cocktail and the adhesive bandage was removed. A 50 µL of MRSA suspension

$(1 \times 10^8$ CFU/ mL) was dropped onto each wound. A dressing (Tegaderm™ film, 3M, Company, St. Louis, MA, USA) was applied to cover the wound immediately to maintain wound moisture. Thirty minutes after bacterial inoculation, 50 μL of 800 μM PS solutions or Fucidin® cream was injected under the dressing by syringe and allowed to spread over the wound. Photoactivation (Biolitec group, Bonn, Germany) was initiated immediately. A single dosage of laser at 0.5 W for 60 s was delivered by an optical fiber 2 mm in diameter, corresponding to 30 J/wound. The dark control (PS alone) groups and the Fucidin® cream (2% fucidic acid) group (positive control) did not receive any laser irradiation but were sham-irradiated under visible light. The animals were returned to individually ventilated cages (IVC) after treatment and thoroughly examined daily. To avoid any possible phototoxicity, all mice were kept in a dark room for 4 h after PDT/sham irradiation. After 2 days of treatment, once daily (Figure 3), the dressings were removed and the wounds were exposed. The wound sizes were recorded every two days. On Day 10, animals were euthanized with an overdose of dorminal pentobarbital solution. The wound (5 × 10 mm) was then excised aseptically.

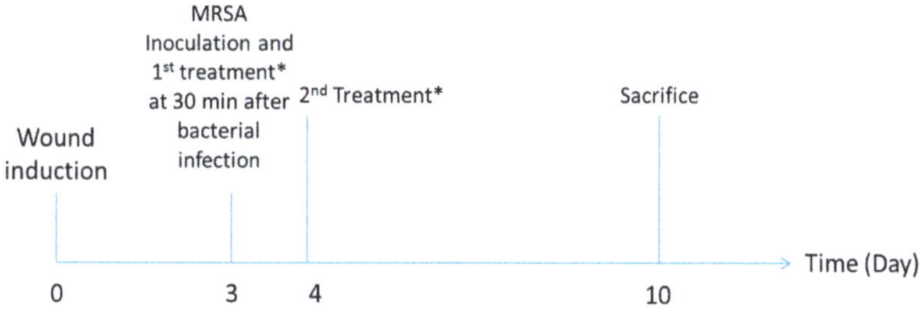

Figure 3. Timeline for the topical treatment.

Each skin sample was divided into two portions. One-piece was used for histological examination to determine the maturity of wound repair, and the second was weighed and homogenized in 0.5 mL of PBS solution for bacterial viability counts. Quantification of viable bacteria was performed by culturing serial dilutions (10 μL) of the bacterial suspension on blood agar plates. For this purpose, all plates were incubated at 37 °C for 24 h and evaluated for the presence of the staphylococcal strain. The bacteria were quantified by counting the number of CFU per plate.

2.3.4. Histological Evaluation

Wound tissues collected from the animal study were initially fixed in 10% buffered formalin, followed by dehydration and paraffin-embedding. Paraffin blocks were cut into 5 μm tissue sections including the epidermis, the dermis, and the subcutaneous panniculus. The sections were stained with hematoxylin and eosin (H&E) and assessed by light microscopy for wound healing.

2.4. In Vitro Cytotoxic MTT (3-(4,5-Dimethylthiazol-2-yl)-2,5-diphenyltetrazolium Bromide) Assay on Human Neutrophils

Human neutrophils were purified from the fresh buffy coat fraction of blood from adult volunteers at the Hong Kong Red Cross Blood Transfusion Service, Hong Kong and separated by the Percoll method which was routinely performed in our laboratory [49].

Our studies showed that human neutrophils exhibited a short lifespan after isolating from buffy coats. Most human neutrophils did not survive 48 h after isolation. Therefore, freshly isolated human neutrophils were used in the present study, and the experiments were done within 24 h after isolation. Freshly isolated human neutrophils were plated in

96-well plates at 10^5 cells/well. Serial dilutions of three photosensitizers, Pa, MB, and Hy were added to the wells. After 24 h at 37 °C incubation, MTT solution (50 μL, 5 mg/mL) were added to each well. Then, the plates were incubated at 37 °C for 3 h. After incubation, 150 μL of DMSO was added to each well. The OD of the wells was determined by a spectrophotometer at 590 nm. Toxicity was represented by the ratio of OD of a well in the presence of compounds with the OD of control wells in the presence of a medium containing DMSO.

3. Results

3.1. Bactericidal Activity Assay on MRSA Strains

It is apparent from Table 1 that the Pa-PDT group showed significantly higher ($p < 0.05$) anti-MRSA activity against MRSA ATCC RN4220/pUL5054, W44, and W46-47 (MBC; 3.125–12.5 μM) than the positive control MB (MBC; 120->160 μM). Similarly, the Hy-PDT group demonstrated significantly higher ($p < 0.05$) anti-MRSA activity against MRSA ATCC RN4220/pUL5054 and a whole panel of CA-MRSA strains (MBC; 0.625–10 μM) compared to MB. However, HA-MRSA was more resistant towards HY-PDT, except HA-232 (MBC; 2.5 μM). Interestingly, Hy-PDT showed the lowest MBC values compared to Pa-PDT or MB-PDT, indicating the importance of further investigations. The dark toxicities of all three PSs were 4-8 times lower than their PDT counterparts (Table 2). Out of these sixteen MRSA strains tested, RN4220/pUL5054 was sensitive to three photosensitizers, especially to Hy and Pa. Therefore, it was selected to establish the in vivo model.

Table 2. The Minimal Bactericidal Concentrations (MBCs) of Pa, Hy and MB against sixteen MRSA strains.

MRSA Type	Strain	MBC Values											
		Hy-PDT		Hy Dark Control		Pa-PDT		Pa Dark Control		MB-PDT		MB Dark Control	
		μM	μg/mL	μM	μg/mL	μM	μg/mL	μM	μg/mL	μM	μg/mL	μM	μg/mL
ATCC	43300	>35	>16	>35	>16	>300	>128	>300	>128	160	32	>160	>32
ATCC	BAA 42	>35	>16	>35	>16	>300	>128	>300	>128	>160	>32	>160	>32
ATCC	BAA 43	>35	>16	>35	>16	>300	>128	>300	>128	80	32	>160	>32
ATCC	BAA 44	>35	>16	>35	>16	>300	>128	>300	>128	>160	>32	>160	>32
Mutant	APH2AAC 6	>35	>16	>35	>16	>300	>128	>300	>128	>160	>32	>160	>32
Mutant	RN4220 /pUL5054	0.625	0.5	5	4	6.25	4	50	32	120	>32	>160	>32
CA [a]	W44	0.625	0.5	3.125	2.5	12.5	8	75	48	140	>32	>160	>32
CA	W45	10	8	>35	>16	>300	>128	>300	>128	>160	>32	>160	>32
CA	W46	1.25	1	7.5	6	6.25	4	31.25	20	>160	>32	>160	>32
CA	W47	5	4	>35	>16	3.125	2	18.75	12	>160	>32	>160	>32
CA	W48	1.25	1	5	4	>300	>128	>300	>128	140	>32	>160	>32
HA [b]	W231	>35	>16	>35	>16	>300	>128	>300	>128	>160	>32	>160	>32
HA	W232	2.5	2	15	12	>300	>128	>300	>128	>160	>32	>160	>32
HA	W233	>35	>16	>35	>16	>300	>128	>300	>128	120	>32	>160	>32
HA	W234	>35	>16	>35	>16	>300	>128	>300	>128	80	32	>160	>32
HA	W235	>35	>16	>35	>16	>300	>128	>300	>128	>160	>32	>160	>32

[a] CA: community associated; [b] HA: hospital associated.

3.2. Animal Studies-Mouse Model of MRSA-Infected Wound

3.2.1. Effect of PDT of Pa and Hy with Intravenous Injection in MRSA-Infected Wound Model

Neither of the PDTs (30 s or 10 min) upon Pa and Hy intravenous (i.v.) injection (2.5 mg/kg) significantly ($p < 0.05$) reduced bacterial load at Day 8 (Figure 4). It was observed that mice receiving 30 s of Pa-PDT and Hy-PDT treatments resulted in slight but insignificant promotion of wound closure. However, this trend could not be observed in the 10 min PDT-treated groups (Figure 5). There was no body weight loss in treatment groups when compared with the control group, implying that the treatments did not cause distress in the mice (Figure 6).

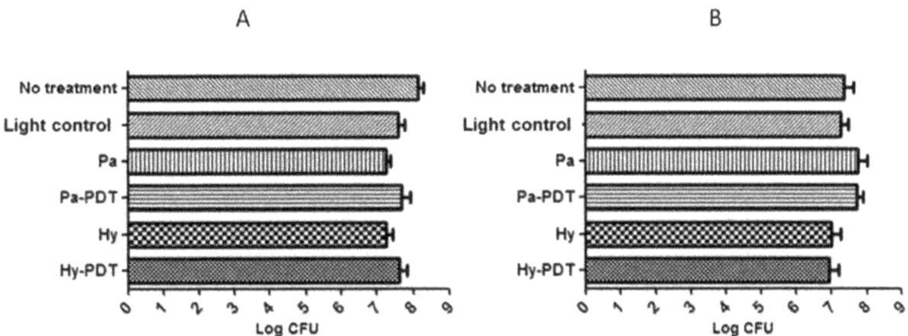

Figure 4. Bacterial load of wounds in different groups of mice after intravenous injection treatment with or without PDT. (**A**) Pa and Hy 2.5 mg/kg, irradiation for 30 s for PDT groups; (**B**) Pa and Hy 2.5 mg/kg, irradiation for 10 min for PDT groups. Data are mean $\pm$ SEM (n = 5).

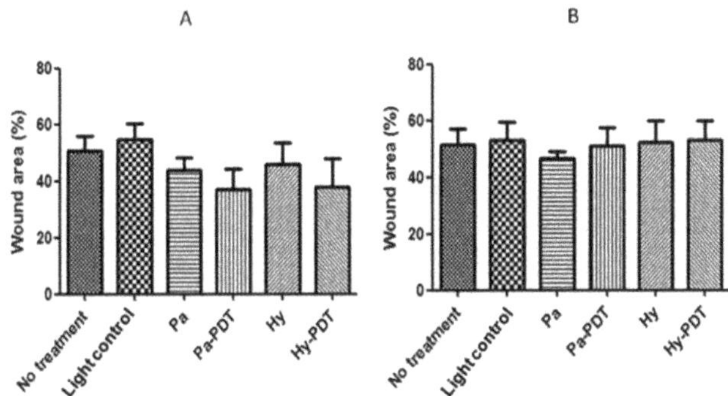

Figure 5. Wound areas in different groups of mice after intravenous injection treatment with or without PDT. (**A**) Pa and Hy 2.5 mg/kg irradiation for 30 s for PDT groups; (**B**) Pa and Hy 2.5 mg/kg, irradiation for 10 min for PDT groups. Data are mean $\pm$ SEM (n = 5).

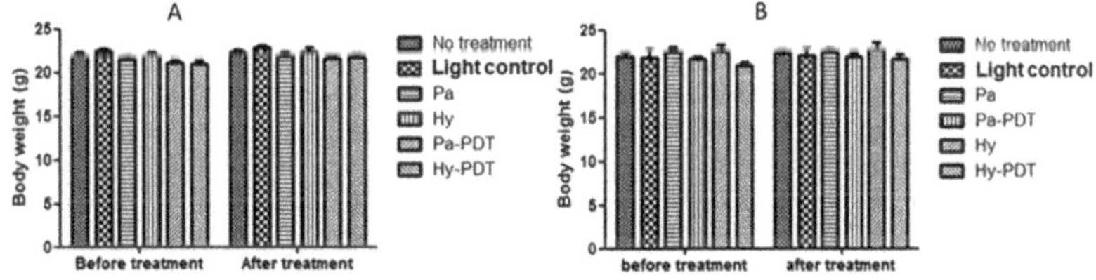

Figure 6. Body weight of different groups of mice after intravenous injection treatment with or without PDT. (**A**) Pa and Hy 2.5 mg/kg, irradiation for30 s for PDT groups; (**B**) Pa and Hy 2.5 mg/kg, irradiation for 10 min for PDT groups. Data are mean $\pm$ SEM (n=5).

3.2.2. Effect of PDT of Topically Applied MB, Pa or Hy in MRSA-Infected Wound Model

Topical application of Fucidin cream eradicated MRSA in the wound. Pa-PDT and Hy-PDT treatment groups showed significant antibacterial effects against MRSA when compared with the no treatment group (1 log decrease of CFU, $p < 0.05$) (Figure 7). The size of Fucidin cream-treated wounds was slightly larger and there was no great difference in wound sizes among all other groups after treatment (Figure 8). There was no body weight loss in the treatment groups when compared with the control group (Figure 9), implying that the treatments did not cause distress in the mice.

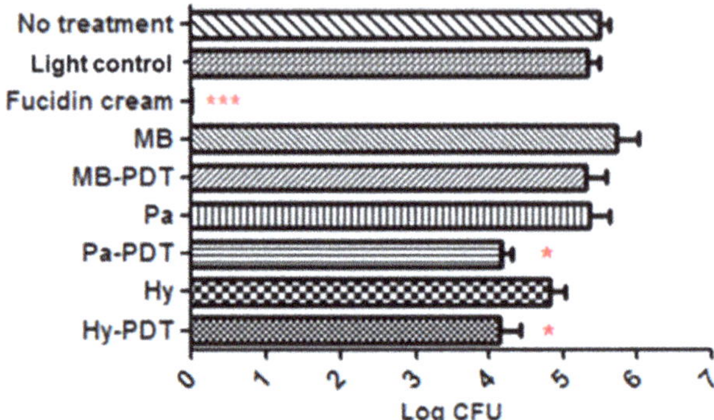

Figure 7. Bacterial load of wounds in different groups of mice after topical treatment with or without PDT. Data are mean $\pm$ SEM ($n = 6$–10).* $p < 0.05$ and *** $p < 0.001$ indicated significant bacterial load difference between No treatment and treatment groups by Student's *t*-test.

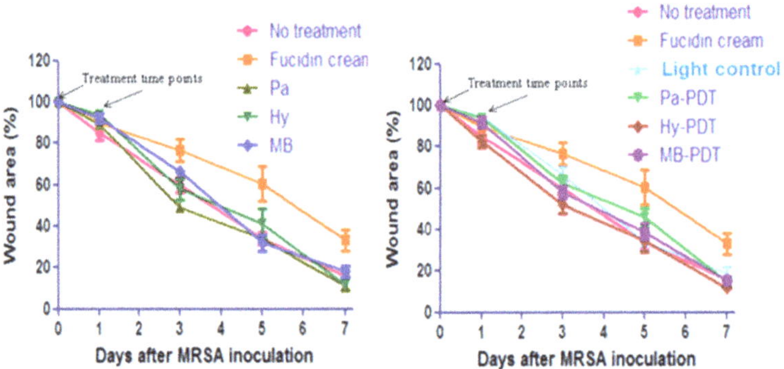

Figure 8. Wound areas of mice in different groups of mice after topical treatment with or without PDT. Data are mean $\pm$ SEM ($n = 6$–10).

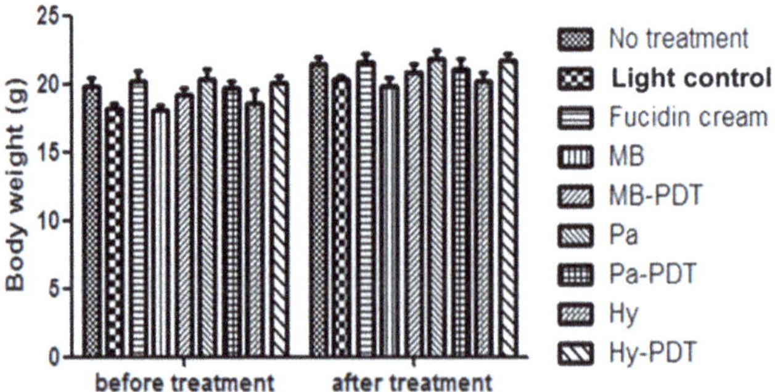

Figure 9. Body weight of different groups of mice before and after treatment with or without PDT. Data are mean $\pm$ SEM ($n = 6$–10).

3.2.3. Histological Evaluation

Histopathological assessment of untreated wounds on day 8 (Figure 10) indicated incomplete epithelialization, loose granulation tissue with areas of poorly stained extracellular matrix where collagen fibers were either immature or lacking as a sign of obvious ulcer formation. It is apparent from Figure 10, Hy and Hy-PDT mediated groups showed rather good wound healing compared to the treatment naïve group by showing epithelial cells and fibrous tissue proliferation. MB-PDT had a minor healing effect in granulation and collagen formation and its re-epithelialization was worse than that of the no treatment group. Wound healing of the Fucidin cream-treated group was worse than the no treatment group (Figure 10).

3.3. Cytotoxicity Effect of Pa-PDT, Hy-PDT or MB-PDT on Human Neutrophils

Three photosensitizers, Pa, MB and Hy, were incubated with human neutrophils for 24 h. No light irradiation was applied to the photosensitizers. Viability of human neutrophils was determined by MTT assay. As shown in Figure 11, Pa and MB were more cytotoxic to human neutrophils at 24-h incubation with LC_{50} at ~10.16 µM and ~11.22 µM, respectively, whereas Hy showed a LC_{50} higher than 50 µM.

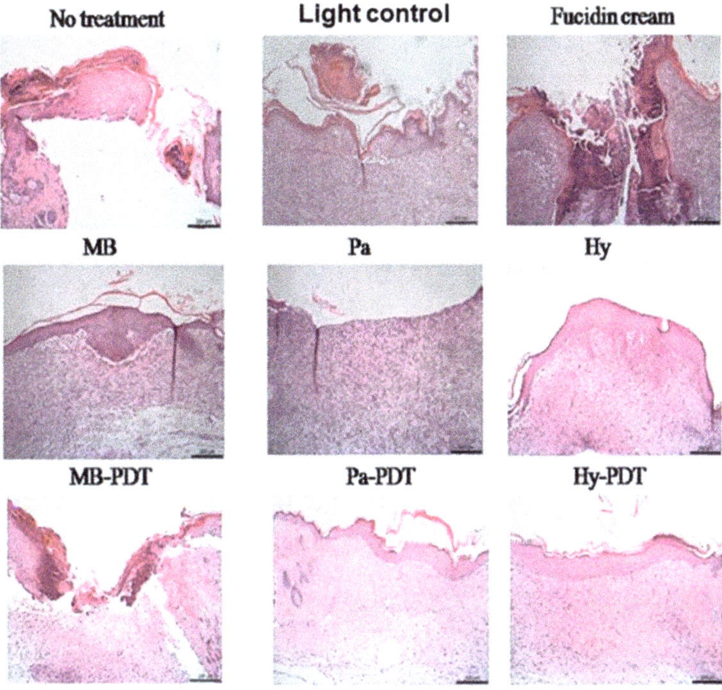

Figure 10. Representative wound section with haematoxylin and eosin stained and examined at ×100 magnification. Scale bar, 200 µm.

Cytotoxicity of the three photosensitizers with light irradiation for 20 min (48 J/cm^2) on human neutrophils was also examined. As shown in Figure 12, Hy-PDT possessed the strongest cytotoxicity with LC_{50} less than 3 µM, whereas LC_{50} of Pa-PDT and MB-PDT were ~4.44 µM and ~4.23 µM, respectively. As predicted, drugs with light irradiation exhibited significantly higher cytotoxicity than drugs without light irradiation. Given the high cytotoxic properties of photosensitizers with light irradiation, we expected that no cytokines would be produced in human neutrophils, as well as in the photosensitizer-administered wound sites when light irradiation was applied.

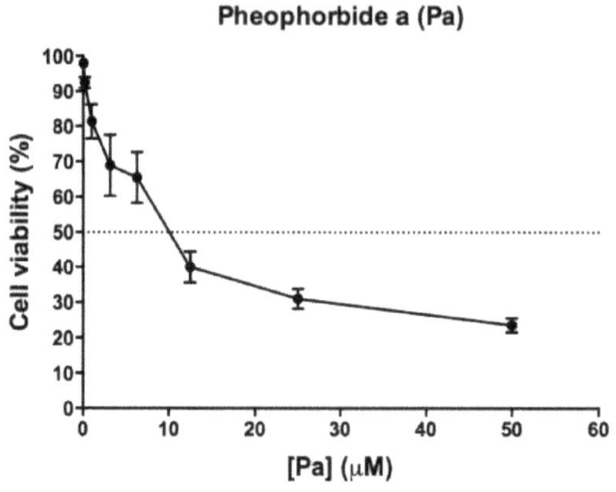

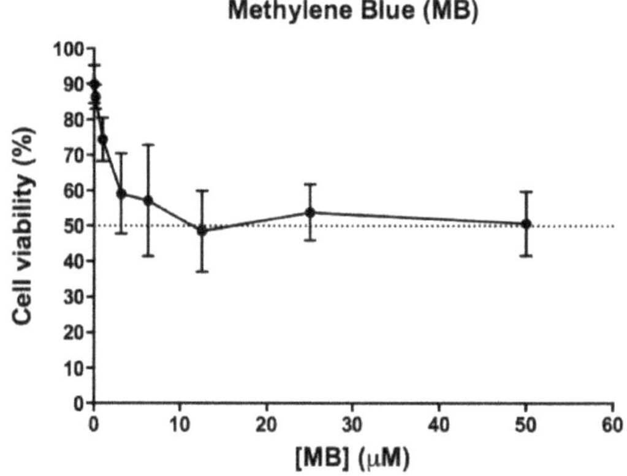

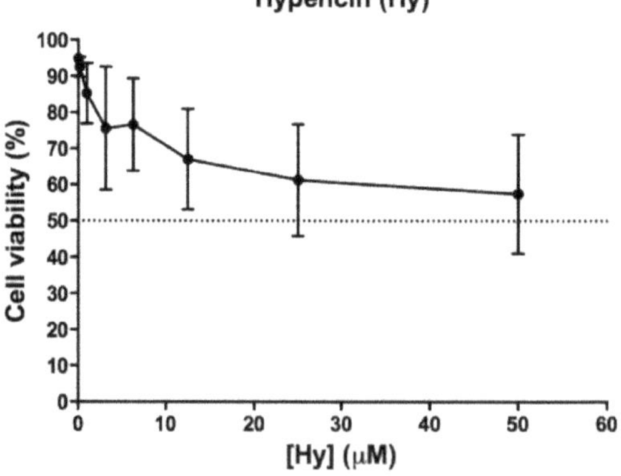

Figure 11. Cytotoxicity of Pa, MB and Hy, without light irradiation on human neutrophils isolated from buffy coat. Data are mean ± SEM ($n = 3$).

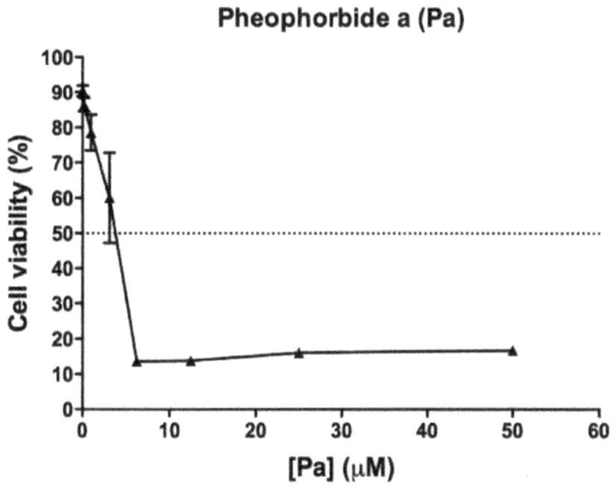

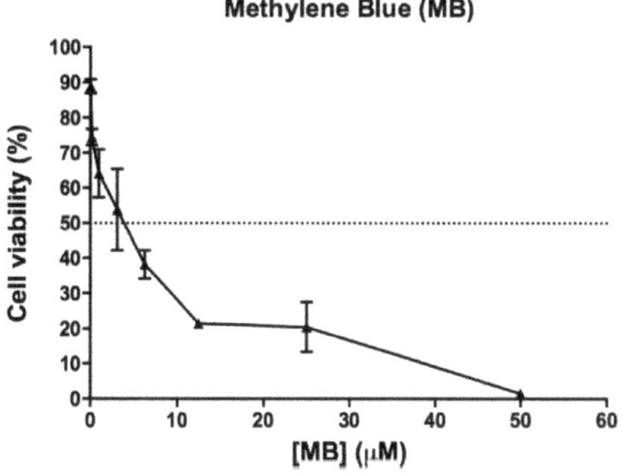

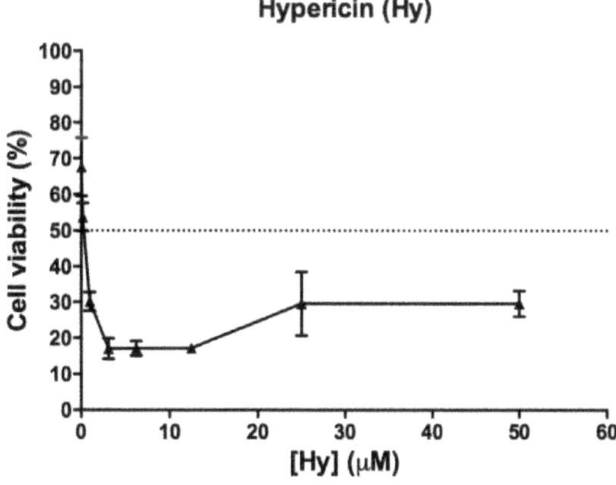

Figure 12. Cytotoxicity of Pa, MB and Hy, with light irradiation, on human neutrophils isolated from buffy coat with. Data are mean $\pm$ SEM ($n = 3$).

4. Discussion

PDT with MB is widely tested to be effective against Gram-positive bacteria, including MRSA, and acts as the gold standard for efficacy comparison for PDTs with Pa and Hy. In vitro results showed that Hy-PDT and Pa-PDT against some MRSA strains had better bactericidal activity than MB-PDT. Among the sixteen MRSA strains tested, the lowest MBC for PDTs with Hy, Pa, and MB were 0.625 µM (0.5 µg/mL), 3.125 µM (2 µg/mL), and 80 µM (32 µg/mL) respectively. Therefore, Hy-PDT and Pa-PDT exhibited potent bactericidal activity on MRSA strains. Hy possesses appropriate photochemical and photobiological properties, such as a high singlet oxygen quantum yield and cytoplasmic membrane localization, which makes it suitable for use as a PS in PDT [29,50]. Furthermore, its absorption maxima (λmax) 570 nm at longer wavelength [51] makes Hy suitable for PDT because of its high penetration ability. The obtained in vitro results for Hy are comparable with the previously published data in Yow et al. [29] for two CA-MRSA strains (W45 and W47, Table 2) where a > 6 $\log_{10}$ CFU reduction of MRSA was obtained at an 8 µM concentration and 30 Jcm^{-2} light dose. However, MBC values for RN4220/pUL5054, W44, W46 and W48 showed significantly lower ($p < 0.05$) MBC values compared to the published report.

In addition, it was found that Pa-PDT inhibited P-glycoprotein-mediated multidrug resistance via c-Jun N-terminal kinase (JNK) activation in human hepatocellular carcinoma [52]. As P-glycoprotein is an important class of efflux pumps that is always associated with a high prevalence of antibiotic resistance, it is hypothesized that Pa can circumvent drug resistance in MRSA as well.

To evaluate the efficacies of photodynamic therapy mediated by Pa and Hy, an MRSA-infected wound-bearing mice model was used. We found that MRSA was far more resistant to PDT in the more complicated environment of the murine dorsal wound than in transparent Petri dishes in vitro. It is possible because the wound tissue provided more layers of organic material to scatter light and to host the bacteria. The open wound of skin tissue, as the natural colony of Staphylococci, might provide a more nutritious matrix for bacterial survival and prosperity, and the aqua dependency of the cytotoxicity of photosensitizers might also partly be impeded in the relatively lower moistures microenvironment of the wounds.

However, the bodyweight of all mice was weighed before and after 8 days of treatment and there was no difference between groups and there was also no significant behavioral change in the mice, indicating that wound induction, MRSA infection, and our treatments had little effect on the general health status of the mice.

It was observed that intravenous injection of photosensitizers with PDT treatments had no antibacterial effect. This may be because none of the photosensitizers had an affinity to wound tissue, so the circulating photosensitizers had little opportunity to aggregate at the wound to generate an adequate amount of ROS by light illumination to kill MRSA. We found that 10 min of PDT resulted in burnt scab formation shortly after light illumination in some cases, while 30 s of PDT did not. The 30 s of PDT treatment also resulted in better wound healing.

The in vivo antibacterial effect of Hy-PDT and Pa-PDT was found to be significant and also stronger than MB-PDT at the same concentration, suggesting encouraging antibacterial effects of them against MRSA wound infection. However, it is interesting that the complete elimination of MRSA by Fucidin cream treatment was accompanied by worse wound healing as reflected by large open wound areas. It was also surprising that the histological appearance of wound healing in MRSA-infected wounds receiving only water and light illumination was significantly higher than that of No treatment group. We also observed more abundant vessel formation in the dermis of some samples than in any other group without significant improvement in open wound area. It has been reported that red light illumination promoted wound healing by promoting ATP release from mitochondria, activating the lymphatic system, increasing blood circulation and forming new capillaries [53]. Since we did not do further cellular nor molecular analyses, we are not in a position to explain the exact reason behind the varied wound closures among different treatment groups.

We tested the wound healing effects of several commercially available antibacterial-agent-free moisturizer skin creams and none of them promoted nor inhibited skin wound healing (data not shown). Pa-PDT, MB-PDT, and Hy-PDT were all toxic to neutrophils in vitro but the neutrophils were more resistant to Pa, MB, and Hy treatments without light illumination. Hy and MB were less toxic, while Pa showed some toxicity even without light illumination. Since our treatments did not include whole body illumination, the topically applied photosensitizers and locally irradiated light illumination could hardly have any toxic effects on the immune system inside the body. In addition, insignificant body weight differences among all groups also indicated the minimum effect of our treatments on the general health of animals.

5. Conclusions

Hy-PDT, Pa-PDT and MB-PDT were all capable of killing MRSA in vitro, and Hy-PDT showed highest efficacy against panel of MRSAs. Both Hy-PDT and Pa-PDT showed better efficacy than MB-PDT in antibacterial and wound healing effects against MRSA-infected wounds in our murine model, and the efficacy of Hy-PDT was the best among all treatments tested.

Author Contributions: B.C.L.C., P.D., B.W., K.M.L., C.C.L., D.W.S.C., J.Y.W.C. and G.G.L.Y. were responsible for conducting the experimental work; C.B.S.L. and C.K.W. were responsible for the study design, K.P.F. and M.I. were responsible for supervision, administration and writing the manuscript. All authors have read and agreed to the published version of the manuscript.

Funding: This work was supported by a grant from the Health and Medical Research Fund (HMRF 13120302).

Institutional Review Board Statement: The study was conducted according to the guidelines of the Declaration of Helsinki, and approved by the Animal Experimentation Ethics Committee of The Chinese University of Hong Kong (Ref No.12/076/MIS and 8 February 2013).

Informed Consent Statement: Not applicable.

Conflicts of Interest: The authors declare no conflict of interest. The funders had no role in the design of the study; in the collection, analyses, or interpretation of data; in the writing of the manuscript, or in the decision to publish the results.

References

1. Verhoeven, P.O.; Gagnaire, J.; Botelho-Nevers, E.; Grattard, F.; Carricajo, A.; Lucht, F. Detection and clinical relevance of *Staphylococcus aureus* nasal carriage: An update. *Expert Rev. Anti-Infect. Ther.* **2014**, *12*, 75–89. [CrossRef]
2. Cheikh, A.; Belefquih, B.; Chajai, Y.; Cheikhaoui, Y.; El Hassani, A.; Benouda, A. Enterobacteriaceae producing extended-spectrum beta-lactamases (ESBLs) colonization as a risk factor for developing ESBL infections in pediatric cardiac surgery patients: Retrospective cohort study. *BMC Infect. Dis.* **2017**, *17*, 237–242. [CrossRef]
3. Dubinsky-Pertzov, B.; Temkin, E.; Harbarth, S.; Fankhauser-Rodriguez, C.; Carevic, B.; Radovanovic, I. Carriage of extended-spectrum beta-lactamase-producing enterobacteriaceae and the risk of surgical site infection after colorectal surgery: A prospective cohort study. *Clin. Infect. Dis.* **2019**, *68*, 1699–1704. [CrossRef]
4. Neyra, R.C.; Frisancho, J.A.; Rinsky, J.L.; Resnick, C.; Carroll, K.C.; Rule, A.M. Multidrug-resistant and methicillin-resistant *Staphylococcus aureus* (MRSA) in hog slaughter and processing plant workers and their community in North Carolina (USA). *Environ. Health Perspect.* **2014**, *122*, 471–477. [CrossRef]
5. Yuen, J.W.; Chung, T.W.; Loke, A.Y. Methicillin-resistant *Staphylococcus aureus* (MRSA) contamination in bedside surfaces of a hospital ward and the potential effectiveness of enhanced disinfection with an antimicrobial polymer surfactant. *Int. J. Environ. Res. Public Health* **2015**, *12*, 3026–3041. [CrossRef] [PubMed]
6. Ho, P.L.; Chiu, S.S.; Chan, M.Y.; Gan, Y.; Chow, K.H.; Lai, E.L. Molecular epidemiology and nasal carriage of *Staphylococcus aureus* and methicillin-resistant *S. aureus* among young children attending day care centers and kindergartens in Hong Kong. *J. Infect.* **2012**, *64*, 500–506. [CrossRef] [PubMed]
7. Wong, J.W.; Ip, M.; Tang, A.; Wei, V.W.; Wong, S.; Riley, S.; Kwok, K.O. Prevalence and risk factors of community-associated methicillin-resistant *Staphylococcus aureus* carriage in Asia-Pacific region from 2000 to 2016: A systematic review and meta-analysis. *Clin. Epidemiol.* **2018**, *10*, 1489–1501. [CrossRef] [PubMed]

8.   Spika, J.; Rud, E.W. Immunization as a tool to combat antimicrobial resistance. *Can. Commun. Dis. Rep.* **2015**, *41*, 7–10. [CrossRef]
9.   Hamblin, M.R.; Mroz, P. History of PDT: The first hundred years. In *Advances in Photodynamic Therapy: Basic, Translational and Clinical Phoenix*; Hamblin, M.R., Mroz, P., Eds.; Artech House, Inc.: Norwood, MA, USA, 2008; pp. 1–12.
10. Dharmaratne, P.; Sapugahawatte, D.N.; Wang, B.; Chan, C.L.; Lau, K.; Lau, C.B.; Fung, K.P.; Ng, D.K.P.; IP, M. Contemporary approaches and future perspectives of antibacterial photodynamic therapy (aPDT) against methicillin-resistant *Staphylococcus aureus* (MRSA): A systematic review. *Eur. J. Med. Chem.* **2020**, *200*, 112341. [CrossRef]
11. Sperandio, F.; Huang, Y.; Hamblin, M. Antimicrobial Photodynamic therapy to kill Gram-negative bacteria. *Recent Pat. Anti-Infect. Drug Discov.* **2013**, *8*, 108–120. [CrossRef]
12. Dharmaratne, P.; Wong, R.C.; Wang, J.; Lo, P.; Wang, B.; Chan, B.C.; Ip, M.; Ng, D.K. Synthesis and in vitro Photodynamic activity of cationic boron dipyrromethene-based photosensitizers against methicillin-resistant *Staphylococcus aureus*. *Biomedicines* **2020**, *8*, 140. [CrossRef]
13. Biel, M.; Sievert, C.; Loebel, N.; Rose, A.; Zimmermann, R. Reduction of endotracheal tube biofilms using antimicrobial photodynamic therapy. *Las. Sur. Med.* **2011**, *43*, 586–590. [CrossRef] [PubMed]
14. Hu, X.; Huang, Y.; Wang, Y.; Wang, X.; Hamblin, M.R. Antimicrobial Photodynamic therapy to control clinically relevant Biofilm infections. *Front. Microbiol.* **2018**, *9*, 1299. [CrossRef]
15. Venturini, M.; Calzavara-Pinton, P. Photodynamic antifungal therapy. *Photodiag. Photod. Ther.* **2017**, *17*, A17. [CrossRef]
16. Lyon, J.P.; Moreira, L.M.; De Moraes, P.C.; Dos Santos, F.V.; De Resende, M.A. Photodynamic therapy for pathogenic fungi. *Mycoses* **2011**, *54*, 265–271. [CrossRef]
17. Pinto, J.G.; Fontana, L.C.; De Oliveira, M.A.; Kurachi, C.; Raniero, L.J.; Ferreira-Strixino, J. In vitro evaluation of photodynamic therapy using curcumin on Leishmania major and Leishmania braziliensis. *Lasers Med. Sci.* **2016**, *31*, 883–890. [CrossRef]
18. Ke, M.R.; Eastel, J.M.; Ngai, K.L.K.; Cheung, Y.Y.; Chan, P.K.S.; Hui, M.; Ng, D.K.P.; Lo, P.-C. Photodynamic inactivation of bacteria and viruses using two monosubstituted zinc(II) phthalocyanines. *Eur. J. Med. Chem.* **2014**, *84*, 278–283. [CrossRef]
19. Garcez, A.S.; Nunez, S.C.; Hamblin, M.R.; Ribeiro, M.S. Antimicrobial effects of photodynamic therapy on patients with necrotic pulps and periapical lesion. *J. Endod.* **2008**, *34*, 138–142. [CrossRef] [PubMed]
20. Pinheiro, S.L.; Schenka, A.A.; Neto, A.A.; De Souza, C.P.; Rodriguez, H.M.; Ribeiro, M.C. Photodynamic therapy in endodontic treatment of deciduous teeth. *Lasers Med. Sci.* **2008**, *4*, 112–117. [CrossRef]
21. Mills, S.J.; Farrar, M.D.; Ashcroft, G.S.; Griffiths, C.E.M.; Hardman, M.J.; Rhodes, L.E. Topical photodynamic therapy following excisional wounding of human skin increases production of transforming growth factor-β3 and matrix metalloproteinases 1 and 9, with associated improvement in dermal matrix organization. *Br. J. Dermatol.* **2014**, *171*, 55–62. [CrossRef]
22. Cappugi, P.; Comacchi, C.; Torchia, D. Photodynamic therapy for chronic venous ulcers. *Acta Dermatovenerol. Croat* **2014**, *22*, 129–131.
23. Widodo, A.; Spratt, D.; Sousa, V.; Petrie, A.; Donos, N. An in vitro study on disinfection of titanium surfaces. *Clin. Oral Implant. Res.* **2016**, *27*, 1227–1232. [CrossRef]
24. Tanaka, M.; Mroz, P.; Dai, T. Photodynamic therapy can induce a protective innate immune response against murine bacterial arthritis via neutrophil accumulation. *PLoS ONE* **2012**, *7*, e39823. [CrossRef]
25. Tanaka, M.; Kinoshita, M.; Yoshihara, Y. Influence of intra-articular neutrophils on the effects of photodynamic therapy for murine MRSA arthritis. *Photochem. Photobiol.* **2010**, *86*, 403–409. [CrossRef]
26. Saddiqe, Z.; Naeem, I.; Maimoona, A. A review of the antibacterial activity of *Hypericum perforatum* L. *J. Ethnopharmacol.* **2010**, *131*, 511–521. [CrossRef]
27. Bugaj, A.M. Targeted photodynamic therapy e a promising strategy of tumor Treatment. *Photochem. Photobiol. Sci.* **2011**, *10*, 1097. [CrossRef] [PubMed]
28. Rook, A.H.; Wood, G.S.; Duvic, M.; Vonderheid, E.C.; Tobia, A.; Cabana, B. A phase II placebo-controlled study of photodynamic therapy with topical hypericin and visible light irradiation in the treatment of cutaneous T-cell lymphoma and psoriasis. *J. Am. Acad. Dermatol.* **2010**, *63*, 984e990. [CrossRef]
29. Yow, C.M.; Tang, H.M.; Chu, E.S.; Huang, Z. Hypericin-mediated photodynamic antimicrobial effect on clinically isolated pathogens. *Photochem. Photobiol.* **2012**, *88*, 626–632. [CrossRef] [PubMed]
30. Suntar, I.P.; Akkol, E.K.; Yilmazer, D. Investigations on the in vivo wound healing potential of *Hypericum perforatum* L. *J. Ethnopharmacol.* **2010**, *127*, 468–477. [CrossRef]
31. Tang, P.M.; Bui-Xuan, N.; Wong, C.; Fong, W.; Fung, K. Pheophorbide a-mediated Photodynamic therapy triggers HLA class I-restricted antigen presentation in human hepatocellular carcinoma. *Transl. Oncol.* **2010**, *3*, 114–122. [CrossRef] [PubMed]
32. Lee, W.Y.; Lim, D.S.; Ko, S.H.; Park, Y.J.; Ryu, K.S.; Ahn, M.Y.; Kim, Y.R.; Lee, D.W.; Cho, C.W. Photoactivation of pheophorbide a induces a mitochondrialmediated apoptosis in Jurkat leukaemia cells. *J. Photochem. Photobiol. B* **2004**, *75*, 119–126. [CrossRef]
33. Hajri, A.; Wack, S.; Meyer, C.; Smith, M.K.; Leberquier, C.; Kedinger, M.; Aprahamian, M. In vitro and in vivo efficacy of photofrin and pheophorbide a, a bacteriochlorin, in photodynamic therapy of colonic cancer cells. *Photochem. Photobiol.* **2002**, *75*, 140–148. [CrossRef]
34. Jin, Z.H.; Miyoshi, N.; Ishiguro, K.; Umemura, S.; Kawabata, K.; Yumita, N.; Sakata, I.; Takaoka, K.; Udagawa, T.; Nakajima, S. Combination effect of photodynamic and sonodynamic therapy on experimental skin squamous cell carcinoma in C3H/HeN mice. *J. Dermatol.* **2000**, *27*, 294–306. [CrossRef] [PubMed]

35. Hajri, A.; Coffy, S.; Vallat, F.; Evrard, S.; Marescaux, J.; Aprahamian, M. Human pancreatic carcinoma cells are sensitive to photodynamic therapy in vitro and in vivo. *J. Br. Surg.* **1999**, *86*, 899–906. [CrossRef]
36. Yamamoto, T.; Iriuchishima, T.; Aizawa, S.; Okano, T.; Goto, B.; Nagai, Y. Bactericidal Effect of Photodynamic Therapy Using Na-Pheophorbide a: Evaluation of Adequate Light Source. *Photomed. Laser Surg.* **2009**, *27*, 849–853. [CrossRef]
37. Kamuhabwa, A.; Agostinis, P.; Ahmed, B.; Landuyt, W.; Van Cleynenbreugel, B.; Van Poppel, H.; De Witte, P. Hypericin as a potential phototherapeutic agent in superficial transitional cell carcinoma of the bladder. *Photochem. Photobiol. Sci.* **2004**, *3*, 668–772. [CrossRef]
38. Junqueira, H.C.; Severino, D.; Dias, L.G.; Gugliotti, M.; Baptista, M.S. Modulation of the methylene blue photochemical properties based on the adsorption at aqueous micelle interfaces. *Phys. Chem. Chem. Phys.* **2002**, *4*, 2320–2328. [CrossRef]
39. Severino, D.; Junqueira, H.C.; Gabrielli, D.S.; Gugliotti, M.; Baptista, M.S. Influence of negatively charged interfaces on the ground and excited state properties of methylene blue. *Photochem. Photobiol.* **2003**, *77*, 459–468. [CrossRef]
40. Wynn, J.L.; Cotton, T.M. Spectroscopic properties of Hypericin in solution and at surfaces. *J. Phys. Chem.* **1995**, *99*, 4317–4323. [CrossRef]
41. Losi, A. Fluorescence and time-resolved Photoacoustics of Hypericin inserted in liposomes: Dependence on pigment concentration and Bilayer phase. *Photochem. Photobiol.* **1997**, *65*, 791–801. [CrossRef]
42. Wang, K.; Li, J.; Kim, B.; Lee, J.; Shin, H.; Ko, S.; Lee, W.; Lee, C.; Jung, S.H.; Kim, Y. Photophysical properties of pheophorbide-a derivatives and their Photodynamic therapeutic effects on a tumor cell Line in vitro. *Int. J. Photoenergy* **2014**, *1*, 1–7. [CrossRef]
43. Eichwurzel, I.; Stiel, H.; Röder, B. Photophysical studies of the pheophorbide a dimer. *J. Photochem. Photobio B Biol.* **2000**, *54*, 194–200. [CrossRef]
44. Dharmaratne, P.; Yu, L.; Wong, R.C.; Chan, B.C.; Lau, K.; Wang, B.; Fung, K.P.; Ng, D.K.P.; Ip, M. A novel Dicationic boron dipyrromethene-based photosensitizer for antimicrobial Photodynamic therapy against methicillin-resistant *Staphylococcus aureus*. *Curr. Med. Chem.* **2020**, *28*, 1–12.
45. Dharmaratne, P.; Wang, B.; Wong, R.C.; Chan, B.C.; Lau, K.; Ke, M.; Fung, K.P.; Ng, D.K.P.; Ip, M. Monosubstituted tricationic Zn(II) phthalocyanine enhances antimicrobial photodynamic inactivation (aPDI) of methicillin-resistant *Staphylococcus aureus* (MRSA) and cytotoxicity evaluation for topical applications: In vitro and in vivo study. *Emerg. Microbes Infect.* **2020**, *9*, 1628–1637. [CrossRef]
46. Simonetti, O.; Cirioni, O.; Orlando, F. Effectiveness of antimicrobial photodynamic therapy with a single treatment of RLP068/Cl in an experimental model of *Staphylococcus aureus* wound infection. *Br. J. Dermatol.* **2011**, *164*, 987–995. [CrossRef]
47. Fila, G.; Kasimova, K.; Arenas, Y. Murine Model Imitating Chronic Wound Infections for Evaluation of Antimicrobial Photodynamic Therapy Efficacy. *Front. Microbiol.* **2016**, *7*, 1258. [CrossRef]
48. Hoi, S.W.; Wong, H.M.; Chan, J.Y.; Yue, G.G.; Tse, G.M.; Law, B.K.; Fung, K.P. Photodynamic therapy of Pheophorbide a inhibits the proliferation of human breast tumour via both caspase-dependent and -independent Apoptotic pathways in in vitro and in vivo models. *Phytother. Res.* **2011**, *26*, 734–742. [CrossRef]
49. Kuhns, D.B.; Priel, D.A.; Chu, J.; Zarember, K.A. Isolation and functional analysis of human neutrophils. *Curr. Protoc. Immunol.* **2015**, *111*, 1–9. [CrossRef]
50. Lopez Chicon, P.; Paz-Cristobal, M.P.; Rezusta, A.; Aspiroz, C.; Royo-Canas, M.; Andres-Ciriano, E. On the mechanism of *Candida* spp. Photoinactivation by hypericin. *Photochem. Photobiol. Sci.* **2012**, *11*, 1099. [CrossRef]
51. Theodossiou, T.A.; Hothersall, J.S.; De Witte, P.A.; Pantos, A.; Agostinis, P. The multifaceted photocytotoxic profile of hypericin. *Mol. Pharm.* **2009**, *6*, 1775–1789. [CrossRef]
52. Tang, P.M.; Zhang, D.M.; Xuan, N.H.; Tsui, S.K.; Waye, M.M.; Kong, S.K.; Fong, W.P.; Fung, K.P. Photodynamic therapy inhibits P-glycoprotein mediated multidrug resistance via JNK activation in human hepatocellular carcinoma using the photosensitizer pheophorbide a. *Mol. Cancer* **2009**, *8*, 56. [CrossRef] [PubMed]
53. Agrawal, T.; Gupta, G.K.; Rai, V. Pre-conditioning with low-level laser (light) therapy: Light before the storm. *Dose-Response* **2014**, *12*, 619–649. [CrossRef]

 *pharmaceutics*

Article

In Vivo Quantification of the Effectiveness of Topical Low-Dose Photodynamic Therapy in Wound Healing Using Two-Photon Microscopy

Hala Zuhayri [1], Viktor V. Nikolaev [1], Anastasia I. Knyazkova [1], Tatiana B. Lepekhina [1], Natalya A. Krivova [1], Valery V. Tuchin [1,2] and Yury V. Kistenev [1,*]

[1] Laboratory of Laser Molecular Imaging and Machine Learning, Tomsk State University, 36 Lenin Av., 634050 Tomsk, Russia; zuhayri.n.hala@gmail.com (H.Z.); vik-nikol@bk.ru (V.V.N.); a_knyazkova@bk.ru (A.I.K.); tatiana_lepekhina@mail.ru (T.B.L.); nakri@res.tsu.ru (N.A.K.); tuchinvv@mail.ru (V.V.T.)

[2] Science Medical Center, Saratov State University, 83 Astrakhanskaya Str., 410012 Saratov, Russia

* Correspondence: yuk@iao.ru

Abstract: The effect of low-dose photodynamic therapy on in vivo wound healing with topical application of 5-aminolevulinic acid and methylene blue was investigated using an animal model for two laser radiation doses (1 and 4 J/cm^2). A second-harmonic-generation-to-auto-fluorescence aging index of the dermis (SAAID) was analyzed by two-photon microscopy. SAAID measured at 60–80 μm depths was shown to be a suitable quantitative parameter to monitor wound healing. A comparison of SAAID in healthy and wound tissues during phototherapy showed that both light doses were effective for wound healing; however, healing was better at a dose of 4 J/cm^2.

Keywords: wound healing; low-dose photodynamic therapy; photosensitizer; two-photon microscopy; second-harmonic-generation-to-auto-fluorescence aging index of the dermis

Citation: Zuhayri, H.; Nikolaev, V.V.; Knyazkova, A.I.; Lepekhina, T.B.; Krivova, N.A.; Tuchin, V.V.; Kistenev, Y.V. In Vivo Quantification of the Effectiveness of Topical Low-Dose Photodynamic Therapy in Wound Healing Using Two-Photon Microscopy. *Pharmaceutics* **2022**, *14*, 287. https://doi.org/10.3390/pharmaceutics14020287

Academic Editors: Udo Bakowsky, Matthias Wojcik, Eduard Preis and Gerhard Litscher

Received: 28 December 2021
Accepted: 21 January 2022
Published: 26 January 2022

Publisher's Note: MDPI stays neutral with regard to jurisdictional claims in published maps and institutional affiliations.

1. Introduction

Wound healing is a complex physiological and dynamic process that occurs in the skin at the cellular level. It involves different overlapping phases of cellular activity that occur in the proper sequence, at a definite time, and for a specific duration [1,2]. Hemostasis begins within minutes of a wound occurring with vascular constriction and ends in the formation of a fibrin clot. The latter is considered essential in promoting the onset of the inflammatory and repair phases [3]. As soon as the clot is formed, pro-inflammatory cytokines and growth factors are released, such as fibroblasts, platelet-derived growth factor and epidermal growth factor. Inflammatory cells migrate into the wound and promote the inflammatory phase, characterized by the sequential infiltration of neutrophils, macrophages, and lymphocytes [3,4]. Neutrophils play an important role in clearing microbes and cellular debris in the wound area. Macrophages are responsible for inducing and removing apoptotic cells, thus paving the way for the resolution of inflammation. As macrophages clear apoptotic cells, they undergo a phenotypic transition to a reparative state that stimulates keratinocytes, fibroblasts, and angiogenesis to promote tissue regeneration. In this way, macrophages promote a transition to the proliferative phase of healing [5]. The latter generally follows and overlaps with the inflammatory phase, beginning approximately on the third or fourth day after injury. It includes granulation tissue formation, re-epithelialization, neoangiogenesis, regeneration of the extracellular matrix (ECM), and wound contraction [6,7].

Within the wound bed, fibroblasts produce collagen, glycosaminoglycans, and proteoglycans, major ECM components. Following proliferation and ECM synthesis, wound healing enters the most prolonged and final remodeling phase, typically beginning one week after injury following collagen deposition in the wound [8]. Collagen, including collagen types I and III, is one of the major skin components [9]. During the wound healing process, type III collagen is replaced by type I collagen, which provides tensile stiffness. Collagen fibers remodel by aligning with the body's tension lines and gaining strength through cross-linking [7–10]. In this final stage of healing, an attempt to recover the normal tissue structure occurs, and the granulation tissue is gradually remodeled, forming scar tissue [10,11].

Photo processes induced by external radiation mediated by photoactive compounds in tissues contribute to skin disease curing, skin rejuvenation, and wound healing [12,13]. Photodynamic therapy (PDT) is widely used to treat skin diseases like acne, viral warts, and skin cancers [14–17]. PDT enables a reduction in treatment time, accelerated tissue repair, and the promotion of wound healing [18,19].

PDT is based on using a photosensitizer (PS), which is accumulated in tissues, followed by irradiation of the tissue with a light source of an appropriate wavelength. The latter causes the generation of reactive oxygen species (ROS) [14,20]. ROS stimulates various cell administration pathways: apoptosis, necrosis, autophagy, depending on the type of cell treated, the concentration of the PS used, the dose of light energy supplied, and PS intracellular location [20–22]. High concentrations of ROS cause cell death, while low concentrations can cause the triggering of cellular repair processes, including proliferation and autophagy, and offer treatment by promoting healing [20].

The first PS generation was mainly constituted by hematoporphyrin and its derivatives, which have low selectivity and high cutaneous phototoxicity [23,24]. The second PS generation included porphyrin, phthalocyanine, benzoporphyrin, thiopurine derivatives, chlorin, and phenothiazines. It is characterized by greater deep light penetration capability due to its maximum absorption in the 630–800 nm spectral range corresponding to the tissue's highest transparency, good solubility, and higher chemical purity [14,25]. For skin treatments, 5-aminolevulinic acid (5-ALA), which is converted into the endogenous PS protoporphyrin IX (PpIX), has an absorption maximum at 630 nm and is widely used [22,26,27]. Methylene blue (MB) is a well-known phenothiazine derivative with great importance in medicine, including antimicrobial and antiviral light-induced activity [25,28,29]. The third PS generation corresponds to enhanced forms of the first and second generations. It is often combined with a carrier such as a lipoprotein, liposomes, nanoparticles, or conjugated antibodies to increase selectivity and contrast in affected areas [30–33].

According to the Arndt–Schultz law, no tissue response will occur when low-level laser light is applied with a low dose. If used with a high dose, it can inhibit tissue response and even induce the proliferation of cancer cells or microorganisms [34,35]. Therefore, there is an optimal dose where a maximal response is obtained [36]. Some studies suggest that open wound healing stimulation occurs in the 0.5–1 J/cm^2 light dose ranges and the 2–4 J/cm^2 range for superficial wound healing stimulation through the skin [37,38]. Alternatively, doses were proposed in the region of 4 J/cm^2 with a range of 1–10 J/cm^2 for superficial targets [38]. However, according to most previous studies, the optimal light doses are in the 1–5 J/cm^2 interval [39–41]. Byrnes et al. established that 632 nm light at an energy density of 4 J/cm^2 activates collagen formation [41]. Prabhu et al. studied the effect of a 632.8 nm laser with different energy doses, including 1 J/cm^2 and 2 J/cm^2, and achieved good results [42]. Another study demonstrates the utility of photobiomodulation (PBM) therapy (810 nm with a total energy of 3 J/cm^2) in mitigating burn injury and provides the biological rationale for its clinical application in wound healing [43]. PBM with various light parameters has been used widely in skincare but can cause certain types of unwanted cells to proliferate in the skin; this can lead to skin tumors, such as papillomas and cancers. H. Goo and his colleagues confirmed that LEDs with a wavelength of 642 nm with a total fluence of 21.6 J/cm^2, increased tumor size, epidermal thickness, and systemic proinflammatory cytokine levels [34]. In an infrared neural stimulation (INS) study, Throckmorton and colleagues evaluated the parameters of light such as wavelength, radiant exposure, and optical spot size using three commonly used wavelengths of INS,

1450 nm, 1875 nm, and 2120 nm. The pulsed diode lasers at 1450 nm and 1875 nm had a consistently higher ($\sim$1.0 J/cm^2) stimulation threshold than that of the Ho:YAG laser at 2120 nm ($\sim$0.7 J/cm^2). An acute histological evaluation of diode-irradiated nerves revealed a safe range of radiant exposures for stimulation [44].

Since two optical methods are mainly used for wound healing—PDT based on exogenous PSs and low-level light therapy (LLLT) without the use of exogenous PSs—it is crucial to find an optimal technology that combines the advantages of both methods—using low doses of radiation, correcting the effects that stimulate healing wounds, and using selective staining of wounds with exogenous photosensitizers with an optimal (relatively low) concentration. This approach is called low-dose PDT (LDPDT).

Traditionally, wounds are observed invasively with a histochemical assessment of biopsies. Similar methods are non-quantitative and may cause tissue damage and healing delay [45]. Recent studies have turned to optical imaging methods. Two-photon microscopy (TPM) is a modern molecular imaging method that enables noninvasive evaluation and monitoring of skin morphological structure and functions at the cellular and subcellular levels with a high spatial resolution [46–48]. TPM, including autofluorescence (AF) and second harmonic generation (SHG), can provide functional and structural imaging of biological tissue and assess cells' in vivo metabolic status [48]. Type I collagen induces SHG due to its noncentrosymmetric molecular structure. Therefore, TPM can be used for collagen disordering control, which occurs in wound healing [49–51]. The SHG-to AF aging index of the dermis (SAAID) is defined as the difference between SHG and AF intensity signals, indicative of type I collagen and elastin, respectively, and normalized to the sum of both signals as follows [52–56]:

$$SAAID = (SHG - AF)/(SHG + AF). \tag{1}$$

SAAID is a suitable quantitative parameter for characterizing a skin condition, which, as a specific parameter for monitoring wound healing, was proposed in [54], but has not been studied in detail.

This study aims to investigate the dynamics of SAAID in vivo using TPM during LDPDT wound healing by employing the topical administration of two different photosensitizers, 5-ALA and MB, and two laser fluences, 1 J/cm^2 and 4 J/cm^2.

2. Materials and Methods

2.1. An Animal Model of a Wound

In vivo experiments were performed using fifteen male CD1 mice aged 6–7 weeks and weighing 25–30 g according to experimental protocol No.4, 10.02.2021, registration No. 6, as approved by the Bioethical Committee of Tomsk State University. Animals were obtained from the Department of Experimental Biological Models of the Research Institute of Pharmacology, TSC SB RAMS. Before the experiment, the mice were kept for 7 days in the standard conditions of a conventional vivarium with free access to water and food, and a 12/12 light regime in a ventilated room at a temperature of 20 ± 2 °C and a humidity of 60%.

The mice were anesthetized by isoflurane using the Ugo Basile gas anesthesia system, where the mice were put in a glass chamber connected to isoflurane. The parameter on the isoflurane cylinder was set to 3–4%. It is recommended that the mouse be placed inside for 10–15 min. The wound area was depilated by Veet cream (France), rinsed with saline solution, and sterilized with chlorhexidine 20%. No additional drugs were used. The paws' skin was folded and raised using forceps. Then, the mouse was placed in a lateral position and using medical scissors the skin layers were completely removed and excisional circular wounds (diameter 5 mm) were created, as shown in Figure 1. This procedure was repeated on both the hind paws of each animal. The operations were carried out entirely under the influence of isoflurane gas attached via a mask to the mouse's nose, but here the isoflurane cylinder was set to 1–1.5%. The experiment was performed in a time-lapsed schedule for wound aging on days 1, 3, 7, and 14. The day of wound formation was denoted as day 0.

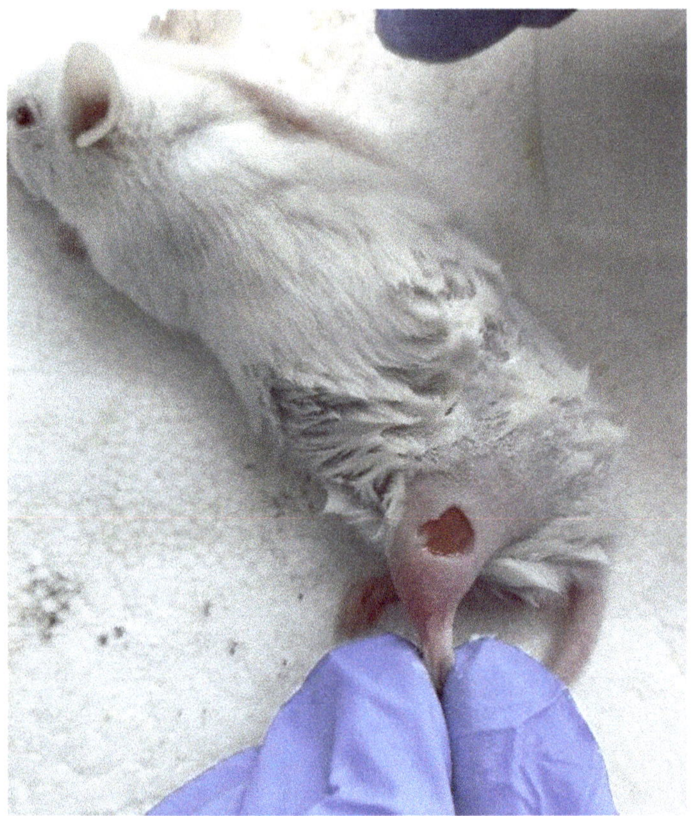

Figure 1. A full-thickness cutaneous wound.

2.2. Wound Healing Assay

The photos for illustrations were taken using a 16 MP camera with a 26 mm focal length, a 2x magnification lens, and an f/1.9 aperture at observational time points until the wounds healed; examples are presented in Figure 2. The wound size was calculated with a digital caliper by measuring its larger A and minor B diameters on every measurement day. The wound area was determined by the formula $S = (A \times B \times \pi)/4$. The digital caliper accuracy was 0.02 mm. Thus, this value makes an insignificant contribution to the measurement error when calculating the area and it was not considered in our analysis. The percentage of wound closure was calculated as follows:

$$[(S_0 - S)/S_0] \times 100, \tag{2}$$

where S_0 is the area of the original wound and S is the area of the current wound.

2.3. Low Dose Photodynamic Therapy Protocol

The photosensitizers were prepared by dissolving the powder 5-ALA and MB in saline solution. The 0.1–0.2 mL of 5-ALA 20%/MB 0.01% saline solutions were topically administered for 30 min of incubation, dripping directly on the wound when the mouse was placed on a base under isoflurane. The wounds were irradiated by an AlGaInP laser ($\lambda = 630$ nm, $P = 5$ mW) with two fluences: 1 J/cm^2 and 4 J/cm^2 for 3 min 45 s and 15 min, respectively. The animals were randomly divided into three basic groups; each group included 5 mice: the control group, LDPDT/1 J/cm^2 group, and LDPDT/4 J/cm^2 group.

In the LDPDT groups, 5-ALA was applied on the wounds on the right hind paws, MB was applied on the left ones. Thus, in total, there were 5 groups: the control, LDPDT-5-ALA/1 J/cm^2, LDPDT-MB/1 J/cm^2, LDPDT-5-ALA/4 J/cm^2, and LDPDT-MB/4 J/cm^2 groups. The LDPDT procedure was repeated once immediately after wound formation.

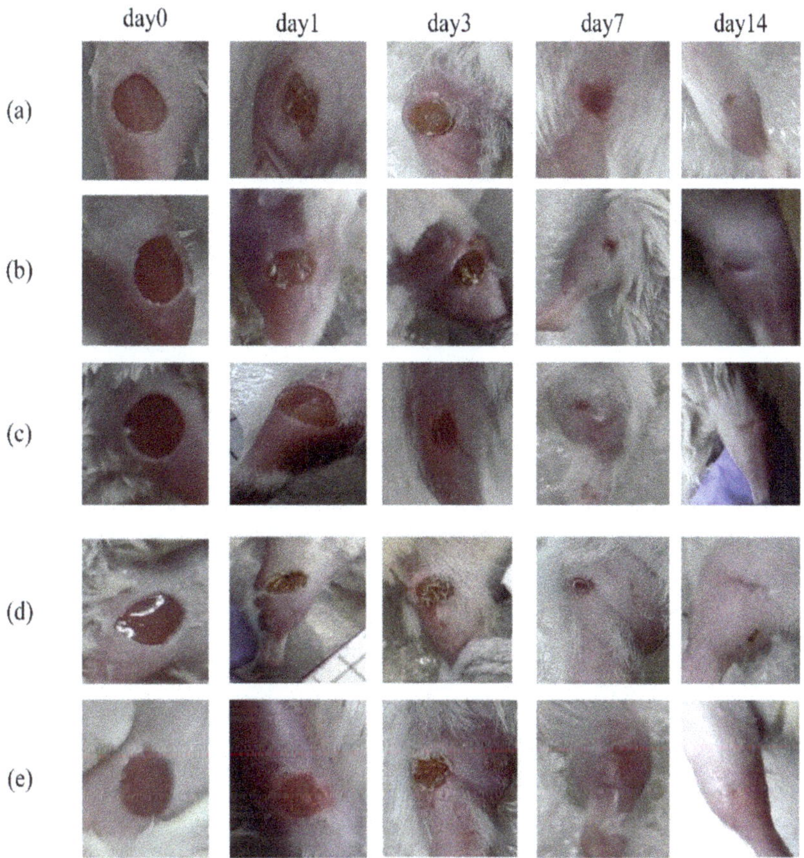

Figure 2. Digital photograph assessment of healing progression from day 0 to day 14: (**a**) control group; (**b**) LDPDT–5-ALA/1 J/cm^2; (**c**) LDPDT–5-ALA/4 J/cm^2; (**d**) LDPDT–MB/1 J/cm^2; (**e**) LDPDT–MB/4 J/cm^2.

2.4. Two-Photon Microscope

The wound was analyzed using a two-photon microscope, MPTflex (Jenlab GmbH, Jena, Germany). The pump laser wavelength was 760 nm and filters at bands 373–387 nm for a SHG signal and at 406–610 nm for an AF signal were used. The repetition frequency was 80 MHz with laser pulse width ~200 fs. The manufacturer's declared spatial resolution <0.5 μm (horizontal); <2 μm (vertical) with focusing optics: magnification 40× NA 1.3. The object was placed directly under the cover glass of a 100–170 μm thickness. A special metal ring was used as a cover glass holder. The space between the glass and the lens was filled with Carl Zeiss™ Immersol™ immersion oil to obtain a better signal. Skin structure was studied at 0–80 μm depth with a 4 μm step, while the pump laser power was varied from 5 mW at a depth of 0 μm (the beginning of the stratum corneum) to 40 mW at a depth of 80 μm. SHG and AF images were recorded on a 512 × 512-pixel matrix; the image size was 70 × 70 μm. The AF and SHG channels were electronically separated using appropriate spectral filters and recorded in digital matrices in "*.tiff" format in two independent channels. RGB color space was used for visualization, where the SHG and AF signals were shown in the red and green channels, respectively. Five stacks for different skin areas were scanned for each mouse during measurement. AF and SHG images were processed using SPCImage and Python software. This study did not perform data preprocessing as areas with acceptable magnification and image quality were selected. The SAAID index was chosen as the main characteristics of the AF and SHG signals. SAAID had been calculated depending on depth for all groups.

2.5. Statistical Analysis

All calculated parameters were expressed as mean $\pm$ standard deviation of the mean (SD). The Mann–Whitney U test was used to analyze the differences between two independent datasets. The level of significance was set at $p < 0.05$.

3. Results

3.1. Visual Observation

The digital photographs of the wounds at different time points were taken for the control and LDPDT groups on the surgery day (0) and successively on days 1, 3, 7, and 14 (Figure 2). The presented photos were only shown to illustrate the process of wound healing in the groups.

During healing, the wound sizes were evaluated according to Equation (2) for five mice from each group (Table 1). The values are presented as mean $\pm$ standard deviation. Figure 3 illustrates the wound healing rate for all groups.

Table 1. The wound size in mm^2 during wound healing (mean $\pm$ standard deviation).

	Day 0	Day 1	Day 3	Day 7	Day 14
control	20.82 $\pm$ 1.37	15.89 $\pm$ 1.06	11.34 $\pm$ 0.88	5.8 $\pm$ 0.72	1.98 $\pm$ 0.35
LDPDT–5-ALA 1 J/cm^2	20.02 $\pm$ 0.75	14.84 $\pm$ 0.89	9.61 $\pm$ 0.65	4.71 $\pm$ 0.42	0.92 $\pm$ 0.14
LDPDT–5-ALA 4 J/cm^2	20.41 $\pm$ 0.54	14.51 $\pm$ 0.92	8.79 $\pm$ 0.78	3.97 $\pm$ 0.57	0.7 $\pm$ 0.09
LDPDT–MB 1 J/cm^2	21.64 $\pm$ 0.48	15.53 $\pm$ 1.04	9.77 $\pm$ 0.43	5.52 $\pm$ 0.22	1.05 $\pm$ 0.16
LDPDT–MB 4 J/cm^2	20.02 $\pm$ 1.09	14.18 $\pm$ 1.11	8.29 $\pm$ 0.64	3.62 $\pm$ 0.13	0.25 $\pm$ 0.07

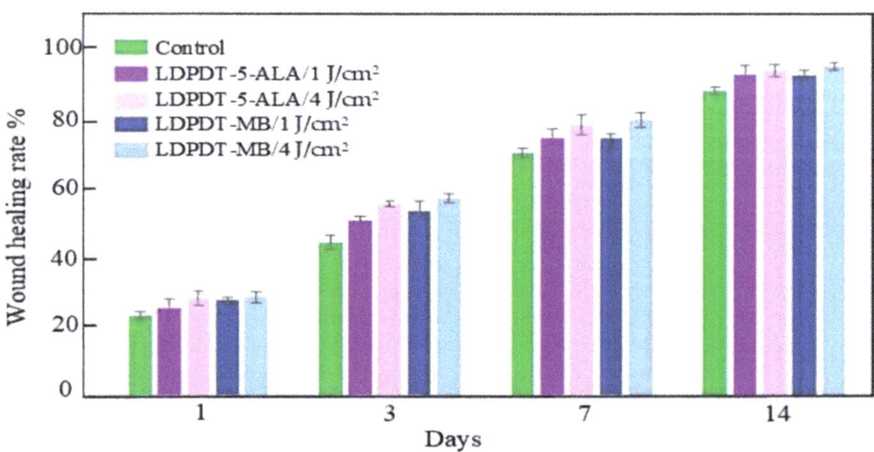

Figure 3. Wound healing rate.

3.2. In Vivo TPM Imaging

In vivo TPM imaging was used to assess differences in wound healing. Figure 4 shows the SHG and AF images of all groups at a depth of 60 μm. Type I collagen fibers indicated by the SHG signal are shown in red and the elastin fibers indicated by AF signals are in green. The white scale bar shows a length of 10 μm.

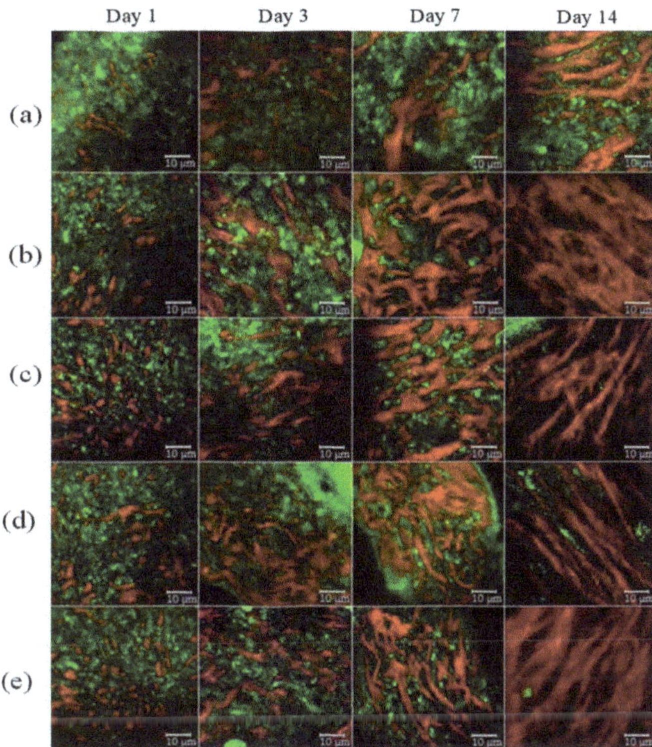

Figure 4. SHG (red) and AF (green) images of wound healing. (**a**) Control group; (**b**) LDPDT–5-ALA/1 J/cm^2; (**c**) LDPDT–5-ALA/4 J/cm^2; (**d**) LDPDT–MB/1 J/cm^2; (**e**) LDPDT–MB/4 J/cm^2.

Different forms of collagen structure could be observed between days 1 and 14: on day 1, the collagen was disorganized and dispersed in a blurry form. Over the following days, organized collagen gradually started to form, which was confirmed by the increase in SHG signal intensity. On day 14, the collagen fibers were rearranged, cross-linked, and aligned (Figure 4). On the seventh day, the SHG signal intensity in LDPDT groups was higher so the collagen formation was better in LDPDT groups. Figure 5 shows the difference between the organized collagen in healthy tissue and the disorganized collagen in a wound at depths from 4 to 80 μm.

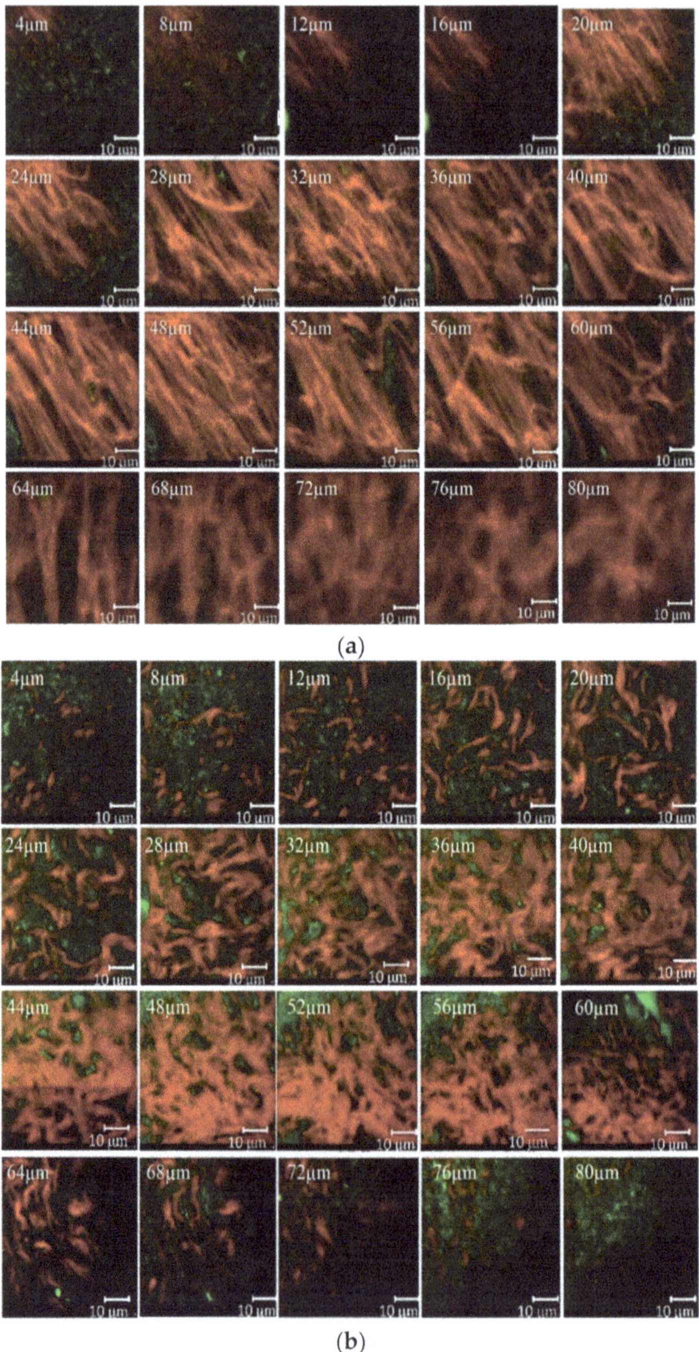

Figure 5. (**a**) SHG (red) and AF (green) of healthy skin at depths from 4 to 80μm; (**b**) SHG (red) and AF (green) of a wound at depths from 4 to 80 μm.

3.3. The SAAID Estimation

For quantification of relative amounts of elastin and collagen, the SAAID was estimated for various depths. The corresponding SAAID curve for healthy skin (comparison area) is shown in Figure 6. The SAAID index for the control group is shown in Figure 7a. The SAAID curve started at slightly negative values and was still negative on day 1. On day 3, the curve did not differ much as the values remained negative. On day 7, the SAAID had a minimum value of −0.4 (a depth of 40 μm). On day 14, the curve shape changed compared to day 1 and day 3, the value of SAAID at 70 μm increased to zero, and then the SAAID reached its maximum of 0.2 at a depth of 80 μm. Figure 7b shows the SAAID index for the LDPDT 5-ALA 4 J/cm^2 group. It was different from the control group; in general the values were larger and on days 7 and 14 they were approximately close to healthy skin values, contrary to the control group. On day 7, the SAAID index started from zero and reached a minimum value of −0.4 (at a depth of 35 μm), rising to zero at 75 μm and a maximum of +0.2 at a depth of 80 μm. With a slight change and larger values, the curve for day 14 is shown. The SAAID reached its minimum at a value of −0.35 (a depth of 25 μm), after which the SAAID curve reached a maximum of +0.3 at a depth of 80 μm. Therefore, the SAAID was found to be substantially affected by the day of the wound healing process. The SAAID of healthy skin vs. control group and LDPDT 5-ALA (1 J/cm^2 and 4 J/cm^2) groups on day 14 are shown in Figure 8a. The difference was evident, especially in the range from 40 μm to 80 μm, with higher values for LDPDT 5-ALA/4 J/cm^2. Additionally, Figure 8b shows the SAAID for healthy skin, the control group, and LDPDT–MB (1 J/cm^2 and 4 J/cm^2) groups on day 14.

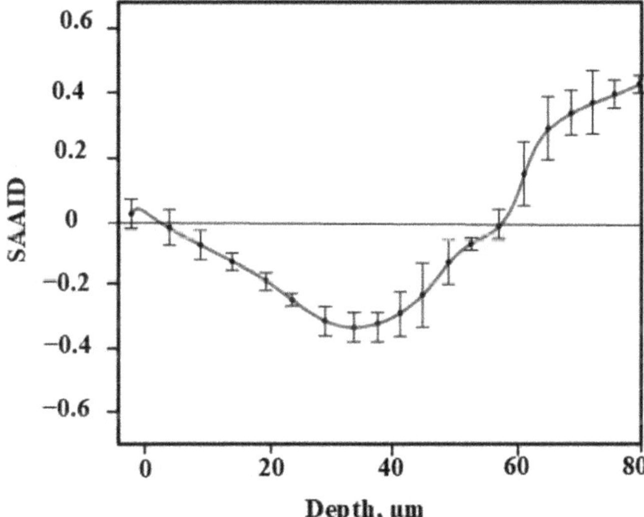

Figure 6. SAAID for healthy skin.

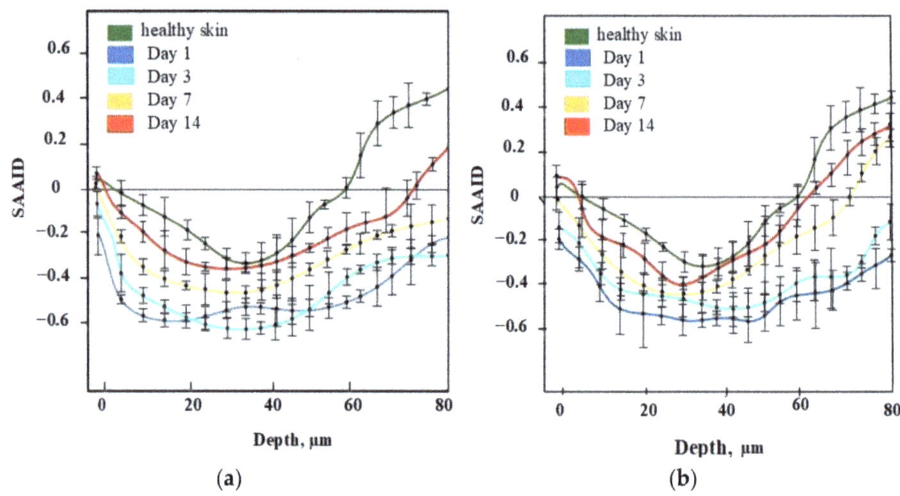

Figure 7. (a) SAAID for control group; (b) SAAID for LDPDT–5-ALA/4 J/cm^2 group.

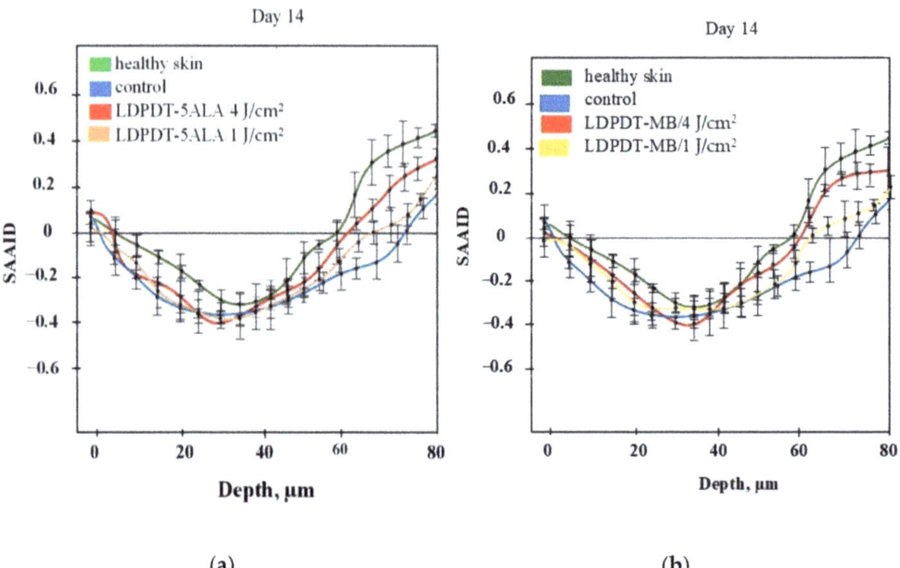

Figure 8. (a) SAAID for LDPDT–5-ALA (4 J/cm^2 and 1 J/cm^2) groups; (b) SAAID for LDPDT–MB (4 J/cm^2 and 1 J/cm^2) groups.

A comparison of LDPDT 5-ALA and MB/4 J/cm^2 vs. healthy skin and the control group on day 14 is presented in Figure 9. When MB was employed, the values were closer to those of healthy skin. The specificity of the SAAID for healthy skin, the control group of wound healing, and the LDPDT groups with 5-ALA and MB (1 J/cm^2 and 4 J/cm^2) was analyzed. The SAAID differences for the studied groups at various depths are shown in Figure 10 and these differences were more evident at depths of 60 μm and 80 μm.

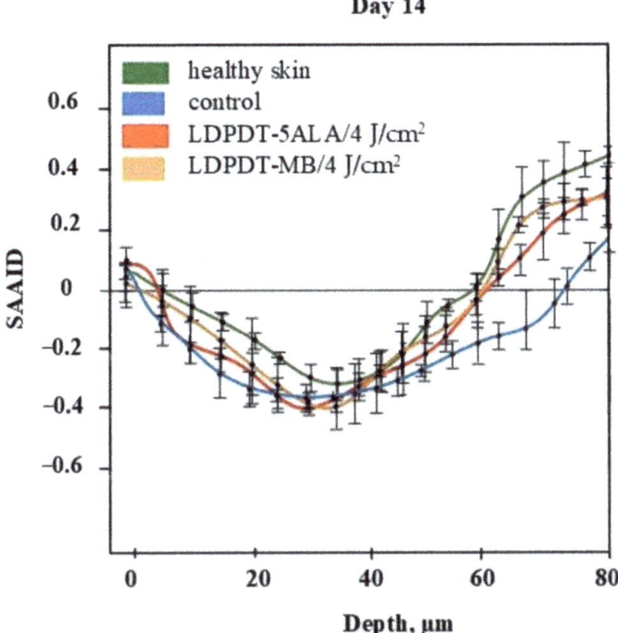

Figure 9. SAAID for LDPDT–5-ALA and LDPDT–MB/4 J/cm² groups.

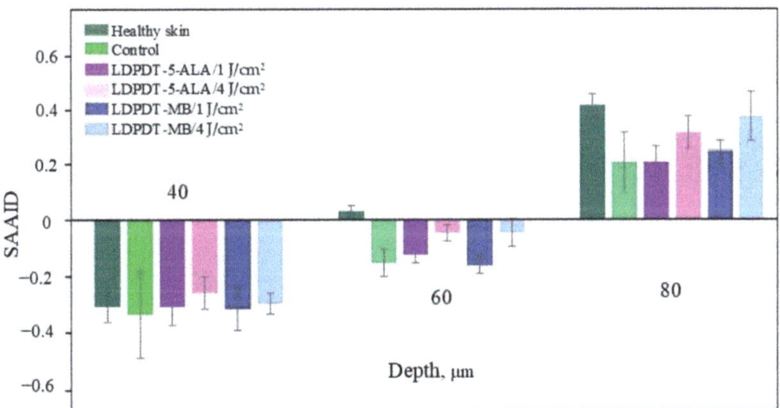

Figure 10. Index SAAID on day 14 at 40 μm, 60 μm and 80 μm for healthy skin, control, LDPDT–5-ALA/1 J/cm², LDPDT–5-ALA/4 J/cm², LDPDT–MB/1 J/cm², and LDPDT–MB/4 J/cm².

The SAAID index at depths from 0 to 40 μm on day 14 described a healthy and regenerated skin epidermis layer. The index values for healthy skin, the control area, and skin after LDPDT exposure did not differ significantly, as demonstrated in Figures 7–10. It indicates that the index did not show significant differences in the epidermis after recovery or that these changes were minor. There were differences in SAAID values deeper than 60 μm (papillary dermis layer) between the control and LDPDT groups. These differences were most pronounced at a depth of 80 μm. It indicates a significant recovery of the papillary layer during LDPDT so it could be concluded that on day 14 after exposure to LDPDT/4 J/cm², both 5-ALA and MB showed accelerated skin regeneration.

3.4. Mann–Whitney U Test

The Mann–Whitney test was applied to assess differences in between the wound skin group and the healthy skin group. SAAID variation p-values for the five groups, depths from 40 to 80 µm, in all days are shown in Tables 2–4. Statistical power was set at 0.95. The higher the p-value, the more minor the difference between a wound skin group and the healthy skin group. Thus, Figure 11 shows the p-value on day 14 by depth. This dependence was calculated with a smaller step in depths than the data presented in Figure 10. Indeed, we see that SAAID measured in the 47–50 µm depth interval is the most sensitive to the wound state.

Table 2. The p-value for the SAAID of the control group relative to healthy skin.

$n = 5$	Control			
Depth, µm	Day 1	Day 3	Day 7	Day 14
40	0.0041 *	0.0041 *	0.0061 *	0.072
44	0.0061 *	0.0061 *	0.0061 *	0.072
48	0.0061 *	0.0061 *	0.0061 *	0.030 *
52	0.0061 *	0.0061 *	0.0059 *	0.0061 *
56	0.0061 *	0.0061 *	0.0061 *	0.0061 *
60	0.0041 *	0.0041 *	0.0041 *	0.030 *
80	0.0061 *	0.0061 *	0.0061 *	0.030 *

* $p < 0.05$.

Table 3. The p-value for the SAAID in LDPDT–5-ALA/4 J/cm^2 and MB/4 J/cm^2 relative to healthy skin.

$n = 5$	LDPDT–5-ALA/4 J/cm^2				LDPDT–MB/4 J/cm^2			
Depth, µm	Day 1	Day 3	Day 7	Day 14	Day 1	Day 3	Day 7	Day 14
40	0.0061 *	0.0061 *	0.0301 *	0.072	0.0061 *	0.0061 *	0.047 *	0.072
44	0.0061 *	0.0061 *	0.0108 *	0.072	0.0061 *	0.0061 *	0.072	0.148
48	0.0061 *	0.0061 *	0.0718	0.105	0.0061 *	0.0061 *	0.202	0.238
52	0.0061 *	0.0061 *	0.105	0.105	0.0061 *	0.0061 *	0.0718	0.165
56	0.0061 *	0.0061 *	0.0108 *	0.072	0.0061 *	0.0061 *	0.0061 *	0.105
60	0.0041 *	0.0041 *	0.0025 *	0.064	0.004 *	0.004 *	0.004 *	0.085
80	0.0059 *	0.0061 *	0.0301 *	0.105	0.0061 *	0.0061 *	0.0718	0.148

* $p < 0.05$.

Table 4. The p-value for the SAAID in LDPDT 5ALA/1 J/cm^2 and MB/1 J/cm^2 relative to healthy skin.

$n = 5$	LDPDT–5-ALA/1 J/cm^2				LDPDT–MB/1 J/cm^2			
Depth, µm	Day 1	Day 3	Day 7	Day 14	Day 1	Day 3	Day 7	Day 14
40	0.0061 *	0.0061 *	0.0108 *	0.072	0.0061 *	0.0061 *	0.0301 *	0.072
44	0.0061 *	0.0061 *	0.0108 *	0.072	0.0061 *	0.0061 *	0.072	0.072
48	0.0061 *	0.0061 *	0.072	0.105	0.0061 *	0.0061 *	0.047 *	0.0301 *
52	0.0061 *	0.0061 *	0.105	0.105	0.0061 *	0.0061 *	0.047 *	0.006 *
56	0.0061 *	0.0061 *	0.0061 *	0.047 *	0.0061 *	0.0061 *	0.0061 *	0.072
60	0.0041 *	0.0041 *	0.0041 *	0.0025 *	0.0041 *	0.0041 *	0.017 *	0.017 *
80	0.0061 *	0.0061 *	0.047 *	0.047 *	0.0061 *	0.0061 *	0.047 *	0.047 *

* $p < 0.05$.

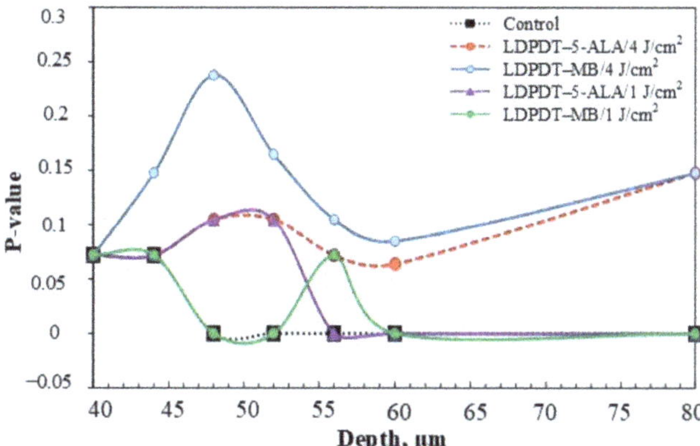

Figure 11. *p*-values of SAAID for all groups at depths from 40 to 60 μm and 80 μm on day 14.

It is worth noting that the *p*-value for LDPDT/4 J/cm^2 was much higher than for the control group, indicating better healing.

4. Discussion

The most common collagens are fibrils (collagens of types I, II, III, V, and XI) and reticular structures (collagens IV, VIII, and X) in the intercellular matrix. Since the type of collagen is directly related to the amino acid sequence underlying its structure, the optical properties of collagens vary depending on its type. As mentioned in the Introduction, collagen I, typical for skin, can be visualized through SHG-microscopy. According to our data, during the wound healing process, especially in the first three days, the collagen was disorganized and disordered. However, after a time, a newly laid collagen matrix gradually formed to fill the wound gap, which was detected by the increase in SHG signal intensity. The disorganized collagen is rearranged, cross-linked, and aligned, which can be observed on day 14 (see Figure 4).

From studying and comparing the five groups (control, LDPDT–5-ALA/4 J/cm^2, LDPDT–5-ALA/1 J/cm^2, LDPDT–MB/4 J/cm^2, and LDPDT–MB/1 J/cm^2), it has been demonstrated that the 4 J/cm^2 laser dose is better in comparison with 1 J/cm^2. For the 4 J/cm^2 dose, MB enables better healing compared to 5-ALA (see Figure 3). The benefits of MB are its low cost and wide availability. In any case, a fourfold fluence increase did not cause a substantial increase in efficiency, which suggests that at the used concentrations of photosensitizers, it is quite acceptable to choose a fluence between 1 J/cm^2 and 4 J/cm^2.

SAAID values have previously been demonstrated for the papillary dermis during skin aging [53]. In [51], SAAID was applied to estimate skin damage by curettage. Every phase of wound healing was studied, as in our research, and similar values were obtained for different depths. Later, the same group of researchers [54] showed similar results for chronic wounds, consistent with our research. The results for estimating wound healing during the photodynamic therapy process using SAAID have not previously been studied in the literature. In this study, variations of the SAAID value during the wound healing process were established. On day 14, in the LDPDT 4 J/cm^2 group, the SAAID curve was practically the same as for healthy skin (see Figures 7 and 8).

SAAID variance during wound healing demonstrates that the healing efficiency does not increase significantly with a fourfold fluence increase. Moreover, the healing process with low fluence is more uniform across the dermis (see Figure 8).

The *p*-value of the difference of SAAID measured in wounds relative to the healthy skin group was calculated using the Mann–Whitney test. On day 14, the *p*-values were higher, so the differences are smaller relative to healthy skin (see Figure 11). This appears significantly and more clearly in both LDPDT groups. Consequently, SAAID is a suitable quantitative parameter for monitoring the wound healing process.

Interestingly, the most essential SAAID variance between healthy and wounded skin was in the 47–55 µm depth interval (Figure 11). It is an area of the papillary dermis with a large amount of collagen and elastin, but there are practically no capillary blood vessels yet. Recent studies showed that liquid or small molecules preferentially colocalize with elastin fibers [57], affecting its fluorescence in an area of inflammation. Therefore, SAAID change during wound healing can be associated with both collagen and elastin transformation.

5. Conclusions

Two-photon microscopy is becoming a conventional tool for noninvasive medical diagnostics of superficial tissues and dynamic monitoring of skin pathology treatment. The latter requires the discovery of specific quantitative criteria of the tissue state. The SAAID index was shown to be a suitable variant of the quantitative criterium for wound healing supervision, including optimal regimes of wound healing low-dose photodynamic therapy.

The development of this approach can be as follows. First of all, fluorescence lifetime imaging combined with SGH and AF channels will give additional information about cell metabolism in a wound area [58]. Other optical modalities, for example, optical coherence tomography, will allow data to be obtained about wound tissue morphology at a lower spatial resolution (the typical value is about 5–7 µm) but with a much larger dimension of tissue area visualization compared to TPM.

The phototherapy of wounds is closely related to the problem of infectious disease treatment, especially in the presence of antibiotic-resistant bacteria in the bacterial biofilms covering the wound surface [35,59]. Since the development of antibacterial drugs is not keeping pace with microbial resistance, it is necessary to use alternative antibacterial approaches, including antimicrobial PDT. In this sense, the effectiveness of wound healing is determined mainly by the combined effect of light on the bacterial flora hidden in the depth of the wound and the restoration of the main protein components of the skin—collagen and elastin. Both processes require efficient staining of pathogens and dermal cells and a homogeneous distribution of excitation light for the effective production of ROS throughout the entire thickness of the wound.

In this case, it is necessary to overcome the limited depth of light penetration into the tissue in order to enhance the therapeutic effect. It can be done by partially suppressing tissue light scattering when exposed to immersion agents using the optical clearing method [60]. This enables not only an increase in the efficiency of therapy (a decrease in fluence) but also makes it possible to improve optical monitoring of the healing process, providing a greater depth of probing [61], which is vital for deep wounds. It is also important that many optical clearing agents, such as glycerol, are good solvents for photodynamic dyes, ensuring the stability of their absorption spectra [15] and possessing bactericidal properties, which can give synergy in the light treatment of wounds. Deeper photodynamic wound therapy can be achieved by using indocyanine green (ICG) as a PDT dye that is effective in the near-infrared range [16].

Author Contributions: Conceptualization, Y.V.K. and V.V.T.; methodology, H.Z. and N.A.K.; software, V.V.N.; validation, H.Z. and V.V.N.; investigation, H.Z., V.V.N., A.I.K., T.B.L. and N.A.K.; writing—original draft preparation, H.Z. and A.I.K.; writing—review and editing, Y.V.K. and V.V.T.; visualization, H.Z.; supervision, Y.V.K. All authors have read and agreed to the published version of the manuscript.

Funding: This research was funded by a grant under the Decree of the Government of the Russian Federation No. 220 of 9 April 2010 (Agreement No. 075-15-2021-615 of 4 June 2021).

Institutional Review Board Statement: The study was approved by the Ethics Committee of Tomsk State University (Protocol No. 4, 10.02.2021), registration No. 6.

Data Availability Statement: The data presented in this study are available on request from the corresponding author. The data are not publicly available due to privacy or ethical restrictions.

Acknowledgments: The authors thank Peter Mitchell, full member of the Institute of Linguists of Great Britain, for his professional editing.

Conflicts of Interest: The authors declare no conflict of interest.

References

1. Deka, G.; Wu, W.-W.; Kao, F.-J. In Vivo Wound Healing Diagnosis with Second Harmonic and Fluorescence Lifetime Imaging. *J. Biomed. Opt.* **2012**, *18*, 061222. [CrossRef] [PubMed]
2. Zuhayri, H.; Knyazkova, A.I.; Nikolaev, V.V.; Borisov, A.V.; Kistenev, Y.V.; Zakharova, O.A.; Dyachenko, P.A.; Tuchin, V.V. Study of Wound Healing by Terahertz Spectroscopy. In Proceedings of the Fourth International Conference on Terahertz and Microwave Radiation: Generation, Detection, and Applications, Tomsk, Russia, 17 November 2020; Romanovskii, O.A., Kistenev, Y.V., Eds.; SPIE: Tomsk, Russia, 2020; p. 63.
3. Beldon, P. Basic Science of Wound Healing. *Surg. Oxf.* **2010**, *28*, 409–412. [CrossRef]
4. Stroncek, J.D.; Bell, N.; Reichert, W.M. Instructional PowerPoint Presentations for Cutaneous Wound Healing and Tissue Response to Sutures. *J. Biomed. Mater. Res. A* **2009**, *90A*, 1230–1238. [CrossRef]
5. Garcia, J.G.M. The Role of Photodynamic Therapy in Wound Healing and Scarring in Human Skin. Ph.D. Thesis, Univ. Manch, Manchester, England, 2015.
6. Guo, S.; Di Pietro, L.A. Factors Affecting Wound Healing. *J. Dent. Res.* **2010**, *89*, 219–229. [CrossRef] [PubMed]
7. Gonzalez, A.C.d.O.; Costa, T.F.; Andrade, Z.A.; Medrado, A.R.A.P. Wound Healing-A Literature Review. *An. Bras. Dermatol.* **2016**, *91*, 614–620. [CrossRef] [PubMed]
8. Sorg, H.; Tilkorn, D.J.; Hager, S.; Hauser, J.; Mirastschijski, U. Skin Wound Healing: An Update on the Current Knowledge and Concepts. *Eur. Surg. Res.* **2017**, *58*, 81–94. [CrossRef] [PubMed]
9. Liu, H.; Liu, H.; Deng, X.; Chen, M.; Han, X.; Yan, W.; Wang, N. Raman Spectroscopy Combined with SHG Gives a New Perspective for Rapid Assessment of the Collagen Status in the Healing of Cutaneous Wounds. *Anal. Methods* **2016**, *8*, 3503–3510. [CrossRef]
10. Mickelson, M.A.; Mans, C.; Colopy, S.A. Principles of Wound Management and Wound Healing in Exotic Pets. *Vet. Clin. North Am. Exot. Anim. Pract.* **2016**, *19*, 33–53. [CrossRef]
11. Oyama, J.; Fernandes Herculano Ramos-Milaré, Á.C.; Lopes Lera-Nonose, D.S.S.; Nesi-Reis, V.; Galhardo Demarchi, I.; Alessi Aristides, S.M.; Juarez Vieira Teixeira, J.; Gomes Verzignassi Silveira, T.; Campana Lonardoni, M.V. Photodynamic Therapy in Wound Healing in Vivo: A Systematic Review. *Photodiagn. Photodyn. Ther.* **2020**, *30*, 101682. [CrossRef]
12. Tedesco, A.; Jesus, P. Low Level Energy Photodynamic Therapy for Skin Processes and Regeneration. In *PhotomediCine-Advances in Clinical Practice*; Tanaka, Y., Ed.; InTech: Sao Paulo, Brazil, 2017; ISBN 978-953-51-3155-7.
13. Shaw, T.J.; Martin, P. Wound Repair at a Glance. *J. Cell Sci.* **2009**, *122*, 3209–3213. [CrossRef]
14. Kwiatkowski, S.; Knap, B.; Przystupski, D.; Saczko, J.; Kędzierska, E.; Knap-Czop, K.; Kotlińska, J.; Michel, O.; Kotowski, K.; Kulbacka, J. Photodynamic Therapy–Mechanisms, Photosensitizers and Combinations. *Biomed. Pharm.* **2018**, *106*, 1098–1107. [CrossRef] [PubMed]
15. Tuchin, V.V.; Genina, E.A.; Bashkatov, A.N.; Simonenko, G.V.; Odoevskaya, O.D.; Altshuler, G.B. A Pilot Study of ICG Laser Therapy of Acne Vulgaris: Photodynamic and Photothermolysis Treatment. *Lasers Surg. Med.* **2003**, *33*, 296–310. [CrossRef] [PubMed]
16. Genina, E.A.; Bashkatov, A.N.; Simonenko, G.V.; Odoevskaya, O.D.; Tuchin, V.V.; Altshuler, G.B. Low-Intensity Indocyanine-Green Laser Phototherapy of Acne Vulgaris: Pilot Study. *J. Biomed. Opt.* **2004**, *9*, 828. [CrossRef] [PubMed]
17. Kim, M.M.; Darafsheh, A. Light Sources and Dosimetry Techniques for Photodynamic Therapy. *Photochem. Photobiol.* **2020**, *96*, 280–294. [CrossRef] [PubMed]
18. Yang, T.; Tan, Y.; Zhang, W.; Yang, W.; Luo, J.; Chen, L.; Liu, H.; Yang, G.; Lei, X. Effects of ALA-PDT on the Healing of Mouse Skin Wounds Infected with Pseudomonas Aeruginosa and Its Related Mechanisms. *Front. Cell Dev. Biol.* **2020**, *8*, 585132. [CrossRef]
19. Algorri, J.F.; Ochoa, M.; Roldán-Varona, P.; Rodríguez-Cobo, L.; López-Higuera, J.M. Photodynamic Therapy: A Compendium of Latest Reviews. *Cancers* **2021**, *13*, 4447. [CrossRef]
20. Peplow, P.V.; Chung, T.-Y.; Baxter, G.D. Photodynamic Modulation of Wound Healing: A Review of Human and Animal Studies. *Photomed. Laser Surg.* **2012**, *30*, 118–148. [CrossRef]
21. Huang, J.; Guo, M.; Wu, M.; Shen, S.; Shi, L.; Cao, Z.; Wang, X.; Wang, H. Effectiveness of a Single Treatment of Photodynamic Therapy Using Topical Administration of 5-Aminolevulinic Acid on Methicillin-Resistant Staphylococcus Aureus-Infected Wounds of Diabetic Mice. *Photodiagn. Photodyn. Ther.* **2020**, *30*, 101748. [CrossRef]
22. Castano, A.P.; Demidova, T.N.; Hamblin, M.R. Mechanisms in Photodynamic Therapy: Part One—Photosensitizers, Photochemistry and Cellular Localization. *Photodiagn. Photodyn. Ther.* **2004**, *1*, 279–293. [CrossRef]
23. Nesi-Reis, V.; Lera-Nonose, D.S.S.L.; Oyama, J.; Silva-Lalucci, M.P.P.; Demarchi, I.G.; Aristides, S.M.A.; Teixeira, J.J.V.; Silveira, T.G.V.; Lonardoni, M.V.C. Contribution of Photodynamic Therapy in Wound Healing: A Systematic Review. *Photodiagn. Photodyn. Ther.* **2018**, *21*, 294–305. [CrossRef]
24. Kou, J.; Dou, D.; Yang, L. Porphyrin Photosensitizers in Photodynamic Therapy and Its Applications. *Oncotarget* **2017**, *8*, 81591–81603. [CrossRef] [PubMed]

25. Allison, R.R.; Downie, G.H.; Cuenca, R.; Hu, X.-H.; Childs, C.J.; Sibata, C.H. Photosensitizers in Clinical PDT. *Photodiagn. Photodyn. Ther.* **2004**, *1*, 27–42. [CrossRef]

26. Morimoto, K.; Ozawa, T.; Awazu, K.; Ito, N.; Honda, N.; Matsumoto, S.; Tsuruta, D. Photodynamic Therapy Using Systemic Administration of 5-Aminolevulinic Acid and a 410-Nm Wavelength Light-Emitting Diode for Methicillin-Resistant Staphylococcus Aureus-Infected Ulcers in Mice. *PLoS ONE* **2014**, *9*, e105173. [CrossRef] [PubMed]

27. Yin, R.; Lin, L.; Xiao, Y.; Hao, F.; Hamblin, M.R. Combination ALA-PDT and Ablative Fractional Er:YAG Laser (2940 Nm) on the Treatment of Severe Acne: COMBINATION ALA-PDT AND ABLATIVE FRACTIONAL Er:YAG LASER. *Lasers Surg. Med.* **2014**, *46*, 165–172. [CrossRef]

28. Kayabaşı, Y.; Erbaş, O. Methylene Blue and Its Importance in Medicine. *J. Med. Sci.* **2020**, *6*, 136–145. [CrossRef]

29. Kozlikina, E.I.; Pominova, D.V.; Ryabova, A.V.; Efendiev, K.T.; Skobeltsin, A.S.; Rudenko, N.S.; Kulik, O.G.; Muhametzyanova, E.I.; Karal-ogly, D.D.; Zhemerikin, G.A.; et al. Spectroscopic Measurement of Methylene Blue Distribution in Organs and Tissues of Hamadryas Baboons during Oral Administration. *Photonics* **2021**, *8*, 294. [CrossRef]

30. Evangelou, G.; Krasagakis, K.; Giannikaki, E.; Kruger-Krasagakis, S.; Tosca, A. Successful Treatment of Cutaneous Leishmaniasis with Intralesional Aminolevulinic Acid Photodynamic Therapy: Successful Treatment of CL with ALA-PDT. *Photodermatol. Photoimmunol. Photomed.* **2011**, *27*, 254–256. [CrossRef]

31. Jiang, C.; Yang, W.; Wang, C.; Qin, W.; Ming, J.; Zhang, M.; Qian, H.; Jiao, T. Methylene Blue-Mediated Photodynamic Therapy Induces Macrophage Apoptosis via ROS and Reduces Bone Resorption in Periodontitis. *Oxid. Med. Cell. Longev.* **2019**, *2019*, 1–15. [CrossRef]

32. Carneiro, V.S.M.; Catao, M.H.C.d.V.; Menezes, R.F.; Araújo, N.C.; Gerbi, M.E.M. *Methylene Blue Photodynamic Therapy in Rats' Wound Healing: 21 Days Follow-Up*; Kurachi, C., Svanberg, K., Tromberg, B.J., Bagnato, V.S., Eds.; SPIE: Rio de Janeiro, Brazil, 2015; p. 95311S.

33. Sperandio, F.F.; Simões, A.; Aranha, A.C.C.; Corrêa, L.; Orsini Machado de Sousa, S.C. Photodynamic Therapy Mediated by Methylene Blue Dye in Wound Healing. *Photomed. Laser Surg.* **2010**, *28*, 581–587. [CrossRef]

34. Goo, H.; Mo, S.; Park, H.J.; Lee, M.Y.; Ahn, J.-C. Treatment with LEDs at a Wavelength of 642 Nm Enhances Skin Tumor Proliferation in a Mouse Model. *Biomed. Opt. Express* **2021**, *12*, 5583. [CrossRef]

35. Tuchin, V.V.; Genina, E.A.; Tuchina, E.S.; Svetlakova, A.V.; Svenskaya, Y.I. Optical Clearing of Tissues: Issues of Antimicrobial Phototherapy and Drug Delivery. *Adv. Drug Deliv. Rev.* **2022**, *180*, 114037. [CrossRef] [PubMed]

36. Keene, S.A. The Science of Light Biostimulation and Low Level Laser Therapy (LLLT). *Hair Transpl. Forum Int.* **2014**, *24*, 201–209. [CrossRef]

37. Lau, P.; Bidin, N.; Krishnan, G.; AnaybBaleg, S.M.; Sum, M.B.M.; Bakhtiar, H.; Nassir, Z.; Hamid, A. Photobiostimulation Effect on Diabetic Wound at Different Power Density of near Infrared Laser. *J. Photochem. Photobiol. B* **2015**, *151*, 201–207. [CrossRef] [PubMed]

38. Zein, R.; Selting, W.; Hamblin, M.R. Review of Light Parameters and Photobiomodulation Efficacy: Dive into Complexity. *J. Biomed. Opt.* **2018**, *23*, 1. [CrossRef] [PubMed]

39. Kilík, R.; Lakyová, L.; Sabo, J.; Kruzliak, P.; Lacjaková, K.; Vasilenko, T.; Vidová, M.; Longauer, F.; Radoňak, J. Effect of Equal Daily Doses Achieved by Different Power Densities of Low-Level Laser Therapy at 635 Nm on Open Skin Wound Healing in Normal and Diabetic Rats. *BioMed Res. Int.* **2014**, *2014*, 1–9. [CrossRef] [PubMed]

40. Hawkins, D.H.; Abrahamse, H. The Role of Laser Fluence in Cell Viability, Proliferation, and Membrane Integrity of Wounded Human Skin Fibroblasts Following Helium-Neon Laser Irradiation. *Lasers Surg. Med.* **2006**, *38*, 74–83. [CrossRef]

41. Byrnes, K.R.; Barna, L.; Chenault, V.M.; Waynant, R.W.; Ilev, I.K.; Longo, L.; Miracco, C.; Johnson, B.; Anders, J.J. Photobiomodulation Improves Cutaneous Wound Healing in an Animal Model of Type II Diabetes. *Photomed. Laser Surg.* **2004**, *24*, 281–290. [CrossRef]

42. Prabhu, V.; Rao, S.B.S.; Chandra, S.; Kumar, P.; Rao, L.; Guddattu, V.; Satyamoorthy, K.; Mahato, K.K. Spectroscopic and Histological Evaluation of Wound Healing Progression Following Low Level Laser Therapy (LLLT). *J. Biophotonics* **2012**, *5*, 168–184. [CrossRef]

43. Khan, I.; Rahman, S.U.; Tang, E.; Engel, K.; Hall, B.; Kulkarni, A.B.; Arany, P.R. Accelerated Burn Wound Healing with Photobiomodulation Therapy Involves Activation of Endogenous Latent TGF-B1. *Sci. Rep.* **2021**, *11*, 13371. [CrossRef]

44. Throckmorton, G.; Cayce, J.; Ricks, Z.; Adams, W.R.; Jansen, E.D.; Mahadevan-Jansen, A. Identifying Optimal Parameters for Infrared Neural Stimulation in the Peripheral Nervous System. *Neurophotonics* **2021**, *8*, 015012. [CrossRef]

45. Deka, G.; Okano, K.; Wu, W.-W.; Kao, F.-J. *Multiphoton Microscopy for Skin Wound Healing Study in Terms of Cellular Metabolism and Collagen Regeneration*; Periasamy, A., So, P.T.C., König, K., Eds.; SPIE: San Francisco, CA, USA, 2014; p. 894820.

46. Nikolaev, V.V.; Kurochkina, O.S.; Vrazhnov, D.A.; Sandykova, E.A.; Kistenev, Y.V. Research on Lymphedema by Method of High-Resolution Multiphoton Microscopy. *J. Phys. Conf. Ser.* **2019**, *1145*, 012043. [CrossRef]

47. Kistenev, Y.V.; Nikolaev, V.; Drozdova, A.; Ilyasova, E.; Sandykova, E. Improvement of the Multiphoton Fluorescence Microscopy Images Quality Using Digital Filtration. In Proceedings of the International Conference on Atomic and Molecular Pulsed Lasers XIII, Tomsk, Russia, 16 April 2018; Kabanov, A.M., Tarasenko, V.F., Eds.; SPIE: Tomsk, Russia, 2018; p. 98.

48. Kistenev, Y.V.; Borisov, A.V.; Knyazkova, A.I.; Nikolaev, V.V.; Samarinova, A.; Navolokin, N.A.; Tuchina, D.K.; Tuchin, V.V. Differential Diagnostics of Paraffin-Embedded Tissues by IR-THz Spectroscopy and Machine Learning. In Proceedings of the Tissue Optics and Photonics, Strasbourg, France, 2 April 2020; Zalevsky, Z., Tuchin, V.V., Blondel, W.C., Eds.; SPIE: Online Only, 2020; p. 20.
49. Zhuo, G.-Y.; KU, S.; KM, S.; Kistenev, Y.V.; Kao, F.-J.; Nikolaev, V.V.; Zuhayri, H.; Krivova, N.A.; Mazumder, N. Label-Free Multimodal Nonlinear Optical Microscopy for Biomedical Applications. *J. Appl. Phys.* **2021**, *129*, 214901. [CrossRef]
50. Kistenev, Y.V.; Vrazhnov, D.A.; Nikolaev, V.V.; Sandykova, E.A.; Krivova, N.A. Analysis of Collagen Spatial Structure Using Multiphoton Microscopy and Machine Learning Methods. *Biochem. Mosc.* **2019**, *84*, 108–123. [CrossRef] [PubMed]
51. Springer, S.; Zieger, M.; Böttcher, A.; Lademann, J.; Kaatz, M. Examination of Wound Healing after Curettage by Multiphoton Tomography of Human Skin in Vivo. *Skin Res. Technol.* **2017**, *23*, 452–458. [CrossRef]
52. Nikolaev, V.; Sandykova, E.A.; Kurochkina, O.S.; Vrazhnov, D.A.; Krivova, N.A.; Kistenev, Y.V.; Sim, E.S. Estimation of the Collagen and Elastin Condition at Lymphedema Using Multiphoton Microscopy. In Proceedings of the 25th International Symposium on Atmospheric and Ocean Optics: Atmospheric Physics, Novosibirsk, Russia, 18 December 2019; Matvienko, G.G., Romanovskii, O.A., Eds.; SPIE: Novosibirsk, Russia, 2019; p. 177.
53. Pittet, J.-C.; Freis, O.; Vazquez-Duchêne, M.-D.; Périé, G.; Pauly, G. Evaluation of Elastin/Collagen Content in Human Dermis in-Vivo by Multiphoton Tomography—Variation with Depth and Correlation with Aging. *Cosmetics* **2014**, *1*, 211–221. [CrossRef]
54. Springer, S.; Zieger, M.; Hipler, U.C.; Lademann, J.; Albrecht, V.; Bueckle, R.; Meß, C.; Kaatz, M.; Huck, V. Multiphotonic Staging of Chronic Wounds and Evaluation of Sterile, Optical Transparent Bacterial Nanocellulose Covering: A Diagnostic Window into Human Skin. *Skin Res. Technol.* **2019**, *25*, 68–78. [CrossRef]
55. Breunig, H.G.; Studier, H.; König, K. Multiphoton Excitation Characteristics of Cellular Fluorophores of Human Skin In Vivo. *Opt. Express* **2010**, *18*, 7857. [CrossRef]
56. Darvin, M.E.; Richter, H.; Ahlberg, S.; Haag, S.F.; Meinke, M.C.; Le Quintrec, D.; Doucet, O.; Lademann, J. Influence of Sun Exposure on the Cutaneous Collagen/Elastin Fibers and Carotenoids: Negative Effects Can Be Reduced by Application of Sunscreen: Sunscreens Prevent the Destruction of Cutaneous Carotenoids and Collagen. *J. Biophotonics* **2014**, *7*, 735–743. [CrossRef]
57. Weiler, M.J.; Cribb, M.T.; Nepiyushchikh, Z.; Nelson, T.S.; Dixon, J.B. A Novel Mouse Tail Lymphedema Model for Observing Lymphatic Pump Failure during Lymphedema Development. *Sci. Rep.* **2019**, *9*, 10405. [CrossRef]
58. Rico-Jimenez, J.; Lee, J.H.; Alex, A.; Musaad, S.; Chaney, E.; Barkalifa, R.; Spillman, D.R., Jr.; Olson, E.; Adams, D.; Marjanovic, M.; et al. Noninvasive Monitoring of Pharmacodynamics during the Skin Wound Healing Process Using Multimodal Optical Microscopy. *BMJ Open Diabetes Res. Care* **2020**, *8*, e000974. [CrossRef]
59. Chan, B.C.L.; Dharmaratne, P.; Wang, B.; Lau, K.M.; Lee, C.C.; Cheung, D.W.S.; Chan, J.Y.W.; Yue, G.G.L.; Lau, C.B.S.; Wong, C.K.; et al. Hypericin and Pheophorbide a Mediated Photodynamic Therapy Fighting MRSA Wound Infections: A Translational Study from In Vitro to In Vivo. *Pharmaceutics* **2021**, *13*, 1399. [CrossRef] [PubMed]
60. Tuchin, V.V.; Zhu, D.; Genina, E.A. *Handbook of Tissue Optical Clearing, New Prospects in Optical Imaging*; Taylor & Francis Group LLC: Abingdon, UK; CRC Press: Boca Raton, FL, USA, 2022.
61. Sdobnov, A.; Darvin, M.E.; Lademann, J.; Tuchin, V. A Comparative Study of *Ex Vivo* Skin Optical Clearing Using Two-Photon Microscopy. *J. Biophotonics* **2017**, *10*, 1115–1123. [CrossRef] [PubMed]

 *pharmaceutics*

Article

The In Vivo Quantitative Assessment of the Effectiveness of Low-Dose Photodynamic Therapy on Wound Healing Using Optical Coherence Tomography

Hala Zuhayri, Viktor V. Nikolaev, Tatiana B. Lepekhina, Ekaterina A. Sandykova, Natalya A. Krivova and Yury V. Kistenev *

Laboratory of Laser Molecular Imaging and Machine Learning, Tomsk State University, Lenin Ave. 36, 634050 Tomsk, Russia; zuhayri.n.hala@gmail.com (H.Z.); vik-nikol@bk.ru (V.V.N.); tatiana_lepekhina@mail.ru (T.B.L.); katrin_nemchenko@mail.ru (E.A.S.); nakri@res.tsu.ru (N.A.K.)
* Correspondence: yuk@iao.ru

Abstract: The effect of low-dose photodynamic therapy on in vivo wound healing was investigated using optical coherence tomography. This work aims to develop an approach to quantitative assessment of the wound's state during wound healing including the effect of low-dose photodynamic therapy using topical application of two different photosensitizers, 5-aminolevulinic acid and methylene blue, and two laser doses of $1\,J/cm^2$ and $4\,J/cm^2$. It was concluded that the laser dose of $4\,J/cm^2$ was better compared to $1\,J/cm^2$ and allowed the wound healing process to accelerate.

Keywords: low dose photodynamic therapy; wound healing; 5-aminolevulinic acid; methylene blue; optical coherence tomography

Citation: Zuhayri, H.; Nikolaev, V.V.; Lepekhina, T.B.; Sandykova, E.A.; Krivova, N.A.; Kistenev, Y.V. The In Vivo Quantitative Assessment of the Effectiveness of Low-Dose Photodynamic Therapy on Wound Healing Using Optical Coherence Tomography. *Pharmaceutics* **2022**, *14*, 399. https://doi.org/10.3390/pharmaceutics14020399

Academic Editors: Udo Bakowsky, Matthias Wojcik, Eduard Preis and Gerhard Litscher

Received: 18 December 2021
Accepted: 9 February 2022
Published: 11 February 2022

Publisher's Note: MDPI stays neutral with regard to jurisdictional claims in published maps and institutional affiliations.

1. Introduction

According to the World Health Organization (WHO), burn wounds result in approximately 180,000 deaths every year and nearly 11 million injuries that require medical treatment worldwide [1]. Cutaneous wounds are widespread and differentiated into acute and chronic wounds [2].

Wound healing is a complex physiological process at the cellular and molecular levels including the extracellular matrix synthesis, the replacement of type III collagen with type I collagen, and scar tissue formation [3–6]. These processes are divided into four overlapping stages: coagulation (hemostasis), inflammation, proliferation, and remodeling [7,8]. Some underlying diseases affect the wound healing process including peripheral arterial and venous disease or diabetes; acute wounds may have impaired healing, which can lead to a chronic stage [9–11]. In developed countries, 1–6% of the population suffers from chronic wounds [12–14].

It is known that a low dose photo process with photoactive compounds promotes the healing of skin diseases and leads to results in rejuvenation and wound healing [15,16]. Low-dose photodynamic therapy (LDPDT) is widely used to treat skin diseases and wound healing where it reduces the treatment time, accelerates tissue repair, and promotes healing [17,18]. The method is based on using a photosensitizer (PS), which accumulates in tissues, followed by irradiation of the tissue with a light source with an appropriate wavelength. The latter causes the formation of reactive oxygen species (ROS) [19,20]. Low concentrations of ROS can trigger cell repair processes including proliferation and offer promising treatments to accelerate healing. Different PSs have been studied in the wound healing process, which has a relevant role in ensuring PDT effectiveness in skin wound healing [21] such as 5-aminolevulinic acid (5-ALA) and methylene blue (MB) [22–25]. MB is a popular PS among the phenothiazinium derivatives that have attracted the attention of different research groups working and achieving good results in wound healing [25–27].

Recently, MB was shown to have an antioxidant role [28]. Additionally, 5-ALA is among the most effective photosensitizers and is widely used to present a better achievement concerning wound healing [17,22,29].

The Arndt–Schultz Law is an appropriate model to demonstrate that low levels of light have a better effect in wound healing than higher levels, which may have an inhibitory or cytotoxic effect [30,31]. Hawkins and Abrahamse studied the behavior in vitro of human skin fibroblasts using different irradiation doses of 0.5, 2.5, 5, 10, and 16 J/cm^2. They demonstrated that higher laser doses (10 and 16 J/cm^2) resulted in increased cellular damage as well as decreased cell viability and proliferation [32]. Results for different energy doses were described for 4 J/cm^2 [33] and for 1 J/cm^2 and 2 J/cm^2 [34]. Basso et al. demonstrated that irradiation of cultured human gingival fibroblasts with energy doses of 0.5 and 3 J/cm^2 resulted in a significant increase in cellular metabolism compared with the non-irradiated control group and the cells irradiated with higher energy doses of 5 and 7 J/cm^2 [35]. The most significant biological effects were seen with predominant dose values (i.e., up to 5 J/cm^2), which were within the Arndt–Schultz curve [36].

Traditionally, wounds have been observed invasively with a histochemical assessment of the biopsies [3,8]. Visual observation is a common tool for wound assessment. Additionally, clinical wound evaluation is a widely used and the least expensive method of assessing wound depth. This method relies on a subjective evaluation of the external features of the wound such as wound appearance, capillary refill, and burn wound sensibility to touch and pinprick, providing diagnostic accuracy at the level of 60–75% [37]. These methods are not quantitative and can lead to additional tissue damage and impair healing. Accordingly, the development of noninvasive and accurate methods of wound analysis is relevant.

Optical coherence tomography (OCT) is a noninvasive 3D imaging method of biological tissues with a spatial resolution of 5–10 μm and a penetration depth of 1–2 mm [38,39]. Epidermal thickness is a critical parameter for assessing epithelialization during wound healing [40,41].

OCT could detect essential morphological changes during wound healing (e.g., epidermis, dermis, adipose tissue, and granulation) that was based primarily on their backscattering characteristics [42–44]. The use of polarization-sensitive OCT revealed higher birefringence in scars compared to healthy skin [45]. OCT-based angiography provides in vivo, three-dimensional vascular information by using flowing red blood cells as intrinsic contrast agents, allowing visualization of functional vessel networks within microcirculatory tissue beds non-invasively, without needing dye injection [46].

This work aims to develop a method for quantitative in vivo evaluation of wounds using OCT during the wound healing process including a quantitative assessment of the effect of LDPDT using the topical application of two different photosensitizers (5-ALA and MB) and two laser doses of 1 J/cm^2 and 4 J/cm^2.

2. Materials and Methods

2.1. Wound Model Protocol

This study used 15 male laboratory CD1 mice, weighing 25–30 g and aged 6–7 weeks, obtained from the Department of Experimental Biological Models of the Research Institute of Pharmacology, TSC SB RAMS. Before the experiment, the mice were kept seven days in the standard conditions of a conventional vivarium with free access to water and food, and a 12/12 light regime, in a ventilated room at a temperature of $20 \pm 2\,^{\circ}C$ and a humidity of 60%. The experimental protocol of this research was approved by the Bioethical Committee of Tomsk State University (Protocol No. 4, 10.02.2021), registration No. 6.

The mice were anesthetized by isoflurane using the Ugo Basile gas anesthesia system, where the mice were put in a glass chamber connected to isoflurane (Figure 1). The wound area was prepared through depilation using Veet cream (made in France), rinsed with saline solution, and sterilized using chlorhexidine 20%. A full-thickness cutaneous wound (diameter 5 mm) was formed by cutting out a whole layer skin flap with scissors on both of the hind paws of each animal under isoflurane anesthesia. The experiment was performed in a time-lapsed schedule for the wound aging on days 1, 3, 7, and 14. Additionally, the day of wound formation was defined as day 0.

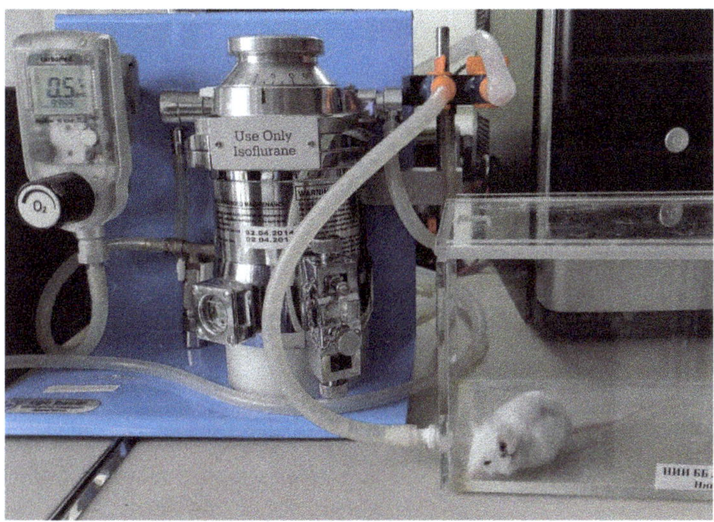

Figure 1. Anesthetized mice by isoflurane using the Ugo Basile gas anesthesia system.

2.2. Low Dose Photodynamic Therapy Protocol

Both of the photosensitizers 5-ALA 20% and MB 0.01% in saline solution were topically administered directly on the wound; after 30 min, the irradiation was started by an AlGaInP laser (λ = 630 nm, P = 5 mW) with two doses: 1 J/cm^2 and 4 J/cm^2, and the procedure was carried out under the influence of isoflurane. 5-ALA was applied on the wounds on the right hind paws, MB was applied on the left ones. The animals were divided according to the laser dose and photosensitizer into five groups: the control group, the LDPDT–5-ALA 1 J/cm^2, LDPDT–MB 1 J/cm^2, LDPDT–5-ALA 4 J/cm^2, and LDPDT–MB 4 J/cm^2. LDPDT procedure was repeated once immediately after wound formation.

2.3. Optical Coherence Tomography Protocol

The experiments were carried out using optical coherence tomography (OCT) on the GANYMEDE–II system (Thorlabs, USA) with the basic scanning module OCTG-900. It is possible to obtain information on the optical characteristics, morphology, and elastic properties of biological tissues using OCT. The GANYMEDE-II system uses a superluminescent diode with an operating wavelength of 930 ± 50 nm. The superluminescent diode allows one to reach a signal penetration depth up to 2.9 mm with an axial resolution of up to 6.0 microns (air/tissue). The width of the spectral band was 100 nm. Figure 2 shows an example of placing a mouse on the substrate of OCT. As a result, B-scans were obtained—two-dimensional images. Data processing was carried out using ThorImageOCT 5.0.1., with the following parameters: size 2469 $\times$ 675 pixel, FOV 4.66 $\times$ 1.94 mm, and pixel size 1.89 $\times$ 2.88 µm, with 20 frames. The experiment was repeated with a 30° rotation around the previous position.

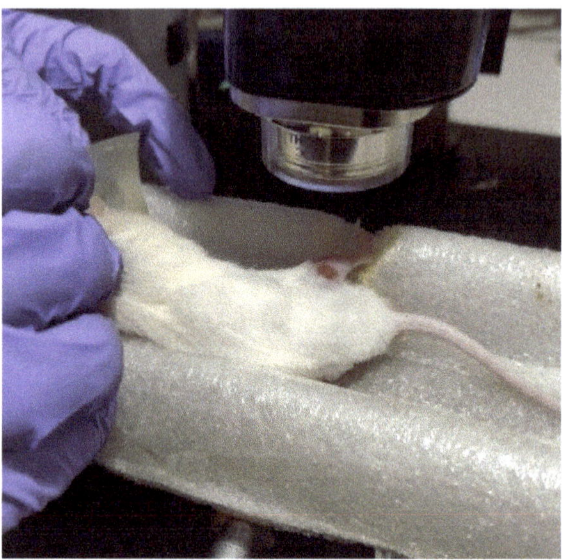

Figure 2. The mouse positioning during OCT imaging.

2.4. Statistical Analysis

The OCT data were exported using the ThorImageOCT 5.0.1 program to files. txt. Statistical analysis and data processing were carried out in Python 3.6 using libraries (numpy, scipy, matplotlib). All calculated parameters were expressed as the mean $\pm$ standard deviation. The Pearson test was applied to assess the level of statistically significant differences among groups under study. Statistical power was used at 0.95 and 0.99. The *p*-values were calculated for all groups on all days.

3. Results

3.1. OCT Imaging

Figure 3a shows a photo of healthy skin, and an OCT B-scan for the area, marked with a red arrow, is shown in Figure 3b. The B-scan data were normalized so that the stratum corneum, corresponding to the region with the highest pixel intensity, was located at the top of the image. After normalization, the signal intensity was calculated at different tissue depths from 0 to 0.8 mm (A-scan) and visualized as shown in Figure 4. The maximum intensity values were at a depth of 0 to 2–4 µm, which corresponded to the stratum corneum, and then the signal intensity gradually decreased to ~20 µm, which corresponded to the epidermis. The dermis starts from 30–40 µm, which was accompanied by a decrease in intensity to the minimum values at a depth of 0.8 mm.

Photos of the observation area and B-scans of the wound at different time points on days 1, 3, 7, and 14 for the control (without LDPDT) are shown in Figure 5.

The wound healed typically without pathologies, and the injury was close to healing on day 14. The difference in signal intensities at different stages of wound healing on different days is shown in Figure 6. On day 1 after the wound forming procedure, the signal had a low intensity, while the signal of the dermis started decreasing from 0.3 µm, so the signal from 0 to 0.3 corresponded to the formed wound scab. The signal intensity increased on days 3 and 7. On day 14, the signal intensity values were close to healthy skin.

OCT images for the LDPDT-5-ALA groups are shown for a laser dose of 1 J/cm² in Figure 7 and 4 J/cm² in Figure 8. The intensity signal for LDPDT-5-ALA 4 J/cm² had the same behavior as the control. On day 1 for LDPDT-5-ALA 1 J/cm², the attenuation signal corresponding to the dermis started from ~0.3 mm. Similar to the control, the signal intensity increased on days 3 and 7, and on day 14, the signal intensity values were close to healthy skin, as shown in Figure 9a. In the same way for LDPDT-5-ALA 4 J/cm^2, the intensity signal started decreasing from ~0.2 mm on day 1. On days 3 and 7, the signal intensity values increased to close to the value of healthy skin on day 14 more than the LDPDT-5-ALA 1 J/cm^2 group, as shown in Figure 9b.

Measurements were similarly carried out for the MB photosensitizer with two laser doses of 1 J/cm^2 (Figure 10) and 4 J/cm^2 (Figure 11). The intensity signal for LDPDT-MB on day 14 was close to the values for healthy skin for different exposure doses, as shown in Figure 12.

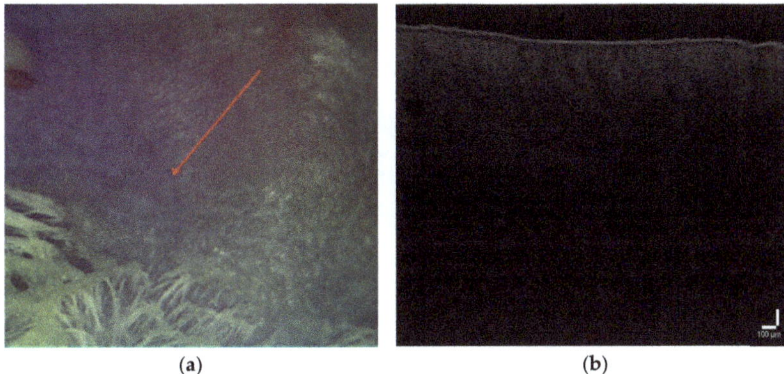

(a)   (b)

Figure 3. (**a**) Visual observation for healthy skin, (**b**) OCT imaging.

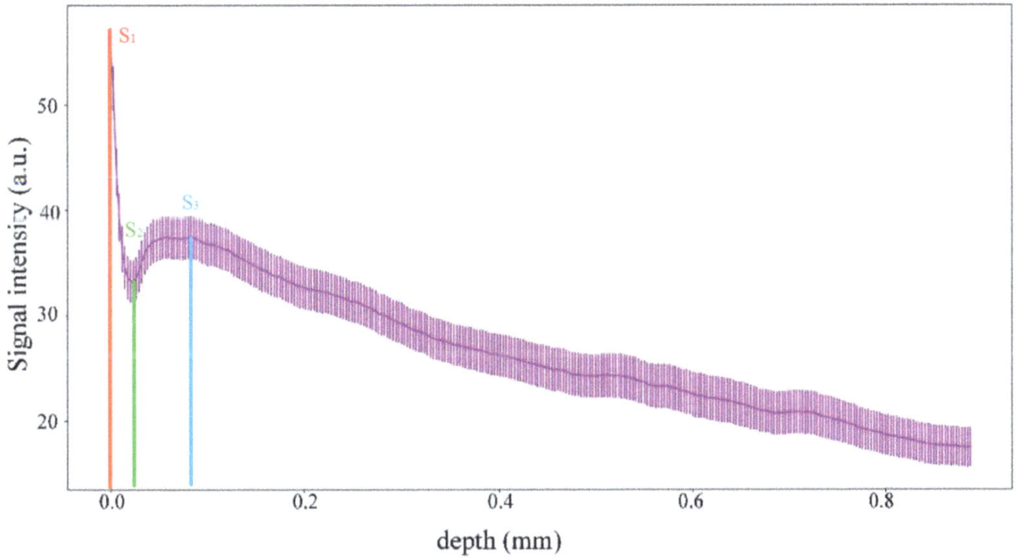

Figure 4. Dependence of signal intensity on depth for healthy skin: S_1 start point of the stratum corneum; S_2—the endpoint of epidermis; S_3—start point of the dermis.

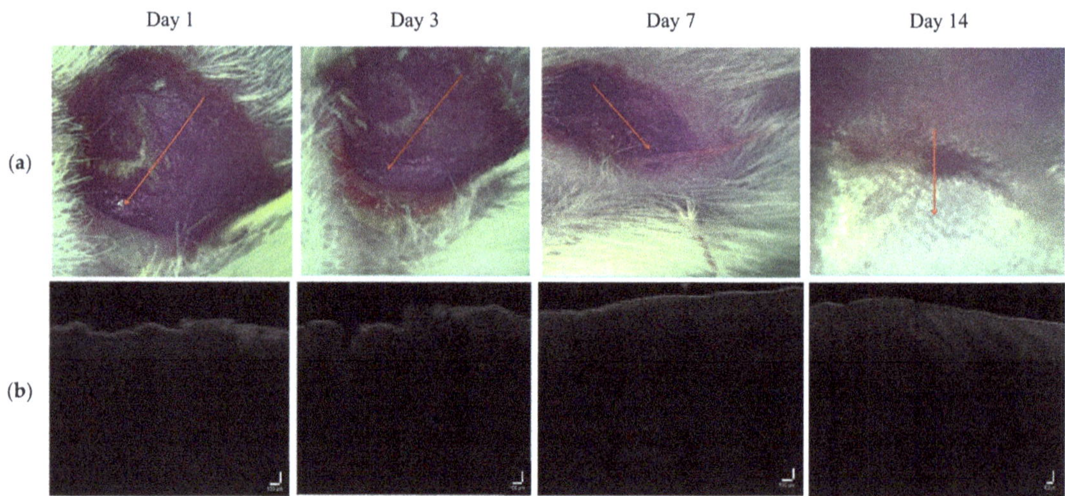

Figure 5. (**a**) Visual observation and (**b**) the corresponding B-scans for the control group.

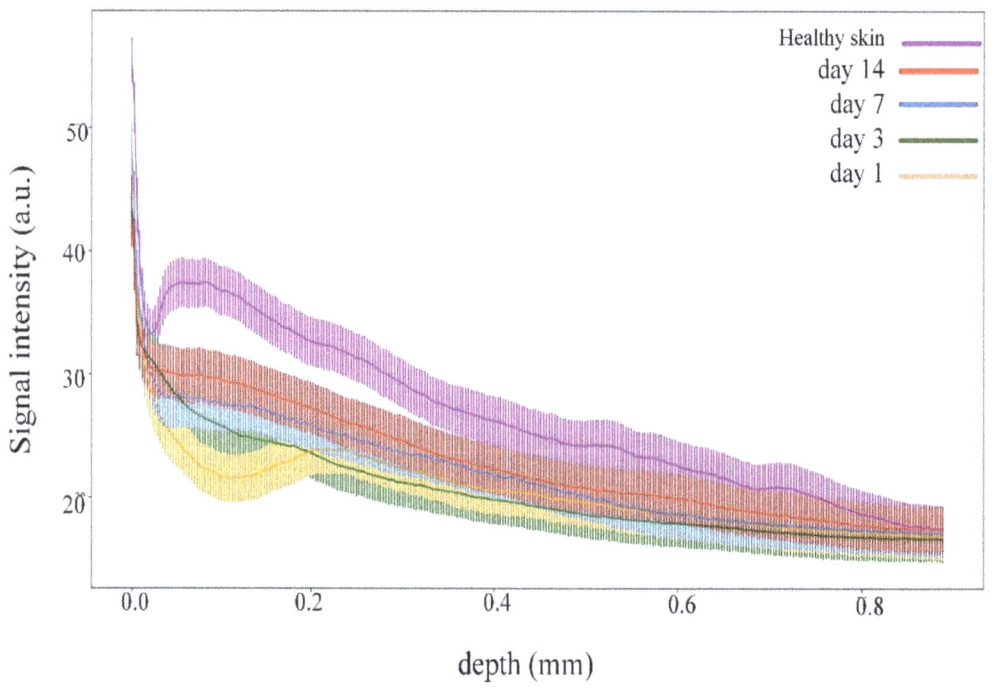

Figure 6. Dependence of signal intensity on depth for the control group during wound healing.

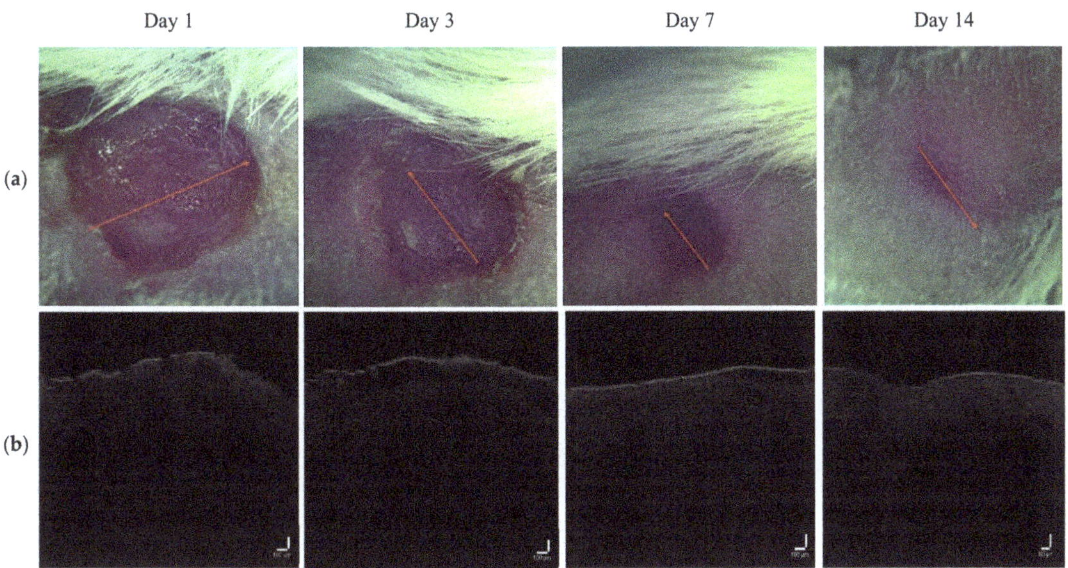

Figure 7. (**a**) Visual observation and (**b**) the corresponding B-scans for the LDPDT-5-ALA 1 J/cm^2 group.

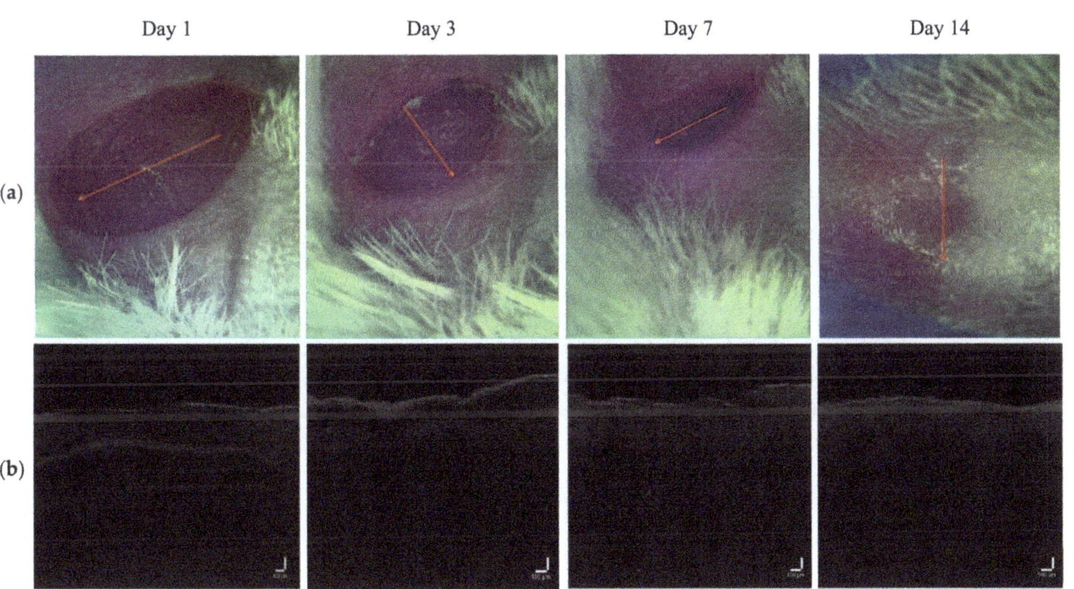

Figure 8. (**a**) Visual observation and (**b**) the corresponding B-scans for the LDPDT-5-ALA 4 J/cm^2 group.

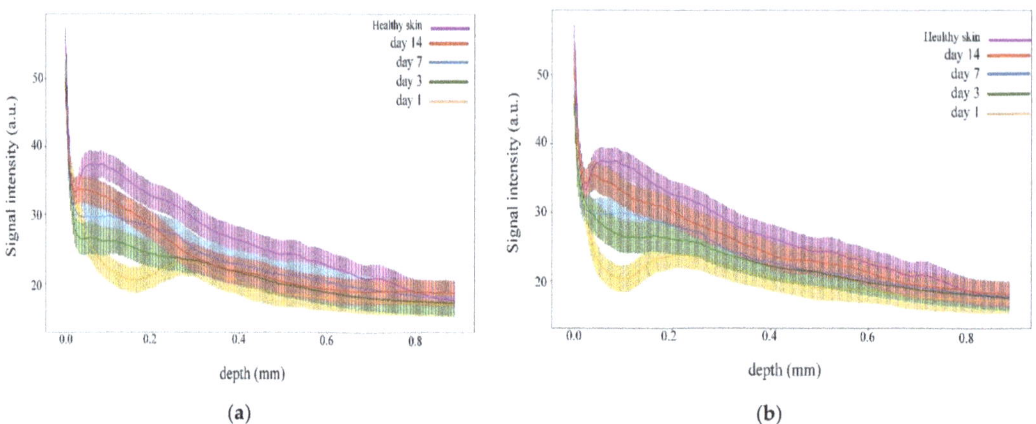

Figure 9. Dependence of signal intensity on depth for the (a) LDPDT-5-ALA 1 J/cm^2 group and (b) LDPDT-5-ALA 4 J/cm^2 group.

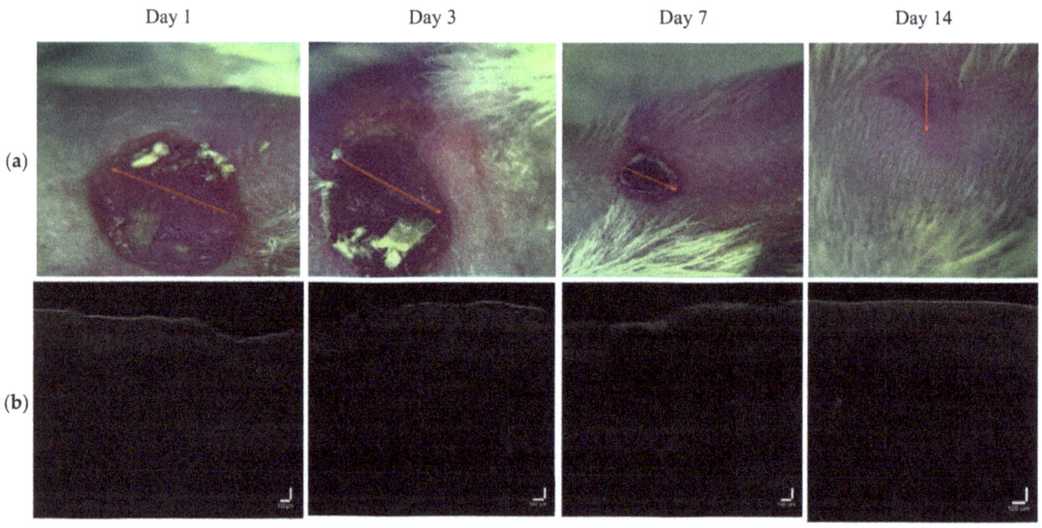

Figure 10. (a) Visual observation and (b) the corresponding B-scans for the LDPDT-MB 1 J/cm^2 group.

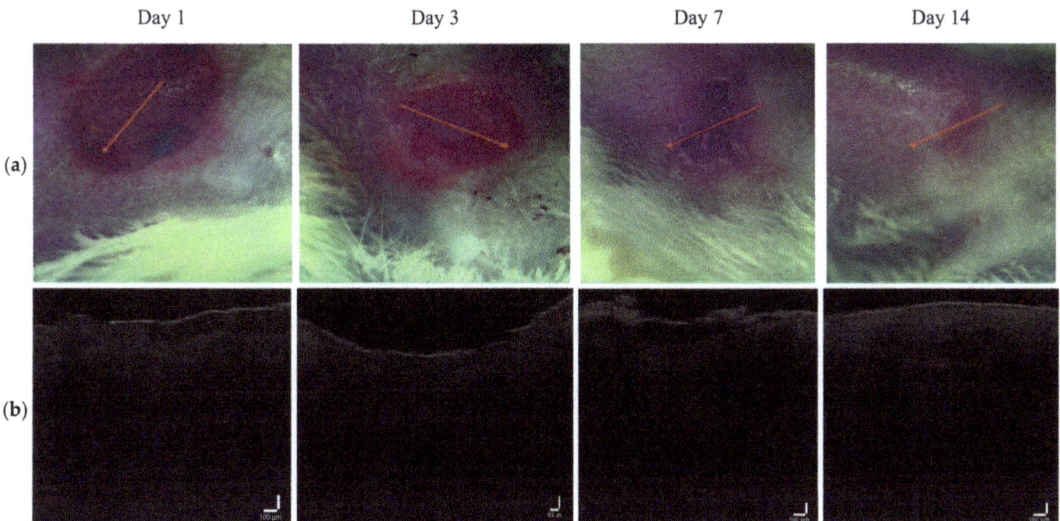

Figure 11. (a) Visual observation and (b) the corresponding B-scans for the LDPDT-MB 4 J/cm² group.

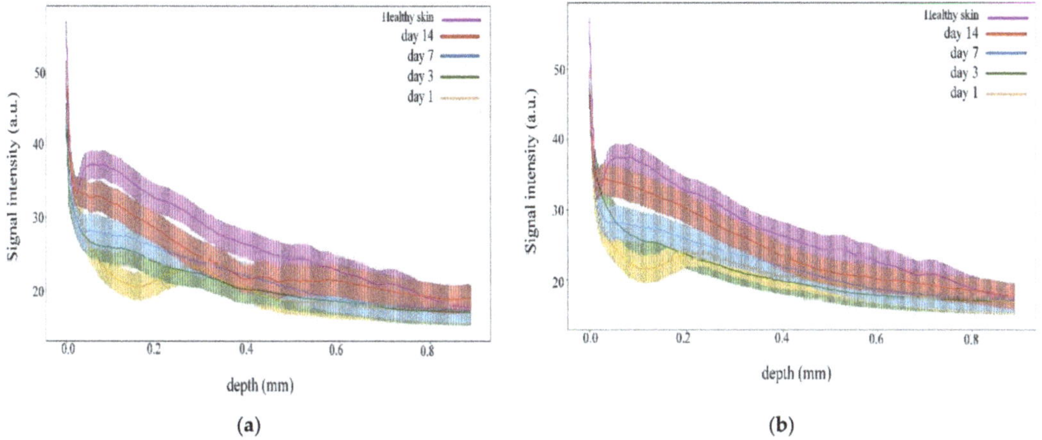

Figure 12. Dependence of signal intensity on depth for the (a) LDPDT-MB 1 J/cm² group and (b) LDPDT-MB 4 J/cm² group.

3.2. Quantitative Comparison of the Spatial Proximity of the OCT Signal Intensity

For a quantitative comparison of spatial profiles, we used the curve proximity factor (CPF) *S*, similar to the Pearson's correlation coefficient, to compare healthy skin and wound curves in all days to all groups [47]:

$$S = \frac{\sum_i |X_i - Y_i|}{\frac{1}{2}\sum_i |X_i + Y_i|},$$ (1)

where X_i, Y_i is the intensity of the OCT signal for a definite depth from the wound and healthy skin, respectively. The higher the CPF value, the closer the wound state to healthy skin. The 0.01 and 0.05 significance levels were used to assess the statistical differences between the wound and healthy skin groups. The CPF values for all groups are shown in Table 1.

Table 1. CPF values for the studied groups (mean $\pm$ standard deviation).

$n = 5$	Control	LDPDT–5-ALA 1 J/cm^2	LDPDT–5-ALA 4 J/cm^2	LDPDT–MB 1 J/cm^2	LDPDT–MB 4 J/cm^2
Day 1	(0.0528 $\pm$ 0.0084) **	(0.0516 $\pm$ 0.0072) **	(0.0517 $\pm$ 0.0045) **	(0.0494 $\pm$ 0.0083) **	(0.0539 $\pm$ 0.0064) **
Day 3	(0.0404 $\pm$ 0.0141) **	(0.0504 $\pm$ 0.0106) *	(0.0480 $\pm$ 0.0087) **	(0.0456 $\pm$ 0.0169) *	(0.0423 $\pm$ 0.0134) *
Day 7	(0.0316 $\pm$ 0.0089) *	(0.0379 $\pm$ 0.0124) *	(0.0315 $\pm$ 0.0111) *	(0.0361 $\pm$ 0.0084) *	(0.0305 $\pm$ 0.0073) *
Day 14	(0.0214 $\pm$ 0.0076) *	(0.0213 $\pm$ 0.0075)	(0.0187 $\pm$ 0.0213)	(0.0201 $\pm$ 0.0054)	(0.0174 $\pm$ 0.0051)

* $p > 0.01$, ** $p > 0.05$.

The CPF (1) was also used to estimate the effectiveness of different laser doses for each photosensitizer on day 14. The CPF values calculated for the OCT signal intensity curves corresponding to 4 J/cm^2 and 1 J/cm^2 for 5-ALA (Figure 9) and MB (Figure 12) are shown in Table 2. These quantitative estimations demonstrated a "proximity" between the curves corresponding to 4 J/cm^2 laser dose and the curves corresponding to the 1 J/cm^2 laser dose for the same photosensitizer.

Table 2. CPF values for the studied groups (mean $\pm$ standard deviation).

$n = 5$	LDPDT–5-ALA	LDPDT–MB
Day 14	(0.0115 $\pm$ 0.0019) **	(0.0142 $\pm$ 0.0011) *

* $p > 0.005$, ** $p > 0.01$.

4. Discussion

The proposed method of wound state quantitative evaluation is based on the OCT visualization of tissue structure transformation. The averaged scatter A-line intensity profile obtained from the horizontal rectangle in the OCT B-scan image of healthy skin is shown in Figure 4. Three areas are highlighted in the figure, representing changes in the attenuation coefficient. The red (S_1), green (S_2), and blue (S_3) lines correspond to the beginning of the stratum corneum, the end of the epidermis, and the beginning of the dermis, respectively. After inflicting a wound in the first days, there are no surface layers of the skin (horny, epidermis); instead, a scab forms on the surface. These changes in the skin are visible on A-scans. Over time, the skin recovers, and on A-scans, we can see the appearance of areas characteristic of the epidermis's end and the dermis's beginning. These changes are reflected in the OCT signal attenuation curve (see Figures 6, 8, 9 and 12).

For wound state quantitative estimation, we used the curve proximity factor, introduced by us earlier [47]. According to Table 1, in the control group, the CPF mean value on day 1 was about 0.053, and decreased to 0.021 on day 14, while in the LDPDT groups, the mean values of this coefficient on day 14 ranged from 0.017 to 0.021. The Pearson test demonstrated that for LDPDT groups on day 14, p-value did not exceed 0.01, while for the control group, this value was equal to 0.03. Therefore, for all LPDT groups, the wound state had no statistically significant difference compared to healthy skin for the used statistical power levels.

The CPF value for LDPDT groups for the 4 J/cm^2 laser dose was smaller than the LDPDT groups for 1 J/cm^2. We also calculated the CPF values for the OCT signal intensity curves corresponding to 4 J/cm^2 and 1 J/cm^2 for 5-ALA (the first column in Table 1) and MB (the second column in Table 2) and conducted a Pearson test to check the statistical significance of these differences. p-value was shown to be larger for 5-ALA.

Therefore, after comparing the CPF parameter for five groups: control, LDPDT 5-ALA 4 J/cm^2, LDPDT 5-ALA 1 J/cm^2, LDPDT-MB 4 J/cm^{2,} and LDPDT-MB 1 J/cm^2, it was concluded that the laser dose of 4 J/cm^2 for LDPDT 5-ALA was definitely better compared to 1 J/cm^2 and probably better for LDPDT MB. It should be noted that the conclusion depends on the volume and quality of the dataset.

A possible reason for the 4 J/cm^2 dose preference relative to the 1 J/cm^2 dose is as follows. According to previous works [33–35], when low-level laser light is applied and a dose is too low, no tissue response will occur. If too a high dose is applied, it can inhibit a tissue response. It has been seen in studies of wound healing where too low a dose did not have an impact, and too high a dose (above $5\,J/cm^2$) prolonged wound healing while the optimal dose resulted in faster healing. In this interval, according to the Arndt–Schultz curve, the larger dose causes a stronger biological effect.

In any case, LDPDT allows for accelerating of the wound healing process, which is consistent with the literature data [32–34,48,49].

5. Conclusions

In this paper, a study of quantitative in vivo evaluation of wounds using OCT during the wound healing process was carried out. 5-ALA and MB were used as photosensitizers for LDPDT, with two laser doses of 1 and $4\,J/cm^2$.

An approach to quantitative estimation of wound state based on the CPF, Equation (1) [47] was proposed. The method was used to quantify the effectiveness of LDPDT to accelerate the wound healing process. CPF parameter estimation allowed us to compare LDPDT regimes quantitatively and to obtain objective arguments about the superiority of one regime over another.

Therefore, the proposed CPF parameter, estimated from OCT data, has demonstrated its feasibility for the quantitative estimation of the human wound state during healing. This approach is noninvasive, simple in implementation, and suitable for continuous monitoring throughout the wound healing process and sufficient resolution to assess both anatomy and pathology. It makes it a promising technique for applications in wound healing and the evaluation of novel therapeutics.

Another approach, which was proven to be effective in monitoring wound healing is two-photon microscopy. Previously, our group analyzed the two-photon microscopy images of the wound healing process and succeeded in quantitatively assessing the state of the wound and studying the effect of low-dose photodynamic therapy using the techniques of two-photon microscopy. The results of this study are completely consistent with the results obtained earlier [50].

Author Contributions: Conceptualization, Y.V.K.; Methodology, H.Z. and E.A.S.; Software, V.V.N.; Validation, H.Z. and V.V.N.; Investigation, H.Z., V.V.N., T.B.L. and N.A.K.; Writing—original draft preparation, E.A.S.; Writing—review and editing, Y.V.K.; Visualization, H.Z.; Supervision, Y.V.K. All authors have read and agreed to the published version of the manuscript.

Funding: This research was funded by a grant under the Decree of the Government of the Russian Federation No. 220 of 9 April 2010 (Agreement No. 075-15-2021-615 of 4 June 2021).

Institutional Review Board Statement: The study was approved by the Ethics Committee of Tomsk State University (Protocol No.4, 10.02.2021), registration No. 6.

Data Availability Statement: The data presented in this study are available on request from the corresponding author.

Conflicts of Interest: The authors declare no conflict of interest.

References

1. World Health Organization. Burns. Available online: https://www.who.int/news-room/fact-sheets/detail/burns (accessed on 15 December 2021).
2. Nussbaum, S.R.; Carter, M.J.; Fife, C.E.; DaVanzo, J.; Haught, R.; Nusgart, M.; Cartwright, D. An Economic Evaluation of the Impact, Cost, and Medicare Policy Implications of Chronic Nonhealing Wounds. *Value Health* **2018**, *21*, 27–32. [CrossRef] [PubMed]
3. De Gonzalez, A.C.O.; Costa, T.F.; de Andrade, Z.A.; Medrado, A.R.A.P. Wound Healing—A Literature Review. *An. Bras. Dermatol.* **2016**, *91*, 614–620. [CrossRef] [PubMed]
4. Sorg, H.; Tilkorn, D.J.; Hager, S.; Hauser, J.; Mirastschijski, U. Skin Wound Healing: An Update on the Current Knowledge and Concepts. *Eur. Surg. Res.* **2017**, *58*, 81–94. [CrossRef] [PubMed]

5.	Liu, H.; Liu, H.; Deng, X.; Chen, M.; Han, X.; Yan, W.; Wang, N. Raman Spectroscopy Combined with SHG Gives a New Perspective for Rapid Assessment of the Collagen Status in the Healing of Cutaneous Wounds. *Anal. Methods* **2016**, *8*, 3503–3510. [CrossRef]

6.	Mickelson, M.A.; Mans, C.; Colopy, S.A. Principles of Wound Management and Wound Healing in Exotic Pets. *Vet. Clin. N. Am. Exot. Anim. Pract.* **2016**, *19*, 33–53. [CrossRef] [PubMed]

7.	Deka, G.; Wu, W.-W.; Kao, F.-J. In Vivo Wound Healing Diagnosis with Second Harmonic and Fluorescence Lifetime Imaging. *J. Biomed. Opt.* **2012**, *18*, 061222. [CrossRef]

8.	Zuhayri, H.; Knyazkova, A.I.; Nikolaev, V.V.; Borisov, A.V.; Kistenev, Y.V.; Zakharova, O.A.; Dyachenko, P.A.; Tuchin, V.V. *Study of Wound Healing by Terahertz Spectroscopy. Proceedings of the Fourth International Conference on Terahertz and Microwave Radiation: Generation, Detection, and Applications; Romanovskii, Tomsk, Russia, 24–26 August 2020*; Zakharova, O.A., Kistenev, Y.V., Eds.; SPIE: Tomsk, Russia, 2020; p. 63.

9.	Bolton, L. Peripheral Arterial Disease: Scoping Review of Patient-centred Outcomes. *Int. Wound J.* **2019**, *16*, 1521–1532. [CrossRef] [PubMed]

10.	Baltzis, D.; Eleftheriadou, I.; Veves, A. Pathogenesis and Treatment of Impaired Wound Healing in Diabetes Mellitus: New Insights. *Adv. Ther.* **2014**, *31*, 817–836. [CrossRef] [PubMed]

11.	Mazumder, N.; Ball, K.N.; Zhuo, K.-Q.; Kistenev, Y.V.; Kumar, R.; Kao, F.-J.; Brasselet, S.; Nikolaev, V.V.; Krivova, N.A. Label-Free Multimodal Nonlinear Optical Microscopy for Biomedical Applications. *J. Appl. Phys.* **2021**, *129*, 214901. [CrossRef]

12.	Phillips, C.J.; Humphreys, I.; Fletcher, J.; Harding, K.; Chamberlain, G.; Macey, S. Estimating the Costs Associated with the Management of Patients with Chronic Wounds Using Linked Routine Data: Costs of Wounds Using Routine Data. *Int. Wound J.* **2016**, *13*, 1193–1197. [CrossRef]

13.	Guest, J.F.; Ayoub, N.; McIlwraith, T.; Uchegbu, I.; Gerrish, A.; Weidlich, D.; Vowden, K.; Vowden, P. Health Economic Burden That Wounds Impose on the National Health Service in the UK. *BMJ Open* **2015**, *5*, e009283. [CrossRef]

14.	Heyer, K.; Herberger, K.; Protz, K.; Glaeske, G.; Augustin, M. Epidemiology of Chronic Wounds in Germany: Analysis of Statutory Health Insurance Data: Epidemiology of Chronic Wounds in Germany. *Wound Rep. Reg.* **2016**, *24*, 434–442. [CrossRef] [PubMed]

15.	Tedesco, A.; Jesus, P. Low Level Energy Photodynamic Therapy for Skin Processes and Regeneration. In *Photomedicine—Advances in Clinical Practice*; Tanaka, Y., Ed.; InTech: London, UK, 2017; ISBN 978-953-51-3155-7.

16.	Shaw, T.J.; Martin, P. Wound Repair at a Glance. *J. Cell Sci.* **2009**, *122*, 3209–3213. [CrossRef] [PubMed]

17.	Yang, T.; Tan, Y.; Zhang, W.; Yang, W.; Luo, J.; Chen, L.; Liu, H.; Yang, G.; Lei, X. Effects of ALA-PDT on the Healing of Mouse Skin Wounds Infected With Pseudomonas Aeruginosa and Its Related Mechanisms. *Front. Cell Dev. Biol.* **2020**, *8*, 585132. [CrossRef]

18.	Algorri, J.F.; Ochoa, M.; Roldán-Varona, P.; Rodríguez-Cobo, L.; López-Higuera, J.M. Photodynamic Therapy: A Compendium of Latest Reviews. *Cancers* **2021**, *13*, 4447. [CrossRef] [PubMed]

19.	Kwiatkowski, S.; Knap, B.; Przystupski, D.; Saczko, J.; Kędzierska, E.; Knap-Czop, K.; Kotlińska, J.; Michel, O.; Kotowski, K.; Kulbacka, J. Photodynamic Therapy—Mechanisms, Photosensitizers and Combinations. *Biomed. PharmacoTher.* **2018**, *106*, 1098–1107. [CrossRef] [PubMed]

20.	Peplow, P.V.; Chung, T.-Y.; Baxter, G.D. Photodynamic Modulation of Wound Healing: A Review of Human and Animal Studies. *Photomed. Laser Surg.* **2012**, *30*, 118–148. [CrossRef]

21.	Vallejo, M.C.S.; Moura, N.M.M.; Ferreira Faustino, M.A.; Almeida, A.; Gonçalves, I.; Serra, V.V.; Neves, M.G.P.M.S. An Insight into the Role of Non-Porphyrinoid Photosensitizers for Skin Wound Healing. *IJMS* **2020**, *22*, 234. [CrossRef]

22.	Morimoto, K.; Ozawa, T.; Awazu, K.; Ito, N.; Honda, N.; Matsumoto, S.; Tsuruta, D. Photodynamic Therapy Using Systemic Administration of 5-Aminolevulinic Acid and a 410-Nm Wavelength Light-Emitting Diode for Methicillin-Resistant Staphylococcus Aureus-Infected Ulcers in Mice. *PLoS ONE* **2014**, *9*, e105173. [CrossRef]

23.	Yin, R.; Lin, L.; Xiao, Y.; Hao, F.; Hamblin, M.R. Combination ALA-PDT and Ablative Fractional Er:YAG Laser (2,940 Nm) on the Treatment of Severe Acne: Combination ala-pdt and ablative fractional er:yag laser. *Lasers Surg. Med.* **2014**, *46*, 165–172. [CrossRef]

24.	Jiang, C.; Yang, W.; Wang, C.; Qin, W.; Ming, J.; Zhang, M.; Qian, H.; Jiao, T. Methylene Blue-Mediated Photodynamic Therapy Induces Macrophage Apoptosis via ROS and Reduces Bone Resorption in Periodontitis. *Oxid. Med. Cell. Longev.* **2019**, *2019*, 1–15. [CrossRef] [PubMed]

25.	Sperandio, F.F.; Simões, A.; Aranha, A.C.C.; Corrêa, L.; Orsini Machado de Sousa, S.C. Photodynamic Therapy Mediated by Methylene Blue Dye in Wound Healing. *Photomed. Laser Surg.* **2010**, *28*, 581–587. [CrossRef] [PubMed]

26.	Pérez, M.; Robres, P.; Moreno, B.; Bolea, R.; Verde, M.T.; Pérez-Laguna, V.; Aspiroz, C.; Gilaberte, Y.; Rezusta, A. Comparison of Antibacterial Activity and Wound Healing in a Superficial Abrasion Mouse Model of Staphylococcus Aureus Skin Infection Using Photodynamic Therapy Based on Methylene Blue or Mupirocin or Both. *Front. Med.* **2021**, *8*, 673408. [CrossRef] [PubMed]

27.	Carneiro, V.S.M.; de Catao, M.H.C.V.; Menezes, R.F.; Araújo, N.C.; Gerbi, M.E.M. *Methylene Blue Photodynamic Therapy in Rats' Wound Healing: 21 Days Follow-Up*; Kurachi, C., Svanberg, K., Tromberg, B.J., Bagnato, V.S., Eds.; SPIE Proceedings: Rio de Janeiro, Brazil, 2015; p. 95311S.

28.	Xiong, Z.-M.; O'Donovan, M.; Sun, L.; Choi, J.Y.; Ren, M.; Cao, K. Anti-Aging Potentials of Methylene Blue for Human Skin Longevity. *Sci. Rep.* **2017**, *7*, 2475. [CrossRef]

29.	Yang, Z.; Hu, X.; Zhou, L.; He, Y.; Zhang, X.; Yang, J.; Ju, Z.; Liou, Y.-C.; Shen, H.-M.; Luo, G.; et al. Photodynamic Therapy Accelerates Skin Wound Healing through Promoting Re-Epithelialization. *Burn. Trauma* **2021**, *9*, tkab008. [CrossRef] [PubMed]

30. Huang, Y.-Y.; Chen, A.C.-H.; Carroll, J.D.; Hamblin, M.R. Biphasic Dose Response in Low Level Light Therapy. *Dose-Resp.* **2009**, *7*, 358–383. [CrossRef] [PubMed]
31. Huang, Y.-Y.; Sharma, S.K.; Carroll, J.; Hamblin, M.R. Biphasic Dose Response in Low Level Light Therapy—an Update. *Dose-Resp.* **2011**, *9*, 602–618. [CrossRef] [PubMed]
32. Hawkins, D.H.; Abrahamse, H. The Role of Laser Fluence in Cell Viability, Proliferation, and Membrane Integrity of Wounded Human Skin Fibroblasts Following Helium-Neon Laser Irradiation. *Lasers Surg. Med.* **2006**, *38*, 74–83. [CrossRef] [PubMed]
33. Byrnes, K.R.; Barna, L.; Chenault, V.M.; Waynant, R.W.; Ilev, I.K.; Longo, L.; Miracco, C.; Johnson, B.; Anders, J.J. Photobiomodulation Improves Cutaneous Wound Healing in an Animal Model of Type II Diabetes. *Photomed. Laser Surg.* **2004**, *22*, 281–290. [CrossRef] [PubMed]
34. Prabhu, V.; Rao, S.B.S.; Chandra, S.; Kumar, P.; Rao, L.; Guddattu, V.; Satyamoorthy, K.; Mahato, K.K. Spectroscopic and Histological Evaluation of Wound Healing Progression Following Low Level Laser Therapy (LLLT). *J. Biophoton.* **2012**, *5*, 168–184. [CrossRef] [PubMed]
35. Basso, F.G.; Pansani, T.N.; Turrioni, A.P.S.; Bagnato, V.S.; Hebling, J.; de Souza Costa, C.A. In Vitro Wound Healing Improvement by Low-Level Laser Therapy Application in Cultured Gingival Fibroblasts. *Int. J. Dent.* **2012**, *2012*, 1–6. [CrossRef] [PubMed]
36. De Chaves, M.E.A.; de Araújo, A.R.; Piancastelli, A.C.C.; Pinotti, M. Effects of Low-Power Light Therapy on Wound Healing: LASER x LED. *An. Bras. Dermatol.* **2014**, *89*, 616–623. [CrossRef] [PubMed]
37. Monstrey, S.; Hoeksema, H.; Verbelen, J.; Pirayesh, A.; Blondeel, P. Assessment of Burn Depth and Burn Wound Healing Potential. *Burns* **2008**, *34*, 761–769. [CrossRef] [PubMed]
38. Huang, D.; Swanson, E.A.; Lin, C.P.; Schuman, J.S.; Stinson, W.G.; Chang, W.; Hee, M.R.; Flotte, T.; Gregory, K.; Puliafito, C.A.; et al. Optical Coherence Tomography. *Science* **1991**, *254*, 1178–1181. [CrossRef]
39. Almasian, M.; Bosschaart, N.; van Leeuwen, T.G.; Faber, D.J. Validation of Quantitative Attenuation and Backscattering Coefficient Measurements by Optical Coherence Tomography in the Concentration-Dependent and Multiple Scattering Regime. *J. Biomed. Opt.* **2015**, *20*, 121314. [CrossRef] [PubMed]
40. Evans, N.D.; Oreffo, R.O.C.; Healy, E.; Thurner, P.J.; Man, Y.H. Epithelial Mechanobiology, Skin Wound Healing, and the Stem Cell Niche. *J. Mech. Behav. Biomed. Mater.* **2013**, *28*, 397–409. [CrossRef] [PubMed]
41. Tan, S.T.; Dosan, R. Lessons From Epithelialization: The Reason Behind Moist Wound Environment. *TODJ* **2019**, *13*, 34–40. [CrossRef]
42. Yuan, Z.; Zakhaleva, J.; Ren, H.; Liu, J.; Chen, W.; Pan, Y. Noninvasive and High-Resolution Optical Monitoring of Healing of Diabetic Dermal Excisional Wounds Implanted with Biodegradable In Situ Gelable Hydrogels. *Tissue Eng. Part C: Methods* **2010**, *16*, 237–247. [CrossRef] [PubMed]
43. Bulygin, A.D.; Vrazhnov, D.A.; Sim, E.S.; Meglinski, I.; Kistenev, Y.V. Imitation of Optical Coherence Tomography Images by Wave Monte Carlo-Based Approach Implemented with the Leontovich–Fock Equation. *Opt. Eng.* **2020**, *59*, 1. [CrossRef]
44. Kistenev, Y.; Buligin, A.D.; Sandykova, E.; Sim, E.; Vrazhnov, D. *Optical Coherence Tomography Modeling Method Based on Leontovich—Fock Equation. Proceedings of the 25th International Symposium on Atmospheric and Ocean Optics: Atmospheric Physics*; Matvienko, G.G., Romanovskii, O.A., Eds.; SPIE: Novosibirsk, Russia, 2019; p. 130.
45. Jaspers, M.E.H.; Feroldi, F.; Vlig, M.; de Boer, J.F.; van Zuijlen, P.P.M. In Vivo Polarization-Sensitive Optical Coherence Tomography of Human Burn Scars: Birefringence Quantification and Correspondence with Histologically Determined Collagen Density. *J. Biomed. Opt.* **2017**, *22*, 1. [CrossRef]
46. Chen, C.-L.; Wang, R.K. Optical Coherence Tomography Based Angiography [Invited]. *Biomed. Opt. Expr.* **2017**, *8*, 1056. [CrossRef] [PubMed]
47. Kistenev, Y.V.; Kuzmin, D.A.; Sandykova, E.A.; Shapovalov, A.V. *Quantitative Comparison of the Absorption Spectra of the Gas Mixtures in Analogy to the Criterion of Pearson*; Romanovskii, O.A., Ed.; SPIE Proceedings: Tomsk, Russia, 2015; p. 96803S.
48. Zein, R.; Selting, W.; Hamblin, M.R. Review of Light Parameters and Photobiomodulation Efficacy: Dive into Complexity. *J. Biomed. Opt.* **2018**, *23*, 1. [CrossRef] [PubMed]
49. Kilík, R.; Lakyová, L.; Sabo, J.; Kruzliak, P.; Lacjaková, K.; Vasilenko, T.; Vidová, M.; Longauer, F.; Radoňak, J. Effect of Equal Daily Doses Achieved by Different Power Densities of Low-Level Laser Therapy at 635 Nm on Open Skin Wound Healing in Normal and Diabetic Rats. *BioMed. Res. Int.* **2014**, *2014*, 1–9. [CrossRef] [PubMed]
50. Zuhayri, H.; Nikolaev, V.V.; Knyazkova, A.I.; Lepekhina, T.B.; Krivova, N.A.; Tuchin, V.V.; Kistenev, Y.V. In Vivo Quantification of the Effectiveness of Topical Low-Dose Photodynamic Therapy in Wound Healing Using Two-Photon Microscopy. *Pharmaceutics* **2022**, *14*, 287. [CrossRef]

 *pharmaceutics*

Article

In Vitro Effect of Photodynamic Therapy with Different Lights and Combined or Uncombined with Chlorhexidine on *Candida* spp.

Vanesa Pérez-Laguna [1,*], Yolanda Barrena-López [2], Yolanda Gilaberte [3,4,†] and Antonio Rezusta [2,3,5,†]

1   Max Planck Institute for Evolutionary Biology, 24306 Plön, Germany
2   Department of Microbiology, Preventive Medicine and Public Health, University of Zaragoza, 50001 Zaragoza, Spain; yolandabl265@gmail.com (Y.B.-L.); arezusta@unizar.es (A.R.)
3   Aragon Health Research Institute (IIS Aragón), 50009 Zaragoza, Spain; ygilaberte@salud.aragon.es
4   Department of Dermatology, Miguel Servet University Hospital, 50009 Zaragoza, Spain
5   Department of Microbiology, Miguel Servet University Hospital, 50009 Zaragoza, Spain
*   Correspondence: perezlaguna@evolbio.mpg.de
†   These authors contributed equally to this work.

Abstract: Candidiasis is very common and complicated to treat in some cases due to increased resistance to antifungals. Antimicrobial photodynamic therapy (aPDT) is a promising alternative treatment. It is based on the principle that light of a specific wavelength activates a photosensitizer molecule resulting in the generation of reactive oxygen species that are able to kill pathogens. The aim here is the in vitro photoinactivation of three strains of *Candida* spp., *Candida albicans* ATCC 10231, *Candida parapsilosis* ATCC 22019 and *Candida krusei* ATCC 6258, using aPDT with different sources of irradiation and the photosensitizer methylene blue (MB), alone or in combination with chlorhexidine (CHX). Irradiation was carried out at a fluence of 18 J/cm^2 with a light-emitting diode (LED) lamp emitting in red (625 nm) or a white metal halide lamp (WMH) that emits at broad-spectrum white light (420–700 nm). After the photodynamic treatment, the antimicrobial effect is evaluated by counting colony forming units (CFU). MB-aPDT produces a 6 log$_{10}$ reduction in the number of CFU/100 µL of *Candida* spp., and the combination with CHX enhances the effect of photoinactivation (effect achieved with lower concentration of MB). Both lamps have similar efficiencies, but the WMH lamp is slightly more efficient. This work opens the doors to a possible clinical application of the combination for resistant or persistent forms of Candida infections.

Keywords: candidiasis; *C. albicans*; antimicrobial photodynamic therapy; methylene blue

Citation: Pérez-Laguna, V.; Barrena-López, Y.; Gilaberte, Y.; Rezusta, A. In Vitro Effect of Photodynamic Therapy with Different Lights and Combined or Uncombined with Chlorhexidine on *Candida* spp. *Pharmaceutics* **2021**, *13*, 1176. https://doi.org/10.3390/pharmaceutics13081176

Academic Editors: Udo Bakowsky, Matthias Wojcik, Eduard Preis and Gerhard Litscher

Received: 18 July 2021
Accepted: 28 July 2021
Published: 30 July 2021

1. Introduction

Candida spp. are commensal fungal species commonly colonizing human mucosal and skin surfaces, but they may become pathogenic in some particular scenarios such as treatment with antibiotics, immunocompromised patients, etc., producing in these cases infections that range from superficial to severe skin and mucosal lesions, to even systemic invasion at its worst degree [1]. For example, oral candidiasis is the most common opportunistic infection affecting the human oral cavity. It is caused by an overgrowth of *Candida* spp., being the most prevalent *Candida albicans* [2,3].

Due to the recurrence of *Candida* spp. infections, high systemic antifungal therapy have been widely used, thereby antifungal resistances are increasing. Moreover, patient-dependent, interactions with other medical regimens and organ toxicity can happen [4].

Therefore, it is necessary to develop new treatments such as antimicrobial photodynamic therapy (aPDT). It is based on the use of photosensitizing molecules that are excited with visible light of the appropriate wavelength and reacts with the oxygen, generating reactive species of oxygen to destroy the target pathogen [5–7].

Superficial wound infections are potentially suitable for treatment by aPDT because of the ready accessibility of these wounds for both topical delivery of the photosensitizer and light, and because of the exposure to oxygen [6,8,9].

Different aPDT studies have demonstrated that *Candida* spp. can be effectively photoinactivated in vitro and in vivo [5,10–13].

Future directions of aPDT include the combination with antimicrobials in order to enhance the microbial inactivation and prevent the regrowth when the light from aPDT is turned off and the photoinactivation ends. This original approach has already shown significant potential. It could help to implement the use of aPDT and reduce the amount of antimicrobials used and, thus, the multidrug resistance problem [14–16].

Chlorhexidine (CHX) is an antiseptic drug, mainly available in over-the-counter products as routine hand hygiene in healthcare personnel, to clean and prepare the skin before surgery, and before injections in order to help reduce the amount of microorganisms that potentially can cause skin infections [17–20]. CHX gluconate is also available as a prescription mouthwash to treat gingivitis and as a prescription oral chip to treat periodontal disease [21–23] and recently against COVID-19 in dentistry [24].

Here, we investigate the aPDT and the CHX uncombined or in combination against *Candida* spp. As a photosensitizing molecule, we use methylene blue (MB), the main member of the phenothiazine family, well known for its ability to produce singlet oxygen when it is irradiated by red light and react with molecular oxygen [6,8]. As a source of irradiation, we use a light-emitting diode (LED) lamp emitting in red or a white metal halide lamp (WMH) that emits at broad-spectrum white light which is comparable to the emission spectrum of daylight.

The aim is to compare the antimicrobial effect of MB-aPDT when a specific irradiation source or a non-specific broad-spectrum source is used to excite different concentrations of MB. Furthermore, the effects of the combination of aPDT with CHX are evaluated.

2. Materials and Methods

The procedure used tried to follow the materials and methods of our previous works and was adapted as follows [25–27]:

2.1. Chemicals, Media, Strains and Light Sources

- Solvent: Distilled water.
- Culture Media: Sabouraud dextrose agar (CM0041 Oxoid®, Thermo Scientific, Waltham, MA, USA) and Columbia blood agar BA (Oxoid®; Madrid, Spain).
- Antiseptic: Chlorhexidine (CHX) (CN162301.0, Miclorbic®, Madrid, Spain). Stock CHX solution was diluted in distilled water. CHX was applied at a concentration of 10 µg/mL.
- Photosensitizer: Methylene blue (MB), (Sigma-Aldrich®; Madrid, Spain). Stock MB solution was diluted in distilled water. All solutions were prepared no more than a week prior to use and handled under light-restricted conditions. The concentration ranges from 640 to 0.03 µg/mL were used.
- Strains: *C. albicans*, *C. parapsilosis* and *C. krusei* were acquired from the American Type Culture Collection (ATCC, Rockville, MD, USA). *C. albicans* ATCC 10231, *C. parapsilosis* ATCC 22019 and *C. krusei* ATCC 6258 were used.
- Light sources: Light-emitting diode (LED) lamp, Showtec LED Par 64 Short 18 × RGB 3-in-1 LED, Highlite International, emitting at 625 ± 10 nm (power density 7 mW/cm^2 at a distance between the LEDs and the microtiter plate with the microbial suspension of 17 cm where the diameter of the light beam is approximately 25 cm) and white metal halide lamp (WMH), made by the Department of Applied Physics of the University of Zaragoza, Spain, emitting at 420–700 nm (power density 90 mW/cm^2 at a distance between the lamp and the 96-well microtiter plate of 10 cm where the diameter of the light beam is approximately 21 cm). Supplementary Figure S1 shows the lamps and their emission spectrums. Both were used at a fluence of 18 J/cm^2. This

fluence corresponds to a 42.86 min ($\approx$43 min) irradiation time for the samples using the red-LED lamp and 3 min and 25 sec for the samples irradiated with WMH lamp.

2.2. In Vitro Photodynamic Treatment of Yeast Suspension

C. albicans, *C. parapsilosis* or *C. krusei* seeded on Sabouraud dextrose agar were cultured aerobically overnight at 35 °C. The inoculum was prepared in distilled water and adjusted to 5 ± 0.03 on the McFarland scale (concentrations in the range of >1×10^6 colony-forming units (CFU) per 100 µL and was deposited into 96-well microtiter plates. Two-fold serial dilutions concentrations from 640 µg/mL to 0.03 µg/mL of the MB were added, in absence or presence of 10 µg/mL of CHX (MB+/CHX−/light+) (MB+/CHX+/light+). The final volume in each well was 100 µL. Irradiation proceeded with no preincubation period; the suspensions were immediately subjected to irradiation with fluence of 18 J/cm^2 using the red-LED lamp or the broad spectrum-WMH lamp. Control samples were subjected to identical treatment, in the absence or presence of the photosensitizer, and were either kept in darkness or irradiated to evaluate the effect of each parameter: negative or initial control (MB−/CHX−/light−), irradiation control (MB−/CHX−/light+), control of photosensitizer in darkness (MB+/CHX−/light−) and antiseptic controls (MB−/CHX+/light−) (MB−/CHX+/light+). After completing the aPDT protocol, samples and controls were assessed in serial dilutions of each suspension and were cultured on blood agar and incubated overnight at 35 °C. The dilutions were made and aliquots were cultured to have blood agar plates with a number of CFUs in the range of 0 to 200 per plate in order to be able to count them reliably.

2.3. Efficacy

The efficacy of aPDT treatment was assessed by counting the number of CFU/100 µL using a Flash and Go automatic colony counter (IUL S.A., Barcelona, Spain). A reduction of 6 log$_{10}$ in the number of CFU/100 µL was considered indicative of fungicidal activity. The minimum concentration of MB that reduced yeast survival by 3 log$_{10}$ was also evaluated. All experiments were carried out at least five times. The results are expressed as mean and standard deviation.

3. Results

3.1. Photoinactivation of Yeasts by MB-aPDT (MB+/CHX−/Light+)

MB-aPDT effectively inactivated *Candida* spp. achieving a reduction of 6 log$_{10}$ in the number of CFU/100 µL in all the studied strains (Figure 1). The minimum concentration of MB required to achieve this effect was 320 µg/mL in all cases except in those irradiated with a WMH lamp in *C. parapsilosis* that required 80 µg/mL and in *C. krusei* between 320–640 µg/mL (Table 1). Analyzing in more detail the sensitivity of each strain to MB-aPDT, *C. krusei* is the most resistant and *C. parapsilosis* and *C. albicans* show a very similar ratio of response, although *C. parapsilosis* is slightly more sensitive to white light than *C. albicans* (Figure 1 and Supplementary Material Figure S2I).

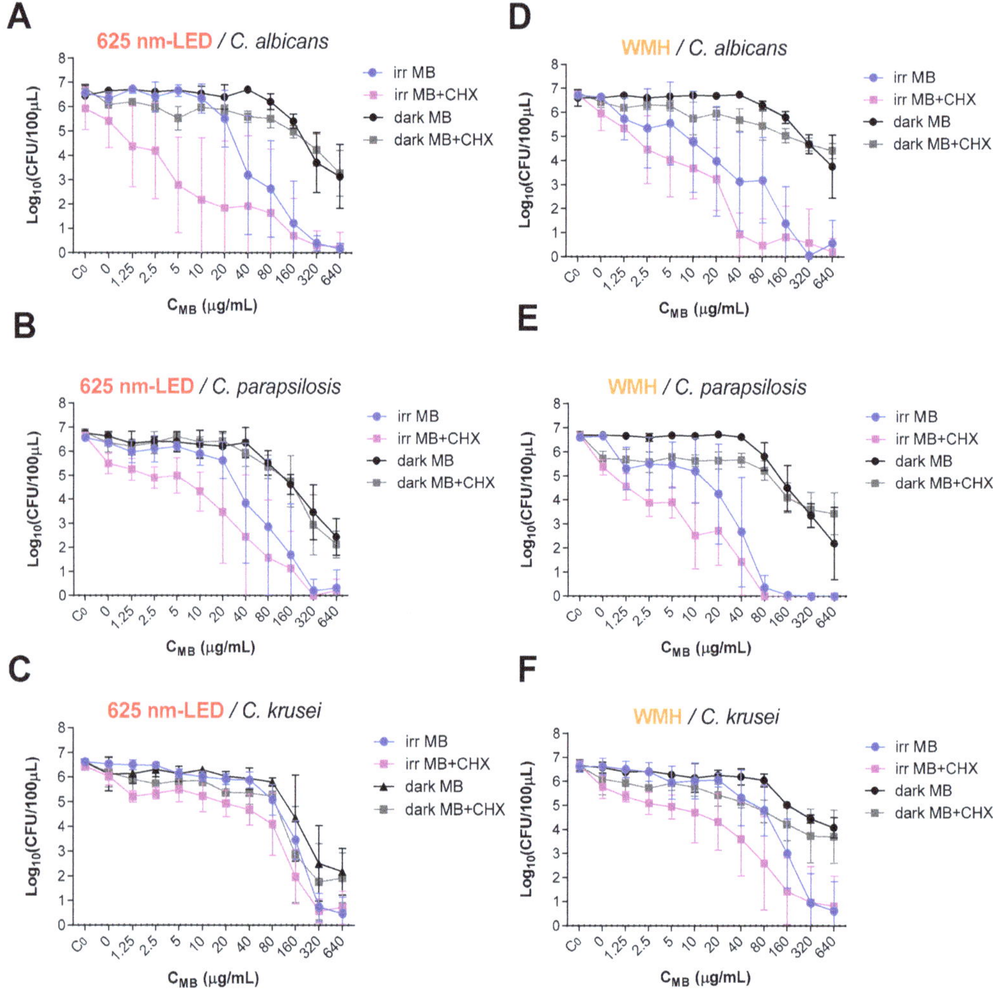

Figure 1. Photoinactivation by antimicrobial photodynamic therapy with methylene blue alone or in combination with chlorhexidine of *C. albicans* (**A**,**D**) *C. parapsilosis* (**B**,**E**) and *C. krusei* (**C**,**F**) using the 625 nm lamp-LED (**A**–**C**) or the WMH lamp (**D**–**F**). The error bars represent the standard deviation calculated for five measurements. C_0, initial inoculum control; CHX, chlorhexidine; LEDs, light-emitting diodes; MB, methylene blue; WMH, white metal halide.

Table 1. Minimum concentrations of methylene blue (µg/mL) required to reduce the number of *Candida* spp. 3 or 6 $\log_{10}$ in the number of colony forming units using the 625 nm-LED lamp or the WMH lamp.

Reduction in the Number of CFU/100 µL	Lamp Used	Treatment	MB Concentration Required for Each Yeast		
			C. albicans	*C. parapsilosis*	*C. krusei*
3 $\log_{10}$	625 nm LED-lamp	MB-aPDT	40	40–80	160
		MB-aPDT + CHX	5	20	80–160
	WMH lamp	MB-aPDT	40	20–40	80–160
		MB-aPDT + CHX	20	5–10	40–80
6 $\log_{10}$	625 nm LED-lamp	MB-aPDT	320	320	320
		MB-aPDT + CHX	320	320	320
	WMH lamp	MB-aPDT	320	80	320–640
		MB-aPDT + CHX	80	80	320–640

aPDT: antimicrobial photodynamic therapy; CHX: chlorhexidine; CFU: colony forming units; LED: light-emitting diodes; MB: methylene blue; WMH: white metal halide.

3.2. Fungicidal Effect of MB-aPDT Combined with CHX (MB+/CHX+/Light+)

The antimicrobial effect of MB-aPDT on *Candida* spp. was maintained in the presence of CHX, as evidenced by the 6 $\log_{10}$ reduction in the number of CFU/100 µL in all experiments. Moreover, the combination of MB-aPDT using the WMH lamp + CHX achieves this degree of reduction on *C. albicans* decreasing 4-fold the required photosensitizer concentration (the necessary concentration is 1/4) (Figure 1 and Table 1).

To achieve a 3 $\log_{10}$ reduction in the number of CFU/100 µL when MB-aPDT is used in combination with CHX, the required concentration of MB needed is at least half compared to the concentration needed using MB-aPDT alone. The greatest reduction of the photosensitizer concentration (8-fold) is achieved against *C. albicans* using the red-LED lamp. On the other hand, the greatest reduction against *C. parapsilosis* occurs with the WMH irradiation (1/4–1/8 of the initial photosensitizer concentration) (Figure 1 and Table 1).

3.3. Control of Inoculum and Toxic Effects of MB (MB+/CHX−/Light−), CHX (MB−/CHX+/Light−) and Irradiation (MB−/CHX−/Light+)

No reduction in the number of CFU/100 µL from the initial inoculums (MB−/CHX−/light−) was observed.

Samples with the different MB concentrations evaluated under the same conditions used in irradiation but keeping it in darkness (MB+/CHX−/light−) (dark MB in Figure 1) show significant reductions at the highest concentrations tested as follows: reductions of up to a maximum of 3.5 $\log_{10}$ in *C. albicans*, 4 $\log_{10}$ in *C. parapsilosis* and 4.5 $\log_{10}$ in *C. krusei* were achieved by 640 µg/mL of MB. In all experiments, the effects of keeping the microbial suspension with the different MB concentrations in dark (light−) for 43 min or 3 min and 25 sec (using the time of the irradiation with the red-LED or the WMH lamp respectively) is similar, except for *C. krusei* (reduction of 4.5 $\log_{10}$ after 43 min vs. 2.5 $\log_{10}$ after 3 min and 25 sec) (Figure 1).

The irradiation with the red-LED lamp or with WMH lamp in the absence of photosensitizer and antimicrobial (MB−/CHX−/light+) did no significantly reduce the number of yeasts (reduction $\leq$0.3 $\log_{10}$, Figure 1).

In the absence of photosensitizer and irradiation, the tested concentration of CHX (10 µg/mL) (MB−/CHX+/light−) (dark MB-CHX with the value of 0 MB concentration in Figure 1) failed to effectively inactivate the yeast. A maximum reduction of 1 $\log_{10}$ was observed against *C. parapsilosis*.

The cumulative effect of CHX and irradiation (MB−/CHX+/light+) (Figure 1) reaches a maximum reduction in the number of CFU/100 µL of 1.3 $\log_{10}$ against *C. parapsilosis* being the most sensitive strain to this effect.

4. Discussion

MB-aPDT is effective in eradicating *Candida* spp. (>6 $\log_{10}$ reduction in the number of CFU/100 μL of *C. albicans*, *C. parapsilosis* or *C. krusei*) and the combination with CHX enhances the photoinactivation, i.e., the effect is achieved with lower aPDT-dose (Figure 1 and Table 1).

Regarding the comparison of MB-aPDT results obtained with those reported by other authors, many variables should be considered. Table 2 summarizes different studies against *Candida* spp. in suspension, specifying the methodology and results. Daliri et al. reported a reduction of 3.43 $\log_{10}$ of *C. albicans* using 200 μg/mL of MB which is notably lower than the one reached in the present study (MB concentration range of 80–160 μg/mL is able to inhibit 4 $\log_{10}$). They used a bigger number of CFU in the inoculum and this could affect but the mismatch may be because they use a laser irradiation source [28]. Application times are usually short when lasers are used and it does not always guarantee adequate oxygenation [6]. Valkov et al. report the absence of effect of MB-aPDT using very low MB concentration (<2 μg/mL) [13]. Ferreira et al. and de Oliveira-Silva et al. achieved different reductions of *C. albicans*, 0.5 $\log_{10}$ and 6 $\log_{10}$ respectively, with 32 μg/mL of MB and a fluence of 30 J/cm^2 for 3–4 min. This shows the variability between experiments [29,30]. The results of Ferreira et al. are closer to those of this work (32 μg/mL of MB does not produce complete photoinactivation) (Table 2).

Focusing on *C. parapsilosis*, Güzel Tunçcan et al. achieved a reduction of 4 $\log_{10}$ with 25 μg/mL of MB. The comparison with our data and the possible explanation is very difficult because the methodology used is dissimilar [31]. Černáková et al. demonstrated that using 9.6 μg/mL of MB inhibited between 1.13–1.27 $\log_{10}$ of *C. parapsilosis* in suspension, similar results to this work (this concentration does not generate complete photoinactivation) [32]. Finally, Ahmed et al. used 100 μg/mL of MB and achieved reductions of 0.58–0.85 $\log_{10}$ at fluences of 90–180 J/cm^2 respectively [33]. Again, the difference may be due to the fact that they used a laser irradiation source and therefore it would be less effective (Table 2).

Against *C. krusei*, concentrations of 16 μg/mL [34] or150 μg/mL of MB [35] even at high fluences only achieves a maximum reduction of 0.65 $\log_{10}$. More similar result to ours was obtained by Souza et al. using 100 μg/mL with a reduction of 1.54 $\log_{10}$ using a fluence of 28 J/cm^2 [36]. All MB-aPDT studies together lead us to conclude that *C. krusei* is the most resistant *Candida* spp. to MB-aPDT as well as it is more resistant to antifungals in general mainly due to the characteristics of its membrane [37] (Table 2).

Regarding the MB-aPDT combination with CHX, it stands out that >6 $\log_{10}$ reduction in the number of CFU/100 μL of *C. albicans* was achieved reducing the concentration of photosensitizer needed from 320 to 80 μg/mL when WMH lamp was used (Figure 1, Table 1 and Supplementary Figure S2). Furthermore, the addition of CHX halved the concentration of MB required to reach a reduction of 3 $\log_{10}$ in *C. albicans* and *C parapsilosis*, and slightly less than half against *C. krusei*. Therefore, a synergistic effect is seen between MB at concentrations unable to achieve complete photoinactivation and CHX. These results are relevant because the presence of CHX could help to avoid the microbial regrowth of those microorganisms not completely destroyed when PDT is finished. This is one of the disadvantages of using aPDT for infections in the clinic, the risk of microbial regrowth after its application. The combination with antimicrobials could play a crucial role to overcome this limitation of aPDT in this context [14,15].

To our knowledge, there are not studies combining aPDT plus CHX in vivo against *Candida* spp. Recently, the effectiveness of MB-aPDT combined with CHX and zinc oxide ointment has been studied on wound healing process after rumenostomy. This study in cattle ratifies the use of aPDT and suggests that it could be performed for other surgical procedures as a complementary approach or an alternative for topical administration of antibiotics [38]. The combination of CHX plus aPDT has been tried against other microorganisms such as *Porphyromonas gingivalis* biofilm on a titanium surface in a dental framework. The application of CHX and subsequent aPDT using toluidine blue O was shown to be an efficient method to reduce *P. gingivalis* in titanium surfaces [39]. Regarding other studies of antimicrobials plus aPDT against *Candida* spp., Giroldo et al. demonstrated that yeasts, both in suspension and in biofilms, were much more susceptible to antifungal treatments after MB-aPDT, explained by the increase of membrane permeability caused by aPDT [40]. Regarding the in vitro combination of MB-aPDT with fluconazole against resistant strains of *C. albicans, C. glabrata* and *C. krusei*, a synergistic effect was found in fluconazole resistant strains of *C. albicans* and *C. glabrata*, but not against *C. krusei*. [34]. These results do not agree with those found by Snell et al. They showed that fluconazole did not increase the aPDT inactivation of *C. albicans* using MB or another photosensitizer of the protoporphyrin family. However, miconazole did enhance the fungicidal activity of aPDT [41]. Moreover, to our knowledge, there are no studies using aPDT in combination with antimicrobials that report antagonistic effects, which support the use of aPDT in combination with CXH due to the possible advantages [15].

Considering clinical practice, MB-aPDT (660 nm and 7.5 J/cm^2) has been tried in HIV patients diagnosed with oral candidiasis comparing it with an antifungal commonly used in candidiasis. After 30 days, the antimicrobial was effective, but there were recurrences except when 450 μg/mL of MB was used [42].

All together indicates that aPDT or antimicrobial alone may not be entirely effective against *Candida* spp. that is characterized for causing highly recurrent infection especially in predisposed or immunosuppressed patients. On the other hand, combined treatments such as aPDT plus antimicrobials may prevent recurrent infections and avoid resistance. In addition, the combination in terms of clinical application would decrease the intensity of blue staining caused when the MB is applied on the skin or mucous membranes, making the aPDT procedure more cosmetically appealing.

Regarding the concentration of 10 μg/mL CHX used for this study, it was chosen by taking into account other protocols for the combination of antimicrobials plus aPDT and considering that by itself produces no effect under experimental conditions [25,27], Figure 1 C_0-CHX. Other studies using other conditions report very different results: e.g., Azizi et al. using 1000 μg/mL of CHX achieved a reduction of 0.71 $\log_{10}$ [43]; Do Vale et al. calculated that the minimum inhibitory concentration for *C. albicans* was 3.74 μg/mL using an exposure time of 12–48 h [44]; Ellepola et al. proved that with 50 μg/mL of CHX applied for 30 min achieved 0.38 and 0.5 $\log_{10}$ of *C. albicans* and *C. krusei* reduction respectively [45].

Regarding the use of the red-LED lamp or the WHM lamp as a source of irradiation for aPDT, the second proved to be more effective in photoinactivating *Candida* spp. with the exception of against *C. krusei* (Figure 1, Table 1 and Supplementary Material Figure S2II). The use of LED lamp emitting in red matching the absorption spectra peak of the MB tends to be more efficient in the sense of not wasting irradiation energy and therefore red emission sources are usually the ones chosen for MB-aPDT studies (e.g., shown in Table 2). In addition, red LEDs lamps have added advantages at the time of transferring the use of aPDT to clinical application because they are available in all the PDT clinical units; in addition, these lamps stimulate cellular repair mechanisms in fibroblasts [46] and are already used to treat acne vulgaris, herpes simplex virus infection, shingles, or severe wound healing [6,47].

It is also worth noting the time factor to facilitate the use in the clinic since the WMH lamp needs 3 min and 25 seconds to photoinactivate *Candida* spp. compared to 43 min for the LED lamp, due to the greater irradiance of the former compare to the latter (90 mW/cm^2 vs. 7 mW/cm^2 respectively). Furthermore, the experiments were performed without preincubation of the photosensitizer MB with *Candida* spp. prior to irradiation. Andrade et al. and Soria-Lozano et al. demonstrated that a pre-incubation time did not produce greater inactivation of the microorganism, so it is not necessary to add more time to the aPDT procedure [12,48].

On the other hand, a broad-spectrum WMH lamp could be a model of daylight, i.e., which could be used as a source of irradiation for aPDT instead of this lamp. The advantages are that the treatment could be carried out at home and it would require less equipment and personnel (cheaper). However, it also has disadvantages, such as the imprecision in the quantification of the dose of light or duration of exposure, considering that intensity of daylight depends on the season of the year, weather conditions, or geographic location [9,49–51]. Another limitation for the use of daylight is the limitation to treat Candida infections not accessible for this light, such as the mouth or the genitalia, which otherwise are the most frequent. Nevertheless, the WMH lamp achieves better results than the LED lamp in this work, demonstrating its efficacy.

Table 2. Representative examples reported in literature of planktonic *C. albicans*, *C. parapsilosis* and *C. krusei* photoinactivation caused by MB-aPDT.

Study	Strain	Concentration (μg/mL)	Media	Source and Wavelength (nm)	Fluence (J/cm^2)	Irradiance (mW/cm^2)	Initial Load (CFU/mL)	Load Reduction ($\log_{10}$)
Güzel Tunçcan et al. (2018) [31]	*C. albicans* ATCC 90028	25	saline	LED-660	0.233	ND	10^6	$3\log_{10}$
de Oliveira-Silva et al. (2019) [29]	*C. albicans* ATCC 10231	32	PBS	LED-660	10 30 60	165	2.5×10^6	$0.5\log_{10}$ $6\log_{10}$ $6\log_{10}$
Ferreira et al. (2016) [30]	*C. albicans* ATCC 90028	32	ND	LED-660	30 60 120	250	6.31×10^5	$0.5\log_{10}$ $6\log_{10}$ $6\log_{10}$
Daliri et al. (2019) [28]	*C. albicans* ATCC 10231	100 200	ND	Laser-660	ND	ND	1.5×10^8	$3.3\log_{10}$ $3.43\log_{10}$
Torres-Hurtado et al. (2019) [52]	*C. albicans*	6.4	PBS	LED-600-650	60	85	2–4×10^5	$>5\log_{10}$
Souza et al. (2010) [53]	*C. albicans* ATCC 18804	100	saline 0.85%	Laser-660	39.5	92	10^6	$3\log_{10}$
Peloi et al. (2008) [54]	*C. albicans* ATCC 90028	22.5	saline 0.85%	LED-663	6	ND	1–2×10^8	$1.31\log_{10}$
Souza et al. (2006) [36]	*C. albicans* ATCC 18804	100	saline 0.85%	Laser-685	28	92	10^6	$1.25\log_{10}$
Valkov et al. (2021) [13]	*C. albicans* ATCC 90028	1.6	saline 0.90%	18 W white luminescent lamp-400–700	27	1.9 ± 0.1	1–3×10^6	0
Soria-Lozano et al. (2015) [12]	*C. albicans* ATCC 10231	160	sterile distilled water	WMH-420-700	37	90	$1 \times 10^{6-7}$	$5\log_{10}$

Table 2. *Cont.*

Study	Strain	Concentration (µg/mL)	Media	Source and Wavelength (nm)	Fluence (J/cm^2)	Irradiance (mW/cm^2)	Initial Load (CFU/mL)	Load Reduction (log$_{10}$)
This work	*C. albicans* ATCC 10231	320	sterile distilled water	LED-625	18	7	>10^6	6 log$_{10}$
		320		WMH-420-700		90		6 log$_{10}$
Güzel Tunçcan et al. (2018) [31]	*C. parapsilosis* ATCC 96142	25	saline	LED-660	0.233	ND	3 × 10^6	4 log$_{10}$
Černáková et al. (2015) [32]	*C. parapsilosis* ATCC 22019	9,6	ND	LED-576-672	15	1.67	ND	1.16 log$_{10}$
	C. parapsilosis 16755/2							1.27 log$_{10}$
	C. parapsilosis 21922/1							1.13 log$_{10}$
Ahmed et al. (2016) [33]	*C. parapsilosis*	100	ND	Laser-660	90	300	350	0.59 log$_{10}$
					180			0.85 log$_{10}$
This work	*C. parapsilosis* ATCC 22019	320	sterile distilled water	LED-625	18	7	>10^6	6 log$_{10}$
		80		WMH-420-700		90		6 log$_{10}$
Lyon et al. (2016) [34]	*C. krusei*	16	ND	ND	ND	200	≈5 × 10^5	0.25 log$_{10}$
Queiroga et al. (2011) [35]	*C. krusei* (ATCC 6258, ATCC 6358, LM08, LM12, LM120)	150	saline 0.85%	Laser-660	60	1000	6 × 10^5	0.18 log$_{10}$
					120			0.40 log$_{10}$
					180			0.65 log$_{10}$
Souza et al. (2006) [36]	*C. krusei* ATCC 6258	100	saline 0.85%	Laser-685	28	92	10^6	1.54 log$_{10}$
This work	*C. krusei* ATCC 6258	320	sterile distilled water	LED-625	18	7	>10^6	6 log$_{10}$
		320–640		WMH-420-700		90		6 log$_{10}$

CFU: colony forming units; MB: methylene blue; ND: no data; PBS: phosphate buffered saline; LED: light-emitting diode lamp; WMH: metal halide lamp.

Overall, our study aims to open the way for the application of this alternative therapy, MB-aPDT alone or better in combination with CHX, either using lamps with a specific or broad- emission spectrum or even daylight as an irradiation source, to deal with cutaneous and mucosal candidiasis. However, it should be borne in mind that the present findings were obtained following in vitro irradiation of *Candida* spp., therefore clinical studies are required to confirm these results.

5. Conclusions

- MB-aPDT is active against *Candida* spp. in water suspension.
- CHX enhances the photoinactivation of *Candida* spp. (aPDT plus CHX increases the photoactivity of MB).
- White light is a suitable light source for aPDT.
- MB-aPDT using a broad-spectrum white light is more efficient than a specific red-LED lamp.
- Transfer of this therapy to the clinic could be very convenient.

Supplementary Materials: The following are available online at https://www.mdpi.com/article/ 10.3390/pharmaceutics13081176/s1, Figure S1: Photos and emission spectra of the lamps used in the photoinactivation experiments. A: graph of the emission spectrum of the red-LED lamp and B: emission spectrum of the white metal halide lamp, Figure S2: Photoinactivation of *Candida* spp. using MB-aPDT. I: Comparison of the response of yeast when they are irradiated with the red-LED lamp (left A) or with the WMH lamp (right B). II: Comparison of the response of each strain to irradiation with the two lamps (A: *C. albicans*; B: *C. parapsilosis*; C: *C. krusei*). The error bars represent the standard deviation calculated for five measurements. C_0, initial inoculum control; LEDs, light-emitting diodes; MB, methylene blue; WMH, white metal halide.

Author Contributions: Conceptualization, V.P.-L., Y.G. and A.R.; methodology, validation and figures, V.P.-L. and Y.B.-L.; analysis, V.P.-L.; resources and funding acquisition, V.P.-L., Y.G. and A.R.; writing—original draft preparation, V.P.-L.; writing—review and editing, V.P.-L., Y.B.-L., Y.G. and A.R.; supervision, V.P.-L. and A.R. All authors have read and agreed to the published version of the manuscript.

Funding: This work was supported by grant CTQ2013-48767-C3-2-R from the Spanish Ministerio de Economía y Competitividad. Antonio Rezusta and Yolanda Gilaberte were funded by the Aragón Government. B10_17R Infectious Diseases of Difficult Diagnosis and Treatment research group and B18_17D Dermatology and Photobiology research group, respectively as recognized by the Government of Aragon.

Institutional Review Board Statement: Not applicable.

Informed Consent Statement: Not applicable.

Data Availability Statement: Not applicable.

Acknowledgments: The authors thank the IIS Aragon for the GIIS-023.

Conflicts of Interest: The authors declare no conflict of interest.

References

1. Rex, J.H.; Walsh, T.J.; Sobel, J.D.; Filler, S.G.; Pappas, P.G.; Dismukes, W.E.; Edwards, J.E. Practice Guidelines for the Treatment of Candidiasis. *Clin. Infect. Dis.* **2000**, *30*, 662–678. [CrossRef] [PubMed]
2. Lalla, R.; Patton, L.; Dongari-Bagtzoglou, A. Oral Candidiasis: Pathogenesis, Clinical Presentation, Diagnosis and Treatment Strategies. *J. Calif. Dent. Assoc.* **2013**, *41*, 263–268.
3. Moran, G.; Coleman, D.; Sullivan, D. An Introduction to the Medically Important Candida Species. In *Candida and Candidiasis*, 2nd ed.; John Wiley & Sons, Ltd.: Hoboken, NJ, USA, 2011; pp. 9–25. [CrossRef]
4. Kashem, S.W.; Kaplan, D.H. Skin Immunity to Candida Albicans. *Trends Immunol.* **2016**, *37*, 440–450. [CrossRef] [PubMed]
5. Gavara, R.; de Llanos, R.; Pérez-Laguna, V.; Arnau del Valle, C.; Miravet, J.F.; Rezusta, A.; Galindo, F. Broad-Spectrum Photo-Antimicrobial Polymers Based on Cationic Polystyrene and Rose Bengal. *Front. Med.* **2021**, *8*. [CrossRef]
6. Perez-Laguna, V.; Garcia-Malinis, A.J.; Aspiroz, C.; Rezusta, A.; Gilaberte, Y. Antimicrobial Effects of Photodynamic Therapy: An Overview. *G. Ital. Dermatol. Venereol. Organo Uff. Soc. Ital. Dermatol. Sifilogr.* **2018**, *153*, 833–846. [CrossRef]

7. Rezusta, A.; López-Chicón, P.; Paz-Cristobal, M.P.; Alemany-Ribes, M.; Royo-Díez, D.; Agut, M.; Semino, C.; Nonell, S.; Revillo, M.J.; Aspiroz, C.; et al. In Vitro Fungicidal Photodynamic Effect of Hypericin on Candida Species. *Photochem. Photobiol.* **2012**, *88*, 613–619. [CrossRef]

8. Kashef, N.; Esmaeeli Djavid, G.; Siroosy, M.; Taghi Khani, A.; Hesami Zokai, F.; Fateh, M. Photodynamic Inactivation of Drug-Resistant Bacteria Isolated from Diabetic Foot Ulcers. *Iran. J. Microbiol.* **2011**, *3*, 36–41. [PubMed]

9. Pérez-Laguna, V.; Rezusta, A.; Ramos, J.J.; Ferrer, L.M.; Gené, J.; Revillo, M.J.; Gilaberte, Y. Daylight Photodynamic Therapy Using Methylene Blue to Treat Sheep with Dermatophytosis Caused by *Arthroderma vanbreuseghemii*. *Small Rumin. Res.* **2017**, *150*, 97–101. [CrossRef]

10. Paz-Cristobal, M.P.; Royo, D.; Rezusta, A.; Andrés-Ciriano, E.; Alejandre, M.C.; Meis, J.F.; Revillo, M.J.; Aspiroz, C.; Nonell, S.; Gilaberte, Y. Photodynamic Fungicidal Efficacy of Hypericin and Dimethyl Methylene Blue against Azole-Resistant *Candida albicans* Strains. *Mycoses* **2014**, *57*, 35–42. [CrossRef] [PubMed]

11. Rodrigues, C.F.; Rodrigues, M.E.; Henriques, M.C.R. Promising Alternative Therapeutics for Oral Candidiasis. *Curr. Med. Chem.* **2019**, *26*, 2515–2528. [CrossRef] [PubMed]

12. Soria-Lozano, P.; Gilaberte, Y.; Paz-Cristobal, M.; Pérez-Artiaga, L.; Lampaya-Pérez, V.; Aporta, J.; Pérez-Laguna, V.; García-Luque, I.; Revillo, M.; Rezusta, A. In Vitro Effect Photodynamic Therapy with Differents Photosensitizers on Cariogenic Microorganisms. *BMC Microbiol.* **2015**, *15*, 187. [CrossRef]

13. Valkov, A.; Zinigrad, M.; Nisnevitch, M. Photodynamic Eradication of Trichophyton Rubrum and *Candida albicans*. *Pathogens* **2021**, *10*, 263. [CrossRef] [PubMed]

14. Pérez-Laguna, V.; García-Luque, I.; Ballesta, S.; Rezusta, A.; Gilaberte, Y. Photodynamic Therapy Combined with Antibiotics or Antifungals against Microorganisms That Cause Skin and Soft Tissue Infections: A Planktonic and Biofilm Approach to Overcome Resistances. *Pharmaceuticals* **2021**, *14*, 603. [CrossRef]

15. Pérez-Laguna, V.; Gilaberte, Y.; Millán-Lou, M.I.; Agut, M.; Nonell, S.; Rezusta, A.; Hamblin, M.R. A Combination of Photodynamic Therapy and Antimicrobial Compounds to Treat Skin and Mucosal Infections: A Systematic Review. *Photochem. Photobiol. Sci.* **2019**, *18*, 1020–1029. [CrossRef]

16. Wozniak, A.; Grinholc, M. Combined Antimicrobial Activity of Photodynamic Inactivation and Antimicrobials-State of the Art. *Front. Microbiol.* **2018**, *9*, 930. [CrossRef] [PubMed]

17. Martin, V.T.; Abdullahi Abdi, M.; Li, J.; Li, D.; Wang, Z.; Zhang, X.; Elodie, W.H.; Yu, B. Preoperative Intranasal Decolonization with Topical Povidone-Iodine Antiseptic and the Incidence of Surgical Site Infection: A Review. *Med. Sci. Monit.* **2020**, *26*, e927052. [CrossRef]

18. Millán-Lou, M.I.; López, C.; Bueno, J.; Pérez-Laguna, V.; Lapresta, C.; Fuertes, M.E.; Rite, S.; Santiago, M.; Romo, M.; Samper, S.; et al. Successful Control of Serratia Marcescens Outbreak in a Neonatal Unit of a Tertiary-Care Hospital in Spain. *Enferm. Infecc. Microbiol. Clínica* **2021**. [CrossRef] [PubMed]

19. Napolitani, M.; Bezzini, D.; Moirano, F.; Bedogni, C.; Messina, G. Methods of Disinfecting Stethoscopes: Systematic Review. *Int. J. Environ. Res. Public Health* **2020**, *17*, 1856. [CrossRef]

20. Yamashita, T.; Takamori, A.; Nakagawachi, A.; Tanigawa, Y.; Hamada, Y.; Aoki, Y.; Sakaguchi, Y. Early Prophylaxis of Central Venous Catheter-Related Thrombosis Using 1% Chlorhexidine Gluconate and Chlorhexidine-Gel-Impregnated Dressings: A Retrospective Cohort Study. *Sci. Rep.* **2020**, *10*, 15952. [CrossRef] [PubMed]

21. Ellepola, A.N.; Samaranayake, L.P. Adjunctive Use of Chlorhexidine in Oral Candidoses: A Review. *Oral Dis.* **2001**, *7*, 11–17. [CrossRef]

22. Fathilah, A.R.; Himratul-Aznita, W.H.; Fatheen, A.R.N.; Suriani, K.R. The Antifungal Properties of Chlorhexidine Digluconate and Cetylpyrinidinium Chloride on Oral Candida. *J. Dent.* **2012**, *40*, 609–615. [CrossRef] [PubMed]

23. Suci, P.A.; Tyler, B.J. Action of Chlorhexidine Digluconate against Yeast and Filamentous Forms in an Early-Stage *Candida albicans* Biofilm. *Antimicrob. Agents Chemother.* **2002**, *46*, 3522–3531. [CrossRef]

24. Vergara-Buenaventura, A.; Castro-Ruiz, C. Use of Mouthwashes against COVID-19 in Dentistry. *Br. J. Oral Maxillofac. Surg.* **2020**, *58*, 924–927. [CrossRef] [PubMed]

25. Pérez-Laguna, V.; García-Luque, I.; Ballesta, S.; Pérez-Artiaga, L.; Lampaya-Pérez, V.; Rezusta, A.; Gilaberte, Y. Photodynamic Therapy Using Methylene Blue, Combined or Not with Gentamicin, against *Staphylococcus aureus* and *Pseudomonas aeruginosa*. *Photodiagnosis Photodyn. Ther.* **2020**, *31*, 101810. [CrossRef]

26. Pérez-Laguna, V.; García-Luque, I.; Ballesta, S.; Pérez-Artiaga, L.; Lampaya-Pérez, V.; Samper, S.; Soria-Lozano, P.; Rezusta, A.; Gilaberte, Y. Antimicrobial Photodynamic Activity of Rose Bengal, Alone or in Combination with Gentamicin, against Planktonic and Biofilm Staphylococcus Aureus. *Photodiagnosis Photodyn. Ther.* **2018**, *21*, 211–216. [CrossRef] [PubMed]

27. Pérez-Laguna, V.; Pérez-Artiaga, L.; Lampaya-Pérez, V.; García-Luque, I.; Ballesta, S.; Nonell, S.; Paz-Cristobal, M.P.; Gilaberte, Y.; Rezusta, A. Bactericidal Effect of Photodynamic Therapy, Alone or in Combination with Mupirocin or Linezolid, on Staphylococcus Aureus. *Front. Microbiol.* **2017**. [CrossRef]

28. Daliri, F.; Azizi, A.; Goudarzi, M.; Lawaf, S.; Rahimi, A. In Vitro Comparison of the Effect of Photodynamic Therapy with Curcumin and Methylene Blue on *Candida albicans* Colonies. *Photodiagnosis Photodyn. Ther.* **2019**, *26*, 193–198. [CrossRef] [PubMed]

29. de Oliveira-Silva, T.; Alvarenga, L.H.; Lima-Leal, C.; Godoy-Miranda, B.; Carribeiro, P.; Suzuki, L.C.; Simões Ribeiro, M.; Tiemy Kato, I.; Pavani, C.; Prates, R.A. Effect of Photodynamic Antimicrobial Chemotherapy on *Candida albicans* in the Presence of Glucose. *Photodiagnosis Photodyn. Ther.* **2019**, *27*, 54–58. [CrossRef]

30. Ferreira, L.R.; Sousa, A.S.; Alvarenga, L.H.; Deana, A.M.; de Santi, M.E.O.S.; Kato, I.T.; Leal, C.R.L.; Ribeiro, M.S.; Prates, R.A. Antimicrobial Photodynamic Therapy on *Candida albicans* Pre-Treated by Fluconazole Delayed Yeast Inactivation. *Photodiagnosis Photodyn. Ther.* **2016**, *15*, 25–27. [CrossRef]

31. Güzel Tunçcan, Ö.; Kalkancı, A.; Unal, E.A.; Abdulmajed, O.; Erdoğan, M.; Dizbay, M.; Çağlar, K. The In Vitro Effect of Antimicrobial Photodynamic Therapy on Candida and Staphylococcus Biofilms. *Turk. J. Med. Sci.* **2018**, *48*, 873–879. [CrossRef]

32. Černáková, L.; Chupáčová, J.; Židlíková, K.; Bujdáková, H. Effectiveness of the Photoactive Dye Methylene Blue versus Caspofungin on the *Candida Parapsilosis* Biofilm In Vitro and Ex Vivo. *Photochem. Photobiol.* **2015**, *91*, 1181–1190. [CrossRef]

33. Ahmed, H.Y.A.; Mekawey, A.A.I.; Morsy, M.E. Cellular Structural Changes in Two Candida Species Caused by Photodynamic Therapy. *Merit Res. J. Microbiol. Biol. Sci.* **2016**, *4*, 39–54.

34. Lyon, J.P.; Carvalho, C.R.; Rezende, R.R.; Lima, C.J.; Santos, F.V.; Moreira, L.M. Synergism between Fluconazole and Methylene Blue-Photodynamic Therapy against Fluconazole-Resistant Candida Strains. *Indian J. Med. Microbiol.* **2016**, *34*, 506–508. [CrossRef]

35. Queiroga, A.S.; Trajano, V.N.; Lima, E.O.; Ferreira, A.F.M.; Queiroga, A.S.; Limeira, F.A. In Vitro Photodynamic Inactivation of *Candida* spp. by Different Doses of Low Power Laser Light. *Photodiagnosis Photodyn. Ther.* **2011**, *8*, 332–336. [CrossRef] [PubMed]

36. de Souza, S.C.; Junqueira, J.C.; Balducci, I.; Koga-Ito, C.Y.; Munin, E.; Jorge, A.O.C. Photosensitization of Different Candida Species by Low Power Laser Light. *J. Photochem. Photobiol. B Biol.* **2006**, *83*, 34–38. [CrossRef]

37. Jamiu, A.T.; Albertyn, J.; Sebolai, O.M.; Pohl, C.H. Update on Candida Krusei, a Potential Multidrug-Resistant Pathogen. *Med. Mycol.* **2021**, *59*, 14–30. [CrossRef]

38. Valandro, P.; Massuda, M.B.; Rusch, E.; Birgel, D.B.; Pereira, P.P.L.; Sellera, F.P.; Ribeiro, M.S.; Pogliani, F.C.; Birgel Junior, E.H. Antimicrobial Photodynamic Therapy Can Be an Effective Adjuvant for Surgical Wound Healing in Cattle. *Photodiagnosis Photodyn. Ther.* **2021**, *33*, 102168. [CrossRef] [PubMed]

39. Cai, Z.; Li, Y.; Wang, Y.; Chen, S.; Jiang, S.; Ge, H.; Lei, L.; Huang, X. Disinfect *Porphyromonas gingivalis* Biofilm on Titanium Surface with Combined Application of Chlorhexidine and Antimicrobial Photodynamic Therapy. *Photochem. Photobiol.* **2019**, *95*, 839–845. [CrossRef]

40. Giroldo, L.M.; Felipe, M.P.; de Oliveira, M.A.; Munin, E.; Alves, L.P.; Costa, M.S. Photodynamic Antimicrobial Chemotherapy (PACT) with Methylene Blue Increases Membrane Permeability in *Candida albicans*. *Lasers Med. Sci.* **2009**, *24*, 109–112. [CrossRef]

41. Snell, S.B.; Foster, T.H.; Haidaris, C.G. Miconazole Induces Fungistasis and Increases Killing of *Candida albicans* Subjected to Photodynamic Therapy. *Photochem. Photobiol.* **2012**, *88*, 596–603. [CrossRef] [PubMed]

42. Scwingel, A.R.; Barcessat, A.R.P.; Núñez, S.C.; Ribeiro, M.S. Antimicrobial Photodynamic Therapy in the Treatment of Oral Candidiasis in HIV-Infected Patients. *Photomed. Laser Surg.* **2012**, *30*, 429–432. [CrossRef]

43. Azizi, A.; Amirzadeh, Z.; Rezai, M.; Lawaf, S.; Rahimi, A. Effect of Photodynamic Therapy with Two Photosensitizers on *Candida albicans*. *J. Photochem. Photobiol. B* **2016**, *158*, 267–273. [CrossRef] [PubMed]

44. do Vale, L.R.; Delbem, A.; Arias, L.S.; Fernandes, R.A.; Vieira, A.; Barbosa, D.B.; Monteiro, D.R. Differential Effects of the Combination of Tyrosol with Chlorhexidine Gluconate on Oral Biofilms. *Oral Dis.* **2017**, *23*, 537–541. [CrossRef] [PubMed]

45. Ellepola, A.N.B.; Chandy, R.; Khan, Z.U. In Vitro Impact of Limited Exposure to Subtherapeutic Concentrations of Chlorhexidine Gluconate on the Adhesion-Associated Attributes of Oral Candida Species. *Med. Princ. Pract.* **2016**, *25*, 355–362. [CrossRef] [PubMed]

46. Niu, T.; Tian, Y.; Ren, Q.; Wei, L.; Li, X.; Cai, Q. Red Light Interferes in UVA-Induced Photoaging of Human Skin Fibroblast Cells. *Photochem. Photobiol.* **2014**, *90*, 1349–1358. [CrossRef]

47. Jagdeo, J.; Austin, E.; Mamalis, A.; Wong, C.; Ho, D.; Siegel, D.M. Light-Emitting Diodes in Dermatology: A Systematic Review of Randomized Controlled Trials. *Lasers Surg. Med.* **2018**, *50*, 613–628. [CrossRef] [PubMed]

48. Andrade, M.C.; Ribeiro, A.P.D.; Dovigo, L.N.; Brunetti, I.L.; Giampaolo, E.T.; Bagnato, V.S.; Pavarina, A.C. Effect of Different Pre-Irradiation Times on Curcumin-Mediated Photodynamic Therapy against Planktonic Cultures and Biofilms of *Candida* spp. *Arch. Oral Biol.* **2013**, *58*, 200–210. [CrossRef]

49. Fitzmaurice, S.; Eisen, D.B. Daylight Photodynamic Therapy: What Is Known and What Is Yet to be Determined. *Dermatol. Surg.* **2016**, *42*, 286–295. [CrossRef]

50. Lee, C.-N.; Hsu, R.; Chen, H.; Wong, T.-W. Daylight Photodynamic Therapy: An Update. *Molecules* **2020**, *25*, 5195. [CrossRef]

51. Nguyen, M.; Sandhu, S.S.; Sivamani, R.K. Clinical Utility of Daylight Photodynamic Therapy in the Treatment of Actinic Keratosis–a Review of the Literature. *Clin. Cosmet. Investig. Dermatol.* **2019**, *12*, 427–435. [CrossRef]

52. Torres-Hurtado, S.A.; Ramírez-Ramírez, J.; Larios-Morales, A.C.; Ramírez-San-Juan, J.C.; Ramos-García, R.; Espinosa-Texis, A.P.; Spezzia-Mazzocco, T. Efficient in Vitro Photodynamic Inactivation Using Repetitive Light Energy Density on *Candida albicans* and *Trichophyton mentagrophytes*. *Photodiagnosis Photodyn. Ther.* **2019**, *26*, 203–209. [CrossRef] [PubMed]

53. Souza, R.C.; Junqueira, J.C.; Rossoni, R.D.; Pereira, C.A.; Munin, E.; Jorge, A.O.C. Comparison of the Photodynamic Fungicidal Efficacy of Methylene Blue, Toluidine Blue, Malachite Green and Low-Power Laser Irradiation Alone against *Candida albicans*. *Lasers Med. Sci.* **2010**, *25*, 385–389. [CrossRef] [PubMed]

54. Peloi, L.S.; Soares, R.R.S.; Biondo, C.E.G.; Souza, V.R.; Hioka, N.; Kimura, E. Photodynamic Effect of Light-Emitting Diode Light on Cell Growth Inhibition Induced by Methylene Blue. *J. Biosci.* **2008**, *33*, 231–237. [CrossRef] [PubMed]

 *pharmaceutics*

Article

Study of Viral Photoinactivation by UV-C Light and Photosensitizer Using a Pseudotyped Model

Mohammad Sadraeian [1,2,*], Fabio Francisco Pinto Junior [1], Marcela Miranda [1], Juliana Galinskas [3], Rafaela Sachetto Fernandes [1], Edgar Ferreira da Cruz [3], Libing Fu [2], Le Zhang [2], Ricardo Sobhie Diaz [3], Gustavo Cabral-Miranda [4,5] and Francisco Eduardo Gontijo Guimarães [1,*,†]

1   Instituto de Física de São Carlos, Universidade de São Paulo, Caixa Postal 369, São Carlos 13560-970, SP, Brazil; fabiojr@ifsc.usp.br (F.F.P.J.); marcelamiranda@usp.br (M.M.); rafaela.fernandes@usp.br (R.S.F.)
2   Institute for Biomedical Materials and Devices (IBMD), Faculty of Science, University of Technology Sydney, Sydney, NSW 2007, Australia; libing.fu@uts.edu.au (L.F.); leo.zhang@uts.edu.au (L.Z.)
3   Laboratório de Retrovirologia, Escola Paulista de Medicina, Universidade Federal de São Paulo, São Paulo 04039-032, SP, Brazil; biomedicaju@hotmail.com (J.G.); edgar.cruz@unifesp.br (E.F.d.C.); rsdiaz@catg.com.br (R.S.D.)
4   Department of Immunology, Institute of Biomedical Sciences, University of São Paulo (ICB/USP), São Paulo 05508-000, SP, Brazil; gcabral.miranda@usp.br
5   Institute of Research and Education in Child Health (PENSI), São Paulo 01228-200, SP, Brazil
*   Correspondence: mohammad.sadraeian@uts.edu.au (M.S.); guimaraes@ifsc.usp.br (F.E.G.G.); Tel.: +55-(16)-33739792 (F.E.G.G.)
†   Current addresses: Av. Trab. São Carlense, 400-Parque Arnold Schimidt, Caixa Postal 369, São Carlos 13566-590, SP, Brazil.

Citation: Sadraeian, M.; Junior, F.F.P.; Miranda, M.; Galinskas, J.; Fernandes, R.S.; da Cruz, E.F.; Fu, L.; Zhang, L.; Diaz, R.S.; Cabral-Miranda, G.; et al. Study of Viral Photoinactivation by UV-C Light and Photosensitizer Using a Pseudotyped Model. *Pharmaceutics* 2022, 14, 683. https://doi.org/10.3390/pharmaceutics14030683

Academic Editor: Maria Nowakowska

Received: 26 February 2022
Accepted: 15 March 2022
Published: 21 March 2022

Publisher's Note: MDPI stays neutral with regard to jurisdictional claims in published maps and institutional affiliations.

Abstract: Different light-based strategies have been investigated to inactivate viruses. Herein, we developed an HIV-based pseudotyped model of SARS-CoV-2 (SC2) to study the mechanisms of virus inactivation by using two different strategies; photoinactivation (PI) by UV-C light and photodynamic inactivation (PDI) by Photodithazine photosensitizer (PDZ). We used two pseudoviral particles harboring the Luciferase-IRES-ZsGreen reporter gene with either a SC2 spike on the membrane or without a spike as a naked control pseudovirus. The mechanism of viral inactivation by UV-C and PDZ-based PDI were studied via biochemical characterizations and quantitative PCR on four levels; free cell viral damage; viral cell entry; DNA integration; and expression of reporter genes. Both UV-C and PDZ treatments could destroy single stranded RNA (ssRNA) and the spike protein of the virus, with different ratios. However, the virus was still capable of binding and entering into the HEK 293T cells expressing angiotensin-converting enzyme 2 (ACE-2). A dose-dependent manner of UV-C irradiation mostly damages the ssRNA, while PDZ-based PDI mostly destroys the spike and viral membrane in concentration and dose-dependent manners. We observed that the cells infected by the virus and treated with either UV-C or PDZ-based PDI could not express the luciferase reporter gene, signifying the viral inactivation, despite the presence of RNA and DNA intact genes.

Keywords: viral inactivation; photodynamic inactivation; SARS-CoV-2 pseudovirus; enveloped virus; UV-C light; photosensitizer

1. Introduction

Novel coronavirus disease (COVID-19), caused by the SC2 virus, was first detected in December 2019 in the Hubei province of China, and has since sparked a global health crisis, with 5.1 million deaths reported by the World Health Organization (WHO) as of November 20 in 2021 [1]. This pandemic situation demands urgent attention toward finding novel strategies that might contribute to the prevention of viral spread via the inactivation of virions on surfaces, aerosols, and the human body.

The UV-C light has been used in healthcare facilities for environmental disinfection (air, liquid, and solid surfaces) [2]. The efficacy of this inactivation may depend not only on the wavelength but also on factors such as the pathogens (e.g., bacterial or viral species), light output, and environmental conditions [3]. UV-C light at 254 nm radiation enables the deposit of the energetic photons during interaction with the coronavirus, damaging the viral genome, and, consequently, the virus replication and proliferation can theoretically be abrupted [4]. In the case of RNA viruses like SC2, UV irradiation forms several RNA photoproducts, primarily from adjacent pyrimidine nucleotides, such as uracil dimers, as well as RNA–protein cross-links [3]. The formation of the uracil dimer potentially leads to frameshift or point mutations in the genome, known as UV-signature mutations of virus [5]. Hence, we should remain vigilant about the long-term effects of irradiation-mediated strategies for viral inactivation. There are several studies on the effects of UV-C for LD90 viral inactivation based on the time and dose of irradiation [2,6,7]; however, the mechanism of action of how UV-C inactivates viruses is still unclear [7,8].

Photodynamic therapy (PDT) is another light-based strategy that has been proposed to treat infections by damaging microorganisms, fungi, parasites, and viral particles. PDT is based on the use of photo-sensitive agents named photosensitizers (PS) which, in light-excited conditions and the presence of molecular oxygen, produce reactive oxygen species (ROS) [9–15]. PDT may damage cells via ROS generation, causing necrosis and apoptosis without harming the neighboring tissues. The advantages of utilizing photosensitizers for photodynamic inactivation (PDI) include its short-term toxicity, the absence of cell genome alterations, and avoiding the development of viral-induced resistance. Hence, the antiviral potential therapeutic effects of PDT and PDI on SC2 have been investigated with promising results [16,17]. Photoditazine photosensitizer (PDZ) is a porphyrin derivative with a chlorine core which allows it a high absorption in the red light spectrum with λmax of between 650–670 nm, as an advantage compared to the first generation of photosensitizers which are porphyrin core-based and which absorbs wavelengths too short for superior tissue penetration [18]. Understanding the mechanism of viral photoinactivation is important in finding and optimizing light-based strategies to battle viral infection. There are several reports on the mechanisms of viral photoinactivation with limited experiments on virion damage and viral propagation [2,19,20] due to the restriction of working with highly pathogenic viruses like HIV and SC2 viruses. Addressing these containment issues, the setting up of pseudotyped models in BSL2 labs can speed up studying the viral–cell mechanism and neutralizing assay towards in vivo studies [21,22]. Herein, we introduced the application of a pseudotyped model for studying the viral mechanism on four levels; virion damage; viral-cell entry; DNA integration; and expression of reporter genes. In this study, we followed the effects of UV-C irradiation and PDI on viral spike proteins and ssRNA in a HIV-based pseudotyped model of SC2 containing the Luciferase-IRES-ZsGreen reporter gene. Finally, we aimed to study the pseudovirus during cell internalization, genome integration, and reporter gene expression, after undergoing treatments by UV-C and PDZ photosensitizer under different concentrations and conditions (Figure 1).

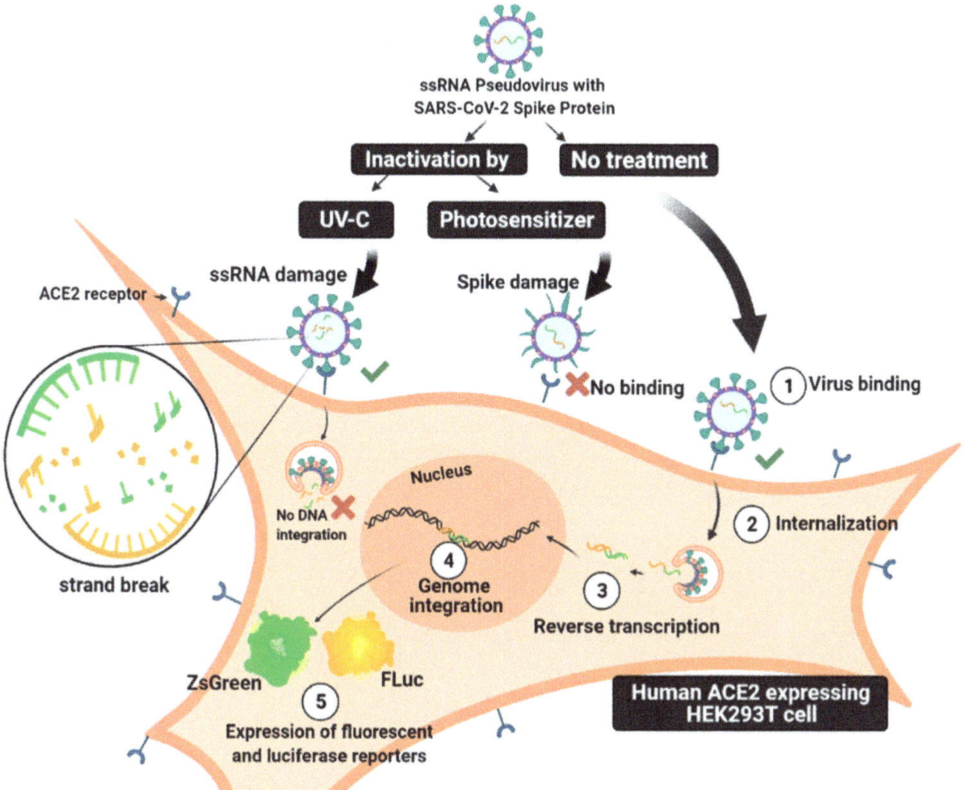

Figure 1. Schematic picture of the mechanism of SARS-CoV-2 pseudovirus infectivity. Unlike SC2 ssRNA virus, which has viral reproduction independent of the host genome, this counterpart pseudovirus carries on reporter ssRNA with LTR, which causes integration into the genome. In this study, the pseudovirus has been treated with either UV-C irradiation or photodynamic inactivation (PDI) by Photodithazine photosensitizer. The mechanism of infectivity of photo-inactivated pseudovirus particles has been compared on four levels; free-cell viral damage; viral cell entry; DNA integration; and expression of reporter genes. The figure was created with BioRender software.

2. Materials and Methods

2.1. Chemical Reagents

All reagents were purchased from Thermo Fisher Scientific (Waltham, MA, USA), unless otherwise stated.

2.2. Cells and Viruses

The HEK 293T cells expressing ACE-2 receptor were gifted from BEI Resources as catalog number NR-52511. ACE-2 enzyme is a critical receptor for virus entrance into the host cell. The HEK 293T cells were used as control cells for assays, and for pseudovirus generation. The cells were maintained at 37 °C in 5% CO_2 in DMEM medium-high glucose (DMEM-HG) with 10% fetal bovine serum (Gibco Invitrogen, Grand Island, NY, USA). HEK 293T is a derivative human cell line isolated from human embryonic kidneys (HEK) and expresses a mutant version of the SV40 large T antigen.

2.3. Plasmids

Vector backbone pMD2.G and Vector backbone psPAX2 were gifts from Didier Trono (Addgene plasmid # 12259 and # 12260, respectively). The other plasmids were donated by BEI Resources and their sequences are available at (https://www.beiresources.org/ (accessed on 11 February 2022)) with the following catalog numbers [23]:

HDM-IDTSpike-fixK (BEI catalog number NR-52514): Plasmid expressing under a CMV promoter the Spike viral entry from SARS-CoV-2 strain Wuhan-Hu-1 (Genbank NC_045512);

pHAGE-CMV-Luc2-IRES-ZsGreen-W (BEI catalog number NR-52516): Lentiviral backbone plasmid that uses a CMV promoter to express luciferase followed by an IRES and ZsGreen;

HDM-Hgpm2 (BEI catalog number NR-52517): lentiviral helper plasmid expressing HIV Gag-Pol under a CMV promoter;

HDM-tat1b (NR-52518): Lentiviral helper plasmid expressing HIV Tat under a CMV promoter.

pRC-CMV-Rev1b (NR-52519): Lentiviral helper plasmid expressing HIV Rev under a CMV promoter.

2.4. UV-Vis Spectroscopy

UV-Vis spectroscopy was used to determine plasmid and protein concentrations by using Nanodrop 1000 UV-Visible spectrophotometer (Thermo Fisher Scientific, Waltham, MA, USA) [24].

2.5. Dynamic Light Scattering

Hydrodynamic radii, electrophoretic mobility, and polydispersity of SC2 Spike-pseudovirus were measured before and after photo inactivation. For UV-C inactivation and PDI inactivation, we followed the inactivation protocols as explained in Sections 2.7 and 2.8. Then, samples with 70 µL volume at 1 mg/mL in UV-transparent 96-well plates were measured using a DLS Wyatt Möbius (Wyatt Technologies, Dernbach, Germany) with incident light at 532 nm, at an angle of 163.5°. Samples were equilibrated at 25 ± 0.1 °C for 600 s before the measurements, and this temperature was held constant throughout the experiments. All samples were measured in triplicate with 10 acquisitions and a 5 s acquisition time. The change in the cumulant-fitted hydrodynamic radius in nanometers was monitored during the storage period. Results were calculated using the Dynamics 7.1.7 software (Wyatt Technologies, Santa Barbara, CA, USA).

2.6. Generation of Pseudovirus with SARS-CoV-2 Spike and Naked Control

SC2 Spike pseudotyped lentiviruses were generated by transfecting 293T cells, adjusted with the protocol explained by Thermo Fisher Scientific (Waltham, MA, USA). Briefly, seed 293T cells to be 95–99% confluent at transfection. At 16–24 h after seeding, the cells were transfected with the plasmids required for lentiviral production by using Lipofectamine 3000 Reagent (Thermo Fisher Scientific, Waltham, MA, USA) following the manufacturer's instructions and using the following plasmid with 1 mL total volume per well of a six-well plate. The 293T cells were transfected with a lentiviral backbone plasmid encoding Firefly luciferase and ZsGreen reporter proteins, a plasmid expressing SC2 Spike, and plasmids expressing HIV-1 *gag*, *pol*, and *tat* proteins, to assemble the membrane of viral particles. The same protocol was followed to generate naked control pseudovirus without adding the viral entry plasmid encoding SC2 Spike. At 8 h post-transfection, the packaging medium was removed and replaced. At 24 h post-transfection, the entire volume of cell supernatant was harvested and stored at 4 °C. Then, 1 mL of fresh medium was replaced. At 52 h post-transfection, the entire volume of the cell supernatant was harvested. The pseudovirus product was aliquoted in small volumes of 400 µL and stored at -80 °C prior to use and underwent a single freeze-thaw.

2.7. Viral Inactivation Using UV-C Irradiation

A total of 40 µL of pseudovirus were diluted in 60 µL of DMEM-HG without supplementation in each well of a 96-well plate, which were exposed to the UV-C lamp 254 nm (HNS G5, OSRAM Germicidal Puritec, Munich, Germany) placed 1 cm above the plate to allow a uniform irradiance over the plate (10 ± 2 mW/cm^2). Light was delivered by 1, 6, and 36 s corresponding to doses of 10, 60, and 360 mJ/cm^2, respectively. Controls were not submitted to irradiation. After irradiation, aliquots of 80 µL were placed into the plates containing the previously seeded 293T/ACE2 cells and incubated for 8 h at 37 °C with 5% CO$_2$ for viral adsorption. Then, 120 µL of DMEM-HG medium containing 12% fetal bovine serum was added, and the plate was incubated for 48 h at 37 °C with 5% CO$_2$. Afterward, the cells were placed into a lysis buffer solution to proceed with either Firefly luciferase assay or proviral DNA assay. Results were normalized in relation to controls for the calculation of viral inhibition rates of each sample.

2.8. Photosensitizer-Based Photodynamic Inactivation

A total of 40 µL of pseudovirus were diluted in 60 µL of DMEM-HG without supplementation in each well of a 96-well plate. The Photodithazine photosensitizer (PDZ) (Photodithazine® Company, Moscow, Russia) with a serial dilution of 10, 50, and 250 µg/mL was added, and incubated in the dark at RT (22 °C) for 15 min, then were irradiated using a homemade LED device at 670 nm (red light). All irradiations were performed with an irradiance rate of 30 mW/cm^2 in a time-dependent manner of 1, 10, and 20 min which equal the light doses of 1.8, 18, and 36 J/cm^2, respectively. Afterwards, the treated ssRNA viruses were either harvested for the viral RNA load and DLS characterization or were incubated with the 293T/ACE-2 cells, as described in Section 2.7. After that, the cells were harvested for previral DNA load and luciferase activity measurement.

2.9. Quantification of Viral RNA and Proviral DNA

The total ssRNA pseudovirus, before and after irradiation, was extracted and purified using the RNeasy Lipid Tissue Mini Kit, according to the manufacturer's (QIAGEN, Hilden, Germany) protocol.

The pseudovirus was treated with either 36 s UV-C irradiation or 10 min PDI in the presence of 10 µg/mL PDZ. The viral RNA load refers to the virus genome of free-cell pseudovirus, before and after treatment. Viral load measurement was carried out using one-step reverse transcriptase (RT) and real-time PCR in a single buffer system using the Abbott Real Time on the automated m2000, over the dynamic range of detection of 40 to 10,000,000 copies/mL (Abbott, IL, USA) [25]. The protocol was followed as described by Kumar et al. for the TaqMan One-Step RT and PCR Master Mix Reagents Kit (Thermo Fisher Scientific, Waltham, MA, USA) with primers and probes for long terminal repeat (LTR) region of 640 bp, with two identical regions located at both ends of the either proviral DNA or RNA viral of SC2 pseudovirus. Briefly, a volume of 5 µL RNA sample and 20 µL Master Mix were used for a one-step RT-qPCR reaction with 20 µM forward primer (5-GCCTCAATAAAGCTTGCCTTGA-3); 20 µM reverse primer (5-GGGCGCCACTGCTAGAGA-3); 10 µM Taqman probe (5-FAM-CCAGAGTCACACAACAG ACGGGCACA-TAMRA-3); and an Applied Biosystems 7500 Fast Real-Time PCR system (Thermo Fisher Scientific, Waltham, MA, USA), as reported previously [25–27].

Quantification of Proviral DNA were completed with the TaqMan Real-Time PCR Assay. The cells were infected with the treated virus (Sections 2.7 and 2.8). The cells were harvested three days after infection, centrifuged, and separated the pellets. The number of infected cells containing proviral DNA of pseudovirus was measured using qPCR. The quantification was executed based on the previously published protocol [27,28] for amplification of proviral DNA of pseudovirus (region LTR) with the primers described in Section 2.9.

2.10. Flow Cytometry

Direct fluorescence detections were applied using flow cytometry (Becton-Dickson Accuri C6, Mountain View, CA, USA) to analyze the expression of ZsGreen in the 293T/ACE-2 cells incubated with treated pseudovirus, as explained before (Sections 2.7 and 2.8). After 48 h, the ZsGreen-positive cells were harvested, fixed by 2% paraformaldehyde (PFA), quantified by blue laser (20 mW) irradiation at 488 nm and analyzed in the channel FL1: 533/30. The acquired data were analyzed by Flow-Jo software version 7.5 (Tree Star Inc., Ashland, OR, USA).

2.11. Luciferase Assay

The infected cells which were harvested before (Sections 2.7 and 2.8) were lysed with 20 µL of Luciferase Cell Culture Lysis Reagent (Promega, Madison, WI, USA), then mixed with 100 µL of Luciferase Assay Reagent (Promega, Madison, WI, USA), and the light emission was measured.

2.12. Titration of Pseudovirus

The pseudovirus particles were titrated using a method similar to SC2 pseudovirus generation. Virus titers were determined by measuring relative luciferase units (RLUs). The HEK293T cells expressing human ACE-2 (293T-ACE-2) were produced to test the correlation between ACE-2 expression and SC2 pseudovirus susceptibility. Particles were generated in two forms; with a SC2 spike and a negative control without a viral entry protein. Both pseudo-typed particles harbored a Luciferase-IRES-ZsGreen backbone. In a mirror plate, the percentage of cell viability was measured during the viral infection with a serial dilution of the virus starting at 50 µL pseudovirus in a total volume of 100 µL (0.5) for the spike pseudovirus. After 8 h of pseudovirus incubation, the media were replaced with 150 µL fresh media. After 48 h incubation, the wells containing 50 µL pseudovirus were studied for cell confluency. Afterwards, the titers of pseudotyped particles were quantified by a Luciferase assay expressed in RLU, to determine the number of transducing particles per mL.

2.13. Confocal Microscopy

One day before UV-C or PDI treatment, 2×10^4 cells per well of 293T/ACE-2 cells were seeded on a multiple-chamber slide (Nalge-Nunc International, Naperville, Ill, USA). The next day, cells were incubated with treated pseudovirus, as explained before (Sections 2.7 and 2.8). After 48 h, the ZsGreen-positive cells were washed four times with PBS, fixed by 2% PFA. Images were obtained with an inverted LSM 780 multiphoton laser scanning confocal microscope (Zeiss, Jena, Germany), a 63×1.2 water immersion objective to couple with the bottom side of the cover slip, and the Zeiss LSM software was used to treat the images. The wavelength of Argon ion laser at 488 nm was used to excite the expressed ZsGreen protein compared to cell autofluorescence. The molecular localization of ZsGreen was analyzed for each image pixel in spectral and channel modes in the ranges 492–700 nm and 492–537 nm, respectively. The cells' autofluorescence were analyzed from 585 to 692 nm.

Each pixel was associated with an emission spectrum which allowed the spatial separation of the expressed ZsGreen fluorescence (bright blue-greenish color) and the cell auto-fluorescence (yellow-orange color). Considering the spectral of the cell autofluorescence (maximum at 575 nm) is almost constant, the expression of the ZsGreen by the active pseudovirus internalization would be promptly signaled by the spectral superposition of the protein emission at around 515 nm.

2.14. Statistical Analyses

Statistical analyses were performed using the GraphPad Prism version 8.0 (GraphPad Software, San Diego, CA, USA). Data are shown as mean and SEM of the indicated number of replicate values. If no error bar appears present, the error bars are smaller than, and obscured by, the symbol. The method for statistical comparison used was unpaired two-tailed Student's *t*-test, unless specifically indicated otherwise.

3. Results and Discussion

3.1. Generation of Pseudovirus with SARS-CoV-2 Spike and Naked Control

The spike-pseudotyped lentiviral particles were generated, which can infect 293T cells expressing the human ACE-2 receptor. In parallel, the naked control pseudovirus was generated, which harbors a backbone plasmid-encoding luciferase-IRES-ZsGreen reporter, but without the SC2 Spike on the membrane (Figure 1).

3.2. Titration of Pseudovirus

The pseudovirus particles were titrated in two forms; particles with a SC2 spike and a negative control without a viral entry protein. Both pseudo-typed particles harbored a Luciferase-IRES-ZsGreen backbone. After 48 h incubation, the cell confluence reached 100% for all wells in a mirror plate containing non-transduced cells, however, the wells containing 50 μL pseudovirus showed 90% cell confluence (Figure 2A). The titers of pseudotyped particles were quantified by a Luciferase assay. Titers of >10^5 RLUs per mL were measured in a 96-well plate (Figure 2B). Unsurprisingly, the ACE2-expressing cells incubated with naked pseudovirus without a spike did not show the expression of luciferase (Figure 2C). The other negative control was the incubation of 293T non-ACE2 control cells with Spike-pseudovirus, which did not show the luciferase expression, as expected (Figure 2D). In previous reports, researchers used polybrene to facilitate the lentiviral infection through minimizing charge-repulsion between the virus and cells [29], but we found this SC2 pseudovirus no need to polybrene for binding to the ACE-2 receptor.

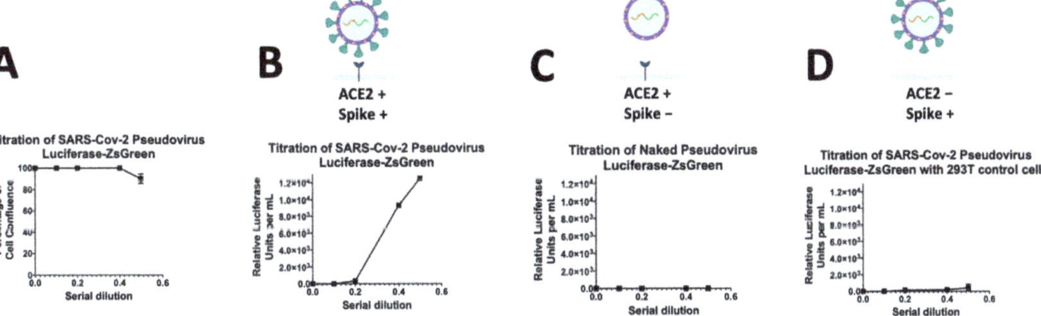
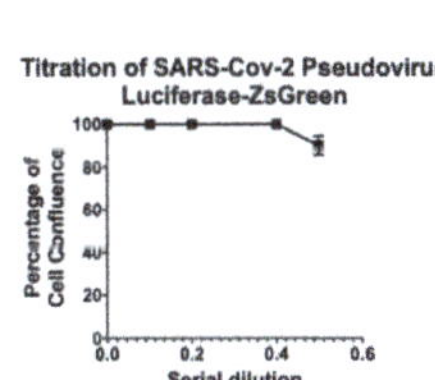

Figure 2. Titration of SARS-CoV-2 spike-pseudovirus particles in 293T cells expressing ACE-2. (**A**) Study of the percentage of cell viability during the viral infection respecting a serial dilution starting at 1:2 (0.5); (**B,C**) The graph shows the titers of the expression of Luciferase reporter as determined by measuring relative luciferase units (RLUs). The RLU data are the average of three-fold serial dilution of virus starting at 50 μL virus in a total volume of 100 μL (0.5) for the Spike-pseudovirus (**B**); naked pseudovirus without spike (**C**); or the Spike-pseudovirus with 293T cells without ACE2 receptor (**D**). After 8 h of pseudovirus incubation, the media were replaced with 150 μL fresh media.

3.3. Viral Inactivation Using UV-C Irradiation or Photosensitizer-Based PDI

A volume of 40 μL of pseudovirus was diluted in 60 μL of DMEM-HG without supplementation in each well of a 96-well plate, which were exposed to the UV-C lamp 254 nm for 1, 6, and 36 s corresponding to doses of 10, 60, and 360 mJ/cm^2, respectively. Figure 3A represents the effect of UV-C irradiation on the photo-inactivation of ssRNA pseudovirus. The results showed that UV-irradiation may inactivate 74%, 93%, and 99.99% of SC2 Spike-pseudovirus particles during 1, 6, and 36 s irradiation, respectively. These results are comparable with the published results on SC2 elsewhere [20], due to the discrepancies of the cell-entry mechanism among virus and pseudovirus. Furthermore, a time-dependent manner of PDI was performed to find the maximum viral inactivation with the minimum time and concentration of Photogem PS (PDZ). As Figure 3B demonstrates, the viral inactivation depended on both time and the PS concentration. We observed that 99.8% of the pseudovirus were inactivated in the presence of 50 µg/mL PDZ with 10 min irradiation. Hence, we selected this time and concentration for further studies. These results indicate that both UV-C irradiation and PDI, as two distinct strategies, are highly effective in inactivating pseudovirus replication, while there could be some differences in the mechanism of infectivity between UV-C irradiation and PDI. Hence, we extended our studies focusing on the viral RNA and proviral DNA loads, as described in Section 3.4.

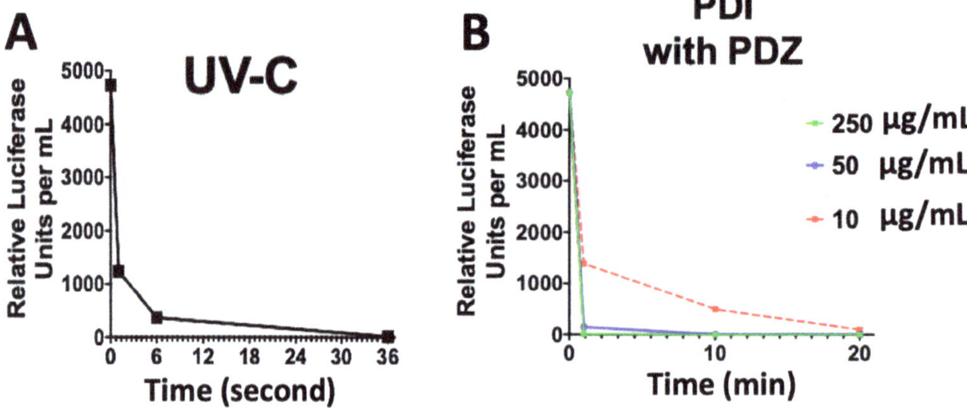

Figure 3. Study of the effect of photo-inactivation of ssRNA pseudovirus considering the relative luciferase units with a time-dependent manner of UV-C irradiation at 1, 6, and 36 s corresponding to doses of 10, 60, and 360 mJ/cm^2, respectively (**A**); and PDZ-based PDI in a serial dilution of 10, 50 and 250 µg/mL in a time-dependent manner of 1, 10 and 20 min which equal the light doses of 1.8, 18 and 36 J/cm^2, respectively (**B**). Data are $\pm$ means S.E.M. (*n* = 3).

3.4. Study the Infectivity Mechanism of UV-C Irradiation and PDI Using qPCR

Viral inactivation could be due to either viral protein or viral genome damage [8,30,31]. We suppose that any damage to the virus spike may lead to loss of the virus binding ability and neutralization of the virus infectivity, while damaging the viral genome may affect the viral and proviral loads of pseudovirus. The results of the viral RNA load showed that both 36 s UV-C irradiation and PDZ-based PDI (10 min irradiation, 50 µg/mL PDZ) can damage ssRNA by 83% and 74%, respectively. The RNA of both control (naked pseudovirus without spike) and spike-positive viral particles were destroyed during irradiation (Figure 4A). By comparing the PDZ-based PDI in two forms of enveloped and non-enveloped (naked) viruses, we found out that PDI may damage the viral genome independently from the virus type.

The results of the proviral DNA assay may interpret the virus's ability to complete the subsequent steps of cell binding, internalization, and genome integration after reverse transcription. The proviral DNA load of 36 s UV-C irradiation was as much as the RNA viral load, signifying that the UV-C based viral inactivation is independent of damaging the spike protein. In parallel, the proviral DNA load of PDZ-based PDI (10 min irradiation, 50 µg/mL PDZ) was decreased by 13%, which is half of the RNA viral load (26%), signifying the PDI-treated pseudovirus may lose cell infectivity due to damaging the spike (Figure 4A,B). Presumably, PDZ-based PDI destroys more of the spike than the viral genome, which leads to losing the binding ability of the virus. Unsurprisingly, the naked control particle showed no DNA load, as the control particle lacks a spike for cell binding. Furthermore, we observed that the cells infected by either UV-C or PDI-treated pseudovirus could not express the luciferase reporter gene (Figure 4C), signifying the total viral inactivation despite the presence of RNA and DNA intact genes.

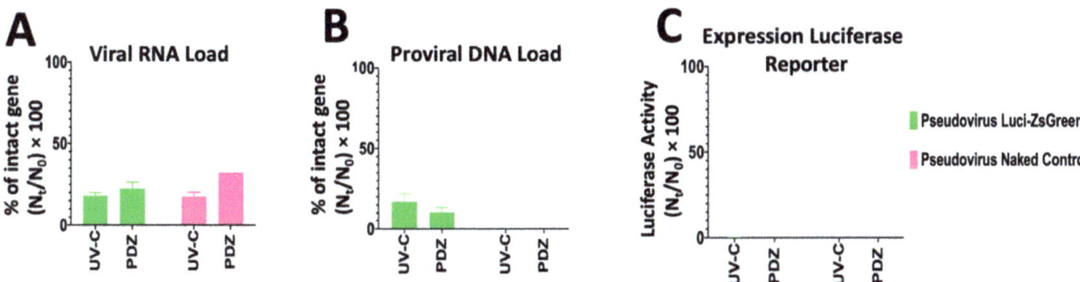

Figure 4. The pseudovirus has been treated with either 36 s UV-C irradiation or 10 min Photodynamic Inactivation in the presence of 50 µg/mL Photodithazine. The mechanism of infectivity of photo-inactivated pseudovirus particles has been compared on three levels; (**A**) viral RNA load referring to the free-cell virus; (**B**) proviral DNA load referring to the ability of the treated virus to complete the subsequent steps of cell internalization, reverse transcription and genome integration; (**C**) luciferase activity referring to the expression of luciferase reporter after DNA integration. n_t/n_0 represents the fraction of the targeted genome region that remained intact after treatment.

In this study, n_t/n_0 represents the fraction of the targeted genome region that remained intact after treatment. In the viral RNA load, the targeted genome region is ssRNA of pseudovirus with LTR sequences. In the proviral DNA load, the targeted genome region is the integrated DNA of pseudovirus genome after reverse transcriptase. Unlike the SC2 virus, the mechanism of infectivity of the SC2 pseudovirus includes DNA integration, which is one of the advantages of utilizing the pseudotyped model. Therefore, we could follow a simple protocol for calculation of RNA and DNA load and compare the qPCR data with luciferase assay results, otherwise to estimate the infectivity based on qPCR data, the infectivity of virus should be assessed by estimation from the qPCR results, according to the protocol published by Sabino et al. [20,32].

3.5. DLS Measurements before and after Irradiation

DLS measurement demonstrated that 36 s UV-C irradiation on pseudovirus with 18 J/cm^2 resulted in a slight decrease in the size distribution compared to the non-irradiated pseudovirus (Figure 5A). On the other hand, in the PDI study, the increase of PDZ concentration from 10 to 50 µg/mL had a significant effect on the size and polydispersity of the virus, and yielded significant aggregated particles (Figure 5B). We assumed that this aggregation may interrupt our results on the cell toxicity therefore we found that PDZ with 10 µg/mL was an appropriate concentration for further studies on flow cytometry and microscopy observations.

3.6. Green Fluorescent Measurement by Flow Cytometry

Furthermore, the cells were infected with viruses, which were treated with either 36 s of UV-C or PDZ (10 µg/mL) of PDI, to measure the expression of the ZsGreen protein. The flow cytometry results showed that 46.3% of the virus-infected cells were emitting green fluorescence, while the cells treated with UV-C or the PDI-treated viruses were not able to express the ZsGreen (Figure 5C).

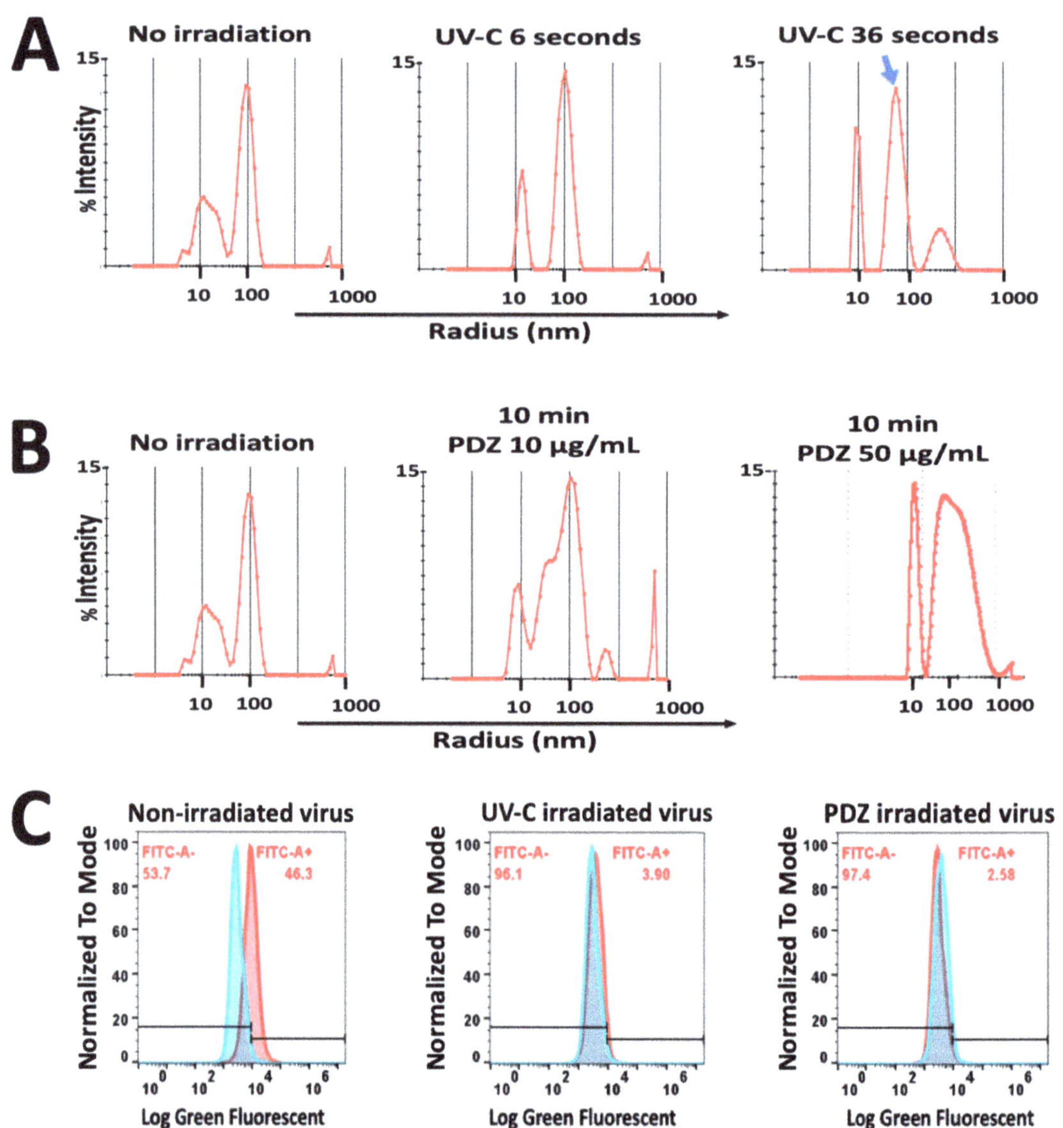

Figure 5. (**A**) Dynamic Light Scattering histograms of hydrodynamic radius (R_h) for pseudovirus showed optimal polydispersity with R_h of 100 nm. During 36 s of UV-C irradiation (360 mJ/cm^2), a slight decrease in the size of pseudovirus was observed (Blue arrow) with no significant aggregation; (**B**) PDZ-based PDI significantly affected the size and polydispersity of the virus, resulting in major aggregated particles in a higher concentration of PDZ (50 µg/mL); (**C**) Flow cytometric diagram on the left demonstrated the percentage of pseudovirus-infected cells expressing ZsGreen Fluorescent protein. The middle and right diagrams represent the cells infected by UV irradiated-virus and PDZ-based PDI virus, which do not express ZsGreen protein. FITC rate indicates the green fluorescent emission from ZsGreen. Data are ± means S.E.M. (*n* = 2).

3.7. Observation of ZsGreen Expression by Confocal Microscopy

Forty-eight h after cell incubation with pseudovirus (with no treatment), the ZsGreen expression was observed using confocal microscopy. Figure 6A shows images of the field in spectral mode (Figure 6A—panel (a)), and in channel mode merged with a wide field transmission image (Figure 6A—panel (c)). The two spectral contributions for both ZsGreen emission and the cell autofluorescence can be separated by taking two regions of interest (ROI) in panel (a) (green and red circles), as depicted as two graphs in Figure 6A—panel (b). In Figure 6A—panel (c) demonstrates that the emission detected between 492 and 532 nm (assigned the bright-blue false color) mainly signals the expression of the ZsGreen protein while the cellular autofluorescence can be differentiated by taking the emission (orange false color) in the spectral range from 585 to 695 nm.

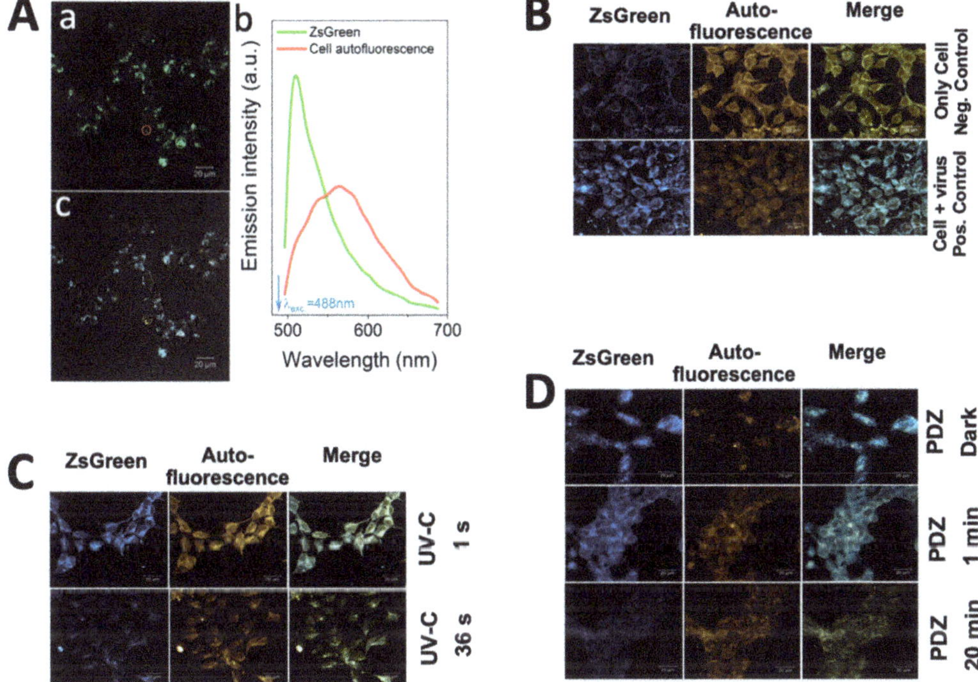

Figure 6. (**A**) Confocal microscopy images showing ZsGreen expression in 293T-ACE2 cells at 48 h after incubation with Spike-pseudotyped lentiviral particles with the ZsGreen backbone. The images are represented in spectral mode (panel (**a**)) and channel mode merged with a wide field transmission image (panel (**c**)). In panel (**b**), the green and red curves represent the regions of interest (ROI) of ZsGreen emission and the cell autofluorescence, respectively; (**B**) The positive control cells incubated with pseudovirus without treatment showed strong green fluorescent emission, indicating the expression of ZsGreen in comparison to negative control cells. The ZsGreen emission was detected between 492 and 532 nm (assigned the bright-blue false color), while the cellular autofluorescence can be differentiated by taking the emission (orange false color) in the spectral range from 585 to 695 nm; (**C**) The results of viruses with 36 s UV-C irradiation (360 mJ/cm^2) did not show green fluorescent emission, while the cells with 1 s UV-C irradiation (10 mJ/cm^2) were still showing somewhat ZsGreen expression; (**D**) The viruses were treated with Photoditazine (PDZ) (10 µg/mL) and irradiated in a time-dependent manner of 0 (so-called dark), 1 and 20 min. The cells infected by PDI virus after 20 min do not express ZsGreen protein. (**A**,**B**) scale bar: 20 µm.

To study the expression of ZsGreen protein by confocal microscopy, the cells were infected with pseudovirus treated with UV-C or PDI. The positive control cells incubated with pseudovirus without treatment showed strong green fluorescent emission indicating the expression of ZsGreen in comparison to negative control cells (Figure 6B). The results of viruses with 36 s UV-C irradiation (360 mJ/cm^2) did not show green fluorescent emission (Figure 6C), while the cells with 1 s UV-C irradiation (10 mJ/cm^2) were still showing slight ZsGreen expression. The results were in agreement with our luciferase assay (Figure 3A), confirming that the 1 s UV-C irradiation is not sufficient to completely inactivate the viruses.

For the PDI study, the viruses were incubated with 10 μg/mL PDZ, and irradiated for 1 or 20 min, which equals the light doses of 1.8 and 36 J/cm^2, respectively (Figure 6D). The dark control groups were submitted to the same procedure, except for light exposure. No green fluorescent emission was observed in the cells after PDI with 20 min irradiation. In contrast, the dark controls showed fluorescent emission of ZsGreen. Neither PDZ irradiated samples nor dark samples showed toxicity on the cell confluency, while an increase of autofluorescence was observable, compared to the negative control cells. These observations confirm our results of luciferase assay (Figure 3B), and are in agreement with our previous studies on PDZ-based PDI, as described elsewhere [15].

In sum, two distinct strategies (UV-C irradiation and PDZ-based PDI) were applied for the inactivation of SC2 pseudovirus produced using HIV-based lentiviral system which specifically infect ACE2-expressing cells. This specificity was demonstrated using luciferase assay compared to the control negative cells and the control naked viruses, which agreed with previous reports [21,23,33,34]. The viral inactivation could be the consequence of either viral protein damage, which affects the cell internalization, or viral genome damage affecting the viral load. Unlike the SC2 RNA virus with viral reproduction independent of the host genome [7,8,30,35,36], this pseudotyped model enabled us to study not only the RNA viral load, but also the DNA integration, as well as the presence or absence of a spike on the viral particle. Several reports demonstrating the results of viral inactivation assays have a high degree of concordance with a clinical isolate of SC2 [33,34]; however, the results cannot be used for the inactivation of the actual SC2 virus unless tested.

4. Conclusions

Considering the advantages of pseudovirus over the actual SC2 virus, which was discussed above, we followed a simple protocol for calculating the RNA and DNA load and compared the qPCR data with luciferase assay results. Hence, we studied the viral inactivation by UV-C and PDI in dose and time-dependent manners via biochemical characterizations and quantitative PCR on four levels; virion damage; viral cell entry; DNA integration; and expression of reporter genes. Both UV-C and PDI treatments could destroy ssRNA and the spike protein of the virus in different ratios; however, the virus was still capable of binding and entering into the ACE-2 expressing 293T cells. UV-C irradiation disinfected the virus mainly through viral genome damage, with no apparent effects on the viral size and virus–cell binding ability. On the other side, PDZ-based PDI mostly destroyed the spike and viral membrane. Ignoring the type of viral destruction (ssRNA or spike), the cells infected by the photo-inactivated virus could not express the luciferase reporter gene. Our findings emphasize the advantages of PDI over UV-C viral inactivation. ROS-mediated damages on the viral envelope may generate debris or the fragments which could stimulate host immune defense. Moreover, viral PDI has affordability compared to other therapeutics like monoclonal antibodies (e.g., Ronapreve), which can be important factors for preventative use at home [37]. Other advantages of PDI include high repeatability without viral resistance or UV-signature mutations, with fast removal of the virus in a very short time.

The other advantage of this model is comparing the viral particles in two forms of enveloped and non-enveloped (naked) viruses, as a matter of importance for side-by-side comparison. Therefore, comparing two viruses with similar genomes but different in their protein envelope enables us to study the effect of each inactivation strategy on damaging the RNA genome in the presence and absence of a spike. Besides, this pseudotyped model can be used for other radiation-based strategies for understanding their mechanism of viral inactivation, with no need to work in BSL-3.

Author Contributions: F.E.G.G., M.M. and M.S. conceptualized and wrote the main manuscript text and prepared figures; F.E.G.G., M.M., R.S.D. and M.S. designed and performed the biochemical and microscopy experiments; F.F.P.J. designed and performed the DLS assay; R.S.D., R.S.F., E.F.d.C., M.S. and J.G. designed and performed the viral assays; L.F., L.Z. and G.C.-M. reviewed and edited the manuscript; M.S., R.S.D. and F.E.G.G. supervised the study. All authors have read and agreed to the published version of the manuscript.

Funding: The authors acknowledge the support provided by: FAPESP (Sao Paulo Research Foundation)–grant numbers: 2017/10910-5 (M. Sadraeian-Pós-Doutorado-Fluxo Contínuo), 2013/07276-1 (CEPOF–CEPID Program), 2019/14526-0 and 2020/05146-7 (G. Cabral-Miranda, JP FAPESP), and CAPES Project (COMBA TE-COVID1673158P) Program (PHYSICS 33002045002P9). The COVID-19 reagents were supported by BEI Resources Repository of ATCC and the NIH.

Institutional Review Board Statement: Not applicable.

Informed Consent Statement: Not applicable.

Data Availability Statement: All data available are reported in the article.

Acknowledgments: We thank BEI Resources Repository of ATCC and the NIH for supporting the reagent related to COVID-19. We thank Seth Pincus, M.D. for his support in the Department of Chemistry and Biochemistry, Montana State University, Bozeman, MT.

Conflicts of Interest: Authors declare no competing interests. Graphical figures were created with BioRender software ((https://biorender.com/ (accessed on 15 July 2021)).

Abbreviations

ACE-2	Angiotensin Converting Enzyme 2
DLS	Dynamic Light Scattering
PDI	Photodynamic Inactivation
PDT	Photodynamic Therapy
PDZ	Photoditazine
PS	Photosensitizer
RLU	Relative Luciferase Unit
SC2	SARS-CoV-2

References

1. World Health Organization. Coronavirus (COVID-19) Dashboard. Available online: https://covid19.who.int/ (accessed on 9 January 2022).
2. Tran, H.D.M.; Boivin, S.; Kodamatani, H.; Ikehata, K.; Fujioka, T. Potential of UV-B and UV-C irradiation in disinfecting microorganisms and removing N-nitrosodimethylamine and 1,4-dioxane for potable water reuse: A review. *Chemosphere* **2021**, *286*, 131682. [CrossRef] [PubMed]
3. Simonet, J.; Gantzer, C. Inactivation of poliovirus 1 and F-specific RNA phages and degradation of their genomes by UV irradiation at 254 nanometers. *Appl. Environ. Microbiol.* **2006**, *72*, 7671–7677. [CrossRef]
4. Nogueira, M.S. Ultraviolet-based biophotonic technologies for control and prevention of COVID-19, SARS and related disorders. *Photodiagn. Photodyn. Ther.* **2020**, *31*, 101890. [CrossRef]
5. Wallace, S.S.; Murphy, D.L.; Sweasy, J.B. Base excision repair and cancer. *Cancer Lett.* **2012**, *327*, 73–89. [CrossRef] [PubMed]
6. Hadi, J.; Dunowska, M.; Wu, S.; Brightwell, G. Control measures for SARS-CoV-2: A review on light-based inactivation of single-stranded rna viruses. *Pathogens* **2020**, *9*, 737. [CrossRef] [PubMed]
7. Biasin, M.; Bianco, A.; Pareschi, G.; Cavalleri, A.; Cavatorta, C.; Fenizia, C.; Galli, P.; Lessio, L.; Lualdi, M.; Tombetti, E.; et al. UV-C irradiation is highly effective in inactivating SARS-CoV-2 replication. *Sci. Rep.* **2021**, *11*, 6260. [CrossRef] [PubMed]
8. Lo, C.-W.; Matsuura, R.; Iimura, K.; Wada, S.; Shinjo, A.; Benno, Y.; Nakagawa, M.; Takei, M.; Aida, Y. UVC disinfects SARS-CoV-2 by induction of viral genome damage without apparent effects on viral morphology and proteins. *Sci. Rep.* **2021**, *11*, 13804. [CrossRef] [PubMed]
9. Bekmukhametova, A.; Ruprai, H.; Hook, J.M.; Mawad, D.; Houang, J.; Lauto, A. Photodynamic therapy with nanoparticles to combat microbial infection and resistance. *Nanoscale* **2020**, *12*, 21034–21059. [CrossRef]

10. Dobre, M.; Boscencu, R.; Neagoe, I.V.; Surcel, M.; Milanesi, E.; Manda, G. Insight into the web of stress responses triggered at gene expression level by porphyrin-pdt in ht29 human colon carcinoma cells. *Pharmaceutics* **2021**, *13*, 1032. [CrossRef]
11. Higuchi, N.; Hayashi, J.-I.; Fujita, M.; Iwamura, Y.; Sasaki, Y.; Goto, R.; Ohno, T.; Nishida, E.; Yamamoto, G.; Kikuchi, T.; et al. Photodynamic Inactivation of an Endodontic Bacteria Using Diode Laser and Indocyanine Green-Loaded Nanosphere. *Int. J. Mol. Sci.* **2021**, *22*, 8384. [CrossRef]
12. Zühlke, M.; Meiling, T.T.; Order, P.; Riebe, D.; Beitz, T.; Bald, I.; Löhmannsröben, H.-G.; Janßen, T.; Erhard, M.; Repp, A. Photodynamic Inactivation of E. coli Bacteria via Carbon Nanodots. *ACS Omega* **2021**, *6*, 23742–23749. [CrossRef] [PubMed]
13. Sadraeian, M.; da Cruz, E.F.; Boyle, R.W.; Bahou, C.; Chudasama, V.; Janini, L.M.R.; Diaz, R.S.; Guimarães, F.E.G. Photoinduced Photosensitizer–Antibody Conjugates Kill HIV Env-Expressing Cells, Also Inactivating HIV. *ACS Omega* **2021**, *6*, 16524–16534. [CrossRef] [PubMed]
14. Sadraeian, M.; Bahou, C.; Da Cruz, E.; Janini, L.; Diaz, R.S.; Boyle, R.; Chudasama, V.; Guimarães, F.E.G. Photoimmunotherapy Using Cationic and Anionic Photosensitizer-Antibody Conjugates against HIV Env-Expressing Cells. *Int. J. Mol. Sci.* **2020**, *21*, 9151. [CrossRef] [PubMed]
15. Romano, R.A.; Pratavieira, S.; Da Silva, A.P.; Kurachi, C.; Guimarães, F.E.G. Light-driven photosensitizer uptake increases Candida albicans photodynamic inactivation. *J. Biophotonics* **2017**, *10*, 1538–1546. [CrossRef] [PubMed]
16. Khorsandi, K.; Fekrazad, S.; Vahdatinia, F.; Farmany, A.; Fekrazad, R. Nano Antiviral Photodynamic Therapy: A Probable Biophysicochemical Management Modality in SARS-CoV-2. *Expert Opin. Drug Deliv.* **2021**, *18*, 265–272. [CrossRef]
17. Almeida, A.; Faustino, M.A.F.; Neves, M.G.P.M.S. Antimicrobial Photodynamic Therapy in the Control of COVID-19. *Antibiotics* **2020**, *9*, 320. [CrossRef]
18. Pires, L.; Bosco, S.D.M.G.; Baptista, M.S.; Kurachi, C. Photodynamic therapy in pythium insidiosum—An in vitro study of the correlation of sensitizer localization and cell death. *PLoS ONE* **2014**, *9*, e85431. [CrossRef]
19. Sharshov, K.; Solomatina, M.; Kurskaya, O.; Kovalenko, I.; Kholina, E.; Fedorov, V.; Meerovich, G.; Rubin, A.; Strakhovskaya, M. The Photosensitizer Octakis(cholinyl)zinc Phthalocyanine with Ability to Bind to a Model Spike Protein Leads to a Loss of SARS-CoV-2 Infectivity In Vitro When Exposed to Far-Red LED. *Viruses* **2021**, *13*, 643. [CrossRef]
20. Sabino, C.P.; Sellera, F.P.; Sales-Medina, D.F.; Machado, R.R.G.; Durigon, E.L.; Freitas-Junior, L.H.; Ribeiro, M.S. UV-C (254 nm) lethal doses for SARS-CoV-2. *Photodiagn. Photodyn. Ther.* **2020**, *32*, 101995. Available online: https://linkinghub.elsevier.com/retrieve/pii/S1572100020303495 (accessed on 14 March 2022). [CrossRef]
21. Ou, X.; Liu, Y.; Lei, X.; Li, P.; Mi, D.; Ren, L.; Guo, L.; Guo, R.; Chen, T.; Hu, J.; et al. Characterization of spike glycoprotein of SARS-CoV-2 on virus entry and its immune cross-reactivity with SARS-CoV. *Nat. Commun.* **2020**, *11*, 1620. [CrossRef]
22. Lam, B.; Kung, Y.J.; Lin, J.; Tseng, S.-H.; Tsai, Y.C.; He, L.; Castiglione, G.; Egbert, E.; Duh, E.J.; Bloch, E.M.; et al. In vivo characterization of emerging SARS-CoV-2 variant infectivity and human antibody escape potential. *Cell Rep.* **2021**, *37*, 109838. [CrossRef]
23. Crawford, K.H.D.; Eguia, R.; Dingens, A.S.; Loes, A.N.; Malone, K.D.; Wolf, C.R.; Chu, H.Y.; Tortorici, M.A.; Veesler, D.; Murphy, M.; et al. Protocol and Reagents for Pseudotyping Lentiviral Particles with SARS-CoV-2 Spike Protein for Neutralization Assays. *Viruses* **2020**, *12*, 513. [CrossRef] [PubMed]
24. Robinson, E.; Nunes, J.P.M.; Vassileva, V.; Maruani, A.; Nogueira, J.C.F.; Smith, M.E.B.; Pedley, R.B.; Caddick, S.; Baker, J.R.; Chudasama, V. Pyridazinediones deliver potent, stable, targeted and efficacious antibody–drug conjugates (ADCs) with a controlled loading of 4 drugs per antibody. *RSC Adv.* **2017**, *7*, 9073–9077. [CrossRef]
25. Marinho, R.D.S.S.; Duro, R.L.S.; Santos, G.L.; Hunter, J.; Teles, M.D.A.R.; Brustulin, R.; Milagres, F.A.D.P.; Sabino, E.C.; Diaz, R.S.; Komninakis, S.V. Detection of coinfection with Chikungunya virus and Dengue virus serotype 2 in serum samples of patients in State of Tocantins, Brazil. *J. Infect. Public Health* **2020**, *13*, 724–729. [CrossRef]
26. Sucupira, M.C.A.; Sanabani, S.; Cortes, R.M.; Giret, M.T.M.; Tomiyama, H.; Sauer, M.M.; Sabino, E.C.; Janini, L.M.; Kallas, E.G.; Diaz, R.S. Faster HIV-1 Disease Progression among Brazilian Individuals Recently Infected with CXCR4-Utilizing Strains. *PLoS ONE* **2012**, *7*, e30292.
27. Kumar, A.M.; Fernandez, J.B.; Singer, E.J.; Commins, D.; Waldrop-Valverde, D.; Ownby, R.L.; Kumar, M. Human immunodeficiency virus type 1 in the central nervous system leads to decreased dopamine in different regions of postmortem human brains. *J. Neuro Virol.* **2009**, *15*, 257–274. [CrossRef] [PubMed]
28. Komninakis, S.V.; Santos, D.E.M.; Santos, C.; Oliveros, M.P.R.; Sanabani, S.; Diaz, R.S. HIV-1 Proviral DNA Loads (as Determined by Quantitative PCR) in Patients Subjected to Structured Treatment Interruption after Antiretroviral Therapy Failure. *J. Clin. Microbiol.* **2012**, *50*, 2132–2133. [CrossRef]
29. Hu, J.; Gao, Q.; He, C.; Huang, A.; Tang, N.; Wang, K. Development of cell-based pseudovirus entry assay to identify potential viral entry inhibitors and neutralizing antibodies against SARS-CoV-2. *Genes Dis.* **2020**, *7*, 551–557. [CrossRef]
30. Svyatchenko, V.A.; Nikonov, S.D.; Mayorov, A.P.; Gelfond, M.L.; Loktev, V.B. Antiviral photodynamic therapy: Inactivation and inhibition of SARS-CoV-2 in vitro using methylene blue and Radachlorin. *Photodiagn. Photodyn. Ther.* **2021**, *33*, 102112. [CrossRef]
31. Alves, F.; Ayala, E.T.P.; Pratavieira, S. Sonophotodynamic Inactivation: The power of light and ultrasound in the battle against microorganisms. *J. Photochem. Photobiol.* **2021**, *7*, 100039. [CrossRef]
32. Sabino, C.P.; Wainwright, M.; dos Anjos, C.; Sellera, F.; Baptista, M.S.; Lincopan, N.; Ribeiro, M. Inactivation kinetics and lethal dose analysis of antimicrobial blue light and photodynamic therapy. *Photodiagn. Photodyn. Ther.* **2019**, *28*, 186–191. [CrossRef]

33. Case, J.B.; Rothlauf, P.W.; Chen, R.E.; Liu, Z.; Zhao, H.; Kim, A.S.; Bloyet, L.-M.; Zeng, Q.; Tahan, S.; Droit, L.; et al. Neutralizing Antibody and Soluble ACE2 Inhibition of a Replication-Competent VSV-SARS-CoV-2 and a Clinical Isolate of SARS-CoV-2. *Cell Host Microbe* **2020**, *28*, 475–485.e5. [CrossRef] [PubMed]
34. Kumar, U.S.; Afjei, R.; Ferrara, K.; Massoud, T.F.; Paulmurugan, R. Gold-Nanostar-Chitosan-Mediated Delivery of SARS-CoV-2 DNA Vaccine for Respiratory Mucosal Immunization: Development and Proof-of-Principle. *ACS Nano* **2021**, *15*, 17582–17601. [CrossRef] [PubMed]
35. Duguay, B.A.; Herod, A.; Pringle, E.S.; Monro, S.M.A.; Hetu, M.; Cameron, C.G.; McFarland, S.A.; McCormick, C. Photodynamic Inactivation of Human Coronaviruses. *Viruses* **2022**, *14*, 110. [CrossRef]
36. Arentz, J.; von der Heide, H.-J. Evaluation of methylene blue based photodynamic inactivation (PDI) against intracellular B-CoV and SARS-CoV2 viruses under different light sources in vitro as a basis for new local treatment strategies in the early phase of a Covid19 infection. *Photodiagn. Photodyn. Ther.* **2022**, *37*, 102642. Available online: https://linkinghub.elsevier.com/retrieve/pii/S1572100021004609 (accessed on 14 March 2022). [CrossRef]
37. Kipshidze, N.; Yeo, N.; Kipshidze, N. Photodynamic therapy for COVID-19. *Nat. Photon.* **2020**, *14*, 651–652. [CrossRef]

Article

Multiple Light-Activated Photodynamic Therapy of Tetraphenylethylene Derivative with AIE Characteristics for Hepatocellular Carcinoma via Dual-Organelles Targeting

Chuxing Chai [1,†], Tao Zhou [2,†], Jianfang Zhu [3,†], Yong Tang [1], Jun Xiong [1], Xiaobo Min [1], Qi Qin [1], Min Li [1,*], Na Zhao [4,5,6] and Chidan Wan [1,*]

[1] Department of Hepatobiliary Surgery, Union Hospital, Tongji Medical College, Huazhong University of Science and Technology, Wuhan 430022, China; chaichuxing@hotmail.com (C.C.); tangyong_2222@126.com (Y.T.); oldxiong@163.com (J.X.); d201981629@hust.edu.cn (X.M.); qinqi8441@163.com (Q.Q.)

[2] Department of Otorhinolaryngology, Union Hospital, Tongji Medical College, Huazhong University of Science and Technology, Wuhan 430022, China; entzt@hust.edu.cn

[3] Central Laboratory, Union Hospital, Tongji Medical College, Huazhong University of Science and Technology, Wuhan 430022, China; 342244204@aliyun.com

[4] Key Laboratory of Macromolecular Science of Shaanxi Province, School of Chemistry & Chemical Engineering, Shaanxi Normal University, Xi'an 710119, China; nzhao@snnu.edu.cn

[5] Key Laboratory of Applied Surface and Colloid Chemistry of Ministry of Education, School of Chemistry & Chemical Engineering, Shaanxi Normal University, Xi'an 710119, China

[6] Key Laboratory of the Ministry of Education for Medicinal Resources and Natural Pharmaceutical Chemistry, Shaanxi Normal University, Xi'an 710119, China

* Correspondence: liminmed@hust.edu.cn (M.L.); chidan_wan@hust.edu.cn (C.W.)

† These authors contributed equally to this work and share first authorship.

Citation: Chai, C.; Zhou, T.; Zhu, J.; Tang, Y.; Xiong, J.; Min, X.; Qin, Q.; Li, M.; Zhao, N.; Wan, C. Multiple Light-Activated Photodynamic Therapy of Tetraphenylethylene Derivative with AIE Characteristics for Hepatocellular Carcinoma via Dual-Organelles Targeting. *Pharmaceutics* 2022, 14, 459. https://doi.org/10.3390/pharmaceutics14020459

Academic Editors: Udo Bakowsky, Matthias Wojcik, Eduard Preis, Gerhard Litscher and Patrick J. Sinko

Received: 8 December 2021
Accepted: 17 January 2022
Published: 21 February 2022

Publisher's Note: MDPI stays neutral with regard to jurisdictional claims in published maps and institutional affiliations.

Abstract: Photodynamic therapy (PDT) has emerged as a promising locoregional therapy of hepatocellular carcinoma (HCC). The utilization of luminogens with aggregation-induced emission (AIE) characteristics provides a new opportunity to design functional photosensitizers (PS). PSs targeting the critical organelles that are susceptible to reactive oxygen species damage is a promising strategy to enhance the effectiveness of PDT. In this paper, a new PS, 1-[2-hydroxyethyl]-4-[4-(1,2,2-triphenylvinyl)styryl]pyridinium bromide (TPE-Py-OH) of tetraphenylethylene derivative with AIE feature was designed and synthesized for PDT. The TPE-Py-OH can not only simultaneously target lipid droplets and mitochondria, but also stay in cells for a long period (more than 7 days). Taking advantage of the long retention ability of TPE-Py-OH in tumor, the PDT effect of TPE-Py-OH can be activated through multiple irradiations after one injection, which provides a specific multiple light-activated PDT effect. We believe that this AIE-active PS will be promising for the tracking and photodynamic ablation of HCC with sustained effectiveness.

Keywords: photodynamic therapy; aggregation-induced emission; organelles targeting; hepatocellular carcinoma

1. Introduction

Locoregional ablation therapy was recommended as an effective treatment for patients with primary hepatocellular carcinoma (HCC) and liver metastases [1,2]. The mechanism of ablation is to cause the destruction of local tumor by physical or chemical damage. Thermal ablation, including radiofrequency ablation and microwave ablation, has been applied as a safe, low-cost, effective alternative of surgery in small and multifocal liver tumor [3]. The ablation procedure often relies on the placement of the needle electrode into target tumor under the guidance of imaging technology which requires additional competence and fails to provide real-time self-monitoring. Multiple treatment sessions and repetitive puncture operation may be required due to difficult tumor conditions and limitation of imaging technology, which may cause additional suffering of patients [4]. Alternatively, photodynamic therapy (PDT) is a potential tumor-ablative locoregional therapy [5–7]. Fluorescent molecules with photodynamic effect can provide precise and safe ablation of tumor at the cellular level while enabling dynamic monitoring of tumor lesion, which has great promise in building cancer-targeted theragnostic platforms [8,9]. PDT can induce both cancer cell death and immune response against the tumor [10,11]. Additionally, intraoperative fluorescence image-guided surgery combined with PDT can eliminate residual tumors and reduce the recurrence of cancer [12–15]. The use of PDT has gained much attention in the treatment of HCC, and has been approved as a feasible approach for HCC treatment in preclinical studies [6,16].

The three crucial elements of PDT are photosensitizer (PS), light irradiation, and oxygen. PS is the core element of PDT. The procedure of PDT involves administration of PSs which accumulate in the tumor tissue, followed by local illumination with specific wavelength to active PSs [17]. After exposure to the light, the PSs can transfer energy from light to molecular oxygen, contribute to their change of energy states, to generate reactive oxygen species (ROS), which then induce apoptosis, necrosis, and autophagy in treated cells and cause cell death [18,19]. However, traditional PSs—such as porphyrin, phthalocyanine, and analogs—are prone to aggregate in the treated cells or aqueous solution, which will lead to a dramatic reduction in ROS production and suppression in fluorescence emission [20,21].

In 2001, Tang's group discovered a novel fluorogen with the opposite characteristics, namely aggregation-induced emission (AIE) [22], which is nonemissive in a molecularly dissolved state, but is highly emissive in their aggregate/solid state due to the mechanism of restriction of intramolecular motions/rotations [23]. Meanwhile, as organic molecules, AIE luminogens (AIEgens) were highly editable. The different additional groups and molecular structures can provide different optical, physical, and chemical properties to AIEgens, which enables broad biological medicine application [24–27]. A series of AIEgens have been developed for translational applications in sensing, imaging, and theranostics with excellent performance than conventional fluorescent probes [28,29]. Recently, some AIEgens were designed to exhibit excellent photosensitization and ROS generation ability, which broadens their potential applications in PDT of cancer. The unique features of AIEgens provide new opportunities for the facile design of PS with special function, high accuracy, and efficacy for image-guided PDT [30–33].

A critical limitation of PDT is that the half-life of ROS (<40 ns) is usually very short so that it can only act on the production site (<20 nm) [34], which means simple transportation PSs into tumor tissues or cancer cells are not efficient enough to achieve expected outcomes [34,35]. Designing PSs to target the critical organelles that are susceptible to ROS damage is a promising strategy to enhance the effectiveness of PDT. The mitochondrion is a promising target for PSs owing to its crucial role in ROS generation, oxidation-reduction status modulation, and cell apoptosis mediation. Mitochondria in tumor cells are sensitive to ROS generation induced by PDT, which ultimately induces tumor cell apoptosis by disrupting the balance of mitochondrial ROS [36]. Recently, several AIEgens have been designed for targeting mitochondria and have exhibited remarkable photostability, high brightness, and excellent PDT effect [37–40]. However, multiple doses and repeated injections of PSs are always required for better outcomes in PDT procedures, which may lead to superfluous side effects and additional toxicity to patients [36]. The destruction of tumor vessels of PDT may also reduce the accumulation of PSs in tumor tissue and influence the efficiency of second PDT. Lipid droplets (LDs), also known as adiposomes or lipid bodies, are intracellular lipid-rich organelles for the long-term storage of lipids [41]. The synthesis of lipid droplets is at a lower level in normal cells, whereas tumor cells actively synthesize lipid droplets owing to their high-metabolism state [42,43]. Due to the suitable metabolic activity and stability, as well as their interactions with other organelles, especially mitochondrial, LDs can be valuable targets for PDT [44,45]. Therefore, AIEgens that could dual-target mitochondria and LDs may provide long-term photodynamic effect and better outcomes in PDT treatment procedure after administration of single-dose PSs.

In this work, for developing a new PS with multiple light-activated photodynamic effect and organelle-targeting abilities, the 1-[2-hydroxyethyl]-4-[4-(1,2,2-triphenylvinyl) styryl]pyridinium bromide (TPE-Py-OH) was designed and synthesized. TPE-Py-OH possesses a typical AIE character and exhibits strong fluorescence emission in the aggregated state. It could simultaneously target LDs and mitochondria in living cells. Due to its lipophilic feature, it can aggregate and store in LDs in a relatively stable state for long-term intracellular retention. Meanwhile, TPE-Py-OH also has efficient ROS generation ability and PDT effect both in vitro and in vivo. All these features suggest its superior performance in long-term tracking of living cancer cells and organelles-targeted PDT effect of cancer cells. Benefiting from the strong fluorescence, good photostability, and efficient ROS generation ability of TPE-Py-OH, the long-term cancer cell tracking and multiple light-activated PDT effect of HCC for more than 7 days were achieved after one PS administration. This AIE-active PS exhibit sustained effectiveness in vitro and in vivo, which suggests that the PS can hopefully be applied in long-term intracellular organelle imaging and multiple light-activated photodynamic ablation of HCC.

2. Materials and Methods

2.1. Materials and Instruments

4-Methylpyridine, 2-bromoethanol, piperidine, anhydrous acetonitrile, and ethanol were purchased from Sigma-Aldrich and used without further purification unless specified otherwise. 4-(1,2,2-Triphenylvinyl)benzaldehyde was synthesized according to the literature method [46,47]. Other chemicals and solvents were all purchased from Aldrich and used as received without further purification. Distilled water was used throughout the experiments. HepG2 cells were obtained from Shanghai Cell Bank of Chinese Academy of Sciences (Shanghai, China). Dulbecco's modified Eagle medium, fetal bovine serum (FBS), BODIPYTM 493/503 Dye, and JC-1 Dye were purchased from Thermo Fisher Scientific (Waltham, MA, USA). Cell Counting Kit-8 (CCK8) was purchased from Dojindo Molecular Technologies, Inc. (Mashikimachi, Japan). Reactive oxygen species (ROS) assay kit, calcein AM/PI double stain kit, Annexin V-FITC/PI double stain kit and MitoTracker Red FM was purchased from Beyotime Biotechnology (Shanghai, China). Annexin V-FITC/PI apoptosis detection kit was purchased from Qcbio Science & Technologies (Shanghai, China). Tissue culture flask, 96-well plates, and 24-well plates were purchased from Corning Incorporated (Corning, NY, USA). Confocal dishes were purchased from Biosharp Life Science (Hefei, China).

^{1}H and ^{13}C NMR spectra were measured on a Bruker ARX 400 NMR spectrometer (Bruker, Tucson, AZ, USA) using DMSO-d_6 as solvents. High-resolution mass spectra (HRMS) were recorded on a Finnigan MAT TSQ 7000 Mass Spectrometer System (Finnigan MAT, San Jose, CA, USA) operating in a MALDI-TOF mode. UV–vis absorption spectra were recorded on a Biochrom UV visible spectrometer (Biochrom, Berlin, Germany). Photoluminescence (PL) spectra were recorded on a PerkinElmer spectrofluorometer LS 55 (PerkinElmer, Singapore, Singapore). The cells were imaged under LSM7 DUO confocal laser scanning microscope (CLSM; Carl Zeiss Microscopy, Jena, Germany) and fluorescence microscope (IX71, Olympus, Tokyo, Japan). The absorbance was measured using a spectrophotometer (Thermo Fisher Scientific, Waltham, MA, USA).

2.2. Synthesis of 1-(2-Hydroxyethyl)-4-methylpyridinium Bromide (Compound 1)

An acetonitrile (20 mL) solution of 4-methylpyridine (0.93 g, 10.00 mmol) had 2-bromoethanol (2.5 g, 20.00 mmol) added to it. The mixture was stirred at reflux temperature (82 °C) under nitrogen for 24 h and then cooled to room temperature. Then, the reaction mixture was precipitated in ethyl acetate (EA) (200 mL). The solid residue was filtrated and washed with excess EA. The filtrate was collected and dried in a vacuum oven. Compound 1 was a light brown powder in 86% yield. ^{1}H NMR (400 MHz, DMSO-$d6$, δ): 8.91 (d, 2H, J = 6.8 Hz), 7.99 (d, 2H, J = 6.4 Hz), 4.64 (t, 2H, J = 15.6 Hz), 3.81 (t, 2H, J = 15.6 Hz), 2.60 (s,

3H). ^{13}C NMR (100 MHz, DMSO-*d6*) δ: 158.5, 143.9, 127.7, 61.9, 59.7, 21.2. HRMS (LDI-TOF) m/z calcd for $[C_8H_{12}NO]^+$, 138.0919 [M-Br]+; found, 138.0920 [M-Br]+.

2.3. 1-[2-Hydroxyethyl]-4-[4-(1,2,2-triphenylvinyl)styryl]pyridinium Bromide (TPE-Py-OH)

A solution of 1 (0.5 g, 2 mmol) and 4-(1,2,2-triphenylvinyl)benzaldehyde (1.01 g, 2.8 mmol) was refluxed at temperature (78 °C) under nitrogen in dry ethanol (20 mL) catalyzed by three drops of piperidine. After cooling to ambient temperature, the solvent was evaporated under reduced pressure. The residue was purified by a silica gel column chromatography (50 mesh) using dichloromethane/methanol mixture (5:1 *v/v*) as eluent to obtain TPE-Py-OH as a yellow solid in 64% yield (0.72 g, 1.29 mmol). ^{1}H NMR (400 MHz, DMSO-*d6*) δ: 8.82 (d, 2H, *J* = 6.8 Hz), 8.16 (d, 2H, *J* = 6.8 Hz), 7.88 (d, 1H, *J* = 16.0 Hz), 7.50 (d, 2H, *J* = 8.4 Hz), 7.42 (d, 1H, *J* = 16.4 Hz), 7.16–6.94 (m, 17H), 4.51 (t, 2H, *J* = 10.0 Hz), 3.82 (m, 2H), 3.5(brs, 1H). ^{13}C NMR (100 MHz, DMSO-*d6*) δ: 152.70, 145.35, 144.50, 142.79, 142.71, 142.56, 141.40, 140.07, 139.72, 133.10, 131.25, 130.53, 130.51, 130.43, 127.79, 127.68, 127.51, 126.69, 126.62, 123.22, 123.02, 61.92, 59.83. HRMS (LDI-TOF) *m/z* calcd for $[C_{15}H_{30}NO]^+$, 480.2327 [M-Br]+; found, 480.2317 [M-Br]+.

2.4. Cellular Uptake and Subcellular Distribution

HepG2 human HCC cells were cultured in DMEM medium supplemented with 10% FBS and 1% penicillin-streptomycin at 37 °C in a 5% CO_2, 90% relative humidity incubator. For cell imaging, the HepG2 cells were seeded in slide chambers at a density of 10^5 and cultured overnight for adhesion. Then, cells were incubated with TPE-Py-OH at various concentrations (2, 5, and 10 μM) for 30 min and continued to incubate with fresh medium for 12 h. Nuclei were labeled with DAPI (1 μg/mL) overnight. After rinsing with PBS, the cells were imaged under a confocal laser scanning microscope (CLSM, Nikon Corporation, Tokyo, Japan). To further explore the subcellular distribution of TPE-Py-OH, 2×10^5 HepG2 cells were plated in the 35-mm confocal dish and cultured overnight, then incubated with 5 μM TPE-Py-OH for 30 min. After rising with PBS, the cells were then stained with 50 nM MitoTracker Red FM, or continuously incubated with fresh culture medium and then stained with 1 μg/mL BODIPY for 30 min, respectively. After three washes, the cell images were captured under CLSM, and the data was analyzed by NIS-Elements Imaging Software (Nikon Corporation, Tokyo, Japan) and Image-J (National Institutes of Health freeware, Bethesda, MD, USA). For TPE-Py-OH: λex = 405 nm and band-pass filter λ = 550–600 nm. For DAPI: λex = 405 nm and band-pass filter λ = 425–475 nm. For MitoTracker Red: λex = 581 nm and band-pass filter λ = 600–650 nm. For BODIPY 493/503: λex = 488 nm and band-pass filter λ = 500–550 nm.

2.5. Cytotoxicity Studies In Vitro

Cell viability after incubation with various concentrations of TPE-Py-OH was evaluated by CCK8 assays. Briefly, the cells at a density of 1×10^4 were seeded into a 96-well plate overnight and then incubated with a series of concentrations of TPE-Py-OH (0, 1, 3, 5, 7, 10, and 15 μM). After pre-treatment, all the cells were incubated with 100 μL fresh medium including with 10 μL CCK8 solution per well and then continued to incubate for 4 h. The absorbance of the solution at 450 nm was measured using a multimode plate reader (Perkin Elmer Pte. Ltd., Singapore, Singapore).

For cytotoxicity effect induced by PDT, the cells were incubated with various concentrations of TPE-Py-OH for 30 min and continued to incubate with fresh medium for 12 h and then exposed to blue laser irradiation with different duration (450 nm, 30 mW/cm^2, 1.8 J/cm^2–18 J/cm^2). The cells were as negative controls without both laser irradiation and TPE-Py-OH treatment. Sequentially, the cells continued to incubate for 24 h and the cell viability was measured according to the above description.

2.6. Intracellular ROS Detection

Intracellular ROS production was measured by 2,7-dichlorodihydrofluorescein diacetate (DCFH-DA) probe according to the manufacturer's instructions. After incubation in the presence or absence of TPE-Py-OH for 30 min, the cells were loaded with 10 μM DCFH-DA at 37 °C for 30 min in dark. The cells were replaced with fresh medium and then exposed to 450 nm blue laser irradiation at the power density of 30 mW/cm^2, then immediately imaged under a fluorescence microscope (IX71, Olympus, Tokyo, Japan). For DCFH-DA: λex = 488 nm and band-pass filter λ = 500–550 nm.

2.7. Calcein-AM and Propidium Iodine (PI) Staining Assay

The live and dead cells were identified using a calcein/PI double stain kit according to the protocol. After treatment, the cells were incubated with 2 μM calcein-AM and 5 μM PI for 30 min at 37 °C. The fluorescence images were captured under a fluorescence microscope (IX71, Olympus, Tokyo, Japan).

2.8. Flow Cytometry Analysis

The apoptotic and necrotic cells were detected by Annexin V-FITC/PI dual staining using flow cytometry. The cells were seeded into a 24-well plate overnight and then incubated with TPE-Py-OH for 30 min. After exposure to laser irradiation (450 nm, 30 mW/cm^2, 4 min, or 8 min), the cells were further incubated for 24 h, then collected and resuspended in 100 μL of binding buffer. According to the manufacturer's instructions, the cells were incubated with 5 μL Annexin V-FITC and 10 μL PI for 15 min in the dark at room temperature and then immediately measured by flow cytometry (BD Falcon, San Jose, CA, USA).

2.9. Mitochondrial Membrane Potential ($\Delta\psi$m) Measurement

The change of mitochondrial membrane potential ($\Delta\psi$m) was measured by JC-1 Dye according to the manufacturer's instructions. Briefly, the cells were incubated with 2 μg/mL of JC-1 for 20 min at 37 °C after different concentrations of TPE-Py-OH administration and different laser irradiation times, washed twice with PBS, and then imaged under CLSM. For JC-1 (monomer): λex = 488 nm and band-pass filter λ = 500–530 nm; (J-aggregate): λex = 585 nm and band-pass filter λ = 590 nm.

2.10. Long-Term Cell Tracking and In Vitro PDT

The cells were first stained with 10 μM TPE-Py-OH for 12 h and then rinsed with PBS to remove the extracellular molecules. The image was captured as a reference. The cells were incubated continuously and one-third of them were sub-cultured into another dish for continual tracking every 2 days. Other cells were seeded into the 35-mm confocal dish. After attachment for 24 h, the fluorescence of TPE-Py-OH in cells was imaged by CLSM, the ROS level in cells was performed according to the above description. The step was repeated every 2 days until non-fluorescence in cells.

To confirm the long-term localization of intracellular TPE-Py-OH, the HepG2 cells were incubated with 10 μM TPE-Py-OH for 12 h and then cultured with fresh DMEM medium, the daughter cells were sequentially co-stained with 1 μg/mL BODIPY for 30 min on days 1, 3, 5, and 7 and their fluorescence images were captured under a fluorescence microscope (IX71, Olympus, Tokyo, Japan).

For long-term PDT in vitro, the HepG2 cells were incubated with 10 μM TPE-Py-OH for 12 h and replaced with fresh medium for following experiments. On days 1, 3, 5, and 7 after TPE-Py-OH treatment, the cells were seeded into 35-mm confocal dishes. After attachment, the cells were exposed to blue laser irradiation with different duration (450 nm, 30 mW/cm^2, 18–45 J/cm^2) with half of the confocal dish covered by foil paper to avoid light irradiation. Then the calcein/PI double stain was used to detect the effect of PDT, and the JC-1 mitochondrial membrane potential probe was used for mitochondrial membrane potential measurement. The fluorescence images were captured under a fluorescence microscope (IX71, Olympus, Tokyo, Japan).

2.11. Animal Model

Male BALB/c mice (4 weeks old, 20 g) were purchased from HFK Bioscience Co, Ltd. (Beijing, China). All animal experiments were approved by the Ethics Committee of the Union Hospital of Huazhong University of Science and Technology and conducted in accordance with the guidelines of the Department of Laboratory Animals of Tongji Medical College. H22 cells (5×10^5) were subcutaneous injection into the right axilla of each mouse to establish xenograft liver tumor model, and the tumor volume reached 60 mm^3 after 5 days.

2.12. In Vivo Multiple Light-Activated PDT

Tumor-bearing mice were randomized into 5 groups of 8 animals per group: (Group 1: TPE-Py-OH 3 IR) 50 ug/100 uL TPE-Py-OH injected intratumorally, and then underwent laser irradiation (450 nm, 100 mw/cm^2, 10 min; 60 J/cm^2) on days 1, 3, and 5 after injection; (Group 2: TPE-Py-OH 1 IR) 50 ug/100 uL TPE-Py-OH injected intratumorally, and then underwent laser irradiation (450nm, 100 mw/cm^2, 10min; 60 J/cm^2) on the first day after injection; (Group 3: TPE-Py-OH non IR) 50 ug/100 uL TPE-Py-OH injected intratumorally without irradiation; (Group 4: 3 IR only) 100 μL of PBS injected intratumorally, and then underwent laser irradiation (450 nm, 100 mw/cm^2, 10 min; 60 J/cm^2) on days 1, 3, and 5 after injection; (Group 5: NC) negative control.

Seven days after injection, 3 mice of every group were sacrificed. Tumors were collected and kept in 4% formalin followed by embedded in paraffin. Slices cut from the paraffin sections were stained by hematoxylin and eosin (H&E) and TUNEL before being scanned with a fluorescence microscope (IX71, Olympus, Tokyo, Japan). The tumor size was measured by a caliper every 2 days and calculated as the volume = (tumor length) $\times$ (tumor width)$^2 \times 0.5$.

3. Results and Discussions

3.1. Synthesis and Characterization

TPE-Py-OH was synthesized according to the synthetic route shown in Figure 1A. Compound 1 was obtained by quaternization of 4-methylpyridine with 2-bromoethanol. The product was purified by precipitation of the reaction mixture in excess EA. The product was obtained by filtrating and washing with EA, which was suitable to be carried forward. 4-(1,2,2-Triphenylvinyl)benzaldehyde (TPE-CHO) was synthesized according to the reported method [46,47]. Finally, compound 1 and TPE-CHO were refluxed in ethanol to give TPE-Py-OH in reasonable yield by aldol condensation. Unexpectedly, TPE-Py-OH can be purified with silica column by using high polarity solvent mixture (DCM: MeOH = 5:1) as eluent. Compound 1 and TPE-Py-OH were fully characterized by HRMS, and ^{1}H and ^{13}C NMR spectroscopies, and gave satisfactory analysis results corresponding to their chemical structures (Figures S1–S4). Due to its ionic character, TPE-Py-OH has good solubility in most of organic solvents like THF, DCM, and DMSO, but it is not soluble in water.

3.2. Aggregation and Micellization of TPE-Py-OH

Figure S5 shows the UV spectrum of TPE-Py-OH (10 μM) in DMSO solutions. The maximum absorption peak of TPE-Py-OH is located at 395 nm. For convenient bio-application, 405 nm was utilized as excitation wavelength for PL measurement. Photoexcitation of its DMSO solution (100 μM) induces a red emission at 650 nm (Figure 1B,C), giving a large Stokes shift of 9932 cm^{-1} due to its extended conjugation as well as the intramolecular charge transfer (ICT) effect from the electron-donating TPE moiety to the electron-accepting pyridinium unit [47]. Due to the positively charged characteristics and pendant hydroxy group, TPE-Py-OH is amphiphilic and shows reversible aggregation behaviors in DMSO/H$_2$O mixture (Figure S6A). Normally, solutions with a concentration of 10^{-5}–10^{-6}

M are used in PL measurements. A highly concentrated solution (100 μM) of TPE-Py-OH is employed for the analysis because TPE-Py-OH only forms aggregates or micelles at higher concentrations. According to previous studies, the aromatic rings on AIEgens have larger twisted angles in the aggregated state than in the solution state [23]. Thus, the AIEgens generally show blue-shifted in the aggregated state compared to that in the solution state. As shown in Figure 1B, TPE-Py-OH emits strongly in aqueous solution at 565 nm due to the activation of the AIE effect by micelles formation. The emission becomes weaker and shifts to 650 nm upon addition of a small amount of DMSO, presumably owing to the disintegration of the micelles, which releases free molecules with a more planar conformation to the solution [48].

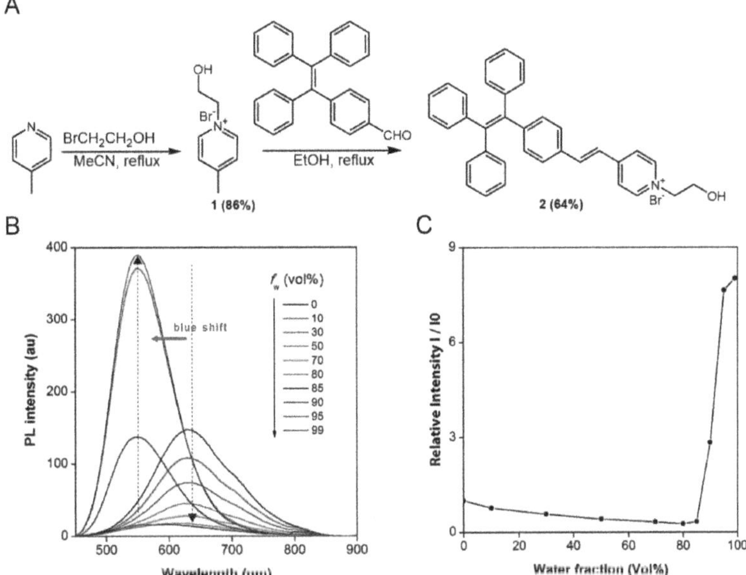

Figure 1. (**A**) Synthesis of molecule 2 (TPE-Py-OH). (**B**) PL spectra of TPE Py OH in DMSO/H₂O mixtures with different water fractions (fw). Concentration: 100 μM; excitation wavelength: 405 nm. (**C**) Plot of PL intensity versus the composition of the DMSO/H₂O mixtures of TPE-Py-OH.

To study the micellization behaviors of TPE-Py-OH, the PL of its solutions at different concentrations were investigated by taking advantage of its AIE property. The PL intensity of TPE-Py-OH increases with increasing the solution concentration with stable emission spectra (Figure S6B). The plot of relative PL intensity (I/I_0) against the logarithm solution concentration generates two lines, the intersection of which determines the critical micelle concentration (CMC) and is found to be 0.3 mM in H₂O/DMSO mixtures (v/v 90%). TPE-Py-OH is molecularly dispersed in solutions with concentrations of below 0.3 mM and displayed weak emission. When the concentration of TPE-Py-OH was more than or equal to 0.3 mM, the FL intensity was remarkably increased due to the AIE feature.

3.3. Dual-Organelles Targeting and Cell Imaging of TPE-Py-OH

We next studied the cellular uptake of TPE-Py-OH in HepG2 cells. The cells were treated with TPE-Py-OH and then observed by CLSM upon excitation at 405 nm with a collection of fluorescent signals above 585 nm. As shown in Figure S7, the bright yellow fluorescence could be detected from the cytoplasm after incubation with TPE-Py-OH for 30 min, which demonstrated that TPE-Py-OH could be internalized with high efficiency by HepG2 cells. Meanwhile, the fluorescence structure in cells was mainly in funicular shape when TPE-Py-OH entered into cells in short time and then fluorescence in punctate shape gradually enhanced, which demonstrated the aggregation of free TPE-Py-OH molecules in cytoplasm as the incubation time prolonged. This phenomenon indicated that the organelles targeting of TPE-Py-OH maybe occur in live cells.

To explore the subcellular location of TPE-Py-OH in living cells, HepG2 cells were co-stained with TPE-Py-OH and commercial fluorescent probes of mitochondria (MitoTracker Red) and LDs (BODIPY 493/503), respectively. As shown in Figure 2A, a large amount of yellow fluorescence from TPE-Py-OH merged with red fluorescence from MitoTracker Red in cells. However, some extra bright yellow fluorescence signals with punctate shape did not overlay the punctate red mitochondrial fluorescence. Sequentially, the co-localization experiment with BODIPY showed that the yellow punctate shape fluorescence from TPE-Py-OH and green fluorescence from BODIPY were co-localized well in live cells (Figure 2B). The Pearson's correlation was further quantitatively determined, which is high as 0.839 (Figure 2C). The intensity profile for the region of interest line across the cells also varied in close synchrony (Figure 2D). All the results indicated that TPE-Py-OH has a dual-targeting ability of mitochondria and LDs with high selectivity due to their structures of membrane phospholipid. Furthermore, it seems that the more dissociative TPE-Py-OH in cytoplasm could be finally inclined to LDs, and then stored in these organelles and kept stable in cells. Additionally, the fluorescent intensity of TPE-Py-OH in cells was dependent on its concentration in culture medium, which demonstrated that stronger fluorescence was captured in cells incubated with higher concentration (Figure 2E).

3.4. In Vitro Anti-Cancer Efficacy of TPE-Py-OH

Biocompatibility is a vital parameter of the PS for cancer therapy in vivo, and thus the biocompatibility of TPE-Py-OH on HepG2 cells was assessed via CCK8 assay before the in vitro PDT experiments. The cytotoxicity of TPE-Py-OH with different concentrations upon incubation with HepG2 cells in dark condition was first evaluated. As shown in Figure 3A, no significant change was observed in cell viability at 24 h after 30 min incubation without light irradiation. Even after 12 h incubation with 15 μM TPE-Py-OH, the cell viability was more than 80%, which indicated minimal cytotoxicity of TPE-Py-OH to cells and excellent biocompatibility. Then the PDT anti-cancer efficiency of TPE-Py-OH against HepG2 cells was further assessed. As shown in Figure 3B,C, the cell viability was significantly decreased in the presence of 450 nm blue laser irradiation. Meanwhile, the amount of cell death was associated with the light duration and the incubation concentration of TPE-Py-OH. The process of ROS generation and mitochondrial-reliant photodynamic killing of TPE-Py-OH was illustrated in Figure 3D. The intracellular ROS level after PDT treatment was measured via ROS indicator 2,7-dichlorodihydrofluorescein diacetate (DCFH-DA), which is highly fluorescent in the presence of ROS in living cells. A JC-1 dye was applied to monitor the change of mitochondrial membrane potential ($\Delta\psi$m). The decline of relative intensity of red to green fluorescence of JC-1 dye could indicate the depolarization of mitochondria, which is commonly used as an indicator of mitochondrial damage [49]. Additionally, after incubation with TPE-Py-OH followed by light irradiation, the appearance change of HepG2 cells was observed. As the concentration of TPE-Py-OH and irradiation time increased, HepG2 cells gradually shrank and turned round, indicating early apoptosis. After staining with DCFH-DA, green fluorescence was negligible without light irradiation or TPE-Py-OH. In contrast, cells incubated with TPE-Py-OH showed bright green fluorescence with light irradiation, which indicated significantly increased ROS production. The results of JC-1 imaging showed that TPE-Py-OH could efficiently promote the decline of $\Delta\psi$m under light irradiation and the decline of $\Delta\psi$m was concentration- and irradiation time-dependent. To further explore the PDT effect of TPE-Py-OH, HepG2 cells were stained with calcein-AM (green fluorescence) and PI (red fluorescence) to label live and dead cells (Figure 3E). The increased red fluorescence in TPE-Py-OH plus light irradiation group indicated good photodynamic cell death effect. An Annexin V-FITC/PI apoptosis detection kit was used to further investigate the cell death modality induced by PDT treatment

(Figure 3E). An increased number of both necrotic cells and apoptotic cells were observed with an elevation of the TPE-Py-OH concentration and irradiation time. All of the above results indicated that TPE-Py-OH could induce cell death through ROS generation and mitochondrial damage in the presence of light irradiation, and the dominant cell death is closely related to the light duration and incubation concentration.

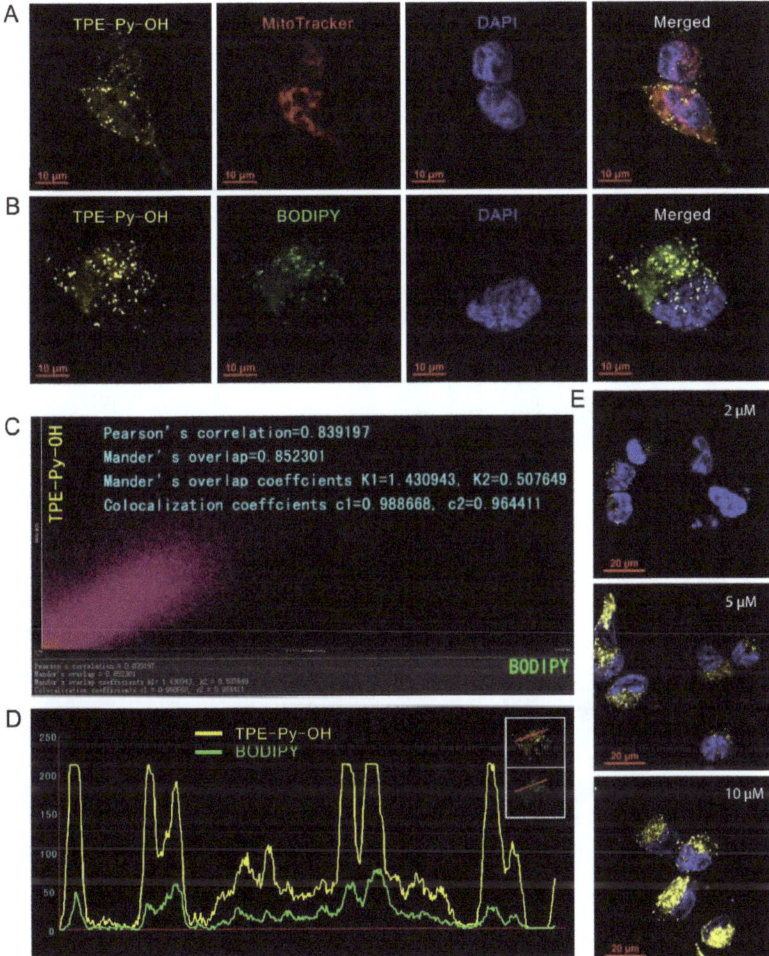

Figure 2. Fluorescence imaging of TPE-Py-OH in HepG2 cells. (**A**) Co-staining of 5 µM TPE-Py-OH (λex = 405 nm) and 50 nM MitoTracker Red (λex = 581 nm) for 30 min. (**B**) Co-staining of cells incubated with 5 µM TPE-Py-OH for 30 min, followed by continuous incubation in fresh medium for 12 h and then staining with 1 µg/L BODIPY (λex = 488 nm) for 30 min. (**C,D**) Quantitation analysis of the cells co-stained with TPE-Py-OH and BODIPY. (**E**) Images of TPE-Py-OH in HepG2 cells incubated at various concentrations for 30 min.

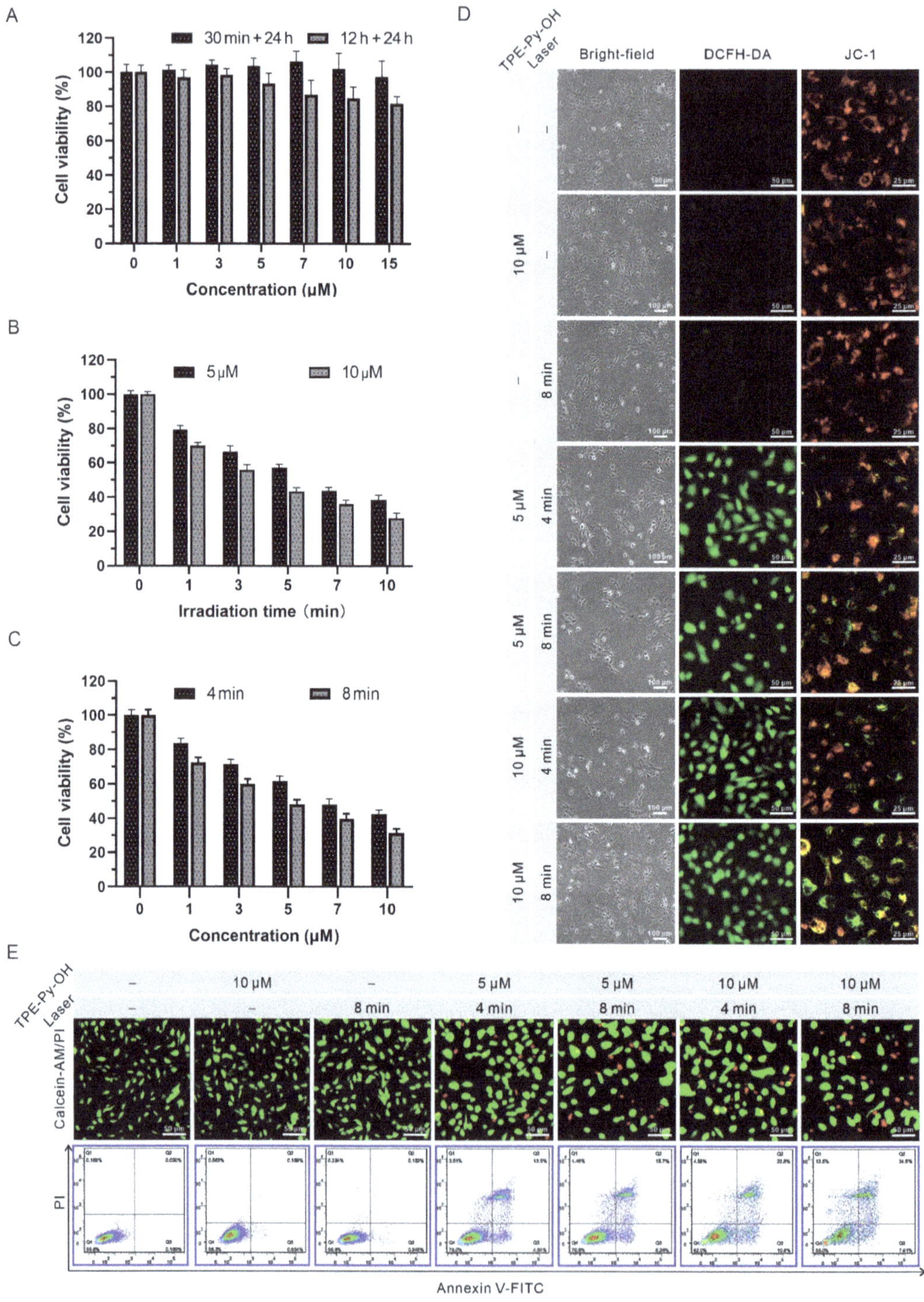

Figure 3. In vitro ROS generation and anti-tumor efficacy of TPE-Py-OH. (**A**) The viability of HepG2

cells pretreated with different concentrations of TPE-Py-OH for different times without light irradiation. (**B**) The viability of HepG2 cells pretreated with TPE-Py-OH, followed by different irradiation times (450 nm, 30 mW/cm^2). (**C**) The viability of HepG2 cells pretreated with different concentrations of TPE-Py-OH, followed by 4 min or 8 min irradiation (450 nm, 30 mW/cm^2; 7.2 J/cm^2 or 14.4 J/cm^2). (**D**) Detection cell morphology, ROS generation, and mitochondrial membrane potential ($\Delta\Psi$m) after different treatment of HepG2 cells. For DCFH-DA: λex = 488 nm and band-pass filter λ = 500–550 nm. For JC-1 (monomer): λex = 488 nm and band-pass filter λ = 500–530 nm; (J-aggregate): λex = 585 nm and band-pass filter λ = 590 nm. (**E**) Fluorescence images stained with calcein-AM/PI (For calcein-AM λex = 490 nm, For PI λex = 545 nm) and flow cytometry stain with Annexin V-FITC/PI after different treatment of HepG2 cells. For flow cytometry: Q1: Necrotic cells; Q2: Late apoptotic cells; Q3: Early apoptotic cells; Q4: Normal cells.

3.5. In Vitro Long-Term Tracking and PDT of TPE-Py-OH

To investigate the long-term retention of TPE-Py-OH in living cells, the HepG2 cells stained with TPE-Py-OH were continuously cultured and their fluorescent signal was monitored under CLSM. As shown in Figure 4A, the yellow fluorescence from TPE-Py-OH was recorded after 12 h of incubation as a reference. Sequentially, the fluorescence signal was captured every 2 days, and the intracellular yellow fluorescence still existed even on the seventh day after incubation with TPE-Py-OH, which demonstrated the long-term intracellular retention ability of TPE-Py-OH. Although the fluorescence intensity decreased as cell divided due to the reduction of the number of the AIEgen in each cell, the residual PSs could still be activated by light irradiation and induce ROS generation, which was confirmed by the DCFH-DA staining. Since previous result showed intracellular storage of TPE-Py-OH, we then explored the location of long-term retention of TPE-Py-OH in live HepG2 cells. Although TPE-Py-OH decreased in single cell as cell division, the yellow fluorescence was still co-localized with BODIPY, indicating that they could keep staying in LDs with high binding characteristics in Figure 4B. To verify long-term photodynamic effect of the remaining TPE-Py-OH in cells, the PDT efficiency was investigated using calcein-AM/PI dual staining. As shown in Figure 4C, the region of light irradiation (right part of the dotted line) showed increased red fluorescence of PI stain even on the seventh day after treatment with TPE-Py-OH in contrast with the region without light irradiation (left part of the dotted line). In the region of light irradiation, cell morphologic changes of the early stage of apoptosis were also observed in the bright field under optical microscope. Although the efficiency of PDT was decreased with the passage of time, the obvious photodynamic killing effect could last for at least 1 week after primary TPE-Py-OH treated with prolonged light irradiation. It seems that TPE-Py-OH can combine with and store in intracellular LDs with stable fluorescence and photodynamic tumor-killing effect. Previous results also showed that TPE-Py-OH could induce mitochondrial damage and dysfunction due to its ability to target mitochondria. Since the TPE-Py-OH could aggregate in LDs after long-term incubation, we further explored the change of damage mechanism. Interestingly, JC-1 staining demonstrated that TPE-Py-OH could still induce dramatically mitochondrial damage in spite of tending to accumulate into LDs in Figure 4D. Singlet oxygen species produced by PDT can oxidize intracellular LDs to form lipid oxide or lipid peroxide [50]. The intracellular lipid peroxidation could initiate cellular ferroptosis, which can induce mitochondrial morphology disruption and depolarization in this process [51]. This may be the mitochondrial damage mechanism of TPE-Py-OH stored in LDs. Hence, the TPE-Py-OH could store in LDs in a relatively stable state and maintain its ROS generation ability for more than 7 days, which allowed multiple-light activated PDT effect and mitochondrial damage.

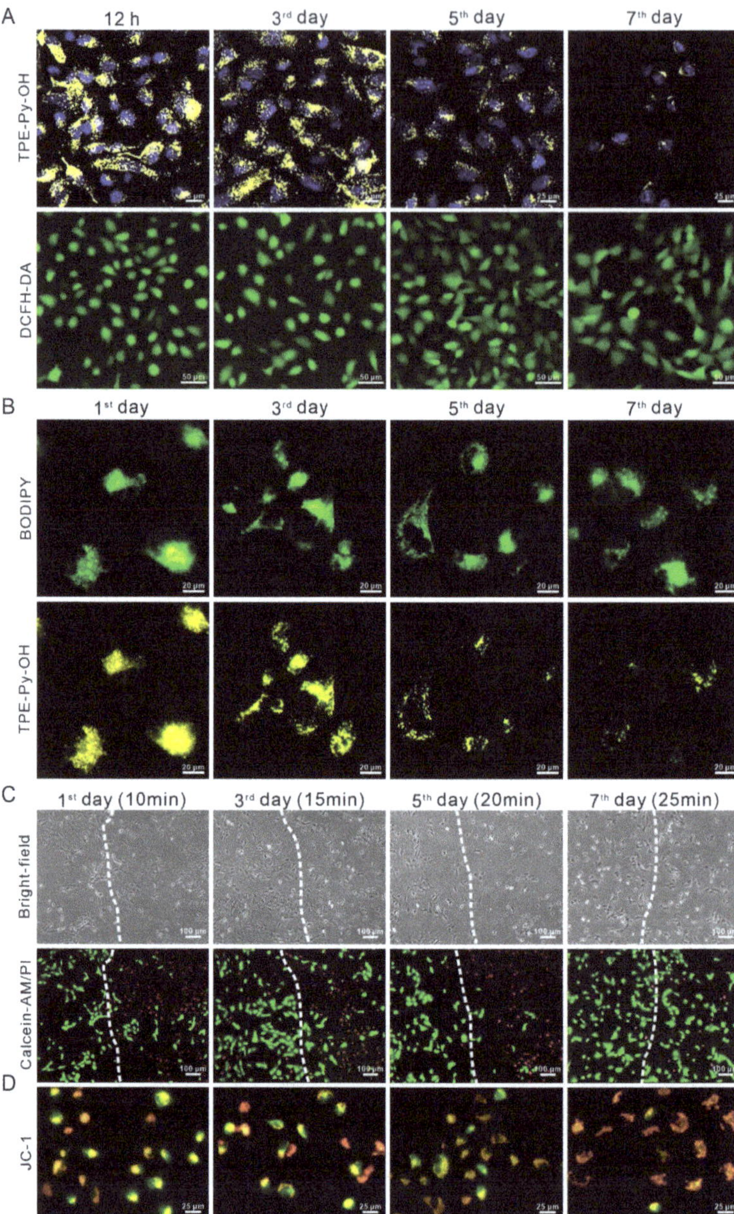

Figure 4. Long-term tracking and PDT effect of HepG2 cells stained with 10 µM TPE-Py-OH for 12 h and continuously cultured for different time. (**A**) Fluorescent images of TPE-Py-OH and intracellular ROS generation after light irradiation (450 nm, 30 mW/cm^2, 10 min). (**B**) Co-staining of HepG2 cells stained with TPE-Py-OH and continuously cultured for different time, followed by staining with 1 µg/mL BODIPY for 30 min. (**C**) Fluorescence images stained with calcein-AM/PI of HepG2 cells cultured for different time. left part of the dotted line: without light irradiation, right part of the dotted line: with light irradiation. (**D**) Fluorescence images stained with JC-1 of HepG2 cells cultured for different time with irradiation.

3.6. In Vivo Multiple Light-Activated PDT Effect of TPE-Py-OH

To further evaluate the potential of PDT performance of TPE-Py-OH in vivo, we carried out a mice H22 tumor model. The tumor model was established by subcutaneous injection of 5×10^5 H22 cells in BALB/c mice. After the tumors had become 60 mm^3 in size, TPE-Py-OH were injected intratumorally. In vivo distribution of TPE-Py-OH in mice over time was monitored by whole animal fluorescence. As shown in Figure S8A, local fluorescence still existed even at 12th day after TPE-Py-OH administration. The compression splice of tumor tissue shows bright yellow fluorescence in tumor cells (Figure S8B). Based on the long-term retention of TPE-Py-OH in tumor foci after administration, we hypothesized that the remaining TPE-Py-OH could kill the tumor cells by repeated irradiation after one administration via PDT effect. Subsequently, to explore the multiple light-activated PDT performance of TPE-Py-OH, H22 tumor-bearing mice were randomized into five groups: (1) TPE-Py-OH with three irradiations; (2) TPE-Py-OH with one irradiation; (3) TPE-Py-OH without irradiation; (4) PBS with irradiation; (5) negative control. TPE-Py-OH was intratumorally injected and further irradiated with laser (450 nm, 100 mW/cm^2, 10 min) and the course of treatment is shown in Figure 5A. As shown in Figure 5B, the mice injected TPE-Py-OH with laser irradiation three times exhibited larger scale local necrosis of tumor and better tumor inhibition than the group treated with irradiation one time. TPE-Py-OH without irradiation group, PBS with irradiation group and negative control group were found to exhibit negligible tumor necrosis and inhibition. The tumor tissues were harvested after the mice were sacrificed on the seventh Day. Hemotoxylin/eosin (H&E) staining of tumor tissue showed a majority of dead cells in TPE-Py-OH with the three-time irradiation group and relative mild tumor cells death in TPE-Py-OH with the one-time irradiation group (Figure 5C). A high apoptosis rate was also observed by TUNEL stain, which demonstrated the multiple light-activated PDT effect of TPE-Py-OH after once injection for more than 1 week. This result was consistent with previous in vitro experiments. Tumor growth in the different groups was monitored by recording the tumor volume every 2 days for a period of 10 days. The tumor volume of TPE-Py-OH with three-time irradiation group was obviously smaller than other groups (Figure 5D). Although TPE-Py-OH with the one-irradiation group showed mild tumor inhibition after treatment, the tumor volume significantly increased after treatment ended. Mice survival was monitored, as shown in Figure 5E. Mice treated with TPE-Py-OH and irradiation three times showed better survival than other groups. All these results demonstrate that TPE-Py-OH could accumulate in the tumor region for a long time and provide multiple light-activated PDT effect for a long time.

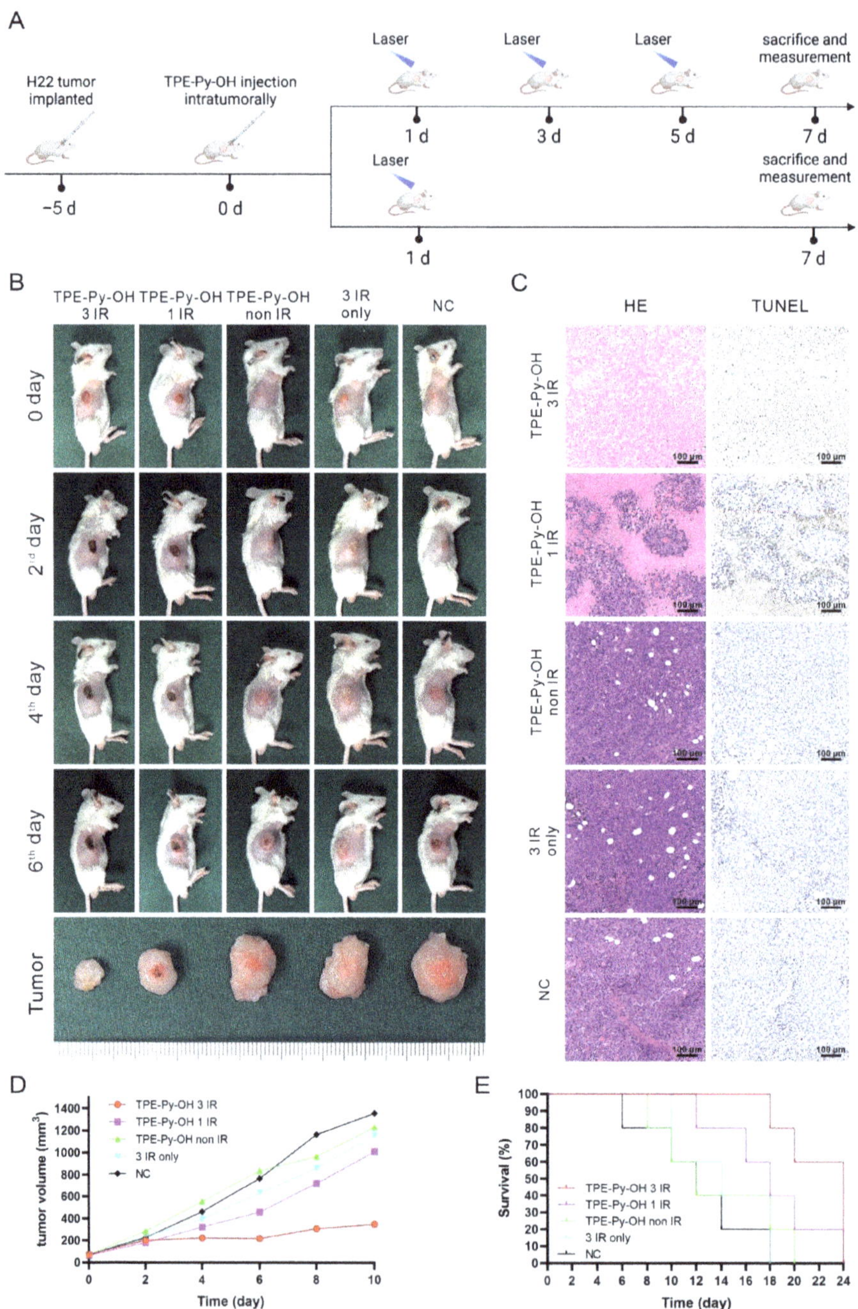

Figure 5. In vivo multiple light-activated PDT effect of TPE-Py-OH in H22 tumor-bearded mice. (**A**) Scheme of in vivo PDT experiment. (**B**) Representative photographs of mice with treatments at different days, the irradiation was performed with 450 nm laser (100 mW/cm^2) for 10 min. (**C**) H&E stained and TUNEL stained tumor tissue slices after 7 days of treatments. (**D**) Tumor growth curves of mice with the various treatments. (**E**) Survival of mice with the various treatments.

4. Conclusions

In summary, we have developed a dual-organelles-targeting AIEgen based on tetraphenylethylene derivative (TPE-Py-OH). TPE-Py-OH can target both mitochondria and LDs, exhibits excellent ability of long-term retention, has stable fluorescence emission and multiple light-activated PDT effect in cancer cells. By storing in intracellular LDs, the fluorescence of TPE-Py-OH in cells can be maintained for a long time, and the photodynamic tumor-killing effect could still be activated by light irradiation at 7 days after a single treatment with TPE-Py-OH in vitro and in vivo. We believe that this AIEgen will be promising for the long-term tracking and multiple light-activated photodynamic ablation of HCC with sustained effectiveness. Although the tumor-targeting ability and the approach of drug administration is less than perfect, the hydroxy group (-OH) of this AIEgen provides excellent molecular modifiability and carrier combinability. Further modification of molecular structure and combination with nanocarrier could improve the cancer selectivity and fluorescence properties of TPE-Py-OH. Based on molecular characteristics of TPE-Py-OH, new PSs with improved function and a better administration approach can be designed to achieve multiple light-activated PDT.

Supplementary Materials: The following are available online at https://www.mdpi.com/article/10.3390/pharmaceutics14020459/s1, Figure S1: HRMS spectra of molecule 1; Figure S2: HRMS spectra of TPE-Py-OH; Figure S3: ^{1}H NMR spectra of (A) TPE-Py-OH and (B) molecular 1 in DMSO-d_6; Figure S4: ^{13}C NMR spectra of (A) TPE-Py-OH and (B) molecular 1 in DMSO-d_6; Figure S5: UV spectra of TPE-Py-OH at the concentration of 10 μM in DMSO; Figure S6: (A) PL spectra of TPE-Py-OH in H$_2$O/DMSO mixtures (v/v 90%) with different concentrations. Excitation wavelength: 405 nm. (B) Plot of PL intensity versus the concentrations of TPE-Py-OH. Inset: Photograph of TPE-Py-OH with various concentrations taken under 365 nm UV irradiation; Figure S7: The dynamic monitoring of the distribution of TPE-Py-OH in living cells. The HepG2 cells were incubated with 5 μM TPE-Py-OH for 30 min and then replaced with fresh medium, Figure S8: (A) The distribution of TPE-Py-OH in vivo after intratumoral injection over time. (B) Imaging of compression slice of H22 tumor on day 12 after TPE-Py-OH administration.

Author Contributions: Conceptualization, M.L. and C.W.; Methodology, M.L., C.C., T.Z. and J.Z.; TPE-Py-OH synthesis and characterization, N.Z.; In vitro and in vivo studies, C.C., T.Z., J.Z. and X.M.; Writing—original draft preparation, C.C.; Writing—review and editing, M.L., Q.Q., J.X. and Y.T.; Supervision, C.W., M.L. and J.X. All authors have read and agreed to the published version of the manuscript.

Funding: This work was financially supported by the National Natural Science Foundation of China (81672919, 81874208, 82172754, 81771005, 81771718).

Institutional Review Board Statement: The animal procedures were conducted at the guidelines of the Department of Laboratory Animals of Tongji Medical College and approved by Institutional animal care and use Committee of Huazhong University of Science and Technology (approval number 2526, date: 12 November 2021).

Informed Consent Statement: Not applicable.

Data Availability Statement: Data sharing not applicable.

Acknowledgments: The authors would like to thank Yilong Chen for his help in the synthesis of TPE-Py-OH. Furthermore, we are grateful to Shunchang Zhou for his help during the animal experiment.

Conflicts of Interest: The authors declare no conflict of interest.

References

1. Vogel, A.; Cervantes, A.; Chau, I.; Daniele, B.; Llovet, J.M.; Meyer, T.; Nault, J.C.; Neumann, U.; Ricke, J.; Sangro, B.; et al. Hepatocellular carcinoma: ESMO Clinical Practice Guidelines for diagnosis, treatment and follow-up. *Ann. Oncol.* **2018**, *29* (Suppl. 4), iv238–iv255. [CrossRef] [PubMed]
2. Yang, G.; Xiong, Y.; Sun, J.; Wang, G.; Li, W.; Tang, T.; Li, J. The efficacy of microwave ablation versus liver resection in the treatment of hepatocellular carcinoma and liver metastases: A systematic review and meta-analysis. *Int. J. Surg.* **2020**, *77*, 85–93. [CrossRef] [PubMed]

3. Makary, M.S.; Khandpur, U.; Cloyd, J.M.; Mumtaz, K.; Dowell, J.D. Locoregional Therapy Approaches for Hepatocellular Carcinoma: Recent Advances and Management Strategies. *Cancers* **2020**, *12*, 1914. [CrossRef] [PubMed]

4. Minami, Y.; Kudo, M. Image Guidance in Ablation for Hepatocellular Carcinoma: Contrast-Enhanced Ultrasound and Fusion Imaging. *Front. Oncol.* **2021**, *11*, 593636. [CrossRef]

5. Castano, A.P.; Mroz, P.; Hamblin, M.R. Photodynamic therapy and anti-tumour immunity. *Nat. Rev. Cancer* **2006**, *6*, 535–545. [CrossRef]

6. Kumar, A.; Morales, O.; Mordon, S.; Delhem, N.; Boleslawski, E. Could Photodynamic Therapy Be a Promising Therapeutic Modality in Hepatocellular Carcinoma Patients? A Critical Review of Experimental and Clinical Studies. *Cancers* **2021**, *13*, 5176. [CrossRef]

7. Broekgaarden, M.; Weijer, R.; van Gulik, T.M.; Hamblin, M.R.; Heger, M. Tumor cell survival pathways activated by photodynamic therapy: A molecular basis for pharmacological inhibition strategies. *Cancer Metastasis Rev.* **2015**, *34*, 643–690. [CrossRef]

8. Ma, Y.; Huang, J.; Song, S.; Chen, H.; Zhang, Z. Cancer-Targeted Nanotheranostics: Recent Advances and Perspectives. *Small* **2016**, *12*, 4936–4954. [CrossRef]

9. Gonzalez-Carmona, M.A.; Bolch, M.; Jansen, C.; Vogt, A.; Sampels, M.; Mohr, R.U.; van Beekum, K.; Mahn, R.; Praktiknjo, M.; Nattermann, J.; et al. Combined photodynamic therapy with systemic chemotherapy for unresectable cholangiocarcinoma. *Aliment. Pharmacol. Ther.* **2019**, *49*, 437–447. [CrossRef]

10. Song, W.; Kuang, J.; Li, C.X.; Zhang, M.; Zheng, D.; Zeng, X.; Liu, C.; Zhang, X.Z. Enhanced Immunotherapy Based on Photodynamic Therapy for Both Primary and Lung Metastasis Tumor Eradication. *ACS Nano* **2018**, *12*, 1978–1989. [CrossRef]

11. Maeding, N.; Verwanger, T.; Krammer, B. Boosting Tumor-Specific Immunity Using PDT. *Cancers* **2016**, *8*, 91. [CrossRef] [PubMed]

12. Jiang, R.; Dai, J.; Dong, X.; Wang, Q.; Meng, Z.; Guo, J.; Yu, Y.; Wang, S.; Xia, F.; Zhao, Z.; et al. Improving Image-Guided Surgical and Immunological Tumor Treatment Efficacy by Photothermal and Photodynamic Therapies Based on a Multifunctional NIR AIEgen. *Adv. Mater.* **2021**, *33*, e2101158. [CrossRef] [PubMed]

13. He, J.; Yang, L.; Yi, W.; Fan, W.; Wen, Y.; Miao, X.; Xiong, L. Combination of Fluorescence-Guided Surgery With Photodynamic Therapy for the Treatment of Cancer. *Mol. Imaging* **2017**, *16*, 1536012117722911. [CrossRef] [PubMed]

14. Akopov, A.; Rusanov, A.; Gerasin, A.; Kazakov, N.; Urtenova, M.; Chistyakov, I. Preoperative endobronchial photodynamic therapy improves resectability in initially irresectable (inoperable) locally advanced non small cell lung cancer. *Photodiagnosis Photodyn. Ther.* **2014**, *11*, 259–264. [CrossRef]

15. Jia, R.; Xu, H.; Wang, C.; Su, L.; Jing, J.; Xu, S.; Zhou, Y.; Sun, W.; Song, J.; Chen, X.; et al. NIR-II emissive AIEgen photosensitizers enable ultrasensitive imaging-guided surgery and phototherapy to fully inhibit orthotopic hepatic tumors. *J. Nanobiotechnology* **2021**, *19*, 419. [CrossRef]

16. Feng, Z.; Birong, W.; Zhengfeng, Z.; Siqin, W.; Chuxing, C.; Dang, S.; Min, L. Photodynamic therapy: A next alternative treatment strategy for hepatocellular carcinoma? *World J. Gastrointest. Surg.* **2021**, *13*, 1523–1525. [CrossRef]

17. Correia, J.H.; Rodrigues, J.A.; Pimenta, S.; Dong, T.; Yang, Z. Photodynamic Therapy Review: Principles, Photosensitizers, Applications, and Future Directions. *Pharmaceutics* **2021**, *13*, 1332. [CrossRef]

18. Zhou, Z.; Song, J.; Nie, L.; Chen, X. Reactive oxygen species generating systems meeting challenges of photodynamic cancer therapy. *Chem. Soc. Rev.* **2016**, *45*, 6597–6626. [CrossRef]

19. Agostinis, P.; Berg, K.; Cengel, K.A.; Foster, T.H.; Girotti, A.W.; Gollnick, S.O.; Hahn, S.M.; Hamblin, M.R.; Juzeniene, A.; Kessel, D.; et al. Photodynamic therapy of cancer: An update. *CA Cancer J. Clin.* **2011**, *61*, 250–281. [CrossRef]

20. Bonnett, R. Photosensitizers of the porphyrin and phthalocyanine series for photodynamic therapy. *Chem. Soc. Rev.* **1995**, *24*, 19–33. [CrossRef]

21. Kang, M.; Zhang, Z.; Song, N.; Li, M.; Sun, P.; Chen, X.; Wang, D.; Tang, B.Z. Aggregation-enhanced theranostics: AIE sparkles in biomedical field. *Aggregate* **2020**, *1*, 80–106. [CrossRef]

22. Luo, J.; Xie, Z.; Lam, J.W.; Cheng, L.; Chen, H.; Qiu, C.; Kwok, H.S.; Zhan, X.; Liu, Y.; Zhu, D.; et al. Aggregation-induced emission of 1-methyl-1,2,3,4,5-pentaphenylsilole. *Chem. Commun.* **2001**, 1740–1741. [CrossRef] [PubMed]

23. Dai, J.; Wu, X.; Ding, S.; Lou, X.; Xia, F.; Wang, S.; Hong, Y. Aggregation-Induced Emission Photosensitizers: From Molecular Design to Photodynamic Therapy. *J. Med. Chem.* **2020**, *63*, 1996–2012. [CrossRef] [PubMed]

24. Gao, Y.; Zheng, Q.C.; Xu, S.D.; Yuan, Y.Y.; Cheng, X.; Jiang, S.; Kenry; Yu, Q.H.; Song, Z.F.; Liu, B.; et al. Theranostic Nanodots with Aggregation-Induced Emission Characteristic for Targeted and Image-Guided Photodynamic Therapy of Hepatocellular Carcinoma. *Theranostics* **2019**, *9*, 1264–1279. [CrossRef]

25. He, W.; Zhang, T.; Bai, H.; Kwok, R.T.K.; Lam, J.W.Y.; Tang, B.Z. Recent Advances in Aggregation-Induced Emission Materials and Their Biomedical and Healthcare Applications. *Adv. Healthc. Mater.* **2021**, e2101055. [CrossRef]

26. Zhao, N.; Li, P.; Zhuang, J.; Liu, Y.; Xiao, Y.; Qin, R.; Li, N. Aggregation-Induced Emission Luminogens with the Capability of Wide Color Tuning, Mitochondrial and Bacterial Imaging, and Photodynamic Anticancer and Antibacterial Therapy. *ACS Appl. Mater. Interfaces* **2019**, *11*, 11227–11237. [CrossRef]

27. Qi, J.; Ou, H.; Liu, Q.; Ding, D. Gathering brings strength: How organic aggregates boost disease phototheranostics. *Aggregate* **2021**, *2*, 95–113. [CrossRef]

28. Cai, X.; Liu, B. Aggregation-Induced Emission: Recent Advances in Materials and Biomedical Applications. *Angew. Chem.* **2020**, *132*, 9952–9970. [CrossRef]

29. Gao, M.; Tang, B.Z. Aggregation-induced emission probes for cancer theranostics. *Drug Discov. Today* **2017**, *22*, 1288–1294. [CrossRef]
30. Li, M.; Gao, Y.; Yuan, Y.Y.; Wu, Y.Z.; Song, Z.F.; Tang, B.Z.; Liu, B.; Zheng, Q.C. One-Step Formulation of Targeted Aggregation-Induced Emission Dots for Image-Guided Photodynamic Therapy of Cholangiocarcinoma. *ACS Nano* **2017**, *11*, 3922–3932. [CrossRef]
31. Wu, W.; Shi, L.; Duan, Y.; Xu, S.; Gao, X.; Zhu, X.; Liu, B. Metabolizable Photosensitizer with Aggregation-Induced Emission for Photodynamic Therapy. *Chem. Mater.* **2021**, *33*, 5974–5980. [CrossRef]
32. Tavakkoli Yaraki, M.; Wu, M.; Middha, E.; Wu, W.; Daqiqeh Rezaei, S.; Liu, B.; Tan, Y.N. Gold Nanostars-AIE Theranostic Nanodots with Enhanced Fluorescence and Photosensitization Towards Effective Image-Guided Photodynamic Therapy. *Nano-Micro Lett.* **2021**, *13*, 58. [CrossRef] [PubMed]
33. Liao, Y.; Wang, R.; Wang, S.; Xie, Y.; Chen, H.; Huang, R.; Shao, L.; Zhu, Q.; Liu, Y. Highly Efficient Multifunctional Organic Photosensitizer with Aggregation-Induced Emission for In Vivo Bioimaging and Photodynamic Therapy. *ACS Appl. Mater. Interfaces* **2021**, *13*, 54783–54793. [CrossRef] [PubMed]
34. Kim, S.; Tachikawa, T.; Fujitsuka, M.; Majima, T. Far-red fluorescence probe for monitoring singlet oxygen during photodynamic therapy. *J. Am. Chem. Soc.* **2014**, *136*, 11707–11715. [CrossRef]
35. Feng, G.; Qin, W.; Hu, Q.; Tang, B.Z.; Liu, B. Cellular and Mitochondrial Dual-Targeted Organic Dots with Aggregation-Induced Emission Characteristics for Image-Guided Photodynamic Therapy. *Adv. Healthc. Mater.* **2015**, *4*, 2667–2676. [CrossRef]
36. Chen, S.; Huang, B.; Pei, W.; Wang, L.; Xu, Y.; Niu, C. Mitochondria-Targeting Oxygen-Sufficient Perfluorocarbon Nanoparticles for Imaging-Guided Tumor Phototherapy. *Int. J. Nanomed.* **2020**, *15*, 8641–8658. [CrossRef]
37. Zhang, L.; Wang, J.L.; Ba, X.X.; Hua, S.Y.; Jiang, P.; Jiang, F.L.; Liu, Y. Multifunction in One Molecule: Mitochondrial Imaging and Photothermal & Photodynamic Cytotoxicity of Fast-Response Near-Infrared Fluorescent Probes with Aggregation-Induced Emission Characteristics. *ACS Appl. Mater. Interfaces* **2021**, *13*, 7945–7954. [CrossRef]
38. Yu, K.; Pan, J.; Husamelden, E.; Zhang, H.; He, Q.; Wei, Y.; Tian, M. Aggregation-induced Emission Based Fluorogens for Mitochondria-targeted Tumor Imaging and Theranostics. *Chem. Asian J.* **2020**, *15*, 3942–3960. [CrossRef]
39. Zhou, T.; Zhu, J.F.; Shang, D.; Chai, C.X.; Li, Y.Z.; Sun, H.Y.; Li, Y.Q.; Gao, M.; Li, M. Mitochondria-anchoring and AIE-active photosensitizer for self-monitored cholangiocarcinoma therapy. *Mater. Chem. Front.* **2020**, *4*, 3201–3208. [CrossRef]
40. Zhang, W.; Huang, Y.; Chen, Y.; Zhao, E.; Hong, Y.; Chen, S.; Lam, J.W.Y.; Chen, Y.; Hou, J.; Tang, B.Z. Amphiphilic Tetraphenylethene-Based Pyridinium Salt for Selective Cell-Membrane Imaging and Room-Light-Induced Special Reactive Oxygen Species Generation. *ACS Appl. Mater. Interfaces* **2019**, *11*, 10567–10577. [CrossRef]
41. Thiam, A.R.; Beller, M. The why, when and how of lipid droplet diversity. *J. Cell Sci.* **2017**, *130*, 315–324. [CrossRef] [PubMed]
42. Cruz, A.L.S.; Barreto, E.A.; Fazolini, N.P.B.; Viola, J.P.B.; Bozza, P.T. Lipid droplets: Platforms with multiple functions in cancer hallmarks. *Cell Death Dis.* **2020**, *11*, 105. [CrossRef] [PubMed]
43. Cheng, C.; Geng, F.; Cheng, X.; Guo, D. Lipid metabolism reprogramming and its potential targets in cancer. *Cancer Commun.* **2018**, *38*, 27. [CrossRef]
44. Tabero, A.; Garcia-Garrido, F.; Prieto-Castaneda, A.; Palao, E.; Agarrabeitia, A.R.; Garcia-Moreno, I.; Villanueva, A.; de la Moya, S.; Ortiz, M.J. BODIPYs revealing lipid droplets as valuable targets for photodynamic theragnosis. *Chem. Commun.* **2020**, *56*, 940–943. [CrossRef] [PubMed]
45. Olzmann, J.A.; Carvalho, P. Dynamics and functions of lipid droplets. *Nat. Rev. Mol. Cell Biol.* **2019**, *20*, 137–155. [CrossRef] [PubMed]
46. Leung, C.W.; Hong, Y.; Chen, S.; Zhao, E.; Lam, J.W.; Tang, B.Z. A photostable AIE luminogen for specific mitochondrial imaging and tracking. *J. Am. Chem. Soc.* **2013**, *135*, 62–65. [CrossRef]
47. Zhao, N.; Li, M.; Yan, Y.; Lam, J.W.Y.; Zhang, Y.L.; Zhao, Y.S.; Wong, K.S.; Tang, B.Z. A tetraphenylethene-substituted pyridinium salt with multiple functionalities: Synthesis, stimuli-responsive emission, optical waveguide and specific mitochondrion imaging. *J. Mater. Chem. C* **2013**, *1*, 4640–4646. [CrossRef]
48. Chen, Y.; Li, M.; Hong, Y.; Lam, J.W.; Zheng, Q.; Tang, B.Z. Dual-modal MRI contrast agent with aggregation-induced emission characteristic for liver specific imaging with long circulation lifetime. *ACS Appl. Mater. Interfaces* **2014**, *6*, 10783–10791. [CrossRef]
49. Sakamuru, S.; Attene-Ramos, M.S.; Xia, M. Mitochondrial Membrane Potential Assay. *Methods Mol. Biol.* **2016**, *1473*, 17–22. [CrossRef]
50. Li, Y.; Zhang, R.; Wan, Q.; Hu, R.; Ma, Y.; Wang, Z.; Hou, J.; Zhang, W.; Tang, B.Z. Trojan Horse-Like Nano-AIE Aggregates Based on Homologous Targeting Strategy and Their Photodynamic Therapy in Anticancer Application. *Adv. Sci.* **2021**, e2102561. [CrossRef]
51. Wu, H.; Wang, F.; Ta, N.; Zhang, T.; Gao, W. The Multifaceted Regulation of Mitochondria in Ferroptosis. *Life* **2021**, *11*, 222. [CrossRef] [PubMed]

pharmaceutics

Insight into the Web of Stress Responses Triggered at Gene Expression Level by Porphyrin-PDT in HT29 Human Colon Carcinoma Cells

Maria Dobre [1], Rica Boscencu [2], Ionela Victoria Neagoe [1], Mihaela Surcel [1], Elena Milanesi [1] and Gina Manda [1,*]

[1] Radiobiology Department, Victor Babes National Institute of Pathology, 99-101 Splaiul Independentei, 050096 Bucharest, Romania; maria_dobre70@yahoo.com (M.D.); neagoevictoria@gmail.com (I.V.N.); msurcel2002@yahoo.com (M.S.); elena.k.milanesi@gmail.com (E.M.)

[2] Faculty of Pharmacy, Carol Davila University of Medicine and Pharmacy, 6 Traian Vuia Street, 020956 Bucharest, Romania; rboscencu@yahoo.com

* Correspondence: gina.manda@gmail.com; Tel.: +40-744246887

Abstract: Photodynamic therapy (PDT), a highly targeted therapy with acceptable side effects, has emerged as a promising therapeutic option in oncologic pathology. One of the issues that needs to be addressed is related to the complex network of cellular responses developed by tumor cells in response to PDT. In this context, this study aims to characterize in vitro the stressors and the corresponding cellular responses triggered by PDT in the human colon carcinoma HT29 cell line, using a new asymmetric porphyrin derivative (P2.2) as a photosensitizer. Besides investigating the ability of P2.2-PDT to reduce the number of viable tumor cells at various P2.2 concentrations and fluences of the activating light, we assessed, using qRT-PCR, the expression levels of 84 genes critically involved in the stress response of PDT-treated cells. Results showed a fluence-dependent decrease of viable tumor cells at 24 h post-PDT, with few cells that seem to escape from PDT. We highlighted following P2.2-PDT the concomitant activation of particular cellular responses to oxidative stress, hypoxia, DNA damage and unfolded protein responses and inflammation. A web of inter-connected stressors was induced by P2.2-PDT, which underlies cell death but also elicits protective mechanisms that may delay tumor cell death or even defend these cells against the deleterious effects of PDT.

Keywords: photodynamic therapy; colon carcinoma cells; stress responses

Citation: Dobre, M.; Boscencu, R.; Neagoe, I.V.; Surcel, M.; Milanesi, E.; Manda, G. Insight into the Web of Stress Responses Triggered at Gene Expression Level by Porphyrin-PDT in HT29 Human Colon Carcinoma Cells. *Pharmaceutics* **2021**, *13*, 1032. https://doi.org/10.3390/pharmaceutics13071032

Academic Editors: Udo Bakowsky, Matthias Wojcik, Eduard Preis and Gerhard Litscher

Received: 27 May 2021
Accepted: 4 July 2021
Published: 7 July 2021

Publisher's Note: MDPI stays neutral with regard to jurisdictional claims in published maps and institutional affiliations.

1. Introduction

Photodynamic therapy (PDT) has lately emerged as a promising targeted therapy for solid tumors. As reviewed by Agostinis et al. [1], PDT consists of the administration of a biocompatible photosensitizer (PS) that is inactive in "dark" conditions and is more or less selectively accumulated by tumor cells. Local activation of PS using visible light in a PS-specific wavelength spectrum triggers a strong singlet oxygen burst that induces locally important oxidative damages and therefore destabilizes the tumor niche. Because PSs should not exhibit "dark" cytotoxicity, and the activating light is highly targeted to the diseased tissue, PDT has minimal effects on healthy tissues. PDT has fewer side-effects than conventional anticancer therapies like radio- and chemotherapy and is sufficiently safe for repeated therapy sessions. Moreover, PDT does not induce immune suppression and can even boost the antitumor immune response which will complement the therapeutic action [2]. Major technical issues raised by PDT, recently summarized [3], are related to PS properties which, besides having to be non-cytotoxic in "dark" conditions, should have convenient amphiphilic properties to load cells and minimal aggregation in physiologic fluids and be activated with light of sufficiently high wavelength to penetrate tissues. Additionally, technical limitations in PDT are mostly related to the devices through which light can be guided to deep-seated tumors for activating PDT locally. An interesting issue for further improving PDT is the characterization of the complex mechanisms underlying its therapeutic effects and the potential therapy-induced resistance [4], both aspects being highly dependent on the tumor type, the structure and properties of the PS and on the PDT settings in terms of light wavelength and fluence. Particular aspects characterizing PDT-induced responses in tumor cells have been already documented, connecting the PDT-induced oxidative stress, cell death, survival mechanisms and danger-associated molecular patterns (DAMPs) production and immunity [5–7], but a detailed picture of the intra-network connections is still missing.

For characterizing at the molecular level the response of tumor cells to PDT, we investigated by qRT-PCR the gene expression pattern in HT29 tumor cells subjected to PDT, addressing 84 genes that are critically involved in cellular responses to stress, aiming to describe the web of stressors elicited by a porphyrin-PDT regimen. As a porphyrinic photosensitizer we used an unsymmetrical meso-tetrasubstituted phenyl porphyrin (4-acetoxy-3-methoxyphenyl)porphyrin, P2.2) that we have previously designed and characterized (Figure 1). As described in [8], P2.2 has remarkable amphyphylic properties, has good solubility in biologically friendly media and long-term stability in polyethylen glycol 200 (PEG 200) and is able to generate PDT-acceptable singlet oxygen yields when activated with light in the spectral domain 600–650 nm. P2.2 was shown to accumulate well into tumor cells following a dose-dependent linear relationship, was significantly less uptaken by blood cells, exhibited good fluorescence for imagistic detection and did not exert in "dark" conditions important in vitro cytotoxicity on cells specific for the tumor niche (tumor colon carcinoma cells and tumorigenic fibroblasts) or on blood cells (peripheral blood mononuclear cells).

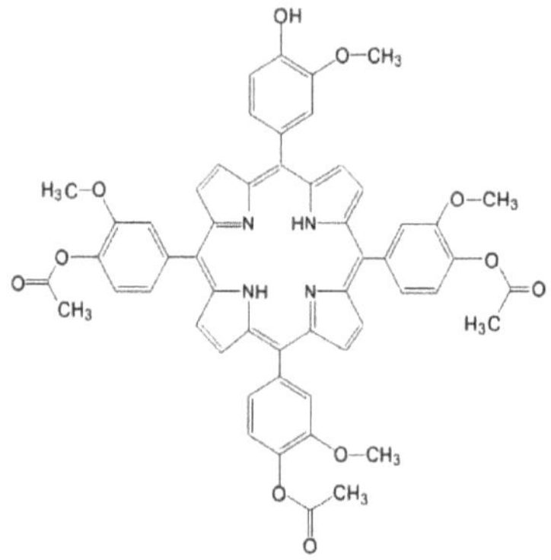

Figure 1. Chemical structure of 5-(4-hydroxy-3-methoxyphenyl)-10, 15, 20-tris-(4-acetoxy-3-methoxyphenyl) porphyrin (P2.2).

Results showed that the new P2.2 photosensitizer produced in vitro PDT in a concentration- and fluence-dependent manner. The gene expression study highlighted that P2.2-PDT generated a particular molecular fingerprint of oxidative stress, hypoxia signaling, DNA damage, endoplasmic reticulum (ER) stress and unfolded protein response (UPR), along with inflammation. The web of inter-connected stress responses elicited by P2.2-PDT in HT29 tumor cells might sustain or limit the therapeutic effect of PDT. The identified genes could represent valuable molecular targets for co-therapies aimed at reinforcing PDT.

2. Results

In vitro PDT was performed on the human colon carcinoma cell line HT29 using the new porphyrinic photosensitizer P2.2 (Figure 1) and activating laser light of 635 nm.

The fluorescent P2.2 photosensitizer was incorporated into HT29 tumor cells following a linear concentration-dependent relationship in the range (2.5–40) μM (Figure 2a). The difference between fluorescence values in samples loaded with consecutive P2.2 concentrations was statistically significant ($p < 0.01$). In the investigated concentration range, P2.2 did not exert "dark" cytotoxicity, according to the lactate dehydrogenase (LDH) release data (not shown). Additionally, the distribution of fluorescence values within a P2.2-treated sample and the corresponding SD value of the distribution (measure of the fluorescence values spread around the mean value in a defined cellular population) showed that cells had incorporated variable amounts of P2.2 (Figure 2b). This may affect the strength of PDT, depending on the cellular P2.2 load.

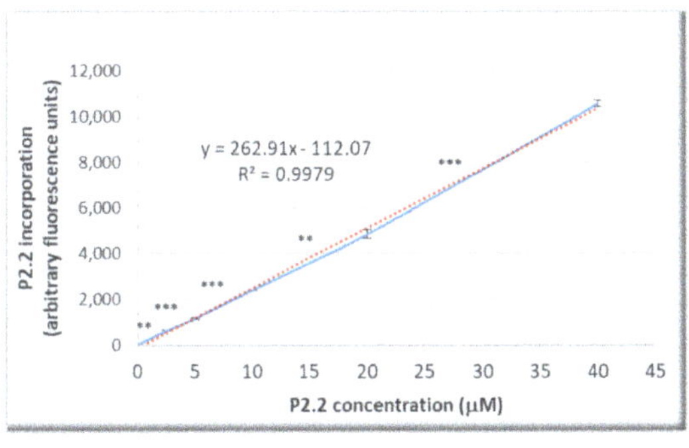

(**a**) Mean cellular fluorescence.

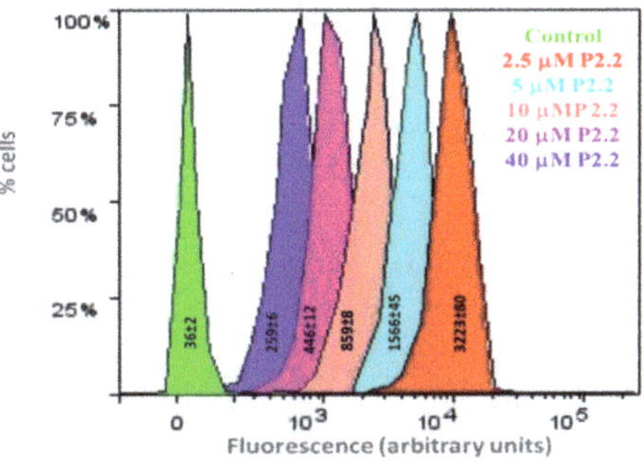

(**b**) Fluorescence distribution in P2.2-loaded cells.

Figure 2. The uptake of the fluorescent P2.2 photosensitizer by HT29 tumor cells. Cells were incubated for 24 h with 2.5–40 µM P2.2. Intracellular P2.2 was evaluated by flow cytometry, based on the red P2.2 fluorescence. Fluorescence data, expressed in arbitrary units, and their median values in each sample were obtained with the BD FACSDiva software (Becton Dickinson). (**a**) Median cellular fluorescence. Data are presented as the mean value of the median fluorescence of each sample of the triplicate (mean $\pm$ SEM). Comparison between samples treated with consecutive P2.2 concentrations was performed using Student's *t*-test considering unequal variances: ** $p < 0.01$, *** $p < 0.001$. (**b**) Fluorescence distribution in samples treated with various P2.2 concentrations. The represented SD values, obtained with the BD FACSDiva 6.1 software, are a measure of the fluorescence values spread around the mean value in a defined cellular population.

The viability of tumor cells following exposure to P2.2-PDT was investigated as a preparative step for the gene expression analysis in PDT-treated samples as compared to non-treated controls.

2.1. PDT-Induced Changes of Cell Viability

2.1.1. PDT-Induced Decrease of Viable Tumor Cells

The dependence of the in vitro PDT outcome (viability of HT29 tumor cells) on P2.2 concentration (2.5–40 µM) was investigated by MTS ([3-(4,5-dimethylthiazol-2-yl)-5-(3-carboxymethoxyphenyl)-2-(4-sulfophenyl)-2H-tetrazolium, inner salt]) reduction at 24 h post-PDT (10 J/cm^2, 50 mW/cm^2). As shown in Figure 3, MTS reduction had a linear decrease in the P2.2 concentration range 2.5–10 µM. In the case of higher P2.2 concentrations (20–40 µM), MTS reduction almost dropped to zero, indicating that only few metabolically active cells remained at 24 h after PDT with high P2.2 concentrations.

An IC50 value of 8.07 $\pm$ 1.69 µM (mean $\pm$ SD) was computed using the Quest Graph™ IC50 Calculator (https://www.aatbio.com/tools/ic50-calculator, accessed on 27 June 2021). A concentration of 10 µM P2.2 was further chosen for performing PDT at various light fluences (5–25 J/cm^2), for further differential investigation of both live and dead cells regarding gene expression.

The number of metabolically active tumor cells in culture dramatically decreased at 24 h post-PDT in comparison with non-treated control cells, as shown by the significant fluence-dependent decrease of MTS reduction (PDT effect < 0.3) at all the tested light fluences (10 J/cm^2, 15 J/cm^2 and 25 J/cm^2, delivered at 50 mW/cm^2) (Figure 4). A subunit PDT effect ($\leq$0.5) following a fluence-dependent curve was also registered at 72 h (Figure 4). A stronger PDT regimen (25 J/cm^2) rendered almost all tumor cells metabolically inactive in the first 24 h after PDT and this effect was maintained until 72 h. Meanwhile, PDT exerted a lower effect on MTS reduction at 72 h than at 24 h post-PDT ($p < 0.001$) at the fluences of 10 J/cm^2 and 15 J/cm^2, indicating that the PDT effect was attenuated over time.

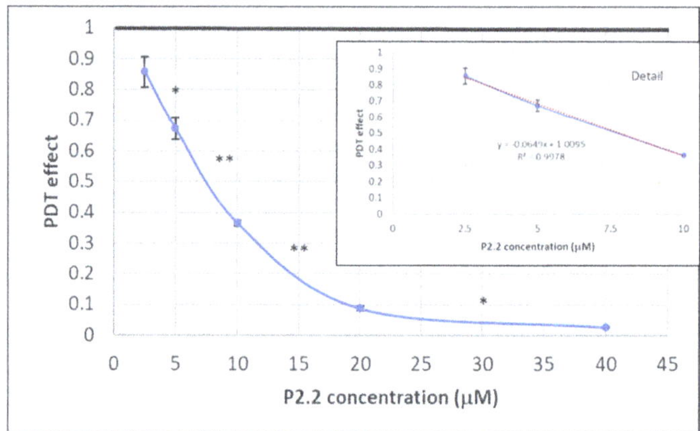

Figure 3. The dependence on P2.2 concentration of MTS reduction by HT29 tumor cells at 24 h post-PDT. Results are presented as PDT effect (mean ± SEM) in triplicate samples. PDT effect was calculated as OD in PDT-treated samples divided by the mean value of OD in control samples. Comparison between samples was performed using Student's *t*-test: paired two sample for means: * $p < 0.05$, ** $p < 0.01$.

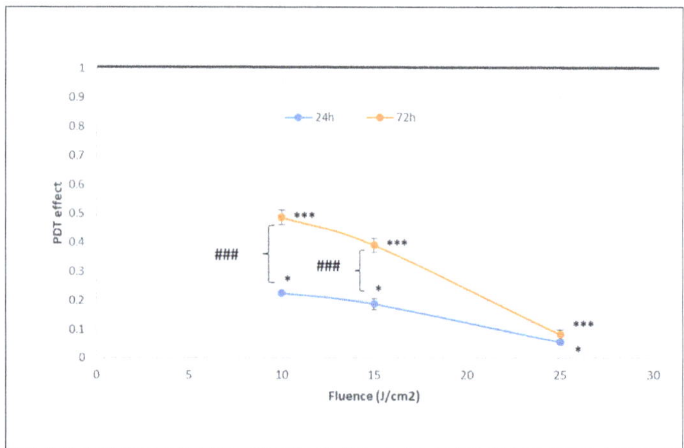

Figure 4. The dependence of MTS reduction by PDT-treated HT29 tumor cells on the fluence of the activation light delivered at a fluence rate of 50 mW/cm^2. Results are presented as PDT effect (mean ± SEM) in triplicate samples. PDT effect was calculated as OD in PDT-treated samples divided by the mean value of OD in control samples. Comparison between samples was performed using Student's *t*-test considering unequal variances: comparison between PDT regimens at a specific time-point: * $p < 0.05$, *** $p < 0.001$; comparison between samples at 24 h and 72 h: ### $p < 0.001$.

This observation was sustained by microscopic investigations that showed the presence of adhered cells at 24 h and 72 h post-PDT, most probably viable cells, even in the case of the stronger PDT regimen of 25 J/cm^2 (Figure 5A,B). While the number of adhered cells decreased over time, a few cells that appeared to be less affected by PDT continued to slowly proliferate and formed cell islets at 120 h post-PDT (Figure 5C).

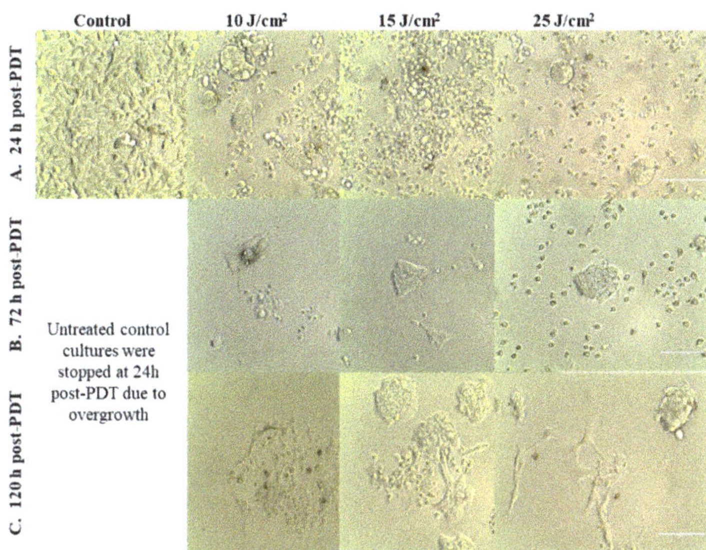

Figure 5. Images of HT29 tumor cells at various time points after PDT performed with various light fluences (10–25 J/cm^2), which were delivered with the fluence rate of 50 mW/cm^2. Cells were visualized by bright-field microscopy at 24 h (**A**), 72 h (**B**) and 120 h (**C**) post-PDT (100 µm scale bar).

The observation that some tumor cells might have escaped from the deleterious action of PDT was also evidenced when cell proliferation was assessed at 72 h post-PDT by flow cytometry with CFDA-SE. Most of the cells exposed to 10 J/cm^2 or 15 J/cm^2 PDT were found at 72 h in lower-order daughter generations as compared to control cells, indicating that their overall proliferation was slower (Figure 6). Concurrently, a low cell percentage was found in higher-order daughter generations at 72 h post-PDT, suggesting that they were not harmed by PDT and actively proliferated at higher rates than control cells. These few cells seem to have gained a proliferation advantage.

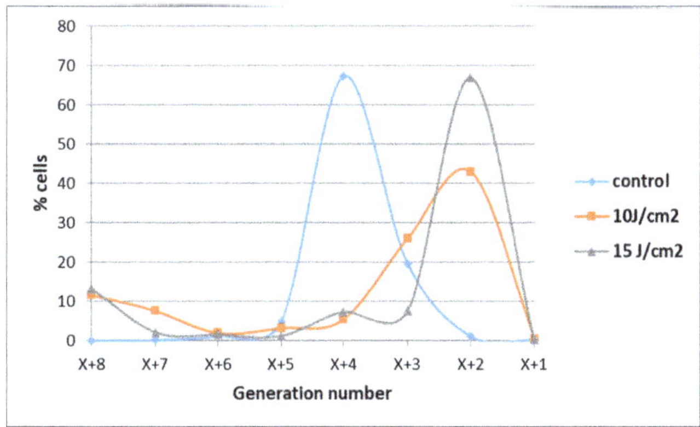

Figure 6. Demonstrative data on the proliferation of HT29 tumor cells subjected to PDT (fluences: 10 J/cm^2 and 15 J/cm^2; fluence rate: 50 mW/cm^2) vs. non-treated control cells at 72 h post-PDT. Results are presented as a percentage of cells in various consecutive daughter generations in comparison with the parent population distanced by X previous generations. Data were obtained by flow cytometry with CFDA-SE and were processed using the ModFit LT software.

2.1.2. PDT-Induced Alteration of Membrane Integrity

The significant decrease of the number of metabolically active tumor cells registered at 24 h post-PDT (Figure 4) was associated with a fluence-dependent linear increase of LDH release, indicating alteration of the plasma membrane (Figure 7). The PDT-induced increase of LDH release was negatively correlated with the decrease of MTS reduction (Pearson r = 0.986, p = 0.106), following a linear regression equation (y = −0.09X + 0.64). Results indicated that, at least partly, the decrease of metabolically active tumor cells in PDT-treated samples was due to membrane alterations generally occurring in necrotic and necroptototic cells [9,10], which allow significant LDH release.

The PDT effect on LDH release was lower at 72 h compared to 24 h for all the investigated light fluences and became fluence-independent in time (Figure 7). Results suggest that PDT induced persistent plasma membrane alterations in the time frame of 24–72 h post-PDT, especially in the first 24 h.

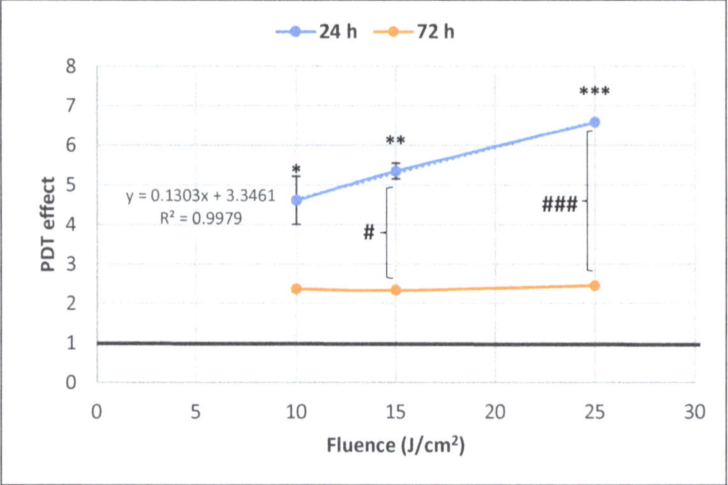

Figure 7. The dependence of LDH release by PDT-treated HT29 tumor cells on the fluence of the activating light delivered at a fluence rate of 50 mW/cm². Analysis was performed at 24 h and 72 h post-PDT. Results are presented as PDT effect (mean ± SEM) in triplicate samples. PDT effect was calculated as OD in samples subjected to PDT divided by the mean value of OD in control samples. Comparison between samples was performed using Student's *t*-test considering unequal variances: comparison between PDT regimens at a specific time-point: * p < 0.05, ** p < 0.01, *** p < 0.001; comparison between samples at 24 h and 72 h: # p < 0.05, ### p < 0.001.

At 24 h post-PDT, when the percentage of metabolically active cells decreased drastically below 50% in a fluence-dependent manner (Figure 4), many cells were found detached and rounded up, as seen in Figure 5A. For clarifying cell death, apoptosis and necrosis were investigated by flow cytometry with annexin V-propidium iodide. An important increase of the percentage of apoptotic cells, more specifically late apoptotic cells, accompanied by a smaller increase of necrotic cells was registered in PDT-treated vs. untreated cultures at 24 h post-PDT (Figure 8).

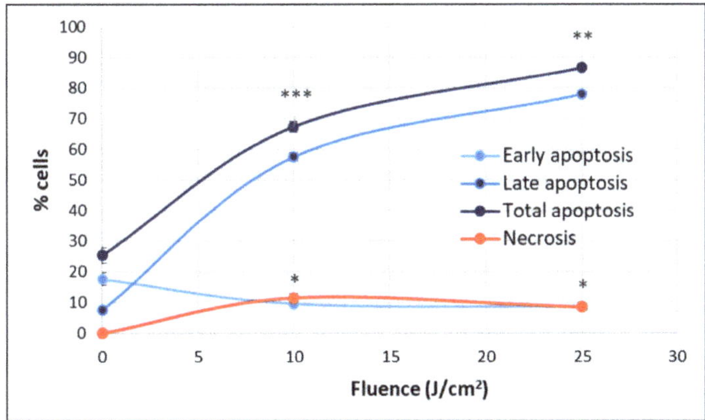

Figure 8. The percentage of apoptotic and necrotic HT29 cells in samples exposed to PDT (10 J/cm^2 and 25 J/cm^2, 50 mW/cm^2) and in non-treated controls. Apoptosis and necrosis were assessed by flow cytometry using FITC-annexin V and propidium iodide at 24 h post-PDT. Results are presented as percentage (mean $\pm$ SEM) of cells in triplicate samples, regarding early, late and total apoptosis, as well as necrosis. Comparison between treated and non-treated samples was performed using Student's *t*-test considering unequal variances. * $p < 0.05$, ** $p < 0.01$, *** $p < 0.001$.

We further investigated whether P2.2-PDT is also efficient in milder PDT conditions with lower light fluences (5–10 J/cm^2) delivered at a lower fluence rate of 10 mW/cm^2. A linear fluence-dependent increase of PDT effect on LDH release was registered at 24 h post-PDT, with significant differences between untreated controls and samples exposed to the higher investigated light fluences of 7.5 J/cm^2 and 10 J/cm^2 (Figure 9). Results emphasized once again that important damages at the level of plasma membrane, resulting in LDH release, were induced by PDT within 24 h post-treatment. Moreover, it appears that LDH release is not influenced by the fluence rate, as demonstrated by similar PDT effects on LDH release at 10 mW/cm^2 and 50 mW/cm^2 for a light fluence of 10 J/cm^2 (PDT effect at 10 mW/cm^2 was 4.2 $\pm$ 0.7 and at 50 mW/cm^2 it was 4.6 $\pm$ 0.6).

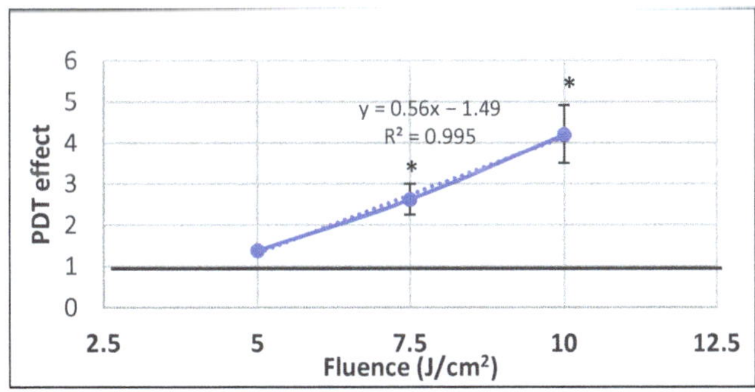

Figure 9. The dependence of LDH release by HT29 tumor cells exposed to milder PDT conditions on the fluence of the activating light (5–7.5–10 J/cm^2) that was delivered at a lower fluence rate of 10 mW/cm^2. Results are presented as PDT effect (mean $\pm$ SEM) in triplicate samples. PDT effect was calculated as OD in samples subjected to PDT divided by the mean OD value in controls. Comparison between samples was performed using Student's *t*-test considering unequal variances: * $p < 0.05$.

Altogether, results showed that, at least from the point of view of LDH release, which provides information on plasma membrane integrity, P2.2-PDT was efficacious on tumor

HT29 cells in the light fluence domain of 7.5–25 J/cm^2, with the highest damaging effects registered at 25 J/cm^2.

2.2. PDT-Induced Gene Expression Changes

We have shown above that in approximately 24 h, more than 50% of HT29 tumor cells were affected by P2.2-PDT, despite the fact that an acute oxidative burst is known to be generated instantaneously in the course of PDT [11]. Possibly, tumor cells develop rescue mechanisms that delay cell death and may even protect some cells against the deleterious action of PDT (Figure 5C).

Therefore, we investigated by qRT-PCR the expression profile of 84 genes (Table 1) that are critically involved in cellular responses to stressors such as oxidative stress, hypoxia, osmotic stress, genotoxic stress, unfolded protein response and inflammation, as well as in cell death by apoptosis, necrosis and autophagy. Briefly, cells exposed in vitro to PDT (10 J/cm^2, 50 mW/cm^2) were harvested 24 h post-PDT. The gene expression pattern was evaluated separately in adhered and detached cells as potentially living and apoptotic/necrotic cells, respectively.

Table 1. The list of analyzed genes, contained in the RT^2 Profiler™ PCR Array Human Stress and Toxicity PathwayFinder (Qiagen, PAHS-003Z).

Oxidative Stress
FTH1, GCLC, GCLM, GSR, GSTP1, HMOX1, NQO1, PRDX1, SQSTM1, TXN, TXNRD1

Hypoxia Signaling
ADM, ARNT, BNIP3L, CA9, EPO, HMOX1, LDHA, MMP9, SERPINE1 (PAI-1), SLC2A1, VEGFA

Osmotic Stress
AKR1B1, AQP1, AQP2, AQP4, CFTR, EDN1, HSPA4L (OSP94), NFAT5, SLC5A3

Cell Death
Apoptosis:
CASP1 (ICE), FAS, MCL1, TNFRSF10A (TRAIL-R), TNFRSF10B (DR5), TNFRSF1A (TNFR1).
AutophagyATG12, ATG5, ATG7, BECN1, FAS, ULK1
Necrosis:
FAS, GRB2, PARP1 (ADPRT1), PVR, RIPK1, TNFRSF10A (TRAIL-R), TNFRSF1A (TNFR1), TXNL4B

DNA Damage and Repair
Cell Cycle Arrest and Checkpoints:
CDKN1A (p21CIP1, WAF1), CHEK1, CHEK2 (RAD53), DDIT3 (GADD153, CHOP), HUS1, MRE11, NBN, RAD17, RAD9A
Other DNA Damage Responses:
ATM, ATR, DDB2, GADD45A, GADD45G, RAD51, TP53 (p53), XPC

Unfolded Protein Response
ATF4, ATF6, ATF6B, BBC3 (PUMA), BID, CALR, DDIT3 (GADD153, CHOP), DNAJC3, HSP90AA1, HSP90B1, HSPA4 (HSP70), HSPA5 (GRP78)

Inflammatory Response
CCL2 (MCP-1), CD40LG, CRP, CXCL8 (IL8), IFNG, IL1A, IL1B, IL6, TLR4, TNF

According to the data presented in Table 2, 43 genes were found differentially expressed in P2.2-treated cells compared to controls (1.5 < FC < −1.5, $p < 0.05$). Among these genes, 38 were upregulated and 5 were downregulated either in adhered or detached cells. Selected genes were classified according to their known association with various types of stress (oxidative stress, hypoxia, genotoxic and proteotoxic stress, along with inflammation), some of them being part of several signaling networks.

Table 2. Genes with statistically significant expression changes ($1.5 < FR < -1.5$, $p < 0.05$) in PDT-treated cells ($10\ J/cm^2$, $50\ mW/cm^2$), either adhered or detached, as compared to untreated controls. Gene expression levels were assessed at 24 h post-PDT. Genes were classified according to the type of stress in which they are involved. Some genes are common to several types of stress. ns = not significant.

a. Oxidative stress

PDT-treated vs. non-treated HT29 tumor cells at 24 h post-PDT

Gene	Adhered cells		Detached cells	
	FR	*p* value	FR	*p* value
HMOX1	8.87	<0.05	7.91	<0.01
FTH1	2.38	<0.05	5.41	<0.01
GCLC	3.36	<0.05	2.74	<0.01
GCLM	2.85	<0.05	3.96	<0.001
GSR	2.01	<0.01	2.85	<0.001
SQSTM1	4.98	<0.001	7.49	<0.001
PRDX1	ns	ns	4.18	<0.001
NQO1	ns	ns	3.02	<0.01
TXN	ns	ns	3.29	0,0.05
GSTP1	ns	ns	2.41	<0.01
TXNRD1	4.16	<0.001	ns	ns

b. Hypoxia signaling

PDT-treated vs. non-treated HT29 tumor cells at 24 h post-PDT

Gene	Adhered cells		Detached cells	
	FR	*p* value	FR	*p* value
HMOX1	8.87	<0.05	7.91	<0.01
SERPINE1	9.78	<0.001	10.74	<0.001
ADM	4.60	<0.01	14.27	<0.001
ARNT	ns	ns	2.36	<0.05
VEGFA	1.81	ns	2.11	≤0.001
BNIP3L	2.02	ns	2.20	<0.05

c. Cell death

PDT-treated vs. non-treated HT29 tumor cells at 24 h post-PDT

Gene	Adhered cells		Detached cells	
	FR	*p* value	FR	*p* value
	Apoptosis			
TNFRSF10B (DR5/TRAILR2)	3.09	<0.001	2.86	<0.001
BID	1.75	ns	3.96	<0.05
BNIP3L	2.02	ns	2.20	<0.05
BBC3	4.19	<0.05	5.30	ns
	Necrosis			
RIPK1	1.78	<0.01	ns	ns

Table 2. *Cont.*

c. Cell death

PDT-treated vs. non-treated HT29 tumor cells at 24 h post-PDT

Gene	Adhered cells		Detached cells	
	FR	*p* value	FR	*p* value
Autophagy				
ATG12	1.85	<0.05	3.29	<0.01
SQSTM1	4.98	<0.001	7.49	<0.001

d. DNA damage responses

PDT-treated vs. non-treated HT29 tumor cells at 24 h post-PDT

Gene	Adhered cells		Detached cells	
	FR	*p* value	FR	*p* value
Cell cycle arrest				
CDKN1A	2.74	<0.05	5.29	<0.01
GADD45A	3.01	<0.001	2.35	<0.001
DDIT3 (GADD153)	4.77	<0.01	12.86	<0.01
GADD45G	3.01	ns	14.77	<0.05
CHEK2	ns	ns	1.71	<0.05
HUS1	1.62	<0.01	ns	ns
Other DNA damage responses				
DDB2	1.51	<0.05	2.63	<0.01
XPC	2.20	<0.05	1.67	ns

e. Unfolded protein response

PDT-treated vs. non-treated HT29 tumor cells at 24 h post-PDT

Gene	Adhered cells		Detached cells	
	FR	*p* value	FR	*p* value
HSP90AA1	16.52	<0.05	9.26	<0.01
DDIT3 (CHOP)	4.77	<0.01	12.86	<0.01
BBC3	4.19	<0.05	5.30	ns
BID	1.75	ns	3.96	<0.05
HSPA4 (HSP70)	2.96	ns	2.95	<0.05
HSPA5 (GRP78)	−2.27	<0.01	−2.42	<0.001
DNAJC3	−1.59	<0.05	−4.07	<0.001
HSP90B1	−1.77	ns	−3.59	<0.001
CALR	−1.89	<0.001	ns	ns

f. Inflammation

PDT-treated vs. non-treated HT29 tumor cells at 24 h post-PDT

Gene	Adhered cells		Detached cells	
	FR	*p* value	FR	*p* value
CXCL8	8.09	≤0.01	6.46	<0.01
IL1B	3.28	<0.001	2.15	ns
IL1A	ns	ns	−2.90	≤0.01

2.2.1. Oxidative Stress

A strong molecular fingerprint of oxidative stress was evidenced in PDT-treated samples by the upregulation of 11 redox genes involved in antioxidant defense (Table 2). The activation of antioxidant mechanisms demonstrated on the one hand that cells were indeed subjected to an oxidative challenge during PDT, as expected considering the PDT-generated burst of singlet oxygen [3,12]. On the other hand, gene expression data pointed out the upregulation of several protective redox genes that aim to counteract or delay cell death induced by the oxidative challenge generated by PDT. The identified antioxidant genes are involved in iron metabolism (*HMOX1, FTH1*) [13], in glutathione (*GCLC, GCLM, GSR, GSTP1*) [14] or thioredoxin (*TXN, TXNRD1*) [15] metabolism, as well as in other cytoprotective pathways (*PRDX1* [16], *NQO1* [17,18] and *SQSTM1* [19,20]). Some of these upregulated genes were common to adhered and detached cells, while other genes were preferentially upregulated either in detached or adhered cells (Figure 10a). Thus, *PRDX1* ($p < 0.001$), *NQO1* ($p < 0.05$), *GSTP1* ($p < 0.05$) and *TXN* ($p < 0.05$) were distinctively expressed in detached cells, while *SQSTM1* ($p < 0.05$) had a higher transcript level in detached cells as compared to adhered cells. Alongside this, *TXNRD1* ($p < 0.05$) was exclusively overexpressed in adhered cells. Altogether, results indicated that a robust antioxidant response is generated by PDT, which triggers antioxidant defense mechanisms that appear to be insufficient to counteract the deleterious action of PDT, resulting in the significant decrease of viable tumor cells (Figures 4, 7 and 8). Surprisingly, the protective antioxidant response was stronger in detached cells than in adhered cells (Table 2). This observation might be explained by a higher photosensitizer load in some cells (Figure 3), that possibly led to a stronger singlet oxygen burst triggered by PDT and, consequently, to robust activation of antioxidant mechanisms.

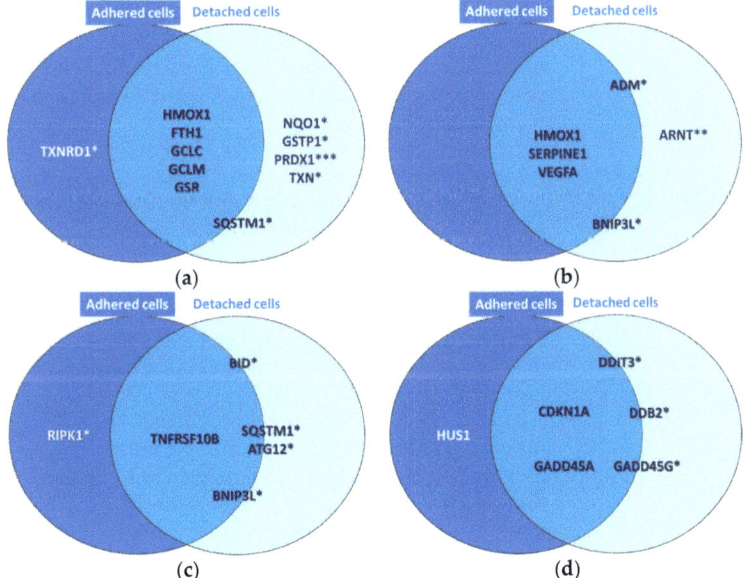

Figure 10. *Cont.*

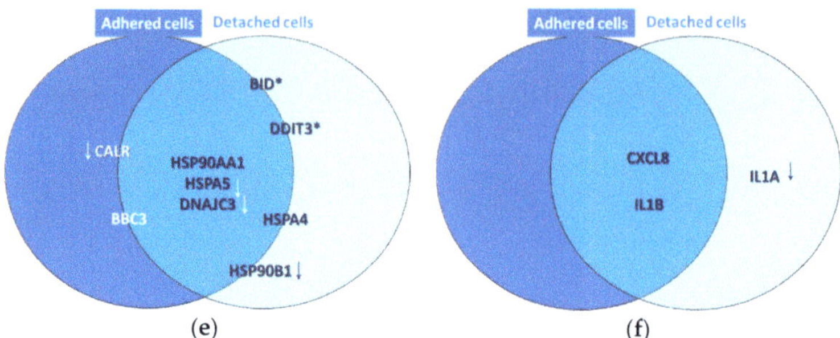

Figure 10. Common and distinctive gene expression patterns in adhered and attached HT29 tumor cells that were exposed to P2.2-PDT (10 J/cm^2, 50 mW/cm^2) at 24 h post-PDT. Some selected genes with modified expression in PDT-treated cells as compared to untreated cells ($1.5 < FR < -1.5$, $p < 0.05$) are represented. These genes are involved in (**a**) oxidative stress, (**b**) Hypoxia signaling, (**c**) cell death, (**d**) DNA damage, (**e**) unfolded protein response and (**f**) inflammation. All the reported genes were upregulated, excepting those marked with downwards arrows. Genes placed at the borders of the two circles are expressed both in adhered and detached cells, but have higher expression in one of the two cell types. This comparison between gene expression levels in adhered and detached cells was performed with a paired samples *t*-test (* $p < 0.05$, ** $p < 0.01$, *** $p < 0.001$).

2.2.2. Hypoxia Signaling

Gene expression data indicated that P2.2-PDT induced the activation of hypoxia signaling, as evidenced by the upregulation of six pathway-relevant genes (Table 2), namely *SERPINE1* [21], *ADM* [22], *ARNT* [23], *VEGFA* [24] and *BNIP3L* [25]. Additionally, *HMOX1* [26], which is at the crossroad of oxidative stress, hypoxia and inflammation [27], was also found highly overexpressed. As shown in Figure 10b, *HMOX1*, *SERPINE1* and *VEGFA* were upregulated both in adhered and detached cells, *ARNT* was upregulated only in detached cells ($p < 0.01$), while the transcript levels of *BNIP3L* and *ADM* were increased in detached vs. adhered cells ($p < 0.05$).

2.2.3. Cell Death

As previously explained, P2.2-PDT was shown to induce significant apoptotic and necrotic cell death within 24 h post-treatment (Figure 8). We also found that several cell death-related genes were upregulated in PDT-exposed cells (Table 2), complementing the functional and phenotypic data on cell death following PDT (Figures 7 and 8). The identified genes are involved in apoptosis (*TNFRSF10B* [28], *BID* [29], *BNIP3L* [25] and *BBC3* [30]), necrosis (*RIPK1* [31]) and autophagy (*ATG12* [32], *SQSTM1* [33]). As shown in Figure 10c, *TNFRSF10B* was found overexpressed both in adhered and detached cells. Meanwhile, *BID*, *ATG12*, *SQSTM1* and *BNIP3L* were preferentially expressed in detached cells ($p < 0.05$) and *RIPK1* ($p < 0.05$) in adhered cells, albeit not exclusively. This gene expression pattern at 24 h post-PDT suggests that distinctive death mechanisms might be activated in detached and adhered cells. Of note, death genes were shown to be overexpressed in adhered cells (presumably living cells), indicating that they were in fact committed to death that may occur sometimes after the first 24 h post-PDT.

In the context of cell death, we identified several genes that were found to be upregulated by P2.2-PDT, which are related to cellular responses to genotoxic or proteotoxic stress, as will be presented below.

2.2.4. DNA Damage

PDT-induced DNA damage was indirectly evidenced by the upregulation of eight genes involved in DNA damage repair mechanisms, such as cell cycle arrest (*CDKN1A* [34], *GADD45A/G* [35], *CHEK2* [36] and *HUS1* [37], or other DNA damage responses (*DDIT3* [38], *DDB2* [39] and *XPC* [40] (Table 2). As shown in Figure 10d, the *CDKN1A* and *GADD45A* genes involved in cell cycle arrest were found overexpressed in both adhered and detached cells. This finding suggests that the observed decrease of the number of metabolically active cells in PDT-treated samples (Figure 4) might be due not only to cell death (Figures 7 and 8, Table 2), but also to a proliferation inhibition (Figure 6). *DDIT3, GADD45G* and *DDB2* had higher transcript levels in detached vs. adhered cells ($p < 0.05$), probably due to increased DNA damage in detached cells, while *HUS1* was moderately overexpressed only in adhered cells (FC = 1.62, $p < 0.05$).

2.2.5. Unfolded Protein Response

P2.2-PDT was shown to induce proteotoxic stress, probably due to the PDT-inflicted oxidative damage of proteins. Proof for ER stress and the consequent UPR was provided by significant expression changes of nine pathway-specific genes, out of which five genes were upregulated and four genes were downregulated (Table 2). Thus, *DDIT3*, which is induced by ER stress [41], along with *BBC3* and *BID*, which link the ER stress response to the mitochondrial apoptosis pathway [42], were concurrently found overexpressed. Some other genes encoding protective ER chaperones (*HSP90AA1* and *HSPA4*) were also found upregulated. They may target misfolded proteins for degradation and even shut down UPR when the stress subsides [43]. Meanwhile, some other chaperone-encoding genes, such as *HSP90AA1* [44], *HSPA5* [45] and *HSP90B1* [46], were found to be downregulated in PDT-treated cells, along with *CALR* [47] and *DNAJC3* [48]. Accordingly, important mechanisms in UPR might be suppressed in cells exposed to PDT, potentially leading in time to cell death. As shown in Figure 10e, while *HSP90AA1* was markedly overexpressed both in adhered and detached cells, *HSPA5* and *DNAJC3* were found concurrently downregulated in both types of cells. Meanwhile, the genes connecting ER stress and apoptosis (*DDIT3* and *BID*) were preferentially overexpressed in detached cells ($p < 0.05$), which also exhibited downregulation of the protective *HSP90B1* gene (Figure 10e). Surprisingly, *CALR* was downregulated by PDT mostly in adherent cells, thus making them more susceptible to ER stress and to its deleterious consequences. In turn, *CALR* downregulation may have beneficial effects in tumors, considering that this can induce inhibition of cell growth, invasion and cell cycle progression [49].

2.2.6. Inflammation

A pro-inflammatory cytokine response was shown to be elicited by PDT in HT29 tumor cells, as seen from the significant overexpression of the pro-inflammatory *CXCL8* and *IL1B* genes in both adhered and detached cells (Table 2, Figure 10f). Surprisingly, *IL1A*, encoding for a danger signal that senses genotoxic stress and is released by cells with damaged plasma membrane [50], was found distinctively downregulated in detached cells ($p < 0.05$) as shown in Table 2 and Figure 10f. Possibly, *IL1A* transcription was inhibited after the activation of the apoptotic machinery, cells having already sensed the death-inducing signals.

3. Discussion

The study brought first experimental in vitro proof on the photosensitizing ability of the new asymmetric porphyrinic compound P2.2 [8] for efficient PDT in the human colon carcinoma HT29 cell line. Following P2.2-PDT, we showed that tumor cells massively died by necrosis and apoptosis within 24 h, especially at high activating light fluences of 25 J/cm^2, as demonstrated using functional and phenotypic tests.

Gene expression data highlighted the molecular mechanisms underlying cell death and survival following a milder P2.2-PDT regimen (10 J/cm^2). Thus, the observed increased apoptosis of PDT treated-cells might be partly mediated by the Death receptor 5 (DR5), also known as TRAIL receptor 2 (TRAILR2) or tumor necrosis factor receptor superfamily member 10B (TNFRSF10B), which triggers the extrinsic apoptotic cascade [51]. DR5 has been found upregulated in various types of tumors, including colorectal carcinomas, and can be exploited for selective apoptotic killing of cancer cells through caspase 8 [52]. Recent studies have shown that PDT extensively sensitizes refractory colon tumors to death signals delivered by long-acting TRAIL [53], possibly by increasing the levels of TRAIL receptors on tumor cells. Furthermore, the increased levels of the *BID* transcripts evidenced by us may sustain the connection of death receptor signaling to the mitochondrial apoptotic machinery [54]. The activation of the mitochondrial apoptotic pathway in PDT-treated cells is also demonstrated by the increased transcript levels of the *BBC3* gene which is under TP53 transcriptional control but can be independently upregulated also by ER stress [55]. A complementary death mechanism, which is highly relevant for tumors and for PDT [56], was evidenced by the overexpression of the *BNIP3L* gene, which is associated with TP53-dependent apoptosis under hypoxia [25].

A particular death pathway bridging apoptosis and necrosis (necroptosis or programmed necrosis) might occur in PDT, as evidenced by *RIPK1* overexpression [57,58]. It is known that necroptosis and necrosis can initiate inflammatory reactions, albeit through distinctive mechanisms, and this may defend or sustain tumor progression, depending on the context [57]. Finally, we have found that autophagy might also be activated by P2.2-PDT, as shown by *ATG12* and *SQSTM1* upregulation, indicating that damaged cells attempt to tolerate the photo-damage or are partly driven to autophagic cell death [59].

We have evidenced that PDT-triggered genotoxic stress may drive tumor cells towards death. In turn, PDT-treated cells responded by upregulating genes involved in cell cycle arrest and in other DNA damage responses (DDR) as defense mechanisms against DNA injury [60]. PDT-induced DNA damage, accompanied by the failure of DNA repair mechanisms, can result in the death of a high percentage of the treated cells [61], as also shown by our cell viability data. Most interesting, among the overexpressed genes involved in DDR, we identified genes with dual roles, which can facilitate DNA repair or induce apoptosis, depending on the context. For instance, the multifunctional DDB2 protein (damaged DNA binding protein 2) is known to contribute to DNA repair through the induction of nucleotide excision repair [62], but can also drive damaged cells towards apoptosis [63,64], for instance by downregulating *CDKN1A* expression [65]. Moreover, it has been shown that DDB2 can regulate the transcription of the antioxidant *SOD2* gene, hence regulating superoxide levels and consequently the redox balance [66].

Molecular analysis of cell death revealed that PDT triggered a complex web of stressors that may lead either to cell death or to resistance of treated tumor cells, as detailed below. We have emphasized that, in time, the few cells surviving PDT continued to grow slowly, probably accounting for disease relapse sometimes after PDT. Therefore, unraveling the rescue mechanisms triggered by PDT may be of highest importance for designing new adjuvant therapeutic strategies aimed at sustaining PDT.

P2.2-PDT induced a strong oxidative burst that was indirectly evidenced in this study through the upregulation of several potent antioxidant genes. Surprisingly, a robust antioxidant response was generated by PDT not only in detached cells (apoptotic cells) but also in adhered cells (presumably living cells), and this protective response was even stronger in detached cells. This observation might be explained by a potentially higher photosensitizer load in some cells, probably resulting in stronger PDT-induced singlet oxygen burst and extensive activation of antioxidant mechanisms. Nonetheless, the elicited antioxidant response was not sufficiently protective to completely counteract the observed PDT-driven decrease of viable cells number. In turn, the remaining viable cells may get shielded against oxidative damage and become resistant to future PDT sessions as well as to other anticancer therapies that rely on oxidative stress for cytotoxicity [3].

The hypoxia signature detected in this study in PDT-treated cells sustains the assumption that the acute oxidative burst triggered by PDT can transiently reduce local oxygen availability due to its utilization for singlet oxygen generation [3]. PDT-induced hypoxia adds to the intrinsic hypoxia of large tumors [67]. Through a feedforward loop, acute hypoxia can induce within minutes a superoxide burst that intensifies the oxidative stress triggered by PDT through the generated singlet oxygen [68]. It has been also shown that reactive oxygen species (ROS) can directly stabilize and activate the transcriptional activity of the hypoxia-inducible factor-1 alpha (HIF1α), which regulates the expression of many hypoxia-responsive genes [69]. Unfortunately, hypoxia signaling is generally associated with increased angiogenesis and epidermal to mesenchymal transition (EMT), both processes favoring tumor progression [70]. Therefore, besides being cytotoxic against most of the treated tumor cells, the PDT regimen investigated by us appears also to support the survival of residual cancer cells that escaped from the therapeutic hit through hypoxia signaling.

The accumulation of oxidatively damaged proteins resulting from the severe oxidative burst triggered by PDT may account for the ER stress signature and the consequent UPR detected by us at transcriptional level in PDT-treated cells [71]. Alternatively, we cannot rule out that ER stress might be induced directly by PDT if the photosensitizer localizes, at least partly, in ER, and a massive amount of ROS is produced locally [72]. Considering the high reactivity of photo-generated ROS, autophagy was shown by us to be initiated to remove damaged organelles [73], if cells were not driven to apoptosis by DDIT3 in conjunction with BID and BBC3 [59,74,75]. Additionally, we evidenced the upregulation of the heat shock protein-encoding genes *HSP90AA1* and *HSPA4*, which may limit the PDT-induced proteotoxic stress. In turn, other protective genes were found downregulated (*HSPA5*, *DNAJC3*, *HSP90B1* and *CALR*), indicating that important cytoprotective mechanisms might have been suppressed by the investigated PDT regimen.

As will be detailed below, several transcription factors regulate the expression of the genes found significantly modified in the present study.

All the investigated redox genes are known targets of the transcription factor NRF2 which controls at the transcriptional level more than 250 cytoprotective genes, most of them having an antioxidant role [76]. It has been shown that singlet oxygen, the peculiar form of ROS generated during PDT, can activate the NRF2 system directly or via derived oxidized products, therefore inducing antioxidant shielding against the PDT-inflicted injuries. Moreover, NRF2 links oxidative conditions with DDR through the p21 protein which is encoded by *CDKN1A* and is mainly involved in cell cycle arrest [34]. NRF2 can induce the transcription of the *CDKN1A* gene which is also under the transcriptional control of TP53 [77]. Through a feed forward loop, enhanced levels of p21 compete with NRF2 for binding to its KEAP1 repressor, hence inducing the activation of the NRF2 pathway [78]. Altogether, NRF2 silencing would be a promising co-therapy for increasing PDT efficacy [79] but, for the moment, there are no NRF2 inhibitors approved for therapeutic use, due to their numerous off-target effects [80].

Most of the proteins encoded by the genes identified in this study are directly or indirectly connected to the redox-sensitive tumor suppressor TP53 (Figure 11). TP53 is one of the most important genome guardians that protects cells against various insults, including oxidative stress, by driving damaged cells towards death or by activating rescue mechanisms [81–83]. Interestingly, it has been shown that TP53 might play a significant role in the death of cells subjected to porphyrin-PDT, through direct interaction with the drug itself, leading to induction of TP53-dependent cell death both in the dark and upon PDT [81]. Nevertheless, our data also point towards a TP53-mediated activation of cytoprotective mechanisms that might render treated cells resistant to future PDT sessions. It is worth mentioning that many genes that are known to be under TP53 control can be also upregulated by other transcription factors, depending on the stressors to which cells were exposed. Therefore, the involvement of TP53 has to be carefully analyzed in PDT for finding the proper therapy regimen that drives tumor cells towards death.

The PDT-mediated upregulation of some gene targets of the redox- and hypoxia-sensitive transcription factor HIF1α, such as *VEGFA*, give evidence that a hypoxia response is indeed triggered by PDT. HIF1α may elicit protective mechanisms that not only defend tumor cells against PDT, but may further drive tumor progression by sustaining angiogenesis and EMT [84,85].

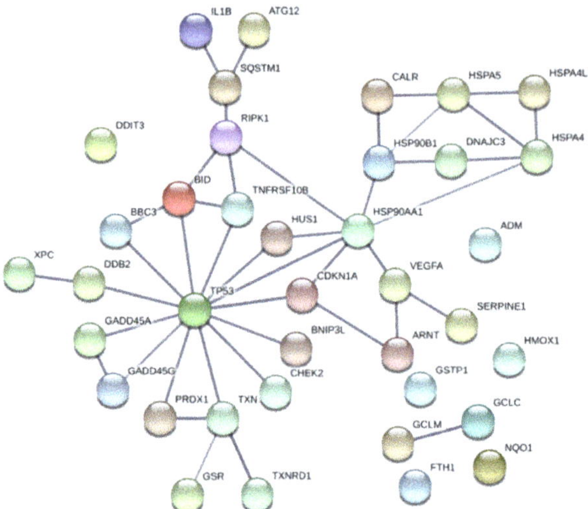

Figure 11. Functional associations between the tumor suppressor TP53 and the proteins encoded by genes with modified expression in PDT-treated HT29 tumor cells, as compared to non-treated cells (Table 2). The network was built using the online tool "STRING: functional protein association network". Only connections deriving from experiments and databases, with a high confidence of 0.7, are represented.

A transcription factor revealed by our results to be transcriptionally upregulated by P2.2-PDT is DDIT3 (CHOP). This stress-induced transcription factor has been shown to integrate the signals from multiple stressors such as DNA damage and ER stress. It mediates a vast array of cellular responses, from UPR to stress-induced cell death, either per se or through interaction with other transcription factors [38]. As evidenced in this study, *DDIT3* overexpression was paralleled by the upregulation of the *TNFRS10B* gene that encodes the death receptor DR5. This might be a functional association between cell death and ER stress, considering that *TNFRS10B* transcription is under the control of DDIT3, in addition to TP53 and NFκB [86,87]. Moreover, it has been shown that DR5 along with other TRAIL receptors serve as stress-associated molecular patterns (SAMPs) to promote ER stress-induced inflammation [88]. Besides the involvement of DDIT3 in apoptotic and autophagic cell death [75], it has been demonstrated that this transcription factor can activate inflammatory responses that generally sustain tumor progression [89]. This can explain, at least partly, some of our results regarding the PDT-induced inflammatory response in HT29 colon carcinoma cells. For instance, we detected in PDT-treated cells a markedly enhanced transcription of the *CXCL8* gene encoding the IL-8 chemokine and this might be mediated by DDIT3 [90]. IL-8 triggers enhanced recruitment of neutrophils in the tumor niche [91], their cytotoxic activity increasing PDT efficacy against tumor cells [92], hence limiting disease progression [93,94]. In turn, increased levels of IL-8 were shown to boost colorectal liver metastasis, hence worsening disease outcome [95]. We do not rule out that enhanced *CXCL8* expression could derive also from NFκB-mediated gene transcription under the PDT pressure [96].

Altogether, molecular data highlighted that the investigated PDT regimen triggered increased transcription of critical genes that underlies the therapeutic effect, but some of the investigated genes may confer a survival advantage to tumor cells. The differential analysis of the gene expression pattern in adhered and detached cells could not provide definite evidence on the particular genes that determine the difference between tumor cells killed by PDT and the surviving ones at 24 h post-PDT. Most of the selected genes were common to adhered and detached cells, indicating that adhered cells might be in fact committed to delayed death. Only *TXNRD1* and *RIPK1* were upregulated specifically in adhered cells, being involved in antioxidant protection and cell death, respectively. The antioxidant genes *NQO1, GSTP1, PRDX1* and *NQO1* were upregulated specifically in detached cells, indicating that antioxidant mechanisms were not fully capable of protecting tumor cells against the PDT-inflicted oxidative injuries.

4. Materials and Methods

4.1. Photosensitizer

The unsymmetrical porphyrin 5-(4-hydroxy-3-methoxyphenyl)-10,15,20-tris-(4-acetoxy-3-methoxyphenyl)porphyrin (Figure 1), abbreviated as P2.2, was obtained according to the method previously described by us [8]. The method is based on the interaction between 4-hydroxy-3-methoxybenzaldehyde, pyrrol and 4-acetoxy-3-methoxybenzaldehyde (3:1 ratio) for approximately 3 h in an acid environment, under strict temperature control (t = 125 °C). P2.2 purification was obtained by column chromatography using Al_2O_3 90 (Merck, 63–200 μm and 70–230 μm mesh) as stationary phase and dichloromethane/diethyl ether (30:1 *v/v*) as eluent. The synthesis yield was 7% and the purity was 100%, as previously shown by NMR analysis. The fluorescence emission and singlet oxygen formation quantum yields, lifetimes and the methods used for quantification are described in [8] and [97]. P2.2 has a Q band with a peak at 628 nm and was therefore activated for PDT using a 635 nm laser in the Modulight ML6600 equipment (Modullight, Tampere, Finland). The obtained P2.2 was dissolved at 10 mM concentration in PEG 200, a biocompatible solvent, and was stored at room temperature in the dark until use.

4.2. Cells

The human colon adenocarcinoma cell line HT29 purchased from the American Tissue and Cell Collection (ATCC® HTB-38™, Manassas, VA, USA) was used. Cells were grown in DMEM-F12 culture medium with GlutaMAX (Gibco), supplemented with 10% fetal bovine serum (FBS, Sigma, Saint Louis, MO, USA), that will be hereinafter referred to as complete culture medium. Twice per week cells were detached with 0.05%/0.02% (*w/v*) Trypsin-EDTA (Biochrom). After trypsin inactivation with two volumes of complete culture medium, cells were washed by centrifugation (1200 rpm, 5 min, 4 °C), were suspended in complete culture medium and were counted by optical microscopy in a Burker–Turk counting chamber, using Trypan blue as dead cells stain. Only cell cultures with a viability >95% were used for experiments. For multiplication, cells were seeded in 25 cm² cell culture flasks (40,000 live cells/cm²) and were cultivated at 37 °C in 5% CO_2 atmosphere.

4.3. Loading of Cells with P2.2

HT29 cells (0.5×10^6 cells) were seeded in 35 mm Petri dishes in 2 mL complete culture medium, in triplicates for each experimental condition. Cells were cultivated for 24 h in 5% CO_2 atmosphere at 37 °C for allowing their adherence. Thereafter, the complete culture medium was discarded and was replaced with 2 mL DMEM-F12 culture medium with GlutaMAX, supplemented with 2% FBS and 10 μM P2.2 (referred as P2.2-loading culture medium). Cells were cultivated for another 24 h to allow P2.2 loading. All the procedures with P2.2 and P2.2-loaded cells were performed in "dark" conditions (no direct light falling on the samples) for avoiding uncontrolled activation of the photosensitizer.

4.4. P2.2 Uptake in HT29 Cells

For measuring P2.2 uptake into HT29 cells, 0.1×10^6 cells were seeded in triplicates in 24 well plates and were cultivated for 24 h in 0.5 mL complete culture medium for allowing their adherence. The culture medium was then discarded and was replaced with P2.2-loading culture medium (see Section 4.3). At 24 h after P2.2 addition to cell cultures, cells were detached with Trypsin-EDTA (see Section 4.2), were washed by centrifugation and were finally suspended in Live Cell Imaging Solution (Thermo Fisher Scientific, Waltham, MA, USA). The uptake control samples were not incubated with P2.2 and were processed exactly as P2.2-treated cells. All the procedures with P2.2-loaded cells were performed in "dark" conditions. P2.2 incorporation into HT29 cells was measured based on P2.2 red fluorescence by flow cytometry on a BD FACSCanto II cytometer with BD FACSDiva 6.1 software (Becton Dickinson, Franklin Lakes, NJ, USA). Data from a minimum of 10,000 events were acquired. Fluorescence was expressed in arbitrary units. Fluorescence data were processed in each sample as median fluorescence value or fluorescence distribution, using the above-mentioned software.

4.5. In Vitro PDT

After loading of cells with P2.2 in 35 mm Petri dishes (see Section 4.3), the P2.2-loading culture medium was discarded and cells were gently washed with complete culture medium at room temperature. Two mL Hank's balanced salt solution supplemented with 2% FBS (PDT culture medium) was added and cells were subjected to in vitro PDT. PDT was performed in test samples at room temperature, using a Modulight ML6600 instrument (Modulight, Tampere, Finland) equipped with a 635 nm laser, illumination chamber for 35 mm Petri dishes and software control of temperature and PDT parameters (light fluence, fluence rate, power and time). The PDT parameters applied in various experimental settings were: 5, 7.5, 10, 15 and 25 J/cm^2. Light was delivered at a fluence rate of 50 mW/cm^2. A comparison between 10 and 50 mW/cm^2 fluence rate was performed for a fluence of 10 J/cm^2. Control samples loaded with P2.2 were prepared by applying the same procedure as described above, were not subjected to PDT and were kept at room temperature during the time when test samples were exposed to PDT. Immediately after PDT, the PDT culture medium was removed and was replaced with 2 mL complete culture medium. Both test samples and controls were further cultivated in 5% CO$_2$ atmosphere at 37 °C for performing various post-PDT investigations.

4.6. Post-PDT Investigations

4.6.1. Preparation of Samples for Post-PDT Investigations

- Cell culture supernatants were harvested from samples cultivated for 24 h post-PDT and were centrifuged for eliminating detached cells. These cell-free culture supernatants were used for the LDH release assay at 24 h post-PDT. Cellular sediments resulting following centrifugation were suspended in a small volume of complete culture medium. Parts of these cells were used for post-PDT investigations at 24 h and parts for cell cultures were analyzed at 72 h post-PDT, as will be described below.
- Adhered cells were detached at 24 h post-PDT with Trypsin-EDTA (see Section 4.2). The resulting cell suspension was centrifuged and the sediment was suspended in complete culture medium. Parts of these cells were used for investigations at 24 h post-PDT and parts were plated for cell cultures to be analyzed at 72 h post-PDT, as will be described below.
- Detached and adhered cells harvested at 24 h post-PDT were mixed. Cells in non-treated samples were counted and the volume containing 10,000 control cells was calculated. For MTS reduction and LDH release (see below Sections 4.6.2 and 4.6.3, respectively), the previously calculated cell suspension volume was collected from all samples, both PDT-treated and non-treated, was placed in 96 well plates and the culture volume was adjusted to 100 µL complete culture medium in each well. Triplicate samples containing only culture medium constituted the background control for these colorimetric tests. Cell cultures were incubated for another 48 h at 37 °C in 5% CO$_2$ atmosphere and were analyzed at 72 h post-PDT.

4.6.2. MTS Reduction

MTS reduction was used for evaluating the relative number of metabolically active cells in PDT-treated and non-treated samples. The method therefore provides information on cell viability and proliferation [8].

Detached and adhered cells harvested at 24 h post-PDT (see Section 4.6.1) were mixed for each sample. Cells in non-treated samples were counted and the volume containing 30,000 cells was calculated. This volume of cell suspension was harvested from all samples, both PDT-treated and non-treated, was adjusted to 100 µL and the resulting samples were analyzed for MTS reduction at 24 h post-PDT. MTS reduction was tested in the cell cultures prepared for 72 h investigations, as follows. A total of 50 µL of supernatant was collected from the cell cultures (see Section 4.6.1) following centrifugation of the 96 well plates at 150 g for 5 min and was used for assessing LDH release (see Section 4.6.3). Then, 50 µL of fresh complete culture medium was added to cell cultures for assessing MTS reduction.

MTS reduction was measured using the colorimetric CellTiter 96® AQueous One Solution Cell Proliferation Assay (MTS) from Promega (Madison, WI, USA), according to the manufacturer's procedure. Briefly, 20 µL of the kit's reagent was added to each well and samples were cultivated 2 h at 37 °C for allowing MTS reduction by metabolically active cells. Finally, the optical density at 490 measured against a 620 nm wavelength reference was measured in each sample using a Sunrise Tecan microplate reader equiped with universal reader control and Magellan data analysis software (Tecan, Männedorf, Switzerland). Data were processed as corrected optical density (OD) obtained by subtrating the mean optical density of the background samples from the the optical density of test samples. Data were processed as PDT effect calculated by dividing the OD of each PDT-treated sample to the mean value of the OD of the corresponding control samples (not treated by PDT).

4.6.3. LDH Release

LDH release was used for evaluating membrane integrity of PDT-treated and non-treated cells [8]. The method provides information on cell death through necrosis/necroptosis [9].

Cell-free supernatants obtained immediately after PDT and at 24 h or 72 h post-PDT were tested for LDH release.

LDH release was measured using the CytoTox 96® Non-Radioactive Cytotoxicity Assay (Promega). The colorimetric test was performed according to the manufacturer's procedure. Briefly, 50 µL of cell-free supernatant and 50 µL of LDH substrate were incubated in the dark at room temperature for 30 min. The reaction was interrupted by the addition of 50 µL stop solution. Finally, the optical density at 490 nm was measured in each sample using a Sunrise Tecan microplate reader equipped with universal reader control and Magellan data analysis software (Tecan). Data were processed as corrected optical density (OD) obtained by subtracting the mean optical density of background samples from the optical density of test samples. Data were processed as PDT effect calculated by dividing the OD of each PDT-treated sample to the mean OD value of control samples (not treated by PDT).

4.6.4. Apoptosis and Necrosis Evaluation by Flow Cytometry

The percentage of apoptotic and necrotic cells in P2.2-treated samples and in untreated controls was evaluated at 24 h post-PDT by flow cytometry using the Annexin A5 Apoptosis Detection Kit (BioLegend, San Diego, CA, USA). The test was performed according to the procedure indicated by the manufacturer.

Briefly, detached and adhered cells harvested at 24 h post-PDT (see Section 4.6.1) were mixed for each samples. Cells were counted and approximately 300,000 cells from each sample were harvested. Cells were washed once with phosphate-buffered saline (PBS), were suspended in 100 µL annexin V binding buffer and were labelled with 5 µL FITC Annexin V and 10 µL of propidium iodide solution. Cell suspensions were incubated for 15 min at room temperature, in the dark. The reaction was stopped by the addition of 400 µL annexin V binding buffer. Samples were analyzed within 30 min by flow cytometry on a BD FACSCanto II cytometer with BD FACSDiva 6.1 software (Becton Dickinson). Data from a minimum of 5000 events were acquired. For PE-FITC the compensation was set at 31, while for FITC-PE it was set at 5. Flow cytometry data were presented as percentage of early apoptotic cells (annexin V^+/PI^-), late apoptotic cells (annexin V^+/PI^+) and necrotic cells (annexin V^-/PI^+). The total pecentage of apoptotic cells was calculated as sum of early and late apoptotic cells percentages.

4.6.5. Cell Proliferation Evaluation

Cell proliferation was assessed by flow cytometry with CFDA-SE (Vybrant® CFDA SE Cell Tracer Kit, Invitrogen, Waltham, MA, USA). According to the information provided by the supplier, CFDA-SE passively diffuses into cells. It is colorless and nonfluorescent until its acetate groups are cleaved by intracellular esterases to yield highly fluorescent, amine-reactive carboxy-fluorescein succinimidyl ester. The succinimidyl ester group reacts with intracellular amines, forming fluorescent conjugates that are well-retained within cells. The dye–protein adducts that form in labeled cells are retained by the cells throughout development, meiosis and in vivo tracing. The label is inherited by daughter cells after cell division and is not transferred to adjacent cells in a population, therefore providing reliable information on cell proliferation.

Adhered cells harvested at 24 h post-PDT (see Section 4.6.1) were counted. From each sample 300,000 cells were collected and were washed twice in cold PBS (1200 rpm, 5 min, 4 °C). Sedimented cells were incubated for 15 min with 5 μM CFDA-SE in 300 μL pre-warmed PBS (37 °C). Cells were washed once in PBS by centrifugation. Sedimented cells were suspended in 500 μL fresh complete culture medium and were incubated for another 30 min at 37 °C to ensure stable loading of cells. The suspension was washed by centrifugation and the cell pellet was suspended in 300 μL complete culture medium. Then, 100 μL of cell suspension from each sample was placed in 24 well plates and the culture volume was adjusted to 1 mL. Cell cultures were incubated for another 48 h at 37 °C in 5% CO_2 atmosphere and were analyzed at 72 h post-PDT. Before investigation, cells were detached with Tripsin-EDTA solution (see Section 4.2) and were washed twice in ice-cold PBS. The intracellular fluorescence of CFDA-SE was measured by flow cytometry using a BD FACSCalibur flow cytometer and CellQuest Pro software (Becton Dickinson, Franklin Lakes, NJ, USA). CFDA-SE fluorescence was activated with a 482 nm laser, while emmission was registered in the FL-1 channel for fluorescein isothiocyanate (FITC). The data from a minimum of 50,000 events were acquired. Flow cytometry data were further processed using the ModFit LT software (Verity Sotware House, Topsham, ME, USA), which provides the distribution of proliferative cells in consequtive daughter generations.

4.6.6. Microscopic Monitoring of Cell Cultures

Cell cultures were monitored in time, up to 120 h post-PDT, by optical microscopy using an EVOS XL Core Cell Imaging System equiped with image acquisition software (ThermoFisher Scientific). Of note is that separate cell cultures were prepared for imaging, equivalent to those described at 4.5, considering that exposure of P2.2-loaded cells to visible light might trigger an artificial activation of PDT and might compromise the results of the post-PDT tests. Due to the same reason, even the samples dedicated to imaging were only rarely subjected to microscopic evaluation (1 time/day, with only short exposure to light).

4.6.7. Gene Expression

Cell cultures in 35 mm Petri dishes were prepared and processed as described in Sections 4.3–4.5.

At 24 h post-PDT, cell supernatants containing detached cells were collected and centrifuged and cellular sediments were finally collected in 1 mL RiboZol reagent (VWR, Radnor, PA, USA). Attached cells were gently washed with warm PBS (37 °C) and 1 mL RiboZol reagent was then added for RNA extraction. Processed samples in RiboZol reagent were stored at −80 °C until use.

Total RNA isolation from cells preserved in RiboZol reagent (VWR Life science) was performed according to the manufacturer's instructions. RNA concentration was quantified using the Nanodrop 2000 (Thermo Fisher Scientific, Waltham, MA, USA). Both the 260/280 nm and 260/230 nm ratios were above 1.8, indicating a high RNA quality. cDNA synthesis was performed using 600 ng of total RNA with the RT^2 First Strand Kit (Qiagen, Hilden, Germany), according to the manufacturer's instructions. The expression of 84 genes involved in stress and toxicity was assessed using the RT^2 Profiler™ PCR Array Human Stress and Toxicity PathwayFinder (PAHS-003Z) from Qiagen Table 1), using the SYBR Green chemistry on ABI7500 Fast PCR System (Thermo Fisher Scientific, Waltham, MA, USA). The expression level of each gene was normalized on the geometric mean values of two housekeeping genes (ACTB and HPRT1), which were selected according to RefFinder algorithm [98] after the analysis of five candidate reference genes (ACTB, B2M, GAPDH, HPRT1 and RPLP0) both in PDT-treated samples and non-treated controls. Gene expression data were analyzed with the RT^2 Profiler PCR Array software package (Qiagen). Gene expression levels were calculated as $2^{-\Delta CT}$ values. Fold change (FC) in gene expression was calculated as the $2^{-\Delta CT}$ mean values in patients divided by $2^{-\Delta CT}$ mean values in controls. Results are presented as fold regulation (FR) as follows: when the FC value was above 1, FR was equal to FC and results were reported as fold upregulation; when the FC value was less than 1, FR was expressed as the negative inverse of FC and results were reported as fold downregulation.

4.7. Data Processing

Whenever possible data were presented as mean value $\pm$ standard error of the mean (SEM) for triplicate samples. The IC50 value of P2.2 was computed using the online Quest Graph™ IC50 Calculator (https://www.aatbio.com/tools/ic50-calculator, accessed on 27 June 2021) and was expressed as mean $\pm$ standard deviation (SD). PDT effect was calculated as a parameter value in PDT-treated samples divided by the mean parameter value in non-treated controls. Supraunit values of PDT effect indicate activation, while subunit values indicate an inhibition. Comparison of the phenotypic and functional data from cells exposed to PDT and from non-treated controls was performed in Excel using the Student's t-test (unequal variances or paired two samples for mean, depending on the case). Comparison of gene expression data between PDT-treated samples and controls was performed with the Student's t-test (equal variances) using the RT^2 Profiler PCR Array software package (Qiagen, Hilden, Germany). Comparison between adhered and attached cells regarding gene expression (FR values) was performed in Excel using the Student's t-test (paired two samples for mean). Differences were considered significant for p values < 0.05. The Pearson test was used for evaluating correlations between parameters, considered as significant for r values > 0.6 and $p < 0.05$.

5. Conclusions

Using an integrative systems biology approach on the expression of stress genes in HT29 human colon carcinoma cells subjected to porphyrin-PDT, the study highlighted the network of molecular mechanisms underlying therapy-induced cell death and other cellular responses to the complex web of stressors triggered concomitantly by P2.2-PDT, encompassing oxidative stress, hypoxia, DNA damage, ER stress and UPR, along with inflammation. The transcription factors potentially responsible for the observed gene expression profile (NRF2, HIF1α, p53, DDIT3 and, possibly, NFκB) were highlighted. Particular cytoprotective mechanisms and upregulated pathway-specific genes were found, which represent promising therapeutic targets for improving PDT efficacy. Nevertheless, this study does not provide information on the impact of the observed gene expression changes at protein and functional level, an aspect that should be addressed in future studies. The identified expression profile of stress genes triggered by the particular P2.2-PDT regimen used in the present study might be further extended to other porphyrinic photosensitizers and PDT regimens.

Author Contributions: M.D. original draft preparation, investigations-gene expression, methodology; R.B. conceptualization-photosensitizers; investigations-P2.2 synthesis, methodology; I.V.N. investigations-cell cultures, in vitro PDT, viability tests; M.S. investigations-flow cytometry, methodology; E.M. methodology, formal analysis, draft revision; G.M. conceptualization, draft preparation and revision. All authors have read and agreed to the published version of the manuscript.

Funding: This research was funded by the Romanian Ministry of Research, Innovation and Digitization, grant number PCCDI 64/2018. Publication charges were funded by the Romanian Ministry of Research, Innovation and Digitization, grant PN 19.29.02.02 (Nucleu Programme). EM is recipient of a post-doctoral contract of the European Regional Development Fund, Competitiveness Operational Program 2014–2020, through the grant P_37_732/2016 REDBRAIN.

Institutional Review Board Statement: Not applicable.

Informed Consent Statement: Not applicable.

Data Availability Statement: The data presented in this study are presented in the paper, and raw data are available on request from the corresponding author.

Conflicts of Interest: The authors declare no conflict of interest.

References

1. Agostinis, P.; Berg, K.; Cengel, K.A.; Foster, T.H.; Girotti, A.W.; Gollnick, S.O.; Hahn, S.M.; Hamblin, M.R.; Juzeniene, A.; Kessel, D.; et al. Photodynamic therapy of cancer: An update. CA. *Cancer J. Clin.* **2011**, *61*, 250–281. [CrossRef] [PubMed]
2. Falk-Mahapatra, R.; Gollnick, S.O. Photodynamic Therapy and Immunity: An Update. *Photochem. Photobiol.* **2020**, *96*, 550–559. [CrossRef] [PubMed]
3. Manda, G.; Hinescu, M.E.; Neagoe, I.V.; Ferreira, L.F.V.; Boscencu, R.; Vasos, P.; Basaga, S.H.; Cuadrado, A. Emerging Therapeutic Targets in Oncologic Photodynamic Therapy. *Curr. Pharm. Des.* **2018**, *24*, 5268–5295. [CrossRef] [PubMed]
4. Li, X.-Y.; Tan, L.-C.; Dong, L.-W.; Zhang, W.-Q.; Shen, X.-X.; Lu, X.; Zheng, H.; Lu, Y.-G. Susceptibility and Resistance Mechanisms During Photodynamic Therapy of Melanoma. *Front. Oncol.* **2020**, *10*, 597. [CrossRef]
5. Donohoe, C.; Senge, M.O.; Arnaut, L.G.; Gomes-da-Silva, L.C. Cell death in photodynamic therapy: From oxidative stress to anti-tumor immunity. *Biochim. Biophys. Acta Rev. Cancer* **2019**, *1872*, 188308. [CrossRef]
6. Garg, A.D.; Krysko, D.V.; Vandenabeele, P.; Agostinis, P. DAMPs and PDT-mediated photo-oxidative stress: Exploring the unknown. *Photochem. Photobiol. Sci.* **2011**, *10*, 670–680. [CrossRef]
7. Bigot, E.; Bataille, R.; Patrice, T. Increased singlet oxygen-induced secondary ROS production in the serum of cancer patients. *J. Photochem. Photobiol. B* **2012**, *107*, 14–19. [CrossRef]
8. Boscencu, R.; Manda, G.; Radulea, N.; Socoteanu, R.P.; Ceafalan, L.C.; Neagoe, I.V.; Ferreira Machado, I.; Basaga, S.H.; Vieira Ferreira, L.F. Studies on the Synthesis, Photophysical and Biological Evaluation of Some Unsymmetrical Meso-Tetrasubstituted Phenyl Porphyrins. *Molecules* **2017**, *22*, 1815. [CrossRef]
9. Chan, F.K.-M.; Moriwaki, K.; De Rosa, M.J. Detection of necrosis by release of lactate dehydrogenase activity. *Methods Mol. Biol.* **2013**, *979*, 65–70. [CrossRef]
10. Ros, U.; Peña-Blanco, A.; Hänggi, K.; Kunzendorf, U.; Krautwald, S.; Wong, W.W.-L.; García-Sáez, A.J. Necroptosis Execution Is Mediated by Plasma Membrane Nanopores Independent of Calcium. *Cell Rep.* **2017**, *19*, 175–187. [CrossRef]
11. Castano, A.P.; Demidova, T.N.; Hamblin, M.R. Mechanisms in photodynamic therapy: Part one-photosensitizers, photochemistry and cellular localization. *Photodiagn. Photodyn. Ther.* **2004**, *1*, 279–293. [CrossRef]
12. Looft, A.; Pfitzner, M.; Preuß, A.; Röder, B. In vivo singlet molecular oxygen measurements: Sensitive to changes in oxygen saturation during PDT. *Photodiagn. Photodyn. Ther.* **2018**, *23*, 325–330. [CrossRef]
13. Shen, Y.; Li, X.; Zhao, B.; Xue, Y.; Wang, S.; Chen, X.; Yang, J.; Lv, H.; Shang, P. Iron metabolism gene expression and prognostic features of hepatocellular carcinoma. *J. Cell. Biochem.* **2018**, *119*, 9178–9204. [CrossRef]
14. Bachhawat, A.K.; Yadav, S. The glutathione cycle: Glutathione metabolism beyond the γ-glutamyl cycle. *IUBMB Life* **2018**, *70*, 585–592. [CrossRef]
15. Mohammadi, F.; Soltani, A.; Ghahremanloo, A.; Javid, H.; Hashemy, S.I. The thioredoxin system and cancer therapy: A review. *Cancer Chemother. Pharmacol.* **2019**, *84*, 925–935. [CrossRef] [PubMed]
16. Li, H.-X.; Sun, X.-Y.; Yang, S.-M.; Wang, Q.; Wang, Z.-Y. Peroxiredoxin 1 promoted tumor metastasis and angiogenesis in colorectal cancer. *Pathol. Res. Pract.* **2018**, *214*, 655–660. [CrossRef] [PubMed]
17. Ji, L.; Wei, Y.; Jiang, T.; Wang, S. Correlation of Nrf2, NQO1, MRP1, cmyc and p53 in colorectal cancer and their relationships to clinicopathologic features and survival. *Int. J. Clin. Exp. Pathol.* **2014**, *7*, 1124–1131. [PubMed]
18. Beaver, S.K.; Mesa-Torres, N.; Pey, A.L.; Timson, D.J. NQO1: A target for the treatment of cancer and neurological diseases, and a model to understand loss of function disease mechanisms. *Biochim. Biophys. Acta Proteins Proteom.* **2019**, *1867*, 663–676. [CrossRef] [PubMed]
19. Kosumi, K.; Masugi, Y.; Yang, J.; Qian, Z.R.; Kim, S.A.; Li, W.; Shi, Y.; da Silva, A.; Hamada, T.; Liu, L.; et al. Tumor SQSTM1 (p62) expression and T cells in colorectal cancer. *Oncoimmunology* **2017**, *6*, e1284720. [CrossRef] [PubMed]

20. Zhang, J.; Yang, S.; Xu, B.; Wang, T.; Zheng, Y.; Liu, F.; Ren, F.; Jiang, J.; Shi, H.; Zou, B.; et al. p62 functions as an oncogene in colorectal cancer through inhibiting apoptosis and promoting cell proliferation by interacting with the vitamin D receptor. *Cell Prolif.* **2019**, *52*, e12585. [CrossRef]

21. Li, S.; Wei, X.; He, J.; Tian, X.; Yuan, S.; Sun, L. Plasminogen activator inhibitor-1 in cancer research. *Biomed. Pharmacother.* **2018**, *105*, 83–94. [CrossRef]

22. Wang, L.; Gala, M.; Yamamoto, M.; Pino, M.S.; Kikuchi, H.; Shue, D.S.; Shirasawa, S.; Austin, T.R.; Lynch, M.P.; Rueda, B.R.; et al. Adrenomedullin is a therapeutic target in colorectal cancer. *Int. J. Cancer* **2014**, *134*, 2041–2050. [CrossRef]

23. Tsai, Y.-P.; Wu, K.-J. Epigenetic regulation of hypoxia-responsive gene expression: Focusing on chromatin and DNA modifications. *Int. J. Cancer* **2014**, *134*, 249–256. [CrossRef] [PubMed]

24. Miyazaki, S.; Kikuchi, H.; Iino, I.; Uehara, T.; Setoguchi, T.; Fujita, T.; Hiramatsu, Y.; Ohta, M.; Kamiya, K.; Kitagawa, K.; et al. Anti-VEGF antibody therapy induces tumor hypoxia and stanniocalcin 2 expression and potentiates growth of human colon cancer xenografts. *Int. J. Cancer* **2014**, *135*, 295–307. [CrossRef] [PubMed]

25. Fei, P.; Wang, W.; Kim, S.; Wang, S.; Burns, T.F.; Sax, J.K.; Buzzai, M.; Dicker, D.T.; McKenna, W.G.; Bernhard, E.J.; et al. Bnip3L is induced by p53 under hypoxia, and its knockdown promotes tumor growth. *Cancer Cell* **2004**, *6*, 597–609. [CrossRef]

26. Chau, L.-Y. Heme oxygenase-1: Emerging target of cancer therapy. *J. Biomed. Sci.* **2015**, *22*, 22. [CrossRef]

27. Alam, J.; Cook, J.L. How many transcription factors does it take to turn on the heme oxygenase-1 gene? *Am. J. Respir. Cell Mol. Biol.* **2007**, *36*, 166–174. [CrossRef]

28. Babinčák, M.; Jendželovský, R.; Košuth, J.; Majerník, M.; Vargová, J.; Mikulášek, K.; Zdráhal, Z.; Fedoročko, P. Death Receptor 5 (TNFRSF10B) Is Upregulated and TRAIL Resistance Is Reversed in Hypoxia and Normoxia in Colorectal Cancer Cell Lines after Treatment with Skyrin, the Active Metabolite of Hypericum spp. *Cancers* **2021**, *13*, 1646. [CrossRef] [PubMed]

29. Gahl, R.F.; Dwivedi, P.; Tjandra, N. Bcl-2 proteins bid and bax form a network to permeabilize the mitochondria at the onset of apoptosis. *Cell Death Dis.* **2016**, *7*, e2424. [CrossRef]

30. Yu, J.; Zhang, L.; Hwang, P.M.; Kinzler, K.W.; Vogelstein, B. PUMA induces the rapid apoptosis of colorectal cancer cells. *Mol. Cell* **2001**, *7*, 673–682. [CrossRef]

31. Wang, L.; Chang, X.; Feng, J.; Yu, J.; Chen, G. TRADD Mediates RIPK1-Independent Necroptosis Induced by Tumor Necrosis Factor. *Front. Cell Dev. Biol.* **2019**, *7*, 393. [CrossRef]

32. Hu, J.L.; He, G.Y.; Lan, X.L.; Zeng, Z.C.; Guan, J.; Ding, Y.; Qian, X.L.; Liao, W.T.; Ding, Y.Q.; Liang, L. Inhibition of ATG12-mediated autophagy by miR-214 enhances radiosensitivity in colorectal cancer. *Oncogenesis* **2018**, *7*, 16. [CrossRef]

33. Bjørkøy, G.; Lamark, T.; Pankiv, S.; Øvervatn, A.; Brech, A.; Johansen, T. Monitoring autophagic degradation of p62/SQSTM1. *Methods Enzymol.* **2009**, *452*, 181–197. [CrossRef] [PubMed]

34. Cazzalini, O.; Scovassi, A.I.; Savio, M.; Stivala, L.A.; Prosperi, E. Multiple roles of the cell cycle inhibitor p21(CDKN1A) in the DNA damage response. *Mutat. Res.* **2010**, *704*, 12–20. [CrossRef] [PubMed]

35. Salvador, J.M.; Brown-Clay, J.D.; Fornace, A.J. Gadd45 in stress signaling, cell cycle control, and apoptosis. *Adv. Exp. Med. Biol.* **2013**, *793*, 1–19. [CrossRef]

36. Xiang, H.; Geng, X.; Ge, W.; Li, H. Meta-analysis of CHEK2 1100delC variant and colorectal cancer susceptibility. *Eur. J. Cancer* **2011**, *47*, 2546–2551. [CrossRef] [PubMed]

37. Zhou, Z.Q.; Zhao, J.-J.; Chen, C.-L.; Liu, Y.; Zeng, J.-X.; Wu, Z.-R.; Tang, Y.; Zhu, Q.; Weng, D.-S.; Xia, J.-C. HUS1 checkpoint clamp component (HUS1) is a potential tumor suppressor in primary hepatocellular carcinoma. *Mol. Carcinog.* **2019**, *58*, 76–87. [CrossRef]

38. Jauhiainen, A.; Thomsen, C.; Strömbom, L.; Grundevik, P.; Andersson, C.; Danielsson, A.; Andersson, M.K.; Nerman, O.; Rörkvist, L.; Ståhlberg, A.; et al. Distinct cytoplasmic and nuclear functions of the stress induced protein DDIT3/CHOP/GADD153. *PLoS ONE* **2012**, *7*, e33208. [CrossRef] [PubMed]

39. Bagchi, S.; Raychaudhuri, P. Damaged-DNA Binding Protein-2 Drives Apoptosis Following DNA Damage. *Cell Div.* **2010**, *5*, 3. [CrossRef] [PubMed]

40. Nemzow, L.; Lubin, A.; Zhang, L.; Gong, F. XPC: Going where no DNA damage sensor has gone before. *DNA Repair* **2015**, *36*, 19–27. [CrossRef]

41. Rashid, H.-O.; Yadav, R.K.; Kim, H.-R.; Chae, H.-J. ER stress: Autophagy induction, inhibition and selection. *Autophagy* **2015**, *11*, 1956–1977. [CrossRef]

42. Reimertz, C.; Kögel, D.; Rami, A.; Chittenden, T.; Prehn, J.H.M. Gene expression during ER stress-induced apoptosis in neurons: Induction of the BH3-only protein Bbc3/PUMA and activation of the mitochondrial apoptosis pathway. *J. Cell Biol.* **2003**, *162*, 587–597. [CrossRef]

43. Ma, Y.; Hendershot, L.M. ER chaperone functions during normal and stress conditions. *J. Chem. Neuroanat.* **2004**, *28*, 51–65. [CrossRef]

44. Theodoraki, M.A.; Caplan, A.J. Quality control and fate determination of Hsp90 client proteins. *Biochim. Biophys. Acta* **2012**, *1823*, 683–688. [CrossRef]

45. Ibrahim, I.M.; Abdelmalek, D.H.; Elfiky, A.A. GRP78: A cell's response to stress. *Life Sci.* **2019**, *226*, 156–163. [CrossRef] [PubMed]

46. Jackson, S.E. Hsp90: Structure and function. *Top. Curr. Chem.* **2013**, *328*, 155–240. [CrossRef] [PubMed]

47. Michalak, M.; Groenendyk, J.; Szabo, E.; Gold, L.I.; Opas, M. Calreticulin, a multi-process calcium-buffering chaperone of the endoplasmic reticulum. *Biochem. J.* **2009**, *417*, 651–666. [CrossRef]

48. Van Huizen, R.; Martindale, J.L.; Gorospe, M.; Holbrook, N.J. P58IPK, a novel endoplasmic reticulum stress-inducible protein and potential negative regulator of eIF2alpha signaling. *J. Biol. Chem.* **2003**, *278*, 15558–15564. [CrossRef]
49. Feng, R.; Ye, J.; Zhou, C.; Qi, L.; Fu, Z.; Yan, B.; Liang, Z.; Li, R.; Zhai, W. Calreticulin down-regulation inhibits the cell growth, invasion and cell cycle progression of human hepatocellular carcinoma cells. *Diagn. Pathol.* **2015**, *10*, 149. [CrossRef]
50. Cohen, I.; Idan, C.; Rider, P.; Peleg, R.; Vornov, E.; Elena, V.; Tomas, M.; Martin, T.; Tudor, C.; Cicerone, T.; et al. IL-1α is a DNA damage sensor linking genotoxic stress signaling to sterile inflammation and innate immunity. *Sci. Rep.* **2015**, *5*, 14756. [CrossRef] [PubMed]
51. Nair, P.; Lu, M.; Petersen, S.; Ashkenazi, A. Apoptosis Initiation Through the Cell-Extrinsic Pathway. *Methods Enzymol.* **2014**, *544*, 99–128. [CrossRef]
52. Oliver, P.G.; LoBuglio, A.F.; Zinn, K.R.; Kim, H.; Nan, L.; Zhou, T.; Wang, W.; Buchsbaum, D.J. Treatment of human colon cancer xenografts with TRA-8 anti-death receptor 5 antibody alone or in combination with CPT-11. *Clin. Cancer Res.* **2008**, *14*, 2180–2189. [CrossRef] [PubMed]
53. She, T.; Shi, Q.; Li, Z.; Feng, Y.; Yang, H.; Tao, Z.; Li, H.; Chen, J.; Wang, S.; Liang, Y.; et al. Combination of long-acting TRAIL and tumor cell-targeted photodynamic therapy as a novel strategy to overcome chemotherapeutic multidrug resistance and TRAIL resistance of colorectal cancer. *Theranostics* **2021**, *11*, 4281–4297. [CrossRef] [PubMed]
54. Billen, L.P.; Shamas-Din, A.; Andrews, D.W. Bid: A Bax-like BH3 protein. *Oncogene* **2008**, *27* (Suppl. S1), S93–S104. [CrossRef] [PubMed]
55. Han, J.; Flemington, C.; Houghton, A.B.; Gu, Z.; Zambetti, G.P.; Lutz, R.J.; Zhu, L.; Chittenden, T. Expression of bbc3, a pro-apoptotic BH3-only gene, is regulated by diverse cell death and survival signals. *Proc. Natl. Acad. Sci. USA* **2001**, *98*, 11318–11323. [CrossRef] [PubMed]
56. Sun, Y.; Zhao, D.; Wang, G.; Wang, Y.; Cao, L.; Sun, J.; Jiang, Q.; He, Z. Recent progress of hypoxia-modulated multifunctional nanomedicines to enhance photodynamic therapy: Opportunities, challenges, and future development. *Acta Pharm. Sin. B* **2020**, *10*, 1382–1396. [CrossRef]
57. Gong, Y.; Fan, Z.; Luo, G.; Yang, C.; Huang, Q.; Fan, K.; Cheng, H.; Jin, K.; Ni, Q.; Yu, X.; et al. The role of necroptosis in cancer biology and therapy. *Mol. Cancer* **2019**, *18*, 100. [CrossRef] [PubMed]
58. Newton, K. RIPK1 and RIPK3: Critical regulators of inflammation and cell death. *Trends Cell Biol.* **2015**, *25*, 347–353. [CrossRef]
59. Martins, W.K.; Belotto, R.; Silva, M.N.; Grasso, D.; Suriani, M.D.; Lavor, T.S.; Itri, R.; Baptista, M.S.; Tsubone, T.M. Autophagy Regulation and Photodynamic Therapy: Insights to Improve Outcomes of Cancer Treatment. *Front. Oncol.* **2020**, *10*, 610472. [CrossRef]
60. Huang, R.-X.; Zhou, P.-K. DNA damage response signaling pathways and targets for radiotherapy sensitization in cancer. *Signal Transduct. Target. Ther.* **2020**, *5*, 60. [CrossRef]
61. Wang, J.Y.J. Cell Death Response to DNA Damage. *Yale J. Biol. Med.* **2019**, *92*, 771–779.
62. Kumar, N.; Raja, S.; Van Houten, B. The involvement of nucleotide excision repair proteins in the removal of oxidative DNA damage. *Nucleic Acids Res.* **2020**, *48*, 11227–11243. [CrossRef]
63. Stoyanova, T.; Roy, N.; Kopanja, D.; Raychaudhuri, P.; Bagchi, S. DDB2 (damaged-DNA binding protein 2) in nucleotide excision repair and DNA damage response. *Cell Cycle* **2009**, *8*, 4067–4071. [CrossRef]
64. Barakat, B.M.; Wang, Q.-E.; Han, C.; Milum, K.; Yin, D.-T.; Zhao, Q.; Wani, G.; Arafa, E.-S.A.; El-Mahdy, M.A.; Wani, A.A. Overexpression of DDB2 enhances the sensitivity of human ovarian cancer cells to cisplatin by augmenting cellular apoptosis. *Int. J. Cancer* **2010**, *127*, 977–988. [CrossRef] [PubMed]
65. Roy, N.; Bagchi, S.; Raychaudhuri, P. Damaged DNA binding protein 2 in reactive oxygen species (ROS) regulation and premature senescence. *Int. J. Mol. Sci.* **2012**, *13*, 11012–11026. [CrossRef] [PubMed]
66. Minig, V.; Kattan, Z.; van Beeumen, J.; Brunner, E.; Becuwe, P. Identification of DDB2 protein as a transcriptional regulator of constitutive SOD2 gene expression in human breast cancer cells. *J. Biol. Chem.* **2009**, *284*, 14165–14176. [CrossRef] [PubMed]
67. Tomioka, Y.; Kushibiki, T.; Awazu, K. Evaluation of oxygen consumption of culture medium and in vitro photodynamic effect of talaporfin sodium in lung tumor cells. *Photomed. Laser Surg.* **2010**, *28*, 385–390. [CrossRef]
68. Hernansanz-Agustín, P.; Izquierdo-Álvarez, A.; Sánchez-Gómez, F.J.; Ramos, E.; Villa-Piña, T.; Lamas, S.; Bogdanova, A.; Martínez-Ruiz, A. Acute hypoxia produces a superoxide burst in cells. *Free Radic. Biol. Med.* **2014**, *71*, 146–156. [CrossRef] [PubMed]
69. Jung, S.-N.; Yang, W.K.; Kim, J.; Kim, H.S.; Kim, E.J.; Yun, H.; Park, H.; Kim, S.S.; Choe, W.; Kang, I.; et al. Reactive oxygen species stabilize hypoxia-inducible factor-1 alpha protein and stimulate transcriptional activity via AMP-activated protein kinase in DU145 human prostate cancer cells. *Carcinogenesis* **2008**, *29*, 713–721. [CrossRef]
70. Muz, B.; de la Puente, P.; Azab, F.; Azab, A.K. The role of hypoxia in cancer progression, angiogenesis, metastasis, and resistance to therapy. *Hypoxia* **2015**, *3*, 83–92. [CrossRef]
71. Lin, Y.; Jiang, M.; Chen, W.; Zhao, T.; Wei, Y. Cancer and ER stress: Mutual crosstalk between autophagy, oxidative stress and inflammatory response. *Biomed. Pharmacother.* **2019**, *118*, 109249. [CrossRef]
72. Moserova, I.; Kralova, J. Role of ER stress response in photodynamic therapy: ROS generated in different subcellular compartments trigger diverse cell death pathways. *PLoS ONE* **2012**, *7*, e32972. [CrossRef]
73. Codogno, P.; Meijer, A.J. Autophagy and signaling: Their role in cell survival and cell death. *Cell Death Differ.* **2005**, *12* (Suppl. S2), 1509–1518. [CrossRef] [PubMed]

74. Sano, R.; Reed, J.C. ER stress-induced cell death mechanisms. *Biochim. Biophys. Acta* **2013**, *1833*, 3460–3470. [CrossRef] [PubMed]
75. Lei, Y.; Wang, S.; Ren, B.; Wang, J.; Chen, J.; Lu, J.; Zhan, S.; Fu, Y.; Huang, L.; Tan, J. CHOP favors endoplasmic reticulum stress-induced apoptosis in hepatocellular carcinoma cells via inhibition of autophagy. *PLoS ONE* **2017**, *12*, e0183680. [CrossRef] [PubMed]
76. Riggs, S.; Alario, A.J.; McHorney, C. Health risk behaviors and attempted suicide in adolescents who report prior maltreatment. *J. Pediatr.* **1990**, *116*, 815–821. [CrossRef]
77. Pereira, E.J.; Burns, J.S.; Lee, C.Y.; Marohl, T.; Calderon, D.; Wang, L.; Atkins, K.A.; Wang, C.-C.; Janes, K.A. Sporadic activation of an oxidative stress-dependent NRF2-p53 signaling network in breast epithelial spheroids and premalignancies. *Sci. Signal.* **2020**, *13*. [CrossRef] [PubMed]
78. Chen, W.; Sun, Z.; Wang, X.-J.; Jiang, T.; Huang, Z.; Fang, D.; Zhang, D.D. Direct interaction between Nrf2 and p21(Cip1/WAF1) upregulates the Nrf2-mediated antioxidant response. *Mol. Cell* **2009**, *34*, 663–673. [CrossRef]
79. Choi, B.; Ryoo, I.; Kang, H.C.; Kwak, M.-K. The sensitivity of cancer cells to pheophorbide a-based photodynamic therapy is enhanced by Nrf2 silencing. *PLoS ONE* **2014**, *9*, e107158. [CrossRef]
80. Robledinos-Antón, N.; Fernández-Ginés, R.; Manda, G.; Cuadrado, A. Activators and Inhibitors of NRF2: A Review of Their Potential for Clinical Development. *Oxid. Med. Cell. Longev.* **2019**, *2019*, 9372182. [CrossRef]
81. Zawacka-Pankau, J.; Krachulec, J.; Grulkowski, I.; Bielawski, K.P.; Selivanova, G. The p53-mediated cytotoxicity of photodynamic therapy of cancer: Recent advances. *Toxicol. Appl. Pharmacol.* **2008**, *232*, 487–497. [CrossRef]
82. Kruiswijk, F.; Labuschagne, C.F.; Vousden, K.H. p53 in survival, death and metabolic health: A lifeguard with a licence to kill. *Nat. Rev. Mol. Cell Biol.* **2015**, *16*, 393–405. [CrossRef] [PubMed]
83. Eriksson, S.E.; Ceder, S.; Bykov, V.J.N.; Wiman, K.G. p53 as a hub in cellular redox regulation and therapeutic target in cancer. *J. Mol. Cell Biol.* **2019**, *11*, 330–341. [CrossRef] [PubMed]
84. Pezzuto, A.; Carico, E. Role of HIF-1 in Cancer Progression: Novel Insights. A Review. *Curr. Mol. Med.* **2018**, *18*, 343–351. [CrossRef] [PubMed]
85. Lamberti, M.J.; Pansa, M.F.; Vera, R.E.; Fernández-Zapico, M.E.; Rumie Vittar, N.B.; Rivarola, V.A. Transcriptional activation of HIF-1 by a ROS-ERK axis underlies the resistance to photodynamic therapy. *PLoS ONE* **2017**, *12*, e0177801. [CrossRef] [PubMed]
86. Li, T.; Su, L.; Lei, Y.; Liu, X.; Zhang, Y.; Liu, X. DDIT3 and KAT2A Proteins Regulate TNFRSF10A and TNFRSF10B Expression in Endoplasmic Reticulum Stress-mediated Apoptosis in Human Lung Cancer Cells. *J. Biol. Chem.* **2015**, *290*, 11108–11118. [CrossRef] [PubMed]
87. Yamaguchi, H.; Wang, H.-G. CHOP is involved in endoplasmic reticulum stress-induced apoptosis by enhancing DR5 expression in human carcinoma cells. *J. Biol. Chem.* **2004**, *279*, 45495–45502. [CrossRef]
88. Sullivan, G.P.; O'Connor, H.; Henry, C.M.; Davidovich, P.; Clancy, D.M.; Albert, M.L.; Cullen, S.P.; Martin, S.J. TRAIL Receptors Serve as Stress-Associated Molecular Patterns to Promote ER-Stress-Induced Inflammation. *Dev. Cell* **2020**, *52*, 714–730.e5. [CrossRef]
89. Greten, F.R.; Grivennikov, S.I. Inflammation and Cancer: Triggers, Mechanisms, and Consequences. *Immunity* **2019**, *51*, 27–41. [CrossRef]
90. Jundi, K.; Greene, C.M. Transcription of Interleukin-8: How Altered Regulation Can Affect Cystic Fibrosis Lung Disease. *Biomolecules* **2015**, *5*, 1386–1398. [CrossRef]
91. Russo, R.C.; Garcia, C.C.; Teixeira, M.M.; Amaral, F.A. The CXCL8/IL-8 chemokine family and its receptors in inflammatory diseases. *Expert Rev. Clin. Immunol.* **2014**, *10*, 593–619. [CrossRef]
92. Galdiero, M.R.; Bianchi, P.; Grizzi, F.; Di Caro, G.; Basso, G.; Ponzetta, A.; Bonavita, E.; Barbagallo, M.; Tartari, S.; Polentarutti, N.; et al. Occurrence and significance of tumor-associated neutrophils in patients with colorectal cancer. *Int. J. Cancer* **2016**, *139*, 446–456. [CrossRef]
93. Berry, R.S.; Xiong, M.-J.; Greenbaum, A.; Mortaji, P.; Nofchissey, R.A.; Schultz, F.; Martinez, C.; Luo, L.; Morris, K.T.; Hanson, J.A. High levels of tumor-associated neutrophils are associated with improved overall survival in patients with stage II colorectal cancer. *PLoS ONE* **2017**, *12*, e0188799. [CrossRef] [PubMed]
94. Arelaki, S.; Arampatzioglou, A.; Kambas, K.; Papagoras, C.; Miltiades, P.; Angelidou, I.; Mitsios, A.; Kotsianidis, I.; Skendros, P.; Sivridis, E.; et al. Gradient Infiltration of Neutrophil Extracellular Traps in Colon Cancer and Evidence for Their Involvement in Tumour Growth. *PLoS ONE* **2016**, *11*, e0154484. [CrossRef] [PubMed]
95. Bie, Y.; Ge, W.; Yang, Z.; Cheng, X.; Zhao, Z.; Li, S.; Wang, W.; Wang, Y.; Zhao, X.; Yin, Z.; et al. The Crucial Role of CXCL8 and Its Receptors in Colorectal Liver Metastasis. *Dis. Markers* **2019**, *2019*, 8023460. [CrossRef] [PubMed]
96. Samuel, T.; Fadlalla, K.; Gales, D.N.; Putcha, B.D.K.; Manne, U. Variable NF-κB pathway responses in colon cancer cells treated with chemotherapeutic drugs. *BMC Cancer* **2014**, *14*, 599. [CrossRef]
97. Ferreira, L.F.V.; Machado, I.F.; Gama, A.; Socoteanu, R.P.; Boscencu, R.; Manda, G.; Calhelha, R.C.; Ferreira, I.C.F.R. Photochemical/Photocytotoxicity Studies of New Tetrapyrrolic Structures as Potential Candidates for Cancer Theranostics. *Curr. Drug Discov. Technol.* **2020**, *17*, 661–669. [CrossRef]
98. Xie, F.; Xiao, P.; Chen, D.; Xu, L.; Zhang, B. miRDeepFinder: A miRNA analysis tool for deep sequencing of plant small RNAs. *Plant Mol. Biol.* **2012**. [CrossRef] [PubMed]

pharmaceutics

Article

Improvement of Pulmonary Photodynamic Therapy: Nebulisation of Curcumin-Loaded Tetraether Liposomes

Jennifer Lehmann [1], Michael R. Agel [1], Konrad H. Engelhardt [1], Shashank R. Pinnapireddy [1], Sabine Agel [2], Lili Duse [1], Eduard Preis [1], Matthias Wojcik [1] and Udo Bakowsky [1,*]

[1] Department of Pharmaceutics and Biopharmaceutics, University of Marburg, Robert-Koch-Str. 4, 35037 Marburg, Germany; jennifer.lehmann@pharmazie.uni-marburg.de (J.L.); michael.agel@pharmazie.uni-marburg.de (M.R.A.); konrad.engelhardt@pharmazie.uni-marburg.de (K.H.E.); shashank.pinnapireddy@pharmazie.uni-marburg.de (S.R.P.); lili.s@hotmail.de (L.D.); eduard.preis@pharmazie.uni-marburg.de (E.P.); matthias_wojcik@web.de (M.W.)
[2] Imaging Unit, Biomedical Research Center (BFS), University of Giessen, Schubertstr. 81, 35392 Giessen, Germany; sabine.agel@bfs.uni-giessen.de
* Correspondence: ubakowsky@aol.com

Abstract: Lung cancer is one of the most common causes for a high number of cancer related mortalities worldwide. Therefore, it is important to improve the therapy by finding new targets and developing convenient therapies. One of these novel non-invasive strategies is the combination of pulmonary delivered tetraether liposomes and photodynamic therapy. In this study, liposomal model formulations containing the photosensitiser curcumin were nebulised via two different technologies, vibrating-mesh nebulisation and air-jet nebulisation, and compared with each other. Particle size and ζ-potential of the liposomes were investigated using dynamic light scattering and laser Doppler anemometry, respectively. Furthermore, atomic force microscopy and transmission electron microscopy were used to determine the morphological characteristics. Using a twin glass impinger, suitable aerodynamic properties were observed, with the fine particle fraction of the aerosols being $\leq 62.7 \pm 1.6\%$. In vitro irradiation experiments on lung carcinoma cells (A549) revealed an excellent cytotoxic response of the nebulised liposomes in which the stabilisation of the lipid bilayer was the determining factor. Internalisation of nebulised curcumin-loaded liposomes was visualised utilising confocal laser scanning microscopy. Based on these results, the pulmonary application of curcumin-loaded tetraether liposomes can be considered as a promising approach for the photodynamic therapy against lung cancer.

Keywords: tetraether liposomes; nebulisation; liposomal stability; photodynamic therapy; curcumin; A549 cells

Citation: Lehmann, J.; Agel, M.R.; Engelhardt, K.H.; Pinnapireddy, S.R.; Agel, S.; Duse, L.; Preis, E.; Wojcik, M.; Bakowsky, U. Improvement of Pulmonary Photodynamic Therapy: Nebulisation of Curcumin-Loaded Tetraether Liposomes. *Pharmaceutics* **2021**, *13*, 1243. https://doi.org/10.3390/pharmaceutics13081243

Academic Editor: Armin Mooranian

Received: 5 July 2021
Accepted: 5 August 2021
Published: 12 August 2021

Publisher's Note: MDPI stays neutral with regard to jurisdictional claims in published maps and institutional affiliations.

1. Introduction

Cancer remains as one of the major challenges in healthcare worldwide and is currently responsible for the majority of global deaths [1]. In recent years, lung cancer was the most common type of cancer and with 18% of the total cancer deaths, it was the leading cause of cancer death in both sexes combined [2]. According to the type of lung cancer and individual risk assessment, standard therapeutic strategies involve chemotherapy, radiotherapy, a combination of both or less frequently, treatment with monoclonal antibodies [3,4]. One of the basic problems in cancer therapy is the high rate of side effects due to undesired tissue destruction and a loss of anatomical and physiological integrity [5]. Therefore, scientists aim to improve the current treatment guidelines as well as to develop treatments with novel mechanisms and targets.

One of these promising mechanisms is the application of light, which itself has a long tradition in the history of medicine, in combination with a photoactive drug collectively termed photodynamic therapy (PDT). The basic principle of PDT is a combination of three otherwise non-toxic components, a photosensitiser, light and oxygen [6,7]. The photosensitiser is a molecule that can be excited by light, mostly due to a conjugated π-electron system. Upon excitation with light of a specific wavelength, the photosensitiser generates reactive oxygen species (ROS) in the presence of oxygen, thereby exhibiting a cytotoxic effect towards surrounding cancer cells [8,9]. The emerged oxygen species has a half-life of only 0.01–0.04 μs and a restricted site of action of 0.01–0.02 μm [8,10]. Due to this, an effect occurs only within a very small range to the irradiation site [11]. Additionally, PDT is able to damage tumour-associated vasculature, resulting in an anaemic infarction of the tumour and can activate the innate immune response against cancer cells. The three mechanisms of PDT can act synergistically and influence each other [12,13]. Other advantages of the PDT are a minor impact towards healthy tissues, its effectiveness against chemo-resistant cancer cell types and being combinable with other therapies such as conventional chemotherapy, microbial adjuvants or cytokines for a T-cell therapy [12,14,15].

There are many different photosensitisers that have been tested in studies, but only very few have reached the stage of advanced human clinical trials or approval for clinical use [14,16]. One such photosensitiser with a huge therapeutic potential is curcumin [17–19]. It is a polyphenol extracted from the rhizomes of turmeric (*Curcuma longa*), naturally occurring in south and southeast Asia [20]. Curcumin has an orange-yellow colour due to its phenylogous π-electron system, and is the most active ingredient of turmeric making approximately 0.5–3% of its weight [14]. Since several decades, it has been an object to pharmaceutical research due to its anti-inflammatory, anti-oxidative, anti-microbial and anti-carcinogenic effects [21–23]. However, curcumin is prone to an extensive first-pass metabolism after oral administration, undergoes a fast metabolic reduction and is subject to biliary excretion [24]. Besides, due to the highly lipophilic properties of this photosensitiser, with a solubility of merely 11 ng/mL in phosphate buffer saline (pH 5), it has a very limited bioavailability [24,25].

These limitations can be overcome by the liposomal encapsulation of curcumin. Within the lipid bilayer of the liposomes, the active ingredient is protected from the physiological degradation on the one hand [9], and the cellular uptake as well as bio-membrane permeability are enhanced on the other hand [9,26]. Altogether, bioavailability, biocompatibility and the circulation time can be improved through liposomal encapsulation, enhancing the overall effectiveness of the photosensitiser [27].

Nebulisers have turned into a trendsetting option in the treatment of a multitude of diseases due to increased patient compliance. For instance with the usage of glycerosomes and hyalurosomes, which are stable towards the process of nebulisation and proved to be good vehicles for a pulmonary application [28,29]. The nebulisation of liposomes can also lead to a considerable improvement in lung cancer disease management.

The major focus of this study was to investigate a non-invasive convenient approach for the treatment of lung cancer by combining nebulised liposomes and PDT. For this purpose, behaviour and stability of three liposomal model formulations were investigated before and after the process of nebulisation and their in vitro efficacy was assessed using A549 lung adenocarcinoma cells. The liposomal formulations had different degrees of membrane stabilisation, amongst others through the usage of bipolar tetraether lipids extracted from the biomass of the archaea *Sulfolobus acidocaldarius*. This plays an important role in membrane integrity and thus the ability to successfully deliver the active ingredient. Curcumin was chosen as it is a commonly used photosensitizer with a well-known impact [14,17,30]. Additionally, the suitability of two different nebulisation techniques, vibrating-mesh and air-jet, was tested and compared for the nebulisation of liposomes in small amounts.

2. Materials and Methods

2.1. Materials

Curcumin (≥80%), cholesterol (≥99%), HEPES (≥99.5%), ethanol (absolute, ≥99.8%), D-(+)-trehalose dehydrate (≥99%) and 3-(4,5-dimethylthiazol-2-yl)-2,5-diphenyltetrazolium bromide (MTT) were obtained from Sigma-Aldrich Chemie GmbH (Taufkirchen, Germany). The lipids 1,2-dipalmitoyl-sn-glycero-3-phosphocholine (DPPC) and 1,2-dioleoyl-sn-glycero-3-phosphoethanolamine (DOPE) were a gift from Lipoid GmbH (Ludwigshafen am Rhein, Germany). Fractions containing the polar tetraether lipids caldarchaeol (GDGT, approx. 10% of the extract) and calditoglycerocaldarchaeol (GDNT, approximately 90% of the extract) were extracted from the biomass of *Sulfolobus acidocaldarius* (Transmit GmbH, Gießen, Germany) [31]. The ends of GDNT contain phosphatidylmyoinositol and β-glucose respectively while GDGT contains phosphatidylmyoinositol and β-D-galactosyl-D-glucose on either ends respectively [32]. Ultrapure water from PURELAB® flex 4 (ELGA LabWater, High Wycombe, UK) was used for all experiments in this study. All solvents used were of analytical or HPLC grade.

2.2. Cell Culture

A549 adenocarcinomic human alveolar basal epithelial cells were obtained from ATCC (American Type Culture Collection, Manassas, VA, USA). The cells were cultured in DMEM (Capricorn Scientific GmbH, Ebsdorfergrund, Germany) supplemented with 10% foetal bovine serum (Capricorn Scientific GmbH). The cells were cultivated at 5% CO_2 and 37 °C under humid conditions. They were grown as monolayers and passaged upon reaching 80% confluency.

2.3. Irradiation Device

For photodynamic activation, a low power prototype LED device consisting of an array of light emitting diodes designed to fit multiwall plates was used. The device was custom manufactured by Lumundus GmbH (Eisenach, Germany), equipped with two different arrays of LEDs capable of emitting light at different wavelengths. Activation took place at 457 nm, the actual fluence used in this study was calculated based on the amperage and irradiation time [26].

2.4. Preparation of Liposomes

Liposomes were prepared via thin-film-hydration method using stock solutions of curcumin in ethanol and lipids dissolved in chloroform:methanol (2:1 (v/v)). The ingredients were mixed in a 10 mL round bottom flask with a curcumin to lipid ratio of 1:10. To gain a thin film, the mixture was evaporated at 40 °C using a Laborota 4000 rotary evaporator (Heidolph Instruments, Schwabach, Germany) fitted with a vacuum pump (KNF Neuberger GmbH, Freiburg im Breisgau, Germany). Afterwards, the film was rehydrated with HEPES (20 mM, pH 7.4) to gain a final concentration of 1 mg/mL. The liposomal formulations were sonicated in a bath sonicator (Elamsonic P30H, Elma Schmidbauer GmbH, Singen, Germany) above the phase transition temperature (T_c) for 30 min.

To homogenise the vesicles, all liposomes were extruded with an Avanti Mini Extruder (Avanti Polar Lipids, INC., Alabaster, AL, USA) using polycarbonate membranes with a pore size of 200 nm. This step was carried out above the T_c. The obtained liposomes were stored at 4 °C.

In total three formulations were compared with each other in this study (Table 1).

Table 1. Lipid compositions of the liposomal formulations. The lipids 1,2-dipalmitoyl-sn-glycero-3-phosphocholine (DPPC), tetraether lipids (TEL), 1,2-dioleoyl-sn-glycero-3-phosphoethanolamine (DOPE) and cholesterol (Chol) were used.

Abbreviations	Formulation	Molar Ratio (%)
DT	DPPC:TEL	90:10
DC	DPPC:Chol	70:30
DD	DPPC:DOPE	75:25

2.5. Lyophilisation

To improve long-term storage stability, the liposomes were lyophilised using an Alpha 1-4 LSC freeze-dryer (Martin Christ, Osterrode am Harz, Germany). In this process, D-(+)-trehalose dihydrate was used to protect the liposomes from collapsing or aggregating during lyophilisation.

Briefly, the liposomal formulations were prepared as stated in Section 2.4 and rehydrated in filtrated HEPES containing 4 mg/mL D-(+)-trehalose dihydrate. A total of 500 µL of the freshly produced liposomes in round bottom flasks were rapidly frozen by shaking in liquid nitrogen to increase the surface and lyophilised with the following parameters: primary drying for 24 h at +15 °C shelf temperature and a partial vacuum of 0.120 mbar, followed by secondary drying for 24 h at +25 °C shelf temperature and 0.100 mbar. The lyophilised residual underwent visual examination. Prior to use and nebulisation with PARI VELOX®, the freeze-dried liposomes were redispersed in ultrapure water.

2.6. Nebulisation, Aerosol Output and Emitted Volume

For the nebulisation of previously produced liposomes, two nebulisers representing two different techniques of nebulisation were used and compared: PARI VELOX® (PARI GmbH, Starnberg, Germany), a metal-based vibrating-mesh nebuliser with a resonance frequency of 160 kHz and PARI BOY® SX utilising compressed air with an operating pressure of approximately 1.6 bar and a nozzle size of 0.48 mm. Briefly, 2 mL of each liposomal formulation were pipetted in the sample reservoir and collected again after nebulisation directly into 5 mL tubes (Sarstedt AG & Co. KG, Nümbrecht, Germany). The nebulisation time and volume of nebulised liposomes were measured for each formulation to calculate the aerosol output rate (L/min) and emitted volume (%). All formulations were nebulised at a total lipid concentration of 1 mg/mL.

2.7. Dynamic Light Scattering and Laser Doppler Anemometry

Particle size (hydrodynamic diameter applying intensity mode) and PdI (polydispersity index) of all liposomal formulations were measured by dynamic light scattering (DLS) using a Zetasizer Nano ZS (Malvern Instruments, Kassel, Germany). All measurements were carried at a wavelength of 633 nm (HeNe laser) and 20.5 °C. For this purpose, a clear disposable folded capillary cell (DTS1060, Malvern Instruments, Kassel, Germany) was utilised. The instrument adjusted attenuation and measurement position of the laser automatically; the detection angle was 173°. Immediately before measurement, 900 µL liposomes were diluted with 100 µL HEPES (20 mM, pH 7.4), thus at the ratio of 1:10. All samples were measured three times with the instrument performing 3 runs with 15 subruns each, all sub-runs lasting 10 s.

Laser Doppler anemometry (LDA) was used to determine the ζ-potential with the Zetasizer Nano ZS. The liposomes were diluted as described above. All samples were measured three times and the instrument performed automatically 15–100 sub-runs. Values from three independent samples were considered.

2.8. Encapsulation Efficiency and Loading Capacity

To extract curcumin from the liposomes, the method of Duse et al. was applied [14,26]. Briefly, 300 µL of each formulation were centrifuged for 90 min at 2000× g (Eppendorf 5418 Centrifuge, Eppendorf AG, Hamburg, Germany) following further centrifugation steps to remove all non-dissolved lipids. The pellet was dissolved in 200 µL ethanol and 200 µL HEPES (20 mM, pH 7.4), 200 µL of the supernatant was mixed with 200 µL ethanol. Subsequently, curcumin in all samples was quantified spectrophotometrically at a wavelength of 425 nm using Multiskan GO plate reader (Thermo Fischer Scientific GmbH, Dreieich, Germany). The required calibration curve was generated with predetermined concentrations of curcumin in the same solvent composition and unloaded liposomes were used as blank. Calculations of the encapsulation efficiency (EE) and loading capacity (LC) were then carried out with the following equations:

$$EE[\%] = \frac{Curcumin_{\,encapsulated}}{Curcumin_{\,total}} \times 100 \tag{1}$$

$$LC[\%] = \frac{Curcumin_{\,encapsulated}}{Ingredients_{\,total}} \times 100 \tag{2}$$

2.9. Atomic Force Microscopy

The morphological characteristics of the liposomal formulations were studied with atomic force microscopy (AFM) using a NanoWizard® 3 (JPK Instruments AG, Berlin, Germany) with silicon cantilevers (HQ:NSC14/AL BS, Mikromasch Europe, Wetzlar, Germany). The measurements were carried out using intermittent contact mode at scan rates between 0.5 to 1 Hz to avoid damaging the lipid membrane [33]. A resonance frequency of 120 kHz and the force constant 5 N/m was used. Different formulations of nebulised liposomes were diluted 1:10 with ultrapure water, 10 μL of the mixture was placed onto glass slides and left to dry at room temperature. To gain the final images, the amplitude signal of the cantilever in the trace direction and the height signal in retrace direction were used and the images were processed using JPKSPM data processing software (JPK Instruments AG).

In addition, AFM images were analysed for size confirmation. For this purpose, the average liposomal size was determined by analysing a representative number of height images using ImageJ software (v1.52a, National Institutes of Health, Bethesda, MD, USA) [33,34].

2.10. Transmission Electron Microscopy

As a second method to examine the liposomal morphology, transmission electron microscopy (TEM) was used. The sample preparation was carried out on 400 mesh copper grids, coated with 1.2% Formvar and carbon (Plano GmbH, Wetzlar, Germany). Briefly, the liposomes were diluted to 1:10 ratio with HEPES (20 mM, pH 7.4) and 10 μL were placed on a grid. After 5 min of incubation, the liquid was withdrawn by suction with a Whatman 1 filter paper (Whatman plc, Maidstone, UK). Then the grids were placed for 5 min on 20 μL 2% uranyl acetate, which was used as contrast agent for the negative staining. The liquid was withdrawn by suction again and the grids were placed on 20 μL H₂O as a washing step. After the remaining liquid was carefully removed, the grids were left to dry. To obtain the images, samples were examined using a Leo 912 AB TEM (Carl Zeiss Microscopy GmbH, Jena, Germany) with different magnifications and an accelerating voltage of 100 kV.

2.11. Aerodynamic Properties

To define the aerodynamic properties, a determination of the fine particle fraction (FPF) was carried out according to monograph 2.9.18 with device A (twin glass impinger) mentioned in the European Pharmacopoeia 9.5. Briefly, 7 mL and 30 mL of water:acetonitrile (1:1 (v/v)) were added to the top and bottom chamber, respectively. The device was assembled and connected to an Erweka vacuum pump VP 1000 (Erweka GmbH, Heusenstamm, Germany). Using a flow meter DFM 2000 (Copley Scientific, Nottingham, UK), the flow rate was adjusted to 60 ± 5 L/min. The liposomal formulations were pipetted into the sample reservoir of a PARI VELOX® nebuliser, which was linked to the glass twin impinger (Hannes & König GmbH, Heusenstamm, Germany) via mouthpiece. The vacuum pump was turned on 10 s before starting the nebulisation. After 60 s, the nebulisation process was stopped, and the vacuum pump was turned off 5 s afterwards. The amount of curcumin in each chamber was quantified spectrophotometrically. FPF was then calculated according to the following equation:

$$\text{FPF}[\%] = \frac{\text{Curcumin}_{\text{bottom chamber}}}{\text{Curcumin}_{\text{total}}} \times 100 \tag{3}$$

2.12. Mucous Membrane Compatibility

The determination of mucosal tolerance for all liposomal formulations was performed by HET-CAM assay [35]. The chorio-allantoic membrane (CAM) is an extraembryonic membrane that results from the fusion of the shell membrane (chorion) with the allantoic cavity (rectal protrusion of the embryo) in a fertilized chicken egg [36]. For preparation, the fertilized chicken eggs had to be thoroughly cleaned with ethanol 70% (*v/v*) after delivery and incubated at 37.8 °C and 60% relative humidity in a Thermo de Luxe 250 incubator (Hemel Brutgeräte GmbH). Eggs were continuously turned 12 times per day until egg development day (EDD) 3. On EDD 3, a round incision (ø 28 mm) was added to the blunt side of the eggs with the aid of an EggPunch pneumatic egg opener (Schuett-Biotec GmbH). An opening was then created by carefully removing the shell as well as the shell membrane with curved Dumostar Style 7 forceps (Manufactures D'Outils Dumont SA). For protection, the opened eggs subsequently had to be covered with a sterile Petri dish and incubated in a vertical position in the incubator.

On EDD 9300 µL of nebulised liposomes were applied to the CAM and observed during 5 min under a Stemi 2000-C stereomicroscope (Carl Zeiss AG) at 13-fold magnification. The study was performed with 6 eggs per sample. HEPES (20 mM, pH 7.4) served as a negative control, sodium dodecyl sulfate (SDS) 1% (m/m) and 0.1 N sodium hydroxide were used as positive control. The determination was based on stereomicroscopic images taken with a Moticam 5 MP digital camera (Motic Deutschland GmbH) and an irritation score (*IS*), which included an assessment of haemolysis (*H*), vessel lysis (*L*), and blood coagulation (*C*) [37]. Events were photographed at the time of their occurrence and the irritation score was calculated as follows according to the literature [38]:

$$IS = \frac{(301 - H[s])}{300} \times 5 + \frac{(301 - L[s])}{300} \times 7 + \frac{(301 - C[s])}{300} \times 9 \tag{4}$$

2.13. In Vitro Cytotoxicity

A549 cells were seeded onto 96-well plates (NUNC, Thermo Fischer Scientific GmbH, Dreieich, Germany) with a seeding density of 1×10^4 cells/0.35 cm^2 (per well) and the plates were incubated for 24 h. For the in vitro cytotoxicity assay, the cells were treated with various concentrations of all liposomal formulations using a serial dilution starting with 100 µM of curcumin. After 4 h of incubation, liposomal formulations were aspirated and replaced with fresh medium. The cells were irradiated at 457 nm using an LED device (Lumundus GmbH) with a fluence of 6.61 J cm^{-2}. After 24 h, the cells were incubated with MTT reagent (0.2 mg/mL) for 4 h and DMSO was then used to dissolve the resulting formazan crystals. The absorbance was determined at a wavelength of 570 nm using FLUOstar® Optima (BMG Labtech, Ortenberg, Germany). Untreated cells (blank) and an unirradiated microtiter plate (dark plate) were used as controls. The viability of blank cells was considered as 100%.

2.14. Cellular Uptake Studies

Liposomal formulations can be absorbed into the cells in different ways and thus develop their effect. For a better understanding of these effects, the uptake pathways were determined. This was done by inhibiting two major pathways respectively, determining the cell survival rate and comparing it to that of untreated cells. Chlorpromazine served as an inhibitor of clathrin-mediated endocytosis, and filipin III inhibited caveolae-mediated endocytosis. As previously described, A549 cells were seeded at a density of 1×10^4 cells per well in 96-well plates and incubated for 24 h. Subsequently, these had to be washed twice with PBS buffer containing Ca^{2+}/Mg^{2+} to ensure removal of serum residues that could interfere with the effect of the inhibitors. Then, preincubation of chlorpromazine and filipin III (5 µg/mL each) for 30 min took place, followed by incubation with liposomes (100 µM each) for 4 h. Afterwards, the liposomes were replaced with fresh medium and irradiation of the cells took place at $\lambda = 457$ nm and with a fluence of 6.61 J/cm². After further incubation of the cells for 24 h, a determination of cell viability was performed by in vitro cytotoxicity assay as described in Section 2.13. Untreated and treated but unirradiated cells as well as cells treated with inhibitor only were used as controls.

2.15. Confocal Laser Scanning Microscopy (CLSM)

A549 cells were seeded onto 15 mm coverslips inside 12-well plates (NUNC, Thermo Fischer Scientific GmbH, Dreieich, Germany) with a seeding density of 9×10^4 cells/3.5 cm² (per well). The cells were treated with liposomal formulations containing 100 µM of curcumin and incubated for 4 h. After irradiation (6.61 J cm⁻² at 457 nm), the cells were washed twice with PBS buffer containing Ca^{2+} and Mg^{2+} (pH 7.4) and fixed with 4% formaldehyde solution. Afterwards, the cells were washed twice with PBS buffer and the cell nuclei were counterstained with 0.1 µg/mL 4′,6-diamidino-2-phenylindole (DAPI). The cells were then washed with PBS buffer and the coverslips were mounted onto slides and sealed using FluorSave™ (Calbiochem Corp., La Jolla, CA, USA). The cells were examined using an LSM700 confocal laser-scanning microscope (Carl Zeiss Microscopy GmbH, Jena, Germany). All micrographs were recorded at a similar adjustment.

2.16. Cellular Migration

The so called "in vitro scratch assay" was performed to determine the growth inhibitory effect of PDT with nebulised liposomes on A549 cells. For this assay, 9×10^4 cells per well were also seeded in 12-well plates and cell growth in the wells was observed until a confluent monolayer was formed (approximately after 48 h). After 4 h of incubation with nebulised liposomes of 100 µM concentration and a subsequent exchange with fresh DMEM medium, it was possible to scratch through the wells with a pipette tip and thus create a gap in the cell monolayer.

2.17. Statistical Analysis

All experiments were carried out in triplicates and the values are presented as mean ± standard deviation unless otherwise stated. The statistical significance of measured values was determined by using a two-tailed t-test and the probability values of $p < 0.05$ were considered statistically significant.

3. Results and Discussion

3.1. Suitability of Nebulisers

In this study, two nebulisers of different technology were compared to identify the device best suited for the nebulisation of liposomes. PARI VELOX® represents vibrating-mesh nebulisation and PARI BOY® SX air-jet nebulisation. Vibrating-mesh nebulisers extrude liquid through a perforated mesh to generate aerosols, whereas air-jet nebulisers use compressed air to press the liquid through a nozzle of defined size (Figure 1a,b) [39,40].

Generally, PARI VELOX® showed a slightly slower output rate compared to PARI BOY® SX, but a much higher emitted dose. On the vibrating-mesh nebuliser, ultrapure water had a low output rate of 0.27 ± 0.04 mL/min due to a lower concentration of electrolytes [41]. However, the liposomal formulations rehydrated in 20 mM HEPES (pH 7.4) showed output rates of 0.48 ± 0.07 mL/min (DT), 0.50 ± 0.01 mL/min (DC) and 0.53 ± 0.03 mL/min (DD). According to the literature, the average output rate of PARI VELOX® for normal saline (0.9% NaCl) is 0.76 ± 0.18 mL/min [42]. Air-jet nebulisation showed an output rate of 0.29 ± 0.03 mL/min for ultrapure water, 0.65 ± 0.05 mL/min for DT, 0.63 ± 0.03 mL/min for DC and 0.69 ± 0.01 mL/min for DD. The emitted dose of vibrating-mesh nebulisation was $93.5 \pm 3.3\%$ (v/v) for ultrapure water, $94.0 \pm 1.5\%$ (v/v) for

DT, 95.5 ± 1.3% (*v/v*) for DC and 94.7 ± 1.0% (*v/v*) for DD. Air-jet nebulisation revealed an emitted dose of 18.5 ± 2.3% (*v/v*) for ultrapure water and 15.3 ± 0.8% (*v/v*) (DT), 17.3 ± 1.3% (*v/v*) (DC), 14.8 ± 1.5% (*v/v*) (DD) for the liposomal formulations. Looking at the liposomal formulations, a slight difference is visible. For both nebulisers, DC had the highest emitted dose. However, the results of all formulations regarding output rate and emitted dose were in the same range for both nebulisers, which was probably because all liposomes were rehydrated with HEPES (20 mM, pH 7.4). The output rate of ultrapure water was much lower and, according to literature, the one of 0.9% NaCl much higher. Thus, an increased conductivity leads to a higher output rate. This confirms previous findings, that the concentration of electrolytes is a significant parameter influencing output rate due to a more efficient liquid breakdown during atomisation [43]. More fluidic liposomes also seem to promote a higher output rate, which can be seen in the case of DD with both nebulisers. It could be explained by smaller and easier separable liposomes which may decrease the liquid resistance.

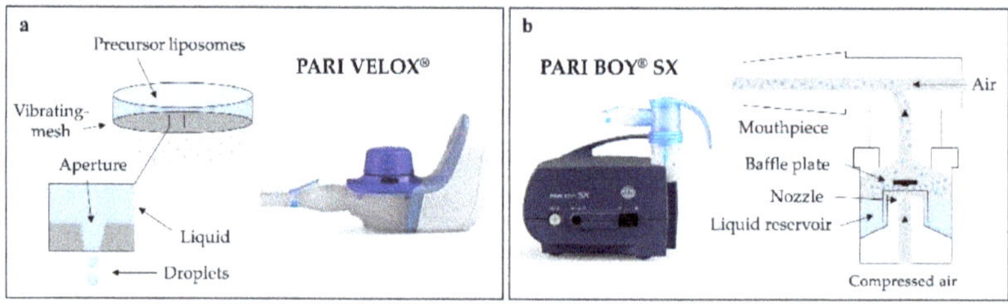

Figure 1. Operating mode of both techniques: (**a**) vibrating-mesh nebulization, (**b**) air-jet nebulisation.

In the recent literature the two nebulisation techniques are described to function with nanoscaled drug carrier systems [44,45] and are suitable for therapies with small volumes since both require a minimum sample volume of only 2 mL. They provide comparable lung deposition profiles and show reportedly no damage of precursor liposomes [39,45]. A comparison of the measured size values for both nebulisation techniques of all formulations (Table 2) revealed no significant difference (*p* > 0.05) between PARI VELOX® and PARI BOY® SX. PdI and ζ-potential likewise showed no major differences for both nebulisation techniques among all formulations, leading to the assumption that both nebulisers are capable of emitting intact liposomes, suitable for deep lung deposition. However, vibrating-mesh nebulisation seemed more suitable than air-jet nebulisation due to a very high emitted dose and a much better handling with similar output rates. Additionally, PARI VELOX® combines other advantages such as being portable, quiet and cost-effective, making the therapy patient-friendly as well. It was therefore used for all further experiments in this study.

3.2. Physicochemical Properties of the Liposomal Formulations

One focus in the course of drug nebulisation was the application of liposomes [39,46]. State of the art in pulmonary drug delivery for the treatment of lung cancer is the nebulisation of different nanocarriers such as nanoparticles, nanoemulsions and liposomes [47–50]. To achieve a more localised drug delivery, different modifications and targets were tested so far [47,51]. The liposomes tested for the treatment of lung cancer in this study have a great advantage, as the used lipids are biocompatible, biodegradable and present in clinical trials or even approved for clinical use [52]. In the field of pulmonary delivery, several liposomal formulations have been tested in animal studies and clinical trials as well [53].

Table 2. DLS analysis of DT (DPPC:TEL), DC (DPPC:Chol) and DD (DPPC:DOPE). Curcumin-loaded liposomes nebulised via vibrating-mesh (PARI VELOX®) and air-jet nebulisation (PARI BOY® SX) were compared with each other, respectively. Unnebulised curcumin-loaded liposomes were used as control. Values of particle size represent the distribution by intensity. Independent formulations were used to measure the triplicates and all results were expressed as mean ± standard deviation.

	Lipid Compositions	Size (nm) ± SD	PdI [1] ± SD	ζ-Potential (mV) ± SD
	DT	129.7 ± 3.2	0.19 ± 0.08	−13.53 ± 2.42
CUR-liposomes	DC	109.4 ± 2.4	0.14 ± 0.09	−5.61 ± 0.79
	DD	94.4 ± 5.9	0.26 ± 0.11	−2.97 ± 0.81
Nebulised	DT	131.1 ± 3.0	0.21 ± 0.06	−16.33 ± 1.95
(PARI VELOX®)	DC	116.4 ± 3.2	0.18 ± 0.07	−3.74 ± 0.64
CUR-liposomes	DD	99.1 ± 5.2	0.30 ± 0.09	−2.86 ± 0.59
Nebulised	DT	129.5 ± 2.8	0.20 ± 0.08	−15.54 ± 2.13
(PARI BOY® SX)	DC	113.3 ± 3.5	0.21 ± 0.09	−5.13 ± 0.92
CUR-liposomes	DD	100.7 ± 4.7	0.32 ± 0.08	−3.11 ± 0.76
Lyophilised &	DT	149.6 ± 4.1	0.46 ± 0.10	−37.89 ± 4.43
nebulised	DC	127.7 ± 4.8	0.47 ± 0.13	−11.47 ± 2.38
(VELOX®) CUR-liposomes	DD	132.1 ± 5.1	0.57 ± 0.13	−9.62 ± 2.09

[1] Polydispersity index.

As there are many suitable lipids for the preparation of liposomes, the present study aimed to compare different representative liposomal compositions. DT (DPPC:TEL 90:10) is the composition with the highest degree of membrane stabilisation as the tetraether lipids are crossing the entire lipid bilayer. This is due to the ring structure of these lipids, originated from four ether bonds (Figure 2b), implementing a tight packing of the lipid bilayer and a rigid behaviour of these liposomes. In addition, TEL reduces liposomal sensitivity towards oxidation [26]. DC (DPPC:Chol 70:30) is a standard medium rigid composition using cholesterol for membrane stabilisation, which is incorporated in the lipid bilayer. The composition DD (DPPC:DOPE 75:25) is more fluidic than the others, with no stabilisation of the membrane (Figure 2a). Upon rehydration of the lipid films, curcumin is packed inside the liposomal bilayer of all formulations having hydrogen bonds to the lipid acyl chains. This encapsulation protects curcumin from degradation and greatly increases its solubility in aqueous media [26,27].

Particle size, PdI and ζ-potential of unnebulised and nebulised liposomal formulations were compared (Table 2), as stable delivery systems are key requirements for a successful application. Looking closer at the results, a few differences between the formulations were noticed. Prior to nebulisation, DT liposomes were slightly larger in diameter than the other formulations due to TEL crossing the lipid bilayer. Particle size and PdI were in the same range before and after nebulisation for this formulation, making it inured to nebulisation. This may be explained by increased packing and stronger hydrogen bonds within the lipid bilayer, reducing the effects of nebulisation on these liposomes [54]. DT also had a more negative ζ-potential compared to the other formulations, which can be explained by the ether bonds of TEL. DC liposomes were smaller than DT, probably due to the smaller cholesterol molecule. Likewise, particle size and PdI were in the same range after nebulisation, revealing them stable to nebulisation as well. Cholesterol can decrease the bilayer fluidity by reducing the movement of the lipid hydrocarbon chains and increase the packing of lipid head groups, which can lead to a smaller particle size [40]. In contrast, DD was a more fluidic formulation due to DOPE, which can destabilise the membrane as it may stimulate a transition from a lamellar to a hexagonal phase [55,56]. This fluidic behaviour could cause the increase in size and PdI after nebulisation, which can be seen in Table 2. Therefore, DD seemed less stable to forces during the process of nebulisation. Liposomal compositions and the total lipid concentration can influence the nebulisation, but conversely, the different nebulisation techniques can have an impact on size, uniformity and integrity of liposomes [57]. It can be seen in this study, that nebulisation with compressed air leads to a lower uniformity of all liposomal formulations. This could be explained by higher shearing forces while pushing the liquid through the nozzle. Both techniques also have a slight impact on liposome sizes and ζ-potentials. The negative ζ-potential of all formulations could be related to the entrapment of curcumin, the different extent seems depending on the liposomal formulation [26].

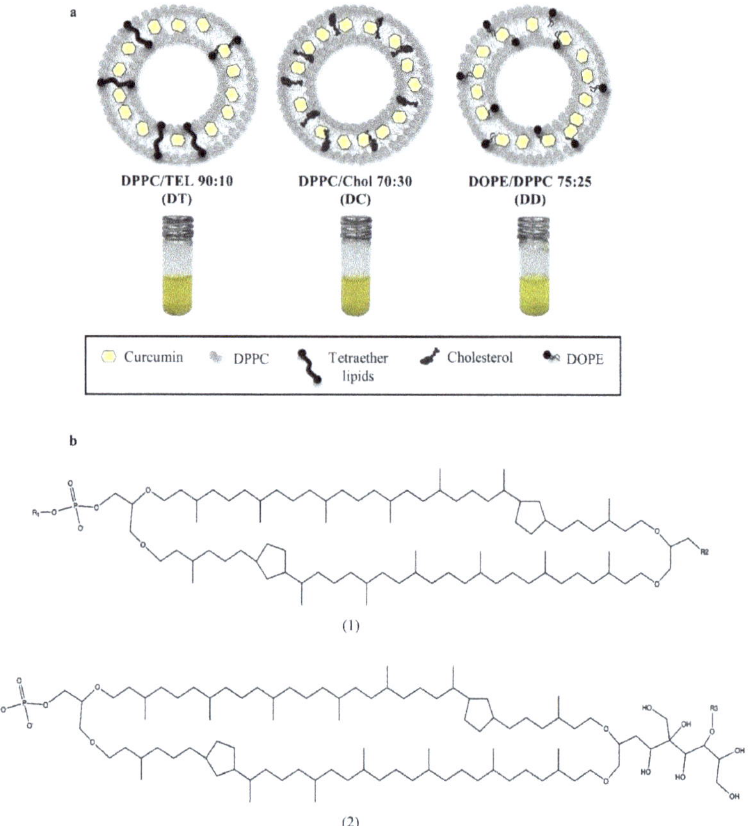

Figure 2. (**a**) Model compositions of liposomes: DT with tetraether lipids providing the highest degree of membrane stabilisation, DC as medium stabilised standard composition with cholesterol and the fluidic composition of DD with no stabilisation of the membrane. (**b**) Structures of the tetraether lipids used in this study. (**1**) GDGT, glycerol dialkylglycerol tetraether and (**2**) GDNT, glycerol dialkylnonitol tetraether with R1: inositol, R2: β-D-glucopyranose and R3: β-D-galactosyl-β-D-glucopyranose.

To increase the storage stability, the three liposomal formulations were lyophilised directly after preparation, rehydrated with ultrapure water and nebulised using PARI VELOX®. It was observed that the ζ-potential of these formulations was more negative, which could be explained by a leakage and attachment of curcumin to the outer surface of the liposomes. This observation is confirmed by results of the encapsulation efficiency (Table 3). A change in the ζ-potential can affect the mucoadhesive properties of the liposomes as they are based on the interaction between charged molecules [58]. Size and PdI of all formulations were also increased. Source of this increase could be the cryoprotectant D-(+)-trehalose dehydrate itself, due to its penetration into the liposomes as it leads to a stabilisation of liposomes from inside and outside [59]. According to literature, lyophilising without cryoprotectant leads usually to intense aggregation or fusion [59,60]. Possible causes for the good protective properties of D-(+)-trehalose dehydrate can be the high glass transition temperature and the low tendency to crystallisation [59]. However, lyophilisation prior to nebulisation seemed to have a negative influence on all three formulations.

Table 3. Encapsulation efficiency (EE) and loading capacity (LC) of DT (DPPC:TEL), DC (DPPC:Chol) and DD (DPPC:DOPE) liposomes. The results in this table were calculated according to Equations (1) and (2) in Section 2.8 for liposomes containing a curcumin concentration of 0.1 mg per 1 mg lipids. Unnebulised curcumin-loaded liposomes were used as control.

	Lipid Compositions	EE (%) ± SD	LC (%) ± SD
	DT	93.9 ± 8.2	1.9 ± 0.1
CUR-liposomes	DC	88.4 ± 12.4	1.7 ± 0.1
	DD	85.1 ± 15.9	1.7 ± 0.1
Nebulised	DT	80.0 ± 12.8	1.6 ± 0.1
(PARI VELOX®)	DC	62.5 ± 12.3	1.3 ± 0.1
CUR-liposomes	DD	57.5 ± 16.6	1.2 ± 0.1
Lyophilised and	DT	74.6 ± 10.9	1.5 ± 0.1
nebulised (VELOX®)	DC	60.3 ± 9.5	1.2 ± 0.1
CUR-liposomes	DD	31.2 ± 11.7	0.6 ± 0.1

Altogether, the membrane stabilised formulations DT and DC appeared stable to nebulisation whereas DD as fluidic formulation did not seem eligible.

Generally, liposomes are very convenient as vehicles for all kinds of drugs due to their relatively high encapsulation efficiency [9,26,46]. The lipophilic drug curcumin was used in this study with 100 µg curcumin per 1 mL liposomes. Typically, curcumin is non-uniformly distributed in the lipid bilayer and entirely located inside the hydrophobic interior, which is important for a high drug loading capacity [27]. Previous research indicated that EE and LC are dependent on the composition of lipid acyl chains forming the liposomal bilayer. Lipids containing shorter acyl chains and hence, a lower Tc, are mostly able to incorporate a larger amount of hydrophobic substances such as curcumin due to altered van der Waals linkages [60]. Additionally, unilamellar vesicles, as they emerge after extrusion through polycarbonate membranes, are considered to have better drug incorporation into the bilayer than multilamellar vesicles [46]. These findings stand in accordance with the results of this study, as Table 3 showed relatively high encapsulation efficiencies for all un-nebulised liposomes. Looking closer at the three formulations, a difference in behaviour was visible. Upon addition of TEL, a high encapsulation efficiency to the extent of 93.9% was achieved. This could be due to the fact that TEL, by crossing the lipid bilayer with its acyl chains containing cyclopentane structures, provided a tight membrane packing through more van der Waals linkages. Resultant in a very good capability of encapsulating hydrophobic substances [61,62]. Both, EE and LC seemed quite stable after the process of nebulisation, which speaks again for the high degree of membrane stability as it makes the liposomes less permeable by preserving membrane integrity. Stabilisation through the bipolar structure of TEL also provides thermal stability [63] and better long-term storage stability rates [64]. Besides, TEL can positively affect the intermembrane exchange of the active ingredient and thereby enhance the delivery [64]. The linkages of Cholesterol inside the lipid bilayer of DC similarly promote the encapsulation of curcumin [60], but with 88.4% (Table 3) to a lower extent. DD, which contains the fluidic lipid DOPE, showed the lowest encapsulation efficiency of 85.1%. Compared to the un-nebulised liposomes, EE values of nebulised curcumin-loaded liposomes were decreased. The slight effect of DT can be explained by leftover curcumin attached to the outer surface of the liposomes, remaining even after extrusion [46]. The residual amount of encapsulated curcumin in DC was also relatively high, as the moderate stabilisation with Cholesterol also leads to decreased leakage rates, whereas for DD a much stronger effect was visible probably due to a disintegration of the fluidic liposomes followed by a loss of curcumin. These findings stand in accordance with the ones of DLS analysis as the changes in size, PdI and ζ-potential were correspondent to the different behaviour of the formulations. This is also the case for lyophilised and nebulised curcumin-liposomes, which had the lowest EE and LC. Again, the cryoprotectant D-(+)-trehalose dehydrate could be the cause, as it can replace curcumin inside the liposomal bilayer [58].

With respect to EE and LC of liposomes, another aspect that must be taken under consideration is the state of entrapped curcumin. Hydrophilic drugs encapsulated in the aqueous core have a high tendency to aggregate which can compromise their therapeutic impact, whereas lipophilic drugs such as curcumin have a much lower tendency to aggregate within the bilayer, probably also due to the presence of van der Waals linkages with the lipid acyl chains building a barrier [60].

Again, the membrane stabilisation of DT and DC seemed to be an advantage. The EE and LC of both formulations were high, and the values of DT were quite stable after nebulisation. However, DD, with no membrane stabilisation, exhibited an unstable behaviour since the EE decreased constantly during lyophilisation and nebulisation. The lack of membrane stabilisation revealed a higher loss of entrapped curcumin making DD not suitable for nebulisation after all.

3.3. AFM Visualisation

AFM images revealed spherically shaped vesicles for all formulations after nebulisation (Figure 3). Smooth surfaces of the vesicles indicated complete incorporation of curcumin. Images of DD (Figure 3e,f), the most fluidic formulation, exhibited several small spherically shaped vesicles, presumably due to a disintegration through the vibrating mesh during the process of nebulisation. This disintegration could be explained by the fluidic behaviour and positive charge, that both lipids have in this formulation. Conductivity greatly affects the process of nebulisation [65], in this case the charge probably led to an interaction with the surface of the metal-based mesh promoting further disintegration. Another explanation can be the high Tc, originated by long hydrocarbon chains. Below this temperature, liposomes act more like solids than as fluids, increasing the probability of decomposition [60]. This stands in accordance with the results of EE and LC, as DD revealed to be not suitable for nebulisation. In contrast, the highly stabilised formulation DT (Figure 3a,b) and the medium stabilised formulation DC (Figure 3c,d) showed constant behaviour in terms of their size. Although the effectiveness of inhalable therapies depends on the aerodynamic properties of the nebulised vehicles, the output of intact liposomes from the nebuliser, which was achieved in this study, is a fundamental requirement for the applicability of the therapy [66]. Liposomes of all formulations appeared slightly bigger in the AFM images. On the one hand, probably due to spreading on the glass slides while they were left to dry as a part of the sample preparation, on the other hand, due to different measuring conditions compared to DLS. DLS results were obtained in aqueous conditions whereas AFM measurements took place under dry conditions. However, size analysis using the height images (Figure 3) stood in agreement with the results obtained with DLS.

3.4. TEM Visualisation

Transmission electron microscopy enabled a precise examination of morphological differences between the liposomal formulations after nebulisation with PARI VELOX®. Through the method of negative staining with uranyl acetate it was possible to image the liposomes in their original environment, which was HEPES (20 mM, pH 7.4) in this study. Likewise seen in the particular AFM images, spherically shaped vesicles were visible for the three formulations. The most rigid formulation DT appeared as well-defined vesicles, which were slightly dented (Figure 4a,b). This was in accordance with AFM measurements. The liposomal diameter determined from TEM images corresponded to the ones obtained via DLS and AFM. These findings indicated the stability of tetraether liposomes during the process of nebulisation. The appearance of the membrane stabilised DC also matched the one in AFM images, as the vesicles were equal round and with nearly uniform sizes

(Figure 4d). Additionally, TEM visualisation revealed the lipid bilayer of DC liposomes in the magnified image (Figure 4c) and hence their intactness. Likewise, the vesicle diameter determined from TEM images corresponded to previous ones indicating DC's stability towards nebulisation. DD, as the most fluidic formulation in this study, showed different behaviour. A few vesicles were identified as intact liposomes due to the visible lipid bilayer (Figure 4e), but the majority was found to be spread on the grid (Figure 4f). For this reason, the diameter determined from TEM images appeared larger compared to the ones obtained by DLS, excluded the intact vesicles. Looking closer at Figure 4f, tiny black spots were visible, matching the results of CLSM for DD. A possible explanation for these spots could be the agglomeration of curcumin after rupture and leakage of the fluidic liposomes during nebulisation. These results indicated the infeasibility of the fluidic formulation DD for nebulisation and were in accordance with the results of AFM, CLSM, DLS and measurements of the encapsulation efficiency.

3.5. Aerodynamic Properties

Aerosols are the products generated from drug solutions or dispersions during the process of nebulisation and their aerodynamic properties determine the location and intensity of the effect [57]. Measurements of the FPF according to monograph 2.9.18 of the 10th Edition of the European Pharmacopoeia were used to specify the aerodynamic characteristics after nebulisation with PARI VELOX®.

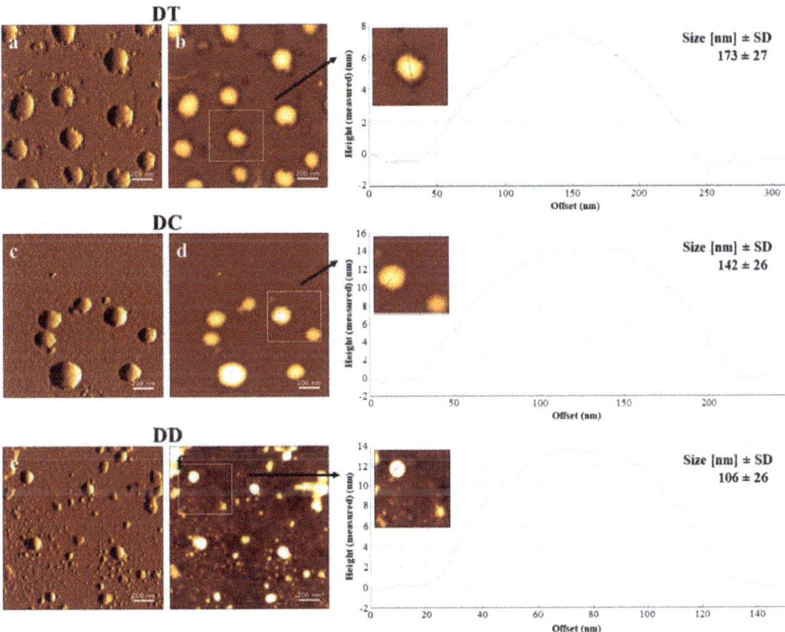

Figure 3. Atomic force microscopy (AFM) visualisation of DT, DC and DD after nebulisation with PARI VELOX®, containing a curcumin concentration of 100 µM. Images (**a,c,e**) are amplitude signals and (**b,d,f**) are height signals. Liposomes in height images together with their cross-sectional profiles were used for a size evaluation along the shown lines and a morphological characterisation. The stated sizes are mean values ± standard deviation gathered by analysing representative height images using ImageJ software. Scale bars represent 200 nm.

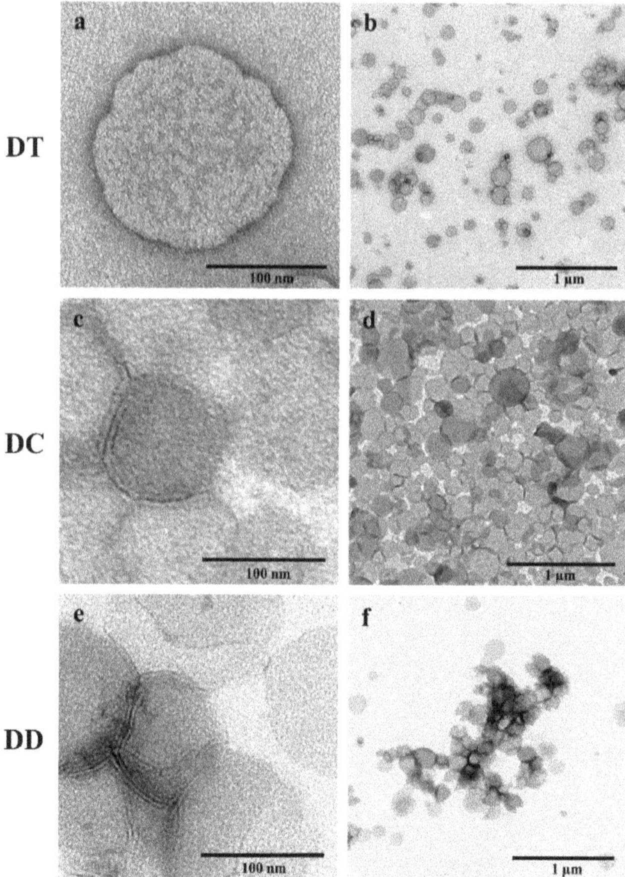

Figure 4. Transmission electron microscope (TEM) images of the liposomal formulations DT, DC and DD after nebulisation, containing a curcumin concentration of 100 µM. The samples were negatively stained with 2% uranyl acetate. Scale bars represent 100 nm for the images (**a,c,e**) and 1 µm for the images (**b,d,f**).

Suitable droplet size distribution, velocity and trajectory of an aerosol optimises the desired pulmonary deposition. Hence, it is important to know which parameters have an impact on these properties and which have none. Higher conductivity of the precursor fluid leads to a more efficient liquid breakdown [65]. Therefore, 20 mM HEPES pH 7.4 seems appropriate for the rehydration of lipid films as it facilitates a good output rate during nebulisation. However, the precise differences between the HEPES buffered liposomal formulations and ultrapure water regarding the output rate are shown in Section 3.1. Surface tension and reservoir volume do not have a relevant impact on aerodynamic properties, as described in the literature [65]. In addition, the vibrating mesh is moving too fast to give surface tension a chance to improve the liquid breakdown [65]. Since nanocarriers are mostly used to encapsulate small amounts of highly potent active ingredients, they are produced in small volumes. To ensure a successful therapy, it has to be certain that these small volumes get nebulised in a reliable manner, which again speaks for the suitability of vibrating-mesh nebulisers in this context, as the hydrostatic pressure on the nebuliser mesh does not affect the aerosol diameter [65]. Previous studies revealed that aerosol droplet sizes have to be between 1 and 6 µm for an optimal effect. Droplets larger than 6 µm may deposit in the mouth or throat, droplets smaller than 1 μm can be exhaled, both cases lead to ineffectiveness [67,68]. According to literature, with PARI VELOX® up to 75% of the produced aerosol droplets are smaller than 6 μm and the average volumetric-median-diameter of the aerosol droplets is 3.6 to 4.4 μm [42]. The FPF measured in this study was defined by monograph 2.9.18 of the European Pharmacopoeia as droplets smaller than 6.4 μm and was stated as percentage ± standard deviation of active ingredient from the total nebulised amount of the latter. The liposomal formulation DC showed the best result with an FPF of 62.7 ± 1.6%, DT showed 59.5 ± 2.4% and DD 58.2 ± 1.4%. For the previously lyophilised liposomes, the FPF was 57.3 ± 3.8% for DT, 55.9 ± 6.2% for DC and 58.0 ± 5.4% for DD.

Overall, the FPF of all three nebulised formulations was relatively high and in the same range. This indicates that the composition and stability of liposomes does not affect the liquid breakdown and that 20 mM HEPES pH 7.4 provides a sufficient conductivity for nebulisation. Among the previously lyophilised liposomes, the stabilisation of DT seemed to be a slight advantage.

3.6. Mucous Membrane Compatibility

The mucous membrane compatibility of the nebulised curcumin-loaded liposomes was tested on the CAM and resulted in an irritation index of 0 for all three formulations. This means that none of the liposomal formulations caused a negative reaction, consequently mucosal intolerance. Therefore, the images were taken at the beginning as well as at the endpoint of the test (5 min) and show no changes (Figure 5). The positive control shows the occurrence of bleeding (Figure 5j).

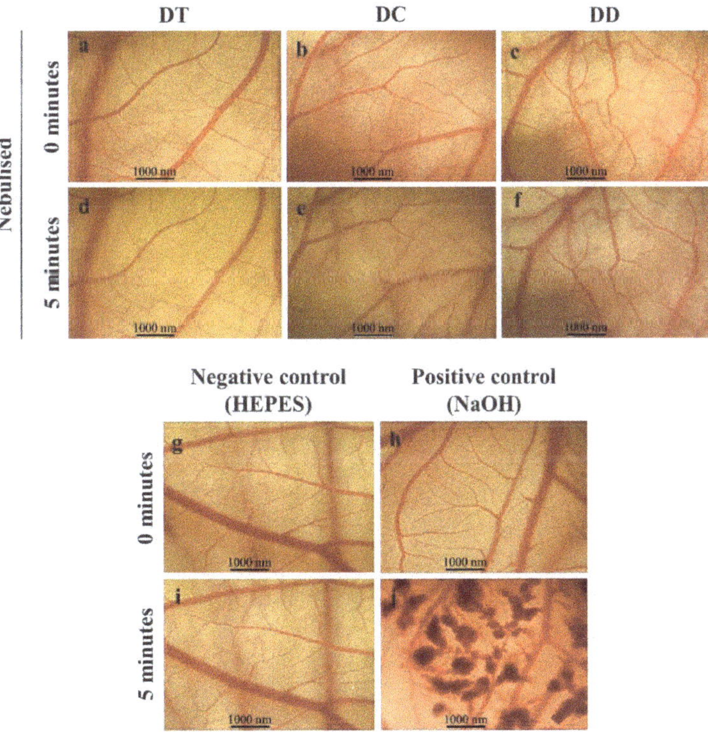

Figure 5. HET-CAM assay for mucous membrane compatibility of nebulised curcumin-loaded liposomes (**a–f**). HEPES 20 mM pH 7.4 served as negative control (**g,i**); 0.1 N NaOH as positive control (**h,j**). Scale bars represent 1000 nm.

3.7. In Vitro Cytotoxicity, Cellular Uptake and Migration

To determine the photodynamic effect of the nebulised curcumin-liposomes, MTT cytotoxicity studies were made. They revealed low dark toxicities of 88.32 ± 4.14% cell viability for DT, 85.03 ± 3.30% for DC and 85.73 ± 1.26% for DD, which was within the acceptable range for all formulations [14,69]. The cell viabilities after treatment with irradiated nebulised liposomes were 10.99 ± 3.04% for DT, 12.76 ± 0.58% for DC and 24.90 ± 2.70% for DD (Figure 6a). All formulations showed a significant difference ($p < 0.05$) between dark control and irradiated nebulised samples, as well as between dark control and irradiated un-nebulised samples, respectively. This indicates an excellent cytotoxic efficacy and an improved nebulisation suitability of liposomal formulations containing tetraether lipids due to the strong adhesion forces within their lipid bilayer [62]. Besides, the ability of TEL to positively affect the intermembrane exchange of the active ingredient, thus enhancing the delivery [64], was also confirmed by these results. The difference between dark control and irradiated lyophilised and nebulised curcumin-liposomes was also significant ($p < 0.05$) for DT and DC, but not for DD. A decreasing cytotoxic effect from DT to DD was visible. Again, these results validate the improved stability of DT and DC liposomes towards nebulisation and the general eligibility of nebulised liposomes for pulmonary PDT. In addition, the results confirm the fact that curcumin is a suitable photosensitiser for a PDT of tumours [70,71].

Qualitative evidence of the internalisation of nebulised curcumin-loaded liposomes into A549 cells was given via visualisation by confocal laser scanning microscopy. After incubation with nebulised formulations containing 100 μM curcumin, a considerable accumulation of the photosensitiser was detected within the cells, which is evident from the z-stack images. Furthermore, for DT and DC a homogenous distribution of curcumin inside the cells was clearly visible (Figure 6b–f). Curcumin from the formulation DD appeared as crystal-like agglomerates (Figure 6b,g,h), probably due to its instability towards nebulisation resulting in a leakage of curcumin. AFM micrographs of DD (Figure 3e,f), recorded after nebulisation of the curcumin-loaded liposomes, showed small liposomal fragments which confirms the thesis of fluidic liposomes being unstable towards nebulisation. Results of the encapsulation efficiency confirm this thesis as well, as the value dropped from 85.1% to 57.5% (Table 3) for DD. Curcumin, originally encapsulated in the lipid bilayer of DD liposomes, was liberated and hence, given the opportunity to aggregate. Similar behaviour of ultra-deformable liposomes regarding fragmentation, decreasing encapsulation efficiency and aggregation of active ingredient occurred in the study of Elhissi et al. [40]. Although no difference in the cell appearance was observable in these micrographs, the assumption that the irradiated nebulised curcumin-loaded liposomes were the source of the cytotoxic effect (Figure 6a), seems very likely and is consistent with the literature [26].

One of the aims of encapsulating a drug in nanocarriers such as liposomes is to transport it to the site of action, in this study adenocarcinoma cells of the lung. For this purpose, the active ingredient must be taken up into the cells as a final step, which can take place by various mechanisms. The most important mechanism is endocytosis, which can play a significant role especially for hydrophobic drugs [72]. This active process can occur without external stimuli or can be induced by ligands [73]. Initially, an insertion is formed in the cell membrane for this purpose, in which extracellular substances or particles are trapped and which is subsequently internalized by constriction. Different endocytosis pathways can be distinguished; of relevance in this study were clathrin-mediated endocytosis on the one hand, and calveolae-mediated endocytosis on the other. Macropinocytosis as well as non-clathrin-non-calveolae-mediated endocytosis can also occur, but were not considered in this work.

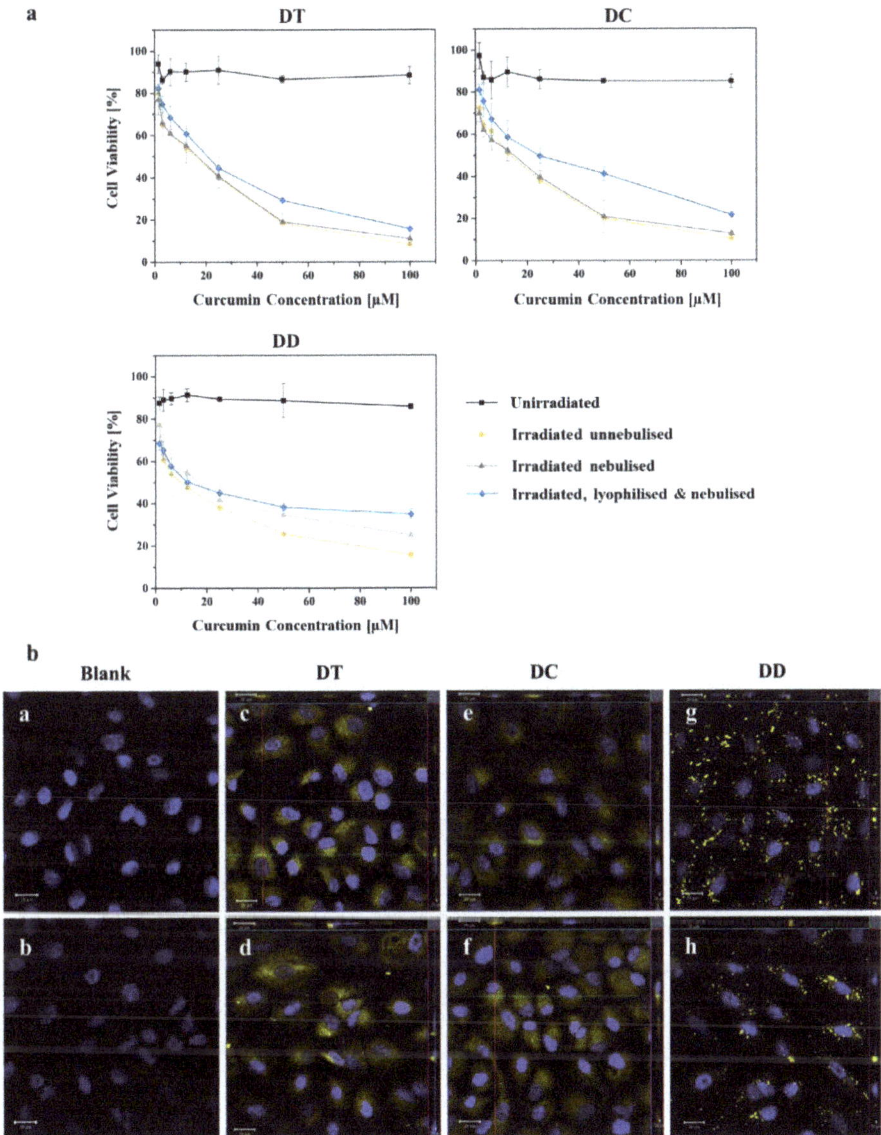

Figure 6. (**a**) MTT in vitro cytotoxicity assay of DT, DC and DD containing a curcumin concentration up to 100 μM. The figure shows a comparison of dark control, irradiated (IR) unnebulised, IR nebulised (PARI VELOX®) and IR lyophilised & nebulised curcumin-loaded liposomes. Irradiation always took place at 457 nm with a fluence of 6.61 J cm^{-2}; (**b**) CLSM micrographs of A549 cells with nebulised curcumin-loaded liposomes containing a curcumin concentration of 100 μM. Dark and irradiated (457 nm, 6.61 J cm^{-2}) samples were compared for the formulations: DT, (**c**,**d**); DC, (**e**,**f**); and DD, (**g**,**h**). Untreated cells were used for the blank micrographs (**a**,**b**). The cell nuclei were counterstained with 0.1 μg/mL DAPI and fixed with a 4% formaldehyde solution for all micrographs. Scale bars represent 20 μM.

To identify the preferred endocytosis pathway of the nebulised curcumin liposomes, appropriate controls were first used to ensure that the inhibitors alone did not reduce cell viability. It can be seen that they have little effect on cell viability (Figure 7). Chlorpromazine served as an inhibitor of clathrin-mediated endocytosis, and filipin III inhibited caveolae-mediated endocytosis. When liposomes were applied after preincubation with filipin III, DT and DC after irradiation showed that the effects without inhibition were comparable to the effects after inhibition of caveolae-mediated endocytosis. This implies that liposome uptake into cells was possible in both cases and suggests that this endocytosis pathway is not the preferred one. The crossmatch after preincubation with chlorpromazine confirms this, as cell viability is significantly increased. Blockade of clathrin-mediated endocytosis inhibited liposome uptake to a large extent. These results agree with previous studies, according to which liposomes with a diameter of up to 200 nm are mainly taken up via the formation of so-called "clathrin-coated pits" and subsequent endocytosis [74]. In contrast, the DD formulation also appears to be taken up via caveolae-mediated to a lesser extent, with clathrin-mediated endocytosis still being preferred (Figure 7). Again, this could be related to the altered size distribution of these fluid liposomes.

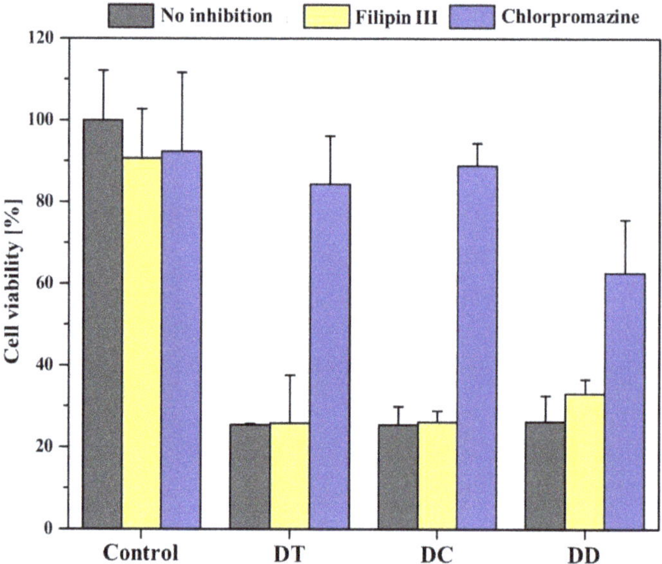

Figure 7. Determination of the endocytotic uptake mechanism of nebulised curcumin liposomes after irradiation with a fluence of 6.61 J cm^{-2} at λ = 457 nm. The endocytosis pathways were either inhibited by filipin III or chlorpromazine or were not inhibited. The incubation time of liposomes was 4 h, and untreated cells represented the 100% value of cell viability.

In addition, the influence of PDT with nebulised liposomes on the migration behaviour of A549 cells was determined in this study. This was done by means of an "in vitro scratch assay", a method based on the observation of an artificially created scratch in the confluent cell monolayer. The cells at the edge of this gap move towards the open area, which happens with the aim of closing the gap as quickly as possible until new cell-cell contacts have been created [75].

The results of this assay show that the natural migration of untreated cells closed the gap in the cell monolayer within 48 h (Figure 8a,e,i,n,r,v). In comparison, it can be seen that the gap in the cells treated with PDT does not close within this time period. The effects on the cell migration behaviour seems to be more distinct for DT and DC, in agreement with the previous results. From the time of irradiation onwards, especially in the case of DT, many cells can be seen to have detached from the bottom of the well (Figure 8f). This can be recognised by the round shape and a bright glow of the cells and is usually a sign of their death. After treatment with the formulation DD, there is still a slight migration or growth of the cells (Figure 8h,m,q,u,y). An explanation for the good migration inhibition of all three formulations could be the fact that curcumin can, in addition to the ROS generation during PDT, inhibit matrix metalloproteinases 2 and 9 in A549 cells, which play a known role in migration, invasion and angiogenesis of these cells [76].

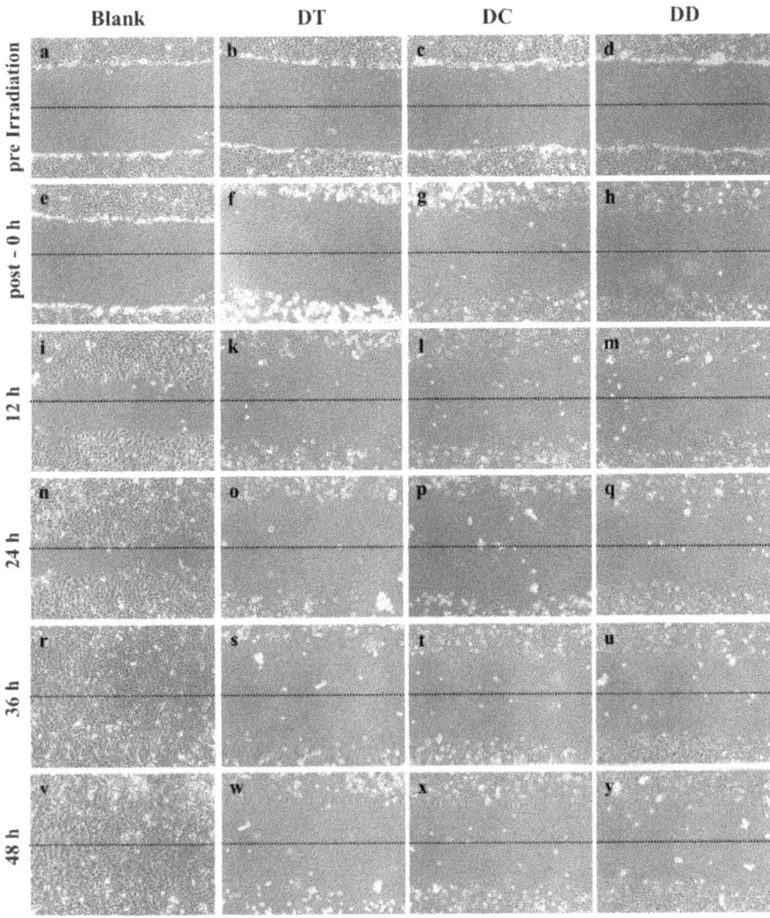

Figure 8. Microscopic images of the A549 cells during 48 h of the "in vitro scratch assay". The images a, e, i, n, r and v show the natural migration of untreated cells (blank) compared to cell migration after PDT treatment with nebulised curcumin liposomes (b,e,d,f,g,h,k,l,m,o,p,q,s,t,u,w,x,y).

4. Conclusions

Summing up, curcumin was taken up via clathrin-coated pits and visualised inside the cells. The good efficacy of nebulised curcumin liposomes with membrane stabilisation on cell viability and migration confirms their eligibility to improve pulmonary PDT.

Author Contributions: Conceptualization, J.L.; methodology, J.L., L.D., K.H.E., S.A. and S.R.P.; software, M.R.A.; validation, J.L.; formal analysis, J.L.; investigation, J.L.; resources, U.B.; data curation, J.L.; writing—original draft preparation, J.L.; writing—review and editing, J.L. and M.R.A.; visualization, J.L.; supervision, E.P., M.W. and U.B.; project administration, U.B.; funding acquisition, U.B. All authors have read and agreed to the published version of the manuscript.

Funding: This research received no external funding.

Institutional Review Board Statement: Not applicable.

Informed Consent Statement: Not applicable.

Data Availability Statement: The data presented in this study are available on request from the corresponding author.

Acknowledgments: The authors would like to express very special thanks to Nikola Schubert for her outstanding support, motivation and technical assistance. Gratitude goes also to Elias Baghdan and Eva Maria Mohr for their support and technical assistance. The authors would also like to thank Lipoid GmbH for providing some of the lipids used in this study and PARI GmbH for kindly providing PARI BOY® SX and PARI VELOX®.

Conflicts of Interest: The authors declare no conflict of interest.

References

1. World Health Organization. *Global Health Observatory*; World Health Organization: Geneva, Switzerland, 2018; Available online: https://who.int/gho/database/en/ (accessed on 3 August 2019).
2. Bray, F.; Ferlay, J.; Soerjomataram, I.; Siegel, R.L.; Torre, L.A.; Jemal, A. Global cancer statistics 2018: GLOBOCAN estimates of incidence and mortality worldwide for 36 cancers in 185 countries. *CA Cancer J. Clin.* **2018**, *68*, 394–424. [CrossRef]
3. Planchard, D.; Popat, S.; Kerr, K.; Novello, S.; Smit, E.; Faivre-Finn, C.; Mok, T.; Reck, M.; Van Schil, P.; Hellmann, M.; et al. Metastatic non-small cell lung cancer: ESMO Clinical Practice Guidelines for diagnosis, treatment and follow-up. *Ann. Oncol.* **2018**, *29*, iv192–iv237. [CrossRef]
4. Postmus, P.E.; Kerr, K.M.; Oudkerk, M.; Senan, S.; Waller, D.A.; Vansteenkiste, J.; Escriu, C.; Peters, S. Early and locally advanced non-small-cell lung cancer (NSCLC): ESMO Clinical Practice Guidelines for diagnosis, treatment and follow-up. *Ann. Oncol.* **2017**, *28*, iv1–iv21. [CrossRef] [PubMed]
5. Forde, P.M.; Chaft, J.E.; Smith, K.N.; Anagnostou, V.; Cottrell, T.R.; Hellmann, M.D.; Zahurak, M.; Yang, S.C.; Jones, D.R.; Broderick, S.; et al. Neoadjuvant PD-1 Blockade in Resectable Lung Cancer. *N. Engl. J. Med.* **2018**, *378*, 1976–1986. [CrossRef]
6. Preis, E.; Anders, T.; Širc, J.; Hobzova, R.; Cocarta, A.-I.; Bakowsky, U.; Jedelská, J. Biocompatible indocyanine green loaded PLA nanofibers for in situ antimicrobial photodynamic therapy. *Mater. Sci. Eng. C* **2020**, *115*, 111068. [CrossRef]
7. Raschpichler, M.; Agel, M.R.; Pinnapireddy, S.R.; Duse, L.; Baghdan, E.; Schäfer, J.; Bakowsky, U. In situ intravenous photodynamic therapy for the systemic eradication of blood stream infections. *Photochem. Photobiol. Sci.* **2018**, *18*, 304–308. [CrossRef] [PubMed]
8. Dougherty, T.J.; Gomer, C.J.; Henderson, B.W.; Jori, G.; Kessel, D.; Korbelik, M.; Moan, J.; Peng, Q. Photodynamic therapy. *J. Natl. Cancer Inst.* **1998**, *90*, 889–905. [CrossRef]
9. Plenagl, N.; Seitz, B.S.; Pinnapireddy, S.R.; Jedelská, J.; Brüßler, J.; Bakowsky, U. Hypericin Loaded Liposomes for Anti-Microbial Photodynamic Therapy of Gram-Positive Bacteria. *Phys. Status Solidi A* **2018**, *215*. [CrossRef]
10. Moan, J.; Berg, K. The photodegradation of porphyrins in cells can be used to estimate the lifetime of singlet oxygen. *Photochem. Photobiol.* **1991**, *53*, 549–553. [CrossRef]
11. Dolmans, D.E.; Fukumura, D.; Jain, R.K. Photodynamic therapy for cancer. *Nat. Rev. Cancer* **2003**, *3*, 380–387. [CrossRef]
12. Banerjee, S.; Dixit, A.; Karande, A.A.; Chakravarty, A.R. Remarkable Selectivity and Photo-Cytotoxicity of an Oxidovanadium(IV) Complex of Curcumin in Visible Light. *Eur. J. Inorg. Chem.* **2014**, *2015*, 447–457. [CrossRef]
13. Juarranz, A.; Jaén, P.; Sanz-Rodríguez, F.; Cuevas, J.; González, S. Photodynamic therapy of cancer. Basic principles and applications. *Clin. Transl. Oncol.* **2008**, *10*, 148–154. [CrossRef] [PubMed]
14. Duse, L.; Agel, M.R.; Pinnapireddy, S.R.; Schäfer, J.; Selo, M.A.; Ehrhardt, C.; Bakowsky, U. Photodynamic Therapy of Ovarian Carcinoma Cells with Curcumin-Loaded Biodegradable Polymeric Nanoparticles. *Pharmaceutics* **2019**, *11*, 282. [CrossRef]
15. Konopka, K.; Goslinski, T. Photodynamic Therapy in Dentistry. *J. Dent. Res.* **2007**, *86*, 694–707. [CrossRef]
16. Brancaleon, L.; Moseley, H. Laser and Non-laser Light Sources for Photodynamic Therapy. *Lasers Med Sci.* **2002**, *17*, 173–186. [CrossRef]
17. Baghdan, E.; Duse, L.; Schüer, J.J.; Pinnapireddy, S.R.; Pourasghar, M.; Schäfer, J.; Schneider, M.; Bakowsky, U. Development of inhalable curcumin loaded Nano-in-Microparticles for bronchoscopic photodynamic therapy. *Eur. J. Pharm. Sci.* **2019**, *132*, 63–71. [CrossRef]
18. Agel, M.R.; Baghdan, E.; Pinnapireddy, S.R.; Lehmann, J.; Schäfer, J.; Bakowsky, U. Curcumin loaded nanoparticles as efficient photoactive formulations against gram-positive and gram-negative bacteria. *Colloids Surf. B Biointerfaces* **2019**, *178*, 460–468. [CrossRef]
19. Duse, L.; Baghdan, E.; Pinnapireddy, S.R.; Engelhardt, K.H.; Jedelská, J.; Schaefer, J.; Quendt, P.; Bakowsky, U. Preparation and Characterization of Curcumin Loaded Chitosan Nanoparticles for Photodynamic Therapy. *Phys. Status Solidi A* **2017**, *215*. [CrossRef]
20. Shishodia, S.; Sethi, G.; Aggarwal, B.B. Curcumin: Getting Back to the Roots. *Ann. N.Y. Acad. Sci.* **2005**, *1056*, 206–217. [CrossRef]
21. Epstein, J.; Sanderson, I.; Macdonald, T.T. Curcumin as a therapeutic agent: The evidence from in vitro, animal and human studies. *Br. J. Nutr.* **2010**, *103*, 1545–1557. [CrossRef]

22. Naksuriya, O.; Okonogi, S.; Schiffelers, R.; Hennink, W.E. Curcumin nanoformulations: A review of pharmaceutical properties and preclinical studies and clinical data related to cancer treatment. *Biomaterials* **2014**, *35*, 3365–3383. [CrossRef]
23. Wilken, R.; Veena, M.S.; Wang, M.B.; Srivatsan, E.S. Curcumin: A review of anti-cancer properties and therapeutic activity in head and neck squamous cell carcinoma. *Mol. Cancer* **2011**, *10*, 12. [CrossRef]
24. Anand, P.; Kunnumakkara, A.B.; Newman, R.A.; Aggarwal, B.B. Bioavailability of Curcumin: Problems and Promises. *Mol. Pharm.* **2007**, *4*, 807–818. [CrossRef] [PubMed]
25. Kaminaga, Y.; Nagatsu, A.; Akiyama, T.; Sugimoto, N.; Yamazaki, T.; Maitani, T.; Mizukami, H. Production of unnatural glucosides of curcumin with drastically enhanced water solubility by cell suspension cultures of Catharanthus roseus. *FEBS Lett.* **2003**, *555*, 311–316. [CrossRef]
26. Duse, L.; Pinnapireddy, S.R.; Strehlow, B.; Jedelská, J.; Bakowsky, U. Low level LED photodynamic therapy using curcumin loaded tetraether liposomes. *Eur. J. Pharm. Biopharm.* **2018**, *126*, 233–241. [CrossRef]
27. Kunwar, A.; Barik, A.; Pandey, R.; Priyadarsini, K.I. Transport of liposomal and albumin loaded curcumin to living cells: An absorption and fluorescence spectroscopic study. *Biochim. Biophys. Acta Gen. Subj.* **2006**, *1760*, 1513–1520. [CrossRef] [PubMed]
28. Melis, V.; Manca, M.L.; Bullita, E.; Tamburini, E.; Castangia, I.; Cardia, M.C.; Valenti, D.; Fadda, A.M.; Peris, J.E.; Manconi, M. Inhalable polymer-glycerosomes as safe and effective carriers for rifampicin delivery to the lungs. *Colloids Surf. B Biointerfaces* **2016**, *143*, 301–308. [CrossRef] [PubMed]
29. Manca, M.L.; Peris, J.E.; Melis, V.; Valenti, D.; Cardia, M.C.; Lattuada, D.; Escribano-Ferrer, E.; Fadda, A.M.; Manconi, M. Nanoincorporation of curcumin in polymer-glycerosomes and evaluation of their in vitro–in vivo suitability as pulmonary delivery systems. *RSC Adv.* **2015**, *5*, 105149–105159. [CrossRef]
30. Preis, E.; Baghdan, E.; Agel, M.R.; Anders, T.; Pourasghar, M.; Schneider, M.; Bakowsky, U. Spray dried curcumin loaded nanoparticles for antimicrobial photodynamic therapy. *Eur. J. Pharm. Biopharm.* **2019**, *142*, 531–539. [CrossRef]
31. Engelhardt, K.H.; Pinnapireddy, S.R.; Baghdan, E.; Jedelská, J.; Bakowsky, U. Transfection Studies with Colloidal Systems Containing Highly Purified Bipolar Tetraether Lipids from Sulfolobus acidocaldarius. *Archaea* **2017**, *2017*, 1–12. [CrossRef]
32. Khan, T.K.; Chong, P.L.-G. Studies of Archaebacterial Bipolar Tetraether Liposomes by Perylene Fluorescence. *Biophys. J.* **2000**, *78*, 1390–1399. [CrossRef]
33. Sitterberg, J.; Özcetin, A.; Ehrhardt, C.; Bakowsky, U. Utilising atomic force microscopy for the characterisation of nanoscale drug delivery systems. *Eur. J. Pharm. Biopharm.* **2010**, *74*, 2–13. [CrossRef]
34. Baghdan, E.; Pinnapireddy, S.R.; Vögeling, H.; Schäfer, J.; Eckert, A.W.; Bakowsky, U. Nano spray drying: A novel technique to prepare well-defined surface coatings for medical implants. *J. Drug Deliv. Sci. Technol.* **2018**, *48*, 145–151. [CrossRef]
35. Luepke, N.; Kemper, F. The HET-CAM test: An alternative to the draize eye test. *Food Chem. Toxicol.* **1986**, *24*, 495–496. [CrossRef]
36. Schüller, M. *Entwicklung Eines Geeigneten Modells zur Untersuchung des selektinvermittelten Zellrollens: Das Chorioallantoismembran (CAM)-Modell*; Universitäts- und Landesbibliothek Sachsen-Anhalt: Halle, Germany, 2005.
37. Goergen, N.; Wojcik, M.; Drescher, S.; Pinnapireddy, S.R.; Brüßler, J.; Bakowsky, U.; Jedelská, J. The Use of Artificial Gel Forming Bolalipids as Novel Formulations in Antimicrobial and Antifungal Therapy. *Pharmaceutics* **2019**, *11*, 307. [CrossRef] [PubMed]
38. European Centre for the Validation of Alternative Methods. *HET-CAM Test INVITTOX n°47: ECVAM DB-ALM: INVITTOX Protocol*; European Centre for the Validation of Alternative Methods: Ispra, Italy, 1992.
39. Nimmano, N.; Somavarapu, S.; Taylor, K.M. Aerosol characterisation of nebulised liposomes co-loaded with erlotinib and genistein using an abbreviated cascade impactor method. *Int. J. Pharm.* **2018**, *512*, 8–17. [CrossRef]
40. Elhissi, A.M.; Giebultowicz, J.; Stec, A.A.; Wroczyński, P.; Ahmed, W.; Alhnan, M.A.; Phoenix, D.; Taylor, K.M. Nebulization of ultradeformable liposomes: The influence of aerosolization mechanism and formulation excipients. *Int. J. Pharm.* **2012**, *436*, 519–526. [CrossRef] [PubMed]
41. Beck-Broichsitter, M.; Oesterheld, N.; Knuedeler, M.-C.; Seeger, W.; Schmehl, T. On the correlation of output rate and aerodynamic characteristics in vibrating-mesh-based aqueous aerosol delivery. *Int. J. Pharm.* **2014**, *461*, 34–37. [CrossRef] [PubMed]
42. Sang, C.H.; Lin, S.C.; Chen, H.F.; Tseng, H.H.; Lo, H.H. Performance of a New Polymer-Based Vibrating Mesh Nebulizer: A Comparison to Metal-Based Mesh Nebulizer. Presented at Drug Delivery to the Lungs 2018, Edinburgh, Scotland, 12–14 December 2018.
43. Beck-Broichsitter, M.; Knuedeler, M.-C.; Seeger, W.; Schmehl, T. Controlling the droplet size of formulations nebulized by vibrating-membrane technology. *Eur. J. Pharm. Biopharm.* **2014**, *87*, 524–529. [CrossRef]
44. Manconi, M.; Manca, M.L.; Valenti, D.; Escribano, E.; Hillaireau, H.; Fadda, A.M.; Fattal, E. Chitosan and hyaluronan coated liposomes for pulmonary administration of curcumin. *Int. J. Pharm.* **2017**, *525*, 203–210. [CrossRef]
45. Pritchard, J.N.; Hatley, R.H.; Denyer, J.; Von Hollen, D. Mesh nebulizers have become the first choice for new nebulized pharmaceutical drug developments. *Ther. Deliv.* **2018**, *9*, 121–136. [CrossRef]
46. Taylor, K.; Taylor, G.; Kellaway, I.; Stevens, J. The stability of liposomes to nebulisation. *Int. J. Pharm.* **1990**, *58*, 57–61. [CrossRef]
47. Singh, A.; Bhatia, S.; Rana, V. Inhalable Nanostructures for Lung Cancer Treatment: Progress and Challenges. *Curr. Nanomed.* **2019**, *9*, 4–29. [CrossRef]
48. Abdelaziz, H.M.; Freag, M.S.; Elzoghby, A.O. Solid Lipid Nanoparticle-Based Drug Delivery for Lung Cancer. In *Nanotechnology-Based Targeted Drug Delivery Systems for Lung Cancer*; Elsevier: Amsterdam, The Netherlands, 2019; pp. 95–121.
49. Sánchez-López, E.; Guerra, M.; Dias-Ferreira, J.; Lopez-Machado, A.; Ettcheto, M.; Cano, A.; Espina, M.; Camins, A.; Garcia, M.L.; Souto, E.B. Current Applications of Nanoemulsions in Cancer Therapeutics. *Nanomaterials* **2019**, *9*, 821. [CrossRef]

50. Akhter, S.; Ahmad, J.; Rizwanullah, M.; Rahman, M.; Ahmad, M.Z.; Alam Rizvi, M.; Ahmad, F.J.; Amin, S.; Kamal, M.A. Nanotechnology-based inhalation treatments for lung cancer: State of the art. *Nanotechnol. Sci. Appl.* **2015**, *8*, 55–66. [CrossRef] [PubMed]

51. Sun, X.; Wang, N.; Yang, L.-Y.; Ouyang, X.-K.; Huang, F. Folic Acid and PEI Modified Mesoporous Silica for Targeted Delivery of Curcumin. *Pharmceutics* **2019**, *11*, 430. [CrossRef] [PubMed]

52. Bulbake, U.; Doppalapudi, S.; Kommineni, N.; Khan, W. Liposomal Formulations in Clinical Use: An Updated Review. *Pharmaceutics* **2017**, *9*, 12. [CrossRef] [PubMed]

53. Rudokas, M.; Najlah, M.; A Alhnan, M.; Elhissi, A. Liposome Delivery Systems for Inhalation: A Critical Review Highlighting Formulation Issues and Anticancer Applications. *Med. Princ. Pr.* **2016**, *25*, 60–72. [CrossRef] [PubMed]

54. Bagatolli, L.; Gratton, E.; Khan, T.K.; Chong, P.L.-G. Two-Photon Fluorescence Microscopy Studies of Bipolar Tetraether Giant Liposomes from Thermoacidophilic Archaebacteria Sulfolobus acidocaldarius. *Biophys. J.* **2000**, *79*, 416–425. [CrossRef]

55. Niven, R.W.; Schreier, H. Nebulization of Liposomes. I. Effects of Lipid Composition. *Pharm. Res.* **1990**, *7*, 1127–1133. [CrossRef] [PubMed]

56. Zhang, Y.; Li, H.; Sun, J.; Gao, J.; Liu, W.; Li, B.; Guo, Y.; Chen, J. DC-Chol/DOPE cationic liposomes: A comparative study of the influence factors on plasmid pDNA and siRNA gene delivery. *Int. J. Pharm.* **2010**, *390*, 198–207. [CrossRef]

57. Yang, S.; Chen, J.; Zhao, D.; Han, D.; Chen, X. Comparative study on preparative methods of DC-Chol/DOPE liposomes and formulation optimization by determining encapsulation efficiency. *Int. J. Pharm.* **2012**, *434*, 155–160. [CrossRef] [PubMed]

58. Beck-Broichsitter, M.; Knuedeler, M.-C.; Oesterheld, N.; Seeger, W.; Schmehl, T. Boosting the aerodynamic properties of vibrating-mesh nebulized polymeric nanosuspensions. *Int. J. Pharm.* **2014**, *459*, 23–29. [CrossRef] [PubMed]

59. Zaru, M.; Mourtas, S.; Klepetsanis, P.; Fadda, A.M.; Antimisiaris, S.G. Liposomes for drug delivery to the lungs by nebulization. *Eur. J. Pharm. Biopharm.* **2007**, *67*, 655–666. [CrossRef]

60. Wolfe, J.; Bryant, G. Freezing, Drying, and/or Vitrification of Membrane–Solute–Water Systems. *Cryobiology* **1999**, *39*, 103–129. [CrossRef]

61. Darwis, Y.; Kellaway, I. Nebulisation of rehydrated freeze-dried beclomethasone dipropionate liposomes. *Int. J. Pharm.* **2001**, *215*, 113–121. [CrossRef]

62. Mahmoud, G.; Jedelská, J.; Strehlow, B.; Bakowsky, U. Bipolar tetraether lipids derived from thermoacidophilic archaeon Sulfolobus acidocaldarius for membrane stabilization of chlorin e6 based liposomes for photodynamic therapy. *Eur. J. Pharm. Biopharm.* **2015**, *95*, 88–98. [CrossRef]

63. Chang, E. Unusual Thermal Stability of Liposomes Made from Bipolar Tetraether Lipids. *Biochem. Biophys. Res. Commun.* **1994**, *202*, 673–679. [CrossRef]

64. Freisleben, H.-J. The Main (Glyco) Phospholipid (MPL) of Thermoplasma acidophilum. *Int. J. Mol. Sci.* **2019**, *20*, 5217. [CrossRef] [PubMed]

65. Ghazanfari, T.; Elhissi, A.M.; Ding, Z.; Taylor, K.M. The influence of fluid physicochemical properties on vibrating-mesh nebulization. *Int. J. Pharm.* **2007**, *339*, 103–111. [CrossRef] [PubMed]

66. Dailey, L.A.; Schmehl, T.; Gessler, T.; Wittmar, M.; Grimminger, F.; Seeger, W.; Kissel, T. Nebulization of biodegradable nanoparticles: Impact of nebulizer technology and nanoparticle characteristics on aerosol features. *J. Control. Release* **2003**, *86*, 131–144. [CrossRef]

67. Olszewski, O.Z.; MacLoughlin, R.; Blake, A.; O'Neill, M.; Mathewson, A.; Jackson, N. A Silicon-based MEMS Vibrating Mesh Nebulizer for Inhaled Drug Delivery. *Procedia Eng.* **2016**, *168*, 1521–1524. [CrossRef]

68. Nikander, K.; Von Hollen, D.; Larhrib, H. The size and behavior of the human upper airway during inhalation of aerosols. *Expert Opin. Drug Deliv.* **2016**, *14*, 621–630. [CrossRef]

69. Mitra, K.; Basu, U.; Khan, I.; Maity, B.; Kondaiah, P.; Chakravarty, A.R. Remarkable anticancer activity of ferrocenyl-terpyridine platinum(ii) complexes in visible light with low dark toxicity. *Dalton Trans.* **2014**, *43*, 751–763. [CrossRef] [PubMed]

70. Pillai, G.R.; Srivastava, A.S.; I Hassanein, T.; Chauhan, D.P.; Carrier, E. Induction of apoptosis in human lung cancer cells by curcumin. *Cancer Lett.* **2004**, *208*, 163–170. [CrossRef] [PubMed]

71. Wu, S.-H.; Hang, L.-W.; Yang, J.-S.; Chen, H.-Y.; Lin, H.-Y.; Chiang, J.-H.; Lu, C.-C.; Yang, J.-L.; Lai, T.-Y.; Ko, Y.-C.; et al. Curcumin induces apoptosis in human non-small cell lung cancer NCI-H460 cells through ER stress and caspase cascade- and mitochondria-dependent pathways. *Anticancer Res.* **2010**, *30*, 2125–2133.

72. Torchilin, V.P. Recent advances with liposomes as pharmaceutical carriers. *Nat. Rev. Drug Discov.* **2005**, *4*, 145–160. [CrossRef] [PubMed]

73. Schifter, C.-R. Untersuchungen zum Einfluss der Größe von Liposomen auf deren Endozytose in verschiedenen Zelllinien. Ph.D. Thesis, Universität Freiburg, Freiburg, Germany, 2008.

74. Fux, C.; Costerton, J.; Stewart, P.; Stoodley, P. Survival strategies of infectious biofilms. *Trends Microbiol.* **2005**, *13*, 34–40. [CrossRef]

75. Liang, C.-C.; Park, A.Y.; Guan, J.-L. In vitro scratch assay: A convenient and inexpensive method for analysis of cell migration in vitro. *Nat. Protoc.* **2007**, *2*, 329–333. [CrossRef]

76. Lin, S.-S.; Lai, K.-C.; Hsu, S.-C.; Yang, J.-S.; Kuo, C.-L.; Lin, J.-P.; Ma, Y.-S.; Wu, C.-C.; Chung, J.-G. Curcumin inhibits the migration and invasion of human A549 lung cancer cells through the inhibition of matrix metalloproteinase-2 and -9 and Vascular Endothelial Growth Factor (VEGF). *Cancer Lett.* **2009**, *285*, 127–133. [CrossRef]

pharmaceutics

Article

Novel Pyropheophorbide Phosphatydic Acids Photosensitizer Combined EGFR siRNA Gene Therapy for Head and Neck Cancer Treatment

Chia-Hsien Yeh [1], Juan Chen [2], Gang Zheng [2,3], Leaf Huang [4,5] and Yih-Chih Hsu [1,5,6,*]

1   Department of Bioscience Technology, Chung Yuan Christian University, Taoyuan 32023, Taiwan; super791031@hotmail.com
2   Princess Margaret Cancer Center, University Health Network (UHN), Toronto, ON M5G 1L7, Canada; juanchen@uhnresearch.ca (J.C.); gang.zheng@uhnresearch.ca (G.Z.)
3   Department of Medical Biophysics, University of Toronto, Toronto, ON M5S 1A1, Canada
4   Division of Pharmacoengineering and Molecular Pharmaceutics, Eshelman School of Pharmacy, University of North Carolina at Chapel Hill, Chapel Hill, NC 27599, USA; leafh@email.unc.edu
5   Center of Cancer Theranostics and Commercialization of Technology, Chung Yuan Christian University, Taoyuan 32023, Taiwan
6   Center for Nanotechnology, Chung Yuan Christian University, Taoyuan 32023, Taiwan
*   Correspondence: ivyhsu@cycu.edu.tw; Tel.: +886-3-265-3522; Fax: +886-3-265-3599

Citation: Yeh, C.-H.; Chen, J.; Zheng, G.; Huang, L.; Hsu, Y.-C. Novel Pyropheophorbide Phosphatydic Acids Photosensitizer Combined EGFR siRNA Gene Therapy for Head and Neck Cancer Treatment. *Pharmaceutics* 2021, 13, 1435. https://doi.org/10.3390/pharmaceutics13091435

Academic Editors: Udo Bakowsky, Matthias Wojcik, Eduard Preis and Gerhard Litscher

Received: 20 July 2021
Accepted: 26 August 2021
Published: 9 September 2021

Publisher's Note: MDPI stays neutral with regard to jurisdictional claims in published maps and institutional affiliations.

Abstract: This study combined two novel nanomedicines, a novel LCP Pyro PA photodynamic therapy (PDT) and LCP EGFR siRNA gene therapy, to treat head and neck cancer. A novel photosensitizer, pyropheophorbide phosphatydic acids (Pyro PA), was first modified into Lipid-Calcium phosphate nanoparticles named LCP Pyro PA NPs, and targeted with aminoethylanisamide as a novel PDT photosensitizer. EGFR siRNA was encapsulated into LCP NPs to silence EGFR expression. Measured sizes of LCP EGFR siRNA NPs and LCP Pyro-PA NPs were 34.9 ± 3.0 and 20 nm respectively, and their zeta potentials were 51.8 ± 1.8 and 52.0 ± 7.6 mV respectively. In vitro studies showed that EGFR siRNA was effectively knocked down after photodynamic therapy (PDT) with significant inhibition of cancer growth. SCC4 or SAS xenografted nude mice were used to verify therapeutic efficacy. The LCP Control siRNA+PDT group of SCC4 and SAS showed significantly reduced tumor volume compared to the phosphate buffered saline (PBS) group. In the LCP-EGFR siRNA+LCP Pyro PA without light group and LCP EGFR siRNA + PBS with light group, SCC4 and SAS tumor volumes were reduced by ~140% and ~150%, respectively, compared to the PBS group. The LCP EGFR siRNA+PDT group of SCC4 and SAS tumor volumes were reduced by ~205% and ~220%, respectively, compared to the PBS group. Combined therapy showed significant tumor volume reduction compared to PBS, control siRNA, or PDT alone. QPCR results showed EGFR expression was significantly reduced after treatment with EGFR siRNA with PDT in SCC4 and SAS compared to control siRNA or PDT alone. Western blot results confirmed decreased EGFR protein expression in the combined therapy group. No toxic results were found in serum biomarkers. No inflammatory factors were found in heart, liver and kidney tissues. Results suggest that the novel LCP Pyro PA mediated PDT combined with LCP siEGFR NPs could be developed in clinical modalities for treating human head and neck cancer in the future.

Keywords: targeted delivery; head and neck cancer; nanoparticles; photodynamic therapy

1. Introduction

Head and neck cancers develop mainly in the human oral cavity [1], larynx, hypopharynx and sinonasal areas. Ninety-five percent of head and neck cancer is caused by squamous cell carcinoma (SCC) [2]. Major risk factors include tobacco, betel nut chewing, alcohol drinking and HPV infections, which are the four major causes for carcinogenesis [2]. Clinical practices to treat human head and neck cancers include surgery, chemotherapy and radiotherapy. Another new modality is photodynamic therapy (PDT), a noninvasive alternative for oral cancer therapy with proven low cumulative side effects after repeated therapies, including little to no observed scarring in the oral cavity [3,4]. PDT uses a specific photosensitizer activated by a specific light wavelength that selectively kills cancerous cells [5]. Photodynamic reactions generate reactive oxygen species that can kill cancer cells. In this study, we are the first to develop a novel Lipid-Calcium phosphate–pyropheophorbide phosphatydic acids nanoparticles (LCP-Pyro PA NPs) technology. LCP Pyro PA NPs consist of LCP encapsulated with a novel photosensitizer, Pyro PA, with targeting ligand aminoethylanisamide (AEAA) on the outer-left layer of LCP NPs. Previous studies have found evidence of EGFR overexpression in human head and neck SCC, and oral cancers have been reported to demonstrate evidence of mRNA expression for EGFR [6,7]. Up-regulation of EGFR by squamous epithelial cells from 24 HNSCC patients with head and neck squamous cell carcinoma (HNSCC) suggests that therapies targeting these genes may be effective in the prevention of head and neck cancer [8]. We delivered small interfering RNA (siRNA) to target genes with specific siRNA to tumor cells without distinct adverse effects. The concept has been proven with transfected HIF1a siRNA [9] and VEGF siRNA [10] and induces in vitro cancer cell apoptosis and in vivo tumor destruction. Therefore, we delivered EGFR siRNA to head and neck tumor cells using lipid-based LCP NPs with AEAA targeting ligands formulated on the outer layer portion of the nanoparticles to target the over-expressed sigma receptors [11,12] on the HNSCC cell surface. The goal of this study was to investigate the therapeutic outcome of integrating a novel targeting photosensitizer, Pyro PA (LCP Pyro PA NPs mediated PDT), with LCP NPs with EGFR siRNA gene therapy and targeting AEAA ligands loaded to treat HNSCC in xenograft animal models (Scheme 1).

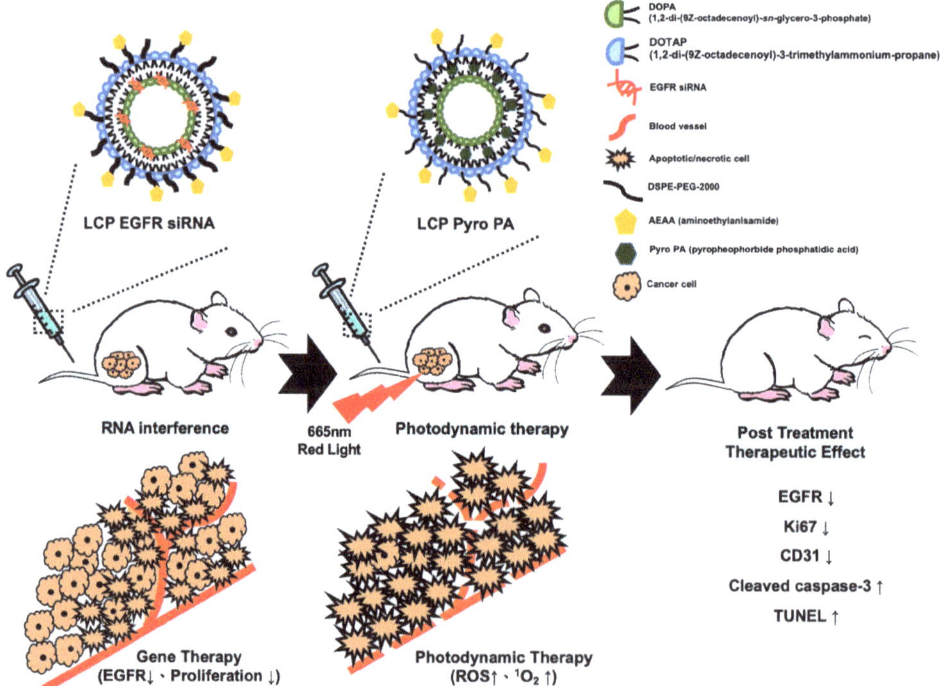

Scheme 1. Schematic illustration of LCP EGFR siRNA and LCP Pyro PA to xenograft models. We used SCC4 and SAS xenografts of Balb/c nude mice to verify the efficacy of the combination therapy. Combined therapy showed greater tumor growth inhibition for both SCC4 and SAS.

2. Materials and Methods

2.1. EGFR siRNA (siEGFR)

The EGFR siRNA (siEGFR) with sense strand 5′-AAG UGC UGG AUG AUA GAC GCA dTdT-3′ was designed using the provided Genome webserver (National Center for Biotechnology Information, Bethesda, MD, USA) [10]. The control siRNA (siControl) was the siGENOME non-targeting siRNA with sense strand 5′-AUGUAUUGGCCUGUAUUAG dTdT-3′ and was purchased from Thermo Scientific (Waltham, MA, USA). Lipofectamine™ RNAiMAX was purchased from Invitrogen (Carlsbad, CA, USA). Dioleoylphosphatydic acid (DOPA), 1,2-dioleoyl-3-trimethylammonium-propane chloride salt (DOTAP), cholesterol and 1,2-distearoryl-*sn*-glycero-3-phosphoethanolamine-*N*-[methoxy(polyethyleneglycol-2000)] ammonium salt (DSPE-PEG2000) were all purchased from Avanti Polar Lipids (Alabaster, AL, USA). DSPE-PEG-AEAA was chemically synthesized in the Hsu lab [12]. Other chemicals used in the study were purchased from Sigma-Aldrich (St. Louis, MO, USA).

2.2. Novel Synthesis Method of Pyro PA Photosensitizer

Pyropheophorbide-phosphatydic acid (Pyro PA) was synthesized and provided by Professor Zheng (Toronto, ON, Canada). Forty mg of pyropheophrobide-phospholipid (Pyro-lipid) was suspended in 3 mL of sodium acetate buffer (40 mM sodium acetate, 80 mM $CaCl_2$, pH 5.0) and sonicated for 5 min for efficient dispersion, after which 3 mL of diethyl ether and 1000 U of phospholipase D obtained from *S. chromofuscus* were added. The mixture was light sensitive and therefore was protected from light and mixed aggressively at 33 °C. The reaction was completed in 5 h to yield Pyro PA, evidenced by HPLC-MS analysis. The HPLC profiles of the Pyro-lipid and Pyro PA and their corresponding UV spectra and mass spectra were further examined and verified to ensure the success of the synthesis.

2.3. Novel Formulation of LCP Pyro PA NP

First, phosphatydic acid was conjugated into Pyro PA for encapsulating into the lipid calcium phosphate nanoparticles. Then, 100 μL of 0.5 M $CaCl_2$ was dispersed in an 8 mL oil phase (cyclohexane/Igepal CO-250 [71/29, *v/v*]) to form a homogenous water-in-oil well-dispersed microemulsion. The phosphate micro-emulsion was prepared by dispersing 100 μL of 100 mM Na_2HPO_4 in a separate 8-mL oil phase (cyclohexane/Igepal CO-250 [71/29, *v/v*]). Concentrations of 17.3 mM DOPA and 17.3 mM Pyro PA were dissolved in chloroform and added to the phosphate micro-emulsion solution. Then, the two micro-emulsions were mixed for 30 min, and ethanol (16 mL) was added to the combined solution. The well-mixed solution was then centrifuged to remove unnecessary ingredients. After three repetitions of the ethanol wash, the CaP core pellets were collected as white precipitates. The final LCP NPs were prepared in a separate step: 50 μL of LCP CaP cores were mixed with 20 μL of 20 mM DOTAP, 20 mM cholesterol, 20 mM DSPE-PEG-AEAA, and 20 mM DSPE-PEG-2000. Finally, residual lipids were evaporated. LCP NPs were formed when rehydrated in 50 μL of 20% glucose solution for tail vein injection into mice. Aminoethylanisamide (AEAA) was modified on the outer leaflet of LCP NPs to target the sigma receptors that are overexpressed on HNSCC cells [13].

2.4. Formulation of LCP Nanoparticles

LCP NPs were formulated as described with slight modifications [13–16]. Sigma receptors are highly expressed on the surface of SCC4 or SAS cell lines, and AEAA is the target moiety against sigma receptors on the surface of LCP NPs [12]. In an oil-phase solution (20 mL) comprising cyclohexane and Igepal CO-250 at 71/29 (*v/v*) ratio, a homogenous water-in-oil microemulsion solution was formed. Subsequently, 300 μL of 2.5 M $CaCl_2$ with 60 μg of siRNA was dispersed into the microemulsion system. A phosphate microemulsion was formulated by adding 300 μL of 12.5 mM Na_2HPO_4 in a 20 mL oil phase bottle. One hundred microliters of 20 mM DOPA dissolved in chloroform solvent was transferred to the phosphate solution. After mixing the two microemulsions homogenously for 20 min, ethanol (40 mL) was added to the final solution and then centrifuged to spin down the LCP NPs. It was crucial to conduct the ethanol wash procedure 3–4 times, after which the CaP core pellets were collected. The final LCP NPs were prepared in a separate step; 50 µL of LCP CaP cores was mixed with 20 µL of 20 mM DOTAP, 20 mM cholesterol, 20 mM DSPE-PEG-AEAA, and 20 mM DSPE-PEG-2000. Finally, residual lipids were evaporated. LCP NPs were formed when rehydrated in 50 µL of 20% glucose solution for tail vein injection into mice.

2.5. Characterization of LCP Nanoparticles

Transmission electron microscope images of LCP were obtained using the Bio-TEM Hitachi HT7700 (Hitachi, Ibaraki, Japan). LCP NP size and zeta potential were determined using the Malvern Zetasizer Nano series (Westborough, MA, USA).

2.6. Human SCC Cell Cultures

The SCC4 cell line was acquired from the Bioresource Collection and Research Center (BCRC, Hsinchu, Taiwan) and the SAS cell line was a gift from Prof. Chun-Ji Liu, Yang Ming University. Both SCC4 and SAS cells were cultured using DMEM/F-12 medium (Invitrogen, Grand Island, NY, USA), supplemented with 10% fetal bovine serum (FBS) (Invitrogen, Grand Island, NY, USA) and incubated at 37 °C with 5% CO_2. SCC4 and SAS cells were removed from culture plates using 0.05% trypsin-EDTA (Invitrogen, Carlsbad, CA, USA) when cancer cells reached 80% confluence on plate.

2.7. In Vitro siRNA Transfection Study

In a six-well plate, 3×10^5 cells per well of SCC4 or SAS cells were cultured on each well and incubated for 24 h. Transfection was conducted with 25 nM siRNA concentration in Opti-MEM Reduced medium (Invitrogen, Grand Island, NY, USA) using Lipofectamine^{TM} RNAiMAX (Invitrogen, Carlsbad, CA, USA) based on manual. SCC4 or SAS cells were then placed into an incubator at 37 °C for 4 h in the Opti-MEM Reduced medium. The medium was then changed to a growth medium containing 10% FBS for 48 h, after which cells were lysated.

2.8. SCC4 and SAS Xenograft Model Establishment

Eight week old male BALB/cAnN.Cg-Foxn1 nude mice were purchased from the National Laboratory Animal Center, Taipei, Taiwan. To establish SCC4 or SAS xenograft mice models, 5×10^5 cells of SCC4 or SAS in DMEM/F12 media were combined gently with matrigel (Corning, Bedford, MA, USA) and infused subcutaneously at the right side of the lower flank of each mouse. SCC4 and SAS xenograft mice were randomly divided into five groups with five mice in each group: (1) PBS; (2) LCP siControl NPs+PDT; (3) LCP Pyro PA NPs without light+LCP siEGFR; (4) LCP siEGFR NPs+PBS+light; and (5) LCP siEGFR NPs+PDT. The five treatments were given intravenously via tail vein infusion. The dosage of LCP siControl NPs and LCP siEGFR NPs was 0.36 mg/kg. The dosage of LCP Pyro PA NPs was 0.78 mg/kg, prepared by solubilizing 0.3 mg Pyro PA in 200 µL PBS for 25 g of weight and given via tail vein. After 55 min, red light at 663 ± 9 nm wavelength with 320 mW/cm^2 power density delivered 100 J/cm^2 energy dosage for 11 min. Tumor size reached $200 \pm 5\%$ mm^3 (190~210 mm^3) and was sufficient for treatment. The formula used to calculate tumor volume was V = (L $\times$ W $\times$ H)/2, where V stands for tumor volume, L stands for length, W stands for width perpendicular to length, and H stands for height of the tumor. Tumor size was determined daily using calipers and all mice were sacrificed on the 13th day for further analysis. Excised tumors and organs were divided and fixed in formalin for H&E and IHC stainings. This animal study was approved on August 6, 2015 (case number #103030) and carried out under strict guidelines based on the recommendations of the Guide for the Care and Use produced by the Institutional Animal Care and Use Committee of Chung Yuan Christian University, Chungli, Taoyuan, Taiwan.

2.9. Western Blot Experiments

Cells were harvested or tumor specimens removed and homogenized using a lysis buffer (PRO-PREPTM^{TM} protein extraction solution, Intron Biotechnology Inc., Seoul, Korea). These samples were assayed. Identical amounts of protein determined using BCA protein assays (Pierce, Thermo Fisher Scientific Inc., Rockford, IL, USA) were assayed for these specimens. Lysated cells or tissues were first heated to denatured soluble proteins in the sample buffer at 100 °C for 5 min. Then, these prepared specimens were loaded in individual wells of 5%/12% Bis-Tris acrylamide gels (stacking/separating gel) with protein markers. SDS-PAGE electrophoresis was conducted at a constant voltage (150 V) at 25 °C. Proteins on the SDS-PAGE gels were transferred to a PVDF membrane (Millipore Corporation, Billerica, MA, USA) electrophoretically at a constant voltage for 2 h at 4 °C. After the protein transfer step, the PVDF membranes were blocked with a blocking buffer (BlockPROTM^{TM}, Visual Protein Biotechnology Corporation, Taipei, Taiwan), then each fraction of the membranes was separated and incubated with rabbit primary antibodies against EGFR (GTX121919; GeneTex, Taipei, Taiwan) (1:1000 dilution) and sigma receptor (SIGMAR1 GTX115389; GeneTex, Taipei, Taiwan) (1:1000 dilution), followed by peroxidase-conjugated goat anti-rabbit IgG (GTX213110 GeneTex, Taipei, Taiwan) (1:10,000 dilution), and then developed in ELC (enhanced chemiluminescence) substrate (PerkinElmer, Inc., Boston, MA, USA). Glyceraldehyde 3-phosphate dehydrogenase (GAPDH GTX100118; GeneTex, Taipei, Taiwan) (1:1000 dilution) was used as the internal control.

2.10. Quantitation of EGFR Gene Knock-Down Using Real Time-PCR

First, RNA was extracted from the tumor tissues. Tumor specimens were divided into small pieces and several pieces of 0.05 g were randomly chosen from each experimental group. Tumor specimens were homogenized for extraction of total tissue RNA using RNAzol® RT solution (Molecular Research Center Inc., Cincinnati, OH, USA). The obtained cDNAs were synthesized using RevertAid cDNA synthesis kits (Fermentas, Thermo Scientific Inc., Waltham, MA, USA). FastStart Universal Master Probe (Roche Applied Science, Mannheim, Germany) was used for quantitative real-time PCR. PCR were operated using a regular cycling schedule: 95 °C for 10 min, 40 cycles of 95 °C for 15 s and 60 °C for 1 min on a real-time PCR system (AB7300, Applied Biosystems, Foster City, CA, USA). The forward primer sequences for human EGFR were 5′-TTCCTCCCAGTGCCTGAA-3′and the reverse primer sequences for human EGFR were 5′ GGGTTCAGAGGCTGATTGTG-3′. The forward primer sequences for human GAPDH were 5′-AGCCACATCGCTCAGACAC-3′ and the reverse primer sequences for human EGFR were 5′-AGCCACATCGCTCAGACAC-3′. The forward and reverse primer pairs were produced by Roche (Roche Applied Science, Mannheim, Germany). Calculated results were based on the relative times of threshold schedules in which the calculated EGFR concentration data were normalized against GAPDH concentration as internal control.

2.11. H&E Staining and Immunohistochemistry (IHC)

Harvested organs and tumor tissues were embedded in paraffin and divided into several connective sections followed by standard H&E staining protocol. Tissue sections of paraffin-embedded SCC4 and SAS tumors were deparaffinized and rehydrated to retrieve epitopes on the antigens for IHC. Tissue slides were treated with either monoclonal or polyclonal antibodies, such as rabbit polyclonal anti-CD31 (1:100, ab28364, Abcam, Cambridge, MA, USA), rabbit monoclonal anti-Ki-67 (1:200, ab16667, Abcam, Cambridge, MA, USA), and rabbit polyclonal anti-EGFR (1:100, GTX121919; GeneTex, Taipei, Taiwan). Sections of tumor tissues were then incubated with HRP-conjugated anti-rabbit antibody (1:200, Santa Cruz Biotechnology, Santa Cruz, CA, USA) for 30 min. Slide observation was conducted using a DAB detection kit (Pierce, Rockland, IL, USA). All specimens were examined using an Olympus light microscope (BX53F model, Olympus, Tokyo, Japan). Intensities of IHC stains were further quantified for comparison at 40× magnification

(10 images per group) using ImageJ software on 21 July 2016 (version 1.8, National Institutes of Health, Bethesda, MD, USA, http://imagej.nih.gov/ij/).

2.12. TUNEL Assays

SCC4 and SAS tumor paraffin-embedded tissue sections were deparaffinized, rehydrated and pretreated for protease. TUNEL assays was executed using in situ Cell Death Detection Kits, POD (Roche, Mannheim, Germany) following the manufacturer's guidelines. Specimens were examined using an Olympus BX53F light microscope (Olympus, Tokyo, Japan). Positive TUNEL cell images were recorded and quantified at $40\times$ magnification using ImageJ software (National Institutes of Health, http://imagej.nih.gov/ij/).

2.13. In Vivo Toxicity Assays

In vivo toxicity assays followed the same protocol as treatment for tail vein infusion of four treated groups including LCP NPs of PBS, siControl, siEGFR, and Pyro PA. Three mice were used in each group. The 7- to 8-week-old C57BL/6JNarl mice were tail vein injected with PBS, LCP siControl NPs, and LCP siEGFR NPs on days 0, 1 and 2, with injections repeated on days 7, 8 and 9, and a 4-day interval between injections (of three experiment groups). The LCP Pyro PA NPs group was also tail vein injected into mice on days 3 and 10. Mice were anesthetized and cardiac punctured to collect blood for toxicity assays and sacrificed on the 14th day. To avoid hemolysis of mouse blood, blood was carefully transferred to a 1.5 mL vial in a slow mode, and blood was set to clot at 25 °C for 20 min prior to centrifugation to separate serum from the supernatant portion (Hermle Z233 MK-2, Hermle, Wehingen, Germany). The collected serum was analyzed for concentrations of liver biomarkers (AST and ALT), kidney biomarkers (CREA and BUN), calcium, and phosphorus. Concentrations of toll-like receptor 3 (TLR3) were also examined using TLR3 ELISA kits (MyBioSource, Inc., San Diego, CA, USA) to ensure no immune activation. The above assays used mouse IL-6 ELISA kits (RayBiotech, Norcross, GA, USA), mouse IL-12 p40/70 ELISA kits (RayBiotech, Norcross, GA, USA), mouse IFN gamma ELISA kits (RayBiotech, Norcross, GA, USA) and TLR3 ELISA kits (MyBioSource Inc., San Diego, CA, USA), respectively.

2.14. Statistical Analysis

All data were presented as mean ± standard deviation. Statistical significance was analyzed using one-way ANOVAs with Tukey's test using SigmaPlot® (Version 12.5, Systat Software, Inc., San Jose, CA, USA).

3. Results

3.1. Results of Novel Synthesis Method of Pyropheophrobide-Phosphatydic Acid (Pyro Pa) from Pyropheophrobide-Phospholipid (Pyro-Lipid)

The innovative synthesis of pyropheophrobide-phosphatydic acid (Pyro Pa) from pyropheophrobide-phospholipid (Pyro-lipid) used phospholipase D (from *S. chromofuscus*) as described in the Section 2 (Figure 1a). The reaction was completed to gain Pyro PA, as evidenced by HPLC-MS analysis (Figure 1b,c). The figures show HPLC profiles of the Pyro-lipid (Figure 1b) and Pyro PA (Figure 1c) and their corresponding UV spectra and mass spectra. The compound absorption spectrum had no significant changes during the reaction. The identified mass signals for Pyro-lipid were: m/z calculated for $C_{57}H_{82}N_5O_9P$ $[M]^+$ 1012.3, found $[M]^+$ 1013.2; $[M]^{2+}$ 507.2; for Pyro PA: m/z calculated for $C_{52}H_{70}N_4O_9P$ $[M]^+$ 927.1, found 928.1.

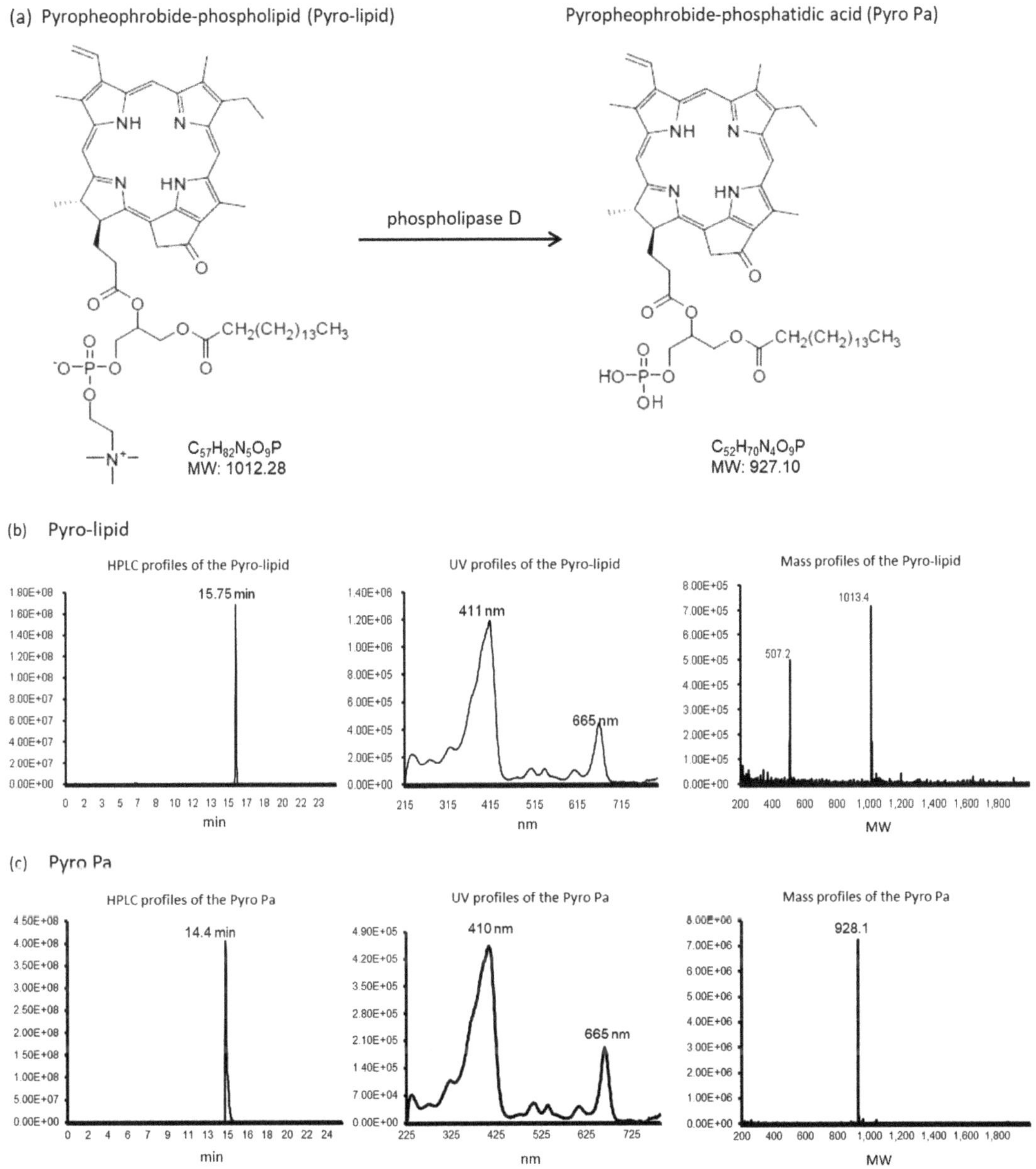

(a) Pyropheophrobide-phospholipid (Pyro-lipid)

Pyropheophrobide-phosphatidic acid (Pyro Pa)

phospholipase D

$C_{57}H_{82}N_5O_9P$
MW: 1012.28

$C_{52}H_{70}N_4O_9P$
MW: 927.10

(b) Pyro-lipid

HPLC profiles of the Pyro-lipid

15.75 min

UV profiles of the Pyro-lipid

411 nm

665 nm

Mass profiles of the Pyro-lipid

1013.4

507.2

(c) Pyro Pa

HPLC profiles of the Pyro Pa

14.4 min

UV profiles of the Pyro Pa

410 nm

665 nm

Mass profiles of the Pyro Pa

928.1

Figure 1. Novel synthesis method of Pyro PA. (**a**) 40 mg of pyropheophrobide-phospholipid (Pyro-lipid) was suspended in 3 mL of sodium acetate buffer (40 mM sodium acetate, 80 mM CaCl₂, pH 5.0) and sonicated for 5 min for efficient dispersion, after which 3 mL of diethyl ether and 1000 unit of phospholipase D obtained from *S. chromofuscus* were added. The mixture was shaded against light and mixed extensively at 33 °C. (**b,c**) The reaction was completed in 5 h to gain pyropheophrobide-phosphatydic acid (Pyro PA) evidenced by HPLC-MS analysis. The HPLC profiles of the Pyro-lipid and Pyro PA and their corresponding UV spectra and mass spectra were examined and verified to ensure success.

3.2. Characterization of Lipid-Calcium-Phosphate Pyropheophrobide Phosphatydic Acid Nanoparticles (LCP Pyro Pa NPs) or Lipid-Calcium-Phosphate EGFR siRNA Nanoparticles (LCP EGFR siRNA NPs)

3.2.1. Preparation of Lipid-Calcium-Phosphate Pyropheophrobide Phosphatydic Acid Nanoparticles (LCP-Pyro Pa NPs)

LCP-Pyro PA NPs were formulated with two different types of phospholipids to form double-layer liposomal nanoparticles that were composed of pH-sensitive calcium phosphate cores stabilized by DOPA and coated with DOTAP. The outer part of the lipid layer was modified with polyethylene glycol (PEG) chains and aminoethylanisamide (AEAA). AEAA is the targeting ligand for sigma receptors on cancer cell surfaces (Figure 2a). TEM photomicrographs indicated that nanoparticles were 15 to 20 nm in size. The zeta potential determined for LCP Pyro Pa NPs was 52.0 ± 7.6 mV with AEAA and 3.5 ± 0.6 mV without AEAA (Figure 2).

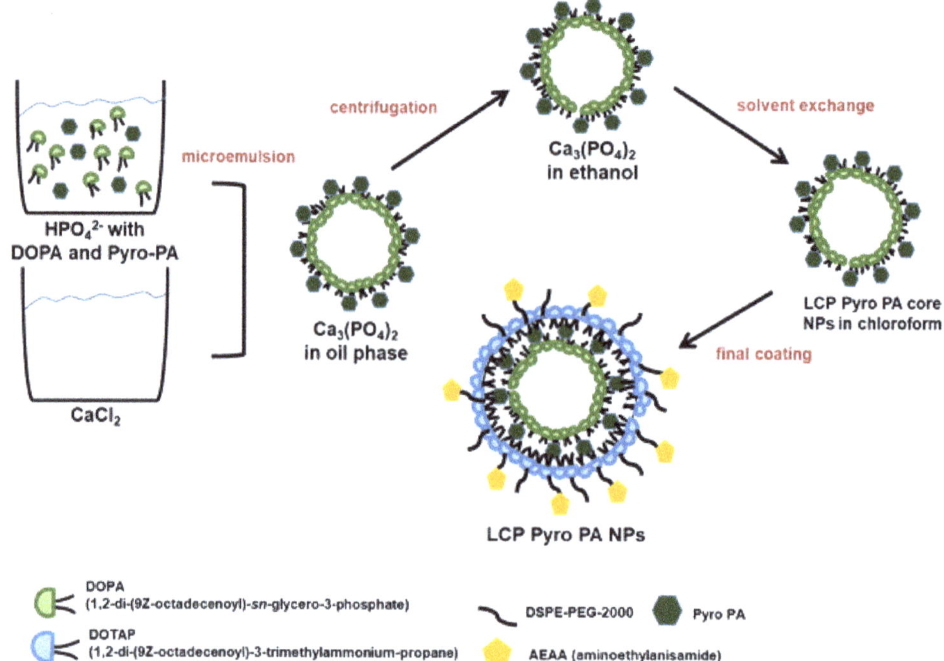

Figure 2. Novel synthesis method of LCP Pyro-Pa nanoparticles. LCP Pyro-Pa NP was composed of a biodegradable calcium phosphate core stabilized using DOPA and DOTAP. The outer layer of lipid was modified with polyethylene glycol (PEG) chains and aminoethylanisamide.

3.2.2. Preparation of Lipid-Calcium Phosphate EGFR siRNA Nanoparticles (LCP EGFR siRNA NPs)

Refer to the similar descriptions shown above for LCP-Pyro PA NPs (Figure 3a). As TEM was conducted in dehydrated conditions, it showed nanoparticle sizes ranging from 25 to 30 nm. The LCP NP sizes detected by dynamic light scattering technology were larger than the TEM-measured NP sizes, from 31 to 37 nm, because of the hydrodynamic diameter. The zeta potential for LCP siEGFR NPs was 50.1 ± 1.8 mV with AEAA and -6.1 ± 0.4 mV without AEAA. Data were presented as mean $\pm$ SD (in triplicate) (Figure 3b–d).

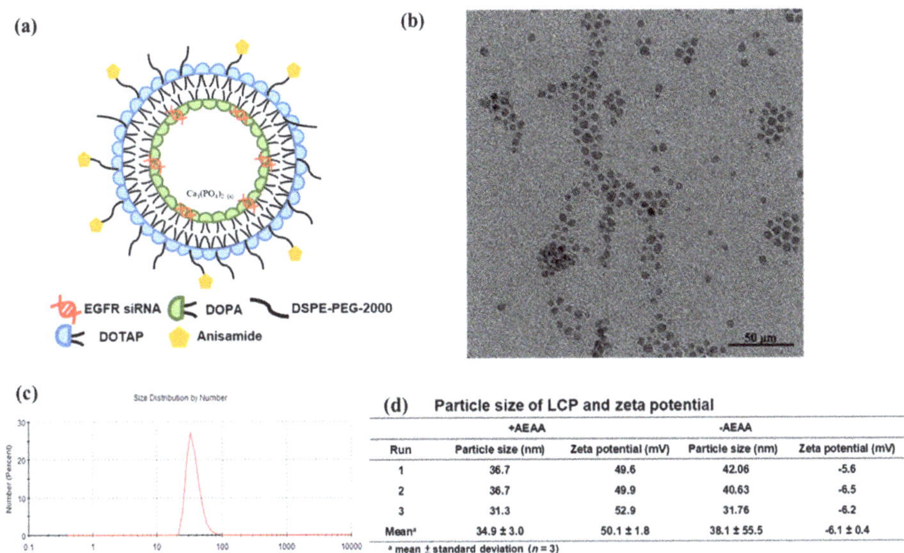

(d) Particle size of LCP and zeta potential

	+AEAA		-AEAA	
Run	Particle size (nm)	Zeta potential (mV)	Particle size (nm)	Zeta potential (mV)
1	36.7	49.6	42.06	-5.6
2	36.7	49.9	40.63	-6.5
3	31.3	52.9	31.76	-6.2
Mean[a]	34.9 ± 3.0	50.1 ± 1.8	38.1 ± 55.5	-6.1 ± 0.4

[a] mean ± standard deviation ($n = 3$)

Figure 3. Characteristic Properties of LCP siEGFR NPs. (**a**) Demonstration of LCP siEGFR NPs. LCP Pyro PA NPs were formulated with two different types of phospholipids to form double-layer liposomal nanoparticles that were composed of pH sensitive calcium phosphate cores stabilized by DOPA and coated with DOTAP. The outer part of the lipid layer was modified with polyethylene glycol (PEG) chains and aminoethylanisamide (AEAA). AEAA is the targeting ligands for the sigma receptors on cancer cell surfaces (**b–d**). Data presented as mean ± SD (in triplicate).

3.3. In Vitro Sigma Receptor and EGFR Protein Expression

Our past studies have reported that sigma receptors are expressed on the surface of HNSCC in vitro [9,10]. We examined the expression of sigma receptors and EGFR proteins at one day after photodynamic therapy. Either SCC4 or SAS cells were then treated with 0~0.25 µg /mL Pyro Pa NPs µg/mL irradiated with 663 ± 9 nm at $10 \, \text{J/cm}^2$ described as the PDT group in the in vitro study. Subsequently, they were incubated in the medium for 24 h. Cell viability was decreased due to the EGFR siRNA knockdown effect and follow-up PDT treatment in SCC4 (Figure 4a) and SAS (Figure 4b) cell lines. Either SCC4 or SAS cells were then transfected with 12.5 nM self-designed EGFR siRNA. After two days' incubation, EGFR protein amount was examined using Western blots. An EGFR silencing effect was observed for SCC4 and SAS cells treated with EGFR siRNA+PDT as compared to the groups treated with phosphate buffered saline (PBS)+PDT or Control siRNA+PDT. This indicates that the novel EGFR siRNA sequence could knockdown EGFR expression (Figure 4c,d).

3.4. In Vivo Treatment Efficacy of Combination Therapy

Two HNSCC xenograft models, SCC4 and SAS models of 204~209 mm^3 tumor volume, were then randomly distributed to five study subgroups: (1) PBS; (2) PDT (LCP siControl+PDT); +LCP Control siRNA, (3) LCP-Pyro PA+LCP EGFR siRNA, (4) PBS+light+LCP EGFR siRNA and (5) PDT (LCP-Pyro PA+light)+LCP EGFR siRNA. The complete therapeutic protocol was conducted for 13 days. All mice received two treatments on days 0, 1, or 2 and 7, 8, or 9. Either photosensitizer or LCP NPs containing siControl or siEGFR were injected into mice via the tail vein. On days 3 and 10, the mice received either PBS or LCP Pyro PA NPs (2 mg/kg) and were given 663 ± 9 nm red light to reach 100 joules after a 55 min metabolic period. For further analysis, all mice were humanely anaesthetized to death on the 13th day to collect major organs and tumors.

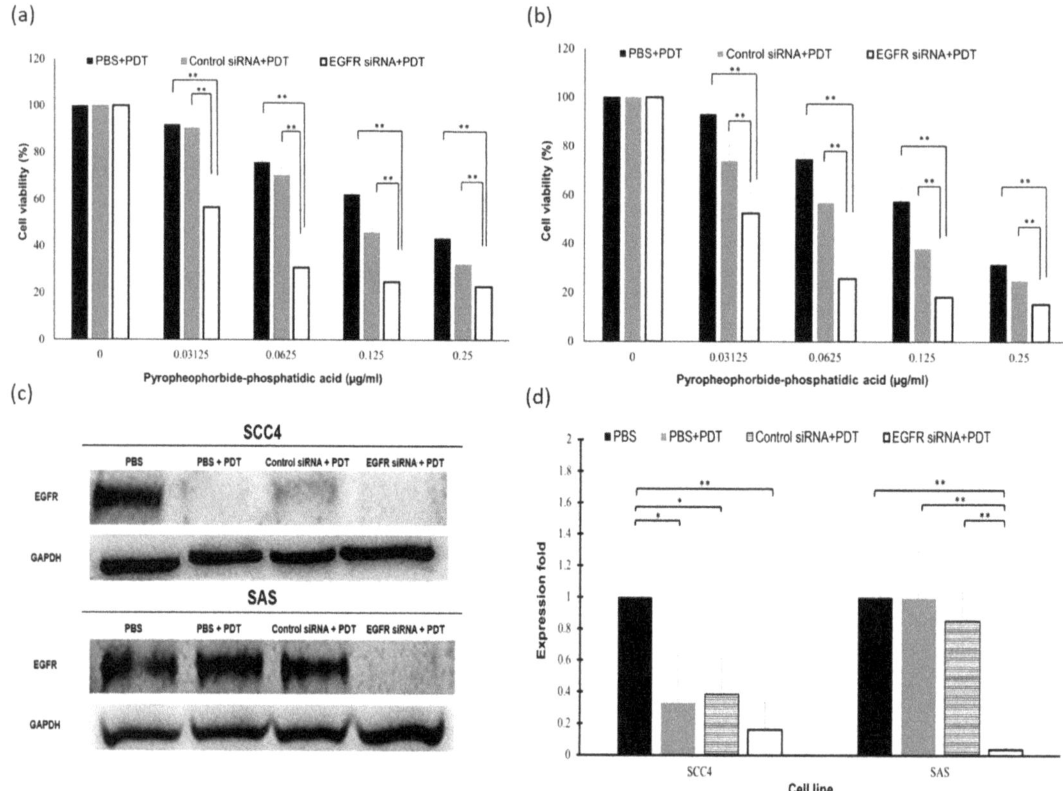

Figure 4. In vitro study of combined therapy. Combined therapy of EGFR siRNA or Control siRNA and photodynamic therapy (PDT) with different concentrations of Photosan photosensitizers in SCC4 (**a**) or SAS (**b**) cell lines. (**c,d**) Western blots of EGFR protein level in siRNA treated SCC4 and SAS cells. Columns, mean ($n = 3$); * stands for $p < 0.05$ compared to PBS group. ** stands for $p < 0.01$ compared to PBS group.

The tumors of the PBS group increased in size by up to 213% and 270% after 13 days in the SCC4 and SAS xenograft models, respectively (Figure 5a,b). On day 13, the LCP siControl+PDT group showed decreased tumor size after the PDT treatments on days 3 and 10 with ~100% and ~145% inhibition of tumor volume in the SCC4 and SAS xenograft models, respectively. Both SCC4 and SAS models demonstrated significant differences ($p < 0.01$) when compared with PBS. On day 13, LCP siEGFR+PS and LCP siEGFR+PBS+light groups showed ~130% and ~175% inhibition of tumor volume in the SCC4 and SAS xenograft models, respectively. Both SCC4 and SAS groups showed tumor volume inhibition resulting from LCP siEGFR in the absence of a complete PDT treatment. The combined group, LCP Pyro PA NPs–mediated PDT, showed significant tumor volume inhibition after LCP siEGFR silencing as compared to the other four groups ($p < 0.01$); combined therapy demonstrated significant tumor volume decrease of ~200% and ~220% in the SCC4 and SAS models, respectively, when compared with untreated PBS groups, with significant differences for both models ($p < 0.01$).

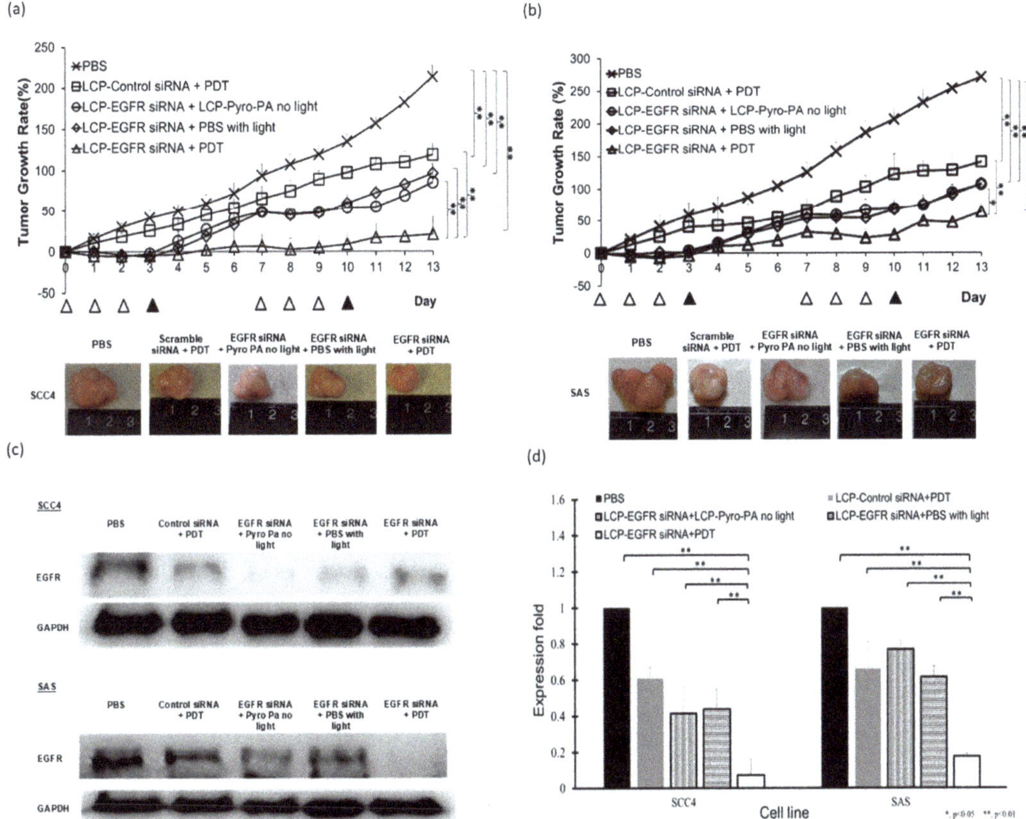

Figure 5. In vivo study of combined therapy. Tumor curves of SCC4 (**a**) and SAS (**b**) xenograft models. The experiments were conducted approximately 3–5 weeks post cell inoculation once the tumor volume reached approximately 200 mm^3. △ indicated IV infusion of LCP with either Control siRNA or EGFR siRNA or PBS; ▲ indicated IV infusion of LCP Pyro PA with or without light. Tumor images of collected SCC4 (**a**) and SAS (**b**) xenograft tumors on day 13. (**c**) Quantified relative EGFR phosphorylated-EGFR (tyr1173) phosphorylated-EGFR (tyr1092) phosphorylated-EGFR (tyr1045), Her2, AKT, phosphorylated-AKT, cleaved caspase-3 and GAPDH protein expression band intensity in SCC4 and SAS cell line. (**d**) QPCR analysis of EGFR protein expression after treatment for SCC4 and SAS xenograft model. Quantified relative EGFR protein expression normalized against GAPDH. GAPDH protein was applied as the internal normalized protein marker with their respective relative EGFR protein expression. Columns, mean (n = 3); *, $p < 0.05$; **, $p < 0.01$.

Quantitative RT-PCR and Western blot tests were conducted to examine SCC4 and SAS tumor tissues obtained on the 13th day. Protein expression of EGFR for LCP siEGFR NPs groups decreased significantly ($p < 0.01$). All data were analyzed and normalized with GAPDH densities when compared with PBS and LCP siControl+PDT groups for SCC4 and SAS xenograft models (Figure 5c). EGFR expression was consistent with assayed siEGFR mRNA expression for SCC4 or SAS xenograft animal models (Figure 5d) and significant differences ($p < 0.01$) were found in LCP siEGFR NPs compared with PBS and the three other treated groups.

3.5. Combination Therapy Does Inhibit HNSCC Tumor Growth Efficiently

Hematoxylin and eosin (H&E) staining assays for SCC4 or SAS tumors tissues obtained from xenografted mice after treatments were conducted to observe tumor, liver and kidney damage. No cell damage was found in livers or kidneys (Figure 6a,b).

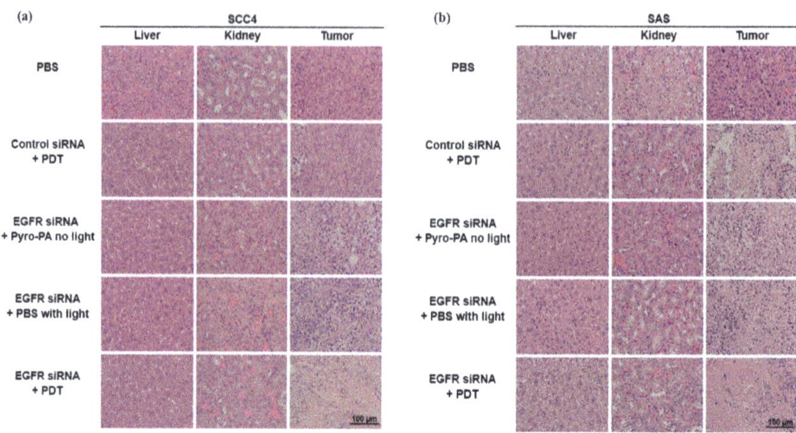

Figure 6. Hematoxylin and eosin staining of SCC4 and SAS Tumors. Hematoxylin and eosin histopathological stains of liver and kidney organs and SCC4 (**a**) and SAS tumors (**b**) of xenograft mice. Original magnification 40×. Scale bar stands for 100 μm.

The following assays were conducted on SCC4 or SAS tumors for the EGFR marker: Ki67 marker was used for cell proliferation staining; CD31 marker was used for microvessel formation; cleaved caspase-3 was used to mark tumor cell apoptosis and TUNEL was used to stain DNA fragmentation. All of these assays were evaluated and quantified using immunohistochemistry staining (IHC) (Figure 7b–f,h–l). EGFR expression was observed at high levels with tumors of the PBS and LCP siControl+PDT groups, suggesting it was highly expressed in SCC4 and SAS cells. Both SCC4 and SAS tumors showed higher EGFR protein expression levels in the PBS and LCP siControl+PDT groups than in the three siEGFR-loaded LCP NPs groups. These three groups treated with LCP siEGFR nanoparticles showed lower EGFR protein expression (Figure 7b,h), which suggests that EGFR genes were silenced. This EGFR gene knockdown effect could lead to tumor cell apoptosis, DNA fragmentation, and reduced microvessel blood vessel formation, especially with the LCP siEGFR NPs+PDT combined therapy. For the LCP Pyro PA NPs–mediated PDT, LCP siEGFR, and both combined treatments groups, there were increased caspase-3 levels, representing apoptotic cells. Significant differences in the expression of TUNEL positive cells and cleaved caspase-3 between all treatment groups compared to the PBS group ($p < 0.01$, $p < 0.05$) were found in both SCC4 and SAS xenograft models (Figure 7c,d,i,j). The combined LCP siEGFR+PDT group showed significantly drastic cell destruction effects that led to 55% and 40% cell apoptosis detected via TUNEL assay (Figure 7c,i) and 35% and 50% cleaved caspase-3 (Figure 7d,j) expression in SCC4 and SAS models, respectively. Moreover, the LCP siEGFR+LCP-Pyro PA–mediated PDT showed a significant difference from LCP siControl+PDT, indicating that the LCP siEGFR NPs enhanced the apoptotic effect on cancer cells when combined with LCP Pyro PA–mediated PDT in SCC4 and SAS xenograft models. The quantified positive field of Ki67 cells indicated a stronger inhibition effect in the LCP siEGFR+PDT group compared to PBS, LCP siControl+PDT, LCP siEGFR+LCP Pyro PA without light, and LCP siEGFR+PBS+light groups ($p < 0.01$) for SCC4 (Figure 7e) and SAS (Figure 7k). This showed that the combination of LCP siEGFR NPs and LCP Pyro PA–mediated PDT enhanced the inhibition effect of the cell proliferation activity significantly. LCP siControl+PDT, LCP siEGFR+LCP-Pyro PA without light, and LCP siEGFR+PBS+light groups demonstrated significant inhibition ($p < 0.01$) in Ki67 amounts compared to PBS group for SCC4 and SAS xenograft models, revealing that LCP-Pyro PA–mediated PDT alone or siEGFR-loaded LCP NPs alone contributed significantly decreasing tumor cell proliferation. The quantified CD31 microvessel intensities of SCC4 and SAS tumors (Figure 7f,l) showed a dramatic decrease in groups treated with LCP siEGFR NPs for SCC4 and SAS models compared with PBS group ($p < 0.01$ for both). These groups exhibited significant differences compared with the LCP siControl+PDT ($p < 0.01$) group, suggesting that LCP siEGFR NPs showed a positive impact on the inhibition of microvasculature formation of the HNSCC tumors.

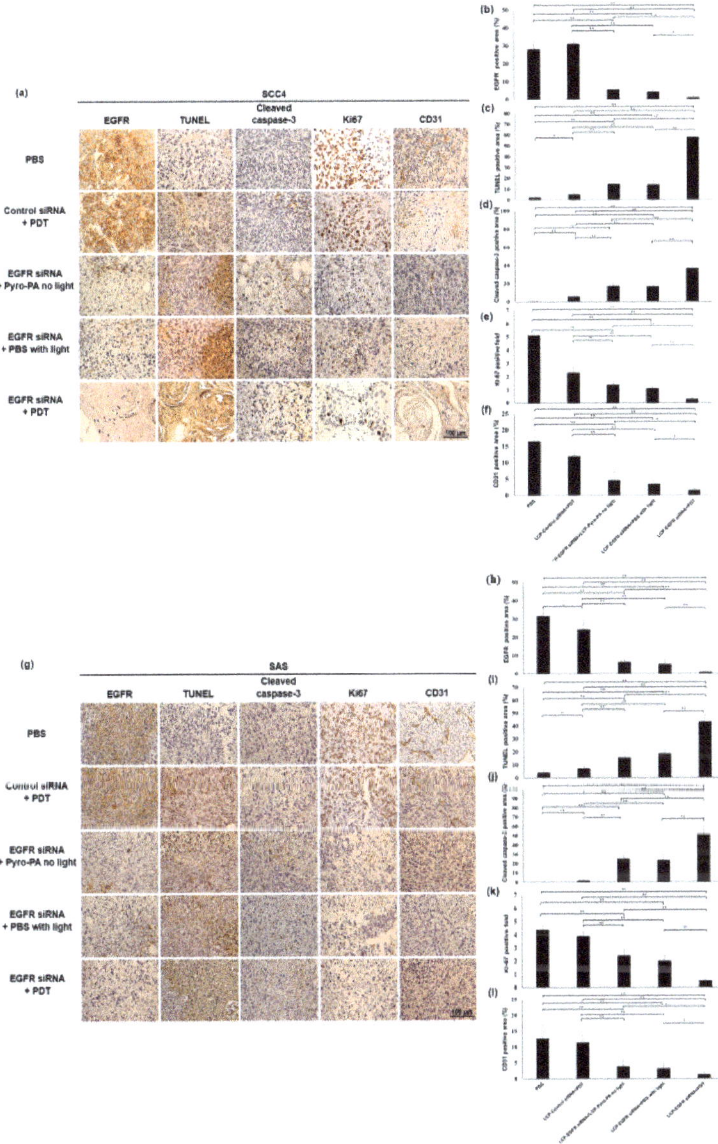

Figure 7. Immunohistochemistry of SCC4 and SAS Tumors; SCC4 tumor cell proliferation and apoptotic cells. Phosphoylated-EGFR (**a**), TUNEL, cleaved caspase-3, Ki-67 and CD31 assays for SCC4 tumor tissue sections. Original magnification 40×. Scale bar stands for 100 μm. Quantitative analysis for Phosphoylated-EGFR (1092) (**b**), TUNEL (**c**), Cleaved Caspase-3 (**d**), Ki-67 (**e**) and CD31 (**f**). SCC4 tumor cell proliferation and apoptotic cells. Phosphoylated-EGFR (**g**), TUNEL, cleaved caspase-3, Ki-67 and CD31 assays for SAS tumor tissue sections. Original magnification 40×. Scale bar stands for 100 μm. Quantitative analysis for phosphoylated-EGFR (1092) (**h**), TUNEL (**i**), cleaved caspase-3 (**j**), Ki-67 (**k**) and CD31 (**l**). Columns, mean (7 images); * $p < 0.05$; ** $p < 0.01$.

3.6. Toxicity and Inflammatory Cytokine Studies In Vivo

The siEGFR-loaded LCP NPs, siControl-loaded LCP NPs and LCP-Pyro PA NPs were investigated for liver and kidney toxicity. C57BL/6 mice were used for the studies. Twelve mice were distributed into four groups that were given three IV infusions of PBS, LCP Control siRNA NPs, LCP EGFR siRNA NPs and LCP-Pyro PA NPs for three days, one injection per day. Mice were sacrificed by cardiac puncture to collect the blood for assays on the fourth day. For liver function, there was no significant difference in aspartate aminotransferase (AST), alanine aminotransferase (ALT), and total-bilirubin (T-BIL) ($p > 0.05$) for LCP siControl, LCP siEGFR, and LCP Pyro PA groups compared to the PBS group. This suggests no liver function damage. Similarly in the kidney index, the creatinine (CREA), blood urea nitrogen (BUN), uric acid (UA), phosphorus (P), and calcium (CA) amounts indicated no significant effect ($p > 0.05$) (Figure 8a). No abnormal effect of LCP NPs on kidney function was noted, even though their cores are made of calcium phosphate. In Figure 8b–e, no significant difference was observed on the levels of interleukin 6 (IL-6), interleukin 12 (IL-12 P40/P70), interferon gamma (INF-γ), and Toll-like receptor 3 (TLR3) in triplicate ($p > 0.05$) for all groups (PBS, LCP siControl, LCP siEGFR, and LCP Pyro Pa). This further indicates no toxicity resulting from significant activation by LCP NPs of IL-6, IL-12, INF-γand TLR3 in mouse serum ($p > 0.05$). These in vivo toxicity assays suggest that both novel LCP Pyro PA NPs and LCP siRNA NPs are safe based on SCC4 and SAS animal model studies (Figure 8b–e).

(a)

	AST (U/L)	ALT (U/L)	T-BIL (µg/L)	CREA(mg/dL)	BUN (mg/dL)	UA (mg/dL)	P (mg/dL)	CA (mg/dL)
PBS	70.9 ± 17.3	34.9 ± 12.1	27.9 ± 6.1	0.2 ± 0.0	31.6 ± 1.6	7.1 ± 1.6	13.6 ± 0.9	13.0 ± 0.7
LCP-Control siRNA	96.3 ± 25.1	31.5 ± 10.0	32.5 ± 18.0	0.3 ± 0.0	33.0 ± 5.4	8.9 ± 3.3	13.3 ± 0.7	13.4 ± 0.9
LCP-EGFR siRNA	65.3 ± 22.9	26.2 ± 3.9	16.7 ± 7.1	0.2 ± 0.0	31.3 ± 1.1	6.7 ± 1.2	12.4 ± 0.5	12.7 ± 0.2
LCP-Pyro-PA	61.0 ± 12.3	29.3 ± 8.0	26.0 ± 2.8	0.3 ± 0.0	29.7 ± 5.5	8.0 ± 2.2	12.9 ± 1.4	13.0 ± 0.7

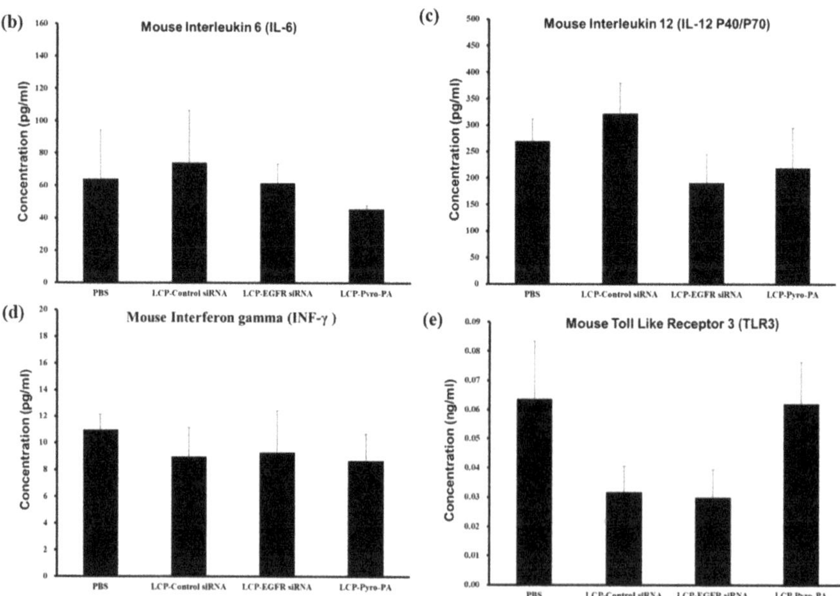

Figure 8. In vivo toxicity and inflammatory response. (**a**) The serum levels of C57BL/6 mice for liver Figure 6. Mouse Interleukin 6 (IL-6) (**b**), interleukin-12 p40/70 (**c**), interferon gamma (**d**), Toll-like receptor 3 (**e**). Data presented as mean ± SD ($n = 3$), * $p > 0.05$ compared to PBS (control) group.

4. Discussion

Combined therapy has become the mainstream present-day strategy for treating head and neck cancer patients. Clinical practices for head and neck cancer therapy include surgery, chemotherapy and radiotherapy. PDT is a noninvasive modality for cancer therapy with minimum side effects [3,4]. In this study, we were the first to design and formulate novel liposomal Pyro PA nanoparticles with targeted ligand AEAA for photodynamic therapy. The combined treatment of targeted LCP Pyro PA with targeted LCP siRNA NPs in vitro and in vivo showed positive preclinical outcomes. Significantly, this is the first design and application of the novel EGFR siRNA sequence in LCP EGFR siRNA targeting EGFR to human HNSCC. The design mechanism of drug release of the LCP NPs was based on cores of calcium phosphate LCP nanoparticles. LCP NPs are pH-sensitive nanoparticles that release drugs locally and transiently. LCP NPs are quickly taken up into endosomes and release encapsulated Pyro PA or encapsulated EGFR siRNA into cell cytoplasmic compartments [11,14]. The asymmetric bilayer structure of LCP NPs was made of anionic lipids (DOPA), cationic lipids (DOTAP) and cholesterol, and LCP NPs were PEGylated to avoid uptake by the immune system and to extend the circulation period in order to enhance the endocytosis of LCP NPs to cancer cells [11,14]. The second leaflet of LCP NPs was PEGylated and also modified with AEAA target ligands, which target the over-expressed sigma receptors in several kinds of cancer cells, including HNSCC cells [9]. LCP Pyro PA NPs and EGFR siRNA NPs were dispersed evenly with similar particle sizes on TEM photomicrographs (Figures 2 and 3). The particle sizes of targeted EGFR siRNA NPs and targeted LCP Pyro PA NPs averaged 34.9 ± 3.0 nm and 15–20 nm, with zeta potentials of 50.1 ± 1.8 mV and 52.0 ± 7.6 mV, respectively. Cabral et al. compared a range of nanoparticle sizes including 30, 50, 70 and 100 nm for accumulation and effectiveness in several poorly and highly penetrable tumor models. Results suggest that 30~50 nm nanomedicines could be the optimal nanoparticle size to penetrate poorly permeable tumors for achieving a better therapeutic effect [16]. Perrault et al. reported that nanoparticle sizes larger than 100 nm exhibited limited permeation into tumors but could be feasible for antiangiogenic therapy; however, the therapeutic effect on malignant tumors was limited due to the larger nanoparticle sizes [17]. The nanoparticle sizes obtained for both LCP Pyro PA NPs and LCP siEGFR NPs were within the suggested 30~50 nm range to enter into tumor cells via endocytosis [16,17]. From our toxicity studies, the CaP cores were shown to be biodegradable and the double lipid bilayer structure of the nanoparticles was relatively functional with regard to safety, with no observed liver and kidney damage caused by LPC NPs (Figure 8). Edmonds et al. reported that EGFR signaling pathway inhibition leads to enhanced PDT cell toxicity with cell apoptosis upregulation. This suggests that targeting EGFR pathways could be a promising solution for PDT clinical trials for patients with cancer metastasis [18]. Protocols of treatment sequence were investigated in vitro to obtain the best feasible order of therapy, which involved conducting LCP siEGFR NPs gene therapy first, followed by PDT (data not shown). We proved that our novel EGFR siRNA sequence did silence EGFR expression. Combined treatment was implemented to apply siEGFR by IV infusion of LCP siRNA NPs for one injection/day for three days, then conducting LCP Pyro PA NPs–mediated PDT on the fourth day. The therapeutic outcome of PDT was demonstrated in the siControl+PDT group. LCP Pyro PA–mediated PDT enhanced inhibition of tumor proliferation and increased tumor apoptosis. In the incomplete PDT treatment groups (LCP siEGFR+LCP Pyro PA without light and LCP siEGFR+PBS+light), the anti-tumorigenic effects of the LCP siEGFR NPs were observed as LCP siEGFR was effectively delivered to cancer cells resulting in lower levels of EGFR mRNA and protein expression. Significant reduction in tumor volume and cell proliferation (Ki67) and upregulation in cell apoptosis (TUNEL and cleaved caspase-3) were observed. LCP siEGFR gene therapy functioned effectively in vivo. Overall, the combination treatment of siEGFR-loaded LCP NPs and LCP Pyro PA mediated PDT demonstrated the greatest inhibition of tumor volume while maintaining inhibition effect. For combined therapy, mRNA and protein expression of EGFR were significantly inhibited, and consistent results in EGFR immunohistochemical stains were documented. The inhibition of EGFR protein expression was consistent with a decrease in micro-vessel density (CD31 biomarker), which may have been caused by the decrease in Ki67 protein tumor proliferation and enhanced tumor apoptosis (cleaved caspase-3 protein and TUNEL data). In a previous study, liposomal porphysomes were previously formed using pyropheophorbide, with a nanoparticle size of 100nm. Porphysomes have large, tunable extinction coefficients and are effective agents for either photothermal or photoacoustic application [17,18]. In the past, porphysomes were intended for use as a photosensitizer for tumors irradiated under PDT conditions. However, porphysomes did not induce PDT–mediated effects on tissue damage, indicating that porphysomes were ineffective for PDT [18]. Cho et al. investigated a nanocomplex containing dextran sulfate and poly-L-arginine-based polyelectrolyte to deliver EGFR siRNA in a HNSCC xenografted mouse model. The size of this polyelectrolyte nanocomplex was less than 200 nm with a positive zeta potential surface charge. Results showed an enhanced EGFR siRNA uptake efficiency in EGFR gene silencing–tumor cells as well as tumor growth inhibition [19]. However, Cabral et al. reported that nanoparticle size is crucial to the therapeutic success of drug delivery to tumor cells [16]. Nanoparticle sizes less than 50 nm can pass through poorly permeable tumors. These findings suggest that LCP NPs are a better nanodelivery system at <50 nm sizes, promoting effective drug delivery and payloads released to targeted cancer cells. In this study, LCP Pyro PA NPs and LCP siEGFR NPs were both within 20–50 nm in size. A PDT effect was observed for the in vivo LCP-control siRNA+PDT group compared to the PBS group ($p < 0.01$). Most importantly, LCP Pyro PA NPs enhanced EGFR siRNA gene therapy, resulting in a significant decrease in tumor volume ($p < 0.01$) compared with the four other experiment groups (Figure 5).

5. Conclusions

In summary, self-designed siEGFR loaded into LCP NPs and combined with novel LCP Pyro PA NPs–mediated PDT promises to improve the treatment effectiveness in HNSCC. This novel nanomedicine combined targeted LCP Pyro PA mediated PDT and gene therapy in order to effectively silence the EGFR oncogene marker.

Author Contributions: All authors discussed the results and contributed to the final manuscript. Conceptualization, L.H. and Y.-C.H.; Methodology, L.H., C.-H.Y., G.Z., J.C. and Y.-C.H.; Writing—original draft preparation, Pyro PA and Pyro lipid synthesis, G.Z. and J.C.; LPC synthesis, characterization, in vitro and in vivo SAS studies, western blots, histology, qPRC, C.-H.Y.; Writing—review & editing, L.H. and Y.-C.H.; Supervision, Y.-C.H.; Funding acquisition, Y.-C.H. All authors have read and agreed to the published version of the manuscript.

Funding: The work conducted in Hsu lab at Departement of Bioscience Technology, Chung Yuan Christian University was supported by Ministry of Science and Technology grant 104-2221-E-033-018-MY3, 108-2119-M-033-001, 109-2221-E-033-008 and 110-2823-8-033-001, Taiwan. The work conducted in Huang lab was supported by NIH grants CA149363, CA151652 and CA149387. The work conducted in Zheng lab was supported by CIHR Foundation Grant #154326, Canada Research Chair Programs and Princess Margaret Cancer Foundation.

Institutional Review Board Statement: These studies were approved (approval number 104011) and carried out in strict accordance with the recommendations in the Guide for the Care and Use produced by the Institutional Animal Care and Use Committee of Chung Yuan Christian University, Chungli, Taoyuan, Taiwan.

Informed Consent Statement: Not applicable.

Data Availability Statement: Not applicable.

Acknowledgments: We thank Thalia Gray and Mary Lin for reviewing, language editing assistance and inputs on manuscript.

Conflicts of Interest: The authors declare no conflict of interest.

References

1. Stransky, N.; Egloff, A.M.; Tward, A.D.; Kostic, A.D.; Cibulskis, K.; Sivachenko, A.; Kryukov, G.V.; Lawrence, M.S.; Sougnez, C.; McKenna, A.; et al. The mutational landscape of head and neck squamous cell carcinoma. *Science* **2011**, *333*, 1157–1160. [CrossRef] [PubMed]
2. Leemans, C.; Braakhuis, B.; Brakenhoff, R. The molecular biology of head and neck cancer. *Nat. Rev. Cancer* **2011**, *11*, 9–22. [CrossRef] [PubMed]
3. Lin, H.P.; Chen, H.M.; Yu, C.H.; Yang, H.; Wang, Y.P.; Chiang, C.P. Topical photodynamic therapy is very effective for oral verrucous hyperplasia and oral erythroleukoplakia. *J. Oral Pathol. Med.* **2010**, *39*, 624–630. [CrossRef] [PubMed]
4. Chiang, C.P.; Huang, W.T.; Lee, J.W.; Hsu, Y.C. Effective treatment of 7,12-dimethylbenz(a)anthracene–induced hamster buccal pouch precancerous lesions by topical photosan-mediated photodynamic therapy. *Head Neck* **2012**, *34*, 505–512. [CrossRef] [PubMed]
5. Dougherty, T.J.; Gomer, C.J.; Henderson, B.W.; Jori, G.; Kessel, D.; Korbelik, M.; Moan, J.; Peng, Q. Photodynamic Therapy. *J Natl. Cancer Inst.* **1998**, *90*, 889–905. [CrossRef] [PubMed]
6. Todd, R.; Donoff, B.R.; Gertz, R.; Chang, A.L.; Chow, P.; Matossian, K.; McBride, J.; Chiang, T.; Gallagher, G.T.; Wong, D.T. TGF-α and EGF-receptor mRNAs in human oral cancer. *Carcinogenesis* **1989**, *10*, 1553–1556. [CrossRef] [PubMed]
7. Todd, R.; Chou, M.Y.; Matossian, K.; Gallagher, G.T.; Donoff, R.B.; Wong, D.T. Cellular Sources of Transforming Growth Factor-Alpha in Human Oral Cancer. *J. Dent. Res.* **1991**, *70*, 917–923. [CrossRef] [PubMed]
8. Grandis, J.R.; Tweardy, D.J. Elevated levels of transforming growth factor α and epidermal growth factor receptor messenger RNA are early markers of carcinogenesis in head and neck cancer. *Cancer Res.* **1993**, *53*, 3579–3584. [PubMed]
9. Chen, W.H.; Lecaros, R.L.; Tseng, Y.C.; Huang, L.; Hsu, Y.C. Nanoparticle delivery of HIF1α siRNA combined with photodynamic therapy as a potential treatment strategy for head-and-neck cancer. *Cancer Lett.* **2015**, *359*, 65–74. [CrossRef] [PubMed]
10. Lecaros, R.L.; Huang, L.; Lee, T.C.; Hsu, Y.C. Nanoparticle Delivered VEGF-A siRNA Enhances Photodynamic Therapy for Head and Neck Cancer Treatment. *Mol. Ther.* **2016**, *24*, 106–116. [CrossRef] [PubMed]
11. Yang, Y.; Hu, Y.; Wang, Y.; Li, J.; Liu, F.; Huang, L. Nanoparticle Delivery of Pooled siRNA for Effective Treatment of Non-Small Cell Lung Cancer. *Mol. Pharm.* **2010**, *9*, 2280–2289. [CrossRef] [PubMed]
12. Banerjee, R.; Tyagi, P.; Li, S.; Huang, L. Anisamide-targeted stealth liposomes: A potent carrier for targeting doxorubicin to human prostate cancer cells. *Int. J. Cancer* **2004**, *112*, 693–700. [CrossRef] [PubMed]
13. Zhang, Y.; Kim, W.Y.; Huang, L. Systemic delivery of gemcitabine triphosphate via LCP nanoparticles for NSCLC and pancreatic cancer therapy. *Biomaterials* **2013**, *34*, 3447–3458. [CrossRef] [PubMed]
14. Li, J.; Chen, Y.C.; Tseng, Y.C.; Mozumdar, S.; Huang, L. Biodegradable calcium phosphate nanoparticle with lipid coating for systemic siRNA delivery. *J. Control. Release* **2010**, *142*, 416–421. [CrossRef] [PubMed]
15. Li, J.; Yang, Y.; Huang, L. Calcium phosphate nanoparticles with an asymmetric lipid bilayer coating for siRNA delivery to the tumor. *J. Control. Release* **2012**, *158*, 108–114. [CrossRef] [PubMed]
16. Cabral, H.; Matsumoto, Y.; Mizuno, K.; Chen, Q.; Murakami, M.; Kimura, M.; Terada, Y.; Kano, M.R.; Miyazono, K.; Uesaka, M.; et al. Accumulation of sub-100 nm polymeric micelles in poorly permeable tumours depends on size. *Nat. Nanotechnol.* **2011**, *6*, 815–823. [CrossRef] [PubMed]
17. Perrault, S.D.; Walkey, C.; Jennings, T.; Fischer, H.C.; Chan, W.C. Mediating Tumor Targeting Efficiency of Nanoparticles through Design. *Nano Lett.* **2009**, *9*, 1909–1915. [CrossRef] [PubMed]
18. Edmonds, C.; Hagan, S.; Gallagher-Colombo, S.M.; Busch, T.M.; Cengel, K.A. Photodynamic therapy activated signaling from epidermal growth factor receptor and STAT3. *Cancer Biol. Ther.* **2012**, *13*, 1463–1470. [CrossRef] [PubMed]
19. Cho, H.J.; Chong, S.; Chung, S.J.; Shim, C.K.; Kim, D.D. Poly-L-arginine and Dextran Sulfate-Based Nanocomplex for Epidermal Growth Factor Receptor (EGFR) siRNA Delivery: Its Application for Head and Neck Cancer Treatment. *Pharm. Res.* **2012**, *29*, 1007–1019. [CrossRef] [PubMed]

Article

Topical Photodynamic Therapy with Different Forms of 5-Aminolevulinic Acid in the Treatment of Actinic Keratosis

Joanna Bartosińska [1,*], Paulina Szczepanik-Kułak [2], Dorota Raczkiewicz [3], Marta Niewiedzioł [2], Agnieszka Gerkowicz [2], Dorota Kowalczuk [4], Mirosław Kwaśny [5] and Dorota Krasowska [2]

[1] Department of Cosmetology and Aesthetic Medicine, Medical University of Lublin, Chodźki 1 St., 20-093 Lublin, Poland

[2] Department of Dermatology, Venereology and Pediatric Dermatology, Medical University of Lublin, Staszica 11 St., 20-081 Lublin, Poland; vpaulinav@gmail.com (P.S.-K.); martan@akcja.pl (M.N.); agnieszka.gerkowicz@umlub.pl (A.G.); dorota.krasowska@umlub.pl (D.K.)

[3] Department of Medical Statistics, School of Public Health, Center of Postgraduate Medical Education, Kleczewska 61/63 St., 01-826 Warsaw, Poland; dorota.bartosinska@gmail.com

[4] Department of Medicinal Chemistry, Faculty of Pharmacy, Medical University of Lublin, Jaczewskiego 4 St., 20-090 Lublin, Poland; dorota.kowalczuk@umlub.pl

[5] Institute of Optoelectronics, The Military University of Technology, Kaliskiego 2 St., 01-476 Warsaw, Poland; miroslaw.kwasny@wat.edu.pl

* Correspondence: joanna.bartosinska@umlub.pl; Tel.: +48-602724298

Citation: Bartosińska, J.; Szczepanik-Kułak, P.; Raczkiewicz, D.; Niewiedzioł, M.; Gerkowicz, A.; Kowalczuk, D.; Kwaśny, M.; Krasowska, D. Topical Photodynamic Therapy with Different Forms of 5-Aminolevulinic Acid in the Treatment of Actinic Keratosis. *Pharmaceutics* **2022**, 14, 346. https://doi.org/10.3390/pharmaceutics14020346

Academic Editors: Udo Bakowsky, Matthias Wojcik, Eduard Preis and Gerhard Litscher

Received: 20 December 2021
Accepted: 29 January 2022
Published: 1 February 2022

Publisher's Note: MDPI stays neutral with regard to jurisdictional claims in published maps and institutional affiliations.

Abstract: Photodynamic therapy (PDT) is safe and effective in the treatment of patients with actinic keratosis (AK). The aim of the study was to assess the efficacy, tolerability and cosmetic outcome of topical PDT in the treatment of AKs with three forms of photosensitizers: 5-Aminolevulinic acid hydrochloride (ALA-HCl), 5-Aminolevulinate methyl ester hydrochloride (MAL-HCl) and 5-Aminolevulinate phosphate (ALA-P). The formulations were applied onto selected scalp/face areas. Fluorescence was assessed with a FotoFinder Dermoscope 800 attachment. Skin areas were irradiated with Red Beam Pro+, Model APRO (MedLight GmbH, Herford, Germany). Applied treatments were assessed during the PDT as well as 7 days and 12 weeks after its completion. Ninety-four percent of patients rated obtained cosmetic effect excellent. The efficacy of applied PSs did not differ significantly. However, pain intensity during the PDT procedure was significantly lower in the area treated with ALA-P (5.8 on average) in comparison to the areas treated with ALA-HCl or MAL-HCl (7.0 on average on 0–10 scale). Obtained results show that ALA-P may undergo more selective accumulation than ALA-HCl and MAL-HCl. Our promising results suggest that PDT with the use of ALA-P in AK treatment may be an advantageous alternative to the already used ALA-HCl and MAL-HCl.

Keywords: photodynamic therapy; actinic keratosis; photosensitizer; 5-aminolevulinic acid

1. Introduction

Melanoma and non-melanoma skin cancer are the most common types of skin malignancies in the Caucasian population. Their incidence rates continue to rise, which is a matter of great concern to the patient and a substantial economic burden for the health care system [1,2]. Squamous cell carcinoma (SCC) deserves particular attention because of its tendency to metastasize. The risk of metastasis in invasive SCC is estimated to be about 4% and up to 2–3 times higher in immunosuppressed patients. Therefore, in addition to prevention, identification and therapy of early SCC is vital to avoid neoplastic progression [3].

One of premalignant conditions, with a malignant transformation rate ranging from 0.025% to 16%, is actinic keratosis (AK), an epidermal keratinocytic disorder induced by chronic exposure to ultraviolet (UV) light. A precise prediction of the risk of AK evolution into invasive SCC is infeasible [4]. However, it could be stated that AK plays an important role in the development of SCC. According to Criscione et al. [5], who studied a high-risk population to estimate the risk of progression of AK to SCC, approximately 65% of all primary squamous cell cancers developed from previously diagnosed AK. Moreover, subclinical and early AK lesions are also capable of direct transformation into invasive malignant disease [6]. All this information is sufficient to conclude that each detected AK lesion needs to undergo appropriate therapy, which is especially relevant to patients with multiple AK lesions in whom the risk of progression to SCC is higher [4,7].

Two separate treatment modalities are distinguished for AK, i.e., lesion-directed therapy and field-directed therapy [4]. The main focus of the latter includes treating subclinical lesions, reducing AK recurrence rates, and potentially lowering the risk of developing SCC. Application of the field-directed therapy is based on the theory of "field cancerization", according to which the skin surrounding the lesions, due to chronic exposure to UV radiation, has the potential to transform to malignancy [8].

Photodynamic therapy (PDT) with the use of 5-aminolevulinic acid hydrochloride (ALA-HCl) as a photosensitizer (PS) and irradiation with blue light for the treatment of AK was approved by the FDA in the year 2000. In 2016, the FDA approved the use of ALA-HCl in combination with red light [8]. At present, due to the high treatment efficacy and a good cosmetic outcome, PDT is a routine, first-line treatment for AK. This type of therapy causes local and selective destruction of the neoplastic skin lesions without damage to the healthy tissue. PDT is a hardly invasive method, is generally well tolerated, is repeatable [9] and is applied either directly onto the lesion or onto the entire field of cancerization [4]. (Figure 1).

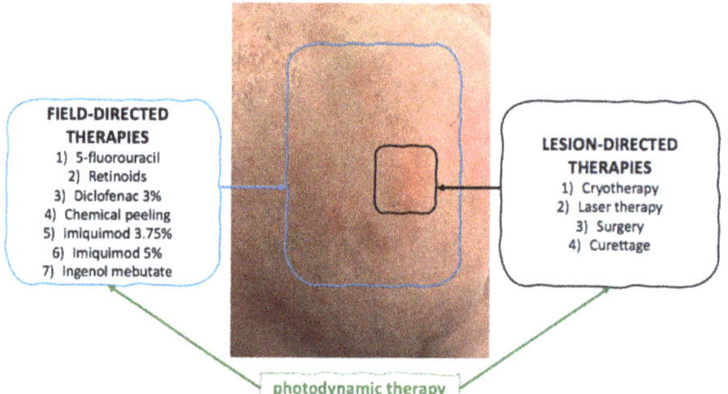

Figure 1. The methods of lesion- and field-directed therapies in actinic keratosis.

PDT has been shown to eliminate AK lesions and prevent their recurrence [8,10]. The action of PDT is based on selective photooxidation of the lesional tissue with the simultaneous involvement of three indispensable components, i.e., a PS, oxygen, and light of appropriate wavelength. The used PS accumulates in the dysplastic cells as well as in all the cells with high proliferation rates, where it is enzymatically metabolized via the heme pathway into the active endogenous photosensitizer protoporphyrin IX (PpIX). When the PS-treated skin is exposed to a light source that spans the absorption spectrum of PpIX (400–730 nm), the compound becomes photoactivated and triggers a photochemical reaction that generates cytotoxic singlet oxygen and free radicals, resulting in subsequent tissue loss [11,12]. The course of PDT and its efficacy are substantially dependent on PpIX production, distribution and depth of its penetration into the skin [13].

ALA-HCl and methyl aminolevulinate (MAL) are the most widely investigated PSs for the treatment of AK lesions. ALA-HCl and MAL are commercially available as Levulan® and Metvix®, respectively [1]. There are reports of some disadvantages associated with these two PSs, especially pain experienced by patients during PDT, which seems to be the main adverse effect and a discouraging factor [14]. Therefore, new substances with more versatile properties are in the pipeline. Unlike ALA-HCl and MAL, the use of 5-aminolevulinic acid phosphate (ALA-P) has not yet been investigated as a PS in PDT of AKs.

The aim of the study was to assess the efficacy, tolerability and cosmetic outcome of topical PDT in the treatment of AK lesions with the use of three different forms of 5-ALA, i.e., 5-Aminolevulinic acid hydrochloride (ALA-HCl), 5-Aminolevulinate methyl ester hydrochloride (MAL-HCl) and 5-Aminolevulinic acid phosphate (ALA-P).

2. Materials and Methods

2.1. Study Group

Twenty-two Caucasian adults (of II and III skin phototypes), within the age range from 60 to 84 years, with multiple mild to severe AK lesions (Grade I-III according to Olsen) localized on the face and/or scalp were included in the study. The patients received PDT treatment at the outpatient clinic of the Department of Dermatology at the Medical University of Lublin, Poland between April 2018 and July 2021.

Inclusion criteria were as follows: multiple AK lesions Grade I-III according to Olsen and willingness to receive therapy and to participate in follow-up visits. Exclusion criteria were: pregnancy, epilepsy, history of photodermatosis, taking photosensitizing medication, and receiving any AK topical therapies at least three months prior to the beginning of the study. Written informed consent was obtained from all the patients before enrollment in the study. The study was approved by the local ethic committee (KE-0254/286/2019).

2.2. Study Formulations

In the study, 5-Aminolevulinic acid hydrochloride (ALA-HCl; 5-amino-4-oxopentanoic acid hydrochloride; $C_5H_{10}ClNO_3$; $[HOOC–CH_2–CH_2–CO–CH_2–NH_3^+]Cl^-$); 5-Aminolevulinic acid phosphate (ALA-P; pentanoic acid, 5-amino-4-oxo-, phosphate (1:1); $C_5H_{12}NO_7P$; $[HOOC–CH_2–CH_2–CO–CH_2–NH_3^+]H_2PO_4^-$); 5-Aminolevulinate methyl ester hydrochloride (MAL-HCl; methyl 5-amino-4-oxopentanoate hydrochloride; $C_6H_{12}ClNO_3$; $H_3COOC–CH_2–CH_2–CO–CH_2–NH_3^+]Cl^-$), obtained from Arisun chempharm Co., Ltd., Xi'an, China, were used PSs. The purity of ALA-HCl was 99.5% and of ALA-P was 99.2%, whereas MAL-HCl (purity of 97.0%) had to be further purified (recrystallization from methyl alcohol in the acidic environment) until its purity reached 99.5%. The purity of the used substances was confirmed with the standards by high-performance liquid chromatography (HPLC). The lipophilicity (LogP value) for the substances to be examined were calculated using the computer programs based on different calculation methods: ACD/LogP (Advanced Chemistry Development Inc., Toronto, ON, Canada; http://www.acdlabs.com (accessed on 17 January 2022)) ALogPs (VCCLAB, Virtual Computational Chemistry Laboratory, German Research Center for Environmental Health, Neuherberg, Munich, Germany; http://www.vcclab.org (accessed on 17 January 2022)) and miLogP (Molinspiration Software; http://www.molinspiration.com/cgi-bin/properties (accessed on 17 January 2022).

In order to obtain the study formulations, each of the three substances was added to the creamy LIPOBAZA. Since each formulation was to contain 10% of pure 5-ALA, the following concentrations of the study formulations were used: ALA-HCl—12.7%, MAL-HCl—12.5%, ALA-P—17.5%. The degree of homogenization was controlled with the use of an optical microscope.

2.3. Treatment Protocol

2.3.1. Application of Study Formulations

In order to avoid the patients' subjective pain assessment, after the scales and crusts were gently removed from a selected face or scalp field designated for the treatment, in each patient, the skin field to be treated was divided into three roughly equal areas onto which a 1 mm-thick layer of the ALA-HCl, MAL-HCl or ALA-P formulation was applied. Caution was taken to avoid overlapping and mixing of the used formulations. Finally, an occlusive plastic dressing and aluminum foil were placed over the treated skin surface. (Figure 2).

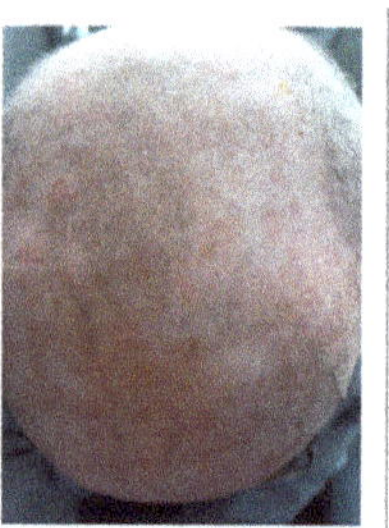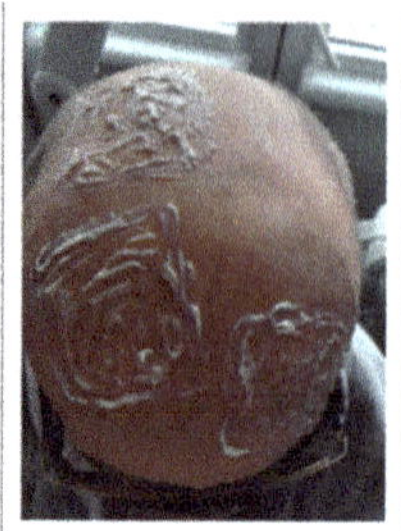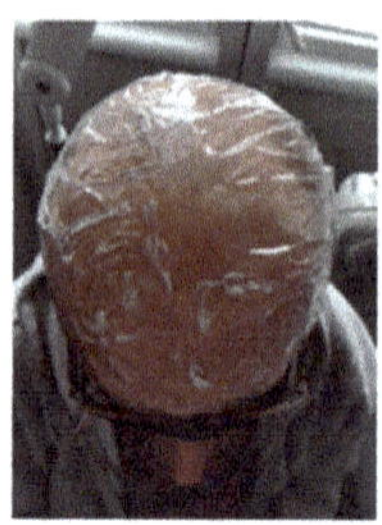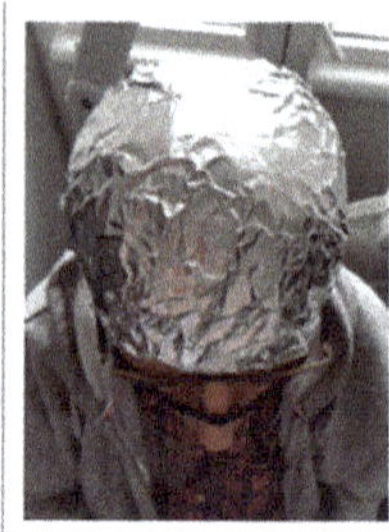

Figure 2. Application of study formulations in PDT of AK lesions.

After a 3 h incubation period, the applied dressing and the remains of the study formulations were removed with 0.9% saline solution swabs.

2.3.2. Assessment of Skin Fluorescence Following Application of Study Formulations

Qualitative absorption of the PSs contained in each of the study formulations was assessed with the use of a special attachment of FotoFinderDermoscope 800 with white and violet LED diodes emitting LED light. The opaque special material of the FotoFinder FD lens was used. (Figure 3a) The fluorescence coming from PpIX formed after application of each PS was rated as high, medium and low, as well as diffuse, confluent and blotchy.

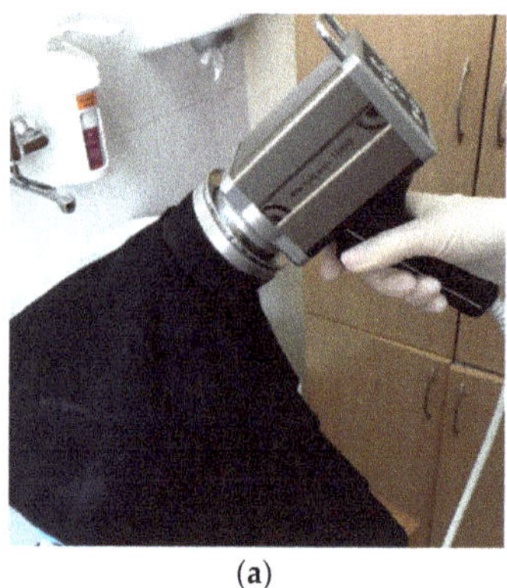

(**a**)

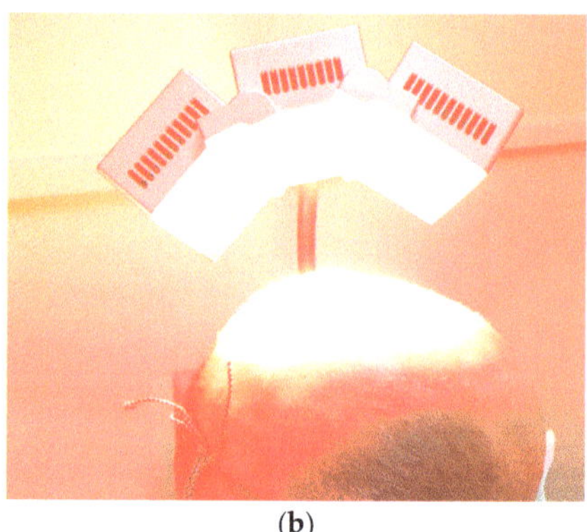

(**b**)

Figure 3. The use of FotoFinderDermoscope 800 to assess the fluorescence (**a**); the use of MedLight GmbH red light 630 ± 5 nm (**b**).

2.3.3. Irradiation with Red Light 630 ± 5 nm

The field-directed PDT was performed on the selected areas of the face and/or scalp. The study participants were instructed to wear protective glasses during irradiation.

Irradiation sessions were conducted with the use of Red Beam Pro+, Model APRO (MedLight GmbH, Herford, Germany), which insures optimal light penetration. (Figure 3b).

The Red Beam Pro+ lamp contains three movable units with 78 high power red light-emitting diodes, which provide red light operating at 630 ± 5 nm [9]. A total dose per session was $37\,\mathrm{J/cm^2}$, and the light power intensity was equal to $68\,\mathrm{mW/cm^2}$. The distance from the face and/or scalp was 10 cm. The patients were instructed to avoid sun exposure for 48 h following the treatment.

2.3.4. PDT with Study Formulations Efficacy Assessment

Assessment of the percent of AK lesion clearance in each treated area was performed 12 weeks after PDT completion.

2.3.5. PDT with Study Formulations Cosmetic Outcome and Patient Satisfaction Assessment

Twelve weeks after the PDT completion, cosmetic outcome assessment was performed with the use of a four-grade scale (Table 1), and the study participants were asked to express their level of satisfaction with the PDT and willingness to repeat the therapy if the need be.

Table 1. Cosmetic outcome of photodynamic therapy assessment.

Grade	Definition
Poor	extensive occurrence of scarring, atrophy, or induration
Fair	slight to moderate occurrence of scarring, atrophy, or induration
Good	no scarring, atrophy, or induration, moderate redness or increase in pigmentation compared with adjacent skin
Excellent	no scarring, atrophy, or induration, slight or no redness or change in pigmentation compared with adjacent skin

2.3.6. PDT with Study Formulations Tolerability Assessment

All the study participants were assessed by the same dermatologists during the procedures, shortly after their completion, and 7 days after the PDT completion.

During the procedure and 7 days after, the study participants were asked to evaluate pain intensity in each treated area on a 10-point VAS scale, and their exact location was determined with a pointer.

Frequency and severity of occurrence of PDT side effects, i.e., erythema, edema, desquamation and crusting, were defined as: none, mild, moderate, or severe.

Erythema and edema were assessed shortly after and 7 days after PDT.

Exfoliation, crusting and pigmentation were assessed 7 days after PDT.

Photographs of the skin lesions were taken at every visit.

2.4. Statistical Methods

The data were analyzed using STATISTICA 13 software (Statsoft, Kraków, Poland). The mean and standard deviation were estimated for numerical variables, as well as for absolute numbers (n) and percentages (%) of the occurrence of items for categorical variables.

Wilcoxon's signed rank test for paired samples was used to compare severity of pain (10 point scale from 0 to 10) or clearance (in percentage) between every pair of two photosensitizers between the first and the second procedures.

The significance level was assumed to be 0.05 in all statistical tests.

3. Results

3.1. Characteristics of the Study Group

Characteristics of the AK patients and AK lesions are presented in Table 2.

Table 2. Characteristics of the studied actinic keratosis patients.

Variable	Category	Parameter	Estimate
Age	years	Min-max	60–84
Sex	male	n (%)	21 (95%)
	female		1 (5%)
Localizations of lesions	face	n (%)	4 (18%)
	scalp		18 (82%)
Thickness grade, according to Olsen et al. [14]	I grade (thin)	n (%)	9 (24%)
	II grade (moderately thick)		18 (49%)
	III grade (thick)		10 (27%)

Since extensive AK lesions are more frequently observed in males, we selected 21 men (18 with AK lesions on the scalp and 3 on the face) as well as 1 woman presenting with facial AK lesions.

Each of the study subjects completed the first PDT procedure, and because of extensive AK lesions, 16 of them qualified for a second PDT session. One patient refused to continue the treatment because of severe pain experienced during the first PDT procedure.

3.2. Fluorescence Assessment Following Application of Study Formulations

Assessment of fluorescence intensity was performed for each of the study subjects before application of irradiation with red light 630 ± 5 nm. In the majority of patients (88% of them in the areas treated with ALA-HCl and 85% of them in the areas treated with MAL-HCl), the fluorescence intensity was high and confluent, while in the majority of patients (82% of them in the areas treated with ALA-P), the fluorescence intensity was lower and blotchy. (Figure 4).

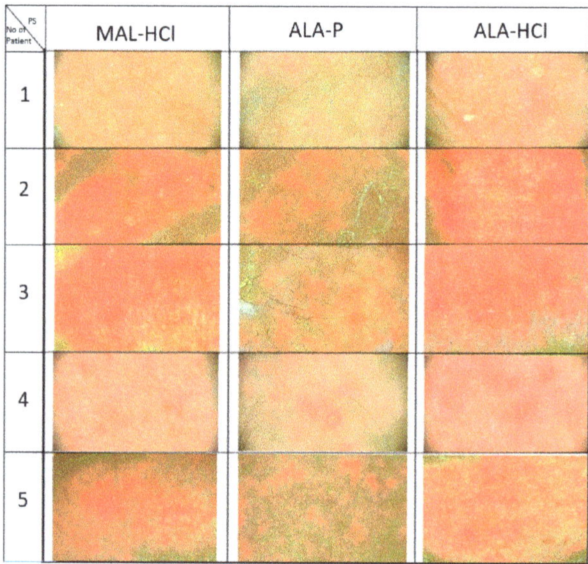

Figure 4. Intensity and distribution of fluorescence after application of ALA-HCl, MAL-HCl, and ALA-P.

3.3. PDT with Study Formulations Efficacy

Clearance 12 weeks after the first and second PDT procedures performed with ALA-HCl, MAL-HCl and ALA-P is presented in Figure 5. The clearance of AK lesions 12 weeks after the first PDT procedure in 22 patients was 83.9 ± 7.7%, 88.2 ± 7.5% and 86.6 ± 7.6% on average for ALA-HCl, MAL-HCl and ALA-P, respectively, while 12 weeks after the second PDT procedure in 15 patients, it was 82.7 ± 5.6%, 86.3 ± 7.2% and 83.7 ± 6.9%, respectively. The efficacy of any of the applied PSs did not differ significantly in the overall treatment of AK lesions ($p > 0.05$) both after the first and second PDT procedure.

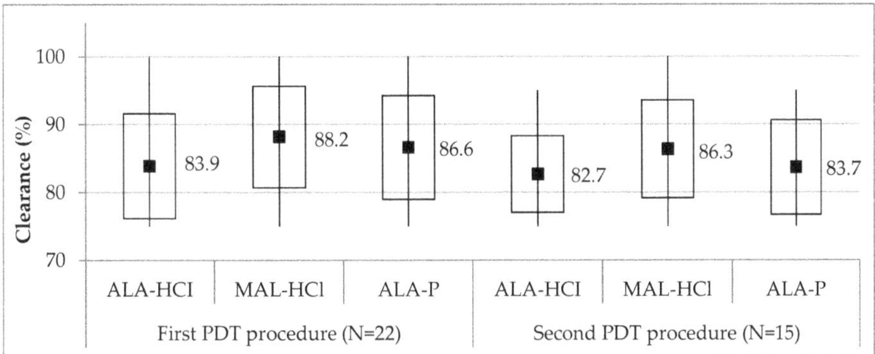

Figure 5. Clearance of actinic keratosis lesions 12 weeks after the first and second PDT procedures with the use of study formulations. Midpoint, mean; box, mean ± standard deviation; whiskers, min–max. Clearance did not significantly differ among three studied PSs as well as between the first and second procedures ($p > 0.05$). p stands for Wilcoxon's signed rank test for paired samples.

3.4. PDT with Study Formulations COSMETIC Outcome and Patient Satisfaction

The vast majority of the study participants evaluated the overall cosmetic effect as excellent (94%), and the remaining 6% of them described it as good. Since scarring, atrophy, or induration were not observed, none of the participants found the PDT cosmetic outcome to be fair or poor. Patient satisfaction with the treatment was high, except for one patient who did not want to repeat the procedure because of excruciating pain experienced during the first PDT procedure.

3.5. PDT with Study Formulations Tolerability

During PDT, all the studied patients reported at least one adverse reaction. Pain was their most frequent complaint, and if it was unbearable, short breaks in irradiation were taken, or a cool dressing was applied.

Pain intensity (on 10-point scale from 0 to 10) during PDT and 7 days after the PDT completion was compared between the PSs grouped in pairs, in both the first and second procedures separately (Figure 6a).

In the 22 studied subjects, pain intensity during the first PDT procedure was significantly lower in the area treated with ALA-P (5.8 on average) in comparison to the areas treated with either ALA-HCl or MAL-HCl (7.0 on average in 0–10 scale). However, pain intensity 7 days after the first PDT procedure was significantly lower for MAL-HCl and ALA-P (1.4 and 1.3 on average, respectively) in comparison to ALA-HCl (1.8 on average).

In 15 patients, pain intensity during the second PDT procedure did not significantly differ between the three used PSs (5.3 on average for ALA-HCl; 5.2 on average for MAL-HCl; 4.7 on average for ALA-P). Pain intensity 7 days after the second PDT procedure was low and did not significantly differ between the three PSs (1.1 on average for ALA-HCl; 1.3 on average for MAL-HCl; 0.9 on average for ALA-P).

Pain intensity (on 10-point scale from 0 to 10) during PDT and 7 days after PDT was compared between the first and second procedures in 15 patients who underwent both procedures (Figure 6b).

Pain intensity was significantly lower during the second PDT procedure than during the first PDT procedure for each PS ($p = 0.001$ for ALA-HCl and MAL-HCl, $p = 0.023$ for ALA-P).

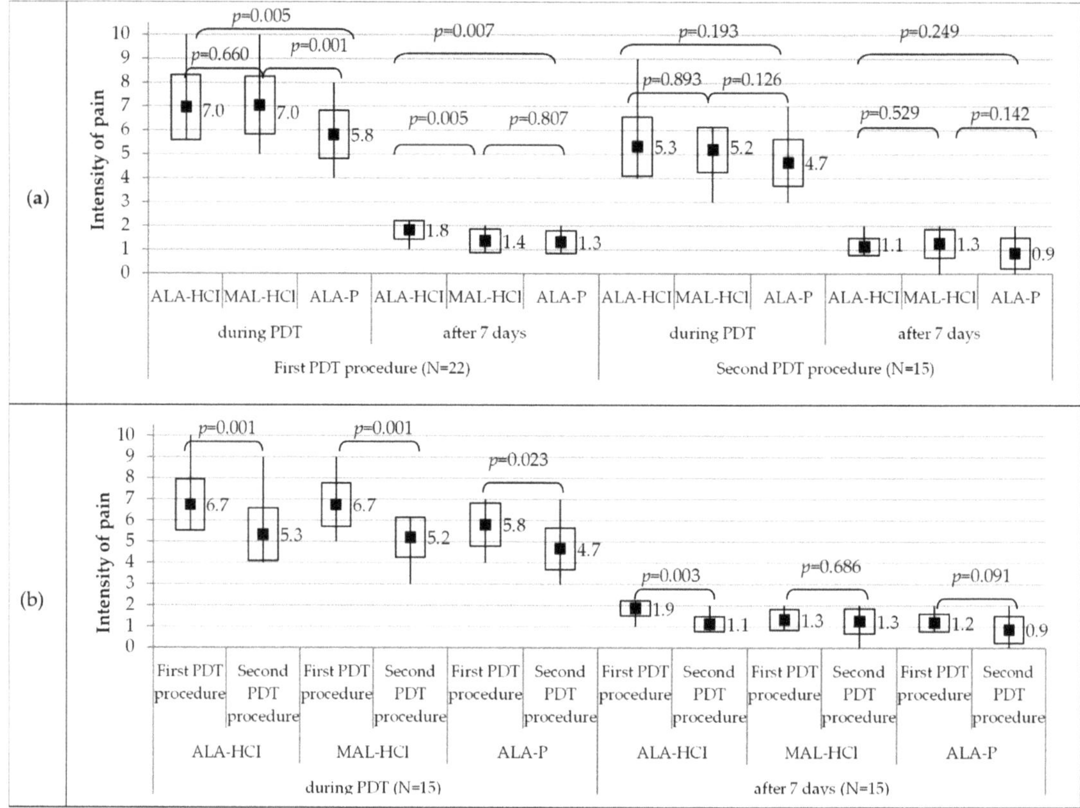

Figure 6. Pain intensity during PDT and 7 days after PDT in the first procedure (*n* = 22 patients) and in the second procedure (*n* = 15 patients): compared between every pair of two photosensitizers (**a**); compared between the first procedure and the second procedure (*n* = 15 patients) (**b**). Pain intensity on 10-point scale from 0 to 10. Midpoint, mean; box, mean ± standard deviation; whiskers, min–max. *p* stands for Wilcoxon's signed rank test for paired samples.

Pain intensity 7 days after the second procedure was significantly lower than 7 days after the first PDT procedure only in ALA-HCl (*p* = 0.003), while it was not significant in MAL-HCl (*p* = 0.686) or ALA-P (*p* = 0.091).

It was noted that the pain experienced during treatment tapered off/subsided together with the time of exposure to the red LED light.

In our study, similar local responses to the three investigated PSs were observed. It was also observed that in the area treated with the ALA-P formulation, the intensity of erythema was mild in over half of the study participants (59%), while in the area where ALA-HCl and MAL-HCl were applied, the erythema intensity was moderate in the vast majority of the studied patients (82% and 91%, respectively). Seven days after the first PDT procedure, the same differences were still observed for the three studied PSs. No differences in erythema intensity were observed between the first and second PDT procedures, both shortly after and 7 days after PDT (Figure 7a).

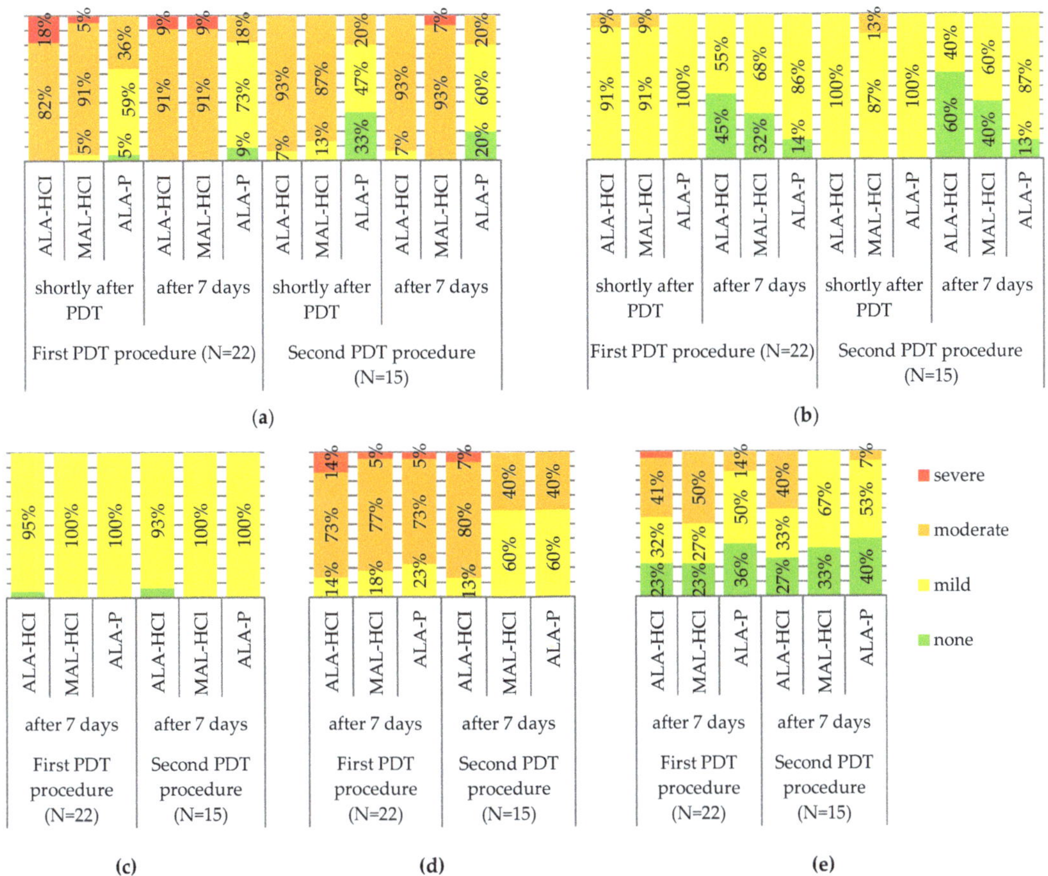

Figure 7. The severity of erythema (**a**), edema (**b**), desquamation (**c**), crusting (**d**), and pigmentation (**e**) in the first and second PDT procedures.

Shortly after and 7 days after PDT completion, edema intensity was mild in all or almost all study subjects regardless of the applied PS during both first and second PDT procedures (Figure 7b).

Seven days after PDT completion, desquamation was mild in all the areas treated with MAL-HCl and ALA-P and in all but one patient treated with ALA-HCl, both in the first and second PDT procedures (Figure 7c).

Seven days after PDT completion, in three-quarters of the study participants, crusting intensity was moderate, regardless of the used formulation in the first PDT procedure, whereas in 80% of the patients, it was moderate in the area treated with ALA-HCl, and in 60% of the patients, it was mild in the areas treated with MAL-HCl and ALA-P (Figure 7d).

Seven days after PDT completion, the lowest pigmentation intensity was observed in the areas treated with ALA-P both in the first and second PDT procedures (Figure 7e).

Some examples of local adverse reactions observed in the studied patients are presented in Figure 8.

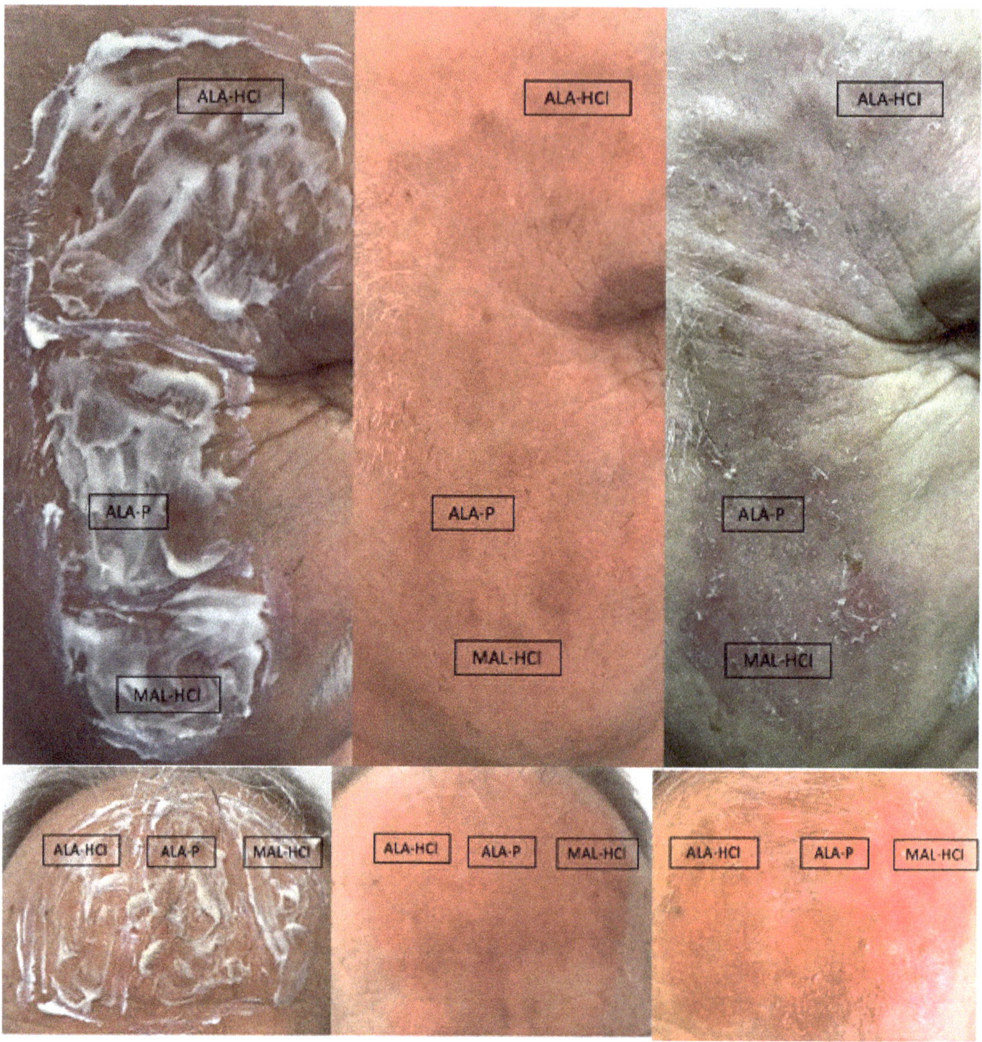

Figure 8. Local adverse reactions observed in the studied patients.

The comparisons of the skin areas in the studied patients before PDT and 12 weeks after procedure completion are shown in Figures 9 and S1 (Supplementary Materials).

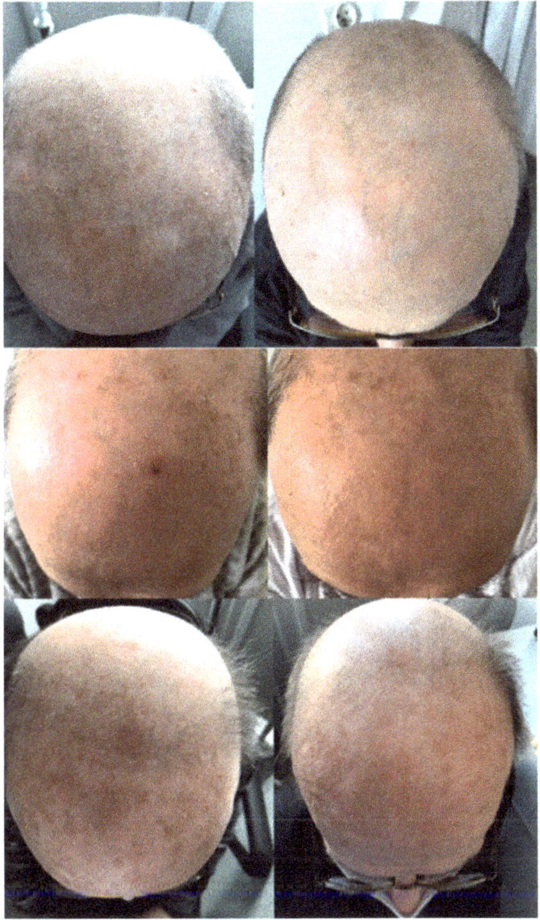

Figure 9. Comparisons of the skin areas in the studied patients before PDT and 12 weeks after procedure completion.

4. Discussion

In actinic keratoses, which are common skin lesions that may progress to invasive SCC, PDT is invaluable because of its minimal invasiveness and high efficacy. This paper presents the first observational, uncontrolled study of the efficacy and tolerability of a novel photosensitizer known as ALA-P and draws comparisons with two other commercially available PSs, i.e., ALA-HCl and MAL-HCl used in PDT of AK lesions. Aminolevulinic acid phosphate ($C_5H_{12}NO_7P$), with a molecular weight of 229.13 g/mol, is a fairly new synthetic chemical compound also known as UNII-FM8DCR39GH; Pentanoic acid, 5-amino-4-oxo-, phosphate (1:1); 868074-65-1; 5-Aminolevulinic acid phosphate. Due to its properties, it has already been approved as a nutritional supplement in Japan and a few other Asian countries [15]. When administered by mouth, ALA-P has been shown to reduce blood glucose levels [16]. Furthermore, Higashikawa et al. observed that ALA-P diminished the severity of negative emotions in individuals continuously feeling physically fatigued [17]. All this made us presume that if orally used ALA-P is safe, effective and well tolerated, it could also be applied onto the skin with no need for in vivo tests to treat AK lesions with the use of PDT.

Thus, the novelty of the method presented in this study consists of the innovative, topical use of ALA-P as a photosensitizer, whereas assessment of the efficacy, tolerability and cosmetic outcome of ALA-P application in comparison with two other photosensitizers, i.e., ALA-HCl and MAL, revealed slightly better tolerance of the PDT procedure with the use of ALA-P.

The use of ALA-HCl and MAL in PDT has proven to be highly effective [18,19]. Nevertheless, the number of papers directly comparing the clinical outcomes of PDTs with ALA-HCl and MAL in AK patients is limited [18]. Moloney et al. performed a randomized, double-blind, prospective study comparing the efficacy and adverse effects of MAL-PDT and ALA-PDT in the treatment of scalp AKs. Their results showed that both ALA-PDT and MAL-PDT caused a significant reduction in the number of AK foci, with no significant difference in efficacy [20]. Fu et al., in a recent meta-analysis investigating the combination of PDT with BF-200 ALA (a 5-aminolevulinic acid nanoemulsion of ALA) versus MAL, indicated that PDT with the former (i.e., BF-200 ALA) had a 9% better chance of complete clearance of AK lesions at 3 months of treatment and a 24% better chance of grade II-III AK lesion clearance in comparison to the results of PDT with the use of the latter (i.e., MAL) [21]. In our study, the clearance assessed 12 weeks after treatment completion demonstrated similarly high PDT efficacy (clearance above 80%), regardless of the applied Ps ($p > 0.05$).

PDT is generally considered to give good cosmetic results with high patient satisfaction. This seems to be of particular importance in the treatment of the lesions localized on the exposed parts of the body [22]. In our study, 94% of the patients found the overall cosmetic effect to be excellent, while the remaining 6% assessed it as good, regardless of the used formulation. A study by Ko et al. compared PDT performed with ALA-HCl as well as MAL in AK patients. The cosmetic outcome after the treatment with the former was rated by 90% of the patients as excellent, while PDT performed with the latter was assessed as excellent by 97% of the study participants, which is in agreement with our results. The excellent result was still observed 12 months after the completion of treatment [23]. Räsänen et al. [24] compared the cosmetic outcome of PDT with the use of either BF-200 ALA or MAL in 69 patients with numerous AK lesions. In their study, 12 months after treatment completion, the cosmetic outcome was excellent or good in >90% of the studied patients, regardless of the used PS.

In order to make sure that PDT is highly effective, the following components need to be present, i.e., proper light energy source, oxygen and a PS with desirable properties such as water solubility, lipophilicity, penetration capability and accumulation in the skin.

In order to demonstrate the accumulation of the investigated PSs in the treated skin area, photodynamic diagnostics with white and violet LED diodes emitting LED light may be used [25]. This type of diagnostic involves the use of PpIX fluorescence, thereby enabling not only determination of PS accumulation in the diseased tissue but also establishing the boundaries between healthy and diseased tissues [26] and the extent and nature of the neoplastic skin lesions, malignant and non-malignant [27].

The intensity of red fluorescence depends on the skin penetration by a selected PS and production of PpIX in the epidermis determined by its lipophilic character. The calculated LogP values for the tested photosensitizers indicate a higher lipophilicity of MAL-HCl (ACD/LogP = -0.57, ALogPs = -1.30, miLogP = -1.62) as the ALA methyl ester compared to ALA-HCl and ALA-P (for both substances: ACD/LogP = -0.93, ALogPs = -2.85, miLogP = -1.93). In our study, confluent, intensive fluorescence seen in the ALA-HCl and MAL-HCl-treated skin areas indicates a remarkably fine accumulation of ALA in the skin and a high amount of PpIX production. However, the lower and blotchy fluorescence we observed in the skin areas treated with ALA-P may be suggestive of a more selective accumulation of ALA-P in the AK lesions. Therefore, because of their lower kinetics of PpIX formation, ALA-P may facilitate obtaining a better contrast between the healthy and diseased tissue [28,29]. It seems that the observed differences in the distribution of ALA-P in relation to ALA-HCl and MAL-HCl may result from the presence of the $H_2PO_4^-$ ion. Available studies suggest that the $H_2PO_4^-$ ion mediates the proton transfer required for the enolization of amino acids by acting simultaneously as both a general base and a general acid. The dihydro-phosphate ion may catalyze various reactions in proteins, including the racemization of amino acid residues. Therefore, it may contribute to better cellular membrane permeability [30], as racemization of the amino acids influences the functions of many intracellular, extracellular and membrane-bound proteins, and it is considered as a critical factor of protein conformation [31]. However, the observations made must be confirmed in further studies.

PDT is not free of local side effects such as pain, erythema, edema, desquamation, crusting, and pustules, which may occur both during the PDT procedure and in the next hours/days. Urticaria, contact dermatitis, erosive pustular dermatosis of the scalp, pigmentary lesions, scarring, and bullous pemphigoid are less frequently reported [32].

One major drawback of PDT is pain, which may be the reason for discontinuation of the treatment altogether and which may discourage the patient from undergoing future treatments [26]. A number of studies have compared the intensity of pain experienced after application of ALA-HCl or MAL. The results, however, are difficult to interpret due to the use of different formulations and study protocols [33]. Therefore, bearing in mind the intensive pain accompanying PDT, the PDT procedure should be divided into a few stages with time intervals. In our study, which consisted of 22 patients, 16 of them qualified for the second PDT procedure because of the extensiveness of their AK lesions. Although all 22 patients completed the first PDT procedure successfully, one of the 16 patients requiring the second PDT procedure refused further treatment because of unbearable pain. We observed that the intensity of pain on a 10-point VAS scale during and shortly after the first PDT procedure performed in the skin areas treated with either ALA-HCl or MAL-HCl was similar (7.0 on average), whereas it was slightly lower in the skin areas treated with ALA-P (5.8 on average). However, 7 days after the first PDT procedure, the intensity of pain was significantly lower in the areas treated with MAL-HCl or ALA-P (1.4 and 1.3 on average, respectively) than in the areas treated with ALA-HCl (1.8 on average). In most of the studies, patients reported more acute pain at the site of ALA-HCl application [20,34–39]. However, in the reports by Yazdanyar et al., who compared the pain response to ALA-HCl and MAL treatment of the scalp and forehead AK lesions, no significant differences in pain intensity between these two formulations were found either during or 30 min after the completion of treatment [40]. Ibotson et al. [41] indicated that it was still unclear which of the tested compounds, i.e., ALA-HCl or MAL, caused more severe pain during irradiation.

In our study, the intensity of pain during the second PDT procedure and 7 days after its completion did not significantly differ regardless of the used PS, and it was slightly lower than during the first PDT procedure (approximately 5 points on average and approximately 1 point on average, respectively). It appears that the pain experienced during the second PDT procedure is better tolerated because of its foreseeable nature.

A limitation of our study is the quality of images showing intensity and distribution of fluorescence in the skin areas after application of ALA-HCl, MAL-HCl, ALA-P. Future studies, e.g., using an animal model and a higher-resolution device are needed.

5. Conclusions

ALA-P, a new PS first tested in our study turned out to be similarly effective as ALA-HCl and MAL-HCl. We suggest that the use of ALA-P in PDT of AK lesions should be further investigated, because despite in a small number of our study subjects, the obtained results are promising enough to acknowledge the fact that this new ALA photosensitizer may be an advantageous alternative to the already used ALA-HCl and MAL since its application appears to be slightly less painful and better tolerated.

Supplementary Materials: The following are available online at https://www.mdpi.com/article/10.3390/pharmaceutics14020346/s1, Figure S1: Comparisons of the skin areas in the studied patients before PDT and 12 weeks after procedure completion.

Author Contributions: Conceptualization, D.K. (Dorota Krasowska), M.K., and J.B.; methodology, D.K. (Dorota Krasowska), M.K., J.B., A.G. and M.N.; software, J.B., P.S.-K. and D.R.; formal analysis, D.R., J.B., and D.K. (Dorota Kowalczuk); investigation, J.B., P.S.-K., M.N. and A.G.; data curation, P.S.-K., J.B., M.N. and D.K. (Dorota Kowalczuk) and D.R.; writing—original draft preparation, J.B., P.S.-K., A.G. and D.R.; writing—review and editing, D.K. (Dorota Krasowska), M.K. and D.K. (Dorota Kowalczuk); visualization, J.B., P.S.-K. and D.R.; supervision, D.K. (Dorota Krasowska) and M.K.; project administration, D.K. (Dorota Krasowska); funding acquisition, D.K. (Dorota Krasowska) and M.K. All authors have read and agreed to the published version of the manuscript.

Funding: This research received no external funding.

Institutional Review Board Statement: The study was conducted according to the guidelines of the Declaration of Helsinki and approved by the Institutional Ethics Committee of Medical University of Lublin, Poland (KE-0254/286/2019).

Informed Consent Statement: Informed consent was obtained from all subjects involved in the study.

Data Availability Statement: Data sharing not applicable.

Conflicts of Interest: The authors declare no conflict of interest.

References

1. Linos, E.; Katz, K.A.; Colditz, G.A. Skin cancer-the importance of prevention. *JAMA Intern. Med.* **2016**, *176*, 1435–1436. [CrossRef] [PubMed]
2. Wu, X.; Elkin, E.E.; Marghoob, A.A. Burden of basal cell carcinoma in USA. *Future Oncol.* **2015**, *11*, 2967–2974. [CrossRef] [PubMed]
3. Combalia, A.; Carrera, C. Squamous cell carcinoma: An update on diagnosis and treatment. *Dermatol. Pract. Concept.* **2020**, *10*, e2020066. [CrossRef] [PubMed]
4. Dianzani, C.; Conforti, C.; Giuffrida, R.; Corneli, P.; di Meo, N.; Farinazzo, E.; Moret, A.; Magaton Rizzi, G.; Zalaudek, I. Current therapies for actinic keratosis. *Int. J. Dermatol.* **2020**, *59*, 677–684. [CrossRef] [PubMed]
5. Criscione, V.D.; Weinstock, M.A.; Naylor, M.F.; Luque, C.; Eide, M.J.; Bingham, S.F. Department of veteran affairs topical tretinoin chemoprevention trial group. Actinic keratoses: Natural history and risk of malignant transformation in the veterans affairs topical tretinoin chemoprevention trial. *Cancer* **2009**, *115*, 2523–2530. [CrossRef] [PubMed]
6. Fernández-Figueras, M.T.; Carrato, C.; Sáenz, X.; Puig, L.; Musulen, E.; Ferrándiz, C.; Ariza, A. Actinic keratosis with atypical basal cells (AK I) is the most common lesion associated with invasive squamous cell carcinoma of the skin. *J. Eur. Acad. Dermatol. Venereol.* **2015**, *29*, 991–997. [CrossRef] [PubMed]
7. Stockfleth, E. From a new vision of actinic keratosis to imiquimod 3.75%, the new treatment standard. *J. Eur. Acad. Dermatol. Venereol.* **2015**, *29*, 1–2. [CrossRef]
8. Jetter, N.; Chandan, N.; Wang, S.; Tsoukas, M. Field cancerization therapies for management of actinic keratosis: A narrative review. *Am. J. Clin. Dermatol.* **2018**, *19*, 543–557. [CrossRef]
9. Kwaśny, M. Photodynamic therapy in dermatology. *Forum. Derm.* **2018**, *4*, 138–147. [CrossRef]
10. Piquero-Casals, J.; Morgado-Carrasco, D.; Gilaberte, Y.; Del Rio, R.; Macaya-Pascual, A.; Granger, C.; López-Estebaranz, J.L. Management pearls on the treatment of actinic keratoses and field cancerization. *Dermatol. Ther.* **2020**, *10*, 903–915. [CrossRef]
11. Ritter, C.G.; Kuhl, I.C.; Lenhardt, C.; Weissbluth, M.L.; Bakos, R.M. Photodynamic therapy with delta-aminolevulinic acid and light-emitting diodes in actinic keratosis. *An. Bras. Dermatol.* **2010**, *85*, 639–645. [CrossRef] [PubMed]
12. Conforti, C.; Corneli, P.; Harwood, C.; Zalaudek, I. Evolving role of systemic therapies in non-melanoma skin cancer. *Clin. Oncol.* **2019**, *31*, 759–768. [CrossRef] [PubMed]
13. Champeau, M.; Vignoud, S.; Mortier, L.; Mordon, S. Photodynamic therapy for skin cancer: How to enhance drug penetration? *J. Photochem. Photobiol. B* **2019**, *197*, 111544. [CrossRef] [PubMed]
14. Yazdanyar, S.; Zarchi, K.; Jemec, G.B.E. Pain during topical photodynamic therapy—Comparing methyl aminolevulinate (Metvix ®) to aminolaevulinic acid (Ameluz ®); an intra-individual clinical study. *Photodiagnosis. Photodyn. Ther.* **2017**, *20*, 6–9. [CrossRef] [PubMed]
15. National Center for Biotechnology Information. PubChem Compound Summary for CID 24737828, Aminolevulinic Acid Phosphate. Available online: https://pubchem.ncbi.nlm.nih.gov/compound/Aminolevulinic-acid-phosphate (accessed on 6 August 2021).
16. Al-Saber, F.; Aldosari, W.; Alselaiti, M.; Khalfan, H.; Kaladari, A.; Khan, G.; Harb, G.; Rehani, R.; Kudo, S.; Koda, A.; et al. The safety and tolerability of 5-aminolevulinic acid phosphate with sodium ferrous citrate in patients with type 2 diabetes mellitus in Bahrain. *J. Diabetes Res.* **2016**, *2016*, 8294805. [CrossRef] [PubMed]
17. Higashikawa, F.; Kanno, K.; Ogata, A.; Sugiyama, M. Reduction of fatigue and anger-hostility by the oral administration of 5-aminolevulinic acid phosphate: A randomized, double-blind, placebo-controlled, parallel study. *Sci. Rep.* **2020**, *10*, 16004. [CrossRef]

18. Tarstedt, M.; Gillstedt, M.; Wennberg Larkö, A.M.; Paoli, J. Aminolevulinic acid and methyl aminolevulinate equally effective in topical photodynamic therapy for non-melanoma skin cancers. *J. Eur. Acad. Dermatol. Venereol.* **2016**, *30*, 420–423. [CrossRef]
19. Ezzedine, K.; Painchault, C.; Brignone, M. Systematic literature review and network meta-analysis of the efficacy and acceptability of interventions in Actinic Keratoses. *Acta Derm.-Venereol.* **2021**, *101*, adv00358. [CrossRef]
20. Moloney, F.J.; Collins, P. Randomized, double-blind, prospective study to compare topical 5-aminolaevulinic acid methylester with topical 5-aminolaevulinic acid photodynamic therapy for extensive scalp actinic keratosis. *Br. J. Dermatol.* **2007**, *157*, 87–91. [CrossRef]
21. Fu, C.; Kuang, B.H.; Qin, L.; Zeng, X.Y.; Wang, B.C. Efficacy and safety of photodynamic therapy with amino-5-laevulinate nanoemulsion versus methyl-5-aminolaevulinate for actinic keratosis: A meta-analysis. *Photodiagnosis. Photodyn. Ther.* **2019**, *27*, 408–414. [CrossRef]
22. Patel, G.; Armstrong, A.W.; Eisen, D.B. Efficacy of photodynamic therapy vs other interventions in randomized clinical trials for the treatment of actinic keratoses: A systematic review and meta-analysis. *JAMA Dermatol.* **2014**, *150*, 1281–1288. [CrossRef] [PubMed]
23. Ko, D.Y.; Kim, K.H.; Song, K.H. Comparative study of photodynamic therapy with topical methyl aminolevulinate versus 5-aminolevulinic acid for facial actinic keratosis with long-term follow-up. *Ann. Dermatol.* **2014**, *26*, 321–331. [CrossRef] [PubMed]
24. Räsänen, J.; Neittaanmäki, N.; Ylitalo, L.; Hagman, J.; Rissanen, P.; Ylianttila, L.; Salmivuori, M.; Snellman, E.; Grönroos, M. DL-PDT for AK: Multicentre trial comparing BF-200 ALA with MAL. *Br. J. Dermatol.* **2019**, *181*, e37. [CrossRef]
25. Menezes, P.F.; Requena, M.B.; Bagnato, V.S. Optimization of photodynamic therapy using negative pressure. *Photomed. Laser. Surg.* **2014**, *32*, 296–301. [CrossRef]
26. Blanco, K.C.; Moriyama, L.T.; Inada, N.M.; Sálvio, A.G.; Menezes, P.F.C.; Leite, E.J.S.; Kurachi, C.; Bagnato, V.S. Fluorescence guided PDT for optimization of the outcome of skin cancer treatment. *Front. Phys.* **2015**, *3*, 30. [CrossRef]
27. Foged, C.; Philipsen, P.A.; Wulf, H.C.; Haedersdal, M.; Togsverd-Bo, K. Skin surface protoporphyrin IX fluorescence is associated with epidermal but not dermal fluorescence intensities. *Photodiagnosis. Photodyn. Ther.* **2020**, *30*, 101681. [CrossRef] [PubMed]
28. Godin, B.; Touitou, E. Transdermal skin delivery: Predictions for humans from in vivo, ex vivo and animal models. *Adv. Drug Deliv. Rev.* **2007**, *59*, 1152–1161. [CrossRef]
29. Rodriguez, L.; de Bruijn, H.S.; Di Venosa, G.; Mamone, L.; Robinson, D.J.; Juarranz, A.; Batlle, A.; Casas, A. Porphyrin synthesis from aminolevulinic acid esters in endothelial cells and its role in photodynamic therapy. *J. Photochem. Photobiol. B* **2009**, *96*, 249–254. [CrossRef]
30. Takahashi, O.; Kirikoshi, R.; Manabe, N. Racemization of serine residues catalyzed by dihydrogen phosphate ion: A computational study. *Catalysts* **2017**, *7*, 363. [CrossRef]
31. Dyakin, V.V.; Wisniewski, T.M.; Lajtha, A. Racemization in post-translational modifications relevance to protein aging, aggregation and neurodegeneration: Tip of the iceberg. *Symmetry* **2021**, *13*, 455. [CrossRef]
32. Borgia, F.; Giuffrida, R.; Caradonna, E.; Vaccaro, M.; Guarneri, F.; Cannavò, S.P. Early and late onset side effects of photodynamic therapy. *Biomedicines* **2018**, *6*, 12. [CrossRef] [PubMed]
33. Warren, C.B.; Karai, L.J.; Vidimos, A.; Maytin, E.V. Pain associated with aminolevulinic acid-photodynamic therapy of skin disease. *J. Am. Acad. Dermatol.* **2009**, *61*, 1033–1043. [CrossRef] [PubMed]
34. Serra-Guillén, C.; Nagore, E.; Bancalari, E.; Kindem, S.; Sanmartín, O.; Llombart, B.; Requena, C.; Serra-Guillén, I.; Calomarde, L.; Diago, A.; et al. A randomized intraindividual comparative study of methyl-5-aminolaevulinate vs. 5-aminolaevulinic acid nanoemulsion (BF-200 ALA) in photodynamic therapy for actinic keratosis of the face and scalp. *Br. J. Dermatol.* **2018**, *179*, 1410–1411. [CrossRef] [PubMed]
35. Räsänen, J.E.; Neittaanmäki, N.; Ylitalo, L.; Hagman, J.; Rissanen, P.; Ylianttila, L.; Salmivuori, M.; Snellman, E.; Grönroos, M. 5-aminolaevulinic acid nanoemulsion is more effective than methyl-5-aminolaevulinate in daylight photodynamic therapy for actinic keratosis: A nonsponsored randomized double-blind multicentre trial. *Br. J. Dermatol.* **2019**, *181*, 265–274. [CrossRef]
36. Szeimies, R.M.; Calzavara-Pinton, P. A randomized, double-blind trial comparing 5-aminolaevulinic acid nanoemulsion with methyl 5-aminolaevulinate in daylight photodynamic therapy for actinic keratosis. *Br. J. Dermatol.* **2019**, *181*, 223–224. [CrossRef]
37. Lee, P.K.; Kloser, A. Current methods for photodynamic therapy in the US: Comparison of MAL/PDT and ALA/PDT. *J. Drugs Dermatol.* **2013**, *12*, 925–930.
38. Dirschka, T.; Radny, P.; Dominicus, R.; Mensing, H.; Brüning, H.; Jenne, L.; Karl, L.; Sebastian, M.; Oster-Schmidt, C.; Klövekorn, W.; et al. AK-CT002 Study Group. Photodynamic therapy with BF-200 ALA for the treatment of actinic keratosis: Results of a multicentre, randomized, observer-blind phase III study in comparison with a registered methyl-5-aminolaevulinate cream and placebo. *Br. J. Dermatol.* **2012**, *166*, 137–146. [CrossRef]
39. Dirschka, T.; Ekanayake-Bohlig, S.; Dominicus, R.; Aschoff, R.; Herrera-Ceballos, E.; Botella-Estrada, R.; Hunfeld, A.; Kremser, M.; Schmitz, B.; Lübbert, H.; et al. A randomized, intraindividual, non-inferiority, Phase III study comparing daylight photodynamic therapy with BF-200 ALA gel and MAL cream for the treatment of actinic keratosis. *J. Eur. Acad. Dermatol. Venereol.* **2019**, *33*, 288–297. [CrossRef]

40. Olsen, E.A.; Abernethy, M.L.; Kulp-Shorten, C.; Callen, J.P.; Glazer, S.D.; Huntley, A.; McCray, M.; Monroe, A.B.; Tschen, E.; Wolf, J.E., Jr. A double-blind, vehicle-controlled study evaluating masoprocol cream in the treatment of actinic keratoses on the head and neck. *J. Am. Acad. Dermatol.* **1991**, *24*, 738–743. [CrossRef]
41. Ibbotson, S.H.; Valentine, R.; Hearn, R. Is the pain of topical photodynamic therapy with methyl aminolevulinate any different from that with 5-aminolaevulinic acid? *Photodermatol. Photoimmunol. Photomed.* **2012**, *28*, 272–273. [CrossRef]

 *pharmaceutics*

Article

Photodynamic Therapy Targeting Macrophages Using IRDye700DX-Liposomes Decreases Experimental Arthritis Development

Daphne N. Dorst [1,2,*,†], Marti Boss [1,†], Mark Rijpkema [1], Birgitte Walgreen [2], Monique M. A. Helsen [2], Desirée L. Bos [1], Louis van Bloois [3], Gerrit Storm [3,4], Maarten Brom [1], Peter Laverman [1], Peter M. van der Kraan [2], Mijke Buitinga [5,6], Marije I. Koenders [2] and Martin Gotthardt [1]

[1] Department of Medical Imaging, Radboudumc, Radboud University, 6525 XZ Nijmegen, The Netherlands; marti.boss@radboudumc.nl (M.B.); Mark.Rijpkema@radboudumc.nl (M.R.); desiree.bos@radboudumc.nl (D.L.B.); maarten.brom@radboudumc.nl (M.B.); peter.laverman@radboudumc.nl (P.L.); martin.gotthardt@radboudumc.nl (M.G.)

[2] Department of Experimental Rheumatology, Radboudumc, Radboud University, 6525 XZ Nijmegen, The Netherlands; birgitte.walgreen@radboudumc.nl (B.W.); monique.helsen@radboudumc.nl (M.M.A.H.); Peter.vanderKraan@radboudumc.nl (P.M.v.d.K.); marije.koenders@radboudumc.nl (M.I.K.)

[3] Department of Pharmaceutics, Utrecht Institute for Pharmaceutical Sciences, Utrecht University, 3508 TC Utrecht, The Netherlands; L.vanBloois@uu.nl (L.v.B.); G.Storm@uu.nl (G.S.)

[4] Department of Biomaterials Science and Technology, University of Twente, 7522 NL Enschede, The Netherlands

[5] Department of Nutrition and Movement Science, Maastricht University, 6229 HX Maastricht, The Netherlands; m.buitinga@maastrichtuniversity.nl

[6] Department of Radiology and Nuclear Medicine, Maastricht University, 6229 HX Maastricht, The Netherlands

* Correspondence: daphne.dorst@radboudumc.nl; Tel.: +31-24-36-66283

† These authors contributed equally to this work.

Citation: Dorst, D.N.; Boss, M.; Rijpkema, M.; Walgreen, B.; Helsen, M.M.A.; Bos, D.L.; van Bloois, L.; Storm, G.; Brom, M.; Laverman, P.; et al. Photodynamic Therapy Targeting Macrophages Using IRDye700DX-Liposomes Decreases Experimental Arthritis Development. *Pharmaceutics* **2021**, *13*, 1868. https://doi.org/10.3390/pharmaceutics13111868

Academic Editors: Udo Bakowsky, Matthias Wojcik, Eduard Preis and Gerhard Litscher

Received: 14 September 2021
Accepted: 1 November 2021
Published: 5 November 2021

Publisher's Note: MDPI stays neutral with regard to jurisdictional claims in published maps and institutional affiliations.

Abstract: Macrophages play a crucial role in the initiation and progression of rheumatoid arthritis (RA). Liposomes can be used to deliver therapeutics to macrophages by exploiting their phagocytic ability. However, since macrophages serve as the immune system's first responders, it is inadvisable to systemically deplete these cells. By loading the liposomes with the photosensitizer IRDye700DX, we have developed and tested a novel way to perform photodynamic therapy (PDT) on macrophages in inflamed joints. PEGylated liposomes were created using the film method and post-inserted with micelles containing IRDye700DX. For radiolabeling, a chelator was also incorporated. RAW 264.7 cells were incubated with liposomes with or without IRDye700DX and exposed to 689 nm light. Viability was determined using CellTiterGlo. Subsequently, biodistribution and PDT studies were performed on mice with collagen-induced arthritis (CIA). PDT using IRDye700DX-loaded liposomes efficiently induced cell death in vitro, whilst no cell death was observed using the control liposomes. Biodistribution of the two compounds in CIA mice was comparable with excellent correlation of the uptake with macroscopic and microscopic arthritis scores. Treatment with 700DX-loaded liposomes significantly delayed arthritis development. Here we have shown the proof-of-principle of performing PDT in arthritic joints using IRDye700DX-loaded liposomes, allowing locoregional treatment of arthritis.

Keywords: photodynamic therapy; liposomes; experimental arthritis

1. Introduction

Rheumatoid arthritis (RA) is a chronic autoimmune disease affecting the synovial joints in 0.5–1% of the general population [1]. The disease is characterized by relapsing and remitting inflammation associated with progressive damage to the joints [2]. Treatment of RA consists of disease-modifying anti-rheumatic drugs (DMARDs). Despite the treatment paradigm having changed, a considerable percentage of patients do not respond adequately to both biological and synthetic DMARD treatment [3].

In RA, the synovial lining, a thin layer of cells providing a barrier between the synovial fluid and the joint tissue, becomes hyperproliferative, infiltrated with immune cells, and forms a tumor-like pannus. A key cell already present in this resident synovial lining is the macrophage. Its phagocytic function is crucial for maintaining joint homeostasis [4]. In RA, the function of the macrophage changes as it adapts a pro-inflammatory phenotype. Through its secretion of pro-inflammatory factors, such as tumor necrosis factor-α (TNF-α), interleukin (IL) −1, −6, −10 and granulocyte-macrophage colony-stimulating factor, it actively contributes to chemotaxis and activation of infiltrating immune cells [5,6]. Furthermore, these cytokines stimulate resident synovial fibroblasts to adapt and propagate an activated and destructive phenotype [6]. Through their production of enzymes such as matrix metalloproteinases, macrophages also directly contribute to joint destruction [6,7]. Due to the crucial role of macrophages in the initiation and propagation of arthritis, they are an interesting therapeutic target. Liposomes can be used to target macrophages since they are preferentially phagocytosed by these cells [8–11]. Depleting macrophages with clodronate-loaded liposomes decreased arthritis in rodents [12–14] and patients [15]. However, systemic administration of clodronate liposomes also resulted in depletion of these cells from the spleen and liver [14]. This may have serious side effects, such as a higher susceptibility to infection and cancer, and is therefore not an advisable strategy.

Instead, the selective depletion of macrophages only in the affected synovial joint would be a preferred treatment method. Therefore, loading the liposomes with a drug that is constantly active, such as clodronate, and that will thus exert its function in all cells that phagocytose it, is not optimal. Instead, a preferable approach would be to load the liposomes used for macrophage targeting with an inert molecule that can selectively be activated at the sites of inflammation. To this end, photodynamic therapy (PDT) may be a promising therapeutic approach. In PDT, a light-sensitive molecule, a so-called photosensitizer (PS), is used. By exposing the region to be treated to light of a specific wavelength, the PS will produce reactive oxygen species (ROS), which are cytotoxic to the cells in its vicinity. Off-target effects are limited, since the photosensitizer only becomes active upon excitation by locally applied light of a PS-specific wave-length (690 nm for the IRDye700DX used in this study), thus also avoiding side effects by daylight exposure. IRDye700DX (Pubchem ID: 102004325) is a phthalocyanine dye with two siloxane chains to improve water solubility. A NHS ester can be attached to the photosensitizer, which facilitates conjugation to free amino groups on macromolecules. This has previously been used by our group to modulate arthritis when conjugated to a monoclonal antibody targeting fibroblast activation protein [16].

Here, we investigate if targeting the macrophages using liposomes loaded with the PS IRDye700DX (700DX-liposomes) can specifically kill macrophages in vitro as well as decrease arthritis severity in the collagen-induced arthritis (CIA) mouse model.

2. Materials and Methods

2.1. Animals

Male DBA/1JRj mice were purchased from Janvier-Elevage at 10–12 weeks of age. Mice were housed in conventional cages on a 12 h day–night cycle and fed standard chow ad libitum. The study protocol was approved by the Radboud University animal ethics committee (RU-DEC-2016-0076, CCD number AVD103002016786). All procedures were performed according to the Institute of Laboratory Animal Research guide for Laboratory Animals.

2.2. Liposome Preparation and Characterization

Liposomes were prepared based on the film method [17]. Briefly, dipalmitoyl phosphatidylcholine (DPPC) and PEG-(2000)-distearoyl phosphatidylethanolamine (PEG-(2000)-DSPE), obtained from Lipoid AG (Steinhausen, Switzerland); DSPE-DTPA, from Avanti Polar Lipids (Birmingham, UK); cholesterol, obtained from Sigma (St. Louis, MO, USA), were used. All other chemicals were of reagent grade. A mix of chloroform and methanol (10:1 *v/v*) containing DPPC, PEG-(2000)-DSPE, cholesterol and DSPE-DTPA was prepared in a molar ratio of 1.85:0.15:1:0.15. The organic phase was evaporated with a rotavapor (BUCHI Labortechnik AG, Flawil, Switzerland), followed by nitrogen flushing to remove residual organic solvent. Liposomes were dispersed in phosphate-buffered saline (PBS). The liposomes were sequentially extruded through two stacked polycarbonate filters with pore sizes of 600, 200, and 100 nm (Nuclepore, Pleaston, CA, USA) under nitrogen pressure, using a Lipex high pressure extruder (Lipex, Nortern Lipids, Vancouver, BC, Canada). The retrieved liposomes had a mean particle size of 120 nm with a PDI (polydispersity index) of 0.02.

700DX-liposomes were prepared by post-insertion of micelles as described previously [18,19]. Briefly, micelles were prepared by mixing DSPE-NH$_2$-PEG-2000 and DSPE-PEG-2000 (both Avanti Polar Lipids, Birmingham, UK) in a 1:1 molar ratio to enable covalent coupling of IRDye-700DX NHS ester (LI-COR Biosciences, Lincoln, NE, USA) to the NH$_2$ terminus of the PEG conjugates. Subsequently, micelles were incubated with the liposomes at 60 °C to allow post-insertion of the PEG conjugates. Unbound IRDye700DX was removed from solution using using a Slide-A-Lyzer Dialysis Cassette (20.000 MWCO, Thermofisher, Waltham, MA, USA) in phosphate-buffered saline (PBS) with 0.5% *w/v* Chelex (Bio-Rad, Hercules, CA, USA).

Fluorescence and absorbance spectra of the liposomes with and without IRDye700DX, as well as those of the free PS, were compared using a TECAN infinite M200 Pro plate reader (PerkinElmer, Groningen, The Netherlands). For fluorescence emission spectrum measurements, an excitation wavelength of 650 nm was used. Results were normalized to the peak absorbance for comparison of the different compounds.

2.3. In Vitro Photodynamic Therapy

RAW 264.7 cells were cultured in Dulbecco's modified Eagle's medium (DMEM) with 4.5 g/L D-glucose and Glutamax, supplemented with 10% fetal calf serum (FCS), 100 IU/mL penicillin G and 10mg/mL streptomycin. Cells were seeded into 24-well plates (Thermo Scientific, Waltham, MA, USA) (150,000 cells/well) and grown overnight ($n = 3$ wells/group). Medium was replaced by binding buffer (medium with 0.1% bovine serum albumin (*w/v*) (BSA)) with 4 µL/mL liposomes with or without IRDye-700DX (200 µg dye/mL liposomes). After incubation at 37 °C for 24 h, cells were washed with binding buffer. Subsequently, cells were irradiated with a NIR light-emitting diode (LED) [20] (peak emission wavelength 690 nm ($\pm$10 nm for minimum and maximum emission), forward voltage: 2.6 V, power output: 490 mW) using 126 individual LED bulbs to ensure homogenous illumination at different radiant exposures (between 0 and 50 J/cm^2) (LED-factory, Leeuwarden, The Netherlands). All experiments were carried out in triplicate. The LED device was calibrated using a FieldMaxII-TO power meter with a PM2 sensor (thermopile head), resolution 1 mW (Coherent, Richmond, CA, USA).

Cell viability was measured 4 h after irradiation. During this time, the cells were kept at 37 °C and 5% CO$_2$. To determine the viability, ATP content was measured using a CellTiter-Glo® luminescent assay (Promega Benelux, Leiden, The Netherlands) according to the instructions of the manufacturer. Luminescence was measured using a TECAN infinite M200 Pro plate reader (PerkinElmer, Groningen, The Netherlands). The ATP content as a measure of cell viability was expressed as a percentage, determined by comparing the luminescent signal with the signal from untreated cells, which were considered 100% viable.

2.4. Collagen-Induced Arthritis

CIA was induced in male DBA/1JRj mice as described previously [21]. Briefly, animals received two injections of 100 µg bovine collagen type II (Radboudumc in-house production, batch 03-04-08). The first dose was emulsified in complete Freund's adjuvant (FCA) and injected intradermally at the base of the tail. The second dose of FCA was administered in PBS intraperitoneally at day 21, with subsequent arthritis developed between days 21 and 25. Mice were scored 3 times a week for development of arthritis on each paw according to a 2-point scale of swelling and redness, as described previously [22]. Mice with a score ≥ 0.5 in one of the hind paws were included in the study. A cumulative inflammation score ≥ 7 was considered a humane endpoint.

2.5. Biodistribution

Male DBA/1JRj mice (starting weight 24.3 ± 0.8 g (n = 10/group)) with CIA were randomly divided between the treatment groups upon development of overt arthritis (macroscopic score of inflammation > 0.5). Upon inclusion, the mice were injected intravenously with 5 µL liposomes (in 200 µL PBS) with or without 1 µg IRDye700DX and labeled with 0.6 MBq ^{111}In for biodistribution analysis. A subset of mice (n = 2 per group) were injected with 18 MBq ^{111}In for SPECT/CT imaging. After 24 h, the mice were sacrificed, and the relevant tissues were dissected and weighed. Tissue uptake of ^{111}In was determined using a γ counter (WIZARD, 2480 Automatic Gamma Counter, Perkin Elmer, Waltham, MA, USA). Results are depicted as percentage of the injected amount per gram tissue (%IA/g). Mice which received a SPECT/CT dose of ^{111}In were scanned for 60 min (4×15 min frames) using a 1-mm-diameter pinhole ultra-high sensitivity mouse collimator (U-SPECT/CT-II, MILabs, Houten, The Netherlands). SPECT scans were followed by CT scans (65 kV, 615 µA). The SPECT scans were reconstructed using software from MILabs using a 0.2 mm voxel size, 1 iteration and 16 subsets.

2.6. Microscopic Scoring of Inflammation

To assess the joint inflammation, microscopically ankle joints were formalin-fixed, decalcified using formic acid and paraffin-embedded. The joints were subsequently cut into 7 µm sections, hematoxylin- and eosin-stained, and scored for the presence of inflammation according to the SMASH recommendations [23].

2.7. In Vivo Photodynamic Therapy

Male DBA/1JRj mice (starting weight 23.4 ± 1.4 g, (n = 8/group)) with CIA were randomly divided between the treatment groups upon development of overt arthritis in one of the hind paws (arbitrary arthritis score of >0.5). Mice were injected intravenously with 5 µL liposomes with or without 1 µg IRDye700DX in 200 µL of PBS, and 24 h after injection, the hind legs of the animals were exposed to 8.8 J/cm^2 or 26.4 J/cm^2 690 nm light. The mice receiving the control liposomes also received the high dose of light. Arthritis development was subsequently scored daily. Mice were sacrificed 5 days after treatment, and the front and hind paws were formalin-fixed, decalcified using formic acid and paraffin-embedded. Histological sections of 7 µm were stained using hematoxylin and eosin, or safranin O, and scored for inflammation and proteoglycan depletion (PG depletion), respectively.

2.8. Statistical Analysis

Results are presented as mean $\pm$ SD. Statistical significance was determined using GraphPad Prism software (Version 5.03; GraphPad Software, San Diego, CA, USA). The tests used were: two-way ANOVA with Bonferroni post-test for the in vitro PDT, biodistribution, and therapy experiment. In the latter, we also accounted for repeated measures. The correlation between arthritis and liposome uptake in inflamed paws was determined using linear regression. A p-value < 0.05 was considered significant.

3. Results

3.1. Fluorescence and Absorption of IRDye700DX-Loaded Liposomes Is Similar to Free IRDye700DX

The fluorescence emission spectrum and absorbance spectrum were measured for liposomes with and without IRDye700DX loading, and these were compared to the respective spectra for free IRDye700DX. Results were normalized to peak fluorescence and absorbance for the corresponding measurement. The absorbance spectrum of 700DX-liposomes was very similar to that of free IRDye700DX, with both showing peak absorbance at 690 nm

(Figure S1A). The fluorescence emission spectra were also very similar, with only a very minor shift in peak emission observed (Figure S1B).

3.2. PDT Using 700DX-Liposomes Induces Cell Death in a Light Dose Dependent Manner

Already at light doses as low as 10 J/cm^2 radiant exposure of 690 nm light, in vitro cell viability of cells incubated with 700DX-liposomes was approximately 50% compared to the untreated control (Figure 1). By increasing the light dose even higher, cell death could be generated until, at a light dose of 50 J/cm^2, only 3% of cells remained. Radiant exposure after incubation with control liposomes without IRDye700DX did not cause cell death, nor did incubation with 700DX-liposomes without light exposure (both $p > 0.5$).

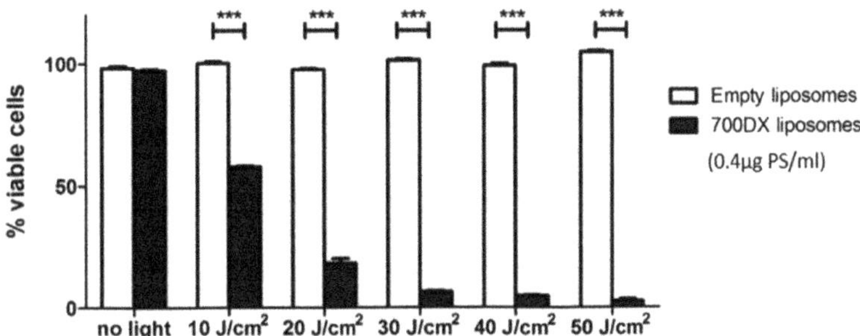

Figure 1. PDT using 700DX-liposomes in RAW 264.7 cells. Cell viability decreased in a light dose dependent manner after incubation with IRDye700DX-liposomes. Cell viability was not affected in cells incubated with control liposomes for any of the tested light doses. Data were normalized to the luminescent values of cells not incubated with liposomes and not exposed to light, and are depicted as mean ±SD ($n = 3$). Two-way ANOVA with Bonferroni post-test, *** = $p < 0.001$.

3.3. Liposomal Accumulation in the Arthritic Joint Is Not Negatively Affected by Loading with IRDye700DX

The biodistribution of 111-indium-radiolabeled 700DX-liposomes in mice with CIA was comparable to that of the control liposomes (Figure 2). No significant difference in uptake was noted, with the exception of the liver, where accumulation of the liposomes with the photosensitizer was significantly increased (24.9 ± 4.9%IA/g versus 14.5 ± 3%IA/g for the liposomes with and without the PS, respectively ($p < 0.01$)). Tracer uptake by the inflamed regions correlated significantly for both constructs with the macroscopic score of arthritis both in the front paw (Figure 3a,b) and ankle joint (Figure 3c,d), as well as with the microscopic score of inflammation (Figure S2). SPECT/CT images that were acquired reflect these results, with clear uptake of the tracer observed in the inflamed joint, and with higher macroscopic inflammation scores also showing more uptake in the scans (Figure 2b,c).

3.4. PDT Using 700DX-Loaded Liposomes Ameliorates Arthritis Progression

The therapeutic effect of PDT using 700DX-liposomes was investigated in male DBA/1JRj mice with CIA. The mice were injected intravenously with the 700DX-liposomes or PBS as a control, and the inflamed paws were exposed to the 690 nm light 24 h post-injection. The arthritis scores for both the 8.8 and 26.4 J/cm^2 PDT treated groups were significantly lower than those for the PBS control at days 1–3 for the 8.8 J/cm^2 and day 1 and 2 for the 26.4 J/cm^2 PDT treated group (Figure 4). This difference became especially clear when accounting for the arthritis development of these mice as a whole using the area under the curve (Figure S3). After this initial delay in arthritis development in the treated paws, no differences in arthritis scores were observed at the later timepoints. In line with this, no differences in histological scores of inflammation or cartilage damage were observed after sacrifice of the animals at day 5 post-therapy (Figure S4).

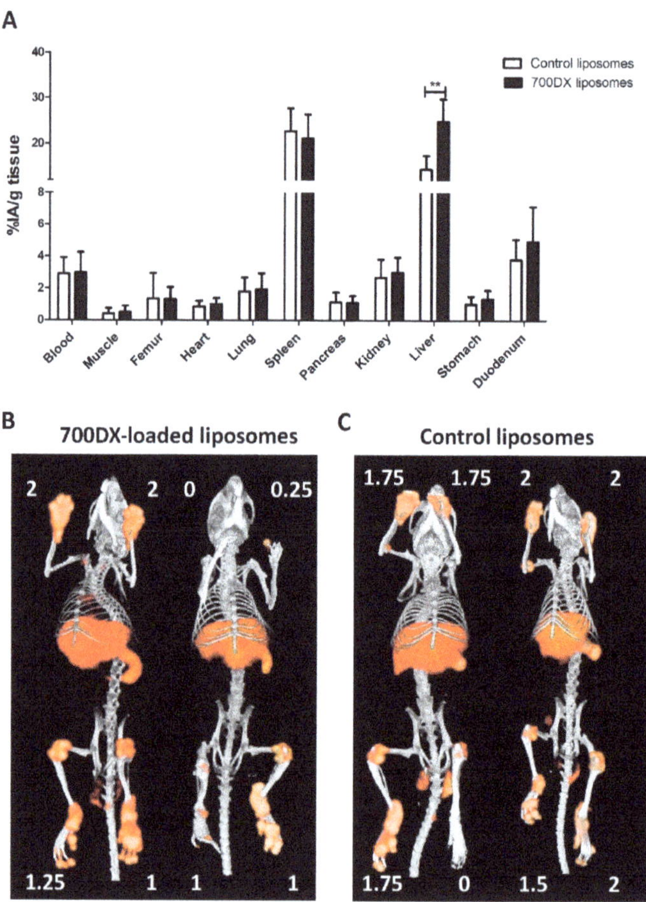

Figure 2. (**A**) Biodistribution of liposomes with or without IRDye700DX (n = 10 mice/group); (**B**) SPECT images of mice after injection of 111In-labeled IRDye700DX-liposomes; (**C**) SPECT images of mice after injection of ^{111}In-labeled control liposomes. Arthritis scores are depicted in white next to the respective joints (** = $p < 0.01$).

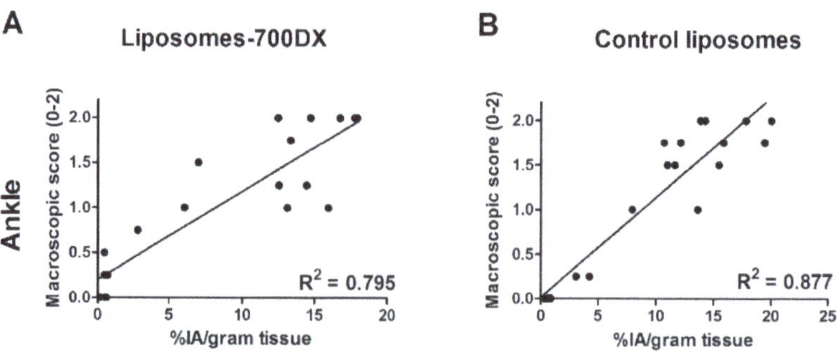

Figure 3. *Conts.*

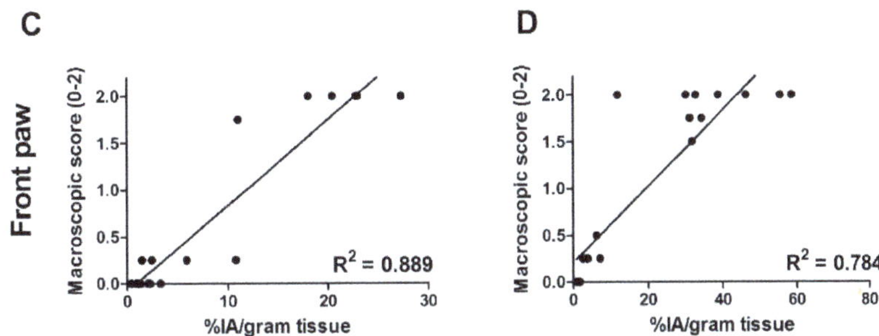

Figure 3. Uptake of 700DX-loaded and control liposomes positively correlated with macroscopic arthritis score in the hind ankle (**A**,**B**) and front paw (**C**,**D**) of mice with CIA ($n = 10$ mice).

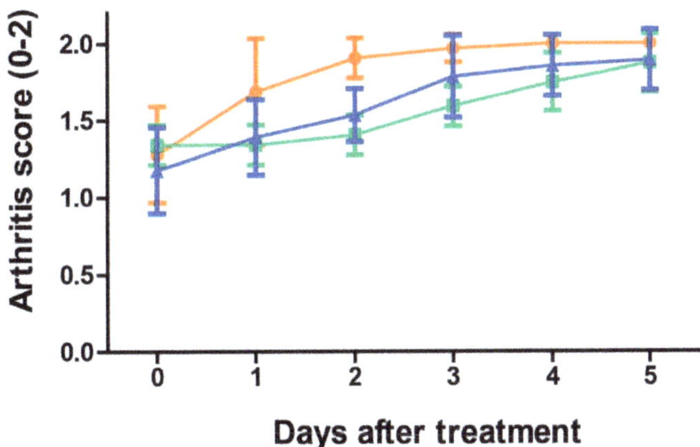

Statistical significance

PBS vs low	ns	**	***	**	ns	ns
PBS vs high	ns	*	**	ns	ns	ns

Legend

			Light dose (J/cm^2)
●	Control	PBS	26.4
■	Treated: low	Liposomes	8.8
▲	Treated: high	Liposomes	26.4

Figure 4. Development of arthritis over time after 700DX-liposome PDT. Initial arthritis development was slower after tPDT. Two-way ANOVA with Bonferroni post-test, * = $p < 0.05$, ** = $p < 0.01$, *** = $p < 0.001$, ns: not significant ($n = 8$ mice/group).

4. Discussion

In this study, we demonstrated the feasibility of depleting macrophages using PS-loaded liposomes for tPDT both in vitro and in vivo in the CIA model of arthritis. Using PS-loaded liposomes, the macrophage cell line RAW264.7 was able to be efficiently depleted in a light dose dependent manner. The construct showed targeting of the arthritic joints in the CIA model of arthritis, with uptake closely correlating with arthritis severity in these mice. Finally, 700DX-liposome PDT was able to ameliorate arthritis development in the initial phase after treatment.

Macrophages are of critical importance for the development of RA. Depletion of macrophages using clodronate-loaded liposomes has previously been shown to be able to decrease the development of arthritis in mice when administered prior to overt arthritis development [12,13]. Additionally, arthritic flares were prevented, and administering the liposomes in the chronic phase of CIA ameliorated the development of the disease [24]. A drawback of the systemic action of clodronate-loaded liposomes is the resulting depletion of macrophages not only in the arthritic joints, but in many tissues, which interferes with the macrophage's normal function of maintaining tissue homeostasis and preventing infection. By using a construct that is inert in the absence of light, we can deplete macrophages specifically in the arthritic joints, eliminating this problem.

Using PDT to treat arthritis in animal models has successfully been employed by several groups. Since these previous studies all used PDT, in which the PS preferentially accumulates in the arthritic joint due to the increased permeability of the vasculature, treatment was accompanied by significant side effects, because accumulation of the photosensitizer in neighboring skin and muscle tissue, which were inadvertently also exposed to light, could not be sufficiently prevented by untargeted approaches, and this resulted in necrosis in the surrounding musculature [25]. By using liposomes to deliver the photosensitizer preferentially to the macrophages, and thus increasing the uptake in the inflamed joints relative to neighboring tissues, these unwanted side-effects were successfully prevented in this study [9].

Current treatment for RA consists of systemic immunosuppression through the administration of biologic or synthetic DMARDS. The advantage of using 700DX-liposome PDT instead of, or complimentary to DMARDS is that the side effects of systemic immunosuppression, such as an increased risk of infectious diseases, can be avoided [26,27]. Furthermore, since 700DX-liposomes are systemically delivered, in contrast to other local adjuvant treatments such as chemical-, surgical- or radio-synovectomy, the inflamed joint is not damaged by the insertion of needles or other instruments.

An important hurdle to achieving clinical translation is the limitation of light penetration. Near infrared light has a penetration depth in tissue between 5 and 10 mm, which is suitable for the treatment of small joints, e.g., interphalangeal joints [28,29]. However, for larger joints, the light will not be able to penetrate deep enough to excite the PS in the whole joint. An alternative to reach the whole synovium could be to deliver the light directly into the joint using endoscopic or fibre-optic systems (reviewed in [28]).

In this study, loading of the liposomes with the photosensitizer IRDye700DX did not significantly change the biodistribution of the compound, with the exception of increased liver accumulation, where the compound is cleared by sinusoidal epithelial cells [30]. This effect has previously been described for other fluorophores and may be due to changes in the charge and lipophilicity of the compounds [31,32]. The time required for elimination of this construct was not determined in this experiment, since biodistribution analysis was only performed 24 h after injection. However, since the liver and spleen are not exposed to light, and the PS has no dark toxicity at the concentrations used in this model, this should not result in any off-target damage.

The limited and transient treatment effect observed in this study may have several explanations. First, using CIA as a model for experimental arthritis to study the effect of PDT using 700DX-loaded liposomes proved challenging in this study. Timing of the treatment and analysis of outcome measures were complicated by the highly variable disease course and incidence, and, since animals were included when they reached inclusion criteria, inclusion was spread over multiple weeks. At inclusion, the cumulative arthritis scores of the animals were variable, and we cannot rule out that the accompanying variation of systemic pro-inflammatory molecules influenced our treatment efficacy. Despite this limitation, we were able to demonstrate an amelioration of arthritis development. The transient nature of our therapy may also indicate that the treatment should be further optimized to give the best possible effect. Further optimization of the therapy could be achieved by performing light dosimetry, as well as by focusing on the timing of administration of the liposomes and subsequent light exposure [28,33]. Combining this therapy with low dose immunosuppressant therapy may also improve the longevity of the PDT effect. This is an especially relevant option for clinical translation since it allows for local treatment of disease flares without increasing systemic immunosuppression.

5. Conclusions

In conclusion, we have shown the proof-of-principle of performing photodynamic therapy in arthritic joints using liposomes loaded with a photosensitizer, which allows for locoregional treatment of RA and thus avoids systemic side effects. To expand on these findings and to assess their therapeutic value with respect to future clinical application, studies into PDT dosimetry and treatment effects in other animal models, as well as RA patient material ex vivo, should be performed.

Supplementary Materials: The following are available online at https://www.mdpi.com/article/10.3390/pharmaceutics13111868/s1, Figure S1: Absorbance and fluorescence spectra of liposome with and without IRDye700DX compared to free IRDye700DX. Both absorbance and fluorescence spectra are similar when comparing IRDye700DX loaded in liposomes or free PS (all in PBS). Fluorescence emission was measured after excitation at 650 nm, Figure S2: Uptake of liposomes with (**a**) and without (**b**) IRDye700DX correlates with histological scores of inflammation in the ankle joints of mice with CIA ($n = 10$ mice), Figure S3: Area under the curve for the development of arthritis over time after 700DX-loaded liposome PDT. The AUC was significantly lower in the treated groups compared to PBS. Two-way ANOVA with Bonferroni post-test, * = $p < 0.05$ ($n = 8$ mice/group), Figure S4: Histological scores of ankle and knee joint inflammation, proteoglycan depletion and knee chondrocyte death and cartilage erosion. Results are depicted as mean with SD ($n = 8$ mice, ($n = 7$ for the 90 s exposed liposomes group)).

Author Contributions: Conceptualization, D.N.D., M.B. (Marti Boss), M.R., G.S., M.B. (Maarten Brom), P.L., P.M.v.d.K., M.I.K. and M.G.; methodology, D.N.D., M.B. (Marti Boss), B.W., M.M.A.H., G.S. and L.v.B.; validation, D.N.D., M.B. (Marti Boss), B.W., M.M.A.H. and L.v.B.; formal analysis, D.N.D., M.B. (Marti Boss); investigation, D.N.D., M.B. (Marti Boss), B.W., M.M.A.H., D.L.B. and L.v.B.; resources, D.N.D., M.B. (Marti Boss), M.R., G.S., M.B. (Maarten Brom), P.L., P.M.v.d.K., M.I.K. and M.G.; data curation, D.N.D., M.B. (Marti Boss), M.R., G.S., M.B. (Maarten Brom), P.L., P.M.v.d.K., M.I.K. and M.G.; writing—original draft preparation, D.N.D. and M.B. (Marti Boss); writing—review and editing, D.N.D., M.B. (Marti Boss), M.R., B.W., M.M.A.H., D.L.B., L.v.B., G.S., M.B. (Maarten Brom), P.L., P.M.v.d.K., M.B. (Mijke Buitinga), M.I.K. and M.G.; visualization, D.N.D. and M.B. (Marti Boss); supervision, P.M.v.d.K., M.B. (Mijke Buitinga), M.I.K. and M.G.; project administration, D.N.D. and M.B. (Marti Boss). All authors have read and agreed to the published version of the manuscript.

Funding: This research received no external funding.

Institutional Review Board Statement: The Radboud University animal ethics committee approved the study protocol (RU-DEC-2016-0076, CCD number AVD103002016786). All procedures were performed according to the Institute of Laboratory Animal Research guide for Laboratory Animals.

Informed Consent Statement: Not applicable.

Data Availability Statement: Data can be made available upon reasonable request.

Conflicts of Interest: The authors declare no conflict of interest.

Abbreviations

%IA/g: percentage of the injected amount per gram tissue; CIA: collagen-induced arthritis; DMARDs: disease-modifying anti-rheumatic drugs; DPPC: dipalmitoyl phosphatidylcholine; FCA: complete Freund's adjuvant; IL-: Interleukin; PDT: photodynamic therapy; PEG-(2000)-DSPE: PEG-(2000)-distearoyl phosphatidylethanolamine; PG depletion: proteoglycan depletion; PS: photosensitizer; RA: rheumatoid arthritis; ROS: reactive oxygen species; SD: standard deviation; SPECT/CT: single photon emission computer tomography/computer tomography; TNF- α: tumor necrosis factor-α.

References

1. Silman, A.J.; Pearson, J.E. Epidemiology and genetics of rheumatoid arthritis. *Arthritis Res.* **2002**, *4* (Suppl. 3), S265–S272. [CrossRef] [PubMed]
2. Smolen, J.S.; Aletaha, D.; McInnes, I.B. Rheumatoid arthritis. *Lancet* **2016**, *388*, 2023–2038. [CrossRef]
3. Buch, M.H. Defining refractory rheumatoid arthritis. *Ann. Rheum. Dis.* **2018**, *77*, 966–969. [CrossRef] [PubMed]
4. Mosser, D.M.; Hamidzadeh, K.; Goncalves, R. Macrophages and the maintenance of homeostasis. *Cell Mol. Immunol.* **2020**, *18*, 579–587. [CrossRef] [PubMed]
5. Kinne, R.W.; Brauer, R.; Stuhlmuller, B.; Palombo-Kinne, E.; Burmester, G.R. Macrophages in rheumatoid arthritis. *Arthritis Res.* **2000**, *2*, 189–202. [CrossRef]
6. Udalova, I.A.; Mantovani, A.; Feldmann, M. Macrophage heterogeneity in the context of rheumatoid arthritis. *Nat. Rev. Rheumatol.* **2016**, *12*, 472–485. [CrossRef]
7. Lam, J.; Takeshita, S.; Barker, J.E.; Kanagawa, O.; Ross, F.P.; Teitelbaum, S.L. TNF-alpha induces osteoclastogenesis by direct stimulation of macrophages exposed to permissive levels of RANK ligand. *J. Clin. Invest.* **2000**, *106*, 1481–1488. [CrossRef] [PubMed]
8. Ahsan, F.; Rivas, I.P.; Khan, M.A.; Torres Suarez, A.I. Targeting to macrophages: Role of Physicochemical properties of particulate carriers—liposomes and microspheres—On the phagocytosis by macrophages. *J. Control. Release* **2002**, *79*, 29–40. [CrossRef]
9. Van Rooijen, N.; Sanders, A. Liposome mediated depletion of macrophages: Mechanism of action, preparation of liposomes and applications. *J. Immunol. Methods* **1994**, *174*, 83–93. [CrossRef]
10. van der Geest, T.; Laverman, P.; Gerrits, D.; Franssen, G.M.; Metselaar, J.M.; Storm, G.; Boerman, O.C. Comparison of three remote radiolabelling methods for long-circulating liposomes. *J. Control. Release* **2015**, *220*, 239–244. [CrossRef] [PubMed]
11. van der Geest, T.; Laverman, P.; Gerrits, D.; Walgreen, B.; Helsen, M.M.; Klein, C.; Nayak, T.K.; Storm, G.; Metselaar, J.M.; Koenders, M.I.; et al. Liposomal Treatment of Experimental Arthritis Can Be Monitored Noninvasively with a Radiolabeled Anti-Fibroblast Activation Protein Antibody. *J. Nucl. Med.* **2017**, *58*, 151–155. [CrossRef]
12. Van Lent, P.L.; Van den Hoek, A.E.; Van den Bersselaar, L.A.; Spanjaards, M.F.; Van Rooijen, N.; Dijkstra, C.D.; Van de Putte, L.B.; Van den Berg, W.B. In vivo role of phagocytic synovial lining cells in onset of experimental arthritis. *Am. J. Pathol.* **1993**, *143*, 1226–1237.
13. van Lent, P.L.; van den Hoek, A.E.; van den Bersselaar, L.; van Rooijen, N.; van den Berg, W.B. Role of phagocytic synovial lining cells in experimental arthritis. *Agents Actions* **1993**, *38*, C92–C94. [CrossRef]
14. Richards, P.J.; Williams, A.S.; Goodfellow, R.M.; Williams, B.D. Liposomal clodronate eliminates synovial macrophages, reduces inflammation and ameliorates joint destruction in antigen-induced arthritis. *Rheumatology* **1999**, *38*, 818–825. [CrossRef]
15. Barrera, P.; Blom, A.; van Lent, P.L.; van Bloois, L.; Beijnen, J.H.; van Rooijen, N.; de Waal Malefijt, M.C.; van de Putte, L.B.; Storm, G.; van den Berg, W.B. Synovial macrophage depletion with clodronate-containing liposomes in rheumatoid arthritis. *Arthritis Rheum.* **2000**, *43*, 1951–1959. [CrossRef]
16. Dorst, D.N.; Rijpkema, M.; Boss, M.; Walgreen, B.; Helsen, M.M.A.; Bos, D.L.; Brom, M.; Klein, C.; Laverman, P.; van der Kraan, P.M.; et al. Targeted photodynamic therapy selectively kills activated fibroblasts in experimental arthritis. *Rheumatology* **2020**, *59*, 3952–3960. [CrossRef]
17. Bartneck, M.; Scheyda, K.M.; Warzecha, K.T.; Rizzo, L.Y.; Hittatiya, K.; Luedde, T.; Storm, G.; Trautwein, C.; Lammers, T.; Tacke, F. Fluorescent cell-traceable dexamethasone-loaded liposomes for the treatment of inflammatory liver diseases. *Biomaterials* **2015**, *37*, 367–382. [CrossRef]
18. Deshantri, A.K.; Fens, M.H.; Ruiter, R.W.J.; Metselaar, J.M.; Storm, G.; van Bloois, L.; Varela-Moreira, A.; Mandhane, S.N.; Mutis, T.; Martens, A.C.M.; et al. Liposomal dexamethasone inhibits tumor growth in an advanced human-mouse hybrid model of multiple myeloma. *J. Control. Release* **2019**, *296*, 232–240. [CrossRef]
19. Theek, B.; Baues, M.; Ojha, T.; Mockel, D.; Veettil, S.K.; Steitz, J.; van Bloois, L.; Storm, G.; Kiessling, F.; Lammers, T. Sonoporation enhances liposome accumulation and penetration in tumors with low EPR. *J. Control. Release* **2016**, *231*, 77–85. [CrossRef]
20. de Boer, E.; Warram, J.M.; Hartmans, E.; Bremer, P.J.; Bijl, B.; Crane, L.M.; Nagengast, W.B.; Rosenthal, E.L.; van Dam, G.M. A standardized light-emitting diode device for photoimmunotherapy. *J. Nucl. Med.* **2014**, *55*, 1893–1898. [CrossRef]
21. Joosten, L.A.B.; Lubberts, E.; Durez, P.; Helsen, M.M.A.; Jacobs, M.J.M.; Goldman, M.; vandenBerg, W.B. Role of interleukin-4 and interleukin-10 in murine collagen-induced arthritis—Protective effect of interleukin-4 and interleukin-10 treatment on cartilage destruction. *Arthritis Rheum.* **1997**, *40*, 249–260. [CrossRef] [PubMed]

22. Laverman, P.; van der Geest, T.; Terry, S.Y.; Gerrits, D.; Walgreen, B.; Helsen, M.M.; Nayak, T.K.; Freimoser-Grundschober, A.; Waldhauer, I.; Hosse, R.J.; et al. Immuno-PET and Immuno-SPECT of Rheumatoid Arthritis with Radiolabeled Anti-Fibroblast Activation Protein Antibody Correlates with Severity of Arthritis. *J. Nucl. Med.* **2015**, *56*, 778–783. [CrossRef] [PubMed]
23. Hayer, S.; Vervoordeldonk, M.J.; Denis, M.C.; Armaka, M.; Hoffmann, M.; Backlund, J.; Nandakumar, K.S.; Niederreiter, B.; Geka, C.; Fischer, A.; et al. 'SMASH' recommendations for standardised microscopic arthritis scoring of histological sections from inflammatory arthritis animal models. *Ann. Rheum. Dis.* **2021**, *80*, 714–726. [CrossRef]
24. Van Lent, P.L.; Van den Bersselaar, L.A.M.; Holthuyzen, A.E.M.; Van Rooijen, N.; Van de Putte, L.B.A.; Van den Berg, W.B. Phagocytic synovial lining cells in experimentally induced chronic arthritis: Down-regulation of synovitis by CL2MDP-liposomes. *Rheumatol. Int.* **1994**, *13*, 221–228. [CrossRef]
25. Funke, B.; Jungel, A.; Schastak, S.; Wiedemeyer, K.; Emmrich, F.; Sack, U. Transdermal photodynamic therapy—A treatment option for rheumatic destruction of small joints? *Lasers Surg. Med.* **2006**, *38*, 866–874. [CrossRef]
26. Ozen, G.; Pedro, S.; England, B.R.; Mehta, B.; Wolfe, F.; Michaud, K. Risk of Serious Infection in Patients With Rheumatoid Arthritis Treated With Biologic Versus Nonbiologic Disease-Modifying Antirheumatic Drugs. *ACR Open Rheumatol.* **2019**, *1*, 424–432. [CrossRef]
27. Doran, M.F.; Crowson, C.S.; Pond, G.R.; O'Fallon, W.M.; Gabriel, S.E. Frequency of infection in patients with rheumatoid arthritis compared with controls: A population-based study. *Arthritis Rheum.* **2002**, *46*, 2287–2293. [CrossRef]
28. Kim, M.M.; Darafsheh, A. Light Sources and Dosimetry Techniques for Photodynamic Therapy. *Photochem. Photobiol.* **2020**, *96*, 280–294. [CrossRef]
29. Ash, C.; Dubec, M.; Donne, K.; Bashford, T. Effect of wavelength and beam width on penetration in light-tissue interaction using computational methods. *Lasers Med. Sci.* **2017**, *32*, 1909–1918. [CrossRef]
30. Kawai, Y.; Smedsrod, B.; Elvevold, K.; Wake, K. Uptake of lithium carmine by sinusoidal endothelial and Kupffer cells of the rat liver: New insights into the classical vital staining and the reticulo-endothelial system. *Cell Tissue Res.* **1998**, *292*, 395–410. [CrossRef]
31. Rijpkema, M.; Bos, D.L.; Cornelissen, A.S.; Franssen, G.M.; Goldenberg, D.M.; Oyen, W.J.; Boerman, O.C. Optimization of Dual-Labeled Antibodies for Targeted Intraoperative Imaging of Tumors. *Mol. Imaging* **2015**, *14*, 348–355. [CrossRef] [PubMed]
32. Boswell, C.A.; Tesar, D.B.; Mukhyala, K.; Theil, F.P.; Fielder, P.J.; Khawli, L.A. Effects of charge on antibody tissue distribution and pharmacokinetics. *Bioconjug. Chem.* **2010**, *21*, 2153–2163. [CrossRef] [PubMed]
33. van Doeveren, T.E.M.; Bouwmans, R.; Wassenaar, N.P.M.; Schreuder, W.H.; van Alphen, M.J.A.; van der Heijden, F.; Tan, I.B.; Karakullukcu, M.B.; van Veen, R.L.P. On the Development of a Light Dosimetry Planning Tool for Photodynamic Therapy in Arbitrary Shaped Cavities: Initial Results. *Photochem. Photobiol.* **2020**, *96*, 405–416. [CrossRef]

 *pharmaceutics*

Article

Application of Photodynamic Therapy with 5-Aminolevulinic Acid to Extracorporeal Photopheresis in the Treatment of Patients with Chronic Graft-versus-Host Disease: A First-in-Human Study

Eidi Christensen [1,2,3,*], Olav A. Foss [4,5], Petter Quist-Paulsen [3,6], Ingrid Staur [1], Frode Pettersen [7], Toril Holien [2,3,6], Petras Juzenas [2] and Qian Peng [2,8]

1 Department of Dermatology, St. Olavs Hospital, Trondheim University Hospital, 7030 Trondheim, Norway
2 Department of Pathology, The Norwegian Radium Hospital, Oslo University Hospital, 0310 Oslo, Norway
3 Department of Clinical and Molecular Medicine, Norwegian University of Science and Technology, 7030 Trondheim, Norway
4 Department of Orthopaedic Surgery, Clinic of Orthopaedy, Rheumatology and Dermatology, St. Olavs Hospital, Trondheim University Hospital, 7030 Trondheim, Norway
5 Department of Neuroscience and Movement Science, Norwegian University of Science and Technology, 7030 Trondheim, Norway
6 Department of Haematology, St. Olavs Hospital, Trondheim University Hospital, 7030 Trondheim, Norway
7 Department of Nephrology, St. Olavs Hospital, Trondheim University Hospital, 7030 Trondheim, Norway
8 Department of Optical Science and Engineering, School of Information Science and Technology, Fudan University, Shanghai 200433, China
* Correspondence: eidi.christensen@ntnu.no

Citation: Christensen, E.; Foss, O.A.; Quist-Paulsen, P.; Staur, I.; Pettersen, F.; Holien, T.; Juzenas, P.; Peng, Q. Application of Photodynamic Therapy with 5-Aminolevulinic Acid to Extracorporeal Photopheresis in the Treatment of Patients with Chronic Graft-versus-Host Disease: A First-in-Human Study. *Pharmaceutics* **2021**, *13*, 1558. https://doi.org/10.3390/pharmaceutics13101558

Academic Editors: Udo Bakowsky, Matthias Wojcik, Eduard Preis and Gerhard Litscher

Received: 22 July 2021
Accepted: 19 September 2021
Published: 26 September 2021

Publisher's Note: MDPI stays neutral with regard to jurisdictional claims in published maps and institutional affiliations.

Abstract: Extracorporeal photopheresis (ECP), an immunomodulatory therapy for the treatment of chronic graft-versus-host disease (cGvHD), exposes isolated white blood cells to photoactivatable 8-methoxypsoralen (8-MOP) and UVA light to induce the apoptosis of T-cells and, hence, to modulate immune responses. However, 8-MOP-ECP kills diseased and healthy cells with no selectivity and has limited efficacy in many cases. The use of 5-aminolevulinic acid (ALA) and light (ALA-based photodynamic therapy) may be an alternative, as ex vivo investigations show that ALA-ECP kills T-cells from cGvHD patients more selectively and efficiently than those treated with 8-MOP-ECP. The purpose of this phase I-(II) study was to evaluate the safety and tolerability of ALA-ECP in cGvHD patients. The study included 82 treatments in five patients. One patient was discharged due to the progression of the haematological disease. No significant persistent changes in vital signs or laboratory values were detected. In total, 62 adverse events were reported. Two events were severe, 17 were moderate, and 43 were mild symptoms. None of the adverse events evaluated by the internal safety review committee were considered to be likely related to the study medication. The results indicate that ALA-ECP is safe and is mainly tolerated well by cGvHD patients.

Keywords: 5-aminolevulinic acid; ALA-based photodynamic therapy; phototherapy; extracorporeal; photopheresis; chronic graft-versus-host disease

1. Introduction

The modification of today's standard extracorporeal photopheresis (ECP) with the introduction of 5-aminolevulinic acid (ALA) based photodynamic therapy (PDT) may improve treatment efficacy. Since the treatment of cutaneous T-cell lymphoma was approved more than 30 years ago, ECP has been used to treat patients with other T-cell-mediated disorders, including chronic graft-versus-host disease (cGvHD) [1–3]. Graft-versus-host disease (GvHD) represents a serious immunologically mediated complication of allogeneic hematopoietic cell transplantation [4]. Donor immune cells, mostly T-lymphocytes (T-cells), are activated and attack the host cells, misrecognising them as foreign. GvHD is classified as chronic based on histological criteria [5]. It can involve multiple organs, of which the skin is the most often affected, and can cause severe morbidity in patients [6]. ECP is widely used in those patients with an initial non or partial response to first-line treatment, including systemic corticosteroids [2].

ECP is a leukapheresis-based therapeutic procedure. The mechanism of its immunomodulatory effect is not fully understood but involves altered T-cell function and the modulation of dendritic cell maturation, with the modification of the cytokine profile and stimulation of regulatory T-cells [2,7,8]. During ECP, leucocytes separated from the whole blood (buffy coat) are exposed to the photosensitising agent 8-methoxypsoralen (8-MOP) and ultraviolet A (UVA) light before reinfusion back into the patient. The combination of 8-MOP-UVA irradiation leads to deoxyribonucleic acid (DNA) crosslinking and, thereby, to the apoptosis of the treated cells. However, a major disadvantage of 8-MOP is that it binds to the DNA of diseased and normal cells with no selectivity, thus inducing the apoptosis of both types of cells after UVA activation. In addition, the treatment is time-consuming, expensive, and often ineffective. This demands the use of an alternative photosensitiser that selectively targets diseased cells to improve the effectiveness of today's standard treatment.

The idea of using 5-aminolevulinic acid (ALA) as a photosensitiser precursor for ECP has emerged from decades of research and experience with photodiagnosis and PDT. ALA-PDT is well established for topical use in the treatment of patients with non-melanoma skin cancer, and the oral intake of ALA is approved for clinical use in the photodetection of glioma and for the treatment of Barrett's oesophagus [9–11]. ALA is a naturally occurring amino acid and a precursor of heme that is metabolised intracellularly to endogenously porphyrin photosensitisers, mainly protoporphyrin IX (PpIX) [12]. It provides a highly selective accumulation of PpIX in the proliferative or activated cells, including tumour cells, due to their higher uptake of ALA and the alteration of the activities of enzymes involved in heme biosynthesis. The photoactivation of ALA-induced PpIX in the presence of oxygen through the formation of reactive oxygen species leads to the apoptosis and necrosis of the targeted cells [13,14]. After illumination, PpIX is partly or completely photobleached (faded). Under experimental conditions, ALA-UVA is shown to kill T-cells in the blood samples of patients with cGvHD more selectively and efficiently than those treated with 8-MOP-UVA [15]; hence, the number and duration of ECP treatments with the use of ALA can be reduced.

The aim of this study was to evaluate the safety and tolerability of multiple treatments with ALA-ECP in patients with cGvHD. In addition, data are presented from assessments of selected organs, performance, and quality of life in patients during the study period.

2. Materials and Methods

2.1. Patients

Patients with cGvHD who responded inadequately to 8-MOP-ECP after a minimum of 3 months of treatment at St. Olavs Hospital, Trondheim University Hospital, Norway, were considered for study inclusion. Inadequate response was defined as the progression of cutaneous cGvHD (>25% worsening of skin score from baseline or insufficient response with a <15% improvement in skin score compared with baseline or a $\leq$25% reduction in corticosteroid dose). Excluded were patients under 18 years of age; with body weight below 40 kg; and those with photosensitive comorbidities such as porphyria or known hypersensitivity to ALA or porphyrins, aphakia, pregnancy or breastfeeding, polyneuropathy, ongoing cardiac and pulmonary diseases, uncontrolled infection or fever, history of heparin-induced thrombocytopenia, an absolute neutrophil count $< 1 \times 10^9$/L, platelet count $< 20 \times 10^9$/L, aspartate transaminase (AST), alanine transaminase (ALT), bilirubin, or an international normalised ratio (INR) value $\geq 3\times$ upper limit of normal or significant ECG findings. Patients considered unlikely to comply with the study procedures were also excluded

2.2. Design and Procedures

This prospective, open, single-centre, phase I–(II) study was approved by the regional committee for medical research ethics (REK-Nord 2014/2316), the Norwegian Medicines Agency (National Regulatory Authority 14/16760-29) and was registered at www.clinicaltrials.gov (accessed on 20 September 2017) as NCT03109353. The study was performed in accordance with the Helsinki Declaration, monitored by the Clinical Research Unit Central Norway, Norwegian University of Science and Technology (NTNU), and patients provided written informed consent.

The CELLEX Photopheresis machine (Therakos, Mallinckrodt Pharmaceuticals, Raritan, NJ, USA) was used. This is a closed system in which leukapheresis, photoactivation, and the reinfusion of white blood cells are achieved sequentially. The patients were connected to the machine through single or double venous access to collect whole blood. Then, through centrifugation, the fraction of white cells (buffy coat) was separated and collected into a treatment bag (buffy bag), while the other blood components, mainly red cells and plasma, were returned to the patient. A commercially available ALA-hydrochloride powder for systemic oral use (Gliolan, photonamic GmbH & Co. KG, Pinneberg, Germany) was used at a dose of 0.168% w/v (10mM). The pharmacy at St. Olavs Hospital reconstituted ALA-hydrochloride in 0.9% sodium chloride (NaCl) physiological saline under sterile condition to a stock solution of 30 mg/mL. Since the amount of buffy coat collected varied on different days for the same patients and between patients, the volume of the ALA solution added to the buffy coat was adjusted to the volume collected on each treatment to reach a dose of 10 mM. After the ALA-solution was added, a 1-h incubation period in the dark was allowed for intracellular PpIX production before UVA light exposure of the cells. Although the most effective light wavelength for PpIX activation is at 400 nm, the use of the UVA light source in the Therakos photopheresis machine can also effectively kill ALA-incubated human T-cell lines [16].

Finally, the treated buffy coat was reinfused into the patient. One treatment-cycle represents two treatments: one treatment given on two consecutive days. Each patient could receive up to 20 ALA-ECP treatments corresponding to 10 treatment-cycles within a year. The duration between two treatment-cycles varied depending on each patient's symptoms and treatment response.

Representative buffy coat samples for the investigation of ALA-induced PpIX were taken after a 1-h ALA incubation at room temperature. The plasma samples for PpIX determination were taken immediately after the reinfusion of the ex vivo ALA-ECP-treated buffy coat and 24-h after treatment. Measurements of ALA-induced PpIX in the buffy coat and plasma samples were performed in an extraction solution [17,18]. The solvent consisted of 1% sodium dodecyl sulfate in 1 N perchloric acid (HClO4) and methanol (CH3OH) (1:1 vol/vol). A 100-µL sample was thoroughly mixed with 900 µL of solvent for approximately 1 min in a 1.5-mL tube (Eppendorf, Hamburg, Germany). The sample homogenates were then kept at $-20\,^{\circ}$C until analysis. Before measurements, the samples were centrifuged for 5 min at 1000 rpm, and the supernatants were carefully transferred to new tubes. The fluorescence of PpIX was measured using a luminescence spectrometer (LS50B, Perkin-Elmer, Norwalk, CT, USA). The PpIX fluorescence in the sample supernatants was measured in a standard 10-mm quartz cuvette placed in the standard holder of the luminescence spectrometer. The fluorescence excitation wavelength was 400 nm, corresponding to the Soret band of the PpIX absorption spectrum. The fluorescence intensity was recorded at 605 nm, corresponding to the PpIX emission peak in the solvent. The amount of PpIX was determined by comparing the PpIX fluorescence intensities with the standard curve made by adding known amounts of PpIX to the solvent. The sensitivity level of our method, that is, the lowest concentration of PpIX that could be detected, was approximately 3 nM in the extraction solution.

2.3. Safety and Tolerability

Safety and tolerability were regularly monitored through clinical and laboratory examinations and patient reports (Table 1). Safety was monitored through frequency, seriousness, and intensity of adverse events, 12 lead electrocardiogram (ECG) recordings, vital signs (blood pressure as seated, pulse at rest, and forehead temperature), physical examinations, and laboratory measurements. The schedules of major events within each treatment-cycle and controls are presented in Table 1. Assessments performed either at screening, if it was less than one week before ALA-ECP or performed just before the first treatment were used as reference (baseline). Any abnormality was considered for clinical significance and for the need for action, for example, repetition of the test to verify the result and/or closer follow-up. Laboratory parameters included haematology (haemoglobin, white blood cell (WBC) count, WBC differential count, eosinophil count, platelet count, mean cell volume, mean cell haemoglobin, and INR), and clinical chemistry (albumin, ALT, alkaline phosphatase, AST, bilirubin, blood urea, cholesterol, C-reactive protein (CRP), calcium cation, creatinine, erythrocyte sedimentation rate (SR), gamma-glutamyl transferase, glycated haemoglobin A1c, lactate dehydrogenase, potassium, sodium, and total protein). Urine was examined with a dipstick for erythrocytes, glucose, ketone, leucocytes, nitrite, pH, and protein.

Table 1. Schedule for the assessment of safety, tolerability and response.

Investigation	Time Point Screening (Baseline)	ECP, Day 1	ECP, Day 2	Post-ECP, Day 4 $\pm$ 1	Post-ECP, Day 16 $\pm$ 2	Controls, Every 3 Months
Vital signs	x	x	x			x
ECG	x					x
Haematology	x	x *				x
Clinical chemistry	x	x *				x
Urine analyses	x	x *				x
Conceivable Adverse Events				x	x	
Adverse Events				x	x	x	x
Organ, performance, and QoL assessments	x	x *				x

* Not needed if less than one week after screening. ECP, extracorporeal photopheresis; ECG, electrocardiogram; QoL, quality of life.

Additional blood tests to analyse AST, ALT, bilirubin, and INR were performed to evaluate the effect of ALA-ECP on the liver. Blood analyses were performed at the Department of Clinical Chemistry, St. Olavs Hospital, and reference values were used. Urine was examined using a dipstick test.

Tolerability was observed through the patients' reported adverse events (AEs). Any event during the study period in which the patients felt unwell or different from usual was recorded as an AE. Conceivable AEs of systemic ALA treatment included nausea, vomiting, headache, photosensitivity, and chills. Patients received a diary in which AEs were noted between visits. The seriousness of all events was graded (grade 1: asymptomatic or mild symptoms to grade 5; death related to AE) according to Common Terminology Criteria for Adverse Events (CTCAE) v4.03. The assessment of study continuation was based on the stated protocol criteria. An internal safety review committee (ISRC) was appointed to assess safety and tolerability data and to evaluate the reported AEs.

2.4. Organ, Performance, and Quality of Life Assessments

Assessments of organ, performance, and quality of life were repeated at approximately 3-month intervals. Investigator-performed evaluations and patient-reported outcome measures were used. The use of immunosuppressive therapy was monitored. Parts of Filipovich and colleagues' clinical organ scoring system were used in the evaluation of skin manifestations [6]. This scoring system combines clinical features with the affected percentage of the body surface area (BSA) (grade 0: no symptoms; grade 1: <18% BSA with disease signs but no sclerotic features; grade 2: 19–50% BSA or superficial sclerotic features; and grade 3: >50% BSA or deep sclerotic features or impaired mobility, ulceration, or severe pruritus). The affected percentage of BSA was additionally registered in the evaluation of skin disease. Pruritus was assessed using a visual analogue scale (VAS) graded from 1 to 10, with an increasing number corresponding to a higher severity of pruritic manifestation. The diagnostic features of mucosal manifestations were not systematically recorded. Schirmer's test was used to evaluate tear production. Sterile paper strips were placed into the inferior-temporal aspect of the conjunctival sac of both eyes without topical anaesthesia, and the wetted length was measured in millimetres after 5 min. A measurement of $\geq$15 mm was considered grade 1 (normal), 14–9 mm as grade 2 (mild), 8–4 as grade 3 (moderate), and less than 4 mm as grade 4 (severe significance). Mouth, gastrointestinal, and performance were assessed using Filipovich's clinical system for scoring the severity and extent of cGvHD using a 4-point scale (grade 0: no symptoms, grade 1: mild symptoms, grade 3: moderate symptoms, and grade 4: severe symptoms) [6]. Functional status was assessed using Karnovsky's performance status scale [19]. The patients were given a score on a linear scale between 0 and 100, summarising their ability to perform daily activities: the lower the score was, the worse the status was. Patients were also asked to complete the Skindex-29 and European Organisation for the Research and Treatment of Cancer Quality of Life Questionnaire Core 30 (EORTC30) and the Functional Assessment of Cancer Therapy–Bone Marrow Transplantation (FACT–BMT) questionnaires [20–22]. Skindex-29 is designed to measure the overall effect of skin disease on health-related quality of life and is sensitive to minor changes over time. EORTC30 is a widely used measure of cancer-specific health-related quality of life. FACT-BMT is a general measure of cancer-specific health-related quality of life in combination with a module developed specifically for evaluating the quality of life of bone marrow transplant patients.

2.5. Statistical Methods

Descriptive data are reported as numbers, percentages (%), and mean (min–max.) and standard deviation ($\pm$), together with line graphs and box plots (median, 25–75% percentile). IBM SPSS software version 27 was used.

3. Results

3.1. Patients and Treatments

Five patients (three women and two men) were included, of which four completed the study. The mean age at entry was 48 years (32–60). Four patients were diagnosed with acute myeloid leukaemia, and one was diagnosed with high-risk myelodysplastic syndrome before undergoing allogeneic bone marrow transplantation. Prior to study inclusion, the patients had received a mean of 25 (10–49) treatment-cycles of 8-MOP-ECP. Three patients received systemic steroids.

During the study period, the patients received a total of 82 single treatments (42 treatment-cycles) with ALA-ECP. One patient missed one treatment due to clotting in the buffy coat. A second patient missed one treatment because of a gastric flu event. The mean volume of the buffy coat collected at treatments was 146 mL (100–251), and the mean irradiation time was 17 min (2–45). As the patients' symptoms of the disease varied, the treatment intervals were individualised. Two patients had an ECP treatment interval of 4 weeks, each receiving 20 treatments within 9 months. The remaining three patients had an interval of 6–8 weeks and received 18, 13, and 11 treatments, respectively. One patient met a protocol-defined discontinuation criterion 8 months into the study when the treatment with systemic prednisolone was increased from 10/5 mg/every other day to 25 mg/day due to the worsening of symptoms. After discharge from the study, the patient received two 8-MOP-ECP treatments before being diagnosed with lung cancer, after which ECP was discontinued.

3.2. Safety and Tolerability

We did not observe any clinically significant changes in the patients' vital signs. The details of the results are presented in Figure 1.

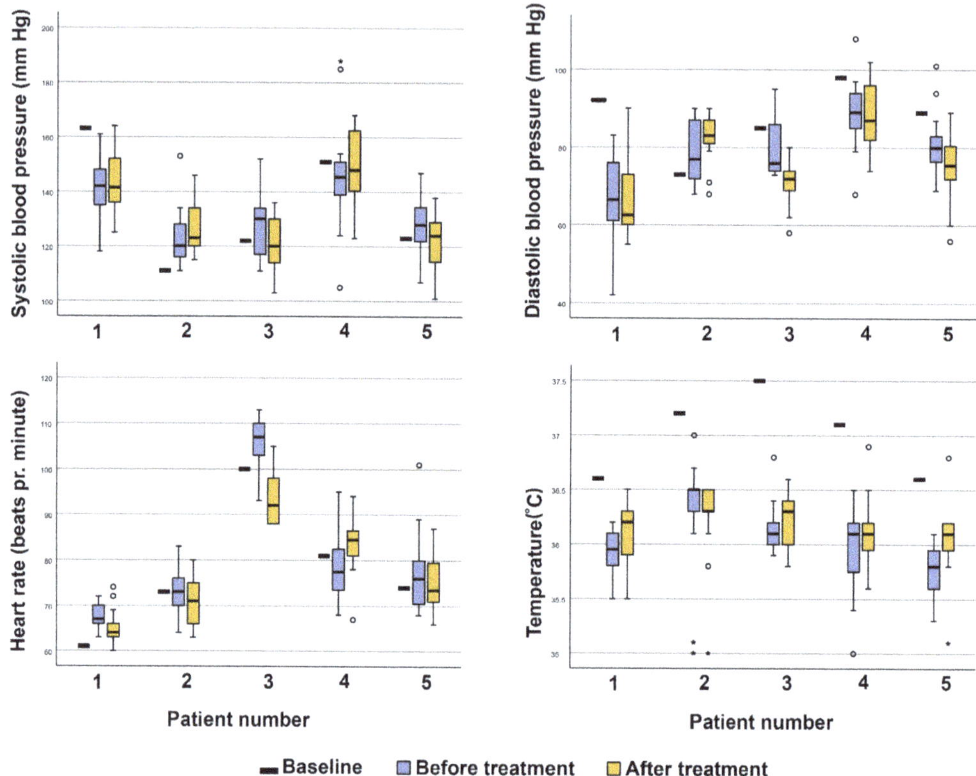

Figure 1. Systolic and diastolic blood pressure, heart rate, and temperature of patients, as measured at baseline and before and after each day in all treatment-cycles. (Boxplot with; median, 25th and 75th percentile, whiskers—value which is not an outlier or an extreme score, ° outlier more than 1.5 box lengths from the box, * extreme case more than 3 box lengths from the box.)

One patient had an increased QTc interval value of 467 ms on ECG at the 3-month control compared to 450 ms at baseline (prolonged QTc defined as a value > 450 ms [23]). Consequently, an ECG was performed before and after three consecutive treatment cycles, and the patient underwent cardiological examination with an echocardiogram and stress ECG with no pathological findings. Both the cardiologist and ISRC concluded that the QTc interval in this patient was unlikely to be affected by the study's medicine.

Few clinically significant changes in laboratory values were observed. Two patients had a transient increase in CRP values (34 and 31 mg/L) and SR values (31 and 21 mm/h), respectively, compared with the baseline. One patient had transient high values of eosinophil count (15%), absolute eosinophil count (1.98×10^9/L), and lactate dehydrogenase level (239 U/L). No evidence of liver toxicity was observed (Figure 2).

Most of the bilirubin and INR measurements showed transiently higher values a few days after treatment. High levels of ALT and AST were confirmed at baseline in one patient (patient two) with a history of elevated liver enzymes. This patient regularly used paracetamol but had no established liver disease. Another patient (patient five) with elevated liver enzyme and bilirubin levels four days after treatment had a known bile disorder. No clinically significant long-term liver effects were observed. Albumin and total protein levels were stable in all the patients throughout the study. Dipstick urinalysis revealed no severe kidney or urinary tract disorders. One test yielded a positive urine culture, resulting in antibiotic treatment. ALA-induced PpIX was determined from 75 buffy coat and 104 plasma samples from the five patients and was detected in all the buffy coat samples with a mean concentration of 122 (123) nM. No PpIX was found in the plasma samples taken immediately or at 24 h after re-infusion.

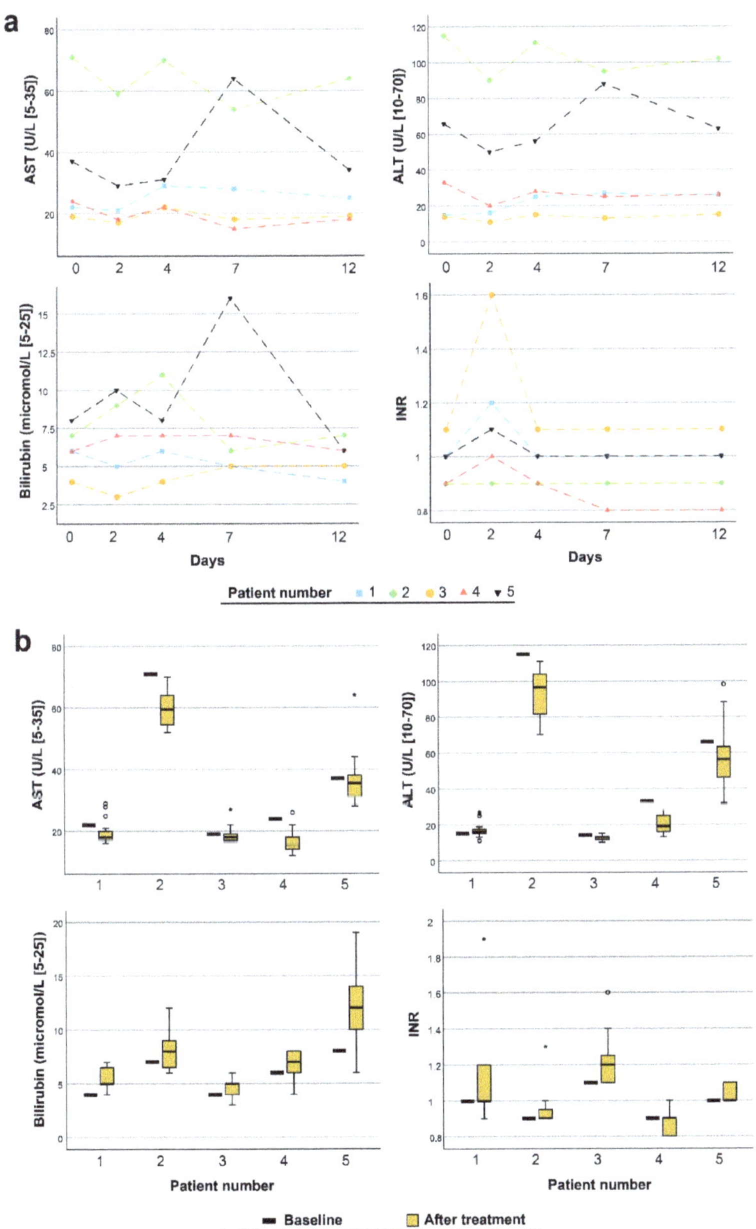

Figure 2. (**a**) Measurements of aspartate transaminase (AST), alkaline phosphatase (ALT), bilirubin, and international normalised ratio (INR) for each patient at baseline and on day 2 of treatment and days 4, 7, and 12 following the first treatment-cycle. (**b**) Measurements of AST, ALT, bilirubin, and INR at baseline and after day 2 of each treatment-cycle. (Boxplot with; median, 25th and 75th percentile, whiskers—value which is not an outlier or an extreme score, ° outlier more than 1.5 box lengths from the box, * extreme case more than 3 box lengths from the box.)

In total, there were 62 cases of AEs (two serious adverse events (SAEs), 20 AEs, and 40 conceivable AEs). Of the 62 events, two (3%) were severe, 17 (27%) were moderate, and 43 (69%) were mild symptoms (Tables 2 and 3). None of the AEs evaluated by the ISRC were considered to be likely related to the study medication (Table 2). The two SAEs were observed in the same patient. The first event was a pulmonary viral infection occurring 12 days after the second treatment-cycle, and the patient was hospitalised for 14 days. The ISRC evaluated the event as possibly related to the study medication. The second SAE was a migraine headache episode that occurred two days after the sixth treatment-cycle and required two days of hospitalisation. The patient had a history of migraines. The ISRC evaluated the SAE as unlikely to be related to the study medication. The most common grade one and two AEs reported were clotting in the treatment system, cold-like symptoms, and dysuria (Table 2). No grade four or five AEs were reported. The most frequently reported conceivable AEs were nausea and headache (Table 3). Of all the cases, 85% had mild grade one, and 15% had grade two symptoms. No grade 3–5 events were reported.

Table 2. Adverse events and Internal Safety Review Committee's assessments of the severity of the events as well as relation to the study medication.

Type of Adverse Event	Number of Patients	Number of Grade 1	Number of Grade 2	Number of Grade 3	Relation to Study Medication
Clotting in buffy	4	0	4	0	Possibly
Cold like symptoms	3	3	1	0	Unlikely
Dysuria	1	0	3	0	Possible
Respiratory infection	2	0	1	1	Possible
Skin worsening	1	0	1	0	Possible
Migraine	1	0	0	1	Unlikely
Flue like symptoms	1	1	0	0	Unlikely
Elevated INR	1	1	0	0	Unlikely
Prickling around mouth	1	1	0	0	Possible
UVI	1	0	1	0	Possible
Malaise	1	1	0	0	Unlikely
Sore throat, pain toes and hip	1	1	0	0	Unlikely
Prolonged QTc interval	1	1	0	0	Unlikely

Grade 1 severity: mild; asymptomatic or mild symptoms; clinical or diagnostic observation only; intervention not indicated; Grade 2 severity: moderate; minimal, local, or non-invasive intervention indicated; limiting age-appropriate instrumental activities of daily living; Grade 3 severity: severe or medically significant but not immediately life-threatening; hospitalisation or prolongation of hospitalisation indicated disabling, limiting self-care activities of daily living.

Table 3. Frequency and severity of conceivable adverse events among patients.

Type of Adverse Event	Number of Patients	Number of Grade 1	Number of Grade 2
Nausea	4	19	2
Vomiting	2	1	1
Headache	4	10	3
Photosensitivity	0	0	0
Chills	2	4	0

Nausea Grade 1: loss of appetite without alteration in eating habits, Grade 2: Oral intake decreased without significant weight loss, dehydration, or malnutrition; *Vomiting* Grade 1: 1–2 episodes in 24 h, Grade 2: 3–5 episodes in 24 h; *Headache* Grade 1: mild pain, Grade 2: moderate pain, limiting instrumental activity of daily living; *Chills* Grade 1: mild sensation of cold, shivering, chatting of teeth.

3.3. Organ, Performance, and Quality of Life Assessments

The assessment scores from the baseline and the patients' last control are shown in Table 4. Most of the conditions showed an improvement in scores with the greatest improvement in skin, BSA involvement, and pruritus. The eye assessment showed the best score at baseline. In one patient, the prednisolone dose of 5 mg/every other day was unchanged, and in another, it was reduced from 10/5 mg/every other day at baseline to 2.5 mg/day at the last control.

Table 4. Results of organ, performance, and quality of life assessments at baseline and at the last control for all patients.

Target	Scoring Tool	Scoring Scale	Baseline (Mean Score, min.- max.)	Last Control (Mean Score, min.- max.)
Skin	Modified organ scoring system	0–3 (0 = no symptom)	2.6 (1–3)	1.6 (1–2)
Skin	Body surface area %	0–100% (0 = no area affected)	15.8 (7–30)	5.0 (3–7)
Pruritus	Visual analogue scale	0–10 (0 = no symptom)	3.3 (0–7)	1.4 (1–2)
Mouth	Modified organ scoring system	0–3 (0 = no symptom)	1.6 (0–3)	1.4 (0–2)
Eye, right	Schirmer's test	1–4 (1 = normal)	3.0 (1–4)	4.0 (4–4)
Eye, left	Schirmer's test	1–4 (1 = normal)	3.6 (3–4)	4.0 (4–4)
Performance	Modified organ scoring system	0–3 (0 = no symptom)	1.2 (1–2)	0.8 (0–1)
Gastrointestinal tract	Modified organ scoring system	0–3 (0 = no symptom)	0.4 (0–1)	0.4 (0–1)
Function	Karnovsky's performance scale	1–100 (0 = low function)	74 (70–90)	82 (70–100)
Skindex, emotions	Questionnaire	0–100 (0 = no symptom)	22.0 (8–48)	10.2 (0–30)
Skindex, symptoms	Questionnaire	0–100 (0 = no symptom)	30.4 (21–39)	24.2 (17–36)
Skindex, function	Questionnaire	0–100 (0 = no symptom)	19.6 (2–56)	13.4 (2–42)
Skindex, single item	Questionnaire	0–100 (0 = no symptom)	10.0 (0–25)	10.0 (0–50)
EORTC30, functional	Questionnaire	0–100% (0 = low function)	76.2 (70–100)	82.0 (57–100)
EORTC30, symptoms	Questionnaire	0–100% (0 = low level of symptom)	23.4 (7–40)	20.0 (3–50)
EORTC30, global health status	Questionnaire	0–100% (0 = low state)	65.2 (42–75)	61.4 (16–83)
FACT-G, total score	Questionnaire	0–108 (0 = low state)	86.8 (63–99)	88.4 (70–97)
FACT-BMT, subscale score	Questionnaire	0–40 (0 = much concern)	27.8 (22–35)	29.6 (18–36)

EORTC30: European Organisation for the Research and Treatment of Cancer Quality of Life Questionnaire Core 30, FACT-G: Functional Assessment of Cancer Therapy—Global, FACT-BMT: FACT—Bone Marrow Transplantation.

4. Discussion

This paper reports the results of 82 ALA-ECP treatments administered to five patients with cGvHD who were considered to respond inadequately to 8-MOP-ECP. We did not observe any clinically significant changes in vital signs or detect any persistent abnormal laboratory findings. Overall, we registered an improvement in the patients' skin scores during the study period.

The rationale for the clinical use of ALA in ECP is multiple. ALA does not induce carcinogenesis, since ALA-induced PpIX localises to the cellular membrane structures outside the nucleus [12,24]. Transformed or activated T-cells produce considerably more ALA-induced PpIX than normal resting cells [15]. As a result, these cells are highly selectively destroyed after light irradiation, while resting normal T-cells remain undamaged. Furthermore, ALA combined with light-based therapy (ALA-based PDT) induces anti-tumour immunity [25]. Besides cell necrosis, it also initiates apoptotic cell death through the mitochondrion and ER-Ca2+-pathways [26], which, in turn, can promote immunosuppressive effects, including regulatory T-cell activation [14]. The highly selective photodamage of transformed or activated T-cells by ALA-PDT with the subsequent immunogenic cell death-mediated induction of an individualised and specific immuno-modulatory response may be a major advantage over non-specific immunosuppressive drugs.

This is the first study of ALA administered to patients treated with ECP. We took the precaution of using a dose of ALA corresponding to 11.2 mg/kg if assuming a 75 kg body weight (10mM). When previously used ex vivo for a 24-h incubation, the same dose gave no dark toxicity to leukocytes from cGvHD patients [16]. Moreover, a single oral dose of 20 mg/kg ALA in drinking water has been approved for the fluorescence-guided resection of glioma [10] in humans, while that of 60 mg/kg has been administered for the PDT treatment of Barrett's oesophagus [11].

Common side effects observed from the use of oral ALA are nausea, vomiting, and a transient rise in some liver enzymes and bilirubin, which typically resolve after 48 h [27,28]. Similar effects were observed in the present study. The additional finding of a modest increase in the INR value on day two of treatment may be seen in relation to the heparinisation of the patients' blood in association with ECP. Phototoxicity was not reported as an adverse reaction after treatment and coincides well with the finding of no PpIX in the patients' plasma.

Conventional ECP with 8-MOP is considered a safe treatment modality with side effects that are commonly sporadic and mild, such as nausea, fever, or headache [29]. Of all the AEs reported in this study, clotting in the buffy coat and episodes of infections were most frequently evaluated to likely be related to the study medicine. We are uncertain whether ALA and/or the 1-h incubation period before the photoactivation of the buffy coat contributed, since abnormal clotting is also recognised with conventional 8-MOP- ECP [30]. Although heparin is part of the standard ECP operational system, the use of anticoagulants varies between hospitals and is adjusted according to the patients' conditions [31]. Clotting during 8-MOP-ECP is commonly managed at our hospital by increasing the heparin dose. In the four study patients, the heparin dose (5000 IE/mL) was increased from 1.0 to 1.5 mL, then no clotting was observed. No definite association between treatment and the increased chance of infection using ALA-ECP was concluded, although three reported cases of infection were considered to possibly be related. There is no evidence of an increased risk of systemic infection after oral ALA or after conventional ECP [2,30]. The occurrence of secondary malignancies in patients after allogeneic stem cell transplantation has been reported, with an incidence of 5.6% [32]. Second malignancy is not associated with traditional ECP [30,33] and is also not expected using systemic ALA [12,34]. However, unexpected events must be monitored in future larger clinical ALA-ECP studies, the main limitations of the present study being the open-label design and the small patient sample size.

Although safety was the primary focus of this study, organ, performance, and quality of life assessments were regularly performed. It should be emphasised that our observations only reflect changes that occurred during the study period and cannot be attributed to ALA-ECP treatment alone. Factors such as the fluctuating course of the diseases, the comedication of the patients, and the lack of investigator-blinded controls may have influenced the results. Overall, the improvement in scores appears promising, especially for skin. In contrast, tear production was measured to be greatest at baseline. The presence of keratoconjunctivitis sicca is well known in cGvHD [6] and four of the patients used artificial teardrops due to dry eyes before entering the study. Information to patients not to use drops before the baseline visit may have been insufficient. This could explain the finding of lower eye moisture at controls. However, we cannot rule out that the treatment may have affected tear production, and this should be further evaluated in future studies.

This study met the aim to evaluate safety and tolerability after multiple treatments with ALA-ECP in patients with cGvHD and lays the basis for designing future clinical ALA-ECP studies.

In conclusion, the results indicate that ALA-ECP is safe. Treatment appears to be tolerated by patients, with most adverse events reported to be in the mild-to-moderate range of severity. Apart from reduced eye moisture, no findings suggestive of organ toxicity were observed. Further research is needed to assess the safety and for the optimisation of the treatment.

Author Contributions: Conceptualization: E.C. and Q.P.; formal analysis: E.C. and O.A.F.; funding acquisition: Q.P.; investigation: E.C., I.S., and F.P.; methodology: E.C., P.Q.-P., T.H., P.J., and Q.P.; project administrator: E.C., I.S., and Q.P.; supervision: E.C.; visualisation: E.C. and O.A.F.; writing—original draft: E.C., O.A.F., and Q.P.; writing—review and editing: E.C., O.A.F., P.Q.-P., I.S., F.P., T.H., P.J., and Q.P. All authors have read and agreed to the published version of the manuscript.

Funding: The South-Eastern Norway Regional Health Authority (project numbers: 2015075, 2016092, 2017058, 2020069) and the Norwegian Cancer Society (project number: 190397).

Institutional Review Board Statement: The study was conducted according to the guidelines of the Declaration of Helsinki, and approved by the regional committee for medical research ethics (REK-Nord 2014/2316) and the Norwegian Medicines Agency (National Regulatory Authority 14/16760-29).

Informed Consent Statement: Oral and written informed consent was obtained from all subjects involved in the study.

Data Availability Statement: No datasets were generated or analysed during the current study.

Acknowledgments: The authors wish to thank the patients for their participation in this study, the staff at the Photopheresis Centre, St. Olavs Hospital and Olav Spigset, Emadoldin Feyzi and Ragnhild Telnes (Norwegian University of Science and Technology (NTNU) and St. Olavs Hospital) for their great efforts in assessing the study safety and tolerability data, Robert Knobler (Medical University of Vienna, Austria) and Trond Warloe for their valuable discussions and photonamic GmbH & Co. KG, Germany, for providing Gliolan for this study, and Daniel DiGirolamo and Erika Odlund (Mallinckrodt Pharmaceuticals, USA) for their suggestions on using the Therakos System.

Conflicts of Interest: Authors Eidi Christensen, Toril Holien, and Qian Peng are co-inventors of patent number 10695371. The additional authors have no conflict of interest to declare.

References

1. Edelson, R.; Berger, C.; Gasparro, F.; Jegasothy, B.; Heald, P.; Wintroub, B.; Vonderheid, E.; Knobler, R.; Wolff, K.; Plewig, G.; et al. Treatment of Cutaneous T-Cell Lymphoma by Extracorporeal Photochemotherapy. *N. Engl. J. Med.* **1987**, *316*, 297–303. [CrossRef]
2. Knobler, R.; Arenberger, P.; Arun, A.; Assaf, C.; Bagot, M.; Berlin, G.; Bohbot, A.; Calzavara-Pinton, P.; Child, F.; Cho, A.; et al. European dermatology forum—Updated guidelines on the use of extracorporeal photopheresis 2020—Part 1. *J. Eur. Acad. Dermatol. Venereol.* **2020**, *34*, 2693–2716. [CrossRef] [PubMed]
3. Nygaard, M.; Wichert, S.; Berlin, G.; Toss, F. Extracorporeal photopheresis for graft-vs-host disease: A literature review and treatment guidelines proposed by the Nordic ECP Quality Group. *Eur. J. Haematol.* **2020**, *104*, 361–375. [CrossRef]
4. Saidu, N.E.B.; Bonini, C.; Dickinson, A.; Grce, M.; Inngjerdingen, M.; Koehl, U.; Toubert, A.; Zeiser, R.; Galimberti, S. New Approaches for the Treatment of Chronic Graft-Versus-Host Disease: Current Status and Future Directions. *Front. Immunol.* **2020**, *11*, 578314. [CrossRef] [PubMed]
5. Shulman, H.M.; Cardona, D.; Greenson, J.K.; Hingorani, S.; Horn, T.; Huber, E.; Kreft, A.; Longerich, T.; Morton, T.; Myerson, D.; et al. NIH Consensus Development Project on Criteria for Clinical Trials in Chronic Graft-versus-Host Disease: II. The 2014 Pathology Working Group Report. *Biol. Blood Marrow Transplant.* **2015**, *21*, 589–603. [CrossRef]
6. Filipovich, A.H.; Weisdorf, D.; Pavletic, S.; Socie, G.; Wingard, J.R.; Lee, S.J.; Martin, P.; Chien, J.; Przepiorka, D.; Couriel, D.; et al. National Institutes of Health Consensus Development Project on Criteria for Clinical Trials in Chronic Graft-versus-Host Disease: I. Diagnosis and Staging Working Group Report. *Biol. Blood Marrow Transplant.* **2005**, *11*, 945–956. [CrossRef]

7. Bruserud, Ø.; Tvedt, T.H.A.; Paulsen, P.Q.; Ahmed, A.B.; Gedde-Dahl, T.; Tjønnfjord, G.E.; Slåstad, H.; Heldal, D.; Reikvam, H. Extracorporeal photopheresis (photochemotherapy) in the treatment of acute and chronic graft versus host disease: Immunological mechanisms and the results from clinical studies. *Cancer Immunol. Immunother.* **2014**, *63*, 757–777. [CrossRef]

8. Szodoray, P.; Papp, G.; Nakken, B.; Harangi, M.; Zeher, M. The molecular and clinical rationale of extracorporeal photochemotherapy in autoimmune diseases, malignancies and transplantation. *Autoimmun. Rev.* **2010**, *9*, 459–464. [CrossRef]

9. Christensen, E.; Warloe, T.; Kroon, S.; Funk, J.; Helsing, P.; Soler, A.M.; Stang, H.J.; Vatne, O.; Mørk, C. Guidelines for practical use of MAL-PDT in non-melanoma skin cancer. *J. Eur. Acad. Dermatol. Venereol.* **2010**, *24*, 505–512. [CrossRef]

10. Stummer, W.; Pichlmeier, U.; Meinel, T.; Wiestler, O.D.; Zanella, F.; Reulen, H.-J. Fluorescence-guided surgery with 5-aminolevulinic acid for resection of malignant glioma: A randomised controlled multicentre phase III trial. *Lancet Oncol.* **2006**, *7*, 392–401. [CrossRef]

11. Dunn, J.; Lovat, L. Photodynamic therapy using 5-aminolaevulinic acid for the treatment of dysplasia in Barrett's oesophagus. *Expert Opin. Pharmacother.* **2008**, *9*, 851–858. [CrossRef] [PubMed]

12. Peng, Q.; Warloe, T.; Berg, K.; Moan, J.; Kongshaug, M.; Giercksky, K.E.; Nesland, J.M. 5-Aminolevulinic acid-based photodynamic therapy. Clinical research and future challenges. *Cancer* **1997**, *79*, 2282–2308. [CrossRef]

13. Darvekar, S.; Juzenas, P.; Oksvold, M.; Kleinauskas, A.; Holien, T.; Christensen, E.; Stokke, T.; Sioud, M.; Peng, Q. Selective Killing of Activated T Cells by 5-Aminolevulinic Acid Mediated Photodynamic Effect: Potential Improvement of Extracorporeal Photopheresis. *Cancers* **2020**, *12*, 377. [CrossRef] [PubMed]

14. Noodt, B.B.; Berg, K.S.; Stokke, T.; Peng, Q.; Nesland, J.M. Apoptosis and necrosis induced with light and 5-aminolaevulinic acid-derived protoporphyrin IX. *Br. J. Cancer* **1996**, *74*, 22–29. [CrossRef] [PubMed]

15. Holien, T.; Gederaas, O.A.; Darvekar, S.R.; Christensen, E.; Peng, Q. Comparison between 8-methoxypsoralen and 5-aminolevulinic acid in killing T cells of photopheresis patients ex vivo. *Lasers Surg. Med.* **2018**, *50*, 469–475. [CrossRef] [PubMed]

16. Čunderlíková, B.; Vasovič, V.; Randeberg, L.; Christensen, E.; Warloe, T.; Nesland, J.; Peng, Q. Modification of extracorporeal photopheresis technology with porphyrin precursors. Comparison between 8-methoxypsoralen and hexaminolevulinate in killing human T-cell lymphoma cell lines in vitro. *Biochim. Biophys. Acta (BBA) Gen. Subj.* **2014**, *1840*, 2702–2708. [CrossRef] [PubMed]

17. Peng, Q.; Moan, J.; Warloe, T.; Nesland, J.M.; Rimington, C. Distribution and photosensitizing efficiency of porphyrins induced by application of exogenous 5-aminolevulinic acid in mice bearing mammary carcinoma. *Int. J. Cancer* **1992**, *52*, 433–443. [CrossRef]

18. Gomer, C.J.; Jester, J.V.; Razum, N.J.; Szirth, B.C.; Murphree, A.L. Photodynamic therapy of intraocular tumors: Examination of hematoporphyrin derivative distribution and long-term damage in rabbit ocular tissue. *Cancer Res.* **1985**, *45*, 3718–3725. [PubMed]

19. Karnofsky, D.A. The clinical evaluation of chemotherapeutic agents in cancer. In *Evaluation of Chemotherapeutic Agents*; MacLeod, C., Ed.; Columbia University Press: New York, NY, USA, 1949; pp. 191–205.

20. Chren, M.-M. The Skindex Instruments to Measure the Effects of Skin Disease on Quality of Life. *Dermatol. Clin.* **2012**, *30*, 231–236. [CrossRef]

21. Kopp, M.; Schweigkofler, H.; Holzner, B.; Nachbaur, D.; Niederwieser, D.; Fleischhacker, W.W.; Kemmler, G.; Sperner-Unterweger, B. EORTC QLQ-C30 and FACT-BMT for the measurement of quality of life in bone marrow transplant recipients: A comparison. *Eur. J. Haematol.* **2000**, *65*, 97–103. [CrossRef] [PubMed]

22. Luckett, T.; King, M.T.; Butow, P.N.; Oguchi, M.; Rankin, N.; Price, M.A.; Hackl, N.A.; Heading, G. Choosing between the EORTC QLQ-C30 and FACT-G for measuring health-related quality of life in cancer clinical research: Issues, evidence and recommendations. *Ann. Oncol.* **2011**, *22*, 2179–2190. [CrossRef]

23. Stallvik, M.; Nordstrand, B.; Kristensen, Ø.; Bathen, J.; Skogvoll, E.; Spigset, O. Corrected QT interval during treatment with methadone and buprenorphine—Relation to doses and serum concentrations. *Drug Alcohol Depend.* **2013**, *129*, 88–93. [CrossRef] [PubMed]

24. Peng, Q.; Berg, K.; Moan, J.; Kongshaug, M.; Nesland, J.M. 5-Aminolevulinic Acid-Based Photodynamic Therapy: Principles and Experimental Research. *Photochem. Photobiol.* **1997**, *65*, 235–251. [CrossRef] [PubMed]

25. Čunderlíková, B.; Vasovič, V.; Sieber, F.; Furre, T.; Borgen, E.; Nesland, J.M.; Peng, Q. Hexaminolevuli-nate-mediated photodynamic purging of marrow grafts with murine breast carcinoma. *Bone Marrow Transplant.* **2010**, *46*, 1118–1127. [CrossRef] [PubMed]

26. Shahzidi, S.; Čunderlíková, B.; Więdłocha, A.; Zhen, Y.; Vasovič, V.; Nesland, J.M.; Peng, Q. Simultaneously targeting mitochondria and endoplasmic reticulum by photodynamic therapy induces apoptosis in human lymphoma cells. *Photochem. Photobiol. Sci.* **2011**, *10*, 1773–1782. [CrossRef]

27. Regula, J.; MacRobert, A.J.; Gorchein, A.; Buonaccorsi, G.A.; Thorpe, S.M.; Spencer, G.M.; Hatfield, A.R.; Bown, S.G. Photosensitisation and photodynamic therapy of oesophageal, duodenal, and colorectal tumours using 5 aminolaevulinic acid in-duced protoporphyrin IX—A pilot study. *Gut* **1995**, *36*, 67–75. [CrossRef]

28. Kim, J.-H.; Yoon, H.-K.; Lee, H.-C.; Park, H.-P.; Park, C.-K.; Dho, Y.-S.; Hwang, J.-W. Preoperative 5-aminolevulinic acid administration for brain tumor surgery is associated with an increase in postoperative liver enzymes: A retrospective cohort study. *Acta Neurochir.* **2019**, *161*, 2289–2298. [CrossRef]

29. Scarisbrick, J.; Taylor, P.; Holtick, U.; Makar, Y.; Douglas, K.; Berlin, G.; Juvonen, E.; Marshall, S.; on behalf of the Photopheresis Expert Group U.K. Consensus statement on the use of extracorporeal photopheresis for treatment of cutaneous T-cell lymphoma and chronic graft-versus-host disease. *Br. J. Dermatol.* **2008**, *158*, 659–678. [CrossRef]

30. Scarisbrick, J. Extracorporeal photopheresis: What is it and when should it be used? *Clin. Exp. Dermatol.* **2009**, *34*, 757–760. [CrossRef]

31. Knobler, R.; Berlin, G.; Calzavara-Pinton, P.; Greinix, H.; Jaksch, P.; Laroche, L.; Ludvigsson, J.; Quaglino, P.; Reinisch, W.; Scarisbrick, J.; et al. Guidelines on the use of extracorporeal photopheresis. *J. Eur. Acad. Dermatol. Venereol.* **2013**, *28*, 1–37. [CrossRef]

32. Shimoni, A.; Shem-Tov, N.; Chetrit, A.; Volchek, Y.; Tallis, E.; Avigdor, A.; Sadetzki, S.; Yerushalmi, R.; Nagler, A. Secondary malignancies after allogeneic stem-cell transplantation in the era of reduced-intensity conditioning; the incidence is not reduced. *Leukemia* **2013**, *27*, 829–835. [CrossRef] [PubMed]

33. Gottlieb, S.L.; Wolfe, J.T.; Fox, F.E.; DeNardo, B.J.; Macey, W.H.; Bromley, P.G.; Lessin, S.R.; Rook, A.H. Treatment of cutaneous T-cell lymphoma with extracorporeal photopheresis monotherapy and in combination with recombinant interferon alfa: A 10-year experience at a single institution. *J. Am. Acad. Dermatol.* **1996**, *35*, 946–957. [CrossRef]

34. Ackroyd, R.; Brown, N.; Vernon, D.; Roberts, D.; Stephenson, T.; Marcus, S.; Stoddard, C.; Reed, M. 5-Aminolevulinic Acid Photosensitization of Dysplastic Barrett's Esophagus: A Pharmacokinetic Study. *Photochem. Photobiol.* **1999**, *70*, 656. [CrossRef] [PubMed]

Review

Photodynamic Therapy Review: Principles, Photosensitizers, Applications, and Future Directions

José H. Correia [1,2], José A. Rodrigues [2,*], Sara Pimenta [2], Tao Dong [1,3] and Zhaochu Yang [1]

[1] Chongqing Key Laboratory of Micro-Nano Systems and Smart Transduction, Collaborative Innovation Center on Micro-Nano Transduction and Intelligent Eco-Internet of Things, Chongqing Key Laboratory of Colleges and Universities on Micro-Nano Systems Technology and Smart Transducing, National Research Base of Intelligent Manufacturing Service, Chongqing Technology and Business University, Nan'an District, Chongqing 400067, China; higino.correia@dei.uminho.pt (J.H.C.); tao.dong@usn.no (T.D.); zhaochu.yang@ctbu.edu.cn (Z.Y.)

[2] CMEMS-UMinho, Department of Industrial Electronics, University of Minho, 4800-058 Guimaraes, Portugal; sara.pimenta@dei.uminho.pt

[3] Department of Microsystems-IMS, Faculty of Technology, Natural Sciences and Maritime Sciences, University of South-Eastern Norway (USN), Postboks 235, 3603 Kongsberg, Norway

* Correspondence: jrodrigues@dei.uminho.pt

Abstract: Photodynamic therapy (PDT) is a minimally invasive therapeutic modality that has gained great attention in the past years as a new therapy for cancer treatment. PDT uses photosensitizers that, after being excited by light at a specific wavelength, react with the molecular oxygen to create reactive oxygen species in the target tissue, resulting in cell death. Compared to conventional therapeutic modalities, PDT presents greater selectivity against tumor cells, due to the use of photosensitizers that are preferably localized in tumor lesions, and the precise light irradiation of these lesions. This paper presents a review of the principles, mechanisms, photosensitizers, and current applications of PDT. Moreover, the future path on the research of new photosensitizers with enhanced tumor selectivity, featuring the improvement of PDT effectiveness, has also been addressed. Finally, new applications of PDT have been covered.

Keywords: PDT mechanisms; new photosensitizers; PDT tumor treatment; antimicrobial PDT; non-oncologic applications of PDT; PDT in medical devices

Citation: Correia, J.H.; Rodrigues, J.A.; Pimenta, S.; Dong, T.; Yang, Z. Photodynamic Therapy Review: Principles, Photosensitizers, Applications, and Future Directions. *Pharmaceutics* **2021**, *13*, 1332. https://doi.org/10.3390/pharmaceutics13091332

Academic Editors: Udo Bakowsky, Matthias Wojcik, Eduard Preis and Gerhard Litscher

Received: 19 July 2021
Accepted: 16 August 2021
Published: 25 August 2021

Publisher's Note: MDPI stays neutral with regard to jurisdictional claims in published maps and institutional affiliations.

1. Introduction

1.1. History of Photodynamic Therapy

Light has been used in the treatment of several diseases since antiquity [1]. The ancient civilizations, Egyptian, Indian, and Chinese, used the sunlight to treat various skin diseases, such as psoriasis, vitiligo, and skin cancer [2]. Herodotus, a famous Greek physician known as the father of heliotherapy, emphasized the importance of whole-body sun exposure for the restoration of health. In the 18th and 19th centuries, in France, sunlight was used in the treatment of several conditions, such as tuberculosis, rickets, scurvy, rheumatism, paralysis, edema, and muscle weakness [3]. At the beginning of the 20th century, the importance of light in the treatment of diseases was recognized, and the 1903 Nobel Prize in Physiology or Medicine was awarded to Niels Finsen for his contribution in this field. Finsen found that sunlight or light from a carbon arc lamp with a heat filter could be used to treat lupus vulgaris, a tubercular condition of the skin. This discovery marked the beginning of modern phototherapy [1,4,5].

Phototherapy describes the use of light in the treatment of a disease. However, photochemotherapy requires the administration of a photosensitizing agent, which is subsequently activated by light in the tissues, where the agent is localized. This form of therapy also dates back over 3000 years, when the Indians and Egyptians used psoralens from natural plants in the treatment of a variety of skin conditions [3,6].

The concept of cell death induced by the interaction of light and chemicals has been recognized for 100 years. In 1900, a German medical student, O. Raab, first reported cell death induced by the interaction of light with chemicals. While working with Professor H. von Tappeiner in Munich, he described the lethal effect that the combination of light and acridine red had on protozoa [3,4,6]. In subsequent experiments, Raab demonstrated that this lethal effect was greater than with acridine red alone, light alone, or acridine red exposed to light and then added the protozoan. He reported that toxicity occurred as a result of fluorescence caused by the transfer of energy from light to the chemical [6,7]. In the same year, the French neurologist J. Prime discovered that patients with epilepsy who were treated with orally administered eosin developed dermatitis in areas exposed to sunlight. Later, in 1903, H. von Tappeiner and A. Jesionek treated skin tumors with topical applications of eosin and white light [3,4,6]. In 1904, H. von Tappeiner and A. Jodlbauer identified that oxygen is an integral component in photosensitization reactions, and in 1907 they introduced the term "photodynamic action" to describe this phenomenon [1–3,6].

In 1841, H. Scherer first produced hematoporphyrin while investigating the nature of blood. However, its fluorescent properties were not described until 1867, and this substance was only named by hematoporphyrin in 1871. In 1911, W. Hausmann reported the effects of hematoporphyrin and light on protozoa and blood cells, describing skin reactions in a mice exposed to light after being administered with hematoporphyrin [3,4,6]. The first report of human photosensitization was in 1913, when F. Meyer-Betz injected himself with 200 mg of hematoporphyrin to determine if similar effects could be induced in humans. He described prolonged pain and swelling in light-exposed areas [2,3,6]. In 1960, R. Lipson and S. Schwartz initiated the concept of photodynamic therapy (PDT) at the Mayo Clinic, discovering the cancer diagnostic and therapeutic effects by injecting a hematoporphyrin derivative (HpD) [1]. In 1975, a significant breakthrough in PDT occurred, when T. Dougherty reported that administration of HpD and its activation with red light completely eradicated the growth of mammary tumor in mice. In the same year, J. F. Kelly proved the elimination of bladder carcinoma in mice, by activating HpD with light [4]. In 1976, another major event in the development of PDT occurred, when J. F. Kelly and M. E. Snell proceeded to the first human study of the effects of PDT in bladder cancer using HpD [3]. The use of this technique in the treatment of pathologies in the gastrointestinal tract was first performed in 1984 by J. S. McCaughan, who used PDT to treat patients with esophageal cancer. One year later, Y. Hayata used PDT to treat patients with gastric carcinoma [3,4]. Dougherty and his coworkers also purified HpD and produced Photofrin, which was the first photosensitizer molecule (PS) approved by the US Food and Drug Administration (FDA) in 1995 for cancer treatment [2]. Since then, PDT has continued to evolve and its clinical application was extended to other areas besides tumors treatment. Dr. M. Weber, known as the pioneer of modern laser therapy, also applies PDT to treat bacterial, viral, and parasitic diseases, named as antimicrobial PDT. The main advantage of this new approach is to combat multiresistant pathogens. Dr. Weber developed the Weberneedle® technology, which allows to apply highly focused and efficient lasers of different wavelengths to intravenous, interstitial, and intra-articular irradiation [8,9].

1.2. Principles of PDT

PDT is based on the dynamic interaction between a PS, light with a specific wavelength, and molecular oxygen, promoting the selective destruction of the target tissue [10]. The PDT treatment consists in the administration of a PS (topically or intravenous), which selectively accumulates in the tumor tissue (during a drug–light interval), followed by subsequent exposition to an appropriate wavelength light (generally in the red spectral region, $\lambda \geq 600$ nm; Figure 1) [11]. The PS itself does not react with biomolecules; however, illumination transfers energy from light to molecular oxygen, to generate reactive oxygen species (ROS), such as singlet oxygen ($^1O_2$), superoxide radical ($O_2^{-\bullet}$), hydroxyl radical ($HO^\bullet$), and hydrogen peroxide (H_2O_2) [1]. These cytotoxic photoproducts start a cascade of biochemical events, which can induce damage and death of the target tissue [11].

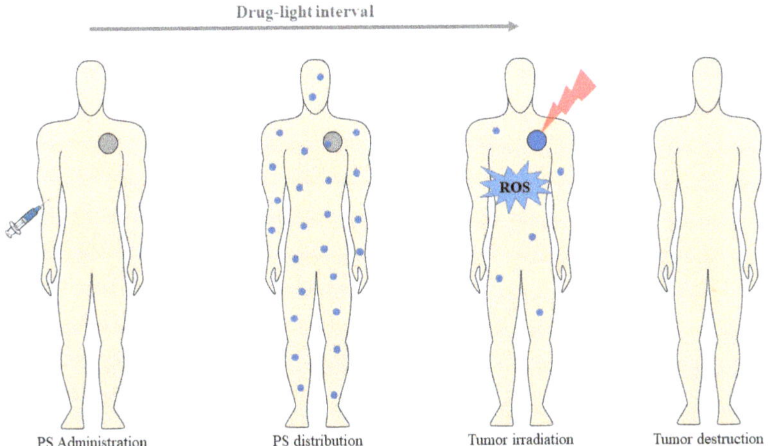

Figure 1. Representation of the clinical application of PDT protocol for the treatment of a solid and localized tumor (based on [12]).

2. Photodynamic Reaction

PDT is a therapeutic modality based on the combination of three factors: PS, light with a specific wavelength, and the presence of molecular oxygen [2]. The photodynamic reaction begins with the absorption of light by the PS in the target tissue, which triggers a series of photochemical reactions that lead to the generation of ROS [10]. After light absorption, the PS is transformed from its ground state (singlet state, ^{1}PS) to a short-lived, electronically excited singlet state (a few nanoseconds or less, ^{1}PS*) [2,4]. This excited state is very unstable and can decay to the ground state, losing the excess of energy through light emission (fluorescence) or heat production (internal conversion) [13]. However, the singlet state can undergo intersystem crossing and progress to a more stable, long-lived, electronically excited state (triplet state, ^{3}PS*), through spin conversion of the electron in the higher energy orbital [13]. This excited state can decay to the ground state through light emission (phosphorescence) or undergo two kinds of reactions [4]. The triplet state has a longer lifetime (up to tens of microseconds), which allows sufficient time for direct transfer of energy to molecular oxygen (O_2). This energy transfer step leads to the formation of singlet oxygen ($^1O_2$) and the fundamental state of the PS, called type II reaction [2,11]. The singlet oxygen is extremely reactive and can interact with a large number of biological substrates, inducing oxidative damage and ultimately cell death [11]. The type I reaction can also occur if the excited state of the PS reacts directly with a substrate, such as cell membrane or a molecule, and undergoes hydrogen atom abstraction or electron transfer reactions, to yield free radicals and radical ions. These radicals react with molecular oxygen, producing ROS, such as $O_2^{-\bullet}$, $HO^\bullet$, and H_2O_2, which produce oxidative damage that can lead to biological lesions [11]. Figure 2 shows the modified Jablonski diagram of the PDT action mechanism.

The products resulting from type I and type II reactions are responsible for the effect of cell death and therapeutic on PDT. Type I and type II reactions can occur simultaneously and the ratio between these processes depends on the PS, substrate, oxygen concentration, and binding affinity of the sensitizer to the substrate [2,4,13]. However, type II reaction is predominant during PDT, and singlet oxygen is the primary cytotoxic agent responsible for the biological effects [11]. The quantum yield of singlet oxygen formation is one of the most important features of a PS and is determined by the quantum yield and lifetime of its triplet excited state [10]. Due to the high reactivity and short half-life of the ROS, only cells close to the area of the ROS production (areas where the PS is localized) are directly affected by PDT. The extent of damage and cytotoxicity resulting from PDT is multifactorial, depending on the type of PS, its extracellular and intracellular location and the total dose administered, the dose of light (light fluence) and the light fluence rate, availability of oxygen, and the time between PS administration and light exposure [4]. PDT of deeper and hypoxic tumors is more difficult due to the low oxygen concentration and low light penetration into the tissue (light absorption by the PS and energy transfer to the oxygen). On the other hand, more superficial and more oxygenated tumors allow greater production of ROS and thus a more effective PDT treatment [14].

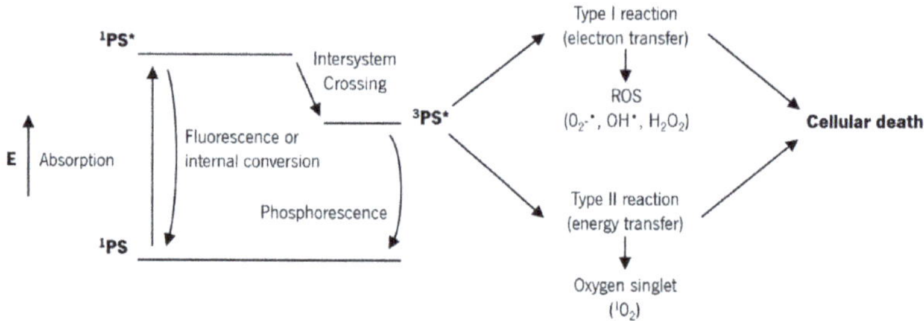

Figure 2. Modified Jablonski diagram of the PDT action mechanism (based on [13]).

3. PDT-Mediated Action Mechanisms

There are three main mechanisms by which PDT mediates tumor destruction (Figure 3) [4]. The ROS produced by PDT photochemical reactions can directly destroy tumor cells by inducing apoptosis and necrosis. PDT can also cause the destruction of the tumor-associated vasculature and the surrounding healthy vessels, leading to an interruption of oxygen and nutrient supply and, consequently, to indirect cell death due to hypoxia. Finally, PDT can induce an inflammatory response that activates an immune response against the tumor cells [4,15]. The outcome of PDT depends on all of these mechanisms, and the contribution of each one is determined by the treatment regime used [4,10].

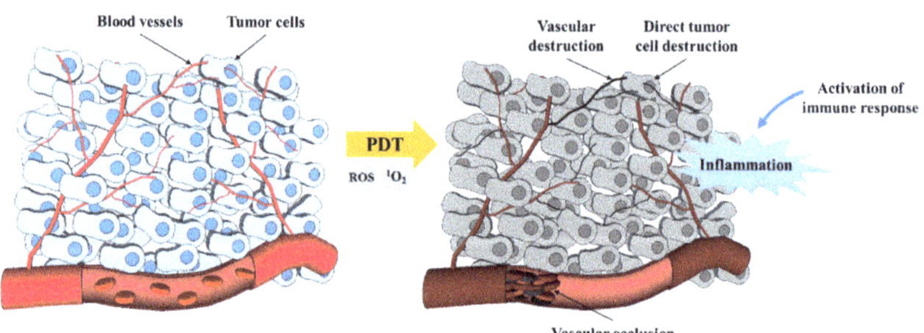

Figure 3. PDT mechanisms for tumor destruction (based on [12]).

3.1. Apoptosis and Necrosis

Tumor destruction from PDT can occur by both programmed (apoptotic) pathways and non-programmed (necrosis) pathways [11,16]. Generally, when high light intensity is employed, the tumor cells are rapidly ablated by necrosis [11]. Necrosis is generally described as a rapid and relatively broad mechanism of cell death, and it is characterized by vacuolization of the cytoplasm and cell membrane breakdown, resulting in a local inflammatory reaction due to the release of cytoplasmic content and pro-inflammatory mediators in the extracellular medium [10].

In contrast, apoptotic death may be initiated by PDT, generally when low light doses are employed [11,16]. Apoptosis is described as a mechanism of programmed cell death that is genetically encoded and energy dependent [10]. Morphologically, it is characterized by chromatin condensation, cleavage of chromosomal DNA into internucleosomal fragments, cell shrinkage, membrane wrinkling, and the formation of apoptotic bodies without plasma membrane breakdown [2,7,10]. No effect or immune response is expected as no toxic chemicals are leaked [11,16].

3.2. Vascular Mechanisms

In addition to the direct destruction of tumor cells, the application of PDT often also leads to the destruction of the tumor microvasculature. Like tumor cells, endothelial cells of the vascular system can concentrate PS to create free radicals when activated by appropriate light. The disrupting of the vascular walls leads to the interruption of the tumor's feeding (i.e., oxygen and nutrients) and, consequently, to the death of the tumor cells. PDT vascular effect has been shown to be very important for the long-term efficacy of PDT [10]. PDT vascular effect can be greatly enhanced by using a short drug–light interval (the time between systemic PS injection and light illumination) when the PS is predominantly localized in the vasculature [1]. Selectivity in vascular PDT protocols is achieved by the precise application of light on the tumor plus a safety margin of the surrounding healthy tissue [10]. Vascular PDT has important advantages in comparison to PDT protocols that require PS accumulation in the tumor cells: it uses photosensitizers that clear rapidly from the organism and minimize skin photosensitivity, gives higher long-term efficacy, and can be performed in one short session [10,17].

3.3. Immunological Mechanisms

For many years, PDT was considered a localized treatment, affecting only tumor cells and tumor vasculature. More recently, numerous studies have demonstrated that PDT can significantly influence the adaptive immune response in disparate ways, either through stimulation or suppression of the immune response. The efficacy of PDT appears to be dependent on the induction of antitumor immunity [10]. Long-term tumor control is a combination of the direct effects of PDT on the lesion and its vasculature with upregulation of the immune system [11].

Under certain conditions, PDT induces immunosuppression, which has been mainly associated with reactions to topical treatments with high fluence rates and in large areas of irradiation [10]. In contrast, non-topical PDT treatments are often described as immunostimulatory. The oxidative damage inflicted by PDT on the tumor stroma will eventually result in cell death. When PDT induces necrosis of tumors and their vasculature, an immune cascade is also initiated. The change in tissue integrity and homeostasis triggers an acute inflammatory response initiated by the release of pro-inflammatory mediators, which include various cytokines, growth factors, and proteins [10,11]. These mediators attract the host's innate immune cells, such as neutrophils, mast cells, macrophages, and dendritic cells, which infiltrate the damaged tissue to restore homeostasis in the affected region [10]. Upon arrival, macrophages phagocytize PDT-damaged cancer cells and present proteins from these tumors to CD4 helper T lymphocytes, which then activate CD8 cytotoxic T lymphocytes [11]. These cytotoxic T cells can recognize and specifically destroy tumor cells and can circulate throughout the body for long periods, ensuring a systemic antitumor immune response [10].

4. PDT Essential Elements

4.1. Photosensitizers

Photosensitizers are key elements for PDT. Ideally, these molecules should accumulate preferentially in the tumors, have a high singlet oxygen quantum yield, have low activity in the absence of light, be quickly eliminated from the patient body, have amphiphilicity, and have a light absorption peak between approximately 600 nm and 800 nm [18,19].

There are a variety of molecular structures of photosensitizers that are currently used in PDT, and it is possible to divide photosensitizers into three generations. The porfimer sodium and the HpD are first-generation photosensitizers. The second-generation photosensitizers arise to overcome some drawbacks of the first-generation ones, related to light absorption at a specific spectral region. Some examples of second-generation photosensitizers are the derivates of chlorins, bacteriochlorins, and phthalocyanines, which can have a stronger action on the tumor regions due to their strong absorbance in the deep red region, and consequently, increased light penetration. Finally, the third-generation photosensitizers are molecules with improved selectivity for tumor regions, due to the conjunction of the PS with targeting molecules or its encapsulation into carriers. Thus, photosensitizers progressed towards the improvement of PDT specificity and efficacy. Today, the functionalization of photosensitizers seems to be the best strategy to achieve a high selectivity to the tumor regions, combining photosensitizers with biomolecules or carriers [18–20]. Photosensitizers can be covalently bonded to several biomolecules, which have affinity to tumors. These biomolecules include antibodies, proteins, carbohydrates, and others. Photosensitizers can also be encapsulated into carriers, such as gold nanoparticles, silica nanoparticles, quantum dots, carbon nanotubes, or others carriers, to guide the photosensitizers to tumors [19,21–24].

Table 1 presents the most used photosensitizers in clinical PDT, including their trade name and class, molecular formula, excitation wavelength, quantum yield, extinction coefficient, and main applications [10,18,25–31]. A variety of photosensitizers under clinical trials for approval in clinical PDT are presented in Table 2 [10,18,26,27,32–34].

Consensus protocol for conventional topical PDT recommends some lesion preparation to increase PS absorption and light penetration. The typical PDT topical treatment protocol follows the next steps [35]:

- Wash the area to be treated with soap and water;
- Remove any residue and remaining oil with a gauze soaked in acetone or alcohol;
- Apply the PS evenly over the entire area to be treated. Apply a second layer of PS after the first one has dried;
- Allow the PS to incubate for 0.5–4 h;
- Activate the PS with the appropriate light source;
- Wash the treated area with soap and water to remove any residual PS;
- Avoid any direct sunlight for 48 h;
- Repeat as needed in 2–3 weeks.

In dermatological indications, PDT is usually performed by topical application of PS, in particular 5-aminolevulinic acid (5-ALA) or its ester methyl aminolevulinate (MAL) [35]. Three photosensitizers are currently approved for topical use in Europe: MAL Metvix®, 5-ALA Ameluz®, and 5-ALA AlaCare®. MAL Metvix® is used along with red light to treat actinic keratosis, Bowen's disease, and superficial basal cell carcinoma (3 h of drug-light interval and 37–75 J/cm^2 of total light dose). 5-ALA Ameluz® is used in combination with red light for the treatment of mild and moderate actinic keratosis and superficial basal cell carcinoma (3 h of drug-light interval and 37–200 J/cm^2 of total light dose). MAL Metvix® and 5-ALA Ameluz® are also used with daylight to treat moderate actinic keratosis (0.5 h of drug-light interval and exposure during 2 h). 5-ALA AlaCare® is approved for the treatment of mild actinic keratosis with red light (4 h of drug-light interval and 37 J/cm^2 of total light dose). A 20% formulation of 5-ALA Levulan® is approved in North America for the treatment of actinic keratosis with blue light (14–18 h of drug-light interval and 10 J/cm^2 of total light dose). Besides, topical PDT is highly recommended for the photorejuvenation and the treatment of acne vulgaris, although these indications currently lack approval for use and the protocols still need to be optimized [35–38].

Table 1. Photosensitizers used in clinical PDT.

Trade Name (Class)	Molecular Formula	Excitation Wavelength (nm)	Quantum Yield	Molar Extinction Coefficient $(M^{-1}\,cm^{-1})$	Main Applications
Photofrin® (porphyrin)	$C_{34}H_{38}N_4NaO_5{}^+$	630	0.01 in PBS	3.0×10^3 in PBS	Esophageal, lung, and endobronchial cancers
Ameluz® (porphyrin)	$C_5H_9NO_3{}^{\bullet}HCl$	630	-	-	Actinic keratosis and basal cell carcinoma
AlaCare® (porphyrin)	$C_5H_9NO_3$	630	-	-	Actinic keratosis
Levulan® (porphyrin)	$C_5H_9NO_3$	635	0.56	5.0×10^3	Actinic keratosis
Hexvix® (porphyrin)	$C_{11}H_{21}NO_3$	635	-	$<1.0 \times 10^3$	Bladder cancer
Foscan® (chlorine)	$C_{44}H_{32}O_4N_4$	652	0.43 in methanol	3.0×10^4 in methanol	Head and neck cancers
Laserphyrin® (chlorine)	$C_{38}H_{37}N_5O_9$	664	0.77 in PBS	4.0×10^4 in PBS	Lung and esophageal cancers and brain tumors
Metvix® (porphyrin)	$C_6H_{11}NO_3$	570–670	-	$<1.0 \times 10^3$	Basal cell carcinoma, Bowen's disease, and actinic keratosis
Visudyne® (porphyrin)	$C_{82}H_{84}N_8O_{16}$	690	0.7 in methanol	3.4×10^4 in methanol	Age-related macular degeneration
Redaporphine® (LUZ11) (bacteriochlorin)	$C_{48}H_{38}F_8N_8O_8S_4$	749	0.43 in ethanol	140×10^3 in ethanol	Biliary tract cancer

Table 2. Photosensitizers under clinical trials for use in PDT.

Trade Name	Molecular Formula	Excitation Wavelength (nm)	Quantum Yield	Molar Extinction Coefficient $(M^{-1}\,cm^{-1})$	Main Applications
Radachlorin® (chlorine)	$C_{34}H_{36}N_4O_6$ $C_{33}H_{34}N_4O_5$ $C_{33}H_{34}N_4O_6$	662	0.52–0.62	3.42×10^4	Skin cancer
Photochlor® (chlorins)	$C_{39}H_{48}N_4O_4$	664	0.48 in CH_2Cl_2	4.75×10^4 in 1% Tween-80 micelles	Head and neck cancer
Purlytin® (purpurin)	$C_{37}H_{42}Cl_2N_4O_2Sn$	664	0.7 in acetonitrile	2.8×10^4	Age-related macular degeneration
Fotolon® (chlorin)	$C_{34}H_{36}N_4O_6$	665	0.63 in dimethylformamide	5.0×10^4 in diethyl ether	Nasopharyngeal sarcoma
Lutrin® (texaphyrin)	$C_{52}H_{72}LuN_5O_{14}$	732	4.2×10^4 in methanol	0.11 in methanol	Coronary artery disease
TOOKAD® (WST09) (bacteriochlorin)	$C_{37}H_{41}K_2N_5O_9PdS$	762	0.99 in organic solvent	8.85×10^4	Prostate cancer

A strategy that seems to be of great interest in the near future is to increase the PDT selectivity through the development of activatable multifunctional photosensitizers, which become active after receiving a biological or physical stimulus. Biological stimuli include the physiological conditions associated with cancer, such as temperature, pH, and enzymatic activity. For example, the peptidic zipper-based PS was designed to react under acidic conditions. Physical stimuli refer to an artificial agent activation, applying magnetic or electric fields, ultrasounds, two-photon excitation, etc. [18,19]. For example, electroporation can be used to support PDT (Figure 4). Electroporation is reported as an effective method that could be used to increase the transport of a PS into the pathological cells, which could lead to the increase of cytotoxicity and PDT efficacy. Several studies have been performed over the years, using different photosensitizers, the clinically approved Photofrin PS being the most relevant. The results, including a study performed with human cancer cells, conclude, undoubtedly, that PDT with electroporation is an attractive approach to cancer treatment, but detailed studies on the mechanisms of this approach are still required [20,39].

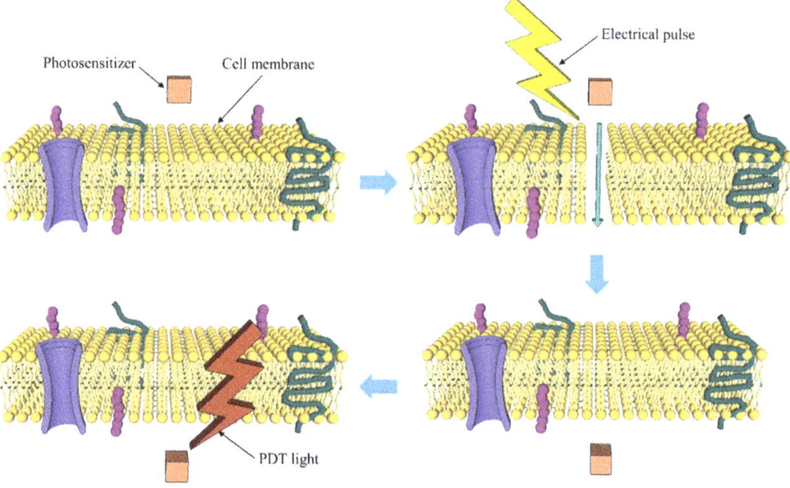

Figure 4. PDT with electroporation (based on [20]).

Very recently, the possibility of using transition metal coordination complexes or organic fluorophores as efficient photosensitizers for PDT has also been reported. The transition metal coordinator complexes, such as ruthenium(II) complexes, iridium(III) complexes, and polymetallic complexes, meet several basic needs for PDT. The most relevant feature is their heavy-atom effect, which mediates strong spin–orbital coupling, providing more time for the excited states to interact with molecular oxygen. Between other features, it can be also be stated that they are easily synthetized, including the possibility of fine-tuning their properties. The organic fluorophores, such as naphthalimides, xanthenes, boron dipyrromethene (BODIPY), and cyanines, can also be designed as photosensitizers for cancer PDT, with high light absorption at relatively long wavelengths and large molar extinction coefficient. Organic fluorophore photosensitizers have low toxicity, good biocompatibility, and long triplet lifetimes, and their fluorescence emission can be used to perform real-time monitoring during PDT treatment [40,41].

4.2. Light

PDT has been performed with various light sources, including lasers, incandescent light, and laser-emitting diodes [42]. Laser light sources are usually expensive and require the use of an optical system to expand the light beam for irradiation of a larger tissue area. Non-laser light sources (e.g., conventional lamps) can be used with optical fibers to specify the light wavelength for tissue irradiation. However, conventional lamps may have thermal effects, which must be avoided in PDT. Finally, light-emitting diodes (LEDs) have also been used in PDT as light sources. LEDs are less expensive, less hazardous, thermally non-destructive, and easily available in flexible arrays [43]. Light penetration into tumor tissue is very complex, as it can be reflected, scattered, or absorbed. The extent of these processes depends on the type of tissue and the wavelength of light [44]. Light absorption is mainly due to endogenous chromophores existing in tissues, such as hemoglobin, myoglobin, and cytochromes, which can reduce the photodynamic process by competing with PS in the absorption process [44,45]. Light absorption by tissues decreases with increasing wavelength, so longer wavelengths of light (red light) penetrate more efficiently through tissue. The region between 600 and 1200 nm is often called the "tissue optical window" [12,44]. Shorter wavelengths (<600 nm) have less tissue penetration and are more absorbed, resulting in increased skin photosensitivity. On the other hand, longer wavelengths (>850 nm) do not have enough energy to excite oxygen in its state of singlet and to produce enough reactive oxygen species. Therefore, the highest tissue permeability occurs between 600 and 850 nm. This range, called the "phototherapeutic window," is predominantly used in PDT [10,12,18,20].

As light is an essential component of PDT, clinical efficacy is highly dependent on the accuracy of its delivery to the target tissue and its dose, which translates into light fluence, light fluence rate, light exposure time, and light delivery mode (single or fractionated) [12]. Light fluence is the total energy of exposed light across a sectional area of irradiated spot and is expressed in J/cm^2. Light fluence rate is the incident energy per second across a sectional area of the irradiated spot and is expressed as W/cm^2 [4,46,47].

Several studies have reported that low light fluence rates are advantageous for PDT [48–50]. The main reason for the lower efficacy of PDT for high light fluence rates is the oxygen depletion in tissues, which leads to a low photo-degradation of the PS. The light fluence rate has also been shown to have an impact on the dominant mechanism of cell death in the PDT. The use of low light fluence rates increases the selective apoptosis of tumor cells, which is more desirable than inflammation and edema that usually occur with the uncontrolled rupturing of cellular content in necrosis [51].

Another relevant light source for PDT is the natural light. The concept of daylight PDT is based on the use of natural light instead of an artificial light source to treat skin lesions, such as actinic keratosis. The main advantages of using daylight PDT are minimal patient discomfort and shorter clinical visits (patients can complete their therapy at home). Moreover, daylight PDT seems to be as effective as conventional PDT for actinic keratosis. For this specific application, the patients expose the sites to daylight for 2 h, after 30 min of PS application. The short PS incubation time in the daylight PDT, compared with conventional PDT (1–3 h required), allows a smaller and more continuous PS activation, leading to lower patient pain intensity associated with PDT [52,53].

4.3. Oxygen

The third key component in the PDT mechanism is molecular oxygen. Oxygen is crucial for the production of ROS during PDT. Oxygen concentration in the tissues truly affects the effectiveness of the PDT treatment. In fact, oxygen concentration can vary significantly between different tumors and even between different regions of the same tumor, depending on the density of the vasculature. Especially in deeper solid tumors, often characterized by their anoxic microenvironment, lack of oxygen can be a limiting factor. As mentioned above, the irradiation of the tumor with a high light fluence rate can lead to a temporary local oxygen depletion. This leads to interruption of ROS production and reduced treatment effectiveness. Oxygen depletion occurs when the rate of oxygen consumption by the photodynamic reaction is greater than the rate of oxygen diffusion in the irradiated area. In addition, PDT can cause occlusion of the tumor vasculature, reducing blood flow to the tumor tissue, further increasing hypoxia [10,26,46].

Real-time measurement of the tissue oxygen levels, before and during PDT, is one of the main challenges in the near future. This allows to optimize the PDT therapeutic result by adjusting the light fluence rate (increasing the irradiation time to maintain the total light dose) or using fractional light dose. Several sensors have been used to monitor the oxygen level in biological media, using imaging agents. However, the combination of these imaging agents with PS has been hardly reported [18]. Other methods to increase oxygen availability in the tumor have been tested: indirect introduction of oxygen and direct introduction of oxygen. One indirect way to increase oxygen concentration in tumor cells is using catalase enzyme to decompose the intracellular hydrogen peroxide into oxygen. The direct delivery of oxygen into tumors is achieved by using oxygen carriers, such as perfluorocarbons and hemoglobin, commonly used to overcome tumor hypoxia in the PDT procedure [37].

5. Advantages and Limitations of PDT

PDT has several advantages over conventional approaches to cancer treatment. First-generation photosensitizers cause increased skin photosensitivity. However, PDT has no long-term side effects when correctly used. It is less invasive than surgical procedures and can be performed on an outpatient basis. In addition to the tumor itself, PDT can also destroy the vasculature associated with it, greatly contributing to tumor death [54]. PDT can be applied directly and accurately in the target tissue, due to its dual selectivity. The two main factors that contribute to the selectivity of PDT are the intrinsic capacity of some photosensitizers to preferentially accumulate in tumor tissue and light irradiation exclusively in the target tissue [10,54]. The selective accumulation of the PS in the tumor is facilitated in the case of topical application, since PS is applied directly and only to the lesions to be treated. When PS is given intravenously, it needs to remain in circulation long enough to reach and accumulate in the tumor [10]. Furthermore, PDT can be repeated several times in the same location, unlike radiation. There is little or no scarring after healing. Finally, it usually costs less than other therapeutic modalities in cancer treatment [54,55].

Like every therapeutic modalities, PDT also has some limitations. The photodynamic effect occurs selectively in the irradiated site, which makes its use in disseminated metastases very difficult with the currently available technology [54]. Tissue oxygenation is crucial for the photodynamic effect to occur, so tumors surrounded by necrotic tissue or dense tumor masses can lead to ineffective PDT. Finally, the accuracy of target tissue irradiation is the most important point when considering PDT as a treatment option. Therefore, deep tumors (not easily accessible without surgical intervention) are difficult to treat due to the low penetration of visible light into the tissue [54,56]. The main advantages and limitations of PDT are summarized in Table 3.

Table 3. Summary of the main advantages and disadvantages of PDT.

Advantages	Limitations
☑Low side effects ☑Less invasive ☑Short treatment time ☑Usable in outpatient basis ☑Cancer selectivity ☑Multiple applications at the same location ☑Excellent cosmetic outcome ☑Lower costs	✗Photosensitivity after treatment ✗Treatment efficacy dependent on accuracy of tumor light irradiation ✗Tissue oxygenation is crucial for the photodynamic effect ✗Very difficult to treat metastatic cancers with current technology

6. Applications of PDT

PDT is a minimally invasive procedure that is clinically used in the treatment of several oncologic human diseases, such as skin, esophageal, head and neck, lung, and bladder cancers [57]. However, PDT also has several non-oncologic applications [58], including the treatment of non-cancerous human diseases, such as dermatologic (acne [59], warts [60], photoaging [61], psoriasis [62], vascular malformations [63], hirsutism [64], keloid [50], and alopecia areata [65]), ophthalmologic (central serous chorioretinopathy [66] and corneal neovascularization [67]), cardiovascular (atherosclerosis [68] and esophageal varix [69]), dental (oral lichen planus [70]), neurologic (Alzheimer's disease [71]), skeletal (rheumatoid arthritis [72]), and gastrointestinal (Crohn's disease [73]).

An extension of PDT procedure can be the inactivation of viruses and microorganisms, including bacteria, yeasts, and fungi, named as photodynamic inactivation of microorganisms (PDI). The viruses or microorganisms are inactivated by combining non-toxic dyes (photosensitizers) with harmless visible light. PDI can be used as an alternative to the use of antibiotics and antiviral drugs that usually cause resistance. The application of PDI is possible in several areas, including human and veterinary medicine, agro-food, wastewater treatment, and biosafety. However, PDI in the treatment of infections is easier to perform in vitro, compared with its clinical applicability, due to the low tissue penetration depth of visible light. Light applied intravenously can be a solution during the clinical treatment of infections. Very recently, the use of the PDT procedure to treat patients with COVID-19 has also been discussed [8,57,74–81].

Dr. M. Weber et al. [8] performed a study to evaluate if the PDT procedure with Riboflavin (also known as vitamin B2) and blue light can be used effectively as a therapy for patients infected with acute COVID-19. The study used COVID-19-positive patients who received PDT therapy and COVID-19-positive patients who received conventional care. The patients that received PDT treatment showed significant improvement in clinical symptoms and viral load within 5 days. The control patients had no significant improvement in clinical symptoms or viral load within 5 days. The results prove the potential of PDT procedure to treat patients infected with COVID-19 virus at an early infection stage and with mild to moderate clinical symptoms. This new application of PDT procedure can prevent hospitalization and intensive care treatments.

Finally, PDT can be implemented in a medical device, e.g., endoscopic capsule. In 2016, G. Tortora et al. [82] developed an ingestible capsule for light delivery in PDT treatment of *Helicobacter pylori* infection. *Helicobacter pylori* bacterium is known to be photosensitive without the exogenous assumption of photosensitizers. This bacterium can be killed by exciting the photosensitizers naturally present in it with the appropriate light wavelength. The capsule with 27 mm length and 14 mm diameter has been equipped with 8 LEDs positioned on an electronic board with a magnetic switch (to turn on the capsule's power) and a battery. The capsule light-emitting module was dimensioned considering the required light necessary to kill the bacterium, with blue (405 nm) and red (625 nm) lights. In 2018, J. A. Rodrigues et al. [26,83] studied the photodynamic activity of the mTHPC (Foscan®) on RKO and HCT-15 cell lines to implement PDT in autonomous medical devices, such as endoscopic capsules for clinical treatment of several gastrointestinal tract tumors. Figure 5 envisages the integration of PDT in the endoscopic capsule. Due to the endoscopic capsule dimensions and battery life, the light fluence and fluence rate of the red light must be minimized to reduce the PDT treatment time. The experimental results showed that a small amount of mTHPC (0.15 mg/kg) and light fluence (5–20 J/cm^2) is sufficient to obtain significant photodynamic activity. An array of LEDs with peak transmittance at 652 nm was used in the in vitro PDT assays. The experimental results show that decreased cell viability (down to 30%) can be obtained for 1–5 μg/mL of mTHPC concentrations and 2.5 J/cm^2 of light fluence. The use of a minimum light fluence (2.5 J/cm^2) and light fluence rate (11 mW/cm^2) allowed to reduce the treatment time to just 3 min and 47 s. The PDT endoscopic capsule was designed with two functional sides. The round side is compound of the conventional optical system of the endoscopic capsule, consisting of the CMOS image sensor, four white LEDs, and focal lenses, and the planar side constitutes the therapeutic module, composed of a red light source (array of 8 SMD red LEDs with total fluence rate of 14 mW/cm^2) and a magnetic switch for turning the red light on and off. This capsule has magnetic locomotion control, for immobilization of the capsule during the treatment time, and is 31 mm in length and 14 mm in diameter. mTHPC-mediated PDT, using a light fluence of 2.5 J/cm^2 and fluence rate of 14 mW/cm^2, reduces the PDT treatment time to approximately 3 min. Faster treatment requires less battery capacity and therefore fewer capsules.

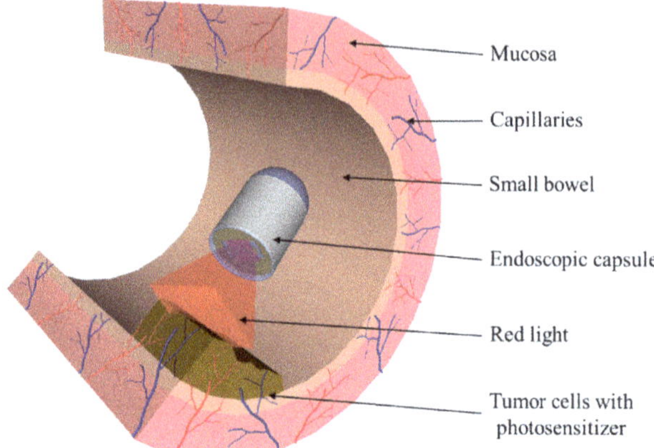

Figure 5. Illustration of PDT integrated in the endoscopic capsule (Adapted with permission from [83], IEEE, 2019).

7. Conclusions

PDT is one of the most interesting and promising approaches to treat various oncologic, non-oncologic, and infectious diseases. This review presented the main principles, mechanisms, and crucial elements of PDT. PDT is based on the dynamic interaction between a PS, light with a specific wavelength, and molecular oxygen, to produce ROS that promote selective destruction of the target tissue. The evolution of photosensitizers was also addressed in this paper, including their future trends. Photosensitizers are evolving towards increasing the PDT efficacy and selectivity, and many possibilities are currently under research. One strategy to increase PDT selectivity involves the development of multifunctional photosensitizers that can be activated by a biological or physical stimulus.

PDT has been increasingly used in many applications, such as destroying tumor tissues, bacteria, fungi, and viruses (including COVID-19). Moreover, PDT can be integrated in medical devices. A light delivery capsule has been developed for mTHPC-mediated PDT treatment of several gastrointestinal tract tumors. The good photodynamic response at low light fluence and low light fluence rate allows to reduce the treatment time to a few minutes and thus integrate the PDT in autonomous medical devices, such as endoscopic capsules of very small dimensions, to provide them with an advanced therapeutic function.

The interdisciplinary uniqueness of PDT inspires physicists, chemists, biologists, and physicians, and its further development and discovery of new applications will only be limited by their enormous imagination.

Author Contributions: The work presented in this paper was a collaboration of all authors. J.A.R. and S.P. made the literature analysis and wrote the first draft of the manuscript. J.H.C., T.D. and Z.Y. corrected and revised the manuscript. J.H.C. supervised and directed the manuscript. All authors have read and agreed to the published version of the manuscript.

Funding: The APC was funded by Chongqing Technology and Business University (CTBU).

Institutional Review Board Statement: Not applicable.

Informed Consent Statement: Not applicable.

Data Availability Statement: Not applicable.

Acknowledgments: This work was supported by FCT with project OpticalBrain reference PTDC/CTM-REF/28406/2017, operation code POCI-01-0145-FEDER-028406, through the COMPETE 2020; CMEMS-UMinho Strategic Project UIDB/04436/2020; Infrastructures Micro&NanoFabs@PT, operation code NORTE 01-0145-FEDER-022090, POR Norte, Portugal 2020; and MPhotonBiopsy, PTDC/FIS-OTI/1259/2020. The authors thank Biolitec research GmbH (Jena, Germany) for providing the photosensitizer Foscan®.

Conflicts of Interest: The authors declare no conflict of interest.

References

1. Hamblin, M.R.; Huang, Y. *Imaging in Photodynamic Therapy*; Taylor & Francis Group: Boca Raton, FL, USA, 2017.
2. Lee, C.-N.; Hsu, R.; Chen, H.; Wong, T.-W. Daylight photodynamic therapy: An update. *Molecules* **2020**, *25*, 5195. [CrossRef]
3. Ackroyd, R.; Kelty, C.; Brown, N.; Reed, M. The history of photodetection and photodynamic therapy. *Photochem. Photobiol.* **2001**, *74*, 656. [CrossRef]
4. Dolmans, D.E.J.G.J.; Fukumura, D.; Jain, R.K. Photodynamic therapy for cancer. *Nat. Rev. Cancer* **2003**, *3*, 380–387. [CrossRef]
5. Li, X.; Lee, S.; Yoon, J. Supramolecular photosensitizers rejuvenate photodynamic therapy. *Chem. Soc. Rev.* **2018**, *47*, 1174–1188. [CrossRef]
6. Mitton, D.; Ackroyd, R. A brief overview of photodynamic therapy in Europe. *Photodiagnosis Photodyn. Ther.* **2008**, *5*, 103–111. [CrossRef] [PubMed]
7. Hamblin, M.R. Photodynamic therapy for cancer: What's past is prologue. *Photochem. Photobiol.* **2020**, *96*, 506–516. [CrossRef] [PubMed]
8. Weber, M.; Mehran, Y.Z.; Orthaber, A.; Saadat, H.H.; Weber, R.; Wojcik, M. Successful reduction of SARS-CoV-2 viral load by photodynamic therapy (PDT) verified by QPCR—A novel approach in treating patients in early infection stages. *Med. Clin. Res.* **2020**, *5*, 311–325.
9. Weber, M. Intravenöse und interstitielle Lasertherapie: Eine neue Option in der Onkologie. *Akupunkt. Aurikulomed.* **2011**, *37*, 32–34.
10. Rocha, L.G.B. Development of a Novel Photosensitizer for Photodynamic Therapy of Cancer. Ph.D. Thesis, University of Coimbra, Coimbra, Portugal, 2015.
11. Fitzgerald, F. *Photodynamic Therapy (PDT): Principles, Mechanisms and Applications*; Nova Science Publishers, Inc.: New York, NY, USA, 2017.
12. Agostinis, P.; Berg, K.; Cengel, K.A.; Foster, T.H.; Girotti, A.W.; Gollnick, S.O.; Hahn, S.M.; Hamblin, M.R.; Juzeniene, A.; Kessel, D.; et al. Photodynamic therapy of cancer: An update. *CA Cancer J. Clin.* **2011**, *61*, 250–281. [CrossRef]
13. Donnelly, R.F.; McCarron, P.A.; Tunney, M.M. Antifungal photodynamic therapy. *Microbiol. Res.* **2008**, *163*, 1–12. [CrossRef] [PubMed]
14. Dąbrowski, J.M. Reactive oxygen species in photodynamic therapy: Mechanisms of their generation and potentiation. *Adv. Inorg. Chem.* **2017**, *70*, 343–394. [CrossRef]
15. Castano, A.P.; Mroz, P.; Hamblin, M.R. Photodynamic therapy and anti-tumour immunity. *Nat. Rev. Cancer* **2006**, *6*, 535–545. [CrossRef]
16. Allison, R.R.; Moghissi, K. Photodynamic therapy (PDT): PDT mechanisms. *Clin. Endosc.* **2013**, *46*, 24. [CrossRef] [PubMed]
17. Dąbrowski, J.M.; Arnaut, L.G. Photodynamic therapy (PDT) of cancer: From local to systemic treatment. *Photochem. Photobiol. Sci.* **2015**, *14*, 1765–1780. [CrossRef] [PubMed]
18. Simões, J.C.S.; Sarpaki, S.; Papadimitroulas, P.; Therrien, B.; Loudos, G. Conjugated photosensitizers for imaging and PDT in cancer research. *J. Med. Chem.* **2020**, *63*, 14119–14150. [CrossRef] [PubMed]
19. Abrahamse, H.; Hamblin, M.R. New photosensitizers for photodynamic therapy. *Biochem. J.* **2016**, *473*, 347–364. [CrossRef] [PubMed]
20. Kwiatkowski, S.; Knap, B.; Przystupski, D.; Saczko, J.; Kędzierska, E.; Knap-Czop, K.; Kotlińska, J.; Michel, O.; Kotowski, K.; Kulbacka, J. Photodynamic therapy—Mechanisms, photosensitizers and combinations. *Biomed. Pharmacother.* **2018**, *106*, 1098–1107. [CrossRef]
21. Duse, L.; Agel, M.R.; Pinnapireddy, S.R.; Schäfer, J.; Selo, M.A.; Ehrhardt, C.; Bakowsky, U. Photodynamic therapy of ovarian carcinoma cells with curcumin-loaded biodegradable polymeric nanoparticles. *Pharmaceutics* **2019**, *11*, 282. [CrossRef]
22. Chizenga, E.P.; Abrahamse, H. Nanotechnology in modern photodynamic therapy of cancer: A review of cellular resistance patterns affecting the therapeutic response. *Pharmaceutics* **2020**, *12*, 632. [CrossRef]
23. Li, T.; Yan, L. Functional polymer nanocarriers for photodynamic therapy. *Pharmaceuticals* **2018**, *11*, 133. [CrossRef]
24. Montaseri, H.; Kruger, C.A.; Abrahamse, H. Inorganic nanoparticles applied for active targeted photodynamic therapy of breast cancer. *Pharmaceutics* **2021**, *13*, 296. [CrossRef] [PubMed]
25. Kataoka, H.; Nishie, H.; Hayashi, N.; Tanaka, M.; Nomoto, A.; Yano, S.; Joh, T. New photodynamic therapy with next-generation photosensitizers. *Ann. Transl. Med.* **2017**, *5*, 183. [CrossRef] [PubMed]
26. Rodrigues, J.A.O. Therapy in Invasive Medical Devices with Image. Ph.D. Thesis, University of Minho, Braga, Portugal, 2019.
27. Yano, S.; Hirohara, S.; Obata, M.; Hagiya, Y.; Ogura, S.; Ikeda, A.; Kataoka, H.; Tanaka, M.; Joh, T. Current states and future views in photodynamic therapy. *J. Photochem. Photobiol. C Photochem. Rev.* **2011**, *12*, 46–67. [CrossRef]
28. Ormond, A.; Freeman, H. Dye sensitizers for photodynamic therapy. *Materials* **2013**, *6*, 817–840. [CrossRef]
29. Chen, J.; Fan, T.; Xie, Z.; Zeng, Q.; Xue, P.; Zheng, T.; Chen, Y.; Luo, X.; Zhang, H. Advances in nanomaterials for photodynamic therapy applications: Status and challenges. *Biomaterials* **2020**, *237*, 119827. [CrossRef] [PubMed]
30. FDA. Available online: https://www.accessdata.fda.gov/drugsatfda_docs/label/2016/208081s000lbl.pdf (accessed on 3 August 2021).

31. Reinhold, U.; Petering, H.; Dirschka, T.; Rozsondai, A.; Gille, J.; Kurzen, H.; Ostendorf, R.; Ebeling, A.; Stocker, M.; Radny, P. Photodynamic therapy with a 5-ALA patch does not increase the risk of conversion of actinic keratoses into squamous cell carcinoma. *Exp. Dermatol.* **2018**, *27*, 1399–1402. [CrossRef]

32. Privalov, V.A.; Lappa, A.V.; Kochneva, E.V. Five years experience of photodynamic therapy with new chlorin photosensitizer. In *Therapeutic Laser Applications and Laser-Tissue Interactions II, Proceedings of the European Conference on Biomedical Optics, Munich, Germany, 12–16 June 2005*; Proceedings SPIE 5863; SPIE: Bellingham, WA, USA, 2005; p. 586310.

33. Nyman, E.S.; Hynninen, P.H. Research advances in the use of tetrapyrrolic photosensitizers for photodynamic therapy. *J. Photochem. Photobiol. B Biol.* **2004**, *73*, 1–28. [CrossRef]

34. O'Connor, A.E.; Gallagher, W.M.; Byrne, A.T. Porphyrin and nonporphyrin photosensitizers in oncology: Preclinical and clinical advances in photodynamic therapy. *Photochem. Photobiol.* **2009**, *85*, 1053–1074. [CrossRef]

35. Ozog, D.M.; Rkein, A.M.; Fabi, S.G.; Gold, M.H.; Goldman, M.P.; Lowe, N.J.; Martin, G.M.; Munavalli, G.S. Photodynamic therapy: A clinical consensus guide. *Dermatol. Surg.* **2016**, *42*, 804–827. [CrossRef]

36. Morton, C.A.; Szeimies, R.-M.; Sidoroff, A.; Braathen, L.R. European guidelines for topical photodynamic therapy part 1: Treatment delivery and current indications—Actinic keratoses, Bowen's disease, basal cell carcinoma. *J. Eur. Acad. Dermatol. Venereol.* **2013**, *27*, 536–544. [CrossRef] [PubMed]

37. Morton, C.A.; Szeimies, R.-M.; Basset-Séguin, N.; Calzavara-Pinton, P.G.; Gilaberte, Y.; Hædersdal, M.; Hofbauer, G.F.L.; Hunger, R.E.; Karrer, S.; Piaserico, S.; et al. European Dermatology Forum guidelines on topical photodynamic therapy 2019 Part 2: Emerging indications—Field cancerization, photorejuvenation and inflammatory/infective dermatoses. *J. Eur. Acad. Dermatol. Venereol.* **2020**, *34*, 17–29. [CrossRef]

38. Morton, C.A.; Szeimies, R.-M.; Basset-Seguin, N.; Calzavara-Pinton, P.; Gilaberte, Y.; Hædersdal, M.; Hofbauer, G.F.L.; Hunger, R.E.; Karrer, S.; Piaserico, S.; et al. European Dermatology Forum guidelines on topical photodynamic therapy 2019 Part 1: Treatment delivery and established indications—Actinic keratoses, Bowen's disease and basal cell carcinomas. *J. Eur. Acad. Dermatol. Venereol.* **2019**, *33*, 2225–2238. [CrossRef] [PubMed]

39. Wezgowiec, J.; Derylo, M.B.; Teissie, J.; Orio, J.; Rols, M.-P.; Kulbacka, J.; Saczko, J.; Kotulska, M. Electric field-assisted delivery of photofrin to human breast carcinoma cells. *J. Membr. Biol.* **2013**, *246*, 725–735. [CrossRef] [PubMed]

40. Zhao, X.; Liu, J.; Fan, J.; Chao, H.; Peng, X. Recent progress in photosensitizers for overcoming the challenges of photodynamic therapy: From molecular design to application. *Chem. Soc. Rev.* **2021**, *50*, 4185–4219. [CrossRef]

41. Soliman, N.; Sol, V.; Ouk, T.-S.; Thomas, C.M.; Gasser, G. Encapsulation of a Ru(II) polypyridyl complex into polylactide nanoparticles for antimicrobial photodynamic therapy. *Pharmaceutics* **2020**, *12*, 961. [CrossRef]

42. Yanovsky, R.L.; Bartenstein, D.W.; Rogers, G.S.; Isakoff, S.J.; Chen, S.T. Photodynamic therapy for solid tumors: A review of the literature. *Photodermatol. Photoimmunol. Photomed.* **2019**, *35*, 295–303. [CrossRef] [PubMed]

43. Chen, D.; Zheng, H.; Huang, Z.; Lin, H.; Ke, Z.; Xie, S.; Li, B. Light-emitting diode-based illumination system for in vitro photodynamic therapy. *Int. J. Photoenergy* **2012**, *2012*, 1–6. [CrossRef]

44. Yoon, I.; Li, J.Z.; Shim, Y.K. Advance in photosensitizers and light delivery for photodynamic therapy. *Clin. Endosc.* **2013**, *46*, 7. [CrossRef]

45. Allison, R.R.; Sibata, C.H. Oncologic photodynamic therapy photosensitizers: A clinical review. *Photodiagnosis Photodyn. Ther.* **2010**, *7*, 61–75. [CrossRef]

46. Kinsella, T.J.; Colussi, V.C.; Oleinick, N.L.; Sibata, C.H. Photodynamic therapy in oncology. *Expert Opin. Pharmacother.* **2001**, *2*, 917–927. [CrossRef]

47. Chilakamarthi, U.; Giribabu, L. Photodynamic therapy: Past, present and future. *Chem. Rec.* **2017**, *17*, 775–802. [CrossRef]

48. Rezzoug, H.; Bezdetnaya, L.; A'amar, O.; Merlin, J.L.; Guillemin, F. Parameters affecting photodynamic activity of Foscan® or metatetra(hydroxyphenyl)chlorin (mTHPC) in vitro and in vivo. *Lasers Med. Sci.* **1998**, *13*, 119–125. [CrossRef]

49. Coutier, S.; Mitra, S.; Bezdetnaya, L.N.; Parache, R.M.; Georgakoudi, I.; Foster, T.H.; Guillemin, F. Effects of fluence rate on cell survival and photobleaching in meta-tetra-(hydroxyphenyl)chlorin–photosensitized colo 26 multicell tumor spheroids. *Photochem. Photobiol.* **2001**, *73*, 297. [CrossRef]

50. Bruscino, N.; Lotti, T.; Rossi, R. Photodynamic therapy for a hypertrophic scarring: A promising choice. *Photodermatol. Photoimmunol. Photomed.* **2011**, *27*, 334–335. [CrossRef] [PubMed]

51. Hartl, B.A.; Hirschberg, H.; Marcu, L.; Cherry, S.R. Characterizing low fluence thresholds for in vitro photodynamic therapy. *Biomed. Opt. Express* **2015**, *6*, 770. [CrossRef] [PubMed]

52. Nguyen, M.; Sandhu, S.; Sivamani, R. Clinical utility of daylight photodynamic therapy in the treatment of actinic keratosis—A review of the literature. *Clin. Cosmet. Investig. Dermatol.* **2019**, *12*, 427–435. [CrossRef]

53. Morton, C.A.; Braathen, L.R. Daylight photodynamic therapy for actinic keratoses. *Am. J. Clin. Dermatol.* **2018**, *19*, 647–656. [CrossRef]

54. Calixto, G.; Bernegossi, J.; de Freitas, L.; Fontana, C.; Chorilli, M. Nanotechnology-based drug delivery systems for photodynamic therapy of cancer: A review. *Molecules* **2016**, *21*, 342. [CrossRef]

55. Dos Santos, A.F.; De Almeida, D.R.Q.; Terra, L.F.; Baptista, M.S.; Labriola, L. Photodynamic therapy in cancer treatment—An update review. *J. Cancer Metastasis Treat.* **2019**, *2019*. [CrossRef]

56. Lange, N.; Szlasa, W.; Saczko, J.; Chwiłkowska, A. Potential of cyanine derived dyes in photodynamic therapy. *Pharmaceutics* **2021**, *13*, 818. [CrossRef]

57. Plaetzer, K.; Berneburg, M.; Kiesslich, T.; Maisch, T. New applications of photodynamic therapy in biomedicine and biotechnology. *Biomed. Res. Int.* **2013**, *2013*, 1–3. [CrossRef]
58. Yoo, S.W.; Oh, G.; Ahn, J.C.; Chung, E. Non-oncologic applications of nanomedicine-based phototherapy. *Biomedicines* **2021**, *9*, 113. [CrossRef] [PubMed]
59. Calzavara-Pinton, P.G.; Rossi, M.T.; Aronson, E.; Sala, R.; The Italian Group for Photodynamic Therapy. A retrospective analysis of real-life practice of off-label photodynamic therapy using methyl aminolevulinate (MAL-PDT) in 20 Italian dermatology departments. Part 1: Inflammatory and aesthetic indications. *Photochem. Photobiol. Sci.* **2013**, *12*, 148. [CrossRef] [PubMed]
60. Stender, I.-M.; Na, R.; Fogh, H.; Gluud, C.; Wulf, H.C. Photodynamic therapy with 5-aminolaevulinic acid or placebo for recalcitrant foot and hand warts: Randomised double-blind trial. *Lancet* **2000**, *355*, 963–966. [CrossRef]
61. Shin, H.T.; Kim, J.H.; Shim, J.; Lee, J.H.; Lee, D.Y.; Lee, J.H.; Yang, J.M. Photodynamic therapy using a new formulation of 5-aminolevulinic acid for wrinkles in Asian skin: A randomized controlled split face study. *J. Dermatol. Treat.* **2015**, *26*, 246–251. [CrossRef]
62. Choi, Y.M.; Adelzadeh, L.; Wu, J.J. Photodynamic therapy for psoriasis. *J. Dermatol. Treat.* **2015**, *26*, 202–207. [CrossRef]
63. Jerjes, W.; Upile, T.; Hamdoon, Z.; Mosse, C.A.; Akram, S.; Morley, S.; Hopper, C. Interstitial PDT for vascular anomalies. *Lasers Surg. Med.* **2011**, *43*, 357–365. [CrossRef]
64. Comacchi, C.; Bencini, P.L.; Galimberti, M.G.; Cappugi, P.; Torchia, D. Topical photodynamic therapy for idiopathic hirsutism and hypertrichosis. *Plast. Reconstr. Surg.* **2012**, *129*, 1012e–1014e. [CrossRef]
65. Linares-González, L.; Ródenas-Herranz, T.; Sáenz-Guirado, S.; Ruiz-Villaverde, R. Successful response to photodynamic therapy with 5-aminolevulinic acid nanoemulsified gel in a patient with universal alopecia areata refractory to conventional treatment. *Dermatol. Ther.* **2020**, *33*, e13416. [CrossRef]
66. van Dijk, E.H.C.; Fauser, S.; Breukink, M.B.; Blanco-Garavito, R.; Groenewoud, J.M.M.; Keunen, J.E.E.; Peters, P.J.H.; Dijkman, G.; Souied, E.H.; MacLaren, R.E.; et al. Half-dose photodynamic therapy versus high-density subthreshold micropulse laser treatment in patients with chronic central serous chorioretinopathy. *Ophthalmology* **2018**, *125*, 1547–1555. [CrossRef]
67. Díaz-Dávalos, C.D.; Carrasco-Quiroz, A.; Rivera-Díez, D. Neovascularization corneal regression in patients treated with photodynamic therapy with verteporfin. *Rev. Med. Inst. Mex. Seguro Soc.* **2016**, *54*, 164–169.
68. Houthoofd, S.; Vuylsteke, M.; Mordon, S.; Fourneau, I. Photodynamic therapy for atherosclerosis. The potential of indocyanine green. *Photodiagn. Photodyn. Ther.* **2020**, *29*, 101568. [CrossRef] [PubMed]
69. Li, C.Z.; Cheng, L.F.; Wang, Z.Q.; Gu, Y. Attempt of photodynamic therapy on esophageal varices. *Lasers Med. Sci.* **2009**, *24*, 167–171. [CrossRef] [PubMed]
70. Cosgarea, R.; Pollmann, R.; Sharif, J.; Schmidt, T.; Stein, R.; Bodea, A.; Auschill, T.; Sculean, A.; Eming, R.; Greene, B.; et al. Photodynamic therapy in oral lichen planus: A prospective case-controlled pilot study. *Sci. Rep.* **2020**, *10*, 1667. [CrossRef] [PubMed]
71. Lee, B., II; Suh, Y.S.; Chung, Y.J.; Yu, K.; Park, C.B. Shedding light on Alzheimer's β-amyloidosis: Photosensitized methylene blue inhibits self-assembly of β-amyloid peptides and disintegrates their aggregates. *Sci. Rep.* **2017**, *7*, 7523. [CrossRef]
72. Gallardo-Villagrán, M.; Leger, D.Y.; Liagre, B.; Therrien, B. Photosensitizers used in the photodynamic therapy of rheumatoid arthritis. *Int. J. Mol. Sci.* **2019**, *20*, 3339. [CrossRef] [PubMed]
73. Favre, L.; Borle, F.; Velin, D.; Bachmann, D.; Bouzourene, H.; Wagnieres, G.; van den Bergh, H.; Ballabeni, P.; Gabrecht, T.; Michetti, P.; et al. Low dose endoluminal photodynamic therapy improves murine T cell-mediated colitis. *Endoscopy* **2011**, *43*, 604–616. [CrossRef]
74. Huang, L.; Dai, T.; Hamblin, M.R. Antimicrobial photodynamic inactivation and photodynamic therapy for infections. *Methods Mol. Biol.* **2010**, *635*, 155–173. [CrossRef]
75. Rapacka-Zdończyk, A.; Woźniak, A.; Michalska, K.; Pierański, M.; Ogonowska, P.; Grinholc, M.; Nakonieczna, J. Factors determining the susceptibility of bacteria to antibacterial photodynamic inactivation. *Front. Med.* **2021**, *8*, 617. [CrossRef]
76. Almeida, A. Photodynamic therapy in the inactivation of microorganisms. *Antibiotics* **2020**, *9*, 138. [CrossRef]
77. Tariq, R.; Khalid, U.A.; Kanwal, S.; Adnan, F.; Qasim, M. Photodynamic therapy: A rational approach toward COVID-19 management. *J. Explor. Res. Pharmacol.* **2021**, *6*, 44–52. [CrossRef]
78. Fekrazad, R. Photobiomodulation and antiviral photodynamic therapy as a possible novel approach in COVID-19 management. *Photobiomodul. Photomed. Laser Surg.* **2020**, *38*, 255–257. [CrossRef]
79. Shen, J.J.; Jemec, G.B.E.; Arendrup, M.C.; Saunte, D.M.L. Photodynamic therapy treatment of superficial fungal infections: A systematic review. *Photodiagn. Photodyn. Ther.* **2020**, *31*, 101774. [CrossRef]
80. Hsieh, Y.-H.; Chuang, W.-C.; Yu, K.-H.; Jheng, C.-P.; Lee, C.-I. Sequential photodynamic therapy with phthalocyanine encapsulated chitosan-tripolyphosphate nanoparticles and flucytosine treatment against *Candida tropicalis*. *Pharmaceutics* **2019**, *11*, 16. [CrossRef] [PubMed]
81. Janeth Rimachi Hidalgo, K.; Cabrini Carmello, J.; Carolina Jordão, C.; Aboud Barbugli, P.; de Sousa Costa, C.A.; Garcia de Oliveira Mima, E.; Pavarina, A.C. Antimicrobial photodynamic therapy in combination with nystatin in the treatment of experimental oral candidiasis induced by candida albicans resistant to fluconazole. *Pharmaceuticals* **2019**, *12*, 140. [CrossRef] [PubMed]

82. Tortora, G.; Orsini, B.; Pecile, P.; Menciassi, A.; Fusi, F.; Romano, G. An ingestible capsule for the photodynamic therapy of helicobacter pylori infection. *IEEE ASME Trans. Mechatron.* **2016**, *21*, 1935–1942. [CrossRef]
83. Rodrigues, J.A.; Amorim, R.; Silva, M.F.; Baltazar, F.; Wolffenbuttel, R.F.; Correia, J.H. Photodynamic therapy at low-light fluence rate: In vitro assays on colon cancer cells. *IEEE J. Sel. Top. Quantum Electron.* **2019**, *25*, 1–6. [CrossRef]

 *pharmaceutics*

Review

Potential of Cyanine Derived Dyes in Photodynamic Therapy

Natalia Lange [1], Wojciech Szlasa [1], Jolanta Saczko [2] and Agnieszka Chwiłkowska [2,*]

[1] Faculty of Medicine, Wroclaw Medical University, Mikulicza-Radeckiego 5, 50-345 Wroclaw, Poland; natalia.lange@student.umed.wroc.pl (N.L.); wojciech.szlasa@outlook.com (W.S.)

[2] Department of Molecular and Cellular Biology, Faculty of Pharmacy, Wroclaw Medical University, Borowska 211A, 50-556 Wroclaw, Poland; jolanta.saczko@umed.wroc.pl

* Correspondence: agnieszka.chwilkowska@umed.wroc.pl

Abstract: Photodynamic therapy (PDT) is a method of cancer treatment that leads to the disintegration of cancer cells and has developed significantly in recent years. The clinically used photosensitizers are primarily porphyrin, which absorbs light in the red spectrum and their absorbance maxima are relatively short. This review presents group of compounds and their derivatives that are considered to be potential photosensitizers in PDT. Cyanine dyes are compounds that typically absorb light in the visible to near-infrared-I (NIR-I) spectrum range (750–900 nm). This meta-analysis comprises the current studies on cyanine dye derivatives, such as indocyanine green (so far used solely as a diagnostic agent), heptamethine and pentamethine dyes, squaraine dyes, merocyanines and phthalocyanines. The wide array of the cyanine derivatives arises from their structural modifications (e.g., halogenation, incorporation of metal atoms or organic structures, or synthesis of lactosomes, emulsions or conjugation). All the following modifications aim to increase solubility in aqueous media, enhance phototoxicity, and decrease photobleaching. In addition, the changes introduce new features like pH-sensitivity. The cyanine dyes involved in photodynamic reactions could be incorporated into sets of PDT agents.

Keywords: cyanine dyes; photodynamic therapy; cancer therapy; irradiation

Citation: Lange, N.; Szlasa, W.; Saczko, J.; Chwiłkowska, A. Potential of Cyanine Derived Dyes in Photodynamic Therapy. *Pharmaceutics* **2021**, *13*, 818. https://doi.org/10.3390/pharmaceutics13060818

Academic Editors: Udo Bakowsky, Matthias Wojcik, Eduard Preis and Gerhard Litscher

Received: 15 April 2021
Accepted: 28 May 2021
Published: 31 May 2021

Publisher's Note: MDPI stays neutral with regard to jurisdictional claims in published maps and institutional affiliations.

1. Introduction

1.1. Basics of Photodynamic Therapy

Photodynamic therapy (PDT) is a low-invasive therapy, that destroys cancer cells through the generation of reactive oxygen species (ROSs). PDT involves a photosensitizer (PS), administered topically or intravenously, a light source and oxygen in the targeted tissue [1]. In the dark, the PS remains in its base energy state. However, the absorption of light at the appropriate wavelength moves the PS to an excited state. An excess of absorbed energy leads to its release and reaction with the oxygen in the tissue, thus inducing the formation of ROS [2]. The photodynamic reaction might propagate in two main ways [3]. The Type I mechanism of photodynamic reaction includes the production of highly reactive intermediates, mainly hydrogen peroxide H_2O_2 and $\bullet OH$. Type II involves the formation of oxygen in the singlet state $^1O_2$ (Figure 1) [4]. The short half-life of free radicals determines the limited diffusion distance of these high cytotoxic molecules [5]. The dysregulation of the homeostasis free radicals induces irreversible oxidation of proteins, nucleic acids, fatty acids, cholesterol and organelles of the tumor cells. The process leads to loss of function and eventually to the death of the irradiated cells depending on the conditions of the therapy, especially the cellular localization of the photosensitizer [5]. Death might also arise from an induced immune response as well as from the destruction of vessels supplying the tumor with nutrition [6].

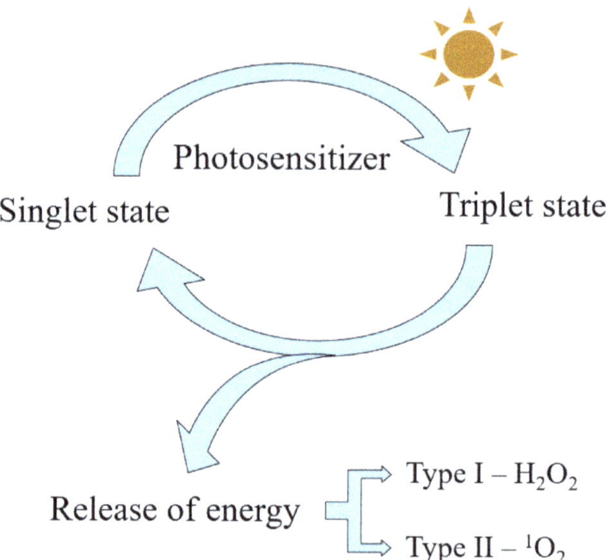

Figure 1. The mechanism of a photosensitizer action in photodynamic therapy.

1.2. Advantages and Disadvantages of PDT

Several advantages of PDT make it a promising treatment against malignancies. Selectivity of the PS towards cancer cells makes it applicable for curative treatment, early-stage tumors, and reducing the size of inoperable tumors. Additionally, as a local low-invasive procedure, PDT has minimal systemic side effects compared with standard chemotherapy or radiotherapy. PDT can also lead to the eradication of the tumor vasculature [5], which will be described in Section 1.5.

Another point in its favor is that PDT may trigger various immune responses to cancerous cells [7]. Primarily, if the irradiated cells undergo necrosis, an acute inflammatory response mediates the removal of the dead tissue [8]. Then, dendritic cells (DCs) absorb the cancer-related antigens through phagocytosis and present it to the immune-system cells, thus triggering the systemic inhibitory response to cancer cells [9].

The penetration of light at an appropriate wavelength to activate the FDA-approved PSs (Porfimer sodium, Temoporfin, Verteporfin [10]) can be suppressed by the tissue, thus putting deep-seated tumors out of reach. To overcome this problem, PSs, that absorb light from the NIR-I could be applied. Another obstacle is the low oxygenation of the tumor microenvironment. Namely, when cancer cells are in hypoxia [11], PDT may be rendered ineffective due to the lack of substrate for ROS generation. The main advantages and disadvantages of PDT are shown in Figure 2, whereas a broad reference to the efficacy and application of the PDT is reviewed by Agostinis et al. [12].

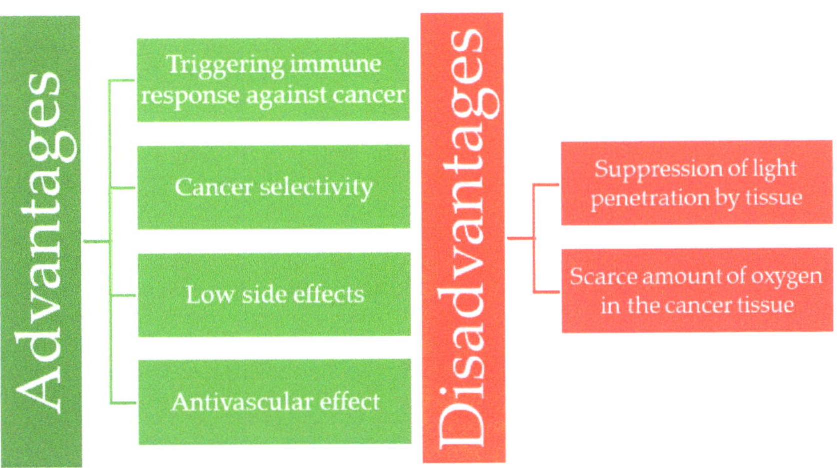

Figure 2. Summary of the main advantages and disadvantages of photodynamic therapy.

1.3. Desired Characteristics of the Photosensitizer

A PS should be characterized by low dark toxicity and selective accumulation in the tumor cells. The compound should be highly susceptible to irradiation and generate as many ROS as possible in the tumor environment [13]. Also, an ideal PS should localize outside the nucleus to avoid the pro-cancerous mutations in the DNA [1]. Preferably, an ROS should be generated in the mitochondria or lysosomes, but an especially effective targeting combination involves both of them [14]. Photodamage to the organelles would result in a controlled cell death, like apoptosis or paraptosis [15], rather than a mutagenic process in the nuclei. Controlled cell death results in the lack of tissue necrosis, which, despite its assets, risks an uncontrolled inflammation response [16]. Moreover, the dye should be characterized by good photostability to prevent photobleaching [13], which is a process of decomposition and loss of fluorescence and occurs when the dye's polymethine chains are oxidized by singlet oxygen species [17]. Interestingly, studies by N.S. James [18] showed no quantitative relation between the photobleaching of the PS and effectiveness of the therapy. The phenomena might be explained by the fact that, aside from the photosensitizer itself, the whole tumor microenvironment undergoes irradiation, leading to the generation of ROS from different sources [19].

1.4. Cellular Effects of PDT

The localization of the PSs is crucial when considering the efficacy of the PDT. There are agents that target mitochondria, lysosomes, the endoplasmic reticulum, Golgi apparatus, plasma membranes or combinations of these sites and condition various cytotoxic mechanisms. A target location in lysosomes and mitochondria was noticed to be associated with higher PDT efficacy [14]. Interestingly, Kessel et al. demonstrated that autophagy can offer cytoprotection after mitochondrial photodamage. This process can be avoided by targeting the lysosomes, which potentiates apoptosis [14], whereas ER photodamage primarily evokes paraptosis [15,20]. Castano et al. summarized the effects of the photosensitizer: organelle interactions and proved photosensitizers induced cytotoxicity in different ways by targeting the specific cellular compartment [21]. Our study revealed that the cytotoxic and genotoxic effects varied depending on the plasmalemmal or intracellular localization of the PSs [22]. Mainly, irradiation of a culture of melanoma cells right after the administration of curcumin as a drug yielded the highest cytotoxicity, yet with the increase in incubation time toxicity to melanoma cells decreased as the drug moved to the intracellular membranous compartments. The mechanism of melanoma cell death, apoptosis, was proved to be mediated by the caspase-12. Surprisingly, even though cytotoxic tendencies were similar among cancerous and non-cancerous cells, the genotoxicity in the culture of normal human fibroblasts was significantly lower than in melanoma cells. Moreover, the cytotoxic effect of cyanine dyes can differ depending on the resistance of a cell culture to certain chemotherapeutics. Kulbacka et al. showed that the doxorubicin-resistant cell line of human breast adenocarcinoma presented a weaker dye distribution than wild-type cell lines, and after irradiation the therapy was significantly more effective among the resistant cells, suggesting that PDT using such dyes may offer an alternative treatment for multidrug-resistant tumors [23].

Characterization of new photosensitizing agents often involves assessing sites of subcellular localization. Cyanine dyes were also studied in case of cellular localization-related death induction. First, Delaey et al. showed that, depending on the partition coefficient, different cyanine dye subgroups concentrate more (~1.5) or less intracellularly [24]. The study proved that hydrophobic cyanine dyes, like indocyanines, localize less intracellularly. Murakami et al. proved that bichromophoric cyanine dyes localize preferentially in mitochondria. In the study, high phototoxicity to melanoma was proven to be the result of the interaction between cyanine and mitochondria [25]. Other studies also proved the potency of cyanine dyes in targeting mitochondria [26].

1.5. Vascular Effects of PDT

A specific and important trigger for cancer cell death mediated by photodynamic therapy is the anti-vascular effect. It is known that the application of ICG in PDT reduces perfusion by photocoagulating blood vessels [27]. This indirect pathway of cancer eradication additionally enhances the efficacy of the therapy. Shafirstein et al. observed how the vasculature surrounding the tumor tissue was impaired after applying ICG in PDT of a murine mammary carcinoma. Both phototoxic and significant photothermal effects of the therapy resulted in necrosis of the endothelial cells and the deposition of fibrin within the blood vessels [28]. Therefore, an anti-vascular effect is not the result of blood-vessel destruction alone. As the endothelial cells are damaged, they release clotting factors that lead to the formation of thrombi, constriction and occlusion of vessels [29]. In case of large tumors, treatment with PDT might lead to hemorrhaging, and additional side effects of the therapy. These processes deprive tumor cells of necessary nutrients and oxygen. Nonetheless, the resulting hypoxia may also have a negative impact on the therapy. Photosensitizers currently in use require oxygen to create free radicals [30]. Insufficient oxygenation, characteristic of the tumor microenvironment (TME) [31] and intensified by oxygen-dependent PDT [30], directly reduces its efficacy. Moreover, hypoxia stimulates the release of angiogenic growth factors that lead to neovascularization. To inhibit such a process, angiogenesis inhibitors could be combined with PDT to minimize the risk of tumor recurrence [32].

1.6. Interstitial PDT

Intratumor light delivery by one or more laser fibers inserted into the target tissue (typically tumor and margins (interstitial PDT, I-PDT)) is applied to activate PSs in deeply seated tumors or tumors more than 10 mm in thickness. If a tumor is too large for the light to be delivered into its entire volume, interstitial PDT (iPDT) could destroy it to the margins of healthy tissue [33]. Moreover, the application of fibers under the guidance of ultrasound or MRI could minimize light scattering by healthy tissue, thus allowing the iPDT to reach the target location and facilitate the eradication of a deep-seated tumor [34] as in pancreatic cancer [35]. According to Chang et al. disulfonated aluminum phthalocyanine (AlS2Pc) application in PDT using fibers placed under ultrasound guidance proved to be effective against prostate cancer by creating lesions up to 12 mm wide. Moreover, the basic connective tissue architecture of the organ and the prostate capsule healed properly, proving that PDT is a less-invasive treatment in comparison to radical surgery [36]. The selective delivery of light could simplify therapies in areas where any excess tissue loss could lead to serious disabilities [37]. An example for such a location is the head and neck, where iPDT using dyes like Foscan [38], or porfimer sodium [37] has already proven to be effective and cause few side effects.

2. Cyanine Derived Dyes

Cyanine dyes consist of two-terminal heterocyclic units linked by a polymethine bridge core structure (Figure 3) [39]. Over the last few years, the cyanine dyes and their derivatives were widely analyzed for the cytotoxic activity against wide spectrum of tumors. It has been reported they meet most of the requirements for being used in PDT against deep-seated cancers. Each of the main subdivisions of the dyes was described in the following paragraphs.

Cyanine derived dyes

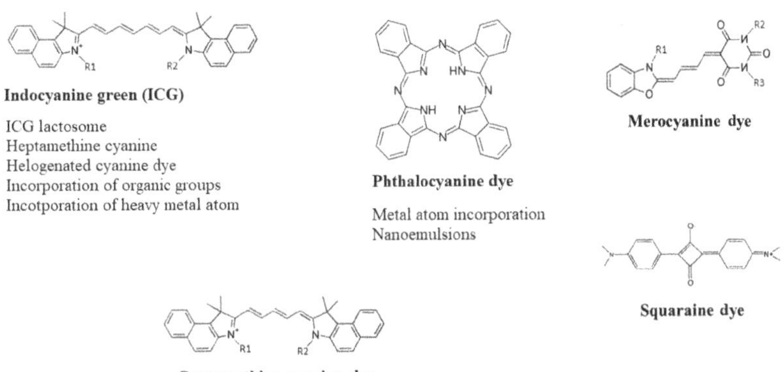

Figure 3. Cyanine-derived dye division and backbone structures. R1, R2 and R3 represent the organic substituent groups.

2.1. Indocyanine Green

One of the most promising cyanine dye is a carbocyanine, widely known as indocyanine green (ICG), the only cyanine dye approved by the Food and Drug Administration (FDA). Due to its absorbance maximum in NIR, prominent fluorescent properties and low dark toxicity, the dye is applied to diagnose liver, cardiovascular and sentinel lymph node pathologies [40,41]. ICG has a high affinity for plasma proteins and is cleared from the body through the biliary pathway, which allows for its use in diagnoses of liver and bile duct functions [42]. Several studies have proven ICG's toxicity towards various cancer cells. For instance, ICG-mediated PDT induced a 38% decrease in choroidal melanoma viability in six months [43]. Here, it is mostly retained in the Golgi apparatus, endoplasmic reticulum, mitochondria and lysosomes [44]. Additionally, ICG proved to be an efficient photothermal agent, suppressing tumor growth under repeated NIR-light irradiation. The process not only includes the induction of oxidative stress but also the production of heat by converting approximately 88% of the absorbed light into radiation [45].

However, the dye also has several disadvantages. These include low photostability, a high level of photobleaching and no specificity towards cancer cells [46]. Of note, the energy yield from the photodynamic reaction is lower than in the PSs characterized by lower absorbance maxima [42,47].

2.2. ICG Lactosomes

Studies show that combining high-mass lactate-derived polymers with ICG in ICG–lactosomes (a core-shell-type polymeric micelle or "nanocarrier") results in greater toxicity towards cancer cells in PDT than by using ICG alone. This tendency was observed in vivo on BALB/c nude mice transfected with human hepatocellular carcinoma (HCC) cell line

HuH-7 [48] and gallbladder cancer NOZ cell lines [49]. In the latest case, ICG–lactosomes were effective not only as photosensitizers but also as fluorescent diagnostic agents because of their selective accumulation in cancer tissue, strengthened cytotoxicity and relatively high fluorescence.

2.3. Heptamethine Cyanine Dye

Heptamethine cyanine dye, Cy7 is an ICG derivative that has maximum absorbance in the near-infrared range (NIR) and shows high selectivity for cancer cells. On a cellular level, the dye is transported through organic anion-transporting polypeptides (OATPs). The expression of OATPs is regulated by hypoxia-inducible factor 1α (HIF1α), which may possibly explain its selectivity for tumor cells [50]. Moreover, Usama et al. proved, that Cy7 cyanine dyes form covalent albumin adducts that can generate long-lasting intratumor fluorescence due to their enhanced permeability and retention in the tumor tissue [51]. Heptamethine dyes aggregate primarily in the mitochondria and lysosomes. Such localization favors inducing apoptosis over necrosis, thereby minimizing the risk of uncontrolled immunological reaction provoked by necrosis of the affected tissue [16].

Cy7 is a promising photosensitizer, diagnostic agent and nanocarrier transporter. Under near-IR light this cyanine derivative might transport anti-cancer drugs exclusively to the tumor [52]. Moreover, Jiang et al. observed that Cy7 conjugated with Gemcitabine displayed relatively high residency time in tumor tissue [53]. Namely, Cy7 remains in a tumor 5–20 days in comparison to ICG, which is removed from the body within 24 h [54].

The disadvantages of heptamethine cyanine derivatives include unfavorable hydrophobicity, which leads to its aggregation in body fluids, extensive photobleaching and insufficiency in ROS production similar to ICG.

2.4. Halogenated Cyanine Dyes

Due to the heavy atom effect, the production of free radicals can be significantly increased by halogenation of the cyanine dyes [55]. Atchison et al. established that iodinated IR-783 derivative presents high efficacy towards BxPC-3 and MIA PaCa-2 pancreatic cancer cell lines [35]. Halogenating the dye resulted in significant suppression of tumor growth. Moreover, the viability of MIA PaCa-2 cells after PDT was less than 10% and BxPc-3 was lower than 40%. Interestingly, after irradiation of the same cell lines with ICG, no photodynamic effect was observed. The potential of the derivative was proven in studies where PDT was more effective than a 5-FU treatment of pancreatic cancer. The compounds may also inhibit the growth of cancer. Namely, in a murine model transfected with a human xenograft of BxPC-3 Luc, the untreated pancreatic tumor grew to about 500% of the initial tumor size, whereas the one irradiated with iodinated cyanine IR-783 increased its volume by only about 39% [35]. The study presents PDT as a prospective neoadjuvant and palliative therapy against highly aggressive tumors.

Cao et al. modified the ICG derivative Cy7 with heavy atom iodine to form the novel NIR dye CyI [56]. They investigated the application of PDT simultaneously with photothermal therapy in the study of HepG2 cancer cells [56]. The iodinated dye enhanced the production of ROS production. In this case, the addition of photothermal therapy (PTT) had a synergistic effect on the treatment and strengthened the suppression of tumor growth.

However, the iodinated ICG derivative CyI proved to have poor solubility and tumor-targeting abilities in clinical application [56]. To overcome the problems the dye was modified by PEGylation and the addition of hyaluronic acid, which increased solubility in water [8].

Another iodidinated cyanine dye, IR-780, accumulates preferentially in tumor tissue after an intravenous injection [57]. The derivative displays a significant in vivo ability to target tumors. The process is dependent on the cancer's energetic metabolism, plasma membrane potential and expression of OATPs [58]. Wang et al. showed no significant difference between laser-irradiated and non-irradiated cells treated with IR-780, which is attributed to its inherent toxicity in the dark [59].

By modifying the heptamethine cyanine dye, Noh et al. developed the mitochondria-targeting photodynamic therapeutic agent MitDt-1 [60]. Bromination of the indoline group of the heptamethine dye proved to increase the production of ROS considerably. MitDt-1 accumulates primarily in mitochondria, thus inducing apoptosis. Also, the derivative containing triphenylphosphonium (TPP) and quaternary ammonium enhanced the dye's solubility and selectivity for the mitochondria of the cancer cells. Moreover, the high toxicity of MitDt-1 toward cancer cells was proven in studies on MCF-7 breast cancer cells in vitro and on NCI-H460 lung cancer both in vitro and in vivo [60].

The analog of IR-780 named IR-808 (MHI-148), was designed, synthesized and screened by Tan et al. [61]. Its photostability and photocytotoxicity was evaluated on the human cervical cancer cell line HeLa and Lewis lung carcinoma (LLC) in mouse xenografts. IR-808 displayed selective aggregation in tumor cells, distinct optical properties and high photostability in serum. Both IR-808 and IR-783 accumulate primarily in mito-chondria and lysosomes [62]. Furthermore, IR-808 showed a significant dose-dependent phototoxic effect and distinct suppression of tumor growth after irradiation [62]. In the histopathological examination of the experimental mice, no aggregation in systemic cir-culation and interstitial fluids were detected. Further investigations into IR-808 proved a high selectivity for cancer and thus potential for imaging gastric [63], prostate [62] and kidney cancer [64].

2.5. Incorporation of Organic Groups

Further modifications of IR-808 aimed to increase its water solubility, which included replacing of one of its side chains with $(CH_2)_4SO_3$ in DZ-1 derivative [65]. To prove the potential of the new derivative, ICG and the newly synthesized fluorophore were compared. ICG itself has high selectivity for HCC with a high tumor to-background ratio (255:1) [66], making it an important tool for identifying HCC lesions during surgery [67]. However, the fluorescence displayed by DZ-1 was significant and lasted longer in contrast to ICG, which allowed for the identification of smaller tumor regions. Moreover, the uptake of the dye in vivo was assessed on HCC Hep3B-Luc cell line xenografts and male New Zealand rabbits. In this case, DZ-1 displayed no accumulation in liver or lung tissue, proving that DZ-1 has a higher specificity for cancer cells than ICG does.

A study by Yang et al., presented the modified heptamethine dye by the addition of 4-amino-2,2,6,6,-tetramethylpiperidine-N-oxyl [68]. The dye was highly effective in ROS production. Moreover, this derivative, first introduced by Jiao et al. [69], has a long triplet-state lifetime and the incorporation of sulfonic acid improves its water-solubility. Additionally, the dye prompted a significant apoptosis of the HepG2 cells after NIR irradiation and showed low dark toxicity.

2.6. Incorporation of a Heavy Metal Atom

Incorporation of a heavy metal into the structure of the dye enhances crossing due to the heavy metal effect [70]. Nevertheless, the introduction of the heavy atom in a cyanine dye poses a risk of enhancing dark toxicity [68] and accumulating in healthy tissues [71]. Conjugation of ICG with Au-based nanomaterials sufficiently enhanced the absorption, emission and stability of the fluorophore [72]. It simultaneously induced PTT and enhanced ROS production in PDT, efficiently killing A549 malignant cells [72]. Similarly, the conjugate of ICG with gold–gold sulfide, where gold also acted as an agent for PTT, presented greater stability and improved cytotoxicity towards a HeLa cell line [73]. In addition, silver was incorporated into ICG nanoparticles, resulting in a synergistic interaction between the photothermal effect and PDT. In the study performed by Tan et al., PEGylated silver nanoparticles with a polyaniline shell acted as nanocarriers for ICG and efficiently induced hyperthermia in HeLa cancer cells. In this case, standalone ICG was responsible for fluorescence and phototoxicity [74].

The platinum (II) complex of heptamethine cyanine, IR797-Platin, proved to have extremely high cytotoxicity under NIR-light conditions towards C-33 A (cervical cancer) and MCF-7 breast cancer cell lines [75]. The drug's cytotoxicity was shown by the photosensitivity of IR797 and the inhibition of DNA transcription and replication by platinum [76].

Zhao et al. synthesized the CYBF2 agent by incorporating: boron difluoride (BF2) into the core structure of cyanine [17]. BF2 reduced the dye's electron density, resulting in the enhanced photostability. The drug accumulated in mitochondria, induced apoptosis and was efficiently absorbed by MCF-7 cells. Furthermore, the compound presented low dark toxicity and a high level of ROS burst induction.

2.7. pH-Sensitive Cyanine Dyes

Heptamethine cyanine dyes were further modified to be pH sensitive, which eventually increased their tumor specificity [39]. Apart from a relatively high cancer cell selectivity the dye induced fluorescence only after it was placed in an acidic environment. Since extracellular cancer fluid has a high concentration of lactic acid, such a modification made it possible to visualize cancer cells and destroy them more accurately. One of the studied dyes, IR2, incorporated a dimethylamine group that worked as an intramolecular charge transporter. Selectivity was proven by a higher uptake in cancer (HepG2 and HeLa) cell lines than normal cells. Also, the apoptosis following irradiation with the drug was much higher among cancer (80%) than normal (HL-7702) cells (9.4%). Interestingly, cell death was mostly the effect of hyperthermia, not photodynamic reaction.

Siriwibool et al. synthesized a pH switchable dye I2-IR783-Mpip [77] composed of IR783 and N-methylpiperazine. In acidic conditions, the color changes from blue to red, but only the red dye can absorb LED light and displays high toxicity towards HepG2 cells. Cancer cell viability in a neutral environment was about 30% and in an acidic environment decreased to 10%. In this case, death was primarily the result of a free radical production, not the photothermal effect.

Another pH switchable agent was presented by Meng et al. who conjugated 5'-carboxyrhodamines (Rho) and heptamethine cyanine IR765 (Cy) [78]. The newly synthesized conjugate (RhoSSCy) had enhanced fluorescence in a decreased pH value and displayed high stability in pH ranging from 5 to 9. The dye accumulated specifically in tumor cells, presenting a significant fluorescence. In the xenograft studies, the phototoxicity towards cancer was high, considerably increasing the survival rate of mice transfected with MCF-7 cells.

2.8. Near-Infrared II Dyes

Further studies of cyanine dyes were aimed at a red-shifted long-wavelength absorption maximum. Unlike the most dyes currently used in PDT, the PSs that absorb energy into the NIR I range show enhanced tissue penetration. Dyes activated with NIR-II light (1000–1700 nm) were analyzed to lessen the scattering of the light by the tissue, enhance the image contrast and improve the deep-seated tumor detection. The NIR-II dyes, which emit light at a wavelength longer than 1000 nm, were first reported by Antaris et al. [79], followed by the discovery of cyanine dye far-red emissions by Zhu et al. [80]. ICG and IRDye800 were shown to possess emission across NIR-I and NIR-II, marking them as promising and highly specific fluorophores for surgical procedures [42]. The photophysical mechanism of NIR-II emission relies on twisted intramolecular charge transfer (TICT). Further, Starolski et al. observed that the contrast-to-noise ratio (CNR) of the ICG window is twice as high in NIR-II than in NIR-I [81]. Ge et al. studied the efficacy of NIR II-emitting polymer nanoparticles: AuNR vesicles (Ru-complex and a cyanine dye (IR 1061)) [82]. Irradiation by NIR II light induced the release of the Ru complex and generated cytotoxic $^1O_2$. The therapy efficiently killed the MCF-7 breast cancer cells both in vitro and in vivo.

3. Pentamethine Cyanine Dyes

Pentamethine cyanine fluorophores were designed to have tissue-specificity and localize primarily in the adrenal and pituitary glands, pancreas and lymph nodes. Their synthesis aimed to give a promising contrast agent for intraoperative imaging of glands [83]. Newly obtained symmetrical penthametine cyanine dyes, based on a benzoindoleninic ring were tried on a human fibrosarcoma cell line (HT-1080) by Ciubini et al. [84]. The dyes remained active at relatively low concentrations (10 nM), and the drugs were rapidly internalized by cancer cells, which led to high ROS burst. Surprisingly, brominating the benzoindolenine did not produce any increase in free radicals.

2-quinolinium pentamethine carbocyanines were analyzed by Ahoulu et al. on an ES2 ovarian carcinoma cell line [85]. The dye, which was brominated at the mesocarbon remained highly phototoxic and reduced cell line viability from $100 \pm 10\%$ to $14 \pm 1\%$ after irradiation at a 694 nm wavelength. The compound was characterized by great stability, little dark toxicity and displayed DNA-cleavage. Dye localized mainly in the cytosol and perinuclear regions, whereupon it generated hydroxyl radicals after irradiation.

4. Carbocyanines against Drug-Resistant Cancer Cells

PDT may act as a prominent treatment for tumors with multiple drug resistance (MDR). Kulbacka et al. investigated four different cyanine dyes: two carbocyanines: HM-118, KF-570, merocyanine FBF-749 and pyridine-thiazolidine ER-139 combined with PDT against malignant breast adenocarcinoma cell lines. One of the latter was resistant to doxorubicin (MCF/DX), but the other wasn't (MCF-WT) [86]. Carbocyanines had the greatest phototoxicity. After irradiation with HM-118, 100% of apoptotic cells were detected in both cell lines. When using KF-570, 98% of the wild-type cells and 95% of the doxorubicin-resistant cells remained apoptotic. Other dyes had a much lower apoptotic effect, and their phototoxicity was insufficient. After irradiation with HM-118, overexpression of AIF, a protein involved in caspase-independent cell death, was detected in MCF-7/WT cells [87]. Conversely, in doxorubicin-resistant cells, the protein was absent.

5. Squaraine Dyes

Squaraine dyes possess several unique properties, such as significant fluorescence, distinct stability and absorbance wavelength maxima of 600–800 nm. The disadvantages of the dyes involve low solubility and low ROS production following PDT. Fernandes et al. incorporated the sulfur atom into the indolenine-based squaraine core and assessed the potency of the drug in combination with PDT on HepG2 and Caco-2 cell lines [88]. ROS production was enhanced after dithiosquaraine dyes aided PDT. Conversely, monothiosquaraine dyes turned out to be ineffective. Despite high phytotoxicity of dithiosquaraine dyes, the compounds degraded easily and aggregated in aqueous media.

5.1. Dicyanomethylene Squaraine Dyes

Martins et al. proved the significant phototoxicity of dicyanomethylene squaraine cyanine dyes against Caco-2 and HepG2 cancer cells in vitro [89]. Despite low singlet oxygen production and moderate light-stability, the chemicals still remained effective. However, further modification of their structure by Wei et al. resulted in dicyanomethylene-substituted benzothiazole squaraines [90]. Among four synthesized dyes, a squaraine derivative with two methyl butyrate sidechains named CSBE showed an excellent phototoxic effect in vitro against seven different cancer cell lines (PC-3, MCF-7, HCT-8, A549, A549T, K562, and LoVo). Negligible dark toxicity was also advantageous. CSBE effectiveness was assessed in xenograft studies, and irradiation following the injection of the dye induced tumor growth suppression. During a histological examination of the liver and kidney, no harm to healthy cells was detected.

Soumya et al. evaluated the potency of symmetrical diiodinated benzothiazolium squaraine (SQDI) dyes in vitro on Ehrlich's Ascites Carcinoma (EAC) cells [91]. The maximum absorbance of the dye was beyond the NIR of 535 nm. A low concentration of the dye (0.2 mg/mL) induced 100% cytotoxicity after irradiation. Moreover, the dye had no dark toxicity. An in vivo study on Swiss albino mice included the measurement of serum biochemical parameters such as SGPT, SGOT, LDH, CK and ALP after the administration of the dye through the intraperitoneal cavity. None of the parameters increased, meaning that the dye did not extend any toxicity to healthy organs.

5.2. Halogenation

Halogenated squaraine dyes were analyzed by Serpe et al. Their efficacy in PDT was tested in vitro on a human fibrosarcoma (HT-1080) tumor cell line [92]. Both brominated and iodinated squaraine dyes proved to induce a significant ROS generation in the first few minutes after irradiation. Despite high initial release of cytochrome c, a drastic reduction was observed 3 h after irradiation. Therefore, in this case necrosis was the main cell death type.

5.3. Aminosquaraine Dyes

A study by Lima et al. evaluated the potency of indolenine-based aminosquaraine cyanine dyes as photosensitizers on several cell lines: Caco-2, MCF-7, PC-3. Non-tumor cell lines (NHDF and N27) were controls [93]. Study revealed that, the zwitterionic dye showed high selectivity for the PC-3 cell line in comparison to the normal human cell line. Almost all aminosquaraine dyes, were specifically cytotoxic towards cancer cells and aggregated in mitochondria.

Magalhães et al. have been evaluated the efficacy of several modified zwitterionic dyes on different cancer cell lines (MCF-7, NCI-H460, HeLa, HepG2) and non-tumor porcine liver primary cell culture (PLP2) in vitro [94]. Modifications to the zwitterionic dyes were designed to increase cellular uptake by enhancing their cationic character, increase the red-shift of the dye's absorption maximum and boost hydrophilicity. All the dyes displayed high phototoxicity towards cancer cell lines, particularly HeLa and MCF-7 cell lines, which showed the highest susceptibility to aminosquaraines. Nevertheless, all dyes showed cytotoxicity against PLP2 cells and a relative inhibition of growth. The use of the aminosquaraine analogues of benzoselenazole was also effective in cancer therapy [95], namely, the inclusion of a heavy metal, selenium, enhanced free radical production while decreasing the dye's fluorescence emission [96]. The absorbance maximum of the modified dyes was in the 665–685 nm range, and its stability improved significantly. The derivatives were more phototoxic than the benzothiazole analogues, but their toxicity in the absence of light was generally higher as well.

To increase the redshift of unsymmetrical squaraine dyes, Lima et al. incorporated quinoline units into the core structure. Some of the new dyes displayed a wavelength of a maximum absorption at in the far-red spectrum (733 nm) [97]. Despite their limitations (i.e., aggregation in aqueous solution and low ROS synthesis) the new derivatives decreased their dark toxicity and presented a higher cellular uptake because of their cationic character. Although production of singlet oxygen was relatively weak, the dyes showed substantial phototherapeutic activity against breast cancer cell lines (MCF-7 and BT-474). These results were comparable to previously studied indolenine-based aminosquaraine dyes. Further modifications to them that would strengthen their singlet oxygen generation and may lead to their successful application in PDT.

6. Merocyanines

Merocyanines raised hope for low-invasive lymphoma [98], leukemia [99] and neuroblastoma [100] treatments as they exhibited high specificity for cancer cells. The compounds are currently undergoing preclinical studies as a treatment for leukemia [99]. The exceptional permeability of the MC540 dye to leukemic leukocytes and immature hemopoietic precursors led to its extensive analysis.

The drawbacks of the group involve maximum absorbance of light outside of the NIR spectrum (556 nm) and preferential peroxidation of phospholipids in the membrane [101].

The latter leads to the induction of necrosis and constraints on the utility of the drugs in vivo. Additionally, their use in PDT against melanoma on the Cloudman S91 cell line turned out to be less effective than the currently applied porphyrin dyes [102]. Nonetheless, the research on merocyanines as a potential hematological cancer treatment hasn't stopped. A rhodamine complex of merocyanine underwent an in vitro study on K562 leukemia cells and revealed a decrease of cancer cell viability [103]. Apart from cancer treatment, merocyanines are also evaluated as an antimicrobial therapy on *Staphylococcus aureus* [104].

Immunoregulatory Agent

It has been shown that merocyanines exhibit immunoregulatory properties. The compound group can regulate an immune response by inhibiting T-lymphocyte proliferation and B-cell differentiation. Also, T-cell helper activity can be stimulated [105]. Therefore, merocyanines reveal the potency against leukemia and lymphoma. The drugs may also may find application in graft-versus-host disease prophylaxis and treatment of several autoimmune diseases. Traul et al. evaluated the application of PDT with MHC540 in reducing GVHD in murine models of allogeneic hematopoietic stem cell transplantation [106]. Prior studies reported, that the sensitivity of cells to merocyanines is determined by the dye's binding to the targeted cells [107]. The binding affinity is reduced in mature lymphocytes and elevated in hematopoietic stem cells. Irradiated lymphocytes displayed no proliferative response after treatment with ConA (concanavalin A), LPS (lipopolysaccharide), PHA (phytohemagglutinin) and IL-2 (interleukin-2). Also the survival of MHC540 treated cells increased by 50–80% [106].

7. Phthalocyanines

Phthalocyanines are successful photosensitizers in the therapy of skin malignancies like basal cell carcinoma and diseases like psoriasis [108]. The compounds belong to the group of second-generation photosensitizers present some similarities to porphyrins. In contrast to the latter, they have absorption maxima in the range of 670–780 nm [109].

7.1. Incorporation of a Metal Atom

The substitution of the phthalocyanine central atom of with zinc (II), aluminum (III), gallium (III) or silicon results in increased cytotoxicity in biological studies of the drugs [110]. The advantages include high stability, fluorescence [111] and low dark toxicity [112]. Their weakness, however, is hydrophobicity, which leads to aggregation in aqueous media [109]. Zinc (II) phthalocyanine Pc13 induced apoptosis and necrosis triggered by free radical production in B16Fo melanoma cells [113]. The level of apoptosis regulators (Bcl-2, Bcl-xL and Bid) increased after irradiation with the drugs. In addition, permeability of the mitochondria towards the inner-derived ROS was detected. The necrotic pathway resulted from an increase in lactate dehydrogenase concentration in extracellular compartments [113].

To enhance water solubility of the zinc phthalocyanine, sulfonic [114], phosphoric [115] and carboxylic group [116] substituents were introduced. The exposure of cervical cancer cells (HeLa) to sulphated zinc (II) phthalocyanines and irradiation with 673 nm diode laser resulted in DNA fragmentation, membrane damage and effective cytotoxicity. This led to a 25% decrease in cell viability [117]. Aniogo et al. combined sulfonated zinc (II) phthalocyanine with doxorubicin and observed a synergistic cytotoxic effect on MCF-7 cell lines [118]. To improve cytotoxicity and penetration of phthalocyanine derivatives into cancer cells, the cyanines might be conjugated with chemotherapeutical drugs. For instance, Al (III) phthalocyanine chloride tetrasulfonic acid (AlPcS4) with different chemotherapeutic agents was studied on gastric cancer cells [119], and zinc phthalocyanine with doxorubicin acted against SK-MEL-3 melanoma cells [120].

7.2. Nanoemulsions

Because phthalocyanines are lipophilic, L.A. Muehlmann developed nanoemulsions that can transport aluminum-phthalocyanine chloride (AlPc) to cancer tissue. The approach prevented aggregation in aqueous media [121]. The nanoemulsions are mostly composed of castor oil, Cremophor ELP® and a monodisperse population of nanodroplets. Cell viability of mammary MCF-7 adenocarcinoma cells significantly decreased and the production of LDH excessively increased. Conversely, without such emulsions AlPc was ineffective. The complex dye aggregated mostly in the cytoplasm and outside the nucleus, thus inducing no damage to the DNA and preventing genome modifications. Subsequently, nanoemulsions of aluminum-phthalocyanine were studied in vivo in a PDT against 4T1 breast adenocarcinoma tumor [122]. The primary breast tumors were eradicated after application of PDT, and contrary to the untreated mice group, metastases to the lungs were not observed.

8. Conclusions

The absorbance maxima of cyanine dyes lie within the NIR-I spectrum and the light used to activate them penetrates deeper into the tissue. They exhibit significant fluorescent properties and thus are applied in photodynamic diagnostics. ICG is an FDA approved dye which has for decades been clinically used as a diagnostic agent. Heptamethine dye is exceptional for cancer cells selectivity. It is transported through organic anion-transporting polypeptides (OATPs) the production of which increases in cancer cells. Certain cyanine dyes (ICG and IRDye800) can emit light at an NIR-II wavelength of 1000–1700 nm. Such a characteristic could provide a better contrast-to-noise ratio (CNR) diagnostic image and higher specificity for tumors [81].

Cyanine dyes do have their weaknesses: poor water solubility and low ROS generation. Nonetheless, the ROS production can be increased by incorporating certain organic groups, heavy atoms, halogenation or metal atoms [55,56,60,72,73,92,96]. This review of the use of cyanines and their derivatives as potential photosensitizers indicates that they could be efficiently activated by light, causing the death of target cells [35,48,49,56,62,68,82,85,89,122]. Cyanine dyes provoke cell death primarily through apoptosis [2,17,69], which benefits the therapy because it prevents an excessive inflammatory response. To overcome the problem of poor water solubility the dyes can be further modified by, for example, incorporating organic [8,69] or other groups [114–116]. Specificity for cancer tissue can be improved by sensitizing cyanine dyes to pH and enhancing their phototoxicity in an acidic environment, which is characteristic for extracellular cancer fluid [39,78].

Moreover, merocyanines, are especially worth mentioning for their unique immunoregulatory properties, namely, the ability to interact with lymphocytes [105]. Currently, merocyanines are in preclinical studies focused on treating leukemia [99].

Current studies involving numerous cyanine dyes show that they may enrich the stock of the permanently used pool of agents in PDT. Considering the very high cost of introducing new agents, it is important to show persuasive evidence of their new and remarkable properties. Further studies to exploit the advantages of cyanine dyes for therapeutic possibilities in low-invasive cancer treatments include absorbance within the NIR, low dark toxicity and fluorescent properties, combined with the application of optical fibers.

Funding: The research was supported Statutory Subsidy Funds of the Department of Molecular and Cellular Biology no. SUB.D260.21.095 (PI: J.Saczko).

Institutional Review Board Statement: Not applicable.

Informed Consent Statement: Not applicable.

Acknowledgments: The authors are very grateful to Julita Kulbacka from Wroclaw Medical University for the scientific impact with this manuscript.

Conflicts of Interest: The authors declare no conflict of interest.

References

1. Van Straten, D.; Mashayekhi, V.; De Bruijn, H.S.; Oliveira, S.; Robinson, D.J. Oncologic Photodynamic Therapy: Basic Principles, Current Clinical Status and Future Directions. *Cancers* **2017**, *9*, 19. [CrossRef]
2. Li, L.; Chen, Y.; Chen, W.; Tan, Y.; Chen, H.; Yin, J. Photodynamic therapy based on organic small molecular fluorescent dyes. *Chin. Chem. Lett.* **2019**, *30*, 1689–1703. [CrossRef]
3. Kwiatkowski, S.; Knap, B.; Przystupski, D.; Saczko, J.; Kędzierska, E.; Knap-Czop, K.; Kotlińska, J.; Michel, O.; Kotowski, K.; Kulbacka, J. Photodynamic therapy—Mechanisms, photosensitizers and combinations. *Biomed. Pharmacother.* **2018**, *106*, 1098–1107. [CrossRef]
4. Ochsner, M. Photophysical and photobiological processes in the photodynamic therapy of tumours. *J. Photochem. Photobiol. B Biol.* **1997**, *39*, 1–18. [CrossRef]
5. Castano, A.P.; Demidova, T.N.; Hamblin, M.R. Mechanisms in photodynamic therapy: Part two—Cellular signaling, cell metabolism and modes of cell death. *Photodiagnosis Photodyn. Ther.* **2005**, *2*, 1–23. [CrossRef]
6. Kaneko, S.; Fujimoto, S.; Yamaguchi, H.; Yamauchi, T.; Yoshimoto, T.; Tokuda, K. Photodynamic Therapy of Malignant Gliomas. In *Progress in Neurological Surgery*; S. Karger AG: Basel, Switzerland, 2018; Volume 32, pp. 1–13.
7. Dolmans, D.E.; Fukumura, D.; Jain, R.K. Photodynamic therapy for cancer. *Nat. Rev. Cancer* **2003**, *3*, 380–387. [CrossRef] [PubMed]
8. Sancho, D.; Joffre, O.; Keller, A.M.; Rogers, N.C.; Martínez, D.; Hernanz-Falcón, P.; Rosewell, I.; Sousa, C.R.E. Identification of a dendritic cell receptor that couples sensing of necrosis to immunity. *Nature* **2009**, *458*, 899–903. [CrossRef] [PubMed]
9. Chi, J.; Ma, Q.; Shen, Z.; Ma, C.; Zhu, W.; Han, S.; Liang, Y.; Cao, J.; Sun, Y. Targeted nanocarriers based on iodinated-cyanine dyes as immunomodulators for synergistic phototherapy. *Nanoscale* **2020**, *12*, 11008–11025. [CrossRef]
10. Baskaran, R.; Lee, J.; Yang, S.-G. Clinical development of photodynamic agents and therapeutic applications. *Biomater. Res.* **2018**, *22*, 25. [CrossRef]
11. Avirah, R.R.; Jayaram, D.T.; Adarsh, N.; Ramaiah, D. Squaraine dyes in PDT: From basic design to in vivo demonstration. *Org. Biomol. Chem.* **2012**, *10*, 911–920. [CrossRef]
12. Agostinis, P.; Berg, K.; Cengel, K.A.; Foster, T.H.; Girotti, A.W.; Gollnick, S.O.; Hahn, S.M.; Hamblin, M.R.; Juzeniene, A.; Kessel, D.; et al. Photodynamic therapy of cancer: An update. *CA Cancer J. Clin.* **2011**, *61*, 250–281. [CrossRef]
13. Usama, S.M.; Thavornpradit, S.; Burgess, K. Optimized Heptamethine Cyanines for Photodynamic Therapy. *ACS Appl. Bio Mater.* **2018**, *1*, 1195–1205. [CrossRef]
14. Kessel, D.; Evans, C.L. Promotion of Proapoptotic Signals by Lysosomal Photodamage: Mechanistic Aspects and Influence of Autophagy. *Photochem. Photobiol.* **2016**, *92*, 620–623. [CrossRef] [PubMed]
15. Kessel, D. Apoptosis, Paraptosis and Autophagy: Death and Survival Pathways Associated with Photodynamic Therapy. *Photochem. Photobiol.* **2018**, *95*, 119–125. [CrossRef]
16. D'Arcy, M.S. Cell death: A review of the major forms of apoptosis, necrosis and autophagy. *Cell Biol. Int.* **2019**, *43*, 582–592. [CrossRef]
17. Zhao, X.; Yang, Y.; Yu, Y.; Guo, S.; Wang, W.; Zhu, S. A cyanine-derivative photosensitizer with enhanced photostability for mitochondria-targeted photodynamic therapy. *Chem. Commun.* **2019**, *55*, 13542–13545. [CrossRef]
18. James, N.S.; Cheruku, R.R.; Missert, J.R.; Sunar, U.; Pandey, R.K. Measurement of Cyanine Dye Photobleaching in Photosensitizer Cyanine Dye Conjugates Could Help in Optimizing Light Dosimetry for Improved Photodynamic Therapy of Cancer. *Molecules* **2018**, *23*, 1842. [CrossRef]
19. Huang, Z.; Xu, H.; Meyers, A.D.; Musani, A.I.; Wang, L.; Tagg, R.; Barqawi, A.B.; Chen, Y.K. Photodynamic Therapy for Treatment of Solid Tumors—Potential and Technical Challenges. *Technol. Cancer Res. Treat.* **2008**, *7*, 309–320. [CrossRef]
20. Kessel, D.J.J.R., Jr. Effects of Combined Lysosomal and Mitochondrial Photodamage in a Non-small-Cell Lung Cancer Cell Line: The Role of Paraptosis. *Photochem. Photobiol.* **2017**, *93*, 1502–1508. [CrossRef]
21. Castano, A.P.; Demidova, T.N.; Hamblin, M.R. Mechanisms in photodynamic therapy: Part one—Photosensitizers, photochemistry and cellular localization. *Photodiagnosis Photodyn. Ther.* **2004**, *1*, 279–293. [CrossRef]
22. Szlasa, W.; Szewczyk, A.; Drąg-Zalesińska, M.; Czapor-Irzabek, H.; Michel, O.; Kiełbik, A.; Cierluk, K.; Zalesińska, A.; Novickij, V.; Tarek, M.; et al. Mechanisms of curcumin-based photodynamic therapy and its effects in combination with electroporation: An in vitro and molecular dynamics study. *Bioelectrochemistry* **2021**, *140*, 107806. [CrossRef]
23. Kulbacka, J.; Pola, A.; Mosiadz, D.; Choromanska, A.; Nowak, P.; Kotulska, M.; Majkowski, M.; Hryniewicz-Jankowska, A.; Purzyc, L.; Saczko, J. Cyanines as efficient photosensitizers in photodynamic reaction: Photophysical properties and in vitro photodynamic activity. *Biochemistry* **2011**, *76*, 473–479. [CrossRef]
24. Delaey, E.; van Laar, F.; De Vos, D.; Kamuhabwa, A.; Jacobs, P.; de Witte, P. A comparative study of the photosensitizing characteristics of some cyanine dyes. *J. Photochem. Photobiol. B Biol.* **2000**, *55*, 27–36. [CrossRef]
25. Murakami, L.; Ferreira, L.; Santos, J.; da Silva, R.; Nomizo, A.; Kuz'Min, V.; Borissevitch, I. Photocytotoxicity of a cyanine dye with two chromophores toward melanoma and normal cells. *Biochim. Biophys. Acta Gen. Subj.* **2015**, *1850*, 1150–1157. [CrossRef]
26. Nödling, A.R.; Mills, E.M.; Li, X.; Cardella, D.; Sayers, E.J.; Wu, S.-H.; Jones, A.T.; Luk, L.Y.P.; Tsai, Y.-H. Cyanine dye mediated mitochondrial targeting enhances the anti-cancer activity of small-molecule cargoes. *Chem. Commun.* **2020**, *56*, 4672–4675. [CrossRef]

27. Babilas, P.; Shafirstein, G.; Baier, J.; Schacht, V.; Szeimies, R.-M.; Landthaler, M.; Bäumler, W.; Abels, C. Photothermolysis of blood vessels using indocyanine green and pulsed diode laser irradiation in the dorsal skinfold chamber model. *Lasers Surg. Med.* **2007**, *39*, 341–352. [CrossRef]

28. Shafirstein, G.; Bäumler, W.; Hennings, L.J.; Siegel, E.R.; Friedman, R.; Moreno, M.A.; Webber, J.; Jackson, C.; Griffin, R.J. Indocyanine green enhanced near-infrared laser treatment of murine mammary carcinoma. *Int. J. Cancer* **2012**, *130*, 1208–1215. [CrossRef]

29. Ben-Hur, E.; Heldman, E.; Crane, S.; Rosenthal, I. Release of clotting factors from photosensitized endothelial cells: A possible trigger for blood vessel occlusion by photodynamic therapy. *FEBS Lett.* **1988**, *236*, 105–108. [CrossRef]

30. Vaupel, M.P.; Thews, O.; Hoeckel, M. Treatment Resistance of Solid Tumors. *Med. Oncol.* **2001**, *18*, 243–260. [CrossRef]

31. Hu, D.; Pan, M.; Yu, Y.; Sun, A.; Shi, K.; Qu, Y.; Qian, Z. Application of nanotechnology for enhancing photodynamic therapy via ameliorating, neglecting, or exploiting tumor hypoxia. *View* **2020**, *1*. [CrossRef]

32. Bhuvaneswari, R.; Gan, Y.Y.; Soo, K.C.; Olivo, M. The effect of photodynamic therapy on tumor angiogenesis. *Cell. Mol. Life Sci.* **2009**, *66*, 2275–2283. [CrossRef] [PubMed]

33. Civantos, F.J.; Karakullukcu, B.; Biel, M.; Silver, C.E.; Rinaldo, A.; Saba, N.F.; Takes, R.P.; Poorten, V.V.; Ferlito, A. A Review of Photodynamic Therapy for Neoplasms of the Head and Neck. *Adv. Ther.* **2018**, *35*, 324–340. [CrossRef]

34. Hester, S.C.; Kuriakose, M.; Nguyen, C.D.; Mallidi, S. Role of Ultrasound and Photoacoustic Imaging in Photodynamic Therapy for Cancer. *Photochem. Photobiol.* **2020**, *96*, 260–279. [CrossRef]

35. Atchison, J.; Kamila, S.; Nesbitt, H.; Logan, K.A.; Nicholas, D.M.; Fowley, C.; Davis, J.; Callan, B.; McHale, A.P.; Callan, J.F. Iodinated cyanine dyes: A new class of sensitisers for use in NIR activated photodynamic therapy (PDT). *Chem. Commun.* **2017**, *53*, 2009–2012. [CrossRef]

36. Chang, S.-C.; Buonaccorsi, G.A.; MacRobert, A.J.; Brown, S.G. Interstitial photodynamic therapy in the canine prostate with disulfonated aluminum phthalocyanine and 5-aminolevulinic acid-induced protoporphyrin IX. *Prostate* **1997**, *32*, 89–98. [CrossRef]

37. Biel, M. Advances in photodynamic therapy for the treatment of head and neck cancers. *Lasers Surg. Med.* **2006**, *38*, 349–355. [CrossRef]

38. Lou, P.-J.; Jäger, H.R.; Jones, L.; Theodossy, T.; Bown, S.G.; Hopper, C. Interstitial photodynamic therapy as salvage treatment for recurrent head and neck cancer. *Br. J. Cancer* **2004**, *91*, 441–446. [CrossRef]

39. Zhang, J.; Liu, Z.; Lian, P.; Qian, J.; Li, X.; Wang, L.; Fu, W.; Chen, L.; Wei, X.; Li, C. Selective imaging and cancer cell death via pH switchable near-infrared fluorescence and photothermal effects. *Chem. Sci.* **2016**, *7*, 5995–6005. [CrossRef]

40. Kitai, T.; Inomoto, T.; Miwa, M.; Shikayama, T. Fluorescence navigation with indocyanine green for detecting sentinel lymph nodes in breast cancer. *Breast Cancer* **2005**, *12*, 211–215. [CrossRef]

41. Bäumler, W.; Abels, C.; Karrer, S.; Weiß, T.; Messmann, H.; Landthaler, M.; Szeimies, R.M. Photo-oxidative killing of human colonic cancer cells using indocyanine green and infrared light. *Br. J. Cancer* **1999**, *80*, 360–363. [CrossRef] [PubMed]

42. Hong, G.; Antaris, A.L.; Dai, H. Near-infrared fluorophores for biomedical imaging. *Nat. Biomed. Eng.* **2017**, *1*, 1–22. [CrossRef]

43. Kubicka-Trzaska, A.; Starzycka, M.; Romanowska-Dixon, B.; Morawski, K. Photodynamic therapy with indocyanine green for choroidal melanoma—A preliminary report. *Klin. Oczna.* **2003**, *105*, 132–135. Available online: https://europepmc.org/article/med/14552169 (accessed on 1 March 2021).

44. Onda, N.; Kimura, M.; Yoshida, T.; Shibutani, M. Preferential tumor cellular uptake and retention of indocyanine green forin vivotumor imaging. *Int. J. Cancer* **2016**, *139*, 673–682. [CrossRef] [PubMed]

45. Shirata, C.; Kaneko, J.; Inagaki, Y.; Kokudo, T.; Sato, M.; Kiritani, S.; Akamatsu, N.; Arita, J.; Sakamoto, Y.; Hasegawa, K.; et al. Near-infrared photothermal/photodynamic therapy with indocyanine green induces apoptosis of hepatocellular carcinoma cells through oxidative stress. *Sci. Rep.* **2017**, *7*, 1–8. [CrossRef]

46. Thavornpradit, S.; Usama, S.M.; Park, G.K.; Shrestha, J.P.; Nomura, S.; Baek, Y.; Choi, H.S.; Burgess, K. QuatCy: A Heptamethine Cyanine Modification with Improved Characteristics. *Theranostics* **2019**, *9*, 2856–2867. [CrossRef] [PubMed]

47. Englman, R.; Jortner, J. The energy gap law for radiationless transitions in large molecules. *Mol. Phys.* **1970**, *18*, 145–164. [CrossRef]

48. Tsuda, T.; Kaibori, M.; Hishikawa, H.; Nakatake, R.; Okumura, T.; Ozeki, E.; Hara, I.; Morimoto, Y.; Yoshii, K.; Kon, M. Near-infrared fluorescence imaging and photodynamic therapy with indocyanine green lactosome has antineoplastic effects for hepatocellular carcinoma. *PLoS ONE* **2017**, *12*, e0183527. [CrossRef]

49. Hishikawa, H.; Kaibori, M.; Tsuda, T.; Matsui, K.; Okumura, T.; Ozeki, E.; Yoshii, K.; Liljedahl, E.; Salford, L.G.; Redebrandt, H.N. Near-infrared fluorescence imaging and photodynamic therapy with indocyanine green lactosomes has antineoplastic effects for gallbladder cancer. *Oncotarget* **2019**, *10*, 5622–5631. [CrossRef] [PubMed]

50. Shi, C.; Wu, J.B.; Chu, G.C.-Y.; Li, Q.; Wang, R.; Zhang, C.; Zhang, Y.; Kim, H.L.; Wang, J.; Zhau, H.E.; et al. Heptamethine carbocyanine dye-mediated near-infrared imaging of canine and human cancers through the HIF-1α/OATPs signaling axis. *Oncotarget* **2014**, *5*, 10114–10126. [CrossRef]

51. Usama, S.M.; Park, G.K.; Nomura, S.; Baek, Y.; Choi, H.S.; Burgess, K. Role of Albumin in Accumulation and Persistence of Tumor-Seeking Cyanine Dyes. *Bioconjugate Chem.* **2020**, *31*, 248–259. [CrossRef]

52. Gorka, A.P.; Nani, R.R.; Zhu, J.; Mackem, S.; Schnermann, M.J. A Near-IR Uncaging Strategy Based on Cyanine Photochemistry. *J. Am. Chem. Soc.* **2014**, *136*, 14153–14159. [CrossRef]

53. Jiang, Z.; Pflug, K.; Usama, S.M.; Kuai, D.; Yan, X.; Sitcheran, R.; Burgess, K. Cyanine–Gemcitabine Conjugates as Targeted Theranostic Agents for Glioblastoma Tumor Cells. *J. Med. Chem.* **2019**, *62*, 9236–9245. [CrossRef]
54. Usama, S.M.; Lin, C.-M.; Burgess, K. On the Mechanisms of Uptake of Tumor-Seeking Cyanine Dyes. *Bioconjugate Chem.* **2018**, *29*, 3886–3895. [CrossRef]
55. Gorman, A.; Killoran, J.; O'Shea, C.; Kenna, T.; Gallagher, A.W.M.; O'Shea, D. In Vitro Demonstration of the Heavy-Atom Effect for Photodynamic Therapy. *J. Am. Chem. Soc.* **2004**, *126*, 10619–10631. [CrossRef] [PubMed]
56. Cao, J.; Chi, J.; Xia, J.; Zhang, Y.; Han, S.; Sun, Y. Iodinated Cyanine Dyes for Fast Near-Infrared-Guided Deep Tissue Synergistic Phototherapy. *ACS Appl. Mater. Interf.* **2019**, *11*, 25720–25729. [CrossRef]
57. Zhang, C.; Liu, T.; Su, Y.; Luo, S.; Zhu, Y.; Tan, X.; Fan, S.; Zhang, L.; Zhou, Y.; Cheng, T.; et al. A near-infrared fluorescent heptamethine indocyanine dye with preferential tumor accumulation for in vivo imaging. *Biomaterials* **2010**, *31*, 6612–6617. [CrossRef]
58. Zhang, E.; Luo, S.; Tan, X.; Shi, C. Mechanistic study of IR-780 dye as a potential tumor targeting and drug delivery agent. *Biomaterials* **2014**, *35*, 771–778. [CrossRef] [PubMed]
59. Wang, K.; Zhang, Y.; Wang, J.; Yuan, A.; Sun, M.; Wu, J.; Hu, Y. Self-assembled IR780-loaded transferrin nanoparticles as an imaging, targeting and PDT/PTT agent for cancer therapy. *Sci. Rep.* **2016**, *6*, 27421. [CrossRef]
60. Noh, I.; Lee, D.; Kim, H.; Jeong, C.; Lee, Y.; Ahn, J.; Hyun, H.; Park, J.; Kim, Y. Enhanced Photodynamic Cancer Treatment by Mitochondria-Targeting and Brominated Near-Infrared Fluorophores. *Adv. Sci.* **2018**, *5*, 1700481. [CrossRef]
61. Tan, X.; Luo, S.; Wang, D.; Su, Y.; Cheng, T.; Shi, C. A NIR heptamethine dye with intrinsic cancer targeting, imaging and photosensitizing properties. *Biomaterials* **2012**, *33*, 2230–2239. [CrossRef]
62. Yuan, J.; Yi, X.; Yan, F.; Wang, F.; Qin, W.; Wu, G.; Yang, X.; Shao, C.; Chung, L.W. Near-infrared fluorescence imaging of prostate cancer using heptamethine carbocyanine dyes. *Mol. Med. Rep.* **2014**, *11*, 821–828. [CrossRef]
63. Zhao, N.; Zhang, C.; Zhao, Y.; Bai, B.; An, J.; Zhang, H.; Wu, J.B.; Shi, C. Optical imaging of gastric cancer with near-infrared heptamethine carbocyanine fluorescence dyes. *Oncotarget* **2016**, *7*, 57277–57289. [CrossRef]
64. Yang, X.; Shao, C.; Wang, R.; Chu, C.-Y.; Hu, P.; Master, V.; Osunkoya, A.O.; Kim, H.L.; Zhau, H.E.; Chung, L.W.K. Optical Imaging of Kidney Cancer with Novel Near Infrared Heptamethine Carbocyanine Fluorescent Dyes. *J. Urol.* **2013**, *189*, 702–710. [CrossRef]
65. Zhang, C.; Zhao, Y.; Zhang, H.; Chen, X.; Zhao, N.; Tan, D.; Zhang, H.; Shi, C. The Application of Heptamethine Cyanine Dye DZ-1 and Indocyanine Green for Imaging and Targeting in Xenograft Models of Hepatocellular Carcinoma. *Int. J. Mol. Sci.* **2017**, *18*, 1332. [CrossRef]
66. Kaneko, J.; Inagaki, Y.; Ishizawa, T.; Gao, J.; Tang, W.; Aoki, T.; Sakamoto, Y.; Hasegawa, K.; Sugawara, Y.; Kokudo, N. Photodynamic therapy for human hepatoma-cell–line tumors utilizing biliary excretion properties of indocyanine green. *J. Gastroenterol.* **2014**, *49*, 110–116. [CrossRef]
67. Ishizawa, T.; Fukushima, N.; Shibahara, J.; Masuda, K.; Tamura, S.; Aoki, T.; Hasegawa, K.; Beck, Y.; Fukayama, M.; Kokudo, N. Real-time identification of liver cancers by using indocyanine green fluorescent imaging. *Cancer* **2009**, *115*, 2491–2504. [CrossRef]
68. Yang, X.; Bai, J.; Qian, Y. The investigation of unique water-soluble heptamethine cyanine dye for use as NIR photosensitizer in photodynamic therapy of cancer cells. *Spectrochim. Acta Part. A Mol. Biomol. Spectrosc.* **2020**, *228*, 117702. [CrossRef]
69. Jiao, L.; Song, F.; Cui, J.; Peng, X. A near-infrared heptamethine aminocyanine dye with a long-lived excited triplet state for photodynamic therapy. *Chem. Commun.* **2018**, *54*, 9198–9201. [CrossRef]
70. Josefsen, L.B.; Boyle, R.W. Photodynamic Therapy and the Development of Metal-Based Photosensitisers. *Met. Drugs* **2008**, *2008*, 1–23. [CrossRef]
71. Mangeolle, T.; Yakavets, I.; Marchal, S.; Debayle, M.; Pons, T.; Bezdetnaya, L.; Marchal, F. Fluorescent Nanoparticles for the Guided Surgery of Ovarian Peritoneal Carcinomatosis. *Nanomaterials* **2018**, *8*, 572. [CrossRef]
72. Kuoabc, W.-S.; Changc, Y.-T.; Chob, K.-C.; Chiub, K.-C.; Lienb, C.-H.; Yehc, C.-S.; Chenab, S.-J. Gold nanomaterials conjugated with indocyanine green for dual-modality photodynamic and photothermal therapy. *Biomaterials* **2012**, *33*, 3270–3278. [CrossRef]
73. Ghorbani, F.; Attaran-Kakhki, N.; Sazgarnia, A. The synergistic effect of photodynamic therapy and photothermal therapy in the presence of gold-gold sulfide nanoshells conjugated Indocyanine green on HeLa cells. *Photodiagnosis Photodyn. Ther.* **2017**, *17*, 48–55. [CrossRef]
74. Tan, X.; Wang, J.; Pang, X.; Liu, L.; Sun, Q.; You, Q.; Tan, F.; Li, N. Indocyanine Green-Loaded Silver Nanoparticle@Polyaniline Core/Shell Theranostic Nanocomposites for Photoacoustic/Near-Infrared Fluorescence Imaging-Guided and Single-Light-Triggered Photothermal and Photodynamic Therapy. *ACS Appl. Mater. Interfaces* **2016**, *8*, 34991–35003. [CrossRef]
75. Mitra, K.; Lyons, C.E.; Hartman, M.C.T. A Platinum(II) Complex of Heptamethine Cyanine for Photoenhanced Cytotoxicity and Cellular Imaging in Near-IR Light. *Angew. Chem. Int. Ed.* **2018**, *57*, 10263–10267. [CrossRef]
76. Bruhn, S.L.; Toney, J.H.; Lippard, S.J. Biological Processing of DNA Modified by Platinum Compounds. *Prog. Inorg. Chem.* **2007**, 477–516. [CrossRef]
77. Siriwibool, S.; Kaekratoke, N.; Chansaenpak, K.; Siwawannapong, K.; Panajapo, P.; Sagarik, K.; Noisa, P.; Lai, R.-Y.; Kamkaew, A. Near-Infrared Fluorescent pH Responsive Probe for Targeted Photodynamic Cancer Therapy. *Sci. Rep.* **2020**, *10*, 1–10. [CrossRef]
78. Meng, X.; Yang, Y.; Zhou, L.; Zhang, L.; Lv, Y.; Li, S.; Wu, Y.; Zheng, M.; Li, W.; Gao, G.; et al. Dual-Responsive Molecular Probe for Tumor Targeted Imaging and Photodynamic Therapy. *Theranostics* **2017**, *7*, 1781–1794. [CrossRef]

79. Antaris, A.L.; Chen, H.; Diao, S.; Ma, Z.; Zhang, Z.; Zhu, S.; Wang, J.; Lozano, A.X.; Fan, Q.; Chew, L.; et al. A high quantum yield molecule-protein complex fluorophore for near-infrared II imaging. *Nat. Commun.* **2017**, *8*, 15269. [CrossRef]

80. Zhu, S.; Hu, Z.; Tian, R.; Yung, B.C.; Yang, Q.; Zhao, S.; Kiesewetter, D.O.; Niu, G.; Sun, H.; Antaris, A.L.; et al. Repurposing Cyanine NIR-I Dyes Accelerates Clinical Translation of Near-Infrared-II (NIR-II) Bioimaging. *Adv. Mater.* **2018**, *30*, e1802546. [CrossRef]

81. Starosolski, Z.; Bhavane, R.; Ghaghada, K.B.; Vasudevan, S.A.; Kaay, A.; Annapragada, A. Indocyanine green fluorescence in second near-infrared (NIR-II) window. *PLoS ONE* **2017**, *12*, e0187563. [CrossRef]

82. Ge, X.; Fu, Q.; Su, L.; Li, Z.; Zhang, W.; Chen, T.; Yang, H.; Song, J. Light-activated gold nanorod vesicles with NIR-II fluorescence and photoacoustic imaging performances for cancer theranostics. *Theranostics* **2020**, *10*, 4809–4821. [CrossRef]

83. Owens, E.A.; Hyun, H.; Tawney, J.G.; Choi, H.S.; Henary, M. Correlating Molecular Character of NIR Imaging Agents with Tissue-Specific Uptake. *J. Med. Chem.* **2015**, *58*, 4348–4356. [CrossRef]

84. Ciubini, B.; Visentin, S.; Serpe, L.; Canaparo, R.; Fin, A.; Barbero, N. Design and synthesis of symmetrical pentamethine cyanine dyes as NIR photosensitizers for PDT. *Dye. Pigment.* **2019**, *160*, 806–813. [CrossRef]

85. Ahoulou, E.O.; Drinkard, K.K.; Basnet, K.; Lorenz, A.S.; Taratula, O.; Henary, M.; Grant, K.B. DNA Photocleavage in the Near-Infrared Wavelength Range by 2-Quinolinium Dicarbocyanine Dyes. *Molecules* **2020**, *25*, 2926. [CrossRef]

86. Kulbacka, J.; Choromańska, A.; Drag-Zalesińska, M.; Nowak, P.; Baczyńska, D.; Kotulska, M.; Bednarz-Misa, I.; Saczko, J.; Chwiłkowska, A. Proapoptotic activity induced by photodynamic reaction with novel cyanine dyes in caspase-3-deficient human breast adenocarcinoma cell lines (MCF/WT and MCF/DX). *Photodiagnosis Photodyn. Ther.* **2020**, *30*, 101775. [CrossRef]

87. Cande, C.; Vahsen, N.; Garrido, C.; Kroemer, G. Apoptosis-inducing factor (AIF): Caspase-independent after all. *Cell Death Differ.* **2004**, *11*, 591–595. [CrossRef] [PubMed]

88. Fernandes, T.C.; Lima, E.; Boto, R.E.; Ferreira, D.; Fernandes, J.R.; Almeida, P.; Ferreira, L.F.; Silva, A.M.; Reis, L.V. In vitro phototherapeutic effects of indolenine-based mono- and dithiosquaraine cyanine dyes against Caco-2 and HepG2 human cancer cell lines. *Photodiagnosis Photodyn. Ther.* **2020**, *31*, 101844. [CrossRef] [PubMed]

89. Martins, T.D.; Lima, E.; Boto, R.E.; Ferreira, D.; Fernandes, J.R.; Almeida, P.; Ferreira, L.F.V.; Silva, A.M.; Reis, L.V. Red and Near-Infrared Absorbing Dicyanomethylene Squaraine Cyanine Dyes: Photophysicochemical Properties and Anti-Tumor Photosensitizing Effects. *Materials* **2020**, *13*, 2083. [CrossRef] [PubMed]

90. Wei, Y.; Hu, X.; Shen, L.; Jin, B.; Liu, X.; Tan, W.; Shangguan, D. Dicyanomethylene Substituted Benzothiazole Squaraines: The Efficiency of Photodynamic Therapy In Vitro and In Vivo. *EBioMedicine* **2017**, *23*, 25–33. [CrossRef]

91. Soumya, M.; Shafeekh, K.; Das, S.; Abraham, A. Symmetrical diiodinated squaraine as an efficient photosensitizer for PDT applications: Evidence from photodynamic and toxicological aspects. *Chem. Interactions* **2014**, *222*, 44–49. [CrossRef]

92. Serpe, L.; Ellena, S.; Barbero, N.; Foglietta, F.; Prandini, F.; Gallo, M.P.; Levi, R.; Barolo, C.; Canaparo, R.; Visentin, S. Squaraines bearing halogenated moieties as anticancer photosensitizers: Synthesis, characterization and biological evaluation. *Eur. J. Med. Chem.* **2016**, *113*, 187–197. [CrossRef] [PubMed]

93. Lima, E.; Ferreira, O.; da Silva, J.F.; dos Santos, A.O.; Boto, R.E.; Fernandes, J.R.; Almeida, P.; Silvestre, S.M.; Reis, L.V. Photodynamic activity of indolenine-based aminosquaraine cyanine dyes: Synthesis and in vitro photobiological evaluation. *Dye. Pigment.* **2020**, *174*, 108024. [CrossRef]

94. Magalhães, Á.F.; Graça, V.C.; Calhelha, R.C.; Ferreira, I.C.; Santos, P.F. Aminosquaraines as potential photodynamic agents: Synthesis and evaluation of in vitro cytotoxicity. *Bioorg. Med. Chem. Lett.* **2017**, *27*, 4467–4470. [CrossRef]

95. Magalhães, Á.F.; Graça, V.C.; Calhelha, R.C.; Machado, I.F.; Ferreira, L.F.V.; Ferreira, I.C.F.R.; Santos, P.F. Synthesis, photochemical and in vitro cytotoxic evaluation of benzoselenazole-based aminosquaraines. *Photochem. Photobiol. Sci.* **2019**, *18*, 336–342. [CrossRef] [PubMed]

96. Ferreira, D.P.; Conceição, D.S.; Ferreira, V.R.A.; Graça, V.C.; Santos, P.F.; Ferreira, L.F.V. Photochemical properties of squarylium cyanine dyes. *Photochem. Photobiol. Sci.* **2013**, *12*, 1948. [CrossRef] [PubMed]

97. Lima, E.; Boto, R.E.; Ferreira, D.; Fernandes, J.R.; Almeida, P.; Ferreira, L.F.V.; Souto, E.B.; Silva, A.M.; Reis, L.V. Quinoline- and Benzoselenazole-Derived Unsymmetrical Squaraine Cyanine Dyes: Design, Synthesis, Photophysicochemical Features and Light-Triggerable Antiproliferative Effects against Breast Cancer Cell Lines. *Materials* **2020**, *13*, 2646. [CrossRef]

98. Itoh, T.; Messner, H.A.; Jamal, N.; Tweeddale, M.; Sieber, F. Merocyanine 540-sensitized photoinactivation of high-grade non-Hodgkin's lymphoma cells: Potential application in autologous BMT. *Bone Marrow Transplant.* **1993**, *12*, 191–196.

99. D'Alessandro, S.; Priefer, R. Non-porphyrin dyes used as photosensitizers in photodynamic therapy. *J. Drug Deliv. Sci. Technol.* **2020**, *60*, 101979. [CrossRef]

100. Sieber, F.; Sieber-Blum, M. Dye-mediated photosensitization of murine neuroblastoma cells. *Cancer Res.* **1986**, *46*, 2072–2076.

101. Zareba, M.; Niziolek, M.; Korytowski, W.; Girotti, A.W. Merocyanine 540-sensitized photokilling of leukemia cells: Role of post-irradiation chain peroxidation of plasma membrane lipids as revealed by nitric oxide protection. *Biochim. Biophys. Acta Gen. Subj.* **2005**, *1722*, 51–59. [CrossRef]

102. Nowak-Sliwinska, P.; Karocki, A.; Elas, M.; Pawlak, A.; Stochel, G.; Urbanska, K. Verteporfin, photofrin II, and merocyanine 540 as PDT photosensitizers against melanoma cells. *Biochem. Biophys. Res. Commun.* **2006**, *349*, 549–555. [CrossRef]

103. Xiang, J.-F.; Liu, Y.-X.; Sun, D.; Zhang, S.-J.; Fu, Y.-L.; Zhang, X.-H.; Wang, L.-Y. Synthesis, spectral properties of rhodanine complex merocyanine dyes as well as their effect on K562 leukemia cells. *Dye. Pigment.* **2012**, *93*, 1481–1487. [CrossRef]

104. Lin, H.-Y.; Chen, C.-T.; Huang, C.-T. Use of Merocyanine 540 for Photodynamic Inactivation of Staphylococcus aureus Planktonic and Biofilm Cells. *Appl. Environ. Microbiol.* **2004**, *70*, 6453–6458. [CrossRef]
105. Lum, L.; Yamagami, M.; Giddings, B.; Joshi, I.; Schober, S.; Sensenbrenner, L.; Sieber, F. The immunoregulatory effects of merocyanine 540 on in vitro human T- and B-lymphocyte functions. *Blood* **1991**, *77*, 2701–2706. [CrossRef]
106. Traul, D.L.; Sieber, F. Inhibitory effects of merocyanine 540-mediated photodynamic therapy on cellular immune functions: A role in the prophylaxis of graft-versus-host disease? *J. Photochem. Photobiol. B Biol.* **2015**, *153*, 153–163. [CrossRef]
107. Del Buono, B.J.; Williamson, P.L.; Schlegel, R.A. Alterations in plasma membrane lipid organization during lymphocyte differentiation. *J. Cell. Physiol.* **1986**, *126*, 379–388. [CrossRef] [PubMed]
108. Miller, J.D.; Baron, E.D.; Scull, H.; Hsia, A.; Berlin, J.C.; McCormick, T.; Colussi, V.; Kenney, M.E.; Cooper, K.D.; Oleinick, N.L. Photodynamic therapy with the phthalocyanine photosensitizer Pc 4: The case experience with preclinical mechanistic and early clinical–translational studies. *Toxicol. Appl. Pharmacol.* **2007**, *224*, 290–299. [CrossRef] [PubMed]
109. Jiang, Z.; Shao, J.; Yang, T.; Wang, J.; Jia, L. Pharmaceutical development, composition and quantitative analysis of phthalocyanine as the photosensitizer for cancer photodynamic therapy. *J. Pharm. Biomed. Anal.* **2014**, *87*, 98–104. [CrossRef]
110. Roguin, L.P.; Chiarante, N.; Vior, M.C.G.; Marino, J. Zinc(II) phthalocyanines as photosensitizers for antitumor photodynamic therapy. *Int. J. Biochem. Cell Biol.* **2019**, *114*, 105575. [CrossRef]
111. Tedesco, A.; Rotta, J.; Lunardi, C.N. Synthesis, Photophysical and Photochemical Aspects of Phthalocyanines for Photodynamic Therapy. *Curr. Org. Chem.* **2003**, *7*, 187–196. [CrossRef]
112. Baldea, I.; Ion, R.M.; Olteanu, D.E.; Nenu, I.; Tudor, D.; Filip, A.G. Photodynamic therapy of melanoma using new, synthetic porphyrins and phthalocyanines as photosensitisers—A comparative study. *Clujul Med.* **2015**, *88*, 175–180. [CrossRef]
113. Valli, F.; Vior, M.C.G.; Roguin, L.P.; Marino, J. Oxidative stress generated by irradiation of a zinc(II) phthalocyanine induces a dual apoptotic and necrotic response in melanoma cells. *Apoptosis* **2019**, *24*, 119–134. [CrossRef] [PubMed]
114. Li, L.; Zhao, J.-F.; Won, N.; Jin, H.; Kim, S.; Chen, J.-Y. Quantum dot-aluminum phthalocyanine conjugates perform photodynamic reactions to kill cancer cells via fluorescence resonance energy transfer. *Nanoscale Res. Lett.* **2012**, *7*, 386. [CrossRef]
115. Neagu, R.-M.; Ion, S.; Popescu, S.; Georgescu, M. Zinc Trisulphonated Phthalocyanine Used in Photodynamic Therapy of Dysplastic Oral Keratinocytes, Rev. Chim. Bucharest Orig. Ed. (n.d.). Available online: https://www.academia.edu/129935 29/Zinc_Trisulphonated_Phthalocyanine_Used_in_Photodynamic_Therapy_of_Dysplastic_Oral_Keratinocytes (accessed on 1 March 2021).
116. García-Iglesias, M.; Yum, J.-H.; Humphry-Baker, R.; Zakeeruddin, S.M.; Péchy, P.; Vazquez, P.; Palomares, E.; Gratzel, M.; Nazeeruddin, M.K.; Torres, T. Effect of anchoring groups in zinc phthalocyanine on the dye-sensitized solar cell performance and stability. *Chem. Sci.* **2011**, *2*, 1145–1150. [CrossRef]
117. Hodgkinson, N.; Kruger, C.A.; Mokwena, M.; Abrahamse, H. Cervical cancer cells (HeLa) response to photodynamic therapy using a zinc phthalocyanine photosensitizer. *J. Photochem. Photobiol. B Biol.* **2017**, *177*, 32–38. [CrossRef]
118. Aniogo, E.C.; George, B.P.A.; Abrahamse, H. In vitro combined effect of Doxorubicin and sulfonated zinc Phthalocyanine–mediated photodynamic therapy on MCF-7 breast cancer cells. *Tumor Biol.* **2017**, *39*, 1–9. [CrossRef] [PubMed]
119. Xin, J.; Wang, S.; Zhang, L.; Xin, B.; He, Y.; Wang, J.; Wang, S.; Shen, L.; Zhang, Z.; Yao, C. Comparison of the synergistic anticancer activity of AlPcS4 photodynamic therapy in combination with different low-dose chemotherapeutic agents on gastric cancer cells. *Oncol. Rep.* **2018**, *40*, 165–178. [CrossRef]
120. Doustvandi, M.A.; Mohammadnejad, F.; Mansoori, B.; Tajalli, H.; Mohammadi, A.; Mokhtarzadeh, A.; Baghbani, E.; Khaze, V.; Hajiasgharzadeh, K.; Moghaddam, M.M.; et al. Photodynamic therapy using zinc phthalocyanine with low dose of diode laser combined with doxorubicin is a synergistic combination therapy for human SK-MEL-3 melanoma cells. *Photodiagnosis Photodyn. Ther.* **2019**, *28*, 88–97. [CrossRef]
121. Muehlmann, L.A.; Rodrigues, M.C.; Longo, J.P.F.; Garcia, M.P.; Py-Daniel, K.R.; Veloso, A.B.; De Souza, P.E.N.; Da Silva, S.W.; Azevedo, R.B. Aluminium-phthalocyanine chloride nanoemulsions for anticancer photodynamic therapy: Development and in vitro activity against monolayers and spheroids of human mammary adenocarcinoma MCF-7 cells. *J. Nanobiotechnol.* **2015**, *13*, 1–11. [CrossRef] [PubMed]
122. Rodrigues, M.C.; Vieira, L.G.; Horst, F.H.; De Araújo, E.C.; Ganassin, R.; Merker, C.; Meyer, T.; Böttner, J.; Venus, T.; Longo, J.P.F.; et al. Photodynamic therapy mediated by aluminium-phthalocyanine nanoemulsion eliminates primary tumors and pulmonary metastases in a murine 4T1 breast adenocarcinoma model. *J. Photochem. Photobiol. B Biol.* **2020**, *204*, 111808. [CrossRef]

 *pharmaceutics*

Review

Latest Innovations and Nanotechnologies with Curcumin as a Nature-Inspired Photosensitizer Applied in the Photodynamic Therapy of Cancer

Laura Marinela Ailioaie [1], **Constantin Ailioaie** [1] and **Gerhard Litscher** [2,*]

[1] Department of Medical Physics, Alexandru Ioan Cuza University, 11 Carol I Boulevard, 700506 Iasi, Romania; lauraailioaie@yahoo.com (L.M.A.); laserail_mail@yahoo.com (C.A.)

[2] President of ISLA (International Society for Medical Laser Applications), Research Unit of Biomedical Engineering in Anesthesia and Intensive Care Medicine, Research Unit for Complementary and Integrative Laser Medicine, and Traditional Chinese Medicine (TCM) Research Center Graz, Medical University of Graz, Auenbruggerplatz 39, 8036 Graz, Austria

* Correspondence: gerhard.litscher@medunigraz.at; Tel.: +43-316-385-83907

Citation: Ailioaie, L.M.; Ailioaie, C.; Litscher, G. Latest Innovations and Nanotechnologies with Curcumin as a Nature-Inspired Photosensitizer Applied in the Photodynamic Therapy of Cancer. *Pharmaceutics* **2021**, *13*, 1562. https://doi.org/10.3390/pharmaceutics13101562

Academic Editor: Nejat Düzgüneş

Received: 29 August 2021
Accepted: 22 September 2021
Published: 26 September 2021

Publisher's Note: MDPI stays neutral with regard to jurisdictional claims in published maps and institutional affiliations.

Abstract: In the context of the high incidence of cancer worldwide, state-of-the-art photodynamic therapy (PDT) has entered as a usual protocol of attempting to eradicate cancer as a minimally invasive procedure, along with pharmacological resources and radiation therapy. The photosensitizer (PS) excited at certain wavelengths of the applied light source, in the presence of oxygen releases several free radicals and various oxidation products with high cytotoxic potential, which will lead to cell death in irradiated cancerous tissues. Current research focuses on the potential of natural products as a superior generation of photosensitizers, which through the latest nanotechnologies target tumors better, are less toxic to neighboring tissues, but at the same time, have improved light absorption for the more aggressive and widespread forms of cancer. Curcumin incorporated into nanotechnologies has a higher intracellular absorption, a higher targeting rate, increased toxicity to tumor cells, accelerates the activity of caspases and DNA cleavage, decreases the mitochondrial activity of cancer cells, decreases their viability and proliferation, decreases angiogenesis, and finally induces apoptosis. It reduces the size of the primary tumor, reverses multidrug resistance in chemotherapy and decreases resistance to radiation therapy in neoplasms. Current research has shown that the use of PDT and nanoformulations of curcumin has a modulating effect on ROS generation, so light or laser irradiation will lead to excessive ROS growth, while nanocurcumin will reduce the activation of ROS-producing enzymes or will determine the quick removal of ROS, seemingly opposite but synergistic phenomena by inducing neoplasm apoptosis, but at the same time, accelerating the repair of nearby tissue. The latest curcumin nanoformulations have a huge potential to optimize PDT, to overcome major side effects, resistance to chemotherapy, relapses and metastases. All the studies reviewed and presented revealed great potential for the applicability of nanoformulations of curcumin and PDT in cancer therapy.

Keywords: light; malignant tumors; nanomedicine; natural photosensitizer; photobiomodulation (PBM); photodynamic therapy (PDT); cancer; curcumin

1. Introduction

Light as a treatment dates back to ancient times, but modern photodynamic therapy (PDT) has advanced following the accidental rediscovery in the early twentieth century of the light-mediated killing effect in the presence of molecular oxygen on acridine-incubated *Paramecium caudatum* [1].

Exactly in the same direction, concerning the use of light in medicine, the first Nobel Prize for outstanding applications of phototherapy was won in 1903 by Niels Finsen for smallpox and skin tuberculosis treatments [2,3].

Herman von Tappeiner defined "photodynamic action" [4] and gave it the name known today as photodynamic therapy, a contemporary and non-invasive form of anti-cancer therapy, well studied in present also for infections, non-oncological disorders, and in some countries, as a standardized protocol, along with radiotherapy and chemotherapy in anti-cancer treatments [5,6].

The first goal of this review was to present and discuss the latest nanotechnologies in relation to curcumin as photosensitizer (PS) to optimize photodynamic therapy (PDT), to overcome major side effects, resistance to chemotherapy, relapses and metastases.

The second objective was to reveal the state of the art of photodynamic therapy as a field of continuous effervescent research and medical applications in direct connection with nanoformulations of curcumin.

The third aim was to analyze the molecular and cellular mechanisms, targeted delivery and localization of nanoparticles in the tumor, as well as the interconnected parts related to PDT, to increase the antiproliferative and apoptotic activity and minimize toxicity in nearby healthy cells.

The fourth goal was to push forward the field of study of natural photosensitizers, especially curcumin- and PDT-related nanotechnologies, providing an up-to-date high-quality evidence base for researchers and scientists to accelerate and rethink new experiments and innovations in cancer.

2. Photosensitizers and Photodynamic Therapy

Photosensitizers (PSs) are molecules unchanged before and after energy exchange, that can absorb electromagnetic radiation from infra-red, visible and UV range and trigger the physicochemical change of a neighboring molecule by yielding an electron to the substrate or by extracting an atom from it, and finally, the PS returns to its ground state, where it remains unchanged until it absorbs radiation again [6,7].

When incident photons are absorbed, PS advances an electron in a single excited state, which, through an intrinsic spin rotation, can pass into an excited triplet state with a longer lifetime, thus increasing the probability of PS interaction with next-door molecules, and finally, conducting to the selective death of diseased cells through the generation of cytotoxic oxidation species (Figure 1). Depending on the internal structure of the PS, they have different efficiencies when interacting with diverse wavelengths [6,8].

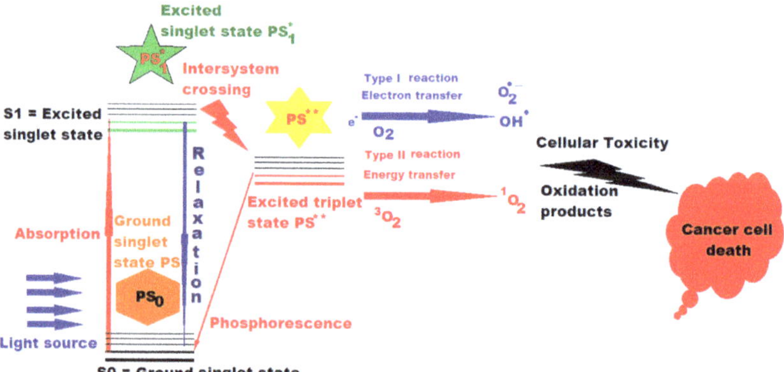

Figure 1. Action of PDT and photosensitizer (PS) in cancer therapy. A single asterisk "*" means "excited singlet state", and two asterisks "**" mean "excited triplet state".

PS had an inherent basis in nature through the green pigment chlorophyll, existing in all green plants and in cyanobacteria, absorbing light to deliver the photosynthesis' energy, as well as other light-sensitive molecules in the plant kingdom.

PDT implies the selective sensitization of tissues to light and the first studies on PS started at the beginning of the last century, when researchers noticed this phenomenon and undertook investigation of it in malignant tumors [9].

In the middle of the last century, Figge et al. showed that exogenous porphyrins accumulated selectively in murine tumors [10] and the results were afterwards applied to cancer patients by first injecting raw hematoporphyrins [11], followed by improvement using a hematoporphyrin "derivative" which highlighted the enhanced selective fluorescence of the neoplasms [12,13].

However, we could consider that the contemporary version of PDT was born in the 1970s in the United States due to the activity of Dr. T.J. Dougherty and collaborators, who later used the more refined version of the "hematoporphyrin derivative", called Photofrin, the most implemented PS worldwide, even nowadays, despite the many difficulties (long-term photosensitivity of the skin of cancer patients treated, reduced absorption in large tumors, due to the limited penetration of photons etc.) [14].

In a paper published 25 years later by Thomas J. Dougherty et al., PDT is well defined as involving the management of a PS, which may also need a metabolic combination (a pro-drug), followed by exposure to light with a certain wavelength for maximum absorption. The effect is an irreversible succession of photochemical and photobiological reactions with the permanent photodeterioration of malignant cells. Preclinical and clinical research has established PDT as a useful care procedure in early or advanced stages for lung, digestive, genitourinary cancer, etc. [15].

The merit of Thomas Dougherty's pre-clinical and clinical trials and efforts, led to FDA approval of PDT in modern clinical practice, and paved the way for current and future advances in PDT, coupled nowadays with nanotechnologies and subsequent drug discoveries [16].

Without Dougherty's endeavor, it is questionable that PDT would not have remained just "a minor biomedical curiosity" [17].

PDT is a therapy that comprises light, a chemical compound that makes cells abnormally sensitive or reactive to light, and in the presence of tissue oxygen induces cell death by generated reactive oxygen species (ROS), i.e., through phototoxicity. Today, this technology is extensively used for certain conditions, incorporating the latest applications in antiviral treatments and cancers that tend to metastasize. For example, in practically applied PDT, the light with specific wavelengths is usually guided by optical fibers to the patient's tumor, which has been given a photoactive drug, which will attach intensely to abnormal cells. The light will stimulate the substance used for treatment, will generate cytotoxic species in the presence of oxygen, and thus, the malignant cells could be destroyed, procedure considered to be invasively negligible and, to a lesser extent, toxic. A disadvantage is the long-term photosensitization, unpleasant and annoying for patients, but counterbalanced by the diminished necessity for fine operations, shortening recovery time and reducing cicatrices or deformities to the smallest possible size [6,18].

PDT practice implies: the PS, the light irradiation system (L) and the molecular oxygen of the tissue (O_2). Wavelength of L must be quantum suited to move PS into action to generate free radicals (type I reaction), as a result of electronic extraction or relocation to an underlying molecule and/or reactive oxygen species, especially singlet oxygen (type II reaction), an extremely reactive species. PDT is composed of several consecutive phases. First and foremost, PS is applied without light, either systemically (intravenous administration), or by topical application. Thereafter, when enough PS is fused into the unhealthy cells to be destroyed, the PS is stimulated by setting the light for a well-defined time interval. The applied dose of electromagnetic radiation provides enough energy to advance PS to higher excitation energy states, but not sufficient to deteriorate the adjacent healthy cells. In the next step of relaxation are generated reactive oxygen species that will put to death the targeted tissue. Unlike other molecules that are normally in a singlet state, molecular oxygen in the atmosphere and in living cells exists in a triplet state, but because quantum physics interdicts reactions between triplets and singlets, this postulates molecular oxygen as inactive in normal states. However, the PS used in medical applications can undergo a process of intersection with oxygen during excitation, a process during which, from an excited singlet state, it will pass to an excited triplet state at the point where the two potential energy curves crosses, as presented in Figure 1 (intersystem crossing), giving rise also to phosphorescence, and consequently, the extremely cytotoxic singlet oxygen will be generated, which will attack any organic substance with which it comes into contact, being meanwhile eliminated very quickly in less than 3 microseconds from the illuminated cells [7,19,20].

When excited in type II reactions, PS provides its excess energy during the process of interacting with molecular triplet oxygen ($^3O_2$) and produces singlet oxygen ($^1O_2$), a highly reactive species that reacts with the substrate to generate other oxidized products that will attack cellular constituents, leading to the targeted killing of cells in the supplied light field.

PSs contain chromophores, i.e., they are photosensitizers. A part of a molecule responsible for its color, is that part of the molecule in which the energy gap between two different molecular orbitals is in the visible area of the spectrum and, when it meets light and absorbs a photon, an electron from its ground state will pass into an excited state, and a conformational change of the molecule will occur. Once stimulated, the photosensitizer passes from the baseline S0 (ground state) into the short-lived excited state, with lots of vibrational sub-levels, it can decrease its energy by rapidly dropping these sub-levels via inner adjustment to populate the first excited singlet state S1, before it quickly relaxes back to S0 (see Figure 1). The transition from S1 to S0 is via fluorescence, with very short lifespans (10^{-9}–10^{-6} s). From S1, it can pass through spin switch via intersystem crossing and can occupy the first excited triplet state, and after that to go downhill to baseline via phosphorescence, with a much longer lifespan (10^{-3} − 1 s); enough to enable the PS in excited triplet state to act on circumambient bio-compounds via type I and type II reactions [7,21].

Singlet oxygen produces effects at 10–55 nm from its origin in about 10–320 ns and can diffuse up to approximately 300 nm in vivo [7,21,22]. All time sequences involved in PS photoactivation and in type I and II reactions play a key role in disrupting the cellular machine, but type II reactions are deemed to be most efficient for cell impair, leading to the ultimate goal of killing unhealthy photo irradiated cells. However, in real practice, PSs with a triplet state life of less than 20 ns could still prove to be productive photodynamic agents [21].

There are a lot of photosensitizers for PDT, classified into porphyrins, chlorins, dyes (e.g., phenothiazinium salts, rose bengal, squaraines), and recently, nature-inspired PSs.

In radiotherapy, the cellular DNA is the target, but almost all PSs will deteriorate other distinct subcellular entities, which makes the distinction between various PSs applied in PDT.

The perfect PS should accumulate, especially in unhealthy cells, and when light is applied, it should produce toxic species that will destroy the target tissue. Characteristics of an ideal PS are the following: high absorption with maximum absorption coefficient in longer wavelengths, close to the red/infrared spectrum with deeper penetration, to make it possible to deal with grand tumors; better adhesion and fixation in unhealthy tissues than in normal ones; reliable and easy dissoluble in biological systems, permitting intravenous management and fast elimination from the organism after PDT, without residual sensitivity of the skin to light or long-term photosensitization; very good chemical balance and insignificant toxic effect on cells in the absence of light or in low darkness; long triplet lifespan and high triplet state effect, as well as high efficiency for triplet formation and singlet oxygen generation; reduced photobleaching and genuine fluorescence, low production price [6,23–26].

During development, there were several generations of PS, such as:

- the first generation (HpD and Photofrin) had poor absorption in the red visible range, limited applications and an unpleasant side effect, the residual sensitivity of the skin;

- the second generation (see some examples depicted in Figure 2) allowed a much more accelerated development of PDT (5-Aminolaevulinic acid (ALA); Benzoporphyrin derivative monoacid ring A (BPD-MA) or Verteporfin; Chlorins sold as Purlytin; Tetra(*m*-hydroxyphenyl)chlorin (*m*THPC) or Foscan; Lutetium texaphyrin with tradename Lutex or Lutrin; 9-Acetoxy-2,7,12,17-tetrakis-(β-methoxyethyl)-porphycene or ATMPn; Zinc phthalocyanine (CGP55847); Naphthalocyanines (NCs) and Porphyrin-type Chromophores (PC) with modified marginal operation by different functional groups, such as nitrophenyl, aminophenyl, hydroxyphenyl, pyridiniumyl derivatives etc. [7,27–29].

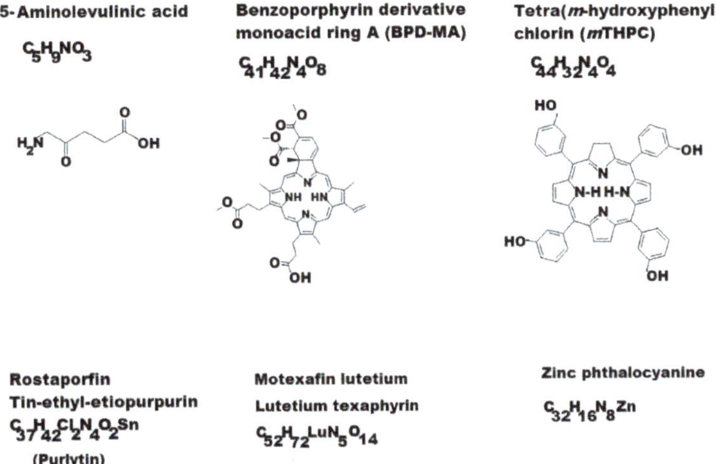

Figure 2. Some examples of second-generation photosensitizers (PSs).

Third generation PSs include antibody-directed photosensitizers, such as, for example, the monoclonal antibodies for rising specificity and enhanced capacity of action, better pharmacokinetics, improved function, selective targeted and delivery etc., and support the hope for a future better featured PDT [28,30].

The use of natural compounds has been approached as a new trend in photodynamic therapy, and the natural photosensitizers investigated coupled with the latest nanotechnologies have been proven to be effective in the clinical practice, as is the case with curcumin.

An illustrative diagram of photodynamic therapy using curcumin as a photosensitizer is shown in Figure 3.

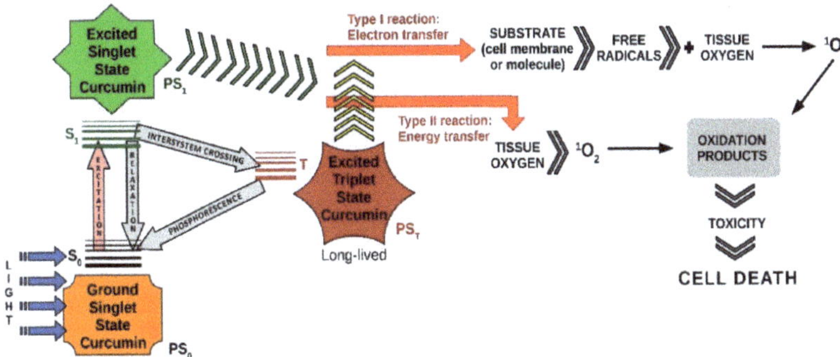

Figure 3. PDT with blue light and curcumin.

The use of nanotechnologies in recent years has brought significant improvements in the pharmaceutical industry through the discovery of nanoparticles, extraordinary innovations have been made for the production and delivery of the active principles of new drugs. Recent studies have shown another way to use curcumin to increase its bioavailability, plasma concentration and ability to penetrate and concentrate inside cells.

There are already many techniques for nano formulations, biomaterials and types of nanoparticles suitable for loading curcumin, to increase its therapeutic efficacy and to avoid its possible side effects, as follows:

2.1. Curcumin-Loaded Liposomes (Lipo-Cur)

Liposomes are spherical vesicles composed of simple or multiple layers that surround aqueous units and have become ideal delivery systems for biologically active substances, because they offer high biocompatibility and biodegradability, high solubility, stability and flexibility, low toxicity, preparation-controlled distribution with specific cell targeting. PEGylated nanoliposomes (attached strands of polyethylene glycol (PEG) to nanoliposomes) were the first nanoparticles approved for nano-drugs by the FDA.

Curcumin is a bioactive agent isolated from *Curcuma longa* rhizomes that, in the last three decades, has reached the peak of research for its biological functions; anti-inflammatory, antioxidant, antimicrobial, antiviral, antimutagenic, antitumor and anti-angiogenic. Liposomes solubilize curcumin and allow its distribution on the aqueous medium and increase its effect for the treatment of various cancers and other diseases.

It has been shown experimentally that blue light stops the multiplication of cancer cells, induces cell death by activating caspase and increasing intracellular reactive oxygen species. A recent study investigated the in vitro effects of curcumin-loaded liposomes at concentrations of 0–100 μmol/L in combination with PDT at 457 nm (blue) and the fluence of 220.2 W/m^2 on three papilloma virus-associated cell lines, and proved that after 24 h, the blue light activation of curcumin-loaded liposomes led to "a significant reduction in colony formation and migratory abilities, as well as to an increase in tumor cell death" [31].

Curcumin-loaded liposomes (Lipo-cur) and blue light distributed by PDT can achieve excellent bioactivity and strong anticancer activity [31–33].

2.2. Cur-Loaded Polymeric Micelles

Polymeric micelles have been used as nanodrugs for over 40 years; they are structured on a hydrophilic shell, and hydrophobic core that can be loaded with hydrophobic drugs [34].

Chang et al. studied how Cur-loaded polymeric micelles influenced endocytosis and exocytosis in colon carcinoma cells and proved that curcumin-loaded micelles had an increased stability compared to the unloaded micelles and were rapidly incorporated by the cells within minutes, becoming cytotoxic after 72 h of exposure.

In this experiment, the micelles applied to encapsulate curcumin had a diameter of about 16–46 nm, and when loading curcumin, their size was approximately doubled, with a loading efficiency of 58%. Results showed the importance of micelle size and loading on drug delivery and cytotoxicity [35].

A wide variety of amphiphilic polymers (diblock, triblock, grafted copolymers, etc.), hydrophobic materials and vitamin E, have been produced for the preparation of mycelium, but the most used are polyethylene glycol (PEG), polyvinylpyrrolidone (PVP) and chitosan [36,37].

Liu et al. have developed curcumin-loaded polymeric micelles to surmount the reduced ability of curcumin to be dissolved in water and to achieve better intravenous administration. Cur-loaded polymeric micelles had a very strong antitumor effect, inhibiting tumor growth and spontaneous lung metastases in a model of breast tumor in mice, proving to be a very good option for breast cancer [37].

In order to easily penetrate the cells and increase the therapeutic intake, the micellar surface was modified by adding ligands (epidermal growth factor receptor, folate, etc.) that more quickly recognize the receptors expressed on the surface by cancer cells [38].

2.3. Cur-Loaded Polymeric NPs

Nanoparticles are designed at the atomic or molecular level, have a diameter 1000 times smaller than a medium-sized cell in our body and are very valuable for the administration of drugs, because their physical, chemical and biological properties are unique. Curcumin can be encapsulated in a wide variety of nanoparticles based on polymers, solid, magnetic, gold and albumin-based lipids, which increase its solubility, pharmacokinetics, controlled release capacity and exact cell strike [39–42].

2.4. Cur-Loaded Mesoporous Silica NPs

Nanotechnology has overcome conventional concepts and ideas in the pharmaceutical industry. More than 30 years ago, Mobil Corporation first produced mesoporous silica nanoparticles (MSNs). The mesoporous structure is unique, provides chemical stability and high drug loading capacity, biocompatibility, large pore volume, large surface area, controlled release at the target, and low toxicity. In cancer, there are large disorders of the normal structures of the lymphatic and vascular system, so that MSN nanoparticles can more easily penetrate cancer cells through the process of phagocytosis and pinocytosis. The nanoencapsulation of curcumin with silica and chitosan increases the stability of curcumin and increases its cytotoxic activity on cell carcinoma cells [43,44].

2.5. Cur-Loaded Protein-Based NPs

Protein-based nanoparticles are widely used as natural biomaterials in the biomedical field because they meet the special qualities of biocompatibility, biodegradability and non-immunogenicity. Albumin from chicken serum or eggs was the most studied protein for obtaining drug-releasing nanostructures. Recently, many types of Cur-loaded protein nanoparticles have emerged, such as bovine serum albumin (BSA), human serum albumin (HSA), ovalbumin (OVA), zein, casein and curcumin–silk fibroin nanoparticles (CM–SF NPs) [32,45,46].

2.6. Cur-Loaded Solid Lipid NPs

Solid lipid nanoparticles (SLNs) are colloidal systems made of biodegradable solid lipids used in the pharmaceutical industry, because they perform through physical stability, high biocompatibility and the controlled release of embedded drugs. Cur-loaded solid lipid NPs were investigated in vitro and in vivo, demonstrating good stability and bioavailability, increased absorption in cells, and high anti-cancer efficacy [47–49].

2.7. Cur-Loaded CDs NPs

Cyclodextrins (CDs) comprise a family of cyclic glucose oligomers, which have excellent biocompatibility properties, complexation with lipophilic structures, very low toxicity, non-immunogenicity and targeted drug release. Cur-loaded CDs NPs have been shown to be highly effective in antioxidant activity, to reduce the oxidative stress associated with various cancer and a preventive role for nosocomial infections [50–52].

2.8. Cur-Loaded Nanogels

The proposed nanogel formulation of curcumin has several advantages over other delivery systems, because it ensures a higher concentration of the drug at the target sites and can reduce the exposure of curcumin to serum proteins and biological degradation after systemic administration. Multifunctional hybrid nanogels, through their ability to emit fluorescence, can be used for imaging, cell monitoring and, because they have high absorption in the near infrared range, are useful in photothermal conversion, and loaded with curcumin, are aimed at suppressing drug-resistant tumors [53–55].

2.9. Cur-Loaded Nanocrystals

Nanocrystals, although small particles have a large surface area for loading the hydrophobic drug, and due to their high solubility and saturation capacity, they increase bioavailability and biodistribution.

For example, in a recent study, Wang et al. have prepared curcumin nanocrystals (CNs) with a mean diameter of 15 nm by the quick emulsion freeze-drying method. CNs proved to be effective in drug delivery to selectively deliver Curcumin to cancer cells. CNs were prepared via the oil-in-water emulsion freeze-dried method: firstly, they prepared the emulsion, the oil phase (O) was dichloromethane solution containing 40 mg/mL curcumin, and the water phase (W) was containing 1 wt% Pluronic®F-127-COOH aqueous solution, at a volume ratio O/W fixed at 1:20, and the blend was mixed through ultrasonic (100 w, 10 min) emulsification to obtain the emulsion (O/W). Secondly, the above emulsion was further added into 20-fold the volume of the water phase, and the mixture was frozen by liquid nitrogen, followed by freeze-drying using a vacuum freeze-dryer. Finally, CNs suspension was obtained when lyophilized products were dispersed into the water, and free surfactants were removed by ultracentrifugation [56].

Curcumin nanocrystals (CNs), by the ability to avoid absorption in the reticuloendothelial system, prolong the circulation time of curcumin, increase the capacity of permeation and retention, accumulating in large quantities in the tumor [56,57].

2.10. Cur-Metal Oxide NPs

Inorganic nanomaterials used as metal nanoparticles (carbon, nanotubes, graphene, minerals and metal oxides) that carry drugs would be more advantageous than organic ones, because it possesses bioavailability, tolerance towards most organic solvents and better stability, increased surface area and porosity, better loading and dispensing capacity of drugs, with lower toxic effects [58].

Nanoconjugated zinc oxide nanoparticles (ZnONP$_{CS}$) with curcumin (ZnONP$_{CS}$-Cur) showed higher cytotoxicity in several cancer cell lines (breast, cervix, osteosarcoma and myeloma) and had anticancer and anti-inflammatory activity in vitro, and therefore, they could be used as possible therapeutic nanoconjugates for future cancer treatments [59].

3. Curcumin and Latest Cancer Applications

On 14 December 2020, International Agency for Research Cancer (IARC) released Globocan 2020, which published the latest data on the incidence of cancer, which rose to 19.3 million new cases and 10 million cancer deaths in 2020. According to Globocan 2020, which is a statistical database for IARC on incidence and mortality in 185 countries for 36 types of cancers, it has been estimated that cancer had 19.3 million new cases per year, of which breast cancer was in first place, with about 11.7% new cases, followed by lung cancer 11.4%, colorectal 10%, prostate 7.3%, and followed by stomach cancer at 5.6% [60,61].

The field of cancer research is a dynamic and evolving domain, with a multitude of models and scenarios proposed for examination over the decades on the hidden cause behind tumors development (genetic mutations, microorganisms, metabolic changes and so on), with fluctuating evidence or achievements, whose major goal remains the discovery of new methods and drugs for stopping disease progression and even eradicative therapeutics.

On this line, the cancer stem cell (CSC) model encompasses unusual immortal cells, such as those existing in tumors or blood cancers, similar to regular stem cells, but capable of generating the full range of cells from a given cancer specimen, which, through self-renewal and differentiation into multiple types of cancer cells, are tumorigenic, i.e., they generate recurrences and metastases, and so, supplementary malignancies [62,63].

In cancers that pursue the CSC model, some intracellular pathways may be attacked with natural compounds, such as curcumin or drugs, to overcome the danger of the development of new tumors at a distance. Promoting appropriate CCS-oriented treatments could improve the survival and quality of the life of patients with metastases [25,63–66].

Very recently, research based on this model shows the effectiveness of curcumin in various forms of cancer [67].

Even though it has reduced bioavailability, being insoluble in water, curcumin has been intensively studied as an authentic polyphenol and practically the main constituent of *Curcuma longa*, for its multiple beneficial effects in the treatment of various inflammatory, auto-immune, degenerative diseases, etc., and going to important applications in cancer, not only for the protective effect, but especially by destroying malignant cells. To overcome this drawback, various water-soluble mixtures have been imagined, such as liposomes or the incorporation of curcumin into micelles at the nanometer scale, with raised assimilations appropriate for cancer studies. The first formulations of curcumin in organic solvents proved to be toxic to living cells, and even with genotoxic capabilities. No investigation with curcumin embedded in the micelles has been planned until not long ago. In a recent experiment, Beltzig et al. comparatively investigated the cytotoxic and genotoxic action of genuine curcumin dissolved in ethanol (Cur-E), or integrated into micelles (Cur-M), and evaluated cell killing, apoptosis, necrosis, senolysis and genotoxicity, on a multitude of elementary and settled cell lines, proving that both formulations reduced viability for all cells in the same dose interval. Cur-E and Cur-M induced apoptosis as a function of dose, without senolytic action. Genotoxic repercussions disappeared in the absence of curcumin, denoting a prompt and full repair of DNA. In every experiment, Cur-E and Cur-M were, to the same extent, dynamic, and had important cytotoxic and genotoxic action, starting with 10 µM. Micelles without curcumin content were fully inoperative. The results proved similar in terms of cytotoxicity and genotoxicity for micellar curcumin as the native one, so the administration of micellar curcumin as a dietary supplement is safe and paves the way for new applications [68].

Major goals of pharmaceutic investigations are the innovative transport/delivery systems of drugs in cancer treatments. Zarrabi et al. have researched the manufacture of a new intelligent biocompatible stealth-nanoliposome to supply curcumin in cancer therapies. Four distinct classes of liposomes (plus or minus pH-sensitive polymeric film) were obtained by the Mozafari process, and then investigated by multiple trials. The embarkation and deliverance of curcumin were assessed at two different pH values, 7.4 and 6.6, but also the cytotoxicity of the specimens. The optimal average size for the smart stealth-liposome was 40 nm, and the efficacy of the drug's catch was about 84%, comparatively with 50 nm and only 74% performance by uncovered liposomes. Nano-carrier discharge from the stealth-liposome was better directed than in the uncovered. Experiments have shown the toxicity of drug's nanocarriers on malignancies. We could conclude that soon, the pH-sensitive intelligent stealth nanoliposome may become a true aspirant in cancer treatments [69].

Resistance to medicine and bad outcome in some cancer cases is often due to the hyperactivation of NRF2, a group of transcription factors, i.e., the nuclear factor erythroid 2 p45, detected in some tumors. It was demonstrated that curcumin can induce either cytoprotection or tumor growth by activating NRF2, as a function of the phase of the malignancy. Garufi et al. highlighted the anticancer effects through manifold molecular processes related to curcumin, and recently explored the fundamental molecular sequence of steps linked to making operative NRF2 by the zinc-curcumin [Zn (II) -curc] complex. Indeed, the therapy with Zn (II)–curc raised the NRF2 proteins concentrations and their connections, the heme oxygenase-1 (HO-1) and p62/SQSTM1, while particularly decreased the levels of Keap1 (Kelch-like ECH-associated protein 1), which stopped the NRF2 in all the investigated malignant cell lines. The inhibition of NRF2 or p62/SQSTM1 with distinctive siRNA proved the existence of the communication channel between the two molecules, and that the easily disassemble of any molecule surged the killing of cancer cells by Zn (II)–curc, a fact that could be implemented in the future to improve the receptivity to tumors treatment by this method [70].

Curcumin displays multiple proven effects on cells; for example, an inhibitive action on thrombocytes, but not known if it is owed to thrombocyte apoptosis or to pro-coagulant platelet organization.

Recently, Rukoyatkina et al. reported that curcumin did not initiate caspase 3—relying on the apoptosis of human thrombocytes, but led to the organization of pro-coagulant thrombocytes. At 5 µM concentration, the effect increased, but at ten times' higher concentration, the thrombocytes apoptosis was stopped by the suppression of ABT-737 (small molecule drug that inhibits Bcl-2 and Bcl-xL) that was interfered with thrombin production.

Curcumin did not alter thrombocytes' ability to survive at low concentrations but decreased it by 17% at higher concentrations. Autophagy caused by curcumin in human thrombocytes was accompanied by the operative configuration of adenosine monophosphate kinase (AMP), and the cessation of protein kinase B function. Curcumin could block the P-glycoprotein (P-gp) in tumors, and therefore defeat the manifold medicines resistance, and likewise could also stop the thrombocytes P-gp action. The effects of curcumin on human thrombocytes are due to complex processes through pro-coagulant thrombocytes organization, and so it can support pro—or against caspase—subordinate thrombocyte's death, but only in distinctive cases [71].

Systematically checking the abnormal functioning of cells, tissues or organs in the inceptive steps of the initiation of malignant processes and the monitorization of the cellular oxygenation is of maximum significance, both for the fundamental applications, but also in the practical medical ones.

A non-invasive modality for both the in vivo and in vitro assessment of cellular oxygenation is evaluating the lifespan of the luminescence of molecular sensors, but still very difficult in the case of increased oxidative stress.

Molecular probes, such as mitochondrial probes or [Ru (Phen)3]$^{2+}$ state-of-the-art, intact cell phosphorescence imaging technologies applied by Huntosova et al., in a model of chorioallantoic membrane (CAM), offer reduced phototoxicities and could also be applied in curcumin cancer therapy in tumors originating from the neuroglia of the brain or spinal cord. These results could be useful and universalized for the evaluation of tissue oxygenation as an advanced and original method, based on the analogies between diverse interacting biological factors, especially in cancer therapies that deal with metabolic or oxygen changes, glucose and lipid loss, and so on [72].

Important attempts to improve the potency of targeted drug carriers in the lung cancer were made, but the prognosis is still very poor, with only 15% survivors, 5 years after identification. The best choice for the direct administration of chemotherapy to the lungs would be the inhalation formulation.

Currently, this type of formulation to accomplish successfully, at the same time, a significant dose of specific chemicals that are selectively destructive to malignant cells and tissues in the solid tumor and to function with reduced local lung toxicity, is still an aim, as only 10–30% of lung chemotherapy nowadays already has the quality of being toxic [73].

Lee et al. imagined a dry powder easy to inhale (DPI) holding a chemotherapeutic agent (paclitaxel, PTX) and the native antioxidant curcumin (CUR) that defends the healthy cells to be damaged during direct lung transfer chemotherapy. Grinding CUR and PTX in co-jet as aerosol formulation, with more than 60% of very fine particles and a fit mass median aerodynamic diameter, exhibits an important cytotoxic effect for lung tumors, giving rise to apoptosis/necrotic cell killing, extending mitochondrial oxidative stress (ROS), depolarizing the mitochondria membranes, and decreasing the ATP in malignant cells. Incorporating CUR is decisive for correcting the cytotoxic effects of PTX against healthy cells and depends on dose, providing an easy and efficient DPI formulation with special discriminating cytotoxicity in lung malignancies [73].

In another experiment concerning lung cancer, Wan Mohd Tajuddin et al. studied the diarylpentanoid (DAP), changed the structure analog from the genuine curcumin, and proved to improve anticancer effects in different forms of malignancies, by comparing the outcomes (toxic impact, proliferative and apoptotic action) on two subtypes of non-small cell lung cancer (NSCLC) cells: the squamous cell carcinoma (NCI-H520) and the adenocarcinoma (NCI-H23). The gene expression to reveal the main signaling pathways, the targeted genes, the cytotoxicity screening, the anti-proliferative action, as well as the apoptosis linked to the rise in caspase-3 activity and decrease in Bcl-2 protein concentration were investigated and proved to be function of dose and time in all studied cells. This new compound, derived from curcumin, should be henceforth investigated as a possible representative anticancer drug for NSCLC cancer treatment [74].

4. Effects of Curcumin and PDT in Various Forms of Cancer

4.1. Breast Cancer

Breast cancer at onset has no clinical symptoms, because it initially affects the glandular epithelium of the ducts or lobe structure, but over the years, it can progress and invade the surrounding breast tissue and lymph nodes or various organs; therefore, if a woman dies, the cause is multiple distant metastases. Breast cancer can occur in women right after puberty, or at different stages of life, and the number of cases has increased year on year, becoming the most common form of cancer worldwide. According to World Health Organization (WHO) publications, as of 2020, 2.3 million women have been confirmed with breast cancer, of which 685,000 deaths have occurred worldwide, and an impressive number have been left with various disabilities in daily life [75].

It is currently known that breast cancer can be classified into three subtypes:

- type 1 breast cancer with estrogen receptor (ER+) positive hormone or progesterone receptor (PR+) positive; this type responds to hormonal treatment.
- type 2 breast cancer with a positive test for the human epidermal growth factor receptor 2 (HER2), a protein which fosters the development of cancer cells, and may respond to HER2-targeted treatments.
- type 3 breast cancer is the one known as triple negative breast cancer (TNBC), because here we do not find ER, PR or HER2. This type of cancer is the most difficult to treat with pharmacological means that have already become classic. Various drugs [Sacituzumab govitecan (Trodelvy)] and immunotherapeutic products, [Pembrolizumab (Keytruda), PARP inhibitors], are being tested in combination with conventional chemotherapy for this type of breast cancer [76].

Although the breast cancer therapy available today (anticancer chemotherapeutic drugs, antihormones, biologics and radiation therapy) has very good results on early detected primary tumor, many deaths still occur due to recurrences and multisystemic metastases. In this pathology, it has been shown that a type of cancer cells, like normal stem cells, has a strong potential for multiplication, self-renewal, differentiation, and their oncogenicity can underlie recurrences and metastases [77].

Radiation therapy is one of the standard treatment methods for breast cancer; applied in the early stages, it can decrease mastectomy surgery; in later stages, it can reduce the risks of recurrence, and in advanced stages, it can prolong the patient's life [75].

However, this method still has many disadvantages, in that the tumor may become radioresistant, and cancer and metastases may recur.

Targeting breast cancer stem cells is the clue for ameliorating the results of breast cancer radiotherapy.

Yang et al. investigated the effects of curcumin combined with glucose nano-gold particles (Glu-GNPs), for a decrease in radiotherapy resistance activity, by targeting MCF-7 and MDA-MB-231 breast cancer stem-like cells (BCSCs), in order to increase apoptotic, colony-forming and antiproliferative activity. Irradiation was carried out with a 6 MV X-ray at a total dose of 0 and 4 Gy; the dose rate was 2 Gy/min, for two minutes, the depth of penetration was 2 cm, and the irradiation distance of 50 cm. The authors demonstrated that curcumin, combined with Glu-GNPs, increased the ROS level of mammals MCF-7 and MDA-MB-231 in hypoxic conditions by inhibiting the factor-1alpha (HIF-1alpha) (an oxygen-sensitive transcriptional activator) and the heat shock protein 90 (HSP90) activities, ameliorating the apoptotic activity of tumor stem cells, and thus, it has increased the sensitivity to radiation therapy [77].

Minafra et al. performed a study to evaluate in vitro the bioavailability and the radiosensitizing effects of curcumin-loaded solid nanoparticles (Cur-SLN) on three breast cell lines, non-tumorigenic MCF10A and tumorigenic cell lines MCF7 and MDA-MB-231 BC, exposed to irradiation. A multi- "omic" assay was used to clarify the radiosensitizing action of Cur-SLN by microarray and metabolomic exploration techniques. The authors demonstrated the antioxidant, radiosensitizing and anti-tumor capacity of Cur-SLN through a transcriptomic and metabolomic analysis [78].

The conventional treatment protocol for breast cancer includes breast removal surgery, X-ray therapy, and the administration of specific drugs, i.e., chemotherapy. Drugs administered for these cancers have many side effects, negatively affecting patients' quality of life, which has required the discovery of new effective, but less toxic treatments. In recent decades there has been a growing trend to find natural remedies with anti-tumor potential, including curcumin and its derivatives.

Research to date has shown that curcumin has anti-neoplastic properties through various mechanisms, including the following: inhibition of endothelial growth factor [79], lipoxygenase enzyme (LOX) pathway [80], blocking the activity of NF-kB and Wnt signaling pathways [81–83], stops cell cycle and p53-dependent apoptosis, disrupts the expression of signaling protein kinase B (Akt) and phosphatidylinositol 3-kinase (PI3K) [84,85].

Experimental studies have shown that curcumin can inhibit EZH2 (enhancer of zest homolog-2), which is a histone methyltransferase that catalyzes the trimethylation of histone H3 in lys 27 (H3K27me3), is found in an increased amount in human cancer, has an oncogenic, metastatic role, and influences drug resistance. It was postulated that, on the EZH2, miR-375, FOXO1 and p53 axis, special processes with direct impact in the proliferation of the breast tumor could take place [86,87].

Gallardo et al. showed, in a study on MCF-10F and MDA-MB-231 cell lines of human breast cancer, that curcumin can modulate the expression of miR-34a and Rho-A, which reduces cancer progression, metastasis, and increases the sensitivity of anti-cancer drugs [88].

The chemical structure of curcumin is responsible for many of its biological and pharmacological activities, as low solubility in aqueous media, poor bioactive absorption, physicochemical instability, rapid metabolism, sensitivity to alkaline environment, which restricts its clinical scope [89,90].

With the patenting of curcumin nanoencapsulations, these impediments have been overcome. There are currently several nanoencapsulation techniques, but ionic gelation and antisolvent precipitation are in vogue. The products used for nanoformulations currently include: liposomes, polymers, nanoparticles, conjugates, solid dispersions, cyclodextrins, micelles, nanospheres and microcapsules and other various nanoformulations [42,90,91].

Due to the maximum wavelength of blue light absorption (408–434 nm), curcumin is used as a natural photosensitizer for PDT applications in various medical, antimicrobial, antiviral and antitumor fields [92,93].

The mechanisms by which PDT can eradicate tumor tissue are synthesized in the following modes of action: initially, it will locate and activate PS in the tumor area, which will release ROS with the potential to destroy neoplastic cells; the second mechanism would be that PDT disrupts the usual supply of oxygen and nutrients by compromising vascularity; the last postulated mechanism is the activation of the immune system, which triggers an inflammatory process against tumor structures.

Sun et al. studied experimentally, on 4T1 mouse breast cells, the effects of carrier-free curcumin nanodrugs (Cur NDs) used in conjunction with PDT administered by a device that emitted blue light at a power of 640 mW on a wavelength of 450 nm. Cur NDs were prepared without using any toxic solvents through an easy and green reprecipitation method, as follows: 1 mg/mL Cur was dissolved in ethyl alcohol; afterwards, 1 mL of the solution was rapidly injected into 20 mL of high-purity water under robust stirring. NDs were purified and concentrated by ultrafiltration, and at the end, the Cur NDs were achieved by lyophilization. In vitro irradiation (λ = 450 nm, 640 mW, 1 min) of 4T1 cells after 24 h of incubation significantly increased the expression of p-JNK, Bax and cleaved caspase-3, with the generation of a large amount of ROS, activation of MAPKs with induction of apoptosis, and reduced the cells viability. The study demonstrates that Cur NDs is a valuable photosensitizer for PDT in eradicating breast cancer, and with very good prospects for use in the clinical practice [94].

Because curcumin has a low solubility, nanoemulsions and microemulsions as administration systems with dimensions between 100–300 nm facilitate an expansion of bioavailability, the therapeutic window, and even a controlled delivery to the desired area [95–97].

Machado et al. investigated the effect of curcumin-nanoemulsion (CNE), a new and well-designed drug delivery system (DDS+) molecule, as a novel photosensitizer in the photodynamic therapy of breast cancer on MCF-7 cell model. CNE was achieved by interfacial pre-polymer deposition and spontaneous nanoemulsification. Curcumin was set up in an oil phase at 0.1 mg/mL. Organic phase (acetone) was obtained from medium-chain-triglycerides, natural soy phospholipids. This was added into the aqueous phase containing an anionic surfactant, poloxamer 188. Organic solvent was taken off through rota-evaporation. The authors revealed that curcumin encapsulation in lipid nanoparticles increased the concentration of this compound, its biological effects, solubility and bioavailability, without substantially affecting the cell viability of HFF-1 and MCF-7 cells. After incubating HFF-1 cells and MCF-7 cells and applying two doses of 80 J/cm^2 with a laser device at 440 nm, the mortality rate of breast carcinoma cells was 65% and 90%, respectively, demonstrating the value of CNE as photosensitizer [98].

Curcumin, a component derived from the dried rhizomes of *Curcuma Longa*, is the most requested phytochemical component for cancer therapy, and is therefore considered the "magic molecule" [99,100].

Remote spread of primary malignancy is the cause of multiple cancer deaths. To date, the entire arsenal of available drugs and treatments are still insufficient to manage the metastases that are forming. Therefore, maximum attention is directed to the prevention of metastases by targeting circulating metastatic tumor cells.

A lot of circulating tumor cells (CTCs) will perish in the blood circulating through the body, but a small number will most likely generate distant metastases, and thus a negative prognosis for the patient. These stem-type CTCs can escape immune surveillance and show a high ability to withstand treatment. The latest strategies should address metastatic CTCs and their progenitors based on their molecular and cellular characteristics, to expand current plans for the successful prevention of metastases [101].

In a recent study in this significant direction, Raschpichler et al. advanced an in vitro exploratory model to mimic the circulation of CTCs, and to render them inactive by PDT and curcumin-loaded poly (lactic-co-glycolic acid) nanoparticles. Researchers obtained after blue laser irradiation (30 min, λ = 447 nm, output power 100 mW), an important decrease in the viability of MDA-MB-231 breast cancer cells under flow conditions. Accumulation of curcumin on cell membranes and high fluorescence signal after irradiation was detected by laser scanning confocal microscopy. PDT determined alterations in the cancer cells, i.e., apoptosis and prompt necrosis, as a function of time, and during laser activation of curcumin nanoparticles, a change in blue absorption spectra and a decrease in total curcumin occurred. Findings are incentive for the extension of these in vitro CTCs studies to in vivo experiments, so that PDT becomes an advanced action plan for future clinical applications for metastasis prevention [102].

The anti-cancer activity of curcumin by formulation as nanostructured lipid carriers (NLCs) using solid lipids (glyceryl monooleate, glyceryl dibehenate, glyceryl distearate) and olive oil as a natural liquid lipid, was tested in vitro on a breast cancer cell line. The curcumin thus prepared has been used as a photosensitizer for PDT in MCF-7 breast cancer cells. The loading of CUR into NLCs increased the penetrating power into cancer cells and the cytotoxic potential in both dark and light conditions [103].

Approaching cancer therapy with current means brings a major drawback, namely the deterioration of healthy tissue in the vicinity of the tumor. To avoid these major problems, the association between PDT and photothermal therapy (PTT) has aroused the interest of researchers in the therapy of a wide range of diseases, including cancer. Very good results with reduced invasive effect and negligible side effects were obtained by combining PTT that releases heat, and PDT that not only produces ROS, but also promotes the release of products with multiple antitumor immunity capacity, in order to dissipate tumors. Phototherapy using nanoparticles can target and destroy undetectable metastatic tumors, restore the anti-tumor capacity of drugs and modulate the immune system in the tumor microenvironment [104–107].

The association between PTT and PDT has the advantage of an excellent curative result by eliminating tumor cells, as well as by immunomodulating the apoptotic and inflammatory responses of tumor cells.

An additional barrier of the immune control point to the photo treatment could increase the antitumor effects, generating more CD4 + and CD8 + T cells in the tumor [108].

An in vivo experiment on Female Balb/c mice tumorized with 4T1 cells has investigated the antitumor effects of curcumin loaded with Fe_3O_4–SiO_2 nanoparticles, i.e., a dual-functioned nanocomposite (NC), and also PDT and PTT. After tumorization, the mice were divided into six groups: control group injected with phosphate buffer saline (PBS); group II with CUR and irradiation with blue diode laser at 450 nm, intensity of 150 mW/cm^2 for 3 min (CUR and PDT group); group III, PBS injection plus irradiation with blue diode laser for 3 min followed by NIR laser with intensity of 0.5 W/cm^2 for 7 min, i.e., the blue + NIR (near-infrared) laser group; group IV, nanocomposite injection (NC group); group V with NC plus irradiation with NIR laser at 808 nm for 7 min (NC and PTT); and the last group with NC + PDT + PTT. The treatment protocol combining NC and PDT together with PTT on the highly invasive triple-negative breast cancer model used in this study could be considered for the replacement of chemotherapy in these types of cancer.

Results of the study revealed that in the group treated with NC and PDT together with PTT, the tumor volume decreased significantly, and the expression of proapoptotic proteins Bax and Caspase 3 increased meaningfully compared to the control group, without weight loss, and no adverse effects on vital organs (liver and lung) in the mice autopsy image [109].

In an in vitro study, Halevas et al. reported that the gallium curcumin complex used as a photosensitizer in photodynamic therapy on human breast MCF-7 adenocarcinoma cells, increased intracellular ROS levels, and significantly decreased the number of cancer cells compared to simple curcumin. New curcumin-Ga complex did not show dark toxicity at low concentrations against MCF-7 breast cancer cells, and the decrease in cell survival was dependent on the laser dose [110].

Table 1 summarizes aspects of recent research that have shown that curcumin nano formulations, together with PDT, can contribute to the treatment of breast cancer by strong penetration into cancer cells, high cytotoxic potential, decreased tumor volume, prevention of metastases, without weight loss, and without adverse effects on vital organs (liver and lungs) in experimental animals, with prospects for future clinical applications [94,98,102,103,109,110].

Table 1. Curcumin nanoformulations and PDT in breast cancer.

References	Type of Study	Type of Light and Curcumin	Total Energy (J) Applied	Analyzed Parameters	Conclusions
[94]	In vitro experimental 4T1 mouse breast cancer cells.	Blue light 450 nm. Curcumin nanodrugs (Cur NDs)	Blue light (P = 640 mW) with a wavelength of 450 nm at a fixed distance of 13 cm for 1 min.	In vitro cytotoxicity. Intracellular ROS production was analyzed using an intracellular ROS kit. c-Jun N-terminal kinase (JNK); p-JNK; mitogen-activated protein kinase (MAPK); caspase-3 (Casp 3); Bcl-2–associated X (Bax); Glyceraldehyde 3-phosphate dehydrogenase (GAPDH).	This study proved that Cur NDs could be a full of promise PS for speeding up the performance and reliability of PDT against breast cancer, with very good prospects for use in clinical practice.

Table 1. *Cont.*

References	Type of Study	Type of Light and Curcumin	Total Energy (J) Applied	Analyzed Parameters	Conclusions
[98]	In vitro breast cancer model, MCF-7 cells.	LED 440 ($\pm$10) nm, with 420 mW output power, 2.52 W total power, with 209 W/cm^2 irradiance. Curcumin-nanoemulsion (CNE).	80 J/cm2 fluency, set at 6.4 s/application	Caspases 3 and 7 activity. Estimation of intracellular reactive oxygen species production by 2'-7'-Dichlorodihydrofluorescein diacetate (DCFH-DA) technique.	Curcumin-nanoemulsion and PDT increased the activity of capsases 3 and 7, had a phototoxic effect with a significant reduction in MCF-7 cell proliferation and stimulated ROS release; this association has great prospects for breast cancer therapy.
[102]	In vitro experimental model of circulating tumor cells (CTCs) with human breast cancer cells (MDA-MB-231, ATCC HTB-26).	Blue light (447 nm, 100 mW). CUR-NPs.	MDA-MB-231 cells were laser irradiated for 30 min under flow conditions (5 cm s^{-1}).	Cell viability assay Morphology of nanoparticles Photodynamic inactivation Scanning electron microscopy of circulating breast cancer cells. CLSM micrographs showing cellular curcumin accumulation and photodynamic effect of curcumin loaded nanoparticles (CUR-NPs) on circulating MDA-MB-231 cells.	Apoptosis and necrosis of metastatic malignant cells were demonstrated by this experimental study in vitro on human breast cancer cells, using CUR-PLGA NPs and 30 min laser irradiation (447 nm and 100 mW) under continuous flow conditions. Results open new perspectives in clinical oncology for targeting metastases.
[103]	In vitro MCF-7 Human Breast Cancer Cell-Line.	LED 430-nm GaAlAs, CW. In vitro release of Curcumin Nanostructured Lipid Carriers (CUR-NLCs) formulas.	Irradiation protocol: blue light (430 nm) for 5 min (power 100 mW), spot size radius 4 cm, irradiance 2 mW/cm^2, fluence 6 J/cm^2).	Determination of encapsulation efficiency and drug loading percentages by spectrophotometry measurements. Morphology changes by transmission electron microscopy. Dark and photo-cytotoxicity studies of MCF-7 cells survival.	Carriers of nanostructured lipids loaded with curcumin and olive oil used in conjunction with PDT have increased the potency of penetrating breast cancer cells and the cytotoxic activity. The results of the study suggest that CUR-NLCs in low doses after exposure to blue light have a significant anticancer effect in breast cancer.
[109]	In vivo on the breast cancer model in Female Balb/c mice (6 to 8 weeks).	Blue diode laser at 450nm, CW for PDT, and 808 nm in NIR range for PTT. An external magnetic field was applied for appropriate delivery of the drug. Curcumin on silica-coated Fe$_3$O$_4$ nanoparticles.	In vivo experiment: with female Balb/c tumorized mice, divided into 6 groups: (I) PBS injection (control group). (II) Curcumin plus irradiation with a blue diode laser at 450 nm with 150 mW/cm^2 for 3 min (CUR + PDT group). (III) Blue diode laser for 3 min followed by NIR laser with 0.5W/cm^2 for 7 min (Blue + NIR lasers group). (IV) injection of 40 μL NC (NC group). (V) injection of 40 μL of NC solution containing 20 μg curcumin (0.46 mg/mL) plus irradiation with NIR laser at 808 nm for 7 min (NC + PTT). (VI) injection of 40 μL of NC containing 20 μg curcumin plus irradiations with two lasers with up-mentioned intensities and exposure times, while a rigid magnet was fixed on the tumor to maintain the injected NC in the tumor position (NC + PDT + PTT group).	Analysis of expression of apoptotic proteins Bax and Caspase 3. In vitro toxicity of Fe$_3$O$_4$/SiO$_2$ NPs after 24 and 48 h. In vitro release of curcumin from the NCs. Antitumor effect of nanocomposite plus PDT and PTT approach in vivo.	In the group treated with NC+PDT+PTT the tumor volume was significantly reduced and the expression of proapoptotic proteins Bax and Caspase 3 increased significantly compared to the control group, without weight loss, no adverse effects on vital organs in the mice autopsy images. Method could replace chemotherapy for triple-negative breast cancers.
[110]	In vitro study of human breast adenocarcinoma MCF-7 cells.	5 min irradiation at 450 nm and 100 mW/cm^2. Ga(III)-curcumin complex.	6 mW/cm^2 and the exposure times were 167 s, 334 s, 501 s, 1002 s, and 6012 s yielding 1, 2, 3, 6 and 10 J/cm^2 fluence, respectively.	Photophysical and photochemical studies (UV-Visible absorption and fluorescence; ROS production; in vitro cytotoxicity assay. Dark cytotoxicity.	Administration of the Ga (III)-curcumin complex studied on MCF-7 breast cancer cells has shown that metal complexation increases its photodynamic effect compared to simple curcumin.

4.2. Gynecologic Cancers

Cancer worldwide is a major health problem, and one of the most common causes of death [111]. According to statistics from the last two decades, the incidence of cancer in women has been steadily increasing by about 0.5% per year for breast cancers, and about 1% for uterine cancer, due to the reduction in the number of fertile women and weight gain [112–114].

Over the past three decades, the survival rate of cancer patients has increased, except for the cancers of the uterine cervix and of the uterine corpus [115].

Cervical cancer, with almost 0.6 million cases and 0.3 million deaths per year in 2018, ranked fourth in the world in terms of incidence and mortality caused by cancer after breast cancer (2.1 million cases), colorectal cancer (0.8 million cases) and lung cancer (0.7 million cases) [116].

Today, special efforts are being made to prevent, detect early and implement new therapeutic strategies to improve the efficacy and safety of chemotherapy and radiotherapy, prevent recurrences, metastases and increase patient survival. PDT, together with various curcumin nanoformulations, are promising treatment modalities for a wide range of cancers, such as cervical cancer.

Nanoemulsion-curcumin behaved as a photosensitizing drug in PDT, generating a high phototoxic effect, with less than 5% of viability in the experiment with cervical carcinoma cell lines (CasKi and SiHa) and human keratinocytes spontaneously immortalized cell line (HaCa). In an experiment done by de Matos et al., increased activity of the enzymes caspase 3 and caspase 4 was observed, suggesting that cell death occurred by apoptosis. The authors propose the use of curcumin-nanoemulsion and PDT through an in situ optic fiber, as an alternative treatment in cervical cancer [117].

A very important issue in PDT is to avoid the accumulation of PS in healthy tissues and prevent unwanted effects, which can be solved by using encapsulated PS in polymeric nanoparticles, in which the active product is protected from degradation by the physiological environment; an example is the case of poly (lactic-co-glycolic acid) (PLGA), recognized as one of the most valuable drug-carrying polymers (DCs) that is used to obtain high quality products. PLGA polymer is approved as a biocompatible and biodegradable polymer by the FDA and the European Medicines Agency (EMA) [118–120].

Duse et al. published the results of in vitro irradiation research (457 nm LED with a radiation flux of 8.6J/cm^2) on the human ovarian adenocarcinoma cell line SK-OV-3, using a photosensitizer with biodegradable PLGA nanoparticles loaded with curcumin (CUR-NP). In this study, we analyzed the effect of nanoformulation on human erythrocytes, by haemolysis test and blood clotting, which showed that there was a slight increase in haemolysis and improved serum stability, assuming that there is an interaction of curcumin with intrinsic proteins or coagulation factors. The authors showed that the nanoformulation allowed the use of higher amounts of curcumin, which showed cytotoxic effects on tumor cells after the administration of PDT at low intensity, thus selectively inhibiting tumor growth, and believe that LEDs provide an economic and technical advantage over laser devices [121].

Curcumin, as a special low-toxicity photosensitizer, has higher pro-apoptotic potency when combined with PDT. However, the mechanisms of action are still poorly understood. He et al. investigated the results of the combination of different concentrations of curcumin mediated PDT with/without Notch receptor blocker (DAPT), after 180 s of irradiation with 445nm laser at a dose of 100 J/cm^2, used together with curcumin, on the survival rate of cervical cancer Me180 cells in female BALB/c mouse. Following the administration of curcumin with PDT plus DAPT, a synergistic interaction occurred that significantly increased the overall rate of cell mortality. The authors claim that this combination could inhibit Notch-1 expression and downstream protein synthesis in in vitro and in vivo cervical cancer. Notch-1, an advanced preserved signaling pathway that controls interactions between adjoining cells and NF-κB, could be the targets of the curcumin-PDT combination in the success of cervical cancer therapy in women. This can be explained by the fact that Notch-1 activation is associated with the onset and development of cervical cancer [122].

Figuratively speaking of Notch, we can consider that it belongs to the group of arbitrators who decide the fate of the cell; in fact, it modulates the balance between differentiation and multiplication of cells. The Notch pathway has a very important role in the evolution of breast, cervix, ovary, and uterine endometrium epithelial tissues, and is frequently involved in the appearance and expansion of cancers [123].

4.3. Skin Cancer

Human skin consists of three layers of cells superimposed from depth to the surface, as follows: epidermis, dermis, and hypodermis. The epidermis consists of five types of overlapping cells, one on top of the other, from the depth to the surface in the basal or germinal layer—spiny, granular, lucidum and horny. All the layers come from the germinal one, whose cells, as they multiply, are pushed to the surface, changing in 26–28 days their shape and structure, and then they will be eliminated as dead cells in the form of barely visible scales. Melanocytes are dermal cells that secrete melanin, a pigment that tans the skin under the action of the sun, and the corpuscles Meissner, Ruffini, Krause, Vater-Pacini and Merkel give the skin the ability to perceive and transmit the proprioceptive information to the brain. If these cells undergo pathological changes, they will give rise to various forms of skin cancer. Each type of skin cancer has a potential for severity, so it must be detected early and treated accordingly. The most common types of skin cancer include the following: basal cell carcinoma, recurrent basal cell carcinoma, squamous cell carcinoma, melanoma, Merkel cell carcinoma, and other types of rare skin cancers [124].

According to data published by the American Cancer Society, about 5.4 million basal and squamous cell skin cancers are diagnosed each year in the US, of which approximately 3.3 million American patients have basal cell carcinoma, meaning that 8 out of 10 have the malignant cells in the basal layer.

Basal cell carcinoma develops in the basal layer of the epidermis, has a tropism for the neck and head regions that are more frequently exposed to the sun, but can evolve elsewhere and over time metastasize to the loco-regional ganglia [125].

Basal cell carcinoma can recur in the same area as before, or in another zone of the body; it has a risk of recurrence of up to 50% after 5 years of diagnosed primary cancer. The risk of recurrence is higher in patients with a personal history of eczema, dry skin, prolonged exposure to the sun or tanning devices with UV light, who have had deep skin carcinoma or a size greater than 2 cm.

Squamous cell skin cancer grows slowly, metastasizes less often, but can invade deep into the skin. This type of skin cancer develops from the flat squamous cells that form the superficial layer of the epidermis, and occurs mainly in the region of the neck, face, ears, external genitalia, dorsal area of the hands, etc. [123].

The exact number of the most common types of basal cell and squamous cell cancer (i.e., keratinocyte carcinoma or KC), known as non-melanoma skin cancer, is not known exactly, because it should not be reported in cancer registries [126].

Among Caucasian populations, squamous cell carcinoma (SCC) of the skin, also known as cutaneous squamous cell carcinoma (cSCC), accounts for 20% of all malignant skin tumors [127].

The incidence of squamous cell carcinoma (SCC) increased in recent decades by 10% per year, so today, the ratio is 1:1 compared to basal cell carcinoma (BCC) in populations in Australia and the US. The most important risk factors for disease or recurrence are chronic sun exposure, ultraviolet A (UVA) radiation, patients using immunosuppressive drugs, those with organ transplants, and the HIV-positive. The disease is more common in the elderly and men. The diagnosis of cSCC can be made especially based on the clinical aspect, which must be confirmed by histopathological examination to anticipate the correct prognosis and the management of therapy. The first intention treatment is the one of complete surgical excision, with histopathological control of the excision edges. When lymph nodes are contained by cSCC, a regional dissection of the lymph nodes will be performed, followed by radiotherapy and various chemotherapeutic agents, and more recently, photodynamic therapy and epidermal growth factor receptor (EGFR) inhibitors [128–132].

Response to treatment is usually good; however, a subtype in the cSCC category needs special monitoring because it has a much higher risk of local recurrence, locoregional metastasis, or distant metastasis and death [133,134].

Although primary cSCC is not a fatal tumor, it can raise major cosmetic problems caused by localization in the face area (surgery can be disfiguring), increased morbidity, and high costs, which will place a significant burden on the public health system [124,131].

The prognosis can be very good, with a 90% survival rate for 10 years, when the disease is in its early stages; if the cancer is extensive or metastasized, survival is reduced to 15.3 months on average, making it the second leading cause of death from cSCC after melanoma in skin cancers [135,136].

Surgical treatment in this type of cancer is the first choice; however, if the tumor is enlarged, its resection will lead to an unsightly scar with a strong psychological impact. Chemotherapy has major disadvantages due to haematological, gastrointestinal, hepatic, or renal side effects, the emergence of drug resistance, and necrosis, fibrosis, secondary tumor formation on long-term administration; after radiotherapy, free radicals can be released that cause local inflammations, dermatitis, ulcers, etc. [137,138].

Given these aspects, it is very important to be found and select effective and safe means of treatment in skin cancer. In recent decades, PDT has established its position in the treatment protocol with predilection for skin cancer, including basal cell carcinoma, recurrent basal cell carcinoma, squamous cell carcinoma, Bowen's disease, actinic keratosis, etc. Depending on the photosensitizer and the light source, PDT triggers oxidative stress by generating ROS in cancer cells and produces their apoptosis [139–144].

Xin Y. et al. investigated the apoptotic effects and molecular mechanisms of action of a treatment combination of ultraviolet B radiation and demethoxycurcumin (DMC) in vitro on A431 and HaCaT cells. The association between demethoxycurcumin as a photosensitizer and ultraviolet B radiation induced apoptosis in vitro in A431 and HaCaT cells, by activating p53 and caspase pathways, increasing Bax and p-p65 expression and suppressing Bcl-2, Mcl-1 and NF-κB pathway. At the same time, high levels of reactive oxygen species were observed, along with the significant depolarization of the mitochondrial membrane [145].

Abdel Fadeel et al. demonstrated the beneficial effects of PDT with violet light (410 nm) and Cur-loaded PEGylated lipid nanoparticles in an in vitro study with increased cytotoxicity in cell culture, human squamous cell carcinoma cell line (A431), and in vivo skin carcinoma in mice [146].

The hypothesis of this study is supported by other researchers [147,148].

Cur can produce higher cytotoxicity when charged on a nanocarrier and, in addition, this quality can be increased by exposure to blue light radiation, therefore, together they will generate ROS that affect the activity of components cell and mitochondrial membrane that will lead to the apoptosis of cancer cells.

In patients with advanced cancer who are malnourished and stay in bed longer, but also in other categories of people who do not mobilize, especially in the elderly, the skin may degrade due to local compression, ischemic and reperfusion disorders [149].

Skin in that area initially becomes erythematous, then necrosis follows and bed-sores or pressure ulcers form. A recently published meta-analysis showed that, in adults, pressure ulcer (PU) has a prevalence of 12.8%, and the incidence rate is 5.4% per 10,000 patients per day, and the rate of PU acquired in hospital was 8.4% [150].

In addition to the high costs for patients and health services, these injuries can even be fatal through local and systemic infections. The conventional treatments with antibiotics, surgical measures, ultrasound, electromagnetic or genetic therapies have failed to completely cure PU in an advanced stage of evolution.

To avoid all these aspects, it seems that a much more beneficial option would be that of the concomitant use of laser radiation, together with curcumin.

The results of studies published so far on the concomitant use of curcumin and laser radiation demonstrate two interesting aspects, namely: laser light [151,152], can excessively increase ROS production, while curcumin is able to inhibit oxidative stress by reducing the activation of ROS-generating enzymes or by eliminating free radicals; these phenomena are very useful for the purpose of the therapy, when it is desired to induce apoptosis in skin cancer or to accelerate the healing process of PU [153,154].

Ebrahiminaseri et al., in an in vitro experiment on mouse MEFs cells using the association between dendrosomal nanocurcumin (DNC) and laser therapy at a dose of 0.95 J/cm^2, demonstrated the significant migration of MEF cells in the denuded area, increase in the S-phase cell population and the growth factors (TGF-β, VEGF) and decrease in the cell population in the GF/G1 phase and of the pro-inflammatory cytokines (TNF-α, IL-6). This combined therapy (DNC + PDT) also highlighted the modulating role of nanocurcumin on the production and excessive accumulation of ROS generated by the action of the laser. In conclusion, the authors consider that additional in vivo studies are needed to confirm the hypothesis that the combined treatment of DNC simultaneously with PDT (450 nm) may favor the wound healing process [155].

Malignant melanoma is an aggressive form of cancer that results from the degeneration of melanocytes, cells of neuroectodermal origin with the role of melanin secretion as a protective response of the skin against the harmful action of ultraviolet sunlight.

Incidence of melanoma [156], mainly metastatic melanoma, is constantly increasing, especially after global warming.

Although current treatment is well established by surgical excision followed by chemotherapy, immunotherapy, recently based on inhibitors of immune checkpoints, small molecule-targeted therapy and oncolytic viral therapy, which have led to an increase in the survival rate of patients at 5 years; however, the incidence of metastases, the high number of deaths and the side effects of drugs require studies to find new therapeutic ways [157,158].

Szlasa et al., in an in vitro study focusing on the anticancer effect of PDT together with curcumin, revealed an increase in the number of apoptotic and necrotic cells by the overexpression of caspase-3, DNA cleavage and reorganization of the actin cytoskeleton, compared to incubation without irradiation. Fibroblasts have been less influenced by this treatment, which seems beneficial for the increased ability to regenerate irradiated skin areas. PDT, together with curcumin, can be an effective way to induce apoptosis in melanoma [159].

Nanoparticles of selenium-polyethylene glycol-curcumin (Se-PEG-Cur) used in an experimental study on melanoma cancer cells as a photosensitizer for phototherapy and sonotherapy greatly increased intracellular ROS levels and cytotoxicity and decreased cell viability [160].

The liposomal formulation of curcumin has been shown to be more effective in killing malignant cells in several studies. The increase in ROS generation and the photo-destructive effects of PDT, together with turmeric tetraether liposomes on microvasculature, were observed by Duse et al. in an in vivo study using the chick chorioallantoic membrane model [121].

Research group led by Vetha et al. reported apoptosis of A549 cancer cells in vitro, by generating singlet oxygen, increasing intracellular ROS levels and activating Caspase-3, by using small CUR molecules, encapsulated in liposome nanoparticles (LIP-CUR) coupled with PDT via emitting diodes blue light (BLED) [161].

In an experimental study, Woźniak et al. demonstrated a significant improvement in the bioavailability and stability of encapsulated liposomal curcumin as a potent apoptotic photosensitizer in squamous cell carcinoma (SCC-25) and melanoma (MugMel2); at the same time, low phototoxicity was observed in normal cutaneous keratinocyte HaCaT cells after PDT treatment [162].

These results promote liposomal curcumin as a potential natural photosensitizer that can improve its absorption, safety and efficacy in photodynamic therapy in human cancers.

Table 2 highlights some of the effects of PDT and curcumin nanoformulations in skin cancers, by significantly improving the bioavailability and stability of nanocurcumin, increasing cytotoxicity on malignant cells, but with low phototoxicity on normal keratinocytes, the apoptotic effects and molecular mechanisms by generating high levels of ROS, along with the significant depolarization of mitochondrial membranes, while other in vitro experiments on mice demonstrated significant cell migration in the denuded area and highlighted the modulatory role of nanocurcumin on excessive production and the accumulation of ROS generated by laser irradiation [145,146,155,159,160,162].

Table 2. Nanocurcumin in skin cancer therapy.

References	Type of Study	Type of Light and Curcumin	Total Energy (J) Applied	Analyzed Parameters	Conclusions
[145]	In vitro A431 -human cell line model (epidermoid carcinoma cell line) and HaCaT cells (spontaneously transformed aneuploid immortal keratinocyte cell line from adult human skin)	Ultraviolet radiation B (UVB) Demethoxycurcumin (DMC)	UVB $(10–100 \text{ mJ/cm}^2)$	Inhibition of tumor cell growth. Enhancement of apoptosis in cells. Apoptosis-associated proteins including Bcl-2, Mcl-1, Bax, nuclear factor-κB (p65), p-p65, p53, caspase-3, caspase-9, and cytochrome C. Measurement of ROS (which increased significantly). Analysis of mitochondrial potential (which decreased: important depolarization occurred).	PDT by ultraviolet B radiation and DMC have experimentally succeeded in causing apoptosis in skin cancer cells. DMC may be a promising photosensitizer for PDT to eradicate skin cancer cells.
[146]	*In-vitro/In-vivo* studies and histopathological examination on a human skin cancer cell line (A431)	Blue light (410 nm); PEGylated lipid nanocarriers (PLN) loaded with curcumin (Cur).	In vitro 300 mW/cm^2 for 4 min by blue light. In vivo LED (420 nm) for 10 min at a fluence of 90 mW/cm^2	Fluorescence intensity measured by confocal laser microscopy. Histopathological studies. In-vitro cytotoxicity.	This in vitro study with Cur-loaded PLN together with blue light proved a significantly higher cytotoxicity than the control sample against human epidermoid squamous cell carcinoma cell line (A431). In vivo study showed a significant improvement in skin carcinoma after photodynamic therapy and Cur-loaded PEGylated lipid nanoparticles. Beneficial effects of this safe and economical method, bring hope in the treatment of cancer.

Table 2. *Cont.*

References	Type of Study	Type of Light and Curcumin	Total Energy (J) Applied	Analyzed Parameters	Conclusions
[155]	In vitro experiments on mouse embryonic fibroblasts (MEFs) cells	Diode laser device with a wavelength of 450 nm and an output power of 75 mW. Dendrosomal Nano-Curcumin (DNC)	Cells were irradiated for 224 s (for getting a dose of 17.9 J, with an energy density of 0.63 J/cm^2), and 337 s (for getting a dose of 26.9 J, with an energy density of 0.95 J/cm^2). For other doses, the time was set in the same way.	- RNA extraction was quantified by spectrophotometry - cDNA synthesis - TGF-β, VEGF, TNF-α, IL-6 and glyceraldehydes 3-phosphate dehydrogenase (GAPDH). In vitro migration assay for cell motility; cell cycle analysis by flow cytometry; quantitation of DNA content stained. Measurements of intracellular ROS.	Results revealed a notable proliferation of mouse embryonic fibroblasts after the combination of DNC + LLLT (450 nm) at a dose of 0.95 J/cm^2. Simultaneous exposure to DNC +LLLT enriched S-phase entry and increased proliferation as well as significant migration of MEF cells in the denuded area, up-regulating growth factors (TGF-β, VEGF) and shortening the inflammatory phase by modulating cytokines (TNF-α, IL-6). Combined therapy (DNC + LLLT) also highlights the modulating role of nanocurcumin on the production and excessive accumulation of ROS generated by laser action.
[159]	In vitro studies on curcumin + PDT on melanotic (A375) and amelanotic melanoma (C32) cell lines.	Lamp with polarized light with power density set to 20 mW/cm^2, blue light (380–500 nm), including maximum absorption of curcumin (410 nm). Curcumin dissolved in dimethyl sulfoxide (DMSO).	Irradiation time = 5 min (6 J/cm^2).	MTT cell viability assay. Cell death evaluation by neutral comet assay (NCA). Fluorescent staining of actin filaments Caspase-3 immunocytochemical staining. Holotomographic microscopy studies. Cell viability and phototoxicity.	PDT + curcumin increased the number of apoptotic and necrotic cells compared to the control without irradiation, it induced overexpression of caspase-3 and DNA cleavage and low cell proliferation due to reorganization of the actin cytoskeleton. PDT together with curcumin can be an effective way to induce apoptosis in melanoma.
[160]	Experimental study on malignant melanoma C540 (B16/F10) cell line.	808-nm laser. Ultrasound (US). Nanoparticles of selenium-polyethylene glycol-curcumin (Se-PEG-Cur).	Output power = 1000 mW Power density = 1.0 W/cm^2. Irradiation time = 10 min. US output power of 1.0 W/cm^2; Frequency of 1MHz; Irradiation time = 1 min.	Detection of intracellular ROS Viability of C540 (B16/F10) cells. Fluorescence intensity (FI).	Se-PEG-Cur can be a very good photosensitizer for phototherapy plus sonotherapy in the destruction of melanoma cancer cells through thermal and ROS-generating effects.
[162]	Experiments on melanoma (MugMel2), squamous cell carcinoma (SCC-25), and normal human keratinocytes (HaCaT) cell lines.	Blue light (380–500 nm); 20 mW/cm^2. Liposomal Curcumin.	Irradiation time = 2 min; 2.5 J/cm^2	Impact of Liposomal Curcumin on Cells Lines' Apoptosis. Bax and Bcl-2 Expression. Cell Viability Assay. Wound-Healing Assay.	Experimental study demonstrated a significant improvement in the bioavailability and stability of liposomal encapsulated curcumin as a potent apoptotic photosensitizer in squamous cell carcinoma (SCC-25) and melanoma (MugMel2). Low phototoxicity was observed in normal cutaneous keratinocyte HaCaT cells after PDT treatment.

4.4. Gastrointestinal Cancers

Incidence of gastrointestinal cancer varies over time, from short intervals of 8 years to longer intervals of 20 years, suggesting that it is under the influence of a predestined epigenetic program. It is assumed that, until the onset of cancer, there are several stages that include accumulations of precursor events, which makes the incidence of cancer higher after the age of 40 years [163].

In the natural process of human evolution from youth to old age, inherited native stem cells age and are replaced by new stem cells that are phenotypically labile and prone to turn into cancer cells. New, unstable stem cells can degenerate malign in a short time through a process of methylation in the so-called early exponential phase of accelerated carcinogenesis. In the colon, stem cells in glandular structures are replaced every 8 years [164–166].

Colorectal cancer has a very high incidence worldwide, ranking third and second in terms of cancer mortality. The incidence varies a lot depending on the level of socio-economic development, geographical regions, food, cultural level, etc. The incidence is lower in the countries of Africa and South-Central Asia and much higher in Europe, Australia, and North America [167].

By expanding and deepening knowledge about the molecular biology and pathogenesis of cancer, the treatment of various tumors has greatly improved today. However, chemoresistance, recurrence rate and mortality are still far from being resolved. Therefore, ways of prevention, and especially new means of treatment with increased efficiency, are constantly being sought. Treatment for colon, gastric and many other cancers usually involves surgery to remove pathological tissue and locoregional lymph nodes, followed by radiotherapy and chemotherapy with conventional drugs and/or immunotherapy.

Vetha et al., in an in vitro experimental study on a CT26 murine colorectal carcinoma cell line, used 450 nm blue light diode-induced photodynamic therapy (BLED-PDT) at a power of 2.4 mW/cm^2 for 30 min (6.3 J/cm^2), together with F127-CUR micelles.

The authors report that cell line carcinoma treated with F127-CUR micelles together with BLED significantly reduced cell density and anticipate that this treatment may be promising for cancer eradication [168].

Şueki et al. have tested curcumin, which is a non-toxic compound that has antitumor properties, to check if it can increase the effectiveness of PDT on resistant cancer cells. For this purpose, PC-3 cancer line of prostate cancer cells and the Caco-2 cell lines of colon cancer were used, which were previously identified as non-resistant and resistant to PDT, respectively. Finally, the authors reported that 5-aminolevulinic acid (5-ALA) -mediated PDT, combined with curcumin, has improved the antitumor efficacy of PDT on cells Caco-2, which is considered a highly resistant cancer cell line [169].

In this scenario, the provision of new drugs and scientific treatment methods based on natural products is part of the research conducted by the group led by de Freitas et al., who studied in vitro effects on Caco-2 intestinal cancer cells treated with curcumin-conjugated silver nanoparticles (CUR-AgNPs) and PDT. The results of the study showed that PDT, in the presence of CUR incorporated into hydrogels consisting of CHT and CS, natural biopolymers, capable of the controlled release of CUR-AgNPs, led to the inhibition of human Caco-2 colon cancer cells [170].

Another attempt to obtain a superior photosensitizer in cancer cell therapy was performed by Tsai et al., which encapsulated curcumin, by crosslinking with chitosan, tripolyphosphate (TPP) and conjugation with epidermal growth factor to target the epidermal growth factor receptor (EGFR), overexpressed on cancer cells. The research targeted two cell lines that were used in this study, including a human gastric cancer cell line (MKN45) and the human gastric (non-cancerous) epithelial mucosa (GHG) cell line, which were irradiated with a light-emitting blue diode (460 nm, 5 ± 0.1 mW, measured on the sample surface) for 30 min, at a dose of 9 J/cm^2. Curcumin-encapsulated chitosan/TPP nanoparticles showed a superior PDT effect in the cancer cells, with a four-fold decrease in IC50. This mode of therapy is promising against cancers that overexpress EGFR [171].

Human hepatoblastoma (HB) is the most common form of liver cancer in infants and children under 5 years of age [172].

To reduce the tumor, but especially to eradicate circulating tumor cells, chemotherapy is used before and after surgery. Despite all the positive results, some cases with extensive or recurrent tumors are a major problem, due to the emergence of drug resistance. Ellerkamp et al. investigated in vitro hepatoblastoma cell lines (HuH6, HepT1) and hepatocellular carcinoma cell lines (HepG2, HC-AFW1) treated with curcumin and exposed to blue light (480 nm, 300 W, for 10 s), and revealed decreased cell viability and a significantly increased level of ROS [173].

The emergence of multidrug resistance (MDR) is a major impediment to the long-term success of therapy against various cancers. It is already known that P-glycoprotein (P-gp) is a membrane transporter, which is ATP-dependent, and it has the function to drain the drug molecules from the cancer cell, so that chemotherapy is less effective. In their adaptive evolution, cancer cells protect themselves by increasing P-gp expression, thus avoiding chemotherapy-induced cellular degradation [174,175].

PDT has become an attractive method of treatment for hepatocellular carcinoma, because it is easy to administer and does not affect normal tissues.

To avoid the effects of chemoresistance, Li et al. have developed a new ICG & Cur @ MoS2 system (ICG and Cur represent indocyanine green and curcumin respectively) nanoplatform, which can perform photothermal-photodynamic therapy and inhibit P-gp efficiently and safely. The researchers used HepG-2 cells (human hepatoma cells) cultured in vitro with ICG and Cur @ MoS2 and irradiated with an 808 nm NIR laser at 2.0 W/cm^2 for 5 min to evaluate the photothermal effect. Acute toxicity was investigated in vivo in female mice, in which the tumor was irradiated with an 808 nm (1.2 W/cm^2) NIR laser for 5 min. Cell apoptosis was significant in the ICG @ MoS2 and NIR group, indicating that it was induced by heat and ROS. MRNA decreased significantly in the ICG and Cur @ MoS2 group, indicating that ICG and Cur @ MoS2 inhibited MDR1 transcription. In conclusion, the ICG and Cur @ MoS2 nanoparticles, under the action of PTT-PDT, have inhibited P-gp effectively, and may have great potential in the treatment of hepatocellular carcinoma [176].

4.5. Lung Cancer

Globally, lung cancer is one of the leading causes of death in both men and women. The small cell form represents about 80% of all types of lung cancer, and the one through which deaths are more common, because the diagnosis is generally made when the disease is quite advanced. After the approval of modern biological therapy, the number of deaths was substantially reduced. However, there are special problems because the disease continues to progress in most cases, due to the development of drug resistance [177–179].

As in other forms of cancer due to these problems, research continues to discover drugs or alternative ways to overcome the phenomenon of drug resistance.

Jiang et al. investigated the effects of the new Cur-SLN photosensitizer on A549 cells of small lung cancer cells irradiated with a 430 nm light-emitting diode (LED; power density 50 mW/cm^2) for 20 min. Cur-SLNs were obtained by emulsification and solidification at low temperature. The organic phase with 0.1 g lecithin, 0.15 g Cur and 0.2 g stearic acid was dissolved in 10 mL of chloroform. Moreover, 0.2 g of polyoxyethylene stearate (40) (Myrj52) dissolved in 30 mL of deionized water formed the aqueous phase. The organic phase was injected into the aqueous phase and stirred until the organic solvent disappeared. Then, in an environment at 0–2 °C, 10 mL of cold water were added and stirred at 1200 rpm. The supernatant was removed by centrifugation and the precipitate was washed twice with deionized water, resuspended in ultrapure water, and refrigerated at −80 °C, and finally lyophilized. The results revealed that Cur-SLN increased the expression of caspase-3, caspase-9, and promoted the Bax/Bcl-2 ratio, which demonstrated that this new Cur photosensitizer significantly induced apoptosis in A549 cells for this type of lung cancer [180].

In the last decade, the delivery agents of photosensitizing nano-carriers have been developed a lot, and those who respond very well to near-infrared (NIR) lasers have received special attention because they absorb light very strongly from 700–900 nm.

Local temperature released from the interaction of NIR light with the photosensitizer deeply destroys cancer cells, which gave high hopes for the beneficial effect of eradicating tumors; however, the damage to nearby tissues is still an unresolved issue.

Its outstanding optical properties and biocompatibility have helped Indocyanine Green (ICG) to be FDA approved as a photosensitizer for NIR in clinical use [25,181,182].

Huang et al. developed an antitumor product consisting of a green photosensitizer with indocyanine (ICG) that was co-encapsulated with curcumin (CUR) in liposomes (LPs), followed by conjugation of the GE11 peptide to provide targeting effects to cancer cells that express on their surface epidermal growth factor receptor (EGFR). A fresh medium containing CUR/ICG, free CUR/ICG-LPs and GE11-CUR/ICG-LPs was added to A549 non-small lung cancer cells line, HeLa human cervical cancer cells and LO2 human normal liver cells, incubated for 24 h, followed by 808 nm NIR laser irradiation, (1W/cm^2) for 0, 5, and 10 min. In this study, after the administration of PDT, the ICG photosensitizer generated an increase in temperature in the tumor area and, on the other hand, the GE11-CUR/ICG-LPs complex slowly released CUR, obtaining strong anticancer effects. The research results showed that GE11-CUR/ICG-LPs, together with PDT, could induce the apoptosis of cancer cells by promoting ROS generation and cell cytoskeleton disruption by the increased stimulation of apoptotic signaling pathways and the inhibition of the EGFR-mediated PI3K/AKT pathway [183].

Another research study to obtain a photosensitizer that is easier to administer and with more effective properties was completed by Baghdan et al. They obtained curcumin-loaded PLGA nanoparticles (PLGA.CUR.NPs) as nano-in-microparticles for inhalation administration. In vitro irradiation results on human lung epithelial carcinoma cells (A549) showed a significant increase in cellular phototoxicity, but dependent on the radiation dose used by the 457 nm LED device. In short, the authors argue that nano-in-microparticles are promising carriers of drugs for the photodynamic therapy of lung cancer [184].

4.6. Other Cancers

Glioblastoma multiforme accounts for approximately 80% of primary malignant brain tumors. Despite current innovations in cancer treatment, this type of tumor has a disappointing prognosis [185].

From the results of several randomized international studies reported so far, it can be seen that they have failed to achieve convincing results with long-term survival after immunotherapy [186,187].

Because there is no reasonable treatment in glioblastoma multiforme, the use of PDT with different CUR nanoparticles is being tested experimentally in vitro. In a recent research study, Kielbik et al. analyzed the cytotoxic effects of CUR alone and as a photosensitizer on glioblastoma SNB-19 cells. After incubation with CUR and irradiation with blue light (6 J/cm^2), over 90% of glioblastoma SNB-19 cells underwent apoptosis [188].

Jamali et al. used biodegradable polymeric nanoparticles (PLGA NPs) conjugated with the anti-EGFRvIII monoclonal antibody (MAb-CUR-PLGA NPs) to enhance the photodynamic action of CUR on overexpressed EGFRvIII cells (DKMG/EGFRvIII cells) of glioblastoma tumors. The results demonstrated in vitro that following PDT, the photocytotoxicity of MAb-CUR-PLGA NPs was significantly higher than that of CUR-PLGA NPs in the DKMG/EGFRvIII glioma cells. The authors propose that anti-EGFRvIII MAb-CUR-PLGA NPs should be used as a PDT-specific photosensitizer in overexpressed EGFRvIII tumor cells [189].

Curcumin has been used as a PS in PDT and has been the subject of numerous studies, due to its valuable antiviral, antimicrobial, and especially antineoplastic effects in various forms of human cancer [190–192].

Kazantzis et al. investigated the natural characteristics of bisdemethoxy curcumin regarding their efficacy together with PDT in vitro on prostate cancer cells. In these experiments, all four curcuminoids released enough ROS that produced cytotoxicity after PDT, with a significant reduction in prostate cancer LNCaP cells [193].

5. Final Remarks and Conclusions

As revealed by all the above-mentioned revised studies, the latest nanoformulations of curcumin applied in PDT illustrate its complex actions on tumor cells, as shown in Figure 4.

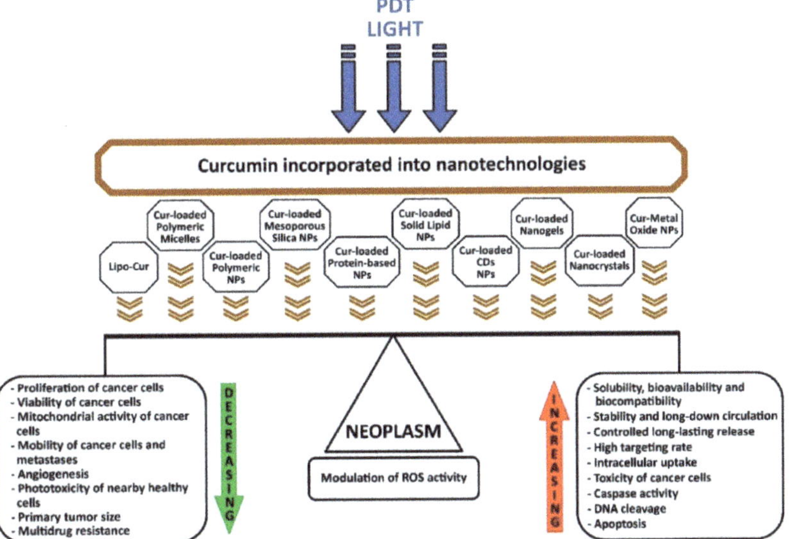

Figure 4. Effects of latest innovations and nanotechnologies with curcumin applied in the photodynamic therapy (PDT) of cancer.

Although curcumin is a polyphenol that has been used in Ayurvedic medicine for many years, as well as a spice for food fragrance, food coloring or adjuvant, its pharmacological complex, anti-inflammatory, antiviral, anti-bacterial, antifungal, anti-neurodegenerative, and its anti-cancer properties have been intensively investigated in recent years. Therapeutic applications of curcumin have been limited, due to its extremely low solubility, instability in body fluids and rapid metabolism. Nanomedicine and the latest nanotechnologies have shown excellent potential to improve the solubility, biocompatibility and therapeutic effects of curcumin.

State-of-the-art research presented here in the photodynamic therapy of various forms of cancer has been able to highlight the significant anticancer potential of curcumin as a valuable photosensitizer when it is combined with nanotechnologies.

In all in vitro or in vivo experiments performed for the study of curcumin nanoformulations, it has been shown to have a high encapsulation efficiency and long-term controlled release, increasing its solubility, bioavailability and biocompatibility, as well as stability and long-down circulation.

Curcumin incorporated into nanotechnologies has a higher intracellular absorption, a superior targeting rate, increased toxicity to cancer cells, accelerates the activity of caspases and DNA cleavage, and ultimately induces apoptosis.

PDT applied together with nanoformulations of curcumin, diminishes the mitochondrial activity of cancer cells, decreases the viability and proliferation of cancer cells, drops the angiogenesis, the cancer cells' mobility and metastases.

The experiments included in this review showed that PDT and nano-curcumin reduce the size of the primary tumor and invert the multidrug resistance in chemotherapy and decrease the resistance to radiotherapy in neoplasms.

PDT applied with curcumin nanoformulations has been shown to have no significant phototoxic effects on healthy cells near the treated tumor.

Regarding breast cancer that came first in terms of incidence, recent experimental research has shown that nanoformulations of curcumin together with PDT can contribute to its treatment by deep infiltration into cancer cells, high cytotoxic action, shrinking tumor volume and stopping metastases.

Current research has shown that the use of PDT and curcumin nanoformulations have a modulating effect on ROS generation, so light or laser irradiation will lead to excessive ROS growth, while nano-curcumin will reduce the activation of ROS-producing enzymes, or will determine the elimination of ROS, seemingly opposite but synergistic phenomena by inducing neoplasm apoptosis, but at the same time, accelerating nearby tissue repair.

All studies reviewed and presented revealed the great potential for applicability of PDT and curcumin nanoformulations in cancer therapy.

This review paves the way for further investigations and advances in this field.

Author Contributions: Conceptualization, L.M.A. and C.A.; writing and original draft preparation, L.M.A.; review and editing, G.L., L.M.A., and C.A.; scientific organizational manuscript issues G.L. All authors (L.M.A., C.A. and G.L.) have read and agreed to the published version of the manuscript. All authors have read and agreed to the published version of the manuscript.

Funding: This research received no external funding.

Institutional Review Board Statement: Not applicable.

Informed Consent Statement: Not applicable.

Data Availability Statement: The references used to support the findings of this study are available from the first (L.M.A.) and the corresponding (G.L.) authors upon request.

Conflicts of Interest: The authors declare no conflict of interest.

Abbreviations

5-Aminolevulinic acid	(5-ALA)
Activated partial thromboplastin time tests	(aPTT)
Adenosine triphosphate	(ATP)
Basal cell carcinoma	(BCC)
Baseline or ground state	(S0)
B-cell lymphoma 2	(Bcl-2)
Bcl-2–associated X	(Bax)
Blue light-emitting diode	(BLED)
Blue-light-emitting diode- photodynamic therapy	(BLED-PDT).
Cervical intraepithelial neoplasia	(CIN)
Chitosan	(CHT)
Cholesteryl hemisuccinate	(CHEMS)
Chondroitin sulfate	(CS)
Circulating tumor cells	(CTCs)
Colorectal cancer	(CRC)
Confocal laser scanning microscopy	(CLSM)
Continuous wave	(CW)
Curcumin	(Cur)
Curcumin conjugated silver nanoparticles	(CUR-AgNPs)
Curcumin loaded PLGA nanoparticles	(CUR-PLGA NPs)
Curcumin nanocrystals	(Cur-NCs)
Curcumin nanoparticles	(CUR-NPs)
Curcumin-loaded liposomes	(Lipo-Cur)
Curcumin-nanoemulsion	(CNE)

Curcumin-loaded solid lipid nanoparticles	(Cur-SLNs)
Cutaneous squamous cell carcinoma	(cSCC)
Cyclodextrins	(CDs)
Demethoxycurcumin	(DMC)
Dendrosomal Nano Curcumin	(DNC)
Deoxyadenosine triphosphate	(dATP)
Dichlorodihydrofluorescein diacetate	(DCFH-DA)
Dioleoyl phosphatidylethanolamine	(DOPE)
Dissolved in dimethyl sulfoxide	(DMSO)
Distearoyl phosphoethanolamine	(DSPE)
EGFRvIII overexpressed human glioblastoma cell line	(DKMG/EGFRvIII cells)
Epidermal growth factor	(EGF)
Epidermal growth factor receptor	(EGFR)
European Medicines Agency	(EMA)
Ferric chloride hexahydrate	(FeCl3·6H2O)
Ferrous sulfate heptahydrate	(FeSO4·7H2O)
Fluorescein isothiocyanate	(FITC)
Fluorescence microscopy imaging system	(FMI)
Food and Drug Administration	(FDA)
Fourier-transform infrared spectroscopy	(FTIR)
Glyceraldehydes 3-phosphate dehydrogenase	(GAPDH)
Immunohistochemistry	(IHC)
Indocyanine Green photosensitizer	(ICG)
Laser irradiation in near-infrared	(NIR)
Light emitting diode	(LED)
Liposome nanocarriers curcumin	(LIP-CUR)
Liposomes	(LPs)
Mesoporous silica nanoparticles	(MSNs)
Monoclonal antibody	(MAb)
Monoclonal antibody against EGFRvIII	(A-EGFRvIII-f)
Mouse embryonic fibroblasts	(MEFs)
Multidrug resistance	(MDR)
Multidrug resistance protein 1	(MDR1)
Myeloid cell leukemia 1	(Mcl-1)
Nanocomposite	(NC)
Nanoparticles	(NPs)
Nanoparticle	(NP)
Nanostructured lipid carriers	(NLCs)
Neutral comet assay	(NCA)
Notch receptor blocker human	(DAPT)
Nuclear Factor-Kappa-B	(NF-κB)
Papillomavirus	(HPV)
Polyethylene glycol	(PEG)
PEGylated lipid nanocarriers	(PLN)
P-glycoprotein	(P-gp)
Phosphate buffer saline	(PBS)
Photodynamic inactivation	(PDI)
Photodynamic therapy	(PDT)
Photosensitizer	(PS)
Photothermal therapy	(PTT)
Poly (lactic-co-glycolic acid) nanoparticles	(PLGA NPs)
Poly (ethylene glycol)	(PEG)
Poly (lactic acid)	(PLA)
Poly (lactic-co-glycolic acid)	(PLGA)
Poly (ε-caprolactone)	(PCL)
Poly (ethylene glycol) polymer	(PEG)
Polyoxyethylene(40)stearate	(Myrj52)
Polyvinylpyrrolidone	(PVP)
Pressure ulcers	(PU)

Reactive oxygen species	(ROS)
Scanning electron microscopy	(SEM)
Silver nanoparticles	(AgNPs)
Excited Singlet state	(S1)
Solid lipid nanoparticles	(SLNs)
Squamous cell carcinoma	(SCC)
Tetraethyl orthosilicate	(TEOS)
The half maximal inhibitory concentration	(IC50)
Transforming growth factor beta	(TGF-β)
Tripolyphosphate	(TPP).
Tumor necrosis factor alpha	(TNF-α)
Tumor Nodes Metastasized	(TNM)
Ultraviolet A	(UVA)
Ultraviolet radiation B	(UVB)
Vascular endothelial growth factor	(VEGF)
Viability measurements	(MTT)
Zinc oxide nanoparticles	(ZnONP$_{CS}$)

References

1. Raab, O. Uber die Wirkung fluoreszierender Stoffe auf Infusorien. *Z. Biol.* **1900**, *39*, 524–526.
2. Dolmans, D.E.; Fukumura, D.; Jain, R.K. Photodynamic therapy for cancer. *Nat. Rev. Cancer* **2003**, *3*, 380–387. [CrossRef]
3. Grzybowski, A.; Sak, J.; Pawlikowski, J. A brief report on the history of phototherapy. *Clin. Dermatol.* **2016**, *34*, 532–537. [CrossRef] [PubMed]
4. Von Tappeiner, H.; Jodlbauer, A. *Die Sensiblilisierende Wirkung Fluoreszierender Substanzen. Gesamte Untersuchungen über die Photodynamische Erscheinung*; FCW Vogel: Leipzig, Germany, 1907.
5. Pröhl, M.; Schubert, U.S.; Weigand, W.; Gottschaldt, M. Metal complexes of curcumin and curcumin derivatives for molecular imaging and anticancer therapy. *Coord. Chem. Rev.* **2016**, *307*, 32–41. [CrossRef]
6. Kwiatkowski, S.; Knap, B.; Przystupski, D.; Saczko, J.; Kędzierska, E.; Knap-Czop, K.; Kotlińska, J.; Michel, O.P.; Kotowski, K.; Kulbacka, J. Photodynamic therapy—Mechanisms, photosensitizers and combinations. *Biomed. Pharmacother.* **2018**, *106*, 1098–1107. [CrossRef] [PubMed]
7. Abrahamse, H.; Hamblin, M.R. New photosensitizers for photodynamic therapy. *Biochem. J.* **2016**, *473*, 347–364. [CrossRef] [PubMed]
8. Zhao, X.; Liu, J.; Fan, J.; Chao, H.; Peng, X. Recent progress in photosensitizers for overcoming the challenges of photodynamic therapy: From molecular design to application. *Chem. Soc. Rev.* **2021**, *50*, 4185. [CrossRef] [PubMed]
9. Huang, Z. A review of progress in clinical photodynamic therapy. *Technol. Cancer Res. Treat.* **2005**, *3*, 283–293. [CrossRef] [PubMed]
10. Figge, F.H.J.; Weiland, G.S.; Nanganiello, L.O.J. Cancer detection and therapy. Affinity of neoplastic, embryonic, and traumatized tissues for porphyrins and metalloporphyrins. *Proc. Soc. Exp. Biol. Med.* **1948**, *68*, 640–641. [CrossRef]
11. Rasmussen-Taxdal, D.S.; Ward, G.E.; Figge, F.H.J. Fluorescence of human lymphatic and cancer tissues following high doses of intravenous hematoporphyrin. *Cancer* **1955**, *8*, 78–81. [CrossRef]
12. Lipson, R.L.; Baldes, E.J.; Olsen, A.M. Hematoporphyrin derivative: A new aid for endoscopic detection of malignant disease. *J. Thorac. Cardiovasc. Surg.* **1961**, *42*, 623–629. [CrossRef]
13. Lipson, R.L.; Baldes, E.J.; Olsen, A.M. The use of a derivative of hematoporphyrin in tumor detection. *J. Natl. Cancer Inst.* **1961**, *26*, 1–11.
14. Agostinis, P.; Berg, K.; Cengel, K.A.; Foster, T.H.; Girotti, A.W.; Gollnick, S.O.; Hahn, S.M.; Hamblin, M.R.; Juzeniene, A.; Kessel, D.; et al. Photodynamic therapy of cancer: An update. *CA Cancer J. Clin.* **2011**, *61*, 250–281. [CrossRef] [PubMed]
15. Dougherty, T.J.; Gomer, C.J.; Henderson, B.W.; Jori, G.; Kessel, D.; Korbelik, M.; Moan, J.; Peng, Q. Photodynamic therapy. *J. Natl. Cancer Inst.* **1998**, *90*, 889–905. [CrossRef]
16. Kessel, D. Photodynamic Therapy: A Brief History. *J. Clin. Med.* **2019**, *8*, 1581. [CrossRef]
17. Kessel, D.; Thomas, J. Dougherty: An Appreciation. *Photochem. Photobiol.* **2020**, *96*, 454–457. [CrossRef]
18. Sivasubramanian, M.; Chuang, Y.C.; Lo, L.-W. Evolution of Nanoparticle-Mediated Photodynamic Therapy: From Superficial to Deep-Seated Cancers. *Molecules* **2019**, *24*, 520. [CrossRef]
19. Skovsen, E.; Snyder, J.W.; Lambert, J.D.C.; Ogilby, P.R. Lifetime and Diffusion of Singlet Oxygen in a Cell. *J. Phys. Chem. B* **2005**, *109*, 8570–8573. [CrossRef]
20. Ożog, L.; David Aebisher, D. Singlet oxygen lifetime and diffusion measurements. *Eur. J. Clin. Exp. Med.* **2018**, *16*, 123–126. [CrossRef]
21. Josefsen, L.B.; Boyle Ross, W. Photodynamic Therapy and the Development of Metal-Based Photosensitisers. *Metal-Based Drugs* **2008**, *2008*, 276109. [CrossRef] [PubMed]

22. Dysart, J.S.; Patterson, M.S. Characterization of Photofrin photobleaching for singlet oxygen dose estimation during photodynamic therapy of MLL cells in vitro. *Phys. Med. Biol.* **2005**, *50*, 2597–2616. [CrossRef]
23. Polat, E.; Kang, K. Natural Photosensitizers in Antimicrobial Photodynamic Therapy. *Biomedicines* **2021**, *9*, 584. [CrossRef]
24. Babu, B.; Mack, J.; Nyokong, T. Sn(iv) N-confused porphyrins as photosensitizer dyes for photodynamic therapy in the near IR region. *Dalton Trans.* **2020**, *49*, 15180–15183. [CrossRef]
25. Wang, Z.; Zhao, J. Bodipy–Anthracene Dyads as Triplet Photosensitizers: Effect of Chromophore Orientation on Triplet-State Formation Efficiency and Application in Triplet–Triplet Annihilation Upconversion. *Org. Lett.* **2017**, *19*, 4492–4495. [CrossRef] [PubMed]
26. Xiao, Q.; Wu, J.; Pang, X.; Jiang, Y.; Wang, P.; Leung, A.W.; Gao, L.; Jiang, S.; Xu, C. Discovery and Development of Natural Products and their Derivatives as Photosensitizers for Photodynamic Therapy. *Curr. Med. Chem.* **2017**, *25*, 839–860. [CrossRef] [PubMed]
27. Allison, R.R.; Sibata, C.H. Oncologic photodynamic therapy photosensitizers: A clinical review. *Photodiagnosis Photodyn. Ther.* **2010**, *7*, 61–75. [CrossRef] [PubMed]
28. Niculescu, A.-G.; Grumezescu, A.M. Photodynamic Therapy—An Up-to-Date Review. *Appl. Sci.* **2021**, *11*, 3626. [CrossRef]
29. Zhang, J.; Zhang, R.; Liu, K.; Li, Y.; Wang, X.; Xie, X.; Jiao, X.; Tang, B. A light-activatable photosensitizer for photodynamic therapy based on a diarylethene derivative. *Chem. Commun.* **2021**, *57*, 8320–8323. [CrossRef]
30. Pye, H.; Stamati, I.; Yahioglu, G.; Butt, M.A.; Deonarain, M. Antibody-Directed Phototherapy (ADP). *Antibodies* **2013**, *2*, 270–305. [CrossRef]
31. Ambreen, G.; Duse, L.; Tariq, I.; Ali, U.; Ali, S.; Pinnapireddy, S.R.; Bette, M.; Bakowsky, U.; Mandic, R. Sensitivity of Papilloma Virus-Associated Cell Lines to Photodynamic Therapy with Curcumin-Loaded Liposomes. *Cancers* **2020**, *12*, 3278. [CrossRef] [PubMed]
32. Moballegh Nasery, M.; Abadi, B.; Poormoghadam, D.; Zarrabi, A.; Keyhanvar, P.; Khanbabaei, H.; Ashrafizadeh, M.; Mohammadinejad, R.; Tavakol, S.; Sethi, G. Curcumin Delivery Mediated by Bio-Based Nanoparticles: A Review. *Molecules* **2020**, *25*, 689. [CrossRef]
33. Lehmann, J.; Agel, M.R.; Engelhardt, K.H.; Pinnapireddy, S.R.; Agel, S.; Duse, L.; Preis, E.; Wojcik, M.; Bakowsky, U. Improvement of Pulmonary Photodynamic Therapy: Nebulisation of Curcumin-Loaded Tetraether Liposomes. *Pharmaceutics* **2021**, *13*, 1243. [CrossRef]
34. Cabral, H.; Kataoka, K. Progress of drug-loaded polymeric micelles into clinical studies. *J. Control. Release* **2014**, *190*, 465–476. [CrossRef]
35. Chang, T.; Trench, D.; Putnam, J.; Stenzel, M.H.; Lord, M.S. Curcumin-loading-dependent stability of PEGMEMA-based micelles affects endocytosis and exocytosis in colon carcinoma cells. *Mol. Pharm.* **2016**, *13*, 924–932. [CrossRef]
36. Kataoka, K.; Harada, A.; Nagasaki, Y. Block copolymer micelles for drug delivery: Design, characterization and biological significance. *Adv. Drug Deliver Rev.* **2012**, *64*, 37–48. [CrossRef]
37. Liu, L.; Sun, L.; Wu, Q.; Guo, W.; Li, L.; Chen, Y.; Li, Y.; Gong, C.; Qian, Z.; Wei, Y. Curcumin loaded polymeric micelles inhibit breast tumor growth and spontaneous pulmonary metastasis. *Int. J. Pharm.* **2013**, *443*, 175–182. [CrossRef] [PubMed]
38. Jin, H.; Pi, J.; Zhao, Y.; Jiang, J.; Li, T.; Zeng, X.; Yang, P.; Evans, C.E.; Cai, J. EGFR-targeting PLGA-PEG nanoparticles as a curcumin delivery system for breast cancer therapy. *Nanoscale* **2017**, *9*, 16365–16374. [CrossRef] [PubMed]
39. Rudramurthy, G.; Swamy, M.; Sinniah, U.; Ghasemzadeh, A. Nanoparticles: Alternatives against drug-resistant pathogenic microbes. *Molecules* **2016**, *21*, 836. [CrossRef]
40. Ferrari, R.; Sponchioni, M.; Morbidelli, M.; Moscatelli, D. Polymer nanoparticles for the intravenous delivery of anticancer drugs: The checkpoints on the road from the synthesis to clinical translation. *Nanoscale* **2018**, *10*, 22701–22719. [CrossRef]
41. Kayani, Z.; Vais, R.D.; Soratijahromi, E.; Mohammadi, S.; Sattarahmady, N. Curcumin-gold-polyethylene glycol nanoparticles as a nanosensitizer for photothermal and sonodynamic therapies: In vitro and animal model studies. *Photodiagn. Photodyn. Ther.* **2021**, *33*, 102139. [CrossRef]
42. Ayubi, M.; Karimi, M.; Abdpour, S.; Rostamizadeh, K.; Parsa, M.; Zamani, M.; Saedi, A. Magnetic nanoparticles decorated with PEGylated curcumin as dual targeted drug delivery: Synthesis, toxicity and biocompatibility study. *Mater. Sci. Eng. C Mater. Biol. Appl.* **2019**, *104*, 109810. [CrossRef]
43. Zhou, Y.; Quan, G.; Wu, Q.; Zhang, X.; Niu, B.; Wu, B.; Huang, Y.; Pan, X.; Wu, C. Mesoporous silica nanoparticles for drug and gene delivery. *Acta Pharm. Sin. B* **2018**, *8*, 165–177. [CrossRef]
44. Kong, Z.L.; Kuo, H.P.; Johnson, A.; Wu, L.C.; Chang, K. Curcumin-Loaded Mesoporous Silica Nanoparticles Markedly Enhanced Cytotoxicity in Hepatocellular Carcinoma Cells. *Int. J. Mol. Sci.* **2019**, *20*, 2918. [CrossRef]
45. Liu, Y.; Ying, D.; Cai, Y.; Le, X. Improved antioxidant activity and physicochemical properties of curcumin by adding ovalbumin and its structural characterization. *Food Hydrocolloid.* **2017**, *72*, 304–311. [CrossRef]
46. Sorolla, A.; Sorolla, A.M.; Wang, E.; Ceña, V. Peptides, proteins and nanotechnology: A promising synergy for breast cancer targeting and treatment. *Expert Opin. Drug Deliv.* **2020**, *11*, 1597–1613. [CrossRef]
47. Sun, J.; Bi, C.; Chan, H.M.; Sun, S.; Zhang, Q.; Zheng, Y. Curcumin-loaded solid lipid nanoparticles have prolonged in vitro antitumour activity, cellular uptake, and improved in vivo bioavailability. *Colloids Surf. B Biointerfaces* **2013**, *111*, 367–375. [CrossRef] [PubMed]

48. Jourghanian, P.; Ghaffari, S.; Ardjmand, M.; Haghighat, S.; Mohammadnejad, M. Sustained release Curcumin loaded Solid Lipid Nanoparticles. *Adv. Pharm. Bull.* **2016**, *6*, 17–21. [CrossRef] [PubMed]
49. Wang, W.; Chen, T.; Xu, H.; Ren, B.; Cheng, X.; Qi, R.; Liu, H.; Wang, Y.; Yan, L.; Chen, S.; et al. Curcumin-Loaded Solid Lipid Nanoparticles Enhanced Anticancer Efficiency in Breast Cancer. *Molecules* **2018**, *23*, 1578. [CrossRef]
50. He, H.; Chen, S.; Zhou, J.; Dou, Y.; Song, L.; Che, L.; Zhou, X.; Chen, X.; Jia, Y.; Zhang, J.; et al. Cyclodextrin-derived pH-responsive nanoparticles for delivery of paclitaxel. *Biomaterials* **2013**, *34*, 5344–5358. [CrossRef] [PubMed]
51. Zhang, L.; Man, S.; Qiu, H.; Liu, Z.; Zhang, M.; Ma, L.; Gao, W. Curcumin-cyclodextrin complexes enhanced the anti-cancer effects of curcumin. *Environ. Toxicol. Pharmacol.* **2016**, *48*, 31–38. [CrossRef]
52. Reddy, D.N.K.; Kumar, R.; Wang, S.P.; Huang, F.Y. Curcumin-C3 Complexed with α-, β-cyclodextrin Exhibits Antibacterial and Antioxidant Properties Suitable for Cancer Treatments. *Curr. Drug Metab.* **2019**, *20*, 988–1001. [CrossRef] [PubMed]
53. Reeves, A.; Vinogradov, S.V.; Morrissey, P.; Chernin, M.; Ahmed, M.M. Curcumin-encapsulating Nanogels as an Effective Anticancer Formulation for Intracellular Uptake. *Mol. Cell Pharmacol.* **2015**, *7*, 25–40. [CrossRef] [PubMed]
54. Howaili, F.; Özliseli, E.; Küçüktürkmen, B.; Razavi, S.M.; Sadeghizadeh, M.; Rosenholm, J.M. Stimuli-Responsive, Plasmonic Nanogel for Dual Delivery of Curcumin and Photothermal Therapy for Cancer Treatment. *Front. Chem.* **2021**, *8*, 602941. [CrossRef] [PubMed]
55. Ghalandarlaki, N.; Alizadeh, A.M.; Ashkani-Esfahani, S. Nanotechnology-applied curcumin for different diseases therapy. *Biomed. Res. Int.* **2014**, *2014*, 394264. [CrossRef] [PubMed]
56. Wang, X.; Peng, Y.; Tan, H.; Li, M.; Li, W. Curcumin nanocrystallites are an ideal nanoplatform for cancer chemotherapy. *Front Nanosci. Nanotech.* **2019**, *5*, 1–4. [CrossRef]
57. Chen, Y.; Lu, Y.; Lee, R.J.; Xiang, G. Nano Encapsulated Curcumin: And Its Potential for Biomedical Applications. *Int. J. Nanomed.* **2020**, *15*, 3099–3120. [CrossRef]
58. Beyene, A.M.; Moniruzzaman, M.; Karthikeyan, A.; Min, T. Curcumin Nanoformulations with Metal Oxide Nanomaterials for Biomedical Applications. *Nanomaterials* **2021**, *1*, 460. [CrossRef]
59. Somu, P.; Paul, S. A biomolecule-assisted one-pot synthesis of zinc oxide nanoparticles and its bioconjugate with curcumin for potential multifaceted therapeutic applications. *New J. Chem.* **2019**, *43*, 11934–11948. [CrossRef]
60. Globocan 2020: New Global Cancer Data. 17 December 2020. Available online: https://www.uicc.org/news/globocan-2020-new-global-cancer-data (accessed on 20 July 2021).
61. Siegel, R.L.; Miller, K.D.; Jemal, A. Cancer statistics, 2020. *CA Cancer J. Clin.* **2020**, *70*, 7–30. [CrossRef]
62. Visvader, J.E.; Lindeman, G.J. Cancer stem cells in solid tumours: Accumulating evidence and unresolved questions. *Nat. Rev. Cancer* **2008**, *8*, 755–768. [CrossRef]
63. Reddy, R.M.; Kakarala, M.; Wicha, M.S. Clinical Trial Design for Testing the Stem Cell Model for the Prevention and Treatment of Cancer. *Cancers* **2011**, *3*, 2696–2708. [CrossRef]
64. Park, C.H.; Hahm, E.R.; Park, S.; Kim, H.K.; Yang, C.H. The inhibitory mechanism of curcumin and its derivative against beta-catenin/Tcf signaling. *FEBS Lett.* **2005**, *579*, 2965–2971. [CrossRef]
65. Kakarala, M.; Brenner, D.E.; Korkaya, H.; Cheng, C.; Tazi, K.; Ginestier, C.; Liu, S.; Dontu, G.; Wicha, M.S. Targeting breast stem cells with the cancer preventive compounds curcumin and piperine. *Breast Cancer Res. Treat.* **2010**, *122*, 77–785. [CrossRef]
66. Carroll, R.E.; Benya, R.V.; Turgeon, D.K.; Vareed, S.; Neuman, M.; Rodriguez, L.; Kakarala, M.; Carpenter, P.M.; McLaren, C.; Meyskens, F.L., Jr.; et al. Phase IIa clinical trial of curcumin for the prevention of colorectal neoplasia. *Cancer Prev. Res.* **2011**, *4*, 354–364. [CrossRef] [PubMed]
67. Mao, X.; Zhang, X.; Zheng, X.; Chen, Y.; Xuan, Z.; Huang, P. Curcumin suppresses LGR5(+) colorectal cancer stem cells by inducing autophagy and via repressing TFAP2A-mediated ECM pathway. *J. Nat. Med.* **2021**, *75*, 590–601. [CrossRef] [PubMed]
68. Beltzig, L.; Frumkina, A.; Schwarzenbach, C.; Kaina, B. Cytotoxic, Genotoxic and Senolytic Potential of Native and Micellar Curcumin. *Nutrients* **2021**, *13*, 2385. [CrossRef]
69. Zarrabi, A.; Zarepour, A.; Khosravi, A.; Alimohammadi, Z.; Thakur, V.K. Synthesis of Curcumin Loaded Smart pH-Responsive Stealth Liposome as a Novel Nanocarrier for Cancer Treatment. *Fibers* **2021**, *9*, 19. [CrossRef]
70. Garufi, A.; Giorno, E.; Gilardini Montani, M.S.; Pistritto, G.; Crispini, A.; Cirone, M.; D'Orazi, G. p62/SQSTM1/Keap1/NRF2 Axis Reduces Cancer Cells Death-Sensitivity in Response to Zn(II)–Curcumin Complex. *Biomolecules* **2021**, *11*, 348. [CrossRef]
71. Rukoyatkina, N.; Shpakova, V.; Sudnitsyna, J.; Panteleev, M.; Makhoul, S.; Gambaryan, S.; Jurk, K. Curcumin at Low Doses Potentiates and at High Doses Inhibits ABT-737-Induced Platelet Apoptosis. *Int. J. Mol. Sci.* **2021**, *22*, 5405. [CrossRef] [PubMed]
72. Huntosova, V.; Horvath, D.; Seliga, R.; Wagnieres, G. Influence of Oxidative Stress on Time-Resolved Oxygen Detection by [Ru(Phen)$_3$]$^{2+}$ In Vivo and In Vitro. *Molecules* **2021**, *26*, 485. [CrossRef] [PubMed]
73. Lee, W.-H.; Loo, C.-Y.; Traini, D.; Young, P.M. Development and Evaluation of Paclitaxel and Curcumin Dry Powder for Inhalation Lung Cancer Treatment. *Pharmaceutics* **2021**, *13*, 9. [CrossRef]
74. Wan Mohd Tajuddin, W.N.B.; Abas, F.; Othman, I.; Naidu, R. Molecular Mechanisms of Antiproliferative and Apoptosis Activity by 1,5-Bis(4-Hydroxy-3-Methoxyphenyl)1,4-Pentadiene-3-one (MS13) on Human Non-Small Cell Lung Cancer Cells. *Int. J. Mol. Sci.* **2021**, *22*, 7424. [CrossRef]
75. World Health Organization. Breast Cancer. 26 March 2021. Available online: https://www.who.int/news-room/fact-sheets/detail/breast-cancer (accessed on 20 July 2021).

76. National Cancer Institute. Advances in Breast Cancer Research. 9 April 2021. Available online: https://www.cancer.gov/types/breast/research (accessed on 20 July 2021).
77. Yang, K.; Liao, Z.; Wu, Y.; Li, M.; Guo, T.; Lin, J.; Li, Y.; Hu, C. Curcumin and Glu-GNPs Induce Radiosensitivity against Breast Cancer Stem-Like Cells. *BioMed Res. Int.* **2020**, *2020*, 3189217. [CrossRef]
78. Minafra, L.; Porcino, N.; Bravatà, V.; Gaglio, D.; Bonanomi, M.; Amore, E.; Cammarata, F.P.; Russo, G.; Militello, C.; Savoca, G.; et al. Radiosensitizing effect of curcumin-loaded lipid nanoparticles in breast cancer cells. *Sci. Rep.* **2019**, *9*, 11134. [CrossRef]
79. Saberi-Karimian, M.; Katsiki, N.; Caraglia, M.; Boccellino, M.; Majeed, M.; Sahebkar, A. Vascular endothelial growth factor: An important molecular target of curcumin. *Crit. Rev. Food Sci. Nutr.* **2019**, *59*, 299–312. [CrossRef]
80. Gouthamchandra, K.; Sudeep, H.V.; Chandrappa, S.; Raj, A.; Naveen, P.; Shyamaprasad, K. Efficacy of a Standardized Turmeric Extract Comprised of 70% Bisdemothoxy-Curcumin (REVERC3) Against LPS-Induced Inflammation in RAW264.7 Cells and Carrageenan-Induced Paw Edema. *J. Inflamm. Res.* **2021**, *14*, 859–868. [CrossRef]
81. Huang, T.; Chen, Z.; Fang, L. Curcumin inhibits LPS-induced EMT through downregulation of NF-κB-Snail signaling in breast cancer cells. *Oncol. Rep.* **2013**, *29*, 117–124. [CrossRef]
82. Wang, Y.; Tang, Q.; Duan, P.; Yang, L. Curcumin as a therapeutic agent for blocking NF-κB activation in ulcerative colitis. *Immunopharmacol. Immunotoxicol.* **2018**, *40*, 476–482. [CrossRef] [PubMed]
83. Ghasemi, F.; Shafiee, M.; Banikazemi, Z.; Pourhanifeh, M.H.; Khanbabaei, H.; Shamshirian, A.; Amiri Moghadam, S.; ArefNezhad, R.; Sahebkar, A.; Avan, A.; et al. Curcumin inhibits NF-kB and Wnt/β-catenin pathways in cervical cancer cells. *Pathol. Res. Pract.* **2019**, *215*, 152556. [CrossRef] [PubMed]
84. Song, X.; Zhang, M.; Dai, E.; Luo, Y. Molecular targets of curcumin in breast cancer. (Review). *Mol. Med. Rep.* **2019**, *19*, 23–29. [CrossRef] [PubMed]
85. Ombredane, A.S.; Silva, V.R.P.; Andrade, L.R.; Pinheiro, W.O.; Simonelly, M.; Oliveira, J.V.; Pinheiro, A.C.; Gonçalves, G.F.; Felice, G.J.; Garcia, M.P.; et al. In Vivo Efficacy and Toxicity of Curcumin Nanoparticles in Breast Cancer Treatment: A Systematic Review. *Front. Oncol.* **2021**, *11*, 612903. [CrossRef]
86. Shahabipour, F.; Caraglia, M.; Majeed, M.; Derosa, G.; Maffioli, P.; Sahebkar, A. Naturally occurring anti-cancer agents targeting EZH2. *Cancer Lett.* **2017**, *400*, 325–335. [CrossRef]
87. Guan, X.; Shi, A.; Zou, Y.; Sun, M.; Zhan, Y.; Dong, Y.; Fan, Z. EZH2-Mediated microRNA-375 Upregulation Promotes Progression of Breast Cancer via the Inhibition of FOXO1 and the p53 Signaling Pathway. *Front. Genet.* **2021**, *12*, 633756. [CrossRef]
88. Gallardo, M.; Kemmerling, U.; Aguayo, F.; Bleak, T.C.; Muñoz, J.P.; Calaf, G.M. Curcumin rescues breast cells from epithelial−mesenchymal transition and invasion induced by anti−miR−34a. *Int. J. Oncol.* **2020**, *56*, 480–493. [CrossRef]
89. Flora, G.; Gupta, D.; Tiwari, A. Nanocurcumin: A promising therapeutic advancement over native curcumin. *Crit. Rev. Ther. Drug Carrier Syst.* **2013**, *30*, 331–368. [CrossRef] [PubMed]
90. Karthikeyan, A.; Senthil, N.; Min, T. Nanocurcumin: A Promising Candidate for Therapeutic Applications. *Front. Pharmacol.* **2020**, *11*, 487. [CrossRef] [PubMed]
91. Silva de Sá, I.; Peron, A.P.; Leimann, F.V.; Bressan, G.N.; Krum, B.N.; Fachinetto, R.; Pinela, J.; Calhelha, R.C.; Barreiro, M.F.; Ferreira, I.C.F.R.; et al. In vitro and in vivo evaluation of enzymatic and antioxidant activity, cytotoxicity and genotoxicity of curcumin loaded solid dispersions. *Food Chem. Toxicol.* **2019**, *125*, 29–37. [CrossRef] [PubMed]
92. Preis, E.; Baghdan, E.; Agel, M.R.; Anders, T.; Pourasghar, M.; Schneider, M.; Bakowsky, U. Spray dried curcumin loaded nanoparticles for antimicrobial photodynamic therapy. *Eur. J. Pharm. Biopharm.* **2019**, *142*, 531–539. [CrossRef]
93. Damyeh, M.S.; Mereddy, R.; Netzel, M.E.; Sultanbawa, Y. An insight into curcumin-based photosensitization as a promising and green food preservation technology. *Compr. Rev. Food Sci. Food Saf.* **2020**, *19*, 1727–1759. [CrossRef]
94. Sun, M.; Zhang, Y.; He, Y.; Xiong, M.; Huang, H.; Pei, S.; Liao, J.; Wang, Y.; Shao, D. Green synthesis of carrier-free curcumin nanodrugs for light-activated breast cancer photodynamic therapy. *Colloids Surf. B Biointerfaces* **2019**, *180*, 313–318. [CrossRef]
95. Singh, Y.; Meher, J.G.; Raval, K.; Khan, F.A.; Chaurasia, M.; Jain, N.K.; Chourasia, M.K. Nanoemulsion: Concepts, development and applications in drug delivery. *J. Control. Release* **2017**, *252*, 28–49. [CrossRef]
96. Barkat, M.A.; Harshita Rizwanullah, M.; Pottoo, F.H.; Beg, S.; Akhter, S.; Ahmad, F.J. Therapeutic Nanoemulsion: Concept to Delivery. *Curr. Pharm. Des.* **2020**, *26*, 1145–1166. [CrossRef] [PubMed]
97. Sarheed, O.; Shouqair, D.; Ramesh, K.V.R.N.S.; Khaleel, T.; Amin, M.; Boateng, J.; Drechsler, M. Formation of stable nanoemulsions by ultrasound-assisted two-step emulsification process for topical drug delivery: Effect of oil phase composition and surfactant concentration and loratadine as ripening inhibitor. *Int. J. Pharm.* **2020**, *576*, 118952. [CrossRef]
98. Machado, F.C.; Adum de Matos, R.P.; Primo, F.L.; Tedesco, A.C.; Rahal, P.; Calmon, M.F. Effect of curcumin-nanoemulsion associated with photodynamic therapy in breast adenocarcinoma cell line. *Bioorg. Med. Chem.* **2019**, *27*, 1882–1890. [CrossRef] [PubMed]
99. Mehanny, M.; Hathout, R.M.; Geneidi, A.S.; Mansour, S. Exploring the use of nanocarrier systems to deliver the magical molecule; Curcumin and its derivatives. *J. Control. Release* **2016**, *225*, 1–30. [CrossRef]
100. Kesharwani, S.S.; Ahmad, R.; Bakkari, M.A.; Rajput, M.; Dachineni, R.; Valiveti, C.K.; Kapur, S.; Jayarama Bhat, G.; Singh, A.B.; Tummala, H. Site-directed non-covalent polymer-drug complexes for inflammatory bowel disease (IBD): Formulation development, characterization and pharmacological evaluation. *J. Control. Release* **2018**, *290*, 165–179. [CrossRef] [PubMed]
101. Menyailo, M.E.; Bokova, U.A.; Ivanyuk, E.E.; Khozyainova, A.A.; Denisov, E.V. Metastasis Prevention: Focus on Metastatic Circulating Tumor Cells. *Mol. Diagn. Ther.* **2021**, *25*, 549–562. [CrossRef]

102. Raschpichler, M.; Preis, E.; Pinnapireddy, S.R.; Baghdan, E.; Pourasghar, M.; Schneider, M.; Bakowsky, U. Photodynamic inactivation of circulating tumor cells: An innovative approach against metastatic cancer. *Eur. J. Pharm. Biopharm.* **2020**, *157*, 38–46. [CrossRef]

103. Kamel, A.E.; Fadel, M.; Louis, D. Curcumin-loaded nanostructured lipid carriers prepared using Peceol™ and olive oil in photodynamic therapy: Development and application in breast cancer cell line. *Int. J. Nanomed.* **2019**, *14*, 5073–5085. [CrossRef]

104. Ghorbani, F.; Attaran-Kakhki, N.; Sazgarnia, A. The synergistic effect of photodynamic therapy and photothermal therapy in the presence of gold-gold sulfide nanoshells conjugated Indocyanine green on HeLa cells. *Photodiagn. Photodyn. Ther.* **2017**, *17*, 48–55. [CrossRef]

105. Hu, J.J.; Cheng, Y.J.; Zhang, X.Z. Recent advances in nanomaterials for enhanced photothermal therapy of tumors. *Nanoscale* **2018**, *10*, 22657–22672. [CrossRef]

106. Hou, Y.J.; Yang, X.X.; Liu, R.Q.; Zhao, D.; Guo, C.N.; Zhu, A.C.; Wen, M.N.; Liu, Z.; Qu, G.F.; Meng, H. Pathological Mechanism of Photodynamic Therapy and Photothermal Therapy Based on Nanoparticles. *Int. J. Nanomed.* **2020**, *15*, 6827–6838. [CrossRef] [PubMed]

107. Nomura, S.; Morimoto, Y.; Tsujimoto, H.; Arake, M.; Harada, M.; Saitoh, D.; Hara, I.; Ozeki, E.; Satoh, A.; Takayama, E.; et al. Highly reliable, targeted photothermal cancer therapy combined with thermal dosimetry using a near-infrared absorbent. *Sci. Rep.* **2020**, *10*, 9765. [CrossRef] [PubMed]

108. Liu, H.; Hu, Y.; Sun, Y.; Wan, C.; Zhang, Z.; Dai, X.; Lin, Z.; He, Q.; Yang, Z.; Huang, P.; et al. Co-delivery of bee venom melittin and a photosensitizer with an organic-inorganic hybrid nanocarrier for photodynamic therapy and immunotherapy. *ACS Nano* **2019**, *13*, 12638–12652. [CrossRef] [PubMed]

109. Ashkbar, A.; Rezaei, F.; Attari, F.; Ashkevarian, S. Treatment of breast cancer in vivo by dual photodynamic and photothermal approaches with the aid of curcumin photosensitizer and magnetic nanoparticles. *Sci. Rep.* **2020**, *10*, 21206. [CrossRef] [PubMed]

110. Halevas, E.; Arvanitidou, M.; Mavroidi, B.; Hatzidimitriou, A.G.; Politopoulos, K.; Alexandratou, E.; Pelecanou, M.; Sagnou, M. A novel curcumin gallium complex as photosensitizer in photodynamic therapy: Synthesis, structural and physicochemical characterization, photophysical properties and in vitro studies against breast cancer cells. *J. Mol. Struct.* **2021**, *1240*, 130485. [CrossRef]

111. Siegel, R.L.; Miller, K.D.; Fuchs, H.E.; Jemal, A. Cancer Statistics, 2021. *CA Cancer J. Clin.* **2021**, *71*, 7–33, Erratum in **2021**, *71*, 359. [CrossRef]

112. Henley, S.J.; Ward, E.M.; Scott, S.; Ma, J.; Anderson, R.N.; Firth, A.U.; Thomas, C.C.; Islami, F.; Weir, H.K.; Lewis, D.R.; et al. Annual Report to the Nation on the Status of Cancer, part I: National cancer statistics. *Cancer* **2020**, *126*, 2225–2249. [CrossRef]

113. Pfeiffer, R.M.; Webb-Vargas, Y.; Wheeler, W.; Gail, M.H. Proportion of U.S. trends in breast cancer incidence attributable to long-term changes in risk factor distributions. *Cancer Epidemiol. Biomark. Prev.* **2018**, *27*, 1214–1222. [CrossRef]

114. Lortet-Tieulent, J.; Ferlay, J.; Bray, F.; Jemal, A. International patterns and trends in endometrial cancer incidence, 1978–2013. *J. Natl. Cancer Inst.* **2018**, *110*, 354–361. [CrossRef]

115. Jemal, A.; Ward, E.M.; Johnson, C.J.; Cronin, K.A.; Ma, J.; Ryerson, B.; Mariotto, A.; Lake, A.J.; Wilson, R.; Sherman, R.L.; et al. Annual Report to the Nation on the Status of Cancer, 1975–2014, featuring survival. *J. Natl. Cancer Inst.* **2017**, *109*, djx030. [CrossRef]

116. Arbyn, M.; Weiderpass, E.; Bruni, L.; de Sanjosé, S.; Saraiya, M.; Ferlay, J.; Bray, F. Estimates of incidence and mortality of cervical cancer in 2018: A worldwide analysis. *Lancet Glob. Health* **2020**, *8*, e191–e203. [CrossRef]

117. De Matos, R.; Calmon, M.F.; Amantino, C.F.; Villa, L.L.; Primo, F.L.; Tedesco, A.C.; Rahal, P. Effect of Curcumin-Nanoemulsion Associated with Photodynamic Therapy in Cervical Carcinoma Cell Lines. *Biomed. Res. Int.* **2018**, *2018*, 4057959. [CrossRef]

118. Makadia, H.K.; Siegel, S.J. Poly Lactic-co-Glycolic Acid (PLGA) as Biodegradable Controlled Drug Delivery Carrier. *Polymers* **2011**, *3*, 1377–1397. [CrossRef]

119. Patel, J.; Amrutiya, J.; Bhatt, P.; Javia, A.; Jain, M.; Misra, A. Targeted delivery of monoclonal antibody conjugated docetaxel loaded PLGA nanoparticles into EGFR overexpressed lung tumour cells. *J. Microencapsul.* **2018**, *35*, 204–217. [CrossRef]

120. Loureiro, J.A.; Pereira, M.C. PLGA Based Drug Carrier and Pharmaceutical Applications: The Most Recent Advances. *Pharmaceutics* **2020**, *12*, 903. [CrossRef] [PubMed]

121. Duse, L.; Agel, M.R.; Pinnapireddy, S.R.; Schäfer, J.; Selo, M.A.; Ehrhardt, C.; Bakowsky, U. Photodynamic Therapy of Ovarian Carcinoma Cells with Curcumin-Loaded Biodegradable Polymeric Nanoparticles. *Pharmaceutics* **2019**, *11*, 282. [CrossRef] [PubMed]

122. He, G.; Mu, T.; Yuan, Y.; Yang, W.; Zhang, Y.; Chen, Q.; Bian, M.; Pan, Y.; Xiang, Q.; Chen, Z.; et al. Effects of Notch Signaling Pathway in Cervical Cancer by Curcumin Mediated Photodynamic Therapy and Its Possible Mechanisms in Vitro and in Vivo. *J. Cancer* **2019**, *10*, 4114–4122. [CrossRef] [PubMed]

123. Orzechowska, M.; Anusewicz, D.; Bednarek, A.K. Functional Gene Expression Differentiation of the Notch Signaling Pathway in Female Reproductive Tract Tissues-A Comprehensive Review with Analysis. *Front. Cell Dev. Biol.* **2020**, *8*, 592616. [CrossRef] [PubMed]

124. Markman, M. Skin Cancer Types. This Page Was Updated on 22 July 2021. Available online: https://www.cancercenter.com/cancer-types/skin-cancer/types (accessed on 8 August 2021).

125. Key Statistics for Basal and Squamous Cell Skin Cancers. American Cancer Society. 12 January 2021. Available online: http://www.cancer.org/cancer/skincancer-basalandsquamouscell/detailedguide/skin-cancer-basal-and-squamous-cell-key-statistics (accessed on 8 August 2021).
126. American Cancer Society. Cancer Facts and Figures. 2021. Available online: https://www.cancer.org/content/dam/cancer-org/research/cancer-facts-and-statistics/annual-cancer-facts-and-figures/2021/cancer-facts-and-figures-2021.pdf (accessed on 8 August 2021).
127. Stratigos, A.; Garbe, C.; Lebbe, C.; Malvehy, J.; del Marmol, V.; Pehamberger, H.; Peris, K.; Becker, J.C.; Zalaudek, I.; Saiag, P.; et al. Diagnosis and treatment of invasive squamous cell carcinoma of the skin: European consensus-based interdisciplinary guideline. *Eur. J. Cancer* **2015**, *5*, 1989–2007. [CrossRef]
128. Kyrgidis, A.; Tzellos, T.G.; Kechagias, N.; Patrikidou, A.; Xirou, P.; Kitikidou, K.; Bourlidou, E.; Vahtsevanos, K.; Antoniades, K. Cutaneous squamous cell carcinoma (SCC) of the head and neck: Risk factors of overall and recurrence-free survival. *Eur. J. Cancer* **2010**, *46*, 1563–1572. [CrossRef]
129. Rogers, H.W.; Weinstock, M.A.; Feldman, S.R.; Coldiron, B.M. Incidence estimate of nonmelanoma skin cancer (keratinocyte carcinomas) in the U.S. population, 2012. *JAMA Dermatol.* **2015**, *151*, 1081–1086. [CrossRef]
130. Maubec, E. Update on the Management of Cutaneous Squamous Cell Carcinoma. *Acta Dermato-Venereol.* **2020**, *100*, adv00143. [CrossRef] [PubMed]
131. Najjar, T.; Meyers, A.D.; Monroe, M.M.; Alam, M.; Baibak, L.M.; Campbell, W.J.; Caputy, G.; de la Torre, J.I.; DeBacker, C.; Dryden, R.M.; et al. Cutaneous Squamous Cell Carcinoma. Medscape. Emedicine, Updated: 8 July 2020. Available online: https://emedicine.medscape.com/article/1965430-overview#a6 (accessed on 8 August 2021).
132. Byeon, H.K.; Ku, M.; Yang, J. Beyond EGFR inhibition: Multilateral combat strategies to stop the progression of head and neck cancer. *Exp. Mol. Med.* **2019**, *51*, 1–14. [CrossRef]
133. Bander, T.S.; Nehal, K.S.; Lee, E.H. Cutaneous Squamous Cell Carcinoma: Updates in Staging and Management. *Derm. Clin.* **2019**, *37*, 241–251. [CrossRef]
134. Stratigos, A.J.; Garbe, C.; Dessinioti, C.; Lebbe, C.; Bataille, V.; Bastholt, L.; Dreno, B.; Fargnoli, M.C.; Forsea, A.M.; Frenard, C.; et al. European interdisciplinary guideline on invasive squamous cell carcinoma of the skin: Part 1. Epidemiology, diagnostics and prevention. *Eur. J. Cancer* **2020**, *128*, 60–82. [CrossRef] [PubMed]
135. Hober, C.; Jamme, P.; Desmedt, E.; Greliak, A.; Mortier, L. Dramatic response of refractory metastatic squamous cell carcinoma of the skin with cetuximab/pembrolizumab. *Ther. Adv. Med. Oncol.* **2021**, *13*, 17588359211015493. [CrossRef]
136. Cowey, C.L.; Robert, N.J.; Espirito, J.L.; Davies, K.; Frytak, J.; Lowy, I.; Fury, M.G. Clinical outcomes among unresectable, locally advanced, and metastatic cutaneous squamous cell carcinoma patients treated with systemic therapy. *Cancer Med.* **2020**, *20*, 7381–7387. [CrossRef] [PubMed]
137. Mei, M.; Chen, Y.H.; Meng, T.; Qu, L.H.; Zhang, Z.Y.; Zhang, X. Comparative efficacy and safety of radiotherapy/cetuximab versus radiotherapy/chemotherapy for locally advanced head and neck squamous cell carcinoma patients: A systematic review of published, primarily non-randomized, data. *Ther. Adv. Med. Oncol.* **2020**, *12*, 1758835920975355. [CrossRef]
138. Liu, S.; Zhao, Q.; Zheng, Z.; Liu, Z.; Meng, L.; Dong, L.; Jiang, X. Status of Treatment and Prophylaxis for Radiation-Induced Oral Mucositis in Patients with Head and Neck Cancer. *Front. Oncol.* **2021**, *11*, 642575. [CrossRef]
139. Wang, X.; Li, J.; Li, L.; Li, X. Photodynamic therapy-induced apoptosis of keloid fibroblasts is mediated by radical oxygen species in vitro. *Clin. Lab.* **2015**, *61*, 1257–1266. [CrossRef]
140. Cohen, D.K.; Lee, P.K. Photodynamic Therapy for Non-Melanoma Skin Cancers. *Cancers* **2016**, *8*, 90. [CrossRef]
141. Griffin, L.L.; Lear, J.T. Photodynamic Therapy and Non-Melanoma Skin Cancer. *Cancers* **2016**, *8*, 98. [CrossRef]
142. Steeb, T.; Schlager, J.G.; Kohl, C.; Ruzicka, T.; Heppt, M.V.; Berking, C. Laser-assisted photodynamic therapy for actinic keratosis: A systematic review and meta-analysis. *J. Am. Acad. Dermatol.* **2019**, *80*, 947–956. [CrossRef]
143. Gu, X.; Zhao, S.; Shen, M.; Su, J.; Chen, X. Laser-assisted photodynamic therapy vs. conventional photodynamic therapy in non-melanoma skin cancers: Systematic review and meta-analysis of randomized controlled trials. *Photodermatol. Photoimmunol. Photomed.* **2021**. [CrossRef]
144. Szeimies, R.M.; Karrer, S. Photodynamische Therapie—Trends und neue Entwicklungen [Photodynamic therapy-trends and new developments]. *Hautarzt* **2021**, *72*, 27–33. [CrossRef]
145. Xin, Y.; Huang, Q.; Zhang, P.; Guo, W.W.; Zhang, L.Z.; Jiang, G. Demethoxycurcumin in combination with ultraviolet radiation B induces apoptosis through the mitochondrial pathway and caspase activation in A431 and HaCaT cells. *Tumour Biol.* **2017**, *39*, 1010428317706216. [CrossRef]
146. Abdel Fadeel, D.A.; Kamel, R.; Fadel, M. PEGylated lipid nanocarrier for enhancing photodynamic therapy of skin carcinoma using curcumin: In-vitro/in-vivo studies and histopathological examination. *Sci. Rep.* **2020**, *10*, 10435. [CrossRef]
147. Luo, H.; Lu, L.; Liu, N.; Li, Q.; Yang, X.; Zhang, Z. Curcumin loaded sub-30 nm targeting therapeutic lipid nanoparticles for synergistically blocking nasopharyngeal cancer growth and metastasis. *J. Nanobiotechnol.* **2021**, *19*, 224. [CrossRef]
148. Fadel, M.; Kassab, K.; Abd El Fadeel, D.A.; Nasr, M.; El Ghoubary, N.M. Comparative enhancement of curcumin cytotoxic photodynamic activity by nanoliposomes and gold nanoparticles with pharmacological appraisal in HepG2 cancer cells and Erlich solid tumor model. *Drug Dev. Ind. Pharm.* **2018**, *44*, 1809–1816. [CrossRef]

149. Yamazaki, S.; Sekiguchi, A.; Uchiyama, A.; Fujiwara, C.; Inoue, Y.; Yokoyama, Y.; Ogino, S.; Torii, R.; Hosoi, M.; Akai, R.; et al. Apelin/APJ signalling suppresses the pressure ulcer formation in cutaneous ischemia-reperfusion injury mouse model. *Sci. Rep.* **2020**, *10*, 1349. [CrossRef]

150. Li, Z.; Lin, F.; Thalib, L.; Chaboyer, W. Global prevalence and incidence of pressure injuries in hospitalized adult patients: A systematic review and meta-analysis. *Int. J. Nurs. Stud.* **2020**, *105*, 103546. [CrossRef]

151. Coelho, V.H.M.; Alvares, L.D.; Carbinatto, F.M.; de Aquino Junior, A.E.; Angarita, D.P.R.; Bagnato, V.S. Phtodynamic Therapy, Laser Therapy and Cellulose Membrane for the Healing of Venous Ulcers: Results of a Pilot Study. *J. Nurs. Care* **2017**, *6*, 387. [CrossRef]

152. Mussttaf, R.A.; Jenkins, D.F.; Jha, A.N. Assessing the impact of low level laser therapy (LLLT) on biological systems: A review. *Int. J. Radiat. Biol.* **2019**, *95*, 120–143. [CrossRef]

153. Tejada, S.; Manayi, A.; Daglia, M.; Nabavi, S.F.; Sureda, A.; Hajheydari, Z.; Gortzi, O.; Pazoki-Toroudi, H.; Nabavi, S.M. Wound healing effects of curcumin: A short review. *Curr. Pharm. Biotechnol.* **2016**, *17*, 1002–1007. [CrossRef]

154. Barchitta, M.; Maugeri, A.; Favara, G.; Magnano San Lio, R.; Evola, G.; Agodi, A.; Basile, G. Nutrition and wound healing: An overview focusing on the beneficial effects of curcumin. *Int. J. Mol. Sci.* **2019**, *20*, 1119. [CrossRef]

155. Ebrahiminaseri, A.; Sadeghizadeh, M.; Moshaii, A.; Asgaritarghi, G.; Safari, Z. Combination treatment of dendrosomal nanocur-cumin and low-level laser therapy develops proliferation and migration of mouse embryonic fibroblasts and alter TGF-β, VEGF, TNF-α and IL-6 expressions involved in wound healing process. *PLoS ONE* **2021**, *16*, e0247098. [CrossRef]

156. Bomar, L.; Senithilnathan, A.; Ahn, C. Systemic Therapies for Advanced Melanoma. *Dermatol. Clin.* **2019**, *37*, 409–423. [CrossRef]

157. Queirolo, P.; Boutros, A.; Tanda, E.; Spagnolo, F.; Quaglino, P. Immune-checkpoint inhibitors for the treatment of metastatic melanoma: A model of cancer immunotherapy. *Semin. Cancer Biol.* **2019**, *59*, 290–297. [CrossRef]

158. Poklepovic, A.S.; Luke, J.J. Considering adjuvant therapy for stage II melanoma. *Cancer* **2020**, *126*, 1166–1174. [CrossRef]

159. Szlasa, W.; Supplitt, S.; Drąg-Zalesińska, M.; Przystupski, D.; Kotowski, K.; Szewczyk, A.; Kasperkiewicz, P.; Saczko, J.; Kulbacka, J. Corrigendum to "Effects of curcumin based PDT on the viability and the organization of actin in melanotic (A375) and amelanotic melanoma (C32)—In vitro studies". *Biomed. Pharmacother.* **2020**, *132*, 110883, Erratum in **2021**, *139*, 111694. [CrossRef]

160. Mohammadi, S.; Soratijahromi, E.; Dehdari Vais, R.; Sattarahmady, N. Phototherapy and Sonotherapy of Melanoma Cancer Cells. Using Nanoparticles of Selenium-Polyethylene Glycol-Curcumin as a Dual-Mode Sensitizer. *J. Biomed. Phys. Eng.* **2020**, *10*, 597–606. [CrossRef] [PubMed]

161. Vetha, S.B.S.; Oh, P.S.; Kim, S.H.; Jeong, H.J. Curcuminoids encapsulated liposome nanoparticles as a blue light emitting diode induced photodynamic therapeutic system for cancer treatment. *J. Photochem. Photobiol. B* **2020**, *205*, 111840. [CrossRef]

162. Woźniak, M.; Nowak, M.; Lazebna, A.; Więcek, K.; Jabłońska, I.; Szpadel, K.; Grzeszczak, A.; Gubernator, J.; Ziółkowski, P. The Comparison of In Vitro Photosensitizing Efficacy of Curcumin-Loaded Liposomes Following Photodynamic Therapy on Melanoma MUG-Mel2, Squamous Cell Carcinoma SCC-25, and Normal Keratinocyte HaCaT Cells. *Pharmaceuticals* **2021**, *14*, 374. [CrossRef]

163. Ferlay, J.; Soerjomataram, I.; Dikshit, R.; Eser, S.; Mathers, C.; Rebelo, M.; Parkin, D.M.; Forman, D.; Bray, F. Cancer incidence and mortality worldwide: Sources, methods and major patterns in GLOBOCAN 2012. *Int. J. Cancer* **2015**, *136*, E359–E386. [CrossRef] [PubMed]

164. Rhyu, M.G.; Oh, J.H.; Hong, S.J. Species-specific role of gene-adjacent retroelements in human and mouse gastric carcinogenesis. *Int. J. Cancer* **2018**, *142*, 1520–1527. [CrossRef]

165. Hung, K.F.; Yang, T.; Kao, S.Y. Cancer stem cell theory: Are we moving past the mist? *J. Chin. Med. Assoc.* **2019**, *82*, 814–818. [CrossRef]

166. Rhyu, M.G.; Oh, J.H.; Kim, T.H.; Kim, J.S.; Rhyu, Y.A.; Hong, S.J. Periodic Fluctuations in the Incidence of Gastrointestinal Cancer. *Front. Oncol.* **2021**, *11*, 558040. [CrossRef]

167. Sung, H.; Ferlay, J.; Siegel, R.L.; Laversanne, M.; Soerjomataram, I.; Jemal, A.; Bray, F. Global Cancer Statistics 2020: GLOBOCAN Estimates of Incidence and Mortality Worldwide for 36 Cancers in 185 Countries. *CA Cancer J. Clin.* **2021**, *71*, 209–249. [CrossRef]

168. Vetha, B.S.S.; Kim, E.M.; Oh, P.S.; Kim, S.H.; Lim, S.T.; Sohn, M.H.; Jeong, H.J. Curcumin Encapsulated Micellar Nanoplatform for Blue Light Emitting Diode Induced Apoptosis as a New Class of Cancer Therapy. *Macromol. Res.* **2019**, *27*, 1179–1184. [CrossRef]

169. Şueki, F.; Ruhi, M.K.; Gülsoy, M. The effect of curcumin in antitumor photodynamic therapy: In vitro experiments with Caco-2 and PC-3 cancer lines. *Photodiagn. Photodyn. Ther.* **2019**, *27*, 95–99. [CrossRef]

170. De Freitas, C.F.; Kimura, E.; Rubira, A.F.; Muniz, E.C. Curcumin and silver nanoparticles carried out from polysaccharide-based hydrogels improved the photodynamic properties of curcumin through metal-enhanced singlet oxygen effect. *Mater. Sci. Eng. C Mater. Biol. Appl.* **2020**, *112*, 110853. [CrossRef]

171. Tsai, W.H.; Yu, K.H.; Huang, Y.C.; Lee, C.I. EGFR-targeted photodynamic therapy by curcumin-encapsulated chitosan/TPP nanoparticles. *Int. J. Nanomed.* **2018**, *13*, 903–916. [CrossRef]

172. Feng, J.; Polychronidis, G.; Heger, U.; Frongia, G.; Mehrabi, A.; Hoffmann, K. Incidence trends and survival prediction of hepatoblastoma in children: A population-based study. *Cancer Commun.* **2019**, *39*, 62. [CrossRef]

173. Ellerkamp, V.; Bortel, N.; Schmid, E.; Kirchner, B.; Armeanu-Ebinger, S.; Fuchs, J. Photodynamic Therapy Potentiates the Effects of Curcumin on Pediatric Epithelial Liver Tumor Cells. *Anticancer Res.* **2016**, *36*, 3363–3372. [PubMed]

174. Robinson, K.; Tiriveedhi, V. Perplexing role of P-glycoprotein in tumor microenvironment. *Front. Oncol.* **2020**, *10*, 265. [CrossRef] [PubMed]

175. Tanaka, K.; Kiguchi, K.; Mikami, M.; Aoki, D.; Iwamori, M. Involvement of the MDR1 gene and glycolipids in anticancer drug-resistance of human ovarian carcinoma-derived cells. *Hum. Cell* **2019**, *32*, 447. [CrossRef] [PubMed]
176. Li, S.; Yang, S.; Liu, C.; He, J.; Li, T.; Fu, C.; Meng, X.; Shao, H. Enhanced Photothermal-Photodynamic Therapy by Indocyanine Green and Curcumin-Loaded Layered MoS$_2$ Hollow Spheres via Inhibition of P-Glycoprotein. *Int. J. Nanomed.* **2021**, *16*, 433–442. [CrossRef] [PubMed]
177. Howlader, N.; Forjaz, G.; Mooradian, M.J.; Meza, R.; Kong, C.Y.; Cronin, K.A.; Cronin, K.A.; Mariotto, A.B.; Lowy, D.R.; Feuer, E.J. The effect of advances in lung-cancer treatment on population mortality. *N. Engl. J. Med.* **2020**, *383*, 640–649. [CrossRef]
178. Lin, J.J.; Shaw, A.T. Resisting Resistance: Targeted Therapies in Lung Cancer. *Trends Cancer* **2016**, *2*, 350–364. [CrossRef]
179. Majeed, U.; Manochakian, R.; Zhao, Y.; Lou, Y. Targeted therapy in advanced non-small cell lung cancer: Current advances and future trends. *J. Hematol. Oncol.* **2021**, *14*, 108. [CrossRef]
180. Jiang, S.; Zhu, R.; He, X.; Wang, J.; Wang, M.; Qian, Y.; Wang, S. Enhanced photocytotoxicity of curcumin delivered by solid lipid nanoparticles. *Int. J. Nanomed.* **2016**, *12*, 167–178. [CrossRef]
181. Yuan, A.; Tang, X.; Qiu, X.; Jiang, K.; Wu, J.; Hu, Y. Activatable photodynamic destruction of cancer cells by NIR dye/photosensitizer loaded liposomes. *Chem. Commun.* **2015**, *51*, 3340–3342. [CrossRef] [PubMed]
182. Feng, L.; Tao, D.; Dong, Z.; Chen, Q.; Chao, Y.; Liu, Z.; Chen, M. Near-infrared light activation of quenched liposomal Ce6 for synergistic cancer phototherapy with effective skin protection. *Biomaterials* **2017**, *127*, 13–24. [CrossRef]
183. Huang, X.; Chen, L.; Zhang, Y.; Zhou, S.; Cai, H.H.; Li, T.; Jin, H.; Cai, J.; Zhou, H.; Pi, J. GE11 Peptide Conjugated Liposomes for EGFR-Targeted and Chemophotothermal Combined Anticancer Therapy. *Bioinorg. Chem. Appl.* **2021**, *2021*, 5534870. [CrossRef]
184. Baghdan, E.; Duse, L.; Schüer, J.J.; Pinnapireddy, S.R.; Pourasghar, M.; Schäfer, J.; Schneider, M.; Bakowsky, U. Development of inhalable curcumin loaded Nano-in-Microparticles for bronchoscopic photodynamic therapy. *Eur. J. Pharm. Sci.* **2019**, *132*, 63–71. [CrossRef] [PubMed]
185. Grech, N.; Dalli, T.; Mizzi, S.; Meilak, L.; Calleja, N.; Zrinzo, A. Rising Incidence of Glioblastoma Multiforme in a Well-Defined Population. *Cureus* **2020**, *12*, e8195. [CrossRef]
186. Tan, A.C.; Ashley, D.M.; López, G.Y.; Malinzak, M.; Friedman, H.S.; Khasraw, M. Management of glioblastoma: State of the art and future directions. *CA Cancer J. Clin.* **2020**, *70*, 299–312. [CrossRef]
187. Lee, A.; Arasaratnam, M.; Chan, D.L.H.; Khasraw, M.; Howell, V.M.; Wheeler, H. Anti-epidermal growth factor receptor therapy for glioblastoma in adults. *Cochrane Database Syst. Rev.* **2020**, *5*, CD013238. [CrossRef]
188. Kielbik, A.; Wawryka, P.; Przystupski, D.; Rossowska, J.; Szewczyk, A.; Saczko, J.; Kulbacka, J.; Chwiłkowska, A. Effects of Photosensitization of Curcumin in Human Glioblastoma Multiforme Cells. *In Vivo* **2019**, *33*, 1857–1864. [CrossRef] [PubMed]
189. Jamali, Z.; Khoobi, M.; Hejazi, S.M.; Eivazi, N.; Abdolahpour, S.; Imanparast, F.; Moradi-Sardareh, H.; Paknejad, M. Evaluation of targeted curcumin (CUR) loaded PLGA nanoparticles for in vitro photodynamic therapy on human glioblastoma cell line. *Photodiagn. Photodyn. Ther.* **2018**, *23*, 190–201. [CrossRef] [PubMed]
190. Bonfim, C.M.D.; Monteleoni, L.F.; Calmon, M.F.; Candido, N.M.; Provazzi, P.J.S.; Lino, V.S.; Rabachini, T.; Sichero, L.; Villa, L.L.; Quintana, S.M.; et al. Antiviral activity of curcumin-nanoemulsion associated with photodynamic therapy in vulvar cell lines transducing different variants of HPV-16. *Artif. Cells Nanomed. Biotechnol.* **2020**, *48*, 515–524. [CrossRef] [PubMed]
191. Trigo-Gutierrez, J.K.; Vega-Chacón, Y.; Soares, A.B.; Mima, E.G.d.O. Antimicrobial Activity of Curcumin in Nanoformulations. A Comprehensive Review. *Int. J. Mol. Sci.* **2021**, *22*, 7130. [CrossRef]
192. Yang, Q.Q.; Farha, A.K.; Kim, G.; Gul, K.; Gan, R.Y.; Corke, H. Antimicrobial and anticancer applications and related mechanisms of curcumin-mediated photodynamic treatments. *Trends Food Sci. Technol.* **2020**, *97*, 341–354. [CrossRef]
193. Kazantzis, K.T.; Koutsonikoli, K.; Mavroidi, B.; Zachariadis, M.; Alexiou, P.; Pelecanou, M.; Politopoulos, K.; Alexandratou, E.; Sagnou, M. Curcumin derivatives as photosensitizers in photodynamic therapy: Photophysical properties and in vitro studies with prostate cancer cells. *Photochem. Photobiol. Sci.* **2020**, *19*, 193–206. [CrossRef] [PubMed]

Review

Liposome Photosensitizer Formulations for Effective Cancer Photodynamic Therapy

Sherif Ashraf Fahmy [1,2], **Hassan Mohamed El-Said Azzazy** [1,*] **and Jens Schaefer** [3,*]

[1] Department of Chemistry, School of Sciences & Engineering, The American University in Cairo, AUC Avenue, P.O. Box 74, New Cairo 11835, Egypt; sheriffahmy@aucegypt.edu

[2] School of Life and Medical Sciences, University of Hertfordshire Hosted by Global Academic Foundation, R5 New Garden City, New Capital AL109AB, Cairo 11835, Egypt

[3] Department of Pharmaceutics and Biopharmaceutics, University of Marburg, Robert-Koch-Street 4, 35037 Marburg, Germany

* Correspondence: hazzazy@aucegypt.edu (H.M.E.-S.A.); j.schaefer@staff.uni-marburg.de (J.S.); Tel.: +2-02-2615-2559 (H.M.E.-S.A.); +49-6421-28-25947 (J.S.)

Abstract: Photodynamic therapy (PDT) is a promising non-invasive strategy in the fight against that which circumvents the systemic toxic effects of chemotherapeutics. It relies on photosensitizers (PSs), which are photoactivated by light irradiation and interaction with molecular oxygen. This generates highly reactive oxygen species (such as $^1O_2$, H_2O_2, O_2, ·OH), which kill cancer cells by necrosis or apoptosis. Despite the promising effects of PDT in cancer treatment, it still suffers from several shortcomings, such as poor biodistribution of hydrophobic PSs, low cellular uptake, and low efficacy in treating bulky or deep tumors. Hence, various nanoplatforms have been developed to increase PDT treatment effectiveness and minimize off-target adverse effects. Liposomes showed great potential in accommodating different PSs, chemotherapeutic drugs, and other therapeutically active molecules. Here, we review the state-of-the-art in encapsulating PSs alone or combined with other chemotherapeutic drugs into liposomes for effective tumor PDT.

Keywords: photodynamic therapy; photosensitizers; liposomes; stealth liposomes; thermosensitive liposomes; tetraether lipids; cancer

Citation: Fahmy, S.A.; Azzazy, H.M.E.-S.; Schaefer, J. Liposome Photosensitizer Formulations for Effective Cancer Photodynamic Therapy. *Pharmaceutics* **2021**, *13*, 1345. https://doi.org/10.3390/pharmaceutics13091345

Academic Editors: Udo Bakowsky, Matthias Wojcik, Eduard Preis and Gerhard Litscher

Received: 20 July 2021
Accepted: 24 August 2021
Published: 27 August 2021

Publisher's Note: MDPI stays neutral with regard to jurisdictional claims in published maps and institutional affiliations.

1. Introduction

One of the fundamental challenges in designing successful tumor-targeting approaches is the selective delivery of anticancer drugs to cancerous cells. Although both the cancerous and healthy tissues are impacted by the cytotoxic effects of anticancer drugs, most targeting approaches depend on the fact that the rapidly proliferating cancer cells would be more affected by chemotherapeutics than healthy ones [1–4]. This warrants the development of novel targeted delivery systems capable of selectively eradicating cancerous cells. Various drug delivery vehicles and nanocarriers have been developed in this context, including polymeric and metal nanoparticles, supramolecular nanocapsules, liposomes, host-guest complexes, and nanofibers [1,2,5,6]. Multimodal systems were designed by exploiting the capability of the nanosystems to co-deliver chemotherapeutic drugs and targeting entities, resulting in more efficient treatment.

Photodynamic therapy (PDT) employs pharmacologically inactive photosensitizers activated upon exposure to light in the presence of oxygen. It is a non-invasive therapeutic approach that does not require sophisticated equipment or setup and has been employed to treat cancer, cardiovascular, and skin diseases [5–15]. This review presents the basic principles and current challenges of using PDT in cancer therapy and state-of-the-art approaches in formulating liposome photosensitizers to improve the therapeutic significance of PDT.

2. Photodynamic Therapy in Treating Cancer

PDT was approved in several countries for treating cancer after its approval for recurrent bladder cancer treatment by the Canadian Health Protection Branch [1]. PDT depends on the interaction of a photosensitizer (PS), delivered to target tissues, and with the light of a specific wavelength in the presence of molecular oxygen dissolved in the cytoplasm [16–22]. Upon light absorption, the PS molecules are transferred from the ground state to an excited singlet state and then to a triplet excited state via the intersystem crossing. In the triplet excited state, the PS undergoes two simultaneous reactions (1) Type I electron transfer reactions which involve the direct reaction of PS with cell components forming anionic or cationic radicals that react with molecular oxygen generating ROS, and (2) Type II energy transfer reactions which involve direct reaction of PS with molecular oxygen producing singlet oxygen ($^1O_2$) (Figure 1). The generated highly reactive ROS (such as $^1O_2$, H_2O_2, O_2, ·OH) exert their cytotoxic effects via irreversible oxidation of the cellular and subcellular organelles and induce apoptosis or necrosis, leading to cell death. ROS can also induce autophagy by several mechanisms, leading to cytoprotective and cell killing responses [23–25].

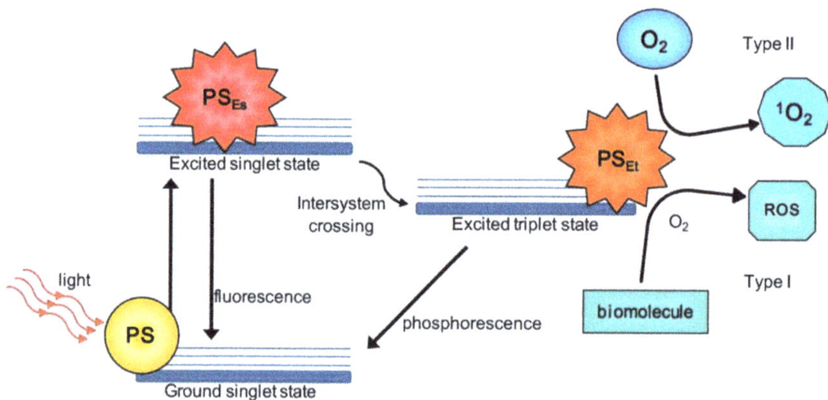

Figure 1. Mechanism of action of PDT demonstrating Type I and Type II reactions. PS_{Es}, PS excited singlet state; PS_{Et}, PS excited triplet state; ROS, reactive oxygen species. Reprinted with permission from ref. [26]. 2016 Calixto et al.

Many parameters influence the effectiveness of PDT. These include the type and dose of PS, route of administration, intensity of light used, type of tumor, and concentration of dissolved cytoplasmic oxygen. The ideal PSs should be pharmacologically inactive in the absence of light irradiation, pure, water-soluble, and selectively present in tumor cells. It should have an absorption spectrum preferably between 650 and 800 nm and rapid elimination rates. Two classes of PSs exist: Porphyrin PSs (three generations) and non-porphyrin PSs [27]. First-generation PSs (such as hematoporphyrins) have been developed and undergone clinical trials for more than 40 years [16]. However, they suffered various drawbacks, including (i) poor tissue penetration, (ii) low chemical stability, (iii) activation when irradiated with wavelengths below 650 nm, (iv) causing skin hypersensitivity reactions, and (v) low elimination rates and long half-lives. Most of the issues attributed to hematoporphyrins refer mainly to Photofrin. This is a complex mixture of porphyrin dimers and higher oligomers, some of which persist in the skin and result in skin photosensitization. Second-generation PSs (such as metalloporphyrins, porphycenes, purpurins, chlorins, protoporphyrins) was developed to overcome most of these shortcomings [17]. 5-Aminolevulinic acid (ALA), methyl aminolevulinate (MAL; Metvix), and Hexvix/Cysview are precursors of protoporphyrin IX, which absorbs at 630 nm [28,29]. They were granted FDA approvals and are used to treat glioblastoma, basal cell carcinoma, Bowen disease, prostate, bladder, and colon cancers [30–32]. Meta-tetrahydroxy phenyl chlorin (m-THPC, temoporfin), excitation wavelength of ~652 nm, is approved in the EU to treat biliary and pancreatic cancers and breast cancer metastases. Verteporfin (Visudyne), a benzo-porphyrin derivative (BPD), with an excitation wavelength of 690 nm, has been granted FDA approval to treat choroidal hemangioma and gastric cancer. Additionally, hypericin (excitation wavelength of 570–650 nm) is an FDA-approved PS for treating skin cancers (Figure 2) [33–35]. Recently, third-generation PSs have been developed by conjugation of second-generation PSs to biological targeting moieties, such as carbohydrates, peptides, or antibodies. This would enhance the selectivity of PSs and minimize undesired adverse effects [29,36]. Non-porphyrin PSs include psoralens, anthracyclines, chalcogenopyrylium dyes, cyanines, and phenothiazinium dyes. Despite the promising effects of PDT in cancer treatment, it still suffers from several shortcomings. These include poor biodistribution of hydrophobic PSs, poor cellular uptake, difficulty applying PDT to deep cancer tissues (hindering light penetration), and low selectivity to cancer cells [37,38]. In addition, cancer cells' oxygenation is essential to achieving effective PDT. Because cancer tissues are bordered by necrotic cells, and compact tumor masses, it is challenging to treat deep cancers using the traditional PDT. Because of the low tissue penetration of visible light, PDT is only effective for treating superficial, and skin tumors [26]. Furthermore, using photofrin may cause long-lasting photosensitivity reactions. Consequently, encapsulation of PSs within different nanocarriers has been proposed to overcome traditional PDT's shortcomings, such as improving water solubility, bioavailability, and selective targeting of PSs [39].

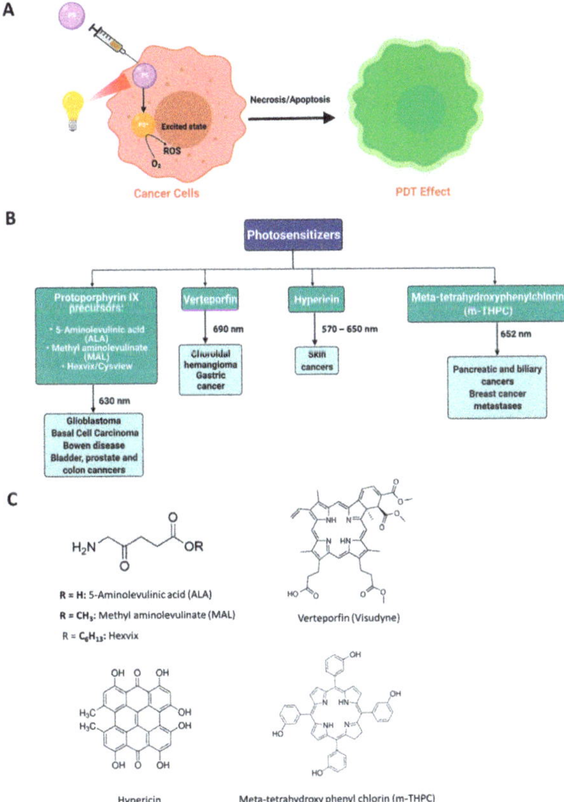

Figure 2. (**A**) General principle of photodynamic therapy; (**B**) FDA and/or EU-approved photosensitizers in cancer photodynamic therapy; (**C**) Chemical structures of approved photosensitizers. PS, photosensitizer; PS*, photosensitizer excited state.

3. Liposomal Photosensitizer Formulations for Tumor Photodynamic Therapy

PSs have entirely or partially been encapsulated, conjugated, or immobilized onto different nanocarriers. Multimodal integration of nanocarriers, such as polymeric nanoparticles, silica nanoparticles, and nanofibers, with PSs, offers several advantages compared to traditional PDT [1–4]. These include: (i) increasing the solubility of hydrophobic PSs, and hence, improving their biodistribution, pharmacokinetics, and cellular uptake; (ii) maintaining constant release rates of PSs at the targeted tumor cells; (iii) the ability to selectively target high loading capacity enabled by the substantial surface-to-volume ratios of nanocarriers which facilitate their surface decoration with particular ligands that can target the overexpressed receptors and proteins in tumor tissues without off-target toxic effects; (iv) boosting the preferential accumulation of PSs into cancer cells via the enhanced permeability and retention (EPR) effect; and (v) expanding the clinical applications of PDT to include additional types of cancer [40–48].

Liposomes are promising nanoplatforms that could be integrated with PDT to enhance the eradication of cancer cells without affecting healthy ones. They are spherical vesicles that consist of a hydrophilic head and a hydrophobic tail. They are self-assembled in an aqueous medium, with the assistance of hydrophobic interaction, developing a sealed structure formed of one or more lamella having the hydrophilic heads oriented towards the outer surfaces of the lamella and the hydrophobic tails founding the lamella interiors. This forms an aqueous core inside the liposomes (Figure 3) [49,50]. Liposomes are prepared using diverse types of naturally occurring phospholipids, which are biocompatible and biodegradable (Table 1). The lipid composition plays a vital role in the stability of liposomes in the systemic circulation, drugs encapsulation efficiencies, and drug release at tumor sites. For instance, the use of DSPC in the liposomal formulation results in improved encapsulation efficiency and stability compared to EPC and DPPC. This is because of the lengthy fatty acid chain of DSPC and the rigidity of the acyl chains of DSPC. Moreover, the use of cholesterol enhances liposomal stability and inhibits the undesirable drug in the systemic circulation. The inclusion of DPPE mPEG5000 in the lipid composition prolongs the blood circulation time of the liposomes, due to the additional steric hindrance created. This reduces liposomal uptake by the reticuloendothelial system. Moreover, the fusogenic features of some lipids (such as DPPG) in the liposomal membrane are reported to improve the ability of liposomes to cross the cancer cell membrane [51,52].

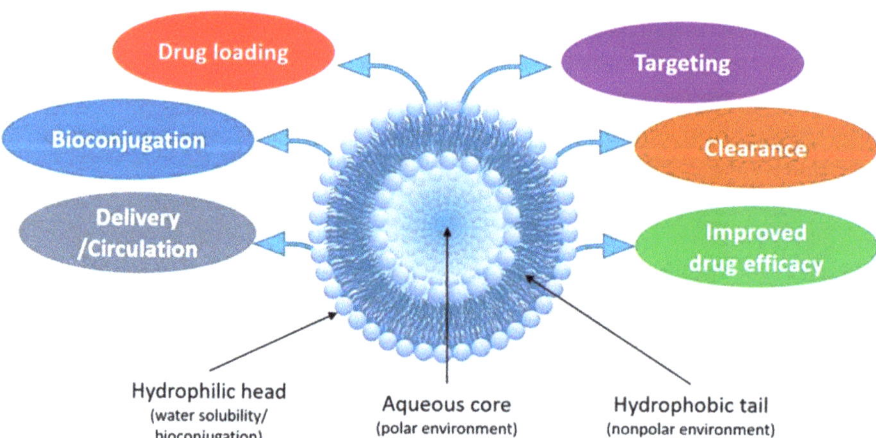

Figure 3. Liposomes are spherical vesicles with an aqueous inner core enclosed by one or more phospholipid bilayers which permit the conjugation of various functional groups. Reprinted with permission from ref. [49]. 2020 Almeida et al.

Table 1. Chemical structure and charge of the most common lipids used in liposomal formulations.

Lipid	Name	Chemical Structure	Charge
DSPC	1,2-Distearoyl-*sn*-glycero-3-phosphocholine		Zwitterion
DSPE	1,2-Distearoyl-*sn*-glycero-3-phosphorylethanolamine		Zwitterion
DPPC	1,2-Dipalmitoyl-*sn*-glycero-3-phosphocholine		Zwitterion
HSPC	L-α-Phosphatidylcholine, hydrogenated (Soy)		Zwitterion
DOPC	1,2-Dioleoyl-*sn*-glycero-3-phosphocholine		Zwitterion
EPC	1,2-Dipalmitoyl-*sn*-glycero-3-ethylphosphocholine (chloride salt)		Zwitterion
DSPG	1,2-Distearoyl-sn-glycero-3-phosphoglycerol, sodium salt		Anionic
Chol	Cholesterol		Neutral
DPPE mPEG5000	1,2-Distearoyl-sn-glycero-3-phosphoethanolamine-N-[methoxy (polyethylene glycol)-2000] (sodium salt)		Anionic

Liposomes have unique biological and physicochemical properties setting them apart from other nanoparticles. They are prepared by the self-assembly of phospholipids and/or tetraethers lipids which are biocompatible and biodegradable [44]. Moreover, liposomes can: (i) Accommodate hydrophilic and hydrophobic agents (such as proteins, nucleic acids, chemotherapeutics, and various PSs); (ii) be decorated and bioconjugated with various functional groups and surface targeting moieties (such as proteins, polymers, and peptides), which would improve their physicochemical properties, drug loading, clearance and/or trigger the release of their cargos into the target tissue; (iii) minimize the opsonization phenomenon by being less identifiable by the reticuloendothelial system leading to prolonging the half-life of the host molecules in the systemic circulation (>48 h) and facilitating the preferential passive accumulation inside tumor tissues; and (iv) widen the therapeutic index for most drugs (Figure 3) [44–49]. Liposomes can enter the cancer cells either by endocytosis or via membrane fusion. For instance, liposomes containing anionic phospholipids display faster endocytosis which augments their intracellular uptake. Moreover, liposomes containing fusogenic lipids demonstrate the ability to fuse and penetrate the cancer cell membrane. A recent publication examined the pathways of cellular internalization of liposomes [50].

It is worth mentioning that proteins in the circulation adsorb to liposomes administered systemically, leading to the formation of a protein corona that interacts with immunoglobulins, complement proteins, and phagocytes in the circulation. This would stimulate cytokine production in the tumor microenvironment leading to adaptive anti-tumor immunity. Moreover, the interaction of liposomes with serum proteins (specially opsonins) plays an essential role in the rapid clearance of liposomal by phagocytes in the blood, liver, and spleen. PEGylation slows the release of liposomes leading to prolongation of their half-lives [53,54].

Various liposomal formulations which improved the physicochemical properties and pharmacokinetics of cancer drugs have received FDA approval (Table 2). Similar liposomal formulations encapsulating PSs can be investigated for improved cancer PDT.

Table 2. FDA-approved liposomal-based formulations in cancer treatment [52].

Name	Approval Year	Lipid Composition	Chemotherapeutic Drug	Clinical Use
Doxil®	1995	HSPC, Cholesterol, and PEG 2000-DSPE	Doxorubicin	Ovarian and Breast cancers
DaunoXome®	1996	DSPC and Cholesterol	Daunorubicin	Kaposi's Sarcoma
Depocyt®	1999	DOPC, DPPG, Cholesterol and Triolein	Cytarabine/Ara-C	Neoplastic meningitis
Myocet®	2000	EPC and Cholesterol	Doxorubicin	Metastatic breast cancer in combination with cyclophosphamide
Mepact®	2004	DOPS and POPC	Mifamurtide	Non-metastatic osteosarcoma
Marqibo®	2012	SM and Cholesterol	Vincristine	Acute leukemia
Onivyde™	2015	DSPC: DPPE mPEG5000	Irinotecan	Metastatic pancreatic cancer

3.1. Liposomes for Photodynamic Tumor Therapy

Several liposomes encapsulating anticancer drugs and carrying targeting moieties (such as mab, glutathione, EndoTAG-1, and transferrin) are currently in clinical trials [55]. In PDT, liposomes can target the loaded PSs to cancer cells via active or passive mechanisms. Active targeting is achieved by decorating the liposomal surface with ligands (such as antibodies) that recognize and bind to overexpressed receptors and proteins in tumor tissues, such as folate, estrogen, spermine, and galactose receptors. Passive targeting takes place via the EPR effect. Blood vessels in healthy tissues are organized and tightly packed, which prevents the extravasation of liposomes. On the other hand, the blood vessels of tumor tissues are disorganized, due to the rapid proliferation of the vascular endothelium inside the tumor tissues [44–56]. Furthermore, the lymphatic drainage is impaired inside the tumor tissue, resulting in the overexpression of permeability mediators, such as bradykinin, nitric oxide, and prostaglandins, which increase EPR [56]. Consequently, nanoliposomes (100–300 nm) can passively cross the loose tumor endothelial barrier through the small open junctions and accumulate inside tumor tissues by the EPR effect [44,56].

Several recent studies have reported the design, optimization, and use of various types of liposomes as nanocarriers for PSs alone or combined with other chemotherapeutic agents, for effective cancer PDT [38,39].

3.1.1. Tetraether Lipid-Based Liposomes

Tetraether lipids have longer lipophilic chains that contain ethers bonds, and hence, show high stability and lower susceptibility to oxidation. Thus, the use of tetraether lipid-based liposomes can increase liposomal membrane stability and integrity [45,56].

Curcumin, with an absorption spectrum of 300–500 nm, is a promising natural PS that could be used safely (at doses up to 12 g/kg/day) in PDT of superficial tumors. Reported studies demonstrate the ability of cancer cell lines to preferentially uptake a curcumin composite compared to normal cell lines. Upon its photoactivation, curcumin (in micromolar concentrations) produces capable of killing cancer cells to treat local superficial infections and cancers [57,58]. Duse et al. developed tetraether lipid-based (TELs) liposomes loaded with curcumin to kill cancer cells selectively [59]. Liposomes were designed by preparing a molar ratio of 90 DSPC:10 TELs using the thin-film hydration method followed by sonication at 56 °C. The curcumin-loaded liposomes had a size of about 208 nm, a zeta potential of $-5.9 \pm$ mV, % encapsulation efficiency (%EE) of 91%, and pronounced stability attributed to the tetraether lipids, which are reported to increase membrane stability and integrity. Furthermore, the curcumin liposomes showed the highest photocytotoxicity on SK-OV-3 ovarian carcinoma cells with a radiation fluence of 13.2 J/cm^2 for the light-induced PDT (IC50 of 8.7 μM). Moreover, the prepared curcumin-loaded liposomes were found to have a minimal hemolytic effect (<40%) and a coagulation time of only 9.7 s compared to 112 s in the case of free curcumin [59].

The in vitro anticancer activities of the prepared curcumin-TEL liposomes were further investigated against cervical cancer cells and papillomavirus-related cancer cell lines [60]. The cancer cell lines were incubated with the designed liposomes at various concentrations ranging from 0 to 100 μmol/L for 4 h followed by LED irradiation at 457 nm for 45, 136, and 227 s at a fluence of 1, 3, and 5 J/cm^2. The cytotoxic effects were then evaluated using the MTT, SYTO9/PI (propidium iodide), Annexin V-FITC (fluorescein isothiocyanate), clonogenic survival, and scratch (wound closure) assays against three different cancer cell lines (HeLa, UD-SCC-2, and VX2). The findings showed enhanced cytotoxicity at a light fluence of 3 J/cm^2 (IC50 values of 9.52 μmol/L, 7.88 μmol/L, and 20.70 μmol/L against HeLa, UD-SCC-2, and VX2 cancer cell lines, respectively), reduced colony formation, proliferation, and cell migration rates. Liposomes loaded with curcumin was suggested as an efficient treatment of papillomavirus-associated cancers.

Plenagl et al. reported tetraether liposomes prepared with a molar ratio of 90 DPPC: 10 TELs encapsulating either hypericin (Hyp-TEL) using thin-film hydration method or hypericin-hydroxypropyl-β-cyclodextrin inclusion complex (HPCD-Hyp-TEL) through dehydration-rehydration vesicle technique [61]. The inclusion of hypericin in hydroxypropyl-β-cyclodextrin in liposomes prevents its photodegradation. The designed Hyp-TEL had an average particle size of 127 ± 14 nm, a zeta potential of -2 ± 1 mV, % encapsulation efficiency (%EE) of 82.5 ± 2.8. At the same time, the HPCD-Hyp-TEL had an average particle size of about 212 ± 13 nm, a zeta potential of -56 ± 1 mV, % encapsulation efficiency (%EE) of 81.6 ± 3.2. The %EE of hypericin was enhanced by increasing the liposomal membrane stability upon the addition of TEL [62]. The phototoxic effect of the prepared liposomes was evaluated on SK-OV-3 cancer cells using an LED fluence of 12.4 J/cm^2. The IC50 values of Hyp-TEL and HPCD-Hyp-TEL were found to be 48 and 136 nM, respectively. Furthermore, considerable amounts of hypericin were uptaken by cancer cells as supported by confocal laser scanning microscopy results. The designed liposomes were hemocompatible and showed a coagulation time range of 5–12 s. These results support the use of integrated liposomal/PDT in cancer-targeted therapy.

Tetraether liposomes were also used to encapsulate protoporphyrin IX PS for vascular targeting and cancer treatment using PDT [63]. The liposomes, designed using 62 mol% TELs, had an average size of 170 nm and an average zeta potential -42 mV. A significant improvement of protoporphyrin IX phototoxicity (IC50 of 5 μM), when loaded in TEL liposomes and tested against SK-OV-3 cancer cells (irradiated with LED light of 672 mJ/cm^2). On the other hand, the protoporphyrin IX-loaded TEL liposomes showed lower phototoxicity (IC50 of 12 μM) when tested on mouse fibroblasts.

Ali et al. reported developing different liposomal formulations accommodating temoporfin (second-generation, synthetic, effective PS) with enhanced phototoxicity against SK-OV-3 cancer cells [64]. Three liposomal formulations loaded with temoporfin (mTHPC) were prepared with molar ratios of 90 DPPC: 10 Cholesterol (DPPC/Chol/T), 95 DPPC: 5 DPPE–mPEG5000 (DPPC/DPPE mPEG5000/T), and 90 DPPC: 10 TEL (DPPC/TEL/T). The prepared liposomes had an average size range of 115 nm and zeta potentials ranging from -6.0 to -13.7 mV. The %EE was 78 ± 4, 81.7 ± 3, and 90 ± 3 % for DPPC/Chol/T, DPPC/DPPE mPEG5000/T and DPPC/TEL/T, respectively. The cell viability of SK-OV-3 cancer cells was reduced to 20% in the three liposomal formulations loaded with temoporfin when exposed to LED light of 10 J/cm^2. Moreover, all the designed liposomal formulations exhibited hemocompatibility (<10% hemolysis) and a coagulation time <40 s.

3.1.2. Stealth Liposomes

Stealth liposomes (PEGylated liposomes) are developed by the adsorption of PEGylated lipids (e.g., DPPE mPEG5000) on the liposomal surface, which prolongs the half-life of liposomes in circulation [65]. Steric hindrance occurs due to PEG reduces liposomal uptake by the reticuloendothelial system and their elimination via renal globular filtration. Moreover, the inclusion of the PEGylated lipids in the liposomal formulations improves their hydrophilicity, and hence, their shelf life [65].

In a different study, curcumin was loaded into liposomes comprising 9.5 HSPC: 0.5 DPPE mPEG5000 HCPC inhibits the leakage of curcumin outside the lipid membranes by imparting rigidity to the liposomal bilayer membrane [66]. The prepared liposomes exhibited significant phototoxic activity against skin cancer melanoma cells (MUG-Mel2) and squamous cell carcinoma (SCC-25) compared to free curcumin after exposure to LED light fluence of 2.5 J/cm^2. The cytotoxicities were 53% (against MUG-Mel2) and 58% (against SCC-25) for curcumin-loaded liposomal mediated PDT compared to 27% (against MUG-Mel2) and 34% (against SCC-25) for free curcumin (10 μM) mediated PDT. The cytotoxic activity of the liposomal preparation against normal keratinocyte cells (HaCaT) was 11%, highlighting the safety of the designed liposomes. Flow cytometry showed that integration of PDT with liposome encapsulating curcumin increased apoptosis to 30% and 40% in MUG-Mel2 and SCC-25 cancer cells, respectively.

A study conducted by Corato et al. loaded iron oxide nanoparticles in the aqueous liposomal core and mTHPC PS in the lipid bilayer [67]. The dual-loaded liposomes were prepared with a molar ratio of 85 DPPC: 10 DSPC: 5 DPPE mPEG5000, which was then mixed with 3.3 mg/mL mTHPC and 0.7 M iron oxide NPs utilizing the reverse-phase evaporation approach. The prepared magneto-photoresponsive liposomes had a spherical structure, an average particle diameter of 150 nm, and one-month stability at 4 °C. The in vitro cell viability was assessed against SK-OV-3 cancer cells after exposure to either magnetic hyperthermia alone, PDT alone, or combined therapies. The % cell viabilities were 10% after exposure to magnetic hyperthermia alone, 5 and 1% after exposure to PDT alone (at 5 and 10 J, respectively), 0.2 and 0% after treatment with the liposomes loaded with iron oxide in the core and mTHPC in the aqueous layer/PDT (at 5 and 10 J, respectively). As supported by proteomic analysis, the dual-loaded liposomes seem to activate intrinsic apoptotic pathways by increasing ROS levels and mitochondrial damage.

Fisher et al. reported the design of lapatinib-loaded PEGylated liposomes for low-dose PDT of the invasive and resistant glioma [68]. Lapatinib is an antineoplastic, clinically approved PS that acts as an epidermal growth factor receptor (EGFR) inhibitor; however, it has poor penetration into glioma cells. PEGylated liposomes were prepared by mixing DPPC: DOTAP (1,2-dioleoyl-3-trimethylammonium-propane): cholesterol: DSPE-mPEG$_{2000}$ at a molar ratio of 0.6:0.079:0.289:0.031. The prepared liposomes had an average particle diameter of 132 ± 9 nm, a zeta potential of 14.3 ± 0.8 mV, and %EE of 64.5 ± 8%. The moderately positive charge aids electrostatic attraction of liposomes to the anionic cells leading to increased lapatinib uptake. The phototoxicity of the nanoformulation loaded with lapatinib and ALA (the latter is metabolized in vivo to protoporphyrin IX) was assessed against human glioblastoma cancer cell lines (U87 and U87vIII) irradiated with 635 nm light provided by a laser at a light dose of 65 mW/cm^2. A remarkable reduction (46%) in the LD50 in the case of lapatinib-loaded liposomes compared to the free lapatinib was observed.

Peng et al. designed long-circulating dual-loaded PEGylated liposomes loaded with Chlorin e6 PS and cisplatin, a first-generation platinum-based chemotherapeutic drug (PL-Ce6-Cis) [69]. In this context, the hydrophilic cisplatin was loaded in the aqueous liposomal core, while hydrophobic Ce6 was incorporated in the outer lipid bilayer. The dual-loaded PEGylated liposomes were prepared with a molar ratio of 10 DSPC:0.2 DSPE-PEG2000:5 cholesterol using the ethanol injection approach. A single dose of the dual-loaded liposomes integrated with light irradiation with a fluence of 100 J/cm^2 could kill 80% of C26 colon cancer cells (compared to only 20% without irradiation), while maintaining minimal adverse toxic effects.

Another recent study reported the use of combined chemotherapy and PDT using dual-loaded PL-Ce6-Cis to treat malignant peripheral nerve sheath cancer (MPNST) [70]. In this regard, the cytotoxicity assessment was conducted against three different MPNST cancer cell lines (T265, ST8814, and S462-TY), derived from NF1 patients using a 662 nm diode laser with a power intensity of 95 mW. The cell viability was reduced significantly (<40%) for PL-Ce6-Cis compared to PL-Ce6 or PL-Cis. Moreover, the PL-Ce6-Cis dual-modal formulation was found to exert minimal neurotoxicity.

3.1.3. Thermosensitive Liposomes (TSLs)

Thermosensitive liposomes (TSLs) are used for triggered release delivery into solid tumors, with at least one formulation currently under phase III clinical trials [71–74]. Exposure to mild hyperthermia (39–43 °C), causes a phase change of the liposomal lipid bilayer from a solid gel well-arranged phase (L$_\beta$) to a fluid tangled phase (L$_\alpha$). This results in enhanced membrane permeability, thus facilitating the release of the liposomal cargo into cancer cells [75–77]. It is of note that the lipid composition of liposomes is key in the TSLs temperature-triggered drug release [78]. Drug release occurs at temperatures equal or lower (by 1–2 °C) to the melting phase transition temperature (Tm) of lipids where structural deformations in the lipid membrane increase the membrane permeability of the lipid bilayer, facilitating the release of the liposomal drug content [79,80]. DPPC, for instance, has a Tm slightly higher than normal body temperature (about 41 °C), and is, thus, used for TSL systems [78].

Shemesh et al. developed an integrated thermosensitive liposomes (TSLs)/PDT encapsulating indocyanine green PS (ICG) to treat triple-negative breast cancer [81]. The TSLs are designed by mixing a molar ratio of 100 DPPC:50 SoyPC (l-α-phosphatidylcholine):30 Chol:0.5 DSPE-PEG 2000 using the thin-film hydration method. The prepared TSLs had an average particle diameter of 71 ± 10 nm and could encapsulate ~500 μM of ICG. The TSLs loaded with ICG (37.5 μM) demonstrated significant cytotoxicity (cell viability < 20%) against MDA-MB-468 and HCC-1806 cells at laser radiation of 14 J/cm^2 compared to free ICG [81].

Meng et al. combined chemotherapy, immunotherapy, and PDT to treat gastric cancer [82]. Thermosensitive liposomes were loaded with IR820 PS, paclitaxel anticancer drug, and imiquimod (R837) (Figure 4). IR820 PS is derived from ICG and is more stable, less expensive, and suitable for photothermal and photodynamic therapies [83,84]. Imiquimod (R837), a dendritic cell activator, is an agonist to TLR7 and has immunomodulatory activities [85]. Mouse fore-stomach carcinoma cell (MFC) treated with the designed liposomes and irradiated with 808 nm light provided by a laser at a light dose of 2.5 W/cm^2 showed that the cell viability decreased to 10% (Figure 5).

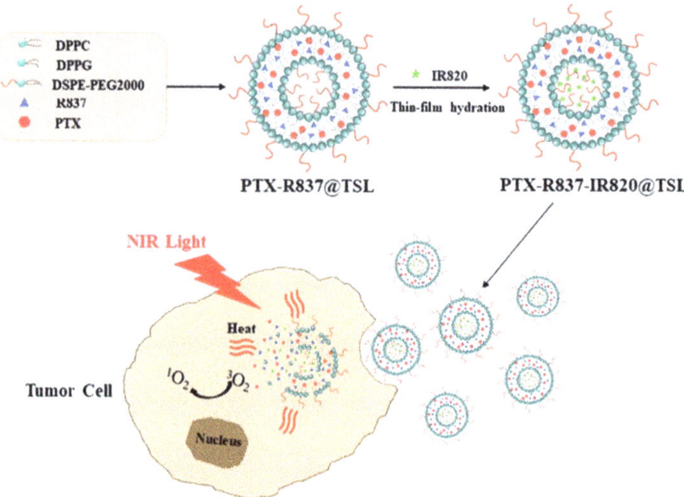

Figure 4. Thermosensitive liposomes loaded with IR820 PS, paclitaxel, and imiquimod (R837) [81]. DPPC, 1,2-Dipalmitoyl-sn-glycero-3-phosphocholine; DPPG, 1,2-Dipalmitoyl-sn-glycero-3-phosphoglycerol, sodium salt; 1,2-distearoyl-sn-glycero-3-phosphoethanolamine-N-[amino(polyethylene glycol)-2000] (ammonium salt); R827, imiquimod; PTX, Paclitaxel; IR820, photosensitizer. Reprinted with permission from ref. [82]. 2019 Meng et al.

3.1.4. Miscellaneous Liposomes

A study reported developing a novel multimodal delivery system, lipopolyplexes (LPPs) loaded with curcumin, combined with PDT for improving gene delivery to SK-OV-3 cancer cells [86]. LPPs used comprised cationic polyethyleneimine and luciferase-expressing pCMV-luc plasmid with anionic liposomes in a ratio of 1:0.5. Liposomes were formulated with a molar ratio of 70 DOPE:15 DPPC:15 cholesterol. The designed LPPs loaded with curcumin were spherical and had a particle size of 200 nm and zeta potential of +8.6 mV ± 1.7 mV. Treatment of SK-OV-3 cells with LPPs loaded with curcumin and irradiation with LED light at 457 nm (irradiation fluence of 1 J/cm^2) resulted in a significant increase in luciferase expression. The designed LPPs had minimal hemolytic effects and plasma coagulation time of 32 s.

Another recent study reported PDT combined with curcumin-loaded magnetic/photoresponsive liposomes (MPLs) in treating papillomavirus-related cancers (Figure 5) [87]. Citric acid-coated iron oxide (CMNPs) magnetic nanoparticles are synthesized employing the chemical precipitation method [88]. Liposomes were prepared using thin-film hydration technique and lipid mixtures in a molar ratio of 5 DSPC: 3 Chol: 1 DDAB (1,2-stearoyl-sn-glycero-3-phosphocholine, dimethyl octadecyl ammonium bromide). In the hydration step, the thin film was hydrated with 0.3 mg/mL CMNPs, 50 µg/mL ICG, producing the ICG-loaded magnetic liposomes. Liposomes were then coated with hyaluronic acid-polyethylene glycol (HA-PEG) PS via sonication. Interactions between anionic HA and the cationic DDAB were responsible for the self-assembly of HA-PEG on the surface of MPLs. HA binds to the over-expressed CD44 receptors on the cell surface of human primary glioblastoma cancer cells (U87MG). The designed HA-PEG MPLs had an average particle size of about 220 nm and exhibited significant cytotoxicity (cell viability of ~30% at a concentration of 2 mg/mL) against U87MG after exposure to near-infrared laser radiation of 2 W/cm^2. Moreover, a xenograft tumor model designed by subcutaneous implantation of U87MG cells in nude mice showed the accumulation of HA-PEG MPLs in cancerous tissues. Following laser irradiation, the tumor size was reduced by 13% compared to the control (Figure 5).

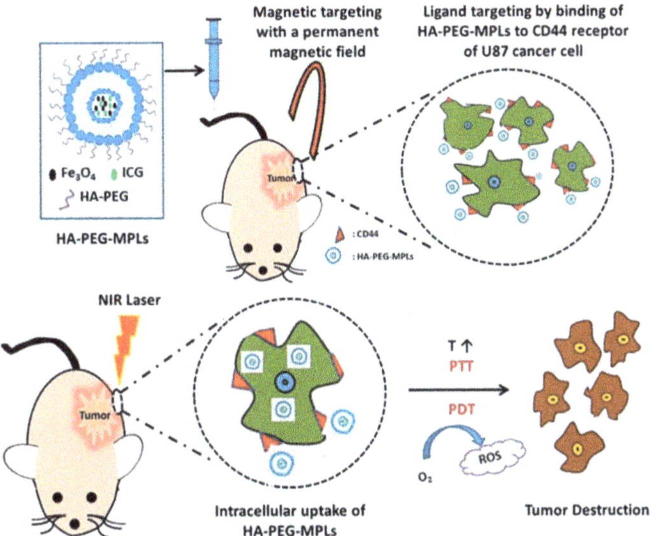

Figure 5. Schematic illustration of HA-PEG coated magnetic/photoresponsive liposomes for combined photothermal/photodynamic cancer therapy. ICG, Indocyanin green PS; HA-PEG, hyaluronic acid-polyethylene glycol; MPLs, magnetic/photoresponsive liposomes; PTT, photothermal therapy; ROS, reactive oxygen species. Reprinted with permission from ref. [87]. 2019 Meng et al.

Another recent study reported the encapsulation of ICG in chitosan-coated liposomes for PDT of melanoma. Chitosan coating has imparted stability to liposomes and improved their cellular uptake into B16-F10 melanoma cancer cells [89]. The liposomes were designed with a molar ratio of 16 DMPC: 3 Chol using the modified freeze-drying method. The positively charged chitosan (0.1%) coated the anionic liposomal surface through electrostatic interactions. The prepared chitosan-coated liposomes loaded with ICG had a mean droplet size of 1983 ± 270 nm, a zeta potential of 43.2 ± 1.2, and a loaded ICG concentration of 1.19 ± 0.03 mg/mL. Cellular uptake of ICG was dramatically increased (owing to the cationic nature of chitosan coats). Moreover, the prepared liposomes showed remarkable phototoxicity against B16-F10 cancer cells after exposure to laser (775 nm) at a power of 0.23 mW for 2.5 min.

A study conducted by Wu et al. reported the preparation of zwitterionic liposomes encapsulating methylene blue (MB) PS that can generate ROS after light exposure and causing cancer cell apoptosis [90]. MB has poor penetration into cancer cells; hence, this study aimed to encapsulate MB into zwitterionic liposomes to protect loaded drugs from degradation, prolong systemic circulation, and enhance cellular uptake. A zwitterionic polymer-lipid poly(12-(methacryloyloxy) dodecyl phosphorylcholine) was self-assembled with DSPC in a molar ratio of 1:4. The prepared zwitterionic liposomes had an average particle size of 150 nm and fast release rates (about 90% of the MB was released over 8 h). A significant increase in the ROS release was observed in the case of MB-loaded liposomes compared to the unloaded MB. Increased cytotoxicity of the MB-loaded zwitterionic liposomes was observed against breast cancer cells (4T1 cells) compared to unloaded MB irradiated with LED light (660 nm) of 165 mW for 6 min. Annexin V-FITC assay showed that cytotoxicity took place via apoptosis.

Finally, Rizvi et al. prepared two liposomal formulations encapsulating Visudyne or a lipid conjugate of BPD. These formulations were combined to localize the PSs and cause photodamage (following irradiation with a 690 nm diode laser at a light dose of 50 mW/cm^2) to mitochondria, endoplasmic reticulum, and lysosomes in a 3D model of ovarian cancer [91]. The recent studies in integrating liposomes with PDT in cancer treatment are summarized in Table 3.

Table 3. Encapsulation of PSs into liposomes in cancer PDT.

Liposomes	Liposomal Formulation	PS	Light Dose	Tumor Type	Outcomes	Ref
Tetraether liposomes	DSPC TELs	Curcumin	13.2 J/cm^2	Ovarian cancer (SK-OV-3)	- Enhanced photocytotoxicity. - Minimal hemolytic effect.	[59]
	DSPC TELs	Curcumin	1, 3, and 5 J/cm^2	Papillomavirus-related cancer cell lines	Enhanced cytotoxicity, reduced colony formation, proliferation, and cell migration rates.	[60]
	DPPC TELs	Hypericin	12.4 J/cm^2	SK-OV-3	- Enhanced photocytotoxicity. - High cancer cell uptake. - Hemocompatibility.	[61]
	TELs	Protoporphyrin IX	672 mJ/cm^2	SK-OV-3	- Improvement of phototoxicity on cancer cells while maintaining lower phototoxicity on normal cells.	[63]
	DPPC TELs	Temoporfin (mTHPC)	10 J/cm^2	SK-OV-3	- Enhanced cytotoxicity. - Hemocompatibility.	[64]
Stealth Liposomes	DPPC DPPE-mPEG5000	Temoporfin (mTHPC)	10 J/cm^2	SK-OV-3	- Enhanced cytotoxicity. - Hemocompatibility.	[64]
	HSPC DPPE mPEG5000	Curcumin	2.5 J/cm^2	Skin cancer melanoma cells (MUG-Mel2) and squamous cell carcinoma (SCC-25)	- Enhanced cytotoxicity on cancer cells. - Lower cytotoxicity on normal keratinocyte cells.	[66]
	DPPC DSPC: DPPE mPEG5000	mTHPC	5 and 10 J/cm^2	SK-OV-3	Enhanced cytotoxicity on cancer cells.	[67]
	DPPC DOTAP	Lapatinib	65 (mW/cm^2)	Human glioblastoma cancer cell lines (U87, U87vIII)	Remarkable reduction in the LD50.	[68]

Table 3. Cont.

Liposomes		Liposomal Formulation	PS	Light Dose	Tumor Type	Outcomes	Ref
		DSPC DSPE-PEG2000 Cholesterol	Chlorin e6	100 J/cm^2	C26 colon cancer cells	- Kills 80% of the C26 colon cancer cells. - Minimum adverse reactions.	[69]
		DSPC DPPE mPEG5000 Cholesterol	Chlorin e6	95 mW/cm^2	Malignant peripheral nerve sheath cancer	- Enhanced cytotoxicity on cancer cells. - Minimal neurotoxicity.	[70]
Thermosensitive liposomes		DPPC SoyPC Chol DSPE-PEG 2000	ICG	14 J/cm^2	Triple-negative breast cancer (MDA-MB-468 and HCC-1806 cells)	Significant cytotoxicity compared to free PS	[81]
		DPPC DPPG DPPE mPEG5000	IR820	2.5 W/cm^2	mouse fore-stomach carcinoma cell (MFC)	Significant cytotoxicity	[82]
Miscellaneous Liposomes	Lipopolyplexes (LPPs)	DOPC DPPC Cholesterol	Curcumin	1 J/cm^2	SK-OV-3	- Significant enhancement in the luciferase expression of SK-OV-3. - Biocompatibility.	[86]
	Magnetic/ photo- responsive liposomes	DSPC Cholesterol DDAB	Indocyanine green (ICG)	2 J/cm^2	Glioblastoma cancer cells (U87MG)	Significant cytotoxicity and accumulation in cancer cells.	[87]
	Chitosan- coated liposomes	DMPC Cholesterol	ICG	250 mW	Melanoma	Improved permeability and phototoxicity.	[89]
	Zwitterionic liposomes	Poly-(12-meth acryloyloxy)- dodecyl phosphoryl- choline DSPC	Methylene blue (MB)	165 mW	Breast cancer cells (4T1 cells)	- Enhanced cytotoxicity on cancer cells. - Improved safety.	[90]

4. Conclusions

In this paper, we reviewed the most recent state-of-the-art studies concerning the use of integrated liposomes/PDT for effective cancer PDT. The ability of liposomes to accommodate hydrophilic and hydrophobic photosensitizers in the aqueous interior and the outer lipid bilayer, respectively, make them top candidates for the delivery of PSs in cancer PDT. Besides, liposomes can target the loaded PSs to cancer cells via active or passive targeting. Moreover, liposome surfaces could be decorated with particular ligands that recognize and bind to overexpressed receptors and proteins in tumor tissues, such as folate, estrogen, spermine, and galactose receptors. Liposomes were also found to improve the selective targeting of PS alone or combined with other chemotherapeutic drugs, hence minimizing their toxic effects. Several studies have reported the enhanced effectiveness of PDT upon encapsulation of PSs in different liposomal formulations. Therefore, these combinatorial/multimodal platforms are likely promising candidates that can improve the effectiveness of cancer PDT.

Author Contributions: Conceptualization, S.A.F., H.M.E.-S.A. and J.S.; methodology, S.A.F.; data curation, S.A.F.; writing—original draft preparation, S.A.F.; writing—review and editing, S.A.F., H.M.E.-S.A. and J.S.; supervision, H.M.E.-S.A. and J.S.; project administration, H.M.E.-S.A. and J.S. All authors have read and agreed to the published version of the manuscript.

Funding: This work has been funded by a grant from the American University in Cairo to H.M.E.-S.A.

Institutional Review Board Statement: Not applicable.

Informed Consent Statement: Not applicable.

Conflicts of Interest: The authors declare no conflict of interest.

References

1. Filipczak, N.; Pan, J.; Yalamarty, S.S.; Torchilin, V.P. Recent advancements in liposome technology. *Adv. Drug Deliv. Rev.* **2020**, *156*, 4–22. [CrossRef]
2. Jhaveri, A.; Deshpande, P.; Torchilin, V. Stimuli-sensitive nanopreparations for combination cancer therapy. *J. Control. Release* **2014**, *190*, 352–370. [CrossRef]
3. Fahmy, S.A.; Ponte, F.; Sicilia, E.; Azzazy, H.M. Experimental and Computational Investigations of Carboplatin Supramolecular Complexes. *ACS Omega* **2020**, *5*, 31456–31466. [CrossRef]
4. Fahmy, S.A.; Ponte, F.; Fawzy, I.M.; Sicilia, E.; Bakowsky, U.; Azzazy, H.S. Host-Guest Complexation of Oxaliplatin and *Para*-Sulfonatocalix[n]Arenes for Potential Use in Cancer Therapy. *Molecules* **2020**, *25*, 5926. [CrossRef]
5. Duse, L.; Baghdan, E.; Pinnapireddy, S.R.; Engelhardt, K.H.; Jedelská, J.; Schaefer, J.; Quendt, P.; Bakowsky, U. Preparation and Characterization of Curcumin Loaded Chitosan Nanoparticles for Photodynamic Therapy. *Phys. Status Solidi A* **2018**, *215*, 1700709. [CrossRef]
6. Fahmy, S.A.; Preis, E.; Bakowsky, U.; Azzazy, H.M. Palladium Nanoparticles Fabricated by Green Chemistry: Promising Chemotherapeutic, Antioxidant and Antimicrobial Agents. *Materials* **2020**, *13*, 3661. [CrossRef] [PubMed]
7. Lim, M.E.; Lee, Y.L.; Zhang, Y.; Chu, J.J. Photodynamic inactivation of viruses using upconversion nanoparticles. *Biomaterials* **2012**, *33*, 1912–1920. [CrossRef]
8. Li, X.; Lee, S.; Yoon, J. Supramolecular photosensitizers rejuvenate photodynamic therapy. *Chem. Soc. Rev.* **2018**, *47*, 1174–1188. [CrossRef] [PubMed]
9. Kwiatkowski, S.; Knap, B.; Przystupski, D.; Saczko, J.; Kędzierska, E.; Knap-Czop, K.; Kotlinska, J.; Michel, O.; Kotowski, K.; Kulbacka, J. Photodynamic therapy—Mechanisms, photosensitizers and combinations. *Biomed. Pharmacother.* **2018**, *106*, 1098–1107. [CrossRef]
10. Sun, J.; Kormakov, S.; Liu, Y.; Huang, Y.; Wu, D.; Yang, Z. Recent Progress in Metal-Based Nanoparticles Mediated Photodynamic Therapy. *Molecules* **2018**, *23*, 1704. [CrossRef]
11. Chizenga, E.P.; Chandran, R.; Abrahamse, H. Photodynamic therapy of cervical cancer by eradication of cervical cancer cells and cervical cancer stem cells. *Oncotarget* **2019**, *10*, 4380–4396. [CrossRef]
12. Dolmans, D.E.; Fukumura, D.; Jain, R.K. Photodynamic therapy for cancer. *Nat. Rev. Cancer* **2003**, *3*, 380–387. [CrossRef]
13. Preis, E.; Baghdan, E.; Agel, M.R.; Anders, T.; Pourasghar, M.; Schneider, M.; Bakowsky, U. Spray dried curcumin loaded nanoparticles for antimicrobial photodynamic therapy. *Eur. J. Pharm. Biopharm.* **2019**, *142*, 531–539. [CrossRef]
14. Preis, E.; Anders, T.; Širc, J.; Hobzova, R.; Cocarta, A.I.; Bakowsky, U.; Jedelská, J. Biocompatible indocyanine green loaded PLA nanofibers for in situ antimicrobial photodynamic therapy. *Mater. Sci. Eng. C Mater. Biol. Appl.* **2020**, *115*, 111068. [CrossRef]
15. Lucky, S.S.; Soo, K.C.; Zhang, Y. Nanoparticles in photodynamic therapy. *Chem. Rev.* **2015**, *115*, 1990–2042. [CrossRef] [PubMed]
16. Polat, E.; Kang, K. Natural Photosensitizers in Antimicrobial Photodynamic Therapy. *Biomedicines* **2021**, *9*, 584. [CrossRef]
17. Park, W.; Cho, S.; Han, J.; Shin, H.; Na, K.; Lee, B.; Kim, D.H. Advanced smart-photosensitizers for more effective cancer treatment. *Biomater. Sci.* **2018**, *6*, 79–90. [CrossRef] [PubMed]
18. Stewart, F.; Baas, P.; Star, W. What does photodynamic therapy have to offer radiation oncologists (or their cancer patients)? *Radiother. Oncol.* **1998**, *48*, 233–248. [CrossRef]
19. Turksoy, A.; Yildiz, D.; Akkaya, E.U. Photosensitization and controlled photosensitization with BODIPY dyes. *Coord. Chem. Rev.* **2019**, *379*, 47–64. [CrossRef]
20. Valkov, A.; Zinigrad, M.; Nisnevitch, M. Photodynamic Eradication of Trichophyton rubrum and Candida albicans. *Pathogens* **2021**, *10*, 263. [CrossRef]
21. Ghate, V.S.; Zhou, W.; Yuk, H.G. Perspectives and Trends in the Application of Photodynamic Inactivation for Microbiological Food Safety. *Compr. Rev. Food Sci. Food Saf.* **2019**, *18*, 402–424. [CrossRef]
22. Bapat, P.; Singh, G.; Nobile, C.J. Visible Lights Combined with Photosensitizing Compounds Are Effective against Candida albicans Biofilms. *Microorganisms* **2021**, *9*, 500. [CrossRef]
23. Sun, H.; Feng, M.; Chen, S.; Wang, R.; Luo, Y.; Yin, B.; Li, J.; Wang, X. Near-infrared photothermal liposomal nanoantagonists for amplified cancer photodynamic therapy. *J. Mater. Chem. B* **2020**, *8*, 7149–7159. [CrossRef] [PubMed]
24. Azad, M.B.; Chen, Y.; Gibson, S.B. Regulation of autophagy by reactive oxygen species (ROS): Implications for cancer progression and treatment. *Antioxid. Redox Signal.* **2009**, *11*, 777–790. [CrossRef] [PubMed]

25. Luby, B.M.; Walsh, C.D.; Zheng, G. Advanced Photosensitizer Activation Strategies for Smarter Photodynamic Therapy Beacons. *Angew. Chem. Int. Ed.* **2019**, *58*, 2558–2569. [CrossRef]
26. Calixto, G.M.; Bernegossi, J.; De Freitas, L.M.; Fontana, C.R.; Chorilli, M. Nanotechnology-Based Drug Delivery Systems for Photodynamic Therapy of Cancer: A Review. *Molecules* **2016**, *21*, 342. [CrossRef]
27. Das, S.; Tiwari, M.; Mondal, D.; Sahoo, B.R.; Tiwari, D.K. Growing tool-kit of photosensitizers for clinical and non-clinical applications. *J. Mater. Chem. B* **2020**, *8*, 10897–10940. [CrossRef]
28. Kleinovink, J.W.; van Driel, P.B.; Snoeks, T.J.; Prokopi, N.; Fransen, M.F.; Cruz, L.J.; Mezzanotte, L.; Chan, A.; Löwik, C.W.; Ossendorp, F. Combination of photodynamic therapy and specific immunotherapy efficiently eradicates established tumors. *Clin. Cancer Res.* **2016**, *22*, 1459–1468. [CrossRef]
29. Mfouo-Tynga, I.S.; Dias, L.D.; Inada, N.M.; Kurachi, C. Features of third generation photosensitizers used in anticancer photodynamic therapy: Review. *Photodiagnosis Photodyn. Ther.* **2021**, *34*, 102091. [CrossRef]
30. Li, Z.; Wang, Y.; Wang, J.; Li, S.; Xiao, Z.; Feng, Y.; Gu, J.; Li, J.; Peng, X.; Li, C.; et al. Evaluation of the efficacy of 5-aminolevulinic acid photodynamic therapy for the treatment of vulvar lichen sclerosus. *Photodiagn. Photodyn. Ther.* **2020**, *29*, 101596. [CrossRef] [PubMed]
31. Malik, Z. Fundamentals of 5-aminolevulinic acid photodynamic therapy and diagnosis: An overview. *Transl. Biophotonics* **2020**, *2*, e201900022. [CrossRef]
32. Alcántara-González, J.; Calzado-Villarreal, L.; Sánchez-Largo, M.E.; Andreu-Barasoain, M.; Ruano-Del Salado, M. Recalcitrant viral warts treated with photodynamic therapy methyl aminolevulinate and red light (630 nm): A case series of 51 patients. *Lasers Med. Sci.* **2020**, *35*, 229–231. [CrossRef] [PubMed]
33. Fukuhara, H.; Yamamoto, S.; Karashima, T.; Inoue, K. Photodynamic diagnosis and therapy for urothelial carcinoma and prostate cancer: New imaging technology and therapy. *Int. J. Clin. Oncol.* **2021**, *26*, 18–25. [CrossRef] [PubMed]
34. Salgado, C.N.; Habigton, P.; Itasaki, N.; Scholz, D. Photodynamic Application of Hexvix for Cancer Therapy. *Cell Mol. Biol. J.* **2016**, *1*, 2.
35. Kiesslich, T.; Berlanda, J.; Plaetzer, K.; Krammer, B.; Berr, F. Comparative characterization of the efficiency and cellular pharmacokinetics of Foscan®- and Foslip®-based photodynamic treatment in human biliary tract cancer cell lines. *Photochem. Photobiol.Sci.* **2007**, *6*, 619–627. [CrossRef]
36. Van Straten, D.; Mashayekhi, V.; De Bruijn, H.S.; Oliveira, S.; Robinson, D.J. Oncologic photodynamic therapy: Basic principles, current clinical status and future directions. *Cancers* **2017**, *9*, 19. [CrossRef]
37. Chatterjee, D.K.; Fong, L.S.; Zhang, Y. Nanoparticles in photodynamic therapy: An emerging paradigm. *Adv. Drug Deliv. Rev.* **2008**, *60*, 1627–1637. [CrossRef]
38. Abrahamse, H.; Kruger, C.A.; Kadanyo, S.; Mishra, A. Nanoparticles for Advanced Photodynamic Therapy of Cancer. *Photomed. Laser Surg.* **2017**, *35*, 581–588. [CrossRef]
39. Yan, K.; Zhang, Y.; Mu, C.; Xu, Q.; Jing, X.; Wang, D.; Dang, D.; Meng, L.; Ma, J. Versatile Nanoplatforms with enhanced Photodynamic Therapy: Designs and Applications. *Theranostics* **2020**, *10*, 7287–7318. [CrossRef]
40. Alsaab, H.O.; Alghamdi, M.S.; Alotaibi, A.S.; Alzhrani, R.; Alwuthaynani, F.; Althobaiti, Y.S.; Almalki, A.H.; Sau, S.; Iyer, A.K. Progress in Clinical Trials of Photodynamic Therapy for Solid Tumors and the Role of Nanomedicine. *Cancers* **2020**, *12*, 2793. [CrossRef]
41. Yano, S.; Hirohara, S.; Obata, M.; Hagiya, Y.; Ogura, S.I.; Ikeda, A.; Kataoka, H.; Tanaka, M.; Joh, T. Current states and future views in photodynamic therapy. *Photochem. Photobiol. C Photochem. Rev.* **2011**, *12*, 46–67. [CrossRef]
42. Debele, T.A.; Peng, S.; Tsai, H.C. Drug carrier for photodynamic cancer therapy. *Int. J. Mol. Sci.* **2015**, *16*, 22094–22136. [CrossRef]
43. Tian, G.; Zhang, X.; Gu, Z.; Zhao, Y. Recent advances in upconversion nanoparticles-based multifunctional nanocomposites for combined cancer therapy. *Adv. Mater.* **2015**, *27*, 7692–7712. [CrossRef]
44. El-Shafie, S.; Fahmy, S.A.; Ziko, L.; Elzahed, N.; Shoeib, T.; Kakarougkas, A. Encapsulation of Nedaplatin in Novel PEGylated Liposomes Increases Its Cytotoxicity and Genotoxicity against A549 and U2OS Human Cancer Cells. *Pharmaceutics* **2020**, *12*, 863. [CrossRef]
45. Torchilin, V.P. Recent advances with liposomes as pharmaceutical carriers. *Nat. Rev. Drug Discov.* **2005**, *4*, 145–160. [CrossRef]
46. Kanasty, R.; Dorkin, J.R.; Vegas, A.; Anderson, D. Delivery materials for siRNA therapeutics. *Nat. Mater.* **2013**, *12*, 967–977. [CrossRef]
47. Yin, H.; Kanasty, R.L.; Eltoukhy, A.A.; Vegas, A.J.; Dorkin, J.R.; Anderson, D.G. Non-viral vectors for gene-based therapy. *Nat. Rev. Genet.* **2014**, *15*, 541–555. [CrossRef]
48. Torchilin, V.P. Multifunctional, stimuli-sensitive nanoparticulate systems for drug delivery. *Nat. Rev. Drug Discov.* **2014**, *13*, 813–827. [CrossRef]
49. Almeida, B.; Nag, O.K.; Rogers, K.E.; Delehanty, J.B. Recent Progress in Bioconjugation Strategies for Liposome-Mediated Drug Delivery. *Molecules* **2020**, *25*, 5672. [CrossRef]
50. Alshehri, A.; Grabowska, A.; Stolnik, S. Pathways of cellular internalisation of liposomes delivered siRNA and effects on siRNA engagement with target mRNA and silencing in cancer cells. *Sci. Rep.* **2018**, *8*, 3748. [CrossRef]
51. Sadeghi, N.; Deckers, R.; Ozbakir, B.; Akthar, S.; Kok, R.J. Lammers T, Storm G. Influence of cholesterol inclusion on the doxorubicin release characteristics of lysolipid-based thermosensitive liposomes. *Int. J. Pharm.* **2018**, *548*, 778–782. [CrossRef]

52. Bulbake, U.; Doppalapudi, S.; Kommineni, N.; Khan, W. Liposomal Formulations in Clinical Use: An Updated Review. *Pharmaceutics* **2017**, *9*, 12. [CrossRef]
53. La-Beck, N.M.; Liu, X.; Wood, L.M. Harnessing Liposome Interactions with the Immune System for the Next Breakthrough in Cancer Drug Delivery. *Front. Pharmacol.* **2019**, *10*, 220. [CrossRef] [PubMed]
54. Inglut, C.T.; Sorrin, A.J.; Kuruppu, T.; Vig, S.; Cicalo, J.; Ahmad, H.; Huang, H.C. Immunological and Toxicological Considerations for the Design of Liposomes. *Nanomaterials* **2020**, *10*, 190. [CrossRef]
55. Li, Y.; Cong, H.; Wang, S.; Yu, B.; Shen, Y. Liposomes modified with bio-substances for cancer treatment. *Biomater. Sci.* **2020**, *8*, 6442–6468. [CrossRef]
56. Shaheen, S.M.; Shakil Ahmed, F.R.; Hossen, M.N.; Ahmed, M.; Amran, M.S.; Ul-Islam, M.A. Liposome as a carrier for advanced drug delivery. *Pak. J. Biol. Sci.* **2006**, *9*, 1181–1191. [CrossRef]
57. Kazantzis, K.T.; Koutsonikoli, K.; Mavroidi, B.; Zachariadis, M.; Alexiou, P.; Pelecanou, M.; Politopoulos, K.; Alexandratou, E.; Sagnou, M. Curcumin derivatives as photosensitizers in photodynamic therapy: Photophysical properties and in vitro studies with prostate cancer cells. *Photochem. Photobiol. Sci.* **2020**, *19*, 193–206. [CrossRef] [PubMed]
58. Dev, A.; Srivastava, A.K.; Choudhury, S.R.; Karmakar, S. Nano-curcumin influences blue light photodynamic therapy for restraining glioblastoma stem cells growth. *RSC Adv.* **2016**, *6*, 95165–95168. [CrossRef]
59. Duse, L.; Pinnapireddy, S.R.; Strehlow, B.; Jedelská, J.; Bakowsky, U. Low level LED photodynamic therapy using curcumin loaded tetraether liposomes. *Eur. J. Pharm. Biopharm.* **2018**, *126*, 233–241. [CrossRef]
60. Ambreen, G.; Duse, L.; Tariq, I.; Ali, U.; Ali, S.; Pinnapireddy, S.R.; Bette, M.; Bakowsky, U.; Mandic, R. Sensitivity of Papilloma Virus-Associated Cell Lines to Photodynamic Therapy with Curcumin-Loaded Liposomes. *Cancers* **2020**, *12*, 3278. [CrossRef] [PubMed]
61. Plenagl, N.; Duse, L.; Seitz, B.S.; Goergen, N.; Pinnapireddy, S.R.; Jedelska, J.; Brüßler, J.; Bakowsky, U. Photodynamic therapy—Hypericin tetraether liposome conjugates and their antitumor and antiangiogenic activity. *Drug Deliv.* **2019**, *26*, 23–33. [CrossRef]
62. Galanou, M.C.; Theodossiou, T.A.; Tsiourvas, D.; Sideratou, Z.; Paleos, C.M. Interactive transport, subcellular relocation and enhanced phototoxicity of hypericin encapsulated in guanidinylated liposomes via molecular recognition. *Photochem. Photobiol.* **2008**, *84*, 1073–1083.
63. Mahmoud, G.; Jedelská, J.; Strehlow, B.; Omar, S.; Schneider, M.; Bakowsky, U. Photo-responsive tetraether lipids based vesicles for prophyrin mediated vascular targeting and direct phototherapy. *Colloids Surf. B Biointerfaces* **2017**, *159*, 720–728. [CrossRef]
64. Ali, S.; Amin, M.U.; Ali, M.Y.; Tariq, I.; Pinnapireddy, S.R.; Duse, L.; Goergen, N.; Wölk, C.; Hause, G.; Jedelská, J.; et al. Wavelength dependent photo-cytotoxicity to ovarian carcinoma cells using temoporfin loaded tetraether liposomes as efficient drug delivery system. *Eur. J. Pharm. Biopharm.* **2020**, *150*, 50–65. [CrossRef]
65. Bakowsky, H.; Richter, T.; Kneuer, C.; Hoekstra, D.; Rothe, U.; Bendas, G.; Ehrhardt, C.; Bakowsky, U. Adhesion characteristics and stability assessment of lectin-modified liposomes for site-specific drug delivery. *Biochim. Biophys. Acta* **2008**, *1778*, 242–249. [CrossRef]
66. Woźniak, M.; Nowak, M.; Lazebna, A.; Więcek, K.; Jabłońska, I.; Szpadel, K.; Grzeszczak, A.; Gubernator, J.; Ziółkowski, P. The Comparison of In Vitro Photosensitizing Efficacy of Curcumin-Loaded Liposomes Following Photodynamic Therapy on Melanoma MUG-Mel2, Squamous Cell Carcinoma SCC-25, and Normal Keratinocyte HaCaT Cells. *Pharmaceuticals* **2021**, *14*, 374. [CrossRef] [PubMed]
67. Corato, R.D.; Béalle, G.; Kolosnjaj-Tabi, J.; Espinosa, A.; Clément, O.; Silva, A.K.; Ménager, C.; Wilhelm, C. Combining Magnetic Hyperthermia and Photodynamic Therapy for Tumor Ablation with Photoresponsive Magnetic Liposomes. *ACS Nano* **2015**, *9*, 2904–2916. [CrossRef]
68. Fisher, C.; Obaid, G.; Niu, C.; Foltz, W.; Goldstein, A.; Hasan, T.; Lilge, L. Liposomal Lapatinib in Combination with Low-Dose Photodynamic Therapy for the Treatment of Glioma. *J. Clin. Med.* **2019**, *8*, 2214. [CrossRef]
69. Peng, P.C.; Hong, R.L.; Tsai, T.; Chen, C.T. Co-Encapsulation of Chlorin e6 and Chemotherapeutic Drugs in a PEGylated Liposome Enhance the Efficacy of Tumor Treatment: Pharmacokinetics and Therapeutic Efficacy. *Pharmaceutics* **2019**, *11*, 617. [CrossRef]
70. Chen, C.T.; Peng, P.C.; Tsai, T.; Chien, H.F.; Lee, M.J. A Novel Treatment Modality for Malignant Peripheral Nerve Sheath Tumor Using a Dual-Effect Liposome to Combine Photodynamic Therapy and Chemotherapy. *Pharmaceutics* **2020**, *12*, 317. [CrossRef]
71. Landon, C.D.; Park, J.Y.; Needham, D.; Dewhirst, M.W. Nanoscale drug delivery and hyperthermia: The materials design and preclinical and clinical testing of low temperature-sensitive liposomes used in combination with mild hyperthermia in the treatment of local cancer. *Open Nanomed. J.* **2011**, *3*, 38–64. [CrossRef]
72. Wust, P.; Hildebrandt, B.; Sreenivasa, G.; Rau, B.; Gellermann, J.; Riess, H.; Felix, R.; Schlag, P.M. Hyperthermia in combined treatment of cancer. *Lancet Oncol.* **2002**, *3*, 487–497. [CrossRef]
73. Huang, S.K.; Stauffer, P.R.; Hong, K.; Guo, J.W.; Phillips, T.L.; Huang, A.; Papahadjopoulos, D. Liposomes and hyperthermia in mice: Increased tumor uptake and therapeutic efficacy of doxorubicin in sterically stabilized liposomes. *Cancer Res.* **1994**, *54*, 2186–2191.
74. Manzoor, A.A.; Lindner, L.H.; Landon, C.D.; Park, J.Y.; Simnick, A.J.; Dreher, M.R.; Das, S.; Hanna, G.; Park, W.; Chilkoti, A.; et al. Overcoming limitations in nanoparticle drug delivery: Triggered, intravascular release to improve drug penetration into tumors. *Cancer Res.* **2012**, *72*, 5566–5575. [CrossRef]
75. Ta, T.; Porter, T.M. Thermosensitive liposomes for localized delivery and triggered release of chemotherapy. *J. Control. Release* **2013**, *169*, 112–125. [CrossRef]

76. Lokerse, W.J.; Kneepkens, E.C.; ten Hagen, T.L.; Eggermont, A.M.; Grüll, H.; Koning, G.A. In depth study on thermosensitive liposomes: Optimizing systems for tumor specific therapy and in vitro to in vivo relations. *Biomaterials* **2016**, *82*, 138–150. [CrossRef]
77. Grüll, H.; Langereis, S. Hyperthermia-triggered drug delivery from temperature-sensitive liposomes using MRI-guided high intensity focused ultrasound. *J. Control. Release* **2012**, *161*, 317–327. [CrossRef]
78. Al-Ahmady, Z.; Kostarelos, K. Chemical components for the design of temperature responsive vesicles as cancer therapeutics. *Chem. Rev.* **2016**, *116*, 3883–3918. [CrossRef]
79. Li, L.; ten Hagen, T.L.; Schipper, D.; Wijnberg, T.M.; van Rhoon, G.C.; Eggermont, A.M.; Lindner, L.H.; Koning, G.A. Triggered content release from optimized stealth thermosensitive liposomes using mild hyperthermia. *J. Control. Release* **2010**, *143*, 274–279. [CrossRef]
80. Li, L.; Ten Hagen, T.L.; Bolkestein, M.; Gasselhuber, A.; Yatvin, J.; Van Rhoon, G.C.; Eggermont, A.M.; Haemmerich, D.; Koning, G.A. Improved intratumoral nanoparticle extravasation and penetration by mild hyperthermia. *J. Control. Release* **2013**, *167*, 130–137. [CrossRef]
81. Shemesh, C.S.; Hardy, C.W.; Yu, D.S.; Fernandez, B.; Zhang, H. Indocyanine green loaded liposome nanocarriers for photodynamic therapy using human triple negative breast cancer cells. *Photodiagn. Photodyn. Ther.* **2014**, *11*, 193–203. [CrossRef]
82. Meng, X.B.; Wang, K.; Lv, L.; Zhao, Y.; Sun, C.; Ma, L.J.; Zhang, B. Photothermal/Photodynamic Therapy with Immune-Adjuvant Liposomal Complexes for Effective Gastric Cancer Therapy. *Part. Part. Syst. Charact.* **2019**, *36*, 1900015. [CrossRef]
83. Wang, C.; Xu, L.; Liang, C.; Xiang, J.; Peng, R.; Liu, Z. Immunological Responses Triggered by Photothermal Therapy with Carbon Nanotubes in Combination with Anti-CTLA-4 Therapy to Inhibit Cancer Metastasis. *Adv. Mater.* **2014**, *26*, 8154. [CrossRef]
84. Feng, X.; Zhang, Y.; Wang, P.; Liu, Q.; Wang, X. Energy metabolism targeted drugs synergize with photodynamic therapy to potentiate breast cancer cell death. *Photochem. Photobiol. Sci.* **2014**, *13*, 1793. [CrossRef]
85. Berghöfer, B.; Frommer, T.; Haley, G.; Fink, L.; Bein, G.; Hackstein, H. TLR7 Ligands Induce Higher IFN-α Production in Females. *J. Immunol.* **2006**, *177*, 2088. [CrossRef]
86. Pinnapireddy, S.R.; Duse, L.; Akbari, D.; Bakowsky, U. Photo-Enhanced Delivery of Genetic Material Using Curcumin Loaded Composite Nanocarriers. *Clin. Oncol.* **2017**, *2*, 1323.
87. Anilkumar, T.S.; Lu, Y.J.; Chen, H.A.; Hsu, H.L.; Jose, G.; Chen, J.P. Dual targeted magnetic photosensitive liposomes for photothermal/photodynamic tumor therapy. *J. Magn. Magn. Mater.* **2018**, *473*, 241–252. [CrossRef]
88. Fahmy, S.A.; Alawak, M.; Brüßler, J.; Bakowsky, U.; El-Sayed, M.M. Nano-enabled Bioseparations: Current developments and future prospects. *BioMed Res. Int. J.* **2019**, *2019*, 4983291.
89. Lee, E.H.; Lim, S.J.; Lee, M.K. Chitosan-coated liposomes to stabilize and enhance transdermal delivery of indocyanine green for photodynamic therapy of melanoma. *Carbohydr. Polym.* **2019**, *224*, 115143. [CrossRef]
90. Wu, P.T.; Lin, C.L.; Lin, C.W.; Chang, N.C.; Tsai, W.B.; Yu, J. Methylene-Blue-Encapsulated Liposomes as Photodynamic Therapy Nano Agents for Breast Cancer Cells. *Nanomaterials* **2019**, *9*, 14. [CrossRef]
91. Rizvi, I.; Nath, S.; Obaid, G.; Ruhi, M.K.; Moore, K.; Bano, S.; Kessel, D.; Hasan, T. A Combination of Visudyne and a Lipid-anchored Liposomal Formulation of Benzoporphyrin Derivative Enhances Photodynamic Therapy Efficacy in a 3D Model for Ovarian Cancer. *Photochem. Photobiol.* **2019**, *95*, 419–429. [CrossRef]

Review

Combination-Based Strategies for the Treatment of Actinic Keratoses with Photodynamic Therapy: An Evidence-Based Review

Stefano Piaserico *, Roberto Mazzetto, Emma Sartor and Carlotta Bortoletti

Dermatology Unit, Department of Medicine, University of Padua, 35121 Padua, Italy
* Correspondence: stefano.piaserico@unipd.it; Tel.: +39-049-8212914; Fax: +39-049-8212502

Abstract: Photodynamic therapy (PDT) is a highly effective and widely adopted treatment strategy for many skin diseases, particularly for multiple actinic keratoses (AKs). However, PDT is ineffective in some cases, especially if AKs occur in the acral part of the body. Several methods to improve the efficacy of PDT without significantly increasing the risks of side effects have been proposed. In this study, we reviewed the combination-based PDT treatments described in the literature for treating AKs; both post-treatment and pretreatment were considered including topical (i.e., diclofenac, imiquimod, adapalene, 5-fluorouracil, and calcitriol), systemic (i.e., acitretin, methotrexate, and polypodium leucotomos), and mechanical–physical (i.e., radiofrequency, thermomechanical fractional injury, microneedling, microdermabrasion, and laser) treatment strategies. Topical pretreatments with imiquimod, adapalene, 5-fluorouracil, and calcipotriol were more successful than PDT alone in treating AKs, while the effect of diclofenac gel was less clear. Both mechanical laser treatment with CO_2 and Er:YAG (Erbium:Yttrium–Aluminum–Garnet) as well as systemic treatment with Polypodium leucotomos were also effective. Different approaches were relatively more effective in particular situations such as in immunosuppressed patients, AKs in the extremities, or thicker AKs. Conclusions: Several studies showed that a combination-based approach enhanced the effectiveness of PDT. However, more studies are needed to further understand the effectiveness of combination therapy in clinical practice and to investigate the role of acitretin, methotrexate, vitamin D, thermomechanical fractional injury, and microdermabrasion in humans.

Keywords: photodynamic therapy (PDT); actinic keratoses; combination; topical; systemic; laser

Citation: Piaserico, S.; Mazzetto, R.; Sartor, E.; Bortoletti, C. Combination-Based Strategies for the Treatment of Actinic Keratoses with Photodynamic Therapy: An Evidence-Based Review. *Pharmaceutics* **2022**, *14*, 1726. https://doi.org/10.3390/pharmaceutics14081726

Academic Editor: Jun Dai

Received: 6 July 2022
Accepted: 12 August 2022
Published: 18 August 2022

Publisher's Note: MDPI stays neutral with regard to jurisdictional claims in published maps and institutional affiliations.

1. Introduction

Actinic keratoses (AKs) are a very common skin disease caused by chronic sun damage. More than 75% of AKs arise in chronically sun-exposed areas such as the face, scalp, neck, and back of the hands/forearms. The rates of malignant transformation vary from 0.025% to 16%, and the risk of progression of squamous cell carcinoma (SCC) increases in patients with multiple AKs [1].

Photodynamic therapy (PDT) is a highly effective and well-tolerated local treatment based on the interaction between a photosensitizer, an appropriate wavelength of light, and oxygen (Figure 1).

It is generally administered with a topical photosensitizer drug, such as 5-aminolevulinic acid (ALA), or derivatives such as methyl aminolevulinate (MAL) [2]. These precursors of the heme biosynthetic pathway are converted within target cells into photoactivatable porphyrins, especially protoporphyrin IX (PpIX). Protoporphyrin IX has its largest absorption peak in the blue region at 410 nm with smaller absorption peaks at 505, 540, 580, and 630 nm². Most light sources for PDT target the 630 nm absorption peak in the red region, associated with greater tissue penetration, although a blue, fluorescent lamp (peak emission 417 nm) is sometimes used with ALA-PDT [3].

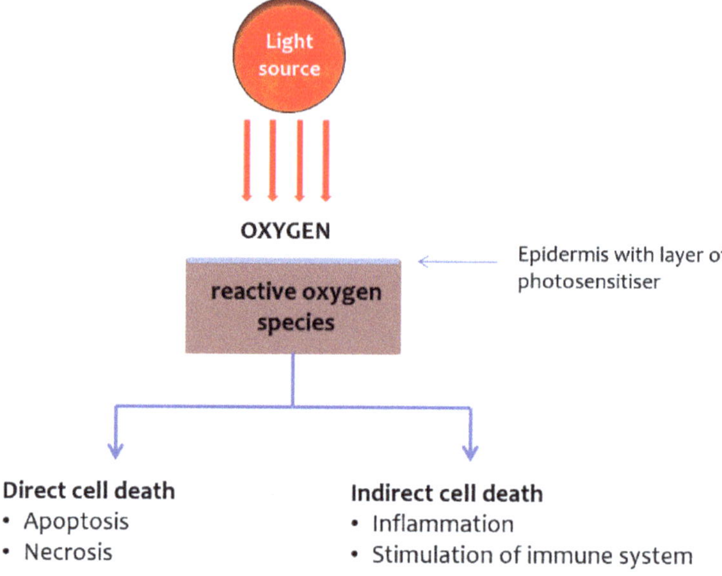

Figure 1. Schematic representation of PDT's mechanism of action.

The antitumor effects of PDT are based on three principal mechanisms: direct cytotoxic effects on cancer cells, indirect effects due to the fact of tumor vasculature damage, and the activation of the immune response (Figure 1) [4]. PDT with 5-ALA or MAL showed lesion clearance rates of 81–92% for thin and moderately thick AKs on the face and scalp [5]. PDT short-term (3 months) and long-term (12 months) efficacy is comparable or superior to imiquimod 5%, 5-fluorouracil 0.5%, and cryosurgery [6–9].

The main limitation of PDT compared to other treatments is pain, which may be the reason for stopping the treatment and may induce the patient to decline further treatment. Nerve block, subcutaneous infiltration anesthesia, and cold analgesia are associated with less PDT-related pain [10]. Lower irradiance (as in daylight PDT) and possibly shorter application times are associated with decreased discomfort without loss of PDT efficacy [3].

Nonetheless, PDT is generally considered to give good cosmetic outcomes [6]. This might be of importance when treating lesions localized on the face or the scalp and may explain patients' preference for and satisfaction with PDT compared to a physically destructive method such as cryotherapy.

However, PDT efficacy may be limited by the reduced ALA/MAL penetration depth and subsequently insufficient PpIX production in lesional skin. Indeed, thicker lesions in some other areas (especially in the upper and lower limbs) are less responsive to this treatment, and some patients show treatment failure or recurrence.

Therefore, AK treatment by combining PDT with other methods of therapy might be more efficacious, although the evidence for such procedures is limited [11].

In this review, we discuss evidence-based in vitro experiments and clinical trials regarding the administration of other therapeutic methods before or after PDT. Specifically, we assessed the effectiveness of topical (i.e., diclofenac, imiquimod, adapalene, 5-fluorouracil, and calcitriol), systemic (i.e., acitretin, methotrexate, vitamin D, and polypodium leucotomos), and mechanical–physical (i.e., radiofrequency, thermomechanical fractional injury ((TMFI)), microdermabrasion, microneedling, and laser) treatment strategies (Figure 2).

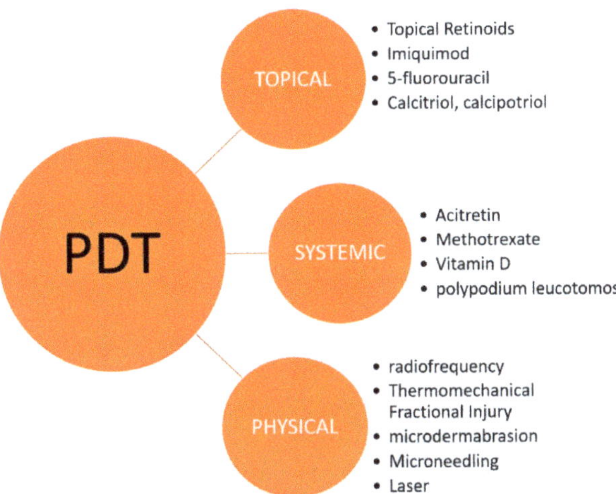

Figure 2. Combination-based strategies for the treatment of actinic keratoses with PDT.

2. Topical Treatments Combined with PDT

2.1. Diclofenac

Diclofenac is a nonsteroidal anti-inflammatory drug that reduces the production of prostaglandins by inhibiting inducible COX-2 (cyclo-oxygenase-2) [12]. It is used for treating AKs, considering that arachidonic acid metabolites promote epithelial tumor growth by stimulating angiogenesis, inhibiting apoptosis, and increasing the invasiveness of tumor cells [12].

Van Der Geer et al. evaluated the effectiveness of 3% diclofenac, applied for four weeks, twice a day, before administering ALA-PDT. The treatment of AKs on the dorsum of the hands of 10 patients had a slightly better long-term outcome (though not significant). Upon evaluating the pain scores during treatment, unbearable pain was detected to a greater extent in the patients of the diclofenac group [13].

However, since 3% diclofenac gel with hyaluronic acid showed better outcomes with longer applications (i.e., 60 and 90 days of administration showed total clearances of 30% and 50%, respectively, in the patients [14]), the effectiveness of PDT treatment in combination with diclofenac and alone should be compared after administration for more than four weeks [15].

2.2. Imiquimod

Imiquimod 5% cream stimulates the innate immune response through the interaction with Toll-like receptor 7 (TLR7).

Shaffelburg [16] compared sequential treatment with ALA-PDT versus ALA-PDT alone. Patients with at least 10 actinic keratoses on the face were treated with photodynamic therapy with aminolevulinic acid (20%) at baseline and month 1. At month 2, imiquimod cream (5%) was applied to half of the face, and the vehicle was applied to the other half twice a week for 16 weeks.

At month 12, the median reduction in the lesion was 89.9% and 74.5% ($p = 0.0023$) on the parts of the face treated with the combination of the imiquimod cream with ALA-PDT and ALA-PDT alone, respectively. Furthermore, tolerance was similar in patients treated with imiquimod and those treated with placebo. Serra et al. [17] compared the effectiveness of PDT alone, imiquimod (5%) alone, and the combined administration in 105 patients (PDT followed by imiquimod twice a week for 16 weeks).

They reported a complete clinic-pathologic response ($p = 0.038$) in 10% of the patients in the PDT-alone group, 27% of the patients in the imiquimod-alone group, and 34% of the patients in the PDT + imiquimod group. Moreover, a significantly greater percentage of patients who received PDT were extremely satisfied with the treatment than those who received only imiquimod (90% PDT vs. 61% imiquimod). Another study with an open-label parallel cohort design investigated the effectiveness of the treatment with imiquimod cream (3.75%), combined with photodynamic therapy (ALA-PDT) vs. imiquimod cream (3.75%) alone [18]. The former group received imiquimod (3.75%) daily for two weeks in a two-week cycle, separated by two weeks of no treatment and followed by one session of photodynamic therapy of the entire face with ALA and blue light. In the latter group, the treatment was administered daily for two weeks in a two-week cycle, separated by two weeks of no treatment. A mean reduction rate of 81% was found for the group treated with PDT, and 83% for the group treated only with imiquimod (the mean difference between the combination treatment and treatment with only imiquimod cream (3.75%) was 2%). However, five out of nine patients who received the combination treatment showed complete clearance of AKs, while only two out of nine patients showed this outcome in the other group. The rates of adverse events observed between the two groups were similar.

2.3. Topical Retinoids

Retinoid drugs are also used widely for the treatment of AKs, as they can disrupt gene expression, modify cell differentiation, and modulate cell proliferation or hyperplasia [19].

Currently, three generations of synthetic retinoids exist. First-generation retinoids include tretinoin (all-trans RA), isotretinoin (13-cis-retinoic acid), and alitretinoin (9-cis RA). Second-generation retinoids include etretinate and acitretin. Third-generation retinoids include adapalene, tazarotene, and bexarotene [1,2]. The application of tazarotene gel (0.1%) twice daily on AKs of the upper extremities one week before the administration of ALA-PDT (20%) increased the reduction in the lesion count by $\geq$50% ($p = 0.0547$) [20].

Galitzer [21] studied the effectiveness of adapalene gel (0.1%) as a pretreatment topical drug for the treatment of AKs on the dorsal hands and forearms by administering a single dose of ALA-PDT. A 10% ALA gel was used instead of a 20% ALA gel, and a red light was used for illumination instead of blue light. The difference between the two groups was significant ($p = 0.0164$), with a median lesion count reduction of 79% in the adapalene pretreated group compared to 57% in the standard therapy group. The discomfort under red light was slightly greater in patients of the standard therapy group, but the difference was not significant.

2.4. 5-Fluorouracil Cream

Topical 5-fluorouracil (5-FU) has greater efficacy than ALA and MAL-PDT [22].

It is converted in the cell to a thymidylate synthase inhibitor that interrupts the synthesis of thymidine, causing growth arrest and apoptosis [23].

In mouse models of SCC (SKH-1 and A431 mice), pretreatment with 5-FU for three days followed by ALA for 4 h caused a considerable, tumor-selective increase in PpIX levels and enhanced cell death upon illumination. The effect might be related to a change in the expression of the heme enzyme (upregulating coproporphyrinogen oxidase and downregulating ferrochelatase) and to a substantial, tumor-selective increase in apoptosis [24].

Likewise, an increase in the PpIX levels (two-fold to three-fold) in 5-FU-pretreated lesions versus controls was also reported in patients with AKs [25].

The clearance rates after treatment with MAL-PDT, with or without six days of 5-FU pretreatment, were 75% and 45% at three months and 67% and 39% at six months, respectively [26].

Niessen et al. [27] found that pretreatment with 5-FU for a week, twice daily before treatment with daylight (DL)-PDT, was more effective for AKs on the dorsal side of hands compared to treatment with DL-PDT alone. At the three-month follow-up, the overall lesion response rate was significantly higher for the combination treatment (62.7%) than for the treatment with DL-PDT alone (51.8%), while no difference was found in the degree of erythema one day after PDT between the treatment groups (pretreatment with 5-FU only caused minimal erythema in a few patients before illumination).

Other studies confirmed that the combined treatment was more effective for AKs on the upper limbs or the face [28–30].

2.5. Topical Calcitriol/Calcipotriol

Vitamin D is a differentiation-promoting hormone, and calcitriol is the most potent and active form of the hormone, the receptor of which is also present in cancer cells [22].

Furthermore, it induces a marked increase in protoporphyrin IX accumulation at the site of the tumor and promotes the activation of the extrinsic apoptotic pathway [31]. Therefore, cell damage after light irradiation is considerably higher in calcitriol-pretreated cells [32–34].

Galimberti first proposed pretreatment with calcipotriol (50 mcg/g) during the treatment of actinic keratoses with MAL DL-PDT [35]. Three months after the treatment, the complete response rate was 85% after calcipotriol pretreatment, while it was 70% for the DL-PDT alone group. Calcipotriol/DL-PDT showed a stronger association with erythema than DL-PDT alone. Some patients preferred DL-PDT alone due to the inconvenience caused by the daily application of calcipotriol and the related erythema and desquamation.

Torezan et al. confirmed these findings and showed improvement in keratinocyte atypia following treatment with MAL-PDT, with slightly more improvement on the parts pretreated with calcipotriol. However, erythema, edema, crusting, and postulation represented the most common adverse events and occurred more frequently in the group treated with the topical vitamin D3 analog [36].

Moreover, a long-term study on treatment with MAL-PDT with prior application of topical calcipotriol or conventional MAL-PDT for AKs on the scalp showed an overall AK clearance of 92% and 82% after three months for CAL-PDT and conventional PDT, respectively ($p < 0.001$). Similar results were found at 6 and 12 months, i.e., 92% vs. 81% ($p < 0.001$) and 90% vs. 77% ($p < 0.001$) for CAL-PDT and conventional PDT, respectively [37].

Another intra-individual, randomized trial [38] was conducted with patients with AKs on the dorsum of the hands or forearms, who were treated with another vitamin D analog, calcitriol (3 mg/g). Calcitriol or a placebo was applied once daily for 14 consecutive days. On day 15, the first administration of MAL DL-PDT was performed, and the second one took place after one week. A higher efficacy was found for grade II and grade III AKs, with a response rate of 55.24% for the group pretreated with calcitriol and 39.58% for the control group ($p - 0.038$).

The administration of topical vitamin D analogs might be more effective than treatment with PDT alone, particularly for thicker and difficult-to-treat AKs on the upper extremities and the scalp. However, the administration of vitamin D analogs might be associated with an increase in local skin reactions.

3. Systemic Treatment Combined with PDT

3.1. Acitretin

Acitretin is a systemic retinoid drug. Although the on-label instruction for acitretin only recommend its use for psoriasis, its off-label use in keratinization disorders is very common [39].

Acitretin induces normalization of differentiation and proliferation as well as the modification of inflammatory responses and neutrophil function in the skin. Moreover, some studies have shown the effectiveness of this drug in preventing non-melanoma skin tumors and AKs [40].

In a nonclinical setting, retinoic acid preconditioning before PDT can induce a moderate (though nonsignificant) increase in ALA–PpIX levels in prostate cancer cells. However, Hasan et al. found a dramatic increase in PDT response after pretreatment with retinoids due to the increase in the production of protoporphyrin IX (PpIX) in the target cells [41,42].

Preclinical studies support the hypothesis that induction of keratinocyte differentiation can increase intracellular PpIX accumulation in cells treated with ALA before light exposure in mice [41]. Treatment with a combination of acitretin and ALA-PDT can significantly increase the apoptosis rate and the mortality rate of SCL-1 cells compared to treatment with acitretin or ALA-PDT alone [43].

However, we found no studies regarding the effectiveness of combining PDT with acitretin in clinical trials or observational studies. This might be an important treatment strategy, considering the encouraging results reported in vitro.

3.2. Methotrexate

Methotrexate (MTX) is widely used for the treatment of psoriasis and other conditions characterized by cell hyperproliferation and lack of differentiation. MTX is a chemotherapeutic agent that inhibits cell proliferation and triggers cellular differentiation [44].

Pretreating human SCC13 cells, HEK1 cells, and normal keratinocytes with MTX for 72 h enhanced the PpIX levels by two-fold to four-fold in cancerous cells relative to the PpIX levels in nontumoral keratinocytes [45]. This was probably because MTX modified the expression of certain enzymes involved in the porphyrin metabolic pathway (a four-fold increase in coproporphyrinogen oxidase and a stable or slight decrease in the expression of ferrochelatase) and induced the expression of differentiation markers (i.e., E-cadherin, involucrin, and filaggrin) in cancer cells [45]. Photodynamic cell killing was thus synergistically enhanced by the combined therapy compared to the effectiveness of PDT alone.

However, the oral administration of MTX before PDT in patients with AKs, specifically in the elderly and those with several comorbidities, might not be possible, since the risks could exceed the benefits compared to the risks with other topical or systemic drugs.

3.3. Vitamin D

We previously discussed the ability of vitamin D derivatives and their prodifferentiating effects to increase protoporphyrin IX accumulation and enhance ALA-PDT-mediated cell death.

Systemic delivery of calcitriol is easily performed in mice and increases the tumoral accumulation of PpIX up to 10-fold, while in humans, high calcitriol levels are associated with the risk of developing hypercalcemia. Interestingly, the deficiency of serum vitamin D is associated with poorer responses of AKs to MAL-PDT [46] and, thus, further studies are needed to investigate its role in the treatment of AKs.

3.4. Polypodium Leucotomos

Polypodium leucotomos (PL) is an effective systemic photoprotective agent that strongly protects the skin against UV radiation [47].

PL is extracted from ferns grown in Central America and has been used for centuries by Native Americans for treating malignant tumors. Studies have shown that PL has antioxidative properties, immunomodulatory properties, and antitumoral activity [48].

Specifically, PL extract supplementation induces p53 activation and reduces acute inflammation by inhibiting the Cox-2 enzyme and increasing the removal of cyclobutane pyrimidine dimers [49,50].

In clinical settings, PL supplementation at a dose of 960 mg per day for one month and then 480 mg per day for five months decreased the recurrence rate of AKs at six months after two sessions of MAL-PDT, administered one week apart. MAL-PDT + PL supplementation showed a median reduction of 87.5% in scalp AKs compared to a reduction of 62.5% in the group treated only with MAL-PDT ($p = 0.040$). Moreover, no major side effects were recorded in either group [51].

Therefore, using PDT-MAL combined with PL might be an effective treatment for AKs due to the fact of its effect on reducing immunosuppression in the human skin not only after UV irradiation but also after PDT treatment, thus reducing the recurrence of AKs in high-risk individuals.

4. Physical and Mechanical Treatments Combined with PDT

Most of the aforementioned topical and systemic drugs improve PDT efficacy by preferentially enhancing PpIX production in AK cells, acting as differentiation-promoting agents.

However, PDT is mostly limited by the poor ALA/MAL penetration within lesional skin. The skin barrier is a lipid layer along which molecules can migrate through diffusion; they can enter through transcellular pathways, intercellular spaces, and the trans-appendageal pathway that includes hair follicles, sebaceous glands, and sweat glands [52,53].

One extensively investigated strategy to enhance drug delivery through the skin is by adding chemical enhancers such as fatty acids, surfactants, esters, and alcohols [54,55].

However, mechanical, physical, and active transport techniques also enhance skin penetration [56]. Some of these treatment methods have been assessed in combination with PDT for AK treatment. The disruption and ablation of the stratum corneum (the primary barrier to the delivery of topical drugs) by using lasers, radiofrequency (RF), thermomechanical fractional injury (TMFI), microdermabrasion (MD), or microneedling can enhance PDT treatment.

4.1. Laser-Assisted PDT

Different types of light sources are used for topical PDT including fluorescent lamps, light-emitting diodes (LEDs), and lasers [2]. However, lasers are not widely used because of their side effects, including pain and discomfort during treatment, as well as erythema, blistering, crusting, and pigmentation after treatment. Moreover, their application is limited while treating larger fields [57].

Due to the advancement in technology, fractional ablative lasers have been developed for promoting topical/transdermal drug delivery. Er:YAG (erbium:yttrium–aluminum–garnet) and CO_2 (carbon dioxide) lasers produce microscopic vertical channels in the skin that enhance the penetration of photosensitizers [58].

Thus, lasers can be used during pretreatment to facilitate the enrichment of ALA or MAL in dysplastic cells, and this approach is also known as laser-assisted drug delivery.

4.1.1. Er:YAG (Erbium:Yttrium–Aluminium–Garnet) Lasers

Shen et al. studied the in vivo kinetics of protoporphyrin IX (PpIX) after topical ALA application, enhanced by an Er:YAG laser; the increase in the ratio of PpIX with laser-treated murine skin ranged from 1.7 to 4.9-times compared to the PpIX levels in the control group [59].

Gellén et al. [60] conducted an intra-patient randomized study and showed that fractional laser pretreatment leads to significantly higher clearance at three months compared to the clearance with PDT alone. However, no difference in recurrence rates was found during follow up after 12 months. In every patient, two random treatment areas received conventional PDT or Er:YAG-ablative fractional laser-assisted PDT (AFL-PT). The number of AKs decreased by $87.5 \pm 17.3\%$ and $82.5 \pm 16.5\%$ ($p = 0.039$) three months after Er:YAG-AFL PDT and cPDT, respectively. Furthermore, upon assessing the immune infiltration, a reduction in Ki67-positive cells and CD8+ T cells was found three months after Er:YAG-assisted PDT [60].

Togsverd-Bo et al. [61] examined difficult-to-treat AKs in organ-transplant recipients (OTRs) using fractional Er:YAG laser-assisted DL-PDT. At three months, the clearance rates were 74% after AFL-dPDT, 46% after dPDT, 50% after cPDT, and 5% after AFL ($p < 0.001$). AFL-dPDT had excellent tolerability, even if the pain was higher in areas treated with AFL-assisted DL-PDT than in areas treated with DL-PDT only.

Another prospective randomized nonblinded trial [62] was conducted to evaluate patients who underwent one session of MAL-PDT using a red light-emitting diode lamp at $37 \, J/cm^2$, and the patients were randomly assigned to receive fractional Er:YAG laser (AFL-PDT) pretreatment. AFL-PDT was significantly more effective than MAL-PDT for treating patients with AKs of all grades (86.9% vs. 61.2%; $p < 0.001$), although the efficacy of AFL-PDT was the most pronounced in treating Olsen grade III AKs (69.4% vs. 32.5%; $p = 0.001$). AFL-PDT also showed a lower lesion recurrence rate than MAL-PDT (9.7% vs. 26.6%; $p = 0.004$), and an excellent or good cosmetic outcome was reported in >90% of the cases. Erythema and hyperpigmentation intensities were higher but not significant in the AFL-PDT group, while the side effects were mild but more frequent in the AFL-PDT group, although the result was not statistically significant [62].

4.1.2. CO_2 (Carbon Dioxide) Laser

Haedersdal et al. evaluated drug delivery of ALA and MAL using CO_2 ablative fractional laser (AFXL). They found that AFXL increased topical uptake of porphyrin precursors when it was followed by the topical application of MAL for 3 h. This pretreatment with laser generated 3 mm microchannels in the skin and, consequently, enabled a more homogeneous distribution of the porphyrins throughout the skin [63].

Syed et al. [64] studied 12 OTRs with multiple (>5) AKs of identical size in two areas on the scalp and/or the forehead. The treatment areas, randomized to AFL-assisted DL-PDT, were first treated with a carbon dioxide laser (eCO_2) targeting single AK lesions, followed by treatment of the whole field with the methyl aminolevulinate (MAL) cream applied to both treatment areas. After 30 min, both treatment areas were exposed to sunlight for 2 h. At the four-month follow up, the overall complete response was 75.5% in areas treated with AFL-assisted DL-PDT and 64.0% in areas treated with DL-PDT. There was a significant interaction between a complete response and the AKs' grade ($p = 0.001$), as well as between a partial response and the AKs grade ($p = 0.007$). Patient-reported pain was significantly higher in areas treated with AFL-assisted PDT in the first two days ($p = 0.008$) but not after five days ($p = 0.11$) [64].

Togsved-Bo et al. [65] found that at the three-month follow up, AFL-PDT was significantly more effective than PDT for all grades of AKs. Complete lesion response of grade II-III AKs was 88% after AFL-PDT and 59% after PDT ($p = 0.02$). In grade I AKs, 100% of the lesions were cleared after AFL-PDT, while 80% of the lesions were cleared after PDT ($p = 0.04$). The AFXL-PDT-treated skin showed significantly fewer new AK lesions ($p = 0.04$) and better photoaging ($p = 0.007$) than the PDT-treated skin. The pain scores during illumination ($p = 0.02$), erythema, and crusting were higher. Inflammatory reactions were more intense in AFL-PDT-treated skin than in PDT-treated skin in 83% of the patients and were of equal intensity in 17% of the patients. PpIX fluorescence was higher in AFL-pretreated skin ($p = 0.003$) [65] and supported the clinical superiority of the combination treatment.

Paasch et al. evaluated the effectiveness and safety of CO_2 AFL-LAD combined with indoor daylight (IDL) ALA-PDT for treating skin field cancerization associated with AKs. All patients showed remission (complete: 71.7%; partial: 28.3%), suggesting that AFL-LAD combined with IDL-PDT is an exceptionally effective treatment approach. However, the high pain score associated with this combined approach is a limiting factor [66].

Jang et al. [67] showed that laser-assisted PDT might reduce the photosensitizer incubation time or the number of sessions required for complete response (efficacy of 70.6% in three sessions).

In conclusion, several studies showed a reduction in the number of AKs and an improvement in the recurrence rate in the approach combining laser treatment and PDT. A recent meta-analysis [30] reported an overall significantly high clearance rates for laser-assisted PDT than for PDT alone (RR: 1.33, 95% CI: 1.24–1.42); however, the evidence for this outcome was graded as low (GRADE ++- -). Furthermore, the pain was similar for both treatments (mean difference of 0.31, 95% CI: 0.12–0.74, low quality of evidence, GRADE ++- -).

4.2. Radiofrequency and Thermomechanical Fractional Injury (TMFI)

Fractional radiofrequency (RF) creates plasma close to the skin and provokes plasma sparks on the skin's surface that induce epidermal ablation and produce microchannels that perforate the dermis, thus improving drug delivery. Additionally, sonophoresis facilitates the movement of molecules through the intact skin under the influence of an ultrasonic perturbation. Low-frequency ultrasound (frequencies below 100 kHz) can highly enhance transdermal transport [68].

Park et al. demonstrated that fractional RF with sonophoresis effectively enhanced ALA penetration in swine skin [69]. Prefractional RF followed by post-treatment with sonophoresis is a promising therapeutic combination for ALA-PDT to enhance ALA uptake [67]. However, no studies on humans have been conducted yet.

Shavit et al. [70] evaluated the efficacy of pretreatment by thermomechanical fractional injury (TMFI) at low-energy settings in five healthy subjects to increase the permeability of the skin to four topical preparations. TMFI, (Tixel®;Novoxel®, Netanya, Israel), is a thermomechanical system developed for providing fractional treatment. The system is designed for treating soft tissues by direct conduction of heat, which allows water to evaporate rapidly while causing low thermal damage to the surrounding tissue.

The authors evaluated the permeability of 20% ALA gel, 10% ALA microemulsion gel, 16.8% MAL cream, and 20% ALA hydroalcoholic solution. Pretreatment with low-energy TMFI at a pulse duration of 6 milliseconds increased the percutaneous permeation of ALA when the 20% gel was used.

Interestingly, after 2 h and 3 h of treatment, the TMFI-treated sites exhibited a higher hourly rate of PpIX fluorescence intensity, which was 156–176% higher than that in the control ($p \leq 0.004$). Thus, TMFI is a powerful method to enhance the transdermal drug delivery of ALA and its derivatives. Further studies are needed to investigate its role in improving PDT treatment.

4.3. Microdermabrasion (MD)

Microdermabrasion (MD) is an easily accessible and cost-effective technique that can be used to increase the efficacy of PDT as a pretreatment method.

Bay et al. compared PpIX accumulation after a range of standardized pretreatments and found significantly higher median intracutaneous PpIX fluorescence intensities after AFL compared to that with alternative physical modalities such as MD [71].

The clinical efficacy of PDT was evaluated after pretreatment with either Er:YAG AFL or MD in a side-by-side trial [72].

MD pretreatment was performed using a pad with particles that had a diameter of 58.5 µm (Ambu® Skin Prep Pads 2121M; Ambu A/S, Ballerup, Denmark). Two large areas were randomly selected in each individual, and a single treatment was administered with AFL + DL-PDT or microdermabrasion + DL-PDT. Interestingly, AFL-dPDT resulted in a significantly higher AK clearance (81% vs. 60%, $p < 0.001$), led to fewer new AKs ($p < 0.001$), and showed higher improvement in dyspigmentation ($p = 0.003$) and skin texture ($p = 0.001$), which was related to microneedling-assisted DL-PDT (MD-dPDT). However, laser-assisted PDT caused more local skin reactions than microdermabrasion pretreatment [72].

4.4. Microneedling-Assisted PDT

Microneedling represents another pretreatment option to improve the effects of PDT. A microneedle consists of tiny needles on a roller or a stamp that punctures the skin and forms microchannels.

In vivo experiments using nude mice showed that microneedle punctures could reduce the application time and ALA dose required to induce high levels of the photosensitizer protoporphyrin IX in the skin. This is beneficial for clinical practice, as shorter application times are more convenient for patients and clinicians [73].

The Microneedling Photodynamic Therapy II (MNPDT-II) study revealed a statistically significant difference in the clearance of AKs for the 20 min incubation arm between the microneedling side and the PDT alone side (76% and 58%, respectively). The microneedle device consisted of a single-use sterile array of microneedles (200 µm long), while the sham treatment consisted of the applicator roller without microneedles. Immediately after sham and microneedle pretreatment, topical ALA was applied to the entire face, followed by exposure to blue-light PDT [74].

Likewise, Spencer and Freeman reported that microneedling pretreatment induced a significantly greater reduction in the AK count compared to that after PDT alone (89.3% and 69.5%, respectively); the occurrence of side effects was similar [75].

Furthermore, a comparative study conducted by Chen et al. to evaluate ALA-PDT revealed that plum-blossom needling (a method of shallow insertion of multiple needles into the skin) caused a broader diffusion of ALA than that with CO_2 AFXL, and the technique had a similar clinical effect at a considerably lower cost. The needle-pretreated-lesion had a stronger surface fluorescence intensity than the laser-pretreated-lesion [76].

However, Lev-Tov et al. [77] conducted a randomized controlled evaluator-blind trial and found no significant differences in the efficacy between microneedling-PDT and ALA-PDT (with a short incubation time of 60 min), while the microneedling side showed significantly higher pain scores. The authors stated that this conflicting result was due to the use of shorter microneedles, which only reached the epidermis (and not the dermis), than those used in other studies.

5. Other Physical and Chemical Treatments

Other methods have been used to increase the effectiveness of PDT. However, no studies have compared them to conventional-PDT or to daylight-PDT for the treatment of AKs.

In vivo studies showed that the incorporation of a thermogenic and vasodilating substances, such as methyl nicotinate, in the MAL cream increased the amount of PpIX produced in the tissue, causing a greater effect on the epidermis after PDT as well as reducing the cream's incubation time [78].

Iontophoresis, a physical treatment aimed at facilitating transport through the skin of ionic species using a voltage-gradient applied to an electrolytic formulation [23], was reported to reduce the incubation time by 1 h in a study comparing iontophoresis-assisted AFL-PDT with AFL-PDT [79].

Photobiomodulation consists of the illumination of tissues with subthermal radiometric conditions (red or near-infrared), stimulating cell metabolism and enhancing the production of PpIX and the effectiveness of PDT in vitro and in vivo tumors (human glioma cells) [80].

Another possible approach to increase 5-ALA penetration and efficiency is the addition of glycolic acid (GA) to 5-ALA for the treatment of patients with SCC [81] and ethylenendiamine-tetra-acetic-acid (EDTA) and dimethylsulphoxid (DMSO) to 5-ALA in order to improve protoporphyrin IX accumulation in certain subtypes of BCC, SCC, and Bowen's disease [82].

Therefore, iontophoresis, photobiomodulation, and the addition of methyl nicotinate, GA, or DMSO in ALA-based topicals represent encouraging possibilities to enhance conventional-PDT or daylight-PDT efficacy and should be further studied to assess their role in treating AKs as well.

6. Conclusions

Our review showed that combination treatment may significantly improve the effectiveness of PDT in reducing AKs (Table 1). Pretreatment with imiquimod, adapalene, 5-FU, and calcipotriol showed greater effectiveness in treating AKs than PDT alone, while evidence supporting the effectiveness of microneedling and diclofenac gel was poor.

Table 1. A list of comparative studies regarding the treatment of AKs with PDT alone and combination-based PDT in the general population.

Combining Therapy	Regimen Adopted	Outcome of Combined Regimen vs. PDT Alone	Adverse Effects of Combined Regimen vs. PDT Alone	Reference
TOPICAL TREATMENTS				
DICLOFENAC 3% gel pKa = 4.2 * logP = 0.7 **	Twice daily for 4 weeks, then one session with ALA-PDT.	12 month decrease in the total number of lesions score of 12.5 in the diclofenac group, while it was 8.8 in the control group. Not significant (p = 0.34).	When looking at the pain scores during treatment, a tendency for a greater, unbearable pain was scored in the diclofenac group.	[13]
IMIQUIMOD 5% pKa = 19.99 § logP = 2.65 §	Treated with ALA-PDT followed by imiquimod (3 times a week for 4 weeks).	Complete clinicopathologic response (p = 0.038) was obtained in 10% of the PDT-alone group, 27% in imiquimod-alone group, and in 34% of the PDT + imiquimod group.	No significant differences were observed among the options and tolerance to treatment.	[17]
	Treated with ALA-PDT at baseline and at month 1. At month 2, imiquimod 5% cream was applied 2 times per week for 16 weeks.	Median lesion reductions were 89.9% versus 74.5% (p = 0.0023), respectively, with a median difference of combination vs. ALA-PDT of 15.5%.	Similar.	[18]
TAZAROTENE 0.1% pKa = 1.23 § logP = 4.2 °	TZ gel 0.1% twice daily on AKs of the upper extremities, 1 week before ALA-PDT with ALA 20% gel.	Lesion count reduction $\geq$ 50% eight weeks after. The significance was borderline (p = 0.0547).	Adverse events were limited to those expected after ALA-PDT. In the pretreated arm five minutes after ALA-PDT, erythema was significantly more severe (p = 0.0029).	[20]
ADAPALENE 0.1% pKa = 3.99 § logP = 6.46 §	Adapalene 0.1% gel twice daily for one week, then one session with ALA-PDT with ALA 10%.	A median lesion count reduction in the adapalene-pretreated group of 79% compared to 57% in the standard therapy group, with a median difference of 22%. (p = 0.0164)	Discomfort during PDT was slightly greater with the standard therapy, but the difference did not achieve significance.	[21]
5-FLUOROURACIL 5% pKa = 8.02 φ logP = 0.89 **	5-FU 5% cream + MAL-PDT. Pretreatment with 5-FU 5% cream for 6 days followed by one session of MAL-PDT. AKs of the face, scalp, and forearms.	Relative clearance rates after PDT with or without 5-FU pretreatment were, respectively, 75% versus 45% at 3 months (mean difference of combinations was 30%) and 67% versus 39% at 6 months (mean difference of combinations was 28%).	5-FU/PDT combination treatment was well tolerated, with no major side effects other than the local inflammatory reaction typically associated with PDT treatment.	[25]
	5-FU 5% cream + dl-PDT. Pretreatment with 5-FU 5% cream twice daily for 7 days followed by one session of dl-PDT. AKs on the dorsal side of hands.	The reduction rate (mean) of the combined treatment group was 62.7%, while it was 51.8% in the PDT-alone group. The difference in combinations vs. monotherapy (mean) was 10.9% (p = 0.001).	No difference was found in the degree of erythema one day after PDT between the 2 treatment groups.	[27]
	Pretreatment with 5-FU 5% cream twice daily for 7 days followed by one session of ALA-PDT. AKs of the face.	A median lesion count reduction in the 5-FU-pretreated group of 100% compared to 66.7% in the standard therapy group with a median difference of 33.5%.	No significant difference in discomfort.	[29]
	5-FU 5% cream twice daily for 7 days followed by one session of ALA-PDT. Unclear regarding the localization of the KAs.	A median lesion count reduction in the 5-FU-pretreated group of 94.6% compared to 68.4% in the standard therapy group, with a median difference of 26.2% (p = 0.001).	Similar.	[30]
CALCIPOTRIOL **50 mcg/g** pKa = 14.39 ° logP = 3.84 °	15 days of treatment with calcipotriol or placebo (once daily) followed by one session of MAL-daylight-PDT.	The complete response rate was 85% while it was 70% for the dl-PDT-alone group; the partial response rate was 12% and 25%, respectively.	Calcipotriol/DL-PDT was associated with more marked erythema than that observed with DL-PDT alone.	[35]
	Calcipotriol was applied daily for 15 days beforehand on the other side.	At three months, overall AK clearance was 92.07% and 82.04% for CAL-PDT and conventional PDT, respectively (p < 0.001). Similar results were found at 6 and 12 months: 92.07% and 81.69% (p < 0.001), and 90.69% and 77.46% (p < 0.001) for CAL-PDT and conventional PDT, respectively.	Slightly superior discomfort after the application of calcipotriol.	[36,37]

Table 1. *Cont.*

Combining Therapy	Regimen Adopted	Outcome of Combined Regimen vs. PDT Alone	Adverse Effects of Combined Regimen vs. PDT Alone	Reference
CALCITRIOL 3 mg/g pKa = 14.39 ° logP = 4.35 °	A layer of calcitriol 3 mg/g or placebo was applied once daily for 14 consecutive days. On day 15 first MAL-DL-PDT was performed, while the second one took place 1 week apart.	A higher efficacy was found for the grade II and grade III AK groups. The response rate was 55.24% for the group pretreated with calcipotriol, whereas it was 39.58% for the control group, with a difference of 15.66% ($p = 0.038$).	Local skin reactions occurred more frequently on the calcitriol DL-PDT-treated sides.	[38]
SYSTEMIC TREATMENT				
POLYPODIUM LEUCOTOMOS	One week after MAL-PDT, PLE supplementation at a dose of 960 mg per day for 1 month and then 480 mg per day for 5 months.	At the 6 month follow up, PDT treatment + PLE supplementation displayed a better clearance rate compared with PDT alone ($p = 0.040$). There was a median reduction in scalp AKs of 87.5% in the combination group, while that of 62.5% in the group treated just with PDT.	No major side effects were recorded in either group.	[51]
PHYSICAL AND MECHANICAL TREATMENT COMBINED WITH PDT				
Er:YAG ABLATIVE FRACTIONAL LASER-ASSISTED PDT (AFL-PT)	The side affected was pretreated with Er:YAG-AFL immediately before ALA application, then PDT was undertaken with 20% ALA, applied for 3 h, then irradiated with water-filtered infrared A light for 20 min.	The number of AKs decreased by $87.56 \pm 17.30\%$ and $82.56 \pm 16.53\%$ ($p = 0.039$) 3 months after Er:YAG-AFL PDT and cPDT, respectively.	Not reported.	[60]
	The side affected was pretreated with Er:YAG-AFL, immediately before MAL application, then PDT was undertaken with a red-light-emitting diode (LED) lamp.	FL-PDT was significantly more effective than MAL-PDT at treating all AK grades (86.9% vs. 61.2%; $p < 0.001$). The efficacy of FL-PDT was most pronounced in treating Olsen grade III AKs (69.4% vs. 32.5%; $p = 0.001$). FL-PDT also showed a lower lesion recurrence rate than MAL-PDT (9.7% vs. 26.6%; $p = 0.004$).	Erythema and hyperpigmentation intensities were higher but not significant in the FL-PDT group, while side effects were mild but more frequent in the FL-PDT group, even though this result was not statistically significant ($p > 0.05$).	[61]
CARBON DIOXIDE LASER (ECO₂)-ASSISTED PDT (AFL-PT)	Carbon dioxide laser (eCO₂), first targeting single AK lesions, followed by treatment of the whole field with methyl aminolaevulinate (MAL) cream applied on both treatment areas. Red-light PDT was used.	At 3 months follow up, the complete lesion response of grade II–III AKs was 88% after AFXL-PDT compared with 59% after PDT ($p = 0.02$). In grade I AKs, 100% of the lesions cleared after AFXL-PDT compared with 80% after PDT ($p = 0.04$). AFXLPDT-treated skin responded with significantly fewer new AK lesions ($p = 0.04$).	Pain during LED illumination was significantly higher in AFXL-PDT-treated areas than in PDT-treated areas. After treatment, patients developed erythema and crusting in both treatment areas.	[65]
	Carbon dioxide laser (eCO₂), first targeting single AK lesions, followed by treatment of the whole field with ALA cream. Red-light PDT was used.	After the study protocol, all patients showed remission (complete: 71.7%; partial: 28.3%).	Higher pain scores were associated with this combined approach.	[66]
	Carbon dioxide laser (eCO₂), first targeting single AK lesions, followed by treatment of the whole field with 20% ALA or MAL. Red-light PDT was used.	70.6% of the lesions showed a complete response (CR) within three sessions of PDT.	No significant side effects were associated with the combination of ablative CO₂ fractional laser and PDT.	[67]

Table 1. *Cont.*

Combining Therapy	Regimen Adopted	Outcome of Combined Regimen vs. PDT Alone	Adverse Effects of Combined Regimen vs. PDT Alone	Reference
MICRONEEDLE-ASSISTED PDT	The microneedle device consisted of a single-use sterile array of microneedles 200 μm in length. Immediately after microneedle pretreatment, each topical ALA was applied to the entire face, and blue-light PDT was used.	Participants experienced significantly superior AK lesion clearance (76% vs. 58%, $p < 0.01$) at 20 min incubation times. While the 10 min group also experienced improvement in AK counts, the clearance rates between the microneedle side and the sham side were not significantly different.	The secondary outcome of pain associated with blue-light exposure during PDT was nominal and not significantly different from the sham side.	[74]
	Microneedling device applied to 1/2 of their face was followed by applying ALA 20% cream. Subsequently, blue-light PDT was used.	The mean percentage reduction in AKs was 89.3% on the microneedling side versus 69.5% on the PDT-alone side. There was a significant difference.	Not different.	[75]
	Microneedling device applied to 1/2 of their face was followed by applying ALA 20% cream for 60 min incubation. Subsequently, blue-light PDT was used.	The average complete response rates for 20, 40, and 60 min microneedling times versus ALA-PDT were 71.4% and 68.3%; 81.1% and 79.9%; 72.1% and 74.2%, respectively. There were no statistically significant differences.	There was statistical significance in pain scores between the microneedling application and the control one, but the absolute difference was small.	[77]

* Settimo, L.; Bellman, K.; Knegte, R.M.A. Comparison of the Accuracy of Experimental and Predicted pKa Values of Basic and Acidic Compounds. *Pharm. Res.* 2014, 31, 1082–1095; https://doi.org/10.1007/s11095-013-1232-z, (accessed on 7 August 2022); ** US Environmental Protection Agency. Comptox Chemicals Dashboard. https://comptox.epa.gov/dashboard/chemical/details/DTXSID3037208 (accessed on 7 August 2022); § https://chemaxon.com/products/calculators-and-predictors#pka, (accessed on 7 August 2022); ° Computed by XLogP3 3.0 (PubChem release 7 May 2021); φ Sangster J; LOGKOW Databank. Sangster Res. Lab., Montreal Quebec, Canada (1994).

Pretreatment with 5-FU, calcitriol, and adapalene can substantially increase the effectiveness of PDT in the extremities (10.9%, 15.7%, and 22% more than PDT alone, respectively). Immunosuppressed patients may benefit from the administration of 5-FU cream pre-PDT or by the combination of PDT with laser. In particular, Er:YAG laser and CO_2 laser-assisted PDT showed greater efficacy than PDT alone (28% and 11.5%, respectively). Treatment of thicker AKs with PDT can be improved by combination with Er:YAG laser-PDT (69.4% effectiveness, 36.9% more than PDT) even though a slight increase in the incidence of side effects was found and should be considered. AKs of higher grade may also be pretreated with higher effectiveness with CO_2 laser-PDT or with calcitriol (3 mg/g).

Therefore, combining treatments may greatly enhance PDT's ability to reduce AKs. However, more studies are needed to further understand the effectiveness of combination therapy in clinical practice and to investigate the role of systemic drugs, such as acitretin, methotrexate, and vitamin D, in humans.

Author Contributions: Conceptualization, S.P.; Methodology, S.P., R.M., E.S. and C.B.; Writing–Original Draft Preparation, S.P., R.M., E.S. and C.B.; Writing—Review and Editing, S.P. and R.M. All authors have read and agreed to the published version of the manuscript.

Funding: This research received no external funding.

Institutional Review Board Statement: Not applicable.

Informed Consent Statement: Not applicable.

Data Availability Statement: Not applicable.

Conflicts of Interest: The authors declare no conflict of interest.

List of Abbreviations

PDT	Photodynamic Therapy
DL-PDT or dPDT	Daylight-Photodynamic Therapy
AKs	Actinic Keratoses
PpIX	Protoporphyrin IX
SCC	Squamous Cell Carcinoma
ALA	5-Aminolevulinic Acid
MAL	Methyl Aminolevulinate
5-FU	Topical 5-Fluorouracil
RF	Radiofrequency
TMFI	Thermomechanical Fractional Injury
MD	Microdermabrasion
LEDs	Light-Emitting Diodes
Er-YAG	laser: Erbium:Yttrium–Aluminum–Garnet
AFL-PT	Er:YAG-ablative fractional laser-assisted PDT
IDL	Indoor Daylight
GA	Glycolic Acid
EDTA	Ethylenendiamine-tetra-acetic-acid
DMSO	Dimethylsulphoxid

References

1. Dianzani, C.; Conforti, C.; Giuffrida, R.; Corneli, P.; Di Meo, N.; Farinazzo, E.; Moret, A.; Magaton Rizzi, G.; Zalaudek, I. Current therapies for actinic keratosis. *Int. J. Dermatol.* **2020**, *59*, 677–684. [CrossRef] [PubMed]
2. Conforti, C.; Corneli, P.; Harwood, C.; Zalaudek, I. Evolving role on systemic therapies in non-melanoma skin cancer. *Clin. Oncol.* **2019**, *31*, 759–768. [CrossRef] [PubMed]
3. Morton, C.A.; Szeimies, R.M.; Basset-Seguin, N.; Calzavara-Pinton, P.; Gilaberte, Y.; Hædersdal, M.; Hofbauer, G.F.L.; Hunger, R.E.; Karrer, S.; Piaserico, S.; et al. European Dermatology Forum guidelines on topical photodynamic therapy 2019 Part 1: Treatment delivery and established indications-Actinic keratoses, Bowen's disease and basal cell carcinomas. *J. Eur. Acad. Dermatol. Venereol.* **2019**, *33*, 2225–2238. [CrossRef] [PubMed]
4. Maytin, E.V.; Hasan, T. Vitamin D and Other Differentiation-promoting Agents as Neoadjuvants for Photodynamic Therapy of Cancer. *Photochem. Photobiol.* **2020**, *96*, 529–538. [CrossRef] [PubMed]
5. Pariser, D.M.; Lowe, N.J.; Stewart, D.M.; Jarratt, M.T.; Lucky, A.W.; Pariser, R.J.; Yamauchi, P.S. Photodynamic therapy with topical methyl aminolevulinate for actinic keratosis: Results of a prospective randomized multicenter trial. *J. Am. Acad. Dermatol.* **2003**, *48*, 227–232. [CrossRef]
6. Patel, G.; Armstrong, A.W.; Eisen, D.B. Efficacy of photodynamic therapy vs other interventions in randomized clinical trials for the treatment of actinic keratoses: A systematic review and meta-analysis. *JAMA Dermatol.* **2014**, *150*, 1281–1288. [CrossRef]
7. Vegter, S.; Tolley, K. A network meta-analysis of the relative efficacy of treatments for actinic keratosis of the face or scalp in Europe. *PLoS ONE* **2014**, *9*, e96829. [CrossRef] [PubMed]
8. Steeb, T.; Wessely, A.; Petzold, A.; Brinker, T.J.; Schmitz, L.; Leiter, U.; Garbe, C.; Schöffski, O.; Berking, C.; Heppt, M.V. Evaluation of long-term clearance rates of interventions for actinic keratosis: A systematic review and network meta-analysis. *JAMA Dermatol.* **2021**, *157*, 1066–1077. [CrossRef] [PubMed]
9. Eisen, D.B.; Asgari, M.M.; Bennett, D.D.; Connolly, S.M.; Dellavalle, R.P.; Freeman, E.E.; Goldenberg, G.; Leffell, D.J.; Peschin, S.; Sligh, J.E.; et al. Guidelines of care for the management of actinic keratosis. *J. Am. Acad. Dermatol.* **2021**, *85*, e209–e233. [CrossRef] [PubMed]
10. Ang, J.M.; Riaz, I.B.; Kamal, M.U.; Paragh, G.; Zeitouni, N.C. Photodynamic therapy and pain: A systematic review. *Photodiagn. Photodyn. Ther.* **2017**, *19*, 308–344. [CrossRef] [PubMed]
11. Peris, K.; Calzavara-Pinton, P.G.; Neri, L.; Girolomoni, G.; Malara, G.; Parodi, A.; Piaserico, S.; Rossi, R.; Pellacani, G. Italian expert consensus for the management of actinic keratosis in immunocompetent patients. *J. Eur. Acad. Dermatol. Venereol.* **2016**, *30*, 1077–1084. [CrossRef]
12. Chiarugi, V.; Magnelli, L.; Gallo, O. Cox-2 iNOS and p53 as play-markers of tumor angiogenesis. *Int. J. Mol. Med.* **1998**, *10*, 715–719. [CrossRef]
13. Van der Geer, S.; Krekels, G.A.M. Treatment of actinic keratoses on the dorsum of the hands: ALA-PDT versus diclofenac 3% gel followed by ALA-PDT. A placebo-controlled, double-blind, pilot study. *J. Dermatol. Treat.* **2009**, *20*, 259–265. [CrossRef] [PubMed]
14. Jarvis, B.; Figgitt, D.P. Topical 3% diclofenac in 2.5% hyaluronic acid gel. A review of its use in patients with actinic keratoses. *Am. J. Clin. Dermatol.* **2003**, *4*, 203–213. [CrossRef] [PubMed]
15. Rivers, J.K.; Arlette, J.; Shear, N.; Guenther, L.; Carey, W.; Poulin, Y. Topical treatment of actinic keratoses with 3% diclofenac in 2.5% hyaluronian gel. *Br. J. Dermatol.* **2002**, *146*, 94–100. [CrossRef] [PubMed]

16. Shaffelburg, M. Treatment of actinic keratoses with sequential use of photodynamic therapy; and imiquimod 5% cream. *J. Drugs Dermatol.* **2009**, *8*, 35–39. [PubMed]
17. Serra-Guillén, C.; Nagore, E.; Hueso, L.; Traves, V.; Messeguer, F.; Sanmartín, O.; Llombart, B.; Requena, C.; Botella-Estrada, R.; Guillén, C. A randomized pilot comparative study of topical methyl aminolevulinate photodynamic therapy versus imiquimod 5% versus sequential application of both therapies in immunocompetent patients with actinic keratosis: Clinical and histologic outcomes. *J. Am. Acad. Dermatol.* **2012**, *39*, E130–E137. [CrossRef] [PubMed]
18. Gold, M.; Cohen, J.; Munavalli, G. Treatment of Actinic Keratoses of the Face With Imiquimod 3.75% Cream Followed by Photodynamic Therapy. 2014. Available online: https://clinicaltrials.gov/ct2/show/NCT01203878 (accessed on 29 April 2022).
19. Wu, C.-S.; Chen, G.-S.; Lin, P.-Y.; Pan, I.-H.; Wang, S.-T.; Hao Lin, S.; Yu, H.-S.; Lin, C.-C. Tazarotene induces apoptosis in human basal cell carcinoma via ativation of caspase-8/t-Bid and the reactive oxygen species-dependent mitochondrial pathway. *DNA Cell Biol.* **2014**, *23*, 652–666. [CrossRef] [PubMed]
20. Galitzer, B.I. Effect of retinoid pretreatment on outcomes of patients treated by photodynamic therapy for actinic keratosis of the hand and forearm. *J. Drugs Dermatol.* **2011**, *10*, 1124–1132.
21. Galitzer, B.I. Photodynamic therapy for actinic keratoses of the upper extremities using 10% aminolevulinic acid gel, red light and adapalene pretreatment. *J. Clin. Aesthet. Dermatol.* **2021**, *14*, 19–24.
22. Gupta, A.K.; Paquet, M. Network meta-analysis of the outcome 'participant complete clearance' in nonimmunosuppressed participants of eight interventions for actinic keratosis: A follow-up on a Cochrane review. *Br. J. Dermatol.* **2013**, *169*, 250–259. [CrossRef] [PubMed]
23. Champeau, M.; Vignoud, S.; Mortier, L.; Mordon, S. Photodynamic therapy for skin cancer: How to enhance drug penetration? *J. Photochem. Photobiol. B.* **2020**, *197*, 111544. [CrossRef] [PubMed]
24. Maytin, E.V.; Anand, S. Combination photodynamic therapy using 5-fluorouracil and aminolevulinate enhances tumor-selective production of protoporphyrin IX and improves treatment efficacy of squamous skin cancers and precancers. *Opt. Methods Tumor Treat. Detect. Mech. Tech. Photodyn. Ther.* **2016**, *16*, 1092–1101. [CrossRef]
25. Maytin, E.V.; Anand, S.; Riha, M.; Lohser, S.; Tellez, A.; Ishak, R.; Karpinski, L.; Sot, J.; Hu, B.; Denisyuk, A.; et al. 5-Fluorouracil Enhances Protoporphyrin IX Accumulation and Lesion Clearance during Photodynamic Therapy of Actinic Keratoses: A Mechanism-Based Clinical Trial. *Clin. Cancer Res.* **2018**, *1*, 3026–3035. [CrossRef] [PubMed]
26. Gilbert, D.J. Treatment for actinic keratosis with sequential combination of 5-fluorouracil and photodynamic therapy. *J. Drugs Dermatol.* **2005**, *4*, 161–163. [PubMed]
27. Nissen, C.V.; Heerfordt, I.M.; Wiegell, S.R.; Mikkelsen, C.S.; Wulf, H.C. Pretreatment with 5-fluotouracil Cream enhances the efficacy of daylight-mediated photodynamic therapy for actinic keratosis. *Acta Derm.-Venereol.* **2017**, *8*, 617–621. [CrossRef] [PubMed]
28. Tyrrell, J.S.; Morton, C.; Campbell, S.M.; Curnow, A. Comparison of protoporphyrin IX accumulation and destruction during methylaminolevulinate photodynamic therapy of skin tumours located at acral and nonacral sites. *Br. J. Dermatol.* **2011**, *164*, 1362–1368. [CrossRef]
29. Tanghetti, E.A.; Hamann, C.; Tanghetti, M. A controlled comparison study of topical fluorouracil 5% cream pretreatment of aminolevulinic acid/photodynamic therapy for actinic keratosis. *J. Drugs Dermatol.* **2015**, *14*, 1241–1244. [PubMed]
30. Pei, S.; Kaminska, E.C.N.; Tsoukas, M.M. Treatment of Actinic Keratoses: A Randomized Split-Site Approach Comparison of Sequential 5-Fluorouracil and 5 Aminolevulinic Acid Photodynamic Therapy to 5-Aminolevulinic Acid Photodynamic Monotherapy. *Dermatol. Surg.* **2017**, *43*, 1170–1175. [CrossRef]
31. Steeb, T.; Wessely, A.; Leiter, U.; French, L.E.; Berking, C.; Heppt, M.V. The more the better? An appraisal of combination therapies for actinic keratosis. *J. Eur. Acad. Dermatol. Venereol.* **2019**, *34*, 727–732. [CrossRef]
32. Anand, S.; Wilson, C.; Hasan, T.; Maytin, E.V. Vitamin D3 enhances the apoptotic response of epithelial tumors to aminolevulinate-based photodynamic therapy. *Cancer Res.* **2012**, *71*, 8980–8984. [CrossRef] [PubMed]
33. Sato, N.; Moore, B.W.; Keevey, S.; Drazba, J.A.; Hasan, T.; Maytin, E.V. Vitamin D enhances ALA-induced protoporphyrin IX production and photodynamic cell death in 3-D organotypic cultures of keratinocytes. *J. Investig. Dermatol.* **2007**, *124*, 925–934. [CrossRef] [PubMed]
34. Cicarma, E.; Tuorkey, M.; Juzeniene, A.; Ma, L.-W.; Moan, J. Calcitriol treatment improves methyl aminolaevulinate-based photodynamic therapy in human squamous cell carcinoma A431 cells. *Br. J. Dermatol.* **2009**, *161*, 413–418. [CrossRef] [PubMed]
35. Galimberti, G.N. Calcipotriol as pretreatment prior to daylight-mediated photodynamic therapy in patients with actinic keratosis: A case series. *Photodiagn. Photodyn. Ther.* **2018**, *21*, 172–175. [CrossRef] [PubMed]
36. Torezan, L.; Grinblat, B.; Haedersdal, M.; Valente, N.; Festa-Neto, C.; Szeimies, R.M. A randomized split-scalp study comparing calcipotriol-assisted methyl aminolaevulinate photodynamic therapy (MAL-PDT) with conventional MAL-PDT for the treatment of actinic keratosis. *Br. J. Dermatol.* **2018**, *179*, 829–835. [CrossRef]
37. Torezan, L.; Grinblat, B.; Haedersdal, M.; Festa-Neto, C.; Szeimies, R.-M. A 12-month follow-up split-scalp study comparing calcipotriol-assisted MAL-PDT with conventional MAL-PDT for the treatment of actinic keratosis: A randomized controlled trial. *Eur. J. Dermatol.* **2021**, *31*, 638–644. [CrossRef]
38. Piaserico, S.; Piccioni, A.; Garcia-Rodrigo, C.G.; Sacco, G.; Pellegrini, C.; Fargnoli, M.C. Sequential treatment with calcitriol and methyl aminolevuliate-daylight photodynamic therapy for patients with multiple actinic keratoses of the upper extremities. *Photodiagn. Photodyn. Ther.* **2021**, *34*, 102325. [CrossRef]

39. Menter, A.; Gelfand, J.M.; Connor, C.; Armstrong, A.W.; Cordoro, K.M.; Davis, D.M.R.; Elewski, B.E.; Gordon, K.B.; Gottlieb, A.B.; Kaplan, D.H.; et al. Joint American Academy of Dermatology-National Psoriasis Foundation guidelines of care for the management of psoriasis with systemic nonbiologic therapies. *J. Am. Acad. Dermatol.* **2020**, *82*, 1445–1486. [CrossRef]

40. Wiegand, U.W.; Chou, R.C. Pharmacokinetics of acitretin and etretinate. *J. Am. Acad. Dermatol.* **1998**, *39*, S25–S33. [CrossRef]

41. Hasan, T. Using cellular mechanisms to develop effective combinations of photodynamic therapy and targeted therapies. *J. Natl. Compr. Cancer Netw.* **2012**, *10*, S23–S26. [CrossRef]

42. Ortel, B.; Chen, N.; Brissette, J.; Dotto, G.P.; Maytin, E.; Hasan, T. Differentiation-specific increase in ALA-induced protoporphyrin IX accumulation in primary mouse keratinocytes. *Br. J. Cancer.* **1998**, *77*, 1744–1751. [CrossRef] [PubMed]

43. Ye, T.; Jiang, B.; Chen, B.; Liu, X.; Yang, L.; Xiong, W.; Yu, B. 5-aminolevulinic acid photodynamic therapy enhance the effect of acitretin on squamous cell carcinoma cells: An in vitro study. *Photodiagn. Photodyn. Ther.* **2020**, *31*, 101887. [CrossRef] [PubMed]

44. Lucena, S.R.; Salazar, N.; Gracia-Cazaña, T.; Zamarrón, A.; González, S.; Juarranz, A.; Gilaberte, Y. Combined Treatments with Photodynamic Therapy for Non-Melanoma Skin Cancer. *Int. J. Mol. Sci.* **2015**, *16*, 25912–25933. [CrossRef] [PubMed]

45. Anand, S.; Honari, G.; Hasan, T.; Elson, P.; Maytin, E.V. Low-dose methotrexate enhances aminolevulinate-based photodynamic therapy in skin carcinoma cells in vitro and in vivo. 2009. *Clin. Cancer Res.* **2009**, *15*, 3333–3343. [CrossRef] [PubMed]

46. Moreno, R.; Nájera, L.; Mascaraque, M.; Juarranz, A.; González, S.; Gilaberte, Y. Influence of serum vitamin D level in the response of actinic keratosis to photodynamic therapy with methylaminolevulinate. *J. Clin. Med.* **2020**, *9*, 398. [CrossRef]

47. Middelkamp-Hup, M.A.; Pathak, M.A.; Parrado, C.; Goukassian, D.; Rius-Díaz, F.; Mihm, M.C.; Fitzpatrick, T.B.; González, S. Oral Polypodium leucotomos extract decreases ultraviolet-induced damage of human skin. *J. Am. Acad. Dermatol.* **2004**, *51*, 910–918. [CrossRef]

48. Horvath, A.; Alvarado, F.; Szöcs, J.; de Alvardo, Z.N.; Padilla, G. Metabolic effects of calagualine, an antitumoral saponine of *Polypodium leucotomos. Nature* **1967**, *214*, 1256–1258. [CrossRef]

49. Zattra, E.; Coleman, C.; Arad, S.; Helms, E.; Levine, D.; Bord, E.; Guillaume, A.; El-Hajahmad, M.; Zwart, E.; van Steeg, H.; et al. Polypodium leucotomos extract decreases UV-induced Cox-2 expression and inflammation, enhances DNA repair, and decreases mutagenesis in hairless mice. *Am. J. Pathol.* **2009**, *175*, 1952–1961. [CrossRef]

50. Siscovick, J.R.; Zapolanski, T.; Magro, C.; Carrington, K.; Prograis, S.; Nussbaum, M.; Gonzalez, S.; Ding, W.; Granstein, R.D. Polypodium leucotomos inhibits ultraviolet B radiation-induced immunosuppression. *Photodermatol. Photoimmunol. Photomed.* **2008**, *12*, 134–141. [CrossRef]

51. Auriemma, M.; Di Nicola, M.; Gonzalez, S.; Piaserico, S.; Capo, A.; Amerio, P. Polypodium leucotomos supplementation in the treatment of scalp actinic keratosis: Could it improve the efficacy of photodynamic therapy? *Dermatol. Surg.* **2015**, *41*, 898–902. [CrossRef]

52. Supe, S.; Takudage, P. Methods for evaluating penetration of drug into the skin: A review. *Ski. Res. Technol.* **2021**, *27*, 299–308. [CrossRef] [PubMed]

53. Senapati, S.; Mahanta, A.K.; Kumar, S.; Maiti, P. Controlled drug delivery vehicles for cancer treatment and their performance. *Signal Transduct. Target. Ther.* **2018**, *7*, 7. [CrossRef] [PubMed]

54. Roberts, M.S.; Cheruvu, H.S.; Mangion, S.E.; Alinaghi, A.; Benson, H.A.E.; Mohammeda, Y.; Holmes, A.; van der Hoek, J.; Pastore, M.; Grice, J.E. Topical drug delivery: History, percutaneous absorption, and product development. *Adv. Drug Deliv. Rev.* **2021**, *177*, 113929. [CrossRef] [PubMed]

55. Mohammeda, Y.; Holmes, A.; Kwok, P.C.L.; Kumeria, T.; Namjoshi, S.; Imran, M.; Matteucci, L.; Ali, M.; Tai, W.; Heather, A.E.B.; et al. Advances and future perspectives in epithelial drug delivery. *Adv. Drug Deliv. Rev.* **2022**, *186*, 114293. [CrossRef] [PubMed]

56. Desai, P.; Patlolla, R.R.; Singh, M. Interaction of nanoparticles and cell-penetrating peptides with skin for transdermal drug delivery. *Mol. Membr. Biol.* **2010**, *27*, 247–259. [CrossRef]

57. Kennedy, J.K.; Marcus, S.L.; Pottier, R.H. Photodynamic therapy and photodiagnosis using endogenous photosensitization induced by 5-aminolevulinic acid: Mechanisms and clinical results. *J. Clin. Laser Med. Surg.* **1996**, *14*, 289–304. [CrossRef]

58. Portugal, I.; Jain, S.; Severino, P.; Priefer, R. Micro-and Nano-Based Transdermal Delivery Systems of Photosensitizing Drugs for the Treatment of Cutaneous Malignancies. *Pharmaceuticals* **2021**, *14*, 772. [CrossRef]

59. Shen, S.-C.; Lee, W.-R.; Fang, Y.-P.; Hu, C.-H.; Fang, J.-Y. In vitro percutaneous absorption and in vivo protoporphyrin IX accumulation in skin and tumors after topical 5-aminolevulinic acid application with enhancement using an erbium: YAG laser. *J. Pharm. Sci.* **2006**, *95*, 929–938. [CrossRef]

60. Fidrus, E.; Janka, E.; Kollár, S.; Paragh, G.; Emri, G.; Remenyik, E. 5-aminolevulinic acid photodynamic therapy with and without Er:YAG laser for actinic keratosis: Changes in immune infiltration. *Photodiagn. Photodyn. Ther.* **2019**, *26*, 270–276. [CrossRef]

61. Togsverd-Bo, K.; Lei, U.; Erlendsson, A.M.; Taudorf, E.H.; Philipsen, P.A.; Wulf, H.C.; Skov, L.; Hædersdal, M. Combination of ablative fractional laser and daylight-mediated photodynamic therapy for actinic keratosis in organ transplant recipients-a randomized controlled trial. *Br. J. Dermatol.* **2015**, *172*, 467–474. [CrossRef]

62. Ko, D.-Y.; Jeon, S.Y.; Kim, K.-H.; Song, K.-H. Fractional erbium: YAG laser-assisted photodynamic therapy for facial actinic keratoses: A randomized, comparative, prospective study. *J. Eur. Acad. Dermatol. Venereol.* **2014**, *28*, 1529–1539. [CrossRef]

63. Haedersdal, M.; Sakamoto, F.H.; Farinelli, W.A.; Doukas, A.G.; Tam, J.; Anderson, R.R. Fractional CO_2 laser-assisted drug delivery. *Lasers Surg. Med.* **2010**, *42*, 113–122. [CrossRef]

64. Mohammad Rizvi, S.; Veierød, M.B.; Mørk, G.; Helsing, P.; Gjersvik, P. Ablative Fractional Laser-assisted Daylight Photodynamic Therapy for Actinic Keratoses of the Scalp and Forehead in Organ Transplant Recipients: A Pilot Study. *Acta Derm.-Venereol.* **2019**, *99*, 1047–1048. [CrossRef]
65. Togsverd-Bo, K.; Haak, C.S.; Thaysen-Petersen, D.; Wulf, H.C.; Anderson, R.R.; Hædersdal, M. Intensified photodynamic therapy of actinic keratoses with fractional CO$_2$ laser: A randomized clinical trial. *Br. J. Dermatol.* **2012**, *166*, 1262–1269. [CrossRef] [PubMed]
66. Paasch, U.; Said, T. Treating field cancerization by ablative fractional laser and indoor daylight: Assessment of efficacy and tolerability. *J. Drugs. Dermatol.* **2020**, *19*, 425–427. [CrossRef] [PubMed]
67. Jang, Y.H.; Lee, D.J.; Shin, J.; Kang, E.J.; Lee, E.S.; Kim, J.C. Photodynamic Therapy with Ablative Carbon Dioxide Fractional Laser in Treatment of Actinic Keratosis. *Ann. Dermatol.* **2013**, *25*, 417–422. [CrossRef] [PubMed]
68. Polat, B.A.; Blankschtein, D.; Langer, R. Low-frequency sonophoresis: Application to the transdermal delivery of macromolecules and hydrophilic drugs. *Expert Opin. Drug Deliv.* **2010**, *7*, 1415–1432. [CrossRef] [PubMed]
69. Park, J.M.; Jeong, K.H.; Bae, M.I.; Lee, S.-J.; Kim, N.-I.; Shin, M.K. Fractional radiofrequency combined with sonophoresis to facilitate skin penetration of 5-aminolevulinic acid. *Lasers Med. Sci.* **2016**, *31*, 113–118. [CrossRef] [PubMed]
70. Shavit, R.; Dierickx, C. A new method for percutaneous drug delivery by thermo-mechanical fractional injury. *Lasers Surg. Med.* **2020**, *52*, 61–69. [CrossRef]
71. Bay, C.; Lerche, C.M.; Ferrick, B.; Philipsen, P.A.; Togsverd-Bo, K.; Haedersdal, M. Comparison of Physical Pretreatment Regimens to Enhance Protoporphyrin IX Uptake in Photodynamic Therapy: A Randomized Clinical Trial. *JAMA Dermatol.* **2017**, *153*, 270–278. [CrossRef]
72. Wenande, E.; Phothong, W.; Bay, C.; Karmisholt, K.E.; Haedersdal, M.; Togsverd-Bo, K. Efficacy and safety of daylight photodynamic therapy after tailored pretreatment with ablative fractional laser or microdermabrasion: A randomized, side-by-side, single-blind trial in patients with actinic keratosis and large-area field cancerization. *Br. J. Dermatol.* **2019**, *180*, 756–764. [CrossRef] [PubMed]
73. Donnelly, R.F.; Morrow, D.I.J.; McCarron, P.A.; Woolfson, A.D.; Morrissey, A.; Juzenas, P.; Juzeniene, A.; Iani, V.; McCarthy, H.O.; Moan, J. Microneedle-mediated intradermal delivery of 5-aminolevulinic acid: Potential for enhanced topical photodynamic therapy. *J. Control Release* **2008**, *129*, 154–162. [CrossRef] [PubMed]
74. Petukhova, T.A.; Hassoun, L.A.; Foolad, N.; Barath, M.; Sivamani, R.K. Effect of Expedited Microneedle-Assisted Photodynamic Therapy for Field Treatment of Actinic Keratoses: A Randomized Clinical Trial. *JAMA Dermatol.* **2017**, *153*, 637–664. [CrossRef]
75. Spencer, J.M.; Freeman, S.A. Microneedling Prior to Levulan PDT for the Treatment of Actinic Keratoses: A Split-Face, Blinded Trial. *J. Drugs Dermatol.* **2016**, *15*, 1072–1074. [PubMed]
76. Chen, J.; Zhang, Y.; Wang, P.; Wang, B.; Zhang, G.; Wang, X. Plum-blossom needling promoted PpIX fluorescence intensity from 5-aminolevulinic acid in porcine skin model and patients with actnic keratosis. *Photodiagn. Photodyn. Ther.* **2016**, *15*, 182–190. [CrossRef]
77. Lev-Tov, H.; Larsen, L.; Zackria, R.; Chahal, H.; Eisen, D.B.; Sivamani, R.K. Microneedle-assisted incubation during aminolaevulinic acid photodynamic therapy of actinic keratoses: A randomized controlled evaluator-blind trial. *Br. J. Dermatol.* **2017**, *176*, 543–545. [CrossRef] [PubMed]
78. Stringasci, M.D.; Ciol, H.; Romano, R.A.; Buzza, H.H.; Leite, I.S.; Inada, N.M.; Bagnato, V.S. MAL-associated methyl nicotinate for topical PDT improvement. *J. Photochem. Photobiol. B* **2020**, *213*, 112071. [CrossRef]
79. Choi, S.-H.; Kim, T.-H.; Song, K.-H. Efficacy of iontophoresis-assisted ablative fractional laser photodynamic therapy with short incubation time for the treatment of actinic keratosis: 12-month follow-up results of a prospective, randomized, comparative trial. *Photodyn. Ther.* **2017**, *18*, 105–110. [CrossRef]
80. Joniov'a, J.; Kazemiraad, K.; Gerelli, E.; Wagnières, G. Stimulation and homogenization of the protoporphyrin IX endogenous production by photobiomodulation to increase the potency of photodynamic therapy. *J. Photochem. Photobiol.* **2021**, *225*, 112347. [CrossRef] [PubMed]
81. Ziolkowski, P.; Osiecka, B.J.; Oremek, G.; Siewinski, M.; Symonowicz, K.; Saleh, Y.; Bronowicz, A. Enhancement of photodynamic therapy by use of aminolevulinic acid/glycolic acid drug mixture. *J. Exp. Ther. Oncol.* **2004**, *4*, 121–129. [PubMed]
82. Harth, Y.; Hirshowitz, B.; Kaplan, B. Modified topical photodynamic therapy of superficial skin tumors, utilizing aminolevulinic acid, penetration enhancers, red light, and hyperthermia. *Dermatol. Surg.* **1998**, *24*, 723–726. [CrossRef] [PubMed]

 *pharmaceutics*

Review

Effectiveness of Antimicrobial Photodynamic Therapy in the Treatment of Periodontitis: A Systematic Review and Meta-Analysis of In Vivo Human Randomized Controlled Clinical Trials

Snehal Dalvi [1,2,*], Stefano Benedicenti [1], Tudor Sălăgean [3,*], Ioana Roxana Bordea [4,†] and Reem Hanna [1,5,†]

1 Department of Surgical Sciences and Integrated Diagnostics, Laser Therapy Centre, University of Genoa, Viale Benedetto XV, 6, 16132 Genoa, Italy; stefano.benedicenti@unige.it (S.B.); reemhanna@hotmail.com (R.H.)
2 Department of Periodontology, Swargiya Dadasaheb Kalmegh Smruti Dental College and Hospital, Nagpur 441110, India
3 Department of Land Measurements and Exact Sciences, University of Agricultural Sciences and Veterinary Medicine Cluj-Napoca, 400372 Cluj-Napoca, Romania
4 Department of Oral Rehabilitation, "Iuliu Hațieganu" University of Medicine and Pharmacy Cluj-Napoca, 400012 Cluj-Napoca, Romania; roxana.bordea@ymail.com
5 Department of Oral Surgery, Dental Institute, King's College Hospital NHS Foundation Trust, London SE5 9RS, UK
* Correspondence: drsnehaldeotale@gmail.com (S.D.); tudor.salagean@usamvcluj.ro (T.S.); Tel.: +39-0-103-537-446 (S.D.); +40-744-707-371 (T.S.)
† These authors contributed equally to the last authorship.

Citation: Dalvi, S.; Benedicenti, S.; Sălăgean, T.; Bordea, I.R.; Hanna, R. Effectiveness of Antimicrobial Photodynamic Therapy in the Treatment of Periodontitis: A Systematic Review and Meta-Analysis of In Vivo Human Randomized Controlled Clinical Trials. *Pharmaceutics* **2021**, *13*, 836. https://doi.org/10.3390/pharmaceutics13060836

Academic Editors: Maria Nowakowska and Nejat Düzgüneş

Received: 17 April 2021
Accepted: 2 June 2021
Published: 4 June 2021

Publisher's Note: MDPI stays neutral with regard to jurisdictional claims in published maps and institutional affiliations.

Abstract: This systematic review and meta-analysis evaluated antimicrobial photodynamic therapy (aPDT) efficacy in periodontitis. The review protocol was conducted in accordance with PRISMA statements, Cochrane Collaboration recommendations and is registered in PROSPERO (CRD 42020161516). Electronic and hand search strategies were undertaken to gather data on in vivo human RCTs followed by qualitative analysis. Differences in probing pocket depth (PPD) and clinical attachment level (CAL) were calculated with 95% confidence intervals and pooled in random effects model at three and six months. Heterogeneity was analyzed, using Q and I^2 tests. Publication bias was assessed by visual examination of the funnel plot symmetry. Sixty percent of 31 eligible studies showed a high risk of bias. Meta-analysis on 18 studies showed no additional benefit in split mouth studies in terms of PPD reduction (SMD 0.166; 95% CI −0.278 to 0.611; $P = 0.463$) and CAL gain (SMD 0.092; 95% CI −0.013 to 0.198; $P = 0.088$). Similar findings noted for parallel group studies; PPD reduction (SMD 0.076; 95% CI −0.420 to 0.573; $P = 0.763$) and CAL gain (SMD 0.056; 95% CI −0.408 to 0.552; $P = 0.745$). Sensitivity analysis minimized heterogeneity for both outcome variables; however, intergroup differences were not statistically significant. Future research should aim for well-designed RCTs in order to determine the effectiveness of aPDT.

Keywords: antimicrobial photodynamic therapy; periodontitis; scaling and root planing; systematic review; meta-analysis

Highlights

1. Limitations of scaling and root planing (SRP) have directed the research to assess alternative comprehensive treatment strategies.
2. Antimicrobial Photodynamic therapy (aPDT) involves photo-excitation of photosensitizer dye upon illumination by a light of a matched wavelength.
3. This systematic review and meta-analysis evaluated the effectiveness of aPDT in the treatment of periodontitis.
4. In spite of the inconsistencies in their findings and methodological bias, the majority of the studies have demonstrated aPDT effectiveness.

5. The efficacy of aPDT in improving treatment outcomes when it is utilized in the non-surgical management of periodontitis remains debatable.

1. Introduction

Antimicrobial Photodynamic therapy (aPDT) involves photo-excitation, which occurs when a photosensitizer (PS) dye is illuminated by a light of a matched wavelength, resulting in its activation and stimulation of a phototoxic response in the presence of ambient oxygen [1]. It has been persistently observed that bacterial recolonizations of *Aggregatibacter actinomycetemcomitans (A.a)* occur in periodontal pockets even after scaling and root planing (SRP) [2]. Aggressive periodontitis (AgP) is frequently associated with fewer local etiologic factors; therefore, it is believed that the affected patients are more likely to benefit from the antimicrobial effect of aPDT [3]. In contrast, chronic periodontitis (CP) patients usually have complex and thick deposits of polymicrobial communities on the affected root surfaces [4]. This may hamper penetration of PS, thereby reducing its effect and leading to an increase in the 'red complex' bacterial counts within a short period of time, resulting in a disease relapse [5]. Hence, the concept of replacing conventional SRP with aPDT is a controversial one with several imperative demerits, as enlisted above.

Utilization of adjunctive aPDT and its comparison with the gold standard SRP is a concept that has been studied extensively in both CP and AgP patients [6–16]. While SRP can quantitively lower the biomass of bacteria, aPDT has a more qualitative approach of a non-invasive nature, by creating alterations in cell membranes or Deoxyribonucleic Acid (DNA) damage [5]. Hence, a combination of these two therapies can be vouched for, since their mechanisms of action on microbiota and role in the periodontal repair process is distinct from the other and thus might have synergistic effects [17].

Distant sites of infection such as tonsils or base of tongue, which are affected due to the spread of tissue penetrating periopathogens, can be successfully reduced with local or systemic antibiotics (AB) [18]. Nonetheless, many clinicians often conduct NSPT without adjunctive AB, which is only used when initial treatment has failed [19]. In AgP, evidence suggests that SRP+AB therapy does not show satisfactory long-term results, unless re-instrumentation of affected sites is performed, as an additional step in the maintenance phase [20]. Furthermore, owing to the development of antibiotic resistant strains, it has been suggested that AB usage should be restricted to those with a highly active disease or a specific microbiological profile [21]. In order to maintain an adequate mean inhibitory concentration (MIC) of any antimicrobial drug, either a sustained-release carrier medium is required or, conversely, a prompt bactericidal approach is needed to overcome the problem of physical displacement from the sulcus [22]. The aPDT falls into the latter category, demonstrating a 4–6-fold logarithmic bacterial reduction within a time frame of 60 s along with repeated applications [23]. A comprehensive assessment to evaluate the impact of these new trials on the role of aPDT in the treatment of periodontitis is unresolved, owing to the diversity in the methodology and results of existing scientific evidence [6–16].

In lieu of the prevailing pertinent literature, the present systematic review and meta-analysis aimed to provide a systematic evaluation of available scientific evidence to determine the efficacy of aPDT in the treatment of periodontitis. The objectives of this critical review were to evaluate the outcomes of this treatment strategy through various PS-laser wavelength combinations, as well as the laser parameters, in order to deduce an ideal PS-laser wavelength combination and treatment protocol for future scientific research.

2. Materials and Methods

2.1. Protocol and Registration

The present systematic review was reported based on the guidelines of Preferred Reporting Items for Systematic Reviews and Meta-Analysis (PRISMA) Statement and Cochrane Collaboration recommendations (Supplementary file 1) [24,25]. The review protocol is published in the Prospective Register of Systematic Reviews (PROSPERO); ref CRD 42020161516.

2.2. Population (P), Intervention (I), Comparison (C) and Outcomes (O)—PICO

- **Population:** Patients diagnosed with Periodontitis (CP or AgP) [26]
- **Intervention:** Utilisation of aPDT as a monotherapy or as an adjunct to SRP
- **Comparison:** Utilisation of SRP alone or SRP with adjunctive AB therapy
- **Outcome:** Evaluation of clinical and/or microbiological and/or immunological profiles

2.3. Focused Research Question

Is aPDT effective as a primary mode of treatment or as an adjunct to SRP compared to SRP alone or in combination with local or systemic antibiotics (AB), in terms of clinical or microbiological or immunological profiles, in patients with Periodontitis?

2.4. Search Strategy

The search strategy only included terms relating to or describing the study's domain and intervention. The use of relevant free text keywords and medical subject heading (Mesh) terms, which were logically connected with the help of Cochrane MEDLINE filters for controlled trials of interventions, was implemented. Individual search algorithms were developed for the following databases: MEDLINE (NCBI PubMed and PMC), Cochrane Central Register of Controlled Trials (CCRCT), Scopus, ScienceDirect, Google Scholar, EMBASE and EBSCO. Electronic search databases were searched thoroughly from their earliest records until 31 December 2019. The following journals were manually searched: Journal of Periodontology, Photomedicine and Laser Surgery, Clinical Oral Investigation, Journal of Clinical Periodontology, Journal of Dental Research, Lasers in Medical Science, Journal of Photochemistry and Photobiology and Photodiagnosis and Photodynamic Therapy. Related review articles and reference lists of all identified articles were searched through for further studies. Abstracts of the American Academy of Periodontology (AAP) and the European Federation of Periodontology (EFP) as well as sources for grey literature were screened to detect unpublished studies. In some instances, an attempt was made to establish a communication with the corresponding author in an attempt to obtain additional information related to the study; however, the attempts were unsuccessful. Search strategy was performed by two blinded, independent reviewers (S.D. and R.H.). In order to assess inter-reviewer reliability analysis, Kappa (κ) statistics were performed and a minimum value of 0.8 was considered acceptable [27]. In case of any disagreements, reviewers would discuss the discrepancies with a third author (S.B.), if necessary.

2.5. Search Algorithms

"Photodynamic therapy" **OR** "photochemotherapy"
AND
"Scaling" **OR** "Root planing" **OR** "non-surgical periodontal therapy"
AND
"Periodontitis" **OR** "Chronic Periodontitis" **OR** "Aggressive Periodontitis" **OR** "Early Onset Periodontitis"

2.6. Eligibility Criteria

2.6.1. Inclusion Criteria

1. Subjects diagnosed with CP or AgP according to 1999 AAP Classification of Periodontal diseases and conditions [26].
2. Studies included: In vivo human RCT's comparing the efficacy of aPDT in CP or AgP as monotherapy or adjunctive to SRP compared to SRP alone or in combination with AB.
3. Parallel group (PG) and split-mouth (SM) studies.
4. Age group >18 years, fit and healthy subjects.
5. No language restrictions for search strategy.
6. Studies that have utilized any PS dye (regardless dose and incubation period) and laser wavelength combination.

7. Studies reporting at least one of the following parameters as an outcome variable: probing pocket depth (PPD), loss of clinical attachment level (CAL), bleeding on probing (BOP), plaque index (PI), gingival index (GI), microbiological profile, or immunological profile.
8. Studies with a minimum follow-up period of at least one month after treatment.

2.6.2. Exclusion Criteria

1. Subjects with systemic diseases or on medications that can influence the outcome variables.
2. Subjects who have undergone any periodontal therapy and/or antibiotic therapy in the last six months prior to RCT enrolment.
3. Studies utilizing low level laser therapy or laser therapy alone, as one of the intervention groups as compared to aPDT.
4. Studies involving utilization of aPDT for residual pockets or in supportive periodontal therapy (SPT).
5. Studies that have utilized light emitting diodes (LEDs) as a light source.
6. No outcome variable of interest.
7. Pregnancy.
8. Smoking.
9. Narrative and systematic reviews, in vitro studies, in vivo animal studies, commentaries, interviews, updates, case series and case reports.

2.7. Systematic Review Outcomes

2.7.1. Primary Outcome Measures

Changes in PPD and CAL from baseline up to the end of follow-up.

2.7.2. Secondary Outcome Measures

Changes in GR, BOP, PI, GI, microbiological and immunological profile from baseline up to the end of follow-up.

2.8. Data Extraction

Two reviewers independently (S.D. and R.H.) selected eligible studies from the search. They performed the review, assessment and data extraction for each eligible study. Each study received an identification with the name of the first author, year of publication and origin. A tabular representation of additional relevant information such as impact factor of journal, study design, sample size, demographics of the participants, baseline characteristics, intervention and comparator groups, type of photosensitizer used and dosage, laser parameters utilized, number of aPDT sessions performed, follow-up duration, statistical tests performed and results and conclusions, were gathered from each eligible study.

2.9. Qualitative Analysis

A qualitative assessment for each study was performed using the Revised Cochrane Risk-of-Bias (RoB) tool for Randomized trials, Version 2.0 (RoB 2) by two independent reviewers (S.D. and R.H.) [28–30]. Detailed assessment under the following headings was performed: 1. Bias arising from the randomization process; 2. Bias due to deviations from intended interventions; 3. Bias due to missing outcome data; 4. Bias in measurement of the outcome; 5. Bias in selection of the reported result. Depending upon fulfilment of above-mentioned criteria, the studies were determined as low, moderate or high RoB. Disagreements between the reviewers were resolved by discussion with a third author (S.B.) as well as use of 'discrepancy check' feature in RoB 2, in order to obtain consensus.

2.10. Statistical Analysis of Data

When appropriate and quantifiable data of interest were extracted from the eligible studies and combined for meta-analyses, using Stata version 15.1 software (StataCorp, Pyrmont, Australia), random effects meta-analyses were conducted to reflect the expected heterogeneity. As continuous outcomes were expected, overall treatment effects were calculated through pooled standardized mean differences (SMDs) with associated 95% confidence intervals (95% CIs) for PPD and CAL. When information was presented in median and inter-quartile ranges, means and SDs were estimated [31]. Results from SM and PG studies were pooled separately at 3 and 6 months, respectively. A pooled overall effect was considered statistically significant when $p < 0.05$. Consequently, statistical heterogeneity to identify outlier studies was performed by visual inspection of forest plots. Additionally, the Cochran Q test was conducted to assess statistical heterogeneity ($p < 0.10$) [32]. I^2 statistics for homogeneity was expressed in a range of 0–100%, with the following interpretation; 0% = no evidence of heterogeneity; 30–60% = moderate heterogeneity; 75–100% = high heterogeneity [33]. Sensitivity analysis was conducted to negate the effect of heterogeneity in between included studies by identifying the outlier studies by visual inspection of forest plots [34]. Publication bias was evaluated by visual assessment of funnel plot symmetry.

3. Results

3.1. Study Selection

Four hundred and sixty-two study titles were obtained from a combined electronic and manual search. Four study titles were obtained from cross-references. Therefore, a total of 466 study titles were included from all databases in the preliminary screening (inter-reviewer agreement, κ = 0.9). Three hundred and eighty-seven articles were excluded, due to duplication and the remaining 79 records were further evaluated (inter-reviewer agreement, κ = 0.94). Twelve articles were excluded based on their titles and abstracts, mainly due to an inappropriate study design (inter-reviewer agreement, κ = 0.92). Thus, 67 articles were assessed for their eligibility. These articles were evaluated based on eligibility criteria. Additionally, 36 studies were excluded due to following reasons: Smokers were included or smoking details were not provided in 12 studies [23,35–45]; Laser or LLLT was utilized, as an adjunct to SRP in eight studies [46–53]; LED-aPDT was performed in seven studies [54–60]; aPDT was used in management of residual pockets in four studies [61–64] and as an adjunct to supportive periodontal therapy in two studies [65,66]; patients with systemic diseases were included in two studies [67,68], whereas one study did not perform a follow-up assessment [69] (inter-reviewer agreement, κ = 1). Hence, out of 67 full text articles, 31 articles were included and analyzed in the present systematic review [2,3,5,17,70–96]. All included articles were in vivo human studies. A meta-analysis on 18 out of 31 studies which assessed efficacy of SRP+aPDT was conducted [17,71,73–76,80,81,84–86,88–93,96] (inter-reviewer agreement, κ = 1). Figure 1 depicts the PRISMA flow diagram for search strategy utilized in the present systematic review and meta-analysis.

3.2. Study Characteristics

3.2.1. Country of Origin

A substantial diversity in the country of origin was noted amongst included papers (Table 1). Distribution of studies was as follows: 11 in Brazil [2,3,5,17,71,82,83,87,92,95,96], 6 in India [78,81,88,89,91,93], 4 in Germany [76,77,80,94], 4 in Iran [70,85,86,90], 3 in Poland [72–74], whereas there is 1 study each, in the following countries; Spain [75], Japan [79], Thailand [84].

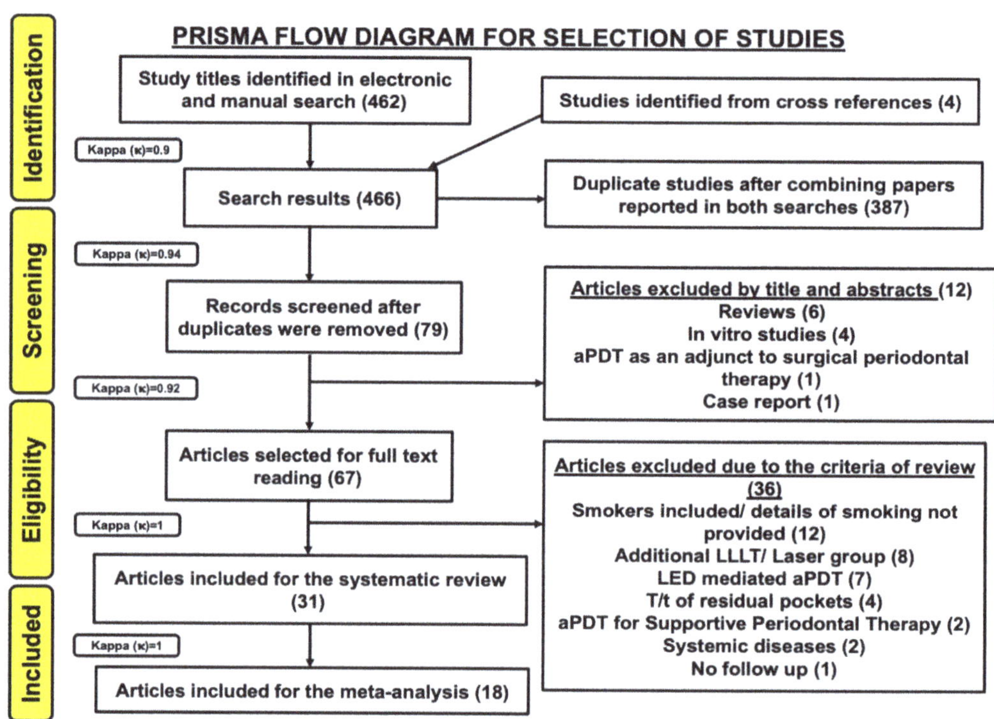

Figure 1. PRISMA flow diagram of the study selection criteria.

3.2.2. Study Design

Twenty studies were conducted using a SM study design [2,3,5,17,70,71,77,80–82, 84–87,89,90,92–95], whereas a PG study design was utilized in the remaining 11 studies [72–76,78,79,83,88,91,96] (Table 1).

3.2.3. Selection Criteria

Several inconsistencies were observed amongst the included studies [2,3,5,17,70–96], which have been outlined in Table 1, in which 21 out of 31 studies included patients with CP [17,70,75–82,84,86,88–96], whereas the remaining 10 studies included patients with AgP [2,3,5,71–74,83,85,87].

Table 1. Tabular representation of eligible in vivo human RCTs in terms of demography, study design, intervention groups, methods of assessment, evaluation period and outcomes. Refer to Supplementary file 2 for list of abbreviations.

Study, Year, Origin and Citation	Journal Name/ Impact Factor (IF)	Study Design	Type of Periodontitis	Sample Size (n)	Gender M/F	Age (Years) (Mean $\pm$ SD)	Intervention Groups		Evaluation Period	Parameters Assessed	Conclusions
De Oliveira et al., 2009 (Brazil) [2]	Journal of Periodontology IF 2020: 3.742 IF 2009: 2.580	SM-RCT	AgP (A minimum of 20 teeth (mean, 26 teeth) with at least one tooth in each posterior sextant and at least one posterior sextant with a minimum of three natural teeth; $\geq$5 mm of attachment loss around at least seven teeth involved, excluding first molars and central incisors)	10	2/8	18–35 Mean: 31.01 $\pm$ 4.43	SRP (Hand instruments) (10 teeth)	aPDT (10 teeth)	−7 (baseline), 0 (immediately after interventions), +1, +7, +30 and +90 days.	TNF-α and RANKL assessment	NSPT with PDT or SRP led to statistically significant reductions in TNF-a level 30 days following treatment ($p < 0.05$) with no statistically significant intergroup differences ($p > 0.5$).
De Oliveira et al., 2007 (Brazil) [3]	Journal of Periodontology IF 2020: 3.742 IF 2007: 2.426	SM-RCT	AgP (A minimum of 20 teeth (mean, 26 teeth) with at least one tooth in each posterior sextant and at least one posterior sextant with a minimum of three natural teeth; $\geq$5 mm of attachment loss around at least seven teeth involved, excluding first molars and central incisors)	10	2/8	18–35 Mean: 31.01 $\pm$ 4.43	SRP (Hand instruments) (10 teeth)	aPDT (10 teeth)	Baseline, 3 months	PD, RCAL, GR, PI, GI, BOP	PDT and SRP showed statistically significant clinical results ($p < 0.05$) in the non-surgical treatment of aggressive periodontitis with no statistically significant differences ($p > 0.5$) in intergroup comparison.
Novaes et al., 2012 (Brazil) [5]	Lasers in Medical Science IF 2019: 2.574 IF 2012: 2.645	SM-RCT	AgP (A minimum of 20 teeth (mean, 26 teeth) with at least one tooth in each posterior sextant, and at least one posterior sextant with a minimum of three natural teeth; $\geq$5 mm of attachment loss around at least seven teeth involved, excluding first molars and central incisors)	10	2/8	18–35 Mean: 31	SRP (Hand instruments)	aPDT	−7, 0 (Baseline), and 3 months	Plaque sample analysis for estimation of 40 subgingival species using DNA-DNA hybridization.	aPDT was more effective in reducing the counts of *A.a* ($p = 0.00$) whereas, SRP reduced red complex bacteria. Combination of both treatment methods would be beneficial for the non-surgical treatment of AgP
Franco et al., 2014 (Brazil) [17]	Photodiagnosis and Photodynamic Therapy IF 2020: 2.894 IF 2014: 2.359	SM-RCT	CP (At least 20 teeth with at least one posterior tooth in each quadrant, and periodontal pockets $\geq$ 5 mm on at least seven teeth)	15	NI	39.5	SRP (Hand instruments)	SRP+aPDT	Baseline and 90 days	BOP, PI, PD, CAL, qPCR gene expression analysis.	Significant improvement in BOP was noted with aPDT group ($p = 0.03$). PDT increased the expression of RANK and OPG, which could indicate a reduction in osteoclastogenesis. Furthermore, the use of PDT in conjunction with conventional treatment significantly increased the expression of FGF2, which has an important role in the periodontal repair process.

Table 1. *Cont.*

Study, Year, Origin and Citation	Journal Name/ Impact Factor (IF)	Study Design	Type of Periodontitis	Sample Size (n)	Gender M/F	Age (Years) (Mean ± SD)	Intervention Groups		Evaluation Period	Parameters Assessed	Conclusions
Pourabbas et al., 2014 (Iran) [70]	Journal of Periodontology IF 2020: 3.742 IF 2014: 2.900	SM-RCT	CP ($\geq$12 natural teeth with a minimum of three in each quadrant; $\geq$3 mm attachment loss in about a minimum of 30% of the existing teeth; $\geq$1 site per quadrant with PPD of $\geq$4 mm and BOP)	24	10/14	46 ± 8	SRP (Sonic and hand instruments)	SRP+aPDT	Baseline and 3 months	PD, BOP, CAL, GR, IL-1β, TNF-α, MMP-8 and MMP-9 analysis	Intragroup comparison showed significant improvements ($p < 0.001$) for all variables in 3-month follow-up compared with baseline. TNF-α was significantly improved in the SRP+aPDT versus SRP group ($p < 0.001$). Total levels of PMNs were reduced for all patients compared with baseline levels ($p < 0.001$).
Moreira et al., 2015 (Brazil) [71]	Journal of Periodontology IF 2020: 3.742 IF 2015: 3.159	SM-RCT	AgP (A minimum of 20 teeth and two pairs of single rooted contralateral teeth with proximal sites presenting PD and CAL $\geq$ 5 mm)	20	2/18	18–35 30.6 ± 4.25	SRP + sham procedure (Hand and ultrasonic instruments) 40 teeth/128 sites	SRP+aPDT 40 teeth/ 135 sites	Baseline, 3 months	PD, CAL, GR, PI, BOP Microbiological analysis for counts of 40 bacterial species using DNA- DNA Hybridization Immunological evaluation for GCF levels of IL-1β, IL- 10 and TNF-α.	In deep periodontal pockets analysis (PD $\geq$ 7 mm at baseline), Test Group presented a decrease in PD and a clinical attachment gain significantly higher than Control Group at 90 days ($p < 0.05$). Test Group also demonstrated significantly less periodontal pathogens of red and orange complexes and a lower ratio IL-1β/IL-10 than Control Group ($p < 0.05$). Four adjunctive sessions of aPDT after SRP have clinical, microbiological and immunological benefits over SRP alone in management of AgP.
Skurska et al., 2015 (Poland) [72]	BMC Oral Health IF 2019: 1.911 IF 2015: 1.605	PG-RCT	AgP (At least 3 sites with PD $\geq$ 6 mm)	35 SRP+AB: 17 SRP+aP DT:18	12/24 SRP+aPDT: 7/10 SRP+AB: 5/13	23–55 SRP+aPDT: 37.3 ± 8.0 SRP+AB: 34.7 ± 9.0	SRP+ AB 141 sites AB: 375 mg of amoxicillin + 250 mg of metronidazole TDS for 7 days, starting on the day of SRP (Hand and ultrasonic instruments)	SRP+aPDT 137 sites	Baseline, 3 and 6 months	MMP-8 and MMP-9 assessment	In the AB group, patients showed a statistically significant ($p = 0.01$) decrease of MMP-8 GCF level at both 3- and 6-months post treatment. In the PDT group, the change of MMP-8 GCF level was not statistically significant. Both groups showed at 3 and 6 months a decrease in MMP-9 levels. However, this change did not reach statistical significance. SRP+AB is more effective in reducing GCF MMP-8 levels compared to SRP+aPDT.

Table 1. *Cont.*

Study, Year, Origin and Citation	Journal Name/ Impact Factor (IF)	Study Design	Type of Periodontitis	Sample Size (n)	Gender M/F	Age (Years) (Mean $\pm$ SD)	Intervention Groups		Evaluation Period	Parameters Assessed	Conclusions
Arweiler et al., 2014 (Poland) [73]	Clinical Oral Investigations IF 2019: 2.903 IF 2014: 2.704	PG-RCT	AgP (At least 3 sites with PD $\geq$ 6 mm)	35 SRP+aPDT: 17 SRP+AB: 18	12/24 SRP+aPDT: 7/10 SRP+AB: 5/13	23–55 SRP+aPDT: 37.3 $\pm$ 8.0 SRP+AB: 34.7 $\pm$ 9.0	SRP+AB 141 sites AB: 375 mg Amoxicillin + 250 mg Metronidazole TDS for 7 days (starting from day of SRP) (Hand and ultrasonic instruments)	SRP+aPDT 137 sites	Baseline, 6 months	PD, CAL, GR, PI, BOP, FMPI, FMBOP	Intragroup comparison revealed statistically significant PD reduction from baseline ($p < 0.001$). SRP+AB showed significant differences in PD reduction and lower number of deep pockets $\geq$ 7 mm ($p < 0.001$) as compared to SRP+aPDT ($p = 0.03$).
Arweiler et al., 2013 (Poland) [74]	Schweiz Monatsschr Zahnmed IF 2020: NA IF 2013: NA	PG-RCT	AgP (At least 3 sites with PD $\geq$ 6 mm)	35 SRP+aPDT: 17 SRP+AB: 18	12/24 SRP+aPDT: 7/10 SRP+AB: 5/13	23–55 SRP+aPDT: 37.3 $\pm$ 8.0 SRP+AB: 34.7 $\pm$ 9.0	SRP+AB 141 sites AB: 375 mg Amoxicillin+250 mg MTZ TDS for 7 days (starting from day of SRP) (Hand and ultrasonic instruments)	SRP+aPDT 137 sites	Baseline, 3 months	PD, CAL, GR, PI, BOP, FMPI, FMBOP	SRP+AB showed significant differences in PD reduction, CAL gain and lower number of deep pockets $\geq$ 7 mm as compared to SRP+aPDT ($p < 0.001$).
Vidal et al., 2017 (Spain) [75]	Journal of Clinical Periodontology IF 2020: 5.241 IF 2017: 4.165	PG-RCT	CP (Four or more periodontal pockets with a PPD $\geq$ 5 mm and BOP)	37	11/26	55 $\pm$ 2	SRP (Hand and ultrasonic instruments)	SRP+aPDT	Baseline, 5, 13 and 25 weeks	PI, PD, GR, CAL, BOP, GCF volume, microbiological and biochemical parameters	RANKL and abundance of *A.a* was significantly decreased in the SRP+aPDT group compared with the SRP group ($p < 0.05$). Except of a reduction in *A.a*, SRP+ aPDT resulted in no additional improvement compared with SRP alone.
Braun et al., 2008 (Germany) [76]	Journal of Clinical Periodontology IF 2020: 5.241 IF 2008: 3.525	SM-RCT	CP (At least one premolar and one molar in every quadrant with a minimum of four teeth each; at least one tooth with an attachment loss of >3 mm in every quadrant)	20	9/11	46.6 $\pm$ 6.1	SRP (Hand and piezo-electric ultrasonic instruments)	SRP+aPDT	Baseline, 1 week, 3 months	SFFR, BOP, RAL PD, GR	Values for RAL, PD, SFFR and BOP decreased significantly 3 months after treatment in the control group with a higher impact on the sites treated with adjunctive aPDT ($p < 0.05$). GR increased 3 months after treatment with and without adjunctive aPDT, with no difference between the groups ($p > 0.05$). In patients with CP, clinical outcomes can be improved by adjunctive aPDT.

Table 1. *Cont.*

Study, Year, Origin and Citation	Journal Name/ Impact Factor (IF)	Study Design	Type of Periodontitis	Sample Size (*n*)	Gender M/F	Age (Years) (Mean ± SD)	Intervention Groups		Evaluation Period	Parameters Assessed	Conclusions
Berakdar et al., 2012 (Germany) [77]	Head and Face Medicine IF 2020: 1.492 IF 2012: 1.519	SM-RCT	CP (At least four teeth with a PPD of ≥5 mm)	22	12/10	59.3 ± 11.7	SRP (Hand instruments)	SRP+aPDT	Baseline, 1, 3 and 6 months	BOP, PI, PD, CAL	At 1, 3 and 6 months after both types of treatment, an improvement in BOP and CAL was observed. The greater reduction of the PD, achieved by a combination of SRP/PDT, was statistically significant after 6 months ($p = 0.007$).
Raut et al., 2018 (India) [78]	Journal of Indian Society of Periodontology IF 2020: 0.460 IF 2018: 0.44	PG-RCT	CP (PPD > 5 mm and CAL > 4 mm)	50	SRP group: 12/13 SRP+aPDT group: 16/9	SRP group: 46.90 ± 4.32 SRP+aPDT group: 51 ± 2.83	SRP+ sham procedure (Hand and ultrasonic instruments)	SRP+aPDT	Baseline and 6 months	PI, BOP, CAL, PD, microbiological analysis	Significant reduction was seen in PD, CAL and BOP in the test group as compared to control group after 6 months ($p < 0.05$). However, intergroup comparison of PI showed nonsignificant results ($p > 0.05$). Anaerobic culture of plaque samples of test group also revealed a significant reduction of microorganisms in comparison with control group.
Hokari et al., 2018 (Japan) [79]	International Journal of Dentistry IF 2019: 0.58 IF 2018: 0.58	PG-RCT	CP (Moderate: 3–4 mm clinical attachment loss, severe: ≥5 mm loss, generalized: >30% of sites affected)	30	aPDT group: 7/8 MO group: 6/9	aPDT group: 61.4 ± 10.2 MO group: 66.7 ± 9.5	SRP+ Minocycline ointment (MO) (Ultrasonic instruments)	SRP+aPDT	Baseline, 1 and 4 weeks	BOP, PD, CAL, PI, GI, microbiological and inflammatory marker analysis	Local MO administration exhibited a significant decrease in scores for clinical parameters ($p < 0.01$) and a significant reduction in bacterial counts ($p < 0.01$) and IL-1β and IF-γ levels at 1 and 4 weeks after treatment ($p < 0.01$). No significant changes were observed in the aPDT group, except in clinical parameters.
Hill et al., 2019 (Germany) [80]	Photodiagnosis and Photodynamic Therapy IF 2020: 2.894 IF 2019: 2.821	SM-RCT	CP (At least one single and one multi-rooted tooth with at least 4 mm PPD in each quadrant)	20	3/17	61.1	SRP (Hand and piezo-electric ultrasonic instruments)	SRP+aPDT	Baseline, 2 week, 3 and 6 months	BOP, SFFR, PD, GR, RAL, Microbiological analysis	Median values for BOP, RAL, PD, decreased significantly in both groups ($p < 0.05$) after three months of treatment without significant difference between the groups ($p > 0.05$). Two weeks after treatment, the SFFR showed significantly lower mean values in the test group (aPDT). With the applied parameters, this study does not conclusively support ICG-based aPDT, though it is promising because no adverse effects occurred.

Table 1. *Cont.*

Study, Year, Origin and Citation	Journal Name/ Impact Factor (IF)	Study Design	Type of Periodontitis	Sample Size (n)	Gender M/F	Age (Years) (Mean $\pm$ SD)	Intervention Groups				Evaluation Period	Parameters Assessed	Conclusions
Ahad et al., 2016 (India) [81]	Journal of Lasers in Medical Sciences IF 2020: 1.570 IF 2016: 0.68	SM-RCT	CP (At least 2 teeth in different quadrants with PD $\geq$ 6 mm, and BOP)	30	21/9	38.67 $\pm$ 10.52	SRP (Hand and ultrasonic instruments)	SRP+aPDT			Baseline, 1 and 3 months	PI, mSBI, PD, CAL	At 1 month follow-up, intergroup difference in mean change was statistically significant in terms of mSBI and PD for the adjunctive aPDT group ($p < 0.05$), at 3 months interval, no statistically significant difference was observed between test and control groups except in terms of mSBI ($p > 0.05$), thus proving that aPDT improved the gingival status in the nonsurgical management of CP.
Balata et al., 2013 (Brazil) [82]	Journal of Applied Oral Science IF 2019: 2.005 IF 2013: 1.153	SM-RCT	CP (Periodontal pockets with CAL $\geq$ 5 mm, BOP and radiographic bone loss; minimum of 2 teeth with PD $\geq$ 7 mm and 2 other teeth with a PD $\geq$ 5 mm, all with BOP and located on opposite sides of the mouth; and $\geq$16 teeth in both jaws)	22	8/14	43.18	SRP (Ultrasonic instruments)	SRP+aPDT			Baseline, 1, 3 and 6 months	PI, GI, BOP, GR, CAL	Both groups revealed statistically significant improvement in the clinical parameters ($p < 0.05$) with no statistically significant differences upon intergroup comparison ($p > 0.05$). aPDT did not provide any additional benefit to those obtained with full-mouth ultrasonic debridement used alone.
Bechara et al., 2018 (Brazil) [83]	Photodiagnosis and Photodynamic Therapy IF 2020: 2.894 IF 2018: 2.624	PG-RCT	AgP (Single-rooted teeth in multiple quadrants, with both PPD and CAL $\geq$ 5 mm, and with BOP)	35 patients (72 sites)	CLM group: 1/17 Placebo group: 1/17	<35 years CLM group: 33.11 $\pm$ 4.26 Placebo group: 31.26 $\pm$ 4.73	CLM group (n = 18) Clarithromycin 500 mg BD for 3 days	Placebo group (n = 18)			Baseline, 3 months and 6 months	PD, CAL, BOP, GR	At 3 months, UPD+aPDT, UPD+CLM and UPD + CLM + aPDT groups all exhibited reduced PD relative to the UPD group ($p < 0.05$). However, at 6 months, the mean PD reduction was greater in the antibiotic groups (UPD+CLM and UPD+CLM+aPDT) than in the UPD and UPD+aPDT groups ($p < 0.05$). Regarding clinical attachment level, only the UPD+CLM+aPDT group presented a significant gain relative to the UPD and UPD+aPDT groups ($p < 0.05$).
							UPD + CLM (18 sites)	UPD+ CLM+ aPDT (18 sites)	UPD (18 sites)	UPD+ aPDT (18 sites)			

432

Table 1. *Cont.*

Study, Year, Origin and Citation	Journal Name/ Impact Factor (IF)	Study Design	Type of Periodontitis	Sample Size (n)	Gender M/F	Age (Years) (Mean $\pm$ SD)	Intervention Groups		Evaluation Period	Parameters Assessed	Conclusions
Bundidpun et al., 2017 (Thailand) [84]	Laser Therapy IF 2020: 0.43 IF 2017: 0.53	SM-RCT	CP (Generalized moderate to severe chronic periodontitis, presence of at least 20 teeth, at least one molar tooth in each quadrant with a minimum of four teeth, at least two teeth and one molar tooth presented with PD > 6 mm in each quadrant)	20	7/13	47.25 ± 8.91	SRP (Piezo-electric ultrasonic instruments)	SRP+aPDT	Baseline, 1, 3 and 6 months	PD, CAL, PI, GBI, GI	All parameters in test group were better than that control group, with statistically significant differences of GBI and GI ($p < 0.05$) at 3 and 6 months after treatment but no statistically significant differences of PD, CAL and PI.
Chitsazi et al., 2014 (Iran) [85]	Journal of Dental Research, Dental Clinics, Dental Prospects IF 2020: 0.69 IF 2014: 1.30	SM-RCT	AgP (Minimum of 12 teeth with at least 3 teeth in each quadrant with ≥ 4 mm of probing depth)	24	9/15	29	SRP (Piezo-electric ultrasonic instruments)	SRP+aPDT	Baseline, 3 months	PD, CAL, GR, PI, GI, BOP, Microbiological analysis for *A.a*	Intragroup comparison showed an improvement in all the clinical parameters and a significant reduction in the counts of *A.a* at 90 days compared to baseline ($p < 0.05$). None of the periodontal parameters exhibited significant differences between the two groups ($p > 0.05$).
Chitsazi et al., 2014 (Iran) [86]	Journal of Advanced Periodontology and Implant Dentistry IF 2020: NA IF 2014: NA	SM-RCT	CP (At least one site per quadrant exhibiting pocket depth of ≥ 4 mm with bleeding on probing)	22	10/12	46.1	SRP (Sonic instruments)	SRP+aPDT	Baseline, 1 and 3 months	PD, CAL, BOP, GR, microbiological analysis	PD values decreased significantly in both groups after 1 month ($p = 0.001$) and 3 months ($p = 0.001$) in the SRP and ($p = 0.001$) in the PDT groups the inter-group differences were not significant after 1 ($p = 0.25$) and 3 months ($p = 0.51$). Clinical measurements showed significant decreases after 1 and 3 months at both sites, without inter-group differences, except for BOP after 1 ($p = 0.004$) and 3 months ($p = 0.0001$).
Garcia et al., 2011 (Brazil) [87]	Revista Periodontia IF 2020: NA IF 2011: NA	SM-RCT	AgP (Bone loss first molars and incisors, and other teeth adjacent, with PPD ≥ 5 mm and loss of CAL ≥ 2 mm)	10	4/6	39.3 ± 5.84	SRP (Hand and ultrasonic instruments)	SRP+aPDT	Baseline, 3 months	PD, RCAL, furcation involvement, tooth mobility	Both groups showed improved clinical results in the nonsurgical treatment of AgP with no statistically significant intergroup differences ($p > 0.05$).

Table 1. *Cont.*

Study, Year, Origin and Citation	Journal Name/ Impact Factor (IF)	Study Design	Type of Periodontitis	Sample Size (n)	Gender M/F	Age (Years) (Mean $\pm$ SD)	Intervention Groups		Evaluation Period	Parameters Assessed	Conclusions
Joseph et al., 2014 (India) [88]	Journal of Clinical Periodontology IF 2020: 5.241 IF 2014: 4.641	PG-RCT	CP (A minimum of 20 teeth; PPD 4–6 mm at least in two different quadrants of the mouth)	90	39/51	39.6 ± 8.7	SRP (Hand and ultrasonic instruments)	SRP+aPDT	Baseline, 2 weeks, 1, 3 and 6 months	PPD, CAL, GI, GBI, PI, halitosis.	PD and CAL showed statistically significant reduction in the test group on evaluation at 3 months and 6 months as compared to the control group ($p < 0.05$). A statistically significant improvement in GI and GBI was seen for the test group after 2 weeks and 1 month of aPDT ($p < 0.01$), whereas the improvement in GI and GBI at 3 months and in plaque index at 2 weeks after aPDT was less ($p < 0.05$). In addition, a significant difference was detected for the test group at 1 month in terms of halitosis, which did not persist for long ($p < 0.05$).
Malgikar et al., 2015 (India) [89]	Journal of Dental Lasers IF 2020: 0.696 IF 2015: NA	SM-RCT	CP (At least one site in each quadrant of the mouth having deep PPD $\geq$ 5 mm and radiographic signs of alveolar bone loss)	24	15/9	M: 36.73 ± 8.46 F: 34.33 ± 6.80	SRP (Hand and piezo-electric ultrasonic instruments)	SRP+aPDT	Baseline, 1, 3 and 6 months.	PI, GI, mSBI, PD, CAL.	A statistically significant decrease in PD, CAL, PI, GI, mSBI scores was seen in SRP+aPDT at the end of 6 months ($p < 0.001$).
Monzavi et al., 2016 (Iran) [90]	Photodiagnosis and Photodynamic Therapy IF 2020: 2.894 IF 2016: 2.503	SM-RCT	CP (At least three teeth exhibiting residual pocket depth of $\geq$ 5 mm with bleeding on probing)	50	25/25	49.6 ± 8.5	SRP (Hand and ultrasonic instruments)	SRP+aPDT	Baseline, 1 and 3 months	BOP, PI, CAL, PPD, FMPS, FMBS	There were no significant differences between two groups at baseline. BOP, PPD and FMBS showed significant improvements in the test group ($p \leq 0.001$). In terms of PI, FMPS and CAL, no significant differences were observed between both groups ($p \geq 0.05$).
Raj et al., 2016 (India) [91]	Indian Journal of Dental Research IF 2020: 0.37 IF 2016: 0.08	PG-RCT	CP (More than 16 natural teeth; PPD $\geq$ 5 mm)	20	8/12	NI	SRP (Type of instruments utilized-NI)	SRP+aPDT	Baseline and 3 months	PI, GI, PD, CAL and microbiological analysis	There was a significant reduction in PI, GI, PD, CAL and microbiologic parameters in test group, following SRP and PDT, when compared with SRP alone in control group ($p < 0.001$). SRP+aPDT has shown additional improvement in periodontal parameters when compared to SRP alone and has a beneficial effect in chronic periodontitis patients.

Table 1. *Cont.*

Study, Year, Origin and Citation	Journal Name/ Impact Factor (IF)	Study Design	Type of Periodontitis	Sample Size (n)	Gender M/F	Age (Years) (Mean $\pm$ SD)	Intervention Groups		Evaluation Period	Parameters Assessed	Conclusions
Sena et al., 2019 (Brazil) [92]	Photobiomodulation, Photomedicine and Laser Surgery IF 2019: 1.913	SM-RCT	CP (At least six sites with PD 5–9 mm; and BOP)	9 (6 sites/ patient: total-54 sites)	NI	NI	SRP+ placebo procedure (Hand and ultrasonic instruments)	SRP+aPDT	Baseline and 3 months	BOP, PD, CAL, VPI	There was a statistically significant decrease in BOP for test group (p = 0.003) and control group (p = 0.001). Intragroup comparison for PD and CAL showed statistically significant differences from baseline (p < 0.05) with no intergroup differences (p > 0.05). Hence, SRP+aPDT did not show any additional benefits over SRP alone.
Shingnapurkar et al., 2016 (India) [93]	Indian Journal of Dental Research IF 2020: 0.37 IF 2016: 0.08	SM-RCT	CP (PD > 5 mm)	60 sites	NI	NI	SRP+ sham procedure (Hand and ultrasonic instruments)	SRP+aPDT	Baseline, 1 and 3 months	PI, GI, PD, RAL	Mean baseline values for PI, GI, PPD and RAL were not different in the test group and control group. Statistically significant difference in PPD and RAL, 3 months after treatment was seen in test group as compared to the control group (p < 0.05).
Sigusch et al., 2010 (Germany) [94]	Journal of Periodontology IF 2020: 3.742 IF 2010: 2.946	PG-RCT	CP (<30% of sites with PPD >3.5 mm)	24 (12 in each group)	PDT group: 4/8 Control group: 3/9	PDT group F: 39.75 M: 45 Control group: F: 44.22 M:42.67	SRP+ sham procedure (Type of instruments utilized- NI)	SRP+aPDT	Baseline, 1, 4 and 12 weeks.	PI, reddening, PD, BOP, CAL, GR Quantitative analysis for *F.n.*	In patients with localized CP who received aPDT treatment, significant reductions in reddening, BOP, and mean PD and CAL were observed during the observation period and with respect to controls (p < 0.001). Four and 12 weeks after aPDT, the mean PD and CAL showed significant differences from baseline values and from those of the control group. In the aPDT group, 12 weeks after treatment, the *F.n.* DNA concentration was found to be significantly reduced compared to the baseline level (p < 0.001) compared to control group.

Table 1. *Cont.*

Study, Year, Origin and Citation	Journal Name/ Impact Factor (IF)	Study Design	Type of Periodontitis	Sample Size (*n*)	Gender M/F	Age (Years) (Mean ± SD)	Intervention Groups			Evaluation Period	Parameters Assessed	Conclusions
Theodoro et al., 2012 (Brazil) [95]	Lasers in Medical Science IF 2019: 2.574 IF 2012: 2.645	SM-RCT	CP (At least three non-adjacent sites with BOP and a PD of 5–9 mm at least 20 teeth in the oral cavity)	33	12/21	43.12 ± 8.2	SRP (Hand instruments)	SRP+ PS (TBO) only	SRP+ aPDT	Baseline, 60, 90 and 180 days	VPI, GI, BOP, PD, CAL, GR, microbiological analysis	All treatment groups showed an improvement in all clinical parameters, and a significant reduction in the proportion of sites positive for periodontopathogens at 60, 90 and 180 days compared to baseline ($p < 0.05$). None of the periodontal parameters showed a significant difference among the groups ($p > 0.05$). At 180 days, PDT treatment led to a significant reduction in the percentage of sites positive for all bacteria compared to SRP alone ($p < 0.05$).
Theodoro et al., 2017 (Brazil) [96]	Journal of Photochemistry and Photobiology B IF 2020: 4.383 IF 2017: 3.438	PG-RCT	CP (Severe generalized CP in at least 6 teeth and with one or several sites with PD $\geq$ 5 mm; a loss of CAL $\geq$ 5 mm; a minimum of 30% of the sites with PD and CAL $\geq$ 4 mm and BOP; and the presence of at least 15 teeth)	34	AB group: 7/7 aPDT group: 9/5	AB group: 46.3 ± 6.8 aPDT group: 48.8 ± 8.3	SRP+ (MTZ+ AMX) MTZ dose: 400mg TDS-7 days AMX dose: 500mg TDS-7 days (Type of instruments utilized for SRP-NI)	SRP +aPDT+ placebo pills		Baseline and 90 days	BOP, PD, CAL	There was a significant improvement in CAL only in the intermediate pocket in the aPDT group com- pared to the MTZ + AMX group between baseline and 90 days post-treatment ($p = 0.01$). There was a reduction of both BOP and the percentage of residual pockets at 90 days after treatment compared with baseline in both groups ($p < 0.05$).

3.2.4. Documentation of Laser Parameters

Table 2 describes various dye laser combinations, as well as laser dosimetry that was utilized to perform aPDT in all eligible studies. Twenty-six out of 31 studies utilized a laser wavelength in the range of 630–690 nm [2,3,5,17,70–77,79,81–88,91,92,94–96] to perform aPDT. While four studies utilized a laser wavelength in the range of 808–810 nm [78,80,90,93], one of the included studies utilized a 980 nm diode laser wavelength to perform aPDT [89] (Table 2) (Figure 2). Emission mode was reported only in five studies [78,88,90,92,93], in which four of them utilized a continuous wave emission mode [78,88,90,92], whilst the remaining one study utilized a gated continuous wave emission mode [93]. Eighteen out of the 31 eligible studies used the laser fibre tip in 'contact mode' with the periodontal pocket in order to perform aPDT [2,3,5,78,81,82,84–86,88–96]. Only 5 studies reported total energy, and it ranged from 1.5–9 J [82,90,92,93,95,96]. Only 19 studies reported power output in the range of 30 mW–1 W [71,75–80,82–85,87,89–96], whereas the use of a power meter to measure the therapeutic power output, reaching the target tissues was not performed in any of the included studies. Spot size was reported in only four studies [5,85,92,96] ranging from 0.02–0.07 cm^2. Ten out of the 31 studies reported the diameter of fibre tip, [2,3,5,80–82,88,89,93,94] ranging from 200–600 μm. The energy density (fluence) was calculated in 18 out of 31 studies [5,17,70–74,78–80,82,83,86,87,92,93,95,96], and its value ranged from 0.01–2829 J/cm^2, whereas the power density (irradiance) values ranged from 60 mW–4 W/cm^2 and were calculated in 13 studies [2,3,5,17,71–74,81,88,92,94,95]. Finally, the exposure time for laser irradiation was mentioned in all included studies except one study [80], and the values ranged from 10–120 s/site amongst included studies.

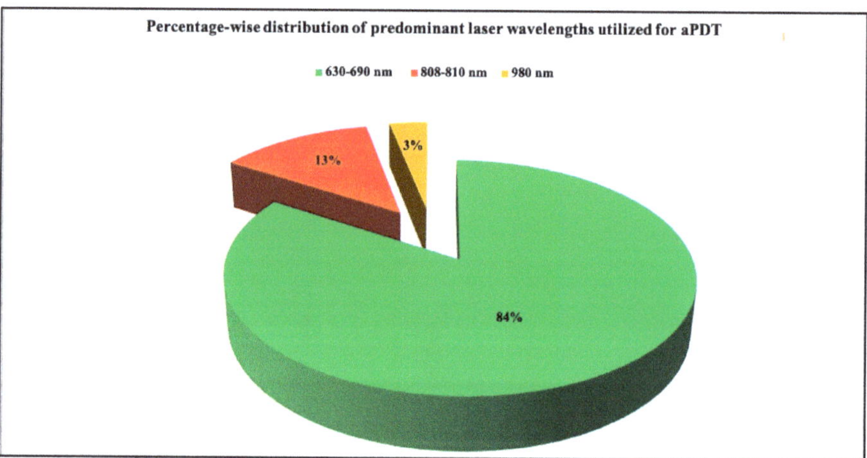

Figure 2. 3D pie diagram illustrating the percentage-wise distribution of predominant laser wavelengths utilized for aPDT in the included studies.

3.2.5. PS Utilized

Type of PS varied amongst eligible clinical trials. Eleven studies utilized phenothiazine chloride [2,3,5,71–74,76,81,84,94] while 10 employed methylene blue [17,75,77,79,82,87–89,96]. Five studies utilized toluidine blue O [70,85,86,91,95], four studies used indocyanine green [78,80,90,93], whereas chloro-aluminum phthalocyanine was utilized in one study [92] (Figure 3). Interestingly, 18 out of 31 studies specified the concentration of the PS [2,3,17,71,75,77–80,82,83,87–90,92,95,96], while 13 studies failed to report the same [5,70,72–74,76,81,84–86,91,93,94] (Table 2).

437

Table 2. Tabular representation of PS dye and laser parameters utilized for aPDT in the selected eligible in vivo human studies. Refer to Supplementary file 2 for list of abbreviations.

Study, Year, Origin and Citation	Photosensitizer (PS) Used and Its Concentration	Pre-Irradiation Exposure Time to PS (min)	Laser Wavelength Utilized	Emission Mode Contact/No Contact Tip Initiation	Energy (J)	Power Output (W)	Pulse Length (Duration), Pulse Interval	Use of Power Meter	Distance from Target	Spot Size/Fibre-Tip Diameter/Spot Diameter	Energy Density [Fluence] (J/cm^2)	Power Density [Irradiance] (W/cm^2)	Exposure Time to Laser Irradiation [Minute (min)/Second (s)]	No. of aPDT Applications
De Oliveira et al., 2009 (Brazil) [2]	Phenothiazine chloride (10 mg/mL)	1 min	660 nm	Contact mode, fibre tip was place at the entrance of the gingival sulcus	NI	NI	NI	NI	NA	Tip diameter: 600 μm	NI	60 mW/cm^2	10 s/site (6 sites = 1 min/tooth)	1
De Oliveira et al., 2007 (Brazil) [3]	Phenothiazine chloride (10 mg/mL)	1 min	660 nm	Contact mode, fibre tip was place at the entrance of the gingival sulcus	NI	NI	NI	NI	NI	Tip diameter: 600 μm	NI	60 mW/cm^2	10 s/site (6 sites = 1 min/tooth)	1
Novaes et al., 2012 (Brazil) [5]	Phenothiazine chloride	NI	660 nm	Contact mode, fibre tip was place at the entrance of the gingival sulcus	NI	NI	NI	NI	NI	Tip diameter: 600 μm [8.5 cm long optic fibre with 60° angulated tip] Spot size: 0.06	212.23 J/cm^2	60 mW/cm^2	10 s/site (6 sites/tooth) 60 s/tooth	1
Franco et al., 2014 (Brazil) [17]	Methylene blue (0.01%)	5 min	660 nm	NI	NI	NI	NI	NI	NI	NI	5.4 J/cm^2	60 mW/cm^2	5 s/site (6 sites/tooth) 90 s/tooth	4
Pourabbas et al., 2014 (Iran) [70]	Toluidine blue	60 s	638 nm	NI	NI	NI	NI	NI	NI	NI	8–10 J/cm^2	NI	120 s	1
Moreira et al., 2015 (Brazil) [71]	Phenothiazine chloride (10 mg/mL)	1 min	670 nm	NI	NI	75 mW	NI	NI	NI	Tip diameter: 600 μm	Fluence/site: 2.49 J/cm^2 Fluence/tooth: 14.94 J/cm^2	0.25 W/cm^2	10 s /site	4 (0, 2nd, 7th and 14th day)
Skurska et al., 2015 (Poland) [72]	Phenothiazine chloride	3 min	660 nm	NI	NI	NI	NI	NI	NI	NI	120 J/cm^2	60 mw/cm^2	60 s/site	2 (0 and 7th day)
Arweiler et al., 2014 (Poland) [73]	Phenothiazine chloride	3 min	660 nm	NI	NI	NI	NI	NI	NI	NI	120 J/cm^2	60 mw/cm^2	60 s/site	2 (0 and 7th day)
Arweiler et al., 2013 (Poland) [74]	Phenothiazine chloride	3 min	660 nm	NI	NI	NI	NI	NI	NI	NI	120 J/cm^2	60 mw/cm^2	60 s/site	2 (0 and 7th day)

Table 2. *Cont.*

Study, Year, Origin and Citation	Photosensitizer (PS) Used and Its Concentration	Pre-Irradiation Exposure Time to PS (min)	Laser Wavelength Utilized	Emission Mode Contact/No Contact Tip Initiation	Energy (J)	Power Output (W)	Pulse Length (Duration), Pulse Interval	Use of Power Meter	Distance from Target	Spot Size/Fibre-Tip Diameter/Spot Diameter	Energy Density [Fluence] (J/cm²)	Power Density [Irradiance] (W/cm²)	Exposure Time to Laser Irradiation [Minute (min)/ Second (s)]	No. of aPDT Applications
Vidal et al., 2017 (Spain) [75]	Methylene blue (0.005%)	NI	670 nm	NI	NI	150 mW	NI	NI	NI	NI	NI	NI	60 s/pocket	3 (1, 5 and 13 weeks)
Braun et al., 2008 (Germany) [76]	Phenothiazine chloride	3 min	660 nm	NI	NI	100 mW	NI	NI	NI	NI	NI	NI	10 s/site (6 sites = 1 min /tooth)	1
Berakdar et al., 2012 (Germany) [77]	Methylene blue 0.005%	NI	670 nm	NI	NI	150 mW	NI	NI	NI	NI	NI	NI	1 min	1
Raut et al., 2018 (India) [78]	Indocyanine green (5 mg/mL)	60 s	810 nm	CW, contact mode	NI	80 mW	NI	NI	NA	NI	5.4 J/cm²	NI	60 s	1
Hokari et al., 2018 (Japan) [79]	Methylene blue dye 0.01%	1 min	670 nm	NI, contact mode	NI	140 mW	NI	NI	NA	NI	21 J/cm²	NI	60 s	2 (0 and 7th day)
Hill et al., 2019 (Germany) [80]	Indocyanine green (0.1 mg/mL)	60 s	808 nm	NI	NI	100 mW	NI	NI	NI	Tip diameter: 300 μm	2829 J/cm²	NI	NI	1
Ahad et al., 2016 (India) [81]	Phenothiazine chloride	3 min	660 nm	Contact mode	NI	NI	NI	NI	NA	Tip diameter: 0.6 μm	NI	100 mW/cm²	10 s/site (6 sites, 1 min/tooth)	1
Balata et al., 2013 (Brazil) [82]	Methylene blue 0.005%	2 min	660 nm	90° angle with the gingival surface and with no contact with the tissues	9 J	100 mW	NI	NI	NI	Tip diameter: 600 μm tip	320 J/cm²	NI	90 s/site	1
Bechara et al., 2018 (Brazil) [83]	Methylene Blue (10 mg/mL)	1 min	660 nm	NI	NI	60 mW	NI	NI	NI	NI	129 J/cm²	NI	60 s/tooth (2 sites/tooth)	1
Bundidpun et al., 2017 (Thailand) [84]	Phenothiazine chloride	1 min	660 nm	Contact mode	NI	100 mW	NI	NI	NA	NI	NI	NI	10 s/site (6 sites) 1 min/tooth	1
Chitsazi et al., 2014 (Iran) [85]	Toluidine Blue	1 min	670–690 nm	Contact mode	NI	75 mW	NI	NI	NA	NI	NI	NI	120 s/site	1

Table 2. *Cont.*

Study, Year, Origin and Citation	Photosensitizer (PS) Used and Its Concentration	Pre-Irradiation Exposure Time to PS (min)	Laser Wavelength Utilized	Emission Mode Contact/No Contact Tip Initiation	Energy (J)	Power Output (W)	Pulse Length (Duration), Pulse Interval	Use of Power Meter	Distance from Target	Spot Size/Fibre-Tip Diameter/Spot Diameter	Energy Density [Fluence] (J/cm^2)	Power Density [Irradiance] (W/cm^2)	Exposure Time to Laser Irradiation [Minute (min)/ Second (s)]	No. of aPDT Applications
Chitsazi et al., 2014 (Iran) [86]	Tolonium chloride (Toluidine Blue O)	60 s	638 nm	Contact mode	NI	NI	NI	NI	NA	NI	$8–10\ J/cm^2$	NI	120 s	1
Garcia et al., 2011 (Brazil) [87]	Methylene blue (0.005%)	5 min	660 nm	NI	NI	40 mW	NI	NI	NI	NI	$120\ J/cm^2$	NI	120 s/site	1
Joseph et al., 2014 (India) [88]	Methylene blue (10 mg/mL)	3 min	655 nm	CW, contact mode, tip was inserted into the gingival sulcus	NI	NI	NI	NI	NA	Tip diameter: 200 μm Probe tip diameter: 0.5 mm	NI	60 mW/ cm^2	60 s/site (4 sites/ tooth)	1
Malgikar et al., 2015 (India) [89]	Methylene blue 1%	3 min	980 nm	Contact mode, tip was initiated	NI	Peak Power: 5 W Average power 1 W	Pulse length: 200 μs, Pulse interval: 200	NI	NA	Tip diameter: 400 μm	NI	NI	30–45 s/site	1
Monzavi et al., 2016 (Iran) [90]	Indocyanine green (1 mg/mL)	NI	810 nm	CW, contact mode	PBM tip: 6 J Bulb tip: 4 J	200 mW	NI	NI	NA	Use of two types of tips: PBM tip was placed on papilla and then the bulb tip was inserted inside the pocket from each buccal or lingual/palatal side, moving from the bottom of the pocket to the coronal aspect.	NI	NI	PBM tip: 30 s Bulb tip: 10 s	4 (0, 7th, 17th and 27th days)
Raj et al., 2016 (India) [91]	Toluidine blue	1 min	635 nm	Contact mode	NI	500 W	NI	NI	NA	NI	NI	NI	60 s	1
Sena et al., 2019 (Brazil) [92]	Chloro-aluminum pthalocyanine (AlClFc) 5 μM	5 min	660 nm	CW, laser optical fiber tip was positioned parallel to the tooth axis in contact with the gingival margin (without penetrating the pocket)	1.5 J	100 mW	NI	NI	NA	Spot size: $0.028\ cm^2$	$54\ J/cm^2$	$4\ W/cm^2$	15 s	1

Table 2. *Cont.*

Study, Year, Origin and Citation	Photosensitizer (PS) Used and Its Concentration	Pre-Irradiation Exposure Time to PS (min)	Laser Wavelength Utilized	Emission Mode Contact/No Contact Tip Initiation	Energy (J)	Power Output (W)	Pulse Length (Duration), Pulse Interval	Use of Power Meter	Distance from Target	Spot Size/Fibre-Tip Diameter/Spot Diameter	Energy Density [Fluence] (J/cm²)	Power Density [Irradiance] (W/cm²)	Exposure Time to Laser Irradiation [Minute (min)/ Second (s)]	No. of aPDT Applications
Shingnapurkar et al., 2016 (India) [93]	Indocyanine green (1 mg/mL)	3 min	810 nm	Gated CW, Contact mode	3 J	200 mW	Pulse duration: 25 μm Duty cycle 50%	NI	NA	Tip diameter: 400 μm	0.0125 J/cm²	NI	30 s/site	1
Sigusch et al., 2010 (Germany) [94]	Phenothiazine chloride	1 min	660 nm	Contact mode	NI	NI	NI	NI	NA	Tip diameter: 600 μm tip	NI	60 mW/cm²	10 s/site (6 sites =1 min /tooth)	1
Theodoro et al., 2012 (Brazil) [95]	Toluidine blue O 100 μg/mL	1 min	660 nm	The laser optical fiber tip was positioned parallel to and in contact with the selected site	4.5 J	30 mW	NI	NI	NA	Spot size: 0.07 cm²	64.28 J/cm²	0.4 W/cm²	150 s	1
Theodoro et al., 2017 (Brazil) [96]	Methylene blue (10 mg/mL)	1 min	660 nm	Contact mode	4.8 J	100 mW	NI	NI	NA	Spot size 0.03 cm²	160 J/cm²	NI	48 s	3 (0, 48 h, 96 h)

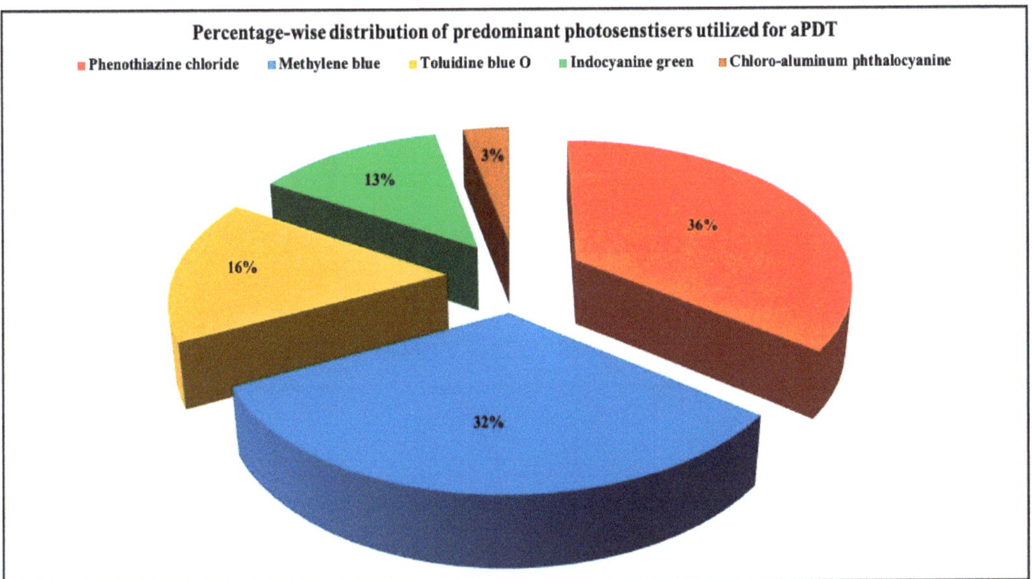

Figure 3. 3D pie diagram illustrating the percentage-wise distribution of predominant photosensitizers utilized for aPDT in the included studies.

3.2.6. Utilization of aPDT as a Mono-Therapeutic or an Adjunctive Therapeutic Agent

While 28 out of the 31 eligible studies utilized SRP+aPDT, aPDT monotherapy was performed in three studies [2,3,5] (Table 1).

3.2.7. Comparison in between SRP+ aPDT versus SRP+AB

Six out of the 31 eligible studies compared efficacy of SRP+aPDT versus SRP+ AB [72–74,79,83,96] (Table 1).

3.2.8. Number of aPDT Sessions

While a single session of aPDT was applied in 22 out of the 31 included studies [2,3,5,70,76–78,80–89,91–95], multiple aPDT sessions were performed in nine studies [17,71–75,79,90,96]. None of the eligible studies compared single versus multiple sessions of aPDT (Table 2).

3.2.9. Follow-Up Assessment

A follow-up assessment at three months from the baseline visit was performed in 18 out of the 31 eligible studies [2,3,5,17,70,71,74,76,81,85–87,90–94,96], whereas 12 studies conducted a longer follow-up assessment at six months [72,73,75,77,78,80,82–84,88,89,95]. Only one study performed a follow-up assessment at one month from the baseline visit [79]. A long-term follow-up of a minimum one year from baseline visit lacked in all eligible studies.

3.3. Qualitative Assessment

Qualitative assessment was performed using the RoB 2 tool, designed for in vivo human RCTs, as depicted in Figures 4 and 5. The most recent version of this tool was utilized to perform a qualitative assessment for both randomized PG and SM human RCTs [29,30]. Figure 4 represents a risk of bias assessment summary of all eligible studies. Figure 5 is a graphical representation of percentage RoB score for each risk domain, which has been evaluated, using the abovementioned tool. Furthermore, 53.1% of included trials were at a high risk of inadequate randomization, whereas 40.6% and 6.3% of included trials were at a low risk or had some concerns, respectively. In addition, 50% of included studies were at a high risk of deviations from intended interventions, whereas 43.7% and 6.3% of them were at a low risk or had some concerns, respectively. All included papers reported substantial evidence (100%) for reporting missing outcome data and, hence, were at a low risk. Although a majority of studies were free of bias arising from reporting outcome measurement (71.9%), 28.1% were at a high risk. In terms of selective reporting of the results, inferences are as follows: 59.4% studies were at a high risk, 37.5% studies were at a low risk, and 3.1% studies had some concerns. Overall, 60% studies reported a high risk of bias, while 35% studies had a low risk of bias, and the final 5% studies had some concerns. It should be noted that information provided in these figures represents the consensual answers verified using the 'Discrepancy check' feature of RoB 2 tool, across two independent reviewers (S.D. and R.H.) (inter-reviewer agreement, κ = 0.94), and, in case of any disagreements, a third author (S.B.) was consulted.

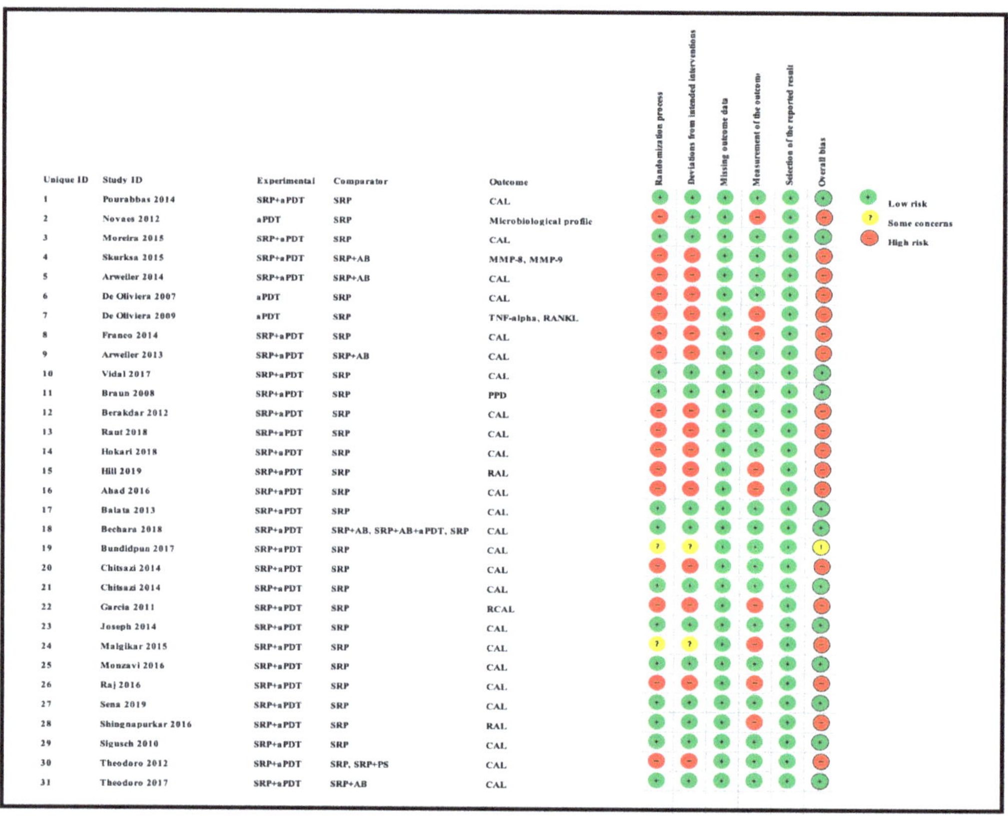

Figure 4. Risk of Bias assessment summary of the included studies based on the consensual answers across two individual assessors (S.D. and R.H.).

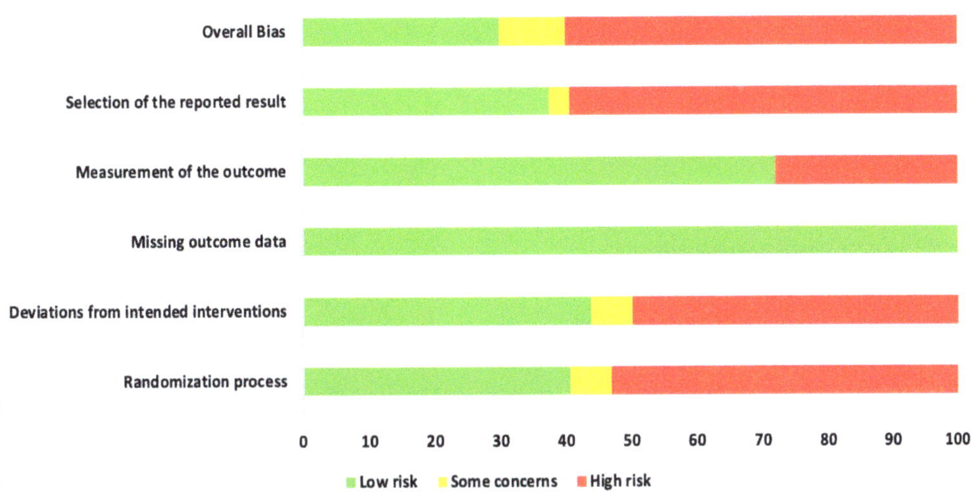

Figure 5. Risk of Bias assessment graph of the included studies expressed as percentages, based on the consensual answers across two individual assessors (S.D. and R.H.).

3.4. Quantitative Assessment

3.4.1. Outcome Variables

Primary outcomes of 18 out of 31 studies, which have assessed efficacy of SRP+aPDT in the management of periodontitis, contributed to this meta-analysis [17,71,73–76,80,81,84–86,88–93,96]. Data were pooled separately for SM and PG studies for differences in PPD and CAL respectively at three and six months, respectively. At three months, the mean difference in PPD reduction was not statistically significant for SM studies (SMD 0.166; 95% CI −0.278 to 0.611; P = 0.463) and PG studies (SMD 0.076; 95% CI −0.420 to 0.573; P = 0.763) along with a high heterogeneity for SM studies (Q = 15.81; P = 0.0001; I^2 = 91.21%) and moderate heterogeneity (Q = 11.87; P = 0.018; I^2 = 66.31%) for PG studies (Table 3). The mean difference in PPD reduction at six months did not show a statistically significant difference for SM studies (SMD 0.005; 95% CI −0.126 to 0.136; P = 0.935) as well as PG studies (SMD 0.141; 95% CI −1.007 to 1.288; P = 0.809) although contrasting findings were noted in terms of level of heterogeneity which was not evident for SM studies (Q = 0.06; P = 0.99; I^2 = 0.00%) and high for PG studies (Q = 18.71; P = 0.0001; I^2 = 89.31%) (Table 4). CAL gain at three months was not statistically significant in SM studies (SMD 0.092; 95% CI −0.013 to 0.198; P = 0.088) with no evident heterogeneity (Q = 8.74; P = 0.655; I^2 = 0.00%) as well as in PG studies (SMD 0.056; 95% CI −0.408 to 0.552; P = 0.745) with moderate heterogeneity (Q = 8.95; P = 0.028; I^2 = 70.31%) (Table 3). At six months, results for SM studies were not statistically significant (SMD −0.013; 95% CI −0.148 to 0.121; P = 0.846) with no evident heterogeneity (Q = 0.03; P = 0.984; I^2 = 0.00%), whereas, for PG studies, the findings were statistically significant (SMD -0.441; 95% CI −0.805 to −0.075; P = 0.018) with no evidence of heterogeneity (Q = 1.70; P = 0.42; I^2 = 0.00%) but favoring control group (Table 4).

Table 3. Forest plots illustrating the overall PPD reduction and CAL gain at three months. Refer to Supplementary file 2 for a list of abbreviations.

Overall PPD Reduction for SM Studies at 3 Months

Study	SMD	SE	95% CI	Weight (%)	
Chitsazi et al., 2014	0.525	0.289	−0.056 to 1.106	8.28	
Moreira et al., 2015	0.205	0.311	−0.425 to 0.834	8.11	
Chitsazi et al., 2014	−1.023	0.316	−1.659 to −0.386	8.07	
Franco et al., 2014	0.631	0.364	−0.115 to 1.378	7.67	
Malgikar et al., 2015	0.119	0.284	−0.453 to 0.691	8.31	
Ahad et al., 2016	0.639	0.204	0.235 to 1.043	8.88	
Monzavi et al., 2016	1.669	0.231	1.211 to 2.127	8.70	$P = 0.463$
Shingnapurkar et al., 2016	0.995	0.271	0.454 to 1.537	8.42	
Bundidpun et al., 2017	0.007	0.310	−0.620 to 0.635	8.11	
Hill et al., 2019	0.039	0.072	−0.103 to 0.181	9.47	
Braun et al., 2008	0.139	0.310	−0.490 to 0.767	8.11	
Sena et al., 2019	−2.169	0.340	−2.851 to −1.487	7.87	
Total (random effects)	0.166	0.227	−0.278 to 0.611	100.00	

Heterogeneity: Q = 15.81; DF = 11; $P = 0.0001$; $I^2 = 91.21\%$

Overall PPD Reduction for PG Studies at 3 Months

Study	SMD	SE	95% CI	Weight (%)	
Arweiler et al., 2013	−0.681	0.340	−1.374 to 0.011	19.80	
Raj et al., 2016	0.669	0.441	−0.258 to 1.595	15.89	
Vidal et al., 2017	−0.060	0.322	−0.714 to 0.593	20.59	$P = 0.763$
Theodoro et al., 2017	−0.127	0.374	−0.897 to 0.643	18.43	
Joseph et al., 2014	0.556	0.215	0.127 to 0.984	25.28	
Total (random effects)	0.076	0.252	−0.420 to 0.573	100.00	

Heterogeneity: Q = 11.87; DF = 4; $P = 0.018$; $I^2 = 66.31\%$

Table 3. *Cont.*

Overall CAL Gain for SM Studies at 3 Months

Study	SMD	SE	95% CI	Weight (%)	
Chitsazi et al., 2014	0.439	0.287	−0.140 to 1.017	3.52	
Moreira et al., 2015	−0.040	0.310	−0.667 to 0.588	3.02	
Chitsazi et al., 2014	0.249	0.297	−0.351 to 0.849	3.29	
Franco et al., 2014	0.601	0.364	−0.144 to 1.346	2.20	
Malgikar et al., 2015	−0.048	0.284	−0.620 to 0.523	3.60	
Ahad et al., 2016	0.158	0.199	−0.237 to 0.552	7.35	
Monzavi et al., 2016	0.080	0.199	−0.314 to 0.474	7.37	$P = 0.088$
Shingnapurkar et al., 2016	0.564	0.260	0.043 to 1.084	4.30	
Bundidpun et al., 2017	0.032	0.310	−0.595 to 0.660	3.02	
Hill et al., 2019	0.019	0.072	−0.123 to 0.161	55.29	
Braun et al., 2008	0.161	0.310	−0.468 to 0.789	3.01	
Sena et al., 2019	0.000	0.268	−0.538 to 0.538	4.04	
Total (random effects)	0.092	0.233	−0.013 to 0.198	100.00	

Heterogeneity: Q = 8.74; DF = 11; P = 0.655; I^2 = 0.00%

Overall CAL Gain for PG Studies at 3 Months

Study	SMD	SE	95% CI	Weight (%)	
Arweiler et al., 2013	−0.662	0.340	−1.356 to 0.010	19.80	
Raj et al., 2016	0.669	0.441	−0.258 to 1.595	15.89	
Vidal et al., 2017	−0.102	0.372	−0.514 to 0.793	20.59	$P = 0.745$
Theodoro et al., 2017	−0.106	0.374	−0.997 to 0.743	18.43	
Joseph et al., 2014	0.456	0.255	−0.120 to 0.673	25.28	
Total (random effects)	0.056	0.358	−0.408 to 0.552	100.00	

Heterogeneity: Q = 8.95; DF = 4; P = 0.028; I^2 = 70.31%

Table 4. Forest plots illustrating the overall PPD reduction and CAL gain at 6 months. Refer to Supplementary file 2 for a list of abbreviations.

Overall PPD Reduction for SM Studies at 6 Months

Study	SMD	SE	95% CI	Weight (%)
Berakdar et al., 2012	0.040	0.296	−0.598 to 0.598	5.08
Malgikar et al., 2015	0.037	0.284	−0.535 to 0.609	5.52
Bundidpun et al., 2017	0.072	0.310	−0.555 to 0.701	4.63
Hill et al., 2019	0.060	0.072	−0.142 to 0.142	84.77
Total (random effects)	0.005	0.066	−0.126 to 0.136	100.00

$P = 0.935$

Heterogeneity: $Q = 0.06$; $DF = 3$; $P = 0.99$; $I^2 = 0.00\%$

Overall PPD Reduction for PG Studies at 6 Months

Study	SMD	SE	95% CI	Weight (%)
Arweiler et al., 2014	−0.722	0.342	−1.417 to −0.027	33.04
Vidal et al., 2017	−0.109	0.322	−0.763 to 0.545	33.47
Raut et al., 2018	1.241	0.321	0.594 to 1.888	33.49
Total (random effects)	0.141	0.579	−1.007 to 1.288	100.00

$P = 0.809$

Heterogeneity: $Q = 18.71$; $DF = 2$; $P = 0.0001$; $I^2 = 89.31\%$

Table 4. *Cont.*

Overall CAL Gain for SM Studies at 6 Months

Study	SMD	SE	95% CI	Weight (%)
Malgikar et al., 2015	−0.057	0.284	−0.629 to 0.515	5.82
Bundidpun et al., 2017	0.025	0.310	−0.602 to 0.653	4.88
Hill et al., 2019	−0.012	0.072	−0.155 to 0.130	89.30
Total (random effects)	−0.013	0.068	−0.148 to 0.121	100.00

$P = 0.846$

Heterogeneity: Q = 0.03; DF = 2; P = 0.984; I^2 = 0.00%

Overall CAL Gain for PG Studies at 6 Months

Study	SMD	SE	95% CI	Weight (%)
Arweiler et al., 2014	−0.539	0.337	−1.224 to 0.146	29.91
Vidal et al., 2017	−0.103	0.322	−0.756 to 0.551	32.69
Raut et al., 2018	−0.658	0.301	−1.265 to −0.050	37.40
Total (random effects)	−0.441	0.184	−0.805 to −0.075	100.00

$P = 0.018$

Heterogeneity: Q = 1.70; DF = 2. P = 0.42; I^2 = 0.00%

Assessment of secondary outcome variables was conducted in the majority of included studies, which are as follows: Changes in GR, BOP, PI and GI in 28 studies [3,17,70,71,73–96], microbiological analysis in 11 studies [5,71,75,78–80,85,86,91,94,95], and immuno-histological in seven studies [2,17,70–72,75,79]. Table 5 provides an overview of clinical parameters which have been assessed in 28 of 31 included studies along with corresponding level of significance, in accordance to data provided in Table 1. Eleven studies performed a microbiological analysis [5,71,75,78–80,85,86,91,94,95], out of which five studies reported that aPDT therapy could significantly reduce periopathogenic burden [71,75,78,91,94] and six studies failed to achieve this outcome [5,79,80,85,86,95]. In terms of immune-histological analysis, seven studies [2,17,70–72,75,79] assessed various pro-inflammatory cytokines and growth factors such as; IL-1β, IL-10, IF-γ, TNF-α, MMP-8, MMP-9, RANK, RANK-L, OPG and FGF-2 (Table 1). Biomarkers for assessment of bone resorption (RANK, RANK-L, OPG) were assessed in three studies [2,17,75]. Two studies [17,75] assessed efficacy of SRP+aPDT in comparison to conventional SRP alone, and showed that SRP+aPDT successfully suppressed the bone resorption process. Levels of IL-1β, IL-10, IF-γ, and TNF-α were assessed in three studies [70,71,79], of which two studies have confirmed immunological benefits of aPDT [70,71], whereas one study [79] failed to show any advantage of aPDT for the same. It should, however, be noted that, while the former two studies [70,71] have compared the efficacy of SRP+aPDT to conventional SRP alone, the latter study [79] has compared SRP+aPDT to SRP+AB and demonstrated the advantages of AB over aPDT. Additionally, SRP+aPDT showed an increased expression of FGF-2, which plays a role in tissue repair as compared to SRP alone, and was assessed in only one study [17]. A meta-analysis on secondary outcomes was not possible due to disparity in scoring methodology, incomplete, or incomparable data.

3.4.2. Sensitivity Analysis

A sensitivity analysis was conducted due to the noteworthy heterogeneity arising from outlier studies which were detected upon visual inspection of Forest plots [74,86,90,92,93] (Tables 3 and 4). This analysis was conducted only for the three-month follow-up due to unavailability of data in included studies (Table 6). In terms of PPD reduction, SM studies (SMD 0.282; 95% CI -0.286 to 0.624; $P = 0.153$) as well as PG studies (SMD 0.257; 95% CI -0.230 to 0.683; $P = 0.361$) did not report a statistically significant improvement. No evident heterogeneity (Q = 9.14; $P = 0.7$; $I^2 = 0.00\%$) in SM studies and in PG studies (Q = 8.87; $P = 0.22$; $I^2 = 0.00\%$) was noted (Table 6). Although improvement in CAL gain was noted after omitting outlier studies, this difference was statistically not significant in both SM (SMD 0.162; 95% CI -0.326 to 0.406; $P = 0.166$) and PG studies (SMD 0.227; 95% CI -0.420 to 0.673; $P = 0.234$) with no evident heterogeneity (Q = 8.40; $P = 0.625$; $I^2 = 0.00\%$) in SM studies as well as in PG studies (Q = 9.7; $P = 0.22$; $I^2 = 0.00\%$) (Table 6).

Table 5. Tabular representation describing the assessment of clinical parameters used for the selected eligible in vivo human studies. Refer to Supplementary file 2 for a list of abbreviations.

Study, Year, Origin and Citation	PPD		CAL		BOP/SBI		PI		GI		GR	
	Statistically Significant Y/N/NI/NS	Not Statistically Significant Y/N/NI/NS	Statistically Significant Y/N/NI/NS	Not Statistically Significant Y/N/NI/NS	Statistically Significant Y/N/NI/NS	Not Statistically Significant Y/N/NI/NS	Statistically Significant Y/N/NI/NS	Not Statistically Significant Y/N/NI/NS	Statistically Significant Y/N/NI/NS	Not Statistically Significant Y/N/NI/NS	Statistically Significant Y/N/NI/NS	Not Statistically Significant Y/N/NI/NS
De Oliveira et al., 2007 (Brazil) [3]	N	Y	N	Y	N	Y	N	Y	N	Y	N	Y
Franco et al., 2014 (Brazil) [17]	N	Y	N	Y	Y	N	N	Y	NS	NS	NS	NS
Pourabbas et al., 2014 (Iran) [70]	N	Y	N	Y	N	Y	NS	NS	NS	NS	N	Y
Moreira et al., 2015 (Brazil) [71]	Y	N	Y	N	Y	N	Y	N	NS	NS	Y	N
Arweiler et al., 2014 (Poland) [73]	N	Y	N	Y	N	Y	N	Y	NS	NS	N	Y
Arweiler et al., 2013 (Poland) [74]	N	Y	N	Y	N	Y	N	Y	NS	NS	N	Y
Vidal et al., 2017 (Spain) [75]	N	Y	N	Y	N	Y	N	Y	NS	NS	N	Y
Braun et al., 2008 (Germany) [76]	Y	N	Y	N	Y	N	NS	NS	NS	NS	N	Y
Berakdar et al., 2012 (Germany) [77]	Y	N	Y	N	N	Y	N	Y	NS	NS	NS	NS
Raut et al., 2018 (India) [78]	Y	N	Y	N	Y	N	N	Y	NS	NS	NS	NS
Hokari et al., 2018 (Japan) [79]	N	Y	N	Y	N	Y	N	Y	N	Y	NS	NS
Hill et al., 2019 (Germany) [80]	N	Y	N	Y	N	Y	NS	NS	NS	NS	N	Y
Ahad et al., 2016 (India) [81]	N	Y	N	Y	Y	N	N	Y	NS	NS	NS	NS
Balata et al., 2013 (Brazil) [82]	N	Y	N	Y	N	Y	N	Y	N	Y	N	Y
Bechara et al., 2018 (Brazil) [83]	Y	N	Y	N	Y	N	NS	NS	NS	NS	Y	N
Bundidpun et al., 2017 (Thailand) [84]	N	Y	N	Y	Y	N	N	Y	Y	N	NS	NS
Chitsazi et al., 2014 (Iran) [85]	N	Y	N	Y	N	Y	N	Y	N	Y	N	Y

Table 5. *Cont.*

Study, Year, Origin and Citation	PPD		CAL		BOP/SBI		PI		GI		GR	
	Statistically Significant Y/N/NI/NS	Not Statistically Significant Y/N/NI/NS	Statistically Significant Y/N/NI/NS	Not Statistically Significant Y/N/NI/NS	Statistically Significant Y/N/NI/NS	Not Statistically Significant Y/N/NI/NS	Statistically Significant Y/N/NI/NS	Not Statistically Significant Y/N/NI/NS	Statistically Significant Y/N/NI/NS	Not Statistically Significant Y/N/NI/NS	Statistically Significant Y/N/NI/NS	Not Statistically Significant Y/N/NI/NS
Chitsazi et al., 2014 (Iran) [86]	N	Y	N	Y	N	Y	NS	NS	NS	NS	N	Y
Garcia et al., 2011 (Brazil) [87]	N	Y	N	Y	NS	NS	NS	NS	NS	NS	NS	NS
Joseph et al., 2014 (India) [88]	Y	N	Y	N	N	Y	N	Y	N	Y	NS	NS
Malgikar et al., 2015 (India) [89]	Y	N	Y	N	Y	N	Y	N	Y	N	NS	NS
Monzavi et al., 2016 (Iran) [90]	Y	N	N	Y	Y	N	N	Y	NS	NS	NS	NS
Raj et al., 2016 (India) [91]	Y	N	N	Y	NS	NS	Y	N	Y	N	NS	NS
Sena et al., 2019 (Brazil) [92]	N	Y	N	Y	N	Y	N	Y	NS	NS	NS	NS
Shingnapurkar et al., 2016 (India) [93]	Y	N	Y	N	NS	NS	N	Y	N	Y	NS	NS
Sigusch et al., 2010 (Germany) [94]	Y	N	Y	N	Y	N	Y	N	NS	NS	Y	N
Theodoro et al., 2012 (Brazil) [95]	N	Y	N	Y	N	Y	N	Y	N	Y	N	Y
Theodoro et al., 2017 (Brazil) [96]	N	Y	N	Y	N	Y	NS	NS	NS	NS	NS	NS

Table 6. Forest plots based on sensitivity analysis illustrating the overall PPD reduction and CAL gain at 3 months without outlier studies. Refer to Supplementary file 2 for a list of abbreviations.

Overall PPD Reduction for SM Studies at 3 Months

Study	SMD	SE	95% CI	Weight (%)
Chitsazi et al., 2014	0.525	0.289	−0.056 to 1.106	12.28
Moreira et al., 2015	0.205	0.311	−0.425 to 0.834	12.11
Franco et al., 2014	0.631	0.364	−0.115 to 1.378	11.67
Malgikar et al., 2015	0.119	0.284	−0.453 to 0.691	12.31
Ahad et al., 2016	0.639	0.204	0.235 to 1.043	12.88
Bundidpun et al., 2017	0.007	0.310	−0.620 to 0.635	12.11
Hill et al., 2019	0.039	0.072	−0.103 to 0.181	14.47
Braun et al., 2008	0.139	0.310	−0.490 to 0.767	12.11
Total (random effects)	0.282	0.234	−0.286 to 0.624	100.00

$P = 0.153$

Heterogeneity: Q = 9.14; DF = 7, $P = 0.71$; $I^2 = 0.00\%$

Overall PPD Reduction for PG Studies at 3 Months

Study	SMD	SE	95% CI	Weight (%)
Raj et al., 2016	0.669	0.441	−0.258 to 1.595	18.89
Vidal et al., 2017	−0.060	0.322	−0.714 to 0.593	25.59
Theodoro et al., 2017	−0.127	0.374	−0.897 to 0.643	22.43
Joseph et al., 2014	0.556	0.215	0.127 to 0.984	33.19
Total (random effects)	0.257	0.278	−0.230 to 0.683	100.00

$P = 0.361$

Heterogeneity: Q = 8.87; DF = 3; $P = 0.22$; $I^2 = 0.00\%$

Table 6. *Cont.*

Overall CAL Gain for SM Studies at 3 Months

Study	SMD	SE	95% CI	Weight (%)
Chitsazi et al., 2014	0.439	0.287	−0.140 to 1.017	5.52
Moreira et al., 2015	−0.040	0.310	−0.667 to 0.588	5.02
Franco et al., 2014	0.601	0.364	−0.144 to 1.346	4.20
Malgikar et al., 2015	−0.048	0.284	−0.620 to 0.523	5.60
Ahad et al., 2016	0.158	0.199	−0.237 to 0.552	10.35
Bundidpun et al., 2017	0.032	0.310	−0.595 to 0.660	5.02
Hill et al., 2019	0.019	0.072	−0.123 to 0.161	59.29
Braun et al., 2008	0.161	0.310	−0.468 to 0.789	5.01
Total (random effects)	0.162	0.253	−0.326 to 0.406	100.00

$P = 0.166$

Heterogeneity: Q = 8.40; DF = 7; P = 0.625; I^2 = 0.00%

Overall CAL Gain for PG Studies at 3 Months

Study	SMD	SE	95% CI	Weight (%)
Raj et al., 2016	0.669	0.441	−0.258 to 1.595	18.89
Vidal et al., 2017	−0.102	0.372	−0.514 to 0.793	25.59
Theodoro et al., 2017	−0.106	0.374	−0.997 to 0.743	22.43
Joseph et al., 2014	0.456	0.255	−0.120 to 0.673	33.19
Total (random effects)	0.227	0.352	−0.420 to 0.673	100.00

$P = 0.234$

Heterogeneity: Q = 9.7; DF = 3; P = 0.22; I^2 = 0.00%

3.4.3. Publication Bias

Visual inspection of funnel plots revealed noticeable asymmetry in SM-study analysis for PPD reduction indicating a probable risk of publication bias in SM studies included in this meta-analysis. However, slight asymmetries for PG studies were noted in corresponding funnel plots suggestive of a low risk of publication bias in the same (Figure 6).

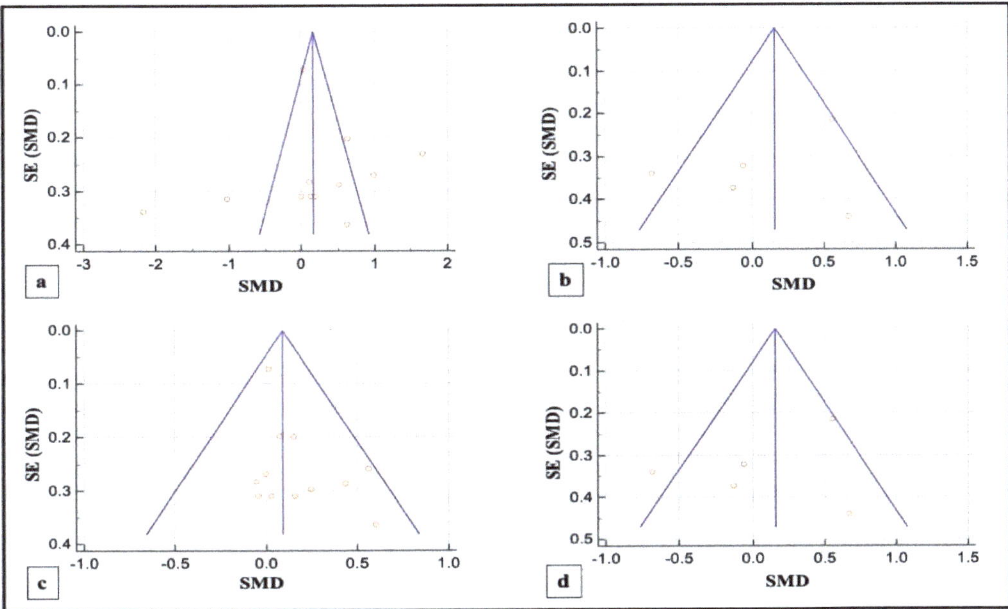

Figure 6. (**a,b**) Funnel plots illustrating the publication bias in overall PPD reduction in SM and PG studies, respectively; (**c,d**) funnel plots illustrating the publication bias in overall CAL gain in SM and PG studies, respectively. Each circle represents a single included study, the y-axis and x-axis represent the standard error of the effect estimate and the results of the study respectively and the graphical plot resembles an inverted funnel with scatter due to sampling variations.

4. Discussion

Based on the hypothesis that aPDT monotherapy or as an adjunct to NSPT can enhance the clinical or microbiological or immunological profile in comparison to conventional SRP, or SRP+AB, a critical appraisal of the available scientific evidence was conducted. After meticulous scrutiny, 31 studies were included in the present systematic review [2,3,5,17,70–96]. Owing to the methodological discrepancies, only 18 out 31 studies were eligible for a meta-analysis [17,71,73–76,80,81,84–86,88–93,96]. This report is the first to evaluate the role of SRP+aPDT compared to SRP alone or SRP+AB in SM and PG studies in AgP as well as in CP patients. The results of this meta-analysis indicated that, in comparison to SRP alone or SRP+AB, SRP+aPDT failed to show any additional benefit in the management of periodontitis up to six months. A significant heterogeneity was reported, arising from confounders in aPDT protocols. Subsequently, after omitting outlier studies [74,86,90,92,93], sensitivity analysis was able to eliminate heterogeneity completely but failed to report statistically significant improvements in primary outcome variables. Furthermore, risk of publication bias was reported indicating a possible selective outcome reporting in eligible published studies. In some instances, missing outcomes could not be detected by comparing the published report with the respective study protocol due to unavailability of the latter. Until now, seven meta-analyses have been reported to assess the role of aPDT in periodontitis [6–12]. The present review protocol is in accordance with the existing reviews [8,10–12]. Azaripour et al. is the only other systematic review and meta-analysis that has assessed the efficacy of SRP+aPDT as compared to SRP alone for SM and PG studies [8]. While three reviews report short-term benefits of aPDT up to 6 months [7–9], four have reported otherwise [6,10–12]. Our findings are in accordance with findings of the latter scientific reports. In order to gain an insight on merits and inadequacies of each included study, a comprehensive and systematic investigation was performed as follows:

4.1. Role of Baseline Characteristics

A key feature of RCTs is the application of balanced baseline characteristics in treatment arms of the trial in order to achieve unbiased treatment outcomes [97]. Most often, researchers provide a tabular representation of relevant variables to confirm an impartial baseline evaluation. In case of missing information on baseline characteristics, a 'selection bias' can be suspected [98]. All eligible studies have provided this vital information and were free from any kind of 'selection bias'. Additionally, evidence-based studies have suggested the potential harmful effects of smoking on the onset and progression of periodontitis, for which smokers were excluded [99,100]. Likewise, the inter-relationship of periodontitis and its systemic manifestations are well-established, resulting in the exclusion of patients with systemic diseases [101,102]. Utilization of 'placebo/sham' procedures to enhance clinical outcomes of the trial is an evidence-based verified concept [103]. In the present systematic review, only six out of 31 studies [71,78,92–94,96] have utilized a 'placebo/sham' procedure as an adjunct to conventional SRP. Furthermore, the role of SRP+aPDT+AB, compared to SRP+AB and SRP+aPDT as well the efficacy of SRP+PS compared to the conventional SRP and SRP+ aPDT have been assessed in this review [83,95]. Differences in study designs were apparent and the majority of clinicians have utilized the SM study design in oral health research [104]. Hence, this meta-analysis included both SM and PG studies in order to assess whether the estimated intervention effect has differed between them.

4.2. Assessment Methods for Various Parameters and Their Inferences to Determine aPDT Efficacy

A decrease in periodontal inflammation is directly proportional to a decrease in the incidence of BOP and detectable plaque levels [105]. Furthermore, the endpoints of a comprehensive periodontal therapy include PPD reduction and CAL gain, both of which are crucial for determining treatment success [106]. Clinical evidence has proven that there is a direct correlation between initial severity of PPD and CAL values and the amount of post-operative differences [107–109]. In terms of disease severity at baseline evaluation, a lack of homogeneity in pre-treatment values of PPD and CAL in the data extracted from various studies was observed (Table 1). Therefore, variations in level of significance across clinical parameters were noted, thus making it difficult to provide a cumulative result (Table 5). Utilization of a narrow laser optic fibre tip in deep periodontal pockets facilitates easy and atraumatic periodontal pocket probing ultimately resulting in the PPD and GR reduction post-operatively [3,88]. Furthermore, evidence-based research has proven that clinical outcome remains unaffected by the type of instrumentation utilized in SRP [76,110], which was observed to vary across all eligible studies (Table 1).

An imbalance amongst the local etiological factors such as dental plaque and calculus, inflammation, and a host defense system can have a great impact on the disease severity and progression [17]. Hence, it is essential to monitor the microbiological and the molecular changes of various growth factors and proinflammatory cytokines. De Oliveira and co-workers have demonstrated, through their studies, the importance of SRP in reducing bacterial load from tooth surface and the failure of aPDT monotherapy in reducing the bacterial counts of *A.a.* periodontal disease activity as well as bone resorption [2,5].

4.3. Representation of the Treatment Outcomes

Positive outcomes of any treatment strategy are governed by several factors such as evaluation of disease status at carefully planned follow-up visits, signs of uneventful healing, role of supportive periodontal therapy (SPT) and patient compliance with treatment [111–113]. The majority of included studies performed follow-up assessment for up to 3 or 6 months, whereas results of a longer follow-up ranging up to 1–2 years was lacking. Collective data obtained from longitudinal studies have confirmed the role of long-term follow up visits in greater reduction of clinical parameters. These findings have been confirmed by three studies included in the present review [73,77,95]. With regard to healing outcomes, 24 out of 31 studies reported no uneventful healing associated with the absence of any postoperative complications, after application of aPDT [2,3,5,70–75,77–81,83–90,92,95]. While six studies have failed to provide any information on healing outcomes [17,76,82,91,93,94], one study [96] has reported a gastrointestinal complication in one patient of the aPDT group. This happens to be unique evidence that has not been registered elsewhere. The presence of residual plaque and calculus resulting in a relapse of periodontal disease severity being inevitable with aPDT monotherapy. Hence, the role of SPT is quintessential. In the present systematic review, apart from the three sequential studies conducted by De Oliveira and co-workers [2,3,5] which utilized aPDT monotherapy, all of the other studies have utilized SRP+aPDT. In the former group of studies, a supragingival professional tooth cleaning was performed 14 days prior to the application of aPDT monotherapy in the treatment sites, which have extensively improved the post-operative clinical findings. Likewise, amongst the studies that have assessed the efficacy of SRP+aPDT, only seven studies [71,75,82,84,90,92,95] have mentioned a planned SPT protocol, which was implemented throughout their study period. Furthermore, information on oral hygiene instructions tailored according to respective study criteria has been specified in all included studies. Co-relation of patient's compliance in adhering to hygiene instructions cannot be overlooked, since it is vital for treatment success and prevention of disease recurrence [114]. Quantitative measurement of plaque by means various indices can aid in monitoring compliance towards therapy. In connection to this, the assessment of plaque levels was performed in 21 studies included in this review [3,17,71,73–75,77–79,81,82,84,85,88–94]. Therefore, an inconsistency in the representation of treatment outcomes in this review is evident.

4.4. Role of Laser Parameters

Apart from study methodology, laser parameters are crucial in determining treatment outcome. Calibration of therapeutic power output with a power-meter can aid in regulating low output power for achieving the desired aPDT effect at treatment sites [115]. This technique can also prevent any inadvertent damage caused by utilization of high output power, resulting in a photothermal effect. However, none of the included studies have utilized a power-meter to calibrate the power output. Furthermore, some other parameters that have been overlooked were: emission mode, contact/non-contact mode, energy/treated site, power output, spot size/fibre diameter, fluence and irradiance. Diode lasers have a high affinity to pigment, which, in the case of aPDT, is the PS. However, inflamed periodontal tissues are rich in blood and high levels of proteins, which are also rich in pigments. Traumatic instrumentation can lacerate the sulcus lining and elicit bleeding [23]. Consequently, the overall aPDT effect, which could be achieved by an effective PS dye-laser wavelength combination, will be compromised [23,116]. Additionally, placement of the fibre tip inside the gingival sulcus needs to be performed judiciously, in order to avoid further trauma to inflamed gingival sulcus, serving as a niche for plaque accumulation and favoring disease relapse [3,88].

Bacterial re-colonization after SRP has been proven to occur after three weeks [117] and, hence, multiple aPDT sessions can help to delay this pathological process. Annaji et al., 2016 [46] have compared the efficacy of three sessions (0, 7th and 21st day) versus a single session (0 day) of aPDT in the management of CP, and concluded that the group receiving multiple sessions demonstrated superior treatment outcomes. Nine studies included in this review conducted multiple aPDT sessions [17,71–75,79,90,96], and have unanimously concluded that the utilization of multiple aPDT sessions has positive healing effects. However, the frequency of aPDT application varies in all these studies (Table 2). As a result, a conclusion on the ideal number of sessions and frequency of application of aPDT cannot be drawn.

Additionally, PS concentration and pre-irradiation time (wash-out or PS incubation time) are governed by the binding capacity of PS to target cells [118]. It was observed in 13 out of 31 studies, whereby the PS concentration was not reported [5,70,72–74,76,81,84–86,91,93,94], whereas a range of PS concentrations have been utilized in the remaining eligible studies. Furthermore, four studies have failed to mention PS pre-irradiation time [5,75,77,90], whereas the remaining 27 studies have reported this time, which ranged from 1–5 min for different PS. An in vitro study by Fumes et al. [119] evaluated the effect of different PS pre-irradiation time periods (1, 2 and 5 min) in aPDT on the biofilms formed by *Streptococcus mutans* and *Candida albicans,* by monitoring the microbial load and have successfully demonstrated that the efficacy of all pre-irradiation times was equal. They emphasized patient discomfort associated with longer pre-irradiation times and thus have advised the use of shorter pre-irradiation times (1 min) in future studies. Up until now, there have been no studies that have determined the minimal duration of PS incubation as well as its role against periopathogens. Hence, further studies on this subject should be sought after. Moreover, discrepancies in the reported data, in terms of the number of sites receiving aPDT application per tooth as well the irradiation time, were noticed amongst the eligible studies (Table 2). These voids have raised concerns regarding the reliability of existing literature, which lacks a reproducible methodology and ultimately hampers the rational use of aPDT in non-surgical management of periodontitis.

4.5. Role of RoB Assessment

A vast majority of bias were raised from inadequate randomization, deviations from intended interventions and selective reporting of results (Figures 4 and 5). Another key finding of this systematic review was the presence of a potential conflict of interest in 10 out of the 31 studies [2,3,5,72–74,76,79,80,94]. Therefore, the results of the included studies are questionable, and their biased methodology cannot be relied upon.

4.6. Limitations of the Present Systematic Review

The majority of the included studies have assessed the efficacy of SRP+aPDT in comparison to aPDT monotherapy resulting in a lack of meta-analysis on the latter. Utilization of a limited number of teeth (mostly anterior teeth) or on specific sites (interproximal sites in case of deep pockets) or on any two quadrants of the dental arch that could facilitate cross contamination from the untreated sites and overshadow the putative benefits of aPDT was noted. In this review, efficacy of aPDT was monitored in systemically healthy non-smokers only, and its benefits in their immunocompromised counterparts were not established. A lack of long-term follow-up in order to determine the stability in healing after aPDT was also observed. The number of studies included in the meta-analysis was nearly half of the studies eligible for a systematic review due to paucity and inconsistency in available literature. Owing to the aforesaid drawbacks, the objective of this review could not be accomplished.

4.7. Future Scope

Future investigations should compare aPDT monotherapy versus SRP+aPDT and provide details of all appropriate laser and PS parameters. Efficacy of aPDT in smokers with various systemic diseases in CP and AgP should be established. The role of patient related outcomes and SPT should be emphasized upon. Nevertheless, future RCTs must have a robust methodology with balanced baseline characteristics, performed by experienced, masked and calibrated clinicians, which will reduce bias. Owing to the evident benefits of a PG-study design, clinicians should prefer its utilization in order to minimize potential 'carry-over' effects. Additionally, researchers should conduct a full mouth study protocol with a long-term follow-up assessment of minimum 6 months duration, consisting of, but not limited to, a vast range of clinical, microbiological, radiographic as well as, immune-histological profiles.

5. Conclusions

Within the limits of the present systematic review and meta-analysis, it can be concluded that the efficacy of aPDT in improving treatment outcomes, when it is utilized in the non-surgical management of periodontitis, remains debatable. However, the results of a majority of included studies have demonstrated the effectiveness of aPDT, and this role is more pronounced for SRP+aPDT rather than aPDT monotherapy. A careful and critical appraisal was performed which helped to obtain a qualitative assessment of eligible studies, thus highlighting the substantial flaws that prevent a reproducible methodology. Data on standardized aPDT study protocol, ideal PS dye-laser combination, optimal laser and PS parameters remain inconsistent and inconclusive amongst the prevalent literature owing to a highly inferior RoB in many studies. Finally, future research should aim for well-designed, robust and preferably PG-RCTs that will overcome the abovementioned limitations and confounders, in order to achieve palpable progress in this field of research while ensuring the use of an appropriate local laser safety protocol.

Supplementary Materials: The following are available online at https://www.mdpi.com/article/10.3390/pharmaceutics13060836/s1, PRISMA Checklist (Supplementary file 1), List of abbreviations (Supplementary file 2).

Author Contributions: Conceptualization, S.D., R.H.; methodology, S.D., R.H.; software, I.R.B., T.S.; validation, S.D., S.B., R.H., I.R.B.; formal analysis, S.B., I.R.B.; investigation, S.D., R.H.; resources, S.D., R.H., S.B., T.S., I.R.B.; data curation, S.D., R.H.; writing: original draft preparation, S.D., R.H.; writing: review and editing, S.D., R.H.; visualization, S.B., I.R.B., T.S.; supervision, S.D., R.H., S.B., T.S., I.R.B.; project administration, T.S., I.R.B.; funding acquisition, T.S. All authors have read and agreed to the published version of the manuscript.

Funding: This research received no external funding.

Institutional Review Board Statement: Not applicable.

Informed Consent Statement: Not applicable.

Conflicts of Interest: The authors declare no conflict of interest.

References

1. Rajesh, S.; Koshi, E.; Philip, K.; Mohan, A. Antimicrobial photodynamic therapy: An overview. *J. Indian Soc. Periodontol.* **2011**, *15*, 323–327. [CrossRef]
2. De Oliveira, R.R.; Schwartz-Filho, H.O.; Novaes, A.B., Jr.; de Souza, R.F.; Taba, M., Jr.; Scombatti de Souza, S.L.; Ribeiro, F.J. Antimicrobial photodynamic therapy in the non-surgical treatment of aggressive periodontitis: Cytokine profile in gingival crevicular fluid, preliminary results. *J. Periodontol.* **2009**, *80*, 98–105. [CrossRef] [PubMed]
3. De Oliveira, R.R.; Schwartz-Filho, H.O.; Novaes, A.B., Jr.; Taba, M., Jr. Antimicrobial photodynamic therapy in the non-surgical treatment of aggressive periodontitis: A preliminary randomized controlled clinical study. *J. Periodontol.* **2007**, *78*, 965–973. [CrossRef] [PubMed]
4. Van der Velden, U. What exactly distinguishes aggressive from chronic periodontitis: Is it mainly a difference in the degree of bacterial invasiveness? *Periodontology 2000* **2017**, *75*, 24–44. [CrossRef] [PubMed]
5. Novaes, A.B., Jr.; Schwartz-Filho, H.O.; de Oliveira, R.R.; Feres, M.; Sato, S.; Figueiredo, L.C. Antimicrobial photodynamic therapy in the non-surgical treatment of aggressive periodontitis: Microbiological profile. *Lasers Med. Sci.* **2012**, *27*, 389–395. [CrossRef] [PubMed]
6. Atieh, M.A. Photodynamic therapy as an adjunctive treatment for chronic periodontitis: A meta-analysis. *Lasers Med. Sci.* **2010**, *4*, 605–613. [CrossRef] [PubMed]
7. Sgolastra, F.; Petrucci, A.; Severino, M.; Graziani, F.; Gatto, R.; Monaco, A. Adjunctive photodynamic therapy to non-surgical treatment of chronic periodontitis: A systemic review and meta-analysis. *J. Clin. Periodontol.* **2013**, *40*, 514–526. [CrossRef] [PubMed]
8. Azaripour, A.; Dittrich, S.; Van Noorden, C.J.F.; Willershausen, C. Efficacy a photodynamic therapy as adjunct treatment of chronic periodontitis: A systematic review and meta-analysis. *Lasers Med. Sci.* **2018**, *33*, 407–423. [CrossRef]
9. Sgolastra, F.; Petrucci, A.; Gatto, R.; Mazzo, G.; Monaco, A. Photodynamic therapy in the treatment of chronic periodontitis: A systematic review and meta-analysis. *Lasers Med. Sci.* **2013**, *28*, 669–682. [CrossRef]

10. Souza, E.; Medeiros, A.C.; Gurgel, B.C.; Sarmento, C. Antimicrobial photodynamic therapy in the treatment of aggressive periodontitis: A systematic review and meta-analysis. *Lasers Med. Sci.* **2016**, *31*, 187–196. [CrossRef] [PubMed]

11. Azarpazhooh, A.; Shah, P.S.; Tenenbaum, H.C.; Goldberg, M.B. The effect of photodynamic therapy for periodontitis: A systematic review and meta-analysis. *J. Periodontol.* **2010**, *81*, 4–14. [CrossRef]

12. Akram, Z.; Hyder, T.; Al-Hammoudi, N.; Binshabaib, M.S.; Alharthi, S.S.; Hanif, A. Efficacy of photodynamic therapy versus antibiotics as an adjunct to scaling and root planing in the treatment of periodontitis: A systematic review and meta-analysis. *Photodiagnosis Photodyn. Ther.* **2017**, *19*, 86–92. [CrossRef] [PubMed]

13. Vohra, F.; Akram, Z.; Safii, S.H.; Vaithilingam, R.D.; Ghanem, A.; Sergis, K.; Javed, F. Role of antimicrobial photodynamic therapy in the treatment of aggressive periodontitis: A systematic review. *Photodiagnosis Photodyn. Ther.* **2016**, *13*, 139–147. [CrossRef] [PubMed]

14. Chatzopoulos, G.S.; Doufexi, A.E. Photodynamic therapy in the treatment of aggressive periodontitis: A systematic review. *Med. Oral Patol. Oral Cir. Bucal* **2016**, *21*, e192–e200. [CrossRef] [PubMed]

15. Reddy, P.; Reddy, C.; Krishna Kumar, R.V.S.; Sudhir, K.M.; Srinivasulu, G. Effect of photodynamic therapy on aggressive periodontitis—A systematic review. *Int. J. Sci. Res. Manag.* **2017**, *5*, 7177–7185.

16. Pal, A.; Paul, S.; Perry, R.; Puryer, J. Is the use of antimicrobial photodynamic therapy or systemic antibiotics more effective in improving periodontal health when used in conjunction with localised non-surgical periodontal therapy? A systematic review. *Dent. J.* **2019**, *7*, 108. [CrossRef]

17. Franco, E.J.; Pogue, R.E.; Sakamoto, L.H.; Cavalcante, L.L.; Carvalho, D.R.; de Andrade, R.V. Increased expression of genes after periodontal treatment with photodynamic therapy. *Photodiagnosis Photodyn. Ther.* **2014**, *11*, 41–47. [CrossRef] [PubMed]

18. Van Assche, N.; Van Essche, M.; Pauwels, M.; Teughels, W.; Quirynen, M. Do periodontopathogens disappear after full-mouth tooth extraction? *J. Clin. Periodontol.* **2009**, *36*, 1043–1047. [CrossRef]

19. Guerrero, A.; Griffiths, G.S.; Nibali, L.; Suvan, J.; Moles, D.R.; Laurell, L.; Tonetti, M.S. Adjunctive benefits of systemic amoxicillin and metronidazole in non-surgical treatment of generalized aggressive periodontitis: A randomized placebo-controlled clinical trial. *J. Clin. Periodontol.* **2005**, *32*, 1096–1107. [CrossRef]

20. Sigusch, B.; Beier, M.; Klinger, G.; Pfister, W.; Glockmann, E. A 2-step non-surgical procedure and systemic antibiotics in the treatment of rapidly progressive periodontitis. *J. Periodontol.* **2001**, *72*, 275–283. [CrossRef]

21. Herrera, D.; Sanz, M.; Jepsen, S.; Needleman, I.; Roldán, S. A systematic review on the effect of systemic antimicrobials as an adjunct to scaling and root planing in periodontitis patients. *J. Clin. Periodontol.* **2002**, *29*, 136–162. [CrossRef]

22. Pallasch, T.J. Antibiotic resistance. *Dent. Clin. N. Am.* **2003**, *47*, 623–639. [CrossRef]

23. Andersen, R.C.; Loebel, N.G. Photodynamic disinfection in the treatment of chronic adult periodontitis: A multicenter clinical trial. *J. Dent. Health Oral Disord. Ther.* **2017**, *8*, 00289. [CrossRef]

24. Moher, D.; Liberati, A.; Tetzlaff, J.; Altman, D.G.; PRISMA Group. Preferred reporting items for systematic reviews and meta-analyses: The PRISMA statement. *BMJ* **2009**, *339*, b2535. [CrossRef] [PubMed]

25. Higgins, J.P.T.; Green, S. Cochrane Handbook for Systematic Reviews of Interventions Version 5.1.0 The Cochrane Collaboration. 2011. Available online: http://www.cochrane-handbook.org (accessed on 12 February 2021).

26. Armitage, G.C. Development of a classification system for periodontal diseases and conditions. *Ann. Periodontol.* **1999**, *4*, 1–6. [CrossRef] [PubMed]

27. McHugh, M.L. Inter-rate reliability: The kappa statistic. *Biochem. Med.* **2012**, *22*, 276–282. [CrossRef]

28. Sterne, J.A.C.; Savović, J.; Page, M.J.; Elbers, R.G.; Blencowe, N.S.; Boutron, I.; Cates, C.J.; Cheng, H.Y.; Corbett, M.S.; Eldridge, S.M.; et al. RoB 2: A revised tool for assessing risk of bias in randomised trials. *BMJ* **2019**, *366*, l4898. [CrossRef] [PubMed]

29. Altman, D.G.; Schulz, K.F.; Moher, D.; Egger, M.; Davidoff, F.; Elbourne, D.; Gøtzsche, P.C.; Lang, T.; CONSORT GROUP (Consolidated Standards of Reporting Trials). The revised CONSORT statement for reporting randomized trials: Explanation and elaboration. *Ann. Intern Med.* **2001**, *134*, 663–694. [CrossRef]

30. Higgins, J.P.T.; Eldridge, S.; Li, T. Chapter 23: Including variants on randomized trials. In *Cochrane Handbook for Systematic Reviews of Interventions Version 6.0*; Higgins, J.P.T., Thomas, J., Chandler, J., Cumpston, M., Li, T., Page, M.J., Welch, V.A., Eds.; (updated July 2019); Cochrane: Oxford, UK, 2019. Available online: www.training.cochrane.org/handbook (accessed on 12 February 2021).

31. Hozo, S.P.; Djulbegovic, B.; Hozo, I. Estimating the mean and variance from the median, range, and the size of a sample. *BMC Med. Res. Methodol.* **2005**, *5*, 13. [CrossRef]

32. Lau, J.; Ioannidis, J.P.; Schmid, C.H. Quantitative synthesis in systematic reviews. *Ann. Intern Med.* **1997**, *127*, 820–826. [CrossRef] [PubMed]

33. Higgins, J.P.T.; Thompson, S.G. Quantifying heterogeneity in a meta-analysis. *Stat. Med.* **2002**, *21*, 1539–1558. [CrossRef] [PubMed]

34. Sterne, J.A.C.; Egger, M. Funnel plots for detecting bias in meta-analysis: Guidelines on choice of axis. *J. Clin. Epidemiol.* **2001**, *54*, 1046–1055. [CrossRef]

35. Theodoro, L.H.; Assem, N.Z.; Longo, M.; Alves, M.L.F.; Duque, C.; Stipp, R.N.; Vizoto, N.L.; Garcia, V.G. Treatment of periodontitis in smokers with multiple sessions of antimicrobial photodynamic therapy or systemic antibiotics: A randomized clinical trial. *Photodiagnosis Photodyn. Ther.* **2018**, *22*, 217–222. [CrossRef] [PubMed]

36. Al-Zahrani, M.S.; Austah, O.N. Photodynamic therapy as an adjunctive to scaling and root planing in treatment of chronic periodontitis in smokers. *Saudi Med. J.* **2011**, *32*, 1183–1188. [PubMed]

37. Tabenski, L.; Moder, D.; Cieplik, F.; Schenke, F.; Hiller, K.A.; Buchalla, W.; Schmalz, G.; Christgau, M. Antimicrobial photodynamic therapy vs. local minocycline in addition to non-surgical therapy of deep periodontal pockets: A controlled randomized clinical trial. *Clin. Oral Investig.* **2017**, *21*, 2253–2264. [CrossRef] [PubMed]
38. Ge, L.; Shu, R.; Li, Y.; Li, C.; Luo, L.; Song, Z.; Xie, Y.; Liu, D. Adjunctive effect of photodynamic therapy to scaling and root planing in the treatment of chronic periodontitis. *Photomed. Laser Surg.* **2011**, *29*, 33–37. [CrossRef] [PubMed]
39. Loebel, N.G.; Galler, C.C.; Andersen, R.C. Antimicrobial photodynamic therapy in the treatment of chronic adult periodontitis. *Oral Health Dental Sci.* **2018**, *2*, 1–7. [CrossRef]
40. Harmouche, L.; Courval, A.; Mathieu, A.; Peti, C.; Huck, O.; Severac, F.; Davideau, J.L. Impact of tooth-related factors on photodynamic therapy effectiveness during active periodontal therapy: A 6-months split-mouth randomized clinical trial. *Photodiagnosis Photodyn. Ther.* **2019**, *27*, 167–172. [CrossRef] [PubMed]
41. Christodoulides, N.; Nikolidakis, D.; Chondros, P.; Becker, J.; Schwarz, F.; Rossler, R.; Sculean, A. Photodynamic therapy as an adjunct to non-surgical periodontal treatment: A randomized, controlled clinical trial. *J. Periodontol.* **2008**, *79*, 1638–1644. [CrossRef]
42. Polansky, R.; Haas, M.; Heschl, A.; Wimmer, G. Clinical effectiveness of photodynamic therapy in the treatment of periodontitis. *J. Clin. Periodontol.* **2009**, *36*, 575–580. [CrossRef]
43. Badea, M.E.; Serbanescu, A.; Hedesiu, M.; Badea, A.F. Photodynamic Laser therapy in patients with periodontitis. *TMJ* **2010**, *60*, 18–22.
44. Andersen, R.; Loebel, N.; Hammond, D.; Wilson, M. Treatment of periodontal disease by photodisinfection compared to scaling and root planing. *J. Clin. Dent.* **2007**, *18*, 34–38.
45. Alwaeli, H.A.; Al-Khateeb, S.N.; Al-Sadi, A. Long-term clinical effect of adjunctive antimicrobial photodynamic therapy in periodontal treatment: A randomized clinical trial. *Lasers Med. Sci.* **2015**, *30*, 801–807. [CrossRef] [PubMed]
46. Annaji, S.; Sarkar, I.; Rajan, P.; Pai, J.; Malagi, S.; Bharmappa, R.; Kamath, V. Efficacy of photodynamic therapy and lasers as an adjunct to scaling and root planing in the treatment of aggressive periodontitis—A clinical and microbiologic short term study. *J. Clin. Diagn. Res.* **2016**, *10*, ZC08–ZC12. [CrossRef] [PubMed]
47. Dilsiz, A.; Canakci, V.; Aydin, T. Clinical effects of potassium-titanyl-phosphate laser and photodynamic therapy on outcomes of treatment of chronic periodontitis: A randomized controlled clinical trial. *J. Periodontol.* **2013**, *84*, 278–286. [CrossRef] [PubMed]
48. Lui, J.; Corbet, E.F.; Jin, L. Combined photodynamic and low-level laser therapies as an adjunct to nonsurgical treatment of chronic periodontitis. *J. Periodontal Res.* **2011**, *46*, 89–96. [CrossRef] [PubMed]
49. Mistry, A.; Pereira, R.; Kini, V.; Padhye, A. Effect of combined therapy using diode laser and photodynamic therapy on levels of IL-17 in gingival crevicular fluid in patients with chronic periodontitis. *J. Lasers Med. Sci.* **2016**, *7*, 250–255. [CrossRef] [PubMed]
50. Srikanth, K.; Chandra, R.V.; Reddy, A.A.; Reddy, B.H.; Reddy, C.; Naveen, A. Effect of a single session of antimicrobial photodynamic therapy using indocyanine green in the treatment of chronic periodontitis: A randomized controlled pilot trial. *Quintessence Int.* **2015**, *46*, 391–400. [CrossRef] [PubMed]
51. Grzech-Leśniak, K.; Matys, J.; Dominiak, M. Comparison of the clinical and microbiological effects of antibiotic therapy in periodontal pockets following laser treatment: An in vivo study. *Adv. Clin. Exp. Med.* **2018**, *27*, 1263–1270. [CrossRef] [PubMed]
52. Kamma, J.J.; Vasdekis, V.G.; Romanos, G.E. The effect of diode laser (980 nm) treatment on aggressive periodontitis: Evaluation of microbial and clinical parameters. *Photomed. Laser Surg.* **2009**, *27*, 11–19. [CrossRef]
53. Matarese, G.; Ramaglia, L.; Cicciù, M.; Cordasco, G.; Isola, G. The effects of diode laser therapy as an adjunct to scaling and root planing in the treatment of aggressive periodontitis: A 1-year randomized controlled clinical trial. *Photomed. Laser Surg.* **2017**, *35*, 702–709. [CrossRef] [PubMed]
54. Annaji, S.; Sarkar, I.; Rajan, P.; Pai, J.; Malagi, S.; Kamath, V.; Barmappa, R. Comparative evaluation of the efficacy of curcumin gel with and without photo activation as an adjunct to scaling and root planing in the treatment of chronic periodontitis: A split mouth clinical and microbiological study. *J. Nat. Sci. Biol. Med.* **2015**, *6*, S102–S109. [CrossRef]
55. Borekci, T.; Meseli, S.E.; Noyan, U.; Kuru, B.E.; Kuru, L. Efficacy of adjunctive photodynamic therapy in the treatment of generalized aggressive periodontitis: A randomized controlled clinical trial. *Lasers Surg. Med.* **2019**, *51*, 167–175. [CrossRef] [PubMed]
56. Pulikkotil, S.J.; Toh, C.G.; Mohandas, K.; Leong, K. Effect of photodynamic therapy adjunct to scaling and root planing in periodontitis patients: A randomized clinical trial. *Aust. Dent. J.* **2016**, *61*, 440–445. [CrossRef]
57. Saitawee, D.; Teerakapong, A.; Morales, N.P.; Jitprasertwong, P.; Hormdee, D. Photodynamic therapy of Curcuma longa extract stimulated with blue light against Aggregatibacter actinomycetemcomitans. *Photodiagnosis Photodyn. Ther.* **2018**, *22*, 101–105. [CrossRef]
58. Bassir, S.H.; Moslemi, N.; Jamali, R.; Mashmouly, S.; Fekrazad, R.; Chiniforush, N.; Shamshiri, A.R.; Nowzari, H. Photoactivated disinfection using light-emitting diode as an adjunct in the management of chronic periodontitis: A pilot double-blind split-mouth randomized clinical trial. *J. Clin. Periodontol.* **2013**, *40*, 65–72. [CrossRef] [PubMed]
59. Husejnagic, S.; Lettner, S.; Laky, M.; Georgopoulos, A.; Moritz, A.; Rausch-Fan, X. Photoactivated disinfection in periodontal treatment: A randomized controlled clinical split-mouth trial. *J. Periodontol.* **2019**, *90*, 1260–1269. [CrossRef] [PubMed]
60. Maria, A.M.; Ursarescu, I.G.; Solomon, S.; Foia, L. Evaluation the effects of led photo-activated disinfection on periodontal clinical parameters in patients with chronic periodontitis. *Balk. J. Dent. Med.* **2016**, *20*, 29–32. [CrossRef]

61. Campos, G.N.; Pimentel, S.P.; Ribeiro, F.V.; Casarin, R.C.; Cirano, F.R.; Saraceni, C.H.; Casati, M.Z. The adjunctive effect of photodynamic therapy for residual pockets in single- rooted teeth: A randomized controlled clinical trial. *Lasers Med. Sci.* **2013**, *28*, 317–324. [CrossRef] [PubMed]

62. Corrêa, M.G.; Oliveira, D.H.; Saraceni, C.H.; Ribeiro, F.V.; Pimentel, S.P.; Cirano, F.R.; Casarin, R.C. Short-term microbiological effects of photodynamic therapy in non-surgical periodontal treatment of residual pockets: A split-mouth RCT. *Lasers Surg. Med.* **2016**, *48*, 944–950. [CrossRef] [PubMed]

63. Petelin, M.; Perkič, K.; Seme, K.; Gašpirc, B. Effect of repeated adjunctive antimicrobial photodynamic therapy on subgingival periodontal pathogens in the treatment of chronic periodontitis. *Lasers Med. Sci.* **2015**, *30*, 1647–1656. [CrossRef] [PubMed]

64. Cappuyns, I.; Cionca, N.; Wick, P.; Giannopoulou, C.; Mombelli, A. Treatment of residual pockets with photodynamic therapy, diode laser, or deep scaling. A randomized, split-mouth controlled clinical trial. *Lasers Med. Sci.* **2012**, *27*, 979–986. [CrossRef] [PubMed]

65. Lulic, M.; Leiggener, G.I.; Salvi, G.E.; Ramseier, C.A.; Mattheos, N.; Lang, N.P. One-year outcomes of repeated adjunctive photodynamic therapy during periodontal maintenance: A proof-of-principle randomized-controlled clinical trial. *J. Clin. Periodontol.* **2009**, *36*, 661–666. [CrossRef] [PubMed]

66. Rühling, A.; Fanghänel, J.; Houshmand, M.; Kuhr, A.; Meisel, P.; Schwahn, C.; Kocher, T. Photodynamic therapy of persistent pockets in maintenance patients-A clinical study. *Clin. Oral Investig.* **2010**, *14*, 637–644. [CrossRef]

67. Barbosa, F.I.; Araújo, P.V.; Machado, L.J.C.; Magalhães, C.S.; Guimarães, M.M.; Moreira, A.N. Effect of photodynamic therapy as an adjuvant to non-surgical periodontal therapy: Periodontal and metabolic evaluation in patients with type 2 diabetes mellitus. *Photodiagnosis Photodyn. Ther.* **2018**, *22*, 245–250. [CrossRef] [PubMed]

68. Ramos, U.D.; Ayub, L.G.; Reino, D.M.; Grisi, M.F.; Taba, M., Jr.; Souza, S.L.; Palioto, D.B.; Novaes, A.B., Jr. Antimicrobial photodynamic therapy as an alternative to systemic antibiotics: Results from a double-blind, randomized, placebo-controlled, clinical study on type 2 diabetics. *J. Clin. Periodontol.* **2016**, *43*, 147–155. [CrossRef]

69. Alvarenga, L.H.; Gomes, A.C.; Carribeiro, P.; Godoy-Miranda, B.; Noschese, G.; Simões Ribeiro, M.; Kato, I.T.; Bussadori, S.K.; Pavani, C.; Geraldo, Y.G.; et al. Parameters for antimicrobial photodynamic therapy on periodontal pocket- Randomized clinical trial. *Photodiagnosis Photodyn. Ther.* **2019**, *27*, 132–136. [CrossRef] [PubMed]

70. Pourabbas, R.; Kashefimehr, A.; Rahmanpour, N.; Babaloo, Z.; Kishen, A.; Tenenbaum, H.C.; Azarpazhooh, A. Effects of photodynamic therapy on clinical and gingival crevicular fluid inflammatory biomarkers in chronic periodontitis: A split-mouth randomized clinical trial. *J. Periodontol.* **2014**, *85*, 1222–1229. [CrossRef] [PubMed]

71. Moreira, A.L.; Novaes, A.B., Jr.; Grisi, M.F.; Taba, M., Jr.; Souza, S.L.; Palioto, D.B.; de Oliveira, P.G.; Casati, M.Z.; Casarin, R.C.; Messora, M.R. Antimicrobial photodynamic therapy as an adjunct to non-surgical treatment of aggressive periodontitis: A split-mouth randomized controlled trial. *J. Periodontol.* **2015**, *86*, 376–386. [CrossRef]

72. Skurska, A.; Dolinska, E.; Pietruska, M.; Pietruski, J.K.; Dymicka, V.; Kemona, H.; Arweiler, N.B.; Milewsk, R.; Sculean, A. Effect of nonsurgical periodontal treatment in conjunction with either systemic administration of amoxicillin and metronidazole or additional photodynamic therapy on the concentration of matrix metalloproteinases 8 and 9 in gingival crevicular fluid in patients with aggressive periodontitis. *BMC Oral Health* **2015**, *15*, 63. [CrossRef]

73. Arweiler, N.B.; Pietruska, M.; Pietruski, J.; Skurska, A.; Dolińsk, E.; Heumann, C.; Auschill, T.M.; Sculean, A. Six-month results following treatment of aggressive periodontitis with antimicrobial photodynamic therapy or amoxicillin and metronidazole. *Clin. Oral Investig.* **2014**, *18*, 2129–2135. [CrossRef] [PubMed]

74. Arweiler, N.B.; Pietruska, M.; Skurska, A.; Dolińska, E.; Pietruski, J.K.; Bläs, M.; Auschill, T.M.; Sculean, A. Nonsurgical treatment of aggressive periodontitis with photodynamic therapy or systemic antibiotics. Three-month results of a randomized, prospective, controlled clinical study. *Schweiz. Monatsschr. Zahnmed.* **2013**, *123*, 532–544. [PubMed]

75. Segarra-Vidal, M.; Guerra-Ojeda, S.; Vallés, L.S.; López-Roldán, A.; Mauricio, M.D.; Aldasoro, M.; Alpiste-Illueca, F.; Vila, J.M. Effects of photodynamic therapy in periodontal treatment: A randomized, controlled clinical trial. *J. Clin. Periodontol.* **2017**, *44*, 915–925. [CrossRef] [PubMed]

76. Braun, A.; Dehn, C.; Krause, F.; Jepsen, S. Short-term clinical effects of adjunctive antimicrobial photodynamic therapy in periodontal treatment: A randomized clinical trial. *J. Clin. Periodontol.* **2008**, *35*, 877–884. [CrossRef] [PubMed]

77. Berakdar, M.; Callaway, A.; Eddin, M.F.; Ross, A.; Willershausen, B. Comparison between scaling-root-planing (SRP) and SRP/Photodynamic therapy: Six-month study. *Head Face Med.* **2012**, *8*, 12. [CrossRef]

78. Raut, C.P.; Sethi, K.S.; Kohale, B.R.; Mamajiwala, A.; Warang, A. Indocyanine green- mediated photothermal therapy in treatment of chronic periodontitis: A clinico- microbiological study. *J. Indian Soc. Periodontol.* **2018**, *22*, 221–227. [CrossRef]

79. Hokari, T.; Morozumi, T.; Komatsu, Y.; Shimizu, T.; Yoshino, T.; Tanaka, M.; Tanaka, Y.; Nohno, K.; Kubota, T.; Yoshie, H. Effects of antimicrobial photodynamic therapy and local administration of minocycline on clinical, microbiological, and inflammatory markers of periodontal pockets: A Pilot Study. *Int. J. Dent.* **2018**, 1748584. [CrossRef] [PubMed]

80. Hill, G.; Dehn, C.; Hinze, A.V.; Frentzen, M.; Meister, J. Indocyanine green-based adjunctive antimicrobial photodynamic therapy for treating chronic periodontitis: A randomized clinical trial. *Photodiagnosis Photodyn. Ther.* **2019**, *26*, 29–35. [CrossRef]

81. Ahad, A.; Lamba, A.K.; Faraz, F.; Tandon, S.; Chawla, K.; Yadav, N. Effect of antimicrobial photodynamic therapy as an adjunct to nonsurgical treatment of deep periodontal pockets: A clinical study. *J. Lasers Med. Sci.* **2016**, *7*, 220–226. [CrossRef]

82. Balata, M.L.; Andrade, L.P.; Santos, D.B.; Cavalcanti, A.N.; Tunes Uda, R.; Ribeiro Édel, P.; Bittencourt, S. Photodynamic therapy associated with full-mouth ultrasonic debridement in the treatment of severe chronic periodontitis: A randomized-controlled clinical trial. *J. Appl. Oral Sci.* **2013**, *21*, 208–214. [CrossRef] [PubMed]

83. Bechara Andere, N.M.R.; Dos Santos, N.C.C.; Araujo, C.F.; Mathias, I.F.; Rossato, A.; de Marco, A.C.; Santamaria, M., Jr.; Jardini, M.A.N.; Santamaria, M.P. Evaluation of the local effect of nonsurgical periodontal treatment with and without systemic antibiotic and photodynamic therapy in generalized aggressive periodontitis. A randomized clinical trial. *Photodiagnosis Photodyn. Ther.* **2018**, *24*, 115–120. [CrossRef] [PubMed]

84. Bundidpun, P.; Srisuwantha, R.; Laosrisin, N. Clinical effects of photodynamic therapy as an adjunct to full-mouth ultrasonic scaling and root planing in treatment of chronic periodontitis. *Laser Ther.* **2018**, *27*, 33–39. [CrossRef] [PubMed]

85. Chitsazi, M.T.; Shirmohammadi, A.; Pourabbas, R.; Abolfazli, N.; Farhoudi, I.; Azar, B.D.; Farhadi, F. Clinical and microbiological effects of photodynamic therapy associated with non- surgical treatment in aggressive periodontitis. *J. Dent. Res. Dent. Clin. Dent. Prospects* **2014**, *8*, 153–159. [CrossRef] [PubMed]

86. Chitsazi, M.T.; Kashefimehr, A.; Pourabbas, R.; Shirmohammadi, A.; Ghasemi Barghi, V.; Daghigh Azar, B. Efficacy of subgingival application of xanthan-based chlorhexidine gel adjunctive to full-mouth root planing assessed by real-time PCR: A Microbiologic and Clinical Study. *J. Dent. Res. Dent. Clin. Dent. Prospects* **2013**, *7*, 95–101. [CrossRef] [PubMed]

87. Garcia, F.B.; Dias, A.T.; Tinoco, B.E.M.; Fischeer, R.G. Evaluation of the efficacy of photodynamic therapy as adjunct to periodontal treatment in patients with aggressive periodontitis. *Rev. Periodontia* **2011**, *21*, 12–19.

88. Betsy, J.; Prasanth, C.S.; Baiju, K.V.; Prasanthila, J.; Subhash, N. Efficacy of antimicrobial photodynamic therapy in the management of chronic periodontitis: A randomized controlled clinical trial. *J. Clin. Periodontol.* **2014**, *4*, 573–581. [CrossRef] [PubMed]

89. Malgikar, S.; Reddy, S.H.; Babu, P.R.; Sagar, S.V.; Kumar, P.S.; Reddy, G.J. A randomized controlled clinical trial on efficacy of photodynamic therapy as an adjunct to nonsurgical treatment of chronic periodontitis. *J. Dent. Lasers* **2015**, *9*, 75–79. [CrossRef]

90. Monzavi, A.; Chinipardaz, Z.; Mousavi, M.; Fekrazad, R.; Moslemi, N.; Azaripour, A.; Bagherpasand, O.; Chiniforush, N. Antimicrobial photodynamic therapy using diode laser activated indocyanine green as an adjunct in the treatment of chronic periodontitis: A randomized clinical trial. *Photodiagnosis Photodyn. Ther.* **2016**, *14*, 93–97. [CrossRef] [PubMed]

91. Raj, K.R.; Musalaiah, S.; Nagasri, M.; Kumar, P.A.; Reddy, P.I.; Greeshma, M. Evaluation of efficacy of photodynamic therapy as an adjunct to nonsurgical periodontal therapy in treatment of chronic periodontitis patients: A clinico-microbiological study. *Indian J. Dent. Res.* **2016**, *27*, 483–487. [CrossRef] [PubMed]

92. Sena, I.A.A.; Silva, D.N.A.; Azevedo, M.L.D.S.; da Silva, N.T.; Longo, J.P.F.; de Moraes, M.; de Aquino Martins, A.R.L. Antimicrobial photodynamic therapy using a chloro-aluminum phthalocyanine adjuvant to nonsurgical periodontal treatment does not improve clinical parameters in patients with chronic periodontitis. *Photobiomodul. Photomed. Laser Surg.* **2019**, *37*, 729–735. [CrossRef]

93. Shingnapurkar, S.H.; Mitra, D.K.; Kadav, M.S.; Shah, R.A.; Rodrigues, S.V.; Prithyani, S.S. The effect of indocyanine green-mediated photodynamic therapy as an adjunct to scaling and root planing in the treatment of chronic periodontitis: A comparative split-mouth randomized clinical trial. *Indian J. Dent. Res.* **2016**, *27*, 609–617. [CrossRef] [PubMed]

94. Sigusch, B.W.; Engelbrecht, M.; Völpel, A.; Holletschke, A.; Pfister, W.; Schütze, J. Full-mouth antimicrobial photodynamic therapy in Fusobacterium nucleatum-infected periodontitis patients. *J. Periodontol.* **2010**, *81*, 975–981. [CrossRef] [PubMed]

95. Theodoro, L.H.; Silva, S.P.; Pires, J.R.; Soares, G.H.G.; Pontes, A.E.F.; Zuza, E.P.; Spolidório, D.M.; de Toledo, B.E.; Garcia, V.G. Clinical and microbiological effects of photodynamic therapy associated with nonsurgical periodontal treatment: A 6-month follow-up. *Lasers Med. Sci.* **2012**, *27*, 687–693. [CrossRef]

96. Theodoro, L.H.; Lopes, A.B.; Nuernberg, M.A.A.; Cláudio, M.M.; Miessi, D.M.J.; Alves, M.L.F.; Duque, C.; Mombelli, A.; Garcia, V.G. Comparison of repeated applications of aPDT with amoxicillin and metronidazole in the treatment of chronic periodontitis: A short-term study. *J. Photochem. Photobiol. B* **2017**, *174*, 364–369. [CrossRef] [PubMed]

97. Festic, E.; Rawal, B.; Gajic, O. How to improve assessment of balance in baseline characteristics of clinical trial participants-example from PROSEVA trial data? *Ann. Transl. Med.* **2016**, *4*, 79. [CrossRef] [PubMed]

98. Tripepi, G.; Jager, K.J.; Dekker, F.W.; Zoccali, C. Selection bias and information bias in clinical research. *Nephron. Clin. Pract.* **2010**, *115*, c94–c99. [CrossRef] [PubMed]

99. Johnson, G.K.; Hill, M. Cigarette smoking and the periodontal patient. *J. Periodontol.* **2004**, *75*, 196–209. [CrossRef]

100. Leite, F.R.M.; Nascimento, G.G.; Scheutz, F.; López, R. Effect of smoking on Periodontitis: A Systematic Review and Meta-regression. *Am. J. Prev. Med.* **2018**, *54*, 831–841. [CrossRef]

101. Arigbede, A.O.; Babatope, B.O.; Bamidele, M.K. Periodontitis and systemic diseases: A literature review. *J. Indian Soc. Periodontol.* **2012**, *16*, 487–491. [CrossRef] [PubMed]

102. Komine-Aizawa, S.; Aizawa, S.; Hayakawa, S. Periodontal diseases and adverse pregnancy outcomes. *J. Obstet.* **2019**, *45*, 5–12. [CrossRef]

103. Blasini, M.; Movsas, S.; Colloca, L. Placebo hypoalgesic effects in pain: Potential applications in dental and orofacial pain management. *Semin. Orthod.* **2018**, *24*, 259–268. [CrossRef] [PubMed]

104. Smaïl-Faugeron, V.; Fron-Chabouis, H.; Courson, F.; Durieux, P. Erratum to: Comparison of intervention effects in split-mouth and parallel-arm randomized controlled trials: A meta-epidemiological study. *BMC Med. Res. Methodol.* **2015**, *15*, 72. [CrossRef] [PubMed]

105. Lang, N.P.; Joss, A.; Orsanic, T.; Gusberti, F.A.; Siegrist, B.E. Bleeding on probing. A predictor for the progression of periodontal disease? *J. Clin. Periodontol.* **1986**, *13*, 590–596. [CrossRef] [PubMed]
106. Greenstein, G. Contemporary interpretation of probing depth assessments: Diagnostic and therapeutic implications. A literature review. *J. Periodontol.* **1997**, *68*, 1194–1205. [CrossRef]
107. Ramfjord, S.P.; Caffesse, R.G.; Morrison, E.C.; Hill, R.W.; Kerry, G.J.; Appleberry, E.A.; Nissle, R.R.; Stults, D.L. 4 modalities of periodontal treatment compared over 5 years. *J. Clin. Periodontol.* **1987**, *14*, 445–452. [CrossRef]
108. Kaldahl, W.B.; Kalkwarf, K.L.; Patil, K.D.; Molvar, M.P.; Dyer, J.K. Long-term evaluation of periodontal therapy: I. Response to 4 therapeutic modalities. *J. Periodontol.* **1996**, *67*, 93–102. [CrossRef] [PubMed]
109. Mailoa, J.; Lin, G.H.; Khoshkam, V.; MacEachern, M.; Chan, H.L.; Wang, H.L. Long-term effect of four surgical periodontal therapies and one non-surgical therapy: A Systematic Review and Meta-Analysis. *J. Periodontol.* **2015**, *86*, 1150–1158. [CrossRef]
110. Singh, S.; Uppoor, A.; Nayak, D. A comparative evaluation of the efficacy of manual, magnetostrictive and piezoelectric ultrasonic instruments- An in vitro profilometric and SEM study. *J. Appl. Oral Sci.* **2012**, *20*, 21–26. [CrossRef] [PubMed]
111. Loos, B.G.; Needleman, I. Endpoints of active periodontal therapy. *J. Clin. Periodontol.* **2020**, *47*, 61–71. [CrossRef] [PubMed]
112. Manresa, C.; Sanz-Miralles, E.C.; Twigg, J.; Bravo, M. Supportive periodontal therapy (SPT) for maintaining the dentition in adults treated for periodontitis. *Cochrane Database Syst. Rev.* **2018**, *1*, CD009376. [CrossRef] [PubMed]
113. Wu, D.; Yang, H.J.; Zhang, Y.; Li, X.E.; Jia, Y.R.; Wang, C.M. Prediction of loss to follow-up in long-term supportive periodontal therapy in patients with chronic periodontitis. *PLoS ONE* **2018**, *13*, e0192221. [CrossRef] [PubMed]
114. Lindhe, J.; Westfelt, E.; Nyman, S.; Socransky, S.S.; Haffajee, A.D. Long-term effect of surgical/non-surgical treatment of periodontal disease. *J. Clin. Periodontol.* **1984**, *11*, 448–458. [CrossRef] [PubMed]
115. De Freitas, L.F.; Hamblin, M.R. Proposed mechanisms of photobiomodulation or low-level light therapy. *IEEE J. Sel. Top Quantum Electron.* **2016**, *22*, 7000417. [CrossRef] [PubMed]
116. Meisel, P.; Kocher, T. Photodynamic therapy for periodontal diseases: State of the art. *J. Photochem. Photobiol. B* **2005**, *79*, 159–170. [CrossRef] [PubMed]
117. Sbordone, L.; Ramaglia, L.; Gulletta, E.; Iacono, V. Recolonization of the subgingival microflora after scaling and root planing in human periodontitis. *J. Periodontol.* **1990**, *61*, 579–584. [CrossRef]
118. Carrera, E.T.; Dias, H.B.; Corti, S.C.T.; Marcantonio, R.A.C.; Bernardi, A.C.A.; Bagnato, V.S. The application of antimicrobial photodynamic therapy (aPDT) in dentistry: A critical review. *Laser Phys.* **2016**, *26*, 1–23. [CrossRef]
119. Fumes, A.C.; Romualdo, P.C.; Monteiro, R.M.; Watanabe, E.; Corona, S.A.M.; Borsatto, M.C. Influence of pre-irradiation time employed in antimicrobial photodynamic therapy with diode laser. *Lasers Med. Sci.* **2018**, *33*, 67–73. [CrossRef] [PubMed]

pharmaceutics

Review

Antimicrobial Photodynamic Therapy: Latest Developments with a Focus on Combinatory Strategies

Raphaëlle Youf [1], Max Müller [2], Ali Balasini [3], Franck Thétiot [4,*], Mareike Müller [2], Alizé Hascoët [1], Ulrich Jonas [3], Holger Schönherr [2,*], Gilles Lemercier [5,*], Tristan Montier [1,6] and Tony Le Gall [1,*]

1   Univ Brest, INSERM, EFS, UMR 1078, GGB-GTCA, F-29200 Brest, France; raphaelle.youf@univ-brest.fr (R.Y.); alz.hascoet@gmail.com (A.H.); tristan.montier@univ-brest.fr (T.M.)
2   Physical Chemistry I & Research Center of Micro- and Nanochemistry and (Bio)Technology of Micro and Nanochemistry and Engineering (Cμ), Department of Chemistry and Biology, University of Siegen, Adolf-Reichwein-Straße 2, 57076 Siegen, Germany; max.mueller@uni-siegen.de (M.M.); m.mueller@chemie-bio.uni-siegen.de (M.M.)
3   Macromolecular Chemistry, Department of Chemistry and Biology, University of Siegen, Adolf-Reichwein-Straße 2, 57076 Siegen, Germany; alibalasini94@gmail.com (A.B.); jonas@chemie.uni-siegen.de (U.J.)
4   Unité Mixte de Recherche (UMR), Centre National de la Recherche Scientifique (CNRS) 6521, Université de Brest (UBO), CS 93837, 29238 Brest, France
5   Coordination Chemistry Team, Unité Mixte de Recherche (UMR), Centre National de la Recherche Scientifique (CNRS) 7312, Institut de Chimie Moléculaire de Reims (ICMR), Université de Reims Champagne-Ardenne, BP 1039, CEDEX 2, 51687 Reims, France
6   CHRU de Brest, Service de Génétique Médicale et de Biologie de la Reproduction, Centre de Référence des Maladies Rares Maladies Neuromusculaires, 29200 Brest, France
*   Correspondence: franck.thetiot@univ-brest.fr (F.T.); schoenherr@chemie.uni-siegen.de (H.S.); gilles.lemercier@univ-reims.fr (G.L.); tony.legall@univ-brest.fr (T.L.G.)

Citation: Youf, R.; Müller, M.; Balasini, A.; Thétiot, F.; Müller, M.; Hascoët, A.; Jonas, U.; Schönherr, H.; Lemercier, G.; Montier, T.; et al. Antimicrobial Photodynamic Therapy: Latest Developments with a Focus on Combinatory Strategies. *Pharmaceutics* **2021**, *13*, 1995. https://doi.org/10.3390/pharmaceutics13121995

Academic Editors: Udo Bakowsky, Matthias Wojcik, Eduard Preis and Gerhard Litscher

Received: 7 October 2021
Accepted: 17 November 2021
Published: 24 November 2021

Abstract: Antimicrobial photodynamic therapy (aPDT) has become a fundamental tool in modern therapeutics, notably due to the expanding versatility of photosensitizers (PSs) and the numerous possibilities to combine aPDT with other antimicrobial treatments to combat localized infections. After revisiting the basic principles of aPDT, this review first highlights the current state of the art of curative or preventive aPDT applications with relevant clinical trials. In addition, the most recent developments in photochemistry and photophysics as well as advanced carrier systems in the context of aPDT are provided, with a focus on the latest generations of efficient and versatile PSs and the progress towards hybrid-multicomponent systems. In particular, deeper insight into combinatory aPDT approaches is afforded, involving non-radiative or other light-based modalities. Selected aPDT perspectives are outlined, pointing out new strategies to target and treat microorganisms. Finally, the review works out the evolution of the conceptually simple PDT methodology towards a much more sophisticated, integrated, and innovative technology as an important element of potent antimicrobial strategies.

Keywords: antimicrobials; ROS; combinatory strategies; photodynamic therapy; multidrug resistance; nanoparticles; photosensitizers

1. Introduction

Antimicrobial resistance (AMR) occurring in bacteria, viruses, fungi, and parasites is a global health and development threat, declared by the WHO as one of the top 10 global public health threats facing humanity. The misuse and overuse of antimicrobials make almost all disease-causing microbes resistant to drugs commonly used to treat them [1]. Multidrug resistance (MDR) to critical classes of antibiotics has gradually increased in nosocomial pathogens, including *Enterococcus faecium*, *Staphylococcus aureus*, *Klebsiella pneumoniae*, *Acinetobacter baumannii*, *Pseudomonas aeruginosa*, and *Enterobacter* spp. (which are gathered in the so-called ESKAPE group) [2]. Currently, in Europe, AMR is estimated to be responsible for 33,000 deaths every year and the annual economic toll, covering extra healthcare costs and productivity losses, amounts to at least EUR 1.5 billion [3,4]. Unless adequately tackled, 10 million people a year will die from drug-resistant infections by 2050, according to the predictions of the government-commissioned O'Neill report [5].

With the decline in the discovery of new antimicrobials since 1970s, the mainstream approach for the development of new drugs to combat emerging and re-emerging resistant pathogens has focused on the modification of existing compounds. However, one key recommendation encourages stimulation of early stage drug discovery [6]. Among emerging antimicrobial therapeutic alternatives, light-based approaches show particular promise [7]. Traditionally, phototherapy was already a common practice in ancient Greece, Egypt, and India to treat skin diseases [8]. At the beginning of the 20th century, Oscar Raab first described the phototoxicity of the dye acridine red against *Paramecium caudatum*, and Tappenier and Jesionek reported the photodynamic effects of eosin suitable for treating diverse cutaneous diseases. Since then, PDT was established as the administration of a non-toxic photosensitizer (PS) followed by exposure to light irradiation at an appropriate wavelength focused on an area to treat [7]. While anti-cancer PDT is a clinical reality for 25 years [9], PDT as an antimicrobial treatment was demonstrated for the first time against drug-resistant infections in the healthcare sector in the early 1990s, leading to a "photo-antimicrobial renaissance era" [7]. Major MDR bacteria have been found susceptible to antimicrobial PDT (aPDT), independently of their drug-resistance profiles [10,11]. To date, resistance to aPDT is rarely reported, indicating that the possibility for microbes to adapt and escape this treatment can occur but is still contained. More effective aPDT systems are continuously developed, notably via combinatory approaches using multiple chemical systems and/or modalities. At the current stage of development, aPDT cannot address systemic infections but it holds great promise for treating localized infections and to fight AMR.

While excellent earlier authoritative reviews provide a detailed description of aPDT [12–15], the present review focuses on most recent developments in the field for the last 5 years, with a focus on aPDT combinatory strategies. It should be noticed that aPDT is sometimes referred to as photodynamic antimicrobial chemotherapy, light-based antimicrobial therapy, photo-controlled antimicrobial therapy, or antimicrobial photo-inactivation. In this review, all these synonyms were considered to present (i) the current state of aPDT applications in preclinical and clinical settings, (ii) a state of the art with recent developments in photosystems, (iii) the implementation of multicomponent nanotechnologies and recent molecular engineering, and (iv) the exploration of combinatory aPDT approaches towards possible future therapeutic innovations.

2. PDT: General Presentation and Features

2.1. Photochemical Pathways and Reactive Oxygen Species Production

Generally speaking, a given PS has the potential to produce reactive oxygen species (ROS) under specific conditions (Figure 1A). Typically, it possesses a stable electronic configuration called ground state level. Following irradiation and absorption of a photon, the PS is converted from a low (fundamental) energy level (^{1}PS) to a "Frank Condon" short-lived, very reactive, excited singlet state ^{1}PS* [16–18]. Subsequently, the PS can lose energy by emitting fluorescence or heat via internal conversion (IC), thereby returning to its initial ground state level; alternatively, it can be converted by a so-called inter-system crossing (ISC) to a longer-living excited triplet state ^{3}PS*. From this state, two types of chemical reaction pathways can occur, known as Type I electron transfer and Type II energy transfer, which can take place simultaneously [19]. In the Type I reaction, the ^{3}PS* captures an electron (e^-) from a reducing molecule (R) in its vicinity, which induces an electron transfer producing the superoxide anion radical ($O_2^{\bullet-}$) and, after the subsequent reduction, leads to the generation of more cytotoxic ROS including hydrogen peroxide (H_2O_2) and hydroxyl radical ($HO^{\bullet}$). In the Type II reaction, a direct energy transfer occurs from the ^{3}PS* to the ground state molecular oxygen ($^3O_2$) that is then converted to singlet oxygen ($^1O_2$). The ROS thus produced encompass $O_2^{\bullet-}$, H_2O_2, $HO^{\bullet}$, and $^1O_2$, the last two being the most reactive and most cytotoxic species but also those with the shortest diffusion distance. One PS molecule can generate thousands of molecules of $^1O_2$, depending notably on its $^1O_2$ quantum yield, the surrounding environment, and the respective occurrence of Type I and Type II mechanisms [12,17,20].

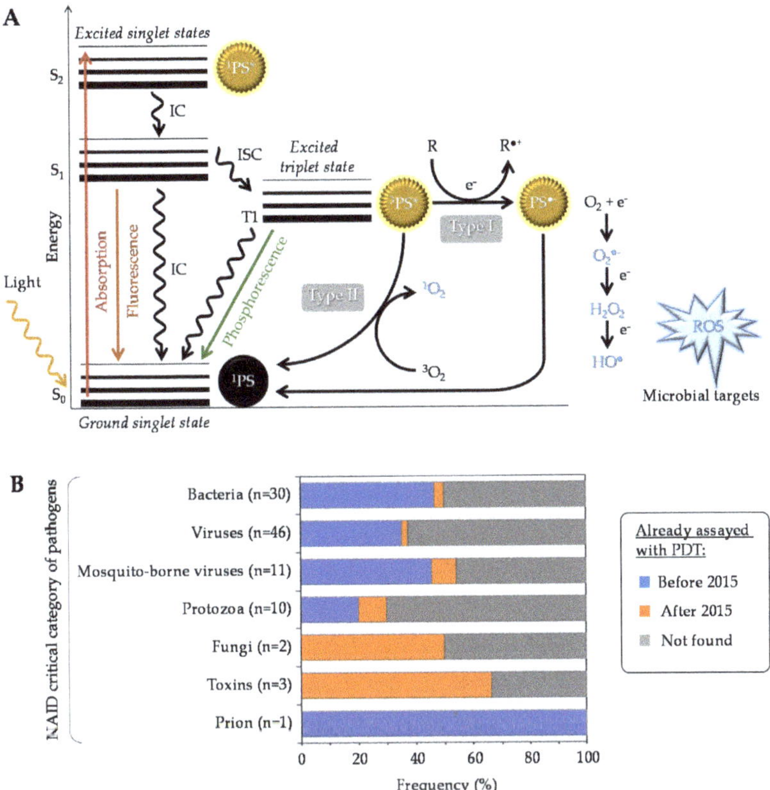

Figure 1. (**A**) Modified Jablonski diagram describing the photochemical and photophysical mechanisms leading to ROS production during PDT. (**B**) Overview of aPDT already applied to the critical category of pathogens, as defined by the NIAID (https://www.niaid.nih.gov/research/emerging-infectious-diseases-pathogens, accessed date: 1 September 2021). For each category, the chart specifies the number and the proportion (percent of pathogens already assayed in at least one aPDT study either before or after 2015).

In addition to the above-mentioned Type I and Type II mechanisms, Hamblin et al. recently proposed introduction of a Type III photochemical pathway, following which the radical anion PS•⁻ and/or inorganic radicals formed in absence of oxygen could also lead to photoinactivation [21]. These authors indeed identified several circumstances in which oxygen-independent photoinactivation of bacteria using specific PSs can be obtained.

2.2. Biological Effects of aPDT: Potential Targets and Related Mechanisms

The main first targets of aPDT are external microbial structures, i.e., the cell wall, cell membrane, or virus capsid and envelope [22,23]. Photodynamic inactivation (PDI) can be obtained against microorganisms growing as planktonic cells and/or in biofilms [13]. In biofilm matrices, the diffusion of PSs can be delayed or PSs can be sequestered, in spite of photodamage induced on various components such as polysaccharides and extracellular

DNA [24,25]. The diffusion potential of ROS depends on (i) the maximal time-limited diffusion length, especially for $^1O_2$ that possesses a shorter half-life compared with other ROS, (ii) the photostability in a given environmental medium, and (iii) the chemical properties of PSs (e.g., molecular size, charge, lipophilicity, stability), which influence the interactions of the latter with target microorganisms [26]. Photoinactivation of Gram(+) bacteria can be obtained with a given PS, irrespective of its charge, whereas that of Gram(−) bacteria generally requires a cationic PS, or a combination of a neutral PS with membrane-disrupting agents [27].

Internalization of PSs in prokaryotic or eukaryotic cells can also occur, thus causing various intracellular oxidative damage (such as in organelles in eukaryotic cells, e.g., nucleus and mitochondria in fungal cells) [28]. To protect their intracellular components, microbial cells can induce the production of antioxidant defenses such as protective enzymes (such as superoxide dismutase (SOD), catalase and glutathion (GSH)-peroxidase) or pigments (such as carotenoids acting as nonphotochemical $^1O_2$ quenchers). Nevertheless, these mechanisms can be insufficient to thwart aPDT-induced oxidative stress because intracellular components (including antioxidant defenses) can be also irreversibly photodamaged by ROS [29]. The latter can act on the DNA level through two mechanisms, i.e., alteration or modulation. Breaks in single-strand and double-strand DNA, and the disappearance of the super-coiled form of plasmid DNA have been reported in both Gram(+) and Gram(−) species. Indeed, PSs can interact with nucleic acids via electrostatic interactions and induce reduction of guanine residues causing DNA cleavage [30]. Again, microorganisms can induce the overproduction of proteins involved in the repair of photodamaged DNA; however, some bacteria, such as *Helicobacter pylori*, possess only a few of such protective repair mechanisms [31]. Upon PDT treatment, ROS and $^1O_2$ can modify bacterial gene expression profiles by modulating (i) the quorum sensing pathway, therefore inhibiting biofilm formation as shown in in vitro models, and/or (ii) the anti-virulence activities by reducing the gene expression of virulence factors in diverse clinical pathogens [32–35].

Given the multi-targeted nature of aPDT, the possibility for microorganisms to develop resistance is supposed to be very limited [36]. However, they may be able to respond to aPDT in different ways. For instance, light response adaptation can occur in some microorganisms, such as *E. coli* upon exposure to blue light inducing the production of a biofilm polysaccharide colonic acid [37]. Moreover, some PSs are substrates of efflux systems that may be overproduced; specific inhibitors of the latter can be used to restore phototoxicity [38–40]. After sublethal aPDT, the biofilm-forming ability of bacteria can increase, making them less susceptible to the same treatment [41]. Each strategy should thus be carefully examined with regard to the ability of target microorganisms to adapt and escape treatments. The latter may be noticeably reduced when using at the same time multiple antimicrobial molecular partners and modalities (read below).

2.3. Important Parameters and Requirements for an "Ideal" aPDT

According to Cieplik et al. [12], an "ideal" aPDT system should meet the following set of general requirements: (i) PS physicochemical properties: efficient PSs for aPDT possess most frequently a high hydrophilicity index and at least one cationic charge promoting interactions with pathogens, especially Gram(−) bacteria; (ii) PS photosensitivity: following irradiation, a good PS produces a high rate of cytotoxic oxygen species (Figure 1A), depending notably on its $^1O_2$ quantum yield, its stability, and the environmental media; (iii) light source parameters: efficient irradiation of PSs must take into account a coherent light exposure ensuring a good transmittance efficiency with no side effects; (iv) safety: the PS has to be specific to target microorganism(s), inducing insignificant or only a few side effects for the host, including no or few immunity responses; and (v) ease for implementation in clinical practice: aPDT has to be relatively easy to use (due to the rapid, non-aggressive, and non-invasive light application), cost-effective, and accessible. It is obvious that the detailed specification of requirements can vary depending on the target applications. The improvement of PSs is an ongoing challenge that implies moving towards more rational chemical engineering and biological investigations [11,42], as discussed in this review.

3. Positioning of aPDT in Current Human Healthcare Treatments

Over the past years, the number of studies dealing with aPDT has dramatically increased, emphasizing the potential of this therapeutic approach to treat a broad spectrum of microorganisms including bacteria, fungi, viruses, and parasites (Figure 1B). In this part, recent in vitro screening studies, preclinical (using animal models) and clinical investigations are briefly reviewed. Details about the PSs involved and their structures are provided in Section 4 "State of the art with recent photo(nano)system developments".

3.1. Curative Preclinical aPDT

3.1.1. Treatment of Bacterial Infections

PDT is a promising alternative approach to antibiotherapy for photoinactivating a broad spectrum of bacterial pathogens, either Gram(+) or Gram(−), responsible for diverse infections in humans. The antibacterial versatility of PDT can be highlighted in different ways, notably by considering lists of critically important human pathogens. First, in recent years, more attention has been paid to the potential of aPDT to fight against bacteria involved in hard-to-treat infections, especially those forming the ESKAPE group [2,43]. Second, other critical pathogen lists can be considered, such as the NIAID emerging infectious diseases/pathogens category that includes biodefense research and additional emerging infectious diseases/pathogens. To our knowledge, the susceptibility of more than 50% of the bacteria in the NIAID critical pathogen list has been considered in at least one aPDT study. In other words, bacteria causing the worst endemic infections including anthrax, botulism, melioidosis, cholerae, plague, and tuberculosis have already been considered. On the opposite, vector-borne diseases transmitted by human parasites, such as *Borrelia mayonii* and *Bartonella henselae* have not been addressed in that regard yet. Multiple experimental settings have been considered to demonstrate the potential of aPDT to photoinactivate pathogenic bacteria, growing as planktonic forms, but also in biofilm matrices, and using diverse animal models [44,45]. Among human pathogens, bacteria implicated in oral infections, especially cariogenic, periodontic, and endodontic injuries, have probably been the most intensely investigated [46]. Although less considered, other indications have also been evaluated with aPDT, including osteomyelitis, meningitis, pneumonia, lung abscess, and emphysema [47,48].

3.1.2. Treatment of Fungal Infections

Fungal infections caused by invasive candidiasis are widely recognized as a major cause of morbidity and mortality in the health care environment [49]. In addition to the opportunistic features of some fungi, resistance to first-line antifungals (such as echinocandins and fluconazole) is spreading, compromising the efficiency of conventional antifungal therapies. Despite the fact that yeasts are naturally more resistant to PDT than bacteria, noticeable in vitro and in vivo effects against fungal infections have been reported, including germ load reduction, biofilm inhibition and clearance, and eradication of persistent colonization [50–53]. Furthermore, antifungal PDT has demonstrated its potential as an adjunctive (potentially synergistic) treatment procedure to the conventional fungicide Nystatin [54]. Chen et al. identified and summarized other fungal infections that may be treated with aPDT, including *onychomycosis*, *tinea cruris*, *pityriasis versicolor*, *chromoblastomycosis*, and the cutaneous *sporothricosis* [55]. Recently, other fungal infections, such as fungi associated to *mucormycosis* (recognized as emerging critical pathogens in the NIAID) were successfully treated with PDT [56].

3.1.3. Treatment of Viral Infections

Although vaccines have drastically reduced the spreading of some of the most virulent viruses around the world, antiviral research and development remain a healthcare priority notably due to emerging viral infectious diseases [57]. PDT holds promises to help treat the latter, as well as other viruses implicated in complications of some cancers. The oldest, but also the most current, application of antiviral PDT concerns the decontamination of blood products potentially containing hepatitis B/C or West Nile virus [23,58]. The PDI of viral infections was explored in many studies considering other various viruses including arbovirus, SV40, poliovirus, encephalitis virus, phages, and HSV [59,60]. In addition, emerging viruses such as Zika, Ebola, or Tickborne hemorrhagic fever viruses have been considered [23], as well as viruses responsible for epidemic/pandemic crises such as influenza virus, SARS-CoV-2 virus, and its mutants/variants [61,62].

3.1.4. Treatment of Parasite Infections

Drug resistance is also rapidly spreading in parasites. For example, resistance to artemisinin (which is used to treat plasmodium infections causing malaria) increases drastically, even when combined with other drugs (WHO, https://www.who.int/news-room/fact-sheets/detail/antimicrobial-resistance, accessed date: 1 September 2021). Antiparasital effect of PDT was demonstrated toward critical parasites in public health such as tropical pathogens including Leishmania, African trypanosoma, and Plasmodium [63,64]. Another way to limit the propagation of the vectors is to rely on photoinductible biolarvicides. Such an approach was investigated in order to control Aedes, Anopheles, and Culex, which are tropical disease-carrying species [65,66]. This was also investigated in Lyme disease using safranin-PDT for reducing the reproduction of ticks [67].

3.1.5. Treatment of Polymicrobial Infections

Quite recently, interest has grown to explore the potential of aPDT against polymicrobial infections involving multispecies pathogens. A study demonstrated that the susceptibility to PDI of *S. aureus* and *C. albicans* growing in mixed biofilms is lower compared with single-species biofilms, which may be due to the difference in the chemical composition and viscosity of the composed matrix [24]. Nevertheless, aPDT applications are of interest regarding hard-to-treat infections due to polymicrobial biofilms colonization, such as chronic rhinosinusitis. They could effectively be eradicated by aPDT in a maxillary sinus cavity model [68]. Moreover, Biel et al. showed that aPDT can eradicate polymicrobial biofilms in the endotracheal tubes, which are factors leading to ventilator-associated pneumonia [69]. More recently, aPDT was shown capable of significantly improving wound healing in mice with polymicrobial infections [70]. However, such an approach remains a challenge due to the respective affinity of PSs to each species being usually reduced in polymicrobial systems.

3.2. Current Clinical aPDT Practices

Many clinical trials have been done for evaluating aPDT approaches in the treatment of bacterial/fungal oral infections. This is facilitated by the development of easy to use light sources in dentistry. On the opposite, it can be compromised by the development of persistent (multispecies) biofilms. Skin infections such as *Acnes vulgaris*, caused by *Propionibacterium acnes*, was one of the first microbial infections to reach the stage of aPDT clinical trial. A few clinical trials demonstrated that onychomycosis, such as *tinea cruris*, *tinea pedis*, and interdigital mycoses, could be treated with aPDT. Results demonstrated that aPDT is effective and well-tolerated, but infections can recur frequently [71,72]. Among cutaneous infections, non-healing chronic wounds in patients with chronic leg and/or foot ulcers were efficiently treated with aPDT, inducing a significant reduction in microbial load (even immediately after the treatment), a better wound healing, and no safety issues [73,74]. Osteomyelitis in patients with chronic ulcers can be treated with aPDT to prevent gangrene and amputations in the extremities of diabetic patients [75]. Clinical studies for treating *H. pylori* in gastric ulcers can be conducted using phototherapy without any PS administration; *H. pylori* naturally accumulates intracellular PSs (porphyrins) and therefore could be inactivated by phototherapy thanks to an ingenious blue/violet light delivery system [76]. A list of recently closed aPDT clinical trials is provided in Table 1; many other trials (not shown here) are still ongoing.

Table 1. List of some recently completed or terminated clinical trials that evaluated aPDT to treat diverse infectious diseases.

Medical Conditions	Target Micro-Organism(s)	Photosensitizer	Trial Phase	Number and Year
Acne	*Propionibacterium acnes*	Butenyl ALA	N.A.	NCT02313467, 2014
		Lemuteporfin	Phase 1/2	NCT01490736, 2011
		5-ALA	Phase 2	NCT01689935, 2012
		Methyl aminolevulinate	Phase 2	NCT00673933, 2013
Dental caries	*Streptococcus mutans, Streptococcus sobrinus, Lactobacillus casei, Fusobacterium nucleatum, and Atopobium rimae*	TBO	Phase 1	NCT02479958, 2015
		MB	Phase 1	NCT02479958, 2015
	Aggregatibacter actinomycetemcomitans, Tannerella forsythia and Porphyromonas gingivalis	N.C.	N.A.	NCT03309748, 2017
Denture-related stomatitis	*Candida albicans*	MB	Phase 4	NCT02642900, 2015
Orthodontic	N.D.	Curcumin	Phase 1	NCT02337192, 2015
Peri-implantitis	N.D.	N.D.	Phase 3	NCT02848482, 2016
Periodontic	*Aggregatibacter Actinomycetemcomitans. Porphyromonas gingivalis, Prevotella intermedia, Tannerella forsythia and Treponema denticola*	MB	N.A.	NCT03750162, 2018
		ICG	Phase 2	NCT02043340, 2014
		Methyl aminolevulinate	Phase 2	NCT00933543, 2013
		MB	N.A.	NCT03262077, 2017
		MB	Phase 2	NCT03074136, 2017
		Phenothiazine hydrochloride	Phase 4	NCT03498404, 2018
		TB	Phase 4	NCT03412331, 2018
Distal subungual onychomycosis	Fungi infecting nails	5-ALA	Phase 2	NCT02355899, 2015
Endodontic	*E. faecalis and C. albicans*	MB	Phase 2	NCT02824601, 2016
HPV infection	*Human Papillomavirus (HPV)*	5-ALA	Phase 2	NCT02631863, 2015
Leg ulcers	Streptococci, anaerobes, coliform, *S. aureus, P. aeruginosa*, yeast, and diphtheroics	PPA904	Phase 2	NCT00825760, 2009

Studies collected from ClinicalTrials.gov (https://clinicaltrials.gov/ct2/results?cond=photodynamic+therapy, Accessed Date: 1 March 2021). ALA, alanine; MB, methylene blue; N.A., not applicable; N.C., not communicated; N.D., not determined; PPA904, 3,7-bis(di-*n*-butylamino)phenothiazin-5-ium bromide; and TB (or TBO), toluidine blue.

3.3. Toward Preventive/Prophylactic Treatments

Beside curative antimicrobial treatments, PDT may be also used to decontaminate medical equipment and tools in hospitals, for preventive/prophylactic aims [27]. For example, photoantimicrobial textiles were reported to efficiently photoinactivate bacteria and viruses, suggesting that self-sterilizing medical gowns could be developed [77]. PDT can also decontaminate medical tools similarly to conventional chemical agents, as demonstrated in a recent comparative study [78]. Its application for the decontamination of routine informatics tools, office equipment, or packing materials demonstrates sterilization potential that could be useful to avoid hospital-acquired infections and to protect healthcare workers. Furthermore, photodisinfection of water and photoinactivation of food-borne pathogens can bring substantial benefits to people's daily lives [65].

4. State of the Art with Recent Photo(nano)System Developments

4.1. Single PSs

4.1.1. Organic PSs and Their Derivatives

Organic PSs used in aPDT have been well described in some recent reviews [79–81] (Figure 2A). Briefly, since the first use of eosin in 1904, various PSs were investigated, especially in the phenothiazinium group, which includes methylene blue (MB) and toluidine blue O (TBO). Thanks to an absorption spectrum in the red region of light, these PSs can be effective in tissues while being less toxic than other PSs. Their aPDT properties are mostly due to a high ROS production following Type I mechanism (Figure 1A). Structural derivatives have been also reported, including new MB and dimethyl-MB [42]. Another group gathers compounds featuring a macrocyclic structure composed of pyrroles, such as porphyrins and its precursor 5-aminolevulinic acid (5-ALA), phthalocyanines, and chlorins. Macrocyclic compounds are generally hydrophilic, positively charged, and they exhibit a good singlet oxygen quantum yield. Modifications of their chemical structure were intensively studied, especially for porphyrins and phthalocyanines [82–87]. The halogenated xanthenes gather PSs with a structure similar to that of fluorescein. Among them, eosin Y, erythrosine, and rose bengal (RB) were the most studied [88–90]. These compounds are anionic, which can limit their interaction with bacterial cells and their aPDT effect in spite of good singlet oxygen quantum yields. Natural compounds, including coumarins, furanocoumarins, benzofurans, anthraquinones, and flavin derivatives are often found in plants and other organisms. They are characterized by an absorption spectrum in white light or UVA. The most used are curcumin, riboflavin, and hypericin [7,80]. Nanostructures such as fullerenes are interesting PSs because of their ability to modulate Type I and Type II reactions (Figure 1A), depending on the near environment and the light source applied [91,92]. In this family, some quantum dots (QDs) can act as photoantimicrobials [80]. Other synthetic fluorescent dyes such as organoboron compounds (e.g., boron–dipyrromethene (BODIPY)), and cyanine dyes (e.g., indocyanine green (ICG)) are known for their high photostability, high extinction coefficients, and high fluorescence quantum yields [93].

Over the past five years, some new organic PSs were reported. Among them, optimized natural PSs such as anthraquinones and diacetylcurcumin can be listed. Others include derivatives of synthetic dyes such as monobrominated neutral red or azure B [79–81] (Figure 2A).

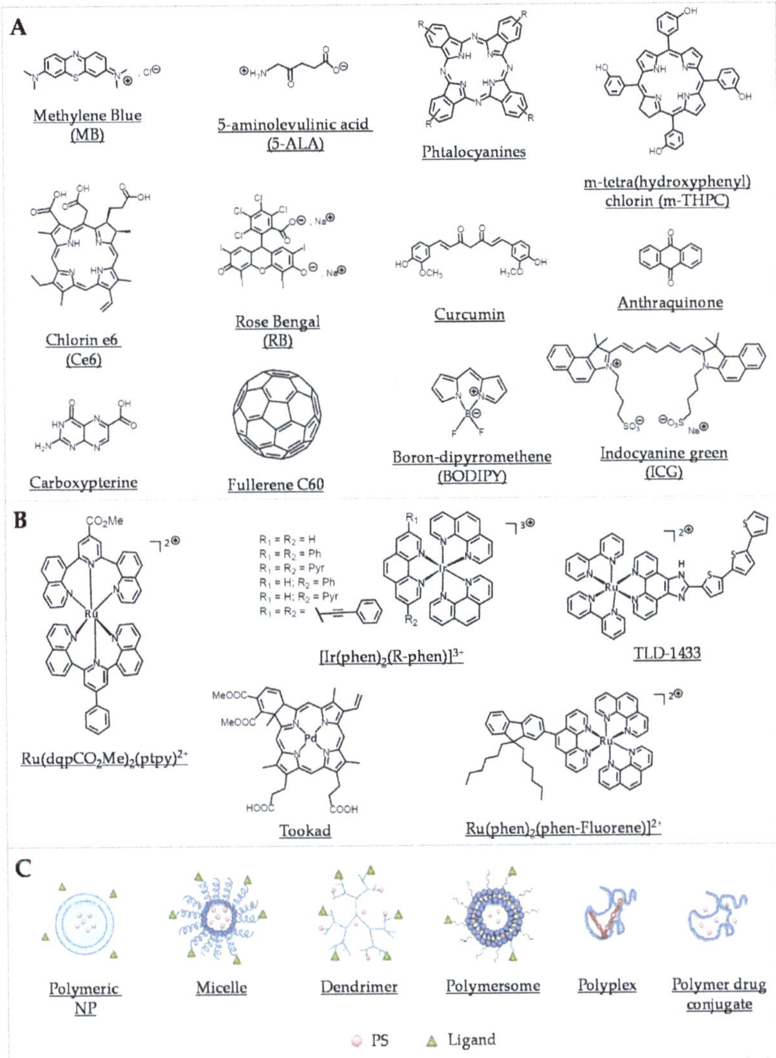

Figure 2. Representative compounds in various classes of PSs used in aPDT. (**A**) Examples of some organic PSs and their derivatives. (**B**) Examples of metallic-based PSs. (**C**) Different types of polymer-based PS carriers, which can be functionalized with ligands for specific target delivery (Adapted from [94], published by MDPI, 2020).

4.1.2. Coordination and Organometallic Complexes-Based PSs

Distinctly from metal nanoparticles (NPs; see Section 4.2.1 "Metal-based systems"), metal complexes, either coordination or organometallic complexes, are of increasing interest as PSs in PDT [95] (Figure 2B). They generally consist of a central metal core combined with ligands, involving coordinate covalent bonds (in coordination compounds) or at least one metal–carbon bond (in organometallic compounds). Compared with organic compounds, metal complexes have been notably much less considered and still remain largely underexploited regarding their potential use as new antibiotics [96]. Frei et al. recently reported on the antimicrobial activity of 906 metal-containing compounds. They showed that, considering antimicrobial activity against critical antibiotic-resistant pathogens, metal-bearing compounds displayed hit rates about 10 times higher than purely organic molecules [97].

Metal complexes display a panel of specific properties that make them promising PSs candidates. The variety of metal ions and ligands can be assembled in scaffolds featuring very diverse geometries [97,98]. Whereas most organic PSs are linear or planar molecules, metal complexes can exhibit much more complex—three-dimensional—geometries, which can improve interaction and molecular recognition with cellular targets, enlarge the spectrum of activity, and impact on biological fate [98–100]. Furthermore, the modulation of the design of metal complexes allows to fine tune their hydrophilic–lipophilic balance, solubility, photophysical properties, and eventual "dark toxicity" (i.e., the toxicity in the absence of specific irradiation). Metal complexes can display many excited-state electronic configurations associated with the central metal, the ligands, or involving both the metal and the ligand(s) in charge-transfer states (metal-to-ligand charge transfer or ligand-to-metal charge transfer). Although it is not always considered, the investigation of excited states of metal complexes (Figure 1A) is however of prime importance. Triplet states can be more easily accessed due to the enhanced spin-orbit coupling induced by the presence of the heavy metallic atom. Compared with natural PSs, metal complexes can act, besides ROS generation, via other mechanisms including redox activation, ligand exchange, and depletion of substrates involved in vital cellular processes [96,97,101].

Besides their first intended development as anticancer compounds, metal complexes have also been envisaged as potential "metallo-antibiotics", benefiting from the knowledge accumulated about their chemical properties and biological behaviors [102]. Quite recently, several metal compounds were characterized for their activity in aPDT. For instance, platinum(II), molybdenum(II), ruthenium(II), cobalt(II), and iridium(III) were proposed as new classes of stable photo-activatable metal complexes capable of combating AMR [11,103–108]. In particular, many mononuclear and polynuclear Ru(II/III) complexes have been considered as potential antibiotics, antifungals, antiparasitics, or antivirals, which have been recently extensively reviewed [107]. It is worth noticing here that, within a series of 906 metal compounds, ruthenium was the most frequent element in active antimicrobial compounds that are nontoxic to eukaryotic cells, followed by silver, palladium, and iridium [97]. Ru(II) polypyridyl complexes were assayed in several aPDT studies. For instance, $Ru(DIP)_2(bdt)$ and $Ru(dqpCO_2Me)_2(ptpy)^{2+}$ (DIP = 4,7-diphenyl-1,10-phenanthroline, bdt = 1,2-benzenedithiolate, $dqpCO_2Me$ = 4-methylcarboxy-2,6-di(quinolin-8-yl)pyridine) and ptpy = 4′-phenyl-2,2′:6′, 2″ -terpyridine) were tested with *S. aureus* and *E. coli* [109]. The complexes were found active against the Gram(+) strain and to a lesser extent against the Gram(−) strain, such difference in susceptibility being commonly reported in studies using other PSs [110] (Figure 2B). This observation was further detailed and rationalized by us when investigating a collection of 17 metal-bearing derivatives; two neutral Ru(II) complexes ($Ru(phen)_2Cl_2$ and $Ru(phen-Fluorene)_2Cl_2$) as well as a mono-cationic Ir(III) complex ($Ir(phen-Fluorene)(ppy)^{2+}$; ppy = phenypyridyl ligand) were found almost inactive, whereas a dicationic Ru(II) ($Ru(phen-Fluorene)(phen)_2^{2+}$) was found to be the most active against a panel of clinical bacterial strains [11]. More recently, Sun et al. described a Ru(II) complex bearing photolabile ligands; they showed its ability to safely photoinactivate intracellular MRSA while inducing only negligible resistance after bacterial exposure for up to 700 generations [111]. Although the precise mechanism(s) of action is not well-established in every case, Ru(II/III) compounds are also increasingly considered for their potential anti-parasitic activity for combating neglected tropical diseases such as malaria, Chagas' disease, and leishmaniasis. Moreover, potent antiviral activities have been noted for the ruthenium complex BOLD-100, particularly against HIV and SARS-CoV-2 [112]; importantly, this compound appears to retain its activity on all mutant strains of the SARS-CoV-2 [107]. All combined, metal complexes—especially ruthenium-based compounds—can display antimicrobial activity via multiple, likely synergistic, mechanisms, involving notably their ability to produce ROS. Therefore, they are of major interest for a wide range of aPDT-related applications.

4.2. Multicomponent PSs and Nanoscale Implementation: Extension to Nanoedifices with PSs

The eventual limitations in the use of singular photoactive molecules as PSs for aPDT applications reside notably in the recurrent lack of solubility and stability in the target media (typically leading to aggregates and/or PS quenching), biocompatibility ("immune stealth") and bioavailability, but also in the relative absence of selectivity for a prospected target (e.g., efficiency in the interactions with a defined target, stimuli-responsive or alternate triggers for controlled release). Thus, the widespread nanoscale implementation into the development progress of upgraded PSs has indubitably provided significant flexibility to first address these drawbacks, and implicitly contribute to the optimization of the aPDT activity via an extensive panel of mainly exclusive features specific to nano entities; the latter typically include an advantageous surface to volume ratio (with a high PS per mass content), access to unique chemical/physical/biological properties (e.g., optical properties with QDs or magnetic ones with superparamagnetic iron oxide NPs (SPIONs)), and almost inexhaustible synthetic options in the design of nanoplatforms [113–115].

Due to their paramount structural diversity and intrinsic variety of properties, the classification of the so-called "PS nanosystems" or "nano-PSs"—which partly include and overlap with "conjugated systems" and "combinatorial strategies"—remains rather challenging; however, we can usually identify the following criteria as the main pillars to rationalize and compare these aPDT nanomaterials [27,80,116]:

(i)   Role(s) and nature of the nanocomponent(s) in the PS nanosystems: the two criteria typically considered for the discrimination of nano-PSs are the role and nature of the nano building block(s) involved. With regards to the role of the latter, we can conventionally discern on the one hand the "PS nanocarriers" (e.g., polymersomes or Au NPs) in which the nanomoiety acts as a delivery system for singular PS molecules (e.g., MB) while either complementing, facilitating, or enhancing the aPDT activity (depending on the nature and eventual intrinsic properties of the nanovector), and on the other hand, the "PS active" nanoagents with the nanocomponent endorsing the role of PS. Among the examples, some versatile nanotemplates may ultimately display a dual role, i.e., "active PS" and "PS conveyor" (e.g., ZnO NPs), while distinct nanomoieties might be simultaneously required for the design of utterly sophisticated hybrid nano-PSs (e.g., Au@AgNP@SiO$_2$@PS) [117,118]. In addition to the chemical composition, the nature of the nano building block(s) will also be defined by the fundamental characteristics of nano-objects, such as size, shape, topology, and crystal structure, which will all incluctably contribute to tailor the biological behavior of the nanomaterials and the interactions with the targets (e.g., with the membranes of the bacteria) [119]. Moreover, for the same nanocore, the nature and role of the eventual surfactant(s) involved (e.g., silica coating or poly(ethylene glycol) (PEG) coating for metal NPs) can drastically alter the overall behaviors of the nanosystems.

(ii)   Type of interactions between the nano entity and the PS, and localization of the PS in the nanosystems: other criteria of relevance when describing nano-PSs—specifically PS nanocarriers—reside in the nature of the interactions between the nanocomponent and the PS molecules involved, but also the location of the PSs. Thus, we can distinguish the common cases of PS molecules "embedded" within a nanovector either by physisorption or functionalized (chemisorption), and alternately the nanoplatforms with surfaces decorated with PSs, again, either by physical or chemical adsorption. The differences between the two types of interactions and distinct localizations of the PSs implicitly imply distinctive chemical engineering and related requirements, and may potentially impact the resulting stability of the nanoedifices, but also the aPDT activity. For instance, in the case of PS molecules located inside the edifice and not released, the selected "nanomatrix" should adequately permit the photoactivation process of the internalized PSs, be sufficiently porous/permeable to both triplet and singlet oxygen and eventual ROS generated by the photosensitizers (i.e., efficient internal diffusion of molecular oxygen to react with the PSs then external diffusion of $^1O_2$ to the targets) while also presenting inertness to the latter to not compromise or quench the aPDT activity. Meanwhile, with surfaces of nano-objects decorated with PSs, the PSs may then contribute to some extent as an interface with the biological medium or the target.

(iii) Biological impacts of the PS nanosystems: in addition to the biocompatibility and aPDT efficiency (including the critical concentrations just as the half-maximal effective concentration EC_{50}, minimum inhibitory concentration MIC, or 50% growth inhibition concentration GIC_{50}), the eventual biodegradability, elimination process, or ecotoxicity of the aPDT nanomaterials can markedly vary from one system to another (based on factors such as composition and size/shape), but are rather difficult to evaluate or compare; ergo, these factors are not systematically addressed in the reports.

(iv) Relative sustainability of the nano-PSs for aPDT applications: the reproducibility, eco-friendliness, and cost-effectiveness parameters of the synthetic protocols and production of aPDT nanomaterials, as well as the ease of storage and use, and the stability over time are also ultimately to be evaluated for any system aiming to be viable and reasonably applied; however, similar to (iii), these parameters are complex and so scarcely investigated.

Thus, in the overview presentation of the different PS nanosystems hereafter, the chemical features (i) and (ii) have been conventionally defined as the main criteria for the classification. Alternatively, the nature of the aPDT applications has also been used as the main criterion for classification in some references [120]. Another approach consists of systemizing all the nanosystems dedicated to a given PS (e.g., curcumin) [121].

Within the extensive collection of aPDT nanomaterials reported to date, the majority belongs to the category labeled as "PS nanocarriers" with the nanocomponent acting as a delivery system for PS molecules; however, increasing examples involving PS-active nano building blocks have emerged as well. Overall, this multipurpose role of the nanomoiety may include avoiding aggregation (e.g., dimerization, trimerization) and correlated PS quenching, enhancing "solubility" (i.e., dispersibility), stability and bioavailability, allowing "biological stealth", on-demand release and target specificity, and ultimately triggering eventual synergistic aPDT activity with the complementary or ameliorative intrinsic properties of the nanocomponent; although irrevocably confirmed in many nano-PSs, the mechanisms involved in the synergy may differ from one system to another, and often remain partly or integrally unresolved due to the complexity of these tacit multiparameter contextures [113]. It is noteworthy that most systems comprise "classic"/"traditional" organic PSs (natural or synthetic, e.g., curcumin, MB) with fewer examples involving metallated PS molecules such as the recent review from Jain et al. dedicated to ruthenium-based photoactive metalloantibiotics [108]. Among aPDT nanomaterials, we can thus identify various families of nanoplatforms based on the nature of the nanocore, starting here with the inorganic vectors followed by the organic templates; as an indication, in the common cases of "multi-component nano-PSs", the classification has been defined hereafter according to the main/prominent nano building block involved in the composition, i.e., metal-based systems, silicon-based systems, carbon-based systems, lipid-based systems, and polymer-based systems. Regarding the following presentation of aPDT nanosystems, it is important to specify that it is not exhaustive, but instead provides a panorama of the main categories of nanosystems—either colloids or surfaces [27,122]—and their related specificities, with an emphasis on recent developments.

4.2.1. Metal-Based Systems

Metal-based nanostructures have been extensively investigated—both as "PS cargo" and PS active entities—through the exploration of the richness and diversity of the respective subcategories related to this class of compounds, as detailed below. Each of the below-mentioned inorganic classes presents distinct specificities of relevance for aPDT applications, with a choice to be defined on a case-by-case basis according to the target, the nature of the PSs involved (with possible preferential affinity), or the anticipated complemental or synergistic role of the selected nano entity (based upon its chemical, physical, and/or biological features, with the eventual PS molecules located either at the surface of the NPs or within, when present). It is important to notice that the nanocores from each subcategory can be either further implemented, with silica or polymer coating for example, or combined (multicomponent nano-PSs) in order to adequately optimize the efficiency of the systems [117]; however, the outcomes of these hybridity processes are complex to anticipate with systematic rationality, with either enhancement or quenching of the properties observed depending on the composition of the combinations.

Metal NPs

Metal NPs—mainly gold, but also silver—maybe considered among the "gold standards" in nanomaterials through the history and expansion of nanosciences in terms of dedicated publications and vastness of related possible applications [123]. As already well documented for Au and Ag NPs, the reasons are numerous and reside notably in their relatively easy accessibility with low-cost and highly reproducible (large scale) biogenic and chemical synthetic routes, in addition to the flexibility to finely tailor the properties via a refined size and shape control (with narrow size distribution and diverse shapes), and the facility for functionalization with various types of molecules. Moreover, gold NPs display biocompatibility, low toxicity, and immunogenicity, almost chemical inertness (distinctly from their inherent catalytic properties), while silver nanomaterials present intrinsic antimicrobial activity against a broad spectrum of microorganisms and related MDR infections (e.g., towards Gram(−) and Gram(+) mature biofilms of MRSA), and disruption of biofilm formations while being safe for mammalian cells [124–126].Ultimately, both Au and Ag NPs share nano features specific to noble metal systems, i.e., localized surface plasmon resonance (LSPR, arising from their resonant oscillation of their free electrons upon light exposure) and resonance energy transfer (RET), with subsequent optical and photothermal properties of enhancing appositeness in aPDT applications and PDI efficacy (e.g., ROS production) [127–129]. For the most part, Au and Ag NPs of various shapes (e.g., spheres, rods, cubes) are combined with organic PS molecules such as RB and MB [128–136], but also with metallated PSs such as ruthenium complexes, metallophthalocyanines, and metalloporphyrins [108,137,138]; the corresponding PS nanovehicles can also be labeled as "conjugates" but they strictly differ from the "mixtures" involving metal NPs and PS molecules [139]. Other noble metal NPs, viz., platinum, have also been employed in aPDT applications due to their multitarget action to inactivate microbes, although to a lower degree up to now owing to synthetic limitations [27,140]. Alternately, redox-active copper NPs are typically less costly and easier to access and present unique features among which the faculty to generate oxidative stress to various microbes through the genesis of ROS [141], such as in the recently developed copper–cysteamine (Cu–Cy) nano-PS that can be activated either by UV, X-ray, microwave, or ultrasound, to produce ROS against cancer cells and bacteria [142,143]. More unwontedly, approaches to treat subcutaneous abscesses lead to the use of acetylcholine (Ach) ruthenium composite NPs (Ach@RuNPs) as an effective appealing PDT/PTT dual-modal phototherapeutic killing agent of pathogenic bacteria, with Ach playing a role in targeting the bacteria and promoting the entry into the bacterial cells [108]. While belonging to the same category, each metal displays particular specificities; consequently, with the objective of optimization, nano-PSs resulting from alloys or multimetallic NPs have been further designed such as the $Au@AgNP@SiO_2@PS$ and $AA@Ru@HA\text{-}MoS_2$ (AA: ascorbic acid, HA: hyaluronic acid) nanocomposites [117,118].

Metal Oxides

Similar to gold and silver, metal oxides such as iron oxide, titanium oxide and zinc oxide have been cornerstone contributors in the global evolution of applied nanomaterials (particularly in medicine), due notably to intrinsic magnetic and optical nanoscale features, with the latter typically available at "room temperature" and commonly finely tunable via shape, size, and crystal structure parameters [144]. Indeed, specific single-domain superparamagnetic iron oxide NPs (SPIONs)—either magnetite (Fe_3O_4) or maghemite

(γ-Fe$_2$O$_3$) of various shapes and sizes, including ferrite or doped derivatives—exhibit an outstanding magnetization behavior with no remanent or coercive responses upon exposure to a magnetic field. As a result, such magnetic NPs have legitimately generated interest and use as magnetic resonance imaging (MRI) and magnetic particle imaging (MPI) agents, as well as magnetic fluid hyperthermia (MFH), magnetic cell separators [145], or drug delivery conveyers with the possibility to guide the NPs to the targeted area via external magnetic fields [146]. More recently, iron oxide nano-objects proved to be also of pertinence for aPDT applications not only as a magnetic "nanocargo" for various organic and inorganic PSs such as curcumin, MB, ICG, BODIPYs, porphyrins, metallophtalocyanines, or ruthenium derivatives among others [27,108,147–155], but also with peroxidase-like activity to enhance the cleavage of biological macromolecules for biofilm elimination [156]. Extension in the design of more elaborate multicomponent architectures involving hybrid iron oxide nanocore led *inter alia* to Ag/Fe$_3$O$_4$, Ag/CuFe$_2$O$_4$, CoFe$_2$O$_4$, and Fe$_3$O$_4$/MnO$_2$ NPs conjugated with different PSs [27,117,147,150,151,157–160].

Other oxides have also drawn heavy attention, in particular zinc oxide and titanium oxide, as the photophysical properties of these wide bandgap semiconducting nanomaterials efficiently translate into a multi-level antimicrobial activity including PS vessel and/or PS active agent (with possible coupled aPDT response), and/or membrane disruptor [161–164]. Thus, ZnO and TiO$_2$ nanoplatforms possess the ability to alter microbes' integrity—through alternate mechanisms involving ROS and/or metallic ions—in the dark or via photoactivation [165]. The latter is customarily triggered by UV or X-ray irradiation, with the eventual possibility to adequately shift to other wavelengths such as visible light irradiation in virtue of diverse modification methods of the oxides encompassing notably: doping or surface alteration (e.g., F-doped ZnO, coatings or oxygen deficiency), coupling with other bandgap semiconductors (e.g., ZnO/TiO$_2$) or sensitizing dyes, and composites [121,156,164,166–171]. Furthermore, beside the size and shape of the NPs, the crystallographic phase appears to be a tuning parameter of particular importance for the antimicrobial effects of some oxides, especially for TiO$_2$ with the distinction between the anatase, rutile, and brookite structures [172–175]. Although reported to a lesser extent, the list of alternate oxides exhibiting potential in aPDT applications comprises CuO/Cu$_2$O, MnO$_2$, and rare earth oxides to mention just a few [141,176–180].

QDs and Metal Chalcogenide Nanomaterials

Aside from zinc/titanium oxides, distinct semiconductors such as QDs and metal chalcogenide (e.g., metal sulfide) nanomaterials (involving elements from different groups in the periodic table) proved to be efficient disruptors against various multi-drug-resistant microorganisms. Due to their smaller size of a few nanometers (ca. up to 10 nm), QDs differ from other nano-objects with physical and optoelectronic properties governed by the rules of quantum mechanics, high chemical stability and resistance to photobleaching, and near-infrared (NIR) emission (typically above 700 nm) notably allowing for deep-tissue imaging [181]. Appositely comparable, metal chalcogenide likewise reveals unorthodox physio and physicochemical properties, accordingly garnering a legit interest for antimicrobial applications [182]. Consequently, not only can these nano building blocks carry PS molecules and alter the integrity of microbial walls/membranes or gene expression, but they may as well act as PSs; when coupled with other PSs, synergistic interactions in the QD-PS edifices might occur resulting from mechanisms such as Förster resonance energy transfer (FRET, non-radiative energy transfer from QD donors to PS acceptors) to generate free radicals and ROS. Among examples of such QDs and metal chalcogenide aPDT systems can be cited CdTe QDs and related CdTe-PS conjugates, CdSe/ZnS QDs combined with PSs, InP and InP-PS, Mn-doped ZnS, MoS$_2$, but also CuS and CdS nanocrystals, with ultimately hybrid systems involving for instance CoZnO/MoS$_2$ or AgBiS$_2$–TiO$_2$ composites [123,183–196].

Metal–Organic Framework (MOF) Nanoscaffolds, Upconversion Nanomaterials and Other Metal Ion Nanostructures

Although less investigated than the above-mentioned alternatives, other original metal-based nanostructures identified as MOF nanostructures, upconversion nanoplatforms, and alternate metal ion nanomaterials tend to further consolidate their promising potential for aPDT applications. Considering their towering surface area and porous ordered structure with substantial loading capacity (e.g., adsorption of O_2 and ensuing photocatalytic production of $^1O_2$ via a heterogeneous process), stable versions of colloidal nano-MOFs—or less common covalent organic frameworks (COFs), i.e., reticulation variants typically defined by non-metal "nodes" instead of metal ones—have emerged as efficient heterogeneous photosensitizers—with frameworks acting as PSs or entrapping PSs—towards antimicrobial applications (e.g., enhanced penetration for bacterial biofilms eradication), with porphyrin-based or porphyrin-containing MOFs and COFs, or Cu-based MOFs embedded with CuS NPs for rapid NIR sterilization among recently reported solutions [197–205].

Moreover, upconversion NPs (UCNPs) generally involve actinide- or lanthanide-doped transition metals and refer to the nonlinear process of photon upconversion, viz., a sequential absorption of two or more photons resulting in an anti-Stokes type emission (i.e., emission of light at a shorter wavelength than the excitation wavelength); when translated into biomedical context, UCNPs can be typically activated by NIR light—characterized by deeper tissue penetration and reduced autofluorescence, phototoxicity, and photodamage when compared with UV or blue light—and produce high energy photons for optical imaging or more recently aPDT when combined with PSs [206–209]. Intrinsically limited by low upconversion quantum yield, the current focus consists of developing hybrid UCNPs to improve aPDT efficiency; thus, auspicious progress has been achieved with examples such as {UCNPs (NaYF4:Mn/Yb/Er)/MB/CuS-chitosan)} multicomponent nanostructured system revealing a superior antibacterial activity with the UCNPs enhancing the energy transfer to MB, the CuS triggering synergistic PDT/PTT effects, and chitosan assuring stability and biocompatibility [156]. Other examples also include silica coating β-NaYF$_4$:Yb, Er@NaYF$_4$ UCNPs loaded with MB as PS and lysozyme as a natural protein-inducing bacterial autolysis, Fe$_3$O$_4$@NaGdF$_4$:Yb:Er combined with the photo/sonosensitizer hematoporphyrin monomethyl ether (HMME), UCNPs@TiO$_2$, N-octyl chitosan (OC) coated UCNP loaded with the photosensitizer zinc phthalocyanine (OC-UCNP-ZnPc), or the UCNPs-CPZ-PVP system (CPZ: β-carboxy-phthalocyanine zinc, PVP: polyvinylpyrrolidone) to name but a few [206,210–213].

Alternatively, more disparate metal-ion aPDT systems have been reported such as PS encapsulated dual-functional metallocatanionic vesicles against drug-resistant bacteria involving copper-based cationic metallosurfactant, or self-assembled porphyrin nanoparticle PSs ZnTPyP@NO using zinc meso-tetra(4-pyridyl)porphyrin (ZnTPyP) and nitric oxide (NO) [214–216].

4.2.2. Silicon-Based NPs

This category will be divided hereafter into two main subclasses—porous silicon (pSi) and (mesoporous) silica (SiO$_2$)—which differ in the oxidation state of the silicon and display distinctive properties, more specifically different quenching behavior and photodynamic activity (singlet oxygen quantum yield under irradiation).

Porous silicon NPs (pSi NPs) are among the most promising types of inorganic nanocarriers for biomedical applications and have been intensively investigated since the first publication by Sailor et al. in 2009 regarding their application for in vivo treatment of ovarian cancer. Composed of pure silicon, pSi NPs indeed exhibit relevant features encompassing not only pores with large capacity for drug loading combined with specific surface area allowing for implemental functionalization, but also degradability in an aqueous environment, and biocompatibility [217]. Moreover, porous silicon particles are known to be photodynamically active with related inherent antimicrobial properties by generation of ROS under irradiation with light of a specific wavelength [218,219]. Because of the low quantum yield of singlet oxygen production from porous silicon itself, particles can be grafted with additional PSs, such as porphyrins, to enhance the yield of singlet oxygen generation and thereby antimicrobial properties for PDT applications [219,220]. Consequently, several silicon-based systems have been reported in recent years, mainly for PDT applications [221,222]. Furthermore, pSi NPs display intrinsic fluorescence, which can be applied for imaging and real-time diagnostics regardless of any surface functionalization [220,223].

As previously mentioned, pSi has a low singlet oxygen quantum yield due to quenching, which makes silica particles in comparison a more suitable substitutional system combining similar biocompatibility with improved optical properties. On the other hand, one of the pivotal advantages to use silica (SiO_2) conjugates with PSs is to achieve a better "solubilization"—or more accurately, dispersibility—of hydrophobic dyes and a better photostabilization, thus limiting the self-photobleaching of PSs. Further advantages to name as the most important features of silica are high biocompatibility, antimicrobial properties, and high surface area for mesoporous silica that can be synthesized easily from commercially available precursors [27,224,225]. Furthermore, SiO_2 exhibits an effective PS-grafting capacity [226]; the latter can be accomplished via adsorption, covalent bonding, binding to the hydroxyl groups from silica surface, and entrapment during formation in silica particles or matrix [27]. Recently, Dube et al. reported about the photo-physicochemical behavior of silica NPs with (3-aminopropyl)triethoxysilane (APTES), and subsequently PS-modified surfaces for aPDT [150]. In addition, silica coatings have also been reported to prevent the degradation of nanocarriers (magnetite) and prolong the stability and functionality of PS systems [227]. Interestingly, coupling PSs to silica or Merrifield resin leads to distinct advantages; indeed, immobilized Ce6 notably displays significantly higher aPDT efficiency in comparison with the free form, which is probably due to an enhancement of the adhesion of PSs to bacterial cells resulting in a stronger cell wall disorganization [228,229]. Unsurprisingly, many approaches with encapsulated PSs in silica NPs for potential aPDT applications have been reported in recent years [229–232].

Distinctly, combinatory approaches involving other nanocomponents, such as silica-containing core-shell particles or silica-coated inorganic NPs, have emerged to further implement the properties of silica with the specific features (e.g., magnetic, photoactive, or antimicrobial properties) from other nanomaterials of relevance [118]. Thus, since the surface of silica can be easily grafted with PSs and is highly biocompatible, the surface modification of silica particles using metallic NPs (e.g., Ag NPs) to enhance the antibacterial photodynamic activity, has been developed for improved aPDT [233] while combinations with carbon quantum dots have been assessed for imaging-guided aPDT [234]. Another example refers to sufficient aPDT/PDI systems, more precisely to mesoporous silica-coated NaYF4:Yb:Er NPs with the PSs (silicon 2,9,16,23-tetra-tert-butyl-29H,31H-phthalocyanine dihydroxide) loaded in the silica shell to enhance bacterial targeting of *E. coli* and *S. aureus* [235].

In addition to silica-based NPs or silica-containing core-shell particles, lesser-known silica nanofibers also proved to be suitable substrates for potential aPDT, PDT, and PDI applications. As an illustration, Mapukata et al. [236] recently reported silver NP-modified silica nanofibers with embedded zinc phthalocyanine as PS for aPDT applications. The nanofiber-based substrates offer the advantage of fast removal after application, which can allow limiting any dark toxicity [236–238].

Silica NPs and fibrous or dendritic fibrous nanosilica have also been reported for the formation of nanocomposites to create antimicrobial photodynamically active surfaces for aPDT or PDI; to create such surfaces, silica NPs can be embedded into polymeric matrices for enhanced biocompatibility and complementary surface properties from the selected polymer [239].

Additionally, silica substrates and nanoconjugates offer a suitable platform for combinatory approaches since they can be easily modified. For example, Zhao et al. described polyelectrolyte-coated silica NPs modified with Ce6 [240]. These complexes could be extracted, by bacteria, from silica NPs to form stable binding on the bacterial surface, changing the aggregation state of Ce6 and leading to both the recovery of PS fluorescence and $^1O_2$ generation. Such bacteria-responsive multifunctional nanomaterials allowed for simultaneous sensing and treating of MRSA. Another approach is illustrated by the photo-induced antibacterial activity of amino- and mannose-decorated silica NPs loaded with MB against *E. coli* and *P. aeruginosa* strains [241]. The modification of silica substrates with mannose led to an increased targeting of *P. aeruginosa* and reduced dark toxicity of the systems.

4.2.3. Carbon-Based Nanomaterials

Akin to silicon-based nano-objects, the notable diversity of allotropic customizable carbon-based nanostructures—either conveyers for traditional PS molecules or intrinsically PS active—legitimizes their distinct consideration in the actual classification of nano-PSs [242–244], with the following subcategories.

Fullerenes, Carbon Nanotubes (CNTs), and Nanodiamonds

In the evolution of carbon-based nano-objects, fullerenes and carbon nanotubes derivatives may be chronologically introduced as the "first generation". Discovered in the mid-1980s, the proper C_n (n = 60–100) spheroidal "soccer ball" π-conjugated structures of the fullerenes yield tremendous chemical modularity and electrochemical and physical properties, including photostability, the propensity to act as a PS via Type I or Type II pathways (Figure 1A) with high ROS quantum yield, and oxygen-independent photo-killing by electron transfer. Despite their intrinsic hydrophobicity typically requiring surface functionalization for biocompatibility and related dispersibility, they have proven even nowadays their effectiveness as broad-spectrum photodynamic antimicrobial agents, with photoactive antimicrobial coating based on a PEDOT-fullerene C_{60} polymeric dyad (PEDOT: poly(3,4-ethylenedioxythiophene)), BODIPY-fullerene C_{60}, diketopyrrolopyrrole–fullerene C_{60}, and cationic fullerene derivatives among recent examples [80,92,156,245–251]. Mainly developed a few years later, carbon nanotubes (CNTs)—either single wall CNTs (SWCNTs) describable simply as a single-layer sheet of a hexagonal arrangement of hybridized carbon atoms (graphene) rolled up into a hollow cylindrical nanostructure, or multiwall CNTs (MWCNTs) consisting of nested SWCNTs—unveil both independent capacities to produce ROS upon irradiation and high surface area for decoration with PS molecules [156]. Neoteric specimen of PS-CNTs encompass toluidine blue, polypyrrole, malachite green, MB, RB, and porphyrins [156,252–257]. Although purportedly older since it was discovered in the 1960s, diamond NPs or nanodiamonds seemingly remain the lesser known carbon-based nanomaterials to date; nevertheless, the latter dispose of legit aPDT arguments with their fluorescence, photostability, proclivity for conjugation with diverse PSs such as porphyrins or metallated phthalocyanines and silver NPs, but also inherent antibacterial activity [120,258–262].

Carbon QDs (CQDs)

As the next momentum in the blossoming of "nano-carbon" era, the carbon QDs were discovered in the early 2000s [263,264]. The physical and chemical properties of these fluorescent particles, commonly quasi-spherical with less than 10 nm in diameter, can be finely tuned upon size/shape variations or doping with heteroatoms (e.g., B, N, O, P, S) [263,265]. By virtue of their biocompatibility and dispersibility, photostability, low toxicity and related eco-friendliness, good quantum yield and conductivity, CQDs have been investigated for various applications, and more recently as antimicrobial agents; withal, their environment-friendly features combined with low cost and rather ecological biogenic or synthetic routes (from natural or synthetic precursors) place them advantageously as a viable scalable photocatalytic disinfection material compared with alternate nano-PSs. Late cases involve doped or hybrid CQDs or more conventional conjugates of CQDs with PSs [121,266–271].

Graphene, Graphene QDs (GQDs), and Graphene Oxide (GO) Nanostructures

In a similar timescale to CQDs is the quantum leap discovery and blooming of graphene and graphene oxide materials. Graphene can be defined as a 2D allotrope of carbon, more accurately a monolayer of atoms with a hexagonal lattice structure (or single-layered graphite) and identifiable as the "building block" for the discrete fullerenes, 1D carbon nanotubes, and 3D graphite. Despite its stunning mechanical/electronic properties and chemical inertness, the limitations of graphene, such as zero bandgap and low absorptivity, lead to the ulterior conversion of the 2D graphene into "0D" GQDs [272,273]. Due to quantum confinement and edge effects, GQDs exhibit different chemical and physical properties when compared with other carbon-based materials, as well as a non-zero bandgap, good dispersibility, and propensity for functionalization and doping. Structurally, GQDs differ from CQDs because they comprise graphene nanosheets with a plane size less than 100 nm [272,274]. Likewise, graphene oxide (GO) is the oxidized form of graphene i.e., a single atomic sheet of graphite with various oxygen-containing moieties either on the basal plane or at the edges. Meanwhile, reduced graphene oxide (rGO) can be summarily described as an "intermediate" structure between graphene and GO, with variable and higher C/O elemental ratios compared with GO, but remaining residual oxygen and structural defects with reference to the pristine graphene structure. Although GO was reported a couple of centuries ago, GO and rGO nanomaterials have mainly emerged for various applications after the discovery of graphene since GO is a precursor to prepare graphene, and both present distinctive physical and chemical properties that differ from graphene. As a result, countless and rising fast illustrations of graphene derivatives for antibacterial applications are regularly reported [121,275–280].

4.2.4. Lipid-Based Systems

Due to their amphiphilic nature (typically hydrophilic "head" and hydrophobic "tail"), some lipids—natural and synthetic—have been extensively studied to develop efficient biocompatible delivery systems—initially for drugs and DNA/RNA, but also for aPDT PSs—with synthetic flexibility and structural diversity. Among the prevalent examples, we can distinguish the micelles (lipid monolayers with polar units at the surface and hydrophobic core) and the liposomes (one or more concentric lipid bilayer with a hydrophilic surface and an internal aqueous compartment). Although sharing similar chemical constituents, the micelles and liposomes present significant differences to be taken into consideration depending on the intended application (nature of the target) and the nature of the PSs. Indeed, the micelles are typically smaller than liposomes (with a diameter starting from a few nanometers for the micelles, and ca. 20 nm for the liposomes), with distinct stability and permeability in biological medium and uptake pathways for the PSs into bacteria. With reference to the nature of the transported PSs, the liposomes display the additional flexibility to carry both hydrophilic PSs (in the core compartment or between the bilayers) or/and hydrophobic PSs (within the lipid bilayer), while the micelles are usually easier and cheaper to prepare [80,156]. As often critical to address for biomedical applications, the surface charge of these nano-objects can be tailored to further optimize the interactions with the bacteria, with cationic modification of liposomes identified as a promising aPDT efficiency "amplifier" [80,156,281]. In addition to recent examples such as the hypericin loaded liposomes against Gram(+) bacteria [282–286], another emerging and promising alternative includes the development of modified liposome-like derivatives labeled either as "ethosomes", "transfersomes", or "invasomes", which can be briefly described as ultra-deformable vesicular carriers with upgraded transdermal penetration and increased permeability into the skin for the PSs compared with conventional liposomes [287–290]. On the other hand, recently reported aPDT micellar systems refer to micelles loaded with various hydrophobic PSs such as curcumin, BODIPY, porphyrins, hypocrellin A, or hypericin among others [121,291–293]. Furthermore, solid lipid NPs (SLN) composed of solid biodegradable lipids have been recently highlighted as delivery systems used for actual mRNA COVID-19 vaccines [294], but they also have been reported as transporters for curcumin for the treatment of oral mucosal infection [121]. Besides, we may also include nanoemulsions in this category of lipid-based nanovehicles since the latter conventionally involve lipids in the oily phase during the formulation process, with recent curcumin/curcuminoid nanoemulsions [121,295].

4.2.5. Polymer-Based Systems

In direct correlation with the above-mentioned lipid-based systems, polymer-based nanocarriers have been positioned as a logical extension with the objective to implement a "degree of freedom" in the synthesis flexibility while expanding the panel of building blocks available in the design of aPDT nanostructures. A categorization of polymeric systems for the delivery of PSs to therapeutically relevant sites can be done using as criteria either the nature of the polymer(s) involved or the type of polymeric (nano)structures. Thus, in the following, we will use a classification primarily based on the structure of the polymeric systems with differentiation between NPs (including hydrogels, biopolymers, and aerogels), polymersomes, polymeric micelles, dendrimers, fibers, and polymeric films and layers (including hybrid systems and nanocomposites). The nanostructure of polymeric systems is implicitly highly correlated to the molecular structure, i.e., composition (nature of the polymer(s), ratios, and distribution and amount of hydrophilic and hydrophobic moieties), charge, and size, as reviewed recently by Osorno et al. [296]. The refined control of these parameters enables the design and fine-tuning of specific polymeric nanostructures spanning from long-ranged ordered lamellar sheets, tubes, and fibers to oval or spherical particles, micelles, and polymersomes [296] (Figure 2C).

Conjugated Polymers as PSs or Polymer-Functionalized PSs

One of the easiest approaches reported to develop polymeric carrier systems is to enhance water solubility and biocompatibility of PS molecules through the use of functionalized polymers, by introducing the PSs in a post-modification reaction to a hydrophilic and/or biocompatible polymer, or to modify the PS with a polymerizable group (e.g., acrylate) to react as a monomer for further polymerization with suitable monomers. Other apt options are conjugated polymers incorporating a backbone with alternating double and single bonds which provide photodynamic and optical properties, and therefore might act as PSs themselves [297,298]. For example, poly[(9,9-bis{6'-[N-(triethylene glycol methyl ether) di(1H imidazolium)-methane]hexyl}-2,7-fluorene)-co-4,7-di-2-thienyl-2,1,3-benzo-thiadiazole] tetrabromide (PFDBT-BIMEG) is a conjugated polymer which affords salt bridges and electrostatic interactions with microorganisms; these interactions enable the simultaneous detection and inhibition of microorganisms [297]. Conjugated polymers were intensively investigated for the development of multifunctional NPs in antibacterial applications because of their generally low-toxicity toward eukaryotic cells, their flexibility, and their high potential to vehicle versatile therapeutic molecules [298]. Improved PSs based on conjugated polymers or polymerized PSs with antimicrobial photodynamic properties have been amply reported in recent years [299–303]. For example, Huang et al. demonstrated the efficiency and selectivity of polyethyleneimine-Ce6 conjugated in potential aPDT/PDI applications [304].

Other polymeric systems also show antimicrobial properties or preferential target infection sites and are thus suitable for aPDT applications when combined with PSs. For example, cationic polymers, which are known to be highly hydrophilic, can be used to target cell walls and thus microbial infections. Hence, such polymers are adequate for potential aPDT applications [305,306]. Exemplary amphiphilic or cationic poly(oxanorbornene)s doped with PSs, which exhibit pronounced antimicrobial activity (99.9999% efficiency) against *E. coli* and *S. aureus* strains, have been reported [307]. By contrast, anionic polymer particles and nanocarriers with negatively charged surfaces and membranes invade the reticuloendothelial system, which leads to elongated blood circulation times [308–310]. Polymers with large fractions of functional groups, such as the biopolymers mentioned above, can be easily modified and offer many loading sites for suitable PSs. Thus, among many suitable materials, hyperbranched and dendrimeric polymers are particularly promising candidates for PS nanocarriers.

Dendrimers

Dendrimers, dendrimeric polymers, or dendrons are highly ordered and highly branched polymers, which may form spherical three-dimensional structures with diameters typically ranging from 1–10 nm [311]. The dimensions of dendrimers are relatively small compared with other drug delivery systems, but as they consist of individual well-defined molecules, the drug loading can be subtly established with reproducibility. The incorporation or encapsulation of drugs may be achieved by covalent binding of the PSs or drug molecules to reactive functional groups along the dendrimeric structure [312,313]. Alternatively, the PS or drug molecule may act as a scaffold from which the dendrimer is synthesized. Finally, drug or PS encapsulation may occur inside the voids of the dendrimer [311,314,315]. Thus, dendrimers offer a substrate for PSs with many reaction sites, and controllable sterical and hydrophobic properties depending on the backbone of the dendrimer. Approaches using hyperbranched polymers for stimuli-responsive release of PSs, such as porphyrins, under acidic or reductive conditions, with improved targeting of bacterial sites have been reported, such as mannose-functionalized polymers by Staegemann et al. [316,317].

Polymeric NPs and Nanocomposites

In many cases, polymeric nano-systems are not clearly classified, but instead typically grouped under the terminology "polymeric NPs", since the exact structure may not be fully characterized or considered less relevant with respect to the application efficiency. Polymeric NPs are commonly defined as physically or chemically crosslinked polymer networks with a size in the range of 1 to 1000 nm, but if not further defined may also include nanocapsules, such as polymersomes, micelles, chitosomes, and even highly branched polymers and dendrimers [318].

In aPDT, polymeric NPs may be either used to encapsulate PS molecules (loading with PSs), or built from inherent photoactive polymers acting as PSs [303]. Various PS molecules (e.g., RB, porphyrin derivatives, or curcumin) encapsulated in biocompatible polymeric NPs, such as polystyrene-, PEG-, polyester- (including poly-beta-amino esters), or polyacrylamide-based NPs, are widely used in aPDT/PDI approaches, as recently reported [319–321]. In addition to the role of nanovehicle, the polymeric particles may also help to address the lack of solubility in the biological medium from the PSs and/or reduce their toxicity, as reported among others by Gualdesi et al. [320,322]. For instance, polystyrene NPs with encapsulated hydrophobic TPP-NP(5,10,15,20-tetraphenylporphyrin) have been reported as nano-PS for efficient aPDI approaches towards multi-resistant bacteria [321]. In addition, biocompatible PLGA (polylactic-co-glycolic acid) NPs were employed to encapsulate curcumin, which could potentially serve as an orthodontic adhesive antimicrobial additive [322]. Alternatively, combinatory approaches such as polymeric NPs merging a polymeric PS, a photothermal polymeric agent, and poly(styrene-co-maleic anhydride) as dispersants have recently been reported for coupled aPDT/PTT nano-platforms in aqueous media [323,324]. Furthermore, Kubát et al. reported an increase in the stability of physically crosslinked polymeric NPs (polystyrene) to comparable nanocapsules (PEG-poly(ε-caprolactone) (PCL) micelles) equipped with identical PS [319].

The term polymeric NP also includes NPs obtained from natural polymers (e.g., chitosan and alginate), hydrogels, or aerogels. Hydrogels are chemically or physically crosslinked polymeric networks, which are able to absorb large amounts of water due to their pronounced hydrophilicity, but do not dissolve because of the crosslinking, whereas aerogels are formed by replacement of liquid with a gas resulting in low-density polymeric structures [325]. Recently, Kirar et al. reported PS-loaded biodegradable NPs fabricated from gelatin, a naturally occurring biopolymer, and hydrogel, for application in aPDT [326]. Moreover, also recently, some polymers such as Carbopol-forming hydrogel matrix entrapping PS molecules attracted attention because of their bio/muco-adhesive property, allowing a prolonged local PS delivery. Nevertheless, the viscosity of such systems can prevent the efficiency of PDT by decreasing the photostability and ROS production, thus calling for further optimizations [327,328]. Furthermore, the incorporation of hydrophobic PSs such as curcumin in polyurethane hydrogel has demonstrated to be an efficient PS release system [329].

Another suitable hydrogel for aPDT is chitosan, which is a polycationic biopolymer with good biocompatibility and antibacterial properties [330,331]. Indeed, cationic polymers, which can be antimicrobials by themselves such as chitosan and poly-Lysine, can assure a better recognition toward bacteria thus improving antimicrobial efficiency. Typically, polycationic chitosan may form nanogels or NPs in the presence of (poly)anionic molecules. Alternatively, another promising aPDT approach for the treatment of *Aggregatibacter actinomycetemcomitans* was recently reported [332,333], in which anionic PS indocyanine green was used to form PS-doped chitosan NPs. These NPs were shown to significantly reduce biofilm growth-related gene expression. Other similar approaches have also been reported, in which PS doped chitosan NPs showed enhanced cellular uptake and improved antimicrobial properties compared with free PS against different bacteria [212,334,335].

The above-mentioned approach has not only been implemented with NPs, but also in thin films and layers to create biocompatible and antimicrobial surfaces [336]. Indeed, in aPDT and especially aPDI, polymers are also used to create antimicrobial surfaces or matrices to embed PS units. Thus, combinatory approaches—using cellulose derivatives, alginates, chitosan, and other polymer-based materials as biocompatible substrates for PSs and nanocomposites to create photoactive antimicrobial surfaces—have recently been reported [337–339]. The development of such surfaces, and in particular antimicrobial membranes, is of considerable interest, especially in numerous and diverse fields in which providing hygienic and sterile surfaces is essential. As an illustration, Müller et al. reported polyethersulfone membranes doped with polycationic PS that provide antimicrobial properties for potential use as filter membranes in water purification or medicine [340]. Other recent distinctive systems for aPDI refer to self-sterilizing and photoactive antimicrobial surfaces made from (i) natural polymers such as chitosan doped with chlorophyll [336], (ii) "bioplastic" poly(lactic acid) surfaces coated with a BODIPY PS [341], or (iii) synthetic polyurethanes doped with curcumin and cationic bacterial biocides [336,342–345].

Alternatively, other combinatory approaches—in which polymeric nanocomposites embedding inorganic nanomaterials with relevant complementary features are conceived—have garnered attention in recent years. Examples include fullerenes or silver NPs that are incorporated as PSs into polymeric matrices [346], phthalocyanine-silver nanoprism conjugates [347], or mesoporous silica NPs loaded into polymer membranes [348]. Additionally, the embedding of zinc-based PS—such as Zn (II) porphyrin [349]—into polymer matrices to create antimicrobial polymers and polymeric surfaces for aPDI and possibly aPDT has also been reported, with pronounced antimicrobial activity against several bacteria strains and viruses [350,351]. Furthermore, protein-based approaches were developed as reported by Ambrósio et al. [352] and Silva et al. [353] using BSA (bovine serum albumin) NPs or BSA conjugated to PSs to improve solubility and biocompatibility.

Lastly, stimuli-responsive polymeric NPs are generally known to be sensitive to an internal or external stimulus (e.g., pH, temperature, light) and so of utmost interest for the controlled release of PSs for PDT and aPDT, whereas the stimuli-responsiveness depends on the structure and properties of the used polymer such as the assembly of polymer chains and linkages [354]. For example, Dolanský et al. reported light- and temperature-triggered ROS and NO release from polystyrene NPs for combinatory aPDT/PTT approaches [355]. Unlike micelles or polymersomes, crosslinked NPs have been reported to be thermodynamically stable, while stimulus-responsive behavior such as pH-responsive release of PSs has been more often achieved using self-assembled micellar or vesicular structures [319,356–358].

Polymersomes

Polymersomes are polymeric vesicles that resemble liposomes, which were previously described in Section 4.2.4 dedicated to lipid-based systems. Polymersomes are formed from amphiphilic copolymers, which self-assemble in aqueous media, resulting in capsules. The lumen of the polymersomes is filled with an aqueous medium and the wall comprises a hydrophobic interior with a hydrophilic corona on both inner and outer interfaces. Polymersomes typically form when the weight fraction of the hydrophilic parts (e.g., PEG in a PEG-b-poly(lactic acid) (PLA) block copolymers) comprises up to 20–40% of the polymer. They also form at comparatively large water fractions in the solvent shift method [359–361]. At higher weight fractions, the system tends to form micellar structures [296,362,363]. The hydrophilic and hydrophobic compartments inside the polymersomes enable the uptake and encapsulation of both hydrophobic and hydrophilic molecules [364–367], making polymersomes a relevant candidate for the development of advanced nanocarriers for PSs [368]. In aPDT and similar approaches, efficient delivery of the PSs to the targeted tissue is essential, not only to minimize toxic side effects and overcome low solubility in body fluids, but also to enable elongated circulation in the blood stream and prevent dimerization and quenching of the PSs. Li et al. reported that PSs encapsulated in polymeric nanocarriers exhibit an increased singlet oxygen $^1O_2$ quantum yield compared with non-encapsulated PSs [308]; by contrast, non-encapsulated PSs may tend to aggregate and lose efficiency [308,369,370]. Additionally, most polymersomes offer certain advantages compared with the established liposome-based systems such as enhanced biocompatibility, lower immune response, controlled membrane properties, stimuli-responsive drug release, biodegradability, and higher stability, with those mainly resulting from the individual design of polymers and polymersomes for the anticipated applications [296,371].

The encapsulation of PSs into specifically designed nanosized polymersomes with stimulus-triggered release of PSs has been reported in recent years, including temperature, pH, and light-induced release of the PSs [296,372,373]. For instance, Lanzilotto et al. recently reported a system used for aPDT consisting of polymersomes of the tri-block-copolymer poly(2-methyl-2-oxazoline)-block-poly-(dimethylsiloxane)-block-poly(2-methyl-2-oxazoline) (PMOXA34–PDMS6–PMOXA34) encapsulating water-soluble porphyrin derivatives [372].

Polymeric Micelles

In contrast to polymersomes or liposomes, micelles are particles that contain a hydrophobic core. Typical micelles range between 10 to 100 nm in size. More specifically, polymeric micelles are formed of amphiphilic polymers, with a higher ratio of hydrophilic parts in the case of block copolymers compared with polymersomes [296,362,363]. They also form at comparatively small water fractions in the solvent shift method [359–361]. Due to their structures, micelles are implicitly used to encapsulate hydrophobic drugs or agents, such as PSs, to convey the latter to the target (e.g., cancer cells or microbial infections) while overcoming their low solubility in aqueous media [374]. Additionally, encapsulation may reduce the toxicity of the PSs [375,376]. When compared with polymersomes, one disadvantage is the lack of flexibility to encapsulate or transport both hydrophilic and hydrophobic molecules. On the other hand, as reported in the review by Kashef et al. [377], polymeric micelles are easier to produce, hence they are more cost-efficient than liposomes and potentially polymersomes, while providing similar applicable features [378,379]. Among examples, an aPDT system of polymeric micelles, fabricated from methoxy-PEG and PCL and loaded with the hydrophobic PS curcumin and ketoconazole for the PDI of fungal biofilms, has been reported with an increased water solubility and controlled release of the PSs on display [380]. Additionally, Caruso et al. recently reported the synthesis of thermodynamically stable PEG-PLA micelles for efficient aPDI of *S. aureus*, suggesting that this delivery system is promising in aPDI applications, which also reduces toxicity compared with pure PSs [291]. Moreover, a significant advantage of polymeric micelles compared with lipid-based micelles resides notably in the adjustability of properties by individual design of the polymer and membrane surface with respect to the selected application. Hence, the polymeric micelles can be equipped with target ligands and/or their morphology can be tuned to increase the cellular uptake of the micelle, PS, or other therapeutic agents in a specific tissue, such as in tumors [381]. Like polymersomes, polymeric micelles can be tuned to release drugs or collapse by application of external stimuli, such as light, temperature or pH [357,358].

Niosomes

Niosomes are closely related to liposomes. They are vesicular structures consisting of non-ionic surfactants, including polymers, and lipids such as cholesterol. The addition of lipids to the non-ionic surfactants leads to increased rigidity of the membrane and the vesicular structure. Niosomes range from 10 to 3000 nm in size, and also may include multi-layered systems typically consisting of more than one bilayer. Niosomes offer enhanced stability and biocompatibility compared with vesicular structures based on ionic surfactants [288]. Due to the bilayer structure and adjustability of the selected polymers for the applications of choice, they exhibit similar properties and offer similar advantages as polymersomes, and can be used for encapsulation of a variety of hydrophobic and hydrophilic molecules [288,358]. They may also be equipped with targeting ligands, such as folic acid that enhances uptake into cancer cells for PDT of cervical cancer [382]. A niosome-based system, using MB as encapsulated PS, has been reported for aPDT treatment of hidradenitis suppurativa [383].

Polymeric Fibers

Another approach to obtain PS carrier systems relies on polymeric or polymer-hybrid fibers with diameters in micro and nano range [384–386]. The fibers can either be loaded after formation with various PSs, such as cationic yellow or RB [387] incorporated into wool/acrylic blended fabrics to obtain antimicrobial properties, or they can be electrospun from PS containing polymer solution to form polymeric fibers loaded with PSs [388]. Since many polymers (such as nylon, cellulose acetate, or polyacrylonitrile) are used in the textile industry, those materials may offer a novel approach to create substrates or textiles with self-sterilizing and antimicrobial properties in the presence of visible light [386,389]. Furthermore, combinatory approaches using various NPs, such as magnetite NPs, to form polymeric fiber based nanocomposites have shown some efficiency for antimicrobial photodynamic chemotherapy [390,391].

To conclude this Section 4.2, the list of above-mentioned aPDT nanosystems is non exhaustive and aims at providing an overview of the diversity and richness in the composition of aPDT nanomaterials. The next section focusing on "combinatorial strategies" partly overlaps with the description of aPDT systems, which further complicates the classification process. Moreover, a distinction must be settled between strict "mixtures" of components and "chemical combinations" of components in the development of aPDT treatments with possible synergistic behaviors.

5. Focus on Combinatory aPDT Approaches

Given its intrinsic characteristics, PDT is highly amenable to versatile combinations with other drugs, treatments, or modalities in view to potentiate therapeutic effects, which include enhanced efficacy, limitation of side effects, and reduction in the risk of resistance emergence. Proof-of-concept for various combinatory aPDT are thus being increasingly recorded, exploiting the additive/synergistic effects arising from single or multiple therapeutic species acting via different mechanistic pathways. The following part aims to review recent works following such strategies (Table 2).

Table 2. Examples of recent studies that combined aPDT with other antimicrobial actives or treatments.

Combination with Antibiotics	Target(s)	In Vitro and/or In Vivo Effect(s)	Reference
5-ALA + Gentamicin	*S. aureus* and *S. epidermidis*	In vitro: antibiofilm synergistic effect	[392]
Photodithazine + Metronidazole	*F. nucleatum* and *P. gingivalis*	In vitro: improvement of antibiofilm effect	[393]
Ce6 NP + Tinidazole	Periodontal pathogenic bacteria	In vitro: synergistic antiperiodontitis effects; in vivo: reduced adsorption of alveolar bone in a rat model of periodontitis	[394]
MB + Clindamycin/Amoxicillin	*E. coli*	In vitro: enhancement of antibiotic susceptibility following aPDT treatment; in vivo: prolonged survival of infected *G. mellonella* larvae	[43]
MB + Gentamicin	*S. aureus* and *P. aeruginosa*	In vitro: synergistic effect on planctonik cultures of both bacteria; positive effect on *P. aeruginosa* biofilm	[395]
MB + Carbapenem	*S. marcescens, K. pneumoniae* and *E. aerogenes*	In vitro: impairment of the enzymatic activity and genetic determinants of carbapenemases; restoration of the susceptibility to Carbapenem	[396]
[Ir(ppy)$_2$(ppdh)]PF$_6$) + Cefotaxime	*K. pneumoniae*	In vitro: synergistic aPDI effect with Cefotaxime	[397]
Combination with other antibacterial compounds	**Target(s)**	**In vitro and/or in vivo effect(s)**	**Reference**
MB or Ce6 + aurein 1.2 monomer or aurein 1.2 C-terminal dimer	*E. faecalis*	In vitro: prevention of biofilm formation with all treatments; improvement of aurein monomer effect when combined with Ce6-PDT	[398]
RB + Concanavalin A	*E. coli*	In vitro: improvement of RB uptake, increased membrane damages and enhanced PDT effect	[399]
MB@GNPDEX-ConA + Carbonyl cyanide m-chlorophenylhydrazone	*K. pneumoniae*	In vitro: enhancement of the MB-NPs mediated phototoxicity with the efflux pump inhibitor CCCP	[40]
Quinine hydrochloride + antimicrobial blue light	MDR *P. aeruginosa* and *A. baumannii*	In vitro: photo-inactivation of planktonic cells and biofilms; in vivo: potentiation of aBL effect in a mouse skin abrasion infection model	[400]
Combination with other antifungal treatment compounds	**Target(s)**	**In vitro and/or in vivo effect(s)**	**Reference**
5-ALA + ITZ, itraconazole; TBF, terbinafine; VOR, voriconazole	*Candida* species, dermatophytes, *A. fumigatus* and *F. monophora*	In vitro: reduction/improvement of lesions, disappearance of plaque	[401]
Photodithazine + Nystatin	Fluconazole-resistant *C. albicans*	In vitro: reduction of fungal viability, decrease in oral lesions and inflammatory reaction; in vivo: decrease in tongue lesions	[54]
5-ALA + Itraconazole	*Trichosporon asahii*	In vitro: better elimination of planktonic and biofilms fungi than single therapy	[402]

Table 2. *Cont.*

Combination with immunotherapy	Target(s)	In vitro and/or in vivo effect(s)	Reference
Schiff base complexes	*E. coli et S. aureus*	In vitro: blockage of the production of inflammatory TNFα cytokine	[403]
Porphyrin + phtalocyanine	HIV-infected cells	In vitro: specific phototoxicity against infected cells	[404]
Combination with sonodynamic therapy (SDT)	**Target(s)**	**In vitro and/or in vivo effect(s)**	**Reference**
Ce6 derivative Photodithazine + RB	*C. albicans*	In vitro: inactivation of biofilm (viability and total biomass)	[405]
UCNPs + hematoporphyrin + SiO_2-RB [1]	Antibiotic-resistant bacteria	In vitro: greater antibacterial effect with SDT and PDT at once	[406]
Combination with electrochemotherapy	**Target(s)**	**In vitro and/or in vivo effect(s)**	**Reference**
Hypericin	*E. coli* and *S. aureus*	In vitro: better bacterial inactivation with combined therapies	[407]
Combination with viral NPs	**Target(s)**	**In vitro and/or in vivo effect(s)**	**Reference**
TVP-A (luminogen) + PAP phage	*P. aeruginosa*	In vitro: synergistic bacterial recognizing and killing; in vivo: acceleration of healing rates	[408]
Pheophorbide A (chlorophyll) + JM-phage	*C. albicans*	In vitro: better specificity of PS targeting	[409]
Ru(bpy2)phen-IA + Cowpea chlorotic mottle virus	*S. aureus*	In vitro: targeted bacterial photodynamic inactivation	[410]
Combination of several PSs	**Target(s)**	**In vitro and/or in vivo effect(s)**	**Reference**
Carboxypterin + MB	*K. pneumoniae*	In vitro: better biofilm eradication	[411]
Phthalocyanines + Graphene QDs	*S. aureus*	In vitro: better bacterial photoinactivation	[412]
ICG + Metformin + Curcumin	*E. faecalis*	In vitro: better biofilm eradication	[35]
Porphyrin + Phthalocyanine	*Leishmania braziliensis*	In vitro: better assimilation of photo-inactivated parasites by macrophages	[413]
Combination with photothermal therapy (PTT)	**Target(s)**	**In vitro and/or in vivo effect(s)**	**Reference**
Ruthenium NPs	Pathogenic bacteria	In vitro: bacterial inhibition; in vivo: reduction of bacterial load and repair of infected wounds	[414]
Graphene oxide	*E. coli* and *S. aureus*	In vitro: efficient vector for both PDT and PTT	[415]
ICG + SPIONs	*E. coli, K. pneumoniae, P. aeruginosa,* and *S. epidermis*	In vitro: antimicrobial and antibiofilm activity at a low dose	[148]
Ag-conjugated graphene QDs	*E. coli* and *S. aureus*	In vitro: efficient photoinactivation by PDT and PTT; in vivo: promoted healing in bacteria-infected rat wounds	[280]
PDPPTT (photothermal agent) + MEH-PPV (PS) [2]	*E. coli*	In vitro: better inhibition rate than PTT/PDT systems used alone	[324]
Mesoporous polydopamine NPs + ICG	*S. aureus*	In vivo: eradication of *S. aureus* biofilm on titanium implant	[416]

Table 2. *Cont.*

Combination with NO phototherapy	Target(s)	In vitro and/or in vivo effect(s)	Reference
N-(3-aminopropyl)-3-(trifluoromethyl)-4-nitrobenzenamine + TMPyP/ZnPc	*E. coli*	In vitro: dual-mode photoantibacterial action	[417]
Sulfonated polystyrene NPs (NO photodonor + porphyrin/phthalocyanine)	*E. coli*	In vitro: strong antibacterial action	[355]
[Ru(bpy)$_3$]Cl$_2$	*P. aeruginosa*	In vitro: PDT/NO synergistic antibiofilm effect	[418]

ALA, alanine; MB, methylene blue; RB, rose bengal; Ce6, chlorin e6; ICG, indocyanine green; SPION, superparamagnetic iron oxide NP. [1], hematoporphyrin monomethyl ether enclosed into yolk-structured up-conversion core and covalently linked RB on SiO2 shell; [2], photothermal agent poly(diketopyrrolopyrrole-thienothiophene) (PDPPTT) and the photosensitizer poly(2-methoxy-5-((2-ethylhexyl)oxy)-p-phenylenevinylene) (MEH-PPV) in the presence of poly(styrene-co-maleic anhydride).

5.1. "Basic" aPDT Combinations

5.1.1. Combination of Several PSs

The simplest combinatory aPDT approach probably consists in combining two or more PSs in a single treatment. Thus, besides aPDT making use of a single PS, aPDT may rely on the simultaneous use of several PSs in an attempt to obtain additive or synergistic antimicrobial effects. The PSs combined may exhibit different photophysical characters showing complementarities. For example, carboxypterin-based aPDT upon sunlight irradiation demonstrates a significant planktonic bacterial load reduction [419]. However, eradication of biofilm formation needs a PS concentration 500 times higher than assays performed with planktonic forms. When combined with MB, Tosato et al. showed that reasonable concentrations of both PSs exert synergistic effect on both biofilm and planktonic MDR bacteria [411]. In other studies, alternative dual-PSs aPDT systems have shown even better antibacterial and antibiofilm properties [35,412].

5.1.2. Addition of Inorganic Salts

The combination of PSs with inorganic salts can modulate the PDT effectiveness, potentiating or inhibiting the antimicrobial activity through the production of additional reactive species or quenching $^1O_2$ [420]. As an example, azide sodium can modulate aPDT effectiveness by promoting or inhibiting the binding of bacteria with PSs, depending on lipophilicity of the latter. Indeed, azide sodium is a $^1O_2$ quencher, but can also produce highly reactive azide radicals, via electron transfer from PS at the excited state. This has been reported with many phenothiazinium dyes and also fullerenes [93,421,422]. Other salts can amplify the bacterial killing mediated by MB-PDT such as potassium iodide (KI), a very versatile salt, involved in the generation of short-lived reactive iodine radicals ($I^{\bullet}/I_2^{\bullet-}$) [423,424]; similar findings have been obtained when using KI with cationic BODIPY derivatives [425] or porphyrin–Schiff base conjugates bearing basic amino groups [426]. Potassium thiocyanate and potassium selenocyanate could also form reactive species, such as the sulfur trioxide radical anion and selenocyanogen $(SeCN)_2$, respectively [427]. In addition, interactions between PSs and target microorganisms could be improved thanks to inorganic salts. For example, calcium and magnesium cations can modulate the electrostatic interaction between PSs and bacterial membranes [428]. It is worth noting here that most of the above-mentioned studies were conducted under in vitro settings. While inorganic salts can be useful to enhance in vitro or ex vivo aPDT effects, the use of such additives in animals or human beings would need to be carefully examined, considering the dose to be used and potential concomitant side effects.

5.2. Combinations of aPDT with Other Antimicrobial Drugs or Antimicrobial Therapies

Various classes of antimicrobial drugs or other antimicrobial therapies, which action does not depend on light, have been considered with regard to their aPDT compatibility.

5.2.1. Antibiotics

Antibiotherapy is an obvious complementary therapy to aPDT, which may allow obtaining stronger antimicrobial effects and/or restore antibiotic susceptibility. Combinations of PSs with antibiotics have been investigated in numerous studies addressing various infectious diseases, such as skin and mucosal infections as recently reviewed [429]. Aminosides are the most common antibiotics used in combination with a PS. For instance, kanamycin, tobramycin, and gentamicin combined respectively to RB, porphyrin, and 5-Alanine have been reported, showing potent effects against bacteria of clinical interest, notably by improving biofilm clearance and reducing microbial loads [392,430–433]. In addition, combinations with other antibiotic classes, such as nitroimidazoles (e.g., metronidazole) or glycopeptides (e.g., vancomycine), also showed some effect against biofilms of *F. nucleatum* and *P. gingivalis* [393], as well as *S. aureus* [434]. Further, PSs can also be useful to combine with antibiotics in order to restore the susceptibility of bacteria to the latter, notably to last-resort antibiotics. As a recent example, Feng et al. reported the photodynamic inactivation of bacterial carbapenemases, both restoring bacterial susceptibility to carbapenems and enhancing the effectiveness of these antibiotics [396]. However, all combinations of PSs and antibiotics may not be effective, since in some cases antibiotics can have an antagonistic effect on PS activity [435]. Finally, it is noteworthy that some antibiotics can act themselves as PSs; Jiang et al. indeed reported light-excited antibiotics for potentiating bacterial killing via ROS generation [436].

5.2.2. Antifungals

The combination of PSs with antifungals is a powerful approach to thwart growing antifungal resistance, especially to fluconazole, which is commonly observed in *C. albicans* strains [437]. Quiroga et al. showed that sublethal photoinactivation mediated by a tetracationic tentacle porphyrin allowed to reduce the MIC of fluconazole in *C. albicans* [438]. Nystatin, a common antifungal used to prevent and treat candidiasis, was combined with photodithazine-based PS and red light in *C. albicans*-infected mice. The combined therapy reduced the fungal viability and decreased the oral lesions and the inflammatory reaction [54]. Moreover, other antifungals (e.g., terbinafine, itraconazole and voriconazole) combined to ALA reported promising results. These combinations could be alternative methods for the treatment of refractory and complex cases of chromoblastomycosis [401,439].

5.2.3. Other Antimicrobial Compounds

Many other antimicrobial compounds may be used in combination with PSs. Antimicrobial peptides (AMPs) are oligopeptides (commonly consisting of 10–50 amino acids) with high affinity for bacterial cells thanks to an overall positive charge. Their antimicrobial spectrum can be modulated through variation of their amino acids sequence. AMPs encompass a large set of natural compounds that may be used in aPDT. For example, de Freitas et al. reported that sub-lethal doses of PSs (either Ce6 or MB) and aurein peptides (either aurein 1.2 monomer or aurein 1.2 C-terminal dimer) were able to prevent biofilm development by *E. faecalis* [398]. In another study, a membrane-anchoring PS, named TBD-anchor, demonstrated both bacterial membrane-anchoring abilities and ROS production [440]. In a similar way to peptide therapy, LPS-binding proteins were used to improve the contact of PSs with the cytoplasmic membrane of bacteria. For instance, the antibacterial efficacy of a complex consisting of a combination of RB with the lectin concanavalin A (ConA) was demonstrated in a planktonic culture of *E. coli*; ConA-RB conjugates increased membrane damages and enhanced the RB efficacy up to 117-fold [399]. In addition, coupling pump efflux inhibitors, such as CCCP EPI (carbonyl cyanide m-chlorophenylhydrazone), to PSs was also investigated. This highlighted the interest to combine PSs with molecules acting on targets susceptible to induce resistance modulations [40]. Other antimicrobial molecules may be good candidates to be used in aPDT strategies. For example, quinine used in combination with antimicrobial blue light was shown efficient to photo-inactivate MDR *P. aeruginosa* and *A. baumannii* [400]. Some cationic molecules, such as cationic lipids, can have a good affinity for bacterial cell membrane and a good antibacterial activity [126,441]. Thus, PS-amphiphiles conjugates would also deserve to be investigated for aPDT applications.

5.2.4. Viral NPs and Phagotherapy

In recent years, viral NPs (VNPs) deriving from phages, animal, or plant viruses have been proposed as biological vehicles for delivery of PSs. Such carriers can exhibit a series of advantageous properties including natural targeting, easy manufacture and good safety profile [442]. Compared with nanomaterials used as PS carriers, VNPs are natural protein-based NPs that may display higher biocompatibility and tissue specificity. In addition, VNPs can be tuned through genetic and synthetic engineering with appropriate biological and chemical modifications (e.g., surface decoration). Alternatively, the combination of aPDT with so-called phagotherapy may be useful for various reasons, following different strategies. One study suggests that, following a PDT treatment, ROS damages can cause quorum sensing and virulence pathway alterations rendering micro-organisms more susceptible to other therapies. Among these, phagotherapy is known to be modulated by multiple virulence factors [443]. Another approach can consist of conjugating phages with PSs, in order to improve interaction/delivery of the latter into target microorganisms. For example, Dong et al. showed that a phage carrying the chlorophyll-based PS pheophorbide efficiently induced apoptosis in *C. albicans*, thus demonstrating the potential of phototherapeutic nanostructures for fungal inactivation [409]. However, such an approach has to be prudently considered regarding the occurrence of phage resistance already reported in numerous investigations [444].

5.3. Combinations of aPDT with Other Light-Based Treatments

Multiple modalities entirely controlled by light stimuli may be combined with aPDT in multifunctional antimicrobial treatments. The following part reports some very recent studies illustrating multiple light-based antimicrobial strategies combined to operate independently, with potential additive or synergistic effects and without reciprocal interferences. This may be obtained with a single PS displaying such multiple properties and/or through combination of PSs with other light-activatable compounds exhibiting complementary properties.

5.3.1. aPDT and Photothermal Therapy

Photothermal therapy (PTT) is a local treatment modality relying on the property of a PS to absorb energy and convert it into heat upon stimulation with an electromagnetic radiation, such as radiofrequency, microwaves, near-infrared irradiation, or visible light. The localized hyperthermia can lead to various damages resulting in microbial inactivation in the treatment area. Following this principle, ruthenium NPs have been used for PDT/PTT dual-modal phototherapeutic killing of pathogenic bacteria [414]. Moreover, GO demonstrated antibacterial effect against *E. coli* and *S. aureus* as a result of both PDT and PTT effects following irradiation with ultra-low doses (65 mW/cm^2) of 630 nm light [415]. Furthermore, combination of sonodynamic, photodynamic, and photothermal therapies with an external controllable source recently reported against breast cancer [445] may also show promising applications for treating bacterial infection. Mai et al. reported a FDA-approved sinoporphyrin sodium (DVDMS) for photo- and sono-dynamic therapy in cancer cells and photoinactivation of *S. aureus* strains, in in vitro and in vivo models. However, no bacterial sonoinactivation by DVDMS was obtained [446].

5.3.2. aPDT and NO Phototherapy

Combination of aPDT with NO phototherapy is gaining increasing interest for antimicrobial applications [447]. For instance, light-responsive dual NO and $^1O_2$ releasing materials showed phototoxicities against *E. coli* [355,417]. More recently, Parisi et al. developed a molecular hybrid based on a BODIPY light-harvesting antenna producing simultaneously NO and $^1O_2$ upon single photon excitation with green light for anticancer applications; according to the authors, this system may also act as an effective PS and NO photodonor antibacterial agent [448].

5.3.3. aPDT and Low Laser Therapy

Photobiomodulation (PBM), also called low-level laser therapy, is a non-destructive process that may alleviate pain and inflammation or promote tissue healing and regeneration. The use of this method coupled to aPDT is a very recent approach. A concomitant use of aPDT and PBM was reported as an adjunct treatment for palatal ulcer [449]. In a clinic-laboratory study, aPDT and PBM showed similar improvement in gingival inflammatory and microbiological parameters compared with conventional treatment [450]. More recently, some benefits of this combined therapy were reported such as the modulation of inflammatory state, pain relief, and acceleration of tissue repair of patients contracting *herpes simplex labialis* virus or orofacial lesions in patients suffering from COVID-19 [451–453].

5.4. Coupling of aPDT with Other Physical Treatments

5.4.1. aPDT and Sonodynamic Therapy

Sonodynamic therapy (SDT), combining so-called sonosensitizers (SS) and ultrasounds (US) is a relatively new approach for treating microbial infections [454,455]. The ultrasonic waves have the property to induce a cavitation phenomenon thus enhancing the efficacy of combined antimicrobial treatments. The rationale for combining PDT with SDT relies on specific advantages of the latter, notably, a deeper propagation of US into the tissue than light; therefore, PDT/SDT may be used to treat deeper lesions in vivo, alleviating the limitations of light propagation and delivery presented by aPDT [405,456]. An approach combining PDT and SDT, called sonophotodynamic therapy (SPDT), has been reported to improve microbial inactivation compared with individual aPDT or SDT [457]. Because of the complicated system of SPDT, its mechanisms have not been clearly revealed yet. Some studies have demonstrated that sonoporation mechanism induced by US improves the transfer of large molecules into the bacteria by forming transient pores. Moreover, US waves could potentiate the microorganisms dispersion in the medium resulting in (i) a better biodisponibility of therapeutic agent and light diffusion and (ii) a reduction of microbial aggregation and networks, such as biofilm [457,458]. The mention of a new class of PS characterized by the dual ability to be activated by both US and light, for SPDT application, has been questioned. Indeed, Harris et al. recently suggested that this specific PS/SS class could be useful for antimicrobial application – beside previously reported anticancer application – with initial investigation using chlorins as dual PS/SS agent [459]. Since this study by Harris et al., a few dual-activated PS/SS have been described that would warrant further investigations [460,461].

5.4.2. aPDT and Electrochemotherapy

Electrochemotherapy, also called pulsed electromagnetic fields (PEMFs) or electropermeabilization, is a method consisting in applying an electrical field to cells in order to enhance their permeability to therapeutic molecules (often chemical drugs or DNA). Combination of PDT with electrochemotherapy has been used many times to treat cancer diseases [14,462,463]. One study showed that, compared with aPDT used alone, hypericin combined with electrochemotherapy allows to achieve more than 2 to 3 $\log_{10}$ CFU reduction in *E. coli* and *S. aureus*, respectively [407]. To our knowledge, no other study combining electrochemotherapy with aPDT was recorded since then. However, combination of aPDT and a cell-permeabilization technique with a controllable toxicity degree may be highly relevant since most PSs can act in extracellular medium without having a specific target. Accordingly, electrochemotherapy may allow boosting aPDT activity by promoting PSs internalization into target microorganisms.

5.5. aPDT and Other Antimicrobial-Related Therapies

Immunotherapeutic effects may be obtained as a result of PDT itself or due to other treatments used in combination. For instance, Schiff base complexes with differential immune-stimulatory and immune-modulatory activities were reported efficient to eliminate both Gram(+) and Gram(−) bacteria. Furthermore, upon photoactivation, these complexes blocked the production of the inflammatory cytokine TNFα, thereby allowing to treat at once bacterial infections associated with damaging inflammation [403]. One recent study proposed the first application of antimicrobial photoimmunotherapy (PIT) by developing a PS-antibody complex, selective to the HIV antigen anchored to the infected cell membranes [404]. Such an approach supports the therapeutic applicability of PDT against antimicrobial infections, especially those mediated by intracellular pathogens. In addition, photodynamic therapy using PSs at sub-lethal concentrations may exhibit interesting properties for inflammatory and infectious conditions [464]. It was shown effective to alter immune cell function and alleviate immune-mediated disease, to hasten the process of wound healing, and to enhance antibacterial immunity. PDT thus appears as a promising therapeutic modality in infectious and chronic inflammatory diseases such as inflammatory bowel disease and arthritis.

6. Other aPDT Perspectives: New Strategies to Efficiently Target Bacteria

Irrespective of the biomedical applications, achieving a precise targeting is crucial to guarantee both efficiency and specificity. For anticancer PDT, many targeting studies have been done, notably for evaluating PSs covalently attached to molecules having affinity for neoplasia or ligands for receptors expressed on tumors. By this way, PSs may be chosen considering primarily their ability to achieve high PDT effects rather than depending on their intrinsic targeting properties. Following the same rationale, aPDT-based combinatory systems can be developed, benefiting from earlier studies performed in multimodal oncology [465].

6.1. Aggregation-Induced Emission (AIE) Luminogens

AIE luminogens exhibit, in the aggregated state, nonradiative decay and show bright fluorescence due to the restriction of intramolecular motions [9]. Recently, their interests for antimicrobial applications have been reported, showing the possibility to simultaneously perform detection and image-guided elimination of bacteria for theranostics applications [466]. In comparison with classical PSs, AIE luminogens in an aggregated state do not exhibit self-quenched fluorescence and ROS production is better. For instance, Gao et al. reported a tetraphenylethylene-based discrete organoplatinum(II) metallacycle electrostatically assembled with a peptide-decorated virus coat protein. This assembly showed strong membrane-intercalating ability, especially in Gram($-$) bacteria, and behaved as a potent AIE-PS upon light irradiation [467].

6.2. Photochemical Internalization (PCI)

PCI may be used to enhance cell internalization of diverse macromolecules. It consists of PDT-induced disruption of endocytic vesicles and lysosomes improving the release of their payloads into the cytoplasm of target cells. Although most PCI applications relate to cancer treatments, PCI could be extended to treat intracellular infections by delivering antimicrobials into infected cells [468]. For instance, Zhang et al. reported PCI as an antibiotic delivery strategy allowing to enhance cytoplasmic release of Gentamicin, to counter intracellular staphylococcal infection in eukaryotic cells and in zebrafish embryos [469].

6.3. Genetically-Encoded PSs

Internalization of PSs inside target microorganisms could be facilitated thanks to their conjugation with adjuvants, as mentioned before. Alternatively, it could be possible to use genetically-encoded ROS-generating proteins (RGPs), also called genetically-encoded PSs. Such an approach represents a powerful way to "completely localize" PSs inside target microorganisms for highly specific antimicrobial phototoxicity. Furthermore, in situ production of RGPs allows to enhance interaction with intracellular targets and better control the biodistribution of PSs, while limiting side-effects for the host tissues and environments [470]. To date, two groups of RGPs have been reported; those that belong to the green fluorescent protein (GFP) family and form their chromophores auto catalytically, and those that use external ubiquitous co-factors (flavins) as chromophores [471]. For example, Endres et al. compared eleven light-oxygen-voltage-based flavin binding fluorescent proteins and showed that most were potent PSs for light-controlled killing of bacteria [472].

6.4. pH-Sensitive aPDT

Some studies have reported smart photoactive systems consisting in PSs assembled in nanoconjugates with acid-cleavable linkers. For instance, Staegemann et al. described porphyrins conjugated with acid-labile benzacetal linkers and demonstrated the cleavage of the active PS agents from the polymer carrier in the acidic bacterial environment [316]. In addition, photoacids may be useful to design pH-sensitive aPDT systems. Upon light irradiation, such agents promote the spatial and temporal control of proton-release processes and could provide a way to convert photoenergy into other types of energy [473]. Thus, proof-of-concept was reported for the use of reversible photo-switchable chemicals as antimicrobials inducing MDR bacteria photoinactivation mediated by the acidification of intercellular environment [474]. To our knowledge, no studies have yet reported the potential of photoacids in combination with aPDT systems. However, some pH-sensitive PSs can induce remarkable variations of antimicrobial photoinactivation levels under different environmental pH [475]. These observations suggest the potential of photoacids as PDT potentiators for enhanced antimicrobial applications.

6.5. DNA Origami as PS Carriers

The quite recent development of DNA origami based on well-established DNA nanotechnology can serve as an excellent scaffold for the functionalization with different kinds of molecules and could be a powerful tool, as described by Yang at al., to study in a real-time conditions the assemble/disassemble of photo-controllable nanostructures [476]. Oligonucleotides organized as DNA origami could thus be used as PS-carrying nanostructures featuring numerous and dense intercalation sites. In addition, the tightly packed double helices can avoid the degradation by DNA hydrolases in the cellular environment. For instance, Zhuang et al. reported the uptake in tumor cells of a PS-loaded DNA origami nanostructure where it generated free radicals, releasing PSs due to DNA photocleavage, and induced cell apoptosis [477]. To our knowledge, such an approach has not yet been investigated for antimicrobial purposes.

7. Discussion

Antimicrobial PDT has the potential to fight against a wide spectrum of infective agents, including those resistant to conventional antimicrobials, under non-clinical and clinical settings. Rather than replacement, aPDT may be a complementary approach to reduce the use of current, especially last resort, antimicrobials. This review aims to give a non-exhaustive overview of the diversity and richness of synthetic, natural, or hybrid single PSs and aPDT nanosystems that were recently reported, with respect to their specific advantages, limitations, and possible evolutions. It is noteworthy that many systems and strategies primarily developed for anticancer PDT have been or could be applied—*per se* or following adaptations—to antimicrobial applications.

Beyond the "chemical space" that can be explored with individual PSs, versatile combinations with other compounds can allow the design of multimodular/multimodal systems. Along this line, the various PSs available may be considered as "basic ingredients". Apart from offering alternative possibilities for overcoming the most common limitations of PSs (i.e., solubility, delivery, and specificity), reasons for implementing PSs in complex systems can be related to (i) the ability to target several types of pathogens at once (extended antimicrobial spectrum), (ii) synergistic antimicrobial effects, (iii) reduction in the dose of each combined component, (iv) beneficial effects in severe poly-pathogenic infections, and (v) reduction in the risk of resistance emergence.

In addition to the many possible variations concerning PSs, optimizations can also consider other critical parameters in PDT, namely light irradiation and oxygenation. The latter was considered for a long time as an indispensable component. However, control of the oxygen level in the aPDT system is questionable. Indeed, additional oxygen-independent phototoxic mechanisms have been reported, for example with psoralens, which can produce more effective aPDT without oxygen [21]. Furthermore, recent studies suggest that various strategies could be used to reduce or bypass the limitations of oxygen and light supplies (read below). All combined, optimizations targeting not only the PSs, but also light irradiation and oxygen supply, could allow to evolve toward integrative, highly sophisticated, antimicrobial photodynamic therapy.

Light irradiation and its various modalities have been reviewed in depth by different authors [12,478–480]. Typically, the irradiation in PDT occurs in the UV (200–400 nm) or in the visible light (400–700 nm) with a power of ≤ 100 mW [479]. However, the low light-penetration depth (around mm) and possible occurrence of tissue photodamage limit the applicability of this spectral range for PDT. This can be circumvented by application of a near-infrared (NIR) irradiation (750–1100 nm), particularly via a two-photon excitation, which is emerging for PDT applications [223,481]. Being a third-order nonlinear optic phenomenon and corresponding to the simultaneous absorption of two photons with half the resonant energy, it allows deeper penetration in biological tissues (around 2 cm), lower scattering losses, and a three-dimensional spatial resolution [482]. Light sources are also constantly improving. Laser is an exceptional source of radiation, capable of producing extremely fine spectral bands, intense, coherent electromagnetic fields ranging from NIR to UV. In comparison, LEDs feature other advantages, notably the possibility to be arranged in many ways, in large quantity, for irradiating wide areas while inducing negligible heating [478,479].

Considering that light-emitted sources can become a brake to any PDT applications, several promising options have been envisaged quite recently. Notably, Blum et al. have identified a series of "self-exiting way" that allows to abolish the need of light to achieve efficient PDT [483]. Among them, chemiluminescence was extensively investigated by using luminol or luciferase energy transfer to induce a chemiexcitation of PS as antibacterial therapeutics [484]. In addition, other methods may be based on Cherenkov radiation, which occurs when an emitted charged particle, such as an electron, moves with high speed through a dielectric medium, such as water. Thus, the polarization of electrons in the medium produces electromagnetic waves in the visible wavelengths that could activate PDT reaction. Another way to induce Cherenkov radiation is to use radioactive isotopes with high beta emissions (e.g., ^{18}F, ^{64}Cu, or ^{68}Ga) as an electronic excitation source [483]. The relevance of such approaches for the design of a self-induced aPDT system remains to be evaluated. In addition, another external source of excitation, such as microwaves, could activate photoactive molecules such as Fe_3O_4 when they are complexed with carbon nanotubes. This was recently evaluated by Qiao et al. for treating MRSA-infected osteomyelitis [48]. Furthermore, X-ray as an activated source can also facilitate the activation of the PDT system by transferring energy harvested from X-ray irradiation to the PS used [485]. Kamanli et al. compared pulse and superpulse radiation modes, showing that the latter is more effective to produce $^1O_2$ and *S. aureus* eradication than the former [486].

Oxygen is also a critical limiting factor that determines PDT efficiency, especially in poorly oxygenated environments. However, it is noteworthy that PDT can be achieved in cancers that typically feature a low oxygenation rate. To alleviate the possible limitations of oxygen supply in PDT systems, several strategies have also been recently considered. One is based on catalase grafting to achieve oxygen self-sufficient NPs in order to convert H_2O_2 into available dissolved oxygen in the tumor environment. The abundant ROS in tumors compared with normal tissues provide a coherent substrate for catalase and thus allows an improvement of PDT activity [487]. Some multifunctional nanomaterials, called nanozymes, can also be used in combination with PSs to achieve a catalase-like activity supplying an oxygen source for the PS functioning [488,489]. In addition, it was also reported that noble metal NPs – such as Ru, Pt, and Au NPs – exhibit catalase-like nanozyme activities [490]. The use of catalases or nanozymes may have the crucial role of oxygen helpers in aPDT for treating deep infections.

The possibilities to design combinatorial aPDT strategies seem unlimited. Those presented in this review partly overlap with the description of aPDT systems, showing the complexity of any classification process. Awareness and caution may be raised about some sort of "paradox" or "dilemma"; indeed, while the seemingly boundless collection of chemical options and modular tools to develop nanoscale aPDT therapeutics implicitly defines extensive design flexibility, it staggeringly complicates and bewilders at the same time the optimization process aiming to compare, rationalize, and identify the "best" option for each application. Along this line, data concerning structure-activity relationships are usually missing, with not many studies yet dedicated to this matter [11,42,91,491]. Similar to the CLSI guidelines defined for antibiotics, uniform research methodologies would be useful to assay aPDT systems under well-defined standards and guidelines, considering notably (i) illumination settings, (ii) positive controls [492], (iii) microorganisms and cell lines relevant for a given application, and (iv) assessment of antimicrobial efficacy; this would guarantee better-conducted preclinical and clinical trials of aPDT systems used as mono or combinatory therapy. Furthermore, a public database compiling the efficacy/side effects of various systems would also be useful, facilitating meta-analyses for delineating (quantitative) structure/activity relationships and computational simulations [493].

Finally, in view of translation to clinical practice, a series of precautions and potential limitations must also be considered, especially when dealing with combinatory strategies. Beside possible reciprocal interferences (antagonisms) between combined partners, safety and specificity parameters must also be carefully examined. One main advantage of aPDT is the possibility to control the production of ROS thanks to the use of nontoxic PSs triggered with inducers (specific light and free oxygen) and/or enhancers. The vast majority of PSs are considered safe and dark cytotoxic side effects toward non-target eukaryotic cells are rarely reported. However, in many cases, more studies are needed for examining—beyond potential side-effects—other parameters such as bioavailability, biodispersion, persistence in host cells and body, and elimination pathways. With regard to combinatory strategies, it is noteworthy that studies conducted to date were mostly based on in vitro evaluation or using animal models bearing well-defined sites of infection (Table 2). Thus, much more data in preclinical and clinical settings are required to support the actual potential of such strategies. Moreover, considering that many microbial infections are systemic, the use of modalities such as sonodynamic therapy or electrochemotherapy is at present not realistically feasible. Therefore, in spite of significant progress and real promise, further important work/innovations are needed to effectively broaden the range of infectious conditions that could be treated via aPDT approaches. Lastly, cost effectiveness of multiplexed aPDT therapies over monomodal conventional antimicrobial agents is another crucial point to be considered in view of clinical applications. Potent, but too expensive solutions may fail to be used in practice, especially in under-developed countries where conventional antimicrobials will continue to be used, allowing microorganisms to increase in resistance.

8. Concluding Remarks

In conclusion, aPDT is a versatile approach that tends to evolve from quite a simple method to reach much higher degrees of complexity, with several expected advantages, but also possible drawbacks or undesirable effects. Among the latter, any direct/indirect impact on AMR should be more thoroughly considered. It is noteworthy that aPDT clinical trials conducted to date evaluated quite simple PSs. Any increase in the complexity of therapeutic systems would lead to an increase in difficulty before being able to reach clinical applications. The development of effective and safe aPDT treatments requires expertise in many fields of research, including biology (microbiology, cell biology, biochemistry, pharmacology), chemistry, physics (optical physics), and engineering. This is even more the case with combinatory strategies involving different modalities as reviewed in this article. Translation to practical applications also implies strong collaborations with the different sectors of health care and pharmaceutical companies. In these conditions, aPDT and its many therapeutic combinations could become a frontline routine treatment to fight against microorganisms possibly responsible for the next healthcare crises [61].

Funding: This work was supported by grants from ANR/BMBF (TARGET-THERAPY; PIs: Tony Le Gall and Holger Schönherr; ANR grant number: ANR-20-AMRB-0009, RPV21103NNA; BMBF Förderkennzeichen: 16GW0342). We also thank the "Association de Transfusion Sanguine et de Biogénétique Gaétan Saleün" (France) and the "Conseil Régional de Bretagne" (France) for their financial support. Raphaëlle Youf is recipient of a PhD fellowship from the French "Ministère de l'Enseignement supérieur, de la Recherche et de l'Innovation" (Paris, France).

Conflicts of Interest: The authors declare no conflict of interest. The funders had no role in the design of the study; in the collection, analyses, or interpretation of data; in the writing of the manuscript, or in the decision to publish the results.

Abbreviations

AIE	aggregation-induced emission
ALA	alanine
AMP	antimicrobial peptide
AMR	antimicrobial resistance
aPDT	antimicrobial PDT
ConA	concanavalin A
CNTs	carbon nanotubes
Ce6	chlorin e6
CFU	colony forming unit
e^-	electron
ESKAPE	*Enterococcus faecium, Staphylococcus aureus, Klebsiella pneumoniae, Acinetobacter baumannii, Pseudomonas aeruginosa,* and *Enterobacter* spp.
IC	internal conversion
ICG	indocyanine green
ISC	inter-system crossing
GO	graphene oxide
H_2O_2	hydrogen peroxide
$HO^\bullet$	hydroxyl radical
MB	methylene blue
MDR	multidrug resistance
MRSA	methicillin resistant *S. aureus*
MOF	metal organic framework
NIR	near infrared
NO	nitic oxide
NP	nanoparticle
O_2	dioxygen
$O_2^{\bullet-}$	superoxide anion radical
$^1O_2$	singlet oxygen
$^3O_2$	ground state molecular oxygen
PDT	photodynamic therapy
PS	photosensitizer
$PS^{\bullet-}$	PS radical anion
^1PS	PS in the ground state
$^1PS^*$	PS in a first excited singlet state
$^3PS^*$	PS in a triplet excited state
PEG	poly(ethylene glycol)
PCI	photochemical internalization
PCL	poly(ε-caprolactone)
PLA	poly(lactic acid)
pSi	porous silicon
PTT	photothermal therapy
QD	quantum dot
R	reduced molecule
$R^{\bullet+}$	oxidized molecule
RB	rose bengal
ROS	reactive oxygen species
SS	sonosensitizer
SDT	sonodynamic therapy
SPDT	sonophotodynamic therapy
SPION	superparamagnetic iron oxide NP
TB(O)	toluidine blue

UCNP   upconversion NP
UV   ultraviolet
WHO   world health organisation

References

1. Laxminarayan, R.; Matsoso, P.; Pant, S.; Brower, C.; Røttingen, J.-A.; Klugman, K.; Davies, S. Access to Effective Antimicrobials: A Worldwide Challenge. *Lancet* **2016**, *387*, 168–175. [CrossRef]
2. Mulani, M.S.; Kamble, E.E.; Kumkar, S.N.; Tawre, M.S.; Pardesi, K.R. Emerging Strategies to Combat ESKAPE Pathogens in the Era of Antimicrobial Resistance: A Review. *Front. Microbiol.* **2019**, *10*, 539. [CrossRef]
3. Antoñanzas, F.; Goossens, H. The Economics of Antibiotic Resistance: A Claim for Personalised Treatments. *Eur. J. Health Econ.* **2019**, *20*, 483–485. [CrossRef]
4. Cassini, A.; Högberg, L.D.; Plachouras, D.; Quattrocchi, A.; Hoxha, A.; Simonsen, G.S.; Colomb-Cotinat, M.; Kretzschmar, M.E.; Devleesschauwer, B.; Cecchini, M.; et al. Attributable Deaths and Disability-Adjusted Life-Years Caused by Infections with Antibiotic-Resistant Bacteria in the EU and the European Economic Area in 2015: A Population-Level Modelling Analysis. *Lancet Infect. Dis.* **2019**, *19*, 56–66. [CrossRef]
5. Dadgostar, P. Antimicrobial Resistance: Implications and Costs. *Infect. Drug Resist.* **2019**, *12*, 3903–3910. [CrossRef] [PubMed]
6. Hutchings, M.I.; Truman, A.W.; Wilkinson, B. Antibiotics: Past, Present and Future. *Curr. Opin. Microbiol.* **2019**, *51*, 72–80. [CrossRef]
7. Wainwright, M.; Maisch, T.; Nonell, S.; Plaetzer, K.; Almeida, A.; Tegos, G.P.; Hamblin, M.R. Photoantimicrobials—Are We Afraid of the Light? *Lancet Infect. Dis.* **2017**, *17*, e49–e55. [CrossRef]
8. Daniell, M.D.; Hill, J.S. A History of Photodynamic Therapy. *Aust. N. Z. J. Surg.* **1991**, *61*, 340–348. [CrossRef]
9. Shi, X.; Zhang, C.Y.; Gao, J.; Wang, Z. Recent Advances in Photodynamic Therapy for Cancer and Infectious Diseases. *WIREs Nanomed. NanoBiotechnol.* **2019**, *11*, e1560. [CrossRef] [PubMed]
10. Sabino, C.P.; Wainwright, M.; Ribeiro, M.S.; Sellera, F.P.; dos Anjos, C.; Baptista, M.d.S.; Lincopan, N. Global Priority Multidrug-Resistant Pathogens Do Not Resist Photodynamic Therapy. *J. Photochem. Photobiol. B Biol.* **2020**, *208*, 111893. [CrossRef] [PubMed]
11. Le Gall, T.; Lemercier, G.; Chevreux, S.; Tücking, K.-S.; Ravel, J.; Thétiot, F.; Jonas, U.; Schönherr, H.; Montier, T. Ruthenium(II) Polypyridyl Complexes as Photosensitizers for Antibacterial Photodynamic Therapy: A Structure-Activity Study on Clinical Bacterial Strains. *ChemMedChem* **2018**, *13*, 2229–2239. [CrossRef]
12. Cieplik, F.; Deng, D.; Crielaard, W.; Buchalla, W.; Hellwig, E.; Al-Ahmad, A.; Maisch, T. Antimicrobial Photodynamic Therapy—What We Know and What We Don't. *Crit. Rev. Microbiol.* **2018**, *44*, 571–589. [CrossRef]
13. Hu, X.; Huang, Y.-Y.; Wang, Y.; Wang, X.; Hamblin, M.R. Antimicrobial Photodynamic Therapy to Control Clinically Relevant Biofilm Infections. *Front. Microbiol.* **2018**, *9*, 1299. [CrossRef]
14. Kwiatkowski, S.; Knap, B.; Przystupski, D.; Saczko, J.; Kędzierska, E.; Knap-Czop, K.; Kotlińska, J.; Michel, O.; Kotowski, K.; Kulbacka, J. Photodynamic Therapy—Mechanisms, Photosensitizers and Combinations. *Biomed. Pharmacother.* **2018**, *106*, 1098–1107. [CrossRef] [PubMed]
15. Wozniak, A.; Grinholc, M. Combined Antimicrobial Activity of Photodynamic Inactivation and Antimicrobials—State of the Art. *Front. Microbiol.* **2018**, *9*, 930. [CrossRef]
16. Baier, J.; Maier, M.; Engl, R.; Landthaler, M.; Bäumler, W. Time-Resolved Investigations of Singlet Oxygen Luminescence in Water, in Phosphatidylcholine, and in Aqueous Suspensions of Phosphatidylcholine or HT29 Cells. *J. Phys. Chem. B* **2005**, *109*, 3041–3046. [CrossRef] [PubMed]
17. Maisch, T.; Baier, J.; Franz, B.; Maier, M.; Landthaler, M.; Szeimies, R.-M.; Bäumler, W. The Role of Singlet Oxygen and Oxygen Concentration in Photodynamic Inactivation of Bacteria. *Proc. Natl. Acad. Sci. USA* **2007**, *104*, 7223–7228. [CrossRef]
18. Ogilby, P.R. Singlet Oxygen: There Is Indeed Something New under the Sun. *Chem. Soc. Rev.* **2010**, *39*, 3181. [CrossRef]
19. Foote, C.S. Definition of Type I and Type II Photosensitized Oxidation. *Photochem. Photobiol.* **1991**, *54*, 659. [CrossRef]
20. Baptista, M.d.S.; Cadet, J.; Di Mascio, P.; Ghogare, A.A.; Greer, A.; Hamblin, M.R.; Lorente, C.; Nunez, S.C.; Ribeiro, M.S.; Thomas, A.H.; et al. Type I and II Photosensitized Oxidation Reactions: Guidelines and Mechanistic Pathways. *Photochem. Photobiol.* **2017**, *93*, 912–919. [CrossRef]
21. Hamblin, M.R.; Abrahamse, H. Oxygen-Independent Antimicrobial Photoinactivation: Type III Photochemical Mechanism? *Antibiotics* **2020**, *9*, 53. [CrossRef]
22. Quiroga, E.D.; Cormick, M.P.; Pons, P.; Alvarez, M.G.; Durantini, E.N. Mechanistic Aspects of the Photodynamic Inactivation of *Candida Albicans* Induced by Cationic Porphyrin Derivatives. *Eur. J. Med. Chem.* **2012**, *58*, 332–339. [CrossRef]
23. Wiehe, A.; O'Brien, J.M.; Senge, M.O. Trends and Targets in Antiviral Phototherapy. *Photochem. Photobiol. Sci.* **2019**, *18*, 2565–2612. [CrossRef]
24. Beirão, S.; Fernandes, S.; Coelho, J.; Faustino, M.A.F.; Tomé, J.P.C.; Neves, M.G.P.M.S.; Tomé, A.C.; Almeida, A.; Cunha, A. Photodynamic Inactivation of Bacterial and Yeast Biofilms with a Cationic Porphyrin. *Photochem. Photobiol.* **2014**, *90*, 1387–1396. [CrossRef]

25. Garcez, A.S.; Núñez, S.C.; Azambuja, N.; Fregnani, E.R.; Rodriguez, H.M.H.; Hamblin, M.R.; Suzuki, H.; Ribeiro, M.S. Effects of Photodynamic Therapy on Gram-Positive and Gram-Negative Bacterial Biofilms by Bioluminescence Imaging and Scanning Electron Microscopic Analysis. *Photomed. Laser Surg.* **2013**, *31*, 519–525. [CrossRef] [PubMed]
26. Alves, E.; Faustino, M.A.; Neves, M.G.; Cunha, A.; Tome, J.; Almeida, A. An Insight on Bacterial Cellular Targets of Photodynamic Inactivation. *Future Med. Chem.* **2014**, *6*, 141–164. [CrossRef]
27. Mesquita, M.Q.; Dias, C.J.; Neves, M.G.P.M.S.; Almeida, A.; Faustino, M.A.F. Revisiting Current Photoactive Materials for Antimicrobial Photodynamic Therapy. *Molecules* **2018**, *23*, 2424. [CrossRef]
28. Lam, M.; Jou, P.C.; Lattif, A.A.; Lee, Y.; Malbasa, C.L.; Mukherjee, P.K.; Oleinick, N.L.; Ghannoum, M.A.; Cooper, K.D.; Baron, E.D. Photodynamic Therapy with Pc 4 Induces Apoptosis of *Candida Albicans*. *Photochem. Photobiol.* **2011**, *87*, 904–909. [CrossRef] [PubMed]
29. Konopka, K.; Goslinski, T. Photodynamic Therapy in Dentistry. *J. Dent. Res.* **2007**, *86*, 694–707. [CrossRef]
30. Hirakawa, K.; Ota, K.; Hirayama, J.; Oikawa, S.; Kawanishi, S. Nile Blue Can Photosensitize DNA Damage through Electron Transfer. *Chem. Res. Toxicol.* **2014**, *27*, 649–655. [CrossRef]
31. Choi, S.S.; Lee, H.K.; Chae, H.S. In Vitro Photodynamic Antimicrobial Activity of Methylene Blue and Endoscopic White Light against *Helicobacter Pylori* 26695. *J. Photochem. Photobiol. B Biol.* **2010**, *101*, 206–209. [CrossRef] [PubMed]
32. Abdulrahman, H.; Misba, L.; Ahmad, S.; Khan, A.U. Curcumin Induced Photodynamic Therapy Mediated Suppression of Quorum Sensing Pathway of *Pseudomonas Aeruginosa*: An Approach to Inhibit Biofilm In Vitro. *Photodiagn. Photodyn. Ther.* **2020**, *30*, 101645. [CrossRef] [PubMed]
33. Boluki, E.; Moradi, M.; Azar, P.S.; Fekrazad, R.; Pourhajibagher, M.; Bahador, A. The Effect of Antimicrobial Photodynamic Therapy against Virulence Genes Expression in Colistin-Resistance *Acinetobacter Baumannii*: APDT in Genes Expression of Colistin-Resistance A. *Baumannii*. *Laser Ther.* **2019**, *28*, 27–33. [CrossRef]
34. Ghorbanzadeh, R.; Assadian, H.; Chiniforush, N.; Parker, S.; Pourakbari, B.; Ehsani, B.; Alikhani, M.Y.; Bahador, A. Modulation of Virulence in *Enterococcus Faecalis* Cells Surviving Antimicrobial Photodynamic Inactivation with Reduced Graphene Oxide-Curcumin: An *Ex Vivo* Biofilm Model. *Photodiagn. Photodyn. Ther.* **2020**, *29*, 101643. [CrossRef] [PubMed]
35. Pourhajibagher, M.; Plotino, G.; Chiniforush, N.; Bahador, A. Dual Wavelength Irradiation Antimicrobial Photodynamic Therapy Using Indocyanine Green and Metformin Doped with Nano-Curcumin as an Efficient Adjunctive Endodontic Treatment Modality. *Photodiagn. Photodyn. Ther.* **2020**, *29*, 101628. [CrossRef] [PubMed]
36. Marasini, S.; Leanse, L.G.; Dai, T. Can Microorganisms Develop Resistance against Light Based Anti-Infective Agents? *Adv. Drug Deliv. Rev.* **2021**, *175*, 113822. [CrossRef] [PubMed]
37. Tschowri, N.; Lindenberg, S.; Hengge, R. Molecular Function and Potential Evolution of the Biofilm-Modulating Blue Light-Signalling Pathway of *Escherichia Coli*. *Mol. Microbiol.* **2012**, *85*, 893–906. [CrossRef]
38. Prates, R.A.; Kato, I.T.; Ribeiro, M.S.; Tegos, G.P.; Hamblin, M.R. Influence of Multidrug Efflux Systems on Methylene Blue-Mediated Photodynamic Inactivation of *Candida Albicans*. *J. Antimicrob. Chemother.* **2011**, *66*, 1525–1532. [CrossRef] [PubMed]
39. Tegos, G.P.; Masago, K.; Aziz, F.; Higginbotham, A.; Stermitz, F.R.; Hamblin, M.R. Inhibitors of Bacterial Multidrug Efflux Pumps Potentiate Antimicrobial Photoinactivation. *Antimicrob. Agents Chemother.* **2008**, *52*, 3202–3209. [CrossRef] [PubMed]
40. Khan, S.; Khan, S.N.; Akhtar, F.; Misba, L.; Meena, R.; Khan, A.U. Inhibition of Multi-Drug Resistant *Klebsiella Pneumonae*: Nanoparticles Induced Photoinactivation in Presence of Efflux Pump Inhibitor. *Eur. J. Pharm. Biopharm.* **2020**, *157*, 165–174. [CrossRef]
41. Kashef, N.; Akbarizare, M.; Kamrava, S.K. Effect of Sub-Lethal Photodynamic Inactivation on the Antibiotic Susceptibility and Biofilm Formation of Clinical *Staphylococcus Aureus* Isolates. *Photodiagn. Photodyn. Ther.* **2013**, *10*, 368–373. [CrossRef] [PubMed]
42. Gollmer, A.; Felgenträger, A.; Bäumler, W.; Maisch, T.; Späth, A. A Novel Set of Symmetric Methylene Blue Derivatives Exhibits Effective Bacteria Photokilling—A Structure–Response Study. *Photochem. Photobiol. Sci.* **2015**, *14*, 335–351. [CrossRef] [PubMed]
43. Garcez, A.S.; Kaplan, M.; Jensen, G.J.; Scheidt, F.R.; Oliveira, E.M.; Suzuki, S.S. Effects of Antimicrobial Photodynamic Therapy on Antibiotic-Resistant *Escherichia Coli*. *Photodiagn. Photodyn. Ther.* **2020**, *32*, 102029. [CrossRef]
44. Paziani, M.H.; Tonani, L.; de Menezes, H.D.; Bachmann, L.; Wainwright, M.; Braga, G.Ú.L.; von Zeska Kress, M.R. Antimicrobial Photodynamic Therapy with Phenothiazinium Photosensitizers in Non-Vertebrate Model *Galleria Mellonella* Infected with *Fusarium Keratoplasticum* and *Fusarium Moniliforme*. *Photodiagn. Photodyn. Ther.* **2019**, *25*, 197–203. [CrossRef] [PubMed]
45. Santezi, C.; Reina, B.D.; Dovigo, L.N. Curcumin-Mediated Photodynamic Therapy for the Treatment of Oral Infections—A Review. *Photodiagn. Photodyn. Ther.* **2018**, *21*, 409–415. [CrossRef] [PubMed]
46. Cieplik, F.; Buchalla, W.; Hellwig, E.; Al-Ahmad, A.; Hiller, K.-A.; Maisch, T.; Karygianni, L. Antimicrobial Photodynamic Therapy as an Adjunct for Treatment of Deep Carious Lesions—A Systematic Review. *Photodiagn. Photodyn. Ther.* **2017**, *18*, 54–62. [CrossRef] [PubMed]
47. Briggs, T.; Blunn, G.; Hislop, S.; Ramalhete, R.; Bagley, C.; McKenna, D.; Coathup, M. Antimicrobial Photodynamic Therapy—a Promising Treatment for Prosthetic Joint Infections. *Lasers Med. Sci.* **2018**, *33*, 523–532. [CrossRef]
48. Qiao, Y.; Liu, X.; Li, B.; Han, Y.; Zheng, Y.; Yeung, K.W.K.; Li, C.; Cui, Z.; Liang, Y.; Li, Z.; et al. Treatment of MRSA-Infected Osteomyelitis Using Bacterial Capturing, Magnetically Targeted Composites with Microwave-Assisted Bacterial Killing. *Nat. Commun.* **2020**, *11*, 4446. [CrossRef] [PubMed]
49. Pappas, P.G.; Lionakis, M.S.; Arendrup, M.C.; Ostrosky-Zeichner, L.; Kullberg, B.J. Invasive Candidiasis. *Nat. Rev. Dis. Primers* **2018**, *4*, 18026. [CrossRef] [PubMed]

50. Baltazar, L.M.; Ray, A.; Santos, D.A.; Cisalpino, P.S.; Friedman, A.J.; Nosanchuk, J.D. Antimicrobial Photodynamic Therapy: An Effective Alternative Approach to Control Fungal Infections. *Front. Microbiol.* **2015**, *6*, 202. [CrossRef] [PubMed]

51. Cabrini Carmello, J.; Alves, F.; Basso, F.G.; de Souza Costa, C.A.; Tedesco, A.C.; Lucas Primo, F.; Mima, E.G.d.O.; Pavarina, A.C. Antimicrobial Photodynamic Therapy Reduces Adhesion Capacity and Biofilm Formation of *Candida Albicans* from Induced Oral Candidiasis in Mice. *Photodiagn. Photodyn. Ther.* **2019**, *27*, 402–407. [CrossRef]

52. Freire, F.; Ferraresi, C.; Jorge, A.O.C.; Hamblin, M.R. Photodynamic Therapy of Oral Candida Infection in a Mouse Model. *J. Photochem. Photobiol. B* **2016**, *159*, 161–168. [CrossRef] [PubMed]

53. Leanse, L.G.; Goh, X.S.; Dai, T. Quinine Improves the Fungicidal Effects of Antimicrobial Blue Light: Implications for the Treatment of Cutaneous Candidiasis. *Lasers Surg. Med.* **2020**, *52*, 569–575. [CrossRef] [PubMed]

54. Janeth Rimachi Hidalgo, K.; Cabrini Carmello, J.; Carolina Jordão, C.; Aboud Barbugli, P.; de Sousa Costa, C.A.; Mima, E.G.d.O.; Pavarina, A.C. Antimicrobial Photodynamic Therapy in Combination with Nystatin in the Treatment of Experimental Oral Candidiasis Induced by *Candida Albicans* Resistant to Fluconazole. *Pharmaceuticals* **2019**, *12*, 140. [CrossRef] [PubMed]

55. Shen, J.J.; Jemec, G.B.E.; Arendrup, M.C.; Saunte, D.M.L. Photodynamic Therapy Treatment of Superficial Fungal Infections: A Systematic Review. *Photodiagn. Photodyn. Ther.* **2020**, *31*, 101774. [CrossRef] [PubMed]

56. Liu, Z.; Tang, J.; Sun, Y.; Gao, L. Effects of Photodynamic Inactivation on the Growth and Antifungal Susceptibility of *Rhizopus Oryzae*. *Mycopathologia* **2019**, *184*, 315–319. [CrossRef]

57. De Clercq, E.; Li, G. Approved Antiviral Drugs over the Past 50 Years. *Clin. Microbiol. Rev.* **2016**, *29*, 695–747. [CrossRef] [PubMed]

58. Mohr, H.; Knüver-Hopf, J.; Gravemann, U.; Redecker-Klein, A.; Müller, T.H. West Nile Virus in Plasma Is Highly Sensitive to Methylene Blue-Light Treatment. *Transfusion* **2004**, *44*, 886–890. [CrossRef] [PubMed]

59. Ohtsuki, A.; Hasegawa, T.; Hirasawa, Y.; Tsuchihashi, H.; Ikeda, S. Photodynamic Therapy Using Light-Emitting Diodes for the Treatment of Viral Warts. *J. Dermatol.* **2009**, *36*, 525–528. [CrossRef]

60. Zhang, W.; Zhang, A.; Sun, W.; Yue, Y.; Li, H. Efficacy and Safety of Photodynamic Therapy for Cervical Intraepithelial Neoplasia and Human Papilloma Virus Infection. *Medicine* **2018**, *97*, e10864. [CrossRef]

61. Almeida, A.; Faustino, M.A.F.; Neves, M.G.P.M.S. Antimicrobial Photodynamic Therapy in the Control of COVID-19. *Antibiotics* **2020**, *9*, 320. [CrossRef]

62. Zarubaev, V.V.; Belousova, I.M.; Kiselev, O.I.; Piotrovsky, L.B.; Anfimov, P.M.; Krisko, T.C.; Muraviova, T.D.; Rylkov, V.V.; Starodubzev, A.M.; Sirotkin, A.C. Photodynamic Inactivation of Influenza Virus with Fullerene C60 Suspension in Allantoic Fluid. *Photodiagn. Photodyn. Ther.* **2007**, *4*, 31–35. [CrossRef] [PubMed]

63. Abada, Z.; Cojean, S.; Pomel, S.; Ferrié, L.; Akagah, B.; Lormier, A.T.; Loiseau, P.M.; Figadère, B. Synthesis and Antiprotozoal Activity of Original Porphyrin Precursors and Derivatives. *Eur. J. Med. Chem.* **2013**, *67*, 158–165. [CrossRef]

64. Espitia-Almeida, F.; Díaz-Uribe, C.; Vallejo, W.; Gómez-Camargo, D.; Romero Bohórquez, A.R. In Vitro Anti-Leishmanial Effect of Metallic Meso-Substituted Porphyrin Derivatives against *Leishmania Braziliensis* and *Leishmania Panamensis* Promastigotes Properties. *Molecules* **2020**, *25*, 1887. [CrossRef]

65. Alves, E.; Faustino, M.A.F.; Neves, M.G.P.M.S.; Cunha, Â.; Nadais, H.; Almeida, A. Potential Applications of Porphyrins in Photodynamic Inactivation beyond the Medical Scope. *J. Photochem. Photobiol. C Photochem. Rev.* **2015**, *22*, 34–57. [CrossRef]

66. de Souza, L.M.; Venturini, F.P.; Inada, N.M.; Iermak, I.; Garbuio, M.; Mezzacappo, N.F.; de Oliveira, K.T.; Bagnato, V.S. Curcumin in Formulations against *Aedes Aegypti*: Mode of Action, Photolarvicidal and Ovicidal Activity. *Photodiagn. Photodyn. Ther.* **2020**, *31*, 101840. [CrossRef]

67. Khater, H.; Hendawy, N.; Govindarajan, M.; Murugan, K.; Benelli, G. Photosensitizers in the Fight against Ticks: Safranin as a Novel Photodynamic Fluorescent Acaricide to Control the Camel Tick *Hyalomma Dromedarii* (Ixodidae). *Parasitol. Res.* **2016**, *115*, 3747–3758. [CrossRef] [PubMed]

68. Biel, M.A.; Pedigo, L.; Gibbs, A.; Loebel, N. Photodynamic Therapy of Antibiotic Resistant Biofilms in a Maxillary Sinus Model. *Int. Forum Allergy Rhinol.* **2013**, *3*, 468–473. [CrossRef] [PubMed]

69. Biel, M.A.; Sievert, C.; Usacheva, M.; Teichert, M.; Wedell, E.; Loebel, N.; Rose, A.; Zimmermann, R. Reduction of Endotracheal Tube Biofilms Using Antimicrobial Photodynamic Therapy. *Lasers Surg. Med.* **2011**, *43*, 586–590. [CrossRef] [PubMed]

70. Karner, L.; Drechsler, S.; Metzger, M.; Hacobian, A.; Schädl, B.; Slezak, P.; Grillari, J.; Dungel, P. Antimicrobial Photodynamic Therapy Fighting Polymicrobial Infections—A Journey from In Vitro to In Vivo. *Photochem. Photobiol. Sci.* **2020**, *19*, 1332–1343. [CrossRef] [PubMed]

71. Calzavara-Pinton, P.G.; Venturini, M.; Capezzera, R.; Sala, R.; Zane, C. Photodynamic Therapy of Interdigital Mycoses of the Feet with Topical Application of 5-Aminolevulinic Acid. *Photodermatol. Photoimmunol. Photomed.* **2004**, *20*, 144–147. [CrossRef] [PubMed]

72. Lin, J.; Wan, M.T. Current Evidence and Applications of Photodynamic Therapy in Dermatology. *Clin. Cosmet. Investig. Dermatol.* **2014**, *7*, 145. [CrossRef] [PubMed]

73. Mannucci, E.; Genovese, S.; Monami, M.; Navalesi, G.; Dotta, F.; Anichini, R.; Romagnoli, F.; Gensini, G. Photodynamic Topical Antimicrobial Therapy for Infected Foot Ulcers in Patients with Diabetes: A Randomized, Double-Blind, Placebo-Controlled Study—The D.A.N.T.E (Diabetic Ulcer Antimicrobial New Topical Treatment Evaluation) Study. *Acta Diabetol.* **2014**, *51*, 435–440. [CrossRef]

74. Morley, S.; Griffiths, J.; Philips, G.; Moseley, H.; O'Grady, C.; Mellish, K.; Lankester, C.L.; Faris, B.; Young, R.J.; Brown, S.B.; et al. Phase IIa Randomized, Placebo-Controlled Study of Antimicrobial Photodynamic Therapy in Bacterially Colonized, Chronic

Leg Ulcers and Diabetic Foot Ulcers: A New Approach to Antimicrobial Therapy. *Br. J. Dermatol.* **2013**, *168*, 617–624. [CrossRef] [PubMed]

75. Tardivo, J.P.; Adami, F.; Correa, J.A.; Pinhal, M.A.S.; Baptista, M.S. A Clinical Trial Testing the Efficacy of PDT in Preventing Amputation in Diabetic Patients. *Photodiagn. Photodyn. Ther.* **2014**, *11*, 342–350. [CrossRef] [PubMed]

76. Hamblin, M.R.; Viveiros, J.; Yang, C.; Ahmadi, A.; Ganz, R.A.; Tolkoff, M.J. Helicobacter Pylori Accumulates Photoactive Porphyrins and Is Killed by Visible Light. *Antimicrob. Agents Chemother.* **2005**, *49*, 2822–2827. [CrossRef]

77. Rahimi, R.; Fayyaz, F.; Rassa, M. The Study of Cellulosic Fabrics Impregnated with Porphyrin Compounds for Use as Photo-Bactericidal Polymers. *Mater. Sci. Eng. C* **2016**, *59*, 661–668. [CrossRef]

78. Fonseca, G.A.M.D.; Dourado, D.C.; Barreto, M.P.; Cavalcanti, M.F.X.B.; Pavelski, M.D.; Ribeiro, L.B.Q.; Frigo, L. Antimicrobial Photodynamic Therapy (APDT) for Decontamination of High-Speed Handpieces: A Comparative Study. *Photodiagn. Photodyn. Ther.* **2020**, *30*, 101686. [CrossRef] [PubMed]

79. Abrahamse, H.; Hamblin, M.R. New Photosensitizers for Photodynamic Therapy. *Biochem. J.* **2016**, *473*, 347–364. [CrossRef] [PubMed]

80. Ghorbani, J.; Rahban, D.; Aghamiri, S.; Teymouri, A.; Bahador, A. Photosensitizers in Antibacterial Photodynamic Therapy: An Overview. *Laser Ther.* **2018**, *27*, 293–302. [CrossRef] [PubMed]

81. Polat, E.; Kang, K. Natural Photosensitizers in Antimicrobial Photodynamic Therapy. *Biomedicines* **2021**, *9*, 584. [CrossRef] [PubMed]

82. Ahmad, A. Phthalocyanines Derivatives as Control Approach for Antimicrobial Photodynamic Therapy. *J. Am. J. Clin. Microbiol. Antimicrob.* **2019**, *2*, 8.

83. Amos-Tautua, B.M.; Songca, S.P.; Oluwafemi, O.S. Application of Porphyrins in Antibacterial Photodynamic Therapy. *Molecules* **2019**, *24*, 2456. [CrossRef]

84. Jeon, Y.-M.; Lee, H.-S.; Jeong, D.; Oh, H.-K.; Ra, K.-H.; Lee, M.-Y. Antimicrobial Photodynamic Therapy Using Chlorin E6 with Halogen Light for Acne Bacteria-Induced Inflammation. *Life Sci.* **2015**, *124*, 56–63. [CrossRef]

85. Li, X.; Lee, D.; Huang, J.-D.; Yoon, J. Phthalocyanine-Assembled Nanodots as Photosensitizers for Highly Efficient Type I Photoreactions in Photodynamic Therapy. *Angew. Chem. Int. Ed.* **2018**, *57*, 9885–9890. [CrossRef]

86. Terra Garcia, M.; Correia Pereira, A.H.; Figueiredo-Godoi, L.M.A.; Jorge, A.O.C.; Strixino, J.F.; Junqueira, J.C. Photodynamic Therapy Mediated by Chlorin-Type Photosensitizers against *Streptococcus Mutans* Biofilms. *Photodiagn. Photodyn. Ther.* **2018**, *24*, 256–261. [CrossRef] [PubMed]

87. Zhang, Y.; Lovell, J.F. Recent Applications of Phthalocyanines and Naphthalocyanines for Imaging and Therapy. *Wiley Interdiscip. Rev. Nanomed. Nanobiotechnol.* **2017**, *9*, e1420. [CrossRef]

88. Hirose, M.; Yoshida, Y.; Horii, K.; Hasegawa, Y.; Shibuya, Y. Efficacy of Antimicrobial Photodynamic Therapy with Rose Bengal and Blue Light against Cariogenic Bacteria. *Arch Oral Biol.* **2021**, *122*, 105024. [CrossRef] [PubMed]

89. Galdino, D.Y.T.; da Rocha Leódido, G.; Pavani, C.; Gonçalves, L.M.; Bussadori, S.K.; Benini Paschoal, M.A. Photodynamic Optimization by Combination of Xanthene Dyes on Different Forms of *Streptococcus Mutans*: An In Vitro Study. *Photodiagnosis Photodyn. Ther.* **2021**, *33*, 102191. [CrossRef]

90. Silva, A.F.; Dos Santos, A.R.; Trevisan, D.A.C.; Bonin, E.; Freitas, C.F.; Batista, A.F.P.; Hioka, N.; Simões, M.; Graton Mikcha, J.M. Xanthene Dyes and Green LED for the Inactivation of Foodborne Pathogens in Planktonic and Biofilm States. *Photochem Photobiol* **2019**, *95*, 1230–1238. [CrossRef]

91. Mizuno, K.; Zhiyentayev, T.; Huang, L.; Khalil, S.; Nasim, F.; Tegos, G.P.; Gali, H.; Jahnke, A.; Wharton, T.; Hamblin, M.R. Antimicrobial Photodynamic Therapy with Functionalized Fullerenes: Quantitative Structure-Activity Relationships. *J. Nanomed. Nanotechnol.* **2011**, *2*, 1–9. [CrossRef] [PubMed]

92. Sharma, S.K.; Chiang, L.Y.; Hamblin, M.R. Photodynamic Therapy with Fullerenes In Vivo: Reality or a Dream? *Nanomedicine* **2011**, *6*, 1813–1825. [CrossRef] [PubMed]

93. Yin, R.; Hamblin, M. Antimicrobial Photosensitizers: Drug Discovery Under the Spotlight. *Curr. Med. Chem.* **2015**, *22*, 2159–2185. [CrossRef] [PubMed]

94. Ali, I.; Alsehli, M.; Scotti, L.; Tullius Scotti, M.; Tsai, S.-T.; Yu, R.-S.; Hsieh, M.F.; Chen, J.-C. Progress in Polymeric Nano-Medicines for Theranostic Cancer Treatment. *Polymers* **2020**, *12*, 598. [CrossRef]

95. Josefsen, L.B.; Boyle, R.W. Photodynamic Therapy and the Development of Metal-Based Photosensitisers. *Met.-Based Drugs* **2008**, *2008*, 1–23. [CrossRef]

96. Frei, A. Metal Complexes, an Untapped Source of Antibiotic Potential? *Antibiotics* **2020**, *9*, 90. [CrossRef] [PubMed]

97. Frei, A.; Zuegg, J.; Elliott, A.G.; Baker, M.; Braese, S.; Brown, C.; Chen, F.; Dowson, C.G.; Dujardin, G.; Jung, N.; et al. Metal Complexes as a Promising Source for New Antibiotics. *Chem. Sci.* **2020**, *11*, 2627–2639. [CrossRef] [PubMed]

98. Morrison, C.N.; Prosser, K.E.; Stokes, R.W.; Cordes, A.; Metzler-Nolte, N.; Cohen, S.M. Expanding Medicinal Chemistry into 3D Space: Metallofragments as 3D Scaffolds for Fragment-Based Drug Discovery. *Chem. Sci.* **2020**, *11*, 1216–1225. [CrossRef] [PubMed]

99. Galloway, W.R.J.D.; Isidro-Llobet, A.; Spring, D.R. Diversity-Oriented Synthesis as a Tool for the Discovery of Novel Biologically Active Small Molecules. *Nat. Commun.* **2010**, *1*, 80. [CrossRef]

100. Hung, A.W.; Ramek, A.; Wang, Y.; Kaya, T.; Wilson, J.A.; Clemons, P.A.; Young, D.W. Route to Three-Dimensional Fragments Using Diversity-Oriented Synthesis. *Proc. Natl. Acad. Sci. USA* **2011**, *108*, 6799–6804. [CrossRef]

101. Claudel, M.; Schwarte, J.V.; Fromm, K.M. New Antimicrobial Strategies Based on Metal Complexes. *Chemistry* **2020**, *2*, 56. [CrossRef]
102. Li, F.; Collins, J.G.; Keene, F.R. Ruthenium Complexes as Antimicrobial Agents. *Chem. Soc. Rev.* **2015**, *44*, 2529–2542. [CrossRef]
103. Deng, T.; Zhao, H.; Shi, M.; Qiu, Y.; Jiang, S.; Yang, X.; Zhao, Y.; Zhang, Y. Photoactivated Trifunctional Platinum Nanobiotics for Precise Synergism of Multiple Antibacterial Modes. *Small* **2019**, *15*, 1902647. [CrossRef] [PubMed]
104. Kirakci, K.; Zelenka, J.; Rumlová, M.; Cvačka, J.; Ruml, T.; Lang, K. Cationic Octahedral Molybdenum Cluster Complexes Functionalized with Mitochondria-Targeting Ligands: Photodynamic Anticancer and Antibacterial Activities. *Biomater. Sci.* **2019**, *7*, 1386–1392. [CrossRef] [PubMed]
105. Sudhamani, C.N.; Bhojya Naik, H.S.; Sangeetha Gowda, K.R.; Girija, D.; Giridhar, M. DNA Binding, Prominent Photonuclease Activity and Antibacterial PDT of Cobalt(II) Complexes of Phenanthroline Based Photosensitizers. *Nucleosides Nucleotides Nucleic Acids* **2018**, *37*, 546–562. [CrossRef] [PubMed]
106. Wang, L.; Monro, S.; Cui, P.; Yin, H.; Liu, B.; Cameron, C.G.; Xu, W.; Hetu, M.; Fuller, A.; Kilina, S.; et al. Heteroleptic Ir(III)N$_6$ Complexes with Long-Lived Triplet Excited States and In Vitro Photobiological Activities. *ACS Appl. Mater. Interfaces* **2019**, *11*, 3629–3644. [CrossRef]
107. Munteanu, A.-C.; Uivarosi, V. Ruthenium Complexes in the Fight against Pathogenic Microorganisms. An Extensive Review. *Pharmaceutics* **2021**, *13*, 874. [CrossRef]
108. Jain, A.; Garrett, N.T.; Malone, Z.P. Ruthenium-based Photoactive Metalloantibiotics. *Photochem. Photobiol.* **2021**, php.13435. [CrossRef] [PubMed]
109. Frei, A.; Rubbiani, R.; Tubafard, S.; Blacque, O.; Anstaett, P.; Felgenträger, A.; Maisch, T.; Spiccia, L.; Gasser, G. Synthesis, Characterization, and Biological Evaluation of New Ru(II) Polypyridyl Photosensitizers for Photodynamic Therapy. *J. Med. Chem.* **2014**, *57*, 7280–7292. [CrossRef] [PubMed]
110. Hamblin, M.R.; Hasan, T. Photodynamic Therapy: A New Antimicrobial Approach to Infectious Disease? *Photochem. Photobiol. Sci.* **2004**, *3*, 436–450. [CrossRef]
111. Sun, W.; Jian, Y.; Zhou, M.; Yao, Y.; Tian, N.; Li, C.; Chen, J.; Wang, X.; Zhou, Q. Selective and Efficient Photoinactivation of Intracellular *Staphylococcus Aureus* and MRSA with Little Accumulation of Drug Resistance: Application of a Ru(II) Complex with Photolabile Ligands. *J. Med. Chem.* **2021**, *64*, 7359–7370. [CrossRef] [PubMed]
112. Ibrahim, I.M.; Abdelmalek, D.H.; Elshahat, M.E.; Elfiky, A.A. COVID-19 Spike-Host Cell Receptor GRP78 Binding Site Prediction. *J. Infect.* **2020**, *80*, 554–562. [CrossRef] [PubMed]
113. Wang, L.; Hu, C.; Shao, L. The Antimicrobial Activity of Nanoparticles: Present Situation and Prospects for the Future. *Int. J. Nanomed.* **2017**, *12*, 1227–1249. [CrossRef]
114. Eleraky, N.E.; Allam, A.; Hassan, S.B.; Omar, M.M. Nanomedicine Fight against Antibacterial Resistance: An Overview of the Recent Pharmaceutical Innovations. *Pharmaceutics* **2020**, *12*, 142. [CrossRef]
115. Khorsandi, K.; Fekrazad, S.; Vahdatinia, F.; Farmany, A.; Fekrazad, R. Nano Antiviral Photodynamic Therapy: A Probable Biophysicochemical Management Modality in SARS-CoV-2. *Expert Opin. Drug Deliv.* **2021**, *18*, 265–272. [CrossRef]
116. Perni, S.; Prokopovich, P.; Pratten, J.; Parkin, I.P.; Wilson, M. Nanoparticles: Their Potential Use in Antibacterial Photodynamic Therapy. *Photochem. Photobiol. Sci.* **2011**, *10*, 712–720. [CrossRef] [PubMed]
117. Basavegowda, N.; Baek, K.-H. Multimetallic Nanoparticles as Alternative Antimicrobial Agents: Challenges and Perspectives. *Molecules* **2021**, *26*, 912. [CrossRef]
118. Zhou, Z.; Peng, S.; Sui, M.; Chen, S.; Huang, L.; Xu, H.; Jiang, T. Multifunctional Nanocomplex for Surface-Enhanced Raman Scattering Imaging and near-Infrared Photodynamic Antimicrobial Therapy of Vancomycin-Resistant Bacteria. *Colloids Surf. B Biointerfaces* **2018**, *161*, 394–402. [CrossRef]
119. Gupta, A.; Landis, R.F.; Rotello, V.M. Nanoparticle-Based Antimicrobials: Surface Functionality Is Critical. *F1000Res* **2016**, *5*. [CrossRef] [PubMed]
120. Alfirdous, R.A.; Garcia, I.M.; Balhaddad, A.A.; Collares, F.M.; Martinho, F.C.; Melo, M.A.S. Advancing Photodynamic Therapy for Endodontic Disinfection with Nanoparticles: Present Evidence and Upcoming Approaches. *Appl. Sci.* **2021**, *11*, 4759. [CrossRef]
121. Trigo-Gutierrez, J.K.; Vega-Chacón, Y.; Soares, A.B.; Mima, E.G.d.O. Antimicrobial Activity of Curcumin in Nanoformulations: A Comprehensive Review. *Int. J. Mol. Sci.* **2021**, *22*, 7130. [CrossRef] [PubMed]
122. Yetisgin, A.A.; Cetinel, S.; Zuvin, M.; Kosar, A.; Kutlu, O. Therapeutic Nanoparticles and Their Targeted Delivery Applications. *Molecules* **2020**, *25*, 2193. [CrossRef] [PubMed]
123. Nair, A.B.; Morsy, M.A.; Shinu, P.; Kotta, S.; Chandrasekaran, M.; Tahir, A. Advances of Non-iron Metal Nanoparticles in Biomedicine. *J. Pharm. Pharm. Sci.* **2021**, *24*, 41–61. [CrossRef] [PubMed]
124. Belekov, E.; Kholikov, K.; Cooper, L.; Banga, S.; Er, A.O. Improved Antimicrobial Properties of Methylene Blue Attached to Silver Nanoparticles. *Photodiagn. Photodyn. Ther.* **2020**, *32*, 102012. [CrossRef]
125. Franci, G.; Falanga, A.; Galdiero, S.; Palomba, L.; Rai, M.; Morelli, G.; Galdiero, M. Silver Nanoparticles as Potential Antibacterial Agents. *Molecules* **2015**, *20*, 8856–8874. [CrossRef] [PubMed]
126. Mottais, A.; Berchel, M.; Sibiril, Y.; Laurent, V.; Gill, D.; Hyde, S.; Jaffrès, P.-A.; Montier, T.; Le Gall, T. Antibacterial Effect and DNA Delivery Using a Combination of an Arsonium-Containing Lipophosphoramide with an N-Heterocyclic Carbene-Silver Complex—Potential Benefits for Cystic Fibrosis Lung Gene Therapy. *Int. J. Pharm.* **2018**, *536*, 29–41. [CrossRef]

127. Ribeiro, M.S.; de Melo, L.S.A.; Farooq, S.; Baptista, A.; Kato, I.T.; Núñez, S.C.; de Araujo, R.E. Photodynamic Inactivation Assisted by Localized Surface Plasmon Resonance of Silver Nanoparticles: In Vitro Evaluation on *Escherichia Coli* and *Streptococcus Mutans*. *Photodiagn. Photodyn. Ther.* **2018**, *22*, 191–196. [CrossRef] [PubMed]
128. Yamamoto, M.; Shitomi, K.; Miyata, S.; Miyaji, H.; Aota, H.; Kawasaki, H. Bovine Serum Albumin-Capped Gold Nanoclusters Conjugating with Methylene Blue for Efficient $^1O_2$ Generation *via* Energy Transfer. *J. Colloid Interface Sci.* **2018**, *510*, 221–227. [CrossRef]
129. Rizzi, V.; Vurro, D.; Placido, T.; Fini, P.; Petrella, A.; Semeraro, P.; Cosma, P. Gold-Chlorophyll a-Hybrid Nanoparticles and Chlorophyll a/Cetyltrimethylammonium Chloride Self-Assembled-Suprastructures as Novel Carriers for Chlorophyll a Delivery in Water Medium: Photoactivity and Photostability. *Colloids Surf. B Biointerfaces* **2018**, *161*, 555–562. [CrossRef] [PubMed]
130. Miyata, S.; Miyaji, H.; Kawasaki, H.; Yamamoto, M.; Nishida, E.; Takita, H.; Akasaka, T.; Ushijima, N.; Iwanaga, T.; Sugaya, T. Antimicrobial Photodynamic Activity and Cytocompatibility of $Au_{25}(Capt)_{18}$ Clusters Photoexcited by Blue LED Light Irradiation. *Int. J. Nanomed.* **2017**, *12*, 2703–2716. [CrossRef] [PubMed]
131. Kubheka, G.; Uddin, I.; Amuhaya, E.; Mack, J.; Nyokong, T. Synthesis and Photophysicochemical Properties of BODIPY Dye Functionalized Gold Nanorods for Use in Antimicrobial Photodynamic Therapy. *J. Porphyr. Phthalocyanines* **2016**, *20*, 1016–1024. [CrossRef]
132. Okamoto, I.; Miyaji, H.; Miyata, S.; Shitomi, K.; Sugaya, T.; Ushijima, N.; Akasaka, T.; Enya, S.; Saita, S.; Kawasaki, H. Antibacterial and Antibiofilm Photodynamic Activities of Lysozyme-Au Nanoclusters/Rose Bengal Conjugates. *ACS Omega* **2021**, *6*, 9279–9290. [CrossRef]
133. Rigotto Caruso, G.; Tonani, L.; Marcato, P.D.; von Zeska Kress, M.R. Phenothiazinium Photosensitizers Associated with Silver Nanoparticles in Enhancement of Antimicrobial Photodynamic Therapy. *Antibiotics* **2021**, *10*, 569. [CrossRef]
134. Morales-de-Echegaray, A.V.; Lin, L.; Sivasubramaniam, B.; Yermembetova, A.; Wang, Q.; Abutaleb, N.S.; Seleem, M.N.; Wei, A. Antimicrobial Photodynamic Activity of Gallium-Substituted Haemoglobin on Silver Nanoparticles. *Nanoscale* **2020**, *12*, 21734–21742. [CrossRef] [PubMed]
135. Parasuraman, P.; Thamanna, R.Y.; Shaji, C.; Sharan, A.; Bahkali, A.H.; Al-Harthi, H.F.; Syed, A.; Anju, V.T.; Dyavaiah, M.; Siddhardha, B. Biogenic Silver Nanoparticles Decorated with Methylene Blue Potentiated the Photodynamic Inactivation of *Pseudomonas Aeruginosa* and *Staphylococcus Aureus*. *Pharmaceutics* **2020**, *12*, 709. [CrossRef] [PubMed]
136. Chen, J.; Yang, L.; Chen, J.; Liu, W.; Zhang, D.; Xu, P.; Dai, T.; Shang, L.; Yang, Y.; Tang, S.; et al. Composite of Silver Nanoparticles and Photosensitizer Leads to Mutual Enhancement of Antimicrobial Efficacy and Promotes Wound Healing. *Chem. Eng. J.* **2019**, *374*, 1373–1381. [CrossRef]
137. Sen, P.; Nyokong, T. Enhanced Photodynamic Inactivation of *Staphylococcus Aureus* with Schiff Base Substituted Zinc Phthalocyanines through Conjugation to Silver Nanoparticles. *J. Mol. Struct.* **2021**, *1232*, 130012. [CrossRef]
138. Shabangu, S.M.; Babu, B.; Soy, R.C.; Managa, M.; Sekhosana, K.E.; Nyokong, T. Photodynamic Antimicrobial Chemotherapy of Asymmetric Porphyrin-Silver Conjugates towards Photoinactivation of *Staphylococcus Aureus*. *J. Coord. Chem.* **2020**, *73*, 593–608. [CrossRef]
139. Wanarska, E.; Maliszewska, I. Gold Nanoparticles in an Enhancement of Antimicrobial Activity. *Physicochem. Probl. Miner. Process.* **2020**, *56*, 269–279. [CrossRef]
140. Managa, M.; Antunes, E.; Nyokong, T. Conjugates of Platinum Nanoparticles with Gallium Tetra—(4-Carboxyphenyl) Porphyrin and Their Use in Photodynamic Antimicrobial Chemotherapy When in Solution or Embedded in Electrospun Fiber. *Polyhedron* **2014**, *76*, 94–101. [CrossRef]
141. Ermini, M.L.; Voliani, V. Antimicrobial Nano-Agents: The Copper Age. *ACS Nano* **2021**, *15*, 6008–6029. [CrossRef]
142. Zhen, X.; Chudal, L.; Pandey, N.K.; Phan, J.; Ran, X.; Amador, E.; Huang, X.; Johnson, O.; Ran, Y.; Chen, W.; et al. A Powerful Combination of Copper-Cysteamine Nanoparticles with Potassium Iodide for Bacterial Destruction. *Mater. Sci. Eng. C Mater. Biol. Appl.* **2020**, *110*, 110659. [CrossRef] [PubMed]
143. Sah, B.; Wu, J.; Vanasse, A.; Pandey, N.K.; Chudal, L.; Huang, Z.; Song, W.; Yu, H.; Ma, L.; Chen, W.; et al. Effects of Nanoparticle Size and Radiation Energy on Copper-Cysteamine Nanoparticles for X-Ray Induced Photodynamic Therapy. *Nanomaterials* **2020**, *10*, 1087. [CrossRef] [PubMed]
144. Shkodenko, L.; Kassirov, I.; Koshel, E. Metal Oxide Nanoparticles Against Bacterial Biofilms: Perspectives and Limitations. *Microorganisms* **2020**, *8*, 1545. [CrossRef] [PubMed]
145. Namanga, J.; Foba, J.; Ndinteh, D.T.; Yufanyi, D.M.; Krause, R.W.M. Synthesis and Magnetic Properties of a Superparamagnetic Nanocomposite "Pectin-Magnetite Nanocomposite". *J. Nanomater.* **2013**, *2013*, 87. [CrossRef]
146. Dadfar, S.M.; Camozzi, D.; Darguzyte, M.; Roemhild, K.; Varvarà, P.; Metselaar, J.; Banala, S.; Straub, M.; Güvener, N.; Engelmann, U.; et al. Size-Isolation of Superparamagnetic Iron Oxide Nanoparticles Improves MRI, MPI and Hyperthermia Performance. *J. Nanobiotechnol.* **2020**, *18*, 22. [CrossRef]
147. Alves, E.; Rodrigues, J.M.M.; Faustino, M.A.F.; Neves, M.G.P.M.S.; Cavaleiro, J.A.S.; Lin, Z.; Cunha, Â.; Nadais, M.H.; Tomé, J.P.C.; Almeida, A. A New Insight on Nanomagnet–Porphyrin Hybrids for Photodynamic Inactivation of Microorganisms. *Dye. Pigment.* **2014**, *110*, 80–88. [CrossRef]
148. Bilici, K.; Atac, N.; Muti, A.; Baylam, I.; Dogan, O.; Sennaroglu, A.; Can, F.; Yagci Acar, H. Broad Spectrum Antibacterial Photodynamic and Photothermal Therapy Achieved with Indocyanine Green Loaded SPIONs under near Infrared Irradiation. *Biomater. Sci.* **2020**, *8*, 4616–4625. [CrossRef]

149. de Santana, W.M.O.S.; Caetano, B.L.; de Annunzio, S.R.; Pulcinelli, S.H.; Ménager, C.; Fontana, C.R.; Santilli, C.V. Conjugation of Superparamagnetic Iron Oxide Nanoparticles and Curcumin Photosensitizer to Assist in Photodynamic Therapy. *Colloids Surf. B Biointerfaces* **2020**, *196*, 111297. [CrossRef] [PubMed]

150. Dube, E.; Soy, R.; Shumba, M.; Nyokong, T. Photophysicochemical Behaviour of Phenoxy Propanoic Acid Functionalised Zinc Phthalocyanines When Grafted onto Iron Oxide and Silica Nanoparticles: Effects in Photodynamic Antimicrobial Chemotherapy. *J. Lumin.* **2021**, *234*, 117939. [CrossRef]

151. Mapukata, S.; Nwahara, N.; Nyokong, T. The Photodynamic Antimicrobial Chemotherapy of *Stapphylococcus Aureus* Using an Asymmetrical Zinc Phthalocyanine Conjugated to Silver and Iron Oxide Based Nanoparticles. *J. Photochem. Photobiol. A Chem.* **2020**, *402*, 112813. [CrossRef]

152. Scanone, A.C.; Gsponer, N.S.; Alvarez, M.G.; Heredia, D.A.; Durantini, A.M.; Durantini, E.N. Magnetic Nanoplatforms for *In Situ* Modification of Macromolecules: Synthesis, Characterization, and Photoinactivating Power of Cationic Nanoiman–Porphyrin Conjugates. *ACS Appl. Bio Mater.* **2020**, *3*, 5930–5940. [CrossRef]

153. Scanone, A.C.; Santamarina, S.C.; Heredia, D.A.; Durantini, E.N.; Durantini, A.M. Functionalized Magnetic Nanoparticles with BODIPYs for Bioimaging and Antimicrobial Therapy Applications. *ACS Appl. Bio Mater.* **2020**, *3*, 1061–1070. [CrossRef]

154. Sun, X.; Wang, L.; Lynch, C.D.; Sun, X.; Li, X.; Qi, M.; Ma, C.; Li, C.; Dong, B.; Zhou, Y.; et al. Nanoparticles Having Amphiphilic Silane Containing Chlorin E6 with Strong Anti-Biofilm Activity against Periodontitis-Related Pathogens. *J. Dent.* **2019**, *81*, 70–84. [CrossRef]

155. Toledo, V.H.; Yoshimura, T.M.; Pereira, S.T.; Castro, C.E.; Ferreira, F.F.; Ribeiro, M.S.; Haddad, P.S. Methylene Blue-Covered Superparamagnetic Iron Oxide Nanoparticles Combined with Red Light as a Novel Platform to Fight Non-Local Bacterial Infections: A Proof of Concept Study against *Escherichia Coli*. *J. Photochem. Photobiol. B* **2020**, *209*, 111956. [CrossRef]

156. Qi, M.; Chi, M.; Sun, X.; Xie, X.; Weir, M.D.; Oates, T.W.; Zhou, Y.; Wang, L.; Bai, Y.; Xu, H.H. Novel Nanomaterial-Based Antibacterial Photodynamic Therapies to Combat Oral Bacterial Biofilms and Infectious Diseases. *Int. J. Nanomed.* **2019**, *14*, 6937–6956. [CrossRef]

157. Collen Makola, L.; Nyokong, T.; Amuhaya, E.K. Impact of Axial Ligation on Photophysical and Photodynamic Antimicrobial Properties of Indium (III) Methylsulfanylphenyl Porphyrin Complexes Linked to Silver-Capped Copper Ferrite Magnetic Nanoparticles. *Polyhedron* **2021**, *193*, 114882. [CrossRef]

158. Makola, L.C.; Managa, M.; Nyokong, T. Enhancement of Photodynamic Antimicrobialtherapy through the Use of Cationic Indium Porphyrin Conjugated to $Ag/CuFe_2O_4$ Nanoparticles. *Photodiagn. Photodyn. Ther.* **2020**, *30*, 101736. [CrossRef]

159. Philip, S.; Kuriakose, S. Photodynamic Antifungal Activity of a Superparamagnetic and Fluorescent Drug Carrier System against Antibiotic-Resistant Fungal Strains. *Cellulose* **2021**, *28*, 9091–9102. [CrossRef]

160. Sun, X.; Sun, J.; Sun, Y.; Li, C.; Fang, J.; Zhang, T.; Wan, Y.; Xu, L.; Zhou, Y.; Wang, L.; et al. Oxygen Self-Sufficient Nanoplatform for Enhanced and Selective Antibacterial Photodynamic Therapy against Anaerobe-Induced Periodontal Disease. *Adv. Funct. Mater.* **2021**, *31*, 2101040. [CrossRef]

161. Siddiqi, K.S.; Ur Rahman, A.; Tajuddin; Husen, A. Properties of Zinc Oxide Nanoparticles and Their Activity Against Microbes. *Nanoscale Res. Lett.* **2018**, *13*, 141. [CrossRef] [PubMed]

162. Gudkov, S.V.; Burmistrov, D.E.; Serov, D.A.; Rebezov, M.B.; Semenova, A.A.; Lisitsyn, A.B. A Mini Review of Antibacterial Properties of ZnO Nanoparticles. *Front. Phys.* **2021**, *9*, 641481. [CrossRef]

163. Da Silva, B.L.; Abuçafy, M.P.; Manaia, E.B.; Junior, J.A.O.; Chiari-Andréo, B.G.; Pietro, R.C.R.; Chiavacci, L.A. Relationship Between Structure and Antimicrobial Activity of Zinc Oxide Nanoparticles: An Overview. *Int. J. Nanomed.* **2019**, *14*, 9395–9410. [CrossRef] [PubMed]

164. Ziental, D.; Czarczynska-Goslinska, B.; Mlynarczyk, D.T.; Glowacka-Sobotta, A.; Stanisz, B.; Goslinski, T.; Sobotta, L. Titanium Dioxide Nanoparticles: Prospects and Applications in Medicine. *Nanomaterials* **2020**, *10*, 387. [CrossRef] [PubMed]

165. Joe, A.; Park, S.-H.; Shim, K.-D.; Kim, D.-J.; Jhee, K.-H.; Lee, H.-W.; Heo, C.-H.; Kim, H.-M.; Jang, E.-S. Antibacterial Mechanism of ZnO Nanoparticles under Dark Conditions. *J. Ind. Eng. Chem.* **2017**, *45*, 430–439. [CrossRef]

166. Liou, J.-W.; Chang, H.-H. Bactericidal Effects and Mechanisms of Visible Light-Responsive Titanium Dioxide Photocatalysts on Pathogenic Bacteria. *Arch. Immunol. Ther. Exp.* **2012**, *60*, 267–275. [CrossRef]

167. Oves, M.; Arshad, M.; Khan, M.S.; Ahmed, A.S.; Azam, A.; Ismail, I.M.I. Anti-Microbial Activity of Cobalt Doped Zinc Oxide Nanoparticles: Targeting Water Borne Bacteria. *J. Saudi Chem. Soc.* **2015**, *19*, 581–588. [CrossRef]

168. Raut, A.V.; Yadav, H.M.; Gnanamani, A.; Pushpavanam, S.; Pawar, S.H. Synthesis and Characterization of Chitosan-TiO_2:Cu Nanocomposite and Their Enhanced Antimicrobial Activity with Visible Light. *Colloids Surf. B Biointerfaces* **2016**, *148*, 566–575. [CrossRef] [PubMed]

169. Xu, X.; Dutta, A.; Khurgin, J.; Wei, A.; Shalaev, V.M.; Boltasseva, A. TiN@TiO_2 Core–Shell Nanoparticles as Plasmon-Enhanced Photosensitizers: The Role of Hot Electron Injection. *Laser Photonics Rev.* **2020**, *14*, 1900376. [CrossRef]

170. Page, K.; Wilson, M.; Parkin, I.P. Antimicrobial Surfaces and Their Potential in Reducing the Role of the Inanimate Environment in the Incidence of Hospital-Acquired Infections. *J. Mater. Chem.* **2009**, *19*, 3819–3831. [CrossRef]

171. Karthikeyan, K.T.; Nithya, A.; Jothivenkatachalam, K. Photocatalytic and Antimicrobial Activities of Chitosan-TiO_2 Nanocomposite. *Int. J. Biol. Macromol.* **2017**, *104*, 1762–1773. [CrossRef]

172. Pantaroto, H.N.; Ricomini-Filho, A.P.; Bertolini, M.M.; Dias da Silva, J.H.; Azevedo Neto, N.F.; Sukotjo, C.; Rangel, E.C.; Barão, V.A.R. Antibacterial Photocatalytic Activity of Different Crystalline TiO_2 Phases in Oral Multispecies Biofilm. *Dent. Mater.* **2018**, *34*, e182–e195. [CrossRef] [PubMed]

173. Suketa, N.; Sawase, T.; Kitaura, H.; Naito, M.; Baba, K.; Nakayama, K.; Wennerberg, A.; Atsuta, M. An Antibacterial Surface on Dental Implants, Based on the Photocatalytic Bactericidal Effect. *Clin. Implant. Dent. Relat. Res.* **2005**, *7*, 105–111. [CrossRef]

174. Kumar, S.G.; Devi, L.G. Review on Modified TiO_2 Photocatalysis under UV/Visible Light: Selected Results and Related Mechanisms on Interfacial Charge Carrier Transfer Dynamics. *J. Phys. Chem. A* **2011**, *115*, 13211–13241. [CrossRef] [PubMed]

175. De Dicastillo, C.L.; Correa, M.G.; Martínez, F.B.; Streitt, C.; Galotto, M.J. *Antimicrobial Effect of Titanium Dioxide Nanoparticles*; IntechOpen: London, UK, 2020; ISBN 978-1-83962-433-9.

176. Farkhonde Masoule, S.; Pourhajibagher, M.; Safari, J.; Khoobi, M. Base-Free Green Synthesis of Copper(II) Oxide Nanoparticles Using Highly Cross-Linked Poly(Curcumin) Nanospheres: Synergistically Improved Antimicrobial Activity. *Res. Chem. Intermed.* **2019**, *45*, 4449–4462. [CrossRef]

177. Varaprasad, K.; López, M.; Núñez, D.; Jayaramudu, T.; Sadiku, E.R.; Karthikeyan, C.; Oyarzúnc, P. Antibiotic Copper Oxide-Curcumin Nanomaterials for Antibacterial Applications. *J. Mol. Liq.* **2020**, *300*, 112353. [CrossRef]

178. Xiu, W.; Gan, S.; Wen, Q.; Qiu, Q.; Dai, S.; Dong, H.; Li, Q.; Yuwen, L.; Weng, L.; Teng, Z.; et al. Biofilm Microenvironment-Responsive Nanotheranostics for Dual-Mode Imaging and Hypoxia-Relief-Enhanced Photodynamic Therapy of Bacterial Infections. *Research* **2020**, *2020*, 9426453. [CrossRef] [PubMed]

179. Sambhaji, R.B.; Vijay, J.S. Folate Tethered Gd_2O_3 Nanoparticles Exhibit Photoactive Antimicrobial Effects and pH Responsive Delivery of 5-Fluorouracil into MCF-7 Cells. *Drug Deliv. Lett.* **2019**, *9*, 58–68.

180. Hossain, M.K.; Khan, M.I.; El-Denglawey, A. A Review on Biomedical Applications, Prospects, and Challenges of Rare Earth Oxides. *Appl. Mater. Today* **2021**, *24*, 101104. [CrossRef]

181. Matea, C.T.; Mocan, T.; Tabaran, F.; Pop, T.; Mosteanu, O.; Puia, C.; Iancu, C.; Mocan, L. Quantum Dots in Imaging, Drug Delivery and Sensor Applications. *Int. J. Nanomed.* **2017**, *12*, 5421–5431. [CrossRef] [PubMed]

182. Tang, Y.; Qin, Z.; Yin, S.; Sun, H. Transition Metal Oxide and Chalcogenide-Based Nanomaterials for Antibacterial Activities: An Overview. *Nanoscale* **2021**, *13*, 6373–6388. [CrossRef] [PubMed]

183. Rajendiran, K.; Zhao, Z.; Pei, D.-S.; Fu, A. Antimicrobial Activity and Mechanism of Functionalized Quantum Dots. *Polymers* **2019**, *11*, 1670. [CrossRef] [PubMed]

184. McCollum, C.R.; Levy, M.; Bertram, J.R.; Nagpal, P.; Chatterjee, A. Photoexcited Quantum Dots as Efficacious and Nontoxic Antibiotics in an Animal Model. *ACS Biomater. Sci. Eng.* **2021**, *7*, 1863–1875. [CrossRef] [PubMed]

185. Nain, A.; Wei, S.-C.; Lin, Y.-F.; Tseng, Y.-T.; Mandal, R.P.; Huang, Y.-F.; Huang, C.-C.; Tseng, F.-G.; Chang, H.-T. Copper Sulfide Nanoassemblies for Catalytic and Photoresponsive Eradication of Bacteria from Infected Wounds. *ACS Appl. Mater. Interfaces* **2021**, *13*, 7865–7878. [CrossRef] [PubMed]

186. Jiang, C.; Scholle, F.; Ghiladi, R.A.; Dai, T.; Wu, M.X.; Popp, J. Mn-Doped Zn/S Quantum Dots as Photosensitizers for Antimicrobial Photodynamic Inactivation. In *Photonic Diagnosis and Treatment of Infections and Inflammatory Diseases II*; SPIE BiOS: San Francisco, CA, USA, 2019; p. 10863. [CrossRef]

187. McCollum, C.R.; Bertram, J.R.; Nagpal, P.; Chatterjee, A. Photoactivated Indium Phosphide Quantum Dots Treat Multidrug-Resistant Bacterial Abscesses In Vivo. *ACS Appl. Mater. Interfaces* **2021**, *13*, 30404–30419. [CrossRef] [PubMed]

188. Tian, X.; Sun, Y.; Fan, S.; Boudreau, M.D.; Chen, C.; Ge, C.; Yin, J.-J. Photogenerated Charge Carriers in Molybdenum Disulfide Quantum Dots with Enhanced Antibacterial Activity. *ACS Appl. Mater. Interfaces* **2019**, *11*, 4858–4866. [CrossRef]

189. Narband, N.; Mubarak, M.; Ready, D.; Parkin, I.P.; Nair, S.P.; Green, M.A.; Beeby, A.; Wilson, M. Quantum Dots as Enhancers of the Efficacy of Bacterial Lethal Photosensitization. *Nanotechnology* **2008**, *19*, 445102. [CrossRef] [PubMed]

190. Nadeau, J.; Chibli, H.; Carlini, L. Photosensitization of InP/ZnS Quantum Dots for Anti-Cancer and Anti-Microbial Applications. In *Proceedings of the Colloidal Nanocrystals for Biomedical Applications VII*; SPIE: San Francisco, CA, USA, 2012; Volume 8232, pp. 87–95.

191. Liu, J.; Cheng, W.; Wang, Y.; Fan, X.; Shen, J.; Liu, H.; Wang, A.; Hui, A.; Nichols, F.; Chen, S. Cobalt-Doped Zinc Oxide Nanoparticle–MoS_2 Nanosheet Composites as Broad-Spectrum Bactericidal Agents. *ACS Appl. Nano Mater.* **2021**, *4*, 4361–4370. [CrossRef]

192. Courtney, C.M.; Goodman, S.M.; McDaniel, J.A.; Madinger, N.E.; Chatterjee, A.; Nagpal, P. Photoexcited Quantum Dots for Killing Multidrug-Resistant Bacteria. *Nat. Mater.* **2016**, *15*, 529–534. [CrossRef]

193. Goodman, S.M.; Levy, M.; Li, F.-F.; Ding, Y.; Courtney, C.M.; Chowdhury, P.P.; Erbse, A.; Chatterjee, A.; Nagpal, P. Designing Superoxide-Generating Quantum Dots for Selective Light-Activated Nanotherapy. *Front. Chem.* **2018**, *6*, 46. [CrossRef] [PubMed]

194. Ganguly, P.; Mathew, S.; Clarizia, L.; Kumar, R.S.; Akande, A.; Hinder, S.; Breen, A.; Pillai, S.C. Theoretical and Experimental Investigation of Visible Light Responsive $AgBiS_2$-TiO_2 Heterojunctions for Enhanced Photocatalytic Applications. *Appl. Catal. B Environ.* **2019**, *253*, 401–418. [CrossRef]

195. Nazir, M.; Aziz, M.I.; Ali, I.; Basit, M.A. Revealing Antimicrobial and Contrasting Photocatalytic Behavior of Metal Chalcogenide Deposited P25-TiO_2 Nanoparticles. *Photonics Nanostruct.-Fundam. Appl.* **2019**, *36*, 100721. [CrossRef]

196. Suganya, M.; Balu, A.R.; Balamurugan, S.; Srivind, J.; Narasimman, V.; Manjula, N.; Rajashree, C.; Nagarethinam, V.S. Photoconductive, Photocatalytic and Antifungal Properties of PbS:Mo Nanoparticles Synthesized *via* Precipitation Method. *Surf. Interfaces* **2018**, *13*, 148–156. [CrossRef]

197. Schlachter, A.; Asselin, P.; Harvey, P.D. Porphyrin-Containing MOFs and COFs as Heterogeneous Photosensitizers for Singlet Oxygen-Based Antimicrobial Nanodevices. *ACS Appl. Mater. Interfaces* **2021**, *13*, 26651–26672. [CrossRef] [PubMed]

198. Qian, S.; Song, L.; Sun, L.; Zhang, X.; Xin, Z.; Yin, J.; Luan, S. Metal-Organic Framework/Poly (ε-Caprolactone) Hybrid Electrospun Nanofibrous Membranes with Effective Photodynamic Antibacterial Activities. *J. Photochem. Photobiol. A Chem.* **2020**, *400*, 112626. [CrossRef]

199. Nie, X.; Wu, S.; Mensah, A.; Wang, Q.; Huang, F.; Wei, Q. FRET as a Novel Strategy to Enhance the Singlet Oxygen Generation of Porphyrinic MOF Decorated Self-Disinfecting Fabrics. *Chem. Eng. J.* **2020**, *395*, 125012. [CrossRef]

200. Yu, P.; Han, Y.; Han, D.; Liu, X.; Liang, Y.; Li, Z.; Zhu, S.; Wu, S. In-Situ Sulfuration of Cu-Based Metal-Organic Framework for Rapid near-Infrared Light Sterilization. *J. Hazard. Mater.* **2020**, *390*, 122126. [CrossRef]

201. Xiong, K.; Li, J.; Tan, L.; Cui, Z.; Li, Z.; Wu, S.; Liang, Y.; Zhu, S.; Liu, X. Ag$_2$S Decorated Nanocubes with Enhanced Near-Infrared Photothermal and Photodynamic Properties for Rapid Sterilization. *Colloid Interface Sci. Commun.* **2019**, *33*, 100201. [CrossRef]

202. Bagchi, D.; Bhattacharya, A.; Dutta, T.; Nag, S.; Wulferding, D.; Lemmens, P.; Pal, S.K. Nano MOF Entrapping Hydrophobic Photosensitizer for Dual-Stimuli-Responsive Unprecedented Therapeutic Action against Drug-Resistant Bacteria. *ACS Appl. Bio Mater.* **2019**, *2*, 1772–1780. [CrossRef]

203. Golmohamadpour, A.; Bahramian, B.; Khoobi, M.; Pourhajibagher, M.; Barikani, H.R.; Bahador, A. Antimicrobial Photodynamic Therapy Assessment of Three Indocyanine Green-Loaded Metal-Organic Frameworks against *Enterococcus Faecalis*. *Photodiagn. Photodyn. Ther.* **2018**, *23*, 331–338. [CrossRef]

204. Wang, M.; Nian, L.; Cheng, Y.; Yuan, B.; Cheng, S.; Cao, C. Encapsulation of Colloidal Semiconductor Quantum Dots into Metal-Organic Frameworks for Enhanced Antibacterial Activity through Interfacial Electron Transfer. *Chem. Eng. J.* **2021**, *426*, 130832. [CrossRef]

205. Hynek, J.; Zelenka, J.; Rathouský, J.; Kubát, P.; Ruml, T.; Demel, J.; Lang, K. Designing Porphyrinic Covalent Organic Frameworks for the Photodynamic Inactivation of Bacteria. *ACS Appl. Mater. Interfaces* **2018**, *10*, 8527–8535. [CrossRef] [PubMed]

206. Liang, G.; Wang, H.; Shi, H.; Wang, H.; Zhu, M.; Jing, A.; Li, J.; Li, G. Recent Progress in the Development of Upconversion Nanomaterials in Bioimaging and Disease Treatment. *J. Nanobiotechnol.* **2020**, *18*, 154. [CrossRef] [PubMed]

207. Zhang, T.; Ying, D.; Qi, M.; Li, X.; Fu, L.; Sun, X.; Wang, L.; Zhou, Y. Anti-Biofilm Property of Bioactive Upconversion Nanocomposites Containing Chlorin E6 against Periodontal Pathogens. *Molecules* **2019**, *24*, 2692. [CrossRef]

208. Chen, B.; Wang, F. Emerging Frontiers of Upconversion Nanoparticles. *Trends Chem.* **2020**, *2*, 427–439. [CrossRef]

209. Yao, J.; Huang, C.; Liu, C.; Yang, M. Upconversion Luminescence Nanomaterials: A Versatile Platform for Imaging, Sensing, and Therapy. *Talanta* **2020**, *208*, 120157. [CrossRef] [PubMed]

210. Li, Z.; Lu, S.; Liu, W.; Dai, T.; Ke, J.; Li, X.; Li, R.; Zhang, Y.; Chen, Z.; Chen, X. Synergistic Lysozyme-Photodynamic Therapy Against Resistant Bacteria Based on an Intelligent Upconversion Nanoplatform. *Angew. Chem. Int. Ed.* **2021**, *60*, 19201–19206. [CrossRef]

211. Wang, Z.; Liu, C.; Zhao, Y.; Hu, M.; Ma, D.; Zhang, P.; Xue, Y.; Li, X. Photomagnetic Nanoparticles in Dual-Modality Imaging and Photo-Sonodynamic Activity against Bacteria. *Chem. Eng. J.* **2019**, *356*, 811–818. [CrossRef]

212. Li, S.; Cui, S.; Yin, D.; Zhu, Q.; Ma, Y.; Qian, Z.; Gu, Y. Dual Antibacterial Activities of a Chitosan-Modified Upconversion Photodynamic Therapy System against Drug-Resistant Bacteria in Deep Tissue. *Nanoscale* **2017**, *9*, 3912–3924. [CrossRef]

213. Zhang, Y.; Huang, P.; Wang, D.; Chen, J.; Liu, W.; Hu, P.; Huang, M.; Chen, X.; Chen, Z. Near-Infrared-Triggered Antibacterial and Antifungal Photodynamic Therapy Based on Lanthanide-Doped Upconversion Nanoparticles. *Nanoscale* **2018**, *10*, 15485–15495. [CrossRef]

214. Sharma, B.; Kaur, G.; Chaudhary, G.R.; Gawali, S.L.; Hassan, P.A. High Antimicrobial Photodynamic Activity of Photosensitizer Encapsulated Dual-Functional Metallocatanionic Vesicles against Drug-Resistant Bacteria *S. Aureus*. *Biomater. Sci.* **2020**, *8*, 2905–2920. [CrossRef] [PubMed]

215. Kaur, G.; Berwal, K.; Sharma, B.; Chaudhary, G.R.; Gawali, S.L.; Hassan, P.A. Enhanced Antimicrobial Photodynamic Activity of Photosensitizer Encapsulated Copper Based Metallocatanionic Vesicles against *E. coli* Using Visible Light. *J. Mol. Liq.* **2021**, *324*, 114688. [CrossRef]

216. Wang, D.; Niu, L.; Qiao, Z.-Y.; Cheng, D.-B.; Wang, J.; Zhong, Y.; Bai, F.; Wang, H.; Fan, H. Synthesis of Self-Assembled Porphyrin Nanoparticle Photosensitizers. *ACS Nano* **2018**, *12*, 3796–3803. [CrossRef] [PubMed]

217. Park, Y.; Yoo, J.; Kang, M.-H.; Kwon, W.; Joo, J. Photoluminescent and Biodegradable Porous Silicon Nanoparticles for Biomedical Imaging. *J. Mater. Chem. B* **2019**, *7*, 6271–6292. [CrossRef]

218. Jabir, M.S.; Nayef, U.M.; Jawad, K.H.; Taqi, Z.J.; Buthenhia, A.H.; Ahmed, N.R. Porous Silicon Nanoparticles Prepared *via* an Improved Method: A Developing Strategy for a Successful Antimicrobial Agent against *Escherichia Coli* and *Staphylococcus Aureus*. *IOP Conf. Ser. Mater. Sci. Eng.* **2018**, *454*, 012077. [CrossRef]

219. Xiao, L.; Gu, L.; Howell, S.B.; Sailor, M.J. Porous Silicon Nanoparticle Photosensitizers for Singlet Oxygen and Their Phototoxicity against Cancer Cells. *ACS Nano* **2011**, *5*, 3651–3659. [CrossRef]

220. Secret, E.; Maynadier, M.; Gallud, A.; Gary-Bobo, M.; Chaix, A.; Belamie, E.; Maillard, P.; Sailor, M.J.; Garcia, M.; Durand, J.-O.; et al. Anionic Porphyrin-Grafted Porous Silicon Nanoparticles for Photodynamic Therapy. *Chem. Commun.* **2013**, *49*, 4202–4204. [CrossRef]

221. Chaix, A.; Cheikh, K.E.; Bouffard, E.; Maynadier, M.; Aggad, D.; Stojanovic, V.; Knezevic, N.; Garcia, M.; Maillard, P.; Morère, A.; et al. Mesoporous Silicon Nanoparticles for Targeted Two-Photon Theranostics of Prostate Cancer. *J. Mater. Chem. B* **2016**, *4*, 3639–3642. [CrossRef]
222. Näkki, S.; Martinez, J.O.; Evangelopoulos, M.; Xu, W.; Lehto, V.-P.; Tasciotti, E. Chlorin E6 Functionalized Theranostic Multistage Nanovectors Transported by Stem Cells for Effective Photodynamic Therapy. *ACS Appl. Mater. Interfaces* **2017**, *9*, 23441–23449. [CrossRef]
223. Knežević, N.Ž.; Stojanovic, V.; Chaix, A.; Bouffard, E.; Cheikh, K.E.; Morère, A.; Maynadier, M.; Lemercier, G.; Garcia, M.; Gary-Bobo, M.; et al. Ruthenium(II) Complex-Photosensitized Multifunctionalized Porous Silicon Nanoparticles for Two-Photon near-Infrared Light Responsive Imaging and Photodynamic Cancer Therapy. *J. Mater. Chem. B* **2016**, *4*, 1337–1342. [CrossRef]
224. Capeletti, L.B.; de Oliveira, L.F.; Gonçalves, K.d.A.; de Oliveira, J.F.A.; Saito, Â.; Kobarg, J.; dos Santos, J.H.Z.; Cardoso, M.B. Tailored Silica–Antibiotic Nanoparticles: Overcoming Bacterial Resistance with Low Cytotoxicity. *Langmuir* **2014**, *30*, 7456–7464. [CrossRef] [PubMed]
225. Guo, Y.; Rogelj, S.; Zhang, P. Rose Bengal-Decorated Silica Nanoparticles as Photosensitizers for Inactivation of Gram-Positive Bacteria. *Nanotechnology* **2010**, *21*, 065102. [CrossRef] [PubMed]
226. Wang, C.; Chen, P.; Qiao, Y.; Kang, Y.; Yan, C.; Yu, Z.; Wang, J.; He, X.; Wu, H. PH Responsive Superporogen Combined with PDT Based on Poly Ce6 Ionic Liquid Grafted on SiO$_2$ for Combating MRSA Biofilm Infection. *Theranostics* **2020**, *10*, 4795–4808. [CrossRef]
227. Scanone, A.C.; Gsponer, N.S.; Alvarez, M.G.; Durantini, E.N. Photodynamic Properties and Photoinactivation of Microorganisms Mediated by 5,10,15,20-Tetrakis(4-Carboxyphenyl)Porphyrin Covalently Linked to Silica-Coated Magnetite Nanoparticles. *J. Photochem. Photobiol. A Chem.* **2017**, *346*, 452–461. [CrossRef]
228. Mesquita, M.Q.; Menezes, J.C.J.M.D.S.; Pires, S.M.G.; Neves, M.G.P.M.S.; Simões, M.M.Q.; Tomé, A.C.; Cavaleiro, J.A.S.; Cunha, Â.; Daniel-da-Silva, A.L.; Almeida, A.; et al. Pyrrolidine-Fused Chlorin Photosensitizer Immobilized on Solid Supports for the Photoinactivation of Gram Negative Bacteria. *Dye. Pigment.* **2014**, *110*, 123–133. [CrossRef]
229. Baigorria, E.; Reynoso, E.; Alvarez, M.G.; Milanesio, M.E.; Durantini, E.N. Silica Nanoparticles Embedded with Water Insoluble Phthalocyanines for the Photoinactivation of Microorganisms. *Photodiagn. Photodyn. Ther.* **2018**, *23*, 261–269. [CrossRef] [PubMed]
230. Amin, A.; Kaduskar, D.V. Comparative Study on Photodynamic Activation of Ortho-Toluidine Blue and Methylene Blue Loaded Mesoporous Silica Nanoparticles Against Resistant Microorganisms. *Recent Pat. Drug Deliv. Formul.* **2018**, *12*, 154–161. [CrossRef] [PubMed]
231. Paramanantham, P.; Siddhardha, B.; Sb, S.L.; Sharan, A.; Alyousef, A.A.; Dosary, M.S.A.; Arshad, M.; Syed, A. Antimicrobial Photodynamic Therapy on *Staphylococcus Aureus* and *Escherichia Coli* Using Malachite Green Encapsulated Mesoporous Silica Nanoparticles: An In Vitro Study. *PeerJ* **2019**, *7*, e7454. [CrossRef]
232. Parasuraman, P.; Antony, A.P.; S.B, S.L.; Sharan, A.; Siddhardha, B.; Kasinathan, K.; Bahkali, N.A.; Dawoud, T.M.S.; Syed, A. Antimicrobial Photodynamic Activity of Toluidine Blue Encapsulated in Mesoporous Silica Nanoparticles against *Pseudomonas Aeruginosa* and *Staphylococcus Aureus*. *Biofouling* **2019**, *35*, 89–103. [CrossRef]
233. Wysocka-Król, K.; Olsztyńska-Janus, S.; Plesch, G.; Plecenik, A.; Podbielska, H.; Bauer, J. Nano-Silver Modified Silica Particles in Antibacterial Photodynamic Therapy. *Appl. Surf. Sci.* **2018**, *461*, 260–268. [CrossRef]
234. Liu, Y.; Liu, X.; Xiao, Y.; Chen, F.; Xiao, F. A Multifunctional Nanoplatform Based on Mesoporous Silica Nanoparticles for Imaging-Guided Chemo/Photodynamic Synergetic Therapy. *RSC Adv.* **2017**, *7*, 31133–31141. [CrossRef]
235. Grüner, M.C.; Arai, M.S.; Carreira, M.; Inada, N.; de Camargo, A.S.S. Functionalizing the Mesoporous Silica Shell of Upconversion Nanoparticles to Enhance Bacterial Targeting and Killing *via* Photosensitizer-Induced Antimicrobial Photodynamic Therapy. *ACS Appl. Bio Mater.* **2018**, *1*, 1028–1036. [CrossRef]
236. Mapukata, S.; Britton, J.; Osifeko, O.L.; Nyokong, T. The Improved Antibacterial Efficiency of a Zinc Phthalocyanine When Embedded on Silver Nanoparticle Modified Silica Nanofibers. *Photodiagn. Photodyn. Ther.* **2021**, *33*, 102100. [CrossRef]
237. Al-Omari, S. Toward a Molecular Understanding of the Photosensitizer-Copper Interaction for Tumor Destruction. *Biophys. Rev.* **2013**, *5*, 305–311. [CrossRef] [PubMed]
238. Eckl, D.B.; Dengler, L.; Nemmert, M.; Eichner, A.; Bäumler, W.; Huber, H. A Closer Look at Dark Toxicity of the Photosensitizer TMPyP in Bacteria. *Photochem. Photobiol.* **2018**, *94*, 165–172. [CrossRef]
239. Ma, J.; Liu, C.; Yan, K. CQDs-MoS2 QDs Loaded on Dendritic Fibrous Nanosilica/Hydrophobic Waterborne Polyurethane Acrylate for Antibacterial Coatings. *Chem. Eng. J.* **2022**, *429*, 132170. [CrossRef]
240. Zhao, Z.; Yan, R.; Wang, J.; Wu, H.; Wang, Y.; Chen, A.; Shao, S.; Li, Y.-Q. A Bacteria-Activated Photodynamic Nanosystem Based on Polyelectrolyte-Coated Silica Nanoparticles. *J. Mater. Chem. B* **2017**, *5*, 3572–3579. [CrossRef] [PubMed]
241. Planas, O.; Bresolí-Obach, R.; Nos, J.; Gallavardin, T.; Ruiz-González, R.; Agut, M.; Nonell, S. Synthesis, Photophysical Characterization, and Photoinduced Antibacterial Activity of Methylene Blue-Loaded Amino- and Mannose-Targeted Mesoporous Silica Nanoparticles. *Molecules* **2015**, *20*, 6284. [CrossRef] [PubMed]
242. Paramanantham, P.; Anju, V.T.; Dyavaiah, M.; Siddhardha, B. Applications of Carbon-Based Nanomaterials for Antimicrobial Photodynamic Therapy. In *Microbial Nanobionics: Volume 2, Basic Research and Applications*; Prasad, R., Ed.; Nanotechnology in the Life Sciences; Springer International Publishing: Cham, Switzerland, 2019; pp. 237–259. ISBN 978-3-030-16534-5.
243. Alavi, M.; Jabari, E.; Jabbari, E. Functionalized Carbon-Based Nanomaterials and Quantum Dots with Antibacterial Activity: A Review. *Expert Rev. Anti-Infect. Ther.* **2021**, *19*, 35–44. [CrossRef] [PubMed]

244. Bacakova, L.; Pajorova, J.; Tomkova, M.; Matejka, R.; Broz, A.; Stepanovska, J.; Prazak, S.; Skogberg, A.; Siljander, S.; Kallio, P. Applications of Nanocellulose/Nanocarbon Composites: Focus on Biotechnology and Medicine. *Nanomaterials* **2020**, *10*, 196. [CrossRef] [PubMed]

245. Agazzi, M.L.; Durantini, J.E.; Gsponer, N.S.; Durantini, A.M.; Bertolotti, S.G.; Durantini, E.N. Light-Harvesting Antenna and Proton-Activated Photodynamic Effect of a Novel BODIPY−Fullerene C60 Dyad as Potential Antimicrobial Agent. *ChemPhysChem* **2019**, *20*, 1110–1125. [CrossRef] [PubMed]

246. Agazzi, M.L.; Almodovar, V.A.S.; Gsponer, N.S.; Bertolotti, S.; Tomé, A.C.; Durantini, E.N. Diketopyrrolopyrrole−Fullerene C60 Architectures as Highly Efficient Heavy Atom-Free Photosensitizers: Synthesis, Photophysical Properties and Photodynamic Activity. *Org. Biomol. Chem.* **2020**, *18*, 1449–1461. [CrossRef] [PubMed]

247. Agazzi, M.L.; Durantini, J.E.; Quiroga, E.D.; Alvarez, M.G.; Durantini, E.N. A Novel Tricationic Fullerene C60 as Broad-Spectrum Antimicrobial Photosensitizer: Mechanisms of Action and Potentiation with Potassium Iodide. *Photochem. Photobiol. Sci.* **2021**, *20*, 327–341. [CrossRef]

248. Huang, Y.-Y.; Sharma, S.K.; Yin, R.; Agrawal, T.; Chiang, L.Y.; Hamblin, M.R. Functionalized Fullerenes in Photodynamic Therapy. *J. Biomed. Nanotechnol.* **2014**, *10*, 1918–1936. [CrossRef] [PubMed]

249. Palacios, Y.B.; Durantini, J.E.; Heredia, D.A.; Martínez, S.R.; González de la Torre, L.; Durantini, A.M. Tuning the Polarity of Fullerene C60 Derivatives for Enhanced Photodynamic Inactivation. *Photochem. Photobiol.* **2021**. [CrossRef] [PubMed]

250. Reynoso, E.; Durantini, A.M.; Solis, C.A.; Macor, L.P.; Otero, L.A.; Gervaldo, M.A.; Durantini, E.N.; Heredia, D.A. Photoactive Antimicrobial Coating Based on a PEDOT-Fullerene C60 Polymeric Dyad. *RSC Adv.* **2021**, *11*, 23519–23532. [CrossRef]

251. Spesia, M.B.; Milanesio, M.E.; Durantini, E.N. Chapter 18—Fullerene Derivatives in Photodynamic Inactivation of Microorganisms. In *Nanostructures for Antimicrobial Therapy*; Ficai, A., Grumezescu, A.M., Eds.; Micro and Nano Technologies; Elsevier: Amsterdam, The Netherlands, 2017; pp. 413–433. ISBN 978-0-323-46152-8.

252. Anju, V.T.; Paramanantham, P.; S.B, S.L.; Sharan, A.; Syed, A.; Bahkali, N.A.; Alsaedi, M.H.; K., K.; Busi, S. Antimicrobial Photodynamic Activity of Toluidine Blue-Carbon Nanotube Conjugate against *Pseudomonas Aeruginosa* and *Staphylococcus Aureus*—Understanding the Mechanism of Action. *Photodiagn. Photodyn. Ther.* **2019**, *27*, 305–316. [CrossRef]

253. Tondro, G.H.; Behzadpour, N.; Keykhaee, Z.; Akbari, N.; Sattarahmady, N. Carbon@polypyrrole Nanotubes as a Photosensitizer in Laser Phototherapy of *Pseudomonas Aeruginosa*. *Colloids Surf. B Biointerfaces* **2019**, *180*, 481–486. [CrossRef] [PubMed]

254. Anju, V.T.; Paramanantham, P.; Siddhardha, B.; Lal, S.S.; Sharan, A.; Alyousef, A.A.; Arshad, M.; Syed, A. Malachite Green-Conjugated Multi-Walled Carbon Nanotubes Potentiate Antimicrobial Photodynamic Inactivation of Planktonic Cells and Biofilms of *Pseudomonas Aeruginosa* and *Staphylococcus Aureus*. *Int. J. Nanomed.* **2019**, *14*, 3861–3874. [CrossRef]

255. Parasuraman, P.; Anju, V.T.; Lal, S.S.; Sharan, A.; Busi, S.; Kaviyarasu, K.; Arshad, M.; Dawoud, T.M.S.; Syed, A. Synthesis and Antimicrobial Photodynamic Effect of Methylene Blue Conjugated Carbon Nanotubes on *E. Coli* and *S. Aureus*. *Photochem. Photobiol. Sci.* **2019**, *18*, 563–576. [CrossRef]

256. Vt, A.; Paramanantham, P.; Sb, S.L.; Sharan, A.; Alsaedi, M.H.; Dawoud, T.M.S.; Asad, S.; Busi, S. Antimicrobial Photodynamic Activity of Rose Bengal Conjugated Multi Walled Carbon Nanotubes against Planktonic Cells and Biofilm of *Escherichia Coli*. *Photodiagn. Photodyn. Ther.* **2018**, *24*, 300–310. [CrossRef]

257. Sah, U.; Sharma, K.; Chaudhri, N.; Sankar, M.; Gopinath, P. Antimicrobial Photodynamic Therapy: Single-Walled Carbon Nanotube (SWCNT)-Porphyrin Conjugate for Visible Light Mediated Inactivation of *Staphylococcus Aureus*. *Colloids Surf. B Biointerfaces* **2018**, *162*, 108–117. [CrossRef]

258. Szunerits, S.; Barras, A.; Boukherroub, R. Antibacterial Applications of Nanodiamonds. *Int. J. Environ. Res Public Health* **2016**, *13*, 413. [CrossRef]

259. Hurtado, C.R.; Hurtado, G.R.; de Cena, G.L.; Queiroz, R.C.; Silva, A.V.; Diniz, M.F.; dos Santos, V.R.; Trava-Airoldi, V.; Baptista, M.d.S.; Tsolekile, N.; et al. Diamond Nanoparticles-Porphyrin MTHPP Conjugate as Photosensitizing Platform: Cytotoxicity and Antibacterial Activity. *Nanomaterials* **2021**, *11*, 1393. [CrossRef]

260. Openda, Y.I.; Nyokong, T. Enhanced Photo-Ablation Effect of Positively Charged Phthalocyanines-Detonation Nanodiamonds Nanoplatforms for the Suppression of *Staphylococcus Aureus* and *Escherichia Coli* Planktonic Cells and Biofilms. *J. Photochem. Photobiol. A Chem.* **2021**, *411*, 113200. [CrossRef]

261. Openda, Y.I.; Nyokong, T. Detonation Nanodiamonds-Phthalocyanine Photosensitizers with Enhanced Photophysicochemical Properties and Effective Photoantibacterial Activity. *Photodiagn. Photodyn. Ther.* **2020**, *32*, 102072. [CrossRef]

262. Openda, Y.I.; Ngoy, B.P.; Nyokong, T. Photodynamic Antimicrobial Action of Asymmetrical Porphyrins Functionalized Silver-Detonation Nanodiamonds Nanoplatforms for the Suppression of *Staphylococcus Aureus* Planktonic Cells and Biofilms. *Front. Chem.* **2021**, *9*, 97. [CrossRef]

263. Gayen, B.; Palchoudhury, S.; Chowdhury, J. Carbon Dots: A Mystic Star in the World of Nanoscience. *J. Nanomater.* **2019**, *2019*, e3451307. [CrossRef]

264. Knoblauch, R.; Geddes, C.D. Carbon Nanodots in Photodynamic Antimicrobial Therapy: A Review. *Materials* **2020**, *13*, 4004. [CrossRef]

265. Moniruzzaman, M.; Lakshmi, B.A.; Kim, S.; Kim, J. Preparation of Shape-Specific (Trilateral and Quadrilateral) Carbon Quantum Dots towards Multiple Color Emission. *Nanoscale* **2020**, *12*, 11947–11959. [CrossRef]

266. Dong, X.; Ge, L.; Abu Rabe, D.I.; Mohammed, O.O.; Wang, P.; Tang, Y.; Kathariou, S.; Yang, L.; Sun, Y.-P. Photoexcited State Properties and Antibacterial Activities of Carbon Dots Relevant to Mechanistic Features and Implications. *Carbon* **2020**, *170*, 137–145. [CrossRef]

267. Knoblauch, R.; Harvey, A.; Geddes, C.D. Metal-Enhanced Photosensitization of Singlet Oxygen (ME$_1$O$_2$) from Brominated Carbon Nanodots on Silver Nanoparticle Substrates. *Plasmonics* **2021**, *16*, 1765–1772. [CrossRef]

268. Kováčová, M.; Špitalská, E.; Markovic, Z.; Špitálský, Z. Carbon Quantum Dots As Antibacterial Photosensitizers and Their Polymer Nanocomposite Applications. *Part. Part. Syst. Charact.* **2020**, *37*, 1900348. [CrossRef]

269. Marković, Z.M.; Kováčová, M.; Humpolíček, P.; Budimir, M.D.; Vajdʼák, J.; Kubát, P.; Mičušík, M.; Švajdlenková, H.; Danko, M.; Capáková, Z.; et al. Antibacterial Photodynamic Activity of Carbon Quantum Dots/Polydimethylsiloxane Nanocomposites against *Staphylococcus Aureus*, *Escherichia Coli* and *Klebsiella Pneumoniae*. *Photodiagn. Photodyn. Ther.* **2019**, *26*, 342–349. [CrossRef]

270. Tejwan, N.; Saini, A.K.; Sharma, A.; Singh, T.A.; Kumar, N.; Das, J. Metal-Doped and Hybrid Carbon Dots: A Comprehensive Review on Their Synthesis and Biomedical Applications. *J. Control. Release* **2021**, *330*, 132–150. [CrossRef] [PubMed]

271. Zhang, J.; Lu, X.; Tang, D.; Wu, S.; Hou, X.; Liu, J.; Wu, P. Phosphorescent Carbon Dots for Highly Efficient Oxygen Photosensitization and as Photo-Oxidative Nanozymes. *ACS Appl. Mater. Interfaces* **2018**, *10*, 40808–40814. [CrossRef]

272. Dong, Y.; Lin, J.; Chen, Y.; Fu, F.; Chi, Y.; Chen, G. Graphene Quantum Dots, Graphene Oxide, Carbon Quantum Dots and Graphite Nanocrystals in Coals. *Nanoscale* **2014**, *6*, 7410–7415. [CrossRef]

273. Tian, P.; Tang, L.; Teng, K.S.; Lau, S.P. Graphene Quantum Dots from Chemistry to Applications. *Mater. Today Chem.* **2018**, *10*, 221–258. [CrossRef]

274. Cui, F.; Ye, Y.; Ping, J.; Sun, X. Carbon Dots: Current Advances in Pathogenic Bacteria Monitoring and Prospect Applications. *Biosens. Bioelectron.* **2020**, *156*, 112085. [CrossRef]

275. Anand, A.; Unnikrishnan, B.; Wei, S.-C.; Chou, C.P.; Zhang, L.-Z.; Huang, C.-C. Graphene Oxide and Carbon Dots as Broad-Spectrum Antimicrobial Agents—A Minireview. *Nanoscale Horiz.* **2019**, *4*, 117–137. [CrossRef] [PubMed]

276. Karahan, H.E.; Wiraja, C.; Xu, C.; Wei, J.; Wang, Y.; Wang, L.; Liu, F.; Chen, Y. Graphene Materials in Antimicrobial Nanomedicine: Current Status and Future Perspectives. *Adv. Healthc. Mater.* **2018**, *7*, 1701406. [CrossRef] [PubMed]

277. Kuo, W.-S.; Shao, Y.-T.; Huang, K.-S.; Chou, T.-M.; Yang, C.-H. Antimicrobial Amino-Functionalized Nitrogen-Doped Graphene Quantum Dots for Eliminating Multidrug-Resistant Species in Dual-Modality Photodynamic Therapy and Bioimaging under Two-Photon Excitation. *ACS Appl. Mater. Interfaces* **2018**, *10*, 14438–14446. [CrossRef]

278. Ristic, B.Z.; Milenkovic, M.M.; Dakic, I.R.; Todorovic-Markovic, B.M.; Milosavljevic, M.S.; Budimir, M.D.; Paunovic, V.G.; Dramicanin, M.D.; Markovic, Z.M.; Trajkovic, V.S. Photodynamic Antibacterial Effect of Graphene Quantum Dots. *Biomaterials* **2014**, *35*, 4428–4435. [CrossRef] [PubMed]

279. Sen, P.; Nwahara, N.; Nyokong, T. Photodynamic Antimicrobial Activity of Benzimidazole Substituted Phthalocyanine When Conjugated to Nitrogen Doped Graphene Quantum Dots against *Staphylococcus Aureus*. *Main Group Chem.* **2021**, *20*, 175–191. [CrossRef]

280. Yu, Y.; Mei, L.; Shi, Y.; Zhang, X.; Cheng, K.; Cao, F.; Zhang, L.; Xu, J.; Li, X.; Xu, Z. Ag-Conjugated Graphene Quantum Dots with Blue Light-Enhanced Singlet Oxygen Generation for Ternary-Mode Highly-Efficient Antimicrobial Therapy. *J. Mater. Chem. B* **2020**, *8*, 1371–1382. [CrossRef] [PubMed]

281. Ferro, S.; Ricchelli, F.; Mancini, G.; Tognon, G.; Jori, G. Inactivation of Methicillin-Resistant *Staphylococcus Aureus* (MRSA) by Liposome-Delivered Photosensitising Agents. *J. Photochem. Photobiol. B Biol.* **2006**, *83*, 98–104. [CrossRef] [PubMed]

282. Jeong, S.; Lee, J.; Im, B.N.; Park, H.; Na, K. Combined Photodynamic and Antibiotic Therapy for Skin Disorder *via* Lipase-Sensitive Liposomes with Enhanced Antimicrobial Performance. *Biomaterials* **2017**, *141*, 243–250. [CrossRef] [PubMed]

283. Miretti, M.; Juri, L.; Cosiansi, M.C.; Tempesti, T.C.; Baumgartner, M.T. Antimicrobial Effects of ZnPc Delivered into Liposomes on Multidrug Resistant (MDR)-*Mycobacterium Tuberculosis*. *ChemistrySelect* **2019**, *4*, 9726–9730. [CrossRef]

284. Plenagl, N.; Seitz, B.S.; Duse, L.; Pinnapireddy, S.R.; Jedelska, J.; Brüßler, J.; Bakowsky, U. Hypericin Inclusion Complexes Encapsulated in Liposomes for Antimicrobial Photodynamic Therapy. *Int. J. Pharm.* **2019**, *570*, 118666. [CrossRef] [PubMed]

285. Sobotta, L.; Dlugaszewska, J.; Ziental, D.; Szczolko, W.; Koczorowski, T.; Goslinski, T.; Mielcarek, J. Optical Properties of a Series of Pyrrolyl-Substituted Porphyrazines and Their Photoinactivation Potential against *Enterococcus Faecalis* after Incorporation into Liposomes. *J. Photochem. Photobiol. A Chem.* **2019**, *368*, 104–109. [CrossRef]

286. Wieczorek, E.; Mlynarczyk, D.T.; Kucinska, M.; Dlugaszewska, J.; Piskorz, J.; Popenda, L.; Szczolko, W.; Jurga, S.; Murias, M.; Mielcarek, J.; et al. Photophysical Properties and Photocytotoxicity of Free and Liposome-Entrapped Diazepinoporphyrazines on LNCaP Cells under Normoxic and Hypoxic Conditions. *Eur. J. Med. Chem.* **2018**, *150*, 64–73. [CrossRef] [PubMed]

287. Babaie, S.; Bakhshayesh, A.R.D.; Ha, J.W.; Hamishehkar, H.; Kim, K.H. Invasome: A Novel Nanocarrier for Transdermal Drug Delivery. *Nanomaterials* **2020**, *10*, 341. [CrossRef] [PubMed]

288. Opatha, S.A.T.; Titapiwatanakun, V.; Chutoprapat, R. Transfersomes: A Promising Nanoencapsulation Technique for Transdermal Drug Delivery. *Pharmaceutics* **2020**, *12*, 855. [CrossRef] [PubMed]

289. Park, H.; Lee, J.; Jeong, S.; Im, B.N.; Kim, M.-K.; Yang, S.-G.; Na, K. Lipase-Sensitive Transfersomes Based on Photosensitizer/Polymerizable Lipid Conjugate for Selective Antimicrobial Photodynamic Therapy of Acne. *Adv. Healthc. Mater.* **2016**, *5*, 3139–3147. [CrossRef]

290. Wang, Y.; Song, J.; Zhang, F.; Zeng, K.; Zhu, X. Antifungal Photodynamic Activity of Hexyl-Aminolevulinate Ethosomes Against *Candida Albicans* Biofilm. *Front. Microbiol.* **2020**, *11*, 2052. [CrossRef]

291. Caruso, E.; Orlandi, V.T.; Malacarne, M.C.; Martegani, E.; Scanferla, C.; Pappalardo, D.; Vigliotta, G.; Izzo, L. Bodipy-Loaded Micelles Based on Polylactide as Surface Coating for Photodynamic Control of *Staphylococcus Aureus*. *Coatings* **2021**, *11*, 223. [CrossRef]

292. Castro, K.A.D.F.; Costa, L.D.; Prandini, J.A.; Biazzotto, J.C.; Tomé, A.C.; Hamblin, M.R.; da Graça, P.M.S. Neves, M.; Faustino, M.A.F.; da Silva, R.S. The Photosensitizing Efficacy of Micelles Containing a Porphyrinic Photosensitizer and KI against Resistant Melanoma Cells. *Chem.—A Eur. J.* **2021**, *27*, 1990–1994. [CrossRef]

293. Dantas Lopes dos Santos, D.; Besegato, J.F.; de Melo, P.B.G.; Oshiro Junior, J.A.; Chorilli, M.; Deng, D.; Bagnato, V.S.; Rastelli, A.N.d.S. Curcumin-Loaded Pluronic®F-127 Micelles as a Drug Delivery System for Curcumin-Mediated Photodynamic Therapy for Oral Application. *Photochem. Photobiol.* **2021**, *97*, 1072–1088. [CrossRef]

294. Schoenmaker, L.; Witzigmann, D.; Kulkarni, J.A.; Verbeke, R.; Kersten, G.; Jiskoot, W.; Crommelin, D.J.A. MRNA-Lipid Nanoparticle COVID-19 Vaccines: Structure and Stability. *Int. J. Pharm.* **2021**, *601*, 120586. [CrossRef]

295. Do Bonfim, C.M.; Monteleoni, L.F.; Calmon, M.d.F.; Cândido, N.M.; Provazzi, P.J.S.; de Souza Lino, V.; Rabachini, T.; Sichero, L.; Villa, L.L.; Quintana, S.M.; et al. Antiviral Activity of Curcumin-Nanoemulsion Associated with Photodynamic Therapy in Vulvar Cell Lines Transducing Different Variants of HPV-16. *Artif. Cells Nanomed. Biotechnol.* **2020**, *48*, 515–524. [CrossRef]

296. Osorno, L.L.; Brandley, A.N.; Maldonado, D.E.; Yiantsos, A.; Mosley, R.J.; Byrne, M.E. Review of Contemporary Self-Assembled Systems for the Controlled Delivery of Therapeutics in Medicine. *Nanomaterials* **2021**, *11*, 278. [CrossRef] [PubMed]

297. Zhu, S.; Wang, X.; Yang, Y.; Bai, H.; Cui, Q.; Sun, H.; Li, L.; Wang, S. Conjugated Polymer with Aggregation-Directed Intramolecular Förster Resonance Energy Transfer Enabling Efficient Discrimination and Killing of Microbial Pathogens. *Chem. Mater.* **2018**, *30*, 3244–3253. [CrossRef]

298. Meng, Z.; Hou, W.; Zhou, H.; Zhou, L.; Chen, H.; Wu, C. Therapeutic Considerations and Conjugated Polymer-Based Photosensitizers for Photodynamic Therapy. *Macromol. Rapid Commun.* **2018**, *39*, 1700614. [CrossRef] [PubMed]

299. Wang, S.; Wu, W.; Manghnani, P.; Xu, S.; Wang, Y.; Goh, C.C.; Ng, L.G.; Liu, B. Polymerization-Enhanced Two-Photon Photosensitization for Precise Photodynamic Therapy. *ACS Nano* **2019**, *13*, 3095–3105. [CrossRef] [PubMed]

300. Wu, W.; Mao, D.; Xu, S.; Kenry; Hu, F.; Li, X.; Kong, D.; Liu, B. Polymerization-Enhanced Photosensitization. *Chem* **2018**, *4*, 1937–1951. [CrossRef]

301. Xu, Q.; He, P.; Wang, J.; Chen, H.; Lv, F.; Liu, L.; Wang, S.; Yoon, J. Antimicrobial Activity of a Conjugated Polymer with Cationic Backbone. *Dye. Pigment.* **2019**, *160*, 519–523. [CrossRef]

302. Zhang, G.; Zhang, D. New Photosensitizer Design Concept: Polymerization-Enhanced Photosensitization. *Chem* **2018**, *4*, 2013–2015. [CrossRef]

303. Zhou, T.; Hu, R.; Wang, L.; Qiu, Y.; Zhang, G.; Deng, Q.; Zhang, H.; Yin, P.; Situ, B.; Zhan, C.; et al. An AIE-Active Conjugated Polymer with High ROS-Generation Ability and Biocompatibility for Efficient Photodynamic Therapy of Bacterial Infections. *Angew. Chem. Int. Ed.* **2020**, *59*, 9952–9956. [CrossRef]

304. Huang, L.; Zhiyentayev, T.; Xuan, Y.; Azhibek, D.; Kharkwal, G.B.; Hamblin, M.R. Photodynamic Inactivation of Bacteria Using Polyethylenimine–Chlorin(E6) Conjugates: Effect of Polymer Molecular Weight, Substitution Ratio of Chlorin(E6) and PH. *Lasers Surg. Med.* **2011**, *43*, 313–323. [CrossRef]

305. Gavara, R.; de Llanos, R.; Pérez-Laguna, V.; Arnau del Valle, C.; Miravet, J.F.; Rezusta, A.; Galindo, F. Broad-Spectrum Photo-Antimicrobial Polymers Based on Cationic Polystyrene and Rose Bengal. *Front. Med.* **2021**, *8*, 641646. [CrossRef]

306. Yang, Y.; Cai, Z.; Huang, Z.; Tang, X.; Zhang, X. Antimicrobial Cationic Polymers: From Structural Design to Functional Control. *Polym. J.* **2018**, *50*, 33–44. [CrossRef]

307. Ahmetali, E.; Sen, P.; Süer, N.C.; Aksu, B.; Nyokong, T.; Eren, T.; Şener, M.K. Enhanced Light-Driven Antimicrobial Activity of Cationic Poly(Oxanorbornene)s by Phthalocyanine Incorporation into Polymer as Pendants. *Macromol. Chem. Phys.* **2020**, *221*, 2000386. [CrossRef]

308. Li, T.; Yan, L. Functional Polymer Nanocarriers for Photodynamic Therapy. *Pharmaceuticals* **2018**, *11*, 133. [CrossRef] [PubMed]

309. Sun, C.-Y.; Shen, S.; Xu, C.-F.; Li, H.-J.; Liu, Y.; Cao, Z.-T.; Yang, X.-Z.; Xia, J.-X.; Wang, J. Tumor Acidity-Sensitive Polymeric Vector for Active Targeted SiRNA Delivery. *J. Am. Chem. Soc.* **2015**, *137*, 15217–15224. [CrossRef]

310. Zhou, J.; Li, T.; Zhang, C.; Xiao, J.; Cui, D.; Cheng, Y. Charge-Switchable Nanocapsules with Multistage PH-Responsive Behaviours for Enhanced Tumour-Targeted Chemo/Photodynamic Therapy Guided by NIR/MR Imaging. *Nanoscale* **2018**, *10*, 9707–9719. [CrossRef] [PubMed]

311. Calixto, G.M.F.; Bernegossi, J.; De Freitas, L.M.; Fontana, C.R.; Chorilli, M. Nanotechnology-Based Drug Delivery Systems for Photodynamic Therapy of Cancer: A Review. *Molecules* **2016**, *21*, 342. [CrossRef] [PubMed]

312. Choudhary, S.; Gupta, L.; Rani, S.; Dave, K.; Gupta, U. Impact of Dendrimers on Solubility of Hydrophobic Drug Molecules. *Front. Pharmacol.* **2017**, *8*, 261. [CrossRef]

313. Svenson, S. Dendrimers as Versatile Platform in Drug Delivery Applications. *Eur. J. Pharm. Biopharm.* **2009**, *71*, 445–462. [CrossRef]

314. Tomalia, D.A.; Khanna, S.N. A Systematic Framework and Nanoperiodic Concept for Unifying Nanoscience: Hard/Soft Nanoelements, Superatoms, Meta-Atoms, New Emerging Properties, Periodic Property Patterns, and Predictive Mendeleev-like Nanoperiodic Tables. *Chem. Rev.* **2016**, *116*, 2705–2774. [CrossRef]

315. Caminade, A.-M.; Turrin, C.-O.; Majoral, J.-P. Biological Properties of Water-Soluble Phosphorhydrazone Dendrimers. *Braz. J. Pharm. Sci.* **2013**, *49*, 33–44. [CrossRef]

316. Staegemann, M.H.; Gräfe, S.; Gitter, B.; Achazi, K.; Quaas, E.; Haag, R.; Wiehe, A. Hyperbranched Polyglycerol Loaded with (Zinc-)Porphyrins: Photosensitizer Release Under Reductive and Acidic Conditions for Improved Photodynamic Therapy. *Biomacromolecules* **2018**, *19*, 222–238. [CrossRef] [PubMed]

317. Staegemann, M.H.; Gitter, B.; Dernedde, J.; Kuehne, C.; Haag, R.; Wiehe, A. Mannose-Functionalized Hyperbranched Polyglycerol Loaded with Zinc Porphyrin: Investigation of the Multivalency Effect in Antibacterial Photodynamic Therapy. *Chem.—A Eur. J.* **2017**, *23*, 3918–3930. [CrossRef] [PubMed]

318. Zielińska, A.; Carreiró, F.; Oliveira, A.M.; Neves, A.; Pires, B.; Venkatesh, D.N.; Durazzo, A.; Lucarini, M.; Eder, P.; Silva, A.M.; et al. Polymeric Nanoparticles: Production, Characterization, Toxicology and Ecotoxicology. *Molecules* **2020**, *25*, 3731. [CrossRef] [PubMed]

319. Kubát, P.; Henke, P.; Raya, R.K.; Štěpánek, M.; Mosinger, J. Polystyrene and Poly(Ethylene Glycol)-b-Poly(ε-Caprolactone) Nanoparticles with Porphyrins: Structure, Size, and Photooxidation Properties. *Langmuir* **2020**, *36*, 302–310. [CrossRef] [PubMed]

320. Gualdesi, M.S.; Vara, J.; Aiassa, V.; Alvarez Igarzabal, C.I.; Ortiz, C.S. New Poly(Acrylamide) Nanoparticles in the Development of Third Generation Photosensitizers. *Dye. Pigment.* **2021**, *184*, 108856. [CrossRef]

321. Malá, Z.; Žárská, L.; Malina, L.; Langová, K.; Večeřová, R.; Kolář, M.; Henke, P.; Mosinger, J.; Kolářová, H. Photodynamic Effect of TPP Encapsulated in Polystyrene Nanoparticles toward Multi-Resistant Pathogenic Bacterial Strains: AFM Evaluation. *Sci. Rep.* **2021**, *11*, 6786. [CrossRef]

322. Ahmadi, H.; Haddadi-Asl, V.; Ghafari, H.-A.; Ghorbanzadeh, R.; Mazlum, Y.; Bahador, A. Shear Bond Strength, Adhesive Remnant Index, and Anti-Biofilm Effects of a Photoexcited Modified Orthodontic Adhesive Containing Curcumin Doped Poly Lactic-Co-Glycolic Acid Nanoparticles: An Ex-Vivo Biofilm Model of *S. Mutans* on the Enamel Slab Bonded Brackets. *Photodiagn. Photodyn. Ther.* **2020**, *30*, 101674. [CrossRef]

323. Liang, Y.; Zhang, H.; Yuan, H.; Lu, W.; Li, Z.; Wang, L.; Gao, L.-H. Conjugated Polymer and Triphenylamine Derivative Codoped Nanoparticles for Photothermal and Photodynamic Antimicrobial Therapy. *ACS Appl. Bio. Mater.* **2020**, *3*, 3494–3499. [CrossRef]

324. Zhang, H.; Liang, Y.; Zhao, H.; Qi, R.; Chen, Z.; Yuan, H.; Liang, H.; Wang, L. Dual-Mode Antibacterial Conjugated Polymer Nanoparticles for Photothermal and Photodynamic Therapy. *Macromol. Biosci.* **2020**, *20*, e1900301. [CrossRef]

325. Ahmed, E.M. Hydrogel: Preparation, Characterization, and Applications: A Review. *J. Adv. Res.* **2015**, *6*, 105–121. [CrossRef]

326. Kirar, S.; Thakur, N.S.; Laha, J.K.; Banerjee, U.C. Porphyrin Functionalized Gelatin Nanoparticle-Based Biodegradable Phototheranostics: Potential Tools for Antimicrobial Photodynamic Therapy. *ACS Appl. Bio Mater.* **2019**, *2*, 4202–4212. [CrossRef]

327. Barbosa, P.M.; Campanholi, K.d.S.S.; Vilsinski, B.H.; Ferreira, S.B.d.S.; Gonçalves, R.S.; de Castro-Hoshino, L.V.; de Oliveira, A.C.V.; Sato, F.; Baesso, M.L.; Bruschi, M.L.; et al. Aluminum Phthalocyanine Hydroxide-Loaded Thermoresponsive Biomedical Hydrogel: A Design for Targeted Photosensitizing Drug Delivery. *J. Mol. Liq.* **2021**, *341*, 117421. [CrossRef]

328. Zheng, Y.; Yu, E.; Weng, Q.; Zhou, L.; Li, Q. Optimization of Hydrogel Containing Toluidine Blue O for Photodynamic Therapy in Treating Acne. *Lasers Med. Sci.* **2019**, *34*, 1535–1545. [CrossRef] [PubMed]

329. Feng, Y.; Xiao, K.; He, Y.; Du, B.; Hong, J.; Yin, H.; Lu, D.; Luo, F.; Li, Z.; Li, J.; et al. Tough and Biodegradable Polyurethane-Curcumin Composited Hydrogel with Antioxidant, Antibacterial and Antitumor Properties. *Mater. Sci. Eng. C* **2021**, *121*, 111820. [CrossRef] [PubMed]

330. de Souza, C.M.; Garcia, M.T.; de Barros, P.P.; Pedroso, L.L.C.; Ward, R.A.d.C.; Strixino, J.F.; Melo, V.M.M.; Junqueira, J.C. Chitosan Enhances the Antimicrobial Photodynamic Inactivation Mediated by Photoditazine®against *Streptococcus Mutans*. *Photodiagn. Photodyn. Ther.* **2020**, *32*, 102001. [CrossRef] [PubMed]

331. Yilmaz Atay, H. Antibacterial Activity of Chitosan-Based Systems. In *Functional Chitosan: Drug Delivery and Biomedical Applications*; Jana, S., Jana, S., Eds.; Springer: Singapore, 2019; pp. 457–489. ISBN 9789811502637.

332. Pourhajibagher, M.; Rokn, A.R.; Rostami-Rad, M.; Barikani, H.R.; Bahador, A. Monitoring of Virulence Factors and Metabolic Activity in *Aggregatibacter Actinomycetemcomitans* Cells Surviving Antimicrobial Photodynamic Therapy *via* Nano-Chitosan Encapsulated Indocyanine Green. *Front. Phys.* **2018**, *6*, 124. [CrossRef]

333. Rad, M.R.; Pourhajibagher, M.; Rokn, A.R.; Barikani, H.R.; Bahador, A. Effect of Antimicrobial Photodynamic Therapy Using Indocyanine Green Doped with Chitosan Nanoparticles on Biofilm Formation-Related Gene Expression of *Aggregatibacter Actinomycetemcomitans*. *Front. Dent.* **2019**, *16*, 187–193. [CrossRef] [PubMed]

334. Shrestha, A.; Kishen, A. Polycationic Chitosan-Conjugated Photosensitizer for Antibacterial Photodynamic Therapy. *Photochem. Photobiol.* **2012**, *88*, 577–583. [CrossRef] [PubMed]

335. Zhang, R.; Li, Y.; Zhou, M.; Wang, C.; Feng, P.; Miao, W.; Huang, H. Photodynamic Chitosan Nano-Assembly as a Potent Alternative Candidate for Combating Antibiotic-Resistant Bacteria. *ACS Appl. Mater. Interfaces* **2019**, *11*, 26711–26721. [CrossRef]

336. Rizzi, V.; Fini, P.; Fanelli, F.; Placido, T.; Semeraro, P.; Sibillano, T.; Fraix, A.; Sortino, S.; Agostiano, A.; Giannini, C.; et al. Molecular Interactions, Characterization and Photoactivity of Chlorophyll a/Chitosan/2-HP-β-Cyclodextrin Composite Films as Functional and Active Surfaces for ROS Production. *Food Hydrocoll.* **2016**, *58*, 98–112. [CrossRef]

337. Sharma, M.; Dube, A.; Majumder, S.K. Antibacterial Photodynamic Activity of Photosensitizer-Embedded Alginate-Pectin-Carboxymethyl Cellulose Composite Biopolymer Films. *Lasers Med. Sci.* **2021**, *36*, 763–772. [CrossRef]

338. Hasanin, M.S.; Abdelraof, M.; Fikry, M.; Shaker, Y.M.; Sweed, A.M.K.; Senge, M.O. Development of Antimicrobial Laser-Induced Photodynamic Therapy Based on Ethylcellulose/Chitosan Nanocomposite with 5,10,15,20-Tetrakis(m-Hydroxyphenyl)Porphyrin. *Molecules* **2021**, *26*, 3551. [CrossRef]

339. Mai, B.; Jia, M.; Liu, S.; Sheng, Z.; Li, M.; Gao, Y.; Wang, X.; Liu, Q.; Wang, P. Smart Hydrogel-Based DVDMS/BFGF Nanohybrids for Antibacterial Phototherapy with Multiple Damaging Sites and Accelerated Wound Healing. *ACS Appl. Mater. Interfaces* **2020**, *12*, 10156–10169. [CrossRef]

340. Müller, A.; Preuß, A.; Bornhütter, T.; Thomas, I.; Prager, A.; Schulze, A.; Röder, B. Electron Beam Functionalized Photodynamic Polyethersulfone Membranes—Photophysical Characterization and Antimicrobial Activity. *Photochem. Photobiol. Sci.* **2018**, *17*, 1346–1354. [CrossRef]

341. Martínez, S.R.; Palacios, Y.B.; Heredia, D.A.; Aiassa, V.; Bartolilla, A.; Durantini, A.M. Self-Sterilizing 3D-Printed Polylactic Acid Surfaces Coated with a BODIPY Photosensitizer. *ACS Appl. Mater. Interfaces* **2021**, *13*, 11597–11608. [CrossRef]

342. Santos, M.R.E.; Mendonça, P.V.; Branco, R.; Sousa, R.; Dias, C.; Serra, A.C.; Fernandes, J.R.; Magalhães, F.D.; Morais, P.V.; Coelho, J.F.J. Light-Activated Antimicrobial Surfaces Using Industrial Varnish Formulations to Mitigate the Incidence of Nosocomial Infections. *ACS Appl. Mater. Interfaces* **2021**, *13*, 7567–7579. [CrossRef]

343. Sautrot-Ba, P.; Contreras, A.; Andaloussi, S.A.; Coradin, T.; Hélary, C.; Razza, N.; Sangermano, M.; Mazeran, P.-E.; Malval, J.-P.; Versace, D.-L. Eosin-Mediated Synthesis of Polymer Coatings Combining Photodynamic Inactivation and Antimicrobial Properties. *J. Mater. Chem. B* **2017**, *5*, 7572–7582. [CrossRef] [PubMed]

344. Kováčová, M.; Kleinová, A.; Vajd'ák, J.; Humpolíček, P.; Kubát, P.; Bodík, M.; Marković, Z.; Špitálský, Z. Photodynamic-Active Smart Biocompatible Material for an Antibacterial Surface Coating. *J. Photochem. Photobiol. B Biol.* **2020**, *211*, 112012. [CrossRef] [PubMed]

345. Wylie, M.P.; Irwin, N.J.; Howard, D.; Heydon, K.; McCoy, C.P. Hot-Melt Extrusion of Photodynamic Antimicrobial Polymers for Prevention of Microbial Contamination. *J. Photochem. Photobiol. B Biol.* **2021**, *214*, 112098. [CrossRef] [PubMed]

346. Moor, K.J.; Osuji, C.O.; Kim, J.-H. Dual-Functionality Fullerene and Silver Nanoparticle Antimicrobial Composites *via* Block Copolymer Templates. *ACS Appl. Mater. Interfaces* **2016**, *8*, 33583–33591. [CrossRef]

347. Mafukidze, D.M.; Sindelo, A.; Nyokong, T. Spectroscopic Characterization and Photodynamic Antimicrobial Chemotherapy of Phthalocyanine-Silver Triangular Nanoprism Conjugates When Supported on Asymmetric Polymer Membranes. *Spectrochim. Acta A Mol. Biomol. Spectrosc.* **2019**, *219*, 333–345. [CrossRef] [PubMed]

348. Sun, J.; Fan, Y.; Zhang, P.; Zhang, X.; Zhou, Q.; Zhao, J.; Ren, L. Self-Enriched Mesoporous Silica Nanoparticle Composite Membrane with Remarkable Photodynamic Antimicrobial Performances. *J. Colloid Interface Sci.* **2020**, *559*, 197–205. [CrossRef]

349. Heredia, D.A.; Martínez, S.R.; Durantini, A.M.; Pérez, M.E.; Mangione, M.I.; Durantini, J.E.; Gervaldo, M.A.; Otero, L.A.; Durantini, E.N. Antimicrobial Photodynamic Polymeric Films Bearing Biscarbazol Triphenylamine End-Capped Dendrimeric Zn(II) Porphyrin. *ACS Appl. Mater. Interfaces* **2019**, *11*, 27574–27587. [CrossRef] [PubMed]

350. Peddinti, B.S.T.; Scholle, F.; Ghiladi, R.A.; Spontak, R.J. Photodynamic Polymers as Comprehensive Anti-Infective Materials: Staying Ahead of a Growing Global Threat. *ACS Appl. Mater. Interfaces* **2018**, *10*, 25955–25959. [CrossRef]

351. Baigorria, E.; Milanesio, M.E.; Durantini, E.N. Synthesis, Spectroscopic Properties and Photodynamic Activity of Zn(II) Phthalocyanine-Polymer Conjugates as Antimicrobial Agents. *Eur. Polym. J.* **2020**, *134*, 109816. [CrossRef]

352. Ambrósio, J.A.R.; Pinto, B.C.D.S.; da Silva, B.G.M.; Passos, J.C.d.S.; Beltrame Junior, M.; Costa, M.S.; Simioni, A.R. BSA Nanoparticles Loaded-Methylene Blue for Photodynamic Antimicrobial Chemotherapy (PACT): Effect on Both Growth and Biofilm Formation by *Candida Albicans*. *J. Biomater. Sci. Polym. Ed.* **2020**, *31*, 2182–2198. [CrossRef]

353. Silva, E.P.O.; Ribeiro, N.M.; Cardoso, M.A.G.; Pacheco-Soares, C.; Jr, M.B. Photodynamic Therapy Using Silicon Phthalocyanine Conjugated with Bovine Serum Albumin as a Drug Delivery System. *Laser Phys.* **2021**, *31*, 075601. [CrossRef]

354. Cheng, R.; Meng, F.; Deng, C.; Klok, H.-A.; Zhong, Z. Dual and Multi-Stimuli Responsive Polymeric Nanoparticles for Programmed Site-Specific Drug Delivery. *Biomaterials* **2013**, *34*, 3647–3657. [CrossRef]

355. Dolanský, J.; Henke, P.; Malá, Z.; Žárská, L.; Kubát, P.; Mosinger, J. Antibacterial Nitric Oxide- and Singlet Oxygen-Releasing Polystyrene Nanoparticles Responsive to Light and Temperature Triggers. *Nanoscale* **2018**, *10*, 2639–2648. [CrossRef]

356. Ren, B.; Li, K.; Liu, Z.; Liu, G.; Wang, H. White Light-Triggered Zwitterionic Polymer Nanoparticles Based on an AIE-Active Photosensitizer for Photodynamic Antimicrobial Therapy. *J. Mater. Chem. B* **2020**, *8*, 10754–10763. [CrossRef]

357. Wang, H.; He, J.; Zhang, M.; Tao, Y.; Li, F.; Tam, K.C.; Ni, P. Biocompatible and Acid-Cleavable Poly(ε-Caprolactone)-Acetal-Poly(Ethylene Glycol)-Acetal-Poly(ε-Caprolactone) Triblock Copolymers: Synthesis, Characterization and PH-Triggered Doxorubicin Delivery. *J. Mater. Chem. B* **2013**, *1*, 6596–6607. [CrossRef] [PubMed]

358. Zhou, Q.; Zhang, L.; Yang, T.; Wu, H. Stimuli-Responsive Polymeric Micelles for Drug Delivery and Cancer Therapy. *IJN* **2018**, *13*, 2921–2942. [CrossRef] [PubMed]

359. Mai, Y.; Eisenberg, A. Self-Assembly of Block Copolymers. *Chem. Soc. Rev.* **2012**, *41*, 5969–5985. [CrossRef]

360. Shen, H.; Eisenberg, A. Morphological Phase Diagram for a Ternary System of Block Copolymer PS310-b-PAA52/Dioxane/H2O. *J. Phys. Chem. B* **1999**, *103*, 9473–9487. [CrossRef]

361. Zhang, L.; Eisenberg, A. Multiple Morphologies of "Crew-Cut" Aggregates of Polystyrene-b-Poly(Acrylic Acid) Block Copolymers. *Science* **1995**, *268*, 1728–1731. [CrossRef]

362. Lombardo, D.; Kiselev, M.A.; Caccamo, M.T. Smart Nanoparticles for Drug Delivery Application: Development of Versatile Nanocarrier Platforms in Biotechnology and Nanomedicine. *J. Nanomater.* **2019**, *2019*, e3702518. [CrossRef]

363. Won, Y.-Y.; Brannan, A.K.; Davis, H.T.; Bates, F.S. Cryogenic Transmission Electron Microscopy (Cryo-TEM) of Micelles and Vesicles Formed in Water by Poly(Ethylene Oxide)-Based Block Copolymers. *J. Phys. Chem. B* **2002**, *106*, 3354–3364. [CrossRef]

364. Haas, S.; Chen, Y.; Fuchs, C.; Handschuh, S.; Steuber, M.; Schönherr, H. Amphiphilic Block Copolymer Vesicles for Active Wound Dressings: Synthesis of Model Systems and Studies of Encapsulation and Release. *Macromol. Symp.* **2013**, *328*, 73–79. [CrossRef]
365. Haas, S.; Hain, N.; Raoufi, M.; Handschuh-Wang, S.; Wang, T.; Jiang, X.; Schönherr, H. Enzyme Degradable Polymersomes from Hyaluronic Acid-Block-Poly(ε-Caprolactone) Copolymers for the Detection of Enzymes of Pathogenic Bacteria. *Biomacromolecules* **2015**, *16*, 832–841. [CrossRef]
366. Handschuh-Wang, S.; Wesner, D.; Wang, T.; Lu, P.; Tücking, K.-S.; Haas, S.; Druzhinin, S.I.; Jiang, X.; Schönherr, H. Determination of the Wall Thickness of Block Copolymer Vesicles by Fluorescence Lifetime Imaging Microscopy. *Macromol. Chem. Phys.* **2017**, *218*, 1600454. [CrossRef]
367. Tücking, K.-S.; Handschuh-Wang, S.; Schönherr, H.; Tücking, K.-S.; Handschuh-Wang, S.; Schönherr, H. Bacterial Enzyme Responsive Polymersomes: A Closer Look at the Degradation Mechanism of PEG-Block-PLA Vesicles. *Aust. J. Chem.* **2014**, *67*, 578–584. [CrossRef]
368. Tang, Q.; Hu, P.; Peng, H.; Zhang, N.; Zheng, Q.; He, Y. Near-Infrared Laser-Triggered, Self-Immolative Smart Polymersomes for in Vivo Cancer Therapy. *Int. J. Nanomed.* **2020**, *15*, 137–149. [CrossRef]
369. Knop, K.; Mingotaud, A.-F.; El-Akra, N.; Violleau, F.; Souchard, J.-P. Monomeric Pheophorbide(a)-Containing Poly(Ethyleneglycol-b-ε-Caprolactone) Micelles for Photodynamic Therapy. *Photochem. Photobiol. Sci.* **2009**, *8*, 396–404. [CrossRef]
370. Zhang, G.-D.; Harada, A.; Nishiyama, N.; Jiang, D.-L.; Koyama, H.; Aida, T.; Kataoka, K. Polyion Complex Micelles Entrapping Cationic Dendrimer Porphyrin: Effective Photosensitizer for Photodynamic Therapy of Cancer. *J. Control. Release* **2003**, *93*, 141–150. [CrossRef] [PubMed]
371. Xiao-ying, Z.; Pei-ying, Z. Polymersomes in Nanomedicine—A Review. *Curr. Nanosci.* **2017**, *13*, 124–129.
372. Lanzilotto, A.; Kyropoulou, M.; Constable, E.C.; Housecroft, C.E.; Meier, W.P.; Palivan, C.G. Porphyrin-Polymer Nanocompartments: Singlet Oxygen Generation and Antimicrobial Activity. *J. Biol. Inorg. Chem.* **2018**, *23*, 109–122. [CrossRef]
373. Li, Z.; Fan, F.; Ma, J.; Yin, W.; Zhu, D.; Zhang, L.; Wang, Z. Oxygen- and Bubble-Generating Polymersomes for Tumor-Targeted and Enhanced Photothermal–Photodynamic Combination Therapy. *Biomater. Sci.* **2021**, *9*, 5841–5853. [CrossRef]
374. Avci, P.; Erdem, S.S.; Hamblin, M.R. Photodynamic Therapy: One Step Ahead with Self-Assembled Nanoparticles. *J. Biomed. Nanotechnol.* **2014**, *10*, 1937–1952. [CrossRef] [PubMed]
375. Gibot, L.; Demazeau, M.; Pimienta, V.; Mingotaud, A.-F.; Vicendo, P.; Collin, F.; Martins-Froment, N.; Dejean, S.; Nottelet, B.; Roux, C.; et al. Role of Polymer Micelles in the Delivery of Photodynamic Therapy Agent to Liposomes and Cells. *Cancers* **2020**, *12*, 384. [CrossRef] [PubMed]
376. Karges, J.; Tharaud, M.; Gasser, G. Polymeric Encapsulation of a Ru(II)-Based Photosensitizer for Folate-Targeted Photodynamic Therapy of Drug Resistant Cancers. *J. Med. Chem.* **2021**, *64*, 4612–4622. [CrossRef] [PubMed]
377. Kashef, N.; Huang, Y.-Y.; Hamblin, M.R. Advances in Antimicrobial Photodynamic Inactivation at the Nanoscale. *Nanophotonics* **2017**, *6*, 853–879. [CrossRef]
378. Vilsinski, B.H.; Gerola, A.P.; Enumo, J.A.; Campanholi, K.d.S.S.; Pereira, P.C.d.S.; Braga, G.; Hioka, N.; Kimura, E.; Tessaro, A.L.; Caetano, W. Formulation of Aluminum Chloride Phthalocyanine in Pluronic(TM) P-123 and F-127 Block Copolymer Micelles: Photophysical Properties and Photodynamic Inactivation of Microorganisms. *Photochem. Photobiol.* **2015**, *91*, 518–525. [CrossRef]
379. Tsai, T.; Yang, Y.-T.; Wang, T. H.; Chien, H.-F.; Chen, C.-T. Improved Photodynamic Inactivation of Gram-Positive Bacteria Using Hematoporphyrin Encapsulated in Liposomes and Micelles. *Lasers Surg. Med.* **2009**, *41*, 316–322. [CrossRef]
380. Teng, F.; Deng, P.; Song, Z.; Zhou, F.; Feng, R. Enhanced Effect in Combination of Curcumin- and Ketoconazole-Loaded Methoxy Poly (Ethylene Glycol)-Poly (ε-Caprolactone) Micelles. *Biomed. Pharmacother.* **2017**, *88*, 43–51. [CrossRef]
381. Yi, X.; Hu, J.-J.; Dai, J.; Lou, X.; Zhao, Z.; Xia, F.; Tang, B.Z. Self-Guiding Polymeric Prodrug Micelles with Two Aggregation-Induced Emission Photosensitizers for Enhanced Chemo-Photodynamic Therapy. *ACS Nano* **2021**, *15*, 3026–3037. [CrossRef]
382. Demir, B.; Barlas, F.B.; Gumus, Z.P.; Unak, P.; Timur, S. Theranostic Niosomes as a Promising Tool for Combined Therapy and Diagnosis: "All-in-One" Approach. *ACS Appl. Nano Mater.* **2018**, *1*, 2827–2835. [CrossRef]
383. Fadel, M.A.; Tawfik, A.A. New Topical Photodynamic Therapy for Treatment of Hidradenitis Suppurativa Using Methylene Blue Niosomal Gel: A Single-Blind, Randomized, Comparative Study. *Clin. Exp. Dermatol.* **2015**, *40*, 116–122. [CrossRef]
384. Belali, S.; Karimi, A.R.; Hadizadeh, M. Novel Nanostructured Smart, Photodynamic Hydrogels Based on Poly(N-Isopropylacrylamide) Bearing Porphyrin Units in Their Crosslink Chains: A Potential Sensitizer System in Cancer Therapy. *Polymer* **2017**, *109*, 93–105. [CrossRef]
385. Preis, E.; Anders, T.; Širc, J.; Hobzova, R.; Cocarta, A.-I.; Bakowsky, U.; Jedelská, J. Biocompatible Indocyanine Green Loaded PLA Nanofibers for *In Situ* Antimicrobial Photodynamic Therapy. *Mater. Sci. Eng. C* **2020**, *115*, 111068. [CrossRef]
386. Stoll, K.R.; Scholle, F.; Zhu, J.; Zhang, X.; Ghiladi, R.A. BODIPY-Embedded Electrospun Materials in Antimicrobial Photodynamic Inactivation. *Photochem. Photobiol. Sci.* **2019**, *18*, 1923–1932. [CrossRef]
387. Chen, W.; Chen, J.; Li, L.; Wang, X.; Wei, Q.; Ghiladi, R.A.; Wang, Q. Wool/Acrylic Blended Fabrics as Next-Generation Photodynamic Antimicrobial Materials. *ACS Appl. Mater. Interfaces* **2019**, *11*, 29557–29568. [CrossRef] [PubMed]
388. Contreras, A.; Raxworthy, M.J.; Wood, S.; Schiffman, J.D.; Tronci, G. Photodynamically Active Electrospun Fibers for Antibiotic-Free Infection Control. *ACS Appl. Bio Mater.* **2019**, *2*, 4258–4270. [CrossRef]
389. Czapka, T.; Winkler, A.; Maliszewska, I.; Kacprzyk, R. Fabrication of Photoactive Electrospun Cellulose Acetate Nanofibers for Antibacterial Applications. *Energies* **2021**, *14*, 2598. [CrossRef]

390. Managa, M.; Amuhaya, E.K.; Nyokong, T. Photodynamic Antimicrobial Chemotherapy Activity of (5,10,15,20-Tetrakis(4-(4-Carboxyphenycarbonoimidoyl)Phenyl)Porphyrinato) Chloro Gallium(III). *Spectrochim. Acta A Mol. Biomol. Spectrosc.* **2015**, *151*, 867–874. [CrossRef] [PubMed]

391. Sindelo, A.; Nyokong, T. Magnetic Nanoparticle—Indium Phthalocyanine Conjugate Embedded in Electrospun Fiber for Photodynamic Antimicrobial Chemotherapy and Photodegradation of Methyl Red. *Heliyon* **2019**, *5*, e02352. [CrossRef]

392. Barra, F.; Roscetto, E.; Soriano, A.A.; Vollaro, A.; Postiglione, I.; Pierantoni, G.M.; Palumbo, G.; Catania, M.R. Photodynamic and Antibiotic Therapy in Combination to Fight Biofilms and Resistant Surface Bacterial Infections. *Int. J. Mol. Sci.* **2015**, *16*, 20417–20430. [CrossRef] [PubMed]

393. Tavares, L.J.; de Avila, E.D.; Klein, M.I.; Panariello, B.H.D.; Spolidório, D.M.P.; Pavarina, A.C. Antimicrobial Photodynamic Therapy Alone or in Combination with Antibiotic Local Administration against Biofilms of *Fusobacterium Nucleatum* and *Porphyromonas Gingivalis*. *J. Photochem. Photobiol. B Biol.* **2018**, *188*, 135–145. [CrossRef] [PubMed]

394. Li, Z.; Pan, W.; Shi, E.; Bai, L.; Liu, H.; Li, C.; Wang, Y.; Deng, J.; Wang, Y. A Multifunctional Nanosystem Based on Bacterial Cell-Penetrating Photosensitizer for Fighting Periodontitis *via* Combining Photodynamic and Antibiotic Therapies. *ACS Biomater. Sci. Eng.* **2021**, *7*, 772–786. [CrossRef] [PubMed]

395. Pérez-Laguna, V.; García-Luque, I.; Ballesta, S.; Pérez-Artiaga, L.; Lampaya-Pérez, V.; Rezusta, A.; Gilaberte, Y. Photodynamic Therapy Using Methylene Blue, Combined or Not with Gentamicin, against *Staphylococcus Aureus* and *Pseudomonas Aeruginosa*. *Photodiagn. Photodyn. Ther.* **2020**, *31*, 101810. [CrossRef] [PubMed]

396. Feng, Y.; Palanisami, A.; Ashraf, S.; Bhayana, B.; Hasan, T. Photodynamic Inactivation of Bacterial Carbapenemases Restores Bacterial Carbapenem Susceptibility and Enhances Carbapenem Antibiotic Effectiveness. *Photodiagn. Photodyn. Ther.* **2020**, *30*, 101693. [CrossRef]

397. Núñez, C.; Palavecino, A.; González, I.A.; Dreyse, P.; Palavecino, C.E. Effective Photodynamic Therapy with Ir(III) for Virulent Clinical Isolates of Extended-Spectrum Beta-Lactamase *Klebsiella Pneumoniae*. *Pharmaceutics* **2021**, *13*, 603. [CrossRef] [PubMed]

398. de Freitas, L.M.; Lorenzón, E.N.; Cilli, E.M.; de Oliveira, K.T.; Fontana, C.R.; Mang, T.S. Photodynamic and Peptide-Based Strategy to Inhibit Gram-Positive Bacterial Biofilm Formation. *Biofouling* **2019**, *35*, 742–757. [CrossRef]

399. Cantelli, A.; Piro, F.; Pecchini, P.; Di Giosia, M.; Danielli, A.; Calvaresi, M. Concanavalin A-Rose Bengal Bioconjugate for Targeted Gram-Negative Antimicrobial Photodynamic Therapy. *J. Photochem. Photobiol. B Biol.* **2020**, *206*, 111852. [CrossRef] [PubMed]

400. Leanse, L.G.; Dong, P.-T.; Goh, X.S.; Lu, M.; Cheng, J.-X.; Hooper, D.C.; Dai, T. Quinine Enhances Photo-Inactivation of Gram-Negative Bacteria. *J. Infect. Dis.* **2020**, *221*, 618–626. [CrossRef] [PubMed]

401. Hu, Y.; Huang, X.; Lu, S.; Hamblin, M.R.; Mylonakis, E.; Zhang, J.; Xi, L. Photodynamic Therapy Combined with Terbinafine Against Chromoblastomycosis and the Effect of PDT on *Fonsecaea Monophora* In Vitro. *Mycopathologia* **2015**, *179*, 103–109. [CrossRef]

402. Lan, Y.; Lu, S.; Zheng, B.; Tang, Z.; Li, J.; Zhang, J. Combinatory Effect of ALA-PDT and Itraconazole Treatment for *Trichosporon Asahii*. *Lasers Surg. Med.* **2020**, *53*, 722–730. [CrossRef]

403. Ayaz, F.; Gonul, I.; Demirbag, B.; Ocakoglu, K. Differential Immunomodulatory Activities of Schiff Base Complexes Depending on Their Metal Conjugation. *Inflammation* **2019**, *42*, 1878–1885. [CrossRef] [PubMed]

404. Sadraeian, M.; Bahou, C.; da Cruz, E.F.; Janini, L.M.R.; Sobhie Diaz, R.; Boyle, R.W.; Chudasama, V.; Eduardo Gontijo Guimarães, F. Photoimmunotherapy Using Cationic and Anionic Photosensitizer-Antibody Conjugates against HIV Env-Expressing Cells. *Int. J. Mol. Sci.* **2020**, *21*, 9151. [CrossRef] [PubMed]

405. Alves, F.; Pavarina, A.C.; Mima, E.G.d.O.; McHale, A.P.; Callan, J.F. Antimicrobial Sonodynamic and Photodynamic Therapies against *Candida Albicans*. *Biofouling* **2018**, *34*, 357–367. [CrossRef]

406. Xu, F.; Hu, M.; Liu, C.; Choi, S.K. Yolk-Structured Multifunctional Up-Conversion Nanoparticles for Synergistic Photodynamic–Sonodynamic Antibacterial Resistance Therapy. *Biomater. Sci.* **2017**, *5*, 678–685. [CrossRef] [PubMed]

407. de Melo, W.d.C.M.A.; Lee, A.N.; Perussi, J.R.; Hamblin, M.R. Electroporation Enhances Antimicrobial Photodynamic Therapy Mediated by the Hydrophobic Photosensitizer, Hypericin. *Photodiagn. Photodyn. Ther.* **2013**, *10*, 647–650. [CrossRef] [PubMed]

408. He, X.; Yang, Y.; Guo, Y.; Lu, S.; Du, Y.; Li, J.-J.; Zhang, X.; Leung, N.L.C.; Zhao, Z.; Niu, G.; et al. Phage-Guided Targeting, Discriminative Imaging, and Synergistic Killing of Bacteria by AIE Bioconjugates. *J. Am. Chem. Soc.* **2020**, *142*, 3959–3969. [CrossRef] [PubMed]

409. Dong, S.; Shi, H.; Zhang, X.; Chen, X.; Cao, D.; Mao, C.; Gao, X.; Wang, L. Difunctional Bacteriophage Conjugated with Photosensitizers for *Candida Albicans*-Targeting Photodynamic Inactivation. *Int. J. Nanomed.* **2018**, *13*, 2199–2216. [CrossRef] [PubMed]

410. Suci, P.A.; Varpness, Z.; Gillitzer, E.; Douglas, T.; Young, M. Targeting and Photodynamic Killing of a Microbial Pathogen Using Protein Cage Architectures Functionalized with a Photosensitizer. *Langmuir* **2007**, *23*, 12280–12286. [CrossRef]

411. Tosato, M.G.; Schilardi, P.; Lorenzo de Mele, M.F.; Thomas, A.H.; Lorente, C.; Miñán, A. Synergistic Effect of Carboxypterin and Methylene Blue Applied to Antimicrobial Photodynamic Therapy against Mature Biofilm of *Klebsiella Pneumoniae*. *Heliyon* **2020**, *6*, e03522. [CrossRef] [PubMed]

412. Openda, Y.I.; Sen, P.; Managa, M.; Nyokong, T. Acetophenone Substituted Phthalocyanines and Their Graphene Quantum Dots Conjugates as Photosensitizers for Photodynamic Antimicrobial Chemotherapy against *Staphylococcus Aureus*. *Photodiagn. Photodyn. Ther.* **2020**, *29*, 101607. [CrossRef] [PubMed]

413. Sharma, R.; Viana, S.M.; Ng, D.K.P.; Kolli, B.K.; Chang, K.P.; de Oliveira, C.I. Photodynamic Inactivation of *Leishmania Braziliensis* Doubly Sensitized with Uroporphyrin and Diamino-Phthalocyanine Activates Effector Functions of Macrophages In Vitro. *Sci. Rep.* **2020**, *10*, 17065. [CrossRef] [PubMed]

414. Huang, X.; Chen, G.; Pan, J.; Chen, X.; Huang, N.; Wang, X.; Liu, J. Effective PDT/PTT Dual-Modal Phototherapeutic Killing of Pathogenic Bacteria by Using Ruthenium Nanoparticles. *J. Mater. Chem. B* **2016**, *4*, 6258–6270. [CrossRef] [PubMed]

415. Romero, M.P.; Marangoni, V.S.; de Faria, C.G.; Leite, I.S.; Silva, C.; de Carvalho Castro e Silva, C.; Maroneze, C.M.; Pereira-da-Silva, M.A.; Bagnato, V.S.; Inada, N.M. Graphene Oxide Mediated Broad-Spectrum Antibacterial Based on Bimodal Action of Photodynamic and Photothermal Effects. *Front. Microbiol.* **2020**, *10*, 2995. [CrossRef]

416. Yuan, Z.; Tao, B.; He, Y.; Mu, C.; Liu, G.; Zhang, J.; Liao, Q.; Liu, P.; Cai, K. Remote Eradication of Biofilm on Titanium Implant *via* Near-Infrared Light Triggered Photothermal/Photodynamic Therapy Strategy. *Biomaterials* **2019**, *223*, 119479. [CrossRef]

417. Dolanský, J.; Henke, P.; Kubát, P.; Fraix, A.; Sortino, S.; Mosinger, J. Polystyrene Nanofiber Materials for Visible-Light-Driven Dual Antibacterial Action *via* Simultaneous Photogeneration of NO and $O_2((1)\Delta g)$. *ACS Appl. Mater. Interfaces* **2015**, *7*, 22980–22989. [CrossRef] [PubMed]

418. Zhao, Z.; Li, H.; Tao, X.; Xie, Y.; Yang, L.; Mao, Z.; Xia, W. Light-Triggered Nitric Oxide Release by a Photosensitizer to Combat Bacterial Biofilm Infections. *Chem. Eur. J.* **2021**, *27*, 5453–5460. [CrossRef]

419. Miñán, A.; Lorente, C.; Ipiña, A.; Thomas, A.H.; Fernández Lorenzo de Mele, M.; Schilardi, P.L. Photodynamic Inactivation Induced by Carboxypterin: A Novel Non-Toxic Bactericidal Strategy against Planktonic Cells and Biofilms of *Staphylococcus Aureus*. *Biofouling* **2015**, *31*, 459–468. [CrossRef] [PubMed]

420. Hamblin, M.R. Potentiation of Antimicrobial Photodynamic Inactivation by Inorganic Salts. *Expert Rev. Anti-Infect. Ther.* **2017**, *15*, 1059–1069. [CrossRef] [PubMed]

421. Huang, L.; St. Denis, T.G.; Xuan, Y.; Huang, Y.-Y.; Tanaka, M.; Zadlo, A.; Sarna, T.; Hamblin, M.R. Paradoxical Potentiation of Methylene Blue-Mediated Antimicrobial Photodynamic Inactivation by Sodium Azide: Role of Ambient Oxygen and Azide Radicals. *Free Radic. Biol. Med.* **2012**, *53*, 2062–2071. [CrossRef] [PubMed]

422. Kasimova, K.R.; Sadasivam, M.; Landi, G.; Sarna, T.; Hamblin, M.R. Potentiation of Photoinactivation of Gram-Positive and Gram-Negative Bacteria Mediated by Six Phenothiazinium Dyes by Addition of Azide Ion. *Photochem. Photobiol. Sci.* **2014**, *13*, 1541–1548. [CrossRef]

423. Vecchio, D.; Gupta, A.; Huang, L.; Landi, G.; Avci, P.; Rodas, A.; Hamblin, M.R. Bacterial Photodynamic Inactivation Mediated by Methylene Blue and Red Light Is Enhanced by Synergistic Effect of Potassium Iodide. *Antimicrob. Agents Chemother.* **2015**, *59*, 5203–5212. [CrossRef]

424. Vieira, C.; Gomes, A.T.P.C.; Mesquita, M.Q.; Moura, N.M.M.; Neves, M.G.P.M.S.; Faustino, M.A.F.; Almeida, A. An Insight Into the Potentiation Effect of Potassium Iodide on APDT Efficacy. *Front. Microbiol.* **2018**, *9*, 2665. [CrossRef]

425. Reynoso, E.; Quiroga, E.D.; Agazzi, M.L.; Ballatore, M.B.; Bertolotti, S.G.; Durantini, E.N. Photodynamic Inactivation of Microorganisms Sensitized by Cationic BODIPY Derivatives Potentiated by Potassium Iodide. *Photochem. Photobiol. Sci.* **2017**, *16*, 1524–1536. [CrossRef]

426. Pérez, M.E.; Durantini, J.E.; Reynoso, E.; Alvarez, M.G.; Milanesio, M.E.; Durantini, E.N. Porphyrin–Schiff Base Conjugates Bearing Basic Amino Groups as Antimicrobial Phototherapeutic Agents. *Molecules* **2021**, *26*, 5877. [CrossRef]

427. Hamblin, M.R.; Abrahamse, H. Inorganic Salts and Antimicrobial Photodynamic Therapy: Mechanistic Conundrums? *Molecules* **2018**, *23*, 3190. [CrossRef]

428. Gsponer, N.S.; Spesia, M.B.; Durantini, E.N. Effects of Divalent Cations, EDTA and Chitosan on the Uptake and Photoinactivation of *Escherichia Coli* Mediated by Cationic and Anionic Porphyrins. *Photodiag. Photodyn. Ther.* **2015**, *12*, 67–75. [CrossRef] [PubMed]

429. Pérez-Laguna, V.; Gilaberte, Y.; Millán-Lou, M.I.; Agut, M.; Nonell, S.; Rezusta, A.; Hamblin, M.R. A Combination of Photodynamic Therapy and Antimicrobial Compounds to Treat Skin and Mucosal Infections: A Systematic Review. *Photochem. Photobiol. Sci.* **2019**, *18*, 1020–1029. [CrossRef] [PubMed]

430. Cahan, R.; Swissa, N.; Gellerman, G.; Nitzan, Y. Photosensitizer–Antibiotic Conjugates: A Novel Class of Antibacterial Molecules. *Photochem. Photobiol.* **2010**, *86*, 418–425. [CrossRef]

431. Lu, Z.; Dai, T.; Huang, L.; Kurup, D.B.; Tegos, G.P.; Jahnke, A.; Wharton, T.; Hamblin, M.R. Photodynamic Therapy with a Cationic Functionalized Fullerene Rescues Mice from Fatal Wound Infections. *Nanomedicine* **2010**, *5*, 1525–1533. [CrossRef] [PubMed]

432. Collins, T.L.; Markus, E.A.; Hassett, D.J.; Robinson, J.B. The Effect of a Cationic Porphyrin on *Pseudomonas Aeruginosa* Biofilms. *Curr. Microbiol.* **2010**, *61*, 411–416. [CrossRef]

433. Almeida, J.; Tomé, J.P.C.; Neves, M.G.P.M.S.; Tomé, A.C.; Cavaleiro, J.A.S.; Cunha, Â.; Costa, L.; Faustino, M.A.F.; Almeida, A. Photodynamic Inactivation of Multidrug-Resistant Bacteria in Hospital Wastewaters: Influence of Residual Antibiotics. *Photochem. Photobiol. Sci.* **2014**, *13*, 626. [CrossRef] [PubMed]

434. Di Poto, A.; Sbarra, M.S.; Provenza, G.; Visai, L.; Speziale, P. The Effect of Photodynamic Treatment Combined with Antibiotic Action or Host Defence Mechanisms on *Staphylococcus Aureus* Biofilms. *Biomaterials* **2009**, *30*, 3158–3166. [CrossRef] [PubMed]

435. Tanaka, M.; Mroz, P.; Dai, T.; Huang, L.; Morimoto, Y.; Kinoshita, M.; Yoshihara, Y.; Shinomiya, N.; Seki, S.; Nemoto, K.; et al. Linezolid and Vancomycin Decrease the Therapeutic Effect of Methylene Blue-Photodynamic Therapy in a Mouse Model of MRSA Bacterial Arthritis. *Photochem. Photobiol.* **2013**, *89*, 679–682. [CrossRef]

436. Jiang, Q.; Fangjie, E.; Tian, J.; Yang, J.; Zhang, J.; Cheng, Y. Light-Excited Antibiotics for Potentiating Bacterial Killing via Reactive Oxygen Species Generation. *ACS Appl. Mater. Interfaces* **2020**, *12*, 16150–16158. [CrossRef] [PubMed]

437. Morschhäuser, J. The Development of Fluconazole Resistance in Candida Albicans—An Example of Microevolution of a Fungal Pathogen. *J. Microbiol.* **2016**, *54*, 192–201. [CrossRef] [PubMed]
438. Quiroga, E.D.; Mora, S.J.; Alvarez, M.G.; Durantini, E.N. Photodynamic Inactivation of *Candida Albicans* by a Tetracationic Tentacle Porphyrin and Its Analogue without Intrinsic Charges in Presence of Fluconazole. *Photodiagn. Photodyn. Ther.* **2016**, *13*, 334–340. [CrossRef]
439. Hu, Y.; Qi, X.; Sun, H.; Lu, Y.; Hu, Y.; Chen, X.; Liu, K.; Yang, Y.; Mao, Z.; Wu, Z.; et al. Photodynamic Therapy Combined with Antifungal Drugs against Chromoblastomycosis and the Effect of ALA-PDT on Fonsecaea In Vitro. *PLoS Negl. Trop. Dis.* **2019**, *13*, e0007849. [CrossRef] [PubMed]
440. Chen, H.; Li, S.; Wu, M.; Kenry; Huang, Z.; Lee, C.; Liu, B. Membrane-Anchoring Photosensitizer with Aggregation-Induced Emission Characteristics for Combating Multidrug-Resistant Bacteria. *Angew. Chem. Int. Ed.* **2020**, *59*, 632–636. [CrossRef] [PubMed]
441. Le Gall, T.; Berchel, M.; Le Hir, S.; Fraix, A.; Salaün, J.Y.; Férec, C.; Lehn, P.; Jaffrès, P.-A.; Montier, T. Arsonium-Containing Lipophosphoramides, Poly-Functional Nano-Carriers for Simultaneous Antibacterial Action and Eukaryotic Cell Transfection. *Adv. Healthc. Mater.* **2013**, *2*, 1513–1524. [CrossRef] [PubMed]
442. Lin, S.; Liu, C.; Han, X.; Zhong, H.; Cheng, C. Viral Nanoparticle System: An Effective Platform for Photodynamic Therapy. *Int. J. Mol. Sci.* **2021**, *22*, 1728. [CrossRef]
443. Hosseini, N.; Pourhajibagher, M.; Chiniforush, N.; Hosseinkhan, N.; Rezaie, P.; Bahador, A. Modulation of Toxin-Antitoxin System Rnl AB Type II in Phage-Resistant Gammaproteobacteria Surviving Photodynamic Treatment. *J. Lasers Med. Sci.* **2018**, *10*, 21–28. [CrossRef]
444. Luong, T.; Salabarria, A.-C.; Roach, D.R. Phage Therapy in the Resistance Era: Where Do We Stand and Where Are We Going? *Clin. Ther.* **2020**, *42*, 1659–1680. [CrossRef]
445. Liu, Z.; Li, J.; Chen, W.; Liu, L.; Yu, F. Light and Sound to Trigger the Pandora's Box against Breast Cancer: A Combination Strategy of Sonodynamic, Photodynamic and Photothermal Therapies. *Biomaterials* **2020**, *232*, 119685. [CrossRef] [PubMed]
446. Mai, B.; Wang, X.; Liu, Q.; Zhang, K.; Wang, P. The Application of DVDMS as a Sensitizing Agent for Sono-/Photo-Therapy. *Front. Pharm.* **2020**, *11*, 19. [CrossRef]
447. Fraix, A.; Sortino, S. Combination of PDT Photosensitizers with NO Photodononors. *Photochem. Photobiol. Sci.* **2018**, *17*, 1709–1727. [CrossRef]
448. Parisi, C.; Failla, M.; Fraix, A.; Rescifina, A.; Rolando, B.; Lazzarato, L.; Cardile, V.; Graziano, A.C.E.; Fruttero, R.; Gasco, A.; et al. A Molecular Hybrid Producing Simultaneously Singlet Oxygen and Nitric Oxide by Single Photon Excitation with Green Light. *Bioorganic Chem.* **2019**, *85*, 18–22. [CrossRef] [PubMed]
449. Maya, R.; Ladeira, L.L.C.; Maya, J.E.P.; Mail, L.M.G.; Bussadori, S.K.; Paschoal, M.A.B. The Combination of Antimicrobial Photodynamic Therapy and Photobiomodulation Therapy for the Treatment of Palatal Ulcers: A Case Report. *J. Lasers Med. Sci.* **2020**, *11*, 228–233. [CrossRef] [PubMed]
450. Alqerban, A. Efficacy of Antimicrobial Photodynamic and Photobiomodulation Therapy against Treponema Denticola, Fusobacterium Nucleatum and Human Beta Defensin-2 Levels in Patients with Gingivitis Undergoing Fixed Orthodontic Treatment: A Clinic-Laboratory Study. *Photodiagn. Photodyn. Ther.* **2020**, *29*, 101659. [CrossRef]
451. Fekrazad, R. Photobiomodulation and Antiviral Photodynamic Therapy as a Possible Novel Approach in COVID-19 Management. *Photobiomodulation Photomed. Laser Surg.* **2020**, *38*, 255–257. [CrossRef] [PubMed]
452. Lago, A.D.N.; Fortes, A.B.C.; Furtado, G.S.; Menezes, C.F.S.; Gonçalves, L.M. Association of Antimicrobial Photodynamic Therapy and Photobiomodulation for *Herpes Simplex* Labialis Resolution: Case Series. *Photodiagn. Photodyn. Ther.* **2020**, *32*, 102070. [CrossRef] [PubMed]
453. Teixeira, I.S.; Leal, F.S.; Tateno, R.Y.; Palma, L.F.; Campos, L. Photobiomodulation Therapy and Antimicrobial Photodynamic Therapy for Orofacial Lesions in Patients with COVID-19: A Case Series. *Photodiagn. Photodyn. Ther.* **2021**, *34*, 102281. [CrossRef]
454. Ma, X.; Pan, H.; Wu, G.; Yang, Z.; Yi, J. Ultrasound May Be Exploited for the Treatment of Microbial Diseases. *Med. Hypotheses* **2009**, *73*, 18–19. [CrossRef] [PubMed]
455. Serpe, L.; Giuntini, F. Sonodynamic Antimicrobial Chemotherapy: First Steps towards a Sound Approach for Microbe Inactivation. *J. Photochem. Photobiol. B Biol.* **2015**, *150*, 44–49. [CrossRef] [PubMed]
456. Costley, D.; Ewan, C.M.; Fowley, C.; McHale, A.P.; Atchison, J.; Nomikou, N.; Callan, J.F. Treating Cancer with Sonodynamic Therapy: A Review. *Int. J. Hyperth.* **2015**, *31*, 107–117. [CrossRef]
457. Alves, F.; Gomes Guimarães, G.; Mayumi Inada, N.; Pratavieira, S.; Salvador Bagnato, V.; Kurachi, C. Strategies to Improve the Antimicrobial Efficacy of Photodynamic, Sonodynamic, and Sonophotodynamic Therapies. *Lasers Surg. Med.* **2021**, lsm.23383. [CrossRef]
458. Erriu, M.; Blus, C.; Szmukler-Moncler, S.; Buogo, S.; Levi, R.; Barbato, G.; Madonnaripa, D.; Denotti, G.; Piras, V.; Orrù, G. Microbial Biofilm Modulation by Ultrasound: Current Concepts and Controversies. *Ultrason. Sonochemistry* **2014**, *21*, 15–22. [CrossRef] [PubMed]
459. Harris, F.; Dennison, S.R.; Phoenix, D.A. Using Sound for Microbial Eradication—Light at the End of the Tunnel? *FEMS Microbiol. Lett.* **2014**, *356*, 20–22. [CrossRef] [PubMed]

460. Pourhajibagher, M.; reza Rokn, A.; reza Barikani, H.; Bahador, A. Photo-Sonodynamic Antimicrobial Chemotherapy *via* Chitosan Nanoparticles-Indocyanine Green against Polymicrobial Periopathogenic Biofilms: *Ex Vivo* Study on Dental Implants. *Photodiagn. Photodyn. Ther.* **2020**, *31*, 101834. [CrossRef]

461. Pourhajibagher, M.; Bahador, A. Attenuation of Aggregatibacter Actinomycetemcomitans Virulence Using Curcumin-Decorated Nanophytosomes-Mediated Photo-Sonoantimicrobial Chemotherapy. *Sci. Rep.* **2021**, *11*, 6012. [CrossRef] [PubMed]

462. Traitcheva, N.; Berg, H. Electroporation and Alternating Current Cause Membrane Permeation of Photodynamic Cytotoxins Yielding Necrosis and Apoptosis of Cancer Cells. *Bioelectrochemistry* **2010**, *79*, 257–260. [CrossRef]

463. Weżgowiec, J.; Kulbacka, J.; Saczko, J.; Rossowska, J.; Chodaczek, G.; Kotulska, M. Biological Effects in Photodynamic Treatment Combined with Electropermeabilization in Wild and Drug Resistant Breast Cancer Cells. *Bioelectrochemistry* **2018**, *123*, 9–18. [CrossRef]

464. Reinhard, A.; Sandborn, W.J.; Melhem, H.; Bolotine, L.; Chamaillard, M.; Peyrin-Biroulet, L. Photodynamic Therapy as a New Treatment Modality for Inflammatory and Infectious Conditions. *Expert Rev. Clin. Immunol.* **2015**, *11*, 637–657. [CrossRef] [PubMed]

465. McFarland, S.A.; Mandel, A.; Dumoulin-White, R.; Gasser, G. Metal-Based Photosensitizers for Photodynamic Therapy: The Future of Multimodal Oncology? *Curr. Opin. Chem. Biol.* **2019**, *56*, 23–27. [CrossRef] [PubMed]

466. Prasad, P.; Gupta, A.; Sasmal, P.K. Aggregation-Induced Emission Active Metal Complexes: A Promising Strategy to Tackle Bacterial Infections. *Chem. Commun.* **2021**, *57*, 174–186. [CrossRef]

467. Gao, S.; Yan, X.; Xie, G.; Zhu, M.; Ju, X.; Stang, P.J.; Tian, Y.; Niu, Z. Membrane Intercalation-Enhanced Photodynamic Inactivation of Bacteria by a Metallacycle and TAT-Decorated Virus Coat Protein. *Proc. Natl. Acad. Sci. USA* **2019**, *116*, 23437–23443. [CrossRef]

468. Jerjes, W.; Theodossiou, T.A.; Hirschberg, H.; Høgset, A.; Weyergang, A.; Selbo, P.K.; Hamdoon, Z.; Hopper, C.; Berg, K. Photochemical Internalization for Intracellular Drug Delivery. From Basic Mechanisms to Clinical Research. *J. Clin. Med.* **2020**, *9*, 528. [CrossRef]

469. Zhang, X.; de Boer, L.; Heiliegers, L.; Man-Bovenkerk, S.; Selbo, P.K.; Drijfhout, J.W.; Høgset, A.; Zaat, S.A.J. Photochemical Internalization Enhances Cytosolic Release of Antibiotic and Increases Its Efficacy against Staphylococcal Infection. *J. Control. Release* **2018**, *283*, 214–222. [CrossRef]

470. Hilgers, F.; Bitzenhofer, N.L.; Ackermann, Y.; Burmeister, A.; Grünberger, A.; Jaeger, K.-E.; Drepper, T. Genetically Encoded Photosensitizers as Light-Triggered Antimicrobial Agents. *Int. J. Mol. Sci.* **2019**, *20*, 4608. [CrossRef] [PubMed]

471. Gorbachev, D.A.; Staroverov, D.B.; Lukyanov, K.A.; Sarkisyan, K.S. Genetically Encoded Red Photosensitizers with Enhanced Phototoxicity. *Int. J. Mol. Sci.* **2020**, *21*, 8800. [CrossRef] [PubMed]

472. Endres, S.; Wingen, M.; Torra, J.; Ruiz-González, R.; Polen, T.; Bosio, G.; Bitzenhofer, N.L.; Hilgers, F.; Gensch, T.; Nonell, S.; et al. An Optogenetic Toolbox of LOV-Based Photosensitizers for Light-Driven Killing of Bacteria. *Sci. Rep.* **2018**, *8*, 15021. [CrossRef] [PubMed]

473. Liao, Y. Design and Applications of Metastable-State Photoacids. *Acc. Chem. Res.* **2017**, *50*, 1956–1964. [CrossRef] [PubMed]

474. Luo, Y.; Wang, C.; Peng, P.; Hossain, M.; Jiang, T.; Fu, W.; Liao, Y.; Su, M. Visible Light Mediated Killing of Multidrug-Resistant Bacteria Using Photoacids. *J. Mater. Chem. B* **2013**, *1*, 997–1001. [CrossRef] [PubMed]

475. Huang, T.-C.; Chen, C.-J.; Ding, S.-J.; Chen, C.-C. Antimicrobial Efficacy of Methylene Blue-Mediated Photodynamic Therapy on Titanium Alloy Surfaces In Vitro. *Photodiagn. Photodyn. Ther.* **2019**, *25*, 7–16. [CrossRef] [PubMed]

476. Yang, Y.; Goetzfried, M.A.; Hidaka, K.; You, M.; Tan, W.; Sugiyama, H.; Endo, M. Direct Visualization of Walking Motions of Photocontrolled Nanomachine on the DNA Nanostructure. *Nano Lett.* **2015**, *15*, 6672–6676. [CrossRef]

477. Zhuang, X.; Ma, X.; Xue, X.; Jiang, Q.; Song, L.; Dai, L.; Zhang, C.; Jin, S.; Yang, K.; Ding, B.; et al. A Photosensitizer-Loaded DNA Origami Nanosystem for Photodynamic Therapy. *ACS Nano* **2016**, *10*, 3486–3495. [CrossRef] [PubMed]

478. Donnelly, R.F.; McCarron, P.A.; Tunney, M.M. Antifungal Photodynamic Therapy. *Microbiol. Res.* **2008**, *163*, 1–12. [CrossRef] [PubMed]

479. Pereira Rosa, L. Antimicrobial Photodynamic Therapy: A New Therapeutic Option to Combat Infections. *J. Med. Microb. Diagn.* **2014**, *3*, 1. [CrossRef]

480. Kim, M.M.; Darafsheh, A. Light Sources and Dosimetry Techniques for Photodynamic Therapy. *Photochem. Photobiol.* **2020**, *96*, 280–294. [CrossRef] [PubMed]

481. Heinemann, F.; Karges, J.; Gasser, G. Critical Overview of the Use of Ru(II) Polypyridyl Complexes as Photosensitizers in One-Photon and Two-Photon Photodynamic Therapy. *Acc. Chem. Res.* **2017**, *50*, 2727–2736. [CrossRef]

482. Boca, S.C.; Four, M.; Bonne, A.; van der Sanden, B.; Astilean, S.; Baldeck, P.L.; Lemercier, G. An Ethylene-Glycol Decorated Ruthenium(II) Complex for Two-Photon Photodynamic Therapy. *Chem. Commun.* **2009**, *30*, 4590–4592. [CrossRef] [PubMed]

483. Blum, N.T.; Zhang, Y.; Qu, J.; Lin, J.; Huang, P. Recent Advances in Self-Exciting Photodynamic Therapy. *Front. Bioeng. Biotechnol.* **2020**, *8*, 594491. [CrossRef] [PubMed]

484. Liu, S.; Yuan, H.; Bai, H.; Zhang, P.; Lv, F.; Liu, L.; Dai, Z.; Bao, J.; Wang, S. Electrochemiluminescence for Electric-Driven Antibacterial Therapeutics. *J. Am. Chem. Soc.* **2018**, *140*, 2284–2291. [CrossRef]

485. Kamkaew, A.; Chen, F.; Zhan, Y.; Majewski, R.L.; Cai, W. Scintillating Nanoparticles as Energy Mediators for Enhanced Photodynamic Therapy. *ACS Nano* **2016**, *10*, 3918–3935. [CrossRef] [PubMed]

486. Kamanli, A.F.; Çetinel, G. Comparison of Pulse and Super Pulse Radiation Modes' Singlet Oxygen Production Effect in Antimicrobial Photodynamic Therapy (AmPDT). *Photodiagn. Photodyn. Ther.* **2020**, *30*, 101706. [CrossRef]

487. Hou, X.; Tao, Y.; Li, X.; Pang, Y.; Yang, C.; Jiang, G.; Liu, Y. CD44-Targeting Oxygen Self-Sufficient Nanoparticles for Enhanced Photodynamic Therapy Against Malignant Melanoma. *Int. J. Nanomed.* **2020**, *15*, 10401–10416. [CrossRef] [PubMed]
488. Liang, M.; Yan, X. Nanozymes: From New Concepts, Mechanisms, and Standards to Applications. *Acc. Chem. Res.* **2019**, *52*, 2190–2200. [CrossRef]
489. Liu, X.; Liu, J.; Chen, S.; Xie, Y.; Fan, Q.; Zhou, J.; Bao, J.; Wei, T.; Dai, Z. Dual-Path Modulation of Hydrogen Peroxide to Ameliorate Hypoxia for Enhancing Photodynamic/Starvation Synergistic Therapy. *J. Mater. Chem. B* **2020**, *8*, 9933–9942. [CrossRef] [PubMed]
490. Wang, D.; Wu, H.; Phua, S.Z.F.; Yang, G.; Qi Lim, W.; Gu, L.; Qian, C.; Wang, H.; Guo, Z.; Chen, H.; et al. Self-Assembled Single-Atom Nanozyme for Enhanced Photodynamic Therapy Treatment of Tumor. *Nat. Commun.* **2020**, *11*, 357. [CrossRef] [PubMed]
491. Nonell, S.; Ferreras, L.R.; Cañete, A.; Lemp, E.; Günther, G.; Pizarro, N.; Zanocco, A.L. Photophysics and Photochemistry of Naphthoxazinone Derivatives. *J. Org. Chem.* **2008**, *73*, 5371–5378. [CrossRef]
492. Lamarque, G.C.C.; Méndez, D.A.C.; Gutierrez, E.; Dionisio, E.J.; Machado, M.A.A.M.; Oliveira, T.M.; Rios, D.; Cruvinel, T. Could Chlorhexidine Be an Adequate Positive Control for Antimicrobial Photodynamic Therapy in-in Vitro Studies? *Photodiagn. Photodyn. Ther.* **2019**, *25*, 58–62. [CrossRef] [PubMed]
493. Pourhajibagher, M.; Bahador, A. Computational Biology Analysis of COVID-19 Receptor-Binding Domains: A Target Site for Indocyanine Green Through Antimicrobial Photodynamic Therapy. *J. Lasers Med. Sci.* **2020**, *11*, 433–441. [CrossRef]

MDPI
St. Alban-Anlage 66
4052 Basel
Switzerland
Tel. +41 61 683 77 34
Fax +41 61 302 89 18
www.mdpi.com

Pharmaceutics Editorial Office
E-mail: pharmaceutics@mdpi.com
www.mdpi.com/journal/pharmaceutics